Calculus I
Differentiation and Integration

Book Title: Calculus I - Differentiation and Integration

Author: Said Hamilton

Editor: Linda Hamilton

Cover design by: Kathleen Myers

First published in the year 2002

Hamilton Education Guides
P.O. Box 681
Vienna, Va. 22183

Library of Congress Catalog Card Number pending
Library of Congress Cataloging-in-Publication Data

ISBN 0-9649954-4-1

This book is dedicated to my wife and children for their support and understanding.

Contents

Chapter 5   Integration (Part II)

Appendix   Exercise Solutions

Introduction and Overview

Similar to the previous books published by the Hamilton Education Guides, the intent of this book is to build a strong foundation by increasing student confidence in solving mathematical problems. To achieve this objective, the author has diligently tried to address each subject in a clear, concise, and easy to understand step-by-step format. A great deal of effort has been made to ensure that the subjects presented in each chapter are explained simply, thoroughly, and adequately. It is the authors hope that this book can fulfill these objectives by building a solid foundation in pursuit of more advanced technical concepts.

The scope of this book is intended for educational levels ranging from the 12th grade to adult. The book can also be used by students in home study programs, parents, teachers, special education programs, tutors, high schools, preparatory schools, and adult educational programs, including colleges and universities as a main text, a thorough reference, or a supplementary book. A thorough knowledge of algebraic concepts in subject areas such as linear equations and inequalities, fractional operations, exponents, radicals, polynomials, factorization, non-linear and quadratic equations is required.

"Calculus I" is divided into five chapters. Sequences and series are introduced in Chapter 1. How to compute and find the limit of arithmetic and geometric sequences and series including expansion and simplification of factorial expressions is discussed in this chapter. Derivatives and its applicable differentiation rules using the Prime and $\frac{d}{dx}$ notations are introduced in Chapter 2. In addition, use of the Chain rule in solving different types of equations, the implicit differentiation method, derivative of functions with fractional exponents, derivative of radical functions, including the steps for solving higher order equations is discussed in this chapter. Differentiation of trigonometric functions, exponential and logarithmic functions, hyperbolic functions, and inverse hyperbolic functions is discussed in Chapter 3. Furthermore, evaluation of expressions referred to as indeterminate forms using a general rule known as L'Hopital's Rule is discussed in Chapter 3. The subject of integration is introduced in Chapter 4. Integration using basic integration formulas and methods such as the substitution method is discussed in this chapter. Additionally, integration of trigonometric functions, inverse trigonometric functions, exponential and logarithmic functions is addressed in Chapter 4. Other integration techniques such as integration by parts, integration using trigonometric substitution, and integration by partial fractions is introduced in Chapter 5. The steps in integrating hyperbolic functions is also discussed in this chapter. Finally, detailed solutions to the exercises are provided in the Appendix. Students are encouraged to solve each problem in the same detailed and step-by-step format as shown in the text.

In keeping with our commitment of excellence in providing clear, easy to follow, and concise educational materials to our readers, I believe this book will again add value to the Hamilton Education Guides series for its clarity and special attention to detail. I hope readers of this book will find it valuable as both a learning tool and as a reference. Any comments or suggestions for improvement of this book will be appreciated.

With best wishes,

Said Hamilton

Chapter 1
Sequences and Series

Quick Reference to Chapter 1 Problems

Chapter 1 - Sequences and Series

The objective of this chapter is to improve the student's ability to solve problems involving sequences and series. Sequences and series are introduced in Sections 1.1 and 1.2. How to solve arithmetic sequences and arithmetic series are discussed in Section 1.3. Solutions to geometric sequences and geometric series are addressed in Section 1.4. The process of identifying convergence or divergence of a sequence or a series, for large values of n, is discussed in Section 1.5. Finally, the factorial notation and its use in expanding binomial expressions is addressed in Section 1.6. Each section is concluded by solving examples with practice problems to further enhance the student's ability.

1.1 Sequences

A **sequence** is a function whose domain contains a set of positive integer terms such as $(1, 2, 3, 4, \cdots)$. Functions generate sequences. For example, the function $s(n) = s_n = n - 2$ whose domain is $(1, 2, 3, 4, 5, 6)$ generates the sequence

$$s(1) = s_1 = 1 - 2 = -1 \qquad s(2) = s_2 = 2 - 2 = 0 \qquad s(3) = s_3 = 3 - 2 = 1$$

$$s(4) = s_4 = 4 - 2 = 2 \qquad s(5) = s_5 = 5 - 2 = 3 \qquad s(6) = s_6 = 6 - 2 = 4$$

where the first six terms of the sequence are $(s_1, s_2, s_3, s_4, s_5, s_6) = (-1, 0, 1, 2, 3, 4)$.

In general, a function $f(x)$ whose domain is the set of positive integers $(1, 2, 3, \cdots, n)$ including a fixed value for n is called a **finite sequence function**. On the other hand, a function whose domain is the set of $(1, 2, 3, \cdots)$ is called an **infinite sequence function**. The elements of the range of a sequence function are called the **terms** of the sequence function. In some instances a sequence is given by presenting its first few terms, followed by its nth term, $s_n = s(n)$, which is commonly referred to as the **general term** of a sequence. For example, the sequence $4, 3, \dfrac{16}{5}, \dfrac{25}{9}, \cdots, \dfrac{(n+1)^2}{2n-1}$ shows the first four terms and the general term of the sequence. In the following examples we will learn how the various terms of a sequence are found:

Example 1.1-1 List the first six terms of the given sequence.

a. $a_n = \dfrac{(-3)^n}{n^3}$ b. $b_k = \dfrac{(-1)^k}{k+1}$ c. $d_n = \dfrac{5}{n(2n-1)}$ d. $c_n = \left(\dfrac{1}{2}\right)^n \cdot \dfrac{(-1)^n}{n}$

Solutions:

a. $\boxed{a_1} = \boxed{\dfrac{(-3)^1}{1^3}} = \boxed{\dfrac{-3}{1}} = \boxed{-3}$ $\boxed{a_2} = \boxed{\dfrac{(-3)^2}{2^3}} = \boxed{\dfrac{9}{8}} = \boxed{1.125}$

$\boxed{a_3} = \boxed{\dfrac{(-3)^3}{3^3}} = \boxed{\dfrac{-27}{27}} = \boxed{-1}$ $\boxed{a_4} = \boxed{\dfrac{(-3)^4}{4^3}} = \boxed{\dfrac{81}{64}} = \boxed{1.265}$

$\boxed{a_5} = \boxed{\dfrac{(-3)^5}{5^3}} = \boxed{\dfrac{-243}{125}} = \boxed{-1.944}$ $\boxed{a_6} = \boxed{\dfrac{(-3)^6}{6^3}} = \boxed{\dfrac{729}{216}} = \boxed{3.375}$

b. $\boxed{b_1} = \boxed{\dfrac{(-1)^1}{1+1}} = \boxed{\dfrac{-1}{2}} = \boxed{-0.5}$ $\boxed{b_2} = \boxed{\dfrac{(-1)^2}{2+1}} = \boxed{\dfrac{1}{3}} = \boxed{0.333}$

$\boxed{b_3} = \boxed{\dfrac{(-1)^3}{3+1}} = \boxed{\dfrac{-1}{4}} = \boxed{-0.25}$ $\boxed{b_4} = \boxed{\dfrac{(-1)^4}{4+1}} = \boxed{\dfrac{1}{5}} = \boxed{0.2}$

$\boxed{b_5} = \boxed{\dfrac{(-1)^5}{5+1}} = \boxed{\dfrac{-1}{6}} = \boxed{-0.167}$ $\boxed{b_6} = \boxed{\dfrac{(-1)^6}{6+1}} = \boxed{\dfrac{1}{7}} = \boxed{0.143}$

c. $\boxed{d_1} = \boxed{\dfrac{5}{1\cdot(2\cdot 1-1)}} = \boxed{\dfrac{5}{2-1}} = \boxed{\dfrac{5}{1}} = \boxed{5}$ $\boxed{d_2} = \boxed{\dfrac{5}{2\cdot(2\cdot 2-1)}} = \boxed{\dfrac{5}{2\cdot 3}} = \boxed{\dfrac{5}{6}} = \boxed{0.833}$

$\boxed{d_3} = \boxed{\dfrac{5}{3\cdot(2\cdot 3-1)}} = \boxed{\dfrac{5}{3\cdot 5}} = \boxed{\dfrac{1}{3}} = \boxed{0.333}$ $\boxed{d_4} = \boxed{\dfrac{5}{4\cdot(2\cdot 4-1)}} = \boxed{\dfrac{5}{4\cdot 7}} = \boxed{\dfrac{5}{28}} = \boxed{0.178}$

$\boxed{d_5} = \boxed{\dfrac{5}{5\cdot(2\cdot 5-1)}} = \boxed{\dfrac{5}{5\cdot 9}} = \boxed{\dfrac{1}{9}} = \boxed{0.111}$ $\boxed{d_6} = \boxed{\dfrac{5}{6\cdot(2\cdot 6-1)}} = \boxed{\dfrac{5}{6\cdot 11}} = \boxed{\dfrac{5}{66}} = \boxed{0.076}$

d. $\boxed{c_1} = \boxed{\left(\dfrac{1}{2}\right)^1\cdot\dfrac{(-1)^1}{1}} = \boxed{\dfrac{1}{2}\cdot -1} = \boxed{-\dfrac{1}{2}} = \boxed{-0.5}$ $\boxed{c_2} = \boxed{\left(\dfrac{1}{2}\right)^2\cdot\dfrac{(-1)^2}{2}} = \boxed{\dfrac{1}{4}\cdot\dfrac{1}{2}} = \boxed{\dfrac{1}{8}} = \boxed{0.125}$

$\boxed{c_3} = \boxed{\left(\dfrac{1}{2}\right)^3\cdot\dfrac{(-1)^3}{3}} = \boxed{\dfrac{1}{8}\cdot -\dfrac{1}{3}} = \boxed{-\dfrac{1}{24}} = \boxed{-0.042}$ $\boxed{c_4} = \boxed{\left(\dfrac{1}{2}\right)^4\cdot\dfrac{(-1)^4}{4}} = \boxed{\dfrac{1}{16}\cdot\dfrac{1}{4}} = \boxed{\dfrac{1}{64}} = \boxed{0.016}$

$\boxed{c_5} = \boxed{\left(\dfrac{1}{2}\right)^5\cdot\dfrac{(-1)^5}{5}} = \boxed{\dfrac{1}{32}\cdot -\dfrac{1}{5}} = \boxed{-\dfrac{1}{160}} = \boxed{-0.006}$ $\boxed{c_6} = \boxed{\left(\dfrac{1}{2}\right)^6\cdot\dfrac{(-1)^6}{6}} = \boxed{\dfrac{1}{64}\cdot\dfrac{1}{6}} = \boxed{\dfrac{1}{384}} = \boxed{0.003}$

Example 1.1-2 Find the indicated terms for the following sequences.

a. Write the third and sixth terms of $s_n = \dfrac{(-2)^{n+1}}{n^3}$ b. Write the tenth term of $a_i = (i-1)^3\cdot 2^{i-4}$

c. Write the third and fourth terms of $a_n = (-1)^{n-2}$ d. Write the seventh term of $a_k = (0.2)^{k-1}$

e. Write the third and twelfth terms of $a_i = (-i)^3$ f. Write the eleventh term of $s_n = (-1)^{n-1}\cdot 2^{n+1}$

Solutions:

a. $\boxed{s_3} = \boxed{\dfrac{(-2)^{3+1}}{3^3}} = \boxed{\dfrac{(-2)^4}{27}} = \boxed{\dfrac{16}{27}} = \boxed{0.593}$ $\boxed{s_6} = \boxed{\dfrac{(-2)^{6+1}}{6^3}} = \boxed{\dfrac{(-2)^7}{216}} = \boxed{\dfrac{-128}{216}} = \boxed{-0.593}$

b. $\boxed{a_{10}} = \boxed{(10-1)^3\cdot 2^{10-4}} = \boxed{9^3\cdot 2^6} = \boxed{729\cdot 64} = \boxed{4.6656\times 10^4}$

c. $\boxed{a_3} = \boxed{(-1)^{3-2}} = \boxed{(-1)^1} = \boxed{-1}$ $\boxed{a_4} = \boxed{(-1)^{4-2}} = \boxed{(-1)^2} = \boxed{1}$

d. $\boxed{a_7} = \boxed{(0.2)^{7-1}} = \boxed{(0.2)^6} = \boxed{0.000064} = \boxed{\mathbf{6.4 \times 10^{-5}}}$

e. $\boxed{a_3} = \boxed{(-3)^3} = \boxed{-3^3} = \boxed{\mathbf{-27}}$ $\qquad\qquad$ $\boxed{a_{12}} = \boxed{(-12)^3} = \boxed{-12^3} = \boxed{\mathbf{-1728}}$

f. $\boxed{s_{11}} = \boxed{(-1)^{11-1} \cdot 2^{11+1}} = \boxed{(-1)^{10} \cdot 2^{12}} = \boxed{1 \cdot 2^{12}} = \boxed{\mathbf{4096}}$

Example 1.1-3 Write s_5, s_6, s_7, and s_{15} for the following sequences.

a. $s_n = \dfrac{(n+1)\pi}{2}$ $\qquad\qquad$ b. $s_n = \dfrac{2^n}{2^{n-1}}$ $\qquad\qquad$ c. $s_n = 2n(n-1)(n-2)$

Solution:

a. $\boxed{s_5} = \boxed{\dfrac{(5+1)\pi}{2}} = \boxed{\dfrac{6\pi}{2}} = \boxed{\mathbf{3\pi}}$ $\qquad\qquad$ $\boxed{s_6} = \boxed{\dfrac{(6+1)\pi}{2}} = \boxed{\dfrac{7\pi}{2}} = \boxed{\mathbf{3.5\pi}}$

$\boxed{s_7} = \boxed{\dfrac{(7+1)\pi}{2}} = \boxed{\dfrac{8\pi}{2}} = \boxed{\mathbf{4\pi}}$ $\qquad\qquad$ $\boxed{s_{15}} = \boxed{\dfrac{(15+1)\pi}{2}} = \boxed{\dfrac{16\pi}{2}} = \boxed{\mathbf{8\pi}}$

b. $\boxed{s_5} = \boxed{\dfrac{2^5}{2^{5-1}}} = \boxed{\dfrac{2^5}{2^4}}; \boxed{2^5 \cdot 2^{-4}}; \boxed{2^{5-4}}; \boxed{2^1} = \boxed{\mathbf{2}}$

$\boxed{s_6} = \boxed{\dfrac{2^6}{2^{6-1}}} = \boxed{\dfrac{2^6}{2^5}} = \boxed{2^6 \cdot 2^{-5}} = \boxed{2^{6-5}} = \boxed{2^1} = \boxed{\mathbf{2}}$

$\boxed{s_7} = \boxed{\dfrac{2^7}{2^{7-1}}} = \boxed{\dfrac{2^7}{2^6}} = \boxed{2^7 \cdot 2^{-6}} = \boxed{2^{7-6}} = \boxed{2^1} = \boxed{\mathbf{2}}$

$\boxed{s_{15}} = \boxed{\dfrac{2^{15}}{2^{15-1}}} = \boxed{\dfrac{2^{15}}{2^{14}}} = \boxed{2^{15} \cdot 2^{-14}} = \boxed{2^{15-14}} = \boxed{2^1} = \boxed{\mathbf{2}}$

c. $\boxed{s_5} = \boxed{2 \cdot 5 \cdot (5-1)(5-2)} = \boxed{10 \cdot 4 \cdot 3} = \boxed{\mathbf{120}}$ $\qquad$ $\boxed{s_6} = \boxed{2 \cdot 6 \cdot (6-1)(6-2)} = \boxed{12 \cdot 5 \cdot 4} = \boxed{\mathbf{240}}$

$\boxed{s_7} = \boxed{2 \cdot 7 \cdot (7-1)(7-2)} = \boxed{14 \cdot 6 \cdot 5} = \boxed{\mathbf{420}}$ $\qquad$ $\boxed{s_{15}} = \boxed{2 \cdot 15 \cdot (15-1)(15-2)} = \boxed{30 \cdot 14 \cdot 13} = \boxed{\mathbf{5460}}$

Example 1.1-4 Find the twelfth term of the following sequences:

a. $0.5, 0.25, 0.125, \cdots, \dfrac{1^{n+1}}{2^n}$ $\quad$ b. $8, 5.063, 4.214, \cdots, \left(1+\dfrac{1}{n}\right)^{n+2}$ $\quad$ c. $4, -12, 32, \cdots, \dfrac{2^n \cdot (n+1)}{(-1)^{n+1}}$

Solutions:

a. $\boxed{s_{12}} = \boxed{\dfrac{1^{12+1}}{2^{12}}} = \boxed{\dfrac{1^{13}}{2^{12}}} = \boxed{\dfrac{1}{4096}} = \boxed{0.000244} = \boxed{\mathbf{2.44 \times 10^{-4}}}$

b. $\boxed{s_{12}} = \boxed{\left(1+\dfrac{1}{12}\right)^{12+2}} = \boxed{\left(\dfrac{12+1}{12}\right)^{14}} = \boxed{\left(\dfrac{13}{12}\right)^{14}} = \boxed{1.0833^{14}} = \boxed{\mathbf{3.066}}$

c. $\boxed{s_{12}} = \boxed{\dfrac{2^{12} \cdot (12+1)}{(-1)^{12+1}}} = \boxed{\dfrac{2^{12} \cdot 13}{(-1)^{13}}} = \boxed{\dfrac{4096 \cdot 13}{-1}} = \boxed{-53248} = \boxed{-5.3248 \times 10^4}$

Example 1.1-5   Given the general term of the sequence $s(n) = s_n = n(n-2) + 5$, write its kth and $k+1$ term.

Solutions:

a. To write the kth term of the sequence simply substitute k in place of n in the general term of

the sequence, i.e., $\boxed{s(k)} = \boxed{s_k} = \boxed{k(k-2)+5} = \boxed{k^2 - 2k + 5}$

b. To write the $k+1$ term of the sequence simply substitute $k+1$ in place of n in the general term

of the sequence, i.e., $\boxed{s(k+1)} = \boxed{s_{k+1}} = \boxed{(k+1)\big[(k+1)-2\big]+5} = \boxed{(k+1)\big[k+1-2\big]+5} = \boxed{(k+1)(k-1)+5}$

$= \boxed{k^2 - k + k - 1 + 5} = \boxed{k^2 + 4}$

Example 1.1-6 For the given domain $(1, 2, 3, 4)$, write the first four terms of the following functions:

a. $f(x) = x^2 + 2x + 1$

b. $s(x) = 3x - 5$

c. $g(x) = \dfrac{2}{3}x$

d. $h(x) = x^{-1} + 1$

Solutions:

a. $\boxed{f(1)} = \boxed{f_1} = \boxed{1^2 + (2 \cdot 1) + 1} = \boxed{1+2+1} = \boxed{4}$ $\boxed{f(2)} = \boxed{f_2} = \boxed{2^2 + (2 \cdot 2) + 1} = \boxed{4+4+1} = \boxed{9}$

$\boxed{f(3)} = \boxed{f_3} = \boxed{3^2 + (2 \cdot 3) + 1} = \boxed{9+6+1} = \boxed{16}$ $\boxed{f(4)} = \boxed{f_4} = \boxed{4^2 + (2 \cdot 4) + 1} = \boxed{16+8+1} = \boxed{25}$

Therefore, the first four terms of the sequence are $(f_1, f_2, f_3, f_4) = (4, 9, 16, 25)$

b. $\boxed{s(1)} = \boxed{s_1} = \boxed{(3 \cdot 1) - 5} = \boxed{3-5} = \boxed{-2}$ $\boxed{s(2)} = \boxed{s_2} = \boxed{(3 \cdot 2) - 5} = \boxed{6-5} = \boxed{1}$

$\boxed{s(3)} = \boxed{s_3} = \boxed{(3 \cdot 3) - 5} = \boxed{9-5} = \boxed{4}$ $\boxed{s(4)} = \boxed{s_4} = \boxed{(3 \cdot 4) - 5} = \boxed{12-5} = \boxed{7}$

Therefore, the first four terms of the sequence are $(s_1, s_2, s_3, s_4) = (-2, 1, 4, 7)$.

c. $\boxed{g(1)} = \boxed{g_1} = \boxed{\dfrac{2}{3} \times 1} = \boxed{\dfrac{2}{3}}$ $\boxed{g(2)} = \boxed{g_2} = \boxed{\dfrac{2}{3} \times 2} = \boxed{\dfrac{4}{3}}$

$\boxed{g(3)} = \boxed{g_3} = \boxed{\dfrac{2}{3} \times 3} = \boxed{\dfrac{2}{1}} = \boxed{2}$ $\boxed{g(4)} = \boxed{g_4} = \boxed{\dfrac{2}{3} \times 4} = \boxed{\dfrac{8}{3}}$

Therefore, the first four terms of the sequence are $(g_1, g_2, g_3, g_4) = \left(\dfrac{2}{3}, \dfrac{4}{3}, 2, \dfrac{8}{3}\right)$

d. $\boxed{h(1)} = \boxed{h_1} = \boxed{1^{-1} + 1} = \boxed{\dfrac{1}{1} + 1} = \boxed{\dfrac{1+1}{1}} = \boxed{2}$ $\boxed{h(2)} = \boxed{h_2} = \boxed{2^{-1} + 1} = \boxed{\dfrac{1}{2} + 1} = \boxed{\dfrac{1+2}{2}} = \boxed{\dfrac{3}{2}}$

$$\boxed{h(3)} = \boxed{h_3} = \boxed{3^{-1}+1} = \boxed{\frac{1}{3}+1} = \boxed{\frac{1+3}{4}} = \boxed{\frac{4}{3}} \qquad\qquad \boxed{h(4)} = \boxed{h_4} = \boxed{4^{-1}+1} = \boxed{\frac{1}{4}+1} = \boxed{\frac{1+4}{4}} = \boxed{\frac{5}{4}}$$

Therefore, the first four terms of the sequence are $(h_1, h_2, h_3, h_4) = \left(2, \dfrac{3}{2}, \dfrac{4}{3}, \dfrac{5}{4}\right)$

In the following section we will discuss series and identify its relation with sequences.

Section 1.1 Practice Problems - Sequences

1. List the first four and tenth terms of the given sequences.

   a. $a_n = \dfrac{2n+1}{-2n}$
   b. $b_k = \dfrac{k(k+1)}{k^2}$
   c. $d_n = 3-(-2)^n$
   d. $k_n = \left(-\dfrac{1}{2}\right)^n \dfrac{(-1)^{n+1}}{n+2}$

2. Write s_3, s_4, s_5, and s_8 for the following sequences.

   a. $s_n = \dfrac{n(n+1)}{2n^{-1}}$
   b. $s_n = (-1)^{n+1}2^{n-2}$
   c. $s_n = \dfrac{(-2)^{n+1}(n-2)}{2n}$

3. Write the first five terms of the following sequences.

   a. $a_n = (-1)^{n+1}(n+2)$
   b. $a_i = 3\left(\dfrac{1}{100}\right)^{i-2}$
   c. $c_i = 3\left(-\dfrac{1}{5}\right)^{i-1}$

   d. $a_n = (3n-5)^2$
   e. $u_k = ar^{k-2}+2$
   f. $b_k = -3\left(\dfrac{2}{3}\right)^{k-2}$

   g. $c_j = \dfrac{j}{j+1}+j$
   h. $y_n = \left(1-\dfrac{1}{n+2}\right)^{n+1}$
   i. $u_k = 1-(-1)^{k+1}$

   j. $y_k = \dfrac{k}{2^{k-1}}$
   k. $y_n = 9^{\frac{1}{k}}(k-2)$
   l. $c_n = \dfrac{n^2-2}{n+1}$

4. Given $n!$ read as "n factorial" which is defined as $n! = n(n-1)(n-2)(n-3)\cdots 5\cdot4\cdot3\cdot2\cdot1$, find

   a. The first eight terms of $n!$.

   b. The first four terms of $a_n = \dfrac{2n+1}{n!}$.

   c. The tenth and twelfth terms of the $c_n = \dfrac{1+3^{n-1}}{(n!)^2}$.

   d. The first, fifth, tenth, and fifteenth terms of $y_n = \dfrac{n!(n-1)}{2+n!}$.

5. Write the first three terms of the following sequences.

   a. $c_n = \dfrac{(2n-3)(n+1)}{(n-4)n}$
   b. $a_n = \left(\dfrac{1}{n-1}\right)\left(\dfrac{n-2}{2+n}\right)$
   c. $s_n = (-1)^{n+1}2^{n+1}$

   d. $y_k = (-1)^{k+1}\dfrac{k(k-1)}{2}$
   e. $b_n = n^2\left(\dfrac{n-1}{2+n}\right)$
   f. $x_a = (5-a)^{a+1}2^a$

1.2 Series

Addition of the terms in any finite sequence result in having the sum of the sequence. The sum of the sequence is referred to as a **series**. For example, the sequence $y_k = \dfrac{1}{2^{k-1}}$ for $k = 1, 2, 3, 4, 5, \text{ and } 6$ can be summed and expressed in the following way:

$$y_1 + y_2 + y_3 + y_4 + y_5 + y_6 = \frac{1}{2^0} + \frac{1}{2^1} + \frac{1}{2^2} + \frac{1}{2^3} + \frac{1}{2^4} + \frac{1}{2^5} = 1 + \frac{1}{2} + \frac{1}{4} + \frac{1}{8} + \frac{1}{16} + \frac{1}{32} = 1.9687$$

The sum of a sequence is generally shown by the Greek letter "$\sum$" (sigma) which is also called summation. Thus, using the sigma notation, the above example can be expressed in the following way $\displaystyle\sum_{i=1}^{6} y_k$ where $y_k = \dfrac{1}{2^{k-1}}$. Note that the variable i is referred to as the *index of summation* and the integer range over which the summation occurs is referred to as the *range of summation*. The following are three properties of summation that students should be familiar with:

$$\sum_{i=1}^{n}\left(a_i + b_i\right) = \sum_{i=1}^{n} a_i + \sum_{i=1}^{n} b_i$$

$$\sum_{i=1}^{n} k a_i = k \sum_{i=1}^{n} a_i$$

$$\sum_{i=1}^{n} k = nk$$

These properties are used extensively in solving the sum of sequences over a specified range as shown in the following examples:

Example 1.2-1 Given $\displaystyle\sum_{i=1}^{n} a_i = 20$ and $\displaystyle\sum_{i=1}^{n} b_i = 40$, find the solution to the following problems using the summation properties.

a. $\displaystyle\sum_{i=1}^{n}\left(2a_i + 3b_i\right) =$
b. $\displaystyle\sum_{i=1}^{n}\left(a_i - b_i\right) =$
c. $\displaystyle\sum_{i=1}^{n}\left(-5a_i + 2b_i\right) =$
d. $\displaystyle\sum_{i=1}^{n}\left(\frac{1}{2}a_i - \frac{1}{4}b_i\right) =$

Solutions:

a. $\boxed{\displaystyle\sum_{i=1}^{n}\left(2a_i + 3b_i\right)} = \boxed{\displaystyle\sum_{i=1}^{n} 2a_i + \sum_{i=1}^{n} 3b_i} = \boxed{2\sum_{i=1}^{n} a_i + 3\sum_{i=1}^{n} b_i} = \boxed{(2 \times 20) + (3 \times 40)} = \boxed{40 + 120} = \boxed{\mathbf{160}}$

b. $\boxed{\displaystyle\sum_{i=1}^{n}\left(a_i - b_i\right)} = \boxed{\displaystyle\sum_{i=1}^{n} a_i - \sum_{i=1}^{n} b_i} = \boxed{20 - 40} = \boxed{\mathbf{-20}}$

c. $\boxed{\displaystyle\sum_{i=1}^{n}\left(-5a_i + 2b_i\right)} = \boxed{\displaystyle\sum_{i=1}^{n} -5a_i + \sum_{i=1}^{n} 2b_i} = \boxed{-5\sum_{i=1}^{n} a_i + 2\sum_{i=1}^{n} b_i} = \boxed{(-5 \times 20) + (2 \times 40)} = \boxed{-100 + 80} = \boxed{\mathbf{-20}}$

d. $\boxed{\sum_{i=1}^{n}\left(\frac{1}{2}a_i - \frac{1}{4}b_i\right)} = \boxed{\sum_{i=1}^{n}\frac{1}{2}a_i - \sum_{i=1}^{n}\frac{1}{4}b_i} = \boxed{\frac{1}{2}\sum_{i=1}^{n}a_i - \frac{1}{4}\sum_{i=1}^{n}b_i} = \boxed{\left(\frac{1}{2}\times 20\right)+\left(-\frac{1}{4}\times 40\right)} = \boxed{10-10} = \boxed{0}$

Example 1.2-2   Solve the following series:

a. Find $\sum_{n=1}^{6} a_n$ where $a_n = \frac{2n+1}{n}$.

b. Find $\sum_{i=1}^{7} x_i$ where $x_i = \left(1+i^2\right)(-2)^i$.

c. Find $\sum_{j=0}^{5}\left(x_j-1\right)^2$ where $x_j = \frac{1}{1+j}$.

d. Find $\sum_{k=0}^{4}\left(u_k\right)^2$ where $u_k = k+1$.

e. Find $\sum_{n=1}^{5}\left(y_n-2\right)^{n+1}$ where $y_n = \frac{n}{1+n}$.

f. Find $\sum_{a=0}^{5}\left(u_a+a\right)^2$ where $u_a = a^2-1$.

Solutions:

a. $\boxed{\sum_{n=1}^{6} a_n \text{ where } a_n = \frac{2n+1}{n}} = \boxed{a_1+a_2+a_3+a_4+a_5+a_6} = \boxed{\frac{2+1}{1}+\frac{4+1}{2}+\frac{6+1}{3}+\frac{8+1}{4}+\frac{10+1}{5}+\frac{12+1}{6}}$

$= \boxed{\frac{3}{1}+\frac{5}{2}+\frac{7}{3}+\frac{9}{4}+\frac{11}{5}+\frac{13}{6}} = \boxed{3+2.5+2.33+2.25+2.2+2.17} = \boxed{\mathbf{14.45}}$

b. $\boxed{\sum_{i=1}^{7} x_i \text{ where } x_i = \left(1+i^2\right)(-2)^i} = \boxed{x_1+x_2+x_3+x_4+x_5+x_6+x_7} = \boxed{\left(1+1^2\right)(-2)^1+\left(1+2^2\right)(-2)^2}$

$\boxed{+\left(1+3^2\right)(-2)^3+\left(1+4^2\right)(-2)^4+\left(1+5^2\right)(-2)^5+\left(1+6^2\right)(-2)^6+\left(1+7^2\right)(-2)^7} = \boxed{(2\cdot-2)+(5\cdot 4)+(10\cdot-8)}$

$\boxed{+(17\cdot 16)+(26\cdot-32)+(37\cdot 64)+(50\cdot-128)} = \boxed{-4+20-80+272-832+2368-6400} = \boxed{\mathbf{-4656}}$

c. $\boxed{\sum_{j=0}^{5}\left(x_j-1\right)^2 \text{ where } x_j = \frac{1}{1+j}} = \boxed{(x_0-1)^2+(x_1-1)^2+(x_2-1)^2+(x_3-1)^2+(x_4-1)^2+(x_5-1)^2}$

$= \boxed{\left(\frac{1}{1}-1\right)^2+\left(\frac{1}{2}-1\right)^2+\left(\frac{1}{3}-1\right)^2+\left(\frac{1}{4}-1\right)^2+\left(\frac{1}{5}-1\right)^2+\left(\frac{1}{6}-1\right)^2} = \boxed{0+\left(-\frac{1}{2}\right)^2+\left(-\frac{2}{3}\right)^2+\left(-\frac{3}{4}\right)^2}$

$\boxed{+\left(\frac{4}{5}\right)^2+\left(-\frac{5}{6}\right)^2} = \boxed{\frac{1}{4}+\frac{4}{9}+\frac{9}{16}+\frac{16}{25}+\frac{25}{36}} = \boxed{0.25+0.4444+0.5625+0.64+0.6944} = \boxed{\mathbf{2.5913}}$

d. $\boxed{\sum_{k=0}^{4}\left(u_k\right)^2 \text{ where } u_k = k+1} = \boxed{(u_0)^2+(u_1)^2+(u_2)^2+(u_3)^2+(u_4)^2} = \boxed{(0+1)^2+(1+1)^2+(2+1)^2}$

$\boxed{+(3+1)^2+(4+1)^2} = \boxed{1^2+2^2+3^2+4^2+5^2} = \boxed{1+4+9+16+25} = \boxed{\mathbf{55}}$

e. $\boxed{\sum_{n=1}^{5}(y_n-2)^{n+1} \text{ where } y_n=\dfrac{n}{1+n}} = \boxed{(y_1-2)^{1+1}+(y_2-2)^{2+1}+(y_3-2)^{3+1}+(y_4-2)^{4+1}+(y_5-2)^{5+1}}$

$= \boxed{\left(\dfrac{1}{1+1}-2\right)^2+\left(\dfrac{2}{1+2}-2\right)^3+\left(\dfrac{3}{1+3}-2\right)^4+\left(\dfrac{4}{1+4}-2\right)^5+\left(\dfrac{5}{1+5}-2\right)^6} = \boxed{(-1.5)^2+(-1.33)^3+(-1.25)^4}$

$\boxed{+(-1.2)^5+(-1.17)^6} = \boxed{2.25-2.35+2.44-2.49+2.56} = \boxed{2.41}$

f. $\boxed{\sum_{a=0}^{5}(u_a+a)^2 \text{ where } u_a=a^2-1} = \boxed{(u_0+0)^2+(u_1+1)^2+(u_2+2)^2+(u_3+3)^2+(u_4+4)^2+(u_5+5)^2}$

$= \boxed{(-1+0)^2+(0+1)^2+(3+2)^2+(8+3)^2+(15+4)^2+(24+5)^2} = \boxed{1+1+25+121+361+841} = \boxed{1350}$

Example 1.2-3 Solve the following series.

a. $\displaystyle\sum_{a=1}^{5}a_n(2a-1) =$

b. $\displaystyle\sum_{a=1}^{5}\dfrac{2a+1}{a} =$

c. $\displaystyle\sum_{i=1}^{5}\dfrac{(-1)^{i+1}}{2i} =$

d. $\displaystyle\sum_{n=0}^{4}(n-1)^2(n+1) =$

e. $\displaystyle\sum_{j=-3}^{3}\dfrac{2^j}{j+5} =$

f. $\displaystyle\sum_{k=1}^{5}\dfrac{(1-k)^{k-1}}{k} =$

Solutions:

a. $\boxed{\sum_{a=1}^{5}a(2a-1)} = \boxed{[1\cdot(2\cdot1-1)]+[2\cdot(2\cdot2-1)]+[3\cdot(2\cdot3-1)]+[4\cdot(2\cdot4-1)]+[5\cdot(2\cdot5-1)]} = \boxed{1+6+15}$

$\boxed{+28+45} = \boxed{95}$

b. $\boxed{\sum_{a=1}^{5}\dfrac{2a+1}{a}} = \boxed{\dfrac{(2\cdot1)+1}{1}+\dfrac{(2\cdot2)+1}{2}+\dfrac{(2\cdot3)+1}{3}+\dfrac{(2\cdot4)+1}{4}+\dfrac{(2\cdot5)+1}{5}} = \boxed{\dfrac{3}{1}+\dfrac{5}{2}+\dfrac{7}{3}+\dfrac{9}{4}+\dfrac{11}{5}} = \boxed{3+2.5}$

$\boxed{+2.33+2.25+2.2} = \boxed{12.28}$

c. $\boxed{\sum_{i=1}^{5}\dfrac{(-1)^{i+1}}{2i}} = \boxed{\dfrac{(-1)^{1+1}}{2\cdot1}+\dfrac{(-1)^{2+1}}{2\cdot2}+\dfrac{(-1)^{3+1}}{2\cdot3}+\dfrac{(-1)^{4+1}}{2\cdot4}+\dfrac{(-1)^{5+1}}{2\cdot5}} = \boxed{\dfrac{(-1)^2}{2}+\dfrac{(-1)^3}{4}+\dfrac{(-1)^4}{6}+\dfrac{(-1)^5}{8}+\dfrac{(-1)^6}{10}}$

$= \boxed{\dfrac{1}{2}-\dfrac{1}{4}+\dfrac{1}{6}-\dfrac{1}{8}+\dfrac{1}{10}} = \boxed{0.5-0.25+0.167-0.125+0.1} = \boxed{0.392}$

d. $\boxed{\sum_{n=0}^{4}(n-1)^2(n+1)} = \boxed{[(0-1)^2(0+1)]+[(1-1)^2(1+1)]+[(2-1)^2(2+1)]+[(3-1)^2(3+1)]+[(4-1)^2(4+1)]}$

$= \boxed{(1\cdot1)+(0\cdot2)+(1\cdot3)+(4\cdot4)+(9\cdot5)} = \boxed{1+0+3+16+45} = \boxed{65}$

e. $\displaystyle\sum_{j=-3}^{3}\frac{2^{j}}{j+5}$ $=$ $\dfrac{2^{-3}}{-3+5}+\dfrac{2^{-2}}{-2+5}+\dfrac{2^{-1}}{-1+5}+\dfrac{2^{0}}{0+5}+\dfrac{2^{1}}{1+5}+\dfrac{2^{2}}{2+5}+\dfrac{2^{3}}{3+5}$ $=$ $\dfrac{2^{-3}}{2}+\dfrac{2^{-2}}{3}+\dfrac{2^{-1}}{4}+\dfrac{2^{0}}{5}+\dfrac{2^{1}}{6}+\dfrac{2^{2}}{7}$

$+\dfrac{2^{3}}{8}$ $=$ $\dfrac{0.125}{2}+\dfrac{0.25}{3}+\dfrac{0.5}{4}+\dfrac{1}{5}+\dfrac{2}{6}+\dfrac{4}{7}+\dfrac{8}{8}$ $=$ $0.0625+0.0833+0.125+0.2+0.3333+0.5714+1$ $=$ **2.3755**

f. $\displaystyle\sum_{k=1}^{5}\frac{(1-k)^{k}}{k}$ $=$ $\dfrac{(1-1)^{1}}{1}+\dfrac{(1-2)^{2}}{2}+\dfrac{(1-3)^{3}}{3}+\dfrac{(1-4)^{4}}{4}+\dfrac{(1-5)^{5}}{5}$ $=$ $\dfrac{0}{1}+\dfrac{1}{2}-\dfrac{8}{3}+\dfrac{81}{4}-\dfrac{1024}{5}$ $=$ $0.5-2.67$

$+20.25-204.8$ $=$ **-186.72**

Example 1.2-4 Prove that both sides of the following series are equal to one another.

a. $\displaystyle\sum_{i=1}^{n}2x_{i}+\sum_{i=1}^{n}4y_{i}=2\sum_{i=1}^{n}(x_{i}+2y_{i})$

b. $\displaystyle\sum_{i=1}^{n}a\,y_{i}^{2}=a\sum_{i=1}^{n}y_{i}^{2}$

c. $\displaystyle\sum_{i=1}^{n}a=n\,a$

d. $\displaystyle\sum_{i=1}^{n}(x_{i}+a)=\sum_{i=1}^{n}x_{i}+na$

Solutions:

a. $\displaystyle\sum_{i=1}^{n}2x_{i}+\sum_{i=1}^{n}4y_{i}$ $=$ $(2x_{1}+2x_{2}+2x_{3}+\cdots+2x_{n})+(4y_{1}+4y_{2}+4y_{3}+\cdots+4y_{n})$ $=$ $(2x_{1}+4y_{1})+(2x_{2}+4y_{2})$

$+(2x_{3}+4y_{3})+\cdots+(2x_{n}+4y_{n})$ $=$ $2(x_{1}+2y_{1})+2(x_{2}+2y_{2})+2(x_{3}+2y_{3})+\cdots+2(x_{n}+2y_{n})$

$=$ $2\displaystyle\sum_{i=1}^{n}(x_{i}+2y_{i})$

b. $\displaystyle\sum_{i=1}^{n}a\,y_{i}^{2}$ $=$ $ay_{1}^{2}+ay_{2}^{2}+ay_{3}^{2}+ay_{4}^{2}+\cdots+ay_{n}^{2}$ $=$ $a\left(y_{1}^{2}+y_{2}^{2}+y_{3}^{2}+y_{4}^{2}+\cdots+y_{n}^{2}\right)$ $=$ $a\displaystyle\sum_{i=1}^{n}y_{i}^{2}$

c. $\displaystyle\sum_{i=1}^{n}a$ $=$ $\underbrace{a+a+a+a+\cdots+a}_{n\ terms}$ $=$ $\boldsymbol{n\,a}$

d. $\displaystyle\sum_{i=1}^{n}(x_{i}+a)$ $=$ $(x_{1}+a)+(x_{2}+a)+(x_{3}+a)+\cdots+(x_{n}+a)$ $=$ $(x_{1}+x_{2}+x_{3}+\cdots+x_{n})+(a+a+a+\cdots+a)$

$=$ $\displaystyle\sum_{i=1}^{n}x_{i}+na$

Example 1.2-5 Use the properties of summation to evaluate the following series.

a. $\displaystyle\sum_{i=1}^{6}2k=$

b. $\displaystyle\sum_{i=1}^{7}(4k-3)=$

c. $\displaystyle\sum_{k=1}^{4}\left(k^{3}-2k\right)=$

d. $\sum_{k=1}^{5}\left(k^2+a\right)=$ e. $\sum_{k=1}^{4}2\cdot\left(-\dfrac{2}{3}\right)^{k+1}=$ f. $\sum_{k=1}^{5}\left(2^k+k\right)=$

Solutions:

a. $\boxed{\sum_{i=1}^{6}2k}=\boxed{2\sum_{i=1}^{6}k}=\boxed{2\cdot6k}=\boxed{12k}$

b. $\boxed{\sum_{i=1}^{7}(4k-3)}=\boxed{\sum_{i=1}^{7}4k+\sum_{i=1}^{7}-3}=\boxed{4\sum_{i=1}^{7}k-\sum_{i=1}^{7}3}=\boxed{4\cdot7k-7\cdot3}=\boxed{28k-21}$

c. $\boxed{\sum_{k=1}^{4}\left(k^3-2k\right)}=\boxed{\sum_{k=1}^{4}k^3+\sum_{k=1}^{4}-2k}=\boxed{\sum_{k=1}^{4}k^3-2\sum_{k=1}^{4}k}=\boxed{\left(1^3+2^3+3^3+4^3\right)-2(1+2+3+4)}=\boxed{1+8+27+64}$

$\boxed{-(2\cdot10)}=\boxed{100-20}=\boxed{80}$

d. $\boxed{\sum_{k=1}^{5}\left(k^2+a\right)}=\boxed{\sum_{k=1}^{5}k^2+\sum_{k=1}^{5}a}=\boxed{\sum_{k=1}^{5}k^2+5a}=\boxed{\left(1^2+2^2+3^2+4^2+5^2\right)+5a}=\boxed{(1+4+9+16+25)+5a}$

$=\boxed{55+5a}=\boxed{5(11+a)}$

e. $\boxed{\sum_{k=1}^{4}2\cdot\left(-\dfrac{2}{3}\right)^{k+1}}=\boxed{2\sum_{k=1}^{4}\left(-\dfrac{2}{3}\right)^{k+1}}=\boxed{2\cdot\left[\left(-\dfrac{2}{3}\right)^2+\left(-\dfrac{2}{3}\right)^3+\left(-\dfrac{2}{3}\right)^4+\left(-\dfrac{2}{3}\right)^5\right]}=\boxed{2\cdot\left(\dfrac{4}{9}-\dfrac{8}{27}+\dfrac{16}{81}-\dfrac{32}{243}\right)}$

$=\boxed{2\cdot(0.4444-0.2963+0.1975-0.1317)}=\boxed{2\cdot0.2139}=\boxed{0.4278}$

f. $\boxed{\sum_{k=1}^{5}\left(2^k+k\right)}=\boxed{\sum_{k=1}^{5}2^k+\sum_{k=1}^{5}k}=\boxed{\left(2^1+2^2+2^3+2^4+2^5\right)+(1+2+3++4+5)}=\boxed{(2+4+8+16+32)+15}$

$=\boxed{62+15}=\boxed{77}$

Section 1.2 Practice Problems - Series

1. Given $\sum_{i=1}^{n}a_i=10$ and $\sum_{i=1}^{n}b_i=25$, find

   a. $\sum_{i=1}^{n}(2a_i+4b_i)=$ b. $\sum_{i=1}^{n}(-a_i+b_i)=$ c. $\sum_{i=1}^{n}(3a_i+5b_i)=$ d. $\sum_{i=1}^{n}\left(\dfrac{1}{2}a_i+\dfrac{1}{5}b_i\right)=$

2. Evaluate each of the following series.

   a. $\sum_{k=1}^{5}y_k\ where\ y_k=2+k$ b. $\sum_{n=0}^{6}x_n\ where\ x_n=\dfrac{1}{(-2)^{n+1}}$ c. $\sum_{n=0}^{4}x_n\ where\ x_n=(-1)^{n+1}$

d. $\displaystyle\sum_{j=-3}^{3} u_j$ where $u_j = j - 3j^2$ e. $\displaystyle\sum_{a=3}^{5} y^a$ where $y = a + 2$ f. $\displaystyle\sum_{i=0}^{5} x_i$ where $x_i = \dfrac{(-1)^{i+1}}{2^i}$

g. $\displaystyle\sum_{k=-2}^{3} y^{k+2}$ where $y = 2k - 3$ h. $\displaystyle\sum_{m=1}^{5} (x_m - 1)^2$ where $x_m = \dfrac{1}{m}$

3. Find the sum of the following series within the specified range.

a. $\displaystyle\sum_{i=-3}^{3} 10^i \;=$ b. $\displaystyle\sum_{n=0}^{6} \dfrac{n-1}{2^n} \;=$ c. $\displaystyle\sum_{a=0}^{4} \dfrac{1}{10^a} \;=$

d. $\displaystyle\sum_{n=1}^{5} \left(n^2 - n\right) \;=$ e. $\displaystyle\sum_{m=0}^{6} (-1)^{m+1} \;=$ f. $\displaystyle\sum_{k=0}^{5} \dfrac{1+(-1)^k}{2^k} \;=$

g. $\displaystyle\sum_{a=1}^{6} 5(a-1) + 3 \;=$ h. $\displaystyle\sum_{k=0}^{5} \left(-\dfrac{1}{3}\right)^{k-1} \;=$ i. $\displaystyle\sum_{j=1}^{5} \left(j - 3j^2\right) \;=$

j. $\displaystyle\sum_{n=1}^{4} \dfrac{n+1}{n} - \sum_{n=1}^{4} \dfrac{n^2}{n+1} \;=$ k. $\displaystyle\sum_{k=1}^{5} 5k^{-1} \;=$ l. $\displaystyle\sum_{i=1}^{4} (-0.1)^{2i-5} \;=$

4. Rewrite the following terms using the sigma notation.

a. $\dfrac{1}{2} + \dfrac{1}{3} + \dfrac{1}{4} + \dfrac{1}{5} + \dfrac{1}{6} + \dfrac{1}{7} =$ b. $\dfrac{1}{2} + \dfrac{2}{3} + \dfrac{3}{4} + \dfrac{4}{5} + \dfrac{5}{6} + \dfrac{6}{7} =$ c. $2 + 4 + 8 + 16 + 32 + 64 =$

d. $1 + \dfrac{1}{2} + \dfrac{1}{3} + \dfrac{1}{4} + \dfrac{1}{5} + \dfrac{1}{6} =$ e. $0 + \dfrac{1}{2} + \dfrac{2}{3} + \dfrac{3}{4} + \dfrac{4}{5} + \dfrac{5}{6} =$ f. $1 - \dfrac{1}{2} + \dfrac{1}{3} - \dfrac{1}{4} + \dfrac{1}{5} - \dfrac{1}{6} =$

1.3   Arithmetic Sequences and Arithmetic Series

An arithmetic sequence is a sequence in which each term, after the first term, is obtained by adding a common number to the preceding term. The common number added to each term can be found by taking the **common difference**, denoted by d, of two successive terms. For example, the sequences $3, 6, 9, 12, 15, \cdots$ and $\frac{1}{2}, 1, \frac{3}{2}, 2, \frac{5}{2}, 3, \frac{7}{2}, \cdots$ are arithmetic sequences because the common difference that is added to each term in order to obtain the next term is $6 - 3 = 3$ and $1 - \frac{1}{2} = \frac{1}{2}$, respectively. Note that the nth term in both examples can easily be stated as $s_n = 3n$ and $s_n = \frac{1}{2}n$. Therefore, the two arithmetic sequences can be written as $3, 6, 9, 12, 15, \cdots, 3n$ and $\frac{1}{2}, 1, \frac{3}{2}, 2, \frac{5}{2}, 3, \frac{7}{2}, \cdots, \frac{1}{2}n$.

To obtain the nth term of an arithmetic sequence, a general form can be developed by letting s_n and d be the nth term and the common difference of an arithmetic sequence. Thus, the first terms can be written as:

$s_1 = a$

$s_2 = s_1 + d$ where a and d are real numbers and n is a positive integer

$s_3 = s_2 + d = (s_1 + d) + d = s_1 + 2d$

$s_4 = s_3 + d = (s_1 + 2d) + d = s_1 + 3d$

$s_5 = s_4 + d = (s_1 + 3d) + d = s_1 + 4d$

$\vdots$

$s_n = s_{n-1} + d = [s_1 + (n-2)d] + d = s_1 + nd - 2d + d = s_1 + nd - d = s_1 + (n-1)d$

$s_{n+1} = s_n + d = [s_1 + (n-1)d] + d = s_1 + nd - d + d = s_1 + nd$

Thus, the nth and $n+1$ term of an arithmetic sequence is equal to

$$s_n = s_1 + (n-1)d \qquad (1)$$

$$s_{n+1} = s_1 + nd \qquad (2)$$

In the following examples the above equations (1) and (2) are used in order to find several terms of arithmetic sequences.

Example 1.3-1   Find the next five terms of the following arithmetic sequences.

a. $s_1 = 5$, $d = 3$ b. $s_1 = -5$, $d = 2$ c. $s_1 = 20$, $d = 0.4$

Solutions:

a. The nth term for an arithmetic sequence is equal to $s_n = s_1 + (n-1)d$. Substituting $s_1 = 5$ and $d = 3$ into the general arithmetic expression for $n = 2, 3, 4, 5, and\ 6$ we obtain

$\boxed{s_2} = \boxed{s_1 + (2-1)d} = \boxed{s_1 + d} = \boxed{5 + 3} = \boxed{\textbf{8}}$

$\boxed{s_3} = \boxed{s_1 + (3-1)d} = \boxed{s_1 + 2d} = \boxed{5 + (2 \times 3)} = \boxed{5 + 6} = \boxed{\textbf{11}}$

$$\boxed{s_4} = \boxed{s_1 + (4-1)d} = \boxed{s_1 + 3d} = \boxed{5 + (3 \times 3)} = \boxed{5+9} = \boxed{\mathbf{14}}$$

$$\boxed{s_5} = \boxed{s_1 + (5-1)d} = \boxed{s_1 + 4d} = \boxed{5 + (4 \times 3)} = \boxed{5+12} = \boxed{\mathbf{17}}$$

$$\boxed{s_6} = \boxed{s_1 + (6-1)d} = \boxed{s_1 + 5d} = \boxed{5 + (5 \times 3)} = \boxed{5+15} = \boxed{\mathbf{20}}$$

Thus, the first six terms of the arithmetic sequence are $(\mathbf{5, 8, 11, 14, 17, 20})$.

b. Substituting $s_1 = -5$ and $d = 2$ into the general arithmetic expression for $n = 2, 3, 4, 5, and\ 6$ we obtain

$$\boxed{s_2} = \boxed{s_1 + (2-1)d} = \boxed{s_1 + d} = \boxed{-5+2} = \boxed{\mathbf{-3}}$$

$$\boxed{s_3} = \boxed{s_1 + (3-1)d} = \boxed{s_1 + 2d} = \boxed{-5 + (2 \times 2)} = \boxed{-5+4} = \boxed{\mathbf{-1}}$$

$$\boxed{s_4} = \boxed{s_1 + (4-1)d} = \boxed{s_1 + 3d} = \boxed{-5 + (3 \times 2)} = \boxed{-5+6} = \boxed{\mathbf{1}}$$

$$\boxed{s_5} = \boxed{s_1 + (5-1)d} = \boxed{s_1 + 4d} = \boxed{-5 + (4 \times 2)} = \boxed{-5+8} = \boxed{\mathbf{3}}$$

$$\boxed{s_6} = \boxed{s_1 + (6-1)d} = \boxed{s_1 + 5d} = \boxed{-5 + (5 \times 2)} = \boxed{-5+10} = \boxed{\mathbf{5}}$$

Thus, the first six terms of the arithmetic sequence are $(\mathbf{-5, -3, -1, 1, 3, 5})$.

c. Substituting $s_1 = 20$ and $d = 0.4$ into the general arithmetic expression for $n = 2, 3, 4, 5, and\ 6$ we obtain

$$\boxed{s_2} = \boxed{s_1 + (2-1)d} = \boxed{s_1 + d} = \boxed{20 + 0.4} = \boxed{\mathbf{20.4}}$$

$$\boxed{s_3} = \boxed{s_1 + (3-1)d} = \boxed{s_1 + 2d} = \boxed{20 + (2 \times 0.4)} = \boxed{20 + 0.8} = \boxed{\mathbf{20.8}}$$

$$\boxed{s_4} = \boxed{s_1 + (4-1)d} = \boxed{s_1 + 3d} = \boxed{20 + (3 \times 0.4)} = \boxed{20 + 1.2} = \boxed{\mathbf{21.2}}$$

$$\boxed{s_5} = \boxed{s_1 + (5-1)d} = \boxed{s_1 + 4d} = \boxed{20 + (4 \times 0.4)} = \boxed{20 + 1.6} = \boxed{\mathbf{21.6}}$$

$$\boxed{s_6} = \boxed{s_1 + (6-1)d} = \boxed{s_1 + 5d} = \boxed{20 + (5 \times 0.4)} = \boxed{20 + 2} = \boxed{\mathbf{22}}$$

Thus, the first six terms of the arithmetic sequence are $(\mathbf{20, 20.4, 20.8, 21.2, 21.6, 22})$.

Example 1.3-2   Find the general term and the fiftieth term of the following arithmetic sequences.

a. $s_1 = 3$, $d = 5$ b. $s_1 = -2$, $d = 4$ c. $s_1 = 10$, $d = -2.5$

Solutions:

a. The nth term for an arithmetic sequence is equal to $s_n = s_1 + (n-1)d$. Substituting $s_1 = 3$ and

$d = 5$ into the general arithmetic expression we obtain

$$\boxed{s_n} = \boxed{3 + (n-1)5} = \boxed{3 + 5n - 5} = \boxed{5n + (3-5)} = \boxed{\mathbf{5n-2}}$$

substituting $n = 50$ into the general equation $s_n = 5n - 2$ we find

$$\boxed{s_{50}} = \boxed{(5 \times 50) - 2} = \boxed{250 - 2} = \boxed{\mathbf{248}}$$

b. Substituting $s_1 = -2$ and $d = 4$ into the general arithmetic expression $s_n = s_1 + (n-1)d$ we obtain

$$\boxed{s_n} = \boxed{-2 + (n-1)4} = \boxed{-2 + 4n - 4} = \boxed{4n + (-2 - 4)} = \boxed{\mathbf{4n - 6}}$$

substituting $n = 50$ into the general equation $s_n = 4n - 6$ we find

$$\boxed{s_{50}} = \boxed{(4 \times 50) - 6} = \boxed{200 - 6} = \boxed{\mathbf{194}}$$

c. Substituting $s_1 = 10$ and $d = -2.5$ into the general arithmetic expression $s_n = s_1 + (n-1)d$ we

obtain $\boxed{s_n} = \boxed{10 + (n-1) \times -2.5} = \boxed{10 - 2.5n + 2.5} = \boxed{-2.5n + (10 + 2.5)} = \boxed{\mathbf{-2.5n + 12.5}}$

substituting $n = 50$ into the general equation $s_n = -2.5n + 12.5$ we find

$$\boxed{s_{50}} = \boxed{(-2.5 \times 50) + 12.5} = \boxed{-125 + 12.5} = \boxed{\mathbf{-112.5}}$$

Example 1.3-3 Find the next four terms in each of the following arithmetic sequences.

a. $6, 10, \cdots$ b. $x, x+2, \cdots$ c. $2x+1, 2x+5, \cdots$ d. $x, x-29, \cdots$

Solutions:

a. The first term s_1 and the common difference d are equal to $s_1 = 6$ and $d = 10 - 6 = 4$. Thus,

using the general arithmetic equation $s_n = s_1 + (n-1)d$ or $s_{n+1} = s_n + d$ the next four terms are as follows: Let's use $s_{n+1} = s_n + d$. Then,

$$\boxed{s_3} = \boxed{s_2 + d} = \boxed{10 + 4} = \boxed{\mathbf{14}} \qquad \boxed{s_4} = \boxed{s_3 + d} = \boxed{14 + 4} = \boxed{\mathbf{18}}$$

$$\boxed{s_5} = \boxed{s_4 + d} = \boxed{18 + 4} = \boxed{\mathbf{22}} \qquad \boxed{s_6} = \boxed{s_5 + d} = \boxed{22 + 4} = \boxed{\mathbf{26}}$$

b. The first term s_1 and the common difference d are equal to $s_1 = x$ and $d = x+2-x = 2$. Thus,

$$\boxed{s_3} = \boxed{s_2 + d} = \boxed{(x+2)+2} = \boxed{\mathbf{x+4}} \qquad \boxed{s_4} = \boxed{s_3 + d} = \boxed{(x+4)+2} = \boxed{\mathbf{x+6}}$$

$$\boxed{s_5} = \boxed{s_4 + d} = \boxed{(x+6)+2} = \boxed{\mathbf{x+8}} \qquad \boxed{s_6} = \boxed{s_5 + d} = \boxed{(x+8)+2} = \boxed{\mathbf{x+10}}$$

c. The first term s_1 and the common difference d are equal to $s_1 = 2x+1$ and $d = (2x+5)-(2x+1)$. $= 2x+5-2x-1 = 4$. Thus,

$$\boxed{s_3} = \boxed{s_2 + d} = \boxed{(2x+1)+4} = \boxed{\mathbf{2x+5}} \qquad \boxed{s_4} = \boxed{s_3 + d} = \boxed{(2x+5)+4} = \boxed{\mathbf{2x+9}}$$

$$\boxed{s_5} = \boxed{s_4 + d} = \boxed{(2x+9)+4} = \boxed{\mathbf{2x+13}} \qquad \boxed{s_6} = \boxed{s_5 + d} = \boxed{(2x+13)+4} = \boxed{\mathbf{2x+17}}$$

d. The first term s_1 and the common difference d are equal to $s_1 = x$ and $d = (x-29)-x = -29$. Thus,

$$\boxed{s_3} = \boxed{s_2 + d} = \boxed{(x-29)-29} = \boxed{\mathbf{x-58}} \qquad \boxed{s_4} = \boxed{s_3 + d} = \boxed{(x-58)-29} = \boxed{\mathbf{x-87}}$$

$$\boxed{s_5} = \boxed{s_4 + d} = \boxed{(x-87)-29} = \boxed{\boldsymbol{x-116}} \qquad\qquad \boxed{s_6} = \boxed{s_5 + d} = \boxed{(x-116)-29} = \boxed{\boldsymbol{x-145}}$$

Example 1.3-4 The first term of an arithmetic sequence is -5 and the fourth term is 10. Find the twentieth term.

Solution:

Since $s_1 = -5$ and $s_4 = 10$ we use the general formula $s_n = s_1 + (n-1)d$ in order to solve for d.

$\boxed{s_4 = s_1 + (4-1)d}$; $\boxed{10 = -5 + (4-1)d}$; $\boxed{10 = -5 + 3d}$; $\boxed{10+5 = 3d}$; $\boxed{15 = 3d}$; $\boxed{d = \dfrac{15}{3}}$; $\boxed{d = 5}$. Then,

$\boxed{s_{20} = s_1 + (20-1)d}$; $\boxed{s_{20} = -5 + 19d}$; $\boxed{s_{20} = -5 + (19 \times 5)}$; $\boxed{s_{20} = -5 + 95}$; $\boxed{s_{20} = 90}$

Having learned about arithmetic sequences and the steps for finding the terms of an arithmetic sequence, we will next learn about arithmetic series and the steps for finding the sum of arithmetic series over a given range.

Addition of the terms in an arithmetic sequence result in having an arithmetic series. To obtain the arithmetic series formula let $s_k = s_1 + (k-1)d$ be an arithmetic sequence and denote the sum of the first n terms by

$$S_n = \sum_{k=1}^{n} s_1 + (k-1)d$$

then,

$$S_n = s_1 + (s_1 + d) + \cdots + [s_1 + (n-2)d] + [s_1 + (n-1)d] \qquad\qquad (a)$$

Let's write the sum in reverse order and add the two series (a) and (b) together.

$$S_n = [s_1 + (n-1)d] + [s_1 + (n-2)d] + \cdots + (s_1 + d) + s_1 \qquad\qquad (b)$$

$$S_n + S_n = \{ s_1 + [s_1 + (n-1)d] \} + \{ (s_1 + d) + [s_1 + (n-2)d] \} + \cdots + \{ [s_1 + (n-2)d] + (s_1 + d) \} + \{ [s_1 + (n-1)d] + s_1 \}$$

$$2S_n = [s_1 + s_1 + (n-1)d] + [s_1 + d + s_1 + (n-2)d] + \cdots + [s_1 + (n-2)d + s_1 + d] + [s_1 + (n-1)d + s_1]$$

$$2S_n = [2s_1 + (n-1)d] + [s_1 + s_1 + (n-2)d + d] + \cdots + [s_1 + s_1 + (n-2)d + d] + [s_1 + s_1 + (n-1)d]$$

$$2S_n = [2s_1 + (n-1)d] + [2s_1 + nd - 2d + d] + \cdots + [2s_1 + nd - 2d + d] + [2s_1 + (n-1)d]$$

$$2S_n = [2s_1 + (n-1)d] + [2s_1 + nd - d] + \cdots + [2s_1 + nd - d] + [2s_1 + (n-1)d]$$

$$2S_n = [2s_1 + (n-1)d] + [2s_1 + (n-1)d] + \cdots + [2s_1 + (n-1)d] + [2s_1 + (n-1)d]$$

$$2S_n = n[2s_1 + (n-1)d] \; ; \; S_n = \frac{n[2s_1 + (n-1)d]}{2} = \frac{n}{2}[2s_1 + (n-1)d]$$

Therefore, the arithmetic series can be written in the following two forms:

$$S_n = \sum_{k=1}^{n} s_1 + (k-1)d \qquad\qquad (1)$$

$$S_n = \frac{n}{2}[2s_1 + (n-1)d] \qquad\qquad (2)$$

Note that equation (2), similar to the nth term of the arithmetic sequence $\left[s_n = s_1 + (n-1)d\right]$, is given in terms of s_1, n, and d.

In the following examples the above equations (1) and (2) are used in order to find the sum of arithmetic series.

Example 1.3-5 Find the sum of the following arithmetic series.

a. $\displaystyle\sum_{i=1}^{20}(2i+1) =$ b. $\displaystyle\sum_{i=1}^{15}(3i-2) =$ c. $\displaystyle\sum_{j=3}^{15}(5j-1) =$

Solutions:

a. **First** - Write the first three terms of the arithmetic series in expanded form, i.e.,

$$\boxed{\sum_{i=1}^{20}(2i+1)} = \boxed{(2+1)+(4+1)+(6+1)+\cdots} = \boxed{3+5+7+\cdots}$$

Second - Identify the first term, s_1, the difference between the two terms, d, and n, i.e., $s_1 = 3$, $d = 5-3 = 2$, and $n = 20$.

Third - Use the arithmetic series formula to obtain the sum of the twenty terms.

$$S_n = \frac{n}{2}\left[2s_1 + (n-1)d\right]$$

$$\boxed{S_{20}} = \boxed{\frac{20}{2}\left[2s_1+(20-1)d\right]} = \boxed{10\left[2s_1+19d\right]} = \boxed{10\left[(2\times3)+(19\times2)\right]} = \boxed{10\left[6+38\right]} = \boxed{10\times44} = \boxed{440}$$

Note that prior to learning the arithmetic series formula the only method that we could use was by summing each term as shown below:

$$\boxed{\sum_{i=1}^{20}(2i+1)} = \boxed{(2+1)+(4+1)+(6+1)+(8+1)+(10+1)+(12+1)+(14+1)+(16+1)+(18+1)+(20+1)}$$

$$\boxed{+(22+1)+(24+1)+(26+1)+(28+1)+(30+1)+(32+1)+(34+1)+(36+1)+(38+1)+(40+1)}$$

$$= \boxed{3+5+7+9+11+13+15+17+19+21+23+25+27+29+31+33+35+37+39+41} = \boxed{440}$$

As you note, it is much easier to use the arithmetic series formula as opposed to the summation of each term which is fairly long and time consuming.

b. **First** - Write the first three terms of the arithmetic series in expanded form, i.e.,

$$\boxed{\sum_{i=1}^{15}(3i-2)} = \boxed{(3-2)+(6-2)+(9-2)+\cdots} = \boxed{1+4+7+\cdots}$$

Second - Identify the first term, s_1, the difference between the two terms, d, and n, i.e., $s_1 = 1$, $d = 4-1 = 3$, and $n = 15$.

Third - Use the arithmetic series formula to obtain the sum of the fifteen terms.

$$S_n = \frac{n}{2}\left[2s_1 + (n-1)d\right]$$

$$S_{15} = \boxed{\dfrac{15}{2}\big[2s_1 + (15-1)d\big]} = \boxed{7.5\big[2s_1 + 14d\big]} = \boxed{7.5\big[(2\times1)+(14\times3)\big]} = \boxed{7.5\big[2+42\big]} = \boxed{7.5\times44} = \boxed{330}$$

or, we can obtain the answer by summing up the first fifteen terms of the series, i.e.,

$$\boxed{\sum_{i=1}^{15}(3i-2)} = \boxed{1+4+7+10+13+16+19+22+25+28+31+34+37+40+43} = \boxed{330}$$

c. **First** - Write the first three terms of the arithmetic series in expanded form, i.e.,

$$\boxed{\sum_{j=3}^{15}(5j-1)} = \boxed{(15-1)+(20-1)+(25-1)+\cdots} = \boxed{14+19+24+\cdots}$$

Second - Identify the first term, s_1, the difference between the two terms, d, and n, i.e.,
$s_1 = 14$, $d = 19-14 = 5$, and $n = 13$.

Third - Use the arithmetic series formula to obtain the sum of the thirteen terms.

$$S_n = \frac{n}{2}\big[2s_1 + (n-1)d\big]$$

$$S_{15} = \boxed{\dfrac{13}{2}\big[2s_1 + (13-1)d\big]} = \boxed{6.5\big[2s_1 + 12d\big]} = \boxed{6.5\big[(2\times14)+(12\times5)\big]} = \boxed{6.5\big[28+60\big]} = \boxed{6.5\times88} = \boxed{572}$$

or, we can obtain the answer by summing up the first thirteen terms of the series, i.e.,

$$\boxed{\sum_{j=3}^{15}(5j-1)} = \boxed{14+19+24+29+34+39+44+49+54+59+64+69+74} = \boxed{572}$$

Example 1.3-6   Given the first term s_1 and d, find S_{80} for each of the following arithmetic sequences.

a. $s_1 = 5$, $d = 2$ b. $s_1 = -10$, $d = 3$ c. $s_1 = 500$, $d = 25$

Solutions:

a. The nth term for an arithmetic series is equal to $S_n = \frac{n}{2}\big[2s_1 + (n-1)d\big]$. Substituting $s_1 = 5$ and $d = 2$ into the general arithmetic expression we obtain

$$\boxed{S_{80}} = \boxed{\dfrac{80}{2}\big[2s_1 + (80-1)d\big]} = \boxed{40\big[2s_1 + 79d\big]} = \boxed{40\big[(2\times5)+(79\times2)\big]} = \boxed{40\big[10+158\big]} = \boxed{40\times168} = \boxed{6720}$$

b. Substituting $s_1 = -10$ and $d = 3$ into $S_n = \frac{n}{2}\big[2s_1 + (n-1)d\big]$ we obtain

$$\boxed{S_{80}} = \boxed{\dfrac{80}{2}\big[2s_1 + (80-1)d\big]} = \boxed{40\big[2s_1 + 79d\big]} = \boxed{40\big[(2\times-10)+(79\times3)\big]} = \boxed{40\big[-20+237\big]} = \boxed{8680}$$

c. Substituting $s_1 = 500$ and $d = 25$ into $s_n = \frac{n}{2}\big[2s_1 + (n-1)d\big]$ we obtain

$$\boxed{S_{80}} = \boxed{\dfrac{80}{2}\big[2s_1 + (80-1)d\big]} = \boxed{40\big[2s_1 + 79d\big]} = \boxed{40\big[(2\times500)+(79\times25)\big]} = \boxed{40\big[1000+1975\big]} = \boxed{119000}$$

Example 1.3-7 Find the sum of the following sequences for the indicated values.

a. S_{35} for the sequence $-5, 3, \cdots$ b. S_{200} for the sequence $-10, 10, \cdots$

Solutions:

a. The first term s_1 and the common difference d are equal to $s_1 = -5$ and $d = 3 - (-5) = 3 + 5 = 8$.

Thus, using the general arithmetic series $S_n = \dfrac{n}{2} [2s_1 + (n-1)d]$ we obtain

$$\boxed{S_{35}} = \boxed{\dfrac{35}{2} [2s_1 + (35-1)d]} = \boxed{17.5 [2s_1 + 34d]} = \boxed{17.5 [(2 \times -5) + (34 \times 8)]} = \boxed{17.5 [-10 + 272]} = \boxed{4585}$$

b. The first term s_1 and the common difference d are equal to $s_1 = -10$ and $d = 10 - (-10) = 20$.

Thus, using the general arithmetic series $S_n = \dfrac{n}{2} [2s_1 + (n-1)d]$ we obtain

$$\boxed{S_{200}} = \boxed{\dfrac{200}{2} [2s_1 + (200-1)d]} = \boxed{100 [2s_1 + 199d]} = \boxed{100 [(2 \times -10) + (199 \times 20)]} = \boxed{100 [-20 + 3980]}$$

$$= \boxed{396000}$$

Section 1.3 Practice Problems - Arithmetic Sequences and Arithmetic Series

1. Find the next seven terms of the following arithmetic sequences.

    a. $s_1 = 3$, $d = 2$ b. $s_1 = -3$, $d = 2$ c. $s_1 = 10$, $d = 0.8$

2. find the general term and the eighth term of the following arithmetic sequences.

    a. $s_1 = 3$, $d = 4$ b. $s_1 = -3$, $d = 5$ c. $s_1 = 8$, $d = -1.2$

3. find the next six terms in each of the following arithmetic sequences.

    a. $5, 8, \cdots$ b. $x, x+4, \cdots$ c. $3x+1, 3x+4, \cdots$ d. $w, w-10, \cdots$

4. Find the sum of the following arithmetic series.

    a. $\displaystyle\sum_{i=10}^{20} (2i - 4) =$ b. $\displaystyle\sum_{k=1}^{1000} k =$ c. $\displaystyle\sum_{k=1}^{100} (2k - 3) =$

    d. $\displaystyle\sum_{i=1}^{15} 3i =$ e. $\displaystyle\sum_{i=1}^{10} (i+1) =$ f. $\displaystyle\sum_{k=5}^{15} (2k - 1) =$

    g. $\displaystyle\sum_{i=4}^{10} (3i + 4) =$ h. $\displaystyle\sum_{j=5}^{13} (3j + 1) =$ i. $\displaystyle\sum_{k=7}^{18} (4k - 3) =$

5. The first term of an arithmetic sequence is 6 and the third term is 24. Find the tenth term.

6. Given the first term s_1 and d, find S_{50} for each of the following arithmetic sequences.

    a. $s_1 = 2$, $d = 5$ b. $s_1 = -5$, $d = 6$ c. $s_1 = 30$, $d = 10$

7. Find the sum of the following sequences for the indicated values.

    a. S_{15} for the sequence $-8, 6, \cdots$ b. S_{100} for the sequence $-20, 20, \cdots$

1.4   Geometric Sequences and Geometric Series

A geometric sequence is a sequence in which each term, after the first term, is obtained by multiplying the preceding term by a common multiplier. This common multiplier is also called the **common ratio** and is denoted by r. The common ratio r is obtained by division of two successive terms in a sequence. For example, the sequences $3, 6, 12, 24, 48, \cdots$ and $\dfrac{1}{2}, \dfrac{1}{4}, \dfrac{1}{8}, \dfrac{1}{16}, \dfrac{1}{32}, \cdots$ are geometric sequences because the common ratio that each term is multiplied by in order to obtain the next term is equal to $\dfrac{6}{3} = 2$ and $\dfrac{\frac{1}{4}}{\frac{1}{2}} = \dfrac{1 \times 2}{4 \times 1} = \dfrac{2}{4} = \dfrac{1}{2}$, respectively.

To obtain the nth term of a geometric sequence, a general form can be developed by letting s_n and r be the nth term and the common multiplier (common ratio) of a geometric sequence. Thus, the first terms can be written as:

$s_1 = a$ where a is a real number and n is a positive integer

$s_2 = s_1 r$

$s_3 = s_2 r = (s_1 r) \cdot r = s_1 r^2$

$s_4 = s_3 r = \left(s_1 r^2\right) \cdot r = s_1 r^3$

$s_5 = s_4 r = \left(s_1 r^3\right) \cdot r = s_1 r^4$

$\vdots$

$s_n = s_{n-1} r = \left(s_1 r^{n-2}\right) \cdot r = s_1 r^{n-2+1} = s_1 r^{n-1}$

$s_{n+1} = s_n r = \left(s_1 r^{n-1}\right) \cdot r = s_1 r^{n-1+1} = s_1 r^n$

Thus, the nth and $n+1$ term of an arithmetic sequence is equal to

$$s_n = s_1 r^{n-1} \qquad (1)$$

$$s_{n+1} = s_1 r^n \qquad (2)$$

In the following examples the above equations (1) and (2) are used in order to find several terms of geometric sequences.

Example 1.4-1   Find the next five terms of the following geometric sequences.

a. $s_1 = 5$, $r = 2$ b. $s_1 = -3$, $r = 3$ c. $s_1 = 10$, $r = 0.5$

Solutions:

a. The nth term for an geometric sequence is equal to $s_n = s_1 r^{n-1}$. Substituting $s_1 = 5$ and $r = 2$ into the general geometric expression for $n = 2, 3, 4, 5,\, and\, 6$ we obtain

$\boxed{s_2} = \boxed{s_1 r^{2-1}} = \boxed{s_1 r} = \boxed{5 \times 2} = \boxed{10}$

$\boxed{s_3} = \boxed{s_1 r^{3-1}} = \boxed{s_1 r^2} = \boxed{5 \times 2^2} = \boxed{5 \times 4} = \boxed{20}$

$\boxed{s_4} = \boxed{s_1 r^{4-1}} = \boxed{s_1 r^3} = \boxed{5 \times 2^3} = \boxed{5 \times 8} = \boxed{40}$

$$\boxed{s_5} = \boxed{s_1 r^{5-1}} = \boxed{s_1 r^4} = \boxed{5 \times 2^4} = \boxed{5 \times 16} = \boxed{\mathbf{80}}$$

$$\boxed{s_6} = \boxed{s_1 r^{6-1}} = \boxed{s_1 r^5} = \boxed{5 \times 2^5} = \boxed{5 \times 32} = \boxed{\mathbf{160}}$$

Thus, the first six terms of the geometric sequence are $(\mathbf{5, 10, 20, 40, 80, 160})$.

b. Substituting $s_1 = -3$ and $r = 3$ into the general geometric expression for $n = 2, 3, 4, 5, and\ 6$ we obtain

$$\boxed{s_2} = \boxed{s_1 r^{2-1}} = \boxed{s_1 r} = \boxed{-3 \times 3} = \boxed{\mathbf{-9}}$$

$$\boxed{s_3} = \boxed{s_1 r^{3-1}} = \boxed{s_1 r^2} = \boxed{-3 \times 3^2} = \boxed{-3 \times 9} = \boxed{\mathbf{-27}}$$

$$\boxed{s_4} = \boxed{s_1 r^{4-1}} = \boxed{s_1 r^3} = \boxed{-3 \times 3^3} = \boxed{-3 \times 27} = \boxed{\mathbf{-81}}$$

$$\boxed{s_5} = \boxed{s_1 r^{5-1}} = \boxed{s_1 r^4} = \boxed{-3 \times 3^4} = \boxed{-3 \times 81} = \boxed{\mathbf{-243}}$$

$$\boxed{s_6} = \boxed{s_1 r^{6-1}} = \boxed{s_1 r^5} = \boxed{-3 \times 3^5} = \boxed{-3 \times 243} = \boxed{\mathbf{-729}}$$

Thus, the first six terms of the geometric sequence are $(\mathbf{-3, -9, -27, -81, -243, -729})$.

c. Substituting $s_1 = 10$ and $r = 0.5$ into the general geometric expression for $n = 2, 3, 4, 5, and\ 6$ we obtain

$$\boxed{s_2} = \boxed{s_1 r^{2-1}} = \boxed{s_1 r} = \boxed{10 \times 0.5} = \boxed{\mathbf{5}}$$

$$\boxed{s_3} = \boxed{s_1 r^{3-1}} = \boxed{s_1 r^2} = \boxed{10 \times 0.5^2} = \boxed{10 \times 0.25} = \boxed{\mathbf{2.5}}$$

$$\boxed{s_4} = \boxed{s_1 r^{4-1}} = \boxed{s_1 r^3} = \boxed{10 \times 0.5^3} = \boxed{10 \times 0.125} = \boxed{\mathbf{1.25}}$$

$$\boxed{s_5} = \boxed{s_1 r^{5-1}} = \boxed{s_1 r^4} = \boxed{10 \times 0.5^4} = \boxed{10 \times 0.0625} = \boxed{\mathbf{0.625}}$$

$$\boxed{s_6} = \boxed{s_1 r^{6-1}} = \boxed{s_1 r^5} = \boxed{10 \times 0.5^5} = \boxed{10 \times 0.03125} = \boxed{\mathbf{0.3125}}$$

Thus, the first six terms of the geometric sequence are $(\mathbf{10, 5, 2.5, 1.25, 0.625, 0.3125})$.

Example 1.4-2   find the general term and the tenth term of the following geometric sequences.

a. $s_1 = 3$, $r = 1.2$ b. $s_1 = -2$, $r = 0.8$ c. $s_1 = 10$, $r = -0.5$

Solutions:

a. The nth term for a geometric sequence is equal to $s_n = s_1 r^{n-1}$. Substituting $s_1 = 3$ and $r = 1.2$ into the general geometric expression we obtain

$$\boxed{s_n} = \boxed{3 \times r^{n-1}} = \boxed{\mathbf{3 \times 1.2^{n-1}}}$$

substituting $n = 10$ into the general equation $s_n = 3 \times 1.2^{n-1}$ we have

$$\boxed{s_{10}} = \boxed{3 \times 1.2^{10-1}} = \boxed{3 \times 1.2^9} = \boxed{3 \times 5.1598} = \boxed{\mathbf{15.479}}$$

b. Substituting $s_1 = -2$ and $r = 0.8$ into the general geometric expression $s_n = s_1 r^{n-1}$ we obtain

$$\boxed{s_n} = \boxed{-2 \times r^{n-1}} = \boxed{-2 \times 0.8^{n-1}}$$

substituting $n = 10$ into the general equation $s_n = -2 \times 0.8^{n-1}$ we have

$$\boxed{s_{10}} = \boxed{-2 \times 0.8^{10-1}} = \boxed{-2 \times 0.8^{9}} = \boxed{-2 \times 0.1342} = \boxed{-0.2684}$$

c. Substituting $s_1 = 10$ and $r = -0.5$ into the general geometric expression $s_n = s_1 r^{n-1}$ we obtain

$$\boxed{s_n} = \boxed{10 \times r^{n-1}} = \boxed{10 \times (-0.5)^{n-1}}$$

substituting $n = 10$ into the general equation $s_n = 10 \times (-0.5)^{n-1}$ we have

$$\boxed{s_{10}} = \boxed{10 \times (-0.5)^{10-1}} = \boxed{10 \times (-0.5)^{9}} = \boxed{10 \times (-0.0019)} = \boxed{-0.019}$$

Example 1.4-3 Find the next four terms and the nth term in each of the following geometric sequences.

a. $1, \dfrac{1}{2}, \cdots$

b. $-\dfrac{1}{3}, \dfrac{1}{9}, \cdots$

c. $\dfrac{1}{2}x, -\dfrac{1}{4}x, \cdots$

Solutions:

a. The first term s_1 and the common ratio r are equal to $s_1 = 1$ and $r = \dfrac{\frac{1}{2}}{1} = \dfrac{\frac{1}{2}}{\frac{1}{1}} = \dfrac{1 \times 1}{2 \times 1} = \dfrac{1}{2}$. Thus,

using the general geometric equation $s_n = s_1 r^{n-1}$ the next four terms are:

$$\boxed{s_3} = \boxed{s_1 r^2} = \boxed{1 \cdot \left(\dfrac{1}{2}\right)^2} = \boxed{\dfrac{1}{2^2}} = \boxed{\dfrac{1}{4}} \qquad \boxed{s_4} = \boxed{s_1 r^3} = \boxed{1 \cdot \left(\dfrac{1}{2}\right)^3} = \boxed{\dfrac{1}{2^3}} = \boxed{\dfrac{1}{8}}$$

$$\boxed{s_5} = \boxed{s_1 r^4} = \boxed{1 \cdot \left(\dfrac{1}{2}\right)^4} = \boxed{\dfrac{1}{2^4}} = \boxed{\dfrac{1}{16}} \qquad \boxed{s_6} = \boxed{s_1 r^5} = \boxed{1 \cdot \left(\dfrac{1}{2}\right)^5} = \boxed{\dfrac{1}{2^5}} = \boxed{\dfrac{1}{32}}$$

Thus, the first six terms of the geometric sequence are $\left(1, \dfrac{1}{2}, \dfrac{1}{4}, \dfrac{1}{8}, \dfrac{1}{16}, \dfrac{1}{32}\right)$ and the nth term is

equal to $\boxed{s_n} = \boxed{1 \cdot \left(\dfrac{1}{2}\right)^{n-1}} = \boxed{\dfrac{1^{n-1}}{2^{n-1}}} = \boxed{\dfrac{1}{2^{n-1}}}$

b. The first term s_1 and the common ratio r are equal to $s_1 = -\dfrac{1}{3}$ and $r = \dfrac{\frac{1}{9}}{-\frac{1}{3}} = -\dfrac{1 \times 3}{9 \times 1} = -\dfrac{1}{3}$. Thus,

using the general geometric equation $s_n = s_1 r^{n-1}$ the next four terms are:

$$\boxed{s_3} = \boxed{s_1 r^2} = \boxed{-\dfrac{1}{3} \cdot \left(-\dfrac{1}{3}\right)^2} = \boxed{-\dfrac{1}{3^3}} = \boxed{-\dfrac{1}{27}} \qquad \boxed{s_4} = \boxed{s_1 r^3} = \boxed{-\dfrac{1}{3} \cdot \left(-\dfrac{1}{3}\right)^3} = \boxed{\dfrac{1}{3^4}} = \boxed{\dfrac{1}{81}}$$

$$\boxed{s_5} = \boxed{s_1 r^4} = \boxed{-\dfrac{1}{3} \cdot \left(-\dfrac{1}{3}\right)^4} = \boxed{-\dfrac{1}{3^5}} = \boxed{-\dfrac{1}{243}} \qquad \boxed{s_6} = \boxed{s_1 r^5} = \boxed{-\dfrac{1}{3} \cdot \left(-\dfrac{1}{3}\right)^5} = \boxed{\dfrac{1}{3^6}} = \boxed{\dfrac{1}{729}}$$

Thus, the first six terms of the geometric sequence are $\left(-\dfrac{1}{3}, \dfrac{1}{9}, -\dfrac{1}{27}, \dfrac{1}{81}, -\dfrac{1}{243}, \dfrac{1}{729}\right)$ and the

n^{th} term is equal to $\boxed{s_n} = \boxed{-\frac{1}{3} \cdot \left(-\frac{1}{3}\right)^{n-1}} = \boxed{-\frac{(-1)^{n-1}}{3 \cdot 3^{n-1}}} = \boxed{-\frac{(-1)^{n-1}}{3^n}}$

c. The first term s_1 and the common ratio r are equal to $s_1 = \frac{1}{2}x$ and $r = -\frac{1}{2}$. Thus, using the general geometric equation $s_n = s_1 r^{n-1}$ the next four terms are:

$\boxed{s_3} = \boxed{s_1 r^2} = \boxed{\frac{1}{2}x \cdot \left(-\frac{1}{2}\right)^2} = \boxed{\frac{1}{2^3}x} = \boxed{\frac{1}{8}x}$ $\boxed{s_4} = \boxed{s_1 r^3} = \boxed{\frac{1}{2}x \cdot \left(-\frac{1}{2}\right)^3} = \boxed{-\frac{1}{2^4}x} = \boxed{-\frac{1}{16}x}$

$\boxed{s_5} = \boxed{s_1 r^4} = \boxed{\frac{1}{2}x \cdot \left(-\frac{1}{2}\right)^4} = \boxed{\frac{1}{2^5}x} = \boxed{\frac{1}{32}x}$ $\boxed{s_6} = \boxed{s_1 r^5} = \boxed{\frac{1}{2}x \cdot \left(-\frac{1}{2}\right)^5} = \boxed{-\frac{1}{2^6}x} = \boxed{-\frac{1}{64}x}$

Thus, the first six terms of the geometric sequence are $\left(\frac{1}{2}x, -\frac{1}{4}x, \frac{1}{8}x, -\frac{1}{16}x, \frac{1}{32}x, -\frac{1}{64}x\right)$

and the nth term is equal to $\boxed{s_n} = \boxed{\frac{1}{2}x \cdot \left(-\frac{1}{2}\right)^{n-1}} = \boxed{\frac{x(-1)^{n-1}}{2 \cdot 2^{n-1}}} = \boxed{\frac{x(-1)^{n-1}}{2^{n-1+1}}} = \boxed{\frac{x(-1)^{n-1}}{2^n}}$

Example 1.4-4   Given the following terms of a geometric sequence, find the common ratio r.

a. $s_1 = 32$ and $s_7 = \frac{1}{2}$ b. $s_1 = 3$ and $s_5 = \frac{1}{27}$ c. $s_1 = 5$ and $s_8 = 1$

Solutions:

a. Substitute $s_1 = 32$ and $s_7 = \frac{1}{2}$ into $s_n = s_1 r^{n-1}$ and solve for r.

$\boxed{s_7 = s_1 r^{7-1}}$; $\boxed{\frac{1}{2} = 32 r^6}$; $\boxed{\frac{\frac{1}{2}}{32} = r^6}$; $\boxed{r^6 = \frac{\frac{1}{2}}{2^5}}$; $\boxed{r^6 = \frac{\frac{1}{2}}{\frac{2^5}{1}}}$; $\boxed{r^6 = \frac{1 \times 1}{2 \times 2^5}}$; $\boxed{r^6 = \frac{1}{2^6}}$; $\boxed{r = \frac{1}{2}}$

b. Substitute $s_1 = 3$ and $s_5 = \frac{1}{27}$ into $s_n = s_1 r^{n-1}$ and solve for r.

$\boxed{s_5 = s_1 r^{5-1}}$; $\boxed{\frac{1}{27} = 3 r^4}$; $\boxed{\frac{\frac{1}{27}}{3} = r^4}$; $\boxed{r^4 = \frac{\frac{1}{3^3}}{3}}$; $\boxed{r^4 = \frac{\frac{1}{3^3}}{\frac{3}{1}}}$; $\boxed{r^4 = \frac{1 \times 1}{3^3 \times 3}}$; $\boxed{r^4 = \frac{1}{3^4}}$; $\boxed{r = \frac{1}{3}}$

c. Substitute $s_1 = 5$ and $s_8 = 1$ into $s_n = s_1 r^{n-1}$ and solve for r.

$\boxed{s_8 = s_1 r^{8-1}}$; $\boxed{1 = 5 r^7}$; $\boxed{\frac{1}{5} = r^7}$; $\boxed{r^7 = \frac{1}{5}}$; $\boxed{r = \frac{1}{\sqrt[7]{5}}}$

Example 1.4-5   Write the first six terms and the nth term of the following geometric sequences.

a. $s_n = \left(\frac{1}{3}\right)^{n-1}$ b. $s_n = \left(\frac{1}{2}\right)^{n+2}$ c. $s_n = \left(-\frac{1}{3}\right)^{2n}$ d. $s_n = \left(-\frac{1}{2}\right)^{n+1}$

Solutions:

a. $\boxed{s_1 = \left(\frac{1}{3}\right)^{1-1}} = \boxed{\left(\frac{1}{3}\right)^0} = \boxed{1}$ $\boxed{s_2 = \left(\frac{1}{3}\right)^{2-1}} = \boxed{\left(\frac{1}{3}\right)^1} = \boxed{\frac{1}{3}}$

$$s_3 = \left(\frac{1}{3}\right)^{3-1} = \left(\frac{1}{3}\right)^2 = \frac{1}{3^2} = \boxed{\frac{1}{9}}$$

$$s_4 = \left(\frac{1}{3}\right)^{4-1} = \left(\frac{1}{3}\right)^3 = \frac{1}{3^3} = \boxed{\frac{1}{27}}$$

$$s_5 = \left(\frac{1}{3}\right)^{5-1} = \left(\frac{1}{3}\right)^4 = \frac{1}{3^4} = \boxed{\frac{1}{81}}$$

$$s_6 = \left(\frac{1}{3}\right)^{6-1} = \left(\frac{1}{3}\right)^5 = \frac{1}{3^5} = \boxed{\frac{1}{243}}$$

Thus, the first six terms of the geometric sequence are $\left(1, \frac{1}{3}, \frac{1}{9}, \frac{1}{27}, \frac{1}{81}, \frac{1}{243}\right)$.

b. $s_1 = \left(\frac{1}{2}\right)^{1+2} = \left(\frac{1}{2}\right)^3 = \frac{1}{2^3} = \boxed{\frac{1}{8}}$

$$s_2 = \left(\frac{1}{2}\right)^{2+2} = \left(\frac{1}{2}\right)^4 = \frac{1}{2^4} = \boxed{\frac{1}{16}}$$

$$s_3 = \left(\frac{1}{2}\right)^{3+2} = \left(\frac{1}{2}\right)^5 = \frac{1}{2^5} = \boxed{\frac{1}{32}}$$

$$s_4 = \left(\frac{1}{2}\right)^{4+2} = \left(\frac{1}{2}\right)^6 = \frac{1}{2^6} = \boxed{\frac{1}{64}}$$

$$s_5 = \left(\frac{1}{2}\right)^{5+2} = \left(\frac{1}{2}\right)^7 = \frac{1}{2^7} = \boxed{\frac{1}{128}}$$

$$s_6 = \left(\frac{1}{2}\right)^{6+2} = \left(\frac{1}{2}\right)^8 = \frac{1}{2^8} = \boxed{\frac{1}{256}}$$

Thus, the first six terms of the geometric sequence are $\left(\frac{1}{8}, \frac{1}{16}, \frac{1}{32}, \frac{1}{64}, \frac{1}{128}, \frac{1}{256}\right)$.

c. $s_1 = \left(-\frac{1}{3}\right)^2 = \frac{1}{3^2} = \boxed{\frac{1}{9.0 \times 10^0}}$

$$s_2 = \left(-\frac{1}{3}\right)^4 = \frac{1}{3^4} = \boxed{\frac{1}{8.1 \times 10^1}}$$

$$s_3 = \left(-\frac{1}{3}\right)^6 = \frac{1}{3^6} = \boxed{\frac{1}{7.29 \times 10^2}}$$

$$s_4 = \left(-\frac{1}{3}\right)^8 = \frac{1}{3^8} = \boxed{\frac{1}{6.56 \times 10^3}}$$

$$s_5 = \left(-\frac{1}{3}\right)^{10} = \frac{1}{3^{10}} = \boxed{\frac{1}{5.9 \times 10^4}}$$

$$s_6 = \left(-\frac{1}{3}\right)^{12} = \frac{1}{3^{12}} = \boxed{\frac{1}{5.31 \times 10^5}}$$

Thus, the six terms are $\left(\frac{1}{9.0 \times 10^0}, \frac{1}{8.1 \times 10^1}, \frac{1}{7.29 \times 10^2}, \frac{1}{6.56 \times 10^3}, \frac{1}{5.9 \times 10^4}, \frac{1}{5.31 \times 10^5}\right)$.

d. $s_1 = \left(-\frac{1}{2}\right)^2 = \frac{1}{2^2} = \boxed{\frac{1}{4}}$

$$s_2 = \left(-\frac{1}{2}\right)^3 = -\frac{1}{2^3} = \boxed{-\frac{1}{8}}$$

$$s_3 = \left(-\frac{1}{2}\right)^4 = \frac{1}{2^4} = \boxed{\frac{1}{16}}$$

$$s_4 = \left(-\frac{1}{2}\right)^5 = -\frac{1}{2^5} = \boxed{-\frac{1}{32}}$$

$$s_5 = \left(-\frac{1}{2}\right)^6 = \frac{1}{2^6} = \boxed{\frac{1}{64}}$$

$$s_6 = \left(-\frac{1}{2}\right)^7 = -\frac{1}{2^7} = \boxed{-\frac{1}{128}}$$

Thus, the first six terms of the geometric sequence are $\left(\frac{1}{4}, -\frac{1}{8}, \frac{1}{16}, -\frac{1}{32}, \frac{1}{64}, -\frac{1}{128}\right)$.

Having learned about geometric sequences and the steps for finding the terms of a geometric sequence, we will next learn about geometric series and the steps for finding the sum of geometric series over a given range.

Similar to arithmetic series, addition of the terms in a geometric sequence result in having a geometric series. To obtain the geometric series formula let $s_k = s_1 r^{k-1}$ be a geometric sequence and denote the sum of the first n terms by

$$S_n = \sum_{k=1}^{n} s_1 r^{k-1}$$

then,

$$S_n = s_1 + s_1 r + s_1 r^2 + \cdots + s_1 r^{n-2} + s_1 r^{n-1} \qquad (a)$$

Let's multiply both sides of the equation (a) by r and subtract (b) from (a).

$$S_n \cdot r = s_1 \cdot r + s_1 r \cdot r + s_1 r^2 \cdot r + \cdots + s_1 r^{n-2} \cdot r + s_1 r^{n-1} \cdot r$$

$$r S_n = s_1 r + s_1 r^2 + s_1 r^3 + \cdots + s_1 r^{n-1} + s_1 r^n \qquad (b)$$

$$S_n - r S_n = \left(s_1 + s_1 r + s_1 r^2 + s_1 r^3 + \cdots + s_1 r^{n-1}\right) - \left(s_1 r + s_1 r^2 + s_1 r^3 + \cdots + s_1 r^{n-1} + s_1 r^n\right)$$

$$S_n(1-r) = \left(s_1 + s_1 r + s_1 r^2 + s_1 r^3 + \cdots + s_1 r^{n-1}\right) + \left(-s_1 r - s_1 r^2 - s_1 r^3 - \cdots - s_1 r^{n-1} - s_1 r^n\right)$$

$$S_n(1-r) = s_1 + \left(s_1 r - s_1 r\right) + \left(s_1 r^2 - s_1 r^2\right) + \left(s_1 r^3 - s_1 r^3\right) + \cdots + \left(s_1 r^{n-1} - s_1 r^{n-1}\right) - s_1 r^n$$

$$S_n(1-r) = s_1 - s_1 r^n \quad ; \quad S_n = \frac{s_1 - s_1 r^n}{1-r} \quad ; \quad S_n = \frac{s_1 \left(1 - r^n\right)}{1-r} \qquad r \neq 1$$

Therefore, the geometric series can be written in the following two forms:

$$S_n = \sum_{k=1}^{n} s_1 r^{k-1} \qquad (1)$$

$$S_n = \frac{s_1 \left(1 - r^n\right)}{1-r} \qquad r \neq 1 \qquad (2)$$

Note that equation (2), similar to the nth term of a geometric sequence $\left(s_n = s_1 r^{n-1}\right)$, is given in terms of s_1, n, and r.

A third alternative way of expressing the geometric series is by substituting $s_1 r^n$ with its equivalent value $s_1 r^n = s_1 r^{n-1} \cdot r = r\left(s_1 r^{n-1}\right) = r s_n$ which result in having

$$S_n = \frac{s_1 - s_1 r^n}{1-r} = \frac{s_1 - r\left(s_1 r^{n-1}\right)}{1-r} = \frac{s_1 - r\left(s_1 r^{n-1}\right)}{1-r} = \frac{s_1 - r s_n}{1-r} \qquad (3)$$

where the geometric series is given in terms of s_1, s_n (the geometric sequence), and r.

In the following examples we will use the above equations (1), (2), and (3) in order to find the sum of geometric series.

Example 1.4-6 Evaluate the sum of the following geometric series.

a. $\displaystyle\sum_{k=1}^{10} 3^{k-2} =$ b. $\displaystyle\sum_{k=1}^{10} (-3)^{k-2} =$ c. $\displaystyle\sum_{k=2}^{6} 8\left(-\frac{1}{2}\right)^{k+1} =$

Solutions:

a. **First** - Write the first few terms of the geometric series in expanded form, i.e.,

$$\boxed{\sum_{k=1}^{10} 3^{k-2}} = \boxed{3^{1-2} + 3^{2-2} + 3^{3-2} + 3^{4-2} + \cdots} = \boxed{3^{-1} + 3^0 + 3^1 + 3^2 + \cdots} = \boxed{3^{-1} + 1 + 3 + 6 + \cdots}$$

Second - Identify the first term, s_1, the common ratio between the two terms, r, and n, i.e.,

$$s_1 = 3^{-1}, \; r = \frac{1}{3^{-1}} = \frac{1}{\frac{1}{3}} = \frac{\frac{1}{1}}{\frac{1}{3}} = \frac{1 \times 3}{1 \times 1} = 3, \text{ and } n = 10.$$

Third - Use the geometric series formula to obtain the sum of the ten terms.

$$S_n = \frac{s_1\left(1 - r^n\right)}{1-r}$$

$$\boxed{S_{10}} = \boxed{\frac{\frac{1}{3}\left(1 - 3^{10}\right)}{1-3}} = \boxed{\frac{\frac{1}{3}\left(1 - 59049\right)}{-2}} = \boxed{\frac{-\frac{59048}{3}}{-2}} = \boxed{\frac{-\frac{59048}{3}}{-\frac{2}{1}}} = \boxed{\frac{59048 \times 1}{3 \times 2}} = \boxed{\frac{59048}{6}} = \boxed{\mathbf{9841.333}}$$

Note that prior to learning the geometric series formula the only method that we could use was by summing each term as shown below:

$$\boxed{\sum_{k=1}^{10} 3^{k-2}} = \boxed{3^{-1} + 3^0 + 3^1 + 3^2 + 3^3 + 3^4 + 3^5 + 3^6 + 3^7 + 3^8} = \boxed{3^{-1} + 1 + 3 + 9 + 27 + 81 + 243 + 729 + 2187}$$

$$\boxed{+ 6561} = \boxed{3^{-1} + 9841} = \boxed{0.333 + 9841} = \boxed{\mathbf{9841.333}}$$

As you note, it is much easier to use the geometric series formula as opposed to the summation of each term which is somewhat long and time consuming.

b. **First** - Write the first few terms of the geometric series in expanded form, i.e.,

$$\boxed{\sum_{k=1}^{10} (-3)^{k-2}} = \boxed{(-3)^{1-2} + (-3)^{2-2} + (-3)^{3-2} + (-3)^{4-2} + \cdots} = \boxed{(-3)^{-1} + (-3)^0 + (-3)^1 + (-3)^2 + \cdots}$$

$$= \boxed{-3^{-1} + 1 - 3 + 9 + \cdots}$$

Second - Identify the first term, s_1, the common ratio between the two terms, r, and n, i.e.,

$$s_1 = -3^{-1}, \; r = \frac{1}{-3^{-1}} = \frac{1}{-\frac{1}{3}} = -\frac{\frac{1}{1}}{\frac{1}{3}} = -\frac{1 \times 3}{1 \times 1} = -3, \text{ and } n = 10.$$

Third - Use the arithmetic series formula to obtain the sum of the ten terms.

$$S_n = \frac{s_1\left(1 - r^n\right)}{1-r}$$

$$\boxed{S_{10}} = \boxed{\frac{-\frac{1}{3}\left[1 - (-3)^{10}\right]}{1-(-3)}} = \boxed{\frac{-\frac{1}{3}\left(1 - 59049\right)}{4}} = \boxed{\frac{\frac{59048}{3}}{4}} = \boxed{\frac{\frac{59048}{3}}{\frac{4}{1}}} = \boxed{\frac{59048 \times 1}{3 \times 4}} = \boxed{\frac{59048}{12}} = \boxed{\mathbf{4920.666}}$$

or, we can obtain the answer by summing up the first ten terms of the series, i.e.,

$$\sum_{k=1}^{10}(-3)^{k-2} = \left[(-3)^{-1}+(-3)^{0}+(-3)^{1}+(-3)^{2}+(-3)^{3}+(-3)^{4}+(-3)^{5}+(-3)^{6}+(-3)^{7}+(-3)^{8}\right]$$

$$= \boxed{-3^{-1}+1-3+9-27+81-243+729-2187+6561} = \boxed{-3^{-1}+4921} = \boxed{-0.333+4921} = \boxed{4920.666}$$

c. **First** - Write the first few terms of the geometric series in expanded form, i.e.,

$$\sum_{k=2}^{6}8\left(-\frac{1}{2}\right)^{k+1} = \left[8\left(-\frac{1}{2}\right)^{2+1}+8\left(-\frac{1}{2}\right)^{3+1}+8\left(-\frac{1}{2}\right)^{4+1}+8\left(-\frac{1}{2}\right)^{5+1}+\cdots\right] = \left[8\left(-\frac{1}{2}\right)^{3}+8\left(-\frac{1}{2}\right)^{4}+8\left(-\frac{1}{2}\right)^{5}\right.$$

$$\left.+8\left(-\frac{1}{2}\right)^{6}+\cdots\right] = \left[8\times-\frac{1}{8}+8\times\frac{1}{16}-8\times\frac{1}{32}+8\times\frac{1}{64}+\cdots\right] = \left[-\frac{8}{8}+\frac{8}{16}-\frac{8}{32}+\frac{8}{64}+\cdots\right] = \left[-1+\frac{1}{2}-\frac{1}{4}+\frac{1}{8}+\cdots\right]$$

Second - Identify the first term, s_1, the common ratio between the two terms, r, and n, i.e.,

$$s_1 = -1, \; r = \frac{\frac{1}{2}}{-1} = -\frac{\frac{1}{2}}{\frac{1}{1}} = -\frac{1\times 1}{2\times 1} = -\frac{1}{2}, \text{ and } n = 5.$$

Third - Use the geometric series formula to obtain the sum of the five terms.

$$S_n = \frac{s_1\left(1-r^n\right)}{1-r}$$

$$\boxed{S_5} = \boxed{\frac{-1\cdot\left[1-\left(-\frac{1}{2}\right)^{5}\right]}{1-\left(-\frac{1}{2}\right)}} = \boxed{\frac{-\left(1+\frac{1}{2^5}\right)}{1+\frac{1}{2}}} = \boxed{\frac{-\left(1+\frac{1}{32}\right)}{\frac{3}{2}}} = \boxed{\frac{-(1+0.03125)}{1.5}} = \boxed{-\frac{1.03125}{1.5}} = \boxed{-0.6875}$$

or, we can obtain the answer by summing up the first five terms of the series, i.e.,

$$\sum_{k=2}^{6}8\left(-\frac{1}{2}\right)^{k+1} = \left[8\cdot\left[\left(-\frac{1}{2}\right)^{3}+\left(-\frac{1}{2}\right)^{4}+\left(-\frac{1}{2}\right)^{5}+\left(-\frac{1}{2}\right)^{6}+\left(-\frac{1}{2}\right)^{7}\right]\right] = \left[8\cdot\left[-\frac{1}{8}+\frac{1}{16}-\frac{1}{32}+\frac{1}{64}-\frac{1}{128}\right]\right]$$

$$= \boxed{8\cdot(-0.125+0.0625-0.03125+0.01563-0.00781)} = \boxed{8\times-0.08593} = \boxed{-0.6875}$$

Example 1.4-7 Given the first term s_1 and r, find S_{10} for each of the following geometric sequences.

a. $s_1 = 5$, $r = 2$ b. $s_1 = -10$, $r = 3$ c. $s_1 = 50$, $r = -2$

Solutions:

a. The nth term for a geometric series is equal to $S_n = \frac{s_1\left(1-r^n\right)}{1-r}$. Substituting $s_1 = 5$ and $r = 2$ into the general geometric expression we obtain

$$\boxed{S_{10}} = \boxed{\frac{5\left(1-2^{10}\right)}{1-2}} = \boxed{\frac{5\left(1-1024\right)}{-1}} = \boxed{5115}$$

b. Substituting $s_1 = -10$ and $r = 3$ into $S_n = \frac{s_1\left(1-r^n\right)}{1-r}$ we obtain

$$S_{10} = \left| \frac{-10\left(1-3^{10}\right)}{1-3} \right| = \left| \frac{-10\left(1-59049\right)}{-2} \right| = \left| 5 \times -59048 \right| = \boxed{-295240}$$

c. Substituting $s_1 = 50$ and $r = -2$ into $S_n = \frac{s_1\left(1-r^n\right)}{1-r}$ we obtain

$$S_{10} = \left| \frac{50\left[1-(-2)^{10}\right]}{1-(-2)} \right| = \left| \frac{50\left[1-1024\right]}{1+2} \right| = \left| \frac{50 \times -1023}{3} \right| = \boxed{-17050}$$

Example 1.4-8 Find the x and y values to the following problems.

a. x if $\displaystyle\sum_{i=1}^{6} ix = 20$

b. x and y if $\displaystyle\sum_{i=1}^{5}(ix+2y) = 30$ and $\displaystyle\sum_{i=2}^{5}(ix+2y) = 10$

c. x if $\displaystyle\sum_{i=1}^{5}(ix+5) = 15$

d. x and y if $\displaystyle\sum_{i=3}^{6}(x+iy) = 50$ and $\displaystyle\sum_{i=4}^{8}(x+iy) = 24$

Solutions:

a. Expanding $\displaystyle\sum_{i=1}^{6} ix = 20$ we obtain $\boxed{x+2x+3x+4x+5x+6x = 20}$; $\boxed{21x = 20}$; $\boxed{x = \dfrac{20}{21}}$; $\boxed{x = 0.952}$

b. Expanding $\displaystyle\sum_{i=1}^{5}(ix+2y) = 30$ we obtain $\boxed{(x+2y)+(2x+2y)+(3x+2y)+(4x+2y)+(5x+2y)=30}$

; $\boxed{(x+2x+3x+4x+5x)+(2y+2y+2y+2y+2y)=30}$; $\boxed{15x+10y=30}$

Expanding $\displaystyle\sum_{i=2}^{5}(ix+2y) = 10$ we obtain $\boxed{(2x+2y)+(3x+2y)+(4x+2y)+(5x+2y)=10}$; $\boxed{14x+8y=10}$

The two linear equations with two unknowns x and y are solved using the substitution method

to obtain $\boxed{x = -7}$ and $\boxed{y = 13.5}$

c. Expanding $\displaystyle\sum_{i=1}^{5}(ix+5) = 15$ we obtain $\boxed{(x+5)+(2x+5)+(3x+5)+(4x+5)+(5x+5)}$; $\boxed{15x+25=15}$

; $\boxed{x = -\dfrac{10}{15}}$; $\boxed{x = -0.667}$

d. Expanding $\displaystyle\sum_{i=3}^{6}(x+iy) = 50$ we obtain $\boxed{(x+3y)+(x+4y)+(x+5y)+(x+6y)=50}$; $\boxed{4x+18y=50}$

Expanding $\displaystyle\sum_{i=4}^{8}(x+iy) = 24$ we obtain $\boxed{(x+4y)+(x+5y)+(x+6y)+(x+7y)+(x+8y)=24}$; $\boxed{5x+30y=24}$

The two linear equations with two unknowns x and y are solved using the substitution method

to obtain $\boxed{x = 35.6}$ and $\boxed{y = -5.133}$

Example 1.4-9 Find the value of x for the following geometric sequences.

a. $2, 4x, 16$.

b. $2^{-1}, 2^{-1}x, 2^{-3}$

c. $5, 5x, 125$

Solutions:

a. Since the common ratio r of a geometric sequence is defined as the ratio of the $(n+1)\,st$ term to the nth term, we can use this principal to solve for x, i.e.,

$$\boxed{r} = \boxed{\dfrac{4x}{2} = \dfrac{16}{4x}} \text{ therefore } \boxed{4x \times 4x = 16 \times 2} ; \boxed{16x^2 = 32} ; \boxed{x^2 = \dfrac{32}{16}} ; \boxed{x^2 = 2} ; \boxed{x = \pm\sqrt{2}}$$

b. Using the common ratio principal we can solve for x in the following way:

$$\boxed{r} = \boxed{\dfrac{2^{-1}x}{2^{-1}} = \dfrac{2^{-3}}{2^{-1}x}} \text{ therefore } \boxed{2^{-1}x \times 2^{-1}x = 2^{-3} \times 2^{-1}} ; \boxed{2^{-2}x^2 = 2^{-4}} ; \boxed{x^2 = \dfrac{2^{-4}}{2^{-2}}} ; \boxed{x^2 = 2^{-4} \cdot 2^2}$$

$$; \boxed{x^2 = 2^{-4+2}} ; \boxed{x^2 = 2^{-2}} ; \boxed{x^2 = \dfrac{1}{2^2}} ; \boxed{x = \pm\dfrac{1}{2}}$$

c. Using the common ratio principal we can solve for x in the following way:

$$\boxed{r} = \boxed{\dfrac{5x}{5} = \dfrac{125}{5x}} \text{ therefore } \boxed{5x \times 5x = 125 \times 5} ; \boxed{25x^2 = 625} ; \boxed{x^2 = \dfrac{625}{25}} ; \boxed{x^2 = 25} ; \boxed{x = \pm 5}$$

Section 1.4 Practice Problems - Geometric Sequences and Geometric Series

1. Find the next four terms of the following geometric sequences.

   a. $s_1 = 3$, $r = 0.5$ b. $s_1 = -5$, $r = 2$ c. $s_1 = 5$, $r = 0.75$

2. Find the eighth and the general term of the following geometric sequences.

   a. $s_1 = 2$, $r = \sqrt{3}$ b. $s_1 = -4$, $r = 1.2$ c. $s_1 = 4$, $r = -2.5$

3. Find the next six terms and the nth term in each of the following geometric sequences.

   a. $1, \dfrac{1}{4}, \cdots$ b. $-\dfrac{1}{2}, \dfrac{1}{4}, \cdots$ c. $\dfrac{1}{3}p, -3p, \cdots$

4. Given the following terms of a geometric sequence, find the common ratio r.

   a. $s_1 = 25$ and $s_4 = \dfrac{1}{5}$ b. $s_1 = 4$ and $s_5 = \dfrac{1}{64}$ c. $s_1 = 3$ and $s_8 = 1$

5. Write the first five terms of the following geometric sequences.

   a. $s_n = \left(-\dfrac{1}{3}\right)^{2n-1}$ b. $s_n = \left(\dfrac{1}{3}\right)^{2n+2}$ c. $s_n = \left(-\dfrac{1}{5}\right)^{2n-3}$ d. $s_n = \left(-\dfrac{1}{2}\right)^{n}$

6. Evaluate the sum of the following geometric series.

   a. $\displaystyle\sum_{k=1}^{6} 3^{k-1} =$ b. $\displaystyle\sum_{k=3}^{10} (-2)^{k-3} =$ c. $\displaystyle\sum_{j=4}^{8} 4\left(-\dfrac{1}{2}\right)^{j+1} =$

d. $\displaystyle\sum_{m=1}^{4}(-2)^{m-3} =$ e. $\displaystyle\sum_{n=5}^{10}(-3)^{n-4} =$ f. $\displaystyle\sum_{k=1}^{5}(-3)^{k-1} =$

g. $\displaystyle\sum_{m=1}^{5}4^{m} =$ h. $\displaystyle\sum_{j=1}^{4}\frac{3^{j}}{27} =$ i. $\displaystyle\sum_{k=3}^{6}6\left(\frac{1}{2}\right)^{k+1} =$

7. Given the first term s_1 and r, find S_8 for each of the following geometric sequences.

   a. $s_1 = 3$, $r = 3$ b. $s_1 = -8$, $r = 0.5$ c. $s_1 = 2$, $r = -2.5$

8. Solve for x and y.

   a. $\displaystyle\sum_{i=3}^{7}(ix+2) = 30$ b. $\displaystyle\sum_{i=1}^{4}(ix+y) = 20$ and $\displaystyle\sum_{i=2}^{6}(ix+y) = 10$

1.5   Limits of Sequences and Series

A sequence $s_1, s_2, s_3, s_4, \cdots, s_n, \cdots$ is said to **converge** to the constant K,

$$\lim_{n \to \infty} s_n = K$$

if and only if, for a large value of n, the absolute value of the difference between the nth term and the constant K is very small. For example, the sequence $2, \dfrac{5}{4}, \dfrac{10}{9}, \dfrac{17}{16}, \cdots, \left(1 + \dfrac{1}{n^2}\right), \cdots$ converges to 1. This is because, the absolute value of the difference between $\left(1 + \dfrac{1}{n^2}\right)$, for large n, and 1 is very small. On the other hand, the sequence $s_1, s_2, s_3, s_4, \cdots, s_n, \cdots$ is said to **diverge,** if and only if, for a large value of n, the sequence approaches to infinity (∞). For example, the sequence $4, 8, 16, 32, \cdots, 2^{n+1}, \cdots$ does not converge. This is because, as n increases, the nth term increases without bound, i.e., it approaches to infinity. In the following examples we will learn how to identify a convergent or a divergent sequence:

Example 1.5-1 State which of the following sequences are convergent.

a. $1, 2, 3, 4, 5, \cdots, n, \cdots =$

b. $2, \dfrac{5}{8}, \dfrac{10}{27}, \dfrac{17}{64}, \cdots, \dfrac{n^2+1}{n^3}, \cdots =$

c. $2, \dfrac{8}{3}, \dfrac{26}{9}, \dfrac{80}{27}, \cdots, \dfrac{3^n-1}{3^{n-1}}, \cdots =$

d. $\dfrac{1}{16}, \dfrac{1}{64}, \dfrac{1}{256}, \dfrac{1}{1024}, \cdots, \dfrac{1}{4^{n+1}}, \cdots =$

e. $3, \dfrac{9}{4}, \dfrac{19}{9}, \dfrac{33}{16}, \cdots, 2+\dfrac{1}{n^2}, \cdots =$

f. $6, \dfrac{7}{4}, \dfrac{8}{9}, \dfrac{9}{16}, \cdots, \dfrac{n+5}{n^2}, \cdots =$

g. $\dfrac{1}{2}, \dfrac{1}{3}, \dfrac{1}{4}, \dfrac{1}{5}, \cdots, \dfrac{1}{n+1}, \cdots =$

h. $3, 6, 9, 12, 15, \cdots, 3n, \cdots =$

Solutions:

In solving this class of problems write the nth term and observe if it converges or diverges as n approaches to infinity.

a. The sequence $1, 2, 3, 4, 5, \cdots, n, \cdots$ continues to increase. $\lim_{n \to \infty} n = \infty$ which is undefined.

Hence, **the sequence diverges or is divergent**.

b. $\boxed{\lim_{n \to \infty} \dfrac{n^2+1}{n^3}} = \boxed{\lim_{n \to \infty} \left(\dfrac{n^2}{n^3} + \dfrac{1}{n^3}\right)} = \boxed{\lim_{n \to \infty} \left(\dfrac{1}{n} + \dfrac{1}{n^3}\right)} = \boxed{\dfrac{1}{\infty} + \dfrac{1}{\infty^3}} = \boxed{0+0} = \boxed{0}$

The sequence converges to 0

c. $\boxed{\lim_{n \to \infty} \dfrac{3^n-1}{3^{n-1}}} = \boxed{\lim_{n \to \infty} \left(\dfrac{3^n}{3^{n-1}} - \dfrac{1}{3^{n-1}}\right)} = \boxed{\lim_{n \to \infty} \left(\dfrac{3^n 3^{-(n-1)}}{1} - \dfrac{1}{3^{n-1}}\right)} = \boxed{\lim_{n \to \infty} \left(\dfrac{3^{n-n+1}}{1} - \dfrac{1}{3^{n-1}}\right)}$

$= \boxed{\lim_{n \to \infty} \left(3 - \dfrac{1}{3^{n-1}}\right)} = \boxed{3 - \dfrac{1}{3^{\infty-1}}} = \boxed{3 - \dfrac{1}{3^\infty}} = \boxed{3 - \dfrac{1}{\infty}} = \boxed{3-0} = \boxed{3}$ **The sequence converges to 3**

d. $\left|\lim_{n\to\infty} \dfrac{1}{4^{n+1}}\right| = \left|\dfrac{1}{4^{\infty+1}}\right| = \left|\dfrac{1}{4^{\infty}}\right| = \left|\dfrac{1}{\infty}\right| = \boxed{0}$ **The sequence converges to** 0

e. $\left|\lim_{n\to\infty} \left(2+\dfrac{1}{n^2}\right)\right| = \left|2+\dfrac{1}{\infty^2}\right| = \left|2+\dfrac{1}{\infty}\right| = \boxed{2+0} = \boxed{2}$ **The sequence converges to** 2

f. $\left|\lim_{n\to\infty} \dfrac{n+5}{n^2}\right| = \left|\lim_{n\to\infty}\left(\dfrac{n}{n^2}+\dfrac{5}{n^2}\right)\right| = \left|\lim_{n\to\infty}\left(\dfrac{1}{n}+\dfrac{5}{n^2}\right)\right| = \left|\dfrac{1}{\infty}+\dfrac{5}{\infty^2}\right| = \left|\dfrac{1}{\infty}+\dfrac{5}{\infty}\right| = \boxed{0+0} = \boxed{0}$

The sequence converges to 0

g. $\left|\lim_{n\to\infty} \dfrac{1}{n+1}\right| = \left|\dfrac{1}{\infty+1}\right| = \left|\dfrac{1}{\infty}\right| = \boxed{0}$

The sequence converges to 0

h. The sequence $3, 6, 9, 12, 15, \cdots, 3n, \cdots$ continues to increase. $\lim_{n\to\infty} 3n = 3\cdot\infty = \infty$ which is undefined. Hence, **the sequence diverges or is divergent**.

Example 1.5-2 State which of the following geometric sequences are convergent.

a. $\dfrac{1}{3}, \dfrac{1}{9}, \dfrac{1}{27}, \dfrac{1}{81}, \cdots, \dfrac{1}{3^n}, \cdots =$ b. $2, 4, 8, 16, 32, \cdots, 2^n, \cdots =$

c. $1, -1, 1, -1, \cdots, (-1)^{n+1}, \cdots =$ d. $10, 1, \dfrac{1}{10}, \dfrac{1}{100}, \dfrac{1}{1000}, \cdots, 10\cdot\left(\dfrac{1}{10}\right)^{n-1}, \cdots =$

e. $0.2, 0.02, 0.0002, \cdots, 2(0.1)^n, \cdots =$ f. $1, \dfrac{4}{3}, \dfrac{16}{9}, \dfrac{64}{27}, \dfrac{256}{81}, \cdots, \left(\dfrac{4}{3}\right)^{n-1}, \cdots =$

g. $-2, 4, -8, 16, -32, \cdots, (-1)^n 2^n, \cdots =$ h. $27, 3, \dfrac{1}{3}, \dfrac{1}{27}, \cdots, 27\cdot\left(\dfrac{1}{9}\right)^{n-1}, \cdots =$

Solutions:

a. The sequence $\dfrac{1}{3}, \dfrac{1}{9}, \dfrac{1}{27}, \dfrac{1}{81}, \cdots, \dfrac{1}{3^n}, \cdots$ **converges to** 0 since, for large value of n, the absolute value of the difference between $\dfrac{1}{3^n}$ and 0 is very small.

b. The sequence $2, 4, 8, 16, 32, \cdots, 2^n, \cdots$ **diverges** since, as n increases, the nth term increases without bound.

c. The sequence $1, -1, 1, -1, \cdots, (-1)^{n+1}, \cdots$ **diverges** since, as n increases, the nth term oscillates back and forth from +1 to −1.

d. The sequence $10, 1, \dfrac{1}{10}, \dfrac{1}{100}, \dfrac{1}{1000}, \cdots, 10\cdot\left(\dfrac{1}{10}\right)^{n-1}, \cdots$ **converges to** 0 since, for large value of n, the absolute value of the difference between $10\cdot\left(\dfrac{1}{10}\right)^{n-1}$ and 0 is very small.

e. The sequence $0.2, 0.02, 0.0002, \cdots, 2(0.1)^n, \cdots$ **converges to** 0 since, for large value of n, the absolute value of the difference between $2(0.1)^n$ and 0 is very small.

f. The sequence $1, \frac{4}{3}, \frac{16}{9}, \frac{64}{27}, \frac{256}{81}, \cdots, \left(\frac{4}{3}\right)^{n-1}, \cdots$ **diverges** since, as n increases, the nth term

increases without bound.

g. The sequence $-2, 4, -8, 16, -32, \cdots, (-1)^n 2^n, \cdots$ **diverges** since, as n increases, the nth term

oscillates back and forth from a large positive number to a large negative number.

h. The sequence $27, 3, \frac{1}{3}, \frac{1}{27}, \cdots, 27 \cdot \left(\frac{1}{9}\right)^{n-1}, \cdots$ **converges to 0** since, for large value of n, the

absolute value of the difference between $27 \cdot \left(\frac{1}{9}\right)^{n-1}$ and 0 is very small.

Example 1.5-3 Discuss the limiting behavior of the following sequences as n approaches ∞.

a. $\frac{1}{n^2} =$

b. $1 - \frac{1}{n^2} =$

c. $\frac{n+5}{n^2} =$

d. $\frac{n^2+5}{n^2} =$

e. $(1)^{-n} \frac{1}{n} =$

f. $\left(1+\frac{1}{2}\right)^{-n} =$

g. $2 + \frac{1}{n^2} =$

h. $100n =$

i. $\frac{5n+10}{n} =$

j. $\frac{2^n - 1}{2^n} =$

Solutions:

a. $\boxed{\lim_{n\to\infty} \frac{1}{n^2}} = \boxed{\frac{1}{\infty^2}} = \boxed{\frac{1}{\infty}} = \boxed{0}$ **converges to 0**

b. $\boxed{\lim_{n\to\infty} 1 - \frac{1}{n^2}} = \boxed{1 - \frac{1}{\infty^2}} = \boxed{1 - \frac{1}{\infty}} = \boxed{1-0} = \boxed{1}$ **converges to 1**

c. $\boxed{\lim_{n\to\infty} \frac{n+5}{n^2}} = \boxed{\lim_{n\to\infty}\left(\frac{n}{n^{2=1}} + \frac{5}{n^2}\right)} = \boxed{\lim_{n\to\infty}\left(\frac{1}{n} + \frac{5}{n^2}\right)} = \boxed{\frac{1}{\infty} + \frac{5}{\infty^2}} = \boxed{\frac{1}{\infty} + \frac{5}{\infty}} = \boxed{0+0} = \boxed{0}$

converges to 0

d. $\boxed{\lim_{n\to\infty} \frac{n^2+5}{n^2}} = \boxed{\lim_{n\to\infty}\left(\frac{n^2}{n^2} + \frac{5}{n^2}\right)} = \boxed{\lim_{n\to\infty}\left(1 + \frac{5}{n^2}\right)} = \boxed{1 + \frac{5}{\infty^2}} = \boxed{1+0} = \boxed{1}$

converges to 1

e. $\boxed{\lim_{n\to\infty} (1)^{-n}\frac{1}{n}} = \boxed{\lim_{n\to\infty}\left(\frac{1}{1^n}\cdot\frac{1}{n}\right)} = \boxed{\frac{1}{1^\infty}\cdot\frac{1}{\infty}} = \boxed{\frac{1}{1}\cdot\frac{1}{\infty}} = \boxed{1\cdot 0} = \boxed{0}$ **converges to 0**

f. $\boxed{\lim_{n\to\infty}\left(1+\frac{1}{2}\right)^{-n}} = \boxed{\lim_{n\to\infty}\left(\frac{3}{2}\right)^{-n}} = \boxed{\lim_{n\to\infty}(1.5)^{-n}} = \boxed{\lim_{n\to\infty}\left(\frac{1}{1.5^n}\right)} = \boxed{\frac{1}{1.5^\infty}} = \boxed{\frac{1}{\infty}} = \boxed{0}$

converges to 0

g. $\boxed{\lim_{n\to\infty} 2 + \dfrac{1}{n^2}} = \boxed{2 + \dfrac{1}{\infty^2}} = \boxed{2 + \dfrac{1}{\infty}} = \boxed{2+0} = \boxed{2}$ **converges to 2**

h. $\boxed{\lim_{n\to\infty} 100n} = \boxed{100\cdot\infty} = \boxed{\infty}$ **diverges**

i. $\boxed{\lim_{n\to\infty} \dfrac{5n+10}{n}} = \boxed{\lim_{n\to\infty}\left(\dfrac{5\not{n}}{\not{n}} + \dfrac{10}{n}\right)} = \boxed{\lim_{n\to\infty}\left(5 + \dfrac{10}{n}\right)} = \boxed{5 + \dfrac{10}{\infty}} = \boxed{5+0} = \boxed{5}$

converges to 5

j. $\boxed{\lim_{n\to\infty} \dfrac{2^n-1}{2^n}} = \boxed{\lim_{n\to\infty}\left(\dfrac{2^n}{2^n} - \dfrac{1}{2^n}\right)} = \boxed{\lim_{n\to\infty}\left(1 - \dfrac{1}{2^n}\right)} = \boxed{1 - \dfrac{1}{2^\infty}} = \boxed{1 - \dfrac{1}{\infty}} = \boxed{1-0} = \boxed{1}$

converges to 1

Note that an easier way of finding the answer to sequences as $n\to\infty$ is by rewriting the sequence in its "almost equivalent" form. This approach is only applicable to cases where n is approaching to infinity. For example, $\lim_{n\to\infty} n+8$ is almost the same as $\lim_{n\to\infty} n$. (This is because $\infty+8$ is the same as ∞. Addition of the number eight to a very large number such as infinity does not significantly change the final answer.) Let's use this approach to solve few of the above problems.

$\boxed{\lim_{n\to\infty} \dfrac{n+5}{n^2}} \approx \boxed{\lim_{n\to\infty} \dfrac{n}{n^2}} = \boxed{\lim_{n\to\infty} \dfrac{1}{n}} = \boxed{\dfrac{1}{\infty}} = \boxed{0}$ which is the same answer as in 1.5-3c.

$\boxed{\lim_{n\to\infty} \dfrac{n^2+5}{n^2}} \approx \boxed{\lim_{n\to\infty} \dfrac{n^2}{n^2}} = \boxed{\lim_{n\to\infty} \dfrac{1}{1}} = \boxed{1}$ which is the same answer as in 1.5-3d.

$\boxed{\lim_{n\to\infty} \dfrac{5n+10}{n}} \approx \boxed{\lim_{n\to\infty} \dfrac{5n}{n}} = \boxed{\lim_{n\to\infty} \dfrac{5}{1}} = \boxed{5}$ which is the same answer as in 1.5-3i.

$\boxed{\lim_{n\to\infty} \dfrac{2^n-1}{2^n}} \approx \boxed{\lim_{n\to\infty} \dfrac{2^n}{2^n}} = \boxed{\lim_{n\to\infty} \dfrac{1}{1}} = \boxed{1}$ which is the same answer as in 1.5-3j.

Example 1.5-4 State whether or not the following sequences are convergent or divergent as n approaches infinity. If the sequence does converge, find its limit.

a. $\dfrac{n^2}{n-1} =$ b. $\dfrac{5n}{\sqrt{n^2+1}} =$ c. $\dfrac{8^n}{2^n+10^5} =$ d. $\dfrac{2^n}{8^n+10^5} =$

e. $\dfrac{8^n}{2^n+1} =$ f. $\dfrac{10^5\sqrt{n}}{1+n} =$ g. $10^n =$ h. $\dfrac{0.5^n}{n^2+1} =$

i. $\left(\dfrac{1}{4}\right)^n =$ j. $\dfrac{n^3+1}{n^3+n+1} =$ k. $\sqrt{25-\dfrac{1}{n}} =$ l. $\dfrac{3^n+1}{3^n} =$

m. $\dfrac{n^6}{12n^4+5} =$ n. $\dfrac{n^4+3n}{n^5+3} =$ o. $\dfrac{n^2}{\sqrt{8n^4+1}} =$ p. $\dfrac{1}{n} - \dfrac{1}{n+3} =$

q. $(0.5)^n =$ r. $(0.5)^{-n} =$ s. $\dfrac{\sqrt{n-1}}{2\sqrt{n}} =$ t. $\dfrac{2\sqrt{n}}{\sqrt{n+1}} =$

Solutions:

a. $\boxed{\lim_{n\to\infty}\dfrac{n^2}{n-1}} \approx \boxed{\lim_{n\to\infty}\dfrac{n^2}{n}} = \boxed{\lim_{n\to\infty}\dfrac{n}{1}} = \boxed{\lim_{n\to\infty} n} = \boxed{\infty}$ **diverges**

b. $\boxed{\lim_{n\to\infty}\dfrac{5n}{\sqrt{n^2+1}}} \approx \boxed{\lim_{n\to\infty}\dfrac{5n}{\sqrt{n^2}}} = \boxed{\lim_{n\to\infty}\dfrac{5n}{n}} = \boxed{\lim_{n\to\infty} 5} = \boxed{5}$ **converges to 5**

c. $\boxed{\lim_{n\to\infty}\dfrac{8^n}{2^n+10^5}} \approx \boxed{\lim_{n\to\infty}\dfrac{8^n}{2^n}} = \boxed{\lim_{n\to\infty}\dfrac{2^{3n}}{2^n}} = \boxed{\lim_{n\to\infty} 2^{3n}\cdot 2^{-n}} = \boxed{\lim_{n\to\infty} 2^{3n-n}} = \boxed{\lim_{n\to\infty} 2^{2n}}$

  $= \boxed{2^\infty} = \boxed{\infty}$ **diverges**

d. $\boxed{\lim_{n\to\infty}\dfrac{2^n}{8^n+10^5}} \approx \boxed{\lim_{n\to\infty}\dfrac{2^n}{8^n}} = \boxed{\lim_{n\to\infty}\dfrac{2^n}{2^{3n}}} = \boxed{\lim_{n\to\infty}\dfrac{1}{2^{3n}\cdot 2^{-n}}} = \boxed{\lim_{n\to\infty}\dfrac{1}{2^{3n-n}}} = \boxed{\lim_{n\to\infty}\dfrac{1}{2^{2n}}}$

  $= \boxed{\dfrac{1}{2^\infty}} = \boxed{\dfrac{1}{\infty}} = \boxed{0}$ **converges to 0**

e. $\boxed{\lim_{n\to\infty}\dfrac{8^n}{2^n+1}} \approx \boxed{\lim_{n\to\infty}\dfrac{8^n}{2^n}} = \boxed{\lim_{n\to\infty}\dfrac{2^{3n}}{2^n}} = \boxed{\lim_{n\to\infty} 2^{3n}\cdot 2^{-n}} = \boxed{\lim_{n\to\infty} 2^{3n-n}} = \boxed{\lim_{n\to\infty} 2^{2n}}$

  $= \boxed{2^\infty} = \boxed{\infty}$ **diverges**

f. $\boxed{\lim_{n\to\infty}\dfrac{10^5\sqrt{n}}{1+n}} \approx \boxed{\lim_{n\to\infty}\dfrac{10^5\sqrt{n}}{n}} = \boxed{\lim_{n\to\infty}\dfrac{10^5 n^{\frac{1}{2}}}{n}} = \boxed{\lim_{n\to\infty}\dfrac{10^5}{n\cdot n^{-\frac{1}{2}}}} = \boxed{\lim_{n\to\infty}\dfrac{10^5}{n^{1-\frac{1}{2}}}} = \boxed{\lim_{n\to\infty}\dfrac{10^5}{n^{\frac{1}{2}}}}$

  $= \boxed{\dfrac{10^5}{\infty^{\frac{1}{2}}}} = \boxed{\dfrac{10^5}{\infty}} = \boxed{0}$ **converges to 0**

g. $\boxed{\lim_{n\to\infty} 10^n} = \boxed{10^\infty} = \boxed{\infty}$ **diverges**

h. $\boxed{\lim_{n\to\infty}\dfrac{0.5^n}{n^2+1}} \approx \boxed{\lim_{n\to\infty}\dfrac{0.5^n}{n^2}} = \boxed{\dfrac{0.5^\infty}{\infty^2}} = \boxed{\dfrac{0}{\infty}} = \boxed{0}$ **converges to 0**

i. $\boxed{\lim_{n\to\infty}\left(\dfrac{1}{4}\right)^n} = \boxed{\left(\dfrac{1}{4}\right)^\infty} = \boxed{0.25^\infty} = \boxed{0}$ **converges to 0**

j. $\boxed{\lim_{n\to\infty}\dfrac{n^3+1}{n^3+n+1}} \approx \boxed{\lim_{n\to\infty}\dfrac{n^3}{n^3}} = \boxed{\lim_{n\to\infty}\dfrac{1}{1}} = \boxed{1}$ **converges to 1**

k. $\boxed{\lim_{n\to\infty}\sqrt{25-\dfrac{1}{n}}} \approx \boxed{\sqrt{25-\dfrac{1}{\infty}}} = \boxed{\sqrt{25-0}} = \boxed{\sqrt{5^2}} = \boxed{5}$ **converges to 5**

l. $\boxed{\lim_{n\to\infty}\dfrac{3^n+1}{3^n}} \approx \boxed{\lim_{n\to\infty}\dfrac{3^n}{3^n}} = \boxed{\lim_{n\to\infty}\dfrac{1}{1}} = \boxed{1}$ **converges to 1**

m. $\boxed{\lim_{n\to\infty}\dfrac{n^6}{12n^4+5}} \approx \boxed{\lim_{n\to\infty}\dfrac{n^6}{n^4}} = \boxed{\lim_{n\to\infty}\dfrac{n^6 n^{-4}}{1}} = \boxed{\lim_{n\to\infty}n^2} = \boxed{\infty^2} = \boxed{\infty}$

                                                                                diverges

n. $\boxed{\lim_{n\to\infty}\dfrac{n^4+3n}{n^5+3}} \approx \boxed{\lim_{n\to\infty}\dfrac{n^4}{n^5}} = \boxed{\lim_{n\to\infty}\dfrac{1}{n^5 n^{-4}}} = \boxed{\lim_{n\to\infty}\dfrac{1}{n}} = \boxed{\dfrac{1}{\infty}} = \boxed{0}$

                                                                                converges to 0

o. $\boxed{\lim_{n\to\infty}\dfrac{n^2}{\sqrt{8n^4+1}}} \approx \boxed{\lim_{n\to\infty}\dfrac{n^2}{\sqrt{8n^4}}} = \boxed{\lim_{n\to\infty}\dfrac{n^2}{2n^2\sqrt{2}}} = \boxed{\lim_{n\to\infty}\dfrac{1}{2\sqrt{2}}} = \boxed{\dfrac{1}{2\sqrt{2}}}$

                                                                                converges to $\dfrac{1}{2\sqrt{2}}$

p. $\boxed{\lim_{n\to\infty}\left(\dfrac{1}{n}-\dfrac{1}{n+3}\right)} \approx \boxed{\lim_{n\to\infty}\left(\dfrac{1}{n}-\dfrac{1}{n}\right)} = \boxed{\lim_{n\to\infty}\left(\dfrac{1-1}{n}\right)} = \boxed{\dfrac{0}{\infty}} = \boxed{0}$ **converges to 0**

q. $\boxed{\lim_{n\to\infty}(0.5)^n} = \boxed{(0.5)^\infty} = \boxed{0}$ **converges to 0**

r. $\boxed{\lim_{n\to\infty}(0.5)^{-n}} = \boxed{\dfrac{1}{(0.5)^n}} = \boxed{\dfrac{1}{(0.5)^\infty}} = \boxed{\dfrac{1}{0}} = \boxed{\infty}$ **diverges**

s. $\boxed{\lim_{n\to\infty}\dfrac{\sqrt{n-1}}{2\sqrt{n}}} \approx \boxed{\lim_{n\to\infty}\dfrac{\sqrt{n}}{2\sqrt{n}}} = \boxed{\lim_{n\to\infty}\dfrac{1}{2}} = \boxed{\dfrac{1}{2}}$ **converges to** $\dfrac{1}{2}$

t. $\boxed{\lim_{n\to\infty}\dfrac{2\sqrt{n}}{\sqrt{n+1}}} \approx \boxed{\lim_{n\to\infty}\dfrac{2\sqrt{n}}{\sqrt{n}}} = \boxed{\lim_{n\to\infty}\dfrac{2}{1}} = \boxed{2}$ **converges to 2**

Infinite Geometric Series:

In section 1.4 we stated that the geometric series can be written in the following two forms:

$$S_n = \sum_{k=1}^{n} s_1 r^{k-1} \qquad\qquad (1)$$

$$S_n = \dfrac{s_1\left(1-r^n\right)}{1-r} \quad \text{where } r \neq 1 \qquad (2)$$

In equation (2), let's consider the criteria where $|r|$ is less than one and n is considerably large.

Then, under these conditions r^n approaches to zero and $S_n = \dfrac{s_1\left(1-r^n\right)}{1-r}$ reduces to $S_n = \dfrac{s_1}{1-r}$, i.e.,

$$S_\infty = \sum_{n=1}^{\infty} s_1 r^{n-1} = \lim_{n\to\infty}\dfrac{s_1\left(1-r^n\right)}{1-r} = \dfrac{s_1(1-0)}{1-r} = \dfrac{s_1}{1-r}.$$

Thus, the sum of an **infinite geometric series** as $n \to \infty$ and if $|r| \langle 1$ is equal to

$$S_\infty = \sum_{n=0}^{\infty} s_1 r^n = \sum_{n=1}^{\infty} s_1 r^{n-1} = \frac{s_1}{1-r}. \qquad\qquad (3)$$

In the following examples we will use the above equation in order to find the sum of infinite geometric series.

Example 1.5-5 Find the sum of the following geometric series.

a. $\sum_{j=0}^{\infty} 2\left(\frac{1}{4}\right)^j =$
	b. $\sum_{j=0}^{\infty} 5\left(-\frac{1}{2}\right)^j =$
	c. $\sum_{k=1}^{\infty} 5\left(\frac{5}{3}\right)^{k-1} =$

d. $\sum_{k=1}^{\infty} \frac{18}{10^{k+1}} =$
	e. $\sum_{j=0}^{\infty} \left(\frac{2}{3}\right)^j =$
	f. $\sum_{j=0}^{\infty} \left(-\frac{1}{8}\right)^j =$

Solutions:

a. $s_1 = 2$ and $r = \frac{1}{4}$. Since $|r| \langle 1$ therefore, we can use equation (3) to obtain the sum

$$\boxed{\sum_{j=0}^{\infty} 2\left(\frac{1}{4}\right)^j} = \boxed{\frac{2}{1-\frac{1}{4}}} = \boxed{\frac{2}{\frac{4-1}{4}}} = \boxed{\frac{2}{\frac{3}{4}}} = \boxed{\frac{\frac{2}{1}}{\frac{3}{4}}} = \boxed{\frac{2\times 4}{1\times 3}} = \boxed{\frac{8}{3}}$$

b. $s_1 = 5$ and $r = -\frac{1}{2}$. Since $|r| \langle 1$ therefore, we can use equation (3) to obtain the sum

$$\boxed{\sum_{j=0}^{\infty} 5\left(-\frac{1}{2}\right)^j} = \boxed{\frac{5}{1-\left(-\frac{1}{2}\right)}} = \boxed{\frac{5}{1+\frac{1}{2}}} = \boxed{\frac{5}{\frac{2+1}{2}}} = \boxed{\frac{5}{\frac{3}{2}}} = \boxed{\frac{\frac{5}{1}}{\frac{3}{2}}} = \boxed{\frac{5\times 2}{1\times 3}} = \boxed{\frac{10}{3}}$$

c. $s_1 = 5$ and $r = \frac{5}{3}$. Since $|r| \rangle 1$ therefore, the geometric series $\sum_{k=1}^{\infty} 5\left(\frac{5}{3}\right)^{k-1}$ **has no finite sum.**

d. $\boxed{\sum_{k=1}^{\infty} \frac{18}{10^{k+1}}} = \boxed{\sum_{k=1}^{\infty} \frac{18}{10^2 \cdot 10^{k-1}}} = \boxed{\sum_{k=1}^{\infty} \frac{18}{100}\left(\frac{1}{10^{k-1}}\right)} = \boxed{\sum_{k=1}^{\infty} \frac{18}{100}\left(\frac{1}{10}\right)^{k-1}}$

$s_1 = \frac{18}{100}$ and $r = \frac{1}{10}$. Since $|r| \langle 1$ therefore, we can use equation (3) to obtain the sum

$$\boxed{\sum_{k=1}^{\infty} \frac{18}{100}\left(\frac{1}{10}\right)^{k-1}} = \boxed{\frac{\frac{18}{100}}{1-\frac{1}{10}}} = \boxed{\frac{\frac{18}{100}}{\frac{10-1}{10}}} = \boxed{\frac{\frac{18}{100}}{\frac{9}{10}}} = \boxed{\frac{18\times 10}{100\times 9}} = \boxed{\frac{180}{900}} = \boxed{\frac{1}{5}}$$

e. $s_1 = 1$ and $r = \frac{2}{3}$. Since $|r| \langle 1$ therefore, we can use equation (3) to obtain the sum

$$\boxed{\sum_{j=0}^{\infty} \left(\frac{2}{3}\right)^j} = \boxed{\frac{1}{1-\frac{2}{3}}} = \boxed{\frac{1}{\frac{3-2}{3}}} = \boxed{\frac{1}{\frac{1}{3}}} = \boxed{\frac{\frac{1}{1}}{\frac{1}{3}}} = \boxed{\frac{1\times 3}{1\times 1}} = \boxed{\frac{3}{1}} = \boxed{3}$$

f. $s_1 = 1$ and $r = -\frac{1}{8}$. Since $|r| \langle 1$ therefore, we can use equation (3) to obtain the sum

$$\sum_{j=0}^{\infty}\left(-\frac{1}{8}\right)^{j} = \boxed{\dfrac{1}{1-\left(-\dfrac{1}{8}\right)}} = \boxed{\dfrac{1}{1+\dfrac{1}{8}}} = \boxed{\dfrac{1}{\dfrac{8+1}{8}}} = \boxed{\dfrac{1}{\dfrac{9}{8}}} = \boxed{\dfrac{\dfrac{1}{1}}{\dfrac{9}{8}}} = \boxed{\dfrac{1\times 8}{1\times 9}} = \boxed{\dfrac{8}{9}}$$

Example 1.5-6   Find the sum of the following infinite geometric series.

a. $2-1+\dfrac{1}{2}-\dfrac{1}{4}+\cdots =$ b. $-\dfrac{1}{3}+1-3+9+\cdots =$

c. $1+\dfrac{1}{9}+\dfrac{1}{81}+\dfrac{1}{729}+\cdots =$ d. $1+\dfrac{1}{2}+\dfrac{1}{4}+\dfrac{1}{16}+\cdots =$

Solutions:

a. Given $2-1+\dfrac{1}{2}-\dfrac{1}{4}+\cdots$, $s_1 = 2$ and $r = -\dfrac{1}{2}$. Since $|r| \langle 1$ therefore, we can use equation (3) to obtain the sum, i.e.,

$$\boxed{2-1+\dfrac{1}{2}-\dfrac{1}{4}+\cdots} = \boxed{\dfrac{2}{1-\left(-\dfrac{1}{2}\right)}} = \boxed{\dfrac{2}{1+\dfrac{1}{2}}} = \boxed{\dfrac{2}{\dfrac{2+1}{2}}} = \boxed{\dfrac{2}{\dfrac{3}{2}}} = \boxed{\dfrac{\dfrac{2}{1}}{\dfrac{3}{2}}} = \boxed{\dfrac{2\times 2}{1\times 3}} = \boxed{\dfrac{4}{3}} = \boxed{1.333}$$

Note that in example 1.4-6c the answer to the same problem when $n = 5$ was 1.375 . However, as $n \to \infty$ the answer approaches to 1.333 .

b. Given $-\dfrac{1}{3}+1-3+9+\cdots$, $s_1 = -\dfrac{1}{3}$ and $r = \dfrac{1}{-\dfrac{1}{3}} = -3$. Since $|r| = |-3| = 3$ is greater than one therefore, the geometric sequence $-\dfrac{1}{3}+1-3+9+\cdots$ **has no finite sum.**

c. Given $1+\dfrac{1}{9}+\dfrac{1}{81}+\dfrac{1}{729}+\cdots$, $s_1 = 1$ and $r = \dfrac{\dfrac{1}{9}}{1} = \dfrac{1}{9}$. Since $|r| \langle 1$ therefore, we can use equation (3) to obtain the sum, i.e.,

$$\boxed{1+\dfrac{1}{9}+\dfrac{1}{81}+\dfrac{1}{729}+\cdots} = \boxed{\dfrac{1}{1-\dfrac{1}{9}}} = \boxed{\dfrac{1}{\dfrac{9-1}{9}}} = \boxed{\dfrac{1}{\dfrac{8}{9}}} = \boxed{\dfrac{\dfrac{1}{1}}{\dfrac{8}{9}}} = \boxed{\dfrac{1\times 9}{1\times 8}} = \boxed{\dfrac{9}{8}}$$

d. Given $1+\dfrac{1}{2}+\dfrac{1}{4}+\dfrac{1}{16}+\cdots$, $s_1 = 1$ and $r = \dfrac{\dfrac{1}{2}}{1} = \dfrac{1}{2}$. Since $|r| \langle 1$ therefore, we can use equation (3) to obtain the sum, i.e.,

$$\boxed{1+\dfrac{1}{2}+\dfrac{1}{4}+\dfrac{1}{16}+\cdots} = \boxed{\dfrac{1}{1-\dfrac{1}{2}}} = \boxed{\dfrac{1}{\dfrac{2-1}{2}}} = \boxed{\dfrac{1}{\dfrac{1}{2}}} = \boxed{\dfrac{\dfrac{1}{1}}{\dfrac{1}{2}}} = \boxed{\dfrac{1\times 2}{1\times 1}} = \boxed{\dfrac{2}{1}} = \boxed{2}$$

Repeating Decimals:

An application of infinite geometric series is in representation of repeating decimals as the quotient of two integers. For example, $0.13\overline{13}\cdots$ and $0.66\overline{66}\cdots$ are repeating decimals. The bar above the repeating numbers denotes that the numbers appearing under it are repeated endlessly. The following examples show the steps as to how repeating decimals are converted to fractional forms:

Example 1.5-7 Write the following repeating decimals as the quotient of two positive integers.

a. $0.131313\cdots =$ b. $5.510510\overline{510}\cdots =$ c. $0.12451\overline{245}\cdots =$

Solutions:

a. **First** - write the decimal number $0.1313\overline{13}\cdots$ in its equivalent form of

$$0.1313\overline{13}\cdots = 0.13 + 0.0013 + 0.000013 + \cdots$$

Second - Since this is a geometric series, write the first term in the series and its ratio, i.e.,

$$s_1 = 0.13 \text{ and } r = \frac{0.0013}{0.13} = 0.01.$$

Third - Since the ratio r is less than one, use the infinite geometric series equation $s_\infty = \dfrac{s_1}{1-r}$

to obtain the sum of the infinite series $0.13 + 0.0013 + 0.000013 + \cdots$, i.e.,

$$\boxed{s_\infty = \frac{s_1}{1-r}} \; ; \; \boxed{s_\infty = \frac{0.13}{1-0.01}} \; ; \; \boxed{s_\infty = \frac{0.13}{0.99}} \; ; \; \boxed{s_\infty = \frac{13}{99}} \text{ thus } \boxed{\mathbf{0.131313\cdots = \frac{13}{99}}}$$

b. **First** - Consider the decimal portion of the number $5.510510\overline{510}\cdots$ and write it in its equivalent

form of $0.510510\overline{510}\cdots = 0.510 + 0.000510 + 0.000000510 + \cdots$

Second - Since this is a geometric series, write the first term in the series and its ratio, i.e.,

$$s_1 = 0.510 \text{ and } r = \frac{0.000510}{0.510} = 0.001.$$

Third - Since the ratio r is less than one, use the infinite geometric series equation $s_\infty = \dfrac{s_1}{1-r}$

to obtain the sum of the infinite series $0.510 + 0.000510 + 0.000000510 + \cdots$, i.e.,

$$\boxed{s_\infty = \frac{s_1}{1-r}} \; ; \; \boxed{s_\infty = \frac{0.510}{1-0.001}} \; ; \; \boxed{s_\infty = \frac{0.510}{0.999}} \; ; \; \boxed{s_\infty = \frac{510}{999}} \text{ thus } \boxed{\mathbf{5.510510\overline{510}\cdots = 5\frac{510}{999}}}$$

c. **First** - write the decimal number $0.12451\overline{245}\cdots$ in its equivalent form of

$$0.12451\overline{245}\cdots = 0.1245 + 0.00001245 + \cdots$$

Second - Since this is a geometric series, write the first term in the series and its ratio, i.e.,

$$s_1 = 0.1245 \text{ and } r = \frac{0.00001245}{0.1245} = 0.0001.$$

Third - Since the ratio r is less than one, use the infinite geometric series equation $s_\infty = \dfrac{s_1}{1-r}$

to obtain the sum of the infinite series $0.1245 + 0.00001245 + \cdots$, i.e.,

$$\boxed{s_\infty = \frac{s_1}{1-r}} \; ; \; \boxed{s_\infty = \frac{0.1245}{1-0.0001}} \; ; \; \boxed{s_\infty = \frac{0.1245}{0.9999}} \; ; \; \boxed{s_\infty = \frac{1245}{9999}} \text{ thus } \boxed{\mathbf{0.12451\overline{245}\cdots = \frac{1245}{9999}}}$$

Section 1.5 Practice Problems - Limits of Sequences and Series

1. State which of the following sequences are convergent.

   a. $1, \dfrac{3}{2}, 2, \dfrac{5}{2}, 3, \cdots, \dfrac{n+1}{2}, \cdots =$

   b. $0, \dfrac{3}{2}, \dfrac{8}{3}, \dfrac{15}{4}, \cdots, \dfrac{n^2-1}{n}, \cdots =$

   c. $4, 8, 16, 32, \cdots, 2^{n+1}, \cdots =$

   d. $\dfrac{1}{16}, \dfrac{1}{64}, \dfrac{1}{256}, \dfrac{1}{1024}, \cdots, \dfrac{1}{4^{n+1}}, \cdots =$

   e. $0, \dfrac{1}{4}, \dfrac{2}{9}, \dfrac{3}{16}, \cdots, \dfrac{n-1}{n^2}, \cdots =$

   f. $\dfrac{1}{25}, \dfrac{1}{125}, \dfrac{1}{625}, \dfrac{1}{3125}, \cdots, \left(\dfrac{1}{5}\right)^{n+1}, \cdots =$

2. State which of the following geometric sequences are convergent.

   a. $\dfrac{1}{4}, \dfrac{1}{16}, \dfrac{1}{64}, \dfrac{1}{256}, \cdots, \dfrac{1}{4^n}, \cdots =$

   b. $-5, 25, -125, 625, -3125, \cdots, (-5)^n, \cdots =$

   c. $2, -2, 2, -2, \cdots, 2(-1)^{n+1}, \cdots =$

   d. $1, \dfrac{1}{2}, \dfrac{1}{4}, \dfrac{1}{8}, \cdots, \left(\dfrac{1}{2}\right)^{n-1}, \cdots =$

   e. $-9, 27, -81, 243, \cdots, (-1)^n 3^{n+1}, \cdots =$

   f. $1, \dfrac{1}{3}, \dfrac{1}{9}, \dfrac{1}{27}, \dfrac{1}{81}, \cdots, \left(\dfrac{1}{3}\right)^{n-1}, \cdots =$

3. State whether or not the following sequences converges or diverges as $n \to \infty$. If it does converge, find the limit.

   a. $\dfrac{n^2}{n^3-4} =$

   b. $\dfrac{5n+1}{\sqrt{n^2+1}} =$

   c. $\dfrac{25^n}{5^{n+1}} =$

   d. $\dfrac{5^n+25}{125^n} =$

   e. $\dfrac{(n+2)^2}{n^2} =$

   f. $\dfrac{2^n}{2^n+1} =$

   g. $\dfrac{\sqrt{n^2+2n}}{\sqrt{n^4+1}} =$

   h. $\dfrac{5}{n^2+1} =$

   i. $\dfrac{n+1}{n-1} =$

   j. $\dfrac{n}{n^3-1} =$

   k. $10^{\frac{1}{n}} =$

   l. $\dfrac{(n-1)^2}{(1-n)(1+n)} =$

   m. $100^{-\frac{1}{n}} =$

   n. $3^{\frac{3}{n}} =$

   o. $\dfrac{n+100}{n^3-10} =$

   p. $\dfrac{100^{\frac{1}{n}}}{n^2+3} =$

   q. $\dfrac{1}{n+1}-1 =$

   r. $(0.25)^{-n} =$

   s. $\dfrac{\sqrt{n+1}}{\sqrt{n}+1} =$

   t. $\dfrac{\sqrt{n}}{\sqrt{n}+1}+2 =$

4. Find the sum of the following geometric series.

   a. $\displaystyle\sum_{j=0}^{\infty} 3\left(\dfrac{1}{8}\right)^j =$

   b. $\displaystyle\sum_{j=0}^{\infty} 3\left(-\dfrac{1}{4}\right)^j =$

   c. $\displaystyle\sum_{k=1}^{\infty} 3\left(\dfrac{3}{2}\right)^{k-1} =$

   d. $\displaystyle\sum_{k=1}^{\infty} \dfrac{5}{100^{k+1}} =$

   e. $\displaystyle\sum_{j=0}^{\infty} \left(\dfrac{1}{3}\right)^j =$

   f. $\displaystyle\sum_{j=0}^{\infty} \left(-\dfrac{1}{5}\right)^j =$

5. Find the sum of the following infinite geometric series.

   a. $5 - 1 + \dfrac{1}{5} - \dfrac{1}{25} + \cdots =$

   b. $-\dfrac{1}{2} + 2 - 8 + 32 + \cdots =$

   c. $1 + \dfrac{1}{6} + \dfrac{1}{36} + \dfrac{1}{216} + \cdots =$

   d. $1 + \dfrac{1}{10} + \dfrac{1}{100} + \dfrac{1}{1000} + \cdots =$

6. Write the following repeating decimals as the quotient of two positive integers.

   a. $0.66666\overline{6} \cdots =$

   b. $3.027027\overline{027} \cdots =$

   c. $0.11111\overline{1} \cdots =$

1.6 The Factorial Notation

As was stated earlier, the product of several consecutive positive integers is usually written using a special symbol $n!$, read as "n factorial," which is defined by

$$n! = n\,(n-1)(n-2)(n-3)(n-4)...(4)(3)(2)(1)$$

For example, 1! (read as "one factorial") through 10! (read as "ten factorial") are written in their equivalent form as:

$1! = 1$

$2! = 2\cdot1 = 2$

$3! = 3\cdot2\cdot1 = 6$

$4! = 4\cdot3\cdot2\cdot1 = 24$

$5! = 5\cdot4\cdot3\cdot2\cdot1 = 120$

$6! = 6\cdot5\cdot4\cdot3\cdot2\cdot1 = 720$

$7! = 7\cdot6\cdot5\cdot4\cdot3\cdot2\cdot1 = 5{,}040$

$8! = 8\cdot7\cdot6\cdot5\cdot4\cdot3\cdot2\cdot1 = 40{,}320$

$9! = 9\cdot8\cdot7\cdot6\cdot5\cdot4\cdot3\cdot2\cdot1 = 362{,}880$

$10! = 10\cdot9\cdot8\cdot7\cdot6\cdot5\cdot4\cdot3\cdot2\cdot1 = 3{,}628{,}800$

Note that, for $n \rangle 1$, since $n! = n\,(n-1)(n-2)(n-3)...4\cdot3\cdot2\cdot1$ and $(n-1)! = (n-1)(n-2)(n-3)\cdots4\cdot3\cdot2\cdot1$

we can rewrite the recursive $n!$ relationship in the following way:

$$n! = n\underbrace{(n-1)(n-2)(n-3)(n-4)..4\cdot3\cdot2\cdot1}_{(n-1)!} = n\,(n-1)!$$

$$n\,(n-1)! = n(n-1)\underbrace{(n-2)(n-3)(n-4)..4\cdot3\cdot2\cdot1}_{(n-2)!} = n\,(n-1)(n-2)!$$

$$n\,(n-1)(n-2)! = n(n-1)(n-2)\underbrace{(n-3)(n-4)..4\cdot3\cdot2\cdot1}_{(n-3)!} = n\,(n-1)(n-2)(n-3)!$$

$$n\,(n-1)(n-2)(n-3)! = n(n-1)(n-2)(n-3)\underbrace{(n-4)..4\cdot3\cdot2\cdot1}_{(n-4)!} = n\,(n-1)(n-2)(n-3)(n-4)!$$

For example,

$7! = 7\cdot6! = 7\cdot6\cdot5! = 7\cdot6\cdot5\cdot4! = 7\cdot6\cdot5\cdot4\cdot3! = 7\cdot6\cdot5\cdot4\cdot3\cdot2!$

$10! = 10\cdot9! = 10\cdot9\cdot8! = 10\cdot9\cdot8\cdot7! = 10\cdot9\cdot8\cdot7\cdot6! = 10\cdot9\cdot8\cdot7\cdot6\cdot5!$

$35! = 35\cdot34! = 35\cdot34\cdot33! = 35\cdot34\cdot33\cdot32! = 35\cdot34\cdot33\cdot32\cdot31! = 35\cdot34\cdot33\cdot32\cdot31\cdot30!$

$(n+4)! = (n+4)(n+3)! = (n+4)(n+3)(n+2)! = (n+4)(n+3)(n+2)(n+1)!$

$(n+8)! = (n+8)(n+7)! = (n+8)(n+7)(n+6)! = (n+8)(n+7)(n+6)(n+5)!$

$(2n+2)! = (2n+2)(2n+1)! = (2n+2)(2n+1)(2n)! = (2n+2)(2n+1)(2n)(2n-1)! = (2n+2)(2n+1)(2n)$

$(2n-1)(2n-2)! = (2n+2)(2n+1)(2n)(2n-1)(2n-2)(2n-3)!$

$$\boxed{(2n+5)!} = \boxed{(2n+5)(2n+4)!} = \boxed{(2n+5)(2n+4)(2n+3)!} = \boxed{(2n+5)(2n+4)(2n+3)(2n+2)!} = \boxed{(2n+5)(2n+4)}$$

$$\boxed{(2n+3)(2n+2)(2n+1)!} = \boxed{\mathbf{(2n+5)(2n+4)(2n+3)(2n+2)(2n+1)(2n)!}}$$

The above principal can be used to prove that $0!=1$. Since $1! = 1\cdot(1-1)! = 1\cdot 0!$ in order for equality on both sides of the equation to hold true $0!$ must be equal to one. Hence, we state that $0!=1$.

Example 1.6-1 Expand and simplify the following factorial expressions.

a. $13! =$ b. $(6-2)! =$ c. $\dfrac{8!}{4!} =$ d. $\dfrac{12!}{11!} =$

e. $\dfrac{10!}{5!4!} =$ f. $\dfrac{8!}{3!(8-2)!} =$ g. $\dfrac{11!8!}{16!} =$ h. $\dfrac{(4-2)!8!}{11!(5-3)!} =$

Solutions:

a. $\boxed{13!} = \boxed{13\cdot12\cdot11\cdot10\cdot9\cdot8\cdot7\cdot6\cdot5\cdot4\cdot3\cdot2\cdot1} = \boxed{\mathbf{6,227,020,800}}$

b. $\boxed{(6-2)!} = \boxed{4!} = \boxed{4\cdot3\cdot2\cdot1} = \boxed{\mathbf{24}}$

c. $\boxed{\dfrac{8!}{4!}} = \boxed{\dfrac{8\cdot7\cdot6\cdot5\cdot4!}{4!}} = \boxed{\dfrac{8\cdot7\cdot6\cdot5\cdot\cancel{4!}}{\cancel{4!}}} = \boxed{8\cdot7\cdot6\cdot5} = \boxed{\mathbf{1680}}$

d. $\boxed{\dfrac{12!}{11!}} = \boxed{\dfrac{12\cdot11!}{11!}} = \boxed{\dfrac{12\cdot\cancel{11!}}{\cancel{11!}}} = \boxed{\mathbf{12}}$

e. $\boxed{\dfrac{10!}{5!4!}} = \boxed{\dfrac{10\cdot9\cdot8\cdot7\cdot6\cdot5!}{5!4!}} = \boxed{\dfrac{10\cdot9\cdot8\cdot7\cdot6\cdot\cancel{5!}}{\cancel{5!}4!}} = \boxed{\dfrac{10\cdot9\cdot8\cdot7\cdot6}{4!}} = \boxed{\dfrac{10\cdot9\cdot\overset{2}{\cancel{8}}\cdot7\cdot6}{4\cdot3\cdot2\cdot1}} = \boxed{\dfrac{10\cdot9\cdot2\cdot7}{1}} = \boxed{\mathbf{1260}}$

f. $\boxed{\dfrac{8!}{3!(8-2)!}} = \boxed{\dfrac{8!}{3!6!}} = \boxed{\dfrac{8\cdot7\cdot6!}{3!6!}} = \boxed{\dfrac{8\cdot7}{3!}} = \boxed{\dfrac{\overset{4}{\cancel{8}}\cdot7}{3\cdot2\cdot1}} = \boxed{\dfrac{4\cdot7}{3}} = \boxed{\dfrac{\mathbf{28}}{\mathbf{3}}}$

g. $\boxed{\dfrac{11!8!}{16!}} = \boxed{\dfrac{11!8!}{16\cdot15\cdot14\cdot13\cdot12\cdot11!}} = \boxed{\dfrac{11!8\cdot7\cdot6\cdot5\cdot4\cdot3\cdot2\cdot1}{16\cdot15\cdot14\cdot13\cdot12\cdot11!}} = \boxed{\dfrac{8\cdot7\cdot6\cdot5\cdot4\cdot3\cdot2\cdot1}{\underset{2}{\cancel{16}}\cdot\underset{3}{\cancel{15}}\cdot\underset{2}{\cancel{14}}\cdot13\cdot\underset{2}{\cancel{12}}}} = \boxed{\dfrac{4\cdot3\cdot2\cdot1}{2\cdot3\cdot2\cdot13\cdot2}} = \boxed{\dfrac{\mathbf{1}}{\mathbf{13}}}$

h. $\boxed{\dfrac{(4-2)!8!}{11!(5-3)!}} = \boxed{\dfrac{2!8!}{11!2!}} = \boxed{\dfrac{8!}{11!}} = \boxed{\dfrac{8!}{11\cdot10\cdot9\cdot8!}} = \boxed{\dfrac{8!}{11\cdot10\cdot9\cdot8!}} = \boxed{\dfrac{\mathbf{1}}{\mathbf{990}}}$

Example 1.6-2 Write the following products in factorial form.

a. $4\cdot3\cdot2\cdot1 =$ b. $10\cdot11\cdot12\cdot13 =$ c. $20\cdot21\cdot22\cdot23\cdot24 =$

d. $9\cdot8\cdot7\cdot6\cdot5 =$ e. $3\cdot4\cdot5\cdot6\cdot7 =$ f. $20 =$

g. $15\cdot14\cdot13\cdot12\cdot11 =$ h. $8 =$ i. $1\cdot2\cdot3\cdot4\cdot5\cdot6 =$

Solutions:

a. $\boxed{4\cdot3\cdot2\cdot1} = \boxed{4!}$

b. $\boxed{10\cdot11\cdot12\cdot13} = \boxed{\dfrac{13!}{9!}}$

c. $\boxed{20\cdot21\cdot22\cdot23\cdot24} = \boxed{\dfrac{24!}{19!}}$

d. $\boxed{9\cdot8\cdot7\cdot6\cdot5} = \boxed{\dfrac{9!}{4!}}$

e. $\boxed{3\cdot4\cdot5\cdot6\cdot7} = \boxed{\dfrac{7!}{2!}}$

f. $\boxed{20} = \boxed{\dfrac{20!}{19!}}$

g. $\boxed{15\cdot14\cdot13\cdot12\cdot11} = \boxed{\dfrac{15!}{10!}}$

h. $\boxed{8} = \boxed{\dfrac{8!}{7!}}$

i. $\boxed{1\cdot2\cdot3\cdot4\cdot5\cdot6} = \boxed{6!}$

Example 1.6-3 Expand the following factorial expressions.

a. $n! =$

b. $(n-1)! =$

c. $(n-3)! =$

d. $(n-5)! =$

e. $(n-8)! =$

f. $(2n)! =$

g. $(2n-1)! =$

h. $(3n-5)! =$

i. $(2n+1)! =$

j. $(n+1)! =$

k. $(n+3)! =$

l. $(n+8)! =$

m. $(2n+2)! =$

n. $(2n+5)! =$

Solutions:

a. $\boxed{n!} = \boxed{(n)(n-1)(n-2)(n-3)(n-4)(n-5)(n-6)(n-7)(n-8)(n-9)\cdots4\cdot3\cdot2\cdot1}$

b. $\boxed{(n-1)!} = \boxed{(n-1)(n-2)(n-3)(n-4)(n-5)(n-6)(n-7)(n-8)(n-9)\cdots4\cdot3\cdot2\cdot1}$

c. $\boxed{(n-3)!} = \boxed{(n-3)(n-4)(n-5)(n-6)(n-7)(n-8)(n-9)\cdots4\cdot3\cdot2\cdot1}$

d. $\boxed{(n-5)!} = \boxed{(n-5)(n-6)(n-7)(n-8)(n-9)(n-10)(n-11)\cdots4\cdot3\cdot2\cdot1}$

e. $\boxed{(n-8)!} = \boxed{(n-8)(n-9)(n-10)(n-11)\cdots4\cdot3\cdot2\cdot1}$

f. $\boxed{(2n)!} = \boxed{(2n)(2n-1)(2n-2)(2n-3)(2n-4)\cdots4\cdot3\cdot2\cdot1}$

g. $\boxed{(2n-1)!} = \boxed{(2n-1)(2n-2)(2n-3)(2n-4)\cdots4\cdot3\cdot2\cdot1}$

h. $\boxed{(3n-5)!} = \boxed{(3n-5)(3n-6)(3n-7)(3n-8)(3n-9)(3n-10)(3n-11)\cdots4\cdot3\cdot2\cdot1}$

i. $\boxed{(2n+1)!} = \boxed{(2n+1)(2n)(2n-1)(2n-2)(2n-3)(2n-4)\cdots4\cdot3\cdot2\cdot1}$

j. $\boxed{(n+1)!} = \boxed{(n+1)(n)(n-1)(n-2)(n-3)(n-4)(n-5)(n-6)(n-7)(n-8)\cdots4\cdot3\cdot2\cdot1}$

k. $\boxed{(n+3)!} = \boxed{(n+3)(n+2)(n+1)(n)(n-1)(n-2)(n-3)(n-4)(n-5)(n-6)(n-7)(n-8)\cdots4\cdot3\cdot2\cdot1}$

l. $\boxed{(n+8)!} = \boxed{(n+8)(n+7)(n+6)(n+5)(n+4)(n+3)(n+2)(n+1)(n)(n-1)(n-2)(n-3)\cdots4\cdot3\cdot2\cdot1}$

m. $\boxed{(2n+2)!}$ = $\boxed{(2n+2)(2n+1)(2n)(2n-1)(2n-2)(2n-3)(2n-4)\cdots 4\cdot 3\cdot 2\cdot 1}$

n. $\boxed{(2n+5)!}$ = $\boxed{(2n+5)(2n+4)(2n+3)(2n+2)(2n+1)(2n)(2n-1)(2n-2)(2n-3)(2n-4)\cdots 4\cdot 3\cdot 2\cdot 1}$

Example 1.6-4 Expand and simplify the following factorial expressions.

a. $\dfrac{(n-1)!}{(n-3)!} =$ b. $\dfrac{(n+2)!}{n!} =$ c. $\dfrac{(n+3)!}{(n-1)!} =$ d. $\dfrac{(n+2)(n+2)!}{(n+3)!} =$

e. $\dfrac{(3n+1)!(3n-1)!}{(3n)!(3n-3)!} =$ f. $\dfrac{(n!)^2}{(n+1)!(n-1)!} =$ g. $\dfrac{(2n-2)!\,2(n!)}{(2n)!(n-1)!} =$

Solutions:

a. $\boxed{\dfrac{(n-1)!}{(n-3)!}} = \boxed{\dfrac{(n-1)!}{(n-3)(n-2)(n-1)!}} = \boxed{\dfrac{(n-1)!}{(n-3)(n-2)(n-1)!}} = \boxed{\dfrac{1}{(n-3)(n-2)}}$

b. $\boxed{\dfrac{(n+2)!}{n!}} = \boxed{\dfrac{(n+2)(n+1)n!}{n!}} = \boxed{\dfrac{(n+2)(n+1)n!}{n!}} = \boxed{(n+2)(n+1)}$

c. $\boxed{\dfrac{(n+3)!}{(n-1)!}} = \boxed{\dfrac{(n+3)(n+2)(n+1)(n)(n-1)!}{(n-1)!}} = \boxed{\dfrac{(n+3)(n+2)(n+1)(n)(n-1)!}{(n-1)!}} = \boxed{(n+3)(n+2)(n+1)n}$

d. $\boxed{\dfrac{(n+2)(n+2)!}{(n+3)!}} = \boxed{\dfrac{(n+2)(n+2)!}{(n+3)(n+2)!}} = \boxed{\dfrac{(n+2)(n+2)!}{(n+3)(n+2)!}} = \boxed{\dfrac{n+2}{n+3}}$

e. $\boxed{\dfrac{(3n+1)!(3n-1)!}{(3n)!(3n-3)!}} = \boxed{\dfrac{\left[(3n+1)(3n)!\right]\left[(3n-1)(3n-2)(3n-3)!\right]}{(3n)!(3n-3)!}} = \boxed{\dfrac{\left[(3n+1)(3n)!\right]\left[(3n-1)(3n-2)(3n-3)!\right]}{(3n)!(3n-3)!}}$

$= \boxed{\dfrac{(3n+1)(3n-1)(3n-2)}{1}} = \boxed{(3n+1)(3n-1)(3n-2)}$

f. $\boxed{\dfrac{(n!)^2}{(n+1)!(n-1)!}} = \boxed{\dfrac{(n!)(n!)}{\left[(n+1)(n)!\right](n-1)!}} = \boxed{\dfrac{(n!)(n!)}{\left[(n+1)(n)!\right](n-1)!}} = \boxed{\dfrac{n!}{(n+1)(n-1)!}} = \boxed{\dfrac{n(n-1)!}{(n+1)(n-1)!}} = \boxed{\dfrac{n}{n+1}}$

g. $\boxed{\dfrac{(2n-2)!\,2(n!)}{(2n)!(n-1)!}} = \boxed{\dfrac{(2n-2)!\left[2n(n-1)!\right]}{\left[(2n)(2n-1)(2n-2)!\right](n-1)!}} = \boxed{\dfrac{(2n-2)!\left[2n(n-1)!\right]}{\left[(2n)(2n-1)(2n-2)!\right](n-1)!}} = \boxed{\dfrac{2n}{(2n)(2n-1)}}$

$= \boxed{\dfrac{2n}{(2n)(2n-1)}} = \boxed{\dfrac{1}{2n-1}}$

The factorial notation is used in expansion of the $\dbinom{n}{r}$ read as "the binomial coefficient n, r" in the following way:

$$\binom{n}{r} = \dfrac{n!}{r!(n-r)!}$$

Example 1.6-5 Write the following expressions in factorial notation form. Simplify the answer.

a. $\binom{6}{4} =$ b. $\binom{6}{2} =$ c. $\binom{5}{0} =$ d. $\binom{5}{5} =$ e. $\binom{7}{5} =$

f. $\binom{10}{5} =$ g. $\binom{8}{4} =$ h. $\binom{n}{n} =$ i. $\binom{n}{n-1} =$ j. $\binom{n}{n-4} =$

Solutions:

a. $\binom{6}{4} = \dfrac{6!}{4!(6-4)!} = \dfrac{6!}{4!\cdot2!} = \dfrac{6\cdot5\cdot4\cdot3\cdot2\cdot1}{(4\cdot3\cdot2\cdot1)\cdot(2\cdot1)} = \dfrac{\overset{3}{6}\cdot5\cdot4\cdot3\cdot2\cdot1}{(4\cdot3\cdot2\cdot1)\cdot(2\cdot1)} = \dfrac{3\cdot5}{1} = \boxed{15}$

b. $\binom{6}{2} = \dfrac{6!}{2!(6-2)!} = \dfrac{6!}{2!\cdot4!} = \dfrac{6\cdot5\cdot4\cdot3\cdot2\cdot1}{(2\cdot1)\cdot(4\cdot3\cdot2\cdot1)} = \dfrac{\overset{3}{6}\cdot5\cdot4\cdot3\cdot2\cdot1}{(2\cdot1)\cdot(4\cdot3\cdot2\cdot1)} = \dfrac{3\cdot5}{1} = \boxed{15}$

c. $\binom{5}{0} = \dfrac{5!}{0!(5-0)!} = \dfrac{5!}{0!\cdot5!} = \dfrac{5!}{1\cdot5!} = \dfrac{1}{1} = \boxed{1}$

d. $\binom{5}{5} = \dfrac{5!}{5!(5-5)!} = \dfrac{5!}{5!\cdot0!} = \dfrac{5!}{5!\cdot1} = \dfrac{1}{1} = \boxed{1}$

e. $\binom{7}{5} = \dfrac{7!}{5!(7-5)!} = \dfrac{7!}{5!\cdot2!} = \dfrac{7\cdot6\cdot5!}{5!\cdot2!} = \dfrac{7\cdot\overset{3}{6}\cdot5!}{5!\cdot2\cdot1} = \dfrac{7\cdot3}{1} = \boxed{21}$

f. $\binom{10}{5} = \dfrac{10!}{5!(10-5)!} = \dfrac{10!}{5!\cdot5!} = \dfrac{10\cdot9\cdot8\cdot7\cdot6\cdot5\cdot4\cdot3\cdot2\cdot1}{(5\cdot4\cdot3\cdot2\cdot1)\cdot(5\cdot4\cdot3\cdot2\cdot1)} = \dfrac{\overset{2}{10}\cdot9\cdot\overset{2}{8}\cdot7\cdot6\cdot5\cdot4\cdot3\cdot2\cdot1}{(5\cdot4\cdot3\cdot2\cdot1)\cdot(5\cdot4\cdot3\cdot2\cdot1)} = \dfrac{2\cdot9\cdot2\cdot7}{1} = \boxed{252}$

g. $\binom{8}{4} = \dfrac{8!}{4!(8-4)!} = \dfrac{8!}{4!\cdot4!} = \dfrac{8\cdot7\cdot6\cdot5\cdot4!}{4!\cdot4\cdot3\cdot2\cdot1} = \dfrac{8\cdot7\cdot\overset{2}{6}\cdot5\cdot4!}{4!\cdot4\cdot3\cdot2\cdot1} = \dfrac{7\cdot2\cdot5}{1} = \boxed{70}$

h. $\binom{n}{n} = \dfrac{n!}{n!(n-n)!} = \dfrac{n!}{n!0!} = \dfrac{n!}{n!\cdot1} = \dfrac{1}{1} = \boxed{1}$

i. $\binom{n}{n-1} = \dfrac{n!}{(n-1)![n-(n-1)]!} = \dfrac{n!}{(n-1)!(n-n+1)!} = \dfrac{n!}{(n-1)!1!} = \dfrac{n\cdot(n-1)!}{(n-1)!} = \dfrac{n\cdot(n-1)!}{(n-1)!} = \boxed{n}$

j. $\binom{n}{n-4} = \dfrac{n!}{(n-4)![n-(n-4)]!} = \dfrac{n!}{(n-4)!(n-n+4)!} = \dfrac{n!}{(n-4)!4!} = \dfrac{n\cdot(n-1)\cdot(n-2)\cdot(n-3)\cdot(n-4)!}{(n-4)!4!}$

$= \dfrac{n\cdot(n-1)\cdot(n-2)\cdot(n-3)\cdot(n-4)!}{(n-4)!4!} = \dfrac{n\cdot(n-1)\cdot(n-2)\cdot(n-3)}{4\cdot3\cdot2\cdot1} = \dfrac{n\cdot(n-1)\cdot(n-2)\cdot(n-3)}{24}$

An application of the binomial coefficient n, r, i.e., $\binom{n}{r}$ is in its use for expansion of binomials of the form $(a+b)^n$, where n is a positive integer. In general, the binomial equation of order n can be expanded in the following form:

$$(a+b)^n = \binom{n}{0}a^n + \binom{n}{1}a^{n-1}b + \binom{n}{2}a^{n-2}b^2 + \cdots + \binom{n}{r-1}a^{n-r+1}b^{r-1} + \cdots + \binom{n}{n}b^n \qquad (1)$$

note that the above equation is used for expanding binomial expressions that are raised to the second, third, fourth, or higher powers. For example, $(x-2)^2$, $(x+3)^3$, $(x-1)^5$, $(x-3)^6$, $(2x+y)^4$, $\left(2x-\frac{y}{2}\right)^5$, $(x-\sqrt{3})^4$, $(x-2y)^6$, $(x-y)^{16}$, etc. can all be expanded using the above equation. The following examples show the steps as to how binomial coefficients are used in expanding binomial equations:

Example 1.6-6 Expand the following binomial expressions.

a. $(x-1)^3 =$ b. $(x+2)^5 =$ c. $(x-3)^4 =$

Solutions:

a. **First** - Identify the a, b, and n terms, i.e., $a = x$, $b = -1$, $n = 3$.

Second - Use the general binomial expansion formula, i.e., equation (1) above to expand $(x-1)^3$.

$$\boxed{(x-1)^3} = \boxed{\binom{3}{0}x^3 + \binom{3}{1}x^{3-1}\cdot(-1) + \binom{3}{2}x^{3-2}(-1)^2 + \binom{3}{3}x^{3-3}(-1)^3} = \boxed{\binom{3}{0}x^3 - \binom{3}{1}x^2 + \binom{3}{2}x - \binom{3}{3}x^0}$$

$$= \boxed{\binom{3}{0}x^3 - \binom{3}{1}x^2 + \binom{3}{2}x - \binom{3}{3}\cdot1} = \boxed{\frac{3!}{0!(3-0)!}x^3 - \frac{3!}{1!(3-1)!}x^2 + \frac{3!}{2!(3-2)!}x - \frac{3!}{3!(3-3)!}}$$

$$= \boxed{\frac{3!}{3!}x^3 - \frac{3!}{2!}x^2 + \frac{3!}{2!}x - \frac{3!}{3!}} = \boxed{\frac{3!}{3!}x^3 - \frac{3\cdot2!}{2!}x^2 + \frac{3\cdot2!}{2!}x - \frac{3!}{3!}} = \boxed{x^3 - 3x^2 + 3x - 1}$$

b. **First** - Identify the a, b, and n terms, i.e., $a = x$, $b = 2$, $n = 5$.

Second - Expand $(x+2)^5$ using equation (1).

$$\boxed{(x+2)^5} = \boxed{\binom{5}{0}x^5 + \binom{5}{1}x^{5-1}\cdot2 + \binom{5}{2}x^{5-2}\cdot2^2 + \binom{5}{3}x^{5-3}\cdot2^3 + \binom{5}{4}x^{5-4}\cdot2^4 + \binom{5}{5}x^{5-5}\cdot2^5}$$

$$= \boxed{\binom{5}{0}x^5 + 2\binom{5}{1}x^4 + 4\binom{5}{2}x^3 + 8\binom{5}{3}x^2 + 16\binom{5}{4}x + 32\binom{5}{5}x^0} = \boxed{\binom{5}{0}x^5 + 2\binom{5}{1}x^4 + 4\binom{5}{2}x^3 + 8\binom{5}{3}x^2}$$

$$\boxed{+16\binom{5}{4}x + 32\binom{5}{5}} = \boxed{\frac{5!}{0!(5-0)!}x^5 + 2\frac{5!}{1!(5-1)!}x^4 + 4\frac{5!}{2!(5-2)!}x^3 + 8\frac{5!}{3!(5-3)!}x^2 + 16\frac{5!}{4!(5-4)!}x}$$

$$\boxed{+32\frac{5!}{5!(5-5)!}} = \boxed{\frac{5!}{5!}x^5 + 2\frac{5!}{4!}x^4 + 4\frac{5!}{2!3!}x^3 + 8\frac{5!}{3!2!}x^2 + 16\frac{5!}{4!}x + 32\frac{5!}{5!}} = \boxed{\frac{5!}{5!}x^5 + 2\frac{5\cdot4!}{4!}x^4}$$

$$\boxed{+4\frac{5\cdot4\cdot3!}{2!3!}x^3 + 8\frac{5\cdot4\cdot3!}{3!2!}x^2 + 16\frac{5\cdot4!}{4!}x + 32\frac{5!}{5!}} = \boxed{x^5 + 10\,x^4 + 40\,x^3 + 80x^2 + 80x + 32}$$

c. **First** - Identify the $a, b,$ and n terms, i.e., $a = x$, $b = -3$, $n = 4$.

Second - Expand $(x-3)^4$ using equation (1).

$$\boxed{(x-3)^4} = \boxed{\binom{4}{0}x^4 + \binom{4}{1}x^{4-1}\cdot(-3) + \binom{4}{2}x^{4-2}(-3)^2 + \binom{4}{3}x^{4-3}(-3)^3 + \binom{4}{4}x^{4-4}(-3)^4} = \boxed{\binom{4}{0}x^4 - 3\binom{4}{1}x^3}$$

$$\boxed{+9\binom{4}{2}x^2 - 27\binom{4}{3}x + 81\binom{4}{4}x^0} = \boxed{\frac{4!}{0!(4-0)!}x^4 - 3\frac{4!}{1!(4-1)!}x^3 + 9\frac{4!}{2!(4-2)!}x^2 - 27\frac{4!}{3!(4-3)!}x}$$

$$\boxed{+81\frac{4!}{4!(4-4)!}} = \boxed{\frac{4!}{4!}x^4 - 3\frac{4!}{3!}x^3 + 9\frac{4!}{2!2!}x^2 - 27\frac{4!}{3!}x + 81\frac{4!}{4!0!}} = \boxed{\frac{4!}{4!}x^4 - 3\frac{4\cdot3!}{3!}x^3 + 9\frac{4\cdot3\cdot2!}{2!2!}x^2}$$

$$\boxed{-27\frac{4\cdot3!}{3!}x + 81\frac{4!}{4!0!}} = \boxed{x^4 - 12\,x^3 + 54\,x^2 - 108x + 81}$$

Example 1.6-7 Use the general equation for binomial expansion to solve the following exponential numbers to the nearest hundredth.

a. $(0.83)^6 =$ b. $(1.05)^8 =$ c. $(1.21)^{10} =$

Solutions:

a. **First** - Write the exponential expression in the form of $(0.83)^6 = (1-0.17)^6$.

Second - Identify the $a, b,$ and n terms, i.e., $a = 1$, $b = -0.17$, $n = 6$.

Third - Use the general binomial expansion formula, i.e., equation (1) above to expand $(1-0.17)^6$.

$$\boxed{(1-0.17)^6} = \boxed{\binom{6}{0}1^6 + \binom{6}{1}1^5\cdot(-0.17) + \binom{6}{2}1^4\cdot(-0.17)^2 + \binom{6}{3}1^3\cdot(-0.17)^3 + \binom{6}{4}1^2\cdot(-0.17)^4 + \binom{6}{5}1\cdot(-0.17)^5}$$

$$\boxed{+\binom{6}{6}(-0.17)^6} = \boxed{\binom{6}{0} - 0.17\binom{6}{1} + 0.0289\binom{6}{2} - 0.0049\binom{6}{3} + \cdots} = \boxed{\frac{6!}{0!(6-0)!} - 0.17\frac{6!}{1!(6-1)!}}$$

$$\boxed{+0.0289\frac{6!}{2!(6-2)!} - 0.0049\frac{6!}{3!(6-3)!} + \cdots} = \boxed{\frac{6!}{6!} - 0.17\frac{6!}{5!} + 0.0289\frac{6!}{2!4!} - 0.0049\frac{6!}{3!3!} + \cdots}$$

$$= \boxed{\frac{6!}{6!} - 0.17\frac{6\cdot5!}{5!} + 0.0289\frac{6\cdot5\cdot4!}{2!4!} - 0.0049\frac{6\cdot5\cdot4\cdot3!}{3!3!} + \cdots} = \boxed{1 - 1.02 + 0.4335 - 0.098 + \cdots} \approx \boxed{0.3155}$$

Therefore, $(1-0.17)^6$, to the nearest hundredth, is equal to $\boxed{0.32}$

b. **First** - Write the exponential expression in the form of $(1.05)^8 = (1+0.05)^8$.

Second - Identify the $a, b,$ and n terms, i.e., $a=1$, $b=0.05$, $n=8$.

Third - Use the general binomial expansion formula, i.e., equation (1) above to expand $(1+0.05)^8$.

$$\boxed{(1+0.05)^8} = \boxed{\binom{8}{0}1^8 + \binom{8}{1}1^7\cdot(0.05) + \binom{8}{2}1^6\cdot(0.05)^2 + \binom{8}{3}1^3\cdot(0.05)^3 + \binom{8}{4}1^2\cdot(0.05)^4 + \binom{8}{5}1\cdot(0.05)^5 + \cdots}$$

$$= \boxed{\binom{8}{0} + 0.05\binom{8}{1} + 0.0025\binom{8}{2} + 0.000125\binom{8}{3} + \cdots} = \boxed{\frac{8!}{0!\,(8-0)!} + 0.05\frac{8!}{1!\,(8-1)!} + 0.0025\frac{8!}{2!\,(8-2)!}}$$

$$\boxed{+0.000125\frac{8!}{3!\,(8-3)!} + \cdots} = \boxed{\frac{8!}{8!} + 0.05\frac{8!}{7!} + 0.0025\frac{8!}{2!\,6!} + 0.000125\frac{8!}{3!\,5!} + \cdots}$$

$$= \boxed{\frac{8!}{8!} + 0.05\frac{8\cdot7!}{7!} + 0.0025\frac{8\cdot7\cdot6!}{2!\,6!} + 0.000125\frac{8\cdot7\cdot6\cdot5!}{3!\,5!} + \cdots} = \boxed{1 + 0.4 + 0.07 + 0.007 + \cdots} \approx \boxed{1.477}$$

Therefore, $(1+0.05)^8$, to the nearest hundredth, is equal to $\boxed{1.48}$

c. **First** - Write the exponential expression in the form of $(1.21)^5 = (1+0.21)^5$.

Second - Identify the $a, b,$ and n terms, i.e., $a=1$, $b=0.21$, $n=5$.

Third - Use the general binomial expansion formula, i.e., equation (1) above to expand $(1+0.21)^5$.

$$\boxed{(1+0.21)^5} = \boxed{\binom{5}{0}1^5 + \binom{5}{1}1^4\cdot(0.21) + \binom{5}{2}1^3\cdot(0.21)^2 + \binom{5}{3}1^2\cdot(0.21)^3 + \binom{5}{4}1\cdot(0.21)^4 + \binom{5}{5}\cdot(0.21)^5}$$

$$= \boxed{\binom{5}{0} + 0.21\binom{5}{1} + 0.0441\binom{5}{2} + 0.00926\binom{5}{3} + 0.0019\binom{5}{4} + 0.0004\binom{5}{5}} = \boxed{\frac{5!}{0!\,(5-0)!} + 0.21\frac{5!}{1!\,(5-1)!} + 0.0441\frac{5!}{2!\,(5-2)!}}$$

$$\boxed{+0.00926\frac{5!}{3!\,(5-3)!} + 0.0019\frac{5!}{5!\,(5-4)!} + 0.0004\frac{5!}{5!\,(5-5)!}} = \boxed{\frac{5!}{5!} + 0.21\frac{5!}{4!} + 0.0441\frac{5!}{2!\,3!} + 0.00926\frac{5!}{3!\,2!} + 0.0019\frac{5!}{4!\,1!} + 0.0004\frac{5!}{5!}}$$

$$= \boxed{\frac{5!}{5!} + 0.21\frac{5\cdot4!}{4!} + 0.0441\frac{5\cdot4\cdot3!}{2!\,3!} + 0.00926\frac{5\cdot4\cdot3!}{3!\,2!} + 0.0019\frac{5\cdot4!}{4!} + 0.0004\frac{5!}{5!}} = \boxed{1 + 1.05 + 0.441 + 0.0926 + 0.0095 + 0.0004}$$

$$= \boxed{2.594}$$

Therefore, $(1+0.21)^5$, to the nearest hundredth, is equal to $\boxed{2.59}$

Note that in equation (1) the rth term in a binomial expansion is given by

$$\binom{n}{r-1}a^{n-r+1}b^{r-1} = \frac{n!}{(r-1)!\,(n-r+1)!}\,a^{n-r+1}b^{r-1} \qquad\qquad (2)$$

this implies that we can use the above equation to find any specific term of a binomial. For example, the sixth term of $(x-3)^8$ is equal to

$$\boxed{\binom{8}{5}x^3(-3)^5} = \boxed{\frac{8!}{5!\,(8-5)!}\,x^3\,(-3)^5} = \boxed{\frac{8!}{5!3!}\,x^3\cdot(-243)} = \boxed{-243\cdot\frac{8\cdot7\cdot6\cdot5!}{5!\cdot3\cdot2\cdot1}\,x^3} = \boxed{-13{,}608\,x^3}$$

and the fourth term of $(x-3)^4$ is equal to

$$\boxed{\binom{4}{3}x^{4-4+1}(-3)^3} = \boxed{\frac{4!}{3!\,(4-3)!}x(-3)^3} = \boxed{\frac{4!}{3!1!}x\cdot(-27)} = \boxed{-27\cdot\frac{4\cdot3!}{3!}\,x} = \boxed{-108\,x}$$

Example 1.6-8 Find the stated term of the following binomial expressions.

a. The sixth term of $(x+2)^{10}$ b. The eighth term of $(x-y)^{12}$

c. The fifth term of $(w-a)^{13}\ =$ d. The tenth term of $(x+1)^{20}$

Solutions:

a. **First** - Identify the a,b,r and n terms, i.e., $a=x$, $b=2$, $r=6$, and $n=10$.

   Second - Use equation (2) above to find the sixth term of $(x+2)^{10}$.

$$\boxed{\frac{10!}{(6-1)!\,(10-6+1)!}\,x^{10-6+1}\cdot2^{6-1}} = \boxed{\frac{10!}{5!5!}\,x^5 2^5} = \boxed{\frac{10\cdot9\cdot8\cdot7\cdot6\cdot5!}{5!5!}\,32x^5} = \boxed{\frac{\overset{3}{10}\cdot9\cdot\overset{2}{8}\cdot7\cdot6}{5\cdot4\cdot3\cdot2\cdot1}\,32x^5} = \boxed{\frac{6\cdot7\cdot6}{1}\,32x^5}$$

$$= \boxed{8064\,x^5}$$

b. **First** - Identify the a,b,r and n terms, i.e., $a=x$, $b=-y$, $r=8$, and $n=12$.

   Second - Use equation (2) above to find the eighth term of $(x-y)^{12}$.

$$\boxed{\frac{12!}{(8-1)!\,(12-8+1)!}\,x^{12-8+1}\cdot b^{8-1}} = \boxed{\frac{12!}{7!5!}\,x^5(-y)^7} = \boxed{-\frac{12\cdot11\cdot10\cdot9\cdot8\cdot7!}{7!5!}\,x^5 y^7} = \boxed{-\frac{12\cdot11\cdot10\cdot9\cdot8}{5\cdot4\cdot3\cdot2\cdot1}\,x^5 y^7}$$

$$= \boxed{-(11\cdot9\cdot8)\,x^5 y^7} = \boxed{-792\,x^5 y^7}$$

c. **First** - Identify the a,b,r and n terms, i.e., $a=w$, $b=-a$, $r=5$, and $n=13$.

   Second - Use equation (2) above to find the fifth term of $(w-a)^{13}$.

$$\boxed{\frac{13!}{(5-1)!\,(13-5+1)!}\,w^{13-5+1}(-a)^{5-1}} = \boxed{\frac{13!}{4!9!}\,w^9(-a)^4} = \boxed{\frac{13\cdot12\cdot11\cdot10\cdot9!}{4!9!}\,w^9 a^4} = \boxed{\frac{13\cdot12\cdot11\cdot\overset{5}{10}}{4\cdot3\cdot2\cdot1}\,w^9 a^4}$$

$$= \boxed{\frac{13\cdot11\cdot5}{1}\,w^9 a^4} = \boxed{715\,w^9 a^4}$$

d. **First** - Identify the a, b, r and n terms, i.e., $a = x$, $b = 1$, $r = 10$, and $n = 20$.

Second - Use equation (2) above to find the tenth term of $(x+1)^{20}$.

$$\boxed{\frac{20!}{(10-1)!\,(20-10+1)!}\, x^{20-10+1}\cdot 1^{10-1}} = \boxed{\frac{20!}{9!\,11!}x^{11}\cdot 1^9} = \boxed{\frac{20\cdot19\cdot18\cdot17\cdot16\cdot15\cdot14\cdot13\cdot12\cdot\cancel{11!}}{9!\,\cancel{11!}}x^{11}}$$

$$= \boxed{\frac{\overset{2}{\cancel{20}}\cdot19\cdot\cancel{18}\cdot17\cdot\cancel{16}\cdot15\cdot14\cdot13\cdot12}{9\cdot8\cdot7\cdot6\cdot5\cdot4\cdot3\cdot2\cdot1}x^{11}} = \boxed{\frac{19\cdot17\cdot2\cdot15\cdot13\cdot12}{9}x^{11}} = \boxed{167{,}960\ x^{11}}$$

Section 1.6 Practice Problems - The Factorial Notation

1. Expand and simplify the following factorial expressions.

   a. $11! =$
   b. $(10-3)! =$
   c. $\dfrac{12!}{5!} =$
   d. $\dfrac{14!}{10!} =$

   e. $\dfrac{15!}{8!\,4!} =$
   f. $\dfrac{10!}{4!(10-2)!} =$
   g. $\dfrac{12!\,6!}{14!} =$
   h. $\dfrac{(7-3)!\,9!}{12!(7-2)!} =$

2. Write the following products in factorial form.

   a. $7\cdot6\cdot5\cdot4\cdot3\cdot2\cdot1 =$
   b. $10\cdot11\cdot12\cdot13\cdot14\cdot15 =$
   c. $22\cdot23\cdot24\cdot25 =$

   d. $8\cdot7\cdot6\cdot5\cdot4 =$
   e. $4\cdot5\cdot6\cdot7\cdot8\cdot9 =$
   f. $35 =$

3. Expand the following factorial expressions.

   a. $5(n!) =$
   b. $(n-7)! =$
   c. $(n+10)! =$
   d. $(5n-5)! =$

   e. $(2n-8)! =$
   f. $(2n+6)! =$
   g. $(2n-5)! =$
   h. $(3n+3)! =$

4. Expand and simplify the following factorial expressions.

   a. $\dfrac{(n-2)!}{(n-4)!} =$
   b. $\dfrac{(n+4)!}{n!} =$
   c. $\dfrac{(n+5)!}{(n-2)!} =$
   d. $\dfrac{(n-1)(n+1)!}{(n+2)!} =$

   e. $\dfrac{(3n)!\,(3n-2)!}{(3n+1)!\,(3n-4)!} =$
   f. $\dfrac{(n-1)!}{(n+2)!\,(n!)^2} =$
   g. $\dfrac{(2n-3)!\,2(n!)}{(2n)!\,(n-2)!} =$

5. Write the following expressions in factorial notation form. Simplify the answer.

   a. $\dbinom{5}{3} =$
   b. $\dbinom{10}{6} =$
   c. $\dbinom{8}{0} =$
   d. $\dbinom{8}{8} =$
   e. $\dbinom{6}{3} =$

   f. $\dbinom{5}{1} =$
   g. $\dbinom{n}{n-5} =$
   h. $\dbinom{2n}{2n-1} =$
   i. $\dbinom{3n}{3n-3} =$
   j. $\dbinom{n}{n-6} =$

6. Expand the following binomial expressions.

   a. $(x-2)^4 =$
   b. $(u+2)^7 =$
   c. $(y-3)^5 =$

7. Use the general equation for binomial expansion to solve the following exponential numbers to the nearest hundredth.

    a. $(0.95)^5 =$ b. $(2.25)^7 =$ c. $(1.05)^4 =$

8. Find the stated term of the following binomial expressions.

    a. The eighth term of $(x+3)^{12}$ b. The ninth term of $(x-y)^{10}$

    c. The seventh term of $(u-2a)^{11}$ d. The twelfth term of $(x-1)^{18}$

Chapter 2
Differentiation (Part I)

Quick Reference to Chapter 2 Problems

Chapter 2 – Differentiation (Part I)

The objective of this chapter is to improve the student's ability to solve problems involving derivatives. In Section 2.1 derivatives are computed using the Difference Quotient method. Various differentiation rules using the prime and $\frac{d}{dx}$ notation are introduced in Sections 2.2 and 2.3, respectively. The use of the Chain Rule in finding the derivative of functions that are being added, subtracted, multiplied, and divided are addressed in Section 2.4. Implicit differentiation is discussed in Section 2.5. Finding the derivative of exponential and radical expressions is addressed in Sections 2.6 and 2.7, respectively. Finally, computation of second, third, fourth, or higher order derivatives are discussed in Section 2.8. Each section is concluded by solving examples with practice problems to further enhance the student's ability.

2.1 The Difference Quotient Method

In this section students learn how to differentiate functions using the difference quotient equation as the limit h approaches zero. The expression $\frac{f(x+h)-f(x)}{h}$ is referred to as the difference quotient equation. A function $f(x)$ is said to be differentiable at x if and only if $\lim_{h \to 0} \frac{f(x+h)-f(x)}{h}$ exists. If the limit exists, then the result is referred to as the derivative of $f(x)$ at x which is denoted by $f'(x)$. It should be noted that this approach is rather long and time consuming and is merely presented in order to show the usefulness of the differentiation rules which are addressed in the subsequent sections. The following examples show the steps in finding derivatives of functions using the Difference Quotient method:

Example 2.1-1: Use the Difference Quotient method to find the derivative of the following functions.

a. $f(x) = 3x + 1$　　　　b. $f(x) = 4x^2$　　　　c. $f(x) = \sqrt{x+3}$

d. $f(x) = (2x-7)^2$　　　e. $f(x) = -\sqrt{x+1}$　　　f. $f(x) = \dfrac{1}{\sqrt{x+1}}$

Solutions:

a. To find the derivative of the function $f(x) = 3x + 1$

　　First - Substitute $\boxed{f(x+h)} = \boxed{3(x+h)+1} = \boxed{3x+3h+1}$ and $\boxed{f(x) = 3x+1}$ into the difference quotient

　　equation, i.e., $\boxed{\dfrac{f(x+h)-f(x)}{h}} = \boxed{\dfrac{(3x+3h+1)-(3x+1)}{h}} = \boxed{\dfrac{3x+3h+1-3x-1}{h}} = \boxed{\dfrac{3h}{h}} = \boxed{3}$

　　Second - Compute $f'(x)$ as the $\lim_{h \to 0}$ in the difference quotient equation, i.e.,

　　$\boxed{f'(x)} = \boxed{\lim_{h \to 0} \dfrac{f(x+h)-f(x)}{h}} = \boxed{\lim_{h \to 0} 3} = \boxed{3}$

b. To find the derivative of the function $f(x) = 4x^2$

　　First - Substitute $\boxed{f(x+h)} = \boxed{4(x+h)^2} = \boxed{4(x^2 + h^2 + 2xh)} = \boxed{4x^2 + 4h^2 + 8xh}$ and $\boxed{f(x)} = \boxed{4x^2}$ into the difference quotient equation, i.e., $\boxed{\dfrac{f(x+h)-f(x)}{h}} = \boxed{\dfrac{\left(4x^2+4h^2+8xh\right)-4x^2}{h}} = \boxed{\dfrac{4x^2+4h^2+8xh-4x^2}{h}}$

$= \boxed{\dfrac{4h(h+2x)}{h}} = \boxed{\dfrac{4(h+2x)}{1}} = \boxed{4h+8x}$

Second - Compute $f'(x)$ as the $\lim_{h\to 0}$ in the difference quotient equation, i.e.,

$\boxed{f'(x)} = \boxed{\lim_{h\to 0}\dfrac{f(x+h)-f(x)}{h}} = \boxed{\lim_{h\to 0}(4h+8x)} = \boxed{(4\cdot 0)+8x} = \boxed{0+8x} = \boxed{\mathbf{8x}}$

c. To find the derivative of the function $f(x)=\sqrt{x+3}$

First - Substitute $\boxed{f(x+h)} = \boxed{\sqrt{(x+h)+3}} = \boxed{\sqrt{x+h+3}}$ and $\boxed{f(x)=\sqrt{x+3}}$ into the difference quotient

equation, i.e., $\boxed{\dfrac{f(x+h)-f(x)}{h}} = \boxed{\dfrac{\sqrt{x+h+3}-\sqrt{x+3}}{h}}$

To remove the radical from the numerator multiply both the numerator and the denominator

by $\sqrt{x+h+3}+\sqrt{x+3}$ to obtain the following:

$\boxed{\dfrac{f(x+h)-f(x)}{h}} = \boxed{\dfrac{\sqrt{x+h+3}-\sqrt{x+3}}{h}} = \boxed{\dfrac{\sqrt{x+h+3}-\sqrt{x+3}}{h}\cdot\dfrac{\sqrt{x+h+3}+\sqrt{x+3}}{\sqrt{x+h+3}+\sqrt{x+3}}} = \boxed{\dfrac{1}{\sqrt{x+h+3}+\sqrt{x+3}}}$

Second - Compute $f'(x)$ as the $\lim_{h\to 0}$ in the difference quotient equation, i.e.,

$\boxed{f'(x)} = \boxed{\lim_{h\to 0}\dfrac{f(x+h)-f(x)}{h}} = \boxed{\lim_{h\to 0}\dfrac{1}{\sqrt{x+h+3}+\sqrt{x+3}}} = \boxed{\dfrac{1}{\sqrt{x+3}+\sqrt{x+3}}} = \boxed{\dfrac{\mathbf{1}}{\mathbf{2\sqrt{x+3}}}}$

d. To find the derivative of the function $f(x)=(2x-7)^2$

First - Substitute $\boxed{f(x+h)} = \boxed{[2(x+h)-7]^2} = \boxed{4(x+h)^2+49-28(x+h)} = \boxed{4x^2+4h^2+8xh+49-28x-28h}$

and $\boxed{f(x)} = \boxed{(2x-7)^2} = \boxed{4x^2+49-28x}$ into the difference quotient equation, i.e., $\boxed{\dfrac{f(x+h)-f(x)}{h}}$

$= \boxed{\dfrac{\left(4x^2+4h^2+8xh+49-28x-28h\right)-\left(4x^2+49-28x\right)}{h}} = \boxed{\dfrac{4x^2+4h^2+8xh+49-28x-28h-4x^2-49+28x}{h}}$

$= \boxed{\dfrac{4h^2+8xh-28h}{h}} = \boxed{\dfrac{h(4h+8x-28)}{h}} = \boxed{4h+8x-28}$

Second - Compute $f'(x)$ as the $\lim_{h\to 0}$ in the difference quotient equation, i.e.,

$\boxed{f'(x)} = \boxed{\lim_{h\to 0}\dfrac{f(x+h)-f(x)}{h}} = \boxed{\lim_{h\to 0}(4h+8x-28)} = \boxed{(4\cdot 0)+8x-28} = \boxed{0+8x-28} = \boxed{\mathbf{8x-28}}$

e. To find the derivative of the function $f(x)=-\sqrt{x+1}$

First - Substitute $\boxed{f(x+h)} = \boxed{-\sqrt{(x+h)+1}} = \boxed{-\sqrt{x+h+1}}$ and $\boxed{f(x)} = \boxed{-\sqrt{x+1}}$ into the difference quotient equation, i.e., $\boxed{\dfrac{f(x+h)-f(x)}{h}} = \boxed{-\dfrac{\sqrt{x+h+1}+\sqrt{x+1}}{h}}$

To remove the radical from the numerator multiply both the numerator and the denominator by

$\sqrt{x+h+1}-\sqrt{x+1}$ to obtain the following:

$\boxed{\dfrac{f(x+h)-f(x)}{h}} = \boxed{-\dfrac{\sqrt{x+h+1}+\sqrt{x+1}}{h}} = \boxed{-\dfrac{\sqrt{x+h+1}+\sqrt{x+1}}{h}\cdot\dfrac{\sqrt{x+h+1}-\sqrt{x+1}}{\sqrt{x+h+1}-\sqrt{x+1}}} = \boxed{-\dfrac{1}{\sqrt{x+h+1}+\sqrt{x+1}}}$

Second - Compute $f'(x)$ as the $\lim_{h\to 0}$ in the difference quotient equation, i.e.,

$\boxed{f'(x)} = \boxed{\lim_{h\to 0}\dfrac{f(x+h)-f(x)}{h}} = \boxed{-\lim_{h\to 0}\dfrac{1}{\sqrt{x+h+1}+\sqrt{x+1}}} = \boxed{-\dfrac{1}{\sqrt{x+1}+\sqrt{x+1}}} = \boxed{-\dfrac{1}{2\sqrt{x+1}}}$

f. To find the derivative of the function $f(x) = \dfrac{1}{\sqrt{x+1}}$

First - Substitute $\boxed{f(x+h)} = \boxed{\dfrac{1}{\sqrt{x+h+1}}}$ and $\boxed{f(x)} = \boxed{\dfrac{1}{\sqrt{x+1}}}$ into the difference quotient equation,

i.e., $\boxed{\dfrac{f(x+h)-f(x)}{h}} = \boxed{\dfrac{\dfrac{1}{\sqrt{x+h+1}} - \dfrac{1}{\sqrt{x+1}}}{h}} = \boxed{\dfrac{\sqrt{x+1}-\sqrt{x+h+1}}{h\sqrt{x+1}\cdot\sqrt{x+h+1}}}$

To remove the radical from the numerator multiply both the numerator and the denominator

by $\sqrt{x+1}+\sqrt{x+h+1}$ to obtain the following:

$\boxed{\dfrac{f(x+h)-f(x)}{h}} = \boxed{\dfrac{\sqrt{x+1}-\sqrt{x+h+1}}{h\sqrt{x+1}\cdot\sqrt{x+h+1}}\cdot\dfrac{\sqrt{x+1}+\sqrt{x+h+1}}{\sqrt{x+1}+\sqrt{x+h+1}}} = \boxed{\dfrac{(x+1)-(x+h+1)}{h(x+1)\sqrt{x+h+1}+h(x+h+1)\sqrt{x+1}}}$

$= \boxed{\dfrac{x+1-x-h-1}{h(x+1)\sqrt{x+h+1}+h(x+h+1)\sqrt{x+1}}} = \boxed{\dfrac{-h}{h\left[(x+1)\sqrt{x+h+1}+(x+h+1)\sqrt{x+1}\right]}} = \boxed{\dfrac{-1}{(x+1)\sqrt{x+h+1}+(x+h+1)\sqrt{x+1}}}$

Second - Compute $f'(x)$ as the $\lim_{h\to 0}$ in the difference quotient equation, i.e.,

$\boxed{f'(x)} = \boxed{\lim_{h\to 0}\dfrac{f(x+h)-f(x)}{h}} = \boxed{\lim_{h\to 0}\dfrac{-1}{(x+1)\sqrt{x+h+1}+(x+h+1)\sqrt{x+1}}} = \boxed{\dfrac{-1}{(x+1)\sqrt{x+1}+(x+1)\sqrt{x+1}}}$

$= \boxed{-\dfrac{1}{2(x+1)\sqrt{x+1}}}$

Example 2.1-2: Given the derivative of the functions in example 2.1-1, find:

a. $f'(2)$ 　　　　　　　　　　b. $f'(3)$ 　　　　　　　　　　c. $f'(1)$

d. $f'(0)$ 　　　　　　　　　　e. $f'(15)$ 　　　　　　　　　f. $f'(0)$

Solutions:

a. Given $f'(x) = 3$ then, $\boxed{f'(2)} = \boxed{3}$

Note that since the derivative is constant $f'(x)$ is independent of the x value. $f'(2)$ can also

be calculated directly by using $\boxed{f'(x)} = \boxed{\lim_{h \to 0} \dfrac{f(x+h) - f(x)}{h}} = \boxed{\lim_{h \to 0} \dfrac{[3(x+h)+1] - (3x+1)}{h}}$

and by replacing x with 2, i.e., $\boxed{f'(2)} = \boxed{\lim_{h \to 0} \dfrac{[3(2+h)+1] - (6+1)}{h}} = \boxed{\lim_{h \to 0} \dfrac{6+3h+1-7}{h}}$

$= \boxed{\lim_{h \to 0} \dfrac{3h}{h}} = \boxed{\lim_{h \to 0} 3} = \boxed{3}$

b. Given $f'(x) = 8x$ then, $\boxed{f'(3)} = \boxed{8 \cdot 3} = \boxed{24}$

$f'(3)$ can also be calculated directly by using $\boxed{f'(x)} = \boxed{\lim_{h \to 0} \dfrac{f(x+h) - f(x)}{h}} = \boxed{\lim_{h \to 0} \dfrac{4(x+h)^2 - 4x^2}{h}}$

and by replacing x with 3, i.e., $\boxed{f'(3)} = \boxed{\lim_{h \to 0} \dfrac{4(3+h)^2 - 4 \cdot 3^2}{h}} = \boxed{\lim_{h \to 0} \dfrac{4(9+h^2+6h) - 36}{h}}$

$= \boxed{\lim_{h \to 0} \dfrac{36+4h^2+24h-36}{h}} = \boxed{\lim_{h \to 0} \dfrac{h(4h+24)}{h}} = \boxed{\lim_{h \to 0} 4h+24} = \boxed{0+24} = \boxed{24}$

c. Given $f'(x) = \dfrac{1}{2\sqrt{x+3}}$ then, $\boxed{f'(1)} = \boxed{\dfrac{1}{2\sqrt{1+3}}} = \boxed{\dfrac{1}{2\sqrt{4}}} = \boxed{\dfrac{1}{2 \cdot 2}} = \boxed{\dfrac{1}{4}}$

Again, $f'(1)$ can also be calculated directly by using the equation $\boxed{f'(x)} = \boxed{\lim_{h \to 0} \dfrac{f(x+h) - f(x)}{h}}$

$= \boxed{\lim_{h \to 0} \dfrac{\sqrt{(x+h)+3} - \sqrt{x+3}}{h}}$ and by replacing x with 1, i.e., $\boxed{f'(1)} = \boxed{\lim_{h \to 0} \dfrac{\sqrt{(1+h)+3} - \sqrt{1+3}}{h}}$

$= \boxed{\lim_{h \to 0} \dfrac{\sqrt{h+4} - \sqrt{4}}{h}} = \boxed{\lim_{h \to 0} \dfrac{\sqrt{h+4} - \sqrt{4}}{h} \cdot \dfrac{\sqrt{h+4} + \sqrt{4}}{\sqrt{h+4} + \sqrt{4}}} = \boxed{\lim_{h \to 0} \dfrac{h}{h(\sqrt{h+4} + 2)}} = \boxed{\lim_{h \to 0} \dfrac{1}{\sqrt{h+4} + 2}}$

$= \boxed{\dfrac{1}{\sqrt{0+4} + 2}} = \boxed{\dfrac{1}{\sqrt{4} + 2}} = \boxed{\dfrac{1}{2+2}} = \boxed{\dfrac{1}{4}}$

d. Given $f'(x) = 8x - 28$ then, $\boxed{f'(0)} = \boxed{(8 \times 0) - 28} = \boxed{-28}$

e. Given $f'(x) = -\dfrac{1}{2\sqrt{x+1}}$ then, $\boxed{f'(15)} = \boxed{-\dfrac{1}{2\sqrt{15+1}}} = \boxed{-\dfrac{1}{2\sqrt{16}}} = \boxed{-\dfrac{1}{2 \cdot 4}} = \boxed{-\dfrac{1}{8}}$

f. Given $f'(x) = -\dfrac{1}{2\sqrt{(x+1)^3}}$ then, $\boxed{f'(0)} = \boxed{-\dfrac{1}{2\sqrt{(0+1)^3}}} = \boxed{-\dfrac{1}{2 \cdot 1}} = \boxed{-\dfrac{1}{2}}$

In problems 2.1-2 d, e, and f students may want to practice finding $f'(x)$ for the specific values of

x by using the general equation $f'(x) = \lim_{h \to 0} \dfrac{f(x+h)-f(x)}{h}$. The answers should agree with the above stated solutions.

Finally, it should be noted that every differentiable function is continuous. However, not every continuous function is differentiable. The proof of this statement is beyond the scope of this book and can be found in a calculus book. In the following section we will learn simpler methods of finding derivative of functions using various differentiation rules.

Section 2.1 Practice Problems – The Difference Quotient Method

1. Find the derivative of the following functions by using the Difference Quotient method.

   a. $f(x) = x^2 - 1$

   b. $f(x) = x^3 + 2x - 1$

   c. $f(x) = \dfrac{x}{x-1}$

   d. $f(x) = -\dfrac{1}{x^2}$

   e. $f(x) = 20x^2 - 3$

   f. $f(x) = \sqrt{x^3}$

   g. $f(x) = \dfrac{10}{\sqrt{x-5}}$

   h. $f(x) = \dfrac{ax+b}{cx}$

2. Compute $f'(x)$ for the specified values by using the difference quotient equation as the $\lim_{h \to 0}$.

   a. $f(x) = x^3$ at $x = 1$

   b. $f(x) = 1 + 2x$ at $x = 0$

   c. $f(x) = x^3 + 1$ at $x = -1$

   d. $f(x) = x^2(x+2)$ at $x = 2$

   e. $f(x) = x^{-2} + x^{-1} + 1$ at $x = 1$

   f. $f(x) = \sqrt{x} + 2$ at $x = 10$

2.2   Differentiation Rules Using the Prime Notation

In the previous section we differentiated several functions by writing the difference quotient equation and taking the limit as h approaches to zero. This process, however - as was mentioned earlier, is rather long and time consuming. Instead, we can establish some general rules that make the calculation of derivatives simpler. These rules are as follows:

Rule No. 1 - *The derivative of a constant function is equal to zero, i.e.,*

$$if \; f(x) = k, \qquad then \; f'(x) = 0$$

For example, the derivative of the functions $f(x) = 10$, $g(x) = -100$, and $s(x) = 250$ is equal to $f'(x) = 0$, $g'(x) = 0$, and $s'(x) = 0$, respectively.

Rule No. 2 - *The derivative of the identity function is equal to one, i.e.,*

$$if \; f(x) = x, \qquad then \; f'(x) = 1$$

For example, the derivative of the functions $f(x) = 5x$, $g(x) = -10x$, and $s(x) = \sqrt{5}x$ is equal to $f'(x) = 5 \cdot 1 = 5$, $g'(x) = -10 \cdot 1 = -10$, and $s'(x) = \sqrt{5} \cdot 1 = \sqrt{5}$, respectively.

Rule No. 3A - *The derivative of the function $f(x) = x^n$ is equal to $f'(x) = nx^{n-1}$, where n is a positive integer.*

For example, the derivative of the functions $f(x) = x$, $f(x) = x^2$, $f(x) = x^3$, and $f(x) = x^4$ is equal to $f'(x) = 1 \cdot x^{1-1} = x^0 = 1$, $f'(x) = 2 \cdot x^{2-1} = 2x^1 = 2x$, $f'(x) = 3 \cdot x^{3-1} = 3x^2$, and $f'(x) = 4 \cdot x^{4-1} = 4x^3$, respectively.

Rule No. 3B - *The derivative of the function $f(x) = x^n$ is equal to $f'(x) = nx^{n-1}$, where n is a negative integer.*

For example, the derivative of the functions $f(x) = x^{-1}$, $f(x) = x^{-2}$, $f(x) = x^{-3}$, $f(x) = x^{-4}$, and $f(x) = x^{-\frac{1}{8}}$ is equal to $f'(x) = -1 \cdot x^{-1-1} = -x^{-2}$, $f'(x) = -2 \cdot x^{-2-1} = -2x^{-3}$, $f'(x) = -3 \cdot x^{-3-1} = -3x^{-4}$, and $f'(x) = -\frac{1}{8} \cdot x^{-\frac{1}{8}-1} = -\frac{1}{8}x^{-\frac{9}{8}}$, respectively.

Note that this rule can also be used to obtain the derivative of functions that are in the form of $f(x) = \frac{1}{x^n}$ by rewriting the function in its equivalent form of $f(x) = x^{-n}$. For example, the derivative of the functions $f(x) = \frac{1}{x} = x^{-1}$, $f(x) = \frac{1}{x^2} = x^{-2}$, $f(x) = -\frac{2}{x^3} = -2x^{-3}$, and $f(x) = \frac{1}{x^8} = x^{-8}$ is equal to $f'(x) = -1 \cdot x^{-1-1} = -x^{-2}$, $f'(x) = -2 \cdot x^{-2-1} = -2x^{-3}$, $f'(x) = (-2 \cdot -3) \cdot x^{-3-1} = 6x^{-4}$, and $f'(x) = -8 \cdot x^{-8-1} = -8x^{-9}$, respectively.

Rule No. 4 - *If the function $f(x)$ is differentiable at x, then a constant k multiplied by $f(x)$ is also differentiable at x, i.e.,*

$$(kf)' x = k f'(x)$$

Note that this rule is referred to as the **scalar rule**.

For example, the derivative of the functions $f(x)=5x$, $g(x)=-10x$, and $s(x)=\sqrt{5}x$ is equal to $f'(x)=5$, $g'(x)=-10$, and $s'(x)=\sqrt{5}$, respectively.

Rule No. 5 - *If the function $f(x)$ and $g(x)$ are differentiable at x, then their sum is also differentiable at x, i.e.,*

$$\left(f+g\right)'(x) = f'(x)+g'(x)$$

In other words, the derivative of the sum of two differentiable functions, $\left(f+g\right)'(x)$, is equal to the derivative of the first function, $f'(x)$, plus the derivative of the second function, $g'(x)$. Note that this rule is referred to as the **summation rule**.

For example, the derivative of the functions $h(x)=\underbrace{(5x-3)}_{f(x)}+\underbrace{\left(2x^2-1\right)}_{g(x)}$ and $s(x)=\underbrace{6x^3}_{f(x)}+\underbrace{(3x+2)}_{g(x)}$ is equal to $h'(x)=f'(x)+g'(x)=5+4x$ and $s'(x)=f'(x)+g'(x)=18x^2+3$.

Rule No. 6 - *If the function $f(x)$ and $g(x)$ are differentiable at x, then their product is also differentiable at x, i.e.,*

$$\left(f\cdot g\right)'(x) = f'(x)g(x)+g'(x)f(x)$$

In other words, the derivative of the product of two differentiable functions, $\left(f\cdot g\right)'(x)$, is equal to the derivative of the first function multiplied by the second function, $f'(x)\cdot g(x)$, plus the derivative of the second function multiplied by the first function, $g'(x)\cdot f(x)$. Note that this rule is referred to as the **product rule**.

For example, the derivative of the functions $f(x)=(3x-5)(6x+1)$ and $g(x)=-10x^2\left(5x^3-2\right)$ is equal to $f'(x) = \left[3\cdot(6x+1)\right]+\left[6\cdot(3x-5)\right] = 18x+3+18x-5 = (18x+18x)+(-5+3)= 36x-2$ and

$g'(x) = \left[-20x\cdot\left(5x^3-2\right)\right]+\left[15x^2\cdot-10x^2\right] = -100x^4+40x-150x^4 = -250x^4+40x$

Rule No. 7 - Using the rules 1, 4, 5, and 6 we can write the formula for differentiating polynomials, i.e.,

$$if\ f(x) = a_nx^n+a_{n-1}x^{n-1}+a_{n-2}x^{n-2}+\cdots+a_3x^3+a_2x^2+a_1x^1+a_0\ ,\ then$$

$$f'(x) = na_nx^{n-1}+(n-1)a_{n-1}x^{n-2}+(n-2)a_{n-2}x^{n-3}+\cdots+3a_3x^2+2a_2x+a_1$$

For example, the derivative of the polynomials $f(x)=6x^4+5x^3-3$, $g(x)=2x^5-3x^2-4x$, and $h(x)=\frac{1}{3}x^2-2x+5$ is equal to $f'(x) = (6\cdot4)x^{4-1}+(5\cdot3)x^{3-1} = 24x^3+15x^2$, $g'(x) = (5\cdot2)x^4-(2\cdot3)x-4$

$= 10x^4-6x-4$, and $h'(x) = \left(2\cdot\frac{1}{3}\right)x-2 = \frac{2}{3}x-2$, respectively.

Rule No. 8 - *If the function $f(x)$ and $g(x)$ are differentiable at x, then their quotient is also differentiable at x, i.e.,*

$$\left(\frac{f}{g}\right)'(x) = \frac{f'(x)g(x) - g'(x)f(x)}{[g(x)]^2}$$

In other words, the derivative of the quotient of two differentiable functions, $\left(\dfrac{f}{g}\right)'(x)$, is equal to the derivative of the function in the numerator multiplied by the function in the denominator, $f'(x) \cdot g(x)$, minus the derivative of the function in the denominator multiplied by the function in the numerator, $g'(x) \cdot f(x)$, all divided by the square of the denominator, $[g(x)]^2$. Note that this rule is referred to as the **quotient rule**.

For example, the derivative of the functions $f(x) = \dfrac{1+3x}{1+x}$ and $g(x) = \dfrac{3x^2+5}{x^3+1}$ is equal to $f'(x)$

$$= \frac{[3 \cdot (1+x)] - [1 \cdot (1+3x)]}{(1+x)^2} = \frac{3 + 3x - 1 - 3x}{(1+x)^2} = \frac{2}{(1+x)^2} \text{ and } g'(x) = \frac{[6x \cdot (x^3+1)] - [3x^2 \cdot (3x^2+5)]}{(x^3+1)^2}$$

$$= \frac{6x^4 + 6x - 9x^4 - 15x^2}{(x^3+1)^2} = \frac{-3x^4 - 9x^2}{(x^3+1)^2}$$

In the following examples the above rules are used in order to find the derivative of various functions:

Example 2.2-1: Differentiate the following functions.

a. $f(x) = 5 + x$

b. $f(x) = x^3 + 3x - 1$

c. $f(x) = 2 - x$

d. $f(x) = 10x^3 + 5x^2 + 5$

e. $f(x) = 3(x^2 + 2x)$

f. $f(x) = ax^3 + bx^2 + c$

g. $f(x) = ax^2 + b$

h. $f(x) = \dfrac{x^5}{10} - \dfrac{x^4}{4} + \dfrac{x^3}{6} - \dfrac{x^2}{4}$

i. $f(x) = \dfrac{10}{x^2}$

j. $f(x) = \dfrac{1}{x} - \dfrac{1}{x^3}$

k. $f(x) = x^{-5} + 3x^{-3} - 2x^{-1} + 10$

l. $f(x) = \dfrac{x^2 + 2}{x^4}$

m. $f(x) = \dfrac{x^4}{1-x}$

n. $f(x) = \dfrac{x^4 + 10}{x^2 + 1}$

o. $f(x) = \dfrac{3+x}{x^3 - 5}$

p. $f(x) = \dfrac{2}{x}\left(\dfrac{2+x^3}{x+1}\right)$

q. $f(x) = (x^2 + 1)(x + 5)$

r. $f(x) = (x+1)(x+2)$

s. $f(x) = \left(1 + \dfrac{1}{x}\right)\left(1 - \dfrac{1}{x^3}\right)$

t. $f(x) = \left(\dfrac{3x^2 + 5}{x}\right)(x+1)$

u. $f(x) = \dfrac{1+x^2}{1-x^2}$

v. $f(x) = \dfrac{2x^2 + 3x + 1}{x^2 + 1}$

w. $f(x) = \dfrac{ax^2 + bx + c}{ax^2 - b}$

x. $f(x) = \dfrac{3-x}{\dfrac{1}{x} - 5}$

Solutions:

a. Given $\boxed{f(x) = 5 + x}$ then $\boxed{f'(x)} = \boxed{0 + x^{1-1}} = \boxed{x^0} = \boxed{1}$

b. Given $\boxed{f(x)=x^3+3x-1}$ then $\boxed{f'(x)}=\boxed{3x^{3-1}+3\cdot x^{1-1}-0}=\boxed{3x^2+3\cdot x^0}=\boxed{3x^2+3\cdot 1}=\boxed{\mathbf{3x^2+3}}$

c. Given $\boxed{f(x)=2-x}$ then $\boxed{f'(x)}=\boxed{0-x^{1-1}}=\boxed{-x^0}=\boxed{\mathbf{-1}}$

d. Given $\boxed{f(x)=10x^3+5x^2+5}$ then $\boxed{f'(x)}=\boxed{10\cdot 3x^{3-1}+5\cdot 2x^{2-1}+0}=\boxed{\mathbf{30x^2+10x}}$

e. Given $\boxed{f(x)=3\left(x^2+2x\right)}$ then $\boxed{f'(x)}=\boxed{3\left(2x^{2-1}+2\cdot x^{1-1}\right)}=\boxed{3\left(2x+2\cdot x^0\right)}=\boxed{3\left(2x+2\cdot 1\right)}=\boxed{\mathbf{6x+6}}$

f. Given $\boxed{f(x)=ax^3+bx^2+c}$ then $\boxed{f'(x)}=\boxed{a\cdot 3x^{3-1}+b\cdot 2x^{2-1}+0}=\boxed{\mathbf{3ax^2+2bx}}$

g. Given $\boxed{f(x)=ax^2+b}$ then $\boxed{f'(x)}=\boxed{a\cdot 2x^{2-1}+0}=\boxed{\mathbf{2a\,x}}$

h. Given $\boxed{f(x)=\dfrac{x^5}{10}-\dfrac{x^4}{4}+\dfrac{x^3}{6}-\dfrac{x^2}{4}}$ then $\boxed{f'(x)}=\boxed{\dfrac{1}{10}\cdot 5x^{5-1}-\dfrac{1}{4}\cdot 4x^{4-1}+\dfrac{1}{6}\cdot 3x^{3-1}-\dfrac{1}{4}\cdot 2x^{2-1}}$

$=\boxed{\dfrac{5}{10}x^4-\dfrac{4}{4}x^3+\dfrac{3}{6}x^2-\dfrac{2}{4}x}=\boxed{\dfrac{x^4}{2}-\dfrac{x^3}{1}+\dfrac{x^2}{2}-\dfrac{x}{2}}$ or, using the quotient rule we obtain

$\boxed{f'(x)}=\boxed{\dfrac{\left(5x^{5-1}\cdot 10\right)-\left(0\cdot x^5\right)}{10^2}-\dfrac{\left(4x^{4-1}\cdot 4\right)-\left(0\cdot x^4\right)}{4^2}+\dfrac{\left(3x^{3-1}\cdot 6\right)-\left(0\cdot x^3\right)}{6^2}-\dfrac{\left(2x^{2-1}\cdot 4\right)-\left(0\cdot x^2\right)}{4^2}}$

$=\boxed{\dfrac{50x^4}{100}-\dfrac{16x^3}{16}+\dfrac{18x^2}{36}-\dfrac{8x}{16}}=\boxed{\dfrac{x^4}{2}-\dfrac{x^3}{1}+\dfrac{x^2}{2}-\dfrac{x}{2}}$

i. Given $\boxed{f(x)=\dfrac{10}{x^2}}$ then $\boxed{f'(x)}=\boxed{\dfrac{\left(0\cdot x^2\right)-\left(2x^{2-1}\cdot 10\right)}{\left(x^2\right)^2}}=\boxed{\dfrac{0-\left(2x\cdot 10\right)}{x^4}}=\boxed{\dfrac{-20x}{x^{4=3}}}=\boxed{-\dfrac{\mathbf{20}}{\mathbf{x^3}}}$

j. Given $\boxed{f(x)=\dfrac{1}{x}-\dfrac{1}{x^3}}$ then $\boxed{f'(x)}=\boxed{\dfrac{\left(0\cdot x\right)-\left(1\cdot 1\right)}{x^2}-\dfrac{\left(0\cdot x^3\right)-\left(3x^{3-1}\cdot 1\right)}{\left(x^3\right)^2}}=\boxed{-\dfrac{1}{x^2}+\dfrac{3x^2}{x^{6=4}}}=\boxed{-\dfrac{\mathbf{1}}{\mathbf{x^2}}+\dfrac{\mathbf{3}}{\mathbf{x^4}}}$

k. Given $\boxed{f(x)=x^{-5}+3x^{-3}-2x^{-1}+10}$ then $\boxed{f'(x)}=\boxed{-5x^{-5-1}+3\cdot-3x^{-3-1}-2\cdot-1x^{-1-1}+0}$

$=\boxed{\mathbf{-5x^{-6}-9x^{-4}+2x^{-2}}}$ or, we can rewrite $\boxed{f(x)}$ as $f(x)=\dfrac{1}{x^5}+\dfrac{3}{x^3}-\dfrac{2}{x}+10$ and then find $f'(x)$

using the quotient rule.

$\boxed{f'(x)}=\boxed{\dfrac{\left(0\cdot x^5\right)-\left(5x^4\cdot 1\right)}{\left(x^5\right)^2}+\dfrac{\left(0\cdot x^3\right)-\left(3x^2\cdot 3\right)}{\left(x^3\right)^2}-\dfrac{\left(0\cdot x\right)-\left(1\cdot 2\right)}{x^2}+0}=\boxed{-\dfrac{5x^4}{x^{10}}-\dfrac{9x^2}{x^6}+\dfrac{2}{x^2}}=\boxed{-\dfrac{5x^4}{x^{10=6}}-\dfrac{9x^2}{x^{6=4}}+\dfrac{2}{x^2}}$

$$= -\frac{5}{x^6} - \frac{9}{x^4} + \frac{2}{x^2} = \boxed{-5x^{-6} - 9x^{-4} + 2x^{-2}}$$

l. Given $\boxed{f(x) = \dfrac{x^2 + 2}{x^4}}$ then $\boxed{f'(x)} = \boxed{\dfrac{\left[\left(2x^{2-1} + 0\right) \cdot x^4\right] - \left[4x^{4-1} \cdot \left(x^2 + 2\right)\right]}{\left(x^4\right)^2}} = \boxed{\dfrac{\left[2x \cdot x^4\right] - \left[4x^3 \cdot \left(x^2 + 2\right)\right]}{x^8}}$

$$= \boxed{\frac{2x^5 - 4x^5 - 8x^3}{x^8}} = \boxed{\frac{-2x^5 - 8x^3}{x^8}} = \boxed{\frac{-2x^3\left(x^2 + 4\right)}{x^{8=5}}} = \boxed{\frac{-2\left(x^2 + 4\right)}{x^5}}$$

m. Given $\boxed{f(x) = \dfrac{x^4}{1-x}}$ then $\boxed{f'(x)} = \boxed{\dfrac{\left[4x^{4-1} \cdot (1-x)\right] - \left[(0-1) \cdot x^4\right]}{(1-x)^2}} = \boxed{\dfrac{4x^3(1-x) + x^4}{(1-x)^2}} = \boxed{\dfrac{4x^3 - 4x^4 + x^4}{(1-x)^2}}$

$$= \boxed{\frac{4x^3 - 3x^4}{(1-x)^2}} = \boxed{\frac{4x^3 - 3x^4}{(1-x)^2}}$$

n. Given $\boxed{f(x) = \dfrac{x^4 + 10}{x^2 + 1}}$ then $\boxed{f'(x)} = \boxed{\dfrac{\left[\left(4x^{4-1} + 0\right)\left(x^2 + 1\right)\right] - \left[\left(2x^{2-1} + 0\right)\left(x^4 + 10\right)\right]}{\left(x^2 + 1\right)^2}} = \boxed{\dfrac{4x^3\left(x^2 + 1\right) - 2x\left(x^4 + 10\right)}{\left(x^2 + 1\right)^2}}$

$$= \boxed{\frac{4x^5 + 4x^3 - 2x^5 - 20x}{\left(x^2 + 1\right)^2}} = \boxed{\frac{2x^5 + 4x^3 - 20x}{\left(x^2 + 1\right)^2}} = \boxed{\frac{2x\left(x^4 + 2x^2 - 10\right)}{\left(x^2 + 1\right)^2}}$$

o. Given $\boxed{f(x) = \dfrac{3+x}{x^3 - 5}}$ then $\boxed{f'(x)} = \boxed{\dfrac{\left[(0+1)\left(x^3 - 5\right)\right] - \left[\left(3x^{3-1} - 0\right)(3+x)\right]}{\left(x^3 - 5\right)^2}} = \boxed{\dfrac{\left(x^3 - 5\right) - 3x^2(3+x)}{\left(x^3 - 5\right)^2}}$

$$= \boxed{\frac{x^3 - 5 - 9x^2 - 3x^3}{\left(x^3 - 5\right)^2}} = \boxed{\frac{-2x^3 - 9x^2 - 5}{\left(x^3 - 5\right)^2}}$$

p. Given $\boxed{f(x) = \dfrac{2}{x}\left(\dfrac{2 + x^3}{x + 1}\right)}$ then $\boxed{f'(x)} = \boxed{\left[\dfrac{(0 \cdot x) - (1 \cdot 2)}{x^2}\right]\left(\dfrac{2 + x^3}{x + 1}\right) + \left\{\dfrac{\left[\left(0 + 3x^{3-1}\right)(x+1)\right] - (1+0)\left(2 + x^3\right)}{(x+1)^2}\right\}\left(\dfrac{2}{x}\right)}$

$$= \boxed{\left\{\frac{\left[3x^2(x+1)\right] - \left(2 + x^3\right)}{(x+1)^2}\right\}} = \boxed{-\frac{2}{x^2}\left(\frac{2 + x^3}{x + 1}\right) + \frac{2}{x}\left[\frac{3x^3 + 3x^2 - x^3 - 2}{(x+1)^2}\right]} = \boxed{-\frac{2}{x^2}\left(\frac{2 + x^3}{x + 1}\right) + \frac{2}{x}\left[\frac{2x^3 + 3x^2 - 2}{(x+1)^2}\right]}$$

q. Given $\boxed{f(x) = \left(x^2 + 1\right)(x + 5)}$ then $\boxed{f'(x)} = \boxed{\left[\left(2x^{2-1} + 0\right)(x + 5)\right] + \left[(1+0)\left(x^2 + 1\right)\right]} = \boxed{2x(x + 5) + \left(x^2 + 1\right)}$

$$= \boxed{2x^2 + 10x + x^2 + 1} = \boxed{3x^2 + 10x + 1}$$

A second method would be to multiply the binomial terms by one another, using the FOIL method, and then taking the derivative of $f(x)$ as follows:

$$\boxed{f(x)} = \boxed{(x^2+1)(x+5)} = \boxed{x^3 + 5x^2 + x + 5} \text{ then } f'(x) = \boxed{3x^{3-1} + (5\cdot2)x^{2-1} + x^{1-1} + 0} = \boxed{3x^2 + 10x + 1}$$

r. Given $\boxed{f(x) = (x+1)(x+2)}$ then $\boxed{f'(x)} = \boxed{[(1+0)(x+2)] + [(1+0)(x+1)]} = \boxed{(x+2) + (x+1)} = \boxed{2x+3}$

An alternative way is by multiplying the terms in parenthesis together and then taking the derivative of the product, i.e.,

$$\boxed{f(x)} = \boxed{(x+1)(x+2)} = \boxed{x^2 + 2x + x + 2} \text{ then } \boxed{f'(x)} = \boxed{2x^{2-1} + 2\cdot1 + 1 + 0} = \boxed{2x+2+1} = \boxed{2x+3}$$

s. Given $\boxed{f(x) = \left(1+\frac{1}{x}\right)\left(1-\frac{1}{x^3}\right)}$ then $\boxed{f'(x)} = \boxed{\left[0 + \frac{(0\cdot x)-(1\cdot1)}{x^2}\right]\left(1-\frac{1}{x^3}\right) + \left[0 - \frac{(0\cdot x^3)-(3x^{3-1}\cdot1)}{x^6}\right]\left(1+\frac{1}{x}\right)}$

$$= \boxed{-\frac{1}{x^2}\left(1-\frac{1}{x^3}\right) + \frac{3x^2}{x^6}\left(1+\frac{1}{x}\right)} = \boxed{-\frac{1}{x^2} + \frac{1}{x^5} + \frac{3x^2}{x^6} + \frac{3x^2}{x^7}} = \boxed{-\frac{1}{x^2} + \frac{1}{x^5} + \frac{3x^2}{x^{6=4}} + \frac{3x^2}{x^{7=5}}}$$

$$= \boxed{-\frac{1}{x^2} + \frac{1}{x^5} + \frac{3}{x^4} + \frac{3}{x^5}} = \boxed{\frac{4}{x^5} + \frac{3}{x^4} - \frac{1}{x^2}}$$

A perhaps simpler way is to write $f(x)$ in the following form:

$$\boxed{f(x)} = \boxed{\left(1+\frac{1}{x}\right)\left(1-\frac{1}{x^3}\right)} = \boxed{(1+x^{-1})(1-x^{-3})} \text{ then } \boxed{f'(x)} = \boxed{[(0-x^{-1-1})\cdot(1-x^{-3})] + [(0+3x^{-3-1})\cdot(1+x^{-1})]}$$

$$= \boxed{[-x^{-2}\cdot(1-x^{-3})] + [3x^{-4}\cdot(1+x^{-1})]} = \boxed{(-x^{-2}+x^{-2-3}) + (3x^{-4}+3x^{-4-1})} = \boxed{-x^{-2} + x^{-5} + 3x^{-4} + 3x^{-5}}$$

$$= \boxed{4x^{-5} + 3x^{-4} - x^{-2}} = \boxed{\frac{4}{x^5} + \frac{3}{x^4} - \frac{1}{x^2}}$$

t. Given $\boxed{f(x) = \left(\frac{3x^2+5}{x}\right)(x+1)}$ then $\boxed{f'(x)} = \boxed{\left\{\frac{[(3\cdot2x^{2-1}+0)x] - [1\cdot(3x^2+5)]}{x^2}\right\}(x+1) + \left[1\cdot\left(\frac{3x^2+5}{x}\right)\right]}$

$$= \boxed{\left[\left(\frac{6x^2-3x^2-5}{x^2}\right)(x+1)\right] + \left(\frac{3x^2+5}{x}\right)} = \boxed{\left[\left(\frac{3x^2-5}{x^2}\right)(x+1)\right] + \left(\frac{3x^2+5}{x}\right)}$$

u. Given $\boxed{f(x) = \frac{1+x^2}{1-x^2}}$ then $\boxed{f'(x)} = \boxed{\frac{[(0+2x^{2-1})(1-x^2)] - [(0-2x^{2-1})(1+x^2)]}{(1-x^2)^2}} = \boxed{\frac{2x(1-x^2) + 2x(1+x^2)}{(1-x^2)^2}}$

$$= \frac{2x - 2x^3 + 2x + 2x^3}{x^4 - 2x^2 + 1} = \frac{(2x + 2x) - 2x^3 + 2x^3}{\left(1 - x^2\right)^2} = \frac{4x}{\left(1 - x^2\right)^2}$$

v. Given $f(x) = \dfrac{2x^2 + 3x + 1}{x^2 + 1}$ then $f'(x) = \dfrac{\left[\left(4x^{2-1} + 3 + 0\right)\left(x^2 + 1\right)\right] - \left[\left(2x^{2-1} + 0\right)\left(2x^2 + 3x + 1\right)\right]}{\left(x^2 + 1\right)^2}$

$$= \frac{\left[(4x + 3)\left(x^2 + 1\right)\right] - \left[2x\left(2x^2 + 3x + 1\right)\right]}{\left(x^2 + 1\right)^2} = \frac{4x^3 + 4x + 3x^2 + 3 - 4x^3 - 6x^2 - 2x}{\left(x^2 + 1\right)^2} = \frac{-3x^2 + 2x + 3}{\left(x^2 + 1\right)^2}$$

w. Given $f(x) = \dfrac{ax^2 + bx + c}{ax^2 - b}$ then $f'(x) = \dfrac{\left[\left(2ax^{2-1} + b\right)\left(ax^2 - b\right)\right] - \left[2ax^{2-1}\left(ax^2 + bx + c\right)\right]}{\left(ax^2 - b\right)^2}$

$$= \frac{\left[(2ax + b)\left(ax^2 - b\right)\right] - \left[2ax\left(ax^2 + bx + c\right)\right]}{\left(ax^2 - b\right)^2} = \frac{2a^2x^3 - 2abx + abx^2 - b^2 - 2a^2x^3 - 2abx^2 - 2acx}{\left(ax^2 - b\right)^2}$$

$$= \frac{-abx^2 - 2abx - 2acx - b^2}{\left(ax^2 - b\right)^2} = \frac{-abx^2 - 2a(b + c)x - b^2}{\left(ax^2 - b\right)^2}$$

x. Given $f(x) = \dfrac{3 - x}{\dfrac{1}{x} - 5}$ then $f'(x) = \dfrac{\left[-1 \cdot \left(\dfrac{1}{x} - 5\right)\right] - \left\{\left[\dfrac{(0 \cdot x) - (1 \cdot 1)}{x^2}\right](3 - x)\right\}}{\left(\dfrac{1}{x} - 5\right)^2} = \dfrac{\left(-\dfrac{1}{x} + 5\right) + \dfrac{1}{x^2}(3 - x)}{\left(\dfrac{1}{x} - 5\right)^2}$

$$= \frac{-\dfrac{1}{x} + \dfrac{3 - x}{x^2} + 5}{\left(\dfrac{1}{x} - 5\right)^2} = \frac{-\dfrac{1}{x} + \dfrac{3}{x^2} - \dfrac{1}{x} + 5}{\left(\dfrac{1}{x} - 5\right)^2} = \frac{\dfrac{3}{x^2} - \dfrac{2}{x} + 5}{\dfrac{1}{x^2} - \dfrac{10}{x} + 25} = \frac{\dfrac{3x - 2x^2}{x^3} + \dfrac{5}{1}}{\dfrac{x - 10x^2}{x^3} + \dfrac{25}{1}} = \frac{\dfrac{3x - 2x^2 + 5x^3}{x^3}}{\dfrac{x - 10x^2 + 25x^3}{x^3}}$$

$$= \frac{5x^3 - 2x^2 + 3x}{25x^3 - 10x^2 + x} = \frac{x\left(5x^2 - 2x + 3\right)}{x\left(25x^2 - 10x + 1\right)} = \frac{5x^2 - 2x + 3}{25x^2 - 10x + 1}$$

Example 2.2-2: Find $f'(0)$, $f'(1)$, and $f'(-2)$ for the following functions.

a. $f(x) = (x + 5)x^2$

b. $f(x) = 3x^2 + 1$

c. $f(x) = x^{-5} - 2x^{-4} - 3x^{-2} + 1$

d. $f(x) = x^{-1}(x + 2)$

e. $f(x) = x^3 + 2x + \dfrac{1}{x}$

f. $f(x) = x^2(x + 1)$

g. $f(x) = \dfrac{x^2 + 4}{3x^2 + 1}$ h. $f(x) = x^5 - 2x^2 + 3x + 10$ i. $f(x) = (x^2 + 3)(x - 1)$

Solutions:

a. Given $f(x) = (x + 5)x^2$, then

$$\boxed{f'(x)} = \boxed{\left[(1+0)\cdot x^2\right] + \left[2x^{2-1}\cdot(x+5)\right]} = \boxed{x^2 + 2x(x+5)} = \boxed{x^2 + 2x^2 + 10x} = \boxed{3x^2 + 10x} \text{ and}$$

$$\boxed{f'(0)} = \boxed{(3\cdot 0^2) + (10\cdot 0)} = \boxed{0}$$

$$\boxed{f'(1)} = \boxed{(3\cdot 1^2) + (10\cdot 1)} = \boxed{3 + 10} = \boxed{13}$$

$$\boxed{f'(-2)} = \boxed{\left[3\cdot(-2)^2\right] + (10\cdot -2)} = \boxed{(3\cdot 4) - 20} = \boxed{12 - 20} = \boxed{-8}$$

b. Given $f(x) = 3x^2 + 1$, then

$$\boxed{f'(x)} = \boxed{(3\cdot 2)x^{2-1} + 0} = \boxed{6x} \text{ and}$$

$$\boxed{f'(0)} = \boxed{6\cdot 0} = \boxed{0} \qquad \boxed{f'(1)} = \boxed{6\cdot 1} = \boxed{6} \qquad \boxed{f'(-2)} = \boxed{6\cdot -2} = \boxed{-12}$$

c. Given $f(x) = x^{-5} - 2x^{-4} - 3x^{-2} + 1$, then

$$\boxed{f'(x)} = \boxed{-5x^{-5-1} + (-2\cdot -4)x^{-4-1} + (-3\cdot -2)x^{-2-1} + 0} = \boxed{-5x^{-6} + 8x^{-5} + 6x^{-3}} = \boxed{-\dfrac{5}{x^6} + \dfrac{8}{x^5} + \dfrac{6}{x^3}} \text{ and}$$

$$\boxed{f'(0)} = \boxed{-\dfrac{5}{0^6} + \dfrac{8}{0^5} + \dfrac{6}{0^3}} = \boxed{-\dfrac{5}{0} + \dfrac{8}{0} + \dfrac{6}{0}} \quad f'(0) \text{ is undefined due to division by zero}$$

$$\boxed{f'(1)} = \boxed{-\dfrac{5}{1^6} + \dfrac{8}{1^5} + \dfrac{6}{1^3}} = \boxed{-\dfrac{5}{1} + \dfrac{8}{1} + \dfrac{6}{1}} = \boxed{9}$$

$$\boxed{f'(-2)} = \boxed{-\dfrac{5}{(-2)^6} + \dfrac{8}{(-2)^5} + \dfrac{6}{(-2)^3}} = \boxed{-\dfrac{5}{64} + \dfrac{8}{-32} + \dfrac{6}{-8}} = \boxed{-0.078 - 0.25 - 0.75} = \boxed{-1.078}$$

d. Given $f(x) = x^{-1}(x + 2)$, then

$$\boxed{f'(x)} = \boxed{\left[-x^{-1-1}(x+2)\right] + \left(1\cdot x^{-1}\right)} = \boxed{\left[-x^{-2}(x+2)\right] + x^{-1}} = \boxed{\left[-\dfrac{x+2}{x^2}\right] + \dfrac{1}{x}} = \boxed{-\dfrac{1}{x} - \dfrac{2}{x^2} + \dfrac{1}{x}} = \boxed{-\dfrac{2}{x^2}} \text{ and}$$

$$\boxed{f'(0)} = \boxed{-\dfrac{2}{0^2}} = \boxed{-\dfrac{2}{0}} \quad f'(0) \text{ is undefined due to division by zero}$$

$$\boxed{f'(1)} = \boxed{-\dfrac{2}{1^2}} = \boxed{-\dfrac{2}{1}} = \boxed{-2}$$

$$\boxed{f'(-2)} = \boxed{-\frac{2}{(-2)^2}} = \boxed{-\frac{2}{4}} = \boxed{-\frac{1}{2}}$$

e. Given $f(x) = x^3 + 2x + \dfrac{1}{x} = x^3 + 2x + x^{-1}$, then

$$\boxed{f'(x)} = \boxed{3x^{3-1} + 2x^{1-1} - x^{-1-1}} = \boxed{3x^2 + 2 - x^{-2}} \text{ and}$$

$$\boxed{f'(0)} = \boxed{3x^2 + 2 - x^{-2}} = \boxed{3x^2 + 2 - \frac{1}{x^2}} = \boxed{3 \cdot 0^2 + 2 - \frac{1}{0^2}} = \boxed{2 - \frac{1}{0}} \; f'(0) \text{ is undefined due to division by zero}$$

$$\boxed{f'(1)} = \boxed{3x^2 + 2 - x^{-2}} = \boxed{3x^2 + 2 - \frac{1}{x^2}} = \boxed{3 \cdot 1^2 + 2 - \frac{1}{1^2}} = \boxed{3 + 2 - \frac{1}{1}} = \boxed{3 + 2 - 1} = \boxed{4}$$

$$\boxed{f'(-2)} = \boxed{3x^2 + 2 - x^{-2}} = \boxed{3 \cdot (-2)^2 + 2 - \frac{1}{(-2)^2}} = \boxed{3 \cdot 4 + 2 - \frac{1}{4}} = \boxed{12 + 2 - 0.25} = \boxed{13.75}$$

f. Given $f(x) = x^2(x+1)$, then

$$\boxed{f'(x)} = \boxed{\left[2x^{2-1}(x+1)\right] + \left(1 \cdot x^2\right)} = \boxed{\left[2x(x+1)\right] + x^2} = \boxed{2x^2 + 2x + x^2} = \boxed{3x^2 + 2x} \text{ and}$$

$$\boxed{f'(0)} = \boxed{\left(3 \cdot 0^2\right) + (2 \cdot 0)} = \boxed{0}$$

$$\boxed{f'(1)} = \boxed{\left(3 \cdot 1^2\right) + (2 \cdot 1)} = \boxed{3 + 2} = \boxed{5}$$

$$\boxed{f'(-2)} = \boxed{\left[3 \cdot (-2)^2\right] + (2 \cdot -2)} = \boxed{(3 \cdot 4) - 4} = \boxed{12 - 4} = \boxed{8}$$

g. Given $f(x) = \dfrac{x^2 + 4}{3x^2 + 1}$, then

$$\boxed{f'(x)} = \boxed{\frac{\left[(3x^2+1) \cdot 2x\right] - \left[(x^2+4) \cdot 6x\right]}{\left(3x^2+1\right)^2}} = \boxed{\frac{(6x^3+2x) - (6x^3+24x)}{\left(3x^2+1\right)^2}} = \boxed{\frac{6x^3+2x-6x^3-24x}{\left(3x^2+1\right)^2}} = \boxed{-\frac{22x}{\left(3x^2+1\right)^2}}$$

$$\boxed{f'(0)} = \boxed{-\frac{22 \cdot 0}{\left(3 \cdot 0^2 + 1\right)^2}} = \boxed{-\frac{0}{(0+1)^2}} = \boxed{-\frac{0}{1^2}} = \boxed{-\frac{0}{1}} = \boxed{0}$$

$$\boxed{f'(1)} = \boxed{-\frac{22 \cdot 1}{\left(3 \cdot 1^2 + 1\right)^2}} = \boxed{-\frac{22}{(3+1)^2}} = \boxed{-\frac{22}{4^2}} = \boxed{-\frac{22}{16}} = \boxed{-1.375}$$

$$\boxed{f'(-2)} = \boxed{-\frac{22 \cdot -2}{\left[3 \cdot (-2)^2 + 1\right]^2}} = \boxed{-\frac{-44}{(12+1)^2}} = \boxed{\frac{44}{13^2}} = \boxed{\frac{44}{169}} = \boxed{0.26}$$

h. Given $f(x) = x^5 - 2x^2 + 3x + 10$, then

$\boxed{f'(x)} = \boxed{5x^{5-1} - 2 \cdot 2x^{2-1} + 3 + 0} = \boxed{5x^4 - 4x + 3}$

$\boxed{f'(0)} = \boxed{5 \cdot 0^4 - 4 \cdot 0 + 3} = \boxed{0 - 0 + 3} = \boxed{3}$

$\boxed{f'(1)} = \boxed{5 \cdot 1^4 - 4 \cdot 1 + 3} = \boxed{5 - 4 + 3} = \boxed{4}$

$\boxed{f'(-2)} = \boxed{5 \cdot (-2)^4 + (-4 \cdot -2) + 3} = \boxed{(5 \cdot 16) + 8 + 3} = \boxed{80 + 11} = \boxed{91}$

i. Given $f(x) = (x^2 + 3)(x - 1)$, then

$\boxed{f'(x)} = \boxed{[2x \cdot (x-1)] + [1 \cdot (x^2 + 3)]} = \boxed{2x^2 - 2x + x^2 + 3} = \boxed{3x^2 - 2x + 3}$

$\boxed{f'(0)} = \boxed{3 \cdot 0^2 - 2 \cdot 0 + 3} = \boxed{0 - 0 + 3} = \boxed{3}$

$\boxed{f'(1)} = \boxed{3 \cdot 1^2 - 2 \cdot 1 + 3} = \boxed{3 - 2 + 3} = \boxed{4}$

$\boxed{f'(-2)} = \boxed{3 \cdot (-2)^2 + (-2 \cdot -2) + 3} = \boxed{3 \cdot 4 + 4 + 3} = \boxed{12 + 7} = \boxed{19}$

Example 2.2-3: Given $g(x) = \dfrac{1}{x} + 1$ and $h(x) = x$, find $f(x)$, $f'(x)$ and $f'(0)$.

a. $f(x) = x\,g(x)$ b. $f(x) = 2x^2 - 5h(x)$ c. $f(x) = g(x) + \dfrac{x}{h(x)}$

d. $h(x) = 3x\,f(x)$ e. $h(x) = 1 - f(x)$ f. $3h(x) = 2x\,f(x) - 1$

Solutions:

a. Given $g(x) = \dfrac{1}{x} + 1$ and $f(x) = x\,g(x)$, then

$\boxed{f(x)} = \boxed{xg(x)} = \boxed{x \cdot \left(\dfrac{1}{x} + 1\right)} = \boxed{1 + x}$ therefore $\boxed{f'(x)} = \boxed{1}$ and $\boxed{f'(0)} = \boxed{1}$

b. Given $h(x) = x$ and $f(x) = 2x^2 - 5h(x)$, then

$\boxed{f(x)} = \boxed{2x^2 - 5h(x)} = \boxed{2x^2 - 5 \cdot x} = \boxed{2x^2 - 5x}$ therefore $\boxed{f'(x)} = \boxed{(2 \cdot 2)x^{2-1} - 5} = \boxed{4x - 5}$ and

$\boxed{f'(0)} = \boxed{(4 \cdot 0) - 5} = \boxed{-5}$

c. Given $g(x) = \dfrac{1}{x} + 1$, $h(x) = x$, and $f(x) = g(x) + \dfrac{x}{h(x)}$, then

$\boxed{f(x)} = \boxed{g(x) + \dfrac{x}{h(x)}} = \boxed{\left(\dfrac{1}{x} + 1\right) + \dfrac{x}{x}} = \boxed{\left(\dfrac{1}{x} + 1\right) + 1} = \boxed{\dfrac{1}{x} + 2} = \boxed{x^{-1} + 2}$ therefore $\boxed{f'(x)} = \boxed{-x^{-1-1} + 0}$

$= \boxed{-x^{-2}} = \boxed{-\dfrac{1}{x^2}}$ and $\boxed{f'(0)} = \boxed{-\dfrac{1}{0^2}} = \boxed{-\dfrac{1}{0}}$ which is undefined due to division by zero.

d. Given $h(x) = x$ and $h(x) = 3x\,f(x)$, then

$\boxed{f(x)} = \boxed{\dfrac{h(x)}{3x}} = \boxed{\dfrac{x}{3x}} = \boxed{\dfrac{1}{3}}$ therefore, $\boxed{f'(x)} = \boxed{0}$ and $\boxed{f'(0)} = \boxed{0}$

e. Given $h(x) = x$ and $h(x) = 1 - f(x)$, then

$\boxed{f(x)} = \boxed{1 - h(x)} = \boxed{1 - x}$ therefore, $\boxed{f'(x)} = \boxed{-1}$ and $\boxed{f'(0)} = \boxed{-1}$

f. Given $h(x) = x$ and $3h(x) = 2x\,f(x) - 1$, then

$\boxed{f(x)} = \boxed{\dfrac{3h(x) + 1}{2x}} = \boxed{\dfrac{3x + 1}{2x}}$ therefore, $\boxed{f'(x)} = \boxed{\dfrac{[2x \cdot 3] - [2 \cdot (3x + 1)]}{(2x)^2}} = \boxed{\dfrac{6x - 6x - 2}{4x^2}} = \boxed{-\dfrac{2}{4x^2}}$

$= \boxed{-\dfrac{1}{2x^2}}$ and $\boxed{f'(0)} = \boxed{-\dfrac{1}{2 \cdot 0^2}} = \boxed{-\dfrac{1}{0}}$ which is undefined due to division by zero.

Section 2.2 Practice Problems - Differentiation Rules Using the Prime Notation

1. Find the derivative of the following functions. Compare your answers with Practice Problem number 1 in Section 2.1.

   a. $f(x) = x^2 - 1$
   b. $f(x) = x^3 + 2x - 1$
   c. $f(x) = \dfrac{x}{x - 1}$
   d. $f(x) = -\dfrac{1}{x^2}$

   e. $f(x) = 20x^2 - 3$
   f. $f(x) = \sqrt{x^3}$
   g. $f(x) = \dfrac{10}{\sqrt{x - 5}}$
   h. $f(x) = \dfrac{ax + b}{cx}$

2. Differentiate the following functions:

   a. $f(x) = x^2 + 10x + 1$
   b. $f(x) = x^8 + 3x^2 - 1$
   c. $f(x) = 3x^4 - 2x^2 + 5$

   d. $f(x) = 2\left(x^5 + 10x^4 + 5x\right)$
   e. $f(x) = a^2x^3 + b^2x + c^2$
   f. $f(x) = x^2(x - 1) + 3x$

   g. $f(x) = \left(x^3 + 1\right)\left(x^2 - 5\right)$
   h. $f(x) = \left(3x^2 + x - 1\right)(x - 1)$
   i. $f(x) = x\left(x^3 + 5x^2\right) - 4x$

   j. $f(x) = \dfrac{x^3 + 1}{x}$
   k. $f(x) = \dfrac{x^5 + 2x^2 - 1}{3x^2}$
   l. $f(x) = \dfrac{x^2}{(x - 1) + 3x}$

   m. $f(x) = x^2\left(2 + \dfrac{1}{x}\right)$
   n. $f(x) = (x + 1) \cdot \dfrac{2x}{x - 1}$
   o. $f(x) = \dfrac{x^3 + 3x - 1}{x^4}$

   p. $f(x) = \left(x^2 - 1\right)\left(\dfrac{2x^3 + 5}{x}\right)$
   q. $f(x) = \dfrac{3x^4 + x^2 + 2}{x - 1}$
   r. $f(x) = x^{-1} + \dfrac{1}{x^{-2}}$

3. Compute $f'(x)$ at the specified value of x. Compare your answers with the practice problem number 2 in Section 2.1.

   a. $f(x) = x^3$ at $x = 1$
   b. $f(x) = 1 + 2x$ at $x = 0$
   c. $f(x) = x^3 + 1$ at $x = -1$

   d. $f(x) = x^2(x + 2)$ at $x = 2$
   e. $f(x) = x^{-2} + x^{-1} + 1$ at $x = 1$
   f. $f(x) = \sqrt{x} + 2$ at $x = 10$

4. Find $f'(0)$ and $f'(2)$ for the following functions:

a. $f(x) = x^3 - 3x^2 + 5$

b. $f(x) = (x^3 + 1)(x - 1)$

c. $f(x) = x(x^2 + 1)$

d. $f(x) = 2x^5 + 10x^4 - 4x$

e. $f(x) = 2x^{-2} - 3x^{-1} + 5x$

f. $f(x) = x^{-2}(x^5 - x^3) + x$

g. $f(x) = \dfrac{x}{1 + x^2}$

h. $f(x) = \dfrac{1}{x} + x^3$

i. $f(x) = \dfrac{ax^2 + bx}{cx - d}$

5. Given $f(x) = x^2 + 1$ and $g(x) = 2x - 5$, find $h(x)$ and $h'(x)$.

a. $h(x) = x^3 f(x)$

b. $f(x) = 3 + h(x)$

c. $2g(x) = h(x) - 1$

d. $3h(x) = 2x \, g(x) - 1$

e. $3[f(x)]^2 - 2h(x) = 1$

f. $h(x) = g(x) \cdot 3f(x)$

g. $3h(x) - f(x) = 0$

h. $2g(x) + h(x) = f(x)$

i. $f(x) = x^3 + 5x^2 + h(x)$

j. $h(x) = \dfrac{x^3 + 1}{x} - f(x)$

k. $h(x) = 2f(x) + g(x)$

l. $[h(x)]^2 - f(x) = 10$

m. $f(x) = \dfrac{2g(x)}{h(x)}$

n. $\dfrac{3f(x)}{h(x)} = \dfrac{1}{x}$

o. $f(x) = \dfrac{1}{h(x) + 4}$

2.3 Differentiation Rules Using the $\dfrac{d}{dx}$ Notation

In the previous section the prime notation was used as a means to show the derivative of a function. For example, derivative of the functions $y = f(x) = x^2 + 3x + 1$ was represented as $y' = f'(x) = 2x + 3$. However, derivatives can also be represented by what is referred to as the "double-d" notation. For example, the derivative of the function $y = f(x) = x^2 + 3x + 1$ can be shown as $\dfrac{dy}{dx} = \dfrac{d}{dx} f(x) = 2x + 3$. Following are the differentiation rules in the double-d notation form:

Rule No. 1 - *The derivative of a constant function is equal to zero, i.e.,*

$$if \ f(x) = k, \qquad then \ \frac{d}{dx} f(x) = 0$$

Rule No. 2 - *The derivative of the identity function is equal to one, i.e.,*

$$if \ f(x) = x, \qquad then \ \frac{d}{dx} f(x) = 1$$

Rule No. 3 - *The derivative of the function $f(x) = x^n$ is equal to $\dfrac{d}{dx} f(x) = n x^{n-1}$, where n is a positive or negative integer.*

Rule No. 4 (scalar rule) - *If the function $f(x)$ is differentiable at x, then a constant k multiplied by $f(x)$ is also differentiable at x, i.e.,*

$$\frac{d}{dx}\left[(k\,f)x\right] = k\left[\frac{d}{dx} f(x)\right]$$

Rule No. 5 (summation rule) - *If the function $f(x)$ and $g(x)$ are differentiable at x, then their sum is also differentiable at x, i.e.,*

$$\frac{d}{dx}\left[(f+g)x\right] = \frac{d}{dx} f(x) + \frac{d}{dx} g(x)$$

Rule No. 6 (product rule) - *If the function $f(x)$ and $g(x)$ are differentiable at x, then their product is also differentiable at x, i.e.,*

$$\left[\frac{d}{dx}(f \cdot g)\right](x) = \left[\frac{d}{dx} f(x)\right] g(x) + \left[\frac{d}{dx} g(x)\right] f(x)$$

Rule No. 7 - Using the rules 1, 4, 5, and 6 we can write the formula for differentiating polynomials, i.e.,

$$if \ f(x) = a_n x^n + a_{n-1} x^{n-1} + a_{n-2} x^{n-2} + \cdots + a_3 x^3 + a_2 x^2 + a_1 x^1 + a_0 \ then,$$

$$\frac{d}{dx} f(x) = n a_n x^{n-1} + (n-1)a_{n-1} x^{n-2} + (n-2)a_{n-2} x^{n-3} + \cdots + 3a_3 x^2 + 2a_2 x + a_1$$

Rule No. 8 (quotient rule) - *If the function $f(x)$ and $g(x)$ are differentiable at x, then their quotient is also differentiable at x, i.e.,*

$$\left[\frac{d}{dx}\left(\frac{f}{g}\right)\right](x) = \frac{\left[\frac{d}{dx}f(x)\right]g(x) - \left[\frac{d}{dx}g(x)\right]f(x)}{\left[g(x)\right]^2}$$

Note 1 - depending on the letter used to express the terms of a function, the double-d notation of a derivative is then shown as $\dfrac{da}{db}$ $\dfrac{\left(\textit{where a is equal to the letter used in the left hand side of the equation}\right)}{\left(\textit{where b is equal to the letter used in the right hand side of the equation}\right)}$.

For example,

- if the function y is represented by $f(x)$, i.e., $y = f(x) = x^2 + 2x$, then its derivative is shown as $\dfrac{dy}{dx} = \dfrac{d}{dx}f(x) = \dfrac{d}{dx}\left(x^2 + 2x\right) = 2x + 2$.

- if the function y is represented by $f(t)$, i.e., $y = f(t) = t^3 + 2t^2 + 4$, then its derivative is shown as $\dfrac{dy}{dt} = \dfrac{d}{dt}f(t) = \dfrac{d}{dt}\left(t^3 + 2t^2 + 4\right) = 3t^2 + 4t$.

- if the function u is represented by $f(v)$, i.e., $u = f(v) = v^3 + 3v$, then its derivative is shown as $\dfrac{du}{dv} = \dfrac{d}{dv}f(v) = \dfrac{d}{dv}\left(v^3 + 3v\right) = 3v^2 + 3$.

- if the function p is represented by $f(r)$, i.e., $p = f(r) = 2r^3 - 2r^2 - 3$, then its derivative is shown as $\dfrac{dp}{dr} = \dfrac{d}{dr}f(r) = \dfrac{d}{dr}\left(2r^3 - 2r^2 - 3\right) = 3r^2 - 4r$.

- if the function y is represented by $f(z)$, i.e., $y = f(z) = z^5 + 3z^2 + 1$, then its derivative is shown as $\dfrac{dy}{dz} = \dfrac{d}{dz}f(z) = \dfrac{d}{dz}\left(z^5 + 3z^2 + 1\right) = 5z^4 + 6z$.

- if the function v is represented by $f(x)$, i.e., $v = f(x) = x^8 + 4$, then its derivative is shown as $\dfrac{dv}{dx} = \dfrac{d}{dx}f(x) = \dfrac{d}{dx}\left(x^8 + 4\right) = 8x^7$, etc.

In the following examples the above rules are used in order to find the derivative of various functions:

Example 2.3-1: Find $\dfrac{dy}{dx}$ for the following functions.

a. $y = x^3 - 2x^2 + 5$ b. $y = 4x^5 - 3x^2 - 1$ c. $y = x^2 + \dfrac{1}{x}$

d. $y = \dfrac{3x^2}{1+x}$ e. $y = 5x + \dfrac{2x}{x^2 + 1}$ f. $y = x^3\left(x^2 + 1\right)$

g. $y = (x+1)\left(x^2 - 3\right)$ h. $y = 5x(x+1)$ i. $y = 5 + \dfrac{1-x}{x}$

j. $y = x(x+1)(x-2)$ k. $y = x^2\left(\dfrac{x-3}{5}\right)$ l. $y = (x+1)(x-1)^{-2}$

Solutions:

a. $\boxed{\dfrac{dy}{dx}} = \dfrac{d}{dx}\left(x^3 - 2x^2 + 5\right) = \dfrac{d}{dx}\left(x^3\right) + \dfrac{d}{dx}\left(-2x^2\right) + \dfrac{d}{dx}(5) = \boxed{3x^2 + (-2\cdot2)x + 0} = \boxed{3x^2 - 4x}$

b. $\boxed{\dfrac{dy}{dx}} = \boxed{\dfrac{d}{dx}\left(4x^5 - 3x^2 - 1\right)} = \boxed{\dfrac{d}{dx}\left(4x^5\right) + \dfrac{d}{dx}\left(-3x^2\right) + \dfrac{d}{dx}(-1)} = \boxed{(4\cdot 5)x^4 + (-3\cdot 2)x + 0} = \boxed{\mathbf{20x^4 - 6x}}$

c. $\boxed{\dfrac{dy}{dx}} = \boxed{\dfrac{d}{dx}\left(x^2 + \dfrac{1}{x}\right)} = \boxed{\dfrac{d}{dx}\left(x^2 + x^{-1}\right)} = \boxed{\dfrac{d}{dx}\left(x^2\right) + \dfrac{d}{dx}\left(x^{-1}\right)} = \boxed{2x + \left(-x^{-2}\right)} = \boxed{2x - x^{-2}} = \boxed{\mathbf{2x - \dfrac{1}{x^2}}}$

d. $\boxed{\dfrac{dy}{dx}} = \boxed{\dfrac{d}{dx}\left(\dfrac{3x^2}{1+x}\right)} = \boxed{\dfrac{\left[(1+x)\dfrac{d}{dx}\left(3x^2\right)\right] - \left[\left(3x^2\right)\dfrac{d}{dx}(1+x)\right]}{(1+x)^2}} = \boxed{\dfrac{[(1+x)\cdot 6x] - \left[\left(3x^2\right)\cdot 1\right]}{(1+x)^2}} = \boxed{\dfrac{\mathbf{3x^2 + 6x}}{\mathbf{(1+x)^2}}}$

e. $\boxed{\dfrac{dy}{dx}} = \boxed{\dfrac{d}{dx}\left(5x + \dfrac{2x}{x^2+1}\right)} = \boxed{\dfrac{d}{dx}(5x) + \dfrac{d}{dx}\left(\dfrac{2x}{x^2+1}\right)} = \boxed{\dfrac{d}{dx}(5x) + \dfrac{\left[\left(x^2+1\right)\dfrac{d}{dx}(2x)\right] - \left[2x\dfrac{d}{dx}\left(x^2+1\right)\right]}{\left(x^2+1\right)^2}}$

$= \boxed{5 + \dfrac{\left[\left(x^2+1\right)\cdot 2\right] - [2x\cdot 2x]}{\left(x^2+1\right)^2}} = \boxed{5 + \dfrac{2x^2 + 2 - 4x^2}{\left(x^2+1\right)^2}} = \boxed{\mathbf{5 + \dfrac{-2x^2 + 2}{\left(x^2+1\right)^2}}}$

f. $\boxed{\dfrac{dy}{dx}} = \boxed{\dfrac{d}{dx}\left[x^3\left(x^2+1\right)\right]} = \boxed{\left[\left(x^2+1\right)\dfrac{d}{dx}\left(x^3\right)\right] + \left[x^3\dfrac{d}{dx}\left(x^2+1\right)\right]} = \boxed{\left(x^2+1\right)\cdot 3x^2 + x^3\cdot 2x} = \boxed{\mathbf{5x^4 + 3x^2}}$ or,

$\boxed{\dfrac{dy}{dx}} = \boxed{\dfrac{d}{dx}\left[x^3\left(x^2+1\right)\right]} = \boxed{\dfrac{d}{dx}\left(x^5 + x^3\right)} = \boxed{\dfrac{d}{dx}x^5 + \dfrac{d}{dx}x^3} = \boxed{5x^{5-1} + 3x^{3-1}} = \boxed{\mathbf{5x^4 + 3x^2}}$

g. $\boxed{\dfrac{dy}{dx}} = \boxed{\dfrac{d}{dx}\left[(x+1)\left(x^2-3\right)\right]} = \boxed{\left[\left(x^2-3\right)\dfrac{d}{dx}(x+1)\right] + \left[(x+1)\dfrac{d}{dx}\left(x^2-3\right)\right]} = \boxed{\left[\left(x^2-3\right)\cdot 1\right] + [(x+1)\cdot 2x]}$

$= \boxed{x^2 - 3 + 2x^2 + 2x} = \boxed{\mathbf{3x^2 + 2x - 3}}$ or,

$\boxed{\dfrac{dy}{dx}} = \boxed{\dfrac{d}{dx}\left[(x+1)\left(x^2-3\right)\right]} = \boxed{\dfrac{d}{dx}\left(x^3 - 3x + x^2 - 3\right)} = \boxed{\dfrac{d}{dx}x^3 + \dfrac{d}{dx}x^2 + \dfrac{d}{dx}(-3x) + \dfrac{d}{dx}(-3)} = \boxed{\mathbf{3x^2 + 2x - 3}}$

h. $\boxed{\dfrac{dy}{dx}} = \boxed{\dfrac{d}{dx}[5x(x+1)]} = \boxed{\left[(x+1)\dfrac{d}{dx}(5x)\right] + \left[(5x)\dfrac{d}{dx}(x+1)\right]} = \boxed{[(x+1)\cdot 5] + [5x\cdot 1]} = \boxed{5x + 5 + 5x} = \boxed{\mathbf{10x + 5}}$ or,

$\boxed{\dfrac{dy}{dx}} = \boxed{\dfrac{d}{dx}[5x(x+1)]} = \boxed{\dfrac{d}{dx}\left(5x^2 + 5x\right)} = \boxed{\dfrac{d}{dx}5x^2 + \dfrac{d}{dx}5x} = \boxed{(5\cdot 2)x^{2-1} + 5x^{1-1}} = \boxed{\mathbf{10x + 5}}$

i. $\boxed{\dfrac{dy}{dx}} = \boxed{\dfrac{d}{dx}\left(5 + \dfrac{1-x}{x}\right)} = \boxed{\dfrac{d}{dx}(5) + \dfrac{d}{dx}\left(\dfrac{1-x}{x}\right)} = \boxed{\dfrac{d}{dx}(5) + \dfrac{\left[x\dfrac{d}{dx}(1-x)\right] - \left[(1-x)\dfrac{d}{dx}(x)\right]}{x^2}} = \boxed{0 + \dfrac{[x\cdot -1] - [(1-x)\cdot 1]}{x^2}}$

$$= \boxed{\dfrac{-x-1+x}{x^2}} = \boxed{-\dfrac{1}{x^2}}$$

j. $\boxed{\dfrac{dy}{dx}} = \boxed{\dfrac{d}{dx}\left[x(x+1)(x-2)\right]} = \boxed{\dfrac{d}{dx}\left[(x^2+x)(x-2)\right]} = \boxed{\left[(x-2)\dfrac{d}{dx}(x^2+x)\right]+\left[(x^2+x)\dfrac{d}{dx}(x-2)\right]}$

$$= \boxed{\left[(x-2)\cdot(2x+1)\right]+\left[(x^2+x)\cdot 1\right]} = \boxed{2x^2+x-4x-2+x^2+x} = \boxed{3x^2-2x-2}\ \text{or,}$$

$\boxed{\dfrac{dy}{dx}} = \boxed{\dfrac{d}{dx}\left[x(x+1)(x-2)\right]} = \boxed{\dfrac{d}{dx}\left[(x^2+x)(x-2)\right]} = \boxed{\dfrac{d}{dx}\left(x^3-2x^2+x^2-2x\right)} = \boxed{\dfrac{d}{dx}\left(x^3-x^2-2x\right)}$

$$= \boxed{\dfrac{d}{dx}x^3+\dfrac{d}{dx}\left(-x^2\right)+\dfrac{d}{dx}(-2x)} = \boxed{3x^{3-1}+-2x^{2-1}-2x^{1-1}} = \boxed{3x^2-2x-2x^0} = \boxed{3x^2-2x-2}$$

k. $\boxed{\dfrac{dy}{dx}} = \boxed{\dfrac{d}{dx}\left[x^2\left(\dfrac{x-3}{5}\right)\right]} = \boxed{\left[\left(\dfrac{x-3}{5}\right)\dfrac{d}{dx}\left(x^2\right)\right]+\left[\left(x^2\right)\dfrac{d}{dx}\left(\dfrac{x-3}{5}\right)\right]} = \boxed{\left[\left(\dfrac{x-3}{5}\right)\cdot 2x\right]+\left[\left(x^2\right)\cdot\left(\dfrac{1\cdot5-0}{5^2}\right)\right]}$

$$= \boxed{\left(\dfrac{2x^2-6x}{5}\right)+\left(x^2\cdot\dfrac{1}{5}\right)} = \boxed{\dfrac{2x^2-6x}{5}+\dfrac{x^2}{5}} = \boxed{\dfrac{2x^2-6x+x^2}{5}} = \boxed{\dfrac{3x^2-6x}{5}} = \boxed{\dfrac{3x(x-2)}{5}}\ \text{or,}$$

$\boxed{\dfrac{dy}{dx}} = \boxed{\dfrac{d}{dx}\left[x^2\left(\dfrac{x-3}{5}\right)\right]} = \boxed{\dfrac{d}{dx}\left(\dfrac{x^3-3x^2}{5}\right)} = \boxed{\dfrac{\left[5\dfrac{d}{dx}\left(x^3-3x^2\right)\right]-\left[\left(x^3-3x^2\right)\dfrac{d}{dx}(5)\right]}{5^2}}$

$$= \boxed{\dfrac{\left[5\left(3x^2-6x\right)\right]-\left[\left(x^3-3x^2\right)\cdot 0\right]}{5^2}} = \boxed{\dfrac{5\left(3x^2-6x\right)-0}{5^2}} = \boxed{\dfrac{5\left(3x^2-6x\right)}{5^{2=1}}} = \boxed{\dfrac{3x^2-6x}{5}} = \boxed{\dfrac{3x(x-2)}{5}}$$

l. $\boxed{\dfrac{dy}{dx}} = \boxed{\dfrac{d}{dx}\left[(x+1)(x-1)^{-2}\right]} = \boxed{\left[(x-1)^{-2}\dfrac{d}{dx}(x+1)\right]+\left[(x+1)\dfrac{d}{dx}(x-1)^{-2}\right]} = \boxed{\left[(x-1)^{-2}\cdot 1\right]+\left[(x+1)\cdot -2(x-1)^{-3}\right]}$

$$= \boxed{(x-1)^{-2}-2(x+1)(x-1)^{-3}} = \boxed{\dfrac{1}{(x-1)^2}-\dfrac{2(x+1)}{(x-1)^3}}\ \text{or,}$$

$\boxed{\dfrac{dy}{dx}} = \boxed{\dfrac{d}{dx}\left[(x+1)(x-1)^{-2}\right]} = \boxed{\dfrac{d}{dx}\left[\dfrac{x+1}{(x-1)^2}\right]} = \boxed{\dfrac{\left[(x-1)^2\dfrac{d}{dx}(x+1)\right]-\left[(x+1)\dfrac{d}{dx}(x-1)^2\right]}{(x-1)^4}} = \boxed{\dfrac{\left[(x-1)^2\cdot 1\right]}{(x-1)^4}}$

$$\boxed{-\dfrac{\left[(x+1)\cdot 2(x-1)\right]}{(x-1)^4}} = \boxed{\dfrac{(x-1)^2-2(x+1)(x-1)}{(x-1)^4}} = \boxed{\dfrac{(x-1)^2}{(x-1)^{4=2}}-\dfrac{2(x+1)(x-1)}{(x-1)^{4=3}}} = \boxed{\dfrac{1}{(x-1)^2}-\dfrac{2(x+1)}{(x-1)^3}}$$

Example 2.3-2: Find the derivative of the following functions.

a. $\dfrac{d}{dx}\left(3x^2+5x-1\right)=$

b. $\dfrac{d}{dx}\left(8x^4+3x^2+x\right)=$

c. $\dfrac{d}{du}\left[\left(u^3+5\right)(u+1)\right]=$

d. $\dfrac{d}{dt}\left(\dfrac{2t^2+3t+1}{t^3}\right)=$

e. $\dfrac{d}{dt}\left[\left(1-t^2\right)(1+t)+t\right]=$

f. $\dfrac{d}{dt}\left(\dfrac{t^2+1}{t^2-1}\right)=$

g. $\dfrac{d}{dt}\left[t^3\left(\dfrac{-2t}{4}\right)\right]=$

h. $\dfrac{d}{du}\left(\dfrac{u}{1-u}+\dfrac{u^2}{1+u}\right)=$

i. $\dfrac{d}{ds}\left(\dfrac{s^3+3s^2+1}{s^3}\right)=$

j. $\dfrac{d}{dw}\left[\left(w^3+1\right)\left(\dfrac{1}{w}\right)\right]=$

k. $\dfrac{d}{dx}\left(\dfrac{2x}{1+2x}\right)=$

l. $\dfrac{d}{ds}\left(\dfrac{s^2}{1+s^2}\right)=$

m. $\dfrac{d}{dt}\left[t^3\left(t^2+1\right)(3t-1)\right]=$

n. $\dfrac{d}{du}\left[\dfrac{4u^3+2}{u^2}\right]=$

o. $\dfrac{d}{dx}\left[\dfrac{x^3}{x^2+1}\right]=$

Solutions:

a. $\boxed{\dfrac{d}{dx}\left(3x^2+5x-1\right)}=\boxed{\dfrac{d}{dx}\left(3x^2\right)+\dfrac{d}{dx}(5x)+\dfrac{d}{dx}(-1)}=\boxed{(3\cdot2)x^{2-1}+5x^{1-1}+0}=\boxed{6x+5x^0}=\boxed{6x+5}$

b. $\boxed{\dfrac{d}{dx}\left(8x^4+3x^2+x\right)}=\boxed{\dfrac{d}{dx}\left(8x^4\right)+\dfrac{d}{dx}\left(3x^2\right)+\dfrac{d}{dx}(x)}=\boxed{(8\cdot4)x^{4-1}+(3\cdot2)x^{2-1}+x^{1-1}}=\boxed{32x^3+6x+1}$

c. $\boxed{\dfrac{d}{du}\left[\left(u^3+5\right)(u+1)\right]}=\boxed{\left[(u+1)\dfrac{d}{du}\left(u^3+5\right)\right]+\left[\left(u^3+5\right)\dfrac{d}{du}(u+1)\right]}=\boxed{\left[(u+1)\cdot3u^2\right]+\left[\left(u^3+5\right)\cdot1\right]}$

$=\boxed{3u^3+3u^2+u^3+5}=\boxed{4u^3+3u^2+5}$

d. $\boxed{\dfrac{d}{dt}\left(\dfrac{2t^2+3t+1}{t^3}\right)}=\boxed{\dfrac{\left[t^3\dfrac{d}{dt}\left(2t^2+3t+1\right)\right]-\left[\left(2t^2+3t+1\right)\dfrac{d}{dt}\left(t^3\right)\right]}{t^6}}=\boxed{\dfrac{\left[t^3\cdot(4t+3)\right]-\left[\left(2t^2+3t+1\right)\cdot3t^2\right]}{t^6}}$

$=\boxed{\dfrac{\left(4t^4+3t^3\right)-\left(6t^4+9t^3+3t^2\right)}{t^6}}=\boxed{\dfrac{4t^4+3t^3-6t^4-9t^3-3t^2}{t^6}}=\boxed{\dfrac{-2t^4-6t^3-3t^2}{t^6}}=\boxed{\dfrac{-t^2\left(2t^2+6t+3\right)}{t^6}}$

$=\boxed{-\dfrac{t^2\left(2t^2+6t+3\right)}{t^{6=4}}}=\boxed{-\dfrac{2t^2+6t+3}{t^4}}$

e. $\boxed{\dfrac{d}{dt}\left[\left(1-t^2\right)(1+t)+t\right]}=\boxed{\dfrac{d}{dt}\left[\left(1-t^2\right)(1+t)\right]+\dfrac{d}{dt}t}=\boxed{\left\{\left[(1+t)\dfrac{d}{dt}\left(1-t^2\right)\right]+\left[\left(1-t^2\right)\dfrac{d}{dt}(1+t)\right]\right\}+1}$

$=\boxed{\left\{\left[(1+t)\cdot-2t\right]+\left[\left(1-t^2\right)\cdot1\right]\right\}+1}=\boxed{\left(-2t-2t^2+1-t^2\right)+1}=\boxed{-2t-2t^2+1-t^2+1}=\boxed{-3t^2-2t+2}$

f. $\dfrac{d}{dt}\left(\dfrac{t^2+1}{t^2-1}\right) = \dfrac{\left[(t^2-1)\dfrac{d}{dt}(t^2+1)\right]-\left[(t^2+1)\dfrac{d}{dt}(t^2-1)\right]}{(t^2-1)^2} = \dfrac{\left[(t^2-1)\cdot 2t\right]-\left[(t^2+1)\cdot 2t\right]}{(t^2-1)^2} = \dfrac{(2t^3-2t)-(2t^3+2t)}{(t^2-1)^2}$

$= \dfrac{2t^3-2t-2t^3-2t}{(t^2-1)^2} = \dfrac{-4t}{(t^2-1)^2}$

g. $\dfrac{d}{dt}\left[t^3\left(-\dfrac{2t}{4}\right)\right] = \left[\left(\dfrac{-2t}{4}\right)\dfrac{d}{dt}t^3\right]+\left[t^3\dfrac{d}{dt}\left(\dfrac{-2t}{4}\right)\right] = \left[\left(\dfrac{-2t}{4}\right)\cdot 3t^2\right]+\left[t^3\left(\dfrac{4\dfrac{d}{dt}(-2t)-(-2t)\dfrac{d}{dt}(4)}{4^2}\right)\right]$

$= -\dfrac{6t^3}{4}+\left[t^3\left(\dfrac{[4\cdot-2]-[(-2t)\cdot 0]}{16}\right)\right] = -\dfrac{6t^3}{4}+\left[t^3\left(\dfrac{-8}{16}\right)\right] = \dfrac{6t^3}{4}-\dfrac{t^3}{2} = \dfrac{-12t^3-4t^3}{8} = -\dfrac{16t^3}{8} = -2t^3$

A second way of solving this problem is to simplify $\dfrac{d}{dt}\left[t^3\left(-\dfrac{2t}{4}\right)\right]$ as follows:

$\dfrac{d}{dt}\left[t^3\left(-\dfrac{2t}{4}\right)\right] = \dfrac{d}{dt}\left(-\dfrac{2t^4}{4}\right) = \dfrac{d}{dt}\left(-\dfrac{2t^4}{\underset{2}{4}}\right) = -\dfrac{1}{2}\dfrac{d}{dt}t^4 = -\dfrac{1}{2}\cdot 4t^{4-1} = -\dfrac{4t^3}{2} = -2t^3$

h. $\dfrac{d}{du}\left(\dfrac{u}{1-u}+\dfrac{u^2}{1+u}\right) = \dfrac{d}{du}\left(\dfrac{u}{1-u}\right)+\dfrac{d}{du}\left(\dfrac{u^2}{1+u}\right) = \dfrac{\left[(1-u)\dfrac{d}{du}u\right]-\left[u\dfrac{d}{du}(1-u)\right]}{(1-u)^2}+\dfrac{\left[(1+u)\dfrac{d}{du}u^2\right]-\left[u^2\dfrac{d}{du}(1+u)\right]}{(1+u)^2}$

$= \dfrac{[(1-u)\cdot 1]-[u\cdot-1]}{(1-u)^2}+\dfrac{[(1+u)\cdot 2u]-[u^2\cdot 1]}{(1+u)^2} = \dfrac{1-u+u}{(1-u)^2}+\dfrac{2u+u^2-u^2}{(1+u)^2} = \dfrac{1}{(1-u)^2}+\dfrac{2u}{(1+u)^2}$

i. $\dfrac{d}{ds}\left(\dfrac{s^3+3s^2+1}{s^3}\right) = \dfrac{\left[s^3\dfrac{d}{ds}(s^3+3s^2+1)\right]-\left[(s^3+3s^2+1)\dfrac{d}{ds}s^3\right]}{s^6} = \dfrac{\left[s^3\cdot(3s^2+6s)\right]-\left[(s^3+3s^2+1)\cdot 3s^2\right]}{s^6}$

$= \dfrac{(3s^5+6s^4)-(3s^5+9s^4+3s^2)}{s^6} = \dfrac{3s^5+6s^4-3s^5-9s^4-3s^2}{s^6} = \dfrac{-3s^4-3s^2}{s^6} = \dfrac{-3s^2(s^2+1)}{s^{6=4}} = \dfrac{-3(s^2+1)}{s^4}$

j. $\dfrac{d}{dw}\left[(w^3+1)\left(\dfrac{1}{w}\right)\right] = \dfrac{d}{dw}\left[(w^3+1)w^{-1}\right] = \dfrac{d}{dw}(w^{3-1}+w^{-1}) = \dfrac{d}{dw}w^2+\dfrac{d}{dw}w^{-1} = 2w-w^{-2} = 2w-\dfrac{1}{w^2}$

k. $\dfrac{d}{dx}\left(\dfrac{2x}{1+2x}\right) = \dfrac{\left[(1+2x)\dfrac{d}{dx}(2x)\right]-\left[(2x)\dfrac{d}{dx}(1+2x)\right]}{(1+2x)^2} = \dfrac{[(1+2x)\cdot 2]-[2x\cdot 2]}{(1+2x)^2} = \dfrac{2+4x-4x}{(1+2x)^2} = \dfrac{2}{(1+2x)^2}$

l. $\dfrac{d}{ds}\left(\dfrac{s^2}{1+s^2}\right) = \dfrac{\left[\left(1+s^2\right)\dfrac{d}{ds}s^2\right]-\left[s^2\dfrac{d}{ds}\left(1+s^2\right)\right]}{\left(1+s^2\right)^2} = \dfrac{\left[\left(1+s^2\right)\cdot 2s\right]-\left[s^2\cdot 2s\right]}{\left(1+s^2\right)^2} = \dfrac{2s+2s^3-2s^3}{\left(1+s^2\right)^2} = \dfrac{2s}{\left(1+s^2\right)^2}$

m. $\dfrac{d}{dt}\left[t^3\left(t^2+1\right)(3t-1)\right] = \dfrac{d}{dt}\left[\left(t^5+t^3\right)(3t-1)\right] = \left[(3t-1)\dfrac{d}{dt}\left(t^5+t^3\right)\right]+\left[\left(t^5+t^3\right)\dfrac{d}{dt}(3t-1)\right]$

$= \left[(3t-1)\cdot\left(5t^4+3t^2\right)\right]+\left[\left(t^5+t^3\right)\cdot 3\right] = \left(15t^5+9t^3-5t^4-3t^2\right)+\left(3t^5+3t^3\right) = 18t^5-5t^4+12t^3-3t^2$

Another way of solving this problem is by multiplication of the binomial terms using the FOIL method prior to taking the derivative of the function as follows.

$\dfrac{d}{dt}\left[t^3\left(t^2+1\right)(3t-1)\right] = \dfrac{d}{dt}\left[t^3\left(3t^3-t^2+3t-1\right)\right] = \dfrac{d}{dt}\left(3t^6-t^5+3t^4-t^3\right) = 18t^5-5t^4+12t^3-3t^2$

n. $\dfrac{d}{du}\left[\dfrac{4u^3+2}{u^2}\right] = \dfrac{\left[u^2\dfrac{d}{du}\left(4u^3+2\right)\right]-\left[\left(4u^3+2\right)\dfrac{d}{du}u^2\right]}{u^4} = \dfrac{\left(u^2\cdot 12u^2\right)-\left(4u^3+2\right)\cdot 2u}{u^4} = \dfrac{12u^4-8u^4-4u}{u^4}$

$= \dfrac{4u^4-4u}{u^4} = \dfrac{4u\left(u^3-1\right)}{u^{4=3}} = \dfrac{4\left(u^3-1\right)}{u^3} = 4\left(\dfrac{u^3-1}{u^3}\right) = 4\left(\dfrac{u^3}{u^3}-\dfrac{1}{u^3}\right) = 4\left(1-\dfrac{1}{u^3}\right)$

o. $\dfrac{d}{dx}\left[\dfrac{x^3}{x^2+1}\right] = \dfrac{\left[\left(x^2+1\right)\dfrac{d}{du}x^3\right]-\left[x^3\dfrac{d}{du}\left(x^2+1\right)\right]}{\left(x^2+1\right)^2} = \dfrac{\left[\left(x^2+1\right)\cdot 3x^2\right]-\left(x^3\cdot 2x\right)}{\left(x^2+1\right)^2} = \dfrac{3x^4+3x^2-2x^4}{\left(x^2+1\right)^2}$

$= \dfrac{x^4+3x^2}{\left(x^2+1\right)^2} = \dfrac{x^2\left(x^2+3\right)}{\left(x^2+1\right)^2}$

Example 2.3-3: Find the derivative of the following functions at the specified value.

a. $\dfrac{d}{dt}\left[(t+1)(t-2)+3t\right]$ at $t=1$

b. $\dfrac{d}{du}\left(\dfrac{u^2+1}{u^3-1}\right)$ at $u=2$

c. $\dfrac{d}{dx}\left[\dfrac{x-1}{(x+1)(2x+1)}\right]$ at $x=0$

d. $\dfrac{d}{dx}\left[\dfrac{\left(x^3+1\right)(x-1)}{2x^2}\right]$ at $x=2$

e. $\dfrac{d}{ds}\left(\dfrac{s^2+3s}{s^2+1}\right)$ at $s=-1$

f. $\dfrac{d}{dz}\left(\dfrac{z^2+3z-5}{z}\right)$ at $z=2$

Solutions:

a. $\dfrac{d}{dt}\left[(t+1)(t-2)+3t\right] = \dfrac{d}{dt}\left[(t+1)(t-2)\right]+\dfrac{d}{dt}(3t) = \left[(t-2)\dfrac{d}{dt}(t+1)+(t+1)\dfrac{d}{dt}(t-2)\right]+\dfrac{d}{dt}(3t)$

$$= \boxed{\left[(t-2)\cdot 1+(t+1)\cdot 1\right]+3} = \boxed{(t-2)+(t+1)+3} = \boxed{t-2+t+1+3} = \boxed{2t+2}$$

Therefore, at $t=1$ $\boxed{\dfrac{d}{dt}\left[(t+1)(t-2)+3t\right]} = \boxed{2t+2} = \boxed{(2\cdot 1)+2} = \boxed{2+2} = \boxed{4}$

b. $\boxed{\dfrac{d}{du}\left(\dfrac{u^2+1}{u^3-1}\right)} = \boxed{\dfrac{\left[(u^3-1)\dfrac{d}{du}(u^2+1)\right]-\left[(u^2+1)\dfrac{d}{du}(u^3-1)\right]}{(u^3-1)^2}} = \boxed{\dfrac{\left[(u^3-1)\cdot 2u\right]-\left[(u^2+1)\cdot 3u^2\right]}{(u^3-1)^2}}$

$$= \boxed{\dfrac{2u^4-2u-3u^4-3u^2}{(u^3-1)^2}} = \boxed{\dfrac{-u^4-3u^2-2u}{(u^3-1)^2}} = \boxed{-\dfrac{u\left(u^3+3u+2\right)}{(u^3-1)^2}}$$

Therefore, at $u=2$ $\boxed{\dfrac{d}{du}\left(\dfrac{u^2+1}{u^3-1}\right)} = \boxed{-\dfrac{u\left(u^3+3u+2\right)}{(u^3-1)^2}} = \boxed{-\dfrac{2\cdot\left[2^3+(3\cdot 2)+2\right]}{(2^3-1)^2}} = \boxed{-\dfrac{2\cdot 16}{7^2}} = \boxed{-\dfrac{32}{49}} = \boxed{-0.653}$

c. $\boxed{\dfrac{d}{dx}\left[\dfrac{x-1}{(x+1)(2x+1)}\right]} = \boxed{\dfrac{d}{dx}\left[\dfrac{x-1}{2x^2+3x+1}\right]} = \boxed{\dfrac{\left[(2x^2+3x+1)\dfrac{d}{dx}(x-1)\right]-\left[(x-1)\dfrac{d}{dx}(2x^2+3x+1)\right]}{(2x^2+3x+1)^2}}$

$$= \boxed{\dfrac{\left[(2x^2+3x+1)\cdot 1\right]-\left[(x-1)\cdot(4x+3)\right]}{(2x^2+3x+1)^2}} = \boxed{\dfrac{(2x^2+3x+1)-(4x^2+3x-4x-3)}{(2x^2+3x+1)^2}} = \boxed{\dfrac{(2x^2+3x+1)-(4x^2-x-3)}{(2x^2+3x+1)^2}}$$

$$= \boxed{\dfrac{2x^2+3x+1-4x^2+x+3}{(2x^2+3x+1)^2}} = \boxed{\dfrac{-2x^2+4x+4}{(2x^2+3x+1)^2}}$$

Therefore, at $x=0$ $\boxed{\dfrac{d}{dx}\left[\dfrac{x-1}{(x+1)(2x+1)}\right]} = \boxed{\dfrac{-2x^2+4x+4}{(2x^2+3x+1)^2}} = \boxed{\dfrac{\left(-2\cdot 0^2\right)+(4\cdot 0)+4}{\left[\left(2\cdot 0^2\right)+[3\cdot 0]+1\right]^2}} = \boxed{\dfrac{4}{1}} = \boxed{4}$

d. $\boxed{\dfrac{d}{dx}\left[\dfrac{(x^3+1)(x-1)}{2x^2}\right]} = \boxed{\dfrac{d}{dx}\left[\dfrac{x^4-x^3+x-1}{2x^2}\right]} = \boxed{\dfrac{\left[(2x^2)\dfrac{d}{dx}(x^4-x^3+x-1)\right]-\left[(x^4-x^3+x-1)\dfrac{d}{dx}(2x^2)\right]}{(2x^2)^2}}$

$$= \boxed{\dfrac{\left[2x^2\left(4x^3-3x^2+1\right)\right]-\left[(x^4-x^3+x-1)\cdot 4x\right]}{4x^4}} = \boxed{\dfrac{\left(8x^5-6x^4+2x^2\right)-\left(4x^5-4x^4+4x^2-4x\right)}{4x^4}}$$

$$\boxed{\dfrac{8x^5 - 6x^4 + 2x^2 - 4x^5 + 4x^4 - 4x^2 + 4x}{4x^4}} = \boxed{\dfrac{4x^5 - 2x^4 - 2x^2 + 4x}{4x^4}} = \boxed{\dfrac{2x\left(2x^4 - x^3 - x + 2\right)}{4x^{4=3}}} = \boxed{\dfrac{2x^4 - x^3 - x + 2}{2x^3}}$$

Therefore, at $x = 2$ $\boxed{\dfrac{d}{dx}\left[\dfrac{\left(x^3+1\right)(x-1)}{2x^2}\right]} = \boxed{\dfrac{2x^4 - x^3 - x + 2}{2x^3}} = \boxed{\dfrac{\left(2\cdot 2^4\right) - 2^3 - 2 + 2}{2\cdot 2^3}} = \boxed{\dfrac{24}{16}} = \boxed{1.5}$

e. $\boxed{\dfrac{d}{ds}\left(\dfrac{s^2+3s}{s^2+1}\right)} = \boxed{\dfrac{\left[\left(s^2+1\right)\dfrac{d}{ds}\left(s^2+3s\right)\right] - \left[\left(s^2+3s\right)\dfrac{d}{ds}\left(s^2+1\right)\right]}{\left(s^2+1\right)^2}} = \boxed{\dfrac{\left[\left(s^2+1\right)(2s+3)\right] - \left[\left(s^2+3s\right)\cdot 2s\right]}{\left(s^2+1\right)^2}}$

$$= \boxed{\dfrac{\left(2s^3 + 3s^2 + 2s + 3\right) - \left(2s^3 + 6s^2\right)}{\left(s^2+1\right)^2}} = \boxed{\dfrac{2s^3 + 3s^2 + 2s + 3 - 2s^3 - 6s^2}{\left(s^2+1\right)^2}} = \boxed{\dfrac{-3s^2 + 2s + 3}{\left(s^2+1\right)^2}}$$

Therefore, at $s = -1$ $\boxed{\dfrac{d}{ds}\left(\dfrac{s^2+3s}{s^2+1}\right)} = \boxed{\dfrac{-3s^2 + 2s + 3}{\left(s^2+1\right)^2}} = \boxed{\dfrac{\left[-3\cdot(-1)^2\right] + (2\cdot -1) + 3}{\left[\left(-1^2\right)+1\right]^2}} = \boxed{\dfrac{-3-2+3}{2^2}} = \boxed{-\dfrac{1}{2}}$

f. $\boxed{\dfrac{d}{dz}\left(\dfrac{z^2+3z-5}{z}\right)} = \boxed{\dfrac{\left[z\dfrac{d}{dz}\left(z^2+3z-5\right)\right] - \left[\left(z^2+3z-5\right)\dfrac{d}{dz}(z)\right]}{z^2}} = \boxed{\dfrac{\left[z\cdot(2z+3)\right] - \left[\left(z^2+3z-5\right)\cdot 1\right]}{z^2}}$

$$= \boxed{\dfrac{2z^2 + 3z - z^2 - 3z + 5}{z^2}} = \boxed{\dfrac{z^2+5}{z^2}}$$

Therefore, at $z = 2$ $\boxed{\dfrac{d}{dz}\left(\dfrac{z^2+3z-5}{z}\right)} = \boxed{\dfrac{z^2+5}{z^2}} = \boxed{\dfrac{2^2+5}{2^2}} = \boxed{\dfrac{9}{4}} = \boxed{2.25}$

Example 2.3-4: Given the functions below, find their derivatives at the specified value.

a. $\dfrac{dy}{dx}$, given $y = \dfrac{x^2 + 2x - 1}{x^3}$ at $x = 2$

b. $\dfrac{dv}{du}$, given $v = \dfrac{u^2}{1-u}$ at $u = 4$

c. $\dfrac{dv}{dx}$, given $v = \left(x^3+1\right)\left(3x^2+5\right)$ at $x = 5$

d. $\dfrac{du}{dx}$, given $u = \dfrac{3x}{(x-1)^2}$ at $x = 2$

Solutions:

a. $\boxed{\dfrac{dy}{dx}} = \boxed{\dfrac{d}{dx}\left(\dfrac{x^2+2x-1}{x^3}\right)} = \boxed{\dfrac{\left[x^3\dfrac{d}{dx}\left(x^2+2x-1\right)\right] - \left[\left(x^2+2x-1\right)\dfrac{d}{dx}\left(x^3\right)\right]}{\left(x^3\right)^2}} = \boxed{\dfrac{\left[x^3(2x+2)\right] - \left[\left(x^2+2x-1\right)\left(3x^2\right)\right]}{x^6}}$

$$= \boxed{\dfrac{2x^4 + 2x^3 - \left(3x^4 + 6x^3 - 3x^2\right)}{x^6}} = \boxed{\dfrac{2x^4 + 2x^3 - 3x^4 - 6x^3 + 3x^2}{x^6}} = \boxed{\dfrac{-x^4 - 4x^3 + 3x^2}{x^6}}$$

Therefore, at $x = 2$ $\boxed{\dfrac{dy}{dx}} = \boxed{\dfrac{d}{dx}\left(\dfrac{x^2 + 2x - 1}{x^3}\right)} = \boxed{\dfrac{-x^4 - 4x^3 + 3x^2}{x^6}} = \boxed{\dfrac{-2^4 - 4 \cdot 2^3 + 3 \cdot 2^2}{2^6}} = \boxed{\dfrac{-36}{64}} = \boxed{-0.56}$

b. $\boxed{\dfrac{dv}{du}} = \boxed{\dfrac{d}{du}\left(\dfrac{u^2}{1-u}\right)} = \boxed{\dfrac{\left[(1-u)\dfrac{d}{du}\left(u^2\right)\right] - \left[u^2 \dfrac{d}{du}(1-u)\right]}{(1-u)^2}} = \boxed{\dfrac{[(1-u)\cdot 2u] - [u^2 \cdot -1]}{(1-u)^2}} = \boxed{\dfrac{2u - 2u^2 + u^2}{(1-u)^2}} = \boxed{\dfrac{-u^2 + 2u}{(1-u)^2}}$

Therefore, at $u = 4$ $\boxed{\dfrac{dv}{du}} = \boxed{\dfrac{d}{du}\left(\dfrac{u^2}{1-u}\right)} = \boxed{\dfrac{-u^2 + 2u}{(1-u)^2}} = \boxed{\dfrac{-(4)^2 + (2\cdot 4)}{(1-4)^2}} = \boxed{\dfrac{-16 + 8}{(-3)^2}} = \boxed{\dfrac{-8}{9}} = \boxed{-0.889}$

c. $\boxed{\dfrac{dv}{dx}} = \boxed{\dfrac{d}{dx}\left[\left(x^3 + 1\right)\left(3x^2 + 5\right)\right]} = \boxed{\left(3x^2 + 5\right)\dfrac{dv}{dx}\left(x^3 + 1\right) + \left(x^3 + 1\right)\dfrac{dv}{dx}\left(3x^2 + 5\right)} = \boxed{\left(3x^2 + 5\right)\cdot 3x^2 + \left(x^3 + 1\right)\cdot 6x}$

$= \boxed{9x^4 + 15x^2 + 6x^4 + 6x} = \boxed{15x^4 + 15x^2 + 6x}$

Thus, at $x = 5$ $\boxed{\dfrac{dv}{dx}} = \boxed{\dfrac{d}{dx}\left(\dfrac{x^2 + 2x - 1}{x^3}\right)} = \boxed{15x^4 + 15x^2 + 6x} = \boxed{15 \cdot 5^4 + 15 \cdot 5^2 + 6 \cdot 5} = \boxed{9375 + 375 + 30} = \boxed{9780}$

d. $\boxed{\dfrac{du}{dx}} = \boxed{\dfrac{d}{dx}\left[\dfrac{3x}{(x-1)^2}\right]} = \boxed{\dfrac{\left[(x-1)^2 \dfrac{d}{du}(3x)\right] - \left[3x \dfrac{d}{du}(x-1)^2\right]}{(x-1)^4}} = \boxed{\dfrac{\left[(x-1)^2 \cdot 3\right] - [3x \cdot 2(x-1)]}{(x-1)^4}} = \boxed{\dfrac{3(x-1)^2 - 6x(x-1)}{(x-1)^4}}$

$= \boxed{\dfrac{3\left(x^2 - 2x + 1\right) - 6x^2 + 6x}{(x-1)^4}} = \boxed{\dfrac{3x^2 - 6x + 3 - 6x^2 + 6x}{(x-1)^4}} = \boxed{\dfrac{-3x^2 + 3}{(x-1)^4}}$

Therefore, at $x = 2$ $\boxed{\dfrac{du}{dx}} = \boxed{\dfrac{d}{dx}\left[\dfrac{3x}{(x-1)^2}\right]} = \boxed{\dfrac{-3x^2 + 3}{(x-1)^4}} = \boxed{\dfrac{\left(-3 \cdot 2^2\right) + 3}{(2-1)^4}} = \boxed{\dfrac{-12 + 3}{1}} = \boxed{-\dfrac{9}{1}} = \boxed{-9}$

Section 2.3 - Differentiation Rules Using the $\frac{d}{dx}$ Notation

1. Find $\dfrac{dy}{dx}$ for the following functions:

a. $y = x^5 + 3x^2 + 1$
b. $y = 3x^2 + 5$
c. $y = x^3 - \dfrac{1}{x}$

d. $y = \dfrac{x^2}{1 - x^3}$
e. $y = 4x^2 + \dfrac{1}{x-1}$
f. $y = \dfrac{x^2 + 2x}{x^3 + 1}$

g. $y = x^3\left(x^2 + 5x - 2\right)$
h. $y = x^2(x+3)(x-1)$
i. $y = 5x - \dfrac{1}{x^3}$

j. $\ y=\dfrac{(x-1)(x+3)}{x^2}$

k. $\ y=x\left(\dfrac{x-1}{3}\right)$

l. $\ y=x^2(x+3)^{-1}$

m. $\ y=\left(\dfrac{x}{1+x}\right)\left(\dfrac{x-3}{5}\right)$

n. $\ y=x^3\left(1+\dfrac{1}{x-1}\right)$

o. $\ y=\dfrac{1}{x}\left(\dfrac{2x-1}{3x+1}\right)$

p. $\ y=\dfrac{ax^2+bx+c}{bx}$

q. $\ y=\dfrac{x^3-2}{x^4-3}$

r. $\ y=\dfrac{5x}{(1+x)^2}$

2. Find the derivative of the following functions:

a. $\ \dfrac{d}{dt}\left(3t^2+5t\right)=$

b. $\ \dfrac{d}{dx}\left(6x^3+5x-2\right)=$

c. $\ \dfrac{d}{du}\left(u^3+2u^2+5\right)=$

d. $\ \dfrac{d}{dt}\left(\dfrac{t^2+2t}{5}\right)=$

e. $\ \dfrac{d}{ds}\left(\dfrac{s^3+3s-1}{s^2}\right)=$

f. $\ \dfrac{d}{dw}\left(w^3+\dfrac{w^2}{1+w}\right)=$

g. $\ \dfrac{d}{dt}\left[t^2(t+1)(t^2-3)\right]=$

h. $\ \dfrac{d}{dx}\left[(x+1)(x^2+5)\right]=$

i. $\ \dfrac{d}{du}\left(\dfrac{u^2}{1-u}-\dfrac{u}{1+u}\right)=$

j. $\ \dfrac{d}{dr}\left(\dfrac{3r^3-2r^2+1}{r}\right)=$

k. $\ \dfrac{d}{ds}\left(\dfrac{3s^2}{s^3+1}-\dfrac{1}{s^2}\right)=$

l. $\ \dfrac{d}{du}\left(\dfrac{u^3}{1-u}-\dfrac{u+1}{u^2}\right)=$

3. Find the derivative of the following functions at the specified value.

a. $\ \dfrac{d}{dx}\left(x^3+3x^2+1\right)$ at $x=2$

b. $\ \dfrac{d}{dx}\left[(x+1)(x^2-1)\right]$ at $x=1$

c. $\ \dfrac{d}{ds}\left[3s^2(s-1)\right]$ at $s=0$

d. $\ \dfrac{d}{dt}\left(\dfrac{t^2+1}{t-1}\right)$ at $t=-1$

e. $\ \dfrac{d}{du}\left[\dfrac{u^3}{(u+1)^2}\right]$ at $u=1$

f. $\ \dfrac{d}{dw}\left[\dfrac{w(w^2+1)}{3w^2}\right]$ at $w=2$

g. $\ \dfrac{d}{dv}\left[(v^2+1)v^3\right]$ at $v=-2$

h. $\ \dfrac{d}{dx}\left(\dfrac{x^3}{x^2+1}\right)$ at $x=0$

i. $\ \dfrac{d}{du}\left[u^3\left(\dfrac{u^2}{1-u}\right)\right]$ at $u=0$

4. Given the functions below, find their derivatives at the specified value.

a. $\ \dfrac{ds}{dt}$, given $s=(t^2-1)+(3t+2)^2$ at $t=2$

b. $\ \dfrac{dy}{dt}$, given $y=\dfrac{t^3+3t^2+1}{2t}$ at $t=1$

c. $\ \dfrac{dw}{dx}$, given $w=(x^2+1)^2+3x$ at $x=-1$

d. $\ \dfrac{dy}{dx}$, given $y=x^2(x^3+2x+1)^2+3x$ at $x=0$

2.4   The Chain Rule

The chain rule is used for finding the derivative of the composition of functions. In general, the chain rule for two and three differentiable functions are defined in the following way:

a. The chain rule for two differentiable functions $f(x)$ and $g(x)$ is defined as:

$$(f \circ g)'(x) = \frac{d}{dx}\{f[g(x)]\} = f'[g(x)] \cdot g'(x)$$

b. The chain rule for three differentiable functions $f(x)$, $g(x)$ and $h(x)$ is defined as:

$$(f \circ g \circ h)'(x) = \frac{d}{dx}[f\{g[h(x)]\}] = f'\{g[h(x)]\} \cdot g'[h(x)] \cdot h'(x)$$

The derivative of four or higher differentiable functions using the chain rule involves addition of additional link(s) to the chain. Note that the pattern in finding the derivative of higher order functions is similar to obtaining the derivative of two or three functions, given that they are differentiable.

One of the most common applications of the chain rule is in taking the derivative of functions that are raised to a power. In general, the derivative of a function to the power of n is defined as:

$$\frac{d}{dx}[f(x)]^n = n[f(x)]^{n-1} \cdot \frac{d}{dx} f(x) = n[f(x)]^{n-1} \cdot f'(x)$$

which means, the derivative of a function raised to an exponent, $[f'(x)]^n$, is equal to the exponent times the function raised to the exponent reduced by one, $n[f(x)]^{n-1}$, multiplied by the derivative of the function, $f'(x)$, i.e., $n[f(x)]^{n-1} \cdot f'(x)$.

Note that the key in using the chain rule is that we always take the derivative of the functions by working our way from outside toward inside. The following examples show in detail the use of chain rule in differentiating different types of functions. Students are encouraged to spend adequate time working these examples.

Example 2.4-1: Find the derivative of the following functions. (It is not necessary to simplify the answer to its lowest level. The objective is to learn how to differentiate using the chain rule.)

a. $f(x) = (3 - 5x)^{-2}$ b. $f(x) = (1 + x)^6$ c. $f(x) = \left(1 + 2x^2\right)^3$

d. $f(x) = \left(x^3 - x^5\right)^8$ e. $f(x) = \left(\dfrac{1}{x^2} + x\right)^3$ f. $f(x) = \left(1 + \dfrac{1}{x}\right)^3$

g. $f(x) = \left(x + x^3\right)^4$ h. $f(x) = \left(\dfrac{1}{1 + x^2}\right)^2$ i. $f(x) = \left(\dfrac{ax + b}{cx - d}\right)^2$

j. $f(t) = \left(\dfrac{1}{1 + t^2}\right)^3$ k. $r(\theta) = \left(\dfrac{\theta^2}{1 + \theta}\right)^3$ l. $p(r) = \left(\dfrac{r^2 + r}{1 + r}\right)^3$

m. $g(u) = \left(u^3 + 3u^2\right)^3$ n. $h(t) = \left(\dfrac{t^3}{t^4 - 1} + t^2\right)^2$ o. $s(t) = \left[\left(1 + t^3\right)^{-1}\right]^3$

p. $f(x) = \left(\dfrac{x^2}{3} - \dfrac{2x}{5}\right)^{-1}$ $\qquad$ q. $r(\theta) = \left(\dfrac{\theta}{1+\theta^2}\right)^{-1}$ $\qquad$ r. $f(t) = \left(\dfrac{t^3}{1+t^2}\right)^4$

s. $f(x) = \left[\left(x^3 + 2x\right)^2 - x^2\right]^4$ $\qquad$ t. $f(x) = \left[\left(x^{-1} + x^{-3}\right)^2 + x\right]^3$ $\qquad$ u. $f(x) = \left(2 - x^{-1}\right)^{-3}$

v. $f(x) = \left[\left(1+2x^2\right)^3 - x^{-2}\right]^5$ $\qquad$ w. $f(x) = \left[\left(2 - x^{-1}\right)^{-3} + 2x^3\right]^{-1}$

Solutions:

a. Given $\boxed{f(x) = (3-5x)^{-2}}$ then $\boxed{f'(x)} = \boxed{-2(3-5x)^{-2-1}\cdot\left(0-5x^{1-1}\right)} = \boxed{-2(3-5x)^{-3}\cdot(-5)} = \boxed{10\,(3-5x)^{-3}}$

b. Given $\boxed{f(x) = (1+x)^6}$ then $\boxed{f'(x)} = \boxed{6(1+x)^{6-1}\cdot\left(0+x^{1-1}\right)} = \boxed{6\,(1+x)^5\cdot 1} = \boxed{6\,(1+x)^5}$

c. Given $\boxed{f(x) = \left(1+2x^2\right)^3}$ then $\boxed{f'(x)} = \boxed{3\left(1+2x^2\right)^{3-1}\cdot\left(0+4x^{2-1}\right)} = \boxed{3\left(1+2x^2\right)^2\cdot 4x} = \boxed{12x\left(1+2x^2\right)^2}$

d. Given $\boxed{f(x) = \left(x^3 - x^5\right)^8}$ then $\boxed{f'(x)} = \boxed{8\left(x^3 - x^5\right)^{8-1}\cdot\left(3x^{3-1} - 5x^{5-1}\right)} = \boxed{8\left(x^3 - x^5\right)^7\left(3x^2 - 5x^4\right)}$

e. Given $\boxed{f(x) = \left(\dfrac{1}{x^2} + x\right)^3} = \boxed{\left(x^{-2} + x\right)^3}$ then $\boxed{f'(x)} = \boxed{3\left(x^{-2} + x\right)^{3-1}\cdot\left(-2x^{-2-1} + 1\right)} = \boxed{3\left(x^{-2} + x\right)^2\left(-2x^{-3} + 1\right)}$

f. Given $\boxed{f(x) = \left(1 + \dfrac{1}{x}\right)^3} = \boxed{f(x) = \left(1 + x^{-1}\right)^3}$ then $\boxed{f'(x)} = \boxed{3\left(1 + x^{-1}\right)^{3-1}\cdot\left(0 - x^{-1-1}\right)} = \boxed{-3x^{-2}\left(1 + x^{-1}\right)^2}$

g. Given $\boxed{f(x) = \left(x + x^3\right)^4}$ then $\boxed{f'(x)} = \boxed{(f \circ g)'(x)} = \boxed{4\left(x + x^3\right)^3\left(1 + 3x^2\right)}$

h. Given $\boxed{f(x) = \left(\dfrac{1}{1+x^2}\right)^2}$ then $\boxed{f'(x)} = \boxed{2\left(\dfrac{1}{1+x^2}\right)^{2-1}\cdot\left\{\dfrac{\left[0\cdot\left(1+x^2\right)\right] - \left[2x\cdot 1\right]}{\left(1+x^2\right)^2}\right\}} = \boxed{\left(\dfrac{2}{1+x^2}\right)\cdot\dfrac{-2x}{\left(1+x^2\right)^2}}$

$= \boxed{\dfrac{2\cdot -2x}{\left(1+x^2\right)\cdot\left(1+x^2\right)^2}} = \boxed{\dfrac{-4x}{\left(1+x^2\right)^{1+2}}} = \boxed{\dfrac{-4x}{\left(1+x^2\right)^3}}$

A second method is to rewrite $\boxed{f(x)} = \boxed{\left(\dfrac{1}{1+x^2}\right)^2} = \boxed{\left[\left(1+x^2\right)^{-1}\right]^2} = \boxed{\left(1+x^2\right)^{-2}}$ and then take the

derivative of $f(x)$. Hence, $\boxed{f'(x)} = \boxed{-2\left(1+x^2\right)^{-2-1}\cdot\left(0+2x\right)} = \boxed{-2\left(1+x^2\right)^{-3}\cdot\left(2x\right)} = \boxed{\dfrac{-4x}{\left(1+x^2\right)^3}}$

i. Given $f(x) = \left(\dfrac{ax+b}{cx-d}\right)^2$ then $f'(x) = 2\left(\dfrac{ax+b}{cx-d}\right)^{2-1} \cdot \left[\dfrac{[a\cdot(cx-d)]-[c\cdot(ax+b)]}{(cx-d)^2}\right]$

$= 2\left(\dfrac{ax+b}{cx-d}\right)\left[\dfrac{(acx-ad)-(acx+bc)}{(cx-d)^2}\right] = 2\left(\dfrac{ax+b}{cx-d}\right)\left[\dfrac{acx-ad-acx-bc}{(cx-d)^2}\right] = 2\left(\dfrac{ax+b}{cx-d}\right)\left[\dfrac{-ad-bc}{(cx-d)^2}\right]$

$= \dfrac{2(ax+b)(-ad-bc)}{(cx-d)(cx-d)^2} = \dfrac{-2(ax+b)(ad+bc)}{(cx-d)^{1+2}} = \dfrac{-2(ax+b)(ad+bc)}{(cx-d)^3}$

j. Given $f(t) = \left(\dfrac{1}{1+t^2}\right)^3$ then $f'(t) = 3\left(\dfrac{1}{1+t^2}\right)^{3-1} \cdot \left[\dfrac{0\cdot(1+t^2)-2t\cdot 1}{(1+t^2)^2}\right] = 3\left(\dfrac{1}{1+t^2}\right)^2\left[\dfrac{-2t}{(1+t^2)^2}\right]$

$= \dfrac{3}{(1+t^2)^2}\cdot\dfrac{-2t}{(1+t^2)^2} = \dfrac{3\cdot-2t}{(1+t^2)^2\cdot(1+t^2)^2} = \dfrac{-6t}{(1+t^2)^{2+2}} = \dfrac{-6t}{(1+t^2)^4}$

A second method is to rewrite $f(t) = \left(\dfrac{1}{1+t^2}\right)^3 = \left[(1+t^2)^{-1}\right]^3 = (1+t^2)^{-3}$ and then take the

derivative of $f(t)$. Hence, $f'(t) = -3(1+t^2)^{-3-1}\cdot(0+2t) = -3(1+t^2)^{-4}2t = \dfrac{-6t}{(1+t^2)^4}$

k. Given $r(\theta) = \left(\dfrac{\theta^2}{1+\theta}\right)^3$ then $r'(\theta) = 3\left(\dfrac{\theta^2}{1+\theta}\right)^{3-1}\cdot\left\{\dfrac{[2\theta\cdot(1+\theta)]-[1\cdot\theta^2]}{(1+\theta)^2}\right\} = 3\left(\dfrac{\theta^2}{1+\theta}\right)^2\left[\dfrac{2\theta+2\theta^2-\theta^2}{(1+\theta)^2}\right]$

$= \dfrac{3\theta^4}{(1+\theta)^2}\cdot\dfrac{\theta^2+2\theta}{(1+\theta)^2} = \dfrac{3\theta^4\cdot(\theta^2+2\theta)}{(1+\theta)^2\cdot(1+\theta)^2} = \dfrac{3\theta^4\cdot\theta(\theta+2)}{(1+\theta)^{2+2}} = \dfrac{3\theta^{4+1}(\theta+2)}{(1+\theta)^4} = \dfrac{3\theta^5(\theta+2)}{(1+\theta)^4}$

l. Given $p(r) = \left(\dfrac{r^2+r}{1+r}\right)^3$ then $p'(r) = 3\left(\dfrac{r^2+r}{1+r}\right)^{3-1}\cdot\left\{\dfrac{[(2r+1)(1+r)]-[1\cdot(r^2+r)]}{(1+r)^2}\right\}$

$= 3\left(\dfrac{r^2+r}{(1+r)}\right)^2\left[\dfrac{2r+2r^2+1+r-r^2-r}{(1+r)^2}\right] = 3\left[\dfrac{r(1+r)}{(1+r)}\right]^2\left[\dfrac{r^2+2r+1}{(1+r)^2}\right] = 3r^2\left[\dfrac{(1+r)^2}{(1+r)^2}\right] = 3r^2\cdot1 = 3r^2$

A simpler way is to note that $\boxed{p(r)} = \left(\boxed{\dfrac{r^2 + r}{1+r}}\right)^3 = \left[\boxed{\dfrac{r(1+r)}{(1+r)}}\right]^3 = \left[\boxed{\dfrac{r(1+\cancel{r})}{(1+\cancel{r})}}\right]^3 = \boxed{r^3}$ then $\boxed{p'(r)} = \boxed{3r^2}$

m. Given $\boxed{g(u) = \left(u^3 + 3u^2\right)^3}$ then $\boxed{g'(u)} = \boxed{3\left(u^3 + 3u^2\right)^{3-1} \cdot \left[3u^2 + (3 \cdot 2)u\right]} = \boxed{3\left(u^3 + 3u^2\right)^2\left(3u^2 + 6u\right)}$

n. Given $\boxed{h(t) = \left(\dfrac{t^3}{t^4 - 1} + t^2\right)^2}$ then $\boxed{h'(t)} = \boxed{2\left(\dfrac{t^3}{t^4-1} + t^2\right)^{2-1} \cdot \left\{\dfrac{\left[\left(t^4 - 1\right)\cdot 3t^2\right] - \left[t^3 \cdot \left(4t^3 - 0\right)\right]}{\left(t^4 - 1\right)^2} + 2t\right\}}$

$= \boxed{2\left(\dfrac{t^3}{t^4-1} + t^2\right)\left[\dfrac{3t^6 - 3t^2 - 4t^6}{\left(t^4 - 1\right)^2} + 2t\right]} = \boxed{2\left(\dfrac{t^3}{t^4-1} + t^2\right)\left[\dfrac{-t^6 - 3t^2}{\left(t^4 - 1\right)^2} + 2t\right]}$

o. Given $\boxed{s(t) = \left[\left(1 + t^3\right)^{-1}\right]^3 = \left(1 + t^3\right)^{-3}}$ then $\boxed{s'(t)} = \boxed{-3\left(1 + t^3\right)^{-3-1} \cdot \left(0 + 3t^2\right)} = \boxed{-3\left(1 + t^3\right)^{-4} \cdot 3t^2}$

$= \boxed{-9t^2\left(1 + t^3\right)^{-4}} = \boxed{-\dfrac{9t^2}{\left(1 + t^3\right)^4}}$

p. Given $\boxed{f(x) = \left(\dfrac{x^2}{3} - \dfrac{2x}{5}\right)^{-1}}$ then $\boxed{f'(x)} = \boxed{-\left(\dfrac{x^2}{3} - \dfrac{2x}{5}\right)^{-1-1} \cdot \left(\dfrac{2}{3}x - \dfrac{2}{5}\right)} = \boxed{-\left(\dfrac{x^2}{3} - \dfrac{2x}{5}\right)^{-2}\left(\dfrac{2}{3}x - \dfrac{2}{5}\right)}$

q. Given $\boxed{r(\theta) = \left(\dfrac{\theta}{1 + \theta^2}\right)^{-1}}$ then $\boxed{r'(\theta)} = \boxed{-\left(\dfrac{\theta}{1 + \theta^2}\right)^{-1-1} \cdot \left[\dfrac{1 \cdot \left(1 + \theta^2\right) - 2\theta \cdot \theta}{\left(1 + \theta^2\right)^2}\right]} = \boxed{-\left(\dfrac{\theta}{1 + \theta^2}\right)^{-2}\left[\dfrac{1 + \theta^2 - 2\theta^2}{\left(1 + \theta^2\right)^2}\right]}$

$= \boxed{-\left(\dfrac{\theta}{1 + \theta^2}\right)^{-2} \cdot \dfrac{1 - \theta^2}{\left(1 + \theta^2\right)^2}} = \boxed{-\left(\dfrac{1 + \theta^2}{\theta}\right)^2 \cdot \dfrac{1 - \theta^2}{\left(1 + \theta^2\right)^2}} = \boxed{-\dfrac{\left(1 + \theta^2\right)^2}{\theta^2} \cdot \dfrac{1 - \theta^2}{\left(1 + \theta^2\right)^2}} = \boxed{-\dfrac{1 - \theta^2}{\theta^2}} = \boxed{\dfrac{\theta^2 - 1}{\theta^2}}$

r. Given $\boxed{f(t) = \left(\dfrac{t^3}{1 + t^2}\right)^4}$ then $\boxed{f'(t)} = \boxed{4\left(\dfrac{t^3}{1 + t^2}\right)^{4-1} \cdot \left[\dfrac{3t^2 \cdot \left(1 + t^2\right) - 2t \cdot t^3}{\left(1 + t^2\right)^2}\right]} = \boxed{4\left(\dfrac{t^3}{1 + t^2}\right)^3\left[\dfrac{3t^2 + 3t^4 - 2t^4}{\left(1 + t^2\right)^2}\right]}$

$= \boxed{\dfrac{4t^9}{\left(1 + t^2\right)^3} \cdot \dfrac{3t^2 + t^4}{\left(1 + t^2\right)^2}} = \boxed{\dfrac{4t^9}{\left(1 + t^2\right)^3} \cdot \dfrac{t^2\left(3 + t^2\right)}{\left(1 + t^2\right)^2}} = \boxed{\dfrac{4t^{9+2}\left(3 + t^2\right)}{\left(1 + t^2\right)^{3+2}}} = \boxed{\dfrac{4t^{11}\left(3 + t^2\right)}{\left(1 + t^2\right)^5}}$

s. Given $f(x) = \left[\left(x^3 + 2x \right)^2 - x^2 \right]^4$ then $f'(x) = 4\left[\left(x^3 + 2x \right)^2 - x^2 \right]^{4-1} \cdot \left\{ \left[2\left(x^3 + 2x \right)^{2-1} \cdot \left(3x^2 + 2 \right) \right] - 2x \right\}$

$= 4\left[\left(x^3 + 2x \right)^2 - x^2 \right]^3 \cdot \left\{ \left[2\left(x^3 + 2x \right)\left(3x^2 + 2 \right) \right] - 2x \right\}$

t. Given $f(x) = \left[\left(x^{-1} + x^{-3} \right)^2 + x \right]^3$ then $f'(x) = 3\left[\left(x^{-1} + x^{-3} \right)^2 + x \right]^{3-1} \left\{ \left[2\left(x^{-1} + x^{-3} \right)^{2-1} \left(-x^{-2} - 3x^{-4} \right) \right] + 1 \right\}$

$= 3\left[\left(x^{-1} + x^{-3} \right)^2 + x \right]^2 \left\{ \left[2\left(x^{-1} + x^{-3} \right)\left(-x^{-2} - 3x^{-4} \right) \right] + 1 \right\}$

u. Given $f(x) = \left(2 - x^{-1} \right)^{-3}$ then $f'(x) = -3\left(2 - x^{-1} \right)^{-3-1} \cdot \left(0 + x^{-2} \right) = -3\left(2 - x^{-1} \right)^{-4} x^{-2} = \dfrac{-3x^{-2}}{\left(2 - x^{-1} \right)^4}$

$= -\dfrac{3}{x^2 \left(2 - x^{-1} \right)^4}$

v. Given $f(x) = \left[\left(1 + 2x^2 \right)^3 - x^{-2} \right]^5$ then $f'(x) = 5\left[\left(1 + 2x^2 \right)^3 - x^{-2} \right]^{5-1} \cdot \left[3\left(1 + 2x^2 \right)^{3-1} \cdot 4x + 2x^{-2-1} \right]$

$= 5\left[\left(1 + 2x^2 \right)^3 - x^{-2} \right]^4 \left[12x\left(1 + 2x^2 \right)^2 + 2x^{-3} \right]$

w. Given $f(x) = \left[\left(2 - x^{-1} \right)^{-3} + 2x^3 \right]^{-1}$ then $f'(x) = -\left[\left(2 - x^{-1} \right)^{-3} + 2x^3 \right]^{-1-1} \cdot \left[-3\left(2 - x^{-1} \right)^{-3-1} \cdot x^{-2} + 6x^2 \right]$

$= -\left[\left(2 - x^{-1} \right)^{-3} + 2x^3 \right]^{-2} \left[-3x^{-2}\left(2 - x^{-1} \right)^{-4} + 6x^2 \right]$

Example 2.4-2: Find the derivative at $x = 0$, $x = -1$, and $x = 1$ in example 2.4-1 for problems a - g.

Solutions:

a. Given $f'(x) = 10(3 - 5x)^{-3}$, then

$f'(0) = 10[3 - (5 \cdot 0)]^{-3} = 10[3 - 0]^{-3} = 10 \cdot 3^{-3} = 10 \cdot \dfrac{1}{3^3} = \dfrac{10}{27} = \boxed{0.37}$

$f'(-1) = 10[3 - (5 \cdot -1)]^{-3} = 10[3 + 5]^{-3} = 10 \cdot 8^{-3} = 10 \cdot \dfrac{1}{8^3} = \dfrac{10}{512} = \boxed{0.019}$

$$\boxed{f'(1)} = \boxed{10\left[3-(5\cdot 1)\right]^{-3}} = \boxed{10\left[3-5\right]^{-3}} = \boxed{10\cdot(-2)^{-3}} = \boxed{10\cdot\dfrac{1}{(-2)^3}} = \boxed{\dfrac{10}{-8}} = \boxed{-\dfrac{10}{8}} = \boxed{\mathbf{-1.25}}$$

b. Given $f'(x) = 6(1+x)^5$, then

$$\boxed{f'(0)} = \boxed{6(1+0)^5} = \boxed{6\cdot 1^5} = \boxed{6\cdot 1} = \boxed{\mathbf{6}}$$

$$\boxed{f'(-1)} = \boxed{6(1-1)^5} = \boxed{6\cdot 0^5} = \boxed{6\cdot 0} = \boxed{\mathbf{0}}$$

$$\boxed{f'(1)} = \boxed{6(1+1)^5} = \boxed{6\cdot 2^5} = \boxed{6\cdot 32} = \boxed{\mathbf{192}}$$

c. Given $f'(x) = 12x\left(1+2x^2\right)^2$, then

$$\boxed{f'(0)} = \boxed{(12\cdot 0)\left[1+\left(2\cdot 0^2\right)\right]^2} = \boxed{0\cdot(1+0)^2} = \boxed{0\cdot 1^2} = \boxed{0\cdot 1} = \boxed{\mathbf{0}}$$

$$\boxed{f'(-1)} = \boxed{(12\cdot -1)\left[1+2\cdot(-1)^2\right]^2} = \boxed{-12\,(1+2)^2} = \boxed{-12\cdot 3^2} = \boxed{-12\cdot 9} = \boxed{\mathbf{-108}}$$

$$\boxed{f'(1)} = \boxed{(12\cdot 1)\left[1+\left(2\cdot 1^2\right)\right]^2} = \boxed{12(1+2)^2} = \boxed{12\cdot 3^2} = \boxed{12\cdot 9} = \boxed{\mathbf{108}}$$

d. Given $f'(x) = 8\left(x^3-x^5\right)^7\left(3x^2-5x^4\right)$, then

$$\boxed{f'(0)} = \boxed{8\left(0^3-0^5\right)^7\left[3\cdot 0^2-5\cdot 0^4\right]} = \boxed{8\cdot 0^7\cdot 0} = \boxed{8\cdot 0\cdot 0} = \boxed{\mathbf{0}}$$

$$\boxed{f'(-1)} = \boxed{8\left[(-1)^3-(-1)^5\right]^7\left[3\cdot(-1)^2-5\cdot(-1)^4\right]} = \boxed{8[-1+1]^7[3\cdot 1-5\cdot 1]} = \boxed{8\cdot 0^7\cdot -2} = \boxed{8\cdot 0\cdot -2} = \boxed{\mathbf{0}}$$

$$\boxed{f'(1)} = \boxed{8\left(1^3-1^5\right)^7\left[3\cdot 1^2-5\cdot 1^4\right]} = \boxed{8(1-1)^7(3\cdot 1-5\cdot 1)} = \boxed{8\cdot 0^7\,(3-5)} = 8\cdot 0\cdot -2 = \boxed{\mathbf{0}}$$

e. Given $f'(x) = 3\left(x^{-2}+x\right)^2\left(-2x^{-3}+1\right) = 3\left(\dfrac{1}{x^2}+x\right)^2\left(-\dfrac{2}{x^3}+1\right)$, then

$$\boxed{f'(0)} = \boxed{3\left(\dfrac{1}{0^2}+0\right)^2\left(-\dfrac{2}{0^3}+1\right)} = \boxed{3\left(\dfrac{1}{0}+0\right)^2\left(-\dfrac{2}{0}+1\right)}\qquad f'(0)\text{ is undefined due to division by zero.}$$

$$\boxed{f'(-1)} = \boxed{3\left(\dfrac{1}{(-1)^2}-1\right)^2\left(-\dfrac{2}{(-1)^3}+1\right)} = \boxed{3(1-1)^2\,(2+1)} = \boxed{3\cdot 0^2\cdot 3} = \boxed{\mathbf{0}}$$

$$\boxed{f'(1)} = \boxed{3\left(\dfrac{1}{1^2}+1\right)^2\left(-\dfrac{2}{1^3}+1\right)} = \boxed{3(1+1)^2(-2+1)} = \boxed{3\cdot 2^2\cdot -1} = \boxed{3\cdot 4\cdot -1} = \boxed{\mathbf{-12}}$$

f. Given $f'(x) = -3x^{-2}\left(1+x^{-1}\right)^2 = -\dfrac{3\left(1+x^{-1}\right)^2}{x^2} = -\dfrac{3\left(1+\dfrac{1}{x}\right)^2}{x^2}$, then

$$f'(0) = \frac{3\left(1+\frac{1}{0}\right)^2}{0^2} = -\frac{3\left(1+\frac{1}{0}\right)^2}{0} \qquad f'(0) \text{ is undefined due to division by zero.}$$

$$f'(-1) = -\frac{3\left(1+\frac{1}{-1}\right)^2}{(-1)^2} = \frac{3(1-1)^2}{1} = \frac{3\cdot 0^2}{1} = -\frac{0}{1} = \boxed{0}$$

$$f'(1) = -\frac{3\left(1+\frac{1}{1}\right)^2}{1^2} = -\frac{3(1+1)^2}{1} = -\frac{3\cdot 2^2}{1} = -\frac{3\cdot 4}{1} = \boxed{-12}$$

g. Given $f'(x) = 4\left(x+x^3\right)^3\left(1+3x^2\right)$, then

$$f'(0) = 4\left(0+0^3\right)^3\left(1+3\cdot 0^2\right) = 4\cdot 0^3\cdot 1 = 4\cdot 0\cdot 1 = \boxed{0}$$

$$f'(-1) = 4\left[-1+(-1)^3\right]^3\left[1+3\cdot(-1)^2\right] = 4\cdot(-1-1)^3\cdot(1+3) = 4\cdot(-2)^3\cdot 4 = 4\cdot(-8)\cdot 4 = \boxed{-128}$$

$$f'(1) = 4\left(1+1^3\right)^3\left(1+3\cdot 1^2\right) = 4\cdot(1+1)^3\cdot(1+3) = 4\cdot 2^3\cdot 4 = 4\cdot 8\cdot 4 = \boxed{128}$$

Example 2.4-3: Use the chain rule to differentiate the following functions. Do not simplify the answer to its lowest term.

a. $\dfrac{d}{dx}\left(x^2+3\right)^5 =$

b. $\dfrac{d}{dx}\left[\left(x^2+5\right)^3+1\right]^4 =$

c. $\dfrac{d}{du}\left[\left(u^2+1\right)^3(u+5)\right] =$

d. $\dfrac{d}{dt}\left[\dfrac{\left(t^2+3\right)^5}{t-1}\right] =$

e. $\dfrac{d}{d\theta}\left[\dfrac{\left(\theta^3+2\theta\right)^3}{(\theta+1)^2}\right] =$

f. $\dfrac{d}{dr}\left[r^2\left(r^2+3\right)^4\right] =$

g. $\dfrac{d}{du}\left[\left(u^2+4\right)^6\left(u^3-1\right)\right] =$

h. $\dfrac{d}{dt}\left[\left(t^3+2t^2+1\right)\left(t^2+t+1\right)^3\right] =$

i. $\dfrac{d}{dx}\left(\dfrac{x^3+3x}{1-x^2}\right)^8 =$

j. $\dfrac{d}{dx}\left(\dfrac{x+2}{x-2}\right)^2 =$

k. $\dfrac{d}{dx}\left[\dfrac{(2x+5)^3}{(1+x)^2}\right] =$

l. $\dfrac{d}{dx}\left[\dfrac{(1-x)^2}{x^3+2x}\right] =$

m. $\dfrac{d}{dx}\left(x^2+\dfrac{1}{x^3+5}\right)^4 =$

n. $\dfrac{d}{dx}\left[\left(2x^3+1\right)^3\left(x^2+1\right)^4\right] =$

Solutions:

a. $\dfrac{d}{dx}\left(x^2+3\right)^5 = 5\left(x^2+3\right)^{5-1}\dfrac{d}{dx}\left(x^2+3\right) = 5\left(x^2+3\right)^4\cdot 2x = \boxed{10x\left(x^2+3\right)^4}$

b. $\dfrac{d}{dx}\left[\left(x^2+5\right)^3+1\right]^4 = 4\left[\left(x^2+5\right)^3+1\right]^{4-1}\dfrac{d}{dx}\left[\left(x^2+5\right)^3+1\right] = 4\left[\left(x^2+5\right)^3+1\right]^3\cdot\left[3\left(x^2+5\right)^{3-1}\dfrac{d}{dx}\left(x^2+5\right)+0\right]$

$$= \boxed{4\left[\left(x^2+5\right)^3+1\right]^3 \cdot \left[3\left(x^2+5\right)^2 \cdot 2x\right]} = \boxed{4\left[\left(x^2+5\right)^3+1\right]^3 \cdot 6x\left(x^2+5\right)^2} = \boxed{24x\left[\left(x^2+5\right)^3+1\right]^3\left(x^2+5\right)^2}$$

c. $\boxed{\dfrac{d}{du}\left[\left(u^2+1\right)^3(u+5)\right]} = \boxed{\left[(u+5)\dfrac{d}{du}\left(u^2+1\right)^3\right]+\left[\left(u^2+1\right)^3\dfrac{d}{du}(u+5)\right]} = \boxed{\left\{(u+5)\cdot\left[3\left(u^2+1\right)^{3-1}\dfrac{d}{du}\left(u^2+1\right)\right]\right\}}$

$+ \boxed{\left[\left(u^2+1\right)^3\cdot 1\right]} = \boxed{\left\{(u+5)\cdot\left[3\left(u^2+1\right)^2\cdot 2u\right]\right\}+\left(u^2+1\right)^3} = \boxed{6u(u+5)\left(u^2+1\right)^2+\left(u^2+1\right)^3}$

d. $\boxed{\dfrac{d}{dt}\left[\dfrac{\left(t^2+3\right)^5}{t-1}\right]} = \boxed{\dfrac{\left[(t-1)\dfrac{d}{dt}\left(t^2+3\right)^5\right]-\left[\left(t^2+3\right)^5\dfrac{d}{dt}(t-1)\right]}{(t-1)^2}} = \boxed{\dfrac{\left[(t-1)\cdot5\left(t^2+3\right)^{5-1}\cdot\dfrac{d}{dt}\left(t^2+3\right)\right]-\left[\left(t^2+3\right)^5\cdot 1\right]}{(t-1)^2}}$

$= \boxed{\dfrac{\left[(t-1)\cdot5\left(t^2+3\right)^4\cdot 2t\right]-\left(t^2+3\right)^5}{(t-1)^2}} = \boxed{\dfrac{10t(t-1)\left(t^2+3\right)^4-\left(t^2+3\right)^5}{(t-1)^2}}$

e. $\boxed{\dfrac{d}{d\theta}\left[\dfrac{\left(\theta^3+2\theta\right)^3}{(\theta+1)^2}\right]} = \boxed{\dfrac{\left[(\theta+1)^2\dfrac{d}{d\theta}\left(\theta^3+2\theta\right)^3\right]-\left[\left(\theta^3+2\theta\right)^3\dfrac{d}{d\theta}(\theta+1)^2\right]}{(\theta+1)^4}} = \boxed{\dfrac{\left[(\theta+1)^2\cdot3\left(\theta^3+2\theta\right)^{3-1}\cdot\dfrac{d}{d\theta}\left(\theta^3+2\theta\right)\right]}{(\theta+1)^4}}$

$= \boxed{\dfrac{-\left[\left(\theta^3+2\theta\right)^3\cdot2(\theta+1)^{2-1}\cdot\dfrac{d}{d\theta}(\theta+1)\right]}{(\theta+1)^4}} = \boxed{\dfrac{\left[(\theta+1)^2\cdot3\left(\theta^3+2\theta\right)^2\cdot\left(3\theta^2+2\right)\right]-\left[\left(\theta^3+2\theta\right)^3\cdot2(\theta+1)\cdot1\right]}{(\theta+1)^4}}$

$= \boxed{\dfrac{\left[3(\theta+1)^2\left(\theta^3+2\theta\right)^2\left(3\theta^2+2\right)\right]-\left[2\left(\theta^3+2\theta\right)^3(\theta+1)\right]}{(\theta+1)^4}}$

f. $\boxed{\dfrac{d}{dr}\left[r^2\left(r^2+3\right)^4\right]} = \boxed{\left[\left(r^2+3\right)^4\dfrac{d}{dr}r^2\right]+\left[r^2\dfrac{d}{dr}\left(r^2+3\right)^4\right]} = \boxed{\left[\left(r^2+3\right)^4\cdot2r\right]+\left[r^2\cdot4\left(r^2+3\right)^{4-1}\cdot\dfrac{d}{dr}\left(r^2+3\right)\right]}$

$= \boxed{2r\left(r^2+3\right)^4+\left[4r^2\left(r^2+3\right)^3\cdot2r\right]} = \boxed{2r\left(r^2+3\right)^4+8r^3\left(r^2+3\right)^3} = \boxed{2r\left(r^2+3\right)^3\left[\left(r^2+3\right)+4r^2\right]}$

g. $\boxed{\dfrac{d}{du}\left[\left(u^2+4\right)^6\left(u^3-1\right)\right]} = \boxed{\left[\left(u^3-1\right)\dfrac{d}{du}\left(u^2+4\right)^6\right]+\left[\left(u^2+4\right)^6\dfrac{d}{du}\left(u^3-1\right)\right]} = \boxed{\left[\left(u^3-1\right)\cdot6\left(u^2+4\right)^{6-1}\cdot\dfrac{d}{du}\left(u^2+4\right)\right]}$

$+ \boxed{\left[\left(u^2+4\right)^6\cdot3u^2\right]} = \boxed{\left[\left(u^3-1\right)\cdot6\left(u^2+4\right)^5\cdot2u\right]+\left[3u^2\left(u^2+4\right)^6\right]} = \boxed{\left[12u\left(u^3-1\right)\left(u^2+4\right)^5\right]+\left[3u^2\left(u^2+4\right)^6\right]}$

$$= \boxed{3u\left(u^2+4\right)^5\left[4\left(u^3-1\right)+u\left(u^2+4\right)\right]}$$

h. $\boxed{\dfrac{d}{dt}\left[\left(t^3+2t^2+1\right)\left(t^2+t+1\right)^3\right]} = \boxed{\left[\left(t^2+t+1\right)^3\dfrac{d}{dt}\left(t^3+2t^2+1\right)\right]+\left[\left(t^3+2t^2+1\right)\dfrac{d}{dt}\left(t^2+t+1\right)^3\right]}$

$= \boxed{\left[\left(t^2+t+1\right)^3\cdot\left(3t^2+4t\right)\right]+\left[\left(t^3+2t^2+1\right)\cdot3\left(t^2+t+1\right)^{3-1}\cdot\dfrac{d}{dt}\left(t^2+t+1\right)\right]} = \boxed{\left[\left(t^2+t+1\right)^3\cdot\left(3t^2+4t\right)\right]}$

$+ \boxed{\left[\left(t^3+2t^2+1\right)\cdot3\left(t^2+t+1\right)^2\cdot(2t+1)\right]} = \boxed{\left(t^2+t+1\right)^2\left\{t\left(t^2+t+1\right)(3t+4)\right]+\left[3\left(t^3+2t^2+1\right)(2t+1)\right]\right\}}$

i. $\boxed{\dfrac{d}{dx}\left(\dfrac{x^3+3x}{1-x^2}\right)^8} = \boxed{8\left(\dfrac{x^3+3x}{1-x^2}\right)^{8-1}\cdot\dfrac{d}{dx}\left(\dfrac{x^3+3x}{1-x^2}\right)} = \boxed{8\left(\dfrac{x^3+3x}{1-x^2}\right)^7\cdot\dfrac{\left[\left(1-x^2\right)\dfrac{d}{dx}\left(x^3+3x\right)\right]-\left[\left(x^3+3x\right)\dfrac{d}{dx}\left(1-x^2\right)\right]}{\left(1-x^2\right)^2}}$

$= \boxed{8\left(\dfrac{x^3+3x}{1-x^2}\right)^7\cdot\dfrac{\left[\left(1-x^2\right)\cdot\left(3x^2+3\right)\right]-\left[\left(x^3+3x\right)\cdot(-2x)\right]}{\left(1-x^2\right)^2}} = \boxed{8\left(\dfrac{x^3+3x}{1-x^2}\right)^7\cdot\dfrac{\left[\left(1-x^2\right)\left(3x^2+3\right)\right]+\left[2x^2\left(x^2+3\right)\right]}{\left(1-x^2\right)^2}}$

j. $\boxed{\dfrac{d}{dx}\left(\dfrac{x+2}{x-2}\right)^2} = \boxed{2\left(\dfrac{x+2}{x-2}\right)^{2-1}\cdot\dfrac{d}{dx}\left(\dfrac{x+2}{x-2}\right)} = \boxed{2\left(\dfrac{x+2}{x-2}\right)\cdot\dfrac{\left[(x-2)\dfrac{d}{dx}(x+2)\right]-\left[(x+2)\dfrac{d}{dx}(x-2)\right]}{(x-2)^2}}$

$= \boxed{2\left(\dfrac{x+2}{x-2}\right)\cdot\dfrac{[(x-2)\cdot1]-[(x+2)\cdot1]}{(x-2)^2}} = \boxed{2\left(\dfrac{x+2}{x-2}\right)\cdot\dfrac{\cancel{x}-2-\cancel{x}-2}{(x-2)^2}} = \boxed{2\left(\dfrac{x+2}{x-2}\right)\dfrac{-4}{(x-2)^2}} = \boxed{-\dfrac{8(x+2)}{(x-2)^3}}$

k. $\boxed{\dfrac{d}{dx}\left[\dfrac{(2x+5)^3}{(1+x)^2}\right]} = \boxed{\dfrac{\left[(1+x)^2\dfrac{d}{dx}(2x+5)^3\right]-\left[(2x+5)^3\dfrac{d}{dx}(1+x)^2\right]}{(1+x)^4}} = \boxed{\dfrac{\left[(1+x)^2\cdot3(2x+5)^{3-1}\dfrac{d}{dx}(2x+5)\right]}{(1+x)^4}}$

$- \boxed{\dfrac{\left[(2x+5)^3\cdot2(1+x)^{2-1}\dfrac{d}{dx}(1+x)\right]}{(1+x)^4}} = \boxed{\dfrac{\left[(1+x)^2\cdot3(2x+5)^2\cdot2\right]-\left[(2x+5)^3\cdot2(1+x)\cdot1\right]}{(1+x)^4}} = \boxed{\dfrac{\left[6(1+x)^2(2x+5)^2\right]}{(1+x)^4}}$

$- \boxed{\dfrac{\left[2(2x+5)^3(1+x)\right]}{(1+x)^4}} = \boxed{\dfrac{2(1+x)(2x+5)^2[3(1+x)-(2x+5)]}{(1+x)^{4=3}}} = \boxed{\dfrac{2(2x+5)^2(3+3x-2x-5)}{(1+x)^3}} = \boxed{\dfrac{2(2x+5)^2(x-2)}{(1+x)^3}}$

l. $\boxed{\dfrac{d}{dx}\left[\dfrac{(1-x)^2}{x^3+2x}\right]} = \boxed{\dfrac{\left[\left(x^3+2x\right)\dfrac{d}{dx}(1-x)^2\right]-\left[(1-x)^2\dfrac{d}{dx}\left(x^3+2x\right)\right]}{\left(x^3+2x\right)^2}} = \boxed{\dfrac{\left[\left(x^3+2x\right)\cdot2(1-x)^{2-1}\cdot\dfrac{d}{dx}(1-x)\right]}{\left(x^3+2x\right)^2}}$

$$-\frac{\left[(1-x)^2\cdot\left(3x^2+2\right)\right]}{\left(x^3+2x\right)^2}=\frac{\left[2\left(x^3+2x\right)(1-x)\cdot-1\right]-\left[(1-x)^2\left(3x^2+2\right)\right]}{\left(x^3+2x\right)^2}=\frac{\left[-2\left(x^3+2x\right)(1-x)\right]-\left[(1-x)^2\left(3x^2+2\right)\right]}{\left(x^3+2x\right)^2}$$

m. $\dfrac{d}{dx}\left(x^2+\dfrac{1}{x^3+5}\right)^4=4\left(x^2+\dfrac{1}{x^3+5}\right)^{4-1}\cdot\dfrac{d}{dx}\left(x^2+\dfrac{1}{x^3+5}\right)=4\left(x^2+\dfrac{1}{x^3+5}\right)^3\cdot\left[\dfrac{d}{dx}\left(x^2\right)+\dfrac{d}{dx}\left(\dfrac{1}{x^3+5}\right)\right]$

$$=4\left(x^2+\frac{1}{x^3+5}\right)^3\cdot\left[2x+\frac{\left[\left(x^3+5\right)\frac{d}{dx}(1)\right]-\left[1\cdot\frac{d}{dx}\left(x^3+5\right)\right]}{\left(x^3+5\right)^2}\right]=4\left(x^2+\frac{1}{x^3+5}\right)^3\left[2x-\frac{3x^2}{\left(x^3+5\right)^2}\right]$$

n. $\dfrac{d}{dx}\left[\left(2x^3+1\right)^3\left(x^2+1\right)^4\right]=\left[\left(x^2+1\right)^4\dfrac{d}{dx}\left(2x^3+1\right)^3\right]+\left[\left(2x^3+1\right)^3\dfrac{d}{dx}\left(x^2+1\right)^4\right]$

$$=\left[\left(x^2+1\right)^4\cdot3\left(2x^3+1\right)^{3-1}\cdot\frac{d}{dx}\left(2x^3+1\right)\right]+\left[\left(2x^3+1\right)^3\cdot4\left(x^2+1\right)^{4-1}\cdot\frac{d}{dx}\left(x^2+1\right)\right]=\left[\left(x^2+1\right)^4\cdot3\left(2x^3+1\right)^2\cdot6x^2\right]$$

$$+\left[\left(2x^3+1\right)^3\cdot4\left(x^2+1\right)^3\cdot2x\right]=\left[18x^2\left(x^2+1\right)^4\left(2x^3+1\right)^2\right]+\left[8x\left(2x^3+1\right)^3\left(x^2+1\right)^3\right]$$

In some instances students are asked to find the derivative of a function y, where y is a function of u and u is a function of x. We can solve this class of problems using one of two methods.

The first method, and perhaps the easiest one, is performed by substituting u into the y equation and taking the derivative of y with respect to x. The second method is to find the derivative of y by using the equation $\dfrac{dy}{dx}=\dfrac{dy}{du}\cdot\dfrac{du}{dx}$. This method is most often used in calculus books and can be time consuming. For example, let's find the derivative of the function $y=u^2+1$ where $u=x+1$ using each of these methods.

First Method: Given the function $y=u^2+1$ where $u=x+1$, substitute u with $x+1$ in the function y and simplify, i.e., $y=u^2+1=(x+1)^2+1=x^2+2x+1+1=x^2+2x+2$. Next, take the derivative of y with respect to x, i.e., $\dfrac{dy}{dx}=\dfrac{dy}{dx}\left(x^2+2x+2\right)=2x^{2-1}+2x^{1-1}+0=\mathbf{2x+2}$.

Second Method: Given the function $y=u^2+1$ where $u=x+1$, find $\dfrac{dy}{du}$ and $\dfrac{du}{dx}$, i.e., $\dfrac{dy}{du}=2u^{2-1}+0=2u$ and $\dfrac{du}{dx}=x^{1-1}+0=x^0=1$. Next, substitute $\dfrac{dy}{du}$ and $\dfrac{du}{dx}$ in the equation $\dfrac{dy}{dx}=\dfrac{dy}{du}\cdot\dfrac{du}{dx}$ to find the derivative $\dfrac{dy}{dx}$, i.e., $\dfrac{dy}{dx}=2u\cdot1=2u$. Substituting $u=x+1$ in place of u we obtain $\dfrac{dy}{dx}=2u=2(x+1)=\mathbf{2x+2}$.

The second method is generally beyond the scope of this book, therefore the first method is used in order to solve this class of problems. Examples 2.4-4 and 2.4-5 below provide additional examples as to how these types of problems are solved.

Example 2.4-4: Find $\dfrac{dy}{dx}$ given:

a. $y = \dfrac{1}{1+u}$ and $u = 3x+1$

b. $y = u^2 + 2u + 1$ and $u = 3x + 2$

c. $y = \dfrac{u}{1+u^2}$ and $u = 5x+1$

d. $y = u^2 + 1$ and $u = \dfrac{x+1}{x^2-1}$

e. $y = \dfrac{1+u}{u^3}$ and $u = x^2 + 1$

f. $y = u^3 + 1$ and $u = \left(x^2+1\right)^{-1}$

Solutions:

a. Given $y = \dfrac{1}{1+u}$ and $u = 3x+1$, then $y = \dfrac{1}{1+(3x+1)} = \dfrac{1}{3x+2}$ and

$$\boxed{\dfrac{dy}{dx}} = \boxed{\dfrac{\left[(3x+2)\dfrac{d}{dx}(1)\right]-\left[1\cdot\dfrac{d}{dx}(3x+2)\right]}{(3x+2)^2}} = \boxed{\dfrac{0-3}{(3x+2)^2}} = \boxed{-\dfrac{3}{(3x+2)^2}}$$

b. Given $y = u^2 + 2u + 1$ and $u = 3x+2$, then $y = (3x+2)^2 + 2(3x+2) + 1 = (3x+2)^2 + 6x + 5$ and

$$\boxed{\dfrac{dy}{dx}} = \boxed{\dfrac{d}{dx}(3x+2)^2 + \dfrac{d}{dx}6x + \dfrac{d}{dx}5} = \boxed{2(3x+2)^{2-1}\cdot\dfrac{d}{dx}(3x+2)+6+0} = \boxed{2(3x+2)\cdot3+6} = \boxed{18(x+1)}$$

c. Given $y = \dfrac{u}{1+u^2}$ and $u = 5x+1$, then $y = \dfrac{(5x+1)}{1+(5x+1)^2}$ and

$$\boxed{\dfrac{dy}{dx}} = \boxed{\dfrac{\left[\left[1+(5x+1)^2\right]\dfrac{d}{dx}(5x+1)\right]-\left[(5x+1)\cdot\dfrac{d}{dx}\left[1+(5x+1)^2\right]\right]}{\left[1+(5x+1)^2\right]^2}} = \boxed{\dfrac{\left\{\left[1+(5x+1)^2\right]\cdot5\right\}-\left[(5x+1)\cdot2\,(5x+1)\cdot5\right]}{\left[1+(5x+1)^2\right]^2}}$$

$$= \boxed{\dfrac{5\left[1+(5x+1)^2\right]-\left[10(5x+1)^2\right]}{\left[1+(5x+1)^2\right]^2}} = \boxed{\dfrac{5+5(5x+1)^2-10(5x+1)^2}{\left[1+(5x+1)^2\right]^2}} = \boxed{\dfrac{5-5(5x+1)^2}{\left[1+(5x+1)^2\right]^2}}$$

d. Given $y = u^2 + 1$ and $u = \dfrac{x+1}{x^2-1}$, then $y = \left(\dfrac{x+1}{x^2-1}\right)^2 + 1$ and

$$\boxed{\dfrac{dy}{dx}} = \boxed{\left[2\left(\dfrac{x+1}{x^2-1}\right)^{2-1}\cdot\dfrac{d}{dx}\left(\dfrac{x+1}{x^2-1}\right)\right]+0} = \boxed{2\left(\dfrac{x+1}{x^2-1}\right)\cdot\dfrac{\left[(x^2-1)\dfrac{d}{dx}(x+1)\right]-\left[(x+1)\cdot\dfrac{d}{dx}(x^2-1)\right]}{(x^2-1)^2}}$$

$$= \boxed{2\left(\dfrac{x+1}{x^2-1}\right)\cdot\dfrac{\left[(x^2-1)\cdot1\right]-\left[(x+1)\cdot2x\right]}{(x^2-1)^2}} = \boxed{2\left(\dfrac{x+1}{x^2-1}\right)\cdot\dfrac{x^2-1-2x^2-2x}{(x^2-1)^2}} = \boxed{2\left(\dfrac{x+1}{x^2-1}\right)\cdot-\dfrac{x^2+2x+1}{(x^2-1)^2}}$$

$$= \boxed{-2\left(\frac{x+1}{x^2-1}\right)\cdot\frac{(x+1)^2}{\left(x^2-1\right)^2}} = \boxed{-2\frac{(x+1)(x+1)^2}{\left(x^2-1\right)\left(x^2-1\right)^2}} = \boxed{-2\frac{(x+1)^3}{\left(x^2-1\right)^3}} = \boxed{-2\left(\frac{x+1}{x^2-1}\right)^3}$$

e. Given $y = \dfrac{1+u}{u^3}$ and $u = x^2+1$, then $y = \dfrac{1+\left(x^2+1\right)}{\left(x^2+1\right)^3} = \dfrac{x^2+2}{\left(x^2+1\right)^3}$

$$\boxed{\frac{dy}{dx}} = \boxed{\frac{\left[\left(x^2+1\right)^3\cdot\frac{d}{dx}\left(x^2+2\right)\right]-\left[\left(x^2+2\right)\cdot\frac{d}{dx}\left(x^2+1\right)^3\right]}{\left(x^2+1\right)^6}} = \boxed{\frac{\left[\left(x^2+1\right)^3\cdot 2x\right]-\left[\left(x^2+2\right)\cdot 3\left(x^2+1\right)^2\frac{d}{dx}\left(x^2+1\right)\right]}{\left(x^2+1\right)^6}}$$

$$= \boxed{\frac{2x\left(x^2+1\right)^3-\left[\left(x^2+2\right)\cdot 3\left(x^2+1\right)^2\cdot 2x\right]}{\left(x^2+1\right)^6}} = \boxed{\frac{2x\left(x^2+1\right)^3-\left[6x\left(x^2+2\right)\left(x^2+1\right)^2\right]}{\left(x^2+1\right)^6}}$$

f. Given $y = u^3+1$ and $u = \left(x^2+1\right)^{-1}$, then $y = \left[\left(x^2+1\right)^{-1}\right]^3+1 = \left(x^2+1\right)^{-3}+1$

$$\boxed{\frac{dy}{dx}} = \boxed{\left[-3\left(x^2+1\right)^{-3-1}\cdot\frac{d}{dx}\left(x^2+1\right)\right]+0} = \boxed{-3\left(x^2+1\right)^{-4}\cdot 2x} = \boxed{-6x\left(x^2+1\right)^{-4}} = \boxed{-\frac{6x}{\left(x^2+1\right)^4}}$$

Example 2.4-5: Find y' given that:

a. $y = 3u^3-1$ and $u = x^2+1$						b. $y = \dfrac{3u^2}{1+u}$ and $u = x^2$

c. $y = \dfrac{u+1}{u-1}$ and $u = x^2+3$						d. $y = \dfrac{1-5u}{u^2}$ and $u = 1-x$

e. $y = \dfrac{u}{u^2+1}$ and $u = 2x-1$						f. $y = u+\dfrac{1}{4}$ and $u = (2x-1)^4$

g. $y = \dfrac{2u}{(u-1)^2}$ and $u = x+2$						h. $y = u^4-1$ and $u = \dfrac{1+x}{1-x}$

Solutions:

a. Given $y = 3u^3-1$ and $u = x^2+1$, then $y = 3\left(x^2+1\right)^3-1$ and

$$\boxed{y'} = \boxed{3\cdot 3\left(x^2+1\right)^{3-1}\cdot 2x-0} = \boxed{9\left(x^2+1\right)^2\cdot 2x} = \boxed{18x\left(x^2+1\right)^2}$$

b. Given $y = \dfrac{3u^2}{1+u}$ and $u = x^2$, then $y = \dfrac{3x^4}{1+x^2}$ and

$$\boxed{y'} = \boxed{\frac{\left[\left(3\cdot 4x^{4-1}\right)\cdot\left(1+x^2\right)\right]-\left[2x\cdot 3x^4\right]}{\left(1+x^2\right)^2}} = \boxed{\frac{12x^3\left(1+x^2\right)-6x^5}{\left(1+x^2\right)^2}} = \boxed{\frac{12x^3+6x^5}{\left(1+x^2\right)^2}} = \boxed{\frac{6x^3\left(2+x^2\right)}{\left(1+x^2\right)^2}}$$

c. Given $y = \dfrac{u+1}{u-1}$ and $u = x^2 + 3$, then $y = \dfrac{\left(x^2+3\right)+1}{\left(x^2+3\right)-1} = \dfrac{x^2+4}{x^2+2}$ and

$$\boxed{y'} = \boxed{\dfrac{\left[2x^{2-1}\cdot\left(x^2+2\right)\right]-\left[2x^{2-1}\cdot\left(x^2+4\right)\right]}{\left(x^2+2\right)^2}} = \boxed{\dfrac{\left[2x\left(x^2+2\right)\right]-\left[2x\left(x^2+4\right)\right]}{\left(x^2+2\right)^2}} = \boxed{\dfrac{2x^3+4x-2x^3-8x}{\left(x^2+2\right)^2}} = \boxed{\dfrac{-4x}{\left(x^2+2\right)^2}}$$

d. Given $y = \dfrac{1-5u}{u^2}$ and $u = 1-x$, then $y = \dfrac{1-5(1-x)}{(1-x)^2} = \dfrac{1-5+5x}{(1-x)^2} = \dfrac{5x-4}{(1-x)^2}$

$$\boxed{y'} = \boxed{\dfrac{\left[5\cdot(...1-x)^2\right]-\left[2(1-x)^{2-1}\cdot-1\cdot(5x-4)\right]}{(1-x)^4}} = \boxed{\dfrac{5(1-x)^2-\left[-2(1-x)(5x-4)\right]}{(1-x)^4}} = \boxed{\dfrac{5(1-x)^2+2(1-x)(5x-4)}{(1-x)^4}}$$

$$= \boxed{\dfrac{5\left(x^2-2x+1\right)+2\left(5x-4-5x^2+4x\right)}{(1-x)^4}} = \boxed{\dfrac{5x^2-10x+5-10x^2+18x-8}{(1-x)^4}} = \boxed{\dfrac{-5x^2+8x-3}{(1-x)^4}}$$

e. Given $y = \dfrac{u}{u^2+1}$ and $u = 2x-1$, then $y = \dfrac{2x-1}{(2x-1)^2+1}$ and

$$\boxed{y'} = \boxed{\dfrac{2\cdot\left[(2x-1)^2+1\right]-\left[2(2x-1)^{2-1}\cdot2\cdot(2x-1)\right]}{\left[(2x-1)^2+1\right]^2}} = \boxed{\dfrac{2\cdot\left[(2x-1)^2+1\right]-\left[4(2x-1)\cdot(2x-1)\right]}{\left[(2x-1)^2+1\right]^2}}$$

$$= \boxed{\dfrac{2(2x-1)^2+2-4(2x-1)^2}{\left[(2x-1)^2+1\right]^2}} = \boxed{\dfrac{2-2(2x-1)^2}{\left[(2x-1)^2+1\right]^2}} = \boxed{\dfrac{2\left[1-(2x-1)^2\right]}{\left[(2x-1)^2+1\right]^2}} = \boxed{\dfrac{-2\left[(2x-1)^2-1\right]}{\left[(2x-1)^2+1\right]^2}}$$

f. Given $y = u + \dfrac{1}{4}$ and $u = (2x-1)^4$, then $y = (2x-1)^4 + \dfrac{1}{4}$ and

$$\boxed{y'} = \boxed{4(2x-1)^{4-1}\cdot2+0} = \boxed{4(2x-1)^3\cdot2} = \boxed{8(2x-1)^3}$$

g. Given $y = \dfrac{2u}{(u-1)^2}$ and $u = x+2$, then $y = \dfrac{2(x+2)}{(x+2-1)^2} = \dfrac{2x+4}{(x+1)^2}$

$$\boxed{y'} = \boxed{\dfrac{\left[2\cdot(x+1)^2\right]-\left[2(x+1)^{2-1}\cdot(2x+4)\right]}{(x+1)^4}} = \boxed{\dfrac{2(x+1)^2-2(x+1)(2x+4)}{(x+1)^4}} = \boxed{\dfrac{2\left(x^2+2x+1\right)-2\left(2x^2+6x+4\right)}{(x+1)^4}}$$

$$= \boxed{\dfrac{2x^2+4x+2-4x^2-12x-8}{(x+1)^4}} = \boxed{\dfrac{-2x^2-8x-6}{(x+1)^4}} = \boxed{\dfrac{-2\left(x^2+4x+3\right)}{(x+1)^4}}$$

h. Given $y = u^4 - 1$ and $u = \dfrac{1+x}{1-x}$, then $y = \left(\dfrac{1+x}{1-x}\right)^4 - 1$ and

$$y' = \boxed{4\left(\dfrac{1+x}{1-x}\right)^{4-1} \cdot \dfrac{[1\cdot(1-x)]-[-1\cdot(1+x)]}{(1-x)^2} - 0} = \boxed{4\left(\dfrac{1+x}{1-x}\right)^3 \cdot \dfrac{(1-x)+(1+x)}{(1-x)^2}} = \boxed{4\left(\dfrac{1+x}{1-x}\right)^3 \cdot \dfrac{2}{(1-x)^2}}$$

$$= \boxed{4\left[\dfrac{(1+x)^3}{(1-x)^3} \cdot \dfrac{2}{(1-x)^2}\right]} = \boxed{\dfrac{8(1+x)^3}{(1-x)^{3+2}}} = \boxed{\dfrac{8(1+x)^3}{(1-x)^5}}$$

Section 2.4 Practice Problems - The Chain Rule

1. Find the derivative of the following functions. Do not simplify the answer to its lowest term.

a. $y = (x^2 + 2)^3$

b. $y = (x^2 + 1)^{-2}$

c. $y = (x^3 - 1)^5$

d. $y = \left(1 - \dfrac{1}{x^2}\right)^2$

e. $y = 2x^3 + \dfrac{1}{3x^2}$

f. $y = \left(\dfrac{1+x^2}{r^3}\right)^4$

g. $y = x^2 \left(\dfrac{x+1}{3}\right)^3$

h. $y = \left[x\,(x+1)^2 + 2x\right]^3$

i. $y = \left(\dfrac{x}{3} - 2x^3\right)^{-1}$

j. $y = (x^3 + 3x^2 + 1)^4$

k. $y = \left(\dfrac{t^2}{1+t^2}\right)^3$

l. $y = (1 + x^{-2})^{-1}$

m. $y = \dfrac{(x+1)^{-2}}{x^3}$

n. $y = \left(\dfrac{1}{1-x^3}\right)^2 + \dfrac{1}{x}$

o. $y = \dfrac{x^3}{x^3+2} - x^2$

2. Find the derivative of the following functions at $x = 0$, $x = 1$, and $x = -1$.

a. $y = (x^3 + 1)^5$

b. $y = (x^3 + 3x^2 - 1)^4$

c. $y = \left(\dfrac{x}{x+1}\right)^2$

d. $y = x\,(x^2 + 1)^2$

e. $y = x^3 + 2(x^2 + 1)^3$

f. $y = \left(\dfrac{x^2}{1+x^2}\right)^3$

g. $y = \left(\dfrac{x}{x^2+1}\right)^5$

h. $y = (x^2 + 1)^3 \cdot \dfrac{1}{x^2}$

i. $y = \left(\dfrac{x^3}{x-1}\right)^2 + 5x$

3. Use the chain rule to differentiate the following functions.

a. $\dfrac{d}{dt}\left[\dfrac{(t+1)^3}{t^2}\right] =$

b. $\dfrac{d}{du}\left[\dfrac{(u^2+1)^3}{3u^4}\right] =$

c. $\dfrac{d}{dx}\left[\dfrac{(2x+1)^3}{(1-x)^2}\right] =$

d. $\dfrac{d}{dx}\left[(x^3-1)^2(2x+1)^3\right] =$

e. $\dfrac{d}{ds}\left[s^3 - \dfrac{1}{s^2+6}\right]^2 =$

f. $\dfrac{d}{dt}\left[\dfrac{(t^2-1)^3}{t^2+1}\right] =$

g. $\dfrac{d}{du}\left[\left(u^2+1\right)^3\left(\dfrac{1}{u+1}\right)^2\right]=$ h. $\dfrac{d}{d\theta}\left[\dfrac{\theta^2+3}{(\theta-1)^3}\right]^2=$ i. $\dfrac{d}{dr}\left[\dfrac{r^7}{\left(r^2+2r\right)^3}\right]=$

4. Given the following y functions in terms of u find y'.

a. $y=2u^2-1$ and $u=x-1$ b. $y=\dfrac{u}{u-1}$ and $u=x^3$ c. $y=\dfrac{u}{1+u^2}$ and $u=x^2+1$

d. $y=u^2-\dfrac{1}{2}$ and $y=x^4$ e. $y=u^4$ and $u=\dfrac{1}{1-x^2}$ f. $y=\dfrac{u^2}{(u+1)^3}$ and $u=x-1$

2.5   Implicit Differentiation

In many instances an equation is explicitly represented in the form where y is the only term in the left-hand side of the equation. In these instances y' is obtained by applying the differentiation rules to the right hand side of the equation. However, for cases where y is not explicitly given, we must either first solve for y (if y can be factored) and then differentiate or use the implicit differentiation method. For example, to differentiate the equation $xy = x^2 + y$ we can either solve for y by bringing the y terms to one side of the equation and then differentiate as follows:

$$\boxed{xy = x^2 + y}\;;\;\boxed{xy - y = x^2}\;;\;\boxed{y(x-1) = x^2}\;;\;\boxed{y = \frac{x^2}{x-1}} \text{ therefore, } y' \text{ is equal to:}$$

$$\boxed{y'} = \boxed{\frac{\left[2x^{2-1}\cdot(x-1)\right]-\left[1\cdot x^2\right]}{(x-1)^2}} = \boxed{\frac{\left[2x(x-1)\right]-x^2}{(x-1)^2}} = \boxed{\frac{2x^2-2x-x^2}{(x-1)^2}} = \boxed{\frac{x^2-2x}{(x-1)^2}}$$

or, we can use the implicit differentiation method as shown below.

$$\boxed{xy = x^2 + y}\;;\;\boxed{(1\cdot y + y'\cdot x) = 2x^{2-1} + y'}\;;\;\boxed{y + y'x = 2x + y'}\;;\;\boxed{y'x - y' = 2x - y}\;;\;\boxed{y'(x-1) = 2x - y}\;;\;\boxed{y' = \frac{2x-y}{x-1}}$$

Substituting $y = \dfrac{x^2}{x-1}$ into the y' equation we obtain:

$$\boxed{y'} = \boxed{\frac{2x-y}{x-1}} = \boxed{\frac{2x - \frac{x^2}{x-1}}{x-1}} = \boxed{\frac{\frac{2x(x-1)-x^2}{x-1}}{x-1}} = \boxed{\frac{\frac{2x(x-1)-x^2}{x-1}}{\frac{x-1}{1}}} = \boxed{\frac{\left[2x(x-1)-x^2\right]\cdot 1}{(x-1)\cdot(x-1)}} = \boxed{\frac{x^2-2x}{(x-1)^2}}$$

Note that the key in using the implicit differentiation method is that the chain rule must be applied each time we come across a term with y in it. Following are additional examples showing the two methods of differentiation when y is not explicitly given:

Example 2.5-1: Given $xy + x = y + 3$, find y'.

Solution:

First Method: Let's solve for y by bringing the y terms to the left-hand side of the equation,

i.e., $\boxed{xy + x = y + 3}\;;\;\boxed{y(x-1) = -x+3}\;;\;\boxed{y = \dfrac{-x+3}{x-1}}$

We can now solve for y' using the differentiation rule for division.

$$\boxed{y'} = \boxed{\frac{\left[-1\cdot(x-1)\right]-\left[1\cdot(-x+3)\right]}{(x-1)^2}} = \boxed{\frac{-x+1+x-3}{(x-1)^2}} = \boxed{\frac{-2}{(x-1)^2}}$$

Second Method: Use the implicit differentiation method to solve for y', i.e., given

$$\boxed{xy + x = y + 3} \text{ then, } \boxed{(1\cdot y + y'\cdot x)+1 = y'+0}\;;\;\boxed{y+1 = y'-y'x}\;;\;\boxed{y'(1-x) = y+1}\;;\;\boxed{y' = \frac{y+1}{1-x}}$$

Substituting $y = \dfrac{-x+3}{x-1}$ into the y' equation we obtain:

$$\boxed{y'} = \boxed{y' = \frac{y+1}{1-x}} = \boxed{\frac{\frac{-x+3}{x-1}+1}{1-x}} = \boxed{\frac{\frac{-x+3}{x-1}+\frac{1}{1}}{1-x}} = \boxed{\frac{\frac{[(-x+3)\cdot 1]+[1\cdot(x-1)]}{(x-1)\cdot 1}}{1-x}} = \boxed{\frac{\frac{-\not x+3+\not x-1}{x-1}}{1-x}} = \boxed{\frac{\frac{2}{x-1}}{1-x}} = \boxed{\frac{\frac{2}{x-1}}{\frac{1-x}{1}}}$$

$$= \boxed{\frac{2\cdot 1}{(x-1)\cdot(1-x)}} = \boxed{\frac{2}{(x-1)\cdot-(x-1)}} = \boxed{-\frac{2}{(x-1)(x-1)}} = \boxed{\frac{-2}{(x-1)^2}}$$

Example 2.5-2: Given $x^2 y+5 = y+2x$, find y'.

Solution:

First Method: Let's solve for y by bringing the y terms to the left-hand side of the equation,

i.e., $\boxed{x^2 y - y = 2x-5}$; $\boxed{y(x^2-1)=2x-5}$; $\boxed{y=\dfrac{2x-5}{x^2-1}}$

We can now solve for y' using the differentiation rule for division.

$$\boxed{y'} = \boxed{\frac{[2\cdot(x^2-1)]-[2x^{2-1}\cdot(2x-5)]}{(x^2-1)^2}} = \boxed{\frac{2x^2-2-4x^2+10x}{(x^2-1)^2}} = \boxed{\frac{-2x^2+10x-2}{(x^2-1)^2}}$$

Second Method: Use the implicit differentiation method to solve for y', i.e., given

$\boxed{x^2 y+5 = y+2x}$ then, $\boxed{(2x\cdot y + y'\cdot x^2)+0 = y'+2x^{1-1}}$; $\boxed{2xy+y'x^2 = y'+2}$; $\boxed{y'x^2 - y' = 2-2xy}$

; $\boxed{y'(x^2-1)=2x-2xy}$; $\boxed{y'=\dfrac{2-2xy}{x^2-1}}$

Substituting $y = \dfrac{2x-5}{x^2-1}$ into the y' equation we obtain:

$$\boxed{y'} = \boxed{\frac{2-2xy}{x^2-1}} = \boxed{\frac{2-2x\cdot\frac{2x-5}{x^2-1}}{x^2-1}} = \boxed{\frac{2-\frac{4x^2-10x}{x^2-1}}{x^2-1}} = \boxed{\frac{\frac{2(x^2-1)-(4x^2-10x)}{x^2-1}}{x^2-1}} = \boxed{\frac{\frac{2x^2-2-4x^2+10x}{x^2-1}}{x^2-1}}$$

$$= \boxed{\frac{\frac{-2x^2+10x-2}{x^2-1}}{x^2-1}} = \boxed{\frac{\frac{-2x^2+10x-2}{x^2-1}}{\frac{x^2-1}{1}}} = \boxed{\frac{(-2x^2+10x-2)\cdot 1}{(x^2-1)\cdot(x^2-1)}} = \boxed{\frac{-2x^2+10x-2}{(x^2-1)^2}}$$

In the previous examples, to find y' we could either first solve for y and then differentiate or use the implicit differentiation rule. However, sometimes we can not simply solve for y by bringing the y terms to the left-hand side of the equation. In these instances, as is shown in the following examples, we can only use implicit differentiation in order to differentiate y.

Example 2.5-3: Given $x^2 y^2 + y = 3y^3 - 1$, find $y' = \dfrac{dy}{dx}$.

Solution:

$\boxed{\dfrac{d}{dx}\left(x^2 y^2 + y\right) = \dfrac{d}{dx}\left(3y^3 - 1\right)}$; $\boxed{\left(2x \cdot y^2 + 2y^{2-1} y' \cdot x^2\right) + y' = (3 \cdot 3)y^{3-1} \cdot y' - 0}$; $\boxed{2xy^2 + 2yy'x^2 + y' = 9y^2 y'}$

; $\boxed{2x^2 y y' - 9y^2 y' + y' = -2xy^2}$; $\boxed{y'\left(2x^2 y - 9y^2 + 1\right) = -2xy^2}$; $\boxed{y' = -\dfrac{2xy^2}{2x^2 y - 9y^2 + 1}}$

Example 2.5-4: Given $xy + x^2 y^2 + y^3 = 10x$, find $\dfrac{dy}{dx}$ in terms of x and y.

Solution:

$\boxed{\dfrac{d}{dx}\left(xy + x^2 y^2 + y^3\right) = \dfrac{d}{dx}(10x)}$; $\boxed{\left(y + x y'\right) + \left(2x y^2 + 2x^2 y y'\right) + 3y^2 y' = 10}$

; $\boxed{x y' + 2x^2 y y' + 3y^2 y' = -2x y^2 - y + 10}$; $\boxed{y'\left(x + 2x^2 y + 3y^2\right) = -2x y^2 - y + 10}$; $\boxed{y' = \dfrac{-2x y^2 - y + 10}{x + 2x^2 y + 3y^2}}$

Example 2.5-5: Given $3x^3 y^3 + 2y^2 = y + 1$, find $\dfrac{dy}{dx}$ in terms of x and y.

Solution:

$\boxed{\dfrac{d}{dx}\left(3x^3 y^3 + 2y^2\right) = \dfrac{d}{dx}(y + 1)}$; $\boxed{3\left(3x^2 y^3 + 3y^2 y'x^3\right) + \left(4yy'\right) = y' + 0}$; $\boxed{9x^2 y^3 + 9x^3 y^2 y' + 4 y y' = y'}$

; $\boxed{9x^3 y^2 y' + 4 y y' - y' = -9x^2 y^3}$; $\boxed{y'\left(9x^3 y^2 + 4 y - 1\right) = -9x^2 y^3}$; $\boxed{y' = -\dfrac{9x^2 y^3}{9x^3 y^2 + 4 y - 1}}$

Example 2.5-6: Given $x y + x^3 y^3 = 5$, find $\dfrac{dy}{dx}$ in terms of x and y.

Solution:

$\boxed{\dfrac{d}{dx}\left(x y + x^3 y^3\right) = \dfrac{d}{dx}(5)}$; $\boxed{\left(1 \cdot y + y' \cdot x\right) + \left(3x^2 \cdot y^3 + 3y^2 y' \cdot x^3\right) = 0}$; $\boxed{y + x y' + 3x^2 y^3 + 3x^3 y^2 y' = 0}$

; $\boxed{x y' + 3x^3 y^2 y' = -3x^2 y^3 - y}$; $\boxed{y'\left(x + 3x^3 y^2\right) = -3x^2 y^3 - y}$; $\boxed{y' = -\dfrac{3x^2 y^3 + y}{x + 3x^3 y^2}}$

Example 2.5-7: Given $3x y + y = \left(x^2 + y^2\right)$, find $\dfrac{dy}{dx}$ in terms of x and y.

Solution:

$\boxed{\dfrac{d}{dx}(3x y + y) = \dfrac{d}{dx}\left(x^2 + y^2\right)}$; $\boxed{3\left(1 \cdot y + y' \cdot x\right) + y' = 2x + 2yy'}$; $\boxed{3y + 3x y' + y' = 2x + 2y y'}$

; $\boxed{3x y' + y' - 2y y' = 2x - 3y}$; $\boxed{y'\left(3x + 1 - 2y\right) = 2x - 3y}$; $\boxed{y' = \dfrac{2x - 3y}{3x - 2y + 1}}$

Example 2.5-8: Given $\dfrac{1}{x} + \dfrac{1}{y^2} = 10x$, find $\dfrac{dy}{dx}$ in terms of x and y.

Solution:

$$\boxed{\frac{d}{dx}\left(\frac{1}{x}+\frac{1}{y^2}\right)=\frac{d}{dx}(10x)} \;;\; \boxed{\frac{d}{dx}\left(x^{-1}+y^{-2}\right)=\frac{d}{dx}(10x)} \;;\; \boxed{-x^{-2}-2y^{-3}y'=10} \;;\; \boxed{-2y^{-3}y'=x^{-2}+10}$$

$$;\; \boxed{y'=\frac{x^{-2}+10}{-2y^{-3}}} \;;\; \boxed{y'=\frac{\frac{1}{x^2}+10}{\frac{-2}{y^3}}} \;;\; \boxed{y'=\frac{\frac{1+10x^2}{x^2}}{\frac{-2}{y^3}}} \;;\; \boxed{y'=\frac{y^3\left(1+10x^2\right)}{-2x^2}} \;;\; \boxed{y'=-\frac{y^3}{2}\left(\frac{1}{x^2}+10\right)}$$

Example 2.5-9: Given $xy^2+yx^2=x^2$, find $\dfrac{dy}{dx}$ in terms of x and y.

Solution:

$$\boxed{\frac{d}{dx}\left(xy^2+yx^2\right)=\frac{d}{dx}\left(x^2\right)} \;;\; \boxed{\left(1\cdot y^2+2y\,y'\cdot x\right)+\left(y'\cdot x^2+2x\cdot y\right)=2x} \;;\; \boxed{y^2+2x\,y\,y'+x^2y'+2x\,y=2x}$$

$$;\; \boxed{2x\,y\,y'+x^2y'=-y^2-2x\,y+2x} \;;\; \boxed{y'\left(x^2+2x\,y\right)=-y^2-2x\,y+2x} \;;\; \boxed{y'=\frac{-y^2-2x\,y+2x}{x^2+2x\,y}}$$

Example 2.5-10: Given $y^{\frac{2}{3}}+x^3y=y$, find $\dfrac{dy}{dx}$ in terms of x and y.

Solution:

$$\boxed{\frac{d}{dx}\left(y^{\frac{2}{3}}+x^3y\right)=\frac{d}{dx}(y)} \;;\; \boxed{\frac{2}{3}y^{\frac{2}{3}-1}\cdot y'+\left(3x^2\cdot y+y'\cdot x^3\right)=y'} \;;\; \boxed{\frac{2}{3}y^{-\frac{1}{3}}y'+3x^2y+x^3y'=y'}$$

$$;\; \boxed{\frac{2}{3}y^{-\frac{1}{3}}y'+x^3y'-y'=-3x^2y} \;;\; \boxed{y'\left(\frac{2}{3}y^{-\frac{1}{3}}+x^3-1\right)=-3x^2y} \;;\; \boxed{y'=\frac{-3x^2y}{\frac{2}{3}y^{-\frac{1}{3}}+x^3-1}}$$

Example 2.5-11: Given $xy+y^2=y^{\frac{1}{8}}$, find $\dfrac{dy}{dx}$ in terms of x and y.

Solution:

$$\boxed{\frac{d}{dx}\left(xy+y^2\right)=\frac{d}{dx}\left(y^{\frac{1}{8}}\right)} \;;\; \boxed{\left[(1\cdot y)+\left(y'\cdot x\right)\right]+2yy'=\frac{1}{8}y^{\frac{1}{8}-1}\cdot y'} \;;\; \boxed{y+x\,y'+2y\,y'=\frac{1}{8}y^{-\frac{7}{8}}y'}$$

$$;\; \boxed{x\,y'+2y\,y'-\frac{1}{8}y^{-\frac{7}{8}}y'=-y} \;;\; \boxed{y'\left(x+2y-\frac{1}{8}y^{-\frac{7}{8}}\right)=-y} \;;\; \boxed{y'=-\frac{y}{x+2y-\frac{1}{8}y^{-\frac{7}{8}}}}$$

Example 2.5-12: Given $xy^2+y=x^2+3$, find $\dfrac{dy}{dx}$ in terms of x and y.

Solution:

$$\boxed{\frac{d}{dx}\left(xy^2+y\right)=\frac{d}{dx}\left(x^2+3\right)} \;;\; \boxed{\left[\left(1\cdot y^2\right)+\left(2y\,y'\cdot x\right)\right]+y'=2x+0} \;;\; \boxed{y^2+2xy\,y'+y'=2x} \;;\; \boxed{2xy\,y'+y'=2x-y^2}$$

$$;\; \boxed{y'(2xy+1)=2x-y^2} \;;\; \boxed{y'=\frac{2x-y^2}{2xy+1}}$$

Example 2.5-13: Given $x^4 y^3 + y^2 = x+4$, find $\dfrac{dy}{dx}$ in terms of x and y.

Solution:

$\boxed{\dfrac{d}{dx}\left(x^4 y^3 + y^2\right) = \dfrac{d}{dx}(x+4)}$; $\boxed{\left[\left(4x^3 \cdot y^3\right)+\left(3y^2 y' \cdot x^4\right)\right]+2y\,y' = 1+0}$; $\boxed{4x^3 y^3 + 3x^4 y^2 y' + 2y\,y' = 1}$

; $\boxed{3x^4 y^2 y' + 2y\,y' = 1 - 4x^3 y^3}$; $\boxed{y'\left(3x^4 y^2 + 2y\right) = 1 - 4x^3 y^3}$; $\boxed{y' = \dfrac{1-4x^3 y^3}{3x^4 y^2 + 2y}}$

Example 2.5-14: Given $y^6 + x^3 y^5 + x^2 = 5$, find $\dfrac{dy}{dx}$ in terms of x and y.

Solution:

$\boxed{\dfrac{d}{dx}\left(y^6 + x^3 y^5 + x^2\right) = \dfrac{d}{dx}(5)}$; $\boxed{6y^5 y' + \left[\left(3x^2 \cdot y^5\right)+\left(5y^4 y' \cdot x^3\right)\right] + 2x = 0}$; $\boxed{6y^5 y' + 3x^2 y^5 + 5x^3 y^4 y' + 2x = 0}$

; $\boxed{6y^5 y' + 5x^3 y^4 y' = -3x^2 y^5 - 2x}$; $\boxed{y'\left(6y^5 + 5x^3 y^4\right) = -3x^2 y^5 - 2x}$; $\boxed{y' = -\dfrac{3x^2 y^5 + 2x}{6y^5 + 5x^3 y^4}}$

Section 2.5 Practice Problems - Implicit Differentiation

Use implicit differentiation method to solve the following functions.

a. $x^2 y + x = y$

b. $xy - 3x^2 + y = 0$

c. $x^2 y^2 + y = 3y^3$

d. $xy + y^3 = 5x$

e. $4x^4 y^4 + 2y^2 = y - 1$

f. $xy + x^2 y^2 - 10 = 0$

g. $xy^2 + y = x^2$

h. $xy^3 + x^3 y = x$

i. $y^{\frac{1}{2}} + x^2 y = x$

j. $x^2 y + y^2 = y^{\frac{1}{4}}$

k. $x + y^2 = x^2 - 3$

l. $x^4 y^2 + y = -3$

m. $y^7 - x^2 y^4 - x = 8$

n. $(x+3)^2 = y^2 - x$

o. $3x^2 y^5 + y^2 = -x$

2.6   The Derivative of Functions with Fractional Exponents

The derivative of a function $f(x)$ with fractional exponent is obtained by applying the chain rule in the following way:

$$\frac{d}{dx}\left[f(x)\right]^{\frac{a}{b}} = \frac{a}{b}\left[f(x)\right]^{\frac{a}{b}-1} \cdot \frac{d}{dx}\left[f(x)\right]$$

For example, the derivative of $f(x) = x^{\frac{a}{b}}$ is equal to

$$\frac{d}{dx}x^{\frac{a}{b}} = \frac{a}{b}x^{\frac{a}{b}-1} \cdot \frac{d}{dx}x = \frac{a}{b}x^{\frac{a}{b}-1} \cdot 1 = \frac{a}{b}x^{\frac{a}{b}-1} = \frac{a}{b}x^{\frac{a-b}{b}}$$

Note that the steps in finding the derivative of a function with fractional exponent is similar to finding the derivative of a function that is raised to a power as discussed in Section 2.4. The following examples illustrate how to obtain the derivative of exponential functions:

Example 2.6-1: Find the derivative for the following exponential expressions.

a. $y = x^{\frac{2}{3}}$

b. $y = \left(3x^2\right)^{\frac{1}{3}}$

c. $y = \left(3x^3 + 2x\right)^{\frac{1}{4}}$

d. $y = \left(3x^2 + 6x\right)^{\frac{2}{5}}$

e. $y = (2x+1)^{\frac{3}{4}}$

f. $y = \left(3x^2 + 8\right)^{\frac{2}{7}}$

g. $y = \left(x^2\right)^{\frac{1}{3}} + (2x+1)^{\frac{3}{5}}$

h. $y = x\left(x^2+1\right)^{\frac{2}{3}}$

i. $y = (x+1)^{\frac{1}{2}}\left(x^2+3\right)^{\frac{1}{3}}$

j. $y = \dfrac{\left(x^2+1\right)^{\frac{1}{2}}}{x}$

k. $y = \dfrac{(x+3)^{\frac{1}{5}}}{x^{\frac{2}{3}}}$

l. $y = \dfrac{x^3}{(x+1)^{\frac{2}{3}}}$

m. $y = (x+1)\cdot\dfrac{1}{x^{\frac{1}{7}}}$

Solutions:

a. Given $\boxed{y = x^{\frac{2}{3}}}$ then $\boxed{y' =}$ $\boxed{\dfrac{2}{3}x^{\frac{2}{3}-1}}$ = $\boxed{\dfrac{2}{3}x^{\frac{2}{3}-\frac{1}{1}}\cdot 1}$ = $\boxed{\dfrac{2}{3}x^{\frac{2-3}{3}}}$ = $\boxed{\dfrac{2}{3}x^{-\frac{1}{3}}}$ = $\boxed{\dfrac{2}{3}\cdot\dfrac{1}{x^{\frac{1}{3}}}}$ = $\boxed{\dfrac{2}{3}\cdot\dfrac{1}{\sqrt[3]{x}}}$ = $\boxed{\dfrac{2}{3\sqrt[3]{x}}}$

Note that the answer does not necessarily need to be in radical form. We can simply stop when $y' = \dfrac{2}{3}x^{-\frac{1}{3}}$. However, for review purposes only, the answer to some of the problems are shown in radical form (see Sections 1.1 and 1.2 from *Mastering Algebra – Advanced Level* to review the subjects of exponents and radicals).

b. Given $\boxed{y = \left(3x^2\right)^{\frac{1}{3}}}$ then $\boxed{y' =}$ $\boxed{\dfrac{1}{3}\left(3x^2\right)^{\frac{1}{3}-1}\cdot 6x}$ = $\boxed{\dfrac{6}{3}x\left(3x^2\right)^{\frac{1}{3}-\frac{1}{1}}}$ = $\boxed{2x\left(3x^2\right)^{\frac{1-3}{3}}}$ = $\boxed{2x\left(3x^2\right)^{-\frac{2}{3}}}$ = $\boxed{2x\dfrac{1}{\left(3x^2\right)^{\frac{2}{3}}}}$

$= \boxed{2x\dfrac{1}{\sqrt[3]{\left(3x^2\right)^2}}}$ = $\boxed{2x\dfrac{1}{\sqrt[3]{9x^4}}}$ = $\boxed{\dfrac{2x}{x\sqrt[3]{9x}}}$ = $\boxed{\dfrac{2}{\sqrt[3]{9x}}}$

c. Given $y = \left(3x^3 + 2x\right)^{\frac{1}{4}}$ then $y' = \frac{1}{4}\left(3x^3 + 2x\right)^{\frac{1}{4}-1}\left(9x^2 + 2\right) = \frac{1}{4}\left(3x^3 + 2x\right)^{\frac{1}{4}-\frac{1}{1}}\left(9x^2 + 2\right)$

$= \frac{1}{4}\left(3x^3 + 2x\right)^{\frac{1-4}{4}}\left(9x^2 + 2\right) = \frac{1}{4}\left(3x^3 + 2x\right)^{-\frac{3}{4}}\left(9x^2 + 2\right) = \frac{9x^2 + 2}{4\left(3x^3 + 2x\right)^{\frac{3}{4}}} = \frac{9x^2 + 2}{4\sqrt[4]{\left(3x^3 + 2x\right)^3}}$

d. Given $y = \left(3x^2 + 6x\right)^{\frac{2}{5}}$ then $y' = \frac{2}{5}\left(3x^2 + 6x\right)^{\frac{2}{5}-1}\left(6x + 6\right) = \frac{12}{5}\left(3x^2 + 6x\right)^{\frac{2}{5}-\frac{1}{1}}\left(x + 1\right) = \frac{12}{5}\left(3x^2 + 6x\right)^{\frac{2-5}{5}}\left(x + 1\right)$

$= \frac{12}{5}\left(3x^2 + 6x\right)^{-\frac{3}{5}}\left(x + 1\right) = \frac{12}{5}\left(x + 1\right) \cdot \frac{1}{\left(3x^2 + 6x\right)^{\frac{3}{5}}} = \frac{12}{5} \cdot \frac{x + 1}{\sqrt[5]{\left(3x^2 + 6x\right)^3}} = \frac{12\left(x + 1\right)}{5\sqrt[5]{\left(3x^2 + 6x\right)^3}}$

e. Given $y = \left(2x + 1\right)^{\frac{3}{4}}$ then $y' = \frac{3}{4}\left(2x + 1\right)^{\frac{3}{4}-1} \cdot 2 = \frac{6}{4}\left(2x + 1\right)^{\frac{3}{4}-\frac{1}{1}} = \frac{3}{2}\left(2x + 1\right)^{\frac{3-4}{4}} = \frac{3}{2}\left(2x + 1\right)^{-\frac{1}{4}}$

$= \frac{3}{2} \cdot \frac{1}{\left(2x + 1\right)^{\frac{1}{4}}} = \frac{3}{2} \cdot \frac{1}{\sqrt[4]{2x + 1}} = \frac{3}{2\sqrt[4]{2x + 1}}$

f. Given $y = \left(3x^2 + 8\right)^{\frac{2}{7}}$ then $y' = \frac{2}{7}\left(3x^2 + 8\right)^{\frac{2}{7}-1} \cdot 6x = \frac{12x}{7}\left(3x^2 + 8\right)^{\frac{2}{7}-\frac{1}{1}} = \frac{12x}{7}\left(3x^2 + 8\right)^{\frac{2-7}{7}}$

$= \frac{12x}{7}\left(3x^2 + 8\right)^{-\frac{5}{7}} = \frac{12x}{7} \cdot \frac{1}{\left(3x^2 + 8\right)^{\frac{5}{7}}} = \frac{12x}{7} \cdot \frac{1}{\sqrt[7]{\left(3x^2 + 8\right)^5}} = \frac{12x}{7\sqrt[7]{\left(3x^2 + 8\right)^5}}$

g. Given $y = \left(x^2\right)^{\frac{1}{3}} + \left(2x + 1\right)^{\frac{3}{5}} = x^{\frac{2}{3}} + \left(2x + 1\right)^{\frac{3}{5}}$ then $y' = \frac{2}{3}x^{\frac{2}{3}-1} + \frac{3}{5}\left(2x + 1\right)^{\frac{3}{5}-1} \cdot 2 = \frac{2}{3}x^{\frac{2}{3}-\frac{1}{1}} + \frac{6}{5}\left(2x + 1\right)^{\frac{3}{5}-\frac{1}{1}}$

$= \frac{2}{3}x^{\frac{2-3}{3}} + \frac{6}{5}\left(2x + 1\right)^{\frac{3-5}{5}} = \frac{2}{3}x^{-\frac{1}{3}} + \frac{6}{5}\left(2x + 1\right)^{-\frac{2}{5}} = \frac{2}{3} \cdot \frac{1}{x^{\frac{1}{3}}} + \frac{6}{5} \cdot \frac{1}{\left(2x + 1\right)^{\frac{2}{5}}} = \frac{2}{3\sqrt[3]{x}} + \frac{6}{5\sqrt[5]{\left(2x + 1\right)^2}}$

h. Given $y = x\left(x^2 + 1\right)^{\frac{2}{3}}$ then $y' = \left[1 \cdot \left(x^2 + 1\right)^{\frac{2}{3}}\right] + \left\{\left[\frac{2}{3}\left(x^2 + 1\right)^{\frac{2}{3}-1} \cdot 2x\right] \cdot x\right\} = \left(x^2 + 1\right)^{\frac{2}{3}} + \frac{4x^2}{3}\left(x^2 + 1\right)^{\frac{2}{3}-\frac{1}{1}}$

$= \left(x^2 + 1\right)^{\frac{2}{3}} + \frac{4x^2}{3}\left(x^2 + 1\right)^{\frac{2-3}{3}} = \left(x^2 + 1\right)^{\frac{2}{3}} + \frac{4x^2}{3}\left(x^2 + 1\right)^{-\frac{1}{3}} = \left(x^2 + 1\right)^{\frac{2}{3}} + \frac{4x^2}{3\left(x^2 + 1\right)^{\frac{1}{3}}} = \sqrt[3]{\left(x^2 + 1\right)^2} + \frac{4x^2}{3\sqrt[3]{x^2 + 1}}$

i. Given $y = \left(x + 1\right)^{\frac{1}{2}}\left(x^2 + 3\right)^{\frac{1}{3}}$ then $y' = \left[\frac{1}{2}\left(x + 1\right)^{\frac{1}{2}-1} \cdot 1 \cdot \left(x^2 + 3\right)^{\frac{1}{3}}\right] + \left[\frac{1}{3}\left(x^2 + 3\right)^{\frac{1}{3}-1} \cdot 2x \cdot \left(x + 1\right)^{\frac{1}{2}}\right]$

$$= \left[\left[\frac{1}{2}(x+1)^{-\frac{1}{2}} \cdot \left(x^2+3\right)^{\frac{1}{3}} \right] + \left[\frac{2x}{3}\left(x^2+3\right)^{-\frac{2}{3}} \cdot (x+1)^{\frac{1}{2}} \right] \right] = \left[\left[\frac{1}{2} \cdot \left(x^2+3\right)^{\frac{1}{3}} \cdot \frac{1}{(x+1)^{\frac{1}{2}}} \right] + \left[\frac{2x}{3} \cdot (x+1)^{\frac{1}{2}} \cdot \frac{1}{\left(x^2+3\right)^{\frac{2}{3}}} \right] \right]$$

$$= \frac{\left(x^2+3\right)^{\frac{1}{3}}}{2(x+1)^{\frac{1}{2}}} + \frac{2x(x+1)^{\frac{1}{2}}}{3\left(x^2+3\right)^{\frac{2}{3}}} = \frac{\sqrt[3]{x^2+3}}{2\sqrt{x+1}} + \frac{2x\sqrt{x+1}}{3\sqrt[3]{\left(x^2+3\right)^2}}$$

j. Given $y = \dfrac{\left(x^2+1\right)^{\frac{1}{2}}}{x}$ then $\boxed{y'} = \dfrac{\left\{ \left[\frac{1}{2}\left(x^2+1\right)^{\frac{1}{2}-1} \cdot 2x \right] \cdot x \right\} - \left[1 \cdot \left(x^2+1\right)^{\frac{1}{2}} \right]}{x^2} = \dfrac{\frac{2x^2}{2}\left(x^2+1\right)^{-\frac{1}{2}} - \left(x^2+1\right)^{\frac{1}{2}}}{x^2}$

$$= \frac{x^2 \frac{1}{\left(x^2+1\right)^{\frac{1}{2}}} - \left(x^2+1\right)^{\frac{1}{2}}}{x^2} = \frac{\frac{x^2}{\sqrt{x^2+1}} - \sqrt{x^2+1}}{x^2} = \frac{\frac{x^2 - \left(x^2+1\right)}{\sqrt{x^2+1}}}{x^2} = \frac{\frac{-1}{\sqrt{x^2+1}}}{x^2} = \frac{-1}{x^2\sqrt{x^2+1}}$$

k. Given $y = \dfrac{(x+3)^{\frac{1}{5}}}{x^{\frac{2}{3}}}$ then $\boxed{y'} = \dfrac{\left[\frac{1}{5}(x+3)^{\frac{1}{5}-1} \cdot x^{\frac{2}{3}} \right] - \left[\frac{2}{3}x^{\frac{2}{3}-1} \cdot (x+3)^{\frac{1}{5}} \right]}{x^{\frac{4}{3}}} = \dfrac{\left[\frac{1}{5}x^{\frac{2}{3}}(x+3)^{-\frac{4}{5}} \right] - \left[\frac{2}{3}x^{-\frac{1}{3}}(x+3)^{\frac{1}{5}} \right]}{x^{\frac{4}{3}}}$

$$= \frac{\frac{1}{5} \cdot \frac{x^{\frac{2}{3}}}{(x+3)^{\frac{4}{5}}} - \frac{2}{3} \cdot \frac{(x+3)^{\frac{1}{5}}}{x^{\frac{1}{3}}}}{x^{\frac{4}{3}}} = \frac{\frac{x^{\frac{2}{3}}}{5(x+3)^{\frac{4}{5}}} - \frac{2(x+3)^{\frac{1}{5}}}{3x^{\frac{1}{3}}}}{x^{\frac{4}{3}}} = \frac{\frac{\sqrt[3]{x^2}}{5\sqrt[5]{(x+3)^4}} - \frac{2\sqrt[5]{x+3}}{3\sqrt[3]{x}}}{\sqrt[3]{x^4}}$$

l. Given $y = \dfrac{x^3}{(x+1)^{\frac{2}{3}}}$ then $\boxed{y'} = \dfrac{\left[3x^{3-1} \cdot (x+1)^{\frac{2}{3}} \right] - \left[\frac{2}{3}(x+1)^{\frac{2}{3}-1} \cdot x^3 \right]}{(x+1)^{\frac{4}{3}}} = \dfrac{\left[3x^2(x+1)^{\frac{2}{3}} \right] - \left[\frac{2x^3}{3}(x+1)^{-\frac{1}{3}} \right]}{(x+1)^{\frac{4}{3}}}$

$$= \frac{3x^2(x+1)^{\frac{2}{3}} - \frac{2x^3}{3(x+1)^{\frac{1}{3}}}}{(x+1)^{\frac{4}{3}}} = \frac{3x^2\sqrt[3]{(x+1)^2} - \frac{2x^3}{3\sqrt[3]{x+1}}}{\sqrt[3]{(x+1)^4}} = \frac{3x^2\sqrt[3]{(x+1)^2} - \frac{2x^3}{3\sqrt[3]{x+1}}}{(x+1)\sqrt[3]{x+1}}$$

m. Given $y = (x+1) \cdot \dfrac{1}{x^{\frac{1}{7}}} = \dfrac{x+1}{x^{\frac{1}{7}}}$ then $\boxed{y'} = \dfrac{\left[1 \cdot x^{\frac{1}{7}} \right] - \left[\frac{1}{7}x^{\frac{1}{7}-1} \cdot (x+1) \right]}{x^{\frac{2}{7}}} = \dfrac{x^{\frac{1}{7}} - \frac{x+1}{7}x^{-\frac{6}{7}}}{x^{\frac{2}{7}}} = \dfrac{x^{\frac{1}{7}} - \frac{x+1}{7} \cdot \frac{1}{x^{\frac{6}{7}}}}{x^{\frac{2}{7}}}$

$$= \frac{\sqrt[7]{x} - \frac{x+1}{7} \cdot \frac{1}{\sqrt[7]{x^6}}}{\sqrt[7]{x^2}} = \frac{\sqrt[7]{x} - \frac{x+1}{7\sqrt[7]{x^6}}}{\sqrt[7]{x^2}}$$

Example 2.6-2: Find $\dfrac{d}{dx}$ of the following functions.

a. $\dfrac{d}{dx}\left(x^{-\frac{2}{3}}\right) =$

b. $\dfrac{d}{dx}(x+1)^{-\frac{1}{4}} =$

c. $\dfrac{d}{dx}\left(x^3+1\right)^{\frac{1}{8}} =$

d. $\dfrac{d}{dx}\left(3x^2+4x\right)^{\frac{7}{8}} =$

e. $\dfrac{d}{dx}x\left(x^2+1\right)^{-\frac{2}{3}} =$

f. $\dfrac{d}{dx}\left(3x^3+4\right)^{-\frac{1}{3}} =$

g. $\dfrac{d}{dx}\left[\dfrac{(x+1)^{-\frac{1}{2}}}{x}\right] =$

h. $\dfrac{d}{dx}\left(x^2+3x+5\right)^{\frac{1}{4}} =$

i. $\dfrac{d}{dx}\left[(x+1)\,(x-1)^{\frac{5}{4}}\right] =$

j. $\dfrac{d}{dx}\left(x^2+5\right)^{-\frac{1}{6}} =$

k. $\dfrac{d}{dx}\left[x^{\frac{3}{5}}\left(x^2+1\right)^{\frac{1}{4}}\right] =$

l. $\dfrac{d}{dx}\left[\dfrac{x^2}{(x+1)^{-\frac{1}{3}}}\right] =$

Solutions:

a. $\boxed{\dfrac{d}{dx}\left(x^{-\frac{2}{3}}\right)} = \boxed{-\dfrac{2}{3}x^{-\frac{2}{3}-1}\cdot\dfrac{d}{dx}(x)} = \boxed{-\dfrac{2}{3}x^{-\frac{2}{3}-1}\cdot1} = \boxed{-\dfrac{2}{3}x^{\frac{-2-3}{3}}} = \boxed{-\dfrac{2}{3}x^{-\frac{5}{3}}} = \boxed{-\dfrac{2}{3}\cdot\dfrac{1}{x^{\frac{5}{3}}}} = \boxed{\dfrac{-2}{3\sqrt[3]{x^5}}} = \boxed{\dfrac{-2}{3x\sqrt[3]{x^2}}}$

b. $\boxed{\dfrac{d}{dx}(x+1)^{-\frac{1}{4}}} = \boxed{-\dfrac{1}{4}(x+1)^{-\frac{1}{4}-1}\cdot\dfrac{d}{dx}(x+1)} = \boxed{-\dfrac{1}{4}(x+1)^{-\frac{1}{4}-1}\cdot1} = \boxed{-\dfrac{1}{4}(x+1)^{\frac{-1-4}{4}}} = \boxed{-\dfrac{1}{4}(x+1)^{-\frac{5}{4}}}$

$= \boxed{-\dfrac{1}{4}\cdot\dfrac{1}{(x+1)^{\frac{5}{4}}}} = \boxed{-\dfrac{1}{4\sqrt[4]{(x+1)^5}}} = \boxed{-\dfrac{1}{4(x+1)\sqrt[4]{x+1}}}$

c. $\boxed{\dfrac{d}{dx}\left(x^3+1\right)^{\frac{1}{8}}} = \boxed{\dfrac{1}{8}\left(x^3+1\right)^{\frac{1}{8}-1}\cdot\dfrac{d}{dx}\left(x^3+1\right)} = \boxed{\dfrac{1}{8}\left(x^3+1\right)^{\frac{1}{8}-1}\cdot3x^2} = \boxed{\dfrac{3}{8}x^2\left(x^3+1\right)^{\frac{1-8}{8}}} = \boxed{\dfrac{3}{8}x^2\left(x^3+1\right)^{-\frac{7}{8}}}$

$= \boxed{\dfrac{3}{8}x^2\cdot\dfrac{1}{\left(x^3+1\right)^{\frac{7}{8}}}} = \boxed{\dfrac{3x^2}{8\left(x^3+1\right)^{\frac{7}{8}}}} = \boxed{\dfrac{3x^2}{8\sqrt[8]{\left(x^3+1\right)^7}}}$

d. $\boxed{\dfrac{d}{dx}\left(3x^2+4x\right)^{\frac{7}{8}}} = \boxed{\dfrac{7}{8}\left(3x^2+4x\right)^{\frac{7}{8}-1}\cdot\dfrac{d}{dx}\left(3x^2+4x\right)} = \boxed{\dfrac{7}{8}\left(3x^2+4x\right)^{\frac{7}{8}-1}\cdot(6x+4)} = \boxed{\dfrac{7}{8}\left(3x^2+4x\right)^{\frac{7-8}{8}}\cdot(6x+4)}$

$= \boxed{\dfrac{7}{8}\left(3x^2+4x\right)^{-\frac{1}{8}}\cdot(6x+4)} = \boxed{\dfrac{7}{8}\cdot\dfrac{6x+4}{\left(3x^2+4x\right)^{\frac{1}{8}}}} = \boxed{\dfrac{7(6x+4)}{8\sqrt[8]{3x^2+4x}}}$

e. $\boxed{\dfrac{d}{dx}x\left(x^2+1\right)^{-\frac{2}{3}}} = \boxed{\left(x^2+1\right)^{-\frac{2}{3}}\cdot\dfrac{d}{dx}x+x\cdot\dfrac{d}{dx}\left(x^2+1\right)^{-\frac{2}{3}}} = \boxed{\left(x^2+1\right)^{-\frac{2}{3}}\cdot1-x\cdot\dfrac{2}{3}\left(x^2+1\right)^{-\frac{2}{3}-1}}$

$= \boxed{\left(x^2+1\right)^{-\frac{2}{3}}-\dfrac{2x}{3}\left(x^2+1\right)^{-\frac{2}{3}-1}} = \boxed{\left(x^2+1\right)^{-\frac{2}{3}}-\dfrac{2x}{3}\left(x^2+1\right)^{\frac{-2-3}{3}}} = \boxed{\left(x^2+1\right)^{-\frac{2}{3}}-\dfrac{2x}{3}\left(x^2+1\right)^{-\frac{5}{3}}}$

$$= \frac{1}{\left(x^2+1\right)^{\frac{2}{3}}} - \frac{2x}{3}\cdot\frac{1}{\left(x^2+1\right)^{\frac{5}{3}}} = \frac{1}{\sqrt[3]{\left(x^2+1\right)^2}} - \frac{2x}{3\sqrt[3]{\left(x^2+1\right)^5}} = \frac{1}{\sqrt[3]{\left(x^2+1\right)^2}} - \frac{2x}{3\left(x^2+1\right)\sqrt[3]{\left(x^2+1\right)^2}}$$

f. $\dfrac{d}{dx}\left(3x^3+4\right)^{-\frac{1}{3}} = -\dfrac{1}{3}\left(3x^3+4\right)^{-\frac{1}{3}-1}\cdot\dfrac{d}{dx}\left(3x^3+4\right) = -\dfrac{1}{3}\left(3x^3+4\right)^{-\frac{1}{3}-\frac{1}{1}}\cdot9x^2 = -\dfrac{9x^2}{3}\left(3x^3+4\right)^{\frac{-1-3}{3}}$

$= -3x^2\left(3x^3+4\right)^{-\frac{4}{3}} = -3x^2\cdot\dfrac{1}{\left(3x^3+4\right)^{\frac{4}{3}}} = -\dfrac{3x^2}{\sqrt[3]{\left(3x^3+4\right)^4}} = -\dfrac{3x^2}{\left(3x^3+4\right)\sqrt[3]{3x^3+4}}$

g. $\dfrac{d}{dx}\left[\dfrac{(x+1)^{-\frac{1}{2}}}{x}\right] = \dfrac{\left[x\cdot\frac{d}{dx}(x+1)^{-\frac{1}{2}}\right]-\left[(x+1)^{-\frac{1}{2}}\cdot\frac{d}{dx}(x)\right]}{x^2} = \dfrac{\left[x\cdot-\frac{1}{2}(x+1)^{-\frac{1}{2}-1}\cdot\frac{d}{dx}(x+1)\right]-\left[(x+1)^{-\frac{1}{2}}\cdot1\right]}{x^2}$

$= \dfrac{\left[-\frac{x}{2}(x+1)^{-\frac{1}{2}-\frac{1}{1}}\cdot1\right]-(x+1)^{-\frac{1}{2}}}{x^2} = \dfrac{-\frac{x}{2}(x+1)^{\frac{-1-2}{2}}-(x+1)^{-\frac{1}{2}}}{x^2} = \dfrac{-\frac{x}{2}(x+1)^{\frac{-3}{2}}-(x+1)^{-\frac{1}{2}}}{x^2}$

h. $\dfrac{d}{dx}\left(x^2+3x+5\right)^{\frac{1}{4}} = \dfrac{1}{4}\left(x^2+3x+5\right)^{\frac{1}{4}-1}\cdot\dfrac{d}{dx}\left(x^2+3x+5\right) = \dfrac{1}{4}\left(x^2+3x+5\right)^{\frac{1}{4}-\frac{1}{1}}\cdot(2x+3) = \dfrac{1}{4}\left(x^2+3x+5\right)^{\frac{1-4}{4}}\cdot(2x+3)$

$= \dfrac{1}{4}\left(x^2+3x+5\right)^{-\frac{3}{4}}(2x+3) = \dfrac{2x+3}{4}\cdot\dfrac{1}{\left(x^2+3x+5\right)^{\frac{3}{4}}} = \dfrac{2x+3}{4}\cdot\dfrac{1}{\sqrt[4]{\left(x^2+3x+5\right)^3}} = \dfrac{2x+3}{4\sqrt[4]{\left(x^2+3x+5\right)^3}}$

i. $\dfrac{d}{dx}\left[(x+1)(x-1)^{\frac{5}{4}}\right] = \left[(x-1)^{\frac{5}{4}}\frac{d}{dx}(x+1)\right]+\left[(x+1)\frac{d}{dx}(x-1)^{\frac{5}{4}}\right] = \left[(x-1)^{\frac{5}{4}}\cdot1\right]+\left[(x+1)\cdot\frac{5}{4}(x-1)^{\frac{5}{4}-1}\cdot\frac{d}{dx}(x-1)\right]$

$= \left[(x-1)^{\frac{5}{4}}+\left[(x+1)\cdot\frac{5}{4}(x-1)^{\frac{5}{4}-\frac{1}{1}}\cdot1\right]\right] = \left[(x-1)^{\frac{5}{4}}+\left[(x+1)\cdot\frac{5}{4}(x-1)^{\frac{5-4}{4}}\right]\right] = (x-1)^{\frac{5}{4}}+\dfrac{5}{4}(x+1)(x-1)^{\frac{1}{4}}$

j. $\dfrac{d}{dx}\left(x^2+5\right)^{-\frac{1}{6}} = -\dfrac{1}{6}\left(x^2+5\right)^{-\frac{1}{6}-1}\cdot\dfrac{d}{dx}\left(x^2+5\right) = -\dfrac{1}{6}\left(x^2+5\right)^{-\frac{1}{6}-\frac{1}{1}}\cdot2x = -\dfrac{2x}{6}\left(x^2+5\right)^{\frac{-1-6}{6}} = -\dfrac{x}{3}\left(x^2+5\right)^{-\frac{7}{6}}$

$= -\dfrac{x}{3}\cdot\dfrac{1}{\left(x^2+5\right)^{\frac{7}{6}}} = -\dfrac{x}{3}\cdot\dfrac{1}{\sqrt[6]{\left(x^2+5\right)^7}} = -\dfrac{x}{3\left(x^2+5\right)\sqrt[6]{x^2+5}}$

k. $\dfrac{d}{dx}\left[x^{\frac{3}{5}}\left(x^2+1\right)^{\frac{1}{4}}\right] = \left[\left(x^2+1\right)^{\frac{1}{4}}\frac{d}{dx}x^{\frac{3}{5}}\right]+\left[x^{\frac{3}{5}}\frac{d}{dx}\left(x^2+1\right)^{\frac{1}{4}}\right] = \left[\left(x^2+1\right)^{\frac{1}{4}}\cdot\frac{3}{5}x^{\frac{3}{5}-1}\right]+\left[x^{\frac{3}{5}}\frac{1}{4}\left(x^2+1\right)^{\frac{1}{4}-1}\cdot\frac{d}{dx}\left(x^2+1\right)\right]$

$= \left[\frac{3}{5}\left(x^2+1\right)^{\frac{1}{4}}\cdot x^{\frac{3}{5}-1}\right]+\left[x^{\frac{3}{5}}\frac{1}{4}\left(x^2+1\right)^{\frac{1}{4}-1}\cdot2x\right] = \left[\frac{3}{5}\left(x^2+1\right)^{\frac{1}{4}}\cdot x^{\frac{3-5}{5}}\right]+\left[\frac{2x\cdot x^{\frac{3}{5}}}{4}\cdot\left(x^2+1\right)^{\frac{1-4}{4}}\right]$

$$= \left[\frac{3x^{-\frac{2}{5}}}{5}\cdot\left(x^2+1\right)^{\frac{1}{4}}\right] + \left[\frac{x^{\frac{8}{5}}}{2}\cdot\left(x^2+1\right)^{-\frac{3}{4}}\right] = \boxed{\frac{3\sqrt[4]{x^2+1}}{5\sqrt[5]{x^2}} + \frac{x\sqrt[5]{x^3}}{2\sqrt[4]{\left(x^2+1\right)^3}}}$$

1. $$\boxed{\frac{d}{dx}\left[\frac{x^2}{(x+1)^{-\frac{1}{3}}}\right]} = \boxed{\frac{d}{dx}\left[x^2(x+1)^{\frac{1}{3}}\right]} = \boxed{\left[(x+1)^{\frac{1}{3}}\frac{d}{dx}x^2\right] + \left[x^2\frac{d}{dx}(x+1)^{\frac{1}{3}}\right]}$$

$$= \boxed{\left[(x+1)^{\frac{1}{3}}\cdot 2x\right] + \left[x^2\cdot\frac{1}{3}(x+1)^{\frac{1}{3}-1}\cdot\frac{d}{dx}(x+1)\right]} = \boxed{2x(x+1)^{\frac{1}{3}} + \left[x^2\cdot\frac{1}{3}(x+1)^{\frac{1}{3}-1}\cdot 1\right]} = \boxed{2x(x+1)^{\frac{1}{3}} + \frac{x^2}{3}(x+1)^{-\frac{2}{3}}}$$

Example 2.6-3: Find $y'(1)$ in example 2.6-1 a through i.

Solutions:
In Example 2.6-1 we obtained the derivative of the exponential functions a through e to be equal
to the following:

a. $y' = \dfrac{2}{3\sqrt[3]{x}}$ then $\boxed{y'(1)} = \boxed{\dfrac{2}{3\sqrt[3]{1}}} = \boxed{\dfrac{2}{3}} = \boxed{0.67}$

b. $y' = \dfrac{2}{\sqrt[3]{9x}}$ then $\boxed{y'(1)} = \boxed{\dfrac{2}{\sqrt[3]{9\cdot 1}}} = \boxed{\dfrac{2}{\sqrt[3]{9}}} = \boxed{\dfrac{2}{9^{0.33}}} = \boxed{\dfrac{2}{2.08}} = \boxed{0.962}$

c. $y' = \dfrac{9x^2+2}{4\sqrt[4]{\left(3x^3+2x\right)^3}}$ then $\boxed{y'(1)} = \boxed{\dfrac{9\cdot 1^2+2}{4\sqrt[4]{\left(3\cdot 1^3+2\cdot 1\right)^3}}} = \boxed{\dfrac{11}{4\sqrt[4]{5^3}}} = \boxed{\dfrac{11}{4\cdot 125^{0.25}}} = \boxed{\dfrac{11}{4\cdot 3.343}} = \boxed{0.823}$

d. $y' = \dfrac{12(x+1)}{5\sqrt[5]{\left(3x^2+6x\right)^3}}$ then $\boxed{y'(1)} = \boxed{\dfrac{12(1+1)}{5\sqrt[5]{\left(3\cdot 1^2+6\cdot 1\right)^3}}} = \boxed{\dfrac{24}{5\sqrt[5]{729}}} = \boxed{\dfrac{24}{5\cdot 729^{0.2}}} = \boxed{\dfrac{24}{5\cdot 3.74}} = \boxed{1.28}$

e. $y' = \dfrac{3}{2\sqrt[4]{2x+1}}$ then $\boxed{y'(1)} = \boxed{\dfrac{3}{2\sqrt[4]{2\cdot 1+1}}} = \boxed{\dfrac{3}{2\sqrt[4]{2+1}}} = \boxed{\dfrac{3}{2\sqrt[4]{3}}} = \boxed{\dfrac{3}{2\cdot 3^{0.25}}} = \boxed{\dfrac{3}{2\cdot 1.32}} = \boxed{1.14}$

Section 2.6 Practice Problems - The Derivative of Functions with Fractional Exponents

1. Find the derivative of the following exponential expressions.

    a. $y = x^{\frac{1}{5}}$ b. $y = \left(4x^3\right)^{\frac{1}{2}}$ c. $y = (2x+1)^{\frac{1}{3}}$

    d. $y = \left(2x^2+1\right)^{\frac{1}{8}}$ e. $y = \left(2x^3+3x\right)^{\frac{3}{5}}$ f. $y = \left(x^3+8\right)^{\frac{2}{3}}$

    g. $y = \left(x^3\right)^{\frac{1}{2}} - (3x-1)^{\frac{1}{3}}$ h. $y = x^2(x+1)^{\frac{1}{8}}$ i. $y = \left(x^3 \right.^{\text{\textbackslash}\,2}$

    j. $y = \dfrac{x+1}{x^{\frac{2}{3}}}$ k. $y = \dfrac{\left(x^2+1\right)^{\frac{1}{2}}}{x^2}$ l. $y =$

2. Use the $\dfrac{d}{dx}$ notation to find the derivative of the following exponential expressions.

a. $\dfrac{d}{dx}\left(x^{\frac{1}{5}}\right)^{2} =$

b. $\dfrac{d}{dx}(x-1)^{\frac{1}{2}} =$

c. $\dfrac{d}{dx}\left(x^{2}+1\right)^{\frac{1}{3}} =$

d. $\dfrac{d}{dx}\left(x^{3}+1\right)^{-\frac{1}{4}} =$

e. $\dfrac{d}{dx}\left[\dfrac{(x-1)^{\frac{1}{2}}}{x^{2}}\right] =$

f. $\dfrac{d}{dx}\left(x^{3}+2x\right)^{\frac{1}{8}} =$

g. $\dfrac{d}{dx}\left[\left(x^{3}+1\right)\left(x^{2}\right)^{\frac{1}{3}}\right] =$

h. $\dfrac{d}{dx}\left[x^{3}\cdot\dfrac{1}{\left(x^{2}+1\right)^{\frac{1}{2}}}\right] =$

i. $\dfrac{d}{dx}\left[\dfrac{x^{5}}{\left(x^{3}+1\right)^{\frac{2}{3}}}\right] =$

j. $\dfrac{d}{dx}\left[(x-1)^{\frac{1}{2}}(x+1)^{\frac{1}{3}}\right] =$

k. $\dfrac{d}{dx}\left[x^{3}\left(x^{2}+1\right)^{\frac{1}{2}}\right]$

l. $\dfrac{d}{dx}\left[x^{3}\left(x^{2}+1\right)^{-\frac{1}{3}}\right] =$

2.7 The Derivative of Radical Functions

In this section finding the derivative of radical expressions and the steps as to how it is calculated is discussed. The derivative of radical functions is found by using the following steps:

First - Write the radical expression in its equivalent fraction form, i.e., write $\sqrt{x^3}$ as $x^{\frac{3}{2}}$.

Second - Apply the differentiation rules to find the derivative of the exponential expression.

Third – Change the answer from an expression with fractional exponent to an expression with radical expression (optional).

The following examples show the steps in solving functions containing radical terms. Students who have difficulty with simplifying radical expressions may want to review *radicals* addressed in Chapter 5 of the *"Mastering Algebra - An Introduction"*.

Example 2.7-1: Find the derivative for the following Radical expressions.

a. $f(x) = \sqrt{x^3 + 1}$

b. $f(x) = \sqrt{x^2 + 3x + 1}$

c. $f(x) = \sqrt{2x^5 + 1}$

d. $f(u) = \sqrt[5]{u^3} + 3u$

e. $f(t) = t^2 + \sqrt{t+1}$

f. $g(x) = x^2\sqrt{x^3 + x - 5}$

g. $h(w) = \sqrt[3]{w^2 + 1}$

h. $f(z) = \sqrt[4]{z^3 - z^2} + z$

i. $f(x) = \dfrac{1}{\sqrt{x^2 + 1}}$

j. $f(x) = \dfrac{x}{\sqrt{x^2 - 1}}$

k. $r(\theta) = \dfrac{\theta^3 + 1}{\sqrt{\theta^2 + 1}}$

l. $p(r) = \dfrac{r^3}{\sqrt{r^3 - 1}}$

m. $g(u) = \dfrac{\sqrt{u-1}}{\sqrt{u+1}}$

n. $h(t) = \dfrac{\sqrt[3]{t^2 + 1}}{\sqrt{t^3}}$

o. $s(r) = \dfrac{r^2 - 1}{\sqrt{r-1}}$

Solutions:

a. Given $f(x) = \sqrt{x^3 + 1} = \left(x^3 + 1\right)^{\frac{1}{2}}$, then

$$\boxed{f'(x)} = \boxed{\frac{1}{2}\left(x^3+1\right)^{\frac{1}{2}-1}\cdot 3x^{3-1}} = \boxed{\frac{1}{2}\left(x^3+1\right)^{-\frac{1}{2}}\cdot 3x^2} = \boxed{\frac{3}{2}x^2\left(x^3+1\right)^{-\frac{1}{2}}} = \boxed{\frac{3}{2}\frac{x^2}{\left(x^3+1\right)^{\frac{1}{2}}}} = \boxed{\frac{3}{2}\frac{x^2}{\sqrt{x^3+1}}}$$

b. Given $f(x) = \sqrt{x^2 + 3x + 1} = \left(x^2 + 3x + 1\right)^{\frac{1}{2}}$, then

$$\boxed{f'(x)} = \boxed{\frac{1}{2}\left(x^2+3x+1\right)^{\frac{1}{2}-1}\cdot\left(2x^{2-1}+3x^{1-1}\right)} = \boxed{\frac{1}{2}\left(x^2+3x+1\right)^{-\frac{1}{2}}\cdot(2x+3)} = \boxed{\frac{2x+3}{2\left(x^2+3x+1\right)^{\frac{1}{2}}}} = \boxed{\frac{2x+3}{2\sqrt{x^2+3x+1}}}$$

c. Given $f(x) = \sqrt{2x^5 + 1} = \left(2x^5 + 1\right)^{\frac{1}{2}}$, then

$$\boxed{f'(x)} = \boxed{\frac{1}{2}\left(2x^5+1\right)^{\frac{1}{2}-1}\cdot\left(2\times5x^{5-1}+0\right)} = \boxed{\frac{1}{2}\left(2x^5+1\right)^{-\frac{1}{2}}\cdot 10x^4} = \boxed{\frac{\overset{5}{\cancel{10}}x^4}{2\left(2x^5+1\right)^{\frac{1}{2}}}} = \boxed{\frac{5x^4}{\sqrt{2x^5+1}}}$$

d. Given $f(u) = \sqrt[5]{u^3} + 3u = u^{\frac{3}{5}} + 3u$, then

$$\boxed{f'(u)} = \boxed{\frac{3}{5}u^{\frac{3}{5}-1}+3u^{1-1}} = \boxed{\frac{3}{5}u^{-\frac{2}{5}}+3u^0} = \boxed{\frac{3}{5}u^{-\frac{2}{5}}+3} = \boxed{\frac{3}{5u^{\frac{2}{5}}}+3} = \boxed{\frac{3}{5\sqrt[5]{u^2}}+3}$$

e. Given $f(t) = t^2 + \sqrt{t+1} = t^2 + (t+1)^{\frac{1}{2}}$, then

$$\boxed{f'(t)} = \boxed{2t^{2-1} + \frac{1}{2}(t+1)^{\frac{1}{2}-1} \cdot t^{1-1}} = \boxed{2t + \frac{1}{2}(t+1)^{-\frac{1}{2}} \cdot t^0} = \boxed{2t + \frac{1}{2(t+1)^{\frac{1}{2}}}} = \boxed{2t + \frac{1}{2\sqrt{t+1}}}$$

f. Given $g(x) = x^2\sqrt{x^3 + x - 5} = x^2\left(x^3 + x - 5\right)^{\frac{1}{2}}$, then

$$\boxed{g'(x)} = \boxed{\left[2x^{2-1}\left(x^3 + x - 5\right)^{\frac{1}{2}}\right] + \left[\frac{1}{2}\left(x^3 + x - 5\right)^{\frac{1}{2}-1} \cdot \left(3x^{3-1} + x^{1-1} + 0\right)\right]x^2} = \boxed{2x\left(x^3 + x - 5\right)^{\frac{1}{2}} + \frac{1}{2}\left(x^3 + x - 5\right)^{-\frac{1}{2}}}$$

$$\cdot \boxed{\left(3x^2 + 1\right)x^2} = \boxed{2x\left(x^3 + x - 5\right)^{\frac{1}{2}} + \frac{\left(3x^2 + 1\right)x^2}{2\left(x^3 + x - 5\right)^{\frac{1}{2}}}} = \boxed{2x\sqrt{x^3 + x - 5} + \frac{x^2\left(3x^2 + 1\right)}{2\sqrt{x^3 + x - 5}}}$$

g. Given $h(w) = \sqrt[3]{w^2 + 1} = \left(w^2 + 1\right)^{\frac{1}{3}}$, then

$$\boxed{h'(w)} = \boxed{\frac{1}{3}\left(w^2 + 1\right)^{\frac{1}{3}-1}\left(2w^{2-1} + 0\right)} = \boxed{\frac{1}{3}\left(w^2 + 1\right)^{-\frac{2}{3}} \cdot 2w} = \boxed{\frac{2w}{3\left(w^2 + 1\right)^{\frac{2}{3}}}} = \boxed{\frac{2w}{3\sqrt[3]{\left(w^2 + 1\right)^2}}}$$

h. Given $f(z) = \sqrt[4]{z^3 - z^2} + z = \left(z^3 - z^2\right)^{\frac{1}{4}} + z$, then

$$\boxed{f'(z)} = \boxed{\frac{1}{4}\left(z^3 - z^2\right)^{\frac{1}{4}-1}\left(3z^{3-1} - 2z^{2-1}\right) + 1} = \boxed{\frac{\left(z^3 - z^2\right)^{-\frac{3}{4}}\left(3z^2 - 2z\right)}{4} + 1} = \boxed{\frac{3\left(3z^2 - 2z\right)}{4\left(z^3 - z^2\right)^{\frac{3}{4}}} + 1} = \boxed{\frac{3z^2 - 2z}{4\sqrt[4]{\left(z^3 - z^2\right)^3}} + 1}$$

i. Given $f(x) = \dfrac{1}{\sqrt{x^2 + 1}} = \dfrac{1}{\left(x^2 + 1\right)^{\frac{1}{2}}} = \left(x^2 + 1\right)^{-\frac{1}{2}}$, then

$$\boxed{f'(x)} = \boxed{-\frac{1}{2}\left(x^2 + 1\right)^{-\frac{1}{2}-1}\left(2x^{2-1} + 0\right)} = \boxed{-\frac{1}{2}\left(x^2 + 1\right)^{-\frac{3}{2}} \cdot 2x} = \boxed{\frac{x}{\left(x^2 - 1\right)^{\frac{1}{2}}}} = \boxed{\frac{-x}{\sqrt[2]{\left(x^2 + 1\right)^3}}} = \boxed{\frac{-x}{\left(x^2 + 1\right)\sqrt{x^2 + 1}}}$$

j. Given $f(x) = \dfrac{x}{\sqrt{x^2 - 1}} = \dfrac{x}{\left(x^2 - 1\right)^{\frac{1}{2}}}$, then

$$\boxed{f'(x)} = \boxed{\frac{\left[1 \cdot \left(x^2 - 1\right)^{\frac{1}{2}}\right] - \left[\frac{1}{2}\left(x^2 - 1\right)^{\frac{1}{2}-1}\left(2x^{2-1} - 0\right) \cdot x\right]}{x^2 - 1}} = \boxed{\frac{\left(x^2 - 1\right)^{\frac{1}{2}} - \left[\frac{1}{2}\left(x^2 - 1\right)^{-\frac{1}{2}} \cdot 2x^2\right]}{x^2 - 1}} = \boxed{\frac{\left(x^2 - 1\right)^{\frac{1}{2}} - \frac{x^2}{\left(x^2 - 1\right)^{\frac{1}{2}}}}{x^2 - 1}}$$

$$= \boxed{\frac{\frac{\left(x^2 - 1\right)^{\frac{1}{2}}}{1} - \frac{x^2}{\left(x^2 - 1\right)^{\frac{1}{2}}}}{x^2 - 1}} = \boxed{\frac{\frac{\left(x^2 - 1\right)^{\frac{1}{2}} \cdot \left(x^2 - 1\right)^{\frac{1}{2}} - x^2}{\left(x^2 - 1\right)^{\frac{1}{2}}}}{x^2 - 1}} = \boxed{\frac{\frac{\left(x^2 - 1\right)^{\frac{1}{2}+\frac{1}{2}} - x^2}{\left(x^2 - 1\right)^{\frac{1}{2}}}}{x^2 - 1}} = \boxed{\frac{\frac{x^2 - 1 - x^2}{\left(x^2 - 1\right)^{\frac{1}{2}}}}{x^2 - 1}}$$

$$= \boxed{-\frac{1}{\left(x^2 - 1\right)^{\frac{1}{2}}\left(x^2 - 1\right)}} = \boxed{-\frac{1}{\left(x^2 - 1\right)^{\frac{1}{2}+1}}} = \boxed{-\frac{1}{\left(x^2 - 1\right)^{\frac{3}{2}}}} = \boxed{-\frac{1}{\sqrt{\left(x^2 - 1\right)^3}}} = \boxed{-\frac{1}{\left(x^2 - 1\right)\sqrt{x^2 - 1}}}$$

k. Given $r(\theta) = \dfrac{\theta^3+1}{\sqrt{\theta^2+1}} = \dfrac{\theta^3+1}{\left(\theta^2+1\right)^{\frac{1}{2}}}$, then

$$\boxed{r'(\theta)} = \frac{\left[\left(3\theta^{3-1}+0\right)\left(\theta^2+1\right)^{\frac{1}{2}}\right]-\left[\frac{1}{2}\left(\theta^2+1\right)^{\frac{1}{2}-1}\left(2\theta^{2-1}+0\right)\left(\theta^3+1\right)\right]}{\theta^2+1} = \frac{\left[3\theta^2\left(\theta^2+1\right)^{\frac{1}{2}}\right]-\frac{1}{2}\left(\theta^2+1\right)^{-\frac{1}{2}}}{\theta^2+1}$$

$$\cdot \frac{2\theta\left(\theta^3+1\right)}{\theta^2+1} = \frac{3\theta^2\left(\theta^2+1\right)^{\frac{1}{2}}-\frac{2\theta\left(\theta^3+1\right)}{2\left(\theta^2+1\right)^{\frac{1}{2}}}}{\theta^2+1} = \frac{\frac{6\theta^2\left(\theta^2+1\right)-2\theta^4-2\theta}{2\left(\theta^2+1\right)^{\frac{1}{2}}}}{\theta^2+1} = \frac{6\theta^4+6\theta^2-2\theta^4-2\theta}{2\left(\theta^2+1\right)^{\frac{1}{2}}\left(\theta^2+1\right)}$$

$$= \frac{4\theta^4+6\theta^2-2\theta}{2\left(\theta^2+1\right)^{\frac{1}{2}+1}} = \frac{2\theta\left(2\theta^3+3\theta-1\right)}{2\left(\theta^2+1\right)^{\frac{3}{2}}} = \frac{\theta\left(2\theta^3+3\theta-1\right)}{\sqrt{\left(\theta^2+1\right)^3}} = \boxed{\frac{\theta\left(2\theta^3+3\theta-1\right)}{\left(\theta^2+1\right)\sqrt{\theta^2+1}}}$$

l. Given $p(r) = \dfrac{r^3}{\sqrt{r^3-1}} = \dfrac{r^3}{\left(r^3-1\right)^{\frac{1}{2}}}$, then

$$\boxed{p'(r)} = \frac{\left[3r^{3-1}\left(r^3-1\right)^{\frac{1}{2}}\right]-\left[\frac{1}{2}\left(r^3-1\right)^{\frac{1}{2}-1}\left(3r^{3-1}-0\right)\right]}{r^3-1} = \frac{\left[3r^2\left(r^3-1\right)^{\frac{1}{2}}\right]-\left[\frac{1}{2}\left(r^3-1\right)^{-\frac{1}{2}}3r^2\right]\cdot r^3}{r^3-1}$$

$$= \frac{3r^2\left(r^3-1\right)^{\frac{1}{2}}-\frac{3r^5}{2\left(r^3-1\right)^{\frac{1}{2}}}}{r^3-1} = \frac{3r^2\left(r^3-1\right)^{\frac{1}{2}}-\frac{3r^5}{2\left(r^3-1\right)^{\frac{1}{2}}}}{r^3-1} = \frac{\frac{6r^2\left(r^3-1\right)-3r^5}{2\left(r^3-1\right)^{\frac{1}{2}}}}{r^3-1} = \frac{\frac{6r^5-6r^2-3r^5}{2\left(r^3-1\right)^{\frac{1}{2}}}}{r^3-1}$$

$$= \frac{3r^5-6r^2}{2\left(r^3-1\right)\left(r^3-1\right)^{\frac{1}{2}}} = \frac{3r^2\left(r^3-2\right)}{2\left(r^3-1\right)^{\frac{1}{2}+1}} = \frac{3r^2\left(r^3-2\right)}{2\left(r^3-1\right)^{\frac{3}{2}}} = \frac{3r^2\left(r^3-2\right)}{2\sqrt{\left(r^3-1\right)^3}} = \boxed{\frac{3r^2\left(r^3-2\right)}{2\left(r^3-1\right)\sqrt{r^3-1}}}$$

m. Given $g(u) = \dfrac{\sqrt{u-1}}{\sqrt{u+1}} = \dfrac{\left(u-1\right)^{\frac{1}{2}}}{\left(u+1\right)^{\frac{1}{2}}}$, then

$$\boxed{g'(u)} = \frac{\left[\frac{1}{2}\left(u-1\right)^{\frac{1}{2}-1}\left(u+1\right)^{\frac{1}{2}}\right]-\left[\frac{1}{2}\left(u+1\right)^{\frac{1}{2}-1}\left(u-1\right)^{\frac{1}{2}}\right]}{u+1} = \frac{\left[\frac{1}{2}\left(u-1\right)^{-\frac{1}{2}}\left(u+1\right)^{\frac{1}{2}}\right]-\left[\frac{1}{2}\left(u+1\right)^{-\frac{1}{2}}\left(u-1\right)^{\frac{1}{2}}\right]}{u+1}$$

$$= \frac{\left[\frac{\left(u+1\right)^{\frac{1}{2}}}{2\left(u-1\right)^{\frac{1}{2}}}\right]-\left[\frac{\left(u-1\right)^{\frac{1}{2}}}{2\left(u+1\right)^{\frac{1}{2}}}\right]}{u+1} = \frac{\frac{2\left(u+1\right)-2\left(u-1\right)}{4\left(u-1\right)^{\frac{1}{2}}\left(u+1\right)^{\frac{1}{2}}}}{u+1} = \frac{\frac{2u+2-2u+2}{4\left(u-1\right)^{\frac{1}{2}}\left(u+1\right)^{\frac{1}{2}}}}{u+1} = \frac{\frac{4}{4\left(u-1\right)^{\frac{1}{2}}\left(u+1\right)^{\frac{1}{2}}}}{u+1}$$

$$= \frac{4}{4\left(u+1\right)\left(u-1\right)^{\frac{1}{2}}\left(u+1\right)^{\frac{1}{2}}} = \frac{1}{\left(u+1\right)\left(u+1\right)^{\frac{1}{2}}\left(u-1\right)^{\frac{1}{2}}} = \boxed{\frac{1}{\left(u+1\right)\sqrt{u+1}\,\sqrt{u-1}}}$$

n. Given $h(t) = \dfrac{\sqrt[3]{t^2+1}}{\sqrt{t^3}} = \dfrac{\left(t^2+1\right)^{\frac{1}{3}}}{t^{\frac{3}{2}}}$, then

$\boxed{h'(t)} = \dfrac{\left[\frac{1}{3}\left(t^2+1\right)^{\frac{1}{3}-1}\cdot\left(2t^{2-1}+0\right)\right]t^{\frac{3}{2}}-\left[\frac{3}{2}t^{\frac{3}{2}-1}\cdot\left(t^2+1\right)^{\frac{1}{3}}\right]}{t^3} = \dfrac{\left[\frac{1}{3}\left(t^2+1\right)^{-\frac{2}{3}}\cdot 2t\right]t^{\frac{3}{2}}-\left[\frac{3}{2}t^{\frac{1}{2}}\left(t^2+1\right)^{\frac{1}{3}}\right]}{t^3}$

$= \dfrac{\left[\frac{2}{3}t^{1+\frac{3}{2}}\left(t^2+1\right)^{-\frac{2}{3}}\right]-\left[\frac{3}{2}t^{\frac{1}{2}}\left(t^2+1\right)^{\frac{1}{3}}\right]}{t^3} = \dfrac{\dfrac{2t^{\frac{5}{2}}}{3\left(t^2+1\right)^{\frac{2}{3}}}-\left[\frac{3}{2}t^{\frac{1}{2}}\left(t^2+1\right)^{\frac{1}{3}}\right]}{t^3} = \dfrac{\dfrac{2t^{\frac{5}{2}}}{3\left(t^2+1\right)^{\frac{2}{3}}}-\dfrac{\frac{3}{2}t^{\frac{1}{2}}\left(t^2+1\right)^{\frac{1}{3}}}{1}}{t^3}$

$= \dfrac{\dfrac{2t^{\frac{5}{2}}-\frac{3}{2}t^{\frac{1}{2}}\left(t^2+1\right)^{\frac{1}{3}}\cdot 3\left(t^2+1\right)^{\frac{2}{3}}}{3\left(t^2+1\right)^{\frac{2}{3}}}}{t^3} = \dfrac{2t^{\frac{5}{2}}-\frac{9}{2}t^{\frac{1}{2}}\left(t^2+1\right)^{\frac{1}{3}+\frac{2}{3}}}{3t^3\left(t^2+1\right)^{\frac{2}{3}}} = \boxed{\dfrac{2t^2\sqrt{t}-\dfrac{9\sqrt{t}\left(t^2+1\right)}{2}}{3t^3\sqrt[3]{\left(t^2+1\right)^2}}}$

o. Given $s(r) = \dfrac{r^2-1}{\sqrt{r-1}} = \dfrac{r^2-1}{(r-1)^{\frac{1}{2}}}$, then

$\boxed{s'(r)} = \dfrac{\left[\left(2r^{2-1}-0\right)(r-1)^{\frac{1}{2}}\right]-\left[\frac{1}{2}(r-1)^{\frac{1}{2}-1}\left(r^2-1\right)\right]}{r-1} = \dfrac{\left[2r(r-1)^{\frac{1}{2}}\right]-\left[\frac{1}{2}(r-1)^{-\frac{1}{2}}\left(r^2-1\right)\right]}{r-1}$

$= \dfrac{\left[2r(r-1)^{\frac{1}{2}}\right]-\left[\dfrac{\left(r^2-1\right)}{2(r-1)^{\frac{1}{2}}}\right]}{r-1} = \dfrac{\left[\dfrac{2r(r-1)^{\frac{1}{2}}}{1}\right]-\left[\dfrac{\left(r^2-1\right)}{2(r-1)^{\frac{1}{2}}}\right]}{r-1} = \dfrac{\dfrac{4r(r-1)-\left(r^2-1\right)}{2(r-1)^{\frac{1}{2}}}}{r-1} = \dfrac{\dfrac{4r^2-4r-r^2+1}{2(r-1)^{\frac{1}{2}}}}{r-1}$

$= \dfrac{3r^2-4r+1}{2(r-1)(r-1)^{\frac{1}{2}}} = \boxed{\dfrac{3r^2-4r+1}{2(r-1)\sqrt{r-1}}}$

Example 2.7-2: Use the chain rule to differentiate the following radical expressions.

a. $\dfrac{d}{dx}\left(\sqrt{x}-\dfrac{1}{\sqrt{x}}\right)$

b. $\dfrac{d}{dx}\left(\sqrt{\dfrac{x+2}{3x+1}}\right)$

c. $\dfrac{d}{dx}\left(\dfrac{x^2}{\sqrt{x^2+1}}\right)$

d. $\dfrac{d}{dx}\left(\dfrac{\sqrt{x^3+1}}{x}\right)$

e. $\dfrac{d}{dx}\left(\sqrt[3]{x^2}+x^{-2}\right)$

f. $\dfrac{d}{dx}\sqrt[5]{x^3+1}$

Solutions:

a. $\boxed{\dfrac{d}{dx}\left(\sqrt{x}-\dfrac{1}{\sqrt{x}}\right)} = \dfrac{d}{dx}\left(x^{\frac{1}{2}}-\dfrac{1}{x^{\frac{1}{2}}}\right) = \dfrac{d}{dx}\left(x^{\frac{1}{2}}-x^{-\frac{1}{2}}\right) = \dfrac{d}{dx}x^{\frac{1}{2}}-\dfrac{d}{dx}x^{-\frac{1}{2}} = \dfrac{1}{2}x^{\frac{1}{2}-1}+\dfrac{1}{2}x^{-\frac{1}{2}-1}$

$$= \boxed{\frac{1}{2}x^{-\frac{1}{2}} + \frac{1}{2}x^{-\frac{3}{2}}} = \boxed{\frac{1}{2x^{\frac{1}{2}}} + \frac{1}{2x^{\frac{3}{2}}}} = \boxed{\frac{1}{2\sqrt{x}} + \frac{1}{2\sqrt{x^3}}} = \boxed{\mathbf{\frac{1}{2\sqrt{x}} + \frac{1}{2x\sqrt{x}}}}$$

b. $\boxed{\dfrac{d}{dx}\left(\sqrt{\dfrac{x+2}{3x+1}}\right)} = \boxed{\dfrac{d}{dx}\left(\dfrac{x+2}{3x+1}\right)^{\frac{1}{2}}} = \boxed{\dfrac{1}{2}\left(\dfrac{x+2}{3x+1}\right)^{\frac{1}{2}-1} \cdot \dfrac{d}{dx}\left(\dfrac{x+2}{3x+1}\right)} = \boxed{\dfrac{1}{2}\left(\dfrac{x+2}{3x+1}\right)^{-\frac{1}{2}}\left(\dfrac{[1\cdot(3x+1)]-[3\cdot(x+2)]}{(3x+1)^2}\right)}$

$$= \boxed{\dfrac{1}{2\left(\dfrac{x+2}{3x+1}\right)^{\frac{1}{2}}}\left(\dfrac{3x+1-3x-6}{(3x+1)^2}\right)} = \boxed{\dfrac{1}{2\left(\dfrac{x+2}{3x+1}\right)^{\frac{1}{2}}}\left(\dfrac{-5}{(3x+1)^2}\right)} = \boxed{-\dfrac{5}{2\left(\dfrac{x+2}{3x+1}\right)^{\frac{1}{2}}(3x+1)^2}} = \boxed{-\dfrac{5(3x+1)^{\frac{1}{2}}}{2(x+2)^{\frac{1}{2}}(3x+1)^2}}$$

$$= \boxed{-\dfrac{5}{2(x+2)^{\frac{1}{2}}(3x+1)^2(3x+1)^{-\frac{1}{2}}}} = \boxed{-\dfrac{5}{2(x+2)^{\frac{1}{2}}(3x+1)^{2-\frac{1}{2}}}} = \boxed{-\dfrac{5}{2(x+2)^{\frac{1}{2}}(3x+1)^{\frac{4-1}{2}}}} = \boxed{-\dfrac{5}{2(x+2)^{\frac{1}{2}}(3x+1)^{\frac{3}{2}}}}$$

$$= \boxed{-\dfrac{5}{2\left(\sqrt{x+2}\right)\sqrt{(3x+1)^3}}} = \boxed{\mathbf{-\dfrac{5}{2\left(\sqrt{x+2}\right)(3x+1)\sqrt{3x+1}}}}$$

c. $\boxed{\dfrac{d}{dx}\left(\dfrac{x^2}{\sqrt{x^2+1}}\right)} = \boxed{\dfrac{d}{dx}\dfrac{x^2}{\left(x^2+1\right)^{\frac{1}{2}}}} = \boxed{\dfrac{\left(x^2+1\right)^{\frac{1}{2}}\dfrac{d}{dx}x^2 - x^2\dfrac{d}{dx}\left(x^2+1\right)^{\frac{1}{2}}}{x^2+1}} = \boxed{\dfrac{2x\left(x^2+1\right)^{\frac{1}{2}} - \left[\dfrac{1}{2}\left(x^2+1\right)^{\frac{1}{2}-1}\left(2x^{2-1}+0\right)\right]x^2}{x^2+1}}$

$$= \boxed{\dfrac{2x\left(x^2+1\right)^{\frac{1}{2}} - \left[\dfrac{1}{2}\left(x^2+1\right)^{-\frac{1}{2}}\cdot 2x\right]\cdot x^2}{x^2+1}} = \boxed{\dfrac{2x\left(x^2+1\right)^{\frac{1}{2}} - x^3\left(x^2+1\right)^{-\frac{1}{2}}}{x^2+1}} = \boxed{\dfrac{2x\left(x^2+1\right)^{\frac{1}{2}} - \dfrac{x^3}{\left(x^2+1\right)^{\frac{1}{2}}}}{x^2+1}}$$

$$= \boxed{\dfrac{\dfrac{2x\left(x^2+1\right)^{\frac{1}{2}}\left(x^2+1\right)^{\frac{1}{2}} - x^3}{\left(x^2+1\right)^{\frac{1}{2}}}}{x^2+1}} = \boxed{\dfrac{\dfrac{2x\left(x^2+1\right) - x^3}{\left(x^2+1\right)^{\frac{1}{2}}}}{x^2+1}} = \boxed{\dfrac{2x^3+2x-x^3}{\left(x^2+1\right)^{\frac{1}{2}}\left(x^2+1\right)}} = \boxed{\dfrac{x^3+2x}{\left(x^2+1\right)^{\frac{1}{2}+1}}} = \boxed{\dfrac{x^3+2x}{\left(x^2+1\right)^{\frac{3}{2}}}}$$

$$= \boxed{\dfrac{x^3+2x}{\sqrt{\left(x^2+1\right)^3}}} = \boxed{\mathbf{\dfrac{x\left(x^2+2\right)}{\left(x^2+1\right)\sqrt{x^2+1}}}}$$

d. $\boxed{\dfrac{d}{dt}\left(\dfrac{\sqrt{x^3}}{x+1}\right)} = \boxed{\dfrac{d}{dt}\left(\dfrac{x^{\frac{3}{2}}}{x+1}\right)} = \boxed{\dfrac{\left[(x+1)\cdot\dfrac{d}{dx}x^{\frac{3}{2}}\right] - \left[x^{\frac{3}{2}}\dfrac{d}{dx}(x+1)\right]}{(x+1)^2}} = \boxed{\dfrac{\left[(x+1)\cdot\dfrac{3}{2}x^{\frac{3}{2}-1}\right] - \left[x^{\frac{3}{2}}\cdot 1\right]}{(x+1)^2}}$

$$= \boxed{\dfrac{\dfrac{3}{2}x^{\frac{1}{2}}(x+1) - x^{\frac{3}{2}}}{(x+1)^2}} = \boxed{\dfrac{\dfrac{3}{2}x^{\frac{1}{2}+1} + \dfrac{3}{2}x^{\frac{1}{2}} - x^{\frac{3}{2}}}{(x+1)^2}} = \boxed{\dfrac{\dfrac{3}{2}x^{\frac{3}{2}} + \dfrac{3}{2}x^{\frac{1}{2}} - x^{\frac{3}{2}}}{(x+1)^2}} = \boxed{\dfrac{\left(\dfrac{3}{2}-1\right)x^{\frac{3}{2}} + \dfrac{3}{2}x^{\frac{1}{2}}}{(x+1)^2}} = \boxed{\dfrac{\dfrac{1}{2}x^{\frac{3}{2}} + \dfrac{3}{2}x^{\frac{1}{2}}}{(x+1)^2}}$$

$$= \boxed{\dfrac{\frac{x^{\frac{3}{2}}+3x^{\frac{1}{2}}}{2}}{(x+1)^2}} = \boxed{\dfrac{x^{\frac{3}{2}}+3x^{\frac{1}{2}}}{2(x+1)^2}} = \boxed{\dfrac{\sqrt{x^3}+3\sqrt{x}}{2(x+1)^2}} = \boxed{\dfrac{x\sqrt{x}+3\sqrt{x}}{2(x+1)^2}} = \boxed{\dfrac{\sqrt{x}\,(x+3)}{2(x+1)^2}}$$

e. $\boxed{\dfrac{d}{dx}\left(\sqrt[3]{x^2}+x^{-2}\right)} = \boxed{\dfrac{d}{dx}\left(x^{\frac{2}{3}}+x^{-2}\right)} = \boxed{\dfrac{d}{dx}x^{\frac{2}{3}}+\dfrac{d}{dx}x^{-2}} = \boxed{\dfrac{2}{3}x^{\frac{2}{3}-1}-2x^{-2-1}} = \boxed{\dfrac{2}{3}x^{\frac{2-3}{3}}-2x^{-3}} = \boxed{\dfrac{2}{3}x^{-\frac{1}{3}}-2x^{-3}}$

$$= \boxed{\dfrac{2}{3x^{\frac{1}{3}}}-\dfrac{2}{x^3}} = \boxed{\dfrac{2}{3\sqrt[3]{x}}-\dfrac{2}{x^3}}$$

f. $\boxed{\dfrac{d}{dx}\sqrt[5]{x^3+1}} = \boxed{\dfrac{d}{dx}\left(x^3+1\right)^{\frac{1}{5}}} = \boxed{\dfrac{1}{5}\left(x^3+1\right)^{\frac{1}{5}-1}\cdot\dfrac{d}{dx}\left(x^3+1\right)} = \boxed{\dfrac{1}{5}\left(x^3+1\right)^{-\frac{4}{5}}\left(3x^{3-1}+0\right)} = \boxed{\dfrac{1}{5}\left(x^3+1\right)^{-\frac{4}{5}}\cdot 3x^2}$

$$= \boxed{\dfrac{3x^2\left(x^3+1\right)^{-\frac{4}{5}}}{5}} = \boxed{\dfrac{3x^2}{5\left(x^3+1\right)^{\frac{4}{5}}}} = \boxed{\dfrac{3x^2}{5\sqrt[5]{\left(x^3+1\right)^4}}}$$

Example 2.7-3: Use the chain rule to differentiate the following radical expressions.

a. $y = x^3\sqrt{x^2-1}$

b. $y = 3x^2+\sqrt{x+1}$

c. $y = \sqrt{x^3+x^2+1}$

d. $y = \sqrt[5]{x^5+x^2-1}$

e. $y = \sqrt{x^2-1}\cdot\sqrt{x+1}$

f. $y = \sqrt{x}(x+1)^3$

g. $y = \sqrt{\left(x^2+5x-1\right)^5}$

h. $y = x^2\sqrt[3]{x+1}$

i. $y = (x-1)^3\sqrt[3]{x^5}$

Solutions:

a. Given $y = x^3\sqrt{x^2-1} = x^3\left(x^2-1\right)^{\frac{1}{2}}$, then

$$\boxed{y'} = \boxed{\left[3x^2\cdot\left(x^2-1\right)^{\frac{1}{2}}\right]+\left[\dfrac{1}{2}\left(x^2-1\right)^{\frac{1}{2}-1}\cdot 2x\cdot x^3\right]} = \boxed{\left[3x^2\left(x^2-1\right)^{\frac{1}{2}}\right]+\left[\dfrac{2x^4}{2}\left(x^2-1\right)^{\frac{1}{2}-\frac{1}{1}}\right]} = \boxed{\left[3x^2\left(x^2-1\right)^{\frac{1}{2}}\right]}$$

$$+\boxed{\left[x^4\left(x^2-1\right)^{\frac{1-2}{2}}\right]} = \boxed{3x^2\left(x^2-1\right)^{\frac{1}{2}}+x^4\left(x^2-1\right)^{-\frac{1}{2}}} = \boxed{3x^2\left(x^2-1\right)^{\frac{1}{2}}+\dfrac{x^4}{\left(x^2-1\right)^{\frac{1}{2}}}} = \boxed{3x^2\sqrt{x^2-1}+\dfrac{x^4}{\sqrt{x^2-1}}}$$

b. Given $y = 3x^2+\sqrt{x+1} = 3x^2+(x+1)^{\frac{1}{2}}$, then

$$\boxed{y'} = \boxed{(3\cdot 2)x^{2-1}+\dfrac{1}{2}(x+1)^{\frac{1}{2}-1}} = \boxed{6x+\dfrac{1}{2}(x+1)^{\frac{1-2}{2}}} = \boxed{6x+\dfrac{1}{2}(x+1)^{-\frac{1}{2}}} = \boxed{6x+\dfrac{1}{2(x+1)^{\frac{1}{2}}}} = \boxed{6x+\dfrac{1}{2\sqrt{x+1}}}$$

c. Given $y = \sqrt{x^3+x^2+1} = \left(x^3+x^2+1\right)^{\frac{1}{2}}$, then

$$\boxed{y'} = \boxed{\dfrac{1}{2}\left(x^3+x^2+1\right)^{\frac{1}{2}-1}\cdot\left(3x^2+2x\right)} = \boxed{\left(\dfrac{3x^2+2x}{2}\right)\left(x^3+x^2+1\right)^{\frac{1}{2}-\frac{1}{1}}} = \boxed{\left(\dfrac{3x^2+2x}{2}\right)\left(x^3+x^2+1\right)^{\frac{1-2}{2}}}$$

$$= \left[\left(\frac{3x^2+2x}{2}\right)\left(x^3+x^2+1\right)^{-\frac{1}{2}}\right] = \left[\frac{3x^2+2x}{2}\cdot\frac{1}{\left(x^3+x^2+1\right)^{\frac{1}{2}}}\right] = \boxed{\frac{3x^2+2x}{2\sqrt{x^3+x^2+1}}}$$

d. Given $y = \sqrt[5]{x^5+x^2-1} = \left(x^5+x^2-1\right)^{\frac{1}{5}}$, then

$$\boxed{y'} = \left[\frac{1}{5}\left(x^5+x^2-1\right)^{\frac{1}{5}-1}\cdot\left(5x^4+2x\right)\right] = \left[\left(\frac{5x^4+2x}{5}\right)\left(x^5+x^2-1\right)^{\frac{1}{5}-\frac{1}{1}}\right] = \left[\left(\frac{5x^4+2x}{5}\right)\left(x^5+x^2-1\right)^{\frac{1-5}{5}}\right]$$

$$= \left[\left(\frac{5x^4+2x}{5}\right)\left(x^5+x^2-1\right)^{-\frac{4}{5}}\right] = \left[\frac{5x^4+2x}{5}\cdot\frac{1}{\left(x^5+x^2-1\right)^{\frac{4}{5}}}\right] = \boxed{\frac{5x^4+2x}{5\sqrt[5]{\left(x^5+x^2-1\right)^4}}}$$

e. Given $y = \sqrt{x^2-1}\cdot\sqrt{x+1} = \left(x^2-1\right)^{\frac{1}{2}}\cdot(x+1)^{\frac{1}{2}}$, then

$$\boxed{y'} = \left[\left[\frac{1}{2}\left(x^2-1\right)^{\frac{1}{2}-1}\cdot2x\cdot(x+1)^{\frac{1}{2}}\right]+\left[(x+1)^{\frac{1}{2}-1}\cdot\left(x^2-1\right)^{\frac{1}{2}}\right]\right] = \left[\left[\frac{2x}{2}\left(x^2-1\right)^{\frac{1}{2}-\frac{1}{1}}\cdot(x+1)^{\frac{1}{2}}\right]+\left[\frac{1}{2}(x+1)^{\frac{1}{2}-\frac{1}{1}}\cdot\left(x^2-1\right)^{\frac{1}{2}}\right]\right]$$

$$= \left[\left[x\left(x^2-1\right)^{-\frac{1}{2}}\cdot(x+1)^{\frac{1}{2}}\right]+\left[\frac{1}{2}(x+1)^{-\frac{1}{2}}\cdot\left(x^2-1\right)^{\frac{1}{2}}\right]\right] = \left[\left[x\cdot\frac{1}{\left(x^2-1\right)^{\frac{1}{2}}}\cdot(x+1)^{\frac{1}{2}}\right]+\left[\frac{1}{2(x+1)^{\frac{1}{2}}}\cdot\left(x^2-1\right)^{\frac{1}{2}}\right]\right]$$

$$= \left[\left[x\cdot\frac{1}{\sqrt{x^2-1}}\cdot\sqrt{x+1}\right]+\left[\frac{1}{2\sqrt{x+1}}\cdot\sqrt{x^2-1}\right]\right] = \boxed{\frac{x\sqrt{x+1}}{\sqrt{x^2-1}}+\frac{\sqrt{x^2-1}}{2\sqrt{x+1}}}$$

f. Given $y = \sqrt{x}(x+1)^3 = x^{\frac{1}{2}}(x+1)^3$, then

$$\boxed{y'} = \left[\left[\frac{1}{2}x^{\frac{1}{2}-1}\cdot(x+1)^3\right]+\left[3(x+1)^2\cdot x^{\frac{1}{2}}\right]\right] = \left[\left[\frac{(x+1)^3}{2}\cdot x^{\frac{1}{2}-\frac{1}{1}}\right]+\left[3(x+1)^2 x^{\frac{1}{2}}\right]\right] = \left[\left[\frac{(x+1)^3}{2}x^{-\frac{1}{2}}\right]+\left[3(x+1)^2 x^{\frac{1}{2}}\right]\right]$$

$$= \left[\left[\frac{(x+1)^3}{2}\cdot\frac{1}{x^{\frac{1}{2}}}\right]+\left[3(x+1)^2 x^{\frac{1}{2}}\right]\right] = \boxed{\frac{(x+1)^3}{2\sqrt{x}}+3\sqrt{x}(x+1)^2}$$

g. Given $y = \sqrt{\left(x^2+5x-1\right)^5} = \left[\left(x^2+5x-1\right)^5\right]^{\frac{1}{2}} = \left(x^2+5x-1\right)^{\frac{5}{2}}$, then

$$\boxed{y'} = \left[\frac{5}{2}\left(x^2+5x-1\right)^{\frac{5}{2}-1}\cdot(2x+5)\right] = \left[\frac{5(2x+5)}{2}\cdot\left(x^2+5x-1\right)^{\frac{5}{2}-\frac{1}{1}}\right] = \left[\frac{5(2x+5)}{2}\cdot\left(x^2+5x-1\right)^{\frac{5-2}{2}}\right]$$

$$= \left[\frac{5(2x+5)}{2}\cdot\left(x^2+5x-1\right)^{\frac{3}{2}}\right] = \left[\frac{5(2x+5)}{2}\cdot\sqrt{\left(x^2+5x-1\right)^3}\right] = \boxed{\frac{5(2x+5)\left(x^2+5x-1\right)\sqrt{x^2+5x-1}}{2}}$$

h. Given $y = x^2\sqrt[3]{x+1} = x^2(x+1)^{\frac{1}{3}}$, then

$$y' = \left[2x\cdot(x+1)^{\frac{1}{3}}\right] + \left[\frac{1}{3}(x+1)^{\frac{1}{3}-1}\cdot x^2\right] = \left[2x(x+1)^{\frac{1}{3}}\right] + \left[\frac{x^2}{3}(x+1)^{\frac{1}{3}-\frac{1}{1}}\right] = 2x(x+1)^{\frac{1}{3}} + \frac{x^2}{3}(x+1)^{\frac{1-3}{3}}$$

$$= 2x(x+1)^{\frac{1}{3}} + \frac{x^2}{3}(x+1)^{-\frac{2}{3}} = 2x(x+1)^{\frac{1}{3}} + \frac{x^2}{3}\cdot\frac{1}{(x+1)^{\frac{2}{3}}} = 2x\sqrt[3]{x+1} + \frac{x^2}{3\sqrt[3]{(x+1)^2}}$$

i. Given $y = (x-1)^3\sqrt[3]{x^5} = (x-1)^3\left(x^5\right)^{\frac{1}{3}} = (x-1)^3 x^{\frac{5}{3}}$, then

$$y' = \left[3(x-1)^2\cdot x^{\frac{5}{3}}\right] + \left[\frac{5}{3}x^{\frac{5}{3}-1}\cdot(x-1)^3\right] = \left[3x^{\frac{5}{3}}(x-1)^2\right] + \left[\frac{5}{3}x^{\frac{5-3}{3}}(x-1)^3\right] = \left[3x^{\frac{5}{3}}(x-1)^2\right] + \left[\frac{5}{3}x^{\frac{2}{3}}(x-1)^3\right]$$

$$= 3x\sqrt[3]{x^2}(x-1)^2 + \frac{5}{3}\sqrt[3]{x^2}(x-1)^3$$

Example 2.7-4: Find $\dfrac{dy}{dx}$ by implicit differentiation.

a. $\sqrt{x} + \sqrt{y} = 10$

b. $\sqrt{x^2 + y^2} = x$

c. $\sqrt[3]{x^2} + \sqrt[5]{y^3} = 2$

d. $\sqrt{2x+1} = y^2$

e. $\sqrt[5]{x^2+1} = y^{\frac{2}{3}}$

f. $\sqrt[4]{x^2 y^2} = x$

g. $\sqrt[7]{xy^2} = 3x^2$

h. $\sqrt{x^3 + y^3} = 2$

i. $\sqrt{x^2-1} = xy$

Solutions:

a. $\dfrac{d}{dx}\left(\sqrt{x}+\sqrt{y}\right) = \dfrac{d}{dx}(10)$; $\dfrac{d}{dx}\left(x^{\frac{1}{2}}+y^{\frac{1}{2}}\right) = 0$; $\dfrac{1}{2}x^{\frac{1}{2}-1}\cdot\dfrac{d}{dx}(x) + \dfrac{1}{2}y^{\frac{1}{2}-1}\cdot\dfrac{d}{dx}(y) = 0$; $\dfrac{1}{2}x^{-\frac{1}{2}}\cdot 1 + \dfrac{1}{2}y^{-\frac{1}{2}}\cdot y' = 0$

; $\dfrac{1}{2}y^{-\frac{1}{2}}y' = -\dfrac{1}{2}x^{-\frac{1}{2}}$; $\dfrac{1}{2}\dfrac{y'}{y^{\frac{1}{2}}} = -\dfrac{1}{2}\cdot\dfrac{1}{x^{\frac{1}{2}}}$; $2y^{\frac{1}{2}}\cdot\dfrac{1}{2}\cdot\dfrac{y'}{y^{\frac{1}{2}}} = 2y^{\frac{1}{2}}\cdot-\dfrac{1}{2}\cdot\dfrac{1}{x^{\frac{1}{2}}}$; $y' = \dfrac{-y^{\frac{1}{2}}}{x^{\frac{1}{2}}}$; $y' = \dfrac{-\sqrt{y}}{\sqrt{x}}$; $y' = -\sqrt{\dfrac{y}{x}}$

b. $\dfrac{d}{dx}\left(\sqrt{x^2+y^2}\right) = \dfrac{d}{dx}(x)$; $\dfrac{d}{dx}\left(x^2+y^2\right)^{\frac{1}{2}} = 1$; $\dfrac{1}{2}\left(x^2+y^2\right)^{\frac{1}{2}-1}\left(\dfrac{d}{dx}x^2+\dfrac{d}{dx}y^2\right) = 1$; $\dfrac{1}{2}\left(x^2+y^2\right)^{-\frac{1}{2}}(2x+2yy') = 1$

; $\dfrac{2x+2yy'}{2\left(x^2+y^2\right)^{\frac{1}{2}}} = 1$; $2x+2yy' = 2\left(x^2+y^2\right)^{\frac{1}{2}}$; $y' = \dfrac{2\left(x^2+y^2\right)^{\frac{1}{2}}-2x}{2y}$; $y' = \dfrac{\sqrt{x^2+y^2}-x}{y}$

c. $\dfrac{d}{dx}\left(\sqrt[3]{x^2}+\sqrt[5]{y^3}\right) = \dfrac{d}{dx}(2)$; $\dfrac{d}{dx}\left[\left(x^2\right)^{\frac{1}{3}}+\left(y^3\right)^{\frac{1}{5}}\right] = 0$; $\dfrac{d}{dx}\left[x^{\frac{2}{3}}+y^{\frac{3}{5}}\right] = 0$; $\dfrac{2}{3}x^{\frac{2}{3}-1}\cdot\dfrac{d}{dx}(x)+\dfrac{3}{5}y^{\frac{3}{5}-1}\cdot\dfrac{d}{dx}(y) = 0$

; $\dfrac{2}{3}x^{-\frac{1}{3}}\cdot 1+\dfrac{3}{5}y^{-\frac{2}{5}}\cdot y' = 0$; $\dfrac{3}{5}y^{-\frac{2}{5}}y' = \dfrac{-2}{3}x^{-\frac{1}{3}}$; $\dfrac{3y'}{5y^{\frac{2}{5}}} = \dfrac{-2}{3x^{\frac{1}{3}}}$; $9x^{\frac{1}{3}}y' = -10y^{\frac{2}{5}}$; $y' = \dfrac{-10y^{\frac{2}{5}}}{9x^{\frac{1}{3}}}$; $y' = -\dfrac{10\sqrt[5]{y^2}}{9\sqrt[3]{x}}$

d. $\boxed{\dfrac{d}{dx}\left(\sqrt{2x+1}\right)=\dfrac{d}{dx}\left(y^2\right)}$; $\boxed{\dfrac{d}{dx}(2x+1)^{\frac{1}{2}}=2y\cdot\dfrac{d}{dx}(y)}$; $\boxed{6x+\dfrac{1}{2}(x+1)^{\frac{1-2}{2}}}$; $\boxed{\dfrac{1}{2}(2x+1)^{-\frac{1}{2}}\cdot 2x=2y\cdot y'}$

; $\boxed{2y\,y'=x(2x+1)^{-\frac{1}{2}}}$; $\boxed{y'=\dfrac{x(2x+1)^{-\frac{1}{2}}}{2y}}$; $\boxed{y'=\dfrac{x}{2y(2x+1)^{\frac{1}{2}}}}$; $\boxed{y'=\dfrac{x}{2y\sqrt{2x+1}}}$

e. $\boxed{\dfrac{d}{dx}\left(\sqrt[5]{x^2+1}\right)=\dfrac{d}{dx}\left(y^{\frac{2}{3}}\right)}$; $\boxed{\dfrac{d}{dx}\left(x^2+1\right)^{\frac{1}{5}}=\dfrac{2}{3}y^{\frac{2}{3}-1}\cdot\dfrac{d}{dx}(y)}$; $\boxed{\dfrac{1}{5}\left(x^2+1\right)^{\frac{1}{5}-1}\cdot\left[\dfrac{d}{dx}\left(x^2\right)+\dfrac{d}{dx}(1)\right]=\dfrac{2}{3}y^{\frac{2}{3}-1}\cdot\dfrac{d}{dx}(y)}$

; $\boxed{\dfrac{1}{5}\left(x^2+1\right)^{-\frac{4}{5}}\cdot 2x=\dfrac{2}{3}y^{-\frac{1}{3}}\cdot y'}$; $\boxed{\dfrac{2}{3}y^{-\frac{1}{3}}y'=\dfrac{2x}{5}\left(x^2+1\right)^{-\frac{4}{5}}}$; $\boxed{\dfrac{2y'}{3y^{\frac{1}{3}}}=\dfrac{2x}{5\left(x^2+1\right)^{\frac{4}{5}}}}$; $\boxed{10\left(x^2+1\right)^{\frac{4}{5}}y'=6x\,y^{\frac{1}{3}}}$

; $\boxed{y'=\dfrac{6x\,y^{\frac{1}{3}}}{10\left(x^2+1\right)^{\frac{4}{5}}}}$; $\boxed{y'=\dfrac{3x\sqrt[3]{y}}{5\sqrt[5]{\left(x^2+1\right)^4}}}$

f. $\boxed{\dfrac{d}{dx}\left(\sqrt[4]{x^2y^2}\right)=\dfrac{d}{dx}(x)}$; $\boxed{\dfrac{d}{dx}\left(x^2y^2\right)^{\frac{1}{4}}=1}$; $\boxed{\dfrac{1}{4}\left(x^2y^2\right)^{\frac{1}{4}-1}\left[\dfrac{d}{dx}\left(x^2y^2\right)\right]=1}$; $\boxed{\dfrac{1}{4}\left(x^2y^2-1\right)^{-\frac{3}{4}}\left(2xy^2+2x^2y\,y'\right)=1}$

; $\boxed{\dfrac{2xy^2+2x^2y\,y'}{4\left(x^2y^2-1\right)^{\frac{3}{4}}}=1}$; $\boxed{\dfrac{2\left(xy^2+x^2y\,y'\right)}{4\left(x^2y^2-1\right)^{\frac{3}{4}}}=1}$; $\boxed{xy^2+x^2y\,y'=2\left(x^2y^2-1\right)^{\frac{3}{4}}}$; $\boxed{x^2y\,y'=2\left(x^2y^2-1\right)^{\frac{3}{4}}-xy^2}$

; $\boxed{y'=\dfrac{2\left(x^2y^2-1\right)^{\frac{3}{4}}}{x^2y}-\dfrac{xy^2}{x^2y}}$; $\boxed{y'=\dfrac{2\left(x^2y^2-1\right)^{\frac{3}{4}}}{x^2y}-\dfrac{y}{x}}$

g. $\boxed{\dfrac{d}{dx}\left(\sqrt[7]{xy^2}\right)=\dfrac{d}{dx}\left(3x^2\right)}$; $\boxed{\dfrac{d}{dx}\left(xy^2\right)^{\frac{1}{7}}=6x}$; $\boxed{\dfrac{1}{7}\left(xy^2\right)^{\frac{1}{7}-1}\cdot\dfrac{d}{dx}\left(xy^2\right)=6x}$; $\boxed{\dfrac{1}{7}\left(xy^2\right)^{-\frac{6}{7}}\left(1\cdot y^2+2y\,y'\cdot x\right)=6x}$

; $\boxed{\dfrac{y^2+2y\,y'x}{7\left(xy^2\right)^{\frac{6}{7}}}=6x}$; $\boxed{y^2+2y\,y'x=6x\cdot 7\left(xy^2\right)^{\frac{6}{7}}}$; $\boxed{2y\,y'x=42x\left(xy^2\right)^{\frac{6}{7}}-y^2}$; $\boxed{y'=\dfrac{42x^{\frac{13}{7}}y^{\frac{12}{7}}-y^2}{2xy}}$

h. $\boxed{\dfrac{d}{dx}\left(\sqrt{x^3+y^3}\right)=\dfrac{d}{dx}(2)}$; $\boxed{\dfrac{d}{dx}\left(x^3+y^3\right)^{\frac{1}{2}}=0}$; $\boxed{\dfrac{1}{2}\left(x^3+y^3\right)^{\frac{1}{2}-1}\cdot\left(\dfrac{d}{dx}x^3+\dfrac{d}{dx}y^3\right)=0}$; $\boxed{\dfrac{1}{2}\left(x^3+y^3\right)^{-\frac{1}{2}}\cdot\left(3x^2+3y^2y'\right)=0}$

$\boxed{\dfrac{3x^2}{2}\left(x^3+y^3\right)^{-\frac{1}{2}}+\dfrac{3y^2y'}{2}\left(x^3+y^3\right)^{-\frac{1}{2}}=0}$; $\boxed{\dfrac{3y^2y'}{2}\left(x^3+y^3\right)^{-\frac{1}{2}}=-\dfrac{3x^2}{2}\left(x^3+y^3\right)^{-\frac{1}{2}}}$; $\boxed{y^2y'=-x^2}$; $\boxed{y'=-\dfrac{x^2}{y^2}}$

i. $\boxed{\dfrac{d}{dx}\left(\sqrt{x^2-1}\right)=\dfrac{d}{dx}(xy)}$; $\boxed{\dfrac{d}{dx}\left(x^2-1\right)^{\frac{1}{2}}=1\cdot y+x\cdot\dfrac{d}{dx}(y)}$; $\boxed{\dfrac{1}{2}\left(x^2-1\right)^{\frac{1}{2}-1}\cdot 2x=y+x\,y'}$; $\boxed{x\left(x^2-1\right)^{-\frac{1}{2}}=y+x\,y'}$

$; \boxed{y + x\,y' = x\left(x^2 - 1\right)^{-\frac{1}{2}}} \; ; \; \boxed{x\,y' = x\left(x^2 - 1\right)^{-\frac{1}{2}} - y} \; ; \; \boxed{y' = \dfrac{x\left(x^2 - 1\right)^{-\frac{1}{2}} - y}{x}}$

Example 2.7-5: Compute the derivative for the following Radical expressions.

a. $\dfrac{d}{dx}\left(2x + \sqrt{x}\right) =$

b. $\dfrac{d}{dx}\left(\sqrt{x+1}\right) =$

c. $\dfrac{d}{dx}\left(\sqrt[3]{x^2 + 1}\right) =$

d. $\dfrac{d}{dx}\left(x + \sqrt{x^3}\right) =$

e. $\dfrac{d}{dx}\left(\sqrt{x^5} - 3\right) =$

f. $\dfrac{d}{dx}\left(x^3\sqrt{x+1}\right) =$

g. $\dfrac{d}{dx}\left(\dfrac{x+1}{\sqrt{x}}\right) =$

h. $\dfrac{d}{dx}\left(\sqrt{\dfrac{1}{x+1}}\right) =$

i. $\dfrac{d}{dx}\left(\dfrac{x^2}{\sqrt[3]{x}}\right) =$

Solutions:

a. $\boxed{\dfrac{d}{dx}\left(2x + \sqrt{x}\right)} = \boxed{\dfrac{d}{dx}2x + \dfrac{d}{dx}\sqrt{x}} = \boxed{\dfrac{d}{dx}2x + \dfrac{d}{dx}x^{\frac{1}{2}}} = \boxed{2 + \dfrac{1}{2}x^{\frac{1}{2}-1}} = \boxed{2 + \dfrac{1}{2}x^{-\frac{1}{2}}} = \boxed{2 + \dfrac{1}{2x^{\frac{1}{2}}}} = \boxed{2 + \dfrac{1}{2\sqrt{x}}}$

b. $\boxed{\dfrac{d}{dx}\left(\sqrt{x+1}\right)} = \boxed{\dfrac{d}{dx}(x+1)^{\frac{1}{2}}} = \boxed{\dfrac{1}{2}(x+1)^{\frac{1}{2}-1}\cdot 1} = \boxed{\dfrac{1}{2}(x+1)^{-\frac{1}{2}}} = \boxed{\dfrac{1}{2(x+1)^{\frac{1}{2}}}} = \boxed{\dfrac{1}{2\sqrt{x+1}}}$

c. $\boxed{\dfrac{d}{dx}\left(\sqrt[3]{x^2+1}\right)} = \boxed{\dfrac{d}{dx}\left(x^2+1\right)^{\frac{1}{3}}} = \boxed{\dfrac{1}{3}\left(x^2+1\right)^{\frac{1}{3}-1}\cdot 2x} = \boxed{\dfrac{2x}{3}\left(x^2+1\right)^{-\frac{2}{3}}} = \boxed{\dfrac{2x}{3\left(x^2+1\right)^{\frac{2}{3}}}} = \boxed{\dfrac{2x}{3\sqrt[3]{\left(x^2+1\right)^2}}}$

d. $\boxed{\dfrac{d}{dx}\left(x + \sqrt{x^3}\right)} = \boxed{\dfrac{d}{dx}\left(x + x^{\frac{3}{2}}\right)} = \boxed{\dfrac{d}{dx}x + \dfrac{d}{dx}x^{\frac{3}{2}}} = \boxed{1 + \dfrac{3}{2}x^{\frac{3}{2}-1}} = \boxed{1 + \dfrac{3}{2}x^{\frac{1}{2}}} = \boxed{1 + \dfrac{3}{2}\sqrt{x}}$

e. $\boxed{\dfrac{d}{dx}\left(\sqrt{x^5} - 3\right)} = \boxed{\dfrac{d}{dx}\left(x^{\frac{5}{2}} - 3\right)} = \boxed{\dfrac{d}{dx}x^{\frac{5}{2}} + \dfrac{d}{dx}(-3)} = \boxed{\dfrac{5}{2}x^{\frac{5}{2}-1} + 0} = \boxed{\dfrac{5}{2}x^{\frac{3}{2}}} = \boxed{\dfrac{5}{2}\sqrt{x^3}} = \boxed{\dfrac{5x\sqrt{x}}{2}}$

f. $\boxed{\dfrac{d}{dx}\left(x^3\sqrt{x+1}\right)} = \boxed{\dfrac{d}{dx}\left[x^3(x+1)^{\frac{1}{2}}\right]} = \boxed{(x+1)^{\frac{1}{2}}\dfrac{d}{dx}x^3 + x^3\dfrac{d}{dx}(x+1)^{\frac{1}{2}}} = \boxed{(x+1)^{\frac{1}{2}}\cdot 3x^2 + x^3\cdot\dfrac{1}{2}(x+1)^{\frac{1}{2}-1}}$

$= \boxed{3x^2(x+1)^{\frac{1}{2}} + \dfrac{x^3}{2}(x+1)^{-\frac{1}{2}}} = \boxed{3x^2(x+1)^{\frac{1}{2}} + \dfrac{x^3}{2(x+1)^{\frac{1}{2}}}} = \boxed{\dfrac{3x^2(x+1)^{\frac{1}{2}}\cdot 2(x+1)^{\frac{1}{2}} + x^3}{2(x+1)^{\frac{1}{2}}}} = \boxed{\dfrac{6x^2(x+1) + x^3}{2\sqrt{x+1}}}$

$= \boxed{\dfrac{6x^3 + 6x^2 + x^3}{2\sqrt{x+1}}} = \boxed{\dfrac{7x^3 + 6x^2}{2\sqrt{x+1}}} = \boxed{\dfrac{x^2(7x+6)}{2\sqrt{x+1}}}$

g. $\boxed{\dfrac{d}{dx}\left(\dfrac{x+1}{\sqrt{x}}\right)} = \boxed{\dfrac{d}{dx}\left(\dfrac{x+1}{x^{\frac{1}{2}}}\right)} = \boxed{\dfrac{\left[x^{\frac{1}{2}}\cdot\dfrac{d}{dx}(x+1)\right] - \left[(x+1)\cdot\dfrac{d}{dx}x^{\frac{1}{2}}\right]}{x}} = \boxed{\dfrac{\left[x^{\frac{1}{2}}\cdot 1\right] - \left[(x+1)\cdot\dfrac{1}{2}x^{\frac{1}{2}-1}\right]}{x}}$

$$= \frac{x^{\frac{1}{2}} - \left[(x+1) \cdot \frac{1}{2} x^{-\frac{1}{2}} \right]}{x} = \frac{x^{\frac{1}{2}} - \frac{x+1}{2x^{\frac{1}{2}}}}{x} = \frac{\frac{x^{\frac{1}{2}} \cdot 2x^{\frac{1}{2}} - (x+1)}{2x^{\frac{1}{2}}}}{x} = \frac{\frac{2x - x - 1}{2\sqrt{x}}}{x} = \frac{\frac{x-1}{2\sqrt{x}}}{x} = \boxed{\frac{x-1}{2x\sqrt{x}}}$$

h. $\boxed{\dfrac{d}{dx}\left(\sqrt{\dfrac{1}{x+1}} \right)} = \boxed{\dfrac{d}{dx}\left(\dfrac{1}{\sqrt{x+1}} \right)} = \boxed{\dfrac{d}{dx}\left(\dfrac{1}{(x+1)^{\frac{1}{2}}} \right)} = \boxed{\dfrac{\left[(x+1)^{\frac{1}{2}} \cdot \frac{d}{dx}(1) \right] - \left[1 \cdot \frac{d}{dx}(x+1)^{\frac{1}{2}} \right]}{x+1}} = \boxed{\dfrac{0 - \frac{1}{2}(x+1)^{\frac{1}{2}-1}}{x+1}}$

$$= \boxed{\frac{-\frac{1}{2}(x+1)^{-\frac{1}{2}}}{x+1}} = \boxed{\frac{-\frac{1}{2(x+1)^{\frac{1}{2}}}}{(x+1)}} = \boxed{-\frac{1}{2(x+1)^{\frac{1}{2}}(x+1)}} = \boxed{-\frac{1}{2(x+1)^{\frac{3}{2}}}} = \boxed{-\frac{1}{2\sqrt{(x+1)^3}}} = \boxed{-\frac{1}{2(x+1)\sqrt{x+1}}}$$

i. $\boxed{\dfrac{d}{dx}\left(\dfrac{x^2}{\sqrt[3]{x}} \right)} = \boxed{\dfrac{d}{dx}\left(\dfrac{x^2}{x^{\frac{1}{3}}} \right)} = \boxed{\dfrac{d}{dx}\left(x^2 \cdot x^{-\frac{1}{3}} \right)} = \boxed{\dfrac{d}{dx} x^{2-\frac{1}{3}}} = \boxed{\dfrac{d}{dx} x^{\frac{6-1}{3}}} = \boxed{\dfrac{d}{dx} x^{\frac{5}{3}}} = \boxed{\dfrac{5}{3} x^{\frac{5}{3}-1}} = \boxed{\dfrac{5}{3} x^{\frac{2}{3}}} = \boxed{\dfrac{5\sqrt[3]{x^2}}{3}}$

Example 2.7-6: Evaluate the derivative of the following equations at the given values.

a. $y = \sqrt{(x^2+1)^3}$ at $x = 2$ b. $y = \sqrt[3]{x} + \sqrt{x} + 2$ at $x = 10$ c. $y = x\sqrt{x^2+1}$ at $x = 5$

d. $y = \sqrt[3]{2x} + \sqrt[3]{x^2}$ at $x = 2$ e. $y = \dfrac{\sqrt{6x+1}}{x^2}$ at $x = 3$ f. $y = \dfrac{1}{\sqrt{x^2+1}}$ at $x = -5$

g. $y = \dfrac{\sqrt{5x+1}}{2x+1}$ at $x = 3$ h. $y = x\sqrt{x^2-10} + 2x$ at $x = 5$ i. $y = x^2\sqrt{(x+1)^3}$ at $x = 1$

j. $y = \sqrt{\dfrac{x^2+1}{1-x^2}}$ at $x = 0$ k. $y = \dfrac{\sqrt{x^5+1}}{2x}$ at $x = 2$ l. $y = \dfrac{x+1}{\sqrt{x^3+1}}$ at $x = 1$

Solutions:

a. Given $y = \sqrt{(x^2+1)^3} = (x^2+1)^{\frac{3}{2}}$, then

$$\boxed{y'} = \boxed{\frac{3}{2}(x^2+1)^{\frac{3}{2}-1}(2x^{2-1}+0)} = \boxed{\frac{3}{2}(x^2+1)^{\frac{3-2}{2}} \cdot 2x} = \boxed{3x(x^2+1)^{\frac{1}{2}}} = \boxed{3x\sqrt{x^2+1}}$$

Substituting $x = 2$ in place of x in the y' equation we obtain the following value:

$$\boxed{y'} = \boxed{(3 \cdot 2)\sqrt{2^2+1}} = \boxed{6\sqrt{5}} = \boxed{\mathbf{13.42}}$$

b. Given $y = \sqrt[3]{x} + \sqrt{x} + 2 = x^{\frac{1}{3}} + x^{\frac{1}{2}} + 2$, then

$$\boxed{y'} = \boxed{\frac{1}{3}x^{\frac{1}{3}-1} + \frac{1}{2}x^{\frac{1}{2}-1} + 0} = \boxed{\frac{1}{3}x^{-\frac{2}{3}} + \frac{1}{2}x^{-\frac{1}{2}}} = \boxed{\frac{1}{3x^{\frac{2}{3}}} + \frac{1}{2x^{\frac{1}{2}}}} = \boxed{\frac{1}{3\sqrt[3]{x^2}} + \frac{1}{2\sqrt{x}}}$$

Substituting $x = 10$ in place of x in the y' equation we obtain the following value:

$$\boxed{y'} = \boxed{\dfrac{1}{3\cdot 10^{\frac{2}{3}}} + \dfrac{1}{2\cdot 10^{\frac{1}{2}}}} = \boxed{\dfrac{1}{3\cdot 4.634} + \dfrac{1}{2\cdot 3.162}} = \boxed{0.072 + 0.158} = \boxed{\mathbf{0.23}}$$

c. Given $y = x\sqrt{x^2+1} = x\left(x^2+1\right)^{\frac{1}{2}}$, then

$$\boxed{y'} = \boxed{\left[1\cdot\left(x^2+1\right)^{\frac{1}{2}}\right] + \left[\frac{1}{2}\left(x^2+1\right)^{\frac{1}{2}-1}\cdot\left(2x^{2-1}+0\right)\cdot x\right]} = \boxed{\left(x^2+1\right)^{\frac{1}{2}} + \left[\frac{1}{2}\left(x^2+1\right)^{-\frac{1}{2}}\cdot 2x^2\right]} = \boxed{\left(x^2+1\right)^{\frac{1}{2}} + x^2\left(x^2+1\right)^{-\frac{1}{2}}}$$

$$= \boxed{\left(x^2+1\right)^{\frac{1}{2}} + \dfrac{x^2}{\left(x^2+1\right)^{\frac{1}{2}}}} = \boxed{\dfrac{\left(x^2+1\right)^{\frac{1}{2}}\left(x^2+1\right)^{\frac{1}{2}}+x^2}{\left(x^2+1\right)^{\frac{1}{2}}}} = \boxed{\dfrac{\left(x^2+1\right)^{\frac{1}{2}+\frac{1}{2}}+x^2}{\left(x^2+1\right)^{\frac{1}{2}}}} = \boxed{\dfrac{x^2+1+x^2}{\left(x^2+1\right)^{\frac{1}{2}}}} = \boxed{\dfrac{2x^2+1}{\sqrt{x^2+1}}}$$

Substituting $x=5$ in place of x in the y' equation we obtain the following value:

$$\boxed{y'} = \boxed{\dfrac{2\cdot 5^2+1}{\left(5^2+1\right)^{\frac{1}{2}}}} = \boxed{\dfrac{2\cdot 25+1}{(25+1)^{\frac{1}{2}}}} = \boxed{\dfrac{50+1}{26^{\frac{1}{2}}}} = \boxed{\dfrac{51}{5.1}} = \boxed{\mathbf{10}}$$

d. Given $y = \sqrt[3]{2x} + \sqrt[3]{x^2} = (2x)^{\frac{1}{3}} + x^{\frac{2}{3}}$, then

$$\boxed{y'} = \boxed{\frac{1}{3}(2x)^{\frac{1}{3}-1} + \frac{2}{3}x^{\frac{2}{3}-1}} = \boxed{\frac{1}{3}(2x)^{-\frac{2}{3}} + \frac{2}{3}x^{-\frac{1}{3}}} = \boxed{\dfrac{1}{3(2x)^{\frac{2}{3}}} + \dfrac{2}{3x^{\frac{1}{3}}}} = \boxed{\dfrac{1}{3\sqrt[3]{(2x)^2}} + \dfrac{2}{3\sqrt[3]{x}}}$$

Substituting $x=2$ in place of x in the y' equation we obtain the following value:

$$\boxed{y'} = \boxed{\dfrac{1}{3\cdot(2\cdot 2)^{\frac{2}{3}}} + \dfrac{2}{3\cdot 2^{\frac{1}{3}}}} = \boxed{\dfrac{1}{3\cdot 4^{\frac{2}{3}}} + \dfrac{2}{3\cdot 2^{\frac{1}{3}}}} = \boxed{\dfrac{1}{3\cdot 2.53} + \dfrac{2}{3\cdot 1.26}} = \boxed{0.132 + 0.53} = \boxed{\mathbf{0.662}}$$

e. Given $y = \dfrac{\sqrt{6x}+1}{x^2} = \dfrac{(6x)^{\frac{1}{2}}+1}{x^2}$, then

$$\boxed{y'} = \boxed{\dfrac{\left\{\left[\frac{1}{2}(6x)^{\frac{1}{2}-1}\cdot 6+0\right]\cdot x^2\right\} - \left\{2x^{2-1}\cdot\left[(6x)^{\frac{1}{2}}+1\right]\right\}}{x^4}} = \boxed{\dfrac{\left\{\left[\frac{1}{2}(6x)^{-\frac{1}{2}}\cdot 6\right]\cdot x^2\right\} - \left\{2x\cdot\left[(6x)^{\frac{1}{2}}+1\right]\right\}}{x^4}}$$

$$= \boxed{\dfrac{3x^2(6x)^{-\frac{1}{2}} - 2x(6x)^{\frac{1}{2}} - 2x}{x^4}}$$

Substituting $x=3$ in place of x in the y' equation we obtain the following value:

$$\boxed{y'} = \boxed{\dfrac{3\cdot 3^2\cdot(6\cdot 3)^{-\frac{1}{2}} - 2\cdot 3\cdot(6\cdot 3)^{\frac{1}{2}} - 2\cdot 3}{3^4}} = \boxed{\dfrac{27\cdot(18)^{-\frac{1}{2}} - 6\cdot(18)^{\frac{1}{2}} - 6}{81}} = \boxed{\dfrac{(27\cdot 0.236) - (6\cdot 4.243) - 6}{81}}$$

$$= \boxed{\dfrac{6.372 - 25.458 - 6}{81}} = \boxed{-\dfrac{25.086}{81}} = \boxed{\mathbf{-0.31}}$$

f. Given $y = \dfrac{1}{\sqrt{x^2+1}} = \dfrac{1}{\left(x^2+1\right)^{\frac{1}{2}}}$, then

$$\boxed{y'} = \boxed{\dfrac{\left[0\cdot\left(x^2+1\right)^{\frac{1}{2}}\right]-\left[\dfrac{1}{2}\left(x^2+1\right)^{\frac{1}{2}-1}\cdot\left(2x^{2-1}+0\right)\cdot 1\right]}{x^2+1}} = \boxed{\dfrac{0-\left[\dfrac{1}{2}\left(x^2+1\right)^{-\frac{1}{2}}\cdot 2x\right]}{x^2+1}} = \boxed{\dfrac{-x\left(x^2+1\right)^{-\frac{1}{2}}}{x^2+1}}$$

$$= \boxed{\dfrac{-x}{\left(x^2+1\right)\left(x^2+1\right)^{\frac{1}{2}}}} = \boxed{\dfrac{-x}{\left(x^2+1\right)^{1+\frac{1}{2}}}} = \boxed{\dfrac{-x}{\left(x^2+1\right)^{\frac{3}{2}}}} = \boxed{\dfrac{-x}{\sqrt{\left(x^2+1\right)^3}}} = \boxed{\dfrac{-x}{\left(x^2+1\right)\sqrt{x^2+1}}}$$

Substituting $x = -5$ in place of x in the y' equation we obtain the following value:

$$\boxed{y'} = \boxed{-\dfrac{-5}{\sqrt{\left[(-5)^2+1\right]^3}}} = \boxed{\dfrac{5}{\sqrt{(25+1)^3}}} = \boxed{\dfrac{5}{\sqrt{17576}}} = \boxed{\dfrac{5}{132.57}} = \boxed{0.038}$$

g. Given $y = \dfrac{\sqrt{5x+1}}{2x+1} = \dfrac{(5x+1)^{\frac{1}{2}}}{2x+1}$, then

$$\boxed{y'} = \boxed{\dfrac{\left[\dfrac{1}{2}(5x+1)^{\frac{1}{2}-1}\cdot\left(5x^{1-1}+0\right)\cdot(2x+1)\right]-\left[\left(2x^{1-1}+0\right)(5x+1)^{\frac{1}{2}}\right]}{(2x+1)^2}} = \boxed{\dfrac{\left[2.5(5x+1)^{-\frac{1}{2}}\cdot(2x+1)\right]-\left[2(5x+1)^{\frac{1}{2}}\right]}{(2x+1)^2}}$$

Substituting $x = 3$ in place of x in the y' equation we obtain the following value:

$$\boxed{y'} = \boxed{\dfrac{\left[2.5(15+1)^{-\frac{1}{2}}\cdot(6+1)\right]-\left[2(15+1)^{\frac{1}{2}}\right]}{(6+1)^2}} = \boxed{\dfrac{\left[17.5\cdot(16)^{-\frac{1}{2}}\right]-\left[2\cdot(16)^{\frac{1}{2}}\right]}{7^2}} = \boxed{\dfrac{[17.5\cdot0.25]-[2\cdot4]}{49}} = \boxed{-0.074}$$

h. Given $y = x\sqrt{x^2-10}+2x = x\left(x^2-10\right)^{\frac{1}{2}}+2x$, then

$$\boxed{y'} = \boxed{\left[1\cdot\left(x^2-10\right)^{\frac{1}{2}}\right]+\left[0.5\left(x^2-10\right)^{\frac{1}{2}-1}\cdot\left(2x^{2-1}-0\right)\cdot x\right]+2x^{1-1}} = \boxed{\left(x^2-10\right)^{\frac{1}{2}}+\left[x^2\left(x^2-10\right)^{-\frac{1}{2}}\right]+2}$$

Substituting $x = 5$ in place of x in the y' equation we obtain the following value:

$$\boxed{y'} = \boxed{\left(5^2-10\right)^{\frac{1}{2}}+\left[5^2\left(5^2-10\right)^{-\frac{1}{2}}\right]+2} = \boxed{15^{\frac{1}{2}}+\left[25\cdot15^{-\frac{1}{2}}\right]+2} = \boxed{3.873+[25\cdot0.258]+2} = \boxed{12.323}$$

i. Given $y = x^2\sqrt{(x+1)^3} = x^2(x+1)^{\frac{3}{2}}$, then

$$\boxed{y'} = \boxed{\left[2x^{2-1}\cdot(x+1)^{\frac{3}{2}}\right]+\left[\dfrac{3}{2}(x+1)^{\frac{3}{2}-1}\cdot\left(x^{1-1}+0\right)\cdot x^2\right]} = \boxed{\left[2x(x+1)^{\frac{3}{2}}\right]+\left[\dfrac{3}{2}x^3(x+1)^{\frac{1}{2}}\right]}$$

Substituting $x = 1$ in place of x in the y' equation we obtain the following value:

$$\boxed{y'} = \boxed{\left[2(1+1)^{\frac{3}{2}}\right]+\left[\dfrac{3}{2}(1+1)^{\frac{1}{2}}\right]} = \boxed{\left(2\cdot2^{\frac{3}{2}}\right)+\left(\dfrac{3}{2}\cdot2^{\frac{1}{2}}\right)} = \boxed{5.657+2.121} = \boxed{7.778}$$

j. Given $y = \sqrt{\dfrac{x^2+1}{1-x^2}} = \left(\dfrac{x^2+1}{1-x^2}\right)^{\frac{1}{2}}$, then

$$\boxed{y'} = \boxed{\frac{1}{2}\left(\frac{x^2+1}{1-x^2}\right)^{\frac{1}{2}-1}\left[\frac{2x\left(1-x^2\right)+2x\left(...x^2+1\right)}{\left(1-x^2\right)^2}\right]} = \boxed{\frac{1}{2}\left(\frac{x^2+1}{1-x^2}\right)^{-\frac{1}{2}}\left[\frac{2x-2x^3+2x^3+2x}{\left(1-x^2\right)^2}\right]} = \boxed{\frac{1}{2}\left(\frac{x^2+1}{1-x^2}\right)^{-\frac{1}{2}}\frac{4x}{\left(1-x^2\right)^2}}$$

Substituting $x = 0$ in place of x in the y' equation we obtain the following value:

$$\boxed{y'} = \boxed{\frac{1}{2}\left(\frac{0+1}{1-0}\right)^{-\frac{1}{2}}\cdot\frac{4\cdot0}{\left(1-0\right)^2}} = \boxed{\frac{1}{2}\cdot1^{-\frac{1}{2}}\cdot0} = \boxed{0}$$

k. Given $y = \dfrac{\sqrt{x^5+1}}{2x} = \dfrac{\left(x^5+1\right)^{\frac{1}{2}}}{2x}$, then

$$\boxed{y'} = \boxed{\frac{\left[\frac{1}{2}\left(x^5+1\right)^{\frac{1}{2}-1}5x^4\cdot2x\right]-\left[2x^{1-1}\cdot\left(x^5+1\right)^{\frac{1}{2}}\right]}{\left(2x\right)^2}} = \boxed{\frac{\left[5x^5\left(x^5+1\right)^{-\frac{1}{2}}\right]-\left[2\cdot\left(x^5+1\right)^{\frac{1}{2}}\right]}{4x^2}}$$

Substituting $x = 2$ in place of x in the y' equation we obtain the following value:

$$\boxed{y'} = \boxed{\frac{\left[160\left(32+1\right)^{-\frac{1}{2}}\right]-\left[2\cdot\left(32+1\right)^{\frac{1}{2}}\right]}{4\cdot4}} = \boxed{\frac{\left[160\cdot0.174\right]-\left[2\cdot5.744\right]}{16}} = \boxed{\frac{27.8-11.5}{16}} = \boxed{1.02}$$

l. Given $y = \dfrac{x+1}{\sqrt{x^3+1}} = \dfrac{x+1}{\left(x^3+1\right)^{\frac{1}{2}}}$, then

$$\boxed{y'} = \boxed{\frac{\left[1\cdot\left(x^3+1\right)^{\frac{1}{2}}\right]-\left\{\left[\frac{1}{2}\left(x^3+1\right)^{\frac{1}{2}-1}\left(3x^2+0\right)\right]\left(x+1\right)\right\}}{x^3+1}} = \boxed{\frac{\left(x^3+1\right)^{\frac{1}{2}}-\left\{\left[\frac{1}{2}\left(x^3+1\right)^{-\frac{1}{2}}3x^2\right]\left(x+1\right)\right\}}{x^3+1}}$$

$$= \boxed{\frac{\left(x^3+1\right)^{\frac{1}{2}}-\frac{\frac{3}{2}x^2\left(x+1\right)}{\left(x^3+1\right)^{\frac{1}{2}}}}{x^3+1}} = \boxed{\frac{\frac{\left(x^3+1\right)^{\frac{1}{2}}}{1}-\frac{3x^3+3x^2}{2\left(x^3+1\right)^{\frac{1}{2}}}}{x^3+1}} = \boxed{\frac{\frac{\left(x^3+1\right)^{\frac{1}{2}}2\left(x^3+1\right)^{\frac{1}{2}}-3x^3-3x^2}{2\left(x^3+1\right)^{\frac{1}{2}}}}{x^3+1}} = \boxed{\frac{\frac{2\left(x^3+1\right)-3x^3-3x^2}{2\left(x^3+1\right)^{\frac{1}{2}}}}{x^3+1}}$$

$$= \boxed{\frac{2x^3+2-3x^3-3x^2}{2\left(x^3+1\right)^{\frac{1}{2}}\left(x^3+1\right)}} = \boxed{\frac{-x^3-3x^2+2}{2\left(x^3+1\right)^{\frac{1}{2}}\left(x^3+1\right)}} = \boxed{\frac{-x^3-3x^2+2}{2\left(x^3+1\right)^{\frac{1}{2}+1}}} = \boxed{\frac{-x^3-3x^2+2}{2\left(x^3+1\right)^{\frac{3}{2}}}}$$

Substituting $x = 1$ in place of x in the y' equation we obtain the following value:

$$\boxed{y'} = \boxed{\frac{-1^3-\left(3\cdot1^2\right)+2}{2\left(1^3+1\right)^{\frac{3}{2}}}} = \boxed{\frac{-1-3+2}{2\cdot2^{\frac{3}{2}}}} = \boxed{\frac{-2}{2\cdot2.828}} = \boxed{\frac{-1}{2.828}} = \boxed{-0.354}$$

$$\boxed{f''(x)} = \boxed{(40\cdot7)x^{7-1} - (9\cdot2)x^{2-1}} = \boxed{280x^6 - 18x}$$

b. Given $f(x) = x^3(x^2 + x + 5)$, then

$$\boxed{f'(x)} = \boxed{\left[3x^{3-1}\cdot(x^2+x+5)\right] + \left[\left(2x^{2-1}+1x^{1-1}+0\right)\cdot x^3\right]} = \boxed{\left[3x^2(x^2+x+5)\right] + \left[(2x+1)\cdot x^3\right]}$$

$$= \boxed{3x^4 + 3x^3 + 15x^2 + 2x^4 + x^3} = \boxed{5x^4 + 4x^3 + 15x^2} \text{ and}$$

$$\boxed{f''(x)} = \boxed{(5\cdot4)x^{4-1} + (4\cdot3)x^{3-1} + (15\cdot2)x^{2-1}} = \boxed{20x^3 + 12x^2 + 30x}$$

c. Given $f(x) = x^2 + \dfrac{1}{x}$, then

$$\boxed{f'(x)} = \boxed{2x^{2-1} + \frac{[0\cdot x] - (1\cdot1)}{x^2}} = \boxed{2x + \frac{0-1}{x^2}} = \boxed{2x - \frac{1}{x^2}}$$

A second way is to rewrite $f(x)$ as $f(x) = x^2 + x^{-1}$ and find its derivative, i.e.,

$$\boxed{f'(x)} = \boxed{2x^{2-1} - 1\cdot x^{-1-1}} = \boxed{2x - x^{-2}} = \boxed{2x - \frac{1}{x^2}} \text{ and}$$

$$\boxed{f''(x)} = \boxed{2x^{1-1} + (-1\cdot-2)x^{-2-1}} = \boxed{2x^0 + 2x^{-3}} = \boxed{2 + 2x^{-3}} = \boxed{2 + \frac{2}{x^3}}$$

d. Given $f(u) = \dfrac{u^3 - 1}{u+1}$, then

$$\boxed{f'(u)} = \boxed{\frac{\left[\left(3u^{3-1}-0\right)\cdot(u+1)\right] - \left[\left(u^{1-1}+0\right)\cdot\left(u^3-1\right)\right]}{(u+1)^2}} = \boxed{\frac{3u^2\cdot(u+1) - \left(u^3-1\right)}{(u+1)^2}} = \boxed{\frac{3u^3+3u^2-u^3+1}{(u+1)^2}} = \boxed{\frac{2u^3+3u^2+1}{(u+1)^2}}$$

$$\boxed{f''(u)} = \boxed{\frac{\left[\left(2\cdot3u^{3-1}+3\cdot2u^{2-1}+0\right)\cdot(u+1)^2\right] - \left[2(u+1)^{2-1}\cdot\left(2u^3+3u^2+1\right)\right]}{(u+1)^4}}$$

$$= \boxed{\frac{\left[\left(6u^2+6u\right)(u+1)^2\right] - \left[2(u+1)\left(2u^3+3u^2+1\right)\right]}{(u+1)^4}} = \boxed{\frac{\left[\left(6u^2+6u\right)(u+1)^2\right] + \left[(-2u-2)\left(2u^3+3u^2+1\right)\right]}{(u+1)^4}}$$

e. Given $g(x) = x^2 + \dfrac{1}{x^3}$, then

$$\boxed{g'(x)} = \boxed{2x^{2-1} + \frac{\left(0\cdot x^3\right) - \left(3x^{3-1}\cdot1\right)}{x^6}} = \boxed{2x + \frac{-3x^2}{x^6}} = \boxed{2x - \frac{3x^2}{x^{6-4}}} = \boxed{2x - \frac{3}{x^4}}$$

A second way is to rewrite $g(x)$ as $g(x) = x^2 + x^{-3}$ and find its derivative, i.e.,

$\boxed{g'(x)} = \boxed{2x^{2-1} - 3 \cdot x^{-1-3}} = \boxed{2x - 3x^{-4}} = \boxed{2x - \dfrac{3}{x^4}}$ and

$\boxed{g''(x)} = \boxed{2x^{1-1} + (-3 \cdot -4)x^{-4-1}} = \boxed{2x^0 + 12x^{-5}} = \boxed{2 + 12x^{-5}} = \boxed{2 + \dfrac{12}{x^5}}$

f. Given $h(x) = \left(a^2 + x^3\right)^2$, then

$\boxed{h'(x)} = \boxed{2\left(a^2 + x^3\right)^{2-1} \cdot \left(0 + 3x^{3-1}\right)} = \boxed{2\left(a^2 + x^3\right) \cdot 3x^2} = \boxed{6x^2\left(a^2 + x^3\right)} = \boxed{6a^2x^2 + 6x^5} = \boxed{6x^5 + 6a^2x^2}$

$\boxed{h''(x)} = \boxed{(6 \cdot 5)x^{5-1} + \left(6a^2 \cdot 2\right)x^{2-1}} = \boxed{30x^4 + 12a^2x}$

g. Given $f(x) = \left(x^2 + 1\right)^{-1}$, then

$\boxed{f'(x)} = \boxed{\left[-1 \cdot \left(x^2 + 1\right)^{-1-1}\right] \cdot \left(2x^{2-1} + 0\right)} = \boxed{\left[-\left(x^2 + 1\right)^{-2}\right] \cdot 2x} = \boxed{-2x\left(x^2 + 1\right)^{-2}}$

$\boxed{f''(x)} = \boxed{-2\left\{1 \cdot \left(x^2 + 1\right)^{-2} + \left[-2\left(x^2 + 1\right)^{-2-1} \cdot \left(2x^{2-1} + 0\right)\right]\right\}} = \boxed{-2\left\{\left(x^2 + 1\right)^{-2} + \left[-2\left(x^2 + 1\right)^{-3} \cdot 2x\right]\right\}}$

$= \boxed{-2\left\{\left(x^2 + 1\right)^{-2} - 4x\left(x^2 + 1\right)^{-3}\right\}} = \boxed{-2\left(x^2 + 1\right)^{-2} + 8x\left(x^2 + 1\right)^{-3}}$

h. Given $r(\theta) = \theta^2 + \dfrac{1}{(\theta+1)^3}$, then

$\boxed{r'(\theta)} = \boxed{2\theta^{2-1} + \dfrac{\left[0 \cdot (\theta+1)^3\right] - \left[3(\theta+1)^{3-1} \cdot 1\right]}{(\theta+1)^6}} = \boxed{2\theta + \dfrac{0 - 3(\theta+1)^2}{(\theta+1)^6}} = \boxed{2\theta - \dfrac{3(\theta+1)^2}{(\theta+1)^{6=4}}} = \boxed{2\theta - \dfrac{3}{(\theta+1)^2}}$ and

$\boxed{r''(\theta)} = \boxed{2\theta^{1-1} - \dfrac{\left[0 \cdot (\theta+1)^2\right] - \left[2(\theta+1)^{2-1} \cdot 3\right]}{(\theta+1)^4}} = \boxed{2 - \dfrac{0 - 6(\theta+1)}{(\theta+1)^4}} = \boxed{2 + \dfrac{6(\theta+1)}{(\theta+1)^{4=3}}} = \boxed{2 + \dfrac{6}{(\theta+1)^3}}$

i. Given $s(r) = r^2\left(r^2 + 1\right)^3$, then

$\boxed{s'(r)} = \boxed{\left[2r^{2-1} \cdot \left(r^2 + 1\right)^3\right] + \left[3\left(r^2 + 1\right)^{3-1} \cdot \left(2r^{2-1} + 0\right)\right] \cdot r^2} = \boxed{\left[2r\left(r^2 + 1\right)^3\right] + \left[3\left(r^2 + 1\right)^2 \cdot 2r\right] \cdot r^2}$

$= \boxed{2r\left(r^2 + 1\right)^3 + 6r^3\left(r^2 + 1\right)^2}$ and

$\boxed{s''(r)} = \boxed{2\left\{\left[1 \cdot \left(r^2 + 1\right)^3\right] + \left[3\left(r^2 + 1\right)^{3-1} \cdot \left(2r^{2-1} + 0\right)\right] \cdot r\right\} + 6\left\{\left[3r^{3-1} \cdot \left(r^2 + 1\right)^2\right] + \left[2\left(r^2 + 1\right)^{2-1} \cdot \left(2r^{2-1} + 0\right)\right] \cdot r^3\right\}}$

$$= \boxed{2\left\{\left(r^2+1\right)^3 + \left[3\left(r^2+1\right)^{3-1} \cdot 2r\right] \cdot r\right\} + 6\left\{\left[3r^2\left(r^2+1\right)^2\right] + \left[2\left(r^2+1\right) \cdot 2r\right] \cdot r^3\right\}}$$

$$= \boxed{2\left\{\left(r^2+1\right)^3 + \left[6r^2\left(r^2+1\right)^2\right]\right\} + 6\left\{\left[3r^2\left(r^2+1\right)^2\right] + \left[4r^4\left(r^2+1\right)\right]\right\}}$$

j. Given $f(t) = \dfrac{t^3 + t^2 + 1}{10}$, then

$$\boxed{f'(t)} = \boxed{\frac{\left[\left(3t^{3-1} + 2t^{2-1} + 0\right) \cdot 10\right] - \left[0 \cdot \left(t^3 + t^2 + 1\right)\right]}{10^2}} = \boxed{\frac{10\left(3t^2 + 2t\right) - 0}{100}} = \boxed{\frac{\cancel{10}\left(3t^2 + 2t\right)}{\cancel{100} = 10}} = \boxed{\frac{3t^2 + 2t}{10}} \text{ and}$$

$$\boxed{f''(t)} = \boxed{\frac{\left[\left(3 \cdot 2t^{2-1} + 2t^{1-1}\right) \cdot 10\right] - \left[0 \cdot \left(3t^2 + 2t\right)\right]}{10^2}} = \boxed{\frac{10\left(6t + 2\right) - 0}{100}} = \boxed{\frac{\cancel{10}\left(6t + 2\right)}{\cancel{100} = 10}} = \boxed{\frac{2\left(3t+1\right)}{\cancel{10} = 5}} = \boxed{\frac{3t+1}{5}}$$

k. Given $p(r) = r^2 - \dfrac{1}{r}$ which is equal to $p(r) = r^2 - r^{-1}$, then

$$\boxed{p'(r)} = \boxed{2r^{2-1} - \left(-1 \cdot r^{-1-1}\right)} = \boxed{2r + r^{-2}} \text{ and}$$

$$\boxed{p''(r)} = \boxed{2r^{1-1} - 2r^{-2-1}} = \boxed{2r^0 - 2r^{-3}} = \boxed{2 - 2r^{-3}}$$

l. Given $f(x) = \dfrac{x^3}{x+1}$, then

$$\boxed{f'(x)} = \boxed{\frac{\left[3x^{3-1} \cdot (x+1)\right] - \left[\left(1 \cdot x^{1-1} + 0\right) \cdot x^3\right]}{(x+1)^2}} = \boxed{\frac{3x^2(x+1) - x^3}{(x+1)^2}} = \boxed{\frac{3x^3 + 3x^2 - x^3}{(x+1)^2}} = \boxed{\frac{2x^3 + 3x^2}{(x+1)^2}} \text{ and}$$

$$\boxed{f''(x)} = \boxed{\frac{\left\{\left[(2 \cdot 3)x^2 + (3 \cdot 2)x\right] \cdot (x+1)^2\right\} - \left\{2(x+1) \cdot \left(2x^3 + 3x^2\right)\right\}}{(x+1)^4}} = \boxed{\frac{\left[\left(6x^2 + 6x\right)(x+1)^2\right] - \left[2(x+1)\left(2x^3 + 3x^2\right)\right]}{(x+1)^4}}$$

Example 2.8-2: Find $\dfrac{d^3 y}{dx^3}$ for the following functions.

a. $y = (1 - 5x)^3$

b. $y = \left(a - bx^2\right)^{-2}$

c. $y = \dfrac{x^2 + 3x + 1}{x+1}$

d. $y = \dfrac{1}{5}x^5 + \dfrac{1}{4}x^4 + x$

e. $y = \dfrac{ax^2 + b}{c}$

f. $y = \dfrac{x^2 + 1}{x^3}$

Solutions:

a. Given $y = (1 - 5x)^3$, then

$$\boxed{y'} = \boxed{3(1-5x)^{3-1} \cdot \left(0 - 5x^{1-1}\right)} = \boxed{3(1-5x)^2 \cdot (-5)} = \boxed{-15(1-5x)^2}$$

$$\boxed{y''} = \boxed{(-15 \cdot 2)(1-5x)^{2-1} \cdot \left(0 - 5x^{1-1}\right)} = \boxed{-30(1-5x) \cdot (-5)} = \boxed{150(1-5x)}$$

$$\boxed{y'''} = \boxed{150\left(0 - 5x^{1-1}\right)} = \boxed{150 \cdot (-5)} = \boxed{-750}$$

b. Given $y = \left(a - bx^2\right)^{-2}$, then

$$\boxed{y'} = \boxed{-2\left(a - bx^2\right)^{-2-1} \cdot (0 - 2bx)} = \boxed{-2\left(a - bx^2\right)^{-3} \cdot (-2bx)} = \boxed{4bx\left(a - bx^2\right)^{-3}}$$

$$\boxed{y''} = \boxed{\left[(4b \cdot 1)\left(a - bx^2\right)^{-3}\right] + \left\{\left[-3\left(a - bx^2\right)^{-3-1} \cdot (0 - 2bx)\right] \cdot (4bx)\right\}} = \boxed{\left[4b\left(a - bx^2\right)^{-3}\right] + \left[6bx\left(a - bx^2\right)^{-4} \cdot (4bx)\right]}$$

$$= \boxed{\left[4b\left(a - bx^2\right)^{-3}\right] + \left[24b^2x^2\left(a - bx^2\right)^{-4}\right]}$$

$$\boxed{y'''} = \boxed{\left[-12b\left(a - bx^2\right)^{-3-1} \cdot (-2bx)\right] + \left[\left(48b^2x\right) \cdot \left(a - bx^2\right)^{-4} + \left(-96b^2x^2\right)\left(a - bx^2\right)^{-4-1} \cdot (-2bx)\right]}$$

$$= \boxed{\left[24b^2x\left(a - bx^2\right)^{-4}\right] + \left[48b^2x\left(a - bx^2\right)^{-4} + 192b^3x^3\left(a - bx^2\right)^{-5}\right]}$$

c. Given $y = \dfrac{x^2 + 3x + 1}{x + 1}$, then

$$\boxed{y'} = \boxed{\frac{\left[\left(2x^{2-1} + 3x^{1-1} + 0\right) \cdot (x + 1)\right] - \left[\left(x^{1-1} + 0\right) \cdot \left(x^2 + 3x + 1\right)\right]}{(x + 1)^2}} = \boxed{\frac{\left[(2x + 3) \cdot (x + 1)\right] - \left[1 \cdot \left(x^2 + 3x + 1\right)\right]}{(x + 1)^2}}$$

$$= \boxed{\frac{\left(2x^2 + 2x + 3x + 3\right) - \left(x^2 + 3x + 1\right)}{(x + 1)^2}} = \boxed{\frac{2x^2 + 5x + 3 - x^2 - 3x - 1}{(x + 1)^2}} = \boxed{\frac{x^2 + 2x + 2}{(x + 1)^2}}$$

$$\boxed{y''} = \boxed{\frac{\left[\left(2x^{2-1} + 2x^{1-1} + 0\right) \cdot (x + 1)^2\right] - \left[2(x + 1)^{2-1} \cdot \left(x^2 + 2x + 2\right)\right]}{(x + 1)^4}} = \boxed{\frac{\left[(2x + 2)(x + 1)^2\right] - \left[2(x + 1)\left(x^2 + 2x + 2\right)\right]}{(x + 1)^4}}$$

$$= \boxed{\frac{\left[(2x + 2)(x + 1)^2\right] - \left[2(x + 1)\left(x^2 + 2x + 2\right)\right]}{(x + 1)^4}} = \boxed{\frac{\left[2x^3 + 6x^2 + 6x + 2\right] - \left[2x^3 + 6x^2 + 8x + 4\right]}{(x + 1)^4}}$$

$$= \boxed{\frac{2x^3 + 6x^2 + 6x + 2 - 2x^3 - 6x^2 - 8x - 4}{(x + 1)^4}} = \boxed{\frac{-2x - 2}{(x + 1)^4}}$$

$$\boxed{y'''} = \boxed{\frac{\left[\left(-2x^{1-1} + 0\right) \cdot (x + 1)^4\right] - \left[4(x + 1)^{4-1} \cdot (-2x - 2)\right]}{(x + 1)^8}} = \boxed{\frac{\left[-2(x + 1)^4\right] + \left[(x + 1)^3(8x + 8)\right]}{(x + 1)^8}}$$

d. Given $y = \frac{1}{5}x^5 + \frac{1}{4}x^4 + x$, then

$$\boxed{y'} = \boxed{\frac{1}{5}\cdot 5x^{5-1} + \frac{1}{4}\cdot 4x^{4-1} + x^{1-1}} = \boxed{x^4 + x^3 + 1} \qquad\qquad \boxed{y''} = \boxed{4x^{4-1} + 3x^{3-1} + 0} = \boxed{4x^3 + 3x^2}\ \text{and}$$

$$\boxed{y'''} = \boxed{(4\cdot 3)x^{3-1} + (3\cdot 2)x^{2-1}} = \boxed{12x^2 + 6x}$$

e. Given $y = \frac{ax^2 + b}{c}$, then

$$\boxed{y'} = \boxed{\frac{\left[(a\cdot 2x^{2-1} + 0)\cdot c\right] - \left[0\cdot (ax^2 + b)\right]}{c^2}} = \boxed{\frac{(2ax\cdot c) - 0}{c^2}} = \boxed{\frac{2a\cancel{c}x}{c^{2=1}}} = \boxed{\frac{2ax}{c}} \qquad \boxed{y''} = \boxed{\frac{2a}{c}}\ \text{and}\ \boxed{y'''} = \boxed{0}$$

f. Given $y = \frac{x^2 + 1}{x^3}$, then

$$\boxed{y'} = \boxed{\frac{\left[(2x^{2-1} + 0)\cdot x^3\right] - \left[3x^{3-1}\cdot (x^2 + 1)\right]}{x^6}} = \boxed{\frac{\left[2x\cdot x^3\right] - \left[3x^2\cdot (x^2 + 1)\right]}{x^6}} = \boxed{\frac{2x^4 - 3x^4 - 3x^2}{x^6}} = \boxed{\frac{-x^4 - 3x^2}{x^6}}$$

$$= \boxed{\frac{-x^2(x^2 + 3)}{x^{6=4}}} = \boxed{-\frac{x^2 + 3}{x^4}}$$

$$\boxed{y''} = \boxed{-\frac{\left[(2x^{2-1} + 0)\cdot x^4\right] - \left[4x^{4-1}\cdot (x^2 + 3)\right]}{x^8}} = \boxed{-\frac{\left[2x\cdot x^4\right] - \left[4x^3\cdot (x^2 + 3)\right]}{x^8}} = \boxed{\frac{2x^5 - 4x^5 - 12x^3}{x^8}}$$

$$= \boxed{-\frac{-2x^5 - 12x^3}{x^8}} = \boxed{\frac{x^3(2x^2 + 12)}{x^{8=5}}} = \boxed{\frac{2x^2 + 12}{x^5}}$$

$$\boxed{y'''} = \boxed{\frac{\left[(4x^{2-1} + 0)\cdot x^5\right] - \left[5x^{5-1}\cdot (2x^2 + 12)\right]}{x^{10}}} = \boxed{\frac{\left[4x\cdot x^5\right] - \left[5x^4\cdot (2x^2 + 12)\right]}{x^{10}}} = \boxed{\frac{4x^6 - 10x^6 - 60x^4}{x^{10}}}$$

$$= \boxed{\frac{-6x^6 - 60x^4}{x^{10}}} = \boxed{\frac{x^4(-6x^2 - 60)}{x^{10=6}}} = \boxed{-\frac{6x^2 + 60}{x^6}}$$

Example 2.8-3: Find $f''(0)$ and $f''(1)$ for the following functions.

a. $f(x) = 6x^7 + 7x^2 - 2$
b. $f(x) = x^5(x-1)^2$
c. $f(x) = x - \frac{1}{x}$

d. $f(x) = \frac{x^3 + 1}{x}$
e. $f(x) = x^3 - \frac{1}{x+1}$
f. $f(x) = (ax + b)^2$

g. $f(x) = (x-1)^{-2}$
h. $f(x) = \frac{(x+1)^2}{x}$
i. $f(x) = (x+1)(x^2 + 1) + 5$

j. $f(x) = (1 + 5x)^3$
k. $f(x) = \frac{1+x}{x^3}$
l. $f(x) = \frac{1}{3}x^3 + \frac{1}{2}x^2 + x + 10$

Solutions:

a. Given $f(x) = 6x^7 + 7x^2 - 2$, then

$$\boxed{f'(x)} = \boxed{6 \cdot 7x^{7-1} + 7 \cdot 2x^{2-1} - 0} = \boxed{42x^6 + 14x} \text{ and } \boxed{f''(x)} = \boxed{42 \cdot 6x^{6-1} + 14 \cdot 1x^{1-1}} = \boxed{252x^5 + 14}$$

Therefore, $\boxed{f''(0)} = \boxed{252 \cdot 0^5 + 14} = \boxed{0 + 14} = \boxed{14}$ and $\boxed{f''(1)} = \boxed{252 \cdot 1^5 + 14} = \boxed{252 + 14} = \boxed{\mathbf{266}}$

b. Given $f(x) = x^5(x-1)^2$, then

$$\boxed{f'(x)} = \boxed{\left(5x^{5-1} \cdot 1\right) \cdot (x-1)^2 + \left[2(x-1) \cdot 1\right] \cdot x^5} = \boxed{5x^4(x-1)^2 + 2x^5(x-1)} = \boxed{5x^4(x-1)^2 + 2x^6 - 2x^5} \text{ and}$$

$$\boxed{f''(x)} = \boxed{\left[5 \cdot 4x^{4-1}(x-1)^2 + 2(x-1) \cdot 5x^4\right] + 2 \cdot 6x^{6-1} - 2 \cdot 5x^{5-1}} = \boxed{20x^3(x-1)^2 + 10x^4(x-1) + 12x^5 - 10x^4}$$

$$= \boxed{20x^3(x^2 - 2x + 1) + 10x^5 - 10x^4 + 12x^5 - 10x^4} = \boxed{20x^5 - 40x^4 + 20x^3 + 22x^5 - 20x^4}$$

$$= \boxed{42x^5 - 60x^4 + 20x^3} \text{ Therefore,}$$

$$\boxed{f''(0)} = \boxed{42 \cdot 0^5 - 60 \cdot 0^4 + 20 \cdot 0^3} = \boxed{0} \text{ and } \boxed{f''(1)} = \boxed{42 \cdot 1^5 - 60 \cdot 1^4 + 20 \cdot 1^3} = \boxed{42 - 60 + 20} = \boxed{2}$$

c. Given $f(x) = x - \dfrac{1}{x}$, then

$$\boxed{f'(x)} = \boxed{1 - \frac{0 \cdot x - 1 \cdot 1}{x^2}} = \boxed{1 + \frac{1}{x^2}} \text{ and } \boxed{f''(x)} = \boxed{0 + \frac{0 \cdot x - 2x \cdot 1}{x^4}} = \boxed{-\frac{2x}{x^4}} = \boxed{-\frac{2}{x^3}}$$

Therefore, $\boxed{f''(0)} = \boxed{-\dfrac{2}{0^3}} = \boxed{-\dfrac{2}{0}}$ which is not defined and $\boxed{f''(1)} = \boxed{-\dfrac{2}{1^3}} = \boxed{-\dfrac{2}{1}} = \boxed{-2}$

d. Given $f(x) = \dfrac{x^3 + 1}{x}$, then

$$\boxed{f'(x)} = \boxed{\frac{\left[3x^2 \cdot x\right] - \left[1 \cdot (x^3 + 1)\right]}{x^2}} = \boxed{\frac{3x^3 - x^3 - 1}{x^2}} = \boxed{\frac{2x^3 - 1}{x^2}} \text{ and}$$

$$\boxed{f''(x)} = \boxed{\frac{\left[6x^2 \cdot x^2\right] - \left[2x \cdot (2x^3 - 1)\right]}{x^4}} = \boxed{\frac{6x^4 - 4x^4 + 2x}{x^4}} = \boxed{\frac{2x^4 + 2x}{x^4}} = \boxed{\frac{2x(x^3 + 1)}{x^{4=3}}} = \boxed{\frac{2(x^3 + 1)}{x^3}}$$

Therefore, $\boxed{f''(0)} = \boxed{\dfrac{2(0^3 + 1)}{0^3}} = \boxed{\dfrac{2}{0}}$ which is not defined and $\boxed{f''(1)} = \boxed{\dfrac{2(1^3 + 1)}{1^3}} = \boxed{\dfrac{4}{1}} = \boxed{4}$

e. Given $f(x) = x^3 - \dfrac{1}{x+1}$, then

$$\boxed{f'(x)} = \boxed{3x^{3-1} - \frac{0 \cdot (x+1) - 1 \cdot 1}{(x+1)^2}} = \boxed{3x^2 + \frac{1}{(x+1)^2}} \text{ and}$$

$$f''(x) = 3 \cdot 2x^{2-1} + \frac{0 \cdot (x+1)^2 - 2(x+1)^{2-1} \cdot 1}{(x+1)^4} = 6x - \frac{2(x+1)}{(x+1)^4} = 6x - \frac{2}{(x+1)^3} \quad \text{Therefore,}$$

$$f''(0) = 6 \cdot 0 - \frac{2}{(0+1)^3} = -\frac{2}{1} = \boxed{-2} \quad \text{and} \quad f''(1) = 6 \cdot 1 - \frac{2}{(1+1)^3} = 6 - \frac{2}{2^3} = 6 - \frac{2}{8} = 6 - \frac{1}{4} = \boxed{5.75}$$

f. Given $f(x) = (ax+b)^2$, then

$$f'(x) = 2(ax+b)^{2-1} \cdot a = 2a(ax+b) = 2a^2x + 2ab \quad \text{and} \quad f''(x) = 2a^2 + 0 = 2a^2$$

Therefore, $f''(0) = 2a^2$ and $f''(1) = 2a^2$

Note that since $f''(x)$ is independent of x, $f''(x)$ is equal to $2a^2$ for all values of x.

g. Given $f(x) = (x-1)^{-2}$, then

$$f'(x) = -2(x-1)^{-2-1} \cdot 1 = -2(x-1)^{-3} \quad \text{and} \quad f''(x) = -2 \cdot \left[-3(x-1)^{-3-1} \cdot 1 \right] = 6(x-1)^{-4} \quad \text{Therefore,}$$

$$f''(0) = 6(0-1)^{-4} = \frac{6}{(-1)^4} = \frac{6}{1} = \boxed{6} \quad \text{and} \quad f''(1) = 6(1-1)^{-4} = \frac{6}{0^4} = \frac{6}{0} \quad \text{which is undefined}$$

h. Given $f(x) = \dfrac{(x+1)^2}{x}$, then

$$f'(x) = \frac{\left[2(x+1)^{2-1} \cdot 1 \cdot x \right] - \left[1 \cdot (x+1)^2 \right]}{x^2} = \frac{2x(x+1) - (x+1)^2}{x^2} = \frac{2x^2 + 2x - x^2 - 2x - 1}{x^2} = \frac{x^2 - 1}{x^2} \quad \text{and}$$

$$f''(x) = \frac{\left[2x \cdot x^2 \right] - \left[2x \cdot (x^2 - 1) \right]}{x^4} = \frac{2x^3 - 2x^3 + 2x}{x^4} = \frac{2x}{x^4} = \frac{2}{x^3}$$

Therefore, $f''(0) = \dfrac{2}{0^3} = \dfrac{2}{0}$ which is undefined and $f''(1) = \dfrac{2}{1^3} = \dfrac{2}{1} = \boxed{2}$

i. Given $f(x) = (x+1)(x^2+1) + 5 = x^3 + x + x^2 + 1 + 5 = x^3 + x^2 + x + 6$, then

$$f'(x) = 3x^2 + 2x + 1 \quad \text{and} \quad f''(x) = 6x + 2$$

Therefore, $f''(0) = 6 \cdot 0 + 2 = \boxed{2}$ and $f''(1) = 6 \cdot 1 + 2 = \boxed{8}$

j. Given $f(x) = (1+5x)^3$, then

$$f'(x) = 3(1+5x)^{3-1} \cdot 5 = 15(1+5x)^2 \quad \text{and} \quad f''(x) = 0 \cdot (1+5x)^2 + 2(1+5x) \cdot 5 \cdot 15 = 150(1+5x)$$

Therefore, $f''(0) = 150(1+5 \cdot 0) = \boxed{150}$ and $f''(1) = 150(1+5 \cdot 1) = 150(1+5) = 150 \cdot 6 = \boxed{900}$

k. Given $f(x) = \dfrac{1+x}{x^3}$, then

$$\boxed{f'(x)} = \boxed{\dfrac{1 \cdot x^3 - 3x^2(1+x)}{x^6}} = \boxed{\dfrac{x^3 - 3x^2 - 3x^3}{x^6}} = \boxed{\dfrac{-2x^3 - 3x^2}{x^6}} = \boxed{\dfrac{-x^2(2x+3)}{x^{6=4}}} = \boxed{\dfrac{-2x-3}{x^4}}$$

$$\boxed{f''(x)} = \boxed{\dfrac{\left[-2 \cdot x^4\right] - \left[4x^3(-2x-3)\right]}{x^8}} = \boxed{\dfrac{-2x^4 + 8x^4 + 12x^3}{x^8}} = \boxed{\dfrac{6x^4 + 12x^3}{x^8}} = \boxed{\dfrac{6x^3(x+2)}{x^{8=5}}} = \boxed{\dfrac{6x+12}{x^5}}$$

Therefore, $\boxed{f''(0)} = \boxed{\dfrac{6 \cdot 0 + 12}{0^5}} = \boxed{\dfrac{12}{0}}$ which is not defined and $\boxed{f''(1)} = \boxed{\dfrac{6 \cdot 1 + 12}{1^5}} = \boxed{\dfrac{18}{1}} = \boxed{18}$

l. Given $f(x) = \dfrac{1}{3}x^3 + \dfrac{1}{2}x^2 + x + 10$, then

$$\boxed{f'(x)} = \boxed{\dfrac{3}{3}x^{3-1} + \dfrac{2}{2}x^{2-1} + x^{1-1} + 0} = \boxed{x^2 + x + 1} \text{ and } \boxed{f''(x)} = \boxed{2x^{2-1} + x^{1-1} + 0} = \boxed{2x+1}$$

Therefore, $\boxed{f''(0)} = \boxed{2 \cdot 0 + 1} = \boxed{1}$ and $\boxed{f''(1)} = \boxed{2 \cdot 1 + 1} = \boxed{3}$

Example 2.8-4: Find $\dfrac{dy}{dx}$, $\dfrac{d^2 y}{dx^2}$, and $\dfrac{d^3 y}{dx^3}$ for the following functions.

a. $y = x^4 + 5x^3 + 6x^2 + 1$ b. $y = x + \dfrac{1}{x}$ c. $y = x(x+1)^3$

d. $y = \left(x^2 + 1\right)^{-2}$ e. $y = x^3 + 3x^2 + 10$ f. $y = \dfrac{1}{1+x}$

g. $y = x - \dfrac{1}{x}$ h. $y = ax^3 + bx$ i. $y = \dfrac{x^3 + 1}{x^2}$

Solutions:

a. Given $y = x^4 + 5x^3 + 6x^2 + 1$, then

$$\boxed{\dfrac{dy}{dx}} = \boxed{\dfrac{d}{dx}x^4 + 5\dfrac{d}{dx}x^3 + 6\dfrac{d}{dx}x^2 + \dfrac{d}{dx}1} = \boxed{4x^{4-1} + (5 \cdot 3)x^{3-1} + (6 \cdot 2)x^{2-1} + 0} = \boxed{\boldsymbol{4x^3 + 15x^2 + 12x}}$$

$$\boxed{\dfrac{d^2 y}{dx^2}} = \boxed{4\dfrac{d}{dx}x^3 + 15\dfrac{d}{dx}x^2 + 12\dfrac{d}{dx}x} = \boxed{(4 \cdot 3)x^{3-1} + (15 \cdot 2)x^{2-1} + (12 \cdot 1)} = \boxed{\boldsymbol{12x^2 + 30x + 12}}$$

$$\boxed{\dfrac{d^3 y}{dx^3}} = \boxed{12\dfrac{d}{dx}x^2 + 30\dfrac{d}{dx}x + \dfrac{d}{dx}12} = \boxed{(12 \cdot 2)x^{2-1} + (30 \cdot 1)x^{1-1} + 0} = \boxed{24x^1 + 30x^0} = \boxed{\boldsymbol{24x + 30}}$$

b. Given $y = x + \dfrac{1}{x}$ which is the same as $y = x + x^{-1}$, then

$$\boxed{\dfrac{dy}{dx}} = \boxed{\dfrac{d}{dx}x + \dfrac{d}{dx}x^{-1}} = \boxed{1 + \left(-1 \cdot x^{-1-1}\right)} = \boxed{\boldsymbol{1 - x^{-2}}}$$

$$\boxed{\dfrac{d^2 y}{dx^2}} = \boxed{\dfrac{d}{dx}1 - \dfrac{d}{dx}x^{-2}} = \boxed{0 - \left(-2 \cdot x^{-2-1}\right)} = \boxed{\boldsymbol{2x^{-3}}}$$

$$\boxed{\frac{d^3y}{dx^3}} = \boxed{2\frac{d}{dx}x^{-3}} = \boxed{(2 \cdot -3)x^{-3-1}} = \boxed{\boldsymbol{-6x^{-4}}}$$

c. Given $y = x(x+1)^3$, then

$$\boxed{\frac{dy}{dx}} = \boxed{\left[(x+1)^3\frac{d}{dx}x\right] + \left[x\frac{d}{dx}(x+1)^3\right]} = \boxed{\left[(x+1)^3 \cdot 1\right] + \left[x \cdot 3(x+1)^{3-1} \cdot 1\right]} = \boxed{\boldsymbol{(x+1)^3 + 3x(x+1)^2}}$$

$$\boxed{\frac{d^2y}{dx^2}} = \boxed{\frac{d}{dx}(x+1)^3 + \left[3(x+1)^2\frac{d}{dx}x\right] + \left[3x\frac{d}{dx}(x+1)^2\right]} = \boxed{\left[3(x+1)^{3-1} \cdot 1\right] + \left[3(x+1)^2 \cdot 1\right] + \left[3x \cdot 2(x+1)^{2-1} \cdot 1\right]}$$

$$= \boxed{3(x+1)^2 + 3(x+1)^2 + 6x(x+1)} = \boxed{\boldsymbol{6(x+1)^2 + 6x(x+1)}}$$

$$\boxed{\frac{d^3y}{dx^3}} = \boxed{6\frac{d}{dx}(x+1)^2 + \left\{\left[6(x+1)\frac{d}{dx}x\right] + \left[6x\frac{d}{dx}(x+1)\right]\right\}} = \boxed{\left[12(x+1)^{2-1} \cdot 1\right] + \left\{\left[6(x+1) \cdot 1\right] + \left[6x \cdot (x+1)^{1-1} \cdot 1\right]\right\}}$$

$$= \boxed{12(x+1) + 6(x+1) + 6x(x+1)^0} = \boxed{18(x+1) + 6x} = \boxed{18x + 18 + 6x} = \boxed{24x + 18} = \boxed{\boldsymbol{6(4x+3)}}$$

d. Given $y = (x+1)^{-2}$, then

$$\boxed{\frac{dy}{dx}} = \boxed{\frac{d}{dx}(x+1)^{-2}} = \boxed{\left[-2(x+1)^{-2-1}\right] \cdot \frac{d}{dx}(x+1)} = \boxed{-2(x+1)^{-3} \cdot 1} = \boxed{\boldsymbol{-2(x+1)^{-3}}}$$

$$\boxed{\frac{d^2y}{dx^2}} = \boxed{-2\frac{d}{dx}(x+1)^{-3}} = \boxed{\left[(-2 \cdot -3)(x+1)^{-3-1}\right] \cdot \frac{d}{dx}(x+1)} = \boxed{6(x+1)^{-4} \cdot 1} = \boxed{\boldsymbol{6(x+1)^{-4}}}$$

$$\boxed{\frac{d^3y}{dx^3}} = \boxed{6\frac{d}{dx}(x+1)^{-4}} = \boxed{\left[(6 \cdot -4)(x+1)^{-4-1}\right] \cdot \frac{d}{dx}(x+1)} = \boxed{-24(x+1)^{-5} \cdot 1} = \boxed{\boldsymbol{-24(x+1)^{-5}}}$$

e. Given $y = x^3 + 3x^2 + 10$, then

$$\boxed{\frac{dy}{dx}} = \boxed{\frac{d}{dx}x^3 + \frac{d}{dx}3x^2 + \frac{d}{dx}10} = \boxed{\frac{d}{dx}x^3 + 3\frac{d}{dx}x^2 + 0} = \boxed{3x^{3-1} + (3 \cdot 2)x^{2-1}} = \boxed{\boldsymbol{3x^2 + 6x}}$$

$$\boxed{\frac{d^2y}{dx^2}} = \boxed{\frac{d}{dx}3x^2 + \frac{d}{dx}6x} = \boxed{3\frac{d}{dx}x^2 + 6\frac{d}{dx}x} = \boxed{(3 \cdot 2)x^{2-1} + (6 \cdot 1)} = \boxed{\boldsymbol{6x+6}}$$

$$\boxed{\frac{d^3y}{dx^3}} = \boxed{\frac{d}{dx}6x + \frac{d}{dx}6} = \boxed{6\frac{d}{dx}x + 0} = \boxed{6 \cdot 1} = \boxed{\boldsymbol{6}}$$

f. Given $y = \dfrac{1}{1+x}$, then

$$\boxed{\frac{dy}{dx}} = \boxed{\frac{\left[(1+x)\frac{d}{dx}1\right] - \left[1 \cdot \frac{d}{dx}(1+x)\right]}{(1+x)^2}} = \boxed{\frac{0 - (1 \cdot 1)}{(1+x)^2}} = \boxed{\boldsymbol{-\frac{1}{(1+x)^2}}}$$

$$\frac{d^2 y}{dx^2} = -\frac{\left[(1+x)^2 \frac{d}{dx} 1\right] - \left[1 \cdot \frac{d}{dx}(1+x)^2\right]}{(1+x)^4} = -\frac{0 - \left[(1 \cdot 2)(1+x)^{2-1}\right]}{(1+x)^4} = \frac{2(1+x)}{(1+x)^{4=3}} = \boxed{\frac{2}{(1+x)^3}}$$

$$\frac{d^3 y}{dx^3} = \frac{\left[(1+x)^3 \frac{d}{dx} 2\right] - \left[2 \cdot \frac{d}{dx}(1+x)^3\right]}{(1+x)^6} = \frac{0 - \left[(2 \cdot 3)(1+x)^{3-1}\right]}{(1+x)^6} = -\frac{6(1+x)^2}{(1+x)^{6=4}} = \boxed{-\frac{6}{(1+x)^4}}$$

g. Given $y = x - \frac{1}{x}$ which is the same as $y = x - x^{-1}$, then

$$\frac{dy}{dx} = \frac{d}{dy} x - \frac{d}{dx} x^{-1} = \boxed{1 + x^{-1-1}} = \boxed{1 + x^{-2}}$$

$$\frac{d^2 y}{dx^2} = \frac{d}{dy} 1 + \frac{d}{dx} x^{-2} = \boxed{0 - 2x^{-2-1}} = \boxed{-2x^{-3}}$$

$$\frac{d^3 y}{dx^3} = -2 \frac{d}{dy} x^{-3} = \boxed{(-2 \cdot -3)x^{-3-1}} = \boxed{6x^{-4}}$$

h. Given $y = ax^3 + bx$, then

$$\frac{dy}{dx} = \frac{d}{dx} ax^3 + \frac{d}{dx} bx = a\frac{d}{dx} x^3 + b\frac{d}{dx} x = \boxed{(a \cdot 3)x^{3-1} + (b \cdot 1)x^{1-1}} = \boxed{3ax^2 + bx^0} = \boxed{3ax^2 + b}$$

$$\frac{d^2 y}{dx^2} = \frac{d}{dx} 3ax^2 + \frac{d}{dx} b = 3a\frac{d}{dx} x^2 + 0 = \boxed{(3a \cdot 2)x^{2-1}} = \boxed{6ax}$$

$$\frac{d^3 y}{dx^3} = \frac{d}{dx} 6ax = 6a\frac{d}{dx} x = \boxed{6a \cdot 1} = \boxed{6a}$$

i. Given $y = \frac{x^3 + 1}{x^2}$, then

$$\frac{dy}{dx} = \frac{\left[x^2 \frac{d}{dx}(x^3 + 1)\right] - \left[(x^3 + 1)\frac{d}{dx} x^2\right]}{x^4} = \frac{x^2\left[\frac{d}{dx} x^3 + \frac{d}{dx} 1\right] - (x^3 + 1) \cdot 2x}{x^4} = \frac{x^2\left[3x^{3-1} + 0\right] - 2x(x^3 + 1)}{x^4}$$

$$= \frac{3x^4 - 2x^4 - 2x}{x^4} = \frac{x^4 - 2x}{x^4} = \frac{x(x^3 - 2)}{x^{4=3}} = \boxed{\frac{x^3 - 2}{x^3}}$$

$$\frac{d^2 y}{dx^2} = \frac{\left[x^3 \frac{d}{dx}(x^3 - 2)\right] - \left[(x^3 - 2)\frac{d}{dx} x^3\right]}{x^6} = \frac{x^3\left[\frac{d}{dx} x^3 - \frac{d}{dx} 2\right] - (x^3 - 2) \cdot 3x^2}{x^6} = \frac{x^3\left[3x^{3-1} - 0\right] - 3x^2(x^3 - 2)}{x^6}$$

$$= \frac{3x^5 - 3x^5 + 6x^2}{x^6} = \frac{6x^2}{x^{6=4}} = \boxed{\frac{6}{x^4}}$$

$$\frac{d^3y}{dx^3} = \frac{x^4 \frac{d}{dx} 6 - 6\frac{d}{dx}x^4}{x^8} = \frac{\left(x^4 \cdot 0\right) - \left(6 \cdot 4\right)x^{4-1}}{x^8} = \frac{0 - 24x^3}{x^8} = \frac{24x^3}{x^{8=5}} = -\frac{24}{x^5}$$

Example 2.8-5: Find y' and y'' for the following functions. Do not simplify the answer to its lowest term.

a. $x^2 + y^2 = 2$ b. $xy + y^2 = 1$ c. $1 + x^2 y^2 = x$ d. $x^3 y + y = 1$

Solutions:

a. Given $x^2 + y^2 = 2$, then y' is equal to $2x^{2-1} + 2y \cdot y' = 0$; $2x + 2yy' = 0$; $2yy' = -2x$; $y' = \dfrac{-2x}{2y}$

; $y' = -\dfrac{x}{y}$ and $y'' = -\dfrac{(1 \cdot y) - (y' \cdot x)}{y^2}$; $y'' = -\dfrac{y - y'x}{y^2}$

b. Given $xy + y^2 = 1$, then y' is equal to $(1 \cdot y + y' \cdot x) + 2y \cdot y' = 0$; $y + y'x + 2yy' = 0$; $y'(x + 2y) = -y$

; $y' = -\dfrac{y}{x + 2y}$ and $y'' = -\dfrac{[y' \cdot (x + 2y)] - [(1 + 2y') \cdot y]}{(x + 2y)^2}$; $y'' = -\dfrac{xy' + 2yy' - y - 2yy'}{(x + 2y)^2}$; $y'' = -\dfrac{xy' - y}{(x + 2y)^2}$

c. Given $1 + x^2 y^2 = x$, then y' is equal to $0 + \left(2x \cdot y^2 + 2yy' \cdot x^2\right) = 1$; $2xy^2 + 2yy'x^2 = 1$

; $2yy'x^2 = 1 - 2xy^2$; $y' = \dfrac{1 - 2xy^2}{2x^2 y}$ and $y'' = \dfrac{\left\{\left[0 - 2\left(1 \cdot y^2 + 2yy' \cdot x\right)\right]\left(2x^2 y\right)\right\} - \left\{2\left(2xy + x^2 y'\right)\left(1 - 2xy^2\right)\right\}}{\left(2x^2 y\right)^2}$

; $y'' = \dfrac{-4x^2 y\left(y^2 + 2xyy'\right) - 2\left(2xy + x^2 y'\right)\left(1 - 2xy^2\right)}{4x^4 y^2}$

d. Given $x^3 y + y = 1$; $y\left(x^3 + 1\right) = 1$; $y = \dfrac{1}{x^3 + 1}$, then $y' = \dfrac{\left[0 \cdot \left(x^3 + 1\right)\right] - \left[3x^2 \cdot 1\right]}{\left(x^3 + 1\right)^2} = \dfrac{-3x^2}{\left(x^3 + 1\right)^2}$ and

$y'' = -\dfrac{\left[-6x \cdot \left(x^3 + 1\right)^2\right] - \left[2\left(x^3 + 1\right) \cdot 3x^2 \cdot -3x^2\right]}{\left(x^3 + 1\right)^4} = \dfrac{-6x\left(x^3 + 1\right)^2 + 18x^4\left(x^3 + 1\right)}{\left(x^3 + 1\right)^4} = \dfrac{6x\left(x^3 + 1\right)\left[-\left(x^3 + 1\right) + 3x^3\right]}{\left(x^3 + 1\right)^4}$

$= \dfrac{6x\left(-x^3 - 1 + 3x^3\right)}{\left(x^3 + 1\right)^3} = \dfrac{6x\left(2x^3 - 1\right)}{\left(x^3 + 1\right)^3}$

Example 2.8-6: Find the first, second, and third derivative of the following functions.

a. $f(x) = 6x^7 + 7x^2 - 2$

b. $f(x) = 3x^4 - 2x^2 + 5x + 9$

c. $f(x) = x^{-4} + 3x^{-3} + x^{-2} + x$

d. $f(x) = x^7 + 6x^5 + 8x + 3x^{-3}$

e. $f(x) = \dfrac{3}{x^3} + \dfrac{2}{x^2} + \dfrac{1}{x}$

f. $f(x) = x(2x+1)^3$

g. $f(x) = x^3 + \dfrac{1}{x^3}$

h. $f(x) = (x+1)^2 - x^3$

Solutions:

a. Given $f(x) = 6x^7 + 7x^2 - 2$, then

$$\boxed{f'(x)} = \boxed{(6\cdot7)x^{7-1} + (7\cdot2)x^{2-1} - 0} = \boxed{42x^6 + 14x}$$

$$\boxed{f''(x)} = \boxed{(42\cdot6)x^{6-1} + 14x^{1-1}} = \boxed{252x^5 + 14}$$

$$\boxed{f'''(x)} = \boxed{(252\cdot5)x^{5-1} + 0} = \boxed{1260x^4}$$

b. Given $f(x) = 3x^4 - 2x^2 + 5x + 9$, then

$$\boxed{f'(x)} = \boxed{(3\cdot4)x^{4-1} - (2\cdot2)x^{2-1} + 5x^{1-1} + 0} = \boxed{12x^3 - 4x + 5}$$

$$\boxed{f''(x)} = \boxed{(12\cdot3)x^{3-1} - 4x^{1-1}} = \boxed{36x^2 - 4}$$

$$\boxed{f'''(x)} = \boxed{(36\cdot2)x^{2-1} - 0} = \boxed{72x}$$

c. Given $f(x) = x^{-4} + 3x^{-3} + x^{-2} + x$, then

$$\boxed{f'(x)} = \boxed{-4x^{-4-1} + (3\cdot-3)x^{-3-1} - 2x^{-2-1} + x^{1-1}} = \boxed{-4x^{-5} - 9x^{-4} - 2x^{-3} + 1}$$

$$\boxed{f''(x)} = \boxed{(-4\cdot-5)x^{-5-1} + (-9\cdot-4)x^{-4-1} + (-2\cdot-3)x^{-3-1} + 0} = \boxed{20x^{-6} + 36x^{-5} + 6x^{-4}}$$

$$\boxed{f'''(x)} = \boxed{(20\cdot-6)x^{-6-1} + (36\cdot-5)x^{-5-1} + (6\cdot-4)x^{-4-1}} = \boxed{-120x^{-7} - 180x^{-6} - 24x^{-5}}$$

d. Given $f(x) = x^7 + 6x^5 + 8x + 3x^{-3}$, then

$$\boxed{f'(x)} = \boxed{7x^{7-1} + (6\cdot5)x^{5-1} + 8x^{1-1} + (3\cdot-3)x^{-3-1}} = \boxed{7x^6 + 30x^4 + 8 - 9x^{-4}}$$

$$\boxed{f''(x)} = \boxed{(7\cdot6)x^{6-1} + (30\cdot4)x^{4-1} + 0 + (-9\cdot-4)x^{-4-1}} = \boxed{42x^5 + 120x^3 + 36x^{-5}}$$

$$\boxed{(42\cdot5)x^{5-1} + (120\cdot3)x^{3-1} + (36\cdot-5)x^{-5-1}} = \boxed{210x^4 + 360x^2 - 180x^{-6}}$$

e. Given $f(x) = \dfrac{3}{x^3} + \dfrac{2}{x^2} + \dfrac{1}{x}$ which is equal to $f(x) = 3x^{-3} + 2x^{-2} + x^{-1}$, then

$$\boxed{f'(x)} = \boxed{(3 \cdot -3)x^{-3-1} + (2 \cdot -2)x^{-2-1} - x^{-1-1}} = \boxed{-9x^{-4} - 4x^{-3} - x^{-2}}$$

$$\boxed{f''(x)} = \boxed{(-9 \cdot -4)x^{-4-1} + (-4 \cdot -3)x^{-3-1} + (-1 \cdot -2)x^{-2-1}} = \boxed{36x^{-5} + 12x^{-4} + 2x^{-3}}$$

$$\boxed{f'''(x)} = \boxed{(36 \cdot -5)x^{-5-1} + (12 \cdot -4)x^{-4-1} + (2 \cdot -3)x^{-3-1}} = \boxed{-180x^{-6} - 48x^{-5} - 6x^{-4}}$$

f. Given $f(x) = x(2x+1)^3$, then

$$\boxed{f'(x)} = \boxed{\left[1 \cdot (2x+1)^3\right] + \left[3(2x+1)^{3-1} \cdot 2\right] \cdot x} = \boxed{(2x+1)^3 + 6x(2x+1)^2}$$

$$\boxed{f''(x)} = \boxed{\left[3(2x+1)^{3-1} \cdot 2\right] + \left[6 \cdot (2x+1)^2 + 2(2x+1)^{2-1} \cdot 2 \cdot 6x\right]} = \boxed{6(2x+1)^2 + 6(2x+1)^2 + 24x(2x+1)}$$

$$= \boxed{12(2x+1)^2 + 24x(2x+1)}$$

$$\boxed{f'''(x)} = \boxed{\left[(12 \cdot 2)(2x+1)^{2-1} \cdot 2\right] + \left[24 \cdot (2x+1) + (2x+1)^{1-1} \cdot 2 \cdot 24x\right]} = \boxed{48(2x+1) + \left[24(2x+1) + 48x\right]}$$

$$= \boxed{72(2x+1) + 48x}$$

g. Given $f(x) = x^3 + \dfrac{1}{x^3}$ which is equal to $f(x) = x^3 + x^{-3}$, then

$$\boxed{f'(x)} = \boxed{3x^{3-1} - 3x^{-3-1}} = \boxed{3x^2 - 3x^{-4}}$$

$$\boxed{f''(x)} = \boxed{(3 \cdot 2)x^{2-1} + (-3 \cdot -4)x^{-4-1}} = \boxed{6x + 12x^{-5}}$$

$$\boxed{f'''(x)} = \boxed{6x^{1-1} + (12 \cdot -5)x^{-5-1}} = \boxed{6x^0 - 60x^{-6}} = \boxed{6 - 60x^{-6}}$$

h. Given $f(x) = (x+1)^2 - x^3$, then

$$\boxed{f'(x)} = \boxed{2(x+1)^{2-1} \cdot 1 - 3x^{3-1}} = \boxed{2(x+1) - 3x^2} = \boxed{2x + 2 - 3x^2} = \boxed{-3x^2 + 2x + 2}$$

$$\boxed{f''(x)} = \boxed{(-3 \cdot 2)x^{2-1} + 2x^{1-1} + 0} = \boxed{-6x + 2x^0} = \boxed{-6x + 2}$$

$$\boxed{f'''(x)} = \boxed{-6x^{1-1} + 0} = \boxed{-6x^0} = \boxed{-6}$$

Section 2.8 Practice Problems - Higher Order Derivatives

1. Find the second derivative of the following functions.

a. $y = x^3 + 3x^2 + 5x - 1$ b. $y = x^2(x+1)^2$ c. $y = 3x^3 + 50x$

d. $y = x^5 + \dfrac{1}{x^2}$

e. $y = \dfrac{x^3}{x+1} - 5x^2$

f. $y = x^3\left(x^2 - 1\right)$

g. $y = x^4 + \dfrac{x^8 - 7x^5 + 5x}{10}$

h. $y = x^2 - \dfrac{1}{x+1}$

i. $y = \dfrac{1}{x^2} - 3x$

2. Find y''' for the following functions.

a. $y = x^5 + 6x^3 + 10$

b. $y = x^2 + \dfrac{1}{x}$

c. $y = 4x^3\left(x - 1\right)^2$

d. $y = \dfrac{x}{x+1}$

e. $y = x^8 - 10x^5 + 5x - 10$

f. $y = \dfrac{x-1}{x^2} + 5x^3$

3. Find $f''(0)$ and $f''(1)$ for the following functions.

a. $f(x) = 6x^5 + 3x^3 + 5$

b. $f(x) = x^3\left(x + 1\right)^2$

c. $f(x) = x + \left(x - 1\right)^2$

d. $f(x) = \left(x - 1\right)^{-3}$

e. $f(x) = \left(x - 1\right)\left(x^2 + 1\right)$

f. $f(x) = \left(x^3 - 1\right)^2 + \sqrt{2x}$

Chapter 3
Differentiation (Part II)

Quick Reference to Chapter 3 Problems

$$\boxed{\frac{d}{dx}\left(\frac{\tan^2 x}{\cot x}\right)} = \; ; \quad \boxed{\frac{d}{dx}\left(\frac{\sin 3x}{\cos 2x}\right)} = \; ; \quad \boxed{\frac{d}{dx}\left(\frac{\sin 4x}{x+3}\right)} =$$

$$\boxed{\frac{d}{dx}\left[\tan^{-1}\left(x^2+3\right)\right]} = \; ; \quad \boxed{\frac{d}{dx}\left(\frac{3x^2}{arc\tan x}\right)} = \; ; \quad \boxed{\frac{d}{dx}\left(\frac{arc\sin x}{x}+5x^3\right)} =$$

$$\boxed{\frac{d}{dx}\left(e^{\ln x}\tan x\right)} = \; ; \quad \boxed{\frac{d}{dx}\left(e^{\cos 5x}e^{-3x}\right)} = \; ; \quad \boxed{\frac{d}{dx}\left[\ln x^3 + \ln(\csc x)\right]} =$$

$$\boxed{\frac{d}{dx}\left(\sinh x^3 + \cosh 3x\right)} = \; ; \quad \boxed{\frac{d}{dx}\left(\sinh x^2 + \cosh x^2\right)} = \; ; \quad \boxed{\frac{d}{dx}\left(\frac{1}{6}\sinh 3x + 10x\right)} =$$

$$\boxed{\frac{d}{dx}\left(\sinh^{-1} x^2 + \ln x^3\right)} = \; ; \quad \boxed{\frac{d}{dx}\left(\tanh^{-1} 2x + \ln e^x\right)} = \; ; \quad \boxed{\frac{d}{dx}\left(\tanh^{-1} x^3 + 5x\right)} =$$

$$\boxed{\lim_{x \to 10}\frac{e^x - e^{10}}{x-10}} = \; ; \quad \boxed{\lim_{x \to 5}\frac{x^3 - 25x}{x^3 - 125}} = \; ; \quad \boxed{\lim_{x \to +\infty}\frac{2x^2 - \ln x}{3x^2 + 3\ln x}} =$$

Chapter 3 – Differentiation (Part II)

The objective of this chapter is to improve the student's ability to solve additional problems involving derivative of functions that was not addressed in the previous chapter. The derivative of trigonometric functions is addressed in Section 3.1. Finding the derivative of inverse trigonometric functions is discussed in Section 3.2. How to obtain the derivative of exponential and logarithmic expressions is addressed in Section 3.3. Differentiation of hyperbolic and inverse hyperbolic functions is addressed in Sections 3.4 and 3.5, respectively. Finally, evaluation of indeterminate forms of the type $\frac{0}{0}$, $\frac{\infty}{\infty}$, $0 \cdot \infty$, $\infty - \infty$, 1^{∞}, ∞^{0}, 0^{0} using L'Hopital's rule is discussed in Section 3.6. Each section is concluded by solving examples with practice problems to further enhance the student's ability.

3.1   Differentiation of Trigonometric Functions

The differential formulas involving trigonometric functions are defined as:

Table 3.1: Differentiation Formulas for Trigonometric Functions

$\dfrac{d}{dx}\sin u = \cos u \cdot \dfrac{du}{dx}$	$\dfrac{d}{dx}\cot u = -\csc^2 u \cdot \dfrac{du}{dx}$
$\dfrac{d}{dx}\cos u = -\sin u \cdot \dfrac{du}{dx}$	$\dfrac{d}{dx}\sec u = \sec u \tan u \cdot \dfrac{du}{dx}$
$\dfrac{d}{dx}\tan u = \sec^2 u \cdot \dfrac{du}{dx}$	$\dfrac{d}{dx}\csc u = -\csc u \cot u \cdot \dfrac{du}{dx}$

In the following examples we will solve problems using the above differentiation formulas:

Example 3.1-1: Find the derivative of the following trigonometric functions:

a. $y = \sin 8x - \cos 3x$ b. $y = \sin 3x^4$ c. $y = \cos x^2 + \tan x^3$

d. $y = \tan\left(1 + 3x^2\right)$ e. $y = \sin\sqrt{x^3} + \cos\sqrt{x}$ f. $y = \cos^3\sqrt{x}$

g. $y = \sec^5\sqrt{x}$ h. $y = \sin^6\sqrt[3]{x^2}$ i. $y = x^3 \sin^2 3x$

j. $y = (x+1)\cot x^2$ k. $y = \sqrt{\sin 5x}$ l. $y = \sqrt{\csc 5x}$

Solutions:

a. Given $\boxed{y = \sin 8x - \cos 3x}$ then $\boxed{\dfrac{dy}{dx}} = \boxed{\dfrac{d}{dx}(\sin 8x - \cos 3x)} = \boxed{\cos 8x \cdot \dfrac{d}{dx}8x + \sin 3x \cdot \dfrac{d}{dx}3x} = \boxed{\mathbf{8\cos 8x + 3\sin 3x}}$

b. Given $\boxed{y = \sin 3x^4}$ then $\boxed{\dfrac{dy}{dx}} = \boxed{\dfrac{d}{dx}\left(\sin 3x^4\right)} = \boxed{\cos 3x^4 \cdot \dfrac{d}{dx}3x^4} = \boxed{\cos 3x^4 \cdot 12x^{4-1}} = \boxed{\mathbf{12x^3 \cos 3x^4}}$

c. Given $\boxed{y = \cos x^2 + \tan x^3}$ then $\boxed{\dfrac{dy}{dx}} = \boxed{\dfrac{d}{dx}\left(\cos x^2 + \tan x^3\right)} = \boxed{-\sin x^2 \cdot \dfrac{d}{dx}x^2 + \sec^2 x^3 \cdot \dfrac{d}{dx}x^3}$

$$= \boxed{-\sin x^2 \cdot 2x + \sec^2 x^3 \cdot 3\,x^2} = \boxed{-2x\sin x^2 + 3x^2 \sec^2 x^3}$$

d. Given $\boxed{y = \tan\left(1+3x^2\right)}$ then $\boxed{\dfrac{dy}{dx}} = \boxed{\dfrac{d}{dx}\tan\left(1+3x^2\right)} = \boxed{\sec^2\left(1+3x^2\right)\cdot\dfrac{d}{dx}\left(1+3x^2\right)} = \boxed{\sec^2\left(1+3x^2\right)\cdot\left(0+6x\right)}$

$$= \boxed{\sec^2\left(1+3x^2\right)\cdot 6x} = \boxed{6x\sec^2\left(1+3x^2\right)}$$

e. Given $\boxed{y = \sin\sqrt{x^3} + \cos\sqrt{x}}$ then $\boxed{\dfrac{dy}{dx}} = \boxed{\dfrac{d}{dx}\left(\sin\sqrt{x^3} + \cos\sqrt{x}\right)} = \boxed{\dfrac{d}{dx}\left(\sin x^{\frac{3}{2}} + \cos x^{\frac{1}{2}}\right)}$

$$= \boxed{\dfrac{d}{dx}\sin x^{\frac{3}{2}} + \dfrac{d}{dx}\cos x^{\frac{1}{2}}} = \boxed{\cos x^{\frac{3}{2}}\cdot\dfrac{d}{dx}x^{\frac{3}{2}} - \sin x^{\frac{1}{2}}\cdot\dfrac{d}{dx}x^{\frac{1}{2}}} = \boxed{\cos x^{\frac{3}{2}}\cdot\dfrac{3}{2}x^{\frac{3}{2}-1} - \sin x^{\frac{1}{2}}\cdot\dfrac{1}{2}x^{\frac{1}{2}-1}}$$

$$= \boxed{\cos x^{\frac{3}{2}}\cdot\dfrac{3}{2}x^{\frac{1}{2}} - \sin x^{\frac{1}{2}}\cdot\dfrac{1}{2}x^{-\frac{1}{2}}} = \boxed{\dfrac{3}{2}x^{\frac{1}{2}}\cos x^{\frac{3}{2}} - \dfrac{1}{2}x^{-\frac{1}{2}}\sin x^{\frac{1}{2}}} = \boxed{\dfrac{3}{2}\sqrt{x}\cos x^{\frac{3}{2}} - \dfrac{1}{2\sqrt{x}}\sin x^{\frac{1}{2}}}$$

f. Given $\boxed{y = \cos^3\sqrt{x}}$ then $\boxed{\dfrac{dy}{dx}} = \boxed{\dfrac{d}{dx}\cos^3\sqrt{x}} = \boxed{\dfrac{d}{dx}\left(\cos\sqrt{x}\right)^3} = \boxed{3\left(\cos\sqrt{x}\right)^2\cdot\dfrac{d}{dx}\left(\cos\sqrt{x}\right)}$

$$= \boxed{3\left(\cos\sqrt{x}\right)^2\cdot -\sin\sqrt{x}\cdot\dfrac{d}{dx}\left(\sqrt{x}\right)} = \boxed{3\left(\cos\sqrt{x}\right)^2\cdot -\sin\sqrt{x}\cdot\dfrac{d}{dx}x^{\frac{1}{2}}} = \boxed{3\left(\cos\sqrt{x}\right)^2\cdot -\sin\sqrt{x}\cdot\dfrac{1}{2}x^{\frac{1}{2}-1}}$$

$$= \boxed{3\left(\cos\sqrt{x}\right)^2\cdot -\dfrac{1}{2}\sin\sqrt{x}\cdot x^{-\frac{1}{2}}} = \boxed{-\dfrac{3}{2}x^{-\frac{1}{2}}\left(\cos\sqrt{x}\right)^2\cdot\sin\sqrt{x}} = \boxed{-\dfrac{3}{2\sqrt{x}}\cos^2\sqrt{x}\,\sin\sqrt{x}}$$

g. Given $\boxed{y = \sec^5\sqrt{x}}$ then $\boxed{\dfrac{dy}{dx}} = \boxed{\dfrac{d}{dx}\sec^5\sqrt{x}} = \boxed{\dfrac{d}{dx}\left(\sec\sqrt{x}\right)^5} = \boxed{5\left(\sec\sqrt{x}\right)^4\cdot\dfrac{d}{dx}\left(\sec\sqrt{x}\right)}$

$$= \boxed{5\sec^4\sqrt{x}\cdot\sec\sqrt{x}\tan\sqrt{x}\cdot\dfrac{d}{dx}\left(\sqrt{x}\right)} = \boxed{5\sec^4\sqrt{x}\cdot\sec\sqrt{x}\tan\sqrt{x}\cdot\dfrac{1}{2}x^{\frac{1}{2}-1}} = \boxed{\dfrac{5}{2}\sec^4\sqrt{x}\cdot\sec\sqrt{x}\tan\sqrt{x}\cdot x^{-\frac{1}{2}}}$$

$$= \boxed{\dfrac{5}{2\sqrt{x}}\sec^4\sqrt{x}\cdot\sec\sqrt{x}\tan\sqrt{x}} = \boxed{\dfrac{5}{2\sqrt{x}}\sec^5\sqrt{x}\tan\sqrt{x}}$$

h. Given $\boxed{y = \sin^6\sqrt[3]{x^2} = \sin^6 x^{\frac{2}{3}}}$ then $\dfrac{dy}{dx} = \boxed{\dfrac{d}{dx}\left(\sin x^{\frac{2}{3}}\right)^6} = \boxed{6\left(\sin x^{\frac{2}{3}}\right)^5\cdot\dfrac{d}{dx}\sin x^{\frac{2}{3}}} = \boxed{6\sin^5 x^{\frac{2}{3}}\cdot\cos x^{\frac{2}{3}}\cdot\dfrac{d}{dx}x^{\frac{2}{3}}}$

$$= \boxed{6\sin^5 x^{\frac{2}{3}}\cdot\cos x^{\frac{2}{3}}\cdot\dfrac{2}{3}x^{\frac{2}{3}-1}} = \boxed{6\sin^5 x^{\frac{2}{3}}\cdot\cos x^{\frac{2}{3}}\cdot\dfrac{2}{3}x^{-\frac{1}{3}}} = \boxed{\dfrac{12}{3\sqrt[3]{x}}\sin^5 x^{\frac{2}{3}}\cdot\cos x^{\frac{2}{3}}} = \boxed{\dfrac{4}{\sqrt[3]{x}}\sin^5 x^{\frac{2}{3}}\cos x^{\frac{2}{3}}}$$

i. Given $\boxed{y = x^3\sin^2 3x}$ then $\boxed{\dfrac{dy}{dx}} = \boxed{\dfrac{d}{dx}\left(x^3\sin^2 3x\right)} = \boxed{\sin^2 3x\cdot\dfrac{d}{dx}x^3 + x^3\cdot\dfrac{d}{dx}\sin^2 3x}$

$$= \boxed{\sin^2 3x \cdot 3x^2 + x^3 \cdot \frac{d}{dx}(\sin 3x)^2} = \boxed{3x^2 \sin^2 3x + x^3 \cdot 2 \sin 3x \cdot \frac{d}{dx} \sin 3x} = \boxed{3x^2 \sin^2 3x + x^3 \cdot 2 \sin 3x \cdot \cos 3x \cdot \frac{d}{dx} 3x}$$

$$= \boxed{3x^2 \sin^2 3x + x^3 \cdot 2 \sin 3x \cdot \cos 3x \cdot 3} = \boxed{\mathbf{3x^2 \sin^2 3x + 6x^3 \sin 3x \cos 3x}}$$

j. Given $\boxed{y = (x+1)\cot x^2}$ then $\boxed{\frac{dy}{dx}} = \boxed{\frac{d}{dx}\left[(x+1)\cot x^2\right]} = \boxed{\cot x^2 \frac{d}{dx}(x+1) + (x+1)\frac{d}{dx}\cot x^2}$

$$= \boxed{\cot x^2 \cdot 1 + (x+1)\cdot -\csc^2 x^2 \cdot \frac{d}{dx}x^2} = \boxed{\cot x^2 - (x+1)\cdot \csc^2 x^2 \cdot 2x} = \boxed{\mathbf{\cot x^2 - 2x\,(x+1)\csc^2 x^2}}$$

k. Given $\boxed{y = \sqrt{\sin 5x} = (\sin 5x)^{\frac{1}{2}}}$ then $\boxed{\frac{dy}{dx}} = \boxed{\frac{d}{dx}(\sin 5x)^{\frac{1}{2}}} = \boxed{\frac{1}{2}(\sin 5x)^{\frac{1}{2}-1}\frac{d}{dx}\sin 5x} = \boxed{\frac{1}{2}(\sin 5x)^{-\frac{1}{2}} \cdot \cos 5x \cdot \frac{d}{dx}5x}$

$$= \boxed{\frac{1}{2}(\sin 5x)^{-\frac{1}{2}} \cdot \cos 5x \cdot 5} = \boxed{\frac{5}{2}\cdot\frac{\cos 5x}{(\sin 5x)^{\frac{1}{2}}}} = \boxed{\mathbf{\frac{5}{2}\cdot\frac{\cos 5x}{\sqrt{\sin 5x}}}}$$

l. Given $\boxed{y = \sqrt{\csc 5x} = (\csc 5x)^{\frac{1}{2}}}$ then $\boxed{\frac{dy}{dx}} = \boxed{\frac{d}{dx}(\csc 5x)^{\frac{1}{2}}} = \boxed{\frac{1}{2}(\csc 5x)^{\frac{1}{2}-1}\frac{d}{dx}\csc 5x}$

$$= \boxed{\frac{1}{2}(\csc 5x)^{-\frac{1}{2}} \cdot -\csc 5x \cot 5x \cdot \frac{d}{dx}5x} = \boxed{\frac{1}{2}(\csc 5x)^{-\frac{1}{2}} \cdot -\csc 5x \cot 5x \cdot 5} = \boxed{-\frac{5}{2}\cdot\frac{\csc 5x \cot 5x}{(\csc 5x)^{\frac{1}{2}}}}$$

$$= \boxed{-\frac{5}{2}\cdot \csc 5x \cdot (\csc 5x)^{-\frac{1}{2}} \cdot \cot 5x} = \boxed{-\frac{5}{2}\cdot (\csc 5x)^{1-\frac{1}{2}} \cdot \cot 5x} = \boxed{-\frac{5}{2}\cdot (\csc 5x)^{\frac{1}{2}} \cdot \cot 5x} = \boxed{\mathbf{-\frac{5}{2}\sqrt{\csc 5x}\ \cot 5x}}$$

Example 3.1-2: Find the derivative of the following trigonometric functions:

a. $y = \dfrac{\csc x}{x^2}$

b. $y = \dfrac{\sin x^2}{x+3}$

c. $y = \sin 5x$

d. $y = \sin x^5$

e. $y = \sin^5 x$

f. $y = \tan\left(x^2 + 1\right)$

g. $y = \tan\left(x^2 + 1\right)^3$

h. $y = \tan^3\left(x^2 + 1\right)$

i. $y = \cos x^3$

j. $y = \cos^3 x$

k. $y = \csc 10x$

l. $y = \csc x^{10}$

a. Given $\boxed{y = \dfrac{\csc x}{x^2}}$ then $\boxed{\dfrac{dy}{dx}} = \boxed{\dfrac{d}{dx}\left(\dfrac{\csc x}{x^2}\right)} = \boxed{\dfrac{\left(x^2 \cdot \frac{d}{dx}\csc x\right) - \left(\csc x \cdot \frac{d}{dx}x^2\right)}{x^4}} = \boxed{\dfrac{x^2 \cdot -\csc x \cot x - \csc x \cdot 2x}{x^4}}$

$$= \boxed{\dfrac{-x^2 \csc x \cot x - 2x\csc x}{x^4}} = \boxed{\dfrac{-x \csc x(x \cot x + 2)}{x^{4=3}}} = \boxed{\mathbf{\dfrac{-\csc x(x\cot x + 2)}{x^3}}}$$

b. Given $\boxed{y=\dfrac{\sin x^2}{x+3}}$ then $\boxed{\dfrac{dy}{dx}}=\boxed{\dfrac{d}{dx}\left(\dfrac{\sin x^2}{x+3}\right)}=\boxed{\dfrac{\left[(x+3)\cdot\frac{d}{dx}\sin x^2\right]-\left[\sin x^2\cdot\frac{d}{dx}(x+3)\right]}{(x+3)^2}}$

$=\boxed{\dfrac{(x+3)\cdot\cos x^2\cdot\frac{d}{dx}x^2-\sin x^2\cdot 1}{(x+3)^2}}=\boxed{\dfrac{(x+3)\cdot\cos x^2\cdot 2x-\sin x^2}{(x+3)^2}}=\boxed{\dfrac{2x\,(x+3)\cos x^2-\sin x^2}{(x+3)^2}}$

c. Given $\boxed{y=\sin 5x}$ then $\boxed{\dfrac{dy}{dx}}=\boxed{\dfrac{d}{dx}\sin 5x}=\boxed{\cos 5x\cdot\dfrac{d}{dx}5x}=\boxed{\cos 5x\cdot 5}=\boxed{\mathbf{5\cos 5x}}$

d. Given $\boxed{y=\sin x^5}$ then $\boxed{\dfrac{dy}{dx}}=\boxed{\dfrac{d}{dx}\sin x^5}=\boxed{\cos x^5\cdot\dfrac{d}{dx}x^5}=\boxed{\cos x^5\cdot 5x^4}=\boxed{\mathbf{5x^4\cos x^5}}$

e. Given $\boxed{y=\sin^5 x}$ then $\boxed{\dfrac{dy}{dx}}=\boxed{\dfrac{d}{dx}\sin^5 x}=\boxed{\dfrac{d}{dx}(\sin x)^5}=\boxed{5(\sin x)^4\cdot\dfrac{d}{dx}\sin x}=\boxed{5\,(\sin x)^4\cdot\cos x\cdot\dfrac{d}{dx}x}$

$=\boxed{5\,(\sin x)^4\cdot\cos x\cdot 1}=\boxed{\mathbf{5\sin^4 x\,\cos x}}$

f. Given $\boxed{y=\tan\left(x^2+1\right)}$ then $\boxed{\dfrac{dy}{dx}}=\boxed{\dfrac{d}{dx}\tan\left(x^2+1\right)}=\boxed{\sec^2\left(x^2+1\right)\cdot\dfrac{d}{dx}\left(x^2+1\right)}=\boxed{\mathbf{2x\,sec^2\!\left(x^2+1\right)}}$

g. Given $\boxed{y=\tan\left(x^2+1\right)^3}$ then $\boxed{\dfrac{dy}{dx}}=\boxed{\dfrac{d}{dx}\tan\left(x^2+1\right)^3}=\boxed{\sec^2\left(x^2+1\right)^3\cdot\dfrac{d}{dx}\left(x^2+1\right)^3}$

$=\boxed{\sec^2\left(x^2+1\right)^3\cdot 3\left(x^2+1\right)^2\cdot\dfrac{d}{dx}\left(x^2+1\right)}=\boxed{\sec^2\left(x^2+1\right)^3\cdot 3\left(x^2+1\right)^2\cdot 2x}=\boxed{\mathbf{6x\left(x^2+1\right)^2 sec^2\!\left(x^2+1\right)^3}}$

h. Given $\boxed{y=\tan^3\!\left(x^2+1\right)=\left[\tan\left(x^2+1\right)\right]^3}$ then $\boxed{\dfrac{dy}{dx}}=\boxed{\dfrac{d}{dx}\left[\tan\left(x^2+1\right)\right]^3}=\boxed{3\left[\tan\left(x^2+1\right)\right]^2\cdot\dfrac{d}{dx}\tan\left(x^2+1\right)}$

$=\boxed{3\tan^2\!\left(x^2+1\right)\cdot\sec^2\left(x^2+1\right)\cdot\dfrac{d}{dx}\left(x^2+1\right)}=\boxed{3\tan^2\!\left(x^2+1\right)\cdot\sec^2\left(x^2+1\right)\cdot 2x}=\boxed{\mathbf{6x\tan^2\!\left(x^2+1\right) sec^2\!\left(x^2+1\right)}}$

i. Given $\boxed{y=\cos x^3}$ then $\boxed{\dfrac{dy}{dx}}=\boxed{\dfrac{d}{dx}\cos x^3}=\boxed{-\sin x^3\cdot\dfrac{d}{dx}x^3}=\boxed{-\sin x^3\cdot 3x^2}=\boxed{\mathbf{-3\,x^2\sin x^3}}$

j. Given $\boxed{y=\cos^3 x=(\cos x)^3}$ then $\boxed{\dfrac{dy}{dx}}=\boxed{\dfrac{d}{dx}(\cos x)^3}=\boxed{3\,(\cos x)^2\cdot\dfrac{d}{dx}\cos x}=\boxed{3\,(\cos x)^2\cdot-\sin x\cdot\dfrac{d}{dx}x}$

$=\boxed{3\,(\cos x)^2\cdot-\sin x\cdot 1}=\boxed{\mathbf{-3\sin x\,(\cos x)^2}}$

k. Given $\boxed{y=\csc 10x}$ then $\boxed{\dfrac{dy}{dx}}=\boxed{\dfrac{d}{dx}\csc 10x}=\boxed{-\csc 10x\cot 10x\cdot\dfrac{d}{dx}10x}=\boxed{\mathbf{-10\csc 10x\cot 10x}}$

l. Given $\boxed{y = \csc x^{10}}$ then $\boxed{\dfrac{dy}{dx}} = \boxed{\dfrac{d}{dx}\csc x^{10}} = \boxed{-\csc x^{10} \cot x^{10} \cdot \dfrac{d}{dx} x^{10}} = \boxed{-10\, x^9 \csc x^{10} \cot x^{10}}$

Example 3.1-3: Find the derivative of the following trigonometric functions.

a. $y = \csc^{10} x$

b. $y = \sec 5x^4$

c. $y = \sec^4 5x^4$

d. $y = \sec x \cdot \cot x$

e. $y = \sin^2 x \cdot \sec (x+1)$

f. $y = \sin\left(x^2 + 1\right)\cdot \cos^2 x$

g. $y = \sin^2 x + \cos^2 x$

h. $y = \sqrt{4 + \cos 4x}$

i. $y = \sin^2 x^2$

j. $y = \sec^2 x - \tan^2 x$

k. $y = \cos\left(1 - x^2\right)$

l. $y = \cos^2 (3x+2)$

Solutions:

a. Given $\boxed{y = \csc^{10} x}$ then $\boxed{\dfrac{dy}{dx}} = \boxed{\dfrac{d}{dx}\left(\csc^{10} x\right)} = \boxed{10\csc^9 x \cdot \dfrac{d}{dx}\csc x} = \boxed{10\csc^9 x \cdot -\csc x \cot x \cdot \dfrac{d}{dx} x}$

$= \boxed{10\csc^9 x \cdot -\csc x \cot x \cdot 1} = \boxed{-10\,\csc^{10} x \cot x}$

b. Given $\boxed{y = \sec 5x^4}$ then $\boxed{\dfrac{dy}{dx}} = \boxed{\dfrac{d}{dx}\left(\sec 5x^4\right)} = \boxed{\sec 5x^4 \cdot \tan 5x^4 \cdot \dfrac{d}{dx}\left(5x^4\right)} = \boxed{\sec 5x^4 \cdot \tan 5x^4 \cdot (5 \cdot 4)x^{4-1}}$

$= \boxed{\sec 5x^4 \cdot \tan 5x^4 \cdot 20x^3} = \boxed{20x^3 \sec 5x^4 \tan 5x^4}$

c. Given $\boxed{y = \sec^4 5x^4 = \left(\sec 5x^4\right)^4}$ then $\boxed{\dfrac{dy}{dx}} = \boxed{\dfrac{d}{dx}\left(\sec 5x^4\right)^4} = \boxed{4\left(\sec 5x^4\right)^3 \cdot \dfrac{d}{dx}\left(\sec 5x^4\right)} = \boxed{4\left(\sec 5x^4\right)^3 \cdot \sec 5x^4}$

$= \boxed{\times \tan 5x^4 \cdot \dfrac{d}{dx}\left(5x^4\right)} = \boxed{4\left(\sec 5x^4\right)^3 \cdot \sec 5x^4 \cdot \tan 5x^4 \cdot 20x^3} = \boxed{80x^3 \left(\sec 5x^4\right)^4 \tan 5x^4}$

d. Given $\boxed{y = \sec x \cdot \cot x}$ then $\boxed{\dfrac{dy}{dx}} = \boxed{\dfrac{d}{dx}(\sec x \cdot \cot x)} = \boxed{\left(\cot x \cdot \dfrac{d}{dx}\sec x\right) + \left(\sec x \cdot \dfrac{d}{dx}\cot x\right)}$

$= \boxed{\left(\cot x \cdot \sec x \tan x \cdot \dfrac{d}{dx} x\right) + \left(\sec x \cdot -\csc^2 x \cdot \dfrac{d}{dx} x\right)} = \boxed{\left(\cot x \cdot \sec x \tan x \cdot 1\right) + \left(\sec x \cdot -\csc^2 x \cdot 1\right)}$

$= \boxed{\cot x \cdot \sec x \tan x + \sec x \cdot -\csc^2 x} = \boxed{\sec x \left(\tan x \cot x - \csc^2 x\right)}$

e. Given $\boxed{y = \sin^2 x \cdot \sec (x+1)}$ then $\boxed{\dfrac{dy}{dx}} = \boxed{\dfrac{d}{dx}\left[\sin^2 x \cdot \sec (x+1)\right]} = \boxed{\sec (x+1)\cdot \dfrac{d}{dx}\sin^2 x + \sin^2 x \cdot \dfrac{d}{dx}\sec (x+1)}$

$= \boxed{\sec (x+1)\cdot 2\sin x \cdot \cos x \cdot \dfrac{d}{dx} x + \sin^2 x \cdot \sec (x+1)\cdot \tan\ (x+1)\cdot \dfrac{d}{dx}(x+1)} = \boxed{\sec (x+1)\cdot 2\sin x \cdot \cos x \cdot 1}$

$$+\sin^2 x \cdot \sec (x+1)\tan (x+1)\cdot 1 = 2\sin x \cos x \sec (x+1)+\sin^2 x \sec (x+1)\tan (x+1)$$

$$= \sin x \sec (x+1)\big[\, 2\cos x + \sin x \tan (x+1)\,\big]$$

f. Given $y = \sin \left(x^2+1\right)\cdot \cos^2 x$ then $\dfrac{dy}{dx} = \dfrac{d}{dx}\left[\sin \left(x^2+1\right)\cdot \cos^2 x\right] = \left[\cos^2 x \cdot \dfrac{d}{dx}\sin \left(x^2+1\right)\right]$

$$+\left[\sin \left(x^2+1\right)\cdot \dfrac{d}{dx}\cos^2 x\right] = \left[\cos^2 x \cdot \cos \left(x^2+1\right)\cdot \dfrac{d}{dx}\left(x^2+1\right)\right] + \left[\sin \left(x^2+1\right)\cdot 2\cos x \cdot \dfrac{d}{dx}\cos x\right]$$

$$= \left[\cos^2 x \cdot \cos \left(x^2+1\right)\cdot 2x\right] + \left[\sin \left(x^2+1\right)\cdot 2\cos x \cdot -\sin x\right] = 2x \cos^2 x \cos \left(x^2+1\right) - 2\sin x \cos x \sin \left(x^2+1\right)$$

g. Given $y = \sin^2 x + \cos^2 x$ then $\dfrac{dy}{dx} = \dfrac{d}{dx}\left(\sin^2 x + \cos^2 x\right) = \dfrac{d}{dx}\sin^2 x + \dfrac{d}{dx}\cos^2 x = 2\sin x \cdot \dfrac{d}{dx}\sin x$

$$= +2\cos x \cdot \dfrac{d}{dx}\cos x = 2\sin x \cdot \cos x \cdot \dfrac{d}{dx}x + 2\cos x \cdot -\sin x \cdot \dfrac{d}{dx}x = 2\sin x \cdot \cos x \cdot 1 + 2\cos x \cdot -\sin x \cdot 1$$

$$= 2\sin x \cos x - 2\sin x \cos x = \boxed{0}$$

A second way of solving this problem is by noting that $\sin^2 x + \cos^2 x = 1$. Therefore, given

$$y = \sin^2 x + \cos^2 x = 1 \ \text{then}\ \dfrac{dy}{dx} = \dfrac{d}{dx}\left(\sin^2 x + \cos^2 x\right) = \dfrac{d}{dx}(1) = \boxed{0}$$

h. Given $y = \sqrt{4+\cos 4x} = (4+\cos 4x)^{\frac{1}{2}}$ then $\dfrac{dy}{dx} = \dfrac{d}{dx}(4+\cos 4x)^{\frac{1}{2}} = \dfrac{1}{2}(4+\cos 4x)^{\frac{1}{2}-1}\cdot \dfrac{d}{dx}(4+\cos 4x)$

$$= \dfrac{1}{2}(4+\cos 4x)^{-\frac{1}{2}}\cdot \left(\dfrac{d}{dx}4 + \dfrac{d}{dx}\cos 4x\right) = \dfrac{1}{2}(4+\cos 4x)^{-\frac{1}{2}}\cdot \left(0-\sin 4x \cdot \dfrac{d}{dx}4x\right) = \dfrac{1}{2}(4+\cos 4x)^{-\frac{1}{2}}\cdot (-4\sin 4x)$$

$$= -\dfrac{\overset{2}{4}\sin 4x}{2}\cdot (4+\cos 4x)^{-\frac{1}{2}} = -2\sin 4x \cdot (4+\cos 4x)^{-\frac{1}{2}} = -\dfrac{2\sin 4x}{(4+\cos 4x)^{\frac{1}{2}}}$$

i. Given $y = \sin^2 x^2$ then $\dfrac{dy}{dx} = \dfrac{d}{dx}\left(\sin^2 x^2\right) = \dfrac{d}{dx}\left(\sin x^2\right)^2 = 2\sin x^2 \cdot \dfrac{d}{dx}\sin x^2 = 2\sin x^2 \cdot \cos x^2 \cdot \dfrac{d}{dx}x^2$

$$= 2\sin x^2 \cdot \cos x^2 \cdot 2x = 4x \sin x^2 \cos x^2$$

j. Given $\boxed{y=\sec^2 x-\tan^2 x}$ then $\boxed{\dfrac{dy}{dx}}=\boxed{\dfrac{d}{dx}\left(\sec^2 x-\tan^2 x\right)}=\boxed{\dfrac{d}{dx}\sec^2 x-\dfrac{d}{dx}\tan^2 x}=\boxed{2\sec x\cdot\dfrac{d}{dx}\sec x}$

$=\boxed{-2\tan x\cdot\dfrac{d}{dx}\tan x}=\boxed{2\sec x\cdot\sec x\tan x\cdot\dfrac{d}{dx}x-2\tan x\cdot\sec^2 x\cdot\dfrac{d}{dx}x}=\boxed{2\sec^2 x\cdot\tan x\cdot 1-2\tan x\cdot\sec^2 x\cdot 1}$

$=\boxed{2\sec^2 x\ \tan x-\sec^2 x\,2\tan x}=\boxed{0}$

A second way of solving this problem is by noting that $\sec^2 x-\tan^2 x=1$. Therefore, given

$\boxed{y=\sec^2 x-\tan^2 x=1}$ then $\boxed{\dfrac{dy}{dx}}=\boxed{\dfrac{d}{dx}\left(\sec^2 x-\tan^2 x\right)}=\boxed{\dfrac{d}{dx}(1)}=\boxed{0}$

k. Given $\boxed{y=\cos\left(1-x^2\right)}$ then $\boxed{\dfrac{dy}{dx}}=\boxed{\dfrac{d}{dx}\left[\cos\left(1-x^2\right)\right]}=\boxed{-\sin\left(1-x^2\right)\cdot\dfrac{d}{dx}\left(1-x^2\right)}=\boxed{\mathbf{2x\sin\left(1-x^2\right)}}$

l. Given $\boxed{y=\cos^2(3x+2)}$ then $\boxed{\dfrac{dy}{dx}}=\boxed{\dfrac{d}{dx}\left[\cos^2(3x+2)\right]}=\boxed{\dfrac{d}{dx}\left[\cos(3x+2)\right]^2}=\boxed{2\cos(3x+2)\cdot\dfrac{d}{dx}\cos(3x+2)}$

$=\boxed{2\cos(3x+2)\cdot-\sin(3x+2)\cdot\dfrac{d}{dx}(3x+2)}=\boxed{2\cos(3x+2)\cdot-\sin(3x+2)\cdot 3}=\boxed{\mathbf{-6\cos(3x+2)\sin(3x+2)}}$

Note that since $\sin 2\alpha=2\sin\alpha\cos\alpha$ the above answer can also be written as:

$\boxed{\dfrac{dy}{dx}}=\boxed{-6\cos(3x+2)\sin(3x+2)}=\boxed{-3\cdot[2\sin(3x+2)\cos(3x+2)]}=\boxed{-3\cdot[\sin 2\cdot(3x+2)]}=\boxed{\mathbf{-3\sin(6x+4)}}$

Example 3.1-4: Find the derivative of the following trigonometric functions:

a. $y=x^2\sin^3 2x$ b. $y=\dfrac{\sin^2 x}{\cos x}$ c. $y=\dfrac{\tan^2 x}{\cot x}$

d. $y=\sec^2 10x$ e. $y=\cos^3 2x$ f. $y=x^3\cot x^5$

g. $y=\dfrac{\sin^2 x}{x}$ h. $y=\cos^2(1+3x)$ i. $y=x^5\cot^3 x$

j. $y=\tan 4x\sin 5x$ k. $y=\dfrac{\sin 3x}{\cos 2x}$ l. $y=\dfrac{\sin 4x}{x+3}$

Solutions:

a. $\boxed{\dfrac{dy}{dx}}=\boxed{\dfrac{d}{dx}\left(x^2\sin^3 2x\right)}=\boxed{\sin^3 2x\cdot\dfrac{d}{dx}x^2+x^2\cdot\dfrac{d}{dx}\sin^3 2x}=\boxed{\sin^3 2x\cdot 2x+x^2\cdot 3\sin^2 2x\cdot\dfrac{d}{dx}\sin 2x}$

$=\boxed{2x\sin^3 2x+x^2\cdot 3\sin^2 2x\cdot\cos 2x\cdot\dfrac{d}{dx}2x}=\boxed{2x\sin^3 2x+x^2\cdot 3\sin^2 2x\cdot\cos 2x\cdot 2}=\boxed{2x\sin^3 2x+6x^2}$

$$\boxed{\times \sin^2 2x \cos 2x} = \boxed{2x\,\sin^2 2x\left(\sin 2x + 3x\cos 2x\right)}$$

b. $\boxed{\dfrac{dy}{dx}} = \boxed{\dfrac{d}{dx}\left(\dfrac{\sin^2 x}{\cos x}\right)} = \boxed{\dfrac{\left(\cos x\cdot\frac{d}{dx}\sin^2 x\right)-\left(\sin^2 x\cdot\frac{d}{dx}\cos x\right)}{\cos^2 x}} = \boxed{\dfrac{\left(\cos x\cdot 2\sin x\cdot\frac{d}{dx}\sin x\right)-\left(\sin^2 x\cdot -\sin x\right)}{\cos^2 x}}$

$= \boxed{\dfrac{\left(\cos x\cdot 2\sin x\cdot\cos x\right)+\sin^3 x}{\cos^2 x}} = \boxed{\dfrac{2\sin x\cos^2 x+\sin^3 x}{\cos^2 x}} = \boxed{\dfrac{\sin x\left(2\cos^2 x+\sin^2 x\right)}{\cos^2 x}}$

c. $\boxed{\dfrac{dy}{dx}} = \boxed{\dfrac{d}{dx}\left(\dfrac{\tan^2 x}{\cot x}\right)} = \boxed{\dfrac{\left(\cot x\cdot\frac{d}{dx}\tan^2 x\right)-\left(\tan^2 x\cdot\frac{d}{dx}\cot x\right)}{\cot^2 x}} = \boxed{\dfrac{\left(\cot x\cdot 2\tan x\cdot\frac{d}{dx}\tan x\right)-\left(\tan^2 x\cdot -\csc^2 x\right)}{\cot^2 x}}$

$= \boxed{\dfrac{\left(\cot x\cdot 2\tan x\cdot\sec^2 x\right)+\tan^2 x\csc^2 x}{\cot^2 x}} = \boxed{\dfrac{\tan x\left(2\cot x\sec^2 x+\tan x\csc^2 x\right)}{\cot^2 x}}$

d. $\boxed{\dfrac{dy}{dx}} = \boxed{\dfrac{d}{dx}\left(\sec^2 10x\right)} = \boxed{2\sec 10x\cdot\dfrac{d}{dx}\left(\sec 10x\right)} = \boxed{2\sec 10x\cdot\sec 10x\tan 10x\cdot\dfrac{d}{dx}\left(10x\right)} = \boxed{2\sec 10x\cdot\sec 10x}$

$\boxed{\times\tan 10x\cdot 10\cdot\dfrac{d}{dx}x} = \boxed{2\sec^2 10x\tan 10x\cdot 10} = \boxed{20\sec^2 10x\tan 10x}$

e. $\boxed{\dfrac{dy}{dx}} = \boxed{\dfrac{d}{dx}\left(\cos^3 2x\right)} = \boxed{3\cos^2 2x\cdot\dfrac{d}{dx}\left(\cos 2x\right)} = \boxed{3\cos^2 2x\cdot -\sin 2x\dfrac{d}{dx}2x} = \boxed{3\cos^2 2x\cdot -\sin 2x\cdot 2\cdot\dfrac{d}{dx}x}$

$= \boxed{3\cos^2 2x\cdot -\sin 2x\cdot 2} = \boxed{-6\cos^2 2x\sin 2x}$

f. $\boxed{\dfrac{dy}{dx}} = \boxed{\dfrac{d}{dx}\left(x^3\cot x^5\right)} = \boxed{\cot x^5\cdot\dfrac{d}{dx}x^3 + x^3\cdot\dfrac{d}{dx}\cot x^5} = \boxed{\cot x^5\cdot 3x^2 + x^3\cdot -\csc^2 x^5\cdot\dfrac{d}{dx}x^5}$

$= \boxed{3x^2\cot x^5 - x^3\cdot\csc^2 x^5\cdot 5\,x^4} = \boxed{3x^2\cot x^5 - 5x^7\csc^2 x^5} = \boxed{x^2\left(3\cot x^5 - 5x^5\csc^2 x^5\right)}$

g. $\boxed{\dfrac{dy}{dx}} = \boxed{\dfrac{d}{dx}\left(\dfrac{\sin^2 x}{x}\right)} = \boxed{\dfrac{\left(x\cdot\frac{d}{dx}\sin^2 x\right)-\left(\sin^2 x\cdot\frac{d}{dx}x\right)}{x^2}} = \boxed{\dfrac{\left(x\cdot 2\sin x\cdot\frac{d}{dx}\sin x\right)-\left(\sin^2 x\cdot 1\right)}{x^2}}$

$= \boxed{\dfrac{\left(x\cdot 2\sin x\cdot\cos x\right)-\sin^2 x}{x^2}} = \boxed{\dfrac{2x\sin x\cos x-\sin^2 x}{x^2}} = \boxed{\dfrac{\sin x\left(2x\cos x-\sin x\right)}{x^2}}$

h. $\dfrac{dy}{dx} = \dfrac{d}{dx}\left[\cos^2(1+3x)\right] = 2\cos(1+3x)\cdot\dfrac{d}{dx}\cos(1+3x) = 2\cos(1+3x)\cdot-\sin(1+3x)\cdot\dfrac{d}{dx}(1+3x)$

$= -2\cos(1+3x)\cdot\sin(1+3x)\cdot\left(\dfrac{d}{dx}1+\dfrac{d}{dx}3x\right) = -2\cos(1+3x)\cdot\sin(1+3x)\cdot(0+3) = \boxed{\mathbf{-6\cos(1+3x)\sin(1+3x)}}$

i. $\dfrac{dy}{dx} = \dfrac{d}{dx}\left(x^5\cot^3 x\right) = \cot^3 x\cdot\dfrac{d}{dx}x^5 + x^5\cdot\dfrac{d}{dx}\cot^3 x = \cot^3 x\cdot 5x^4 + x^5\cdot 3\cot^2 x\cdot\dfrac{d}{dx}\cot x$

$= 5x^4\cot^3 x + x^5\cdot 3\cot^2 x\cdot-\csc^2 x\cdot\dfrac{d}{dx}x = 5x^4\cot^3 x - x^5\cdot 3\cot^2 x\cdot\csc^2 x\cdot 1 = \boxed{\mathbf{x^4\cot^2 x\left(5\cot x - 3x\csc^2 x\right)}}$

j. $\dfrac{dy}{dx} = \dfrac{d}{dx}(\tan 4x\sin 5x) = \tan 4x\dfrac{d}{dx}\sin 5x + \sin 5x\cdot\dfrac{d}{dx}\tan 4x = \tan 4x\cdot\cos 5x\cdot\dfrac{d}{dx}5x + \sin 5x\cdot\sec^2 4x\cdot\dfrac{d}{dx}4x$

$= \tan 4x\cdot\cos 5x\cdot 5 + \sin 5x\cdot\sec^2 4x\cdot 4 = \boxed{\mathbf{5\tan 4x\cos 5x + 4\sin 5x\sec^2 4x}}$

k. $\dfrac{dy}{dx} = \dfrac{d}{dx}\left(\dfrac{\sin 3x}{\cos 2x}\right) = \dfrac{\left(\cos 2x\cdot\frac{d}{dx}\sin 3x\right)-\left(\sin 3x\cdot\frac{d}{dx}\cos 2x\right)}{\cos^2 2x} = \dfrac{\left(\cos 2x\cdot\cos 3x\cdot\frac{d}{dx}3x\right)-\left(\sin 3x\cdot-\sin 2x\cdot\frac{d}{dx}2x\right)}{\cos^2 2x}$

$= \dfrac{(\cos 2x\cdot\cos 3x\cdot 3)-(\sin 3x\cdot-\sin 2x\cdot 2)}{\cos^2 2x} = \boxed{\dfrac{\mathbf{3\cos 3x\cos 2x + 2\sin 3x\sin 2x}}{\mathbf{\cos^2 2x}}}$

l. $\dfrac{dy}{dx} = \dfrac{d}{dx}\left(\dfrac{\sin 4x}{x+3}\right) = \dfrac{\left[(x+3)\cdot\frac{d}{dx}\sin 4x\right]-\left[\sin 4x\cdot\frac{d}{dx}(x+3)\right]}{(x+3)^2} = \dfrac{\left[(x+3)\cdot\cos 4x\cdot\frac{d}{dx}4x\right]-\left[\sin 4x\cdot 1\right]}{(x+3)^2}$

$= \dfrac{\left[(x+3)\cdot\cos 4x\cdot 4\right]-\sin 4x}{(x+3)^2} = \boxed{\dfrac{\mathbf{4(x+3)\cos 4x - \sin 4x}}{\mathbf{(x+3)^2}}}$

Example 3.1-5: Find the derivative of the following trigonometric functions:

a. $\sin y = \cot 3x$

b. $\cos 3y = \tan 5x$

c. $\sin(y+1) = \cos x^2$

d. $\sin y = \cos 10x$

e. $\tan(3y+2) = \sin 5x$

f. $\sec^2 y = \csc x$

g. $x\cos y = \sin(x+y)$

h. $x^2\sin 2y = \cos(3x+5y)$

i. $x\tan y = \cot x$

a. Given $\sin y = \cot 3x$, let's take the derivative on both sides of the equation to obtain $\dfrac{dy}{dx} = y'$.

$\boxed{\cos y\cdot\dfrac{d}{dx}y = -\csc^2 3x\cdot\dfrac{d}{dx}3x}$; $\boxed{\cos y\cdot y' = -\csc^2 3x\cdot 3}$; $\boxed{y' = -\dfrac{\mathbf{3\csc^2 3x}}{\mathbf{\cos y}}}$

b. Given $\cos 3y = \tan 5x$, let's take the derivative on both sides of the equation to obtain $\frac{dy}{dx} = y'$.

$\boxed{-\sin 3y \cdot \frac{d}{dx} 3y = \sec^2 5x \cdot \frac{d}{dx} 5x}$; $\boxed{-\sin 3y \cdot 3y' = \sec^2 5x \cdot 5}$; $\boxed{-3\sin 3y \cdot y' = 5\sec^2 5x}$; $\boxed{y' = -\frac{5\sec^2 5x}{3\sin 3y}}$

c. Given $\sin(y+1) = \cos x^2$, let's take the derivative on both sides of the equation to obtain $\frac{dy}{dx} = y'$.

$\boxed{\cos(y+1) \cdot \frac{d}{dx}(y+1) = -\sin x^2 \cdot \frac{d}{dx} x^2}$; $\boxed{\cos(y+1) \cdot y' = -\sin x^2 \cdot 2x}$; $\boxed{y' = -\frac{2x\sin x^2}{\cos(y+1)}}$

d. Given $\sin y = \cos 10x$, let's take the derivative on both sides of the equation to obtain $\frac{dy}{dx} = y'$.

$\boxed{\cos y \cdot \frac{d}{dx} y = -\sin 10x \cdot \frac{d}{dx} 10x}$; $\boxed{\cos y \cdot y' = -\sin 10x \cdot 10}$; $\boxed{y' = -\frac{10\sin 10x}{\cos y}}$

e. Given $\tan(3y+2) = \sin 5x$, let's take the derivative on both sides of the equation to obtain $\frac{dy}{dx} = y'$.

$\boxed{\sec^2(3y+2) \cdot \frac{d}{dx}(3y+2) = \cos 5x \cdot \frac{d}{dx} 5x}$; $\boxed{\sec^2(3y+2) \cdot 3y' = \cos 5x \cdot 5}$; $\boxed{y' = \frac{5\cos 5x}{3\sec^2(3y+2)}}$

f. Given $\sec^2 y = \csc x$, let's take the derivative on both sides of the equation to obtain $\frac{dy}{dx} = y'$.

$\boxed{2\sec y \cdot \frac{d}{dx}\sec y = -\csc x \cot x \cdot \frac{d}{dx} x}$; $\boxed{2\sec y \cdot \sec y \tan y \cdot \frac{d}{dx} y = -\csc x \cot x \cdot 1}$; $\boxed{2\sec^2 y \tan y \cdot y' = -\csc x \cot x}$

; $\boxed{y' = \frac{-\csc x \cot x}{2\sec^2 y \tan y}}$; $\boxed{y' = -\frac{\csc x \cot x}{2\sec^2 y \tan y}}$

g. Given $x\cos y = \sin(x+y)$, let's take the derivative on both sides of the equation to obtain $\frac{dy}{dx} = y'$.

$\boxed{\cos y \cdot \frac{d}{dx} x + x \cdot \frac{d}{dx}\cos y = \cos(x+y) \cdot \frac{d}{dx}(x+y)}$; $\boxed{\cos y \cdot 1 + x \cdot -\sin y \cdot \frac{d}{dx} y = \cos(x+y) \cdot \left(\frac{d}{dx} x + \frac{d}{dx} y\right)}$

; $\boxed{\cos y \cdot 1 + x \cdot -\sin y \cdot y' = \cos(x+y) \cdot (1+y')}$; $\boxed{\cos y - x\sin y\, y' = \cos(x+y) + y'\cos(x+y)}$

; $\boxed{-x\sin y\, y' - y'\cos(x+y) = \cos(x+y) - \cos y}$; $\boxed{x\sin y\, y' + y'\cos(x+y) = \cos y - \cos(x+y)}$

; $\boxed{-y'[x\sin y + \cos(x+y)] = \cos y - \cos(x+y)}$; $\boxed{y' = -\frac{\cos y - \cos(x+y)}{x\sin y + \cos(x+y)}}$

h. Given $x^2 \sin 2y = \cos(3x+5y)$, let's take the derivative on both sides of the equation to obtain $\frac{dy}{dx} = y'$.

$\boxed{\sin 2y \cdot \frac{d}{dx} x^2 + x^2 \cdot \frac{d}{dx}\sin 2y = -\sin(3x+5y) \cdot \frac{d}{dx}(3x+5y)}$; $\boxed{\sin 2y \cdot 2x + x^2 \cdot \cos 2y \cdot \frac{d}{dx} 2y = -\sin(3x+5y)}$

$$\times \left(\frac{d}{dx} 3x + \frac{d}{dx} 5y \right) \;;\; \sin 2y \cdot 2x + x^2 \cdot \cos 2y \cdot 2y' = -\sin(3x+5y) \cdot (3+5y') \;;\; 2x \sin 2y + 2y' x^2 \cos 2y$$

$$= -3\sin(3x+5y) - 5y' \sin(3x+5y) \;;\; 2y' x^2 \cos 2y + 5y' \sin(3x+5y) = -3\sin(3x+5y) - 2x \sin 2y$$

$$;\; y' \left[2x^2 \cos 2y + 5 \sin(3x+5y) \right] = -3\sin(3x+5y) - 2x \sin 2y \;;\; y' = -\frac{3\sin(3x+5y) + 2x \sin 2y}{\left[2x^2 \cos 2y + 5 \sin(3x+5y) \right]}$$

i. Given $x \tan y = \cot x$, let's take the derivative on both sides of the equation to obtain $\frac{dy}{dx} = y'$.

$$\tan y \cdot \frac{d}{dx} x + x \cdot \frac{d}{dx} \tan y = -\csc^2 x \cdot \frac{d}{dx} x \;;\; \tan y \cdot 1 + x \cdot \sec^2 y \cdot \frac{d}{dx} y = -\csc^2 x \cdot 1 \;;\; \tan y + x \cdot \sec^2 y \cdot y' = -\csc^2 x$$

$$x \cdot y' \cdot \sec^2 y = -\csc^2 x - \tan y \;;\; y' = -\frac{\csc^2 x + \tan y}{x \sec^2 y}$$

Example 3.1-6: Find the first and second derivative of the following trigonometric functions:

a. $y = \sin 5x$ b. $y = \sin^2 5x$ c. $y = \cos(10x+3)$

d. $y = \tan 2x$ e. $y = \tan x^2$ f. $y = \tan^2 x$

g. $y = \sin 3x + \cos 3x$ h. $y = x \sin x$ i. $y = \tan(5x+1)$

j. $y = \sin x - x \cos x$ k. $y = \sqrt{2 + \sin 2x}$ l. $y = \left(\sin x^2 \right)^2$

Solutions:

a. $\dfrac{dy}{dx} = \dfrac{d}{dx}(\sin 5x) = \cos 5x \cdot \dfrac{d}{dx}(5x) = \cos 5x \cdot 5 = \boxed{5\cos 5x}$

$\dfrac{d^2 y}{dx^2} = \dfrac{d}{dx}(5\cos 5x) = -5\sin 5x \cdot \dfrac{d}{dx} 5x = -5\sin 5x \cdot 5 = \boxed{-25\sin 5x}$

b. $\dfrac{dy}{dx} = \dfrac{d}{dx}(\sin^2 5x) = \dfrac{d}{dx}(\sin 5x)^2 = 2\sin 5x \cdot \dfrac{d}{dx}(\sin 5x) = 2\sin 5x \cdot \cos 5x \cdot \dfrac{d}{dx}(5x) = \boxed{10\sin 5x \cos 5x}$

$\dfrac{d^2 y}{dx^2} = \dfrac{d}{dx}(10\sin 5x \cos 5x) = 10\dfrac{d}{dx}(\sin 5x \cos 5x) = 10\left[\cos 5x \cdot \dfrac{d}{dx}(\sin 5x) + \sin 5x \cdot \dfrac{d}{dx}(\cos 5x) \right]$

$= 10(\cos 5x \cdot 5\cos 5x - 5\sin 5x \cdot \sin 5x) = 10\left(5\cos^2 5x - 5\sin^2 5x\right) = \boxed{50\left(\cos^2 5x - \sin^2 5x\right)}$

Note that since $\sin 2\alpha = 2\sin \alpha \cos \alpha$ the first derivative can also be written as:

$\dfrac{dy}{dx} = 5\left(2\sin 5x \cos 5x\right) = 5\left(\sin 2\cdot 5x\right) = 5\sin 10x$. Thus, $\dfrac{d^2 y}{dx^2} = 5\cdot\cos 10x \cdot \dfrac{d}{dx}10x = 50\cos 10x$

c. $\boxed{\dfrac{dy}{dx}} = \boxed{\dfrac{d}{dx}\left[\cos\left(10x+3\right)\right]} = \boxed{-\sin\left(10x+3\right)\cdot\dfrac{d}{dx}\left(10x+3\right)} = \boxed{-\sin\left(10x+3\right)\cdot 10} = \boxed{\mathbf{-10\sin\left(10x+3\right)}}$

$\boxed{\dfrac{d^2 y}{dx^2}} = \boxed{-\dfrac{d}{dx}\left[10\sin\left(10x+3\right)\right]} = \boxed{-10\cos\left(10x+3\right)\cdot\dfrac{d}{dx}\left(10x+3\right)} = \boxed{-10\cos\left(10x+3\right)\cdot 10} = \boxed{\mathbf{-100\cos\left(10x+3\right)}}$

d. $\boxed{\dfrac{dy}{dx}} = \boxed{\dfrac{d}{dx}\left(\tan 2x\right)} = \boxed{\sec^2 2x\cdot\dfrac{d}{dx}(2x)} = \boxed{\sec^2 2x\cdot 2} = \boxed{\mathbf{2\sec^2 2x}}$

$\boxed{\dfrac{d^2 y}{dx^2}} = \boxed{\dfrac{d}{dx}\left(2\sec^2 2x\right)} = \boxed{2\cdot\dfrac{d}{dx}\left(\sec 2x\right)^2} = \boxed{2\cdot 2\sec 2x\cdot\dfrac{d}{dx}\sec 2x} = \boxed{4\sec 2x\cdot\sec 2x\tan 2x\cdot\dfrac{d}{dx}2x}$

$= \boxed{4\sec 2x\cdot\sec 2x\tan 2x\cdot 2} = \boxed{8\sec 2x\sec 2x\tan 2x} = \boxed{\mathbf{8\sec^2 2x\tan 2x}}$

e. $\boxed{\dfrac{dy}{dx}} = \boxed{\dfrac{d}{dx}\left(\tan x^2\right)} = \boxed{\sec^2 x^2\cdot\dfrac{d}{dx}\left(x^2\right)} = \boxed{\sec^2 x^2\cdot 2x} = \boxed{\mathbf{2x\sec^2 x^2}}$

$\boxed{\dfrac{d^2 y}{dx^2}} = \boxed{\dfrac{d}{dx}\left(2x\sec^2 x^2\right)} = \boxed{\dfrac{d}{dx}2x\left(\sec x^2\right)^2} = \boxed{\left(\sec x^2\right)^2\cdot\dfrac{d}{dx}2x + 2x\cdot\dfrac{d}{dx}\left(\sec x^2\right)^2}$

$= \boxed{\left(\sec x^2\right)^2\cdot 2 + 2x\cdot 2\left(\sec x^2\right)\cdot\dfrac{d}{dx}\left(\sec x^2\right)} = \boxed{\left(\sec x^2\right)^2\cdot 2 + 2x\cdot 2\left(\sec x^2\right)\cdot\sec x^2\tan x^2\cdot\dfrac{d}{dx}x^2}$

$= \boxed{\sec^2 x^2\cdot 2 + 2x\cdot 2\sec^2 x^2\tan x^2\cdot 2x} = \boxed{2\sec^2 x^2 + 8x^2\sec^2 x^2\tan x^2} = \boxed{\mathbf{2\sec^2 x^2\left(1+4x^2\tan x^2\right)}}$

f. $\boxed{\dfrac{dy}{dx}} = \boxed{\dfrac{d}{dx}\left(\tan^2 x\right)} = \boxed{\dfrac{d}{dx}\left(\tan x\right)^2} = \boxed{2\tan x\cdot\dfrac{d}{dx}\tan x} = \boxed{2\tan x\cdot\sec^2 x\cdot\dfrac{d}{dx}x} = \boxed{2\tan x\cdot\sec^2 x\cdot 1} = \boxed{\mathbf{2\tan x\sec^2 x}}$

$\boxed{\dfrac{d^2 y}{dx^2}} = \boxed{\dfrac{d}{dx}\left(2\tan x\sec^2 x\right)} = \boxed{2\left(\sec^2 x\cdot\dfrac{d}{dx}\tan x + \tan x\cdot\dfrac{d}{dx}\sec^2 x\right)} = \boxed{2\left(\sec^2 x\cdot\sec^2 x + \tan x\cdot 2\sec x\cdot\dfrac{d}{dx}\sec x\right)}$

$= \boxed{2\left(\sec^4 x + \tan x\cdot 2\sec x\cdot\sec x\tan x\right)} = \boxed{2\left(\sec^4 x + 2\sec^2 x\,\tan^2 x\right)} = \boxed{\mathbf{2\sec^2 x\left(\sec^2 x + 2\tan^2 x\right)}}$

g. $\boxed{\dfrac{dy}{dx}} = \boxed{\dfrac{d}{dx}\left(\sin 3x + \cos 3x\right)} = \boxed{\dfrac{d}{dx}\sin 3x + \dfrac{d}{dx}\cos 3x} = \boxed{\cos 3x\cdot\dfrac{d}{dx}3x - \sin 3x\cdot\dfrac{d}{dx}3x} = \boxed{\mathbf{3\cos 3x - 3\sin 3x}}$

$\boxed{\dfrac{d^2 y}{dx^2}} = \boxed{\dfrac{d}{dx}\left(3\cos 3x - 3\sin 3x\right)} = \boxed{\dfrac{d}{dx}3\cos 3x - \dfrac{d}{dx}3\sin 3x} = \boxed{-3\sin 3x\cdot\dfrac{d}{dx}3x - 3\cos 3x\cdot\dfrac{d}{dx}3x}$

$$= \boxed{-3\sin 3x \cdot 3 - 3\cos 3x \cdot 3} = \boxed{-9\sin 3x - 9\cos 3x} = \boxed{-9\big(\sin 3x + \cos 3x\big)}$$

h. $\boxed{\dfrac{dy}{dx}} = \boxed{\dfrac{d}{dx}\big(x\sin x\big)} = \boxed{\sin x \cdot \dfrac{d}{dx}x + x \cdot \dfrac{d}{dx}\sin x} = \boxed{\sin x \cdot 1 + x \cdot \cos x} = \boxed{\boldsymbol{\sin x + x\cos x}}$

$\boxed{\dfrac{d^2 y}{dx^2}} = \boxed{\dfrac{d}{dx}\big(\sin x + x\cos x\big)} = \boxed{\dfrac{d}{dx}\sin x + \dfrac{d}{dx}\big(x\cos x\big)} = \boxed{\cos x \cdot \dfrac{d}{dx}x + \left(\cos x \cdot \dfrac{d}{dx}x + x \cdot \dfrac{d}{dx}\cos x\right)}$

$= \boxed{\cos x \cdot 1 + \big(\cos x \cdot 1 - x \cdot \sin x\big)} = \boxed{\cos x + \cos x - x\sin x} = \boxed{\boldsymbol{2\cos x - x\sin x}}$

i. $\boxed{\dfrac{dy}{dx}} = \boxed{\dfrac{d}{dx}\big[\tan\big(5x+1\big)\big]} = \boxed{\sec^2\big(5x+1\big) \cdot \dfrac{d}{dx}\big(5x+1\big)} = \boxed{\sec^2\big(5x+1\big) \cdot 5} = \boxed{\boldsymbol{5\sec^2\big(5x+1\big)}}$

$\boxed{\dfrac{d^2 y}{dx^2}} = \boxed{\dfrac{d}{dx}\big[5\sec^2\big(5x+1\big)\big]} = \boxed{5 \cdot \dfrac{d}{dx}\big[\sec\big(5x+1\big)^2\big]} = \boxed{5 \cdot 2\sec\big(5x+1\big) \cdot \dfrac{d}{dx}\sec\big(5x+1\big)}$

$= \boxed{10\sec\big(5x+1\big) \cdot \sec\big(5x+1\big)\tan\big(5x+1\big) \cdot \dfrac{d}{dx}\big(5x+1\big)} = \boxed{10\sec^2\big(5x+1\big)\tan\big(5x+1\big) \cdot 5} = \boxed{\boldsymbol{50\sec^2\big(5x+1\big)\tan\big(5x+1\big)}}$

j. $\boxed{\dfrac{dy}{dx}} = \boxed{\dfrac{d}{dx}\big(\sin x - x\cos x\big)} = \boxed{\dfrac{d}{dx}\sin x - \dfrac{d}{dx}\big(x\cos x\big)} = \boxed{\cos x \cdot \dfrac{d}{dx}x - \left(\cos x \cdot \dfrac{d}{dx}x + x \cdot \dfrac{d}{dx}\cos x\right)}$

$= \boxed{\cos x \cdot 1 - \big(\cos x \cdot 1 - x \cdot \sin x\big)} = \boxed{\cos x - \cos x + x\sin x} = \boxed{\boldsymbol{x\sin x}}$

$\boxed{\dfrac{d^2 y}{dx^2}} = \boxed{\dfrac{d}{dx}\big(x\sin x\big)} = \boxed{\sin x \cdot \dfrac{d}{dx}x + x \cdot \dfrac{d}{dx}\sin x} = \boxed{\sin x \cdot 1 + x \cdot \cos x} = \boxed{\boldsymbol{\sin x + x\cos x}}$

k. $\boxed{\dfrac{dy}{dx}} = \boxed{\dfrac{d}{dx}\left(\sqrt{2+\sin 2x}\,\right)} = \boxed{\dfrac{d}{dx}\big(2+\sin 2x\big)^{\frac{1}{2}}} = \boxed{\dfrac{1}{2}\big(2+\sin 2x\big)^{\frac{1}{2}-1} \cdot \dfrac{d}{dx}\big(2+\sin 2x\big)} = \boxed{\dfrac{1}{2}\big(2+\sin 2x\big)^{-\frac{1}{2}}}$

$\boxed{\times\left(\dfrac{d}{dx}2 + \dfrac{d}{dx}\sin 2x\right)} = \boxed{\dfrac{1}{2}\big(2+\sin 2x\big)^{-\frac{1}{2}} \times \big(0+2\cos 2x\big)} = \boxed{\dfrac{1}{2} \cdot \dfrac{1}{\big(2+\sin 2x\big)^{\frac{1}{2}}} \cdot 2\cos 2x} = \boxed{\dfrac{\boldsymbol{\cos 2x}}{\boldsymbol{\sqrt{2+\sin 2x}}}}$

$\boxed{\dfrac{d^2 y}{dx^2}} = \boxed{\dfrac{d}{dx}\left(\dfrac{\cos 2x}{\sqrt{2+\sin 2x}}\right)} = \boxed{\dfrac{\big(2+\sin 2x\big)^{\frac{1}{2}} \cdot \dfrac{d}{dx}\cos 2x - \left(\cos 2x \cdot \dfrac{d}{dx}\big(2+\sin 2x\big)^{\frac{1}{2}}\right)}{2+\sin 2x}}$

$= \boxed{\dfrac{\big(2+\sin 2x\big)^{\frac{1}{2}} \cdot -\sin 2x \cdot \dfrac{d}{dx}2x - \left(\cos 2x \cdot \dfrac{1}{2}\big(2+\sin 2x\big)^{\frac{1}{2}-1} \cdot \dfrac{d}{dx}\big(2+\sin 2x\big)\right)}{2+\sin 2x}} = \boxed{\dfrac{\big(2+\sin 2x\big)^{\frac{1}{2}} \cdot -\sin 2x \cdot 2}{2+\sin 2x}}$

$$= \boxed{-2\sin\theta\cos\theta} \quad \text{Therefore} \quad \boxed{\dfrac{dy/d\theta}{dx/d\theta}} = \boxed{\dfrac{dy}{dx}} = \boxed{\dfrac{3\sin^2\theta\cos\theta}{-2\sin\theta\cos\theta}} = \boxed{-\dfrac{3}{2}\sin\theta}$$

d. Given $\boxed{y=\sin^3\alpha}$ and $\boxed{x=\cos^3\alpha}$ then $\boxed{\dfrac{dy}{d\alpha}} = \boxed{\dfrac{d}{d\alpha}\sin^3\alpha} = \boxed{\dfrac{d}{d\alpha}(\sin\alpha)^3} = \boxed{3\sin^2\alpha\cdot\dfrac{d}{d\alpha}\sin\alpha}$

$$= \boxed{3\sin^2\alpha\cos\alpha} \text{ and } \boxed{\dfrac{dx}{d\alpha}} = \boxed{\dfrac{d}{d\alpha}\cos^3\alpha} = \boxed{\dfrac{d}{d\alpha}(\cos\alpha)^3} = \boxed{3\cos^2\alpha\cdot\dfrac{d}{d\alpha}\cos\alpha} = \boxed{3\cos^2\alpha\cdot-\sin\alpha}$$

$$= \boxed{-3\sin\alpha\cos^2\alpha} \quad \text{Therefore} \quad \boxed{\dfrac{dy/d\alpha}{dx/d\alpha}} = \boxed{\dfrac{dy}{dx}} = \boxed{\dfrac{3\sin^2\alpha\cos\alpha}{-3\sin\alpha\cos^2\alpha}} = \boxed{-\dfrac{\sin\alpha}{\cos\alpha}} = \boxed{-\tan\alpha}$$

e. Given $\boxed{y=t\,e^t}$ and $\boxed{x=e^t\cos t}$ then $\boxed{\dfrac{dy}{dt}} = \boxed{\dfrac{d}{dt}t\,e^t} = \boxed{e^t\cdot\dfrac{d}{dt}t+t\cdot\dfrac{d}{dt}e^t} = \boxed{e^t\cdot1+t\cdot e^t} = \boxed{e^t+t\,e^t}$

$$= \boxed{e^t(1+t)} \text{ and } \boxed{\dfrac{dx}{dt}} = \boxed{\dfrac{d}{dt}e^t\cos t} = \boxed{\cos t\cdot\dfrac{d}{dt}e^t+e^t\cdot\dfrac{d}{dt}\cos t} = \boxed{\cos t\cdot e^t+e^t\cdot-\sin t} = \boxed{e^t\cos t-e^t\sin t}$$

$$= \boxed{e^t(\cos t-\sin t)} \quad \text{Therefore} \quad \boxed{\dfrac{dy/dt}{dx/dt}} = \boxed{\dfrac{dy}{dx}} = \boxed{\dfrac{e^t(1+t)}{e^t(\cos t-\sin t)}} = \boxed{\dfrac{1+t}{\cos t-\sin t}}$$

f. Given $\boxed{y=5\ln 3t^2}$ and $\boxed{x=\ln t^3}$ then $\dfrac{dy}{dt} = \boxed{\dfrac{d}{dt}5\ln 3t^2} = \boxed{5\cdot\dfrac{1}{3t^2}\cdot\dfrac{d}{dt}3t^2} = \boxed{5\cdot\dfrac{1}{3t^2}\cdot6t} = \boxed{\dfrac{30t}{3t^2}} = \boxed{\dfrac{10}{t}}$ and

$$\boxed{\dfrac{dx}{dt}} = \boxed{\dfrac{d}{dt}\ln t^3} = \boxed{\dfrac{1}{t^3}\cdot\dfrac{d}{dt}t^3} = \boxed{\dfrac{1}{t^3}\cdot3t^2} = \boxed{\dfrac{3t^2}{t^3}} = \boxed{\dfrac{3}{t}} \quad \text{Therefore} \quad \boxed{\dfrac{dy/dt}{dx/dt}} = \boxed{\dfrac{dy}{dx}} = \boxed{\dfrac{10/t}{3/t}} = \boxed{\dfrac{10}{3}}$$

g. Given $\boxed{y=e^t\cos t}$ and $\boxed{x=t\ln t}$ then $\boxed{\dfrac{dy}{dt}} = \boxed{\dfrac{d}{dt}e^t\cos t} = \boxed{\cos t\cdot\dfrac{d}{dt}e^t+e^t\cdot\dfrac{d}{dt}\cos t}$

$$= \boxed{\cos t\cdot e^t+e^t\cdot-\sin t} = \boxed{e^t\cos t-e^t\sin t} = \boxed{e^t(\cos t-\sin t)} \text{ and } \boxed{\dfrac{dx}{dt}} = \boxed{\dfrac{d}{dt}t\ln t} = \boxed{\ln t\cdot\dfrac{d}{dt}t+t\cdot\dfrac{d}{dt}\ln t}$$

$$= \boxed{\ln t\cdot1+t\cdot\dfrac{1}{t}} = \boxed{\ln t+1} \quad \text{Therefore} \quad \boxed{\dfrac{dy/dt}{dx/dt}} = \boxed{\dfrac{dy}{dx}} = \boxed{\dfrac{e^t(\cos t-\sin t)}{\ln t+1}}$$

h. Given $\boxed{y=\sin^2\sqrt{t}}$ and $\boxed{x=\sec^4\sqrt{t}}$ then $\boxed{\dfrac{dy}{dt}} = \boxed{\dfrac{d}{dt}\sin^2\sqrt{t}} = \boxed{\dfrac{d}{dt}(\sin\sqrt{t})^2} = \boxed{2\sin\sqrt{t}\cdot\dfrac{d}{dt}\sin\sqrt{t}}$

$$= \boxed{2\sin\sqrt{t}\cdot\cos\sqrt{t}\cdot\dfrac{d}{dt}\sqrt{t}} = \boxed{2\sin\sqrt{t}\cdot\cos\sqrt{t}\cdot\dfrac{1}{2\sqrt{t}}} = \boxed{\dfrac{\sin\sqrt{t}\cos\sqrt{t}}{\sqrt{t}}} \text{ and } \boxed{\dfrac{dx}{dt}} = \boxed{\dfrac{d}{dt}\sec^4\sqrt{t}}$$

$$= \boxed{\dfrac{d}{dt}\left(\sec\sqrt{t}\right)^4} = \boxed{4\sec^3\sqrt{t}\cdot\dfrac{d}{dt}\sec\sqrt{t}} = \boxed{4\sec^3\sqrt{t}\cdot\sec\sqrt{t}\tan\sqrt{t}\cdot\dfrac{d}{dt}\sqrt{t}} = \boxed{4\sec^4\sqrt{t}\cdot\tan\sqrt{t}\cdot-\dfrac{1}{2\sqrt{t}}}$$

$$= \boxed{-\dfrac{2\sec^4\sqrt{t}\tan\sqrt{t}}{\sqrt{t}}}\quad \text{Therefore } \boxed{\dfrac{dy/dt}{dx/dt}} = \boxed{\dfrac{dy}{dx}} = \boxed{\dfrac{-\sin\sqrt{t}\cos\sqrt{t}/\sqrt{t}}{-2\sec^4\sqrt{t}\tan\sqrt{t}/\sqrt{t}}} = \boxed{\mathbf{\dfrac{\sin\sqrt{t}\cos\sqrt{t}}{2\sec^4\sqrt{t}\tan\sqrt{t}}}}$$

i. Given $\boxed{y=\cos^2(\theta+1)}$ and $\boxed{x=\sin\theta^2}$ then $\boxed{\dfrac{dy}{d\theta}} = \boxed{\dfrac{d}{d\theta}\cos^2(\theta+1)} = \boxed{\dfrac{d}{d\theta}\left[\cos(\theta+1)\right]^2}$

$$= \boxed{2\cos(\theta+1)\cdot\dfrac{d}{d\theta}\cos(\theta+1)} = \boxed{2\cos(\theta+1)\cdot-\sin(\theta+1)\cdot\dfrac{d}{d\theta}(\theta+1)} = \boxed{-2\cos(\theta+1)\cdot\sin(\theta+1)\cdot 1}$$

$$= \boxed{-2\sin(\theta+1)\cos(\theta+1)}\quad \text{and}\quad \boxed{\dfrac{dx}{d\theta}} = \boxed{\dfrac{d}{dt}\sin\theta^2} = \boxed{\cos\theta^2\cdot\dfrac{d}{dt}\theta^2} = \boxed{\cos\theta^2\cdot 2\theta} = \boxed{2\theta\cos\theta^2}$$

$$\text{Therefore } \boxed{\dfrac{dy/d\theta}{dx/d\theta}} = \boxed{\dfrac{dy}{dx}} = \boxed{\dfrac{-2\sin(\theta+1)\cos(\theta+1)}{2\theta\cos\theta^2}} = \boxed{\mathbf{-\dfrac{\sin(\theta+1)\cos(\theta+1)}{\theta\cos\theta^2}}}$$

j. Given $\boxed{y=3\sin\varphi}$ and $\boxed{x=\cos3\varphi^2}$ then $\boxed{\dfrac{dy}{d\varphi}} = \boxed{\dfrac{d}{d\varphi}3\sin\varphi} = \boxed{3\cos\varphi}$ and $\boxed{\dfrac{dx}{d\varphi}} = \boxed{\dfrac{d}{dt}\cos3\varphi^2}$

$$= \boxed{-\sin3\varphi^2\cdot\dfrac{d}{dt}3\varphi^2} = \boxed{-\sin3\varphi^2\cdot6\varphi} = \boxed{-6\varphi\sin3\varphi^2}\quad \text{So } \boxed{\dfrac{dy/d\varphi}{dx/d\varphi}} = \boxed{\dfrac{dy}{dx}} = \boxed{\dfrac{3\cos\varphi}{-6\varphi\sin3\varphi^2}} = \boxed{\mathbf{-\dfrac{\cos\varphi}{2\varphi\sin3\varphi^2}}}$$

k. Given $\boxed{y=\sqrt[3]{t^2}}$ and $\boxed{x=1+\sqrt{t}}$ then $\boxed{\dfrac{dy}{dt}} = \boxed{\dfrac{d}{dt}\sqrt[3]{t^2}} = \boxed{\dfrac{d}{dt}t^{\frac{2}{3}}} = \boxed{\dfrac{2}{3}t^{\frac{2}{3}-1}} = \boxed{\dfrac{2}{3}t^{-\frac{1}{3}}} = \boxed{\dfrac{2}{3t^{\frac{1}{3}}}} = \boxed{\dfrac{2}{3\sqrt[3]{t}}}$ and

$$\boxed{\dfrac{dx}{dt}} = \boxed{\dfrac{d}{dt}\left(1+\sqrt{t}\right)} = \boxed{\dfrac{d}{dt}1+\dfrac{d}{dt}\sqrt{t}} = \boxed{0+\dfrac{1}{2\sqrt{t}}} = \boxed{\dfrac{1}{2\sqrt{t}}}\quad \text{So } \boxed{\dfrac{dy/dt}{dx/dt}} = \boxed{\dfrac{dy}{dx}} = \boxed{\dfrac{2/3\sqrt[3]{t}}{1/2\sqrt{t}}} = \boxed{\dfrac{2\cdot2\sqrt{t}}{1\cdot3\sqrt[3]{t}}} = \boxed{\mathbf{\dfrac{4\sqrt{t}}{3\sqrt[3]{t}}}}$$

l. Given $\boxed{y=\theta-\sqrt{\theta}}$ and $\boxed{x=\theta^2\cos\theta}$ then $\boxed{\dfrac{dy}{d\theta}} = \boxed{\dfrac{d}{d\theta}\left(\theta-\sqrt{\theta}\right)} = \boxed{\dfrac{d}{d\theta}\theta-\dfrac{d}{d\theta}\sqrt{\theta}} = \boxed{1-\dfrac{1}{2\sqrt{\theta}}} = \boxed{\dfrac{2\sqrt{\theta}-1}{2\sqrt{\theta}}}$

$$\text{and } \boxed{\dfrac{dx}{d\theta}} = \boxed{\dfrac{d}{dt}\left(\theta^2\cos\theta\right)} = \boxed{\cos\theta\cdot\dfrac{d}{dt}\theta^2+\theta^2\cdot\dfrac{d}{dt}\cos\theta} = \boxed{\cos\theta\cdot2\theta+\theta^2\cdot-\sin\theta} = \boxed{2\theta\cos\theta-\theta^2\sin\theta}$$

$$\text{Therefore } \boxed{\dfrac{dy/d\theta}{dx/d\theta}} = \boxed{\dfrac{dy}{dx}} = \boxed{\dfrac{2\sqrt{\theta}-1/2\sqrt{\theta}}{2\theta\cos\theta-\theta^2\sin\theta}} = \boxed{\mathbf{\dfrac{2\sqrt{\theta}-1}{2\sqrt{\theta}\left(2\theta\cos\theta-\theta^2\sin\theta\right)}}}$$

1. Find the derivative of the following trigonometric functions:

   a. $y = \sin(3x+1)$

   b. $y = 5\cos x^3$

   c. $y = x^3 \cos x^2$

   d. $y = \sin 5x \cdot \tan 3x$

   e. $y = \tan^2 x^3$

   f. $y = \cot(x+3)^3$

   g. $y = x^5 \cos x^7$

   h. $y = \sec^4 \sqrt{x^3}$

   i. $y = \sec 3x^2 + x^3$

   j. $y = \tan x^5$

   k. $y = \tan^5 x$

   l. $y = \csc(x^3+1)$

2. Find the derivative of the following trigonometric functions:

   a. $y = \dfrac{\tan x}{\csc x}$

   b. $y = \dfrac{\sin(x^3+1)}{x^2}$

   c. $y = \dfrac{\sec x}{\csc x^3}$

   d. $y = x^5 \tan x^3$

   e. $y = x^5 \sin x^2$

   f. $y = (x+5)^2 \cos x$

   g. $y = x^2 \tan^3 x^5$

   h. $y = \sqrt{x + \sin x^3}$

   i. $y = \sin(1+x^5)$

   j. $y = \cot^2 x^3$

   k. $y = \sin^3(1+5x)$

   l. $y = x^5 \csc x^3$

3. Find the derivative of the following trigonometric functions:

   a. $y = \sin 7x$

   b. $y = \cos x^3$

   c. $y = \cot x^3$

   d. $y = x^3 \tan x^2$

   e. $y = \cot(x+9)$

   f. $y = \sin^2(x^3+5x+2)$

   g. $y = \sin \sqrt{x+3}$

   h. $y = \sin x^2 + \cos x^3$

   i. $y = x^2 \sin x^3$

3.2   Differentiation of Inverse Trigonometric Functions

The differential formulas involving inverse trigonometric functions are defined as:

Table 3.2-1: Differentiation Formulas for Inverse Trigonometric Functions

$\dfrac{d}{dx}arc\sin u = \dfrac{d}{dx}\sin^{-1}u = \dfrac{1}{\sqrt{1-u^2}}\cdot\dfrac{du}{dx}$	$\dfrac{d}{dx}arc\cot u = \dfrac{d}{dx}\cot^{-1}u = -\dfrac{1}{1+u^2}\cdot\dfrac{du}{dx}$
$\dfrac{d}{dx}arc\cos u = \dfrac{d}{dx}\cos^{-1}u = -\dfrac{1}{\sqrt{1-u^2}}\cdot\dfrac{du}{dx}$	$\dfrac{d}{dx}arc\sec u = \dfrac{d}{dx}\sec^{-1}u = \dfrac{1}{u\sqrt{u^2-1}}\cdot\dfrac{du}{dx}$
$\dfrac{d}{dx}arc\tan u = \dfrac{d}{dx}\tan^{-1}u = \dfrac{1}{1+u^2}\cdot\dfrac{du}{dx}$	$\dfrac{d}{dx}arc\csc u = \dfrac{d}{dx}\csc^{-1}u = -\dfrac{1}{u\sqrt{u^2-1}}\cdot\dfrac{du}{dx}$

Let's differentiate some inverse trigonometric functions using the above differentiation formulas.

Example 3.2-1: Find the derivative of the following inverse trigonometric functions:

a. $y = arc\sin(3x-4)$ b. $y = arc\sin x^2$ c. $y = \tan^{-1}(x^2+1)$

d. $y = x^2 arc\sin 2x$ e. $y = \dfrac{\sin^{-1}x}{3x}$ f. $y = \cos x^2 + arc\cos x^2$

g. $y = (1+x)\cot^{-1}x$ h. $y = arc\tan\dfrac{5}{x}$ i. $y = \dfrac{\cos^{-1}x}{\cos x}$

j. $y = \dfrac{arc\sin x}{\cos x}$ k. $y = x^5 + arc\tan\sqrt{x}$ l. $y = arc\cot\sqrt[3]{x^2}$

Solutions:

a. Given $\boxed{y = arc\sin(3x-4)}$ then $\boxed{\dfrac{dy}{dx} = \dfrac{d}{dx}[arc\sin(3x-4)]} = \boxed{\dfrac{1}{\sqrt{1-(3x-4)^2}}\cdot\dfrac{d}{dx}(3x-4)} = \boxed{\dfrac{3}{\sqrt{1-(3x-4)^2}}}$

$= \boxed{\dfrac{3}{\sqrt{1-(9x^2+16-24x)}}} = \boxed{\dfrac{3}{\sqrt{1-9x^2-16+24x}}} = \boxed{\dfrac{3}{\sqrt{-9x^2+24x-15}}}$

b. Given $\boxed{y = arc\sin x^2}$ then $\boxed{\dfrac{dy}{dx} = \dfrac{d}{dx}(arc\sin x^2)} = \boxed{\dfrac{1}{\sqrt{1-x^4}}\cdot\dfrac{d}{dx}x^2} = \boxed{\dfrac{1}{\sqrt{1-x^4}}\cdot 2x} = \boxed{\dfrac{2x}{\sqrt{1-x^4}}}$

c. Given $\boxed{y = \tan^{-1}(x^2+1)}$ then $\boxed{\dfrac{dy}{dx} = \dfrac{d}{dx}[\tan^{-1}(x^2+1)]} = \boxed{\dfrac{1}{1+(x^2+1)^2}\cdot\dfrac{d}{dx}(x^2+1)} = \boxed{\dfrac{1}{1+(x^2+1)^2}\cdot 2x}$

$\boxed{\dfrac{1}{?x+1}\cdot 2x} = \boxed{\dfrac{2x}{x^4+2x+2}}$

$\boxed{2x}$ then $\boxed{\dfrac{dy}{dx} = \dfrac{d}{dx}(x^2 arc\sin 2x)} = \boxed{arc\sin 2x\cdot\dfrac{d}{dx}x^2 + x^2\cdot\dfrac{d}{dx}arc\sin 2x}$

$$= \boxed{arc \sin 2x \cdot 2x + x^2 \cdot \frac{1}{\sqrt{1-(2x)^2}} \cdot \frac{d}{dx} 2x} = \boxed{2x \, arc \sin 2x + x^2 \cdot \frac{1}{\sqrt{1-4x^2}} \cdot 2} = \boxed{\mathbf{2x \, arc \sin 2x + \frac{2x^2}{\sqrt{1-4x^2}}}}$$

e. Given $\boxed{y = \dfrac{\sin^{-1} x}{3x}}$ then $\boxed{\dfrac{dy}{dx}} = \boxed{\dfrac{d}{dx}\left(\dfrac{\sin^{-1} x}{3x}\right)} = \boxed{\dfrac{\left(3x \cdot \frac{d}{dx}\sin^{-1} x\right)-\left(\sin^{-1} x \cdot \frac{d}{dx} 3x\right)}{(3x)^2}} = \boxed{\dfrac{3x \cdot \frac{1}{\sqrt{1-x^2}} - \sin^{-1} x \cdot 3}{9x^2}}$

$$= \boxed{\dfrac{\frac{3x}{\sqrt{1-x^2}} - 3\sin^{-1} x}{9x^2}} = \boxed{\dfrac{\frac{3x-3\sqrt{1-x^2}\,\sin^{-1} x}{\sqrt{1-x^2}}}{9x^2}} = \boxed{\dfrac{3x - 3\sqrt{1-x^2}\,\sin^{-1} x}{9x^2\sqrt{1-x^2}}} = \boxed{\dfrac{x - \sqrt{1-x^2}\,\sin^{-1} x}{3x^2\sqrt{1-x^2}}}$$

f. Given $\boxed{y = \cos x^2 + arc \cos x^2}$ then $\boxed{\dfrac{dy}{dx}} = \boxed{\dfrac{d}{dx}\left(\cos x^2 + arc \cos x^2\right)} = \boxed{\dfrac{d}{dx}\cos x^2 + \dfrac{d}{dx} arc \cos x^2}$

$$= \boxed{-\sin x^2 \cdot \frac{d}{dx} x^2 + \frac{-1}{\sqrt{1-x^4}} \cdot \frac{d}{dx} x^2} = \boxed{-\sin x^2 \cdot 2x - \frac{1}{\sqrt{1-x^4}} \cdot 2x} = \boxed{\mathbf{-2x\left(\sin x^2 + \frac{1}{\sqrt{1-x^4}}\right)}}$$

g. Given $\boxed{y = (1+x)\cot^{-1} x}$ then $\boxed{\dfrac{dy}{dx}} = \boxed{\dfrac{d}{dx}\left[(1+x)\cot^{-1} x\right]} = \boxed{\cot^{-1} x \cdot \frac{d}{dx}(1+x) + (1+x)\cdot \frac{d}{dx}\cot^{-1} x}$

$$= \boxed{\cot^{-1} x \cdot 1 + (1+x)\cdot \frac{-1}{1+x^2}} = \boxed{\mathbf{\cot^{-1} x - \frac{1+x}{1+x^2}}}$$

h. Given $\boxed{y = arc \tan \dfrac{5}{x}}$ then $\boxed{\dfrac{dy}{dx}} = \boxed{\dfrac{d}{dx}\left(arc \tan \dfrac{5}{x}\right)} = \boxed{\dfrac{1}{1+\left(\frac{5}{x}\right)^2} \cdot \dfrac{d}{dx}\dfrac{5}{x}} = \boxed{\dfrac{1}{1+\frac{25}{x^2}} \cdot \dfrac{-5}{x^2}} = \boxed{\dfrac{1}{\frac{x^2+25}{x^2}} \cdot \dfrac{-5}{x^2}}$

$$= \boxed{\frac{x^2}{x^2+25} \cdot \frac{-5}{x^2}} = \boxed{-\frac{5x^2}{x^2\left(x^2+25\right)}} = \boxed{\mathbf{-\frac{5}{x^2+25}}}$$

i. Given $\boxed{y = \dfrac{\cos^{-1} x}{\cos x}}$ then $\boxed{\dfrac{dy}{dx}} = \boxed{\dfrac{d}{dx}\left(\dfrac{\cos^{-1} x}{\cos x}\right)} = \boxed{\dfrac{\left(\cos x \cdot \frac{d}{dx}\cos^{-1} x\right)-\left(\cos^{-1} x \cdot \frac{d}{dx}\cos x\right)}{(\cos x)^2}}$

$$= \boxed{\frac{\cos x \cdot \frac{-1}{\sqrt{1-x^2}} - \left(\cos^{-1} x \cdot -\sin x\right)}{\cos^2 x}} = \boxed{\frac{\frac{-\cos x}{\sqrt{1-x^2}} + \sin x \cos^{-1} x}{\cos^2 x}} = \boxed{\mathbf{\frac{-\cos x + \sqrt{1-x^2}\,\sin x \cos^{-1} x}{\sqrt{1-x^2}\,\cos^2 x}}}$$

j. Given $\boxed{y = \dfrac{arc \sin x}{\cos x}}$ then $\boxed{\dfrac{dy}{dx}} = \boxed{\dfrac{d}{dx}\left(\dfrac{arc \sin x}{\cos x}\right)} = \boxed{\dfrac{\left(\cos x \cdot \frac{d}{dx} arc \sin x\right)-\left(arc \sin x \cdot \frac{d}{dx}\cos x\right)}{(\cos x)^2}}$

$$= \frac{\cos x \cdot \dfrac{1}{\sqrt{1-x^2}} - (arc \sin x \cdot -\sin x)}{\cos^2 x} = \frac{\dfrac{\cos x}{\sqrt{1-x^2}} + \sin x \; arc \sin x}{\cos^2 x} = \boxed{\frac{\cos x + \sqrt{1-x^2}\,\sin x \; arc \sin x}{\sqrt{1-x^2}\,\cos^2 x}}$$

k. Given $\boxed{y = x^5 + arc \tan \sqrt{x}}$ then $\boxed{\dfrac{dy}{dx}} = \boxed{\dfrac{d}{dx}\left(x^5 + arc \tan \sqrt{x}\right)} = \boxed{\dfrac{d}{dx}x^5 + \dfrac{d}{dx} arc \tan \sqrt{x}}$

$$= \boxed{5x^4 + \frac{1}{1+\left(\sqrt{x}\right)^2}\cdot\frac{d}{dx}\sqrt{x}} = \boxed{5x^4 + \frac{1}{1+x}\cdot\frac{d}{dx}x^{\frac{1}{2}}} = \boxed{5x^4 + \frac{1}{1+x}\cdot\frac{1}{2}x^{\frac{1}{2}-1}} = \boxed{5x^4 + \frac{x^{-\frac{1}{2}}}{2(1+x)}} = \boxed{5x^4 + \frac{1}{2\sqrt{x}\,(1+x)}}$$

l. Given $\boxed{y = arc \cot \sqrt[3]{x^2}}$ then $\boxed{\dfrac{dy}{dx}} = \boxed{\dfrac{d}{dx} arc \cot \sqrt[3]{x^2}} = \boxed{\dfrac{d}{dx} arc \cot x^{\frac{2}{3}}} = \boxed{-\dfrac{1}{1+x^{\frac{4}{3}}}\cdot\dfrac{d}{dx}x^{\frac{2}{3}}}$

$$= \boxed{-\frac{1}{1+x^{\frac{4}{3}}}\cdot\frac{2}{3}x^{\frac{2}{3}-1}} = \boxed{-\frac{1}{1+x^{\frac{4}{3}}}\cdot\frac{2}{3}x^{-\frac{1}{3}}} = \boxed{-\frac{1}{1+x^{\frac{4}{3}}}\cdot\frac{2}{3\sqrt[3]{x}}} = \boxed{-\frac{1}{1+\sqrt[3]{x^4}}\cdot\frac{2}{3\sqrt[3]{x}}} = \boxed{-\frac{2}{3\sqrt[3]{x}\left(1+x\sqrt[3]{x}\right)}}$$

Example 3.2-2: Find the derivative of the following inverse trigonometric functions:

a. $y = arc \cos \dfrac{x^2}{a}$

b. $y = \dfrac{1}{3} arc \cot \dfrac{x}{5}$

c. $y = \cot^{-1} x - \tan x$

d. $y = \tan^{-1}\left(x^2 + 3\right)$

e. $y = \dfrac{arc \sin x}{x} + 5x^3$

f. $y = \cos x + \tan^{-1} x$

g. $y = \dfrac{3x^2}{arc \tan x}$

h. $y = \sin x + \sin^{-1} x$

i. $\sin 3y = arc \tan 5x$

j. $\cos y = \sin^{-1} x$

k. $\sin (y+1) = arc \cos x^2$

l. $\cot y = \tan^{-1} x$

Solutions:

a. Given $\boxed{y = arc \cos \dfrac{x^2}{a}}$ then $\boxed{\dfrac{dy}{dx}} = \boxed{\dfrac{d}{dx} arc \cos \dfrac{x^2}{a}} = \boxed{-\dfrac{1}{\sqrt{1-\left(\frac{x^2}{a}\right)^2}}\cdot\dfrac{d}{dx}\dfrac{x^2}{a}} = \boxed{-\dfrac{1}{\sqrt{1-\left(\frac{x^2}{a}\right)^2}}\cdot\dfrac{2x}{a}}$

$$= \boxed{-\frac{1}{\sqrt{1-\frac{x^4}{a^2}}}\cdot\frac{2x}{a}} = \boxed{-\frac{1}{\sqrt{\frac{a^2-x^4}{a^2}}}\cdot\frac{2x}{a}} = \boxed{-\frac{1}{\frac{1}{a}\sqrt{a^2-x^4}}\cdot\frac{2x}{a}} = \boxed{-\frac{a}{\sqrt{a^2-x^4}}\cdot\frac{2x}{a}} = \boxed{-\frac{2x}{\sqrt{a^2-x^4}}}$$

b. Given $\boxed{y = \dfrac{1}{3} arc \cot \dfrac{x}{5}}$ then $\boxed{\dfrac{dy}{dx}} = \boxed{\dfrac{d}{dx}\left(\dfrac{1}{3} arc \cot \dfrac{x}{5}\right)} = \boxed{\dfrac{1}{3}\dfrac{d}{dx}\left(arc \cot \dfrac{x}{5}\right)} = \boxed{\dfrac{1}{3}\dfrac{-1}{1+\left(\frac{x}{5}\right)^2}\cdot\dfrac{d}{dx}\dfrac{x}{5}}$

$$= \left| \frac{1}{3} \frac{-1}{1+\frac{x^2}{25}} \cdot \frac{1}{5} \right| = \left| \frac{1}{15} \frac{-1}{\frac{25+x^2}{25}} \right| = \left| \frac{1}{15} \frac{-25}{25+x^2} \right| = \boxed{-\frac{5}{3\left(25+x^2\right)}}$$

c. Given $\boxed{y = \cot^{-1} x - \tan x}$ then $\boxed{\frac{dy}{dx}} = \boxed{\frac{d}{dx}\left(\cot^{-1} x - \tan x\right)} = \boxed{\frac{d}{dx}\cot^{-1} x - \frac{d}{dx}\tan x} = \boxed{-\frac{1}{1+x^2} - \sec^2 x}$

d. Given $\boxed{y = \tan^{-1}\left(x^2+3\right)}$ then $\boxed{\frac{dy}{dx}} = \boxed{\frac{d}{dx}\tan^{-1}\left(x^2+3\right)} = \boxed{\frac{1}{1+\left(x^2+3\right)^2} \cdot \frac{d}{dx}\left(x^2+3\right)} = \boxed{\frac{2x}{1+\left(x^2+3\right)^2}}$

$$= \boxed{\frac{2x}{1+\left(x^4+9+6x^2\right)}} = \boxed{\frac{2x}{x^4+6x^2+10}}$$

e. Given $\boxed{y = \dfrac{arc \sin x}{x} + 5x^3}$ then $\boxed{\dfrac{dy}{dx}} = \boxed{\dfrac{d}{dx}\left(\dfrac{arc \sin x}{x} + 5x^3\right)} = \boxed{\dfrac{\left(\frac{d}{dx}arc\sin x \cdot x\right)-\left(\frac{d}{dx}x\cdot arc\sin x\right)}{x^2}+15x^2}$

$$= \boxed{\dfrac{\left(x\cdot\frac{1}{\sqrt{1-x^2}}\right)-\left(arc\sin x\cdot 1\right)}{x^2}+15x^2} = \boxed{\dfrac{\frac{x}{\sqrt{1-x^2}}-arc\sin x-}{x^2}+15x^2} = \boxed{\dfrac{x-arc\sin x\sqrt{1-x^2}}{x^2\sqrt{1-x^2}}+15x^2}$$

f. Given $\boxed{y = \cos x + \tan^{-1} x}$ then $\boxed{\frac{dy}{dx}} = \boxed{\frac{d}{dx}\left(\cos x + \tan^{-1} x\right)} = \boxed{\frac{d}{dx}\cos x + \frac{d}{dx}\tan^{-1} x} = \boxed{-\sin x + \frac{1}{1+x^2}}$

g. Given $\boxed{y = \dfrac{3x^2}{arc \tan x}}$ then $\boxed{\dfrac{dy}{dx}} = \boxed{\dfrac{d}{dx}\left(\dfrac{3x^2}{arc\tan x}\right)} = \boxed{\dfrac{\left(arc\tan x\cdot\frac{d}{dx}3x^2\right)-\left(3x^2\cdot\frac{d}{dx}arc\tan x\right)}{arc\tan^2 x}}$

$$= \boxed{\dfrac{\left(arc\tan x\cdot 6x\right)-\left(3x^2\cdot\frac{1}{1+x^2}\right)}{arc\tan^2 x}} = \boxed{\dfrac{6x\,arc\tan x-\frac{3x^2}{1+x^2}}{arc\tan^2 x}} = \boxed{\dfrac{6x\left(1+x^2\right)arc\tan x-3x^2}{\left(1+x^2\right)arc\tan^2 x}}$$

h. Given $\boxed{y = \sin x + \sin^{-1} x}$ then $\boxed{\frac{dy}{dx}} = \boxed{\frac{d}{dx}\left(\sin x + \sin^{-1} x\right)} = \boxed{\frac{d}{dx}\sin x + \frac{d}{dx}\sin^{-1} x} = \boxed{\cos x + \frac{1}{\sqrt{1-x^2}}}$

i. Given $\sin 3y = arc\tan 5x$ let's take the derivative of both sides of the equation to obtain:

$$\boxed{\cos 3y\cdot\frac{d}{dx}3y = \frac{1}{1+\left(5x\right)^2}\cdot\frac{d}{dx}5x}\; ; \;\boxed{\cos 3y\cdot 3y' = \frac{1}{1+25x^2}\cdot 5}\; ; \;\boxed{3\cos 3y\cdot y' = \frac{5}{1+25x^2}}\; ; \;\boxed{y' = \frac{\frac{5}{1+25x^2}}{3\cos 3y}}$$

$; \boxed{y' = \dfrac{\frac{5}{1+25x^2}}{\frac{3\cos 3y}{1}}}$ $; \boxed{y' = \dfrac{5 \cdot 1}{\left(1+25x^2\right) \cdot 3\cos 3y}}$ $; \boxed{\mathbf{y' = \dfrac{5}{3\left(1+25x^2\right)\,\cos 3y}}}$

j. Given $\cos y = \sin^{-1} x$ let's take the derivative of both sides of the equation to obtain:

$\boxed{-\sin y \cdot \dfrac{d}{dx} y = \dfrac{1}{\sqrt{1-x^2}} \cdot \dfrac{d}{dx} x}$ $; \boxed{-\sin y \cdot y' = \dfrac{1}{\sqrt{1-x^2}} \cdot 1}$ $; \boxed{-\sin y \cdot y' = \dfrac{1}{\sqrt{1-x^2}}}$ $; \boxed{y' = \dfrac{\frac{1}{\sqrt{1-x^2}}}{-\sin y}}$ $; \boxed{y' = \dfrac{\frac{1}{\sqrt{1-x^2}}}{\frac{-\sin y}{1}}}$

$; \boxed{y' = \dfrac{1 \cdot 1}{\sqrt{1-x^2} \cdot -\sin y}}$ $; \boxed{\mathbf{y' = -\dfrac{1}{\sqrt{1-x^2}\,\sin\,y}}}$

k. Given $\sin\left(y+1\right) = arc\cos x^2$ let's take the derivative of both sides of the equation to obtain:

$\boxed{\cos\left(y+1\right) \cdot \dfrac{d}{dx}\left(y+1\right) = \dfrac{-1}{\sqrt{1-x^4}} \cdot \dfrac{d}{dx} x^2}$ $; \boxed{\cos\left(y+1\right) \cdot y' = \dfrac{-1}{\sqrt{1-x^4}} \cdot 2x}$ $; \boxed{\cos\left(y+1\right) \cdot y' = \dfrac{-2x}{\sqrt{1-x^4}}}$

$; \boxed{y' = \dfrac{\frac{-2x}{\sqrt{1-x^4}}}{\cos\left(y+1\right)}}$ $; \boxed{y' = \dfrac{\frac{-2x}{\sqrt{1-x^4}}}{\frac{\cos\left(y+1\right)}{1}}}$ $; \boxed{y' = \dfrac{-2x \cdot 1}{\sqrt{1-x^4} \cdot \cos\left(y+1\right)}}$ $; \boxed{\mathbf{y' = -\dfrac{2x}{\sqrt{1-x^4}\,\cos\left(y+1\right)}}}$

l. Given $\cot y = \tan^{-1} x$ let's take the derivative of both sides of the equation to obtain:

$\boxed{-\csc^2 y \cdot \dfrac{d}{dx} y = \dfrac{1}{1+x^2} \cdot \dfrac{d}{dx} x}$ $; \boxed{-\csc^2 y \cdot y' = \dfrac{1}{1+x^2} \cdot 1}$ $; \boxed{-\csc^2 y \cdot y' = \dfrac{1}{1+x^2}}$ $; \boxed{y' = \dfrac{\frac{1}{1+x^2}}{-\csc^2 y}}$ $; \boxed{y' = \dfrac{\frac{1}{1+x^2}}{\frac{-\csc^2 y}{1}}}$

$; \boxed{y' = \dfrac{1 \cdot 1}{\left(1+x^2\right) \cdot -\csc^2 y}}$ $; \boxed{\mathbf{y' = -\dfrac{1}{\left(1+x^2\right)\,\csc^2\,y}}}$

Example 3.2-3: Find the derivative of the following inverse trigonometric functions:

a. $y = x^3 \sin^{-1} x$

b. $y = arc\sin^2 x$

c. $y = \sin^{-1} x^2$

d. $y = arc\tan^2\left(x^3 + 1\right)$

e. $y = arc\cot^3 2x$

f. $y = \cot x \cdot \cot^{-1} x$

g. $y = arc\csc 10x$

h. $y = arc\csc x^{10}$

i. $y = arc\csc^{10} x$

Solutions:

a. $\boxed{\dfrac{dy}{dx}} = \boxed{\dfrac{d}{dx} x^3 \sin^{-1} x} = \boxed{\sin^{-1} x \cdot \dfrac{d}{dx} x^3 + x^3 \cdot \dfrac{d}{dx}\sin^{-1} x} = \boxed{\sin^{-1} x \cdot 3x^2 + x^3 \cdot \dfrac{1}{\sqrt{1-x^2}}} = \boxed{\mathbf{3x^2 \sin^{-1} x + \dfrac{x^3}{\sqrt{1-x^2}}}}$

b. $\dfrac{dy}{dx} = \dfrac{d}{dx} arc \sin^2 x = \dfrac{d}{dx}\left(arc \sin x\right)^2 = 2arc \sin x \cdot \dfrac{d}{dx}\left(arc \sin x\right) = 2arc \sin x \cdot \dfrac{1}{\sqrt{1-x^2}} = \dfrac{2arc \sin x}{\sqrt{1-x^2}}$

c. $\dfrac{dy}{dx} = \dfrac{d}{dx} \sin^{-1} x^2 = \dfrac{1}{\sqrt{1-x^4}} \cdot \dfrac{d}{dx} x^2 = \dfrac{1}{\sqrt{1-x^4}} \cdot 2x = \dfrac{2x}{\sqrt{1-x^4}}$

d. $\dfrac{dy}{dx} = \dfrac{d}{dx}\left[arc \tan^2\left(x^3+1\right)\right] = \dfrac{d}{dx}\left[arc \tan\left(x^3+1\right)\right]^2 = 2\,arc \tan\left(x^3+1\right)\cdot \dfrac{d}{dx} arc \tan\left(x^3+1\right)$

$= 2\,arc \tan\left(x^3+1\right)\cdot \dfrac{1}{1+\left(x^3+1\right)^2}\cdot \dfrac{d}{dx}\left(x^3+1\right) = 2\,arc \tan\left(x^3+1\right)\cdot \dfrac{1}{1+\left(x^3+1\right)^2}\cdot 3x^2 = \dfrac{6x^2 arc \tan\left(x^3+1\right)}{1+\left(x^3+1\right)^2}$

e. $\dfrac{dy}{dx} = \dfrac{d}{dx} arc \cot^3 2x = \dfrac{d}{dx}\left(arc \cot 2x\right)^3 = 3\left(arc \cot 2x\right)^2 \cdot \dfrac{d}{dx} arc \cot 2x = 3\left(arc \cot 2x\right)^2 \cdot \dfrac{-1}{1+(2x)^2}\cdot \dfrac{d}{dx} 2x$

$= 3\left(arc \cot 2x\right)^2 \cdot \dfrac{-1}{1+4x^2}\cdot 2 = \dfrac{-6\left(arc \cot 2x\right)^2}{1+4x^2}$

f. $\dfrac{dy}{dx} = \dfrac{d}{dx}\left(\cot x \cdot \cot^{-1} x\right) = \cot^{-1} x \cdot \dfrac{d}{dx}\cot x + \cot x \cdot \dfrac{d}{dx}\cot^{-1} x = \cot^{-1} x \cdot -\csc^2 x + \cot x \cdot \dfrac{-1}{1+x^2}$

$= -\csc^2 x \cot^{-1} x - \dfrac{\cot x}{1+x^2} = -\dfrac{\left(1+x^2\right)\csc^2 x \cot^{-1} x + \cot x}{1+x^2}$

g. $\dfrac{dy}{dx} = \dfrac{d}{dx}\left(arc \csc 10x\right) = -\dfrac{1}{10x\sqrt{(10x)^2-1}}\cdot \dfrac{d}{dx} 10x = -\dfrac{1}{10x\sqrt{100x^2-1}}\cdot 10 = -\dfrac{1}{x\sqrt{100x^2-1}}$

h. $\dfrac{dy}{dx} = \dfrac{d}{dx}\left(arc \csc x^{10}\right) = -\dfrac{1}{x^{10}\sqrt{x^{20}-1}}\cdot \dfrac{d}{dx} x^{10} = -\dfrac{1}{x^{10}\sqrt{x^{20}-1}}\cdot 10x^9 = -\dfrac{10}{x\sqrt{x^{20}-1}}$

i. $\dfrac{dy}{dx} = \dfrac{d}{dx}\left(arc \csc^{10} x\right) = \dfrac{d}{dx}\left(arc \csc x\right)^{10} = 10\left(arc \csc x\right)^9 \cdot \dfrac{d}{dx} arc \csc x = 10\left(arc \csc x\right)^9 \cdot \dfrac{-1}{x\sqrt{x^2-1}}$

$= \dfrac{10\left(arc \csc x\right)^9}{1}\cdot \dfrac{-1}{x\sqrt{x^2-1}} = \dfrac{-10\left(arc \csc x\right)^9}{x\sqrt{x^2-1}}$

Example 3.2-4: Given that the inverse sine function is given by $y = \sin^{-1} x \Leftrightarrow x = \sin y$ where $-\dfrac{\pi}{2} \le y \le \dfrac{\pi}{2}$ and $-1 \le x \le 1$, show that $\dfrac{d}{dx} \sin^{-1} x = \dfrac{1}{\sqrt{1-x^2}}$.

Solution:

Given $\sin y = x$ then $\dfrac{d}{dx} \sin y = \dfrac{d}{dx} x$; $\cos y \cdot \dfrac{d}{dx} y = 1$; $\dfrac{dy}{dx} = \dfrac{1}{\cos y}$

Since $y = \sin^{-1} x$ we can state that $\dfrac{dy}{dx} = \dfrac{1}{\cos y}$; $\dfrac{d}{dx} \sin^{-1} x = \dfrac{1}{\cos y}$

To express $\cos y$ in terms of x we can use the relation $\sin^2 y + \cos^2 y = 1$; $\cos^2 y = 1 - \sin^2 y$; $\cos^2 y = 1 - x^2$. Therefore, $\cos y = \sqrt{1-x^2}$ and we have shown that $\dfrac{d}{dx} \sin^{-1} x = \dfrac{1}{\cos y} = \dfrac{1}{\sqrt{1-x^2}}$

Example 3.2-5: Given that the inverse sine function is given by $y = \cos^{-1} x \Leftrightarrow x = \cos y$ where $0 \le y \le \pi$ and $-1 \le x \le 1$, show that $\dfrac{d}{dx} \cos^{-1} x = \dfrac{-1}{\sqrt{1-x^2}}$.

Solution:

Given $\cos y = x$ then $\dfrac{d}{dx} \cos y = \dfrac{d}{dx} x$; $-\sin y \cdot \dfrac{d}{dx} y = 1$; $\dfrac{dy}{dx} = \dfrac{1}{-\sin y} = \dfrac{-1}{\sin y}$

Since $y = \cos^{-1} x$ we can state that $\dfrac{dy}{dx} = \dfrac{-1}{\sin y}$; $\dfrac{d}{dx} \cos^{-1} x = \dfrac{-1}{\sin y}$

To express $\sin y$ in terms of x we can use the relation $\sin^2 y + \cos^2 y = 1$; $\sin^2 y = 1 - \cos^2 y$; $\sin^2 y = 1 - x^2$. Therefore, $\sin y = \sqrt{1-x^2}$ and we have shown that $\dfrac{d}{dx} \cos^{-1} x = \dfrac{-1}{\sin y} = \dfrac{-1}{\sqrt{1-x^2}}$

Example 3.2-6: Given that the inverse tangent function is given by $y = \tan^{-1} x \Leftrightarrow x = \tan y$ where $-\dfrac{\pi}{2} \langle y \langle \dfrac{\pi}{2}$ for all x, show that $\dfrac{d}{dx} \tan^{-1} x = \dfrac{1}{1+x^2}$.

Solution:

Given $\tan y = x$ then $\dfrac{d}{dx} \tan y = \dfrac{d}{dx} x$; $\sec^2 y \cdot \dfrac{d}{dx} y = 1$; $\dfrac{dy}{dx} = \dfrac{1}{\sec^2 y}$

Since $y = \tan^{-1} x$ we can state that $\dfrac{dy}{dx} = \dfrac{1}{\sec^2 y}$; $\dfrac{d}{dx} \tan^{-1} x = \dfrac{1}{\sec^2 y}$

To express $\sec^2 y$ in terms of x we can use the relation $\sec^2 y - \tan^2 y = 1$; $\sec^2 y = 1 + \tan^2 y$; $\sec^2 y = 1 + x^2$. Therefore, we have shown that $\dfrac{d}{dx} \tan^{-1} x = \dfrac{1}{\sec^2 y} = \dfrac{1}{1+x^2}$

Example 3.2-7: Given that the inverse tangent function is given by $y = \cot^{-1} x \Leftrightarrow x = \cot y$ where $0 \langle y \langle \pi$ for all x, show that $\dfrac{d}{dx} \cot^{-1} x = \dfrac{-1}{1+x^2}$.

Solution:

Given $\cot y = x$ then $\dfrac{d}{dx}\cot y = \dfrac{d}{dx}x$; $-\csc^2 y \cdot \dfrac{d}{dx}y = 1$; $\dfrac{dy}{dx} = \dfrac{1}{-\csc^2 y} = \dfrac{-1}{\csc^2 y}$

Since $y = \cot^{-1}x$ we can state that $\dfrac{dy}{dx} = \dfrac{-1}{\csc^2 y}$; $\dfrac{d}{dx}\cot^{-1}x = \dfrac{-1}{\csc^2 y}$

To express $\csc^2 y$ in terms of x we can use the relation $\csc^2 y - \cot^2 y = 1$; $\csc^2 y = 1 + \cot^2 y$

; $\csc^2 y = 1 + x^2$. Therefore, we have shown that $\dfrac{d}{dx}\cot^{-1}x = \dfrac{-1}{\csc^2 y} = -\dfrac{1}{1+x^2}$

Note: Similar steps can be taken to show that:

1. $\dfrac{d}{dx}\sec^{-1}x = \dfrac{1}{x\sqrt{x^2-1}}$ where the inverse secant function is given by $y = \sec^{-1}x \Leftrightarrow x = \sec y$ for

   $0 \le y \langle \dfrac{\pi}{2}$ or $\pi \le y \langle \dfrac{3\pi}{2}$ and $x \ge 1$ or $x \le -1$ and

2. $\dfrac{d}{dx}\csc^{-1}x = \dfrac{-1}{x\sqrt{x^2-1}}$ where the inverse cosecant function is given by $y = \csc^{-1}x \Leftrightarrow x = \csc y$ for

   $0 \langle y \langle \dfrac{\pi}{2}$ or $\pi \langle y \le \dfrac{3\pi}{2}$ and $x \ge 1$ or $x \le -1$.

Section 3.2 Practice Problems – Differentiation of Inverse Trigonometric Functions

1. Find the derivative of the following inverse trigonometric functions:

   a. $y = \sin^{-1}3x$

   b. $y = x^3 + arc\cos 5x$

   c. $y = x^5 arc\tan x^4$

   d. $y = arc\sin\left(x^3 + 2\right)$

   e. $y = arc\cot\dfrac{1}{x^3}$

   f. $y = \dfrac{arc\sin 3x}{x^2}$

   g. $y = \dfrac{\tan^{-1}x}{x}$

   h. $y = \dfrac{x^3}{x+5} - arc\sin x$

   i. $y = x + \cos^{-1}x$

2. Find the derivative of the following inverse trigonometric functions:

   a. $y = x^2 + arc\sin ax$

   b. $y = \cos^{-1}6x$

   c. $y = x + arc\sin x^3$

   d. $y = arc\tan^2 x$

   e. $y = x - arc\tan x^2$

   f. $y = x^2 - arc\sin x$

   g. $y = \tan^{-1}x^3$

   h. $y = \tan^{-1}5x$

   i. $y = arc\sin 5x^2$

3.3   Differentiation of Logarithmic and Exponential Functions

When differentiating logarithmic functions the following two rules should be kept in mind:

1. The derivative of a logarithmic function to the base e, i.e., $\log_e x = \ln x$ is equal to:

$$\frac{dy}{dx} = \frac{d}{dx}\left(\log_e x\right) = \frac{1}{x}\cdot\log_e e = \frac{1}{x}\cdot 1 = \frac{1}{x}$$

or,

$$\frac{dy}{dx} = \frac{d}{dx}\left(\ln x\right) = \frac{1}{x} \qquad\qquad (1)$$

2. The derivative of a logarithmic function to the base other than e, i.e., $\log_a x$ is equal to:

$$\frac{dy}{dx} = \frac{d}{dx}\left(\log_a x\right) = \frac{1}{x}\log_a x \qquad\qquad (2)$$

Let's differentiate some logarithmic functions using the above differentiation formulas.

Example 3.3-1:   Find the derivative of the following logarithmic functions:

a. $y = \ln 5x$

b. $y = \ln x^3$

c. $y = 5\ln 3x^2$

d. $y = \ln\left(x^2 + x + 3\right)$

e. $y = \ln\left(x^2 + x + 1\right)^4$

f. $y = x^2 \ln\left(x+1\right)$

g. $y = x\ln x - x$

h. $y = x^2 \ln x - x^3$

i. $y = \ln\left(\ln x + 1\right)$

j. $y = \dfrac{\ln x^2}{x^3}$

k. $y = \ln\left(\dfrac{x+1}{x-3}\right)$

l. $y = \ln\left(\sin 3x + \cos 5x\right)$

Solutions:

a.   Given $\boxed{y = \ln 5x}$ then $\boxed{\dfrac{dy}{dx}} = \boxed{\dfrac{d}{dx}\left(\ln 5x\right)} = \boxed{\dfrac{1}{5x}\cdot\dfrac{d}{dx}\left(5x\right)} = \boxed{\dfrac{1}{5x}\cdot 5} = \boxed{\dfrac{5}{5x}} = \boxed{\dfrac{1}{x}}$

b.   Given $\boxed{y = \ln x^3}$ then $\boxed{\dfrac{dy}{dx}} = \boxed{\dfrac{d}{dx}\left(\ln x^3\right)} = \boxed{\dfrac{1}{x^3}\cdot\dfrac{d}{dx}\left(x^3\right)} = \boxed{\dfrac{1}{x^3}\cdot 3x^2} = \boxed{\dfrac{3x^2}{x^{3=1}}} = \boxed{\dfrac{3}{x}}$

c.   Given $\boxed{y = 5\ln 3x^2}$ then $\boxed{\dfrac{dy}{dx}} = \boxed{\dfrac{d}{dx}\left(5\ln 3x^2\right)} = \boxed{5\cdot\dfrac{1}{3x^2}\cdot\dfrac{d}{dx}\left(3x^2\right)} = \boxed{\dfrac{5}{3x^2}\cdot 6x} = \boxed{\dfrac{30x}{3x^{2=1}}} = \boxed{\dfrac{30}{3x}} = \boxed{\dfrac{10}{x}}$

d.   Given $\boxed{y = \ln\left(x^2 + x + 3\right)}$ then $\boxed{\dfrac{dy}{dx}} = \boxed{\dfrac{d}{dx}\left[\ln\left(x^2 + x + 3\right)\right]} = \boxed{\dfrac{1}{x^2+x+3}\cdot\dfrac{d}{dx}\left(x^2+x+3\right)} = \boxed{\dfrac{2x+1}{x^2+x+3}}$

e.   Given $\boxed{y = \ln\left(x^2 + x + 1\right)^4}$ then $\boxed{\dfrac{dy}{dx}} = \boxed{\dfrac{d}{dx}\left[\ln\left(x^2 + x + 1\right)^4\right]} = \boxed{\dfrac{1}{\left(x^2+x+1\right)^4}\cdot\dfrac{d}{dx}\left(x^2+x+1\right)^4}$

$= \boxed{\dfrac{1}{\left(x^2+x+1\right)^4}\cdot\left[4\left(x^2+x+1\right)^3\cdot\left(2x+1\right)\right]} = \boxed{\dfrac{4\left(x^2+x+1\right)^3\cdot\left(2x+1\right)}{\left(x^2+x+1\right)^{4=1}}} = \boxed{\dfrac{4\left(2x+1\right)}{x^2+x+1}} = \boxed{\dfrac{8x+4}{x^2+x+1}}$

f. Given $\boxed{y = x^2 \ln(x+1)}$ then $\boxed{\dfrac{dy}{dx}} = \boxed{\dfrac{d}{dx}\left[x^2 \ln(x+1)\right]} = \boxed{\left[\ln(x+1)\cdot\dfrac{d}{dx}\left(x^2\right)\right] + \left[x^2\cdot\dfrac{d}{dx}\left[\ln(x+1)\right]\right]}$

$= \boxed{\left[\ln(x+1)\cdot 2x\right] + \left[x^2\cdot\dfrac{1}{x+1}\cdot\dfrac{d}{dx}(x+1)\right]} = \boxed{\left[\ln(x+1)\cdot 2x\right] + \left[x^2\cdot\dfrac{1}{x+1}\cdot 1\right]} = \boxed{\mathbf{2x \ln(x+1) + \dfrac{x^2}{x+1}}}$

g. Given $\boxed{y = x \ln x - x}$ then $\boxed{\dfrac{dy}{dx}} = \boxed{\dfrac{d}{dx}(x\ln x - x)} = \boxed{\dfrac{d}{dx}(x\ln x) - \dfrac{d}{dx}(x)} = \boxed{\left[x\cdot\dfrac{d}{dx}(\ln x) + \ln x\dfrac{d}{dx}(x)\right] - \dfrac{d}{dx}(x)}$

$= \boxed{\left[x\cdot\dfrac{1}{x}\cdot\dfrac{d}{dx}(x) + \ln x\cdot 1\right] - 1} = \boxed{\left[\dfrac{\not{x}}{\not{x}} + \ln x\right] - 1} = \boxed{1 + \ln x - 1} = \boxed{\mathbf{\ln x}}$

h. Given $\boxed{y = x^2 \ln x - x^3}$ then $\boxed{\dfrac{dy}{dx}} = \boxed{\dfrac{d}{dx}\left(x^2 \ln x - x^3\right)} = \boxed{\dfrac{d}{dx}\left(x^2 \ln x\right) - \dfrac{d}{dx}\left(x^3\right)} = \boxed{\left[\ln x\cdot\dfrac{d}{dx}\left(x^2\right) + x^2\cdot\dfrac{d}{dx}(\ln x)\right]}$

$\boxed{-\dfrac{d}{dx}\left(x^3\right)} = \boxed{\left[\ln x\cdot 2x + x^2\cdot\dfrac{1}{x}\cdot\dfrac{d}{dx}(x)\right] - 3x^2} = \boxed{\left[2x\ln x + x^2\cdot\dfrac{1}{x}\cdot 1\right] - 3x^2} = \boxed{\left[2x\ln x + \dfrac{x^{2=1}}{\not{x}}\right] - 3x^2}$

$= \boxed{2x\ln x + x - 3x^2} = \boxed{\mathbf{-3x^2 + x(2\ln x + 1)}}$

i. Given $\boxed{y = \ln(\ln x + 1)}$ then $\boxed{\dfrac{dy}{dx}} = \boxed{\dfrac{1}{\ln x + 1}\cdot\dfrac{d}{dx}(\ln x + 1)} = \boxed{\dfrac{1}{\ln x + 1}\cdot\left[\dfrac{d}{dx}(\ln x) + \dfrac{d}{dx}(1)\right]} = \boxed{\dfrac{1}{\ln x + 1}\cdot\left[\dfrac{1}{x}\dfrac{d}{dx}(x) + 0\right]}$

$= \boxed{\dfrac{1}{\ln x + 1}\cdot\left[\dfrac{1}{x}\cdot 1 + 0\right]} = \boxed{\dfrac{1}{\ln x + 1}\cdot\dfrac{1}{x}} = \boxed{\mathbf{\dfrac{1}{x(\ln x + 1)}}}$

j. Given $\boxed{y = \dfrac{\ln x^2}{x^3}}$ then $\boxed{\dfrac{dy}{dx}} = \boxed{\dfrac{d}{dx}\left(\dfrac{\ln x^2}{x^3}\right)} = \boxed{\dfrac{\left[x^3\cdot\dfrac{d}{dx}\left(\ln x^2\right)\right] - \left[\ln x^2\cdot\dfrac{d}{dx}\left(x^3\right)\right]}{x^6}} = \boxed{\dfrac{\left[x^3\cdot\dfrac{1}{x^2}\cdot 2x\right] - \left[\ln x^2\cdot 3x^2\right]}{x^6}}$

$= \boxed{\dfrac{\dfrac{2x^{4=2}}{x^2} - \left(3x^2\ln x^2\right)}{x^6}} = \boxed{\dfrac{2x^2 - \left(3x^2\ln x^2\right)}{x^6}} = \boxed{\dfrac{x^2\left(2 - 3\ln x^2\right)}{x^{6=4}}} = \boxed{\mathbf{\dfrac{2 - 3\ln x^2}{x^4}}}$

k. Given $\boxed{y = \ln\left(\dfrac{x+1}{x-3}\right)}$ then $\boxed{\dfrac{dy}{dx}} = \boxed{\dfrac{d}{dx}\ln\left(\dfrac{x+1}{x-3}\right)} = \boxed{\dfrac{1}{\dfrac{x+1}{x-3}}\cdot\dfrac{d}{dx}\left(\dfrac{x+1}{x-3}\right)} = \boxed{\dfrac{1}{\dfrac{x+1}{x-3}}\cdot\dfrac{\left[(x-3)\cdot\dfrac{d}{dx}(x+1)\right] - \left[(x+1)\cdot\dfrac{d}{dx}(x-3)\right]}{(x-3)^2}}$

$= \boxed{\dfrac{1}{\dfrac{x+1}{x-3}}\cdot\dfrac{\left[(x-3)\cdot 1\right] - \left[(x+1)\cdot 1\right]}{(x-3)^2}} = \boxed{\dfrac{1}{\dfrac{x+1}{x-3}}\cdot\dfrac{(x-3) - (x+1)}{(x-3)^2}} = \boxed{\dfrac{1\cdot(x-3)}{(x+1)\cdot 1}\cdot\dfrac{(x-3) - (x+1)}{(x-3)^2}} = \boxed{\dfrac{x-3}{x+1}\cdot\dfrac{\not{x}-3-\not{x}-1}{(x-3)^2}}$

$$= \left| \frac{(x-3)}{x+1} \cdot \frac{-4}{(x-3)^{2}=1} \right| = \left| -\frac{4}{(x+1)(x-3)} \right|$$

l. Given $\boxed{y = \ln(\sin 3x + \cos 5x)}$ then $\boxed{\dfrac{dy}{dx}} = \boxed{\dfrac{1}{\sin 3x + \cos 5x} \cdot \dfrac{d}{dx}(\sin 3x + \cos 5x)} = \boxed{\dfrac{1}{\sin 3x + \cos 5x} \cdot [\; \dfrac{d}{dx}\sin 3x}$

$\boxed{+ \dfrac{d}{dx}(\cos 5x)]} = \boxed{\dfrac{1}{\sin 3x + \cos 5x} \cdot (3\cos 3x - 5\sin 5x)} = \boxed{\dfrac{3\cos 3x - 5\sin 5x}{\sin 3x + \cos 5x}}$

Example 3.3-2: Find the derivative of the following logarithmic expressions:

a. $y = (x^3 + 1) \cdot \ln x^2$

b. $y = \sin(\ln 3x) - \cos(\ln 5x)$

c. $y = x^2 \ln x - x$

d. $y = \ln(x^3 - 1) - \ln(\sec x)$

e. $y = \dfrac{\ln x^3}{x^5} + x$

f. $y = \ln(\sec 5x + \tan 5x)$

g. $y = \sin(\ln x) + x$

h. $y = \sin(\ln x) + \cos(\ln x)$

i. $y = \dfrac{1}{3}x^5(\ln x + 3)$

j. $y = \ln x^3 + \ln(\csc x)$

k. $y = \ln\left(\dfrac{x^2}{x+1}\right)$

l. $y = \ln(\sec x + \csc x)$

Solutions:

a. Given $\boxed{y = (x^3 + 1) \cdot \ln x^2}$ then $\boxed{\dfrac{dy}{dx}} = \boxed{\left[\dfrac{d}{dx}(x^3+1)\cdot\ln x^2\right] + \left[\dfrac{d}{dx}(\ln x^2)\cdot(x^3+1)\right]} = \boxed{3x^2 \ln x^2}$

$\boxed{+\left[\dfrac{1}{x^2}\cdot\dfrac{d}{dx}(x^2)\cdot(x^3+1)\right]} = \boxed{3x^2 \ln x^2 + \left[\dfrac{1}{x^{2}=1}\cdot 2x\cdot(x^3+1)\right]} = \boxed{3x^2 \ln x^2 + \dfrac{2(x^3+1)}{x}}$

b. Given $\boxed{y = \sin(\ln 3x) - \cos(\ln 5x)}$ then $\boxed{\dfrac{dy}{dx}} = \boxed{\dfrac{d}{dx}\left[\sin(\ln 3x) - \cos(\ln 5x)\right]} = \boxed{\dfrac{d}{dx}\sin(\ln 3x) - \dfrac{d}{dx}\cos(\ln 5x)}$

$= \boxed{\cos(\ln 3x)\cdot\dfrac{d}{dx}\ln 3x + \sin(\ln 5x)\cdot\dfrac{d}{dx}\ln 5x} = \boxed{\cos(\ln 3x)\cdot\dfrac{1}{3x}\cdot\dfrac{d}{dx}(3x) + \sin(\ln 5x)\cdot\dfrac{1}{5x}\cdot\dfrac{d}{dx}(5x)}$

$= \boxed{\cos(\ln 3x)\cdot\dfrac{1}{3x}\cdot 3 + \sin(\ln 5x)\cdot\dfrac{1}{5x}\cdot 5} = \boxed{\dfrac{3\cos(\ln 3x)}{3x} + \dfrac{5\sin(\ln 5x)}{5x}} = \boxed{\dfrac{\cos(\ln 3x)}{x} + \dfrac{\sin(\ln 5x)}{x}}$

c. Given $\boxed{y = x^2 \ln x - x}$ then $\boxed{\dfrac{dy}{dx}} = \boxed{\left[\dfrac{d}{dx}(x^2)\cdot\ln x + \dfrac{d}{dx}(\ln x)\cdot x^2\right] - \dfrac{d}{dx}(x)} = \boxed{\left[2x\cdot\ln x + \dfrac{1}{x}\cdot\dfrac{d}{dx}(x)\cdot x^2\right] - 1}$

$= \boxed{\left[2x\cdot\ln x + \dfrac{1}{x}\cdot 1\cdot x^2\right] - 1} = \boxed{\left[2x\ln x + \dfrac{x^{2}=1}{\cancel{x}}\right] - 1} = \boxed{2x\ln x + x - 1}$

d. Given $y = \ln\left(x^3 - 1\right) - \ln\left(\sec x\right)$ then $\dfrac{dy}{dx} = \dfrac{d}{dx}\left[\ln\left(x^3 - 1\right) - \ln\left(\sec x\right)\right] = \dfrac{d}{dx}\ln\left(x^3 - 1\right) - \dfrac{d}{dx}\ln\left(\sec x\right)$

$= \dfrac{1}{x^3 - 1} \cdot \dfrac{d}{dx}\left(x^3 - 1\right) - \dfrac{1}{\sec x} \cdot \dfrac{d}{dx}\left(\sec x\right) = \dfrac{1}{x^3 - 1} \cdot 3x^2 - \dfrac{1}{\sec x} \cdot \sec x \tan x = \boxed{\dfrac{3x^2}{x^3 - 1} - \tan x}$

e. Given $y = \dfrac{\ln x^3}{x^5} + x$ then $\dfrac{dy}{dx} = \dfrac{d}{dx}\left(\dfrac{\ln x^3}{x^5} + x\right) = \dfrac{d}{dx}\left(\dfrac{\ln x^3}{x^5}\right) + \dfrac{d}{dx}x = \dfrac{\left(x^5 \cdot \frac{d}{dx}\ln x^3\right) - \left(\ln x^3 \cdot \frac{d}{dx}x^5\right)}{x^{10}} + 1$

$= \dfrac{\left(x^5 \cdot \frac{1}{x^3} \cdot \frac{d}{dx}x^3\right) - \left(\ln x^3 \cdot 5x^4\right)}{x^{10}} + 1 = \dfrac{\left(x^5 \cdot \frac{1}{x^3} \cdot 3x^2\right) - \left(5x^4 \ln x^3\right)}{x^{10}} + 1 = \dfrac{\left(3\frac{x^{7=4}}{x^3}\right) - \left(5x^4 \ln x^3\right)}{x^{10}} + 1$

$= \dfrac{\left(3x^4\right) - \left(5x^4 \ln x^3\right)}{x^{10}} + 1 = \dfrac{x^4\left(3 - 5\ln x^3\right)}{x^{10=6}} + 1 = \boxed{\dfrac{3 - 5\ln x^3}{x^6} + 1}$

f. Given $y = \ln\left(\sec 5x + \tan 5x\right)$ then $\dfrac{dy}{dx} = \dfrac{d}{dx}\left[\ln\left(\sec 5x + \tan 5x\right)\right] = \dfrac{1}{\sec 5x + \tan 5x} \cdot \dfrac{d}{dx}\left[\left(\sec 5x + \tan 5x\right)\right]$

$= \dfrac{1}{\sec 5x + \tan 5x} \cdot \left(\dfrac{d}{dx}\sec 5x + \dfrac{d}{dx}\tan 5x\right) = \dfrac{1}{\sec 5x + \tan 5x} \cdot \left(\sec 5x \tan 5x \cdot \dfrac{d}{dx}5x + \sec^2 5x \cdot \dfrac{d}{dx}5x\right)$

$= \dfrac{\sec 5x \tan 5x \cdot 5 + \sec^2 5x \cdot 5}{\sec 5x + \tan 5x} = \dfrac{5\sec 5x\left(\tan 5x + \sec 5x\right)}{\left(\sec 5x + \tan 5x\right)} = \boxed{5\sec 5x}$

g. Given $y = \sin\left(\ln x\right) + x$ then $\dfrac{dy}{dx} = \dfrac{d}{dx}\left[\sin\left(\ln x\right) + x\right] = \dfrac{d}{dx}\sin\left(\ln x\right) + \dfrac{d}{dx}x = \cos\left(\ln x\right) \cdot \dfrac{d}{dx}\left(\ln x\right) + 1$

$= \cos\left(\ln x\right) \cdot \dfrac{1}{x} \cdot \dfrac{d}{dx}x + 1 = \cos\left(\ln x\right) \cdot \dfrac{1}{x} \cdot 1 + 1 = \boxed{\dfrac{\cos\left(\ln x\right)}{x} + 1}$

h. Given $y = \sin\left(\ln x\right) + \cos\left(\ln x\right)$ then $\dfrac{dy}{dx} = \dfrac{d}{dx}\left[\sin\left(\ln x\right) + \cos\left(\ln x\right)\right] = \dfrac{d}{dx}\sin\left(\ln x\right) + \dfrac{d}{dx}\cos\left(\ln x\right)$

$= \dfrac{d}{dx}\sin\left(\ln x\right) + \dfrac{d}{dx}\cos\left(\ln x\right) = \cos\left(\ln x\right) \cdot \dfrac{d}{dx}\left(\ln x\right) - \sin\left(\ln x\right) \cdot \dfrac{d}{dx}\left(\ln x\right) = \cos\left(\ln x\right) \cdot \dfrac{1}{x} \cdot \dfrac{d}{dx}x - \sin\left(\ln x\right) \cdot \dfrac{1}{x} \cdot \dfrac{d}{dx}x$

$= \cos\left(\ln x\right) \cdot \dfrac{1}{x} \cdot 1 - \sin\left(\ln x\right) \cdot \dfrac{1}{x} \cdot 1 = \dfrac{\cos\left(\ln x\right)}{x} - \dfrac{\sin\left(\ln x\right)}{x} = \boxed{\dfrac{1}{x}\left[\cos\left(\ln x\right) - \sin\left(\ln x\right)\right]}$

i. Given $y = \dfrac{1}{3}x^5\left(\ln x + 3\right)$ then $\dfrac{dy}{dx} = \dfrac{1}{3}\left[\dfrac{d}{dx}x^5 \cdot \left(\ln x + 3\right) + \dfrac{d}{dx}\left(\ln x + 3\right) \cdot x^5\right] = \dfrac{1}{3}\left[5x^4\left(\ln x + 3\right) + \left(\dfrac{d}{dx}\ln x + \dfrac{d}{dx}3\right)x^5\right]$

$$= \boxed{\frac{1}{3}\left[5x^4(\ln x+3)+\left(\frac{1}{x}\cdot\frac{d}{dx}(x)+0\right)\cdot x^5\right]} = \boxed{\frac{1}{3}\left[5x^4(\ln x+3)+\left(\frac{1}{x}\cdot1\right)\cdot x^5\right]} = \boxed{\frac{1}{3}\left[5x^4(\ln x+3)+\frac{x^{5=4}}{x}\right]}$$

$$= \boxed{\frac{1}{3}\left[5x^4(\ln x+3)+x^4\right]} = \boxed{\frac{5}{3}x^4(\ln x+3)+\frac{x^4}{3}} = \boxed{\frac{5x^4\ln x}{3}+\frac{15x^4}{3}+\frac{x^4}{3}} = \boxed{\mathbf{\frac{5x^4\ln x}{3}+\frac{16x^4}{3}}} \qquad \text{or,}$$

Given $\boxed{y=\frac{1}{3}x^5(\ln x+3)=\frac{1}{3}x^5\ln x+x^5}$ then $\boxed{\frac{dy}{dx}}=\boxed{\frac{1}{3}\left[\frac{d}{dx}x^5\cdot(\ln x)+\frac{d}{dx}(\ln x)\cdot x^5\right]+\frac{d}{dx}x^5}$

$$= \boxed{\frac{1}{3}\left[5x^4\cdot(\ln x)+\frac{1}{x}\cdot\frac{d}{dx}(x)\cdot x^5\right]+5x^4} = \boxed{\frac{1}{3}\left[5x^4\cdot(\ln x)+\frac{1}{x}\cdot1\cdot x^5\right]+5x^4} = \boxed{\frac{1}{3}\left[5x^4\ln x+\frac{x^{5=4}}{x}\right]+5x^4}$$

$$= \boxed{\frac{1}{3}\left[5x^4\ln x+x^4\right]+5x^4} = \boxed{\frac{5x^4\ln x}{3}+\frac{x^4}{3}+5x^4} = \boxed{\frac{5x^4\ln x}{3}+\frac{x^4}{3}+\frac{5x^4}{1}} = \boxed{\frac{5x^4\ln x}{3}+\frac{(x^4\cdot1)+(5x^4\cdot3)}{3\cdot1}}$$

$$= \boxed{\frac{5x^4\ln x}{3}+\frac{x^4+15x^4}{3}} = \boxed{\mathbf{\frac{5x^4\ln x}{3}+\frac{16x^4}{3}}}$$

j. Given $\boxed{y=\ln x^3+\ln(\csc x)}$ then $\boxed{\frac{dy}{dx}}=\boxed{\frac{d}{dx}\left[\ln x^3+\ln(\csc x)\right]}=\boxed{\frac{d}{dx}\ln x^3+\frac{d}{dx}\ln(\csc x)}$

$$= \boxed{\frac{1}{x^3}\cdot\frac{d}{dx}(x^3)+\frac{1}{\csc x}\cdot\frac{d}{dx}(\csc x)} = \boxed{\frac{1}{x^3}\cdot3x^2+\frac{1}{\csc x}\cdot-\csc x\cot x} = \boxed{\frac{3x^2}{x^{3=2}}-\frac{\csc x\cot x}{\csc x}} = \boxed{\mathbf{\frac{3}{x}-\cot x}}$$

k. Given $\boxed{y=\ln\left(\frac{x^2}{x+1}\right)}$ then $\boxed{\frac{dy}{dx}}=\boxed{\frac{1}{\frac{x^2}{x+1}}\cdot\frac{d}{dx}\left(\frac{x^2}{x+1}\right)}=\boxed{\frac{1}{\frac{x^2}{x+1}}\cdot\frac{\left[(x+1)\cdot\frac{d}{dx}(x^2)\right]-\left[x^2\cdot\frac{d}{dx}(x+1)\right]}{(x+1)^2}}$

$$= \boxed{\frac{1\cdot(x+1)}{x^2\cdot1}\cdot\frac{[(x+1)\cdot2x]-[x^2\cdot1]}{(x+1)^2}} = \boxed{\frac{(x+1)}{x^2}\cdot\frac{2x^2+2x-x^2}{(x+1)^2}} = \boxed{\frac{(x+1)}{x^2}\cdot\frac{x^2+2x}{(x+1)^{2=1}}} = \boxed{\frac{x(x+2)}{x^{2=1}(x+1)}} = \boxed{\mathbf{\frac{x+2}{x(x+1)}}}$$

l. Given $\boxed{y=\ln(\sec x+\csc x)}$ then $\boxed{\frac{dy}{dx}}=\boxed{\frac{d}{dx}\left[\ln(\sec x+\csc x)\right]}=\boxed{\frac{1}{\sec x+\csc x}\cdot\frac{d}{dx}(\sec x+\csc x)}$

$$= \boxed{\frac{1}{\sec x+\csc x}\cdot(\sec x\tan x-\csc x\cot x)} = \boxed{\mathbf{\frac{\sec x\tan x-\csc x\cot x}{\sec x+\csc x}}}$$

Example 3.3-3: Find the derivative of the following logarithmic functions:

a. $y = \dfrac{\ln(\sec 5x)}{x^2}$

b. $y = \ln \csc x^2$

c. $y = \ln \dfrac{x+1}{\sin x}$

d. $y = \tan\left(\ln x^2\right)$

e. $y = \cot\left(\ln 3x^5\right)$

f. $y = \tan(\ln x) - \cot(\ln x)$

g. $y = \sin 5x + \ln(\sin 5x)$

h. $y = \ln\sqrt{x^2+1}$

i. $y = \sqrt{x}\cdot\ln\left(x^2+3\right)$

j. $y = \ln\sqrt[3]{x-1}$

k. $y = \ln\left(1+x^2\right) + 2\,arc\tan x$

l. $y = \ln\left(arc\tan 3x\right) + 3x^2$

Solutions:

a. $\boxed{\dfrac{dy}{dx}} = \boxed{\dfrac{d}{dx}\dfrac{\ln \sec 5x}{x^2}} = \boxed{\dfrac{\left[x^2\cdot\frac{d}{dx}\ln\sec 5x\right] - \left[\ln\sec 5x\cdot\frac{d}{dx}x^2\right]}{x^4}} = \boxed{\dfrac{\left[x^2\cdot\frac{1}{\sec 5x}\cdot\frac{d}{dx}(\sec 5x)\right] - \left[\ln(\sec 5x)\cdot 2x\right]}{x^4}}$

$= \dfrac{\left[\frac{x^2}{\sec 5x}\cdot\sec 5x\cdot\tan 5x\cdot\frac{d}{dx}5x\right] - 2x\ln\sec 5x}{x^4} = \boxed{\dfrac{\left(x^2\cdot\tan 5x\cdot 5\right) - 2x\ln\sec 5x}{x^4}} = \boxed{\dfrac{\left(5x^2\tan 5x\right) - 2x\ln\sec 5x}{x^4}}$

$= \boxed{\dfrac{\not{x}\left(5x\tan 5x - 2\ln\sec 5x\right)}{x^{4=3}}} = \boxed{\dfrac{\mathbf{5x\tan 5x - 2\ln(\sec 5x)}}{x^3}}$

b. $\boxed{\dfrac{dy}{dx}} = \boxed{\dfrac{d}{dx}\left(\ln\csc x^2\right)} = \boxed{\dfrac{1}{\csc x^2}\cdot\dfrac{d}{dx}\csc x^2} = \boxed{\dfrac{1}{\csc x^2}\cdot -\csc x^2\cdot\cot x^2\cdot\dfrac{d}{dx}x^2} = \boxed{\dfrac{1}{\csc x^2}\cdot -\csc x^2\cdot\cot x^2\cdot 2x}$

$= \boxed{\dfrac{1}{\not{\csc}\, x^2}\cdot -\not{\csc}\, x^2\cdot\cot x^2\cdot 2x} = \boxed{\mathbf{-2x\cot x^2}}$

c. $\boxed{\dfrac{dy}{dx}} = \boxed{\dfrac{d}{dx}\left(\ln\dfrac{x+1}{\sin x}\right)} = \boxed{\dfrac{1}{\frac{x+1}{\sin x}}\cdot\dfrac{d}{dx}\left(\dfrac{x+1}{\sin x}\right)} = \boxed{\dfrac{\frac{1}{1}}{\frac{x+1}{\sin x}}\cdot\dfrac{\left[\sin x\cdot\frac{d}{dx}(x+1)\right] - \left[(x+1)\cdot\frac{d}{dx}\sin x\right]}{(\sin x)^2}} = \boxed{\dfrac{1\cdot\sin x}{(x+1)\cdot 1}}$

$\boxed{\times\dfrac{\left[\sin x\cdot 1\right] - \left[(x+1)\cdot\cos x\right]}{(\sin x)^2}} = \boxed{\dfrac{\not{\sin}\, x}{x+1}\cdot\dfrac{\sin x - (x+1)\cos x}{(\sin x)^{2=1}}} = \boxed{\dfrac{1}{x+1}\cdot\dfrac{\sin x - (x+1)\cos x}{\sin x}} = \boxed{\dfrac{\mathbf{\sin x - (x+1)\cos\ x}}{(x+1)\sin x}}$

d. $\boxed{\dfrac{dy}{dx}} = \boxed{\dfrac{d}{dx}\left[\tan\left(\ln x^2\right)\right]} = \boxed{\sec^2\left(\ln x^2\right)\cdot\dfrac{d}{dx}\ln x^2} = \boxed{\sec^2\left(\ln x^2\right)\cdot\dfrac{1}{x^2}\cdot\dfrac{d}{dx}x^2} = \boxed{\sec^2\left(\ln x^2\right)\cdot\dfrac{1}{x^2}\cdot 2x}$

$= \boxed{\sec^2\left(\ln x^2\right)\cdot\dfrac{2\not{x}}{x^{2=1}}} = \boxed{\dfrac{\mathbf{2\sec^2\left(\ln x^2\right)}}{x}}$

e. $\boxed{\dfrac{dy}{dx}} = \boxed{\dfrac{d}{dx}\left[\cot\left(\ln 3x^5\right)\right]} = \boxed{-\csc^2\left(\ln 3x^5\right)\cdot\dfrac{d}{dx}\left(\ln 3x^5\right)} = \boxed{-\csc^2\left(\ln 3x^5\right)\cdot\dfrac{1}{3x^5}\cdot\dfrac{d}{dx}\left(3x^5\right)}$

$$= \boxed{-\csc^2\left(\ln 3x^5\right)\cdot\frac{1}{3x^5}\cdot15x^4} = \boxed{-\frac{\overset{5}{\cancel{15}}\,x^4\csc^2\left(\ln 3x^5\right)}{3x^{\cancel{5}=1}}} = \boxed{-\frac{5\csc^2\left(\ln 3x^5\right)}{x}}$$

f. $\boxed{\dfrac{dy}{dx}} = \boxed{\dfrac{d}{dx}\left[\tan(\ln x)-\cot(\ln x)\right]} = \boxed{\dfrac{d}{dx}\tan(\ln x)-\dfrac{d}{dx}\cot(\ln x)} = \boxed{\sec^2(\ln x)\cdot\dfrac{d}{dx}(\ln x)+\csc^2(\ln x)\dfrac{d}{dx}(\ln x)}$

$$= \boxed{\sec^2(\ln x)\cdot\frac{1}{x}\cdot\frac{d}{dx}x+\csc^2(\ln x)\cdot\frac{1}{x}\cdot\frac{d}{dx}x} = \boxed{\sec^2(\ln x)\cdot\frac{1}{x}\cdot1+\csc^2(\ln x)\cdot\frac{1}{x}\cdot1} = \boxed{\frac{1}{x}\left[\sec^2(\ln x)+\csc^2(\ln x)\right]}$$

g. $\boxed{\dfrac{dy}{dx}} = \boxed{\dfrac{d}{dx}\left[\sin 5x+\ln(\sin 5x)\right]} = \boxed{\dfrac{d}{dx}\sin 5x+\dfrac{d}{dx}\ln(\sin 5x)} = \boxed{\cos 5x\dfrac{d}{dx}5x+\dfrac{1}{\sin 5x}\cdot\dfrac{d}{dx}(\sin 5x)}$

$$= \boxed{\cos 5x\cdot5+\frac{1}{\sin 5x}\cdot\cos 5x\cdot\frac{d}{dx}5x} = \boxed{5\cos 5x+\frac{1}{\sin 5x}\cdot\cos 5x\cdot5} = \boxed{5\cos 5x+\frac{5\cos 5x}{\sin 5x}} = \boxed{5\cos 5x\left(1+\frac{1}{\sin 5x}\right)}$$

h. $\boxed{\dfrac{dy}{dx}} = \boxed{\dfrac{d}{dx}\left(\ln\sqrt{x^2+1}\right)} = \boxed{\dfrac{1}{\sqrt{x^2+1}}\cdot\dfrac{d}{dx}\sqrt{x^2+1}} = \boxed{\dfrac{1}{\sqrt{x^2+1}}\cdot\dfrac{d}{dx}\left(x^2+1\right)^{\frac12}} = \boxed{\dfrac{1}{\sqrt{x^2-1}}\cdot\dfrac12\left(x^2+1\right)^{\frac12-1}\cdot\dfrac{d}{dx}(2x+1)}$

$$= \boxed{\frac{1}{\sqrt{x^2+1}}\cdot\frac12\left(x^2+1\right)^{\frac12-1}\cdot2x} = \boxed{\frac{1}{\sqrt{x^2+1}}\cdot\frac{2x}{2}\left(x^2+1\right)^{-\frac12}} = \boxed{\frac{x}{\sqrt{x^2+1}}\cdot\frac{1}{\left(x^2+1\right)^{\frac12}}} = \boxed{\frac{x}{\left(x^2+1\right)^{\frac12}}\cdot\frac{1}{\left(x^2+1\right)^{\frac12}}}$$

$$= \boxed{\frac{x}{\left(x^2+1\right)^{\frac12+\frac12}}} = \boxed{\frac{x}{x^2+1}}$$

i. $\boxed{\dfrac{dy}{dx}} = \boxed{\dfrac{d}{dx}\left[\sqrt{x}\cdot\ln\left(x^2+3\right)\right]} = \boxed{\dfrac{d}{dx}\left[x^{\frac12}\cdot\ln\left(x^2+3\right)\right]} = \boxed{\left[\ln\left(x^2+3\right)\cdot\dfrac{d}{dx}x^{\frac12}\right]+\left[x^{\frac12}\cdot\dfrac{d}{dx}\ln\left(x^2+3\right)\right]}$

$$= \boxed{\left[\ln\left(x^2+3\right)\cdot\frac12 x^{\frac12-1}\right]+\left[x^{\frac12}\cdot\frac{1}{x^2+3}\cdot\frac{d}{dx}\left(x^2+3\right)\right]} = \boxed{\left[\ln\left(x^2+3\right)\cdot\frac12 x^{-\frac12}\right]+\left[x^{\frac12}\cdot\frac{1}{x^2+3}\cdot2x\right]}$$

$$= \boxed{\frac{\ln\left(x^2+3\right)}{2x^{\frac12}}+\frac{2x\cdot x^{\frac12}}{x^2+3}} = = \boxed{\frac{\ln\left(x^2+3\right)}{2\sqrt{x}}+\frac{2x\sqrt{x}}{x^2+3}}$$

j. $\boxed{\dfrac{dy}{dx}} = \boxed{\dfrac{d}{dx}\left(\ln\sqrt[3]{x-1}\right)} = \boxed{\dfrac{1}{\sqrt[3]{x-1}}\cdot\dfrac{d}{dx}\sqrt[3]{x-1}} = \boxed{\dfrac{1}{\sqrt[3]{x-1}}\cdot\dfrac{d}{dx}(x-1)^{\frac13}} = \boxed{\dfrac{1}{\sqrt[3]{x-1}}\cdot\dfrac13(x-1)^{\frac13-1}} = \boxed{\dfrac{1}{\sqrt[3]{x-1}}\cdot\dfrac13(x-1)^{-\frac23}}$

$$= \boxed{\frac{1}{\sqrt[3]{x-1}}\cdot\frac{1}{3(x-1)^{\frac23}}} = \boxed{\frac{1}{(x-1)^{\frac13}}\cdot\frac{1}{3(x-1)^{\frac23}}} = \boxed{\frac{1}{3(x-1)^{\frac13+\frac23}}} = \boxed{\frac{1}{3(x-1)^{\frac{1+2}{3}}}} = \boxed{\frac{1}{3(x-1)^{\frac33}}} = \boxed{\frac{1}{3(x-1)}}$$

k. $\boxed{\dfrac{dy}{dx}} = \boxed{\dfrac{d}{dx}\left[\ln\left(1+x^2\right)+2\arctan x\right]} = \boxed{\dfrac{d}{dx}\ln\left(1+x^2\right)+\dfrac{d}{dx}\left(2\arctan x\right)} = \boxed{\dfrac{1}{1+x^2}\cdot\dfrac{d}{dx}\left(1+x^2\right)+2\cdot\dfrac{1}{1+x^2}\cdot\dfrac{d}{dx}x}$

$= \boxed{\dfrac{1}{1+x^2}\cdot 2x+2\cdot\dfrac{1}{1+x^2}\cdot 1} = \boxed{\dfrac{2x}{1+x^2}+\dfrac{2}{1+x^2}} = \boxed{\dfrac{2x+2}{1+x^2}} = \boxed{\dfrac{2\left(1+x\right)}{1+x^2}}$

l. $\boxed{\dfrac{dy}{dx}} = \boxed{\dfrac{d}{dx}\left[\ln\left(\arctan 3x\right)+3x^2\right]} = \boxed{\dfrac{d}{dx}\ln\left(\arctan 3x\right)+\dfrac{d}{dx}\left(3x^2\right)} = \boxed{\dfrac{1}{\arctan 3x}\cdot\dfrac{d}{dx}\left(\arctan 3x\right)+6x}$

$= \boxed{\dfrac{1}{\arctan 3x}\cdot\dfrac{1}{1+(3x)^2}\cdot\dfrac{d}{dx}3x+6x} = \boxed{\dfrac{1}{\arctan 3x}\cdot\dfrac{1}{1+9x^2}\cdot 3+6x} = \boxed{\dfrac{3}{\left(1+9x^2\right)\arctan 3x}+6x}$

When differentiating exponential functions the following two rules should be kept in mind:

$$\text{Given } y=e^u, \text{ then } \frac{dy}{dx}=e^u\cdot\frac{du}{dx}$$

$$\text{Given } y=a^u, \text{ then } \frac{dy}{dx}=a^u\cdot\ln a\cdot\frac{du}{dx}$$

Let's differentiate some exponential functions using the above differentiation formulas.

Example 3.3-4: Find the derivative of the following exponential functions:

a. $y=e^{10x}$

b. $y=x\,e^{2x}$

c. $y=x^2 e^{5x}$

d. $y=(x+8)e^{3x^2}$

e. $y=e^{\sin 5x}$

f. $y=e^{5x}\sin 2x$

g. $y=e^{2x}\cos 3x$

h. $y=e^{\ln x}\tan x$

i. $y=\sqrt{e^x}+\arcsin x$

j. $y=e^{5x}\arccos x$

k. $y=e^{e^x}$

l. $y=e^{\cos 5x}e^{-3x}$

Solutions:

a. Given $\boxed{y=e^{10x}}$ then $\boxed{\dfrac{dy}{dx}} = \boxed{\dfrac{d}{dx}\left(e^{10x}\right)} = \boxed{e^{10x}\cdot\dfrac{d}{dx}(10x)} = \boxed{e^{10x}\cdot 10} = \boxed{10\,e^{10x}}$

b. Given $\boxed{y=x\,e^{2x}}$ then $\boxed{\dfrac{dy}{dx}} = \boxed{\dfrac{d}{dx}\left(x\,e^{2x}\right)} = \boxed{e^{2x}\cdot\dfrac{d}{dx}(x)+x\cdot\dfrac{d}{dx}\left(e^{2x}\right)} = \boxed{e^{2x}\cdot 1+x\cdot e^{2x}\cdot\dfrac{d}{dx}2x}$

$= \boxed{e^{2x}+x\cdot e^{2x}\cdot 2} = \boxed{e^{2x}+2x\,e^{2x}} = \boxed{e^{2x}\left(1+2x\right)}$

c. Given $\boxed{y=x^2 e^{5x}}$ then $\boxed{\dfrac{dy}{dx}} = \boxed{\dfrac{d}{dx}\left(x^2 e^{5x}\right)} = \boxed{e^{5x}\cdot\dfrac{d}{dx}\left(x^2\right)+x^2\cdot\dfrac{d}{dx}\left(e^{5x}\right)} = \boxed{e^{5x}\cdot 2x+x^2\cdot e^{5x}\cdot\dfrac{d}{dx}5x}$

$= \boxed{e^{5x}\cdot 2x+x^2\cdot e^{5x}\cdot 5} = \boxed{2x\,e^{5x}+5x^2 e^{5x}} = \boxed{x\,e^{5x}\left(2+5x\right)}$

d. Given $y=(x+8)e^{3x^2}$ then $\dfrac{dy}{dx}=\dfrac{d}{dx}\left[(x+8)e^{3x^2}\right]=\left[e^{3x^2}\cdot\dfrac{d}{dx}(x+8)+\left[(x+8)\cdot\dfrac{d}{dx}\left(e^{3x^2}\right)\right]\right]$

$=\left[e^{3x^2}\cdot1\right]+\left[(x+8)\cdot e^{3x^2}\cdot\dfrac{d}{dx}3x^2\right]=e^{3x^2}+(x+8)\cdot e^{3x^2}\cdot6x=e^{3x^2}\left[1+6x(x+8)\right]=e^{3x^2}\left(6x^2+48x+1\right)$

e. Given $y=e^{\sin 5x}$ then $\dfrac{dy}{dx}=\dfrac{d}{dx}\left(e^{\sin 5x}\right)=e^{\sin 5x}\cdot\dfrac{d}{dx}(\sin 5x)=e^{\sin 5x}\cdot5\cos 5x=5\cos 5x\,e^{\sin 5x}$

f. Given $y=e^{5x}\sin 2x$ then $\dfrac{dy}{dx}=\dfrac{d}{dx}\left(e^{5x}\sin 2x\right)=\left[\sin 2x\cdot\dfrac{d}{dx}\left(e^{5x}\right)\right]+\left[e^{5x}\cdot\dfrac{d}{dx}(\sin 2x)\right]$

$=\left[\sin 2x\cdot e^{5x}\cdot\dfrac{d}{dx}5x\right]+\left[e^{5x}\cdot\cos 2x\cdot\dfrac{d}{dx}2x\right]=\left[\sin 2x\cdot e^{5x}\cdot5\right]+\left[e^{5x}\cdot\cos 2x\cdot2\right]=e^{5x}\left(5\sin 2x+2\cos 2x\right)$

g. Given $y=e^{2x}\cos 3x$ then $\dfrac{dy}{dx}=\dfrac{d}{dx}\left(e^{2x}\cos 3x\right)=\left[\cos 3x\cdot\dfrac{d}{dx}\left(e^{2x}\right)\right]+\left[e^{2x}\cdot\dfrac{d}{dx}(\cos 3x)\right]$

$=\left[\cos 3x\cdot e^{2x}\cdot\dfrac{d}{dx}2x\right]+\left[e^{2x}\cdot-\sin 3x\cdot\dfrac{d}{dx}3x\right]=\left[\cos 3x\cdot e^{2x}\cdot2\right]+\left[e^{2x}\cdot-\sin 3x\cdot3\right]=e^{2x}\left(2\cos 3x-3\sin 3x\right)$

h. Given $y=e^{\ln x}\tan x$ then $\dfrac{dy}{dx}=\dfrac{d}{dx}\left(e^{\ln x}\tan x\right)=\left[\tan x\cdot\dfrac{d}{dx}\left(e^{\ln x}\right)\right]+\left[e^{\ln x}\cdot\dfrac{d}{dx}(\tan x)\right]$

$=\left[\tan x\cdot e^{\ln x}\cdot\dfrac{d}{dx}(\ln x)\right]+\left[e^{\ln x}\cdot\sec^2 x\right]=\left[\tan x\cdot e^{\ln x}\cdot\dfrac{1}{x}\right]+\left[e^{\ln x}\cdot\sec^2 x\right]=e^{\ln x}\left(\dfrac{\tan x}{x}+\sec^2 x\right)$

i. Given $y=\sqrt{e^x}+arc\sin x$ then $\dfrac{dy}{dx}=\dfrac{d}{dx}\left(\sqrt{e^x}+arc\sin x\right)=\dfrac{d}{dx}\sqrt{e^x}+\dfrac{d}{dx}arc\sin x=\dfrac{d}{dx}e^{\frac{x}{2}}+\dfrac{d}{dx}arc\sin x$

$=e^{\frac{x}{2}}\cdot\dfrac{d}{dx}\dfrac{x}{2}+\dfrac{1}{\sqrt{1-x^2}}\cdot\dfrac{d}{dx}x=e^{\frac{x}{2}}\cdot\dfrac{1}{2}+\dfrac{1}{\sqrt{1-x^2}}\cdot1=\dfrac{e^{\frac{x}{2}}}{2}+\dfrac{1}{\sqrt{1-x^2}}$

j. Given $y=e^{5x}arc\cos x$ then $\dfrac{dy}{dx}=\dfrac{d}{dx}\left(e^{5x}arc\cos x\right)=\left[arc\cos x\cdot\dfrac{d}{dx}\left(e^{5x}\right)\right]+\left[e^{5x}\cdot\dfrac{d}{dx}(arc\cos x)\right]$

$=\left(arc\cos x\cdot e^{5x}\cdot\dfrac{d}{dx}5x\right)+\left(e^{5x}\cdot-\dfrac{1}{\sqrt{1-x^2}}\cdot\dfrac{d}{dx}x\right)=\left(arc\cos x\cdot e^{5x}\cdot5\right)+\left(e^{5x}\cdot-\dfrac{1}{\sqrt{1-x^2}}\cdot1\right)$

$$= \boxed{arc \cos x \cdot 5e^{5x} - e^{5x} \cdot \frac{1}{\sqrt{1-x^2}}} = \boxed{5e^{5x} arc \cos x - \frac{e^{5x}}{\sqrt{1-x^2}}} = \boxed{e^{5x}\left(\mathbf{5}arc \, \mathbf{cos} \, x - \frac{1}{\sqrt{1-x^2}} \right)}$$

k. Given $\boxed{y = e^{e^x}}$ then $\boxed{\dfrac{dy}{dx}} = \boxed{\dfrac{d}{dx} e^{e^x}} = \boxed{e^{e^x} \cdot \dfrac{d}{dx} e^x} = \boxed{e^{e^x} \cdot e^x \cdot \dfrac{d}{dx} x} = \boxed{e^{e^x} \cdot e^x \cdot 1} = \boxed{e^{e^x} e^x} = \boxed{e^{e^x + x}}$

l. Given $\boxed{y = e^{\cos 5x} e^{-3x}}$ then $\boxed{\dfrac{dy}{dx}} = \boxed{\dfrac{d}{dx}\left(e^{\cos 5x} e^{-3x} \right)} = \boxed{\left[e^{-3x} \cdot \dfrac{d}{dx}\left(e^{\cos 5x} \right) \right] + \left[e^{\cos 5x} \cdot \dfrac{d}{dx}\left(e^{-3x} \right) \right]}$

$$= \boxed{\left[e^{-3x} \cdot e^{\cos 5x} \cdot \frac{d}{dx}(\cos 5x) \right] + \left[e^{\cos 5x} \cdot e^{-3x} \cdot \frac{d}{dx}(-3x) \right]} = \boxed{\left[e^{-3x} \cdot e^{\cos 5x} \cdot -\sin 5x \cdot \frac{d}{dx} 5x \right] + \left[e^{\cos 5x} \cdot e^{-3x} \cdot -3 \right]}$$

$$= \boxed{\left[e^{-3x} \cdot e^{\cos 5x} \cdot -\sin 5x \cdot 5 \right] + \left[e^{\cos 5x} \cdot e^{-3x} \cdot -3 \right]} = \boxed{-5 \sin 5x \, e^{-3x} e^{\cos 5x} - 3e^{\cos 5x} e^{-3x}}$$

$$= \boxed{-e^{\cos 5x} e^{-3x} (5 \sin 5x + 3)} = \boxed{-e^{\cos 5x - 3x}(\mathbf{5 \sin 5x + 3})}$$

Example 3.3-5: Differentiate the following exponential functions

a. $y = e^{\ln x^2} - e^{-x}$

b. $y = \dfrac{e^{3x}}{\cot 5x}$

c. $y = x^3 e^{-5x}$

d. $y = e^{3x} \cdot arc \cos x^2$

e. $y = \ln e^{x^2} + arc \tan x + x^3$

f. $y = e^{x^2} - arc \sin x$

g. $y = e^{\ln x^2} + x^3$

h. $y = arc \sin e^x$

i. $y = x \, arc \sin e^{3x}$

Solutions:

a. Given $\boxed{y = e^{\ln x^2} + e^{-x}}$ then $\boxed{\dfrac{dy}{dx}} = \boxed{\dfrac{d}{dx}\left(e^{\ln x^2} - e^{-x} \right)} = \boxed{\dfrac{d}{dx} e^{\ln x^2} - \dfrac{d}{dx} e^{-x}} = \boxed{e^{\ln x^2} \cdot \dfrac{d}{dx} \ln x^2 - e^{-x} \cdot \dfrac{d}{dx}(-x)}$

$$= \boxed{e^{\ln x^2} \cdot \frac{1}{x^2} \cdot \frac{d}{dx} x^2 - e^{-x} \cdot -1} = \boxed{e^{\ln x^2} \cdot \frac{1}{x^{2=1}} \cdot 2x + e^{-x}} = \boxed{\frac{2e^{\ln x^2}}{x} + e^{-x}}$$

b. Given $\boxed{y = \dfrac{e^{3x}}{\cot 5x}}$ then $\boxed{\dfrac{dy}{dx}} = \boxed{\dfrac{d}{dx}\left(\dfrac{e^{3x}}{\cot 5x} \right)} = \boxed{\dfrac{\left[\cot 5x \cdot \frac{d}{dx}\left(e^{3x} \right) \right] - \left[e^{3x} \cdot \frac{d}{dx}(\cot 5x) \right]}{\cot^2 5x}}$

$$= \boxed{\frac{\left[\cot 5x \cdot e^{3x} \cdot \frac{d}{dx} 3x \right] - \left[e^{3x} \cdot -\csc^2 5x \cdot \frac{d}{dx} 5x \right]}{\cot^2 5x}} = \boxed{\frac{\cot 5x \cdot e^{3x} \cdot 3 + e^{3x} \cdot \csc^2 5x \cdot 5}{\cot^2 5x}} = \boxed{\frac{e^{3x}\left(\mathbf{3 \cot 5x + 5 \csc^2 5x} \right)}{\mathbf{\cot^2 5x}}}$$

c. Given $\boxed{y = x^3 e^{-5x}}$ then $\boxed{\dfrac{dy}{dx}} = \boxed{\dfrac{d}{dx}\left(x^3 e^{-5x} \right)} = \boxed{\left(x^3 \cdot \dfrac{d}{dx} e^{-5x} \right) + \left(e^{-5x} \cdot \dfrac{d}{dx} x^3 \right)} = \boxed{\left[x^3 \cdot e^{-5x} \cdot \dfrac{d}{dx}(-5x) \right]}$

$$+\left[e^{-5x}\cdot 3x^2\right]\Big]=\left[\left(x^3\cdot e^{-5x}\cdot -5\right)+\left(e^{-5x}\cdot 3x^2\right)\right]=\boxed{-5x^3e^{-5x}+3x^2e^{-5x}}=\boxed{x^2e^{-5x}(-5x+3)}$$

d. Given $\boxed{y=e^{3x}\,arc\cos x^2}$ then $\boxed{\dfrac{dy}{dx}}=\boxed{\dfrac{d}{dx}\left(e^{3x}\,arc\cos x^2\right)}=\boxed{arc\cos x^2\cdot\dfrac{d}{dx}e^{3x}+e^{3x}\cdot\dfrac{d}{dx}arc\cos x^2}$

$=\boxed{arc\cos x^2\cdot e^{3x}\cdot\dfrac{d}{dx}3x+e^{3x}\cdot -\dfrac{1}{\sqrt{1-x^4}}\cdot\dfrac{d}{dx}x^2}=\boxed{arc\cos x^2\cdot e^{3x}\cdot 3+e^{3x}\cdot -\dfrac{1}{\sqrt{1-x^4}}\cdot 2x}$

$=\boxed{3e^{3x}arc\cos x^2-\dfrac{2xe^{3x}}{\sqrt{1-x^4}}}=\boxed{e^{3x}\left(3\,arc\cos x^2-\dfrac{2x}{\sqrt{1-x^4}}\right)}$

e. Given $\boxed{y=\ln e^{x^2}+arc\tan x+x^3}$ then $\boxed{\dfrac{dy}{dx}}=\boxed{\dfrac{d}{dx}\left(\ln e^{x^2}+arc\tan x+x^3\right)}=\boxed{\dfrac{d}{dx}\ln e^{x^2}+\dfrac{d}{dx}arc\tan x+\dfrac{d}{dx}x^3}$

$=\boxed{\dfrac{1}{e^{x^2}}\cdot\dfrac{d}{dx}e^{x^2}+\dfrac{1}{1+x^2}+3x^2}=\boxed{\dfrac{1}{e^{x^2}}\cdot e^{x^2}\cdot\dfrac{d}{dx}x^2+\dfrac{1}{1+x^2}+3x^2}=\boxed{\dfrac{e^{x^2}}{e^{x^2}}\cdot 2x+\dfrac{1}{1+x^2}+3x^2}=\boxed{3x^2+2x+\dfrac{1}{1+x^2}}$

f. Given $\boxed{y=e^{x^2}-arc\sin x}$ then $\boxed{\dfrac{dy}{dx}}=\boxed{\dfrac{d}{dx}\left(e^{x^2}-arc\sin x\right)}=\boxed{\dfrac{d}{dx}e^{x^2}-\dfrac{d}{dx}arc\sin x}=\boxed{e^{x^2}\cdot\dfrac{d}{dx}x^2-\dfrac{1}{\sqrt{1-x^2}}}$

$=\boxed{e^{x^2}\cdot 2x-\dfrac{1}{\sqrt{1-x^2}}}=\boxed{2xe^{x^2}-\dfrac{1}{\sqrt{1-x^2}}}$

g. Given $\boxed{y=e^{\ln x^2}+x^3}$ then $\boxed{\dfrac{dy}{dx}}=\boxed{\dfrac{d}{dx}\left(e^{\ln x^2}+x^3\right)}=\boxed{\dfrac{d}{dx}e^{\ln x^2}+\dfrac{d}{dx}x^3}=\boxed{e^{\ln x^2}\cdot\dfrac{d}{dx}\ln x^2+3x^2}$

$=\boxed{e^{\ln x^2}\cdot\dfrac{1}{x^2}\cdot\dfrac{d}{dx}x^2+3x^2}=\boxed{e^{\ln x^2}\cdot\dfrac{1}{x^{2-1}}\cdot 2x+3x^2}=\boxed{e^{\ln x^2}\cdot\dfrac{2}{x}+3x^2}=\boxed{\dfrac{2e^{\ln x^2}}{x}+3x^2}$

h. Given $\boxed{y=arc\sin e^x}$ then $\boxed{\dfrac{dy}{dx}}=\boxed{\dfrac{d}{dx}\left(arc\sin e^x\right)}=\boxed{\dfrac{1}{\sqrt{1-e^{2x}}}\cdot\dfrac{d}{dx}e^x}=\boxed{\dfrac{1}{\sqrt{1-e^{2x}}}\cdot e^x\cdot\dfrac{d}{dx}x}=\boxed{\dfrac{e^x}{\sqrt{1-e^{2x}}}}$

i. Given $\boxed{y=x\,arc\sin e^{3x}}$ then $\boxed{\dfrac{dy}{dx}}=\boxed{\dfrac{d}{dx}\left(x\,arc\sin e^{3x}\right)}=\boxed{arc\sin e^{3x}\cdot\dfrac{d}{dx}x+x\cdot\dfrac{d}{dx}arc\sin e^{3x}}$

$=\boxed{arc\sin e^{3x}\cdot 1+x\cdot\dfrac{1}{\sqrt{1-e^{6x}}}\cdot\dfrac{d}{dx}e^{3x}}=\boxed{arc\sin e^{3x}+x\cdot\dfrac{1}{\sqrt{1-e^{6x}}}\cdot e^{3x}\cdot\dfrac{d}{dx}3x}=\boxed{arc\sin e^{3x}+\dfrac{3xe^{3x}}{\sqrt{1-e^{6x}}}}$

- In order to find the derivative of the functions of the form $y = x^{g(x)}$ first multiply both sides of the equation by natural logarithm ($\log_e = \ln$) and then apply the logarithmic rules prior to taking the derivative. The following examples show how to differentiate this class of functions:

Example 3.3-6: Find the derivative of the following functions:

a. $y = x^x$, $x \rangle 0$
b. $y = x^{\sin x}$, $x \rangle 0$
c. $y = (\sin x)^x$, $x \rangle 0$

d. $y = x^{\frac{1}{x}}$, $x \rangle 0$
e. $y = x^{\ln x}$, $x \rangle 0$
f. $y = x^{e^x}$

Solutions:

a. The function $y = x^x$ is equivalent to $\boxed{\ln y = \ln x^x}$; $\boxed{\ln y = x \ln x}$ thus $\boxed{\dfrac{d}{dx} \ln y = \dfrac{d}{dx} x \ln x}$

; $\boxed{\dfrac{1}{y} \cdot \dfrac{dy}{dx} = \ln x \cdot \dfrac{d}{dx} x + x \cdot \dfrac{d}{dx} \ln x}$; $\boxed{\dfrac{1}{y} \cdot y' = \ln x \cdot 1 + x \cdot \dfrac{1}{x}}$; $\boxed{\dfrac{1}{y} \cdot y' = \ln x + 1}$; $\boxed{y' = y(\ln x + 1)}$; $\boxed{\mathbf{y' = x^x(\ln x + 1)}}$

b. The function $y = x^{\sin x}$ is equivalent to $\boxed{\ln y = \ln x^{\sin x}}$; $\boxed{\ln y = \sin x \ln x}$ thus:

$\boxed{\dfrac{d}{dx} \ln y = \dfrac{d}{dx} \sin x \ln x}$; $\boxed{\dfrac{1}{y} \cdot \dfrac{dy}{dx} = \ln x \cdot \dfrac{d}{dx} \sin x + \sin x \cdot \dfrac{d}{dx} \ln x}$; $\boxed{\dfrac{1}{y} \cdot y' = \ln x \cdot \cos x + \sin x \cdot \dfrac{1}{x}}$

; $\boxed{\dfrac{1}{y} \cdot y' = \ln x \cos x + \dfrac{\sin x}{x}}$; $\boxed{y' = y\left(\ln x \cos x + \dfrac{\sin x}{x}\right)}$; $\boxed{\mathbf{y' = x^{\sin x}\left(\ln x \cos x + \dfrac{\sin x}{x}\right)}}$

c. The function $y = (\sin x)^x$ is equivalent to $\boxed{\ln y = \ln(\sin x)^x}$; $\boxed{\ln y = x \ln(\sin x)}$ thus:

$\boxed{\dfrac{d}{dx} \ln y = \dfrac{d}{dx} x \ln(\sin x)}$; $\boxed{\dfrac{1}{y} \cdot \dfrac{dy}{dx} = \ln(\sin x) \cdot \dfrac{d}{dx} x + x \cdot \dfrac{d}{dx} \ln(\sin x)}$; $\boxed{\dfrac{1}{y} \cdot y' = \ln(\sin x) \cdot 1 + x \cdot \dfrac{1}{\sin x} \cdot \cos x}$

; $\boxed{\dfrac{1}{y} \cdot y' = \ln(\sin x) + \dfrac{x \cos x}{\sin x}}$; $\boxed{y' = y[\ln(\sin x) + x \cot x]}$; $\boxed{\mathbf{y' = (\sin x)^x[\ln(\sin x) + x \cot x]}}$

d. The function $y = x^{\frac{1}{x}}$ is equivalent to $\boxed{\ln y = \ln x^{\frac{1}{x}}}$; $\boxed{\ln y = \dfrac{1}{x} \ln x}$ thus:

$\boxed{\dfrac{d}{dx} \ln y = \dfrac{d}{dx}\left(\dfrac{1}{x} \cdot \ln x\right)}$; $\boxed{\dfrac{1}{y} \cdot \dfrac{dy}{dx} = \ln x \cdot \dfrac{d}{dx}\dfrac{1}{x} + \dfrac{1}{x} \cdot \dfrac{d}{dx} \ln x}$; $\boxed{\dfrac{1}{y} \cdot y' = \ln x \cdot -\dfrac{1}{x^2} + \dfrac{1}{x} \cdot \dfrac{1}{x}}$; $\boxed{\dfrac{1}{y} \cdot y' = -\dfrac{\ln x}{x^2} + \dfrac{1}{x^2}}$

; $\boxed{y' = y\left(\dfrac{1 - \ln x}{x^2}\right)}$; $\boxed{y' = x^{\frac{1}{x}}\left(\dfrac{1 - \ln x}{x^2}\right)} = \boxed{\mathbf{y' = \dfrac{x^{\frac{1}{x}}}{x^2}(1 - \ln x)}}$

e. The function $y = x^{\ln x}$ is equivalent to $\boxed{\ln y = \ln x^{\ln x}}$; $\boxed{\ln y = \ln x \cdot \ln x} = \boxed{\ln y = \ln^2 x}$ thus:

$$\boxed{\frac{d}{dx}\ln y = \frac{d}{dx}\ln^2 x} \; ; \; \boxed{\frac{1}{y}\cdot\frac{dy}{dx} = 2\ln x \cdot \frac{d}{dx}\ln x} \; ; \; \boxed{\frac{1}{y}\cdot y' = 2\ln x \cdot \frac{1}{x}\cdot\frac{d}{dx}x} = \boxed{\frac{1}{y}\cdot y' = 2\ln x \cdot \frac{1}{x}\cdot 1} = \boxed{\frac{1}{y}\cdot y' = 2\ln x \cdot \frac{1}{x}}$$

$$; \; \boxed{\frac{1}{y}\cdot y' = \frac{2\ln x}{x}} \; ; \; \boxed{y' = y\left(\frac{2\ln x}{x}\right)} \; ; \; \boxed{y' = 2x^{\ln x}\left(\frac{\ln x}{x}\right)}$$

f. The function $y = x^{e^x}$ is equivalent to $\boxed{\ln y = \ln x^{e^x}}$; $\boxed{\ln y = e^x \ln x}$ thus:

$$\boxed{\frac{d}{dx}\ln y = \frac{d}{dx}e^x \ln x} \; ; \; \boxed{\frac{1}{y}\cdot\frac{dy}{dx} = \ln x \cdot \frac{d}{dx}e^x + e^x \cdot \frac{d}{dx}\ln x} \; ; \; \boxed{\frac{1}{y}\cdot y' = \ln x \cdot e^x \cdot \frac{d}{dx}x + e^x \cdot \frac{1}{x}\cdot\frac{d}{dx}x}$$

$$; \; \boxed{\frac{1}{y}\cdot y' = \ln x \cdot e^x \cdot 1 + e^x \cdot \frac{1}{x}\cdot 1} \; ; \; \boxed{\frac{1}{y}\cdot y' = e^x \ln x + \frac{e^x}{x}} \; ; \; \boxed{y' = y\,e^x\left(\ln x + \frac{1}{x}\right)} \; ; \; \boxed{y' = x^{e^x}\,e^x\left(\ln x + \frac{1}{x}\right)}$$

Example 3.3-7: Find the derivative of the following functions:

a. $y = x^{3e}$

b. $y = (\sin x)^\pi$

c. $y = x^{\sqrt{5}}$

d. $y = x^{2\pi}$

e. $y = x^{\ln a}$

f. $y = x^{\sin\theta} + x^3$

Solutions:

a. Given $\boxed{y = x^{3e}}$ then $\boxed{\frac{dy}{dx}} = \boxed{3e\,x^{3e-1}}$

Note that we could have solved this problem, and other problems in this example, by first multiplying both sides of the equation by natural logarithm ($\log_e = \ln$) and taking the derivative as shown below. However, this is a more difficult way of solving this class of functions and is not recommended.

$$\boxed{y = x^{3e}} \; ; \; \boxed{\ln y = \ln x^{3e}} \text{ then } \boxed{\frac{d}{dx}\ln y = \frac{d}{dx}\ln x^{3e}} \; ; \; \boxed{\frac{1}{y}\cdot\frac{dy}{dx} = \frac{1}{x^{3e}}\cdot\frac{d}{dx}x^{3e}} \; ; \; \boxed{\frac{1}{y}\cdot y' = \frac{1}{x^{3e}}\cdot 3e\,x^{3e-1}}$$

$$; \; \boxed{y' = y\left(3e\,x^{3e-1}\cdot x^{-3e}\right)} \; ; \; \boxed{y' = y\left(3e\,x^{3e-1-3e}\right)} \; ; \; \boxed{y' = y\left(3e\,x^{-1}\right)} \; ; \; \boxed{y' = x^{3e}\left(3e\,x^{-1}\right)} \; ; \; \boxed{y' = 3e\,x^{3e-1}}$$

b. Given $\boxed{y = (\sin x)^\pi}$ then $\boxed{\frac{dy}{dx}} = \boxed{\pi\,(\sin x)^{\pi-1}\cdot\frac{d}{dx}\sin x} = \boxed{\pi\,(\sin x)^{\pi-1}\cos x}$

c. Given $\boxed{y = x^{\sqrt{5}} = x^{\frac{5}{2}}}$ then $\boxed{\frac{dy}{dx}} = \boxed{\frac{5}{2}x^{\frac{5}{2}-1}} = \boxed{\frac{5}{2}x^{\frac{3}{2}}} = \boxed{\frac{5}{2}\sqrt{x^3}} = \boxed{\frac{5}{2}x\sqrt{x}}$

d. Given $y = x^{2\pi}$ then $\dfrac{dy}{dx} = \dfrac{d}{dx} x^{2\pi} = 2\pi\, x^{2\pi - 1}$

e. Given $y = x^{\ln a}$ then $\dfrac{dy}{dx} = \dfrac{d}{dx} x^{\ln a} = \ln a\; x^{\ln a - 1}$

f. Given $y = x^{\sin \theta} + x^3$ then $\dfrac{dy}{dx} = \dfrac{d}{dx}\left(x^{\sin \theta} + x^3\right) = \dfrac{d}{dx} x^{\sin \theta} + \dfrac{d}{dx} x^3 = \sin \theta\; x^{\sin \theta - 1} + 3x^2$

Note: The problems in this example can be written in the standard form of $y = x^a$ where $\dfrac{dy}{dx}$ is equal to $y' = a\,x^{a-1}$. Therefore, students should not get confused if the constant a is replaced with numbers such as $\sqrt{3}$, $\dfrac{1}{3}$, e, $e^{\frac{1}{2}}$, α, π, η, $\sin \theta$, etc. The process of finding the derivative remains the same.

• The derivative of the functions of the form $a^u = e^{u \ln a}$ is equal to the following:

$$\dfrac{d}{dx} a^u = \dfrac{d}{dx} e^{u \ln a} = e^{u \ln a} \cdot \dfrac{d}{dx} u \ln a = e^{u \ln a} \cdot \ln a \cdot \dfrac{d}{dx} u = e^{\ln a^u} \cdot \ln a \cdot \dfrac{du}{dx} = a^u \ln a \dfrac{du}{dx} \quad \text{Thus,}$$

$$\dfrac{d}{dx} a^u = a^u \ln a \dfrac{du}{dx}$$

The following examples show how to differentiate this class of functions:

Example 3.3-8: Find the derivative of the following functions:

a. $y = 2^x$

b. $y = 5^{\sin x}$

c. $y = \pi^{\,x}$

d. $y = 10^{\frac{1}{x}}$

e. $y = 10^{\ln x}$

f. $y = 5^{e^{x^3}}$

Solutions:

a. Given $y = 2^x$ then $\dfrac{dy}{dx} = \dfrac{d}{dx} 2^x = 2^x \cdot \ln 2 \cdot \dfrac{d}{dx} x = 2^x \cdot \ln 2 \cdot 1 = (\ln 2) 2^x$

b. Given $y = 5^{\sin x}$ then $\dfrac{dy}{dx} = \dfrac{d}{dx} 5^{\sin x} = 5^{\sin x} \cdot \ln 5 \cdot \dfrac{d}{dx} \sin x = 5^{\sin x} \cdot \ln 5 \cdot \cos x = (\ln 5) 5^{\sin x} \cos x$

c. Given $y = \pi^{\,x}$ then $\dfrac{dy}{dx} = \dfrac{d}{dx} \pi^{\,x} = \pi^{\,x} \cdot \ln \pi \cdot \dfrac{d}{dx} x = \pi^{\,x} \cdot \ln \pi \cdot 1 = (\ln \pi) \pi^{\,x}$

d. Given $y = 10^{\frac{1}{x}}$ then $\dfrac{dy}{dx} = \dfrac{d}{dx} 10^{\frac{1}{x}} = 10^{\frac{1}{x}} \cdot \ln 10 \cdot \dfrac{d}{dx} \dfrac{1}{x} = 10^{\frac{1}{x}} \cdot \ln 10 \cdot -\dfrac{1}{x^2} = -\dfrac{(\ln 10) 10^{\frac{1}{x}}}{x^2}$

e. Given $y = 10^{\ln x}$ then $\dfrac{dy}{dx} = \dfrac{d}{dx} 10^{\ln x} = 10^{\ln x} \cdot \ln 10 \cdot \dfrac{d}{dx} \ln x = 10^{\ln x} \cdot \ln 10 \cdot \dfrac{1}{x} = \dfrac{(\ln 10) 10^{\ln x}}{x}$

f. Given $y = 5^{e^{x^3}}$ then $\dfrac{dy}{dx} = \dfrac{d}{dx} 5^{e^{x^3}} = 5^{e^{x^3}} \cdot \ln 5 \cdot \dfrac{d}{dx} e^{x^3} = 5^{e^{x^3}} \cdot \ln 5 \cdot e^{x^3} \cdot \dfrac{d}{dx} x^3 = 5^{e^{x^3}} \cdot \ln 5 \cdot e^{x^3} \cdot 3x^2$

$= (\ln 5) 3x^2 \, 5^{e^{x^3}} e^{x^3}$

Section 3.3 Practice Problems – Differentiation of Logarithmic and Exponential Functions

1. Find the derivative of the following logarithmic functions:

   a. $y = \ln 10x$

   b. $y = 10 \ln 5x^3$

   c. $y = \ln\left(x^2 + 3\right)$

   d. $y = x^3 \ln x$

   e. $y = x \ln x - 5x^2$

   f. $y = \left(x^3 + 2\right) \ln x^2$

   g. $y = \sin\left(\ln x^2\right)$

   h. $y = \ln \csc x$

   i. $y = \ln \sqrt{x^3}$

   j. $y = \cos\left(\ln x^2\right)$

   k. $y = \ln \dfrac{x+1}{x-1}$

   l. $y = x^3 \ln x + 5x$

2. Find the derivative of the following exponential and trigonometric functions:

   a. $y = x^3 e^{2x}$

   b. $y = \left(x^2 + 3\right) e^{3x}$

   c. $y = e^{\sin 3x}$

   d. $y = e^{\ln x^2}$

   e. $y = e^{9x} \sin 5x$

   f. $y = e^{2x} arc \sin x$

   g. $y = (x-5) e^{-x}$

   h. $y = e^{\ln (x+1)}$

   i. $y = \dfrac{e^x}{\tan x}$

   j. $y = 3^x \cdot e^x$

   k. $y = x^3 arc \sin x$

   l. $y = e^{5x} \cos (5x+1)$

3.4   Differentiation of Hyperbolic Functions

Hyperbolic functions are defined as:

Table 3.4-1: Hyperbolic Functions

$\sinh x = \dfrac{e^x - e^{-x}}{2}$	$\coth x = \dfrac{\cosh x}{\sinh x} = \dfrac{e^x + e^{-x}}{e^x - e^{-x}}$ where $x \neq 0$
$\cosh x = \dfrac{e^x + e^{-x}}{2}$	$\sec h\, x = \dfrac{1}{\cosh x} = \dfrac{2}{e^x + e^{-x}}$
$\tanh x = \dfrac{\sinh x}{\cosh x} = \dfrac{e^x - e^{-x}}{e^x + e^{-x}}$	$\csc h\, x = \dfrac{1}{\sinh x} = \dfrac{2}{e^x - e^{-x}}$ where $x \neq 0$

The differential formulas involving hyperbolic functions are defined as:

Table 3.4-2: Differentiation Formulas for Hyperbolic Functions

$\dfrac{d}{dx}\sinh u = \cosh u \cdot \dfrac{du}{dx}$	$\dfrac{d}{dx}\coth u = -\csc h^2 u \cdot \dfrac{du}{dx}$
$\dfrac{d}{dx}\cosh u = \sinh u \cdot \dfrac{du}{dx}$	$\dfrac{d}{dx}\sec h\, u = -\sec h\, u \tanh u \cdot \dfrac{du}{dx}$
$\dfrac{d}{dx}\tanh u = \sec h^2 u \cdot \dfrac{du}{dx}$	$\dfrac{d}{dx}\csc h\, u = -\csc h\, u \coth u \cdot \dfrac{du}{dx}$

Let's differentiate some hyperbolic functions using the above formulas:

Example 3.4-1: Find the derivative of the following hyperbolic functions:

a. $y = 5\sinh 6x$ b. $y = \sinh x^3 + \cosh 3x$ c. $y = \cosh^2 5x^3$

d. $y = x^3 \sinh \sqrt{x}$ e. $y = \sinh x^2 + \cosh x^2$ f. $y = \tanh\left(3x^2 - 1\right)$

g. $y = \dfrac{1}{6}\sinh 3x + 10x$ h. $y = \ln \cosh 3x$ i. $y = \ln \tanh 3x^3$

Solutions:

a. Given $\boxed{y = 5\sinh 6x}$ then $\boxed{\dfrac{dy}{dx}} = \boxed{\dfrac{d}{dx}(5\sinh 6x)} = \boxed{5\cosh 6x \cdot \dfrac{d}{dx}(6x)} = \boxed{5\cosh 6x \cdot 6} = \boxed{\mathbf{30\cosh 6x}}$

b. Given $\boxed{y = \sinh x^3 + \cosh 3x}$ then $\boxed{\dfrac{dy}{dx}} = \boxed{\dfrac{d}{dx}\left(\sinh x^3 + \cosh 3x\right)} = \boxed{\dfrac{d}{dx}\left(\sinh x^3\right) + \dfrac{d}{dx}(\cosh 3x)}$

$= \boxed{\cosh x^3 \cdot \dfrac{d}{dx}x^3 + \sinh 3x \cdot \dfrac{d}{dx}3x} = \boxed{\cosh x^3 \cdot 3x^2 + \sinh 3x \cdot 3} = \boxed{\mathbf{3x^2\cosh x^3 + 3\sinh 3x}}$

c. Given $\boxed{y = \cosh^2 5x^3}$ then $\boxed{\dfrac{dy}{dx}} = \boxed{\dfrac{d}{dx}\left(\cosh^2 5x^3\right)} = \boxed{\dfrac{d}{dx}\left(\cosh 5x^3\right)^2} = \boxed{2\cosh 5x^3 \cdot \dfrac{d}{dx}\cosh 5x^3}$

$= \boxed{2\cosh 5x^3 \cdot \sinh 5x^3 \cdot \dfrac{d}{dx}5x^3} = \boxed{2\cosh 5x^3 \cdot \sinh 5x^3 \cdot 15x^2} = \boxed{\mathbf{30x^2\sinh 5x^3\cosh 5x^3}}$

d. Given $\boxed{y = x^3 \sinh \sqrt{x}}$ then $\boxed{\dfrac{dy}{dx}} = \boxed{\dfrac{d}{dx}\left(x^3 \sinh \sqrt{x} \right)} = \boxed{\sinh \sqrt{x} \cdot \dfrac{d}{dx} x^3 + x^3 \cdot \dfrac{d}{dx} \sinh \sqrt{x}}$

$= \boxed{\sinh \sqrt{x} \cdot 3x^2 + x^3 \cdot \cosh \sqrt{x} \cdot \dfrac{d}{dx}\sqrt{x}} = \boxed{3x^2 \sinh \sqrt{x} + x^3 \cosh \sqrt{x} \cdot \dfrac{1}{2} x^{-\frac{1}{2}}} = \boxed{\mathbf{3x^2 \sinh \sqrt{x} + \dfrac{x^3 \cosh \sqrt{x}}{2\sqrt{x}}}}$

e. Given $\boxed{y = \sinh x^2 + \cosh x^2}$ then $\boxed{\dfrac{dy}{dx}} = \boxed{\dfrac{d}{dx}\left(\sinh x^2 + \cosh x^2 \right)} = \boxed{\dfrac{d}{dx}\sinh x^2 + \dfrac{d}{dx}\cosh x^2}$

$= \boxed{\cosh x^2 \cdot \dfrac{d}{dx} x^2 + \sinh x^2 \cdot \dfrac{d}{dx} x^2} = \boxed{\cosh x^2 \cdot 2x + \sinh x^2 \cdot 2x} = \boxed{\mathbf{2x \cosh x^2 + 2x \sinh x^2}}$

f. Given $\boxed{y = \tanh\left(3x^2 - 1\right)}$ then $\boxed{\dfrac{dy}{dx}} = \boxed{\dfrac{d}{dx}\left[\tanh\left(3x^2 - 1\right)\right]} = \boxed{\sec h^2\left(3x^2 - 1\right) \cdot \dfrac{d}{dx}\left(3x^2 - 1\right)} = \boxed{\mathbf{6x \sec h^2\left(3x^2 - 1\right)}}$

g. Given $\boxed{y = \dfrac{1}{6}\sinh 3x + 10x}$ then $\boxed{\dfrac{dy}{dx}} = \boxed{\dfrac{d}{dx}\left(\dfrac{1}{6}\sinh 3x + 10x \right)} = \boxed{\dfrac{d}{dx}\left(\dfrac{1}{6}\sinh 3x \right) + \dfrac{d}{dx} 10x}$

$= \boxed{\dfrac{1}{6}\cosh 3x \cdot \dfrac{d}{dx} 3x + 10} = \boxed{\dfrac{3}{6}\cosh 3x + 10} = \boxed{\mathbf{\dfrac{1}{2}\cosh 3x + 10}}$

h. Given $\boxed{y = \ln \cosh 3x}$ then $\boxed{\dfrac{dy}{dx}} = \boxed{\dfrac{d}{dx}\left(\ln \cosh 3x \right)} = \boxed{\dfrac{1}{\cosh 3x} \cdot \dfrac{d}{dx}\left(\cosh 3x \right)} = \boxed{\dfrac{1}{\cosh 3x} \cdot \sinh 3x \cdot \dfrac{d}{dx} 3x}$

$= \boxed{\dfrac{1}{\cosh 3x} \cdot \sinh 3x \cdot 3} = \boxed{3 \cdot \dfrac{\sinh 3x}{\cosh 3x}} = \boxed{\mathbf{3 \tanh 3x}}$

i. Given $\boxed{y = \ln \tanh 3x^3}$ then $\boxed{\dfrac{dy}{dx}} = \boxed{\dfrac{d}{dx}\left(\ln \tanh 3x^3 \right)} = \boxed{\dfrac{1}{\tanh 3x^3} \cdot \dfrac{d}{dx}\left(\tanh 3x^3 \right)} = \boxed{\dfrac{\sec h^2\, 3x^3}{\tanh 3x^3} \cdot \dfrac{d}{dx} 3x^3}$

$= \boxed{\dfrac{\sec h^2\, 3x^3}{\tanh 3x^3} \cdot 9x^2} = \boxed{\mathbf{\dfrac{9x^2 \sec h^2\, 3x^3}{\tanh 3x^3}}}$

Example 3.4-2: Find the derivative of the following hyperbolic functions:

a. $y = \coth \dfrac{1}{x^2}$

b. $y = \dfrac{\sinh x^2}{\cosh(x+1)}$

c. $y = \dfrac{\sinh 10x^2}{\cosh 3x^5}$

d. $y = \dfrac{\sinh e^{3x}}{\cosh x^3}$

e. $y = x^2 \sinh x^2$

f. $y = (x+1)\cosh x$

g. $y = \sinh \sqrt{x^3} + \cosh \sqrt{x}$

h. $y = x^3 \sinh^2 5x$

i. $y = \csc h\, x + \sqrt[3]{x^2}$

Solutions:

a. Given $\boxed{y = \coth \dfrac{1}{x^2}}$ then $\boxed{\dfrac{dy}{dx}} = \boxed{\dfrac{d}{dx}\left(\coth \dfrac{1}{x^2} \right)} = \boxed{-\csc h^2 \dfrac{1}{x^2} \cdot \dfrac{d}{dx}\dfrac{1}{x^2}} = \boxed{-\csc h^2 \dfrac{1}{x^2} \cdot \dfrac{-2}{x^3}} = \boxed{\mathbf{\dfrac{2}{x^3}\csc h^2 \dfrac{1}{x^2}}}$

b. Given $y = \dfrac{\sinh x^2}{\cosh (x+1)}$ then $\dfrac{dy}{dx} = \dfrac{d}{dx}\left(\dfrac{\sinh x^2}{\cosh (x+1)} \right) = \dfrac{\left[\cosh (x+1) \cdot \frac{d}{dx} \sinh x^2 \right] - \left[\sinh x^2 \cdot \frac{d}{dx} \cosh (x+1) \right]}{\cosh^2 (x+1)}$

$= \dfrac{\left[\cosh (x+1)\cosh x^2 \cdot \frac{d}{dx} x^2 \right] - \left[\sinh x^2 \sinh (x+1) \cdot \frac{d}{dx} (x+1) \right]}{\cosh^2 (x+1)} = \dfrac{\left[\cosh (x+1)\cosh x^2 \cdot 2x \right] - \left[\sinh x^2 \sinh (x+1) \cdot 1 \right]}{\cosh^2 (x+1)}$

$= \dfrac{2x \cosh x^2 \cosh (x+1) - \sinh (x+1) \sinh x^2}{\cosh^2 (x+1)}$

c. Given $y = \dfrac{\sinh 10x^2}{\cosh 3x^5}$ then $\dfrac{dy}{dx} = \dfrac{d}{dx}\left(\dfrac{\sinh 10x^2}{\cosh 3x^5} \right) = \dfrac{\left[\cosh 3x^5 \cdot \frac{d}{dx} \sinh 10x^2 \right] - \left[\sinh 10x^2 \cdot \frac{d}{dx} \cosh 3x^5 \right]}{\cosh^2 3x^5}$

$= \dfrac{\left[\cosh 3x^5 \cdot \cosh 10x^2 \frac{d}{dx} 10x^2 \right] - \left[\sinh 10x^2 \cdot \sinh 3x^5 \frac{d}{dx} 3x^5 \right]}{\cosh^2 3x^5} = \dfrac{\left[\cosh 3x^5 \cdot \cosh 10x^2 \cdot 20x \right]}{\cosh^2 3x^5}$

$= \dfrac{-\left[\sinh 10x^2 \cdot \sinh 3x^5 \cdot 15x^4 \right]}{\cosh^2 3x^5} = \dfrac{20x \cosh 10x^2 \cosh 3x^5 - 15x^4 \sinh 3x^5 \sinh 10x^2}{\cosh^2 3x^5}$

d. Given $y = \dfrac{\sinh e^{3x}}{\cosh x^3}$ then $\dfrac{dy}{dx} = \dfrac{d}{dx}\left(\dfrac{\sinh e^{3x}}{\cosh x^3} \right) = \dfrac{\left[\cosh x^3 \cdot \frac{d}{dx} \sinh e^{3x} \right] - \left[\sinh e^{3x} \cdot \frac{d}{dx} \cosh x^3 \right]}{\cosh^2 x^3}$

$= \dfrac{\cosh x^3 \cdot \cosh e^{3x} \cdot \frac{d}{dx} e^{3x} - \sinh e^{3x} \cdot \sinh x^3 \cdot \frac{d}{dx} x^3}{\cosh^2 x^3} = \dfrac{\cosh x^3 \cdot \cosh e^{3x} \cdot 3 e^{3x} - \sinh e^{3x} \cdot \sinh x^3 \cdot 3 x^2}{\cosh^2 x^3}$

$= \dfrac{3 e^{3x} \cosh e^{3x} \cosh x^3 - 3 x^2 \sinh x^3 \sinh e^{3x}}{\cosh^2 x^3}$

e. Given $y = x^2 \sinh x^2$ then $\dfrac{dy}{dx} = \dfrac{d}{dx}\left(x^2 \sinh x^2 \right) = \sinh x^2 \cdot \dfrac{d}{dx} x^2 + x^2 \cdot \dfrac{d}{dx} \sinh x^2$

$= \sinh x^2 \cdot 2x + x^2 \cdot \cosh x^2 \cdot \dfrac{d}{dx} x^2 = 2x \sinh x^2 + x^2 \cdot \cosh x^2 \cdot 2x = 2x \sinh x^2 + 2x^3 \cosh x^2$

f. Given $y = (x+1)\cosh x$ then $\dfrac{dy}{dx} = \dfrac{d}{dx}\left[(x+1)\cosh x \right] = \cosh x \cdot \dfrac{d}{dx}(x+1) + (x+1)\cdot \dfrac{d}{dx}\cosh x$

$= \cosh x \cdot 1 + (x+1)\cdot \sinh x = \cosh x + (x+1)\sinh x$

g. Given $\boxed{y = \sinh \sqrt{x^3} + \cosh \sqrt{x}}$ then $\boxed{\dfrac{dy}{dx}} = \boxed{\dfrac{d}{dx}\left(\sinh \sqrt{x^3} + \cosh \sqrt{x} \right)} = \boxed{\dfrac{d}{dx}\sinh \sqrt{x^3} + \dfrac{d}{dx}\cosh \sqrt{x}}$

$= \boxed{\dfrac{d}{dx}\sinh x^{\frac{3}{2}} + \dfrac{d}{dx}\cosh x^{\frac{1}{2}}} = \boxed{\cosh x^{\frac{3}{2}}\cdot \dfrac{d}{dx} x^{\frac{3}{2}} + \sinh x^{\frac{1}{2}}\cdot \dfrac{d}{dx} x^{\frac{1}{2}}} = \boxed{\cosh x^{\frac{3}{2}}\cdot \dfrac{3}{2} x^{\frac{3}{2}-1} + \sinh x^{\frac{1}{2}}\cdot \dfrac{1}{2} x^{\frac{1}{2}-1}}$

$= \boxed{\cosh x^{\frac{3}{2}}\cdot \dfrac{3}{2} x^{\frac{1}{2}} + \sinh x^{\frac{1}{2}}\cdot \dfrac{1}{2} x^{-\frac{1}{2}}} = \boxed{\dfrac{3}{2} x^{\frac{1}{2}} \cosh x^{\frac{3}{2}} + \dfrac{1}{2} x^{-\frac{1}{2}} \sinh x^{\frac{1}{2}}} = \boxed{\dfrac{3\sqrt{x}}{2} \cosh \sqrt{x^3} + \dfrac{1}{2\sqrt{x}} \sinh \sqrt{x}}$

h. Given $\boxed{y = x^3 \sinh^2 5x}$ then $\boxed{\dfrac{dy}{dx}} = \boxed{\dfrac{d}{dx}\left(x^3 \sinh^2 5x \right)} = \boxed{\sinh^2 5x\cdot \dfrac{d}{dx} x^3 + x^3 \cdot \dfrac{d}{dx}\sinh^2 5x}$

$= \boxed{\sinh^2 5x \cdot 3x^2 + x^3 \cdot 2\sinh 5x\cdot \dfrac{d}{dx}\sinh 5x} = \boxed{\sinh^2 5x \cdot 3x^2 + x^3 \cdot 2\sinh 5x\cdot \cosh 5x\cdot \dfrac{d}{dx} 5x}$

$= \boxed{\sinh^2 5x \cdot 3x^2 + x^3 \cdot 2\sinh 5x\cdot \cosh 5x\cdot 5} = \boxed{3x^2 \sinh^2 5x + 10x^3 \sinh 5x \cosh 5x}$

i. Given $\boxed{y = \csc h\, x + \sqrt[3]{x^2}}$ then $\boxed{\dfrac{dy}{dx}} = \boxed{\dfrac{d}{dx}\left(\csc h\, x + \sqrt[3]{x^2} \right)} = \boxed{\dfrac{d}{dx}\csc h\, x + \dfrac{d}{dx}\sqrt[3]{x^2}} = \boxed{\dfrac{d}{dx}\csc h\, x + \dfrac{d}{dx} x^{\frac{2}{3}}}$

$= \boxed{-\csc h\, x \coth x\cdot \dfrac{d}{dx} x + \dfrac{2}{3} x^{\frac{2}{3}-1}} = \boxed{-\csc h\, x \coth x\cdot 1 + \dfrac{2}{3} x^{-\frac{1}{3}}} = \boxed{-\csc h\, x \coth x + \dfrac{2}{3\sqrt[3]{x}}}$

Example 3.4-3: Find the first and second derivative of the following hyperbolic functions:

a. $y = \sinh 8x$

b. $y = \sinh^2 3x$

c. $y = \cosh (10x + 3)$

d. $y = \cosh \sqrt{x}$

e. $y = \sinh \left(x^2 + 1 \right)$

f. $y = x \sinh x$

g. $y = \cosh 3x - x^2$

h. $y = \tanh 5x$

i. $y = \coth x^2$

Solutions:

a. $\boxed{\dfrac{dy}{dx}} = \boxed{\dfrac{d}{dx}(\sinh 8x)} = \boxed{\cosh 8x\cdot \dfrac{d}{dx}(8x)} = \boxed{\cosh 8x\cdot 8} = \boxed{8\cosh 8x}$

$\boxed{\dfrac{d^2 y}{dx^2}} = \boxed{\dfrac{d}{dx}(8\cosh 8x)} = \boxed{8\sinh 8x\cdot \dfrac{d}{dx}(8x)} = \boxed{8\sinh 8x\cdot 8} = \boxed{64\sinh 8x}$

b. $\boxed{\dfrac{dy}{dx}} = \boxed{\dfrac{d}{dx}\left(\sinh^2 3x \right)} = \boxed{\dfrac{d}{dx}(\sinh 3x)^2} = \boxed{2\sinh 3x\cdot \dfrac{d}{dx}\sinh 3x} = \boxed{2\sinh 3x\cdot \cosh 3x\cdot \dfrac{d}{dx} 3x}$

$= \boxed{2\sinh 3x\cdot \cosh 3x\cdot 3} = \boxed{6\sinh 3x \cosh 3x}$

$$\dfrac{d^2 y}{dx^2} = \dfrac{d}{dx}\left(6\sinh 3x\cosh 3x\right) = 6\dfrac{d}{dx}\left(\sinh 3x\cosh 3x\right) = 6\left(\cosh 3x\cdot\dfrac{d}{dx}\sinh 3x + \sinh 3x\cdot\dfrac{d}{dx}\cosh 3x\right)$$

$$= 6\left(\cosh 3x\cdot\cosh 3x\cdot\dfrac{d}{dx}3x + \sinh 3x\cdot\sinh 3x\cdot\dfrac{d}{dx}3x\right) = 6\left(\cosh 3x\cdot\cosh 3x\cdot 3 + \sinh 3x\cdot\sinh 3x\cdot 3\right)$$

$$= 6\left(3\cosh^2 3x + 3\sinh^2 3x\right) = \boxed{\mathbf{18\left(\cosh^2 3x + \sinh^2 3x\right)}}$$

c. $$\dfrac{dy}{dx} = \dfrac{d}{dx}\cosh(10x+3) = \sinh(10x+3)\cdot\dfrac{d}{dx}(10x+3) = \sinh(10x+3)\cdot 10 = \boxed{\mathbf{10\sinh(10x+3)}}$$

$$\dfrac{d^2 y}{dx^2} = \dfrac{d}{dx}\left[10\sinh(10x+3)\right] = 10\cosh(10x+3)\cdot\dfrac{d}{dx}(10x+3) = 10\cosh(10x+3)\cdot 10 = \boxed{\mathbf{100\cosh(10x+3)}}$$

d. $$\dfrac{dy}{dx} = \dfrac{d}{dx}\left(\cosh\sqrt{x}\right) = \sinh\sqrt{x}\cdot\dfrac{d}{dx}\left(\sqrt{x}\right) = \sinh\sqrt{x}\cdot\dfrac{d}{dx}x^{\frac{1}{2}} = \sinh\sqrt{x}\cdot\dfrac{1}{2}x^{-\frac{1}{2}} = \boxed{\dfrac{\sinh\sqrt{x}}{2\sqrt{x}}}$$

$$\dfrac{d^2 y}{dx^2} = \dfrac{d}{dx}\left(\dfrac{\sinh\sqrt{x}}{2\sqrt{x}}\right) = \dfrac{1}{2}\cdot\dfrac{d}{dx}\dfrac{\sinh x^{\frac{1}{2}}}{x^{\frac{1}{2}}} = \dfrac{1}{2}\cdot\dfrac{x^{\frac{1}{2}}\cdot\dfrac{d}{dx}\sinh x^{\frac{1}{2}} - \sinh x^{\frac{1}{2}}\cdot\dfrac{d}{dx}x^{\frac{1}{2}}}{x} = \dfrac{1}{2}\cdot\dfrac{x^{\frac{1}{2}}\cdot\cosh x^{\frac{1}{2}}\cdot\dfrac{d}{dx}x^{\frac{1}{2}}}{x}$$

$$\dfrac{-\sinh x^{\frac{1}{2}}\cdot\dfrac{1}{2}x^{-\frac{1}{2}}}{1} = \dfrac{1}{2}\cdot\dfrac{x^{\frac{1}{2}}\cdot\cosh x^{\frac{1}{2}}\cdot\dfrac{1}{2}x^{-\frac{1}{2}} - \sinh x^{\frac{1}{2}}\cdot\dfrac{1}{2}x^{-\frac{1}{2}}}{x} = \dfrac{1}{2}\cdot\dfrac{\sqrt{x}\cdot\cosh x^{\frac{1}{2}}\cdot\dfrac{1}{2\sqrt{x}} - \sinh x^{\frac{1}{2}}\cdot\dfrac{1}{2\sqrt{x}}}{x}$$

$$= \dfrac{1}{2}\cdot\dfrac{\dfrac{1}{2}\cdot\cosh x^{\frac{1}{2}} - \sinh x^{\frac{1}{2}}\cdot\dfrac{1}{2\sqrt{x}}}{x} = \dfrac{1}{4}\cdot\dfrac{\cosh\sqrt{x} - \dfrac{\sinh\sqrt{x}}{\sqrt{x}}}{x} = \dfrac{\dfrac{\sqrt{x}\cosh\sqrt{x} - \sinh\sqrt{x}}{\sqrt{x}}}{4x} = \boxed{\dfrac{\boldsymbol{\sqrt{x}\cosh\sqrt{x} - \sinh\sqrt{x}}}{\mathbf{4x\sqrt{x}}}}$$

e. $$\dfrac{dy}{dx} = \dfrac{d}{dx}\sinh\left(x^2+1\right) = \cosh\left(x^2+1\right)\cdot\dfrac{d}{dx}\left(x^2+1\right) = \cosh\left(x^2+1\right)\cdot 2x = \boxed{\mathbf{2x\cosh\left(x^2+1\right)}}$$

$$\dfrac{d^2 y}{dx^2} = \dfrac{d}{dx}\left[2x\cosh\left(x^2+1\right)\right] = 2\dfrac{d}{dx}\left[x\cosh\left(x^2+1\right)\right] = 2\left[\cosh\left(x^2+1\right)\cdot\dfrac{d}{dx}x + x\cdot\dfrac{d}{dx}\cosh\left(x^2+1\right)\right]$$

$$2\left[\cosh\left(x^2+1\right)\cdot 1 + x\cdot\sinh\left(x^2+1\right)\cdot\dfrac{d}{dx}\left(x^2+1\right)\right] = \boxed{\mathbf{2\left[\cosh\left(x^2+1\right) + 2x^2\sinh\left(x^2+1\right)\right]}}$$

f. $$\dfrac{dy}{dx} = \dfrac{d}{dx}x\sinh x = \sinh x\cdot\dfrac{d}{dx}x + \dfrac{d}{dx}\sinh x\cdot x = \sinh x\cdot 1 + \cosh x\cdot\dfrac{d}{dx}x\cdot x = \boxed{\mathbf{\sinh x + x\cosh x}}$$

$$\frac{d^2 y}{dx^2} = \frac{d}{dx}\left(\sinh x + x \cosh x\right) = \frac{d}{dx}\sinh x + \frac{d}{dx}x \cosh x = \cosh x \cdot \frac{d}{dx}x + \cosh x \cdot \frac{d}{dx}x + \frac{d}{dx}\cosh x \cdot x$$

$$= \cosh x \cdot 1 + \cosh x \cdot 1 + \sinh x \cdot x = \cosh x + \cosh x + x \sinh x = \boxed{2\cosh x + x \sinh x}$$

g. $\dfrac{dy}{dx} = \dfrac{d}{dx}\left(\cosh 3x - x^2\right) = \dfrac{d}{dx}\cosh 3x - \dfrac{d}{dx}x^2 = \sinh 3x \cdot \dfrac{d}{dx}3x - 2x = \sinh 3x \cdot 3 - 2x = \boxed{3\sinh 3x - 2x}$

$$\frac{d^2 y}{dx^2} = \frac{d}{dx}\left(3\sinh 3x - 2x\right) = \frac{d}{dx}3\sinh 3x - \frac{d}{dx}2x = 3\cosh 3x \cdot \frac{d}{dx}3x - 2 = 3\cosh 3x \cdot 3 - 2 = \boxed{9\cosh 3x - 2}$$

h. $\dfrac{dy}{dx} = \dfrac{d}{dx}\left(\tanh 5x\right) = \sec h^2 5x \cdot \dfrac{d}{dx}(5x) = \sec h^2 5x \cdot 5 = \boxed{5\sec h^2 5x}$

$$\frac{d^2 y}{dx^2} = \frac{d}{dx}\left(5\sec h^2 5x\right) = 5 \cdot 2\sec h\, 5x \cdot \frac{d}{dx}\sec h\, 5x = 5 \cdot 2\sec h\, 5x \cdot -\sec h\, 5x \ \tanh 5x \cdot \frac{d}{dx}5x$$

$$= -5 \cdot 2\sec h\, 5x \cdot \sec h\, 5x \ \tanh 5x \cdot 5 = \boxed{-50\sec h^2\, 5x \tanh 5x}$$

i. $\dfrac{dy}{dx} = \dfrac{d}{dx}\left(\coth x^2\right) = -\csc h\, x^2 \cdot \dfrac{d}{dx}x^2 = -\csc h\, x^2 \cdot 2x = \boxed{-2x\csc h\, x^2}$

$$\frac{d^2 y}{dx^2} = \frac{d}{dx}\left(-2x\csc h\, x^2\right) = -2\frac{d}{dx}\left(x\csc h\, x^2\right) = -2\left(\csc h\, x^2 \cdot \frac{d}{dx}x + x \cdot \frac{d}{dx}\csc h\, x^2\right)$$

$$= -2\left(\csc h\, x^2 \cdot 1 - x \cdot \csc h\, x^2 \coth x^2 \cdot 2x\right) = -2\left(\csc h\, x^2 - 2x^2 \coth x^2\right) = \boxed{-2\csc h\, x^2\left(1 - 2x^2 \coth x^2\right)}$$

Section 3.4 Practice Problems – Differentiation of Hyperbolic Functions

Find the derivative of the following hyperbolic functions:

a. $y = \cosh\left(x^3 + 5\right)$ b. $y = x^3 \sinh x$ c. $y = (x + 6)\sinh x^3$

d. $y = \ln\left(\sinh x\right)$ e. $y = e^{\sinh x}$ f. $y = \cosh 3x^5$

g. $y = \dfrac{\tanh^2 x}{x}$ h. $y = \coth \dfrac{1}{x^3}$ i. $y = \left(x^2 + 9\right)\tanh x$

j. $y = \sinh^3 x^2$ k. $y = \tanh^5 x$ l. $y = x^5 \coth\left(x^3 + 1\right)$

3.5   Differentiation of Inverse Hyperbolic Functions

Inverse hyperbolic functions are defined as:

Table 3.5-1: Inverse Hyperbolic Functions

$\sinh^{-1} x = \ln\left(x + \sqrt{x^2 + 1} \right)$ for all x	$\coth^{-1} x = \dfrac{1}{2}\ln\left(\dfrac{x+1}{x-1} \right)$ $x^2 \rangle 1$	
$\cosh^{-1} x = \ln\left(x + \sqrt{x^2 - 1} \right)$ $x \geq 1$	$\sec h^{-1} x = \ln\left(\dfrac{1 + \sqrt{1 - x^2}}{x} \right)$ $0 \langle x \leq 1$	
$\tanh^{-1} x = \dfrac{1}{2}\ln\left(\dfrac{1+x}{1-x} \right)$ $x^2 \langle 1$	$\csc h^{-1} x = \ln\left(\dfrac{1}{x} + \dfrac{\sqrt{1 + x^2}}{\lvert x \rvert} \right)$ $x \neq 0$	

The differential formulas involving inverse hyperbolic functions are defined as:

Table 3.5-2: Differentiation Formulas for Inverse Hyperbolic Functions

$\dfrac{d}{dx}\sinh^{-1} u = \dfrac{1}{\sqrt{u^2 + 1}} \cdot \dfrac{du}{dx}$	$\dfrac{d}{dx}\coth^{-1} u = \dfrac{1}{1 - u^2} \cdot \dfrac{du}{dx}$ $\lvert u \rvert \rangle 1$	
$\dfrac{d}{dx}\cosh^{-1} u = \dfrac{1}{\sqrt{u^2 - 1}} \cdot \dfrac{du}{dx}$ $u \rangle 1$	$\dfrac{d}{dx}\sec h^{-1} u = \dfrac{-1}{u\sqrt{1 - u^2}} \cdot \dfrac{du}{dx}$ $0 \langle u \langle 1$	
$\dfrac{d}{dx}\tanh^{-1} u = \dfrac{1}{1 - u^2} \cdot \dfrac{du}{dx}$ $\lvert u \rvert \langle 1$	$\dfrac{d}{dx}\csc h^{-1} u = \dfrac{-1}{\lvert u \rvert \sqrt{1 + u^2}} \cdot \dfrac{du}{dx}$ $u \neq 0$	

Let's differentiate some inverse hyperbolic functions using the above formulas:

Example 3.5-1: Find the derivative of the following inverse hyperbolic functions:

a. $y = \sinh^{-1} 10x$

b. $y = \cosh^{-1} x^2$

c. $y = \sinh^{-1} \sqrt{x}$

d. $y = x\tanh^{-1} x^3$

e. $y = \sinh^{-1} e^{2x}$

f. $y = x^3 + \coth^{-1} x$

g. $y = \cosh^{-1} x + e^{2x}$

h. $y = \dfrac{\sinh^{-1} x}{3x}$

i. $y = \sec h^{-1} e^{x}$

Solutions:

a. Given $\boxed{y = \sinh^{-1} 10x}$ then $\boxed{\dfrac{dy}{dx}} = \boxed{\dfrac{d}{dx}\sinh^{-1} 10x} = \boxed{\dfrac{1}{\sqrt{(10x)^2 + 1}} \cdot \dfrac{d}{dx} 10x} = \boxed{\dfrac{1}{\sqrt{100x^2 + 1}} \cdot 10} = \boxed{\dfrac{10}{\sqrt{100x^2 + 1}}}$

b. Given $\boxed{y = \cosh^{-1} x^2}$ then $\boxed{\dfrac{dy}{dx}} = \boxed{\dfrac{d}{dx}\cosh^{-1} x^2} = \boxed{\dfrac{1}{\sqrt{\left(x^2\right)^2 - 1}} \cdot \dfrac{d}{dx} x^2} = \boxed{\dfrac{1}{\sqrt{x^4 - 1}} \cdot 2x} = \boxed{\dfrac{2x}{\sqrt{x^4 - 1}}}$

c. Given $\boxed{y = \sinh^{-1} \sqrt{x}}$ then $\boxed{\dfrac{dy}{dx}} = \boxed{\dfrac{d}{dx}\sinh^{-1} \sqrt{x}} = \boxed{\dfrac{1}{\sqrt{\left(\sqrt{x}\right)^2 + 1}} \cdot \dfrac{d}{dx}\sqrt{x}} = \boxed{\dfrac{1}{\sqrt{x+1}} \cdot \dfrac{d}{dx} x^{\frac{1}{2}}} = \boxed{\dfrac{1}{\sqrt{x+1}} \cdot \dfrac{1}{2} x^{\frac{1}{2}-1}}$

$$= \boxed{\frac{1}{\sqrt{x+1}} \cdot \frac{1}{2} x^{-\frac{1}{2}}} = \boxed{\frac{1}{\sqrt{x+1}} \cdot \frac{1}{2\sqrt{x}}} = \boxed{\frac{1}{2\sqrt{x}\,\sqrt{x+1}}}$$

d. Given $\boxed{y = x \tanh^{-1} x^3}$ then $\boxed{\frac{dy}{dx}} = \boxed{\frac{d}{dx}\left(x \tanh^{-1} x^3\right)} = \boxed{\tanh^{-1} x^3 \cdot \frac{d}{dx} x + x \cdot \frac{d}{dx} \tanh^{-1} x^3}$

$$= \boxed{\tanh^{-1} x^3 \cdot 1 + x \cdot \frac{1}{1-x^6} \cdot \frac{d}{dx} x^3} = \boxed{\tanh^{-1} x^3 + x \cdot \frac{1}{1-x^6} \cdot 3x^2} = \boxed{\mathbf{\tanh^{-1} x^3 + \frac{3x^3}{1-x^6}}}$$

e. Given $\boxed{y = \sinh^{-1} e^{2x}}$ then $\boxed{\frac{dy}{dx}} = \boxed{\frac{d}{dx} \sinh^{-1} e^{2x}} = \boxed{\frac{1}{\sqrt{\left(e^{2x}\right)^2 + 1}} \cdot \frac{d}{dx} e^{2x}} = \boxed{\frac{1}{\sqrt{e^{4x}+1}} \cdot e^{2x} \cdot \frac{d}{dx} 2x}$

$$= \boxed{\frac{1}{\sqrt{e^{4x}+1}} \cdot e^{2x} \cdot 2} = \boxed{\mathbf{\frac{2e^{2x}}{\sqrt{e^{4x}+1}}}}$$

f. Given $\boxed{y = x^3 + \coth^{-1} x}$ then $\boxed{\frac{dy}{dx}} = \boxed{\frac{d}{dx}\left(x^3 + \coth^{-1} x\right)} = \boxed{\frac{d}{dx} x^3 + \frac{d}{dx} \coth^{-1} x} = \boxed{\mathbf{3x^2 + \frac{1}{1-x^2}}}$

g. Given $\boxed{y = \cosh^{-1} x + e^{2x}}$ then $\boxed{\frac{dy}{dx}} = \boxed{\frac{d}{dx}\left(\cosh^{-1} x + e^{2x}\right)} = \boxed{\frac{d}{dx} \cosh^{-1} x + \frac{d}{dx} e^{2x}} = \boxed{\frac{1}{\sqrt{x^2-1}} + \frac{d}{dx} e^{2x}}$

$$= \boxed{\frac{1}{\sqrt{x^2-1}} + \frac{d}{dx} e^{2x}} = \boxed{\frac{1}{\sqrt{x^2-1}} + e^{2x} \cdot \frac{d}{dx} 2x} = \boxed{\frac{1}{\sqrt{x^2-1}} + e^{2x} \cdot 2} = \boxed{\mathbf{\frac{1}{\sqrt{x^2-1}} + 2e^{2x}}}$$

h. Given $\boxed{y = \frac{\sinh^{-1} x}{3x}}$ then $\boxed{\frac{dy}{dx}} = \boxed{\frac{d}{dx}\left(\frac{\sinh^{-1} x}{3x}\right)} = \boxed{\frac{3x \cdot \frac{d}{dx} \sinh^{-1} x - \sinh^{-1} x \cdot \frac{d}{dx} 3x}{9x^2}} = \boxed{\frac{3x \cdot \frac{1}{\sqrt{x^2+1}} - \sinh^{-1} x \cdot 3}{9x^2}}$

$$= \boxed{\frac{\frac{3x}{\sqrt{x^2+1}} - 3\sinh^{-1} x}{9x^2}} = \boxed{\frac{3x - 3\sqrt{x^2+1}\,\sinh^{-1} x}{9x^2 \sqrt{x^2+1}}} = \boxed{\mathbf{\frac{x - \sqrt{x^2+1}\,\sinh^{-1} x}{3x^2 \sqrt{x^2+1}}}}$$

i. Given $\boxed{y = \sec h^{-1} e^x}$ then $\boxed{\frac{dy}{dx}} = \boxed{-\frac{1}{e^x \sqrt{1-\left(e^x\right)^2}} \cdot \frac{d}{dx} e^x} = \boxed{-\frac{1}{e^x \sqrt{1-e^{2x}}} \cdot e^x} = \boxed{\mathbf{-\frac{1}{\sqrt{1-e^{2x}}}}}$

Example 3.5-2: Find the derivative of the following inverse hyperbolic functions:

a. $y = \sinh^{-1} x^2 + \ln x^3$

b. $y = \tanh^{-1} \sin x$

c. $y = \coth^{-1} \cos x$

d. $y = \sec h^{-1} x \cdot \sec x$

e. $y = \sinh^{-1} 3e^x$

f. $y = \coth^{-1}(10x+3)$

g. $y = \cosh^{-1} \ln x$

h. $y = \tanh^{-1} 2x + \ln e^x$

i. $y = \tanh^{-1} x + x$

j. $y = e^x \sinh^{-1} x$

k. $y = x \cosh^{-1} x$

l. $y = \tanh^{-1} x^3 + 5x$

Solutions:

a. $\dfrac{dy}{dx} = \dfrac{d}{dx}\left(\sinh^{-1} x^2 + \ln x^3\right) = \dfrac{d}{dx}\sinh^{-1} x^2 + \dfrac{d}{dx}\ln x^3 = \dfrac{1}{\sqrt{x^4+1}} \cdot \dfrac{d}{dx} x^2 + \dfrac{1}{x^3} \cdot \dfrac{d}{dx} x^3 = \dfrac{1}{\sqrt{x^4+1}} \cdot 2x$

$+\dfrac{1}{x^3} \cdot 3\,x^2 = \dfrac{2x}{\sqrt{x^4+1}} + \dfrac{3x^2}{x^{3=1}} = \dfrac{2x}{\sqrt{x^4+1}} + \dfrac{3}{x}$

b. $\dfrac{dy}{dx} = \dfrac{d}{dx}\tanh^{-1}\sin x = \dfrac{1}{1-\sin^2 x} \cdot \dfrac{d}{dx}\sin x = \dfrac{1}{1-\sin^2 x} \cdot \cos x = \dfrac{\cos x}{1-\sin^2 x} = \dfrac{\cos x}{\cos^2 x} = \dfrac{1}{\cos x}$

c. $\dfrac{dy}{dx} = \dfrac{d}{dx}\coth^{-1}\cos x = \dfrac{1}{1-\cos^2 x} \cdot \dfrac{d}{dx}\cos x = \dfrac{1}{1-\cos^2 x} \cdot -\sin x = \dfrac{-\sin x}{1-\cos^2 x} = -\dfrac{\sin x}{\sin^2 x} = -\dfrac{1}{\sin x}$

d. $\dfrac{dy}{dx} = \dfrac{d}{dx}\sec h^{-1} x \cdot \sec x = \sec x \cdot \dfrac{d}{dx}\sec h^{-1} x + \sec h^{-1} x \cdot \dfrac{d}{dx}\sec x = \sec x \cdot \dfrac{-1}{x\sqrt{1-x^2}} + \sec h^{-1} x \cdot \sec x \tan x$

$= -\dfrac{\sec x}{x\sqrt{1-x^2}} + \sec x \tan x \sec h^{-1} x = \dfrac{-\sec x + x\sqrt{1-x^2}\,\sec x \tan x \sec h^{-1} x}{x\sqrt{1-x^2}}$

e. $\dfrac{dy}{dx} = \dfrac{d}{dx}\sinh^{-1} 3e^x = \dfrac{1}{\sqrt{\left(3e^x\right)^2+1}} \cdot \dfrac{d}{dx} 3e^x = \dfrac{1}{\sqrt{9e^{2x}+1}} \cdot 3e^x \cdot \dfrac{d}{dx} x = \dfrac{1}{\sqrt{9e^{2x}+1}} \cdot 3e^x \cdot 1 = \dfrac{3e^x}{\sqrt{9e^{2x}+1}}$

f. $\dfrac{dy}{dx} = \dfrac{d}{dx}\coth^{-1}(x+3) = \dfrac{1}{1-(x+3)^2} \cdot \dfrac{d}{dx}(x+3) = \dfrac{1}{1-\left(x^2+9+6x\right)} \cdot 1 = \dfrac{1}{-x^2-6x-8} = -\dfrac{1}{x^2+6x+8}$

g. $\dfrac{dy}{dx} = \dfrac{d}{dx}\cosh^{-1}\ln x = \dfrac{1}{\sqrt{(\ln x)^2-1}} \cdot \dfrac{d}{dx}\ln x = \dfrac{1}{\sqrt{\ln^2 x-1}} \cdot \dfrac{1}{x} \cdot \dfrac{d}{dx} x = \dfrac{1}{\sqrt{\ln^2 x-1}} \cdot \dfrac{1}{x} \cdot 1 = \dfrac{1}{x\sqrt{\ln^2 x-1}}$

h. $\dfrac{dy}{dx} = \dfrac{d}{dx}\left(\tanh^{-1} 2x + \ln e^x\right) = \dfrac{d}{dx}\tanh^{-1} 2x + \dfrac{d}{dx}\ln e^x = \dfrac{1}{1-(2x)^2} \cdot \dfrac{d}{dx} 2x + \dfrac{1}{e^x} \cdot \dfrac{d}{dx} e^x = \dfrac{1}{1-4x^2} \cdot 2 + \dfrac{1}{e^x} \cdot e^x$

$= \dfrac{2}{1-4x^2} + \dfrac{e^x}{e^x} = \dfrac{2}{1-4x^2} + 1 = \dfrac{2+1-4x^2}{1-2x^2} = \dfrac{-4x^2+3}{-2x^2+1}$

i. $\dfrac{dy}{dx} = \dfrac{d}{dx}\left(\tanh^{-1} x + x\right) = \dfrac{d}{dx}\tanh^{-1} x + \dfrac{d}{dx} x = \dfrac{1}{1-x^2} \cdot \dfrac{d}{dx} x + 1 = \dfrac{1}{1-x^2} \cdot 1 + 1 = \dfrac{1}{1-x^2} + 1$

j. $\dfrac{dy}{dx} = \dfrac{d}{dx}\left(e^x \sinh^{-1} x\right) = \boxed{\sinh^{-1} x \cdot \dfrac{d}{dx} e^x + e^x \cdot \dfrac{d}{dx}\sinh^{-1} x} = \boxed{\sinh^{-1} x \cdot e^x \cdot \dfrac{d}{dx} x + e^x \cdot \dfrac{1}{\sqrt{x^2+1}}\cdot \dfrac{d}{dx} x}$

$= \boxed{\sinh^{-1} x \cdot e^x \cdot 1 + e^x \cdot \dfrac{1}{\sqrt{x^2+1}}\cdot 1} = \boxed{e^x \sinh^{-1} x + \dfrac{e^x}{\sqrt{x^2+1}}} = \boxed{e^x\left(\sinh^{-1} x + \dfrac{1}{\sqrt{x^2+1}}\right)}$

k. $\dfrac{dy}{dx} = \dfrac{d}{dx}\left(x\cosh^{-1} x\right) = \boxed{\cosh^{-1} x \cdot \dfrac{d}{dx} x + x\cdot \dfrac{d}{dx}\cosh^{-1} x} = \boxed{\cosh^{-1} x \cdot 1 + x\cdot \dfrac{1}{\sqrt{x^2-1}}} = \boxed{\cosh^{-1} x + \dfrac{x}{\sqrt{x^2-1}}}$

l. $\dfrac{dy}{dx} = \dfrac{d}{dx}\left(\tanh^{-1} x^3 +5x\right) = \boxed{\dfrac{d}{dx}\tanh^{-1} x^3 + \dfrac{d}{dx}5x} = \boxed{\dfrac{1}{1-x^6}\cdot \dfrac{d}{dx} x^3 +5} = \boxed{\dfrac{1}{1-x^6}\cdot 3x^2 +5} = \boxed{\dfrac{3x^2}{1-x^6}+5}$

Example 3.5-3: In the following examples find $\dfrac{dy}{dx}$:

a. $x = t + \sinh^{-1} t$ and $y = \cos t$

b. $x = \cosh^{-1} t$ and $y = t^3 +3t^2 +t$

c. $x = \sinh^{-1} t^2$ and $y = \sin^2 t$

d. $x = \tanh^{-1} \alpha$ and $y = \sin \alpha$

e. $x = \theta \coth^{-1} \theta$ and $y = \theta$

f. $x = t^2 \sinh^{-1} t$ and $y = t\, e^t$

g. $x = \cosh^{-1} e^t$ and $y = e^{2t}$

h. $x = t^3 +3t$ and $y = \cosh^{-1} \sqrt{t}$

Solutions:

a. Given $\boxed{x = t+\sinh^{-1} t}$ and $\boxed{y = \cos t}$ then $\dfrac{dx}{dt} = \dfrac{d}{dt}\left(t+\sinh^{-1} t\right) = \boxed{1+\dfrac{1}{\sqrt{t^2+1}}} = \boxed{\dfrac{1+\sqrt{t^2+1}}{\sqrt{t^2+1}}}$ and

$\dfrac{dy}{dt} = \dfrac{d}{dt}\cos t = \boxed{-\sin t}$ So $\dfrac{dy/dt}{dx/dt} = \dfrac{-\sin t}{\frac{1+\sqrt{t^2+1}}{\sqrt{t^2+1}}} = \dfrac{-\frac{\sin t}{1}}{\frac{1+\sqrt{t^2+1}}{\sqrt{t^2+1}}} = -\dfrac{\sin t\cdot\sqrt{t^2+1}}{1\cdot 1+\sqrt{t^2+1}} = \boxed{-\dfrac{\sin t\,\sqrt{t^2+1}}{1+\sqrt{t^2+1}}}$

b. Given $\boxed{x = \cosh^{-1} t}$ and $\boxed{y = t^3 +3t^2 +t}$ then $\dfrac{dx}{dt} = \dfrac{d}{dt}\cosh^{-1} t = \boxed{\dfrac{1}{\sqrt{t^2-1}}}$ and $\dfrac{dy}{dt} = \boxed{\dfrac{d}{dt}\left(t^3 +3t^2 +t\right)}$

$= \boxed{3t^2 +6t+1}$ So $\dfrac{dy/dt}{dx/dt} = \dfrac{3t^2 +6t+1}{\frac{1}{\sqrt{t^2-1}}} = \dfrac{\frac{3t^2+6t+1}{1}}{\frac{1}{\sqrt{t^2-1}}} = \dfrac{\left(3t^2 +6t+1\right)\cdot\sqrt{t^2-1}}{1\cdot1} = \boxed{\left(3t^2 +6t+1\right)\sqrt{t^2-1}}$

c. Given $\boxed{x = \sinh^{-1} t^2}$ and $\boxed{y = \sin^2 t}$ then $\dfrac{dx}{dt} = \dfrac{d}{dt}\sinh^{-1} t^2 = \boxed{\dfrac{1}{\sqrt{t^4+1}}}$ and $\dfrac{dy}{dt} = \dfrac{d}{dt}\sin^2 t$

$$= \boxed{\dfrac{d}{dt}(\sin t)^2} = \boxed{2\sin t \cdot \dfrac{d}{dt}\sin t} = \boxed{2\sin t \cos t} \quad \text{Therefore} \quad \boxed{\dfrac{dy/dt}{dx/dt}} = \boxed{\dfrac{2\sin t \cos t}{\dfrac{1}{\sqrt{t^4+1}}}} = \boxed{\dfrac{\dfrac{2\sin t \cos t}{1}}{\dfrac{1}{\sqrt{t^4+1}}}}$$

$$= \boxed{\dfrac{2\sin t \cos t \cdot \sqrt{t^4+1}}{1\cdot 1}} = \boxed{\dfrac{2\sin t \cos t \sqrt{t^4+1}}{1}} = \boxed{2\sqrt{t^4+1}\,\sin t \cos t}$$

d. Given $\boxed{x = \tanh^{-1}\alpha}$ and $\boxed{y = \sin\alpha}$ then $\dfrac{dx}{d\alpha} = \boxed{\tanh^{-1}\alpha} = \boxed{\dfrac{1}{1-\alpha^2}}$ and $\dfrac{dy}{d\alpha} = \boxed{\dfrac{d}{d\alpha}\sin\alpha} = \boxed{\cos\alpha}$

Therefore $\boxed{\dfrac{dy/d\alpha}{dx/d\alpha}} = \boxed{\dfrac{\cos\alpha}{\dfrac{1}{1-\alpha^2}}} = \boxed{\dfrac{\dfrac{\cos\alpha}{1}}{\dfrac{1}{1-\alpha^2}}} = \boxed{\dfrac{\cos\alpha\cdot(1-\alpha^2)}{1\cdot 1}} = \boxed{\dfrac{\cos\alpha\,(1-\alpha^2)}{1}} = \boxed{(1-\alpha^2)\cos\alpha}$

e. Given $\boxed{x = \theta\coth^{-1}\theta}$ and $\boxed{y = \theta}$ then $\dfrac{dx}{d\theta} = \boxed{\dfrac{d}{d\theta}\left(\theta\coth^{-1}\theta\right)} = \boxed{\coth^{-1}\theta\cdot\dfrac{d}{d\theta}\theta + \theta\cdot\dfrac{d}{d\theta}\coth^{-1}\theta}$

$$= \boxed{\coth^{-1}\theta\cdot 1 + \theta\cdot\dfrac{1}{1-\theta^2}\cdot\dfrac{d}{d\theta}\theta} = \boxed{\coth^{-1}\theta + \dfrac{\theta}{1-\theta^2}} = \boxed{\dfrac{(1-\theta^2)\coth^{-1}\theta + \theta}{1-\theta^2}} \text{ and } \dfrac{dy}{d\theta} = \boxed{\dfrac{d}{d\theta}\theta} = \boxed{1}$$

Therefore $\boxed{\dfrac{dy/d\theta}{dx/d\theta}} = \boxed{\dfrac{1}{\left(1-\theta^2\right)\coth^{-1}\theta + \theta\big/1-\theta^2}} = \boxed{\dfrac{1\cdot\left(1-\theta^2\right)}{\left[\left(1-\theta^2\right)\coth^{-1}\theta + \theta\right]\cdot 1}} = \boxed{\dfrac{1-\theta^2}{\left(1-\theta^2\right)\coth^{-1}\theta + \theta}}$

f. Given $\boxed{x = t^2\sinh^{-1}t}$ and $\boxed{y = t\,e^t}$ then $\dfrac{dx}{dt} = \boxed{\dfrac{d}{dt}\left(t^2\sinh^{-1}t\right)} = \boxed{\sinh^{-1}t\cdot\dfrac{d}{dt}t^2 + t^2\cdot\dfrac{d}{dt}\sinh^{-1}t}$

$$= \boxed{\sinh^{-1}t\cdot 2t + t^2\cdot\dfrac{1}{\sqrt{t^2+1}}} = \boxed{2t\sinh^{-1}t + \dfrac{t^2}{\sqrt{t^2+1}}} = \boxed{\dfrac{2t\sqrt{t^2+1}\,\sinh^{-1}t + t^2}{\sqrt{t^2+1}}} \text{ and } \dfrac{dy}{dt} = \boxed{\dfrac{d}{dt}t\,e^t}$$

$$= \boxed{e^t\cdot\dfrac{d}{dt}t + t\cdot\dfrac{d}{dt}e^t} = \boxed{e^t\cdot 1 + t\cdot e^t} = \boxed{e^t(1+t)} \quad \text{Therefore} \quad \boxed{\dfrac{dy/dt}{dx/dt}} = \boxed{\dfrac{e^t(1+t)}{2t\sqrt{t^2+1}\,\sinh^{-1}t + t^2\big/\sqrt{t^2+1}}}$$

$$= \boxed{\dfrac{e^t(1+t)\big/1}{2t\sqrt{t^2+1}\,\sinh^{-1}t + t^2\big/\sqrt{t^2+1}}} = \boxed{\dfrac{e^t(1+t)\sqrt{t^2+1}}{2t\sqrt{t^2+1}\,\sinh^{-1}t + t^2}}$$

g. Given $\boxed{x = \cosh^{-1}e^t}$ and $\boxed{y = e^{2t}}$ then $\dfrac{dx}{dt} = \boxed{\dfrac{d}{dt}\cosh^{-1}e^t} = \boxed{\dfrac{1}{\sqrt{e^{2t}-1}}\cdot\dfrac{d}{dt}e^t} = \boxed{\dfrac{1}{\sqrt{e^{2t}-1}}\cdot e^t}$

$$= \boxed{\dfrac{e^t}{\sqrt{e^{2t}-1}}} \text{ and } \boxed{\dfrac{dy}{dt}} = \boxed{\dfrac{d}{dt}e^{2t}} = \boxed{e^{2t}\cdot\dfrac{d}{dt}2t} = \boxed{e^{2t}\cdot 2} = \boxed{2e^{2t}} \text{ Therefore } \boxed{\dfrac{dy/dt}{dx/dt}} = \boxed{\dfrac{2e^{2t}}{e^t/\sqrt{e^{2t}-1}}}$$

$$= \boxed{\dfrac{2e^{2t}\sqrt{e^{2t}-1}}{e^t}} = \boxed{\dfrac{2e^{2t}\cdot e^{-t}\sqrt{e^{2t}-1}}{1}} = \boxed{2e^{2t-t}\sqrt{e^{2t}-1}} = \boxed{\mathbf{2e^{\,t}\sqrt{e^{2t}-1}}}$$

h. Given $\boxed{x=t^3+3t}$ and $\boxed{y=\cosh^{-1}\sqrt{t}}$ then $\dfrac{dx}{dt} = \boxed{\dfrac{d}{dt}\left(t^3+3t\right)} = \boxed{\dfrac{d}{dt}t^3+\dfrac{d}{dt}3t} = \boxed{3t^2+3}$ and $\boxed{\dfrac{dy}{dt}}$

$$= \boxed{\dfrac{d}{dt}\cosh^{-1}\sqrt{t}} = \boxed{\dfrac{1}{\sqrt{t-1}}\cdot\dfrac{d}{dt}\sqrt{t}} = \boxed{\dfrac{1}{\sqrt{t-1}}\cdot\dfrac{1}{2\sqrt{t}}} = \boxed{\dfrac{1}{2\sqrt{t}\sqrt{t-1}}} \text{ Therefore } \boxed{\dfrac{dy/dt}{dx/dt}} = \boxed{\dfrac{1/2\sqrt{t}\sqrt{t-1}}{3t^2+3}}$$

$$= \boxed{\dfrac{1}{\left(3t^2+3\right)\cdot 2\sqrt{t}\sqrt{t-1}}} = \boxed{\mathbf{\dfrac{1}{6\left(t^2+1\right)\sqrt{t}\sqrt{t-1}}}}$$

Section 3.5 Practice Problems – Differentiation of Inverse Hyperbolic Functions

Find the derivative of the following inverse hyperbolic functions:

a. $y = x\sinh^{-1}3x$

b. $y = \cosh^{-1}\left(x+5\right)$

c. $y = \tanh^{-1}\sqrt{x}$

d. $y = \dfrac{\sinh^{-1}x}{x^3}$

e. $y = \dfrac{\sinh^{-1}x}{\cosh 3x}$

f. $y = \dfrac{\cosh^{-1}x}{x^5}$

g. $y = \left(x^2+3\right)\coth^{-1}x$

h. $y = e^{3x}\cosh^{-1}x$

i. $y = x^3+\tanh^{-1}x^5$

j. $y = \sinh^{-1}7x+\ln e^{x^2}$

k. $y = \tanh^{-1}\left(e^{2x}\right)$

l. $y = \coth^{-1}(3x+5)$

3.6   Evaluation of Indeterminate Forms Using L'Hopital's Rule

In Chapter 1 the concept of limits and how they are used in finding a solution to an expression as the limit of a variable, within the expression, approaches a constant or infinity was addressed. In Chapter 2 and this Chapter finding the derivative of various class of functions was discussed. In this section, we will use what we have learned in the previous chapters in order to solve expressions of the form referred to as indeterminate forms.

We stated earlier that the derivative of a differentiable function $f(x)$ is defined as

$$\lim_{\Delta x \to 0} \frac{f(x+\Delta x)-f(x)}{(x+\Delta x)-x} \tag{1}$$

As the limit $\Delta x \to 0$ the above expression reduces to:

$$\lim_{\Delta x \to 0} \frac{f(x+\Delta x)-f(x)}{(x+\Delta x)-x} = \frac{f(x+0)-f(x)}{(x+0)-x} = \frac{f(x)-f(x)}{x-x} = \frac{0}{0}$$

Since the limit on both the numerator and the denominator is zero, equation (1) is referred to as *indeterminate form* of the type $\frac{0}{0}$. Similarly, when both the numerator and the denominator are equal to ∞, i.e., $\frac{\infty}{\infty}$ the resulting expression is also referred to as *indeterminate form*. Note that indeterminate forms of the type $0\cdot\infty$, $\infty-\infty$, 0^0, ∞^0, and 1^∞ need to be transformed to one of the types $\frac{0}{0}$ or $\frac{\infty}{\infty}$ first. These types of indeterminate forms will be discussed later in this section.

To solve expressions that result in indeterminate forms of the type $\frac{0}{0}$ or $\frac{\infty}{\infty}$ we apply a rule that is referred to as the L'Hopital's rule.

L'Hopital's Rule – Given that $f(x)$ and $g(x)$ are differentiable and $g(x)\neq 0$ and assuming that $\lim_{x\to a} f(x)=f(a)=0$ and $\lim_{x\to a} g(x)=g(a)=0$, then

$$\lim_{x\to a} \frac{f(x)}{g(x)} = \lim_{x\to a} \frac{f'(x)}{g'(x)}$$

is referred to as the L'Hopital's rule provided that $f'(a)$ and $g'(a)$ exist and $g'(a)\neq 0$

Note that we proceed to differentiate $f(x)$ and $g(x)$, i.e., continue to apply the L'Hopital's rule, as long as we still obtain the indeterminate forms of the type $\frac{0}{0}$ or $\frac{\infty}{\infty}$ as the limit x approaches to a. We stop differentiation as soon as either the numerator or the denominator has a finite nonzero limit. The following examples show application of the L'Hopital's rule.

Example 3.6-1: Evaluate the limit of the following functions:

a. $\lim_{x\to 0} \frac{e^x -1}{x^3} =$

b. $\lim_{x\to\infty} \frac{\ln x}{x}$

c. $\lim_{x\to 1} \frac{x^4 -x^3 +x^2 -1}{x^4 -2x^2 +x}$

d. $\lim_{x\to 0} \frac{e^x +e^{-x} -2}{\sin x}$

e. $\lim_{x\to 3} \frac{x^3 -x^2 -x-15}{x^3 -2x^2 -9}$

f. $\lim_{x\to 0} \frac{x-\sin x}{x^3}$

g. $\lim_{x \to \frac{\pi}{2}} \dfrac{2x-\pi}{\cos x}$ 　　　　 h. $\lim_{t \to 1} \dfrac{t^3-1}{4t^3-t-3}$ 　　　　 i. $\lim_{x \to 0} \dfrac{\sin 3x}{x}$

Solutions:

a. $\boxed{\lim_{x \to 0} \dfrac{e^x-1}{x^3}} = \boxed{\dfrac{e^0-1}{0^3}} = \boxed{\dfrac{1-1}{0}} = \boxed{\dfrac{0}{0}}$ which is an indeterminate function. Let's apply the

$\text{L'Hopital's rule } \boxed{\lim_{x \to 0} \dfrac{e^x-1}{x^3}} = \boxed{\lim_{x \to 0} \dfrac{\frac{d}{dx}\left(e^x-1\right)}{\frac{d}{dx}x^3}} = \boxed{\lim_{x \to 0} \dfrac{\frac{d}{dx}e^x-\frac{d}{dx}1}{3x^2}} = \boxed{\lim_{x \to 0} \dfrac{e^x \cdot \frac{d}{dx}x-0}{3x^2}}$

$= \boxed{\lim_{x \to 0} \dfrac{e^x \cdot 1}{3x^2}} = \boxed{\lim_{x \to 0} \dfrac{e^x}{3x^2}} = \boxed{\dfrac{e^0}{3 \cdot 0^2}} = \boxed{\dfrac{1}{0}} = \boxed{\infty}$

b. $\boxed{\lim_{x \to \infty} \dfrac{\ln x}{x}} = \boxed{\dfrac{\ln \infty}{\infty}} = \boxed{\dfrac{\infty}{\infty}}$ which is an indeterminate function. Let's apply the L'Hopital's rule

$\boxed{\lim_{x \to \infty} \dfrac{\ln x}{x}} = \boxed{\lim_{x \to \infty} \dfrac{\frac{d}{dx}\ln x}{\frac{d}{dx}x}} = \boxed{\lim_{x \to \infty} \dfrac{\frac{1}{x} \cdot \frac{d}{dx}x}{1}} = \boxed{\lim_{x \to \infty} \dfrac{\frac{1}{x} \cdot 1}{1}} = \boxed{\lim_{x \to \infty} \dfrac{\frac{1}{x}}{1}} = \boxed{\lim_{x \to \infty} \dfrac{1}{x}} = \boxed{\dfrac{1}{\infty}} = \boxed{0}$

c. $\boxed{\lim_{x \to 1} \dfrac{x^4-x^3+x^2-1}{x^4-2x^2+x}} = \boxed{\dfrac{1^4-1^3+1^2-1}{1^4-2 \cdot 1^2+1}} = \boxed{\dfrac{1-1+1-1}{1-2+1}} = \boxed{\dfrac{2-2}{2-2}} = \boxed{\dfrac{0}{0}}$ Apply the L'Hopital's rule

$\boxed{\lim_{x \to 1} \dfrac{x^4-x^3+x^2-1}{x^4-2x^2+x}} = \boxed{\lim_{x \to 1} \dfrac{\frac{d}{dx}\left(x^4-x^3+x^2-1\right)}{\frac{d}{dx}\left(x^4-2x^2+x\right)}} = \boxed{\lim_{x \to 1} \dfrac{\frac{d}{dx}x^4-\frac{d}{dx}x^3+\frac{d}{dx}x^2-\frac{d}{dx}1}{\frac{d}{dx}x^4-2\frac{d}{dx}x^2+\frac{d}{dx}x}}$

$= \boxed{\lim_{x \to 1} \dfrac{4x^3-3x^2+2x-0}{4x^3-4x+1}} = \boxed{\dfrac{4 \cdot 1^3-3 \cdot 1^2+2 \cdot 1}{4 \cdot 1^3-4 \cdot 1+1}} = \boxed{\dfrac{4-3+2}{4-4+1}} = \boxed{\dfrac{6-3}{1}} = \boxed{\dfrac{3}{1}} = \boxed{3}$

d. $\boxed{\lim_{x \to 0} \dfrac{e^x+e^{-x}-2}{\sin x}} = \boxed{\lim_{x \to 0} \dfrac{e^x+\frac{1}{e^x}-2}{\sin x}} = \boxed{\dfrac{e^0+\frac{1}{e^0}-2}{\sin 0}} = \boxed{\dfrac{1+\frac{1}{1}-2}{0}} = \boxed{\dfrac{1+1-2}{0}} = \boxed{\dfrac{2-2}{0}} = \boxed{\dfrac{0}{0}}$ Apply

$\text{the L'Hopital's rule } \boxed{\lim_{x \to 0} \dfrac{e^x+e^{-x}-2}{\sin x}} = \boxed{\lim_{x \to 0} \dfrac{\frac{d}{dx}\left(e^x+e^{-x}-2\right)}{\frac{d}{dx}\sin x}} = \boxed{\lim_{x \to 0} \dfrac{\frac{d}{dx}e^x+\frac{d}{dx}e^{-x}-\frac{d}{dx}2}{\frac{d}{dx}\sin x}}$

$= \boxed{\lim_{x \to 0} \dfrac{e^x+e^{-x} \cdot \frac{d}{dx}(-x)-0}{\cos x}} = \boxed{\lim_{x \to 0} \dfrac{e^x+e^{-x} \cdot -1}{\cos x}} = \boxed{\lim_{x \to 0} \dfrac{e^x-e^{-x}}{\cos x}} = \boxed{\lim_{x \to 0} \dfrac{e^x-\frac{1}{e^x}}{\cos x}}$

$$= \frac{\left| \frac{e^0 - \frac{1}{e^0}}{\cos 0} \right|}{} = \left| \frac{1 - \frac{1}{1}}{1} \right| = \left| \frac{1-1}{1} \right| = \left| \frac{0}{1} \right| = \boxed{0}$$

e. $\left| \lim_{x \to 3} \frac{x^3 - x^2 - x - 15}{x^3 - 2x^2 - 9} \right| = \left| \frac{3^3 - 3^2 - 3 - 15}{3^3 - 2 \cdot 3^2 - 9} \right| = \left| \frac{27 - 9 - 3 - 15}{27 - 18 - 9} \right| = \left| \frac{27 - 27}{27 - 27} \right| = \left| \frac{0}{0} \right|$ Apply the L'Hopital's

rule $\left| \lim_{x \to 3} \frac{x^3 - x^2 - x - 15}{x^3 - 2x^2 - 9} \right| = \left| \lim_{x \to 3} \frac{\frac{d}{dx}\left(x^3 - x^2 - x - 15\right)}{\frac{d}{dx}\left(x^3 - 2x^2 - 9\right)} \right| = \left| \lim_{x \to 3} \frac{\frac{d}{dx}x^3 - \frac{d}{dx}x^2 - \frac{d}{dx}x - \frac{d}{dx}15}{\frac{d}{dx}x^3 - 2\frac{d}{dx}x^2 - \frac{d}{dx}9} \right|$

$= \left| \lim_{x \to 3} \frac{3x^2 - 2x - 1 - 0}{3x^2 - 4x - 0} \right| = \left| \lim_{x \to 3} \frac{3x^2 - 2x - 1}{3x^2 - 4x} \right| = \left| \frac{3 \cdot 3^2 - 2 \cdot 3 - 1}{3 \cdot 3^2 - 4 \cdot 3} \right| = \left| \frac{27 - 6 - 1}{27 - 12} \right| = \left| \frac{27 - 7}{15} \right| = \left| \frac{20}{15} \right| = \boxed{\frac{4}{3}}$

f. $\left| \lim_{x \to 0} \frac{x - \sin x}{x^3} \right| = \left| \frac{0 - \sin 0}{0^3} \right| = \left| \frac{0 - 0}{0} \right| = \left| \frac{0}{0} \right|$ Apply the L'Hopital's rule $\left| \lim_{x \to 0} \frac{x - \sin x}{x^3} \right|$

$= \left| \lim_{x \to 0} \frac{\frac{d}{dx}\left(x - \sin x\right)}{\frac{d}{dx}x^3} \right| = \left| \lim_{x \to 0} \frac{\frac{d}{dx}x - \frac{d}{dx}\sin x}{\frac{d}{dx}x^3} \right| = \left| \lim_{x \to 0} \frac{1 - \cos x}{3x^2} \right| = \left| \frac{1 - \cos 0}{3 \cdot 0^2} \right| = \left| \frac{1-1}{0} \right| = \left| \frac{0}{0} \right|$

Apply the L'Hopital's rule again $\left| \lim_{x \to 0} \frac{1 - \cos x}{3x^2} \right| = \left| \lim_{x \to 0} \frac{\frac{d}{dx}\left(1 - \cos x\right)}{\frac{d}{dx}3x^2} \right| = \left| \lim_{x \to 0} \frac{\frac{d}{dx}1 - \frac{d}{dx}\cos x}{\frac{d}{dx}3x^2} \right|$

$= \left| \lim_{x \to 0} \frac{0 + \sin x}{6x} \right| = \left| \lim_{x \to 0} \frac{\sin x}{6x} \right| = \left| \frac{\sin 0}{6 \cdot 0} \right| = \left| \frac{0}{0} \right|$ Apply the L'Hopital's rule again $\left| \lim_{x \to 0} \frac{\sin x}{6x} \right|$

$= \left| \lim_{x \to 0} \frac{\frac{d}{dx}\sin x}{\frac{d}{dx}6x} \right| = \left| \lim_{x \to 0} \frac{\cos x}{6} \right| = \left| \frac{\cos 0}{6} \right| = \boxed{\frac{1}{6}}$

g. $\left| \lim_{x \to \frac{\pi}{2}} \frac{2x - \pi}{\cos x} \right| = \left| \frac{2 \cdot \frac{\pi}{2} - \pi}{\cos \frac{\pi}{2}} \right| = \left| \frac{\pi - \pi}{0} \right| = \left| \frac{0}{0} \right|$ Apply the rule $\left| \lim_{x \to \frac{\pi}{2}} \frac{2x - \pi}{\cos x} \right| = \left| \lim_{x \to \frac{\pi}{2}} \frac{\frac{d}{dx}\left(2x - \pi\right)}{\frac{d}{dx}\cos x} \right|$

$= \left| \lim_{x \to \frac{\pi}{2}} \frac{\frac{d}{dx}2x - \frac{d}{dx}\pi}{\frac{d}{dx}\cos x} \right| = \left| \lim_{x \to \frac{\pi}{2}} \frac{2 - 0}{-\sin x} \right| = \left| \frac{2}{-\sin \frac{\pi}{2}} \right| = \left| \frac{2}{-1} \right| = \boxed{-2}$

h. $\left| \lim_{t \to 1} \frac{t^3 - 1}{4t^3 - t - 3} \right| = \left| \frac{1^3 - 1}{4 \cdot 1^3 - 1 - 3} \right| = \left| \frac{1-1}{4-1-3} \right| = \left| \frac{1-1}{4-4} \right| = \left| \frac{0}{0} \right|$ Apply the L'Hopital's rule

$$\boxed{\lim_{t \to 1} \frac{t^3-1}{4t^3-t-3}} = \boxed{\lim_{t \to 1} \frac{\frac{d}{dt}\left(t^3-1\right)}{\frac{d}{dt}\left(4t^3-t-3\right)}} = \boxed{\lim_{t \to 1} \frac{\frac{d}{dt}t^3 - \frac{d}{dt}1}{\frac{d}{dt}4t^3 - \frac{d}{dt}t - \frac{d}{dt}3}} = \boxed{\lim_{t \to 1} \frac{3t^2-0}{12t^2-1-0}}$$

$$= \boxed{\lim_{t \to 1} \frac{3t^2}{12t^2-1}} = \boxed{\frac{3 \cdot 1^2}{12 \cdot 1^2 -1}} = \boxed{\frac{3}{12-1}} = \boxed{\mathbf{\frac{3}{11}}}$$

i. $\boxed{\lim_{x \to 0} \frac{\sin 3x}{x}} = \boxed{\frac{\sin 0}{0}} = \boxed{\frac{0}{0}}$ Apply the L'Hopital's rule $\boxed{\lim_{x \to 0} \frac{\sin 3x}{x}} = \boxed{\lim_{x \to 0} \frac{\frac{d}{dx}\sin 3x}{\frac{d}{dx}x}}$

$$= \boxed{\lim_{x \to 0} \frac{\cos 3x \cdot \frac{d}{dx}3x}{1}} = \boxed{\lim_{x \to 0} \frac{\cos 3x \cdot 3}{1}} = \boxed{\lim_{x \to 0} 3\cos 3x} = \boxed{3\lim_{x \to 0}\cos 3x} = \boxed{3\cos\left(3 \cdot 0\right)}$$

$$= \boxed{3\cos 0} = \boxed{3 \cdot 1} = \boxed{\mathbf{3}}$$

Example 3.6-2: Evaluate the limit of the following functions:

a. $\lim_{\theta \to \frac{\pi}{2}} \frac{\sin\theta}{\pi-\theta}$

b. $\lim_{x \to 0} \frac{x}{x+\sqrt{x}}$

c. $\lim_{t \to 5} \frac{t-5}{t^2-5}$

d. $\lim_{x \to 0} \frac{x}{e^x-1}$

e. $\lim_{x \to 0}\left(\csc x - \cot x\right)$

f. $\lim_{x \to 0} \frac{e^{2x}-1}{\tan 3x}$

g. $\lim_{x \to 0} \frac{\tan 5x}{\tan 7x}$

h. $\lim_{x \to 0} \frac{\sin x^2}{x}$

i. $\lim_{x \to 0} \frac{x\left(\cos x -1\right)}{\sin x - x}$

Solutions:

a. $\boxed{\lim_{\theta \to \frac{\pi}{2}} \frac{\sin\theta}{\pi-\theta}} = \boxed{\frac{\sin\frac{\pi}{2}}{\pi-\frac{\pi}{2}}} = \boxed{\frac{1}{\frac{\pi}{2}}} = \boxed{\mathbf{\frac{2}{\pi}}}$

b. $\boxed{\lim_{x \to 0} \frac{x}{x+\sqrt{x}}} = \boxed{\frac{0}{0+\sqrt{0}}} = \boxed{\frac{0}{0}}$ Apply the L'Hopital's rule $\boxed{\lim_{x \to 0} \frac{x}{x+\sqrt{x}}} = \boxed{\lim_{x \to 0} \frac{\frac{d}{dx}x}{\frac{d}{dx}\left(x+\sqrt{x}\right)}}$

$$= \boxed{\lim_{x \to 0} \frac{\frac{d}{dx}x}{\frac{d}{dx}x + \frac{d}{dx}\sqrt{x}}} = \boxed{\lim_{x \to 0} \frac{1}{1+\frac{1}{2\sqrt{x}}}} = \boxed{\frac{1}{1+\frac{1}{2 \cdot \sqrt{0}}}} = \boxed{\frac{1}{1+\frac{1}{0}}} = \boxed{\frac{1}{1+\frac{1}{0}}} = \boxed{\frac{1}{1+\infty}} = \boxed{\frac{1}{\infty}} = \boxed{\mathbf{0}}$$

c. $\boxed{\lim_{t \to 5} \frac{t-5}{t^2-5}} = \boxed{\frac{5-5}{5^2-5}} = \boxed{\frac{0}{25-5}} = \boxed{\frac{0}{20}} = \boxed{\mathbf{0}}$

d. $\boxed{\lim_{x \to 0} \frac{x}{e^x-1}} = \boxed{\frac{0}{e^0-1}} = \boxed{\frac{0}{1-1}} = \boxed{\frac{0}{0}}$ Apply the L'Hopital's rule $\boxed{\lim_{x \to 0} \frac{x}{e^x-1}} = \boxed{\lim_{x \to 0} \frac{\frac{d}{dx}x}{\frac{d}{dx}\left(e^x-1\right)}}$

$$= \lim_{x \to 0} \frac{\frac{d}{dx} x}{\frac{d}{dx} e^x - \frac{d}{dx} 1} = \lim_{x \to 0} \frac{1}{e^x - 0} = \lim_{x \to 0} \frac{1}{e^x} = \frac{1}{e^0} = \frac{1}{1} = \boxed{1}$$

e. $\lim_{x \to 0} (\csc x - \cot x) = \lim_{x \to 0} \left(\frac{1}{\sin x} - \frac{\cos x}{\sin x} \right) = \lim_{x \to 0} \frac{1 - \cos x}{\sin x} = \frac{1 - \cos 0}{\sin 0} = \frac{1-1}{0} = \frac{0}{0}$ Apply

the L'Hopital's rule to the expression $\lim_{x \to 0} \frac{1 - \cos x}{\sin x} = \lim_{x \to 0} \frac{\frac{d}{dx}(1 - \cos x)}{\frac{d}{dx} \sin x} = \lim_{x \to 0} \frac{\frac{d}{dx} 1 - \frac{d}{dx} \cos x}{\frac{d}{dx} \sin x}$

$$= \lim_{x \to 0} \frac{0 + \sin x}{\cos x} = \lim_{x \to 0} \frac{\sin x}{\cos x} = \lim_{x \to 0} \tan x = \tan 0 = \boxed{0}$$

f. $\lim_{x \to 0} \frac{e^{2x} - 1}{\tan 3x} = \frac{e^{2 \cdot 0} - 1}{\tan 3 \cdot 0} = \frac{e^0 - 1}{\tan 0} = \frac{1-1}{0} = \frac{0}{0}$ Apply the L'Hopital's rule $\lim_{x \to 0} \frac{e^{2x} - 1}{\tan 3x}$

$$= \lim_{x \to 0} \frac{\frac{d}{dx}\left(e^{2x} - 1\right)}{\frac{d}{dx} \tan 3x} = \lim_{x \to 0} \frac{\frac{d}{dx} e^{2x} - \frac{d}{dx} 1}{\frac{d}{dx} \tan 3x} = \lim_{x \to 0} \frac{2e^{2x} - 0}{\sec^2 3x \cdot \frac{d}{dx} 3x} = \lim_{x \to 0} \frac{2e^{2x}}{\sec^2 3x \cdot 3}$$

$$= \lim_{x \to 0} \frac{2e^{2x}}{3 \cdot \frac{1}{\cos^2 3x}} = \lim_{x \to 0} \frac{2e^{2x}}{\frac{3}{\cos^2 3x}} = \lim_{x \to 0} \frac{2e^{2x} \cos^2 3x}{3} = \frac{2e^{2 \cdot 0} \cdot \cos^2 3 \cdot 0}{3} = \frac{2e^0 \cdot \cos^2 0}{3} = \boxed{\frac{2}{3}}$$

g. $\lim_{x \to 0} \frac{\tan 5x}{\tan 7x} = \frac{\tan 5 \cdot 0}{\tan 7 \cdot 0} = \frac{\tan 0}{\tan 0} = \frac{0}{0}$ Apply the L'Hopital's rule $\lim_{x \to 0} \frac{\tan 5x}{\tan 7x}$

$$= \lim_{x \to 0} \frac{\frac{d}{dx} \tan 5x}{\frac{d}{dx} \tan 7x} = \lim_{x \to 0} \frac{\sec^2 5x \cdot \frac{d}{dx} 5x}{\sec^2 7x \cdot \frac{d}{dx} 7x} = \lim_{x \to 0} \frac{\sec^2 5x \cdot 5}{\sec^2 7x \cdot 7} = \lim_{x \to 0} \frac{5 \sec^2 5x}{7 \sec^2 7x} = \lim_{x \to 0} \frac{\frac{5}{\cos^2 5x}}{\frac{7}{\cos^2 7x}}$$

$$= \lim_{x \to 0} \frac{5 \cdot \cos^2 7x}{7 \cdot \cos^2 5x} = \frac{5 \cdot \cos^2 7 \cdot 0}{7 \cdot \cos^2 5 \cdot 0} = \frac{5 \cdot \cos^2 0}{7 \cdot \cos^2 0} = \frac{5 \cdot (\cos 0)^2}{7 \cdot (\cos 0)^2} = \frac{5 \cdot 1^2}{7 \cdot 1^2} = \frac{5 \cdot 1}{7 \cdot 1} = \boxed{\frac{5}{7}}$$

h. $\lim_{x \to 0} \frac{\sin x^2}{x} = \frac{\sin 0^2}{0} = \frac{\sin 0}{0} = \frac{0}{0}$ Apply the L'Hopital's rule $\lim_{x \to 0} \frac{\sin x^2}{x} = \lim_{x \to 0} \frac{\frac{d}{dx} \sin x^2}{\frac{d}{dx} x}$

$$= \lim_{x \to 0} \frac{\cos x^2 \cdot \frac{d}{dx} x^2}{\frac{d}{dx} x} = \lim_{x \to 0} \frac{\cos x^2 \cdot 2x}{1} = \frac{\cos 0^2 \cdot 2 \cdot 0}{1} = \frac{1 \cdot 2 \cdot 0}{1} = \frac{0}{1} = \boxed{0}$$

i. $\lim_{x \to 0} \frac{x(\cos x - 1)}{\sin x - x} = \lim_{x \to 0} \frac{x \cos x - x}{\sin x - x} = \frac{0 \cdot \cos 0 - 0}{\sin 0 - 0} = \frac{0 \cdot 1 - 0}{0 - 0} = \frac{0 - 0}{0 - 0} = \frac{0}{0}$ Apply the L'Hopital's rule $\left| \lim_{x \to 0} \dfrac{\frac{d}{dx}(x\cos x - x)}{\frac{d}{dx}(\sin x - x)} \right| = \left| \lim_{x \to 0} \dfrac{\left(\cos x \cdot \frac{d}{dx}x + x \cdot \frac{d}{dx}\cos x\right) - \frac{d}{dx}x}{\frac{d}{dx}\sin x - \frac{d}{dx}x} \right| = \left| \lim_{x \to 0} \dfrac{(\cos x \cdot 1 + x \cdot -\sin x) - 1}{\cos x - 1} \right|$

$= \left| \lim_{x \to 0} \dfrac{(\cos x - x\sin x) - 1}{\cos x - 1} \right| = \left| \dfrac{(\cos 0 - 0 \cdot \sin 0) - 1}{\cos 0 - 1} \right| = \left| \dfrac{(1 - 0 \cdot 0) - 1}{1 - 1} \right| = \left| \dfrac{1-1}{1-1} \right| = \left| \dfrac{0}{0} \right|$ Apply the

L'Hopital's rule again to the expression $\left| \lim_{x \to 0} \dfrac{(\cos x - x\sin x) - 1}{\cos x - 1} \right| = \left| \lim_{x \to 0} \dfrac{\frac{d}{dx}[(\cos x - x\sin x) - 1]}{\frac{d}{dx}(\cos x - 1)} \right|$

$= \left| \lim_{x \to 0} \dfrac{\frac{d}{dx}\cos x - \frac{d}{dx}x\sin x - \frac{d}{dx}1}{\frac{d}{dx}\cos x - \frac{d}{dx}1} \right| = \left| \lim_{x \to 0} \dfrac{\frac{d}{dx}\cos x - \frac{d}{dx}x\sin x - 0}{\frac{d}{dx}\cos x - 0} \right| = \left| \lim_{x \to 0} \dfrac{\frac{d}{dx}\cos x - \frac{d}{dx}x\sin x}{\frac{d}{dx}\cos x} \right|$

$= \left| \lim_{x \to 0} \dfrac{-\sin x - \left(\sin x \cdot \frac{d}{dx}x + x \cdot \frac{d}{dx}\sin x\right)}{-\sin x} \right| = \left| \lim_{x \to 0} \dfrac{-\sin x - (\sin x + x\cos x)}{-\sin x} \right| = \left| \lim_{x \to 0} \dfrac{-2\sin x - x\cos x}{-\sin x} \right|$

$= \left| \lim_{x \to 0} \dfrac{2\sin x + x\cos x}{\sin x} \right| = \left| \dfrac{2 \cdot 0 + 0 \cdot 1}{0} \right| = \left| \dfrac{0}{0} \right|$ Apply the L'Hopital's rule again to the expression

$\left| \lim_{x \to 0} \dfrac{2\sin x + x\cos x}{\sin x} \right| = \left| \lim_{x \to 0} \dfrac{\frac{d}{dx}(2\sin x + x\cos x)}{\frac{d}{dx}\sin x} \right| = \left| \lim_{x \to 0} \dfrac{2\cos x + \cos x \cdot \frac{d}{dx}x + x \cdot \frac{d}{dx}\cos x}{\cos x} \right|$

$= \left| \lim_{x \to 0} \dfrac{2\cos x + \cos x \cdot 1 - x \cdot \sin x}{\cos x} \right| = \left| \lim_{x \to 0} \dfrac{2\cos x + \cos x - x\sin x}{\cos x} \right| = \left| \lim_{x \to 0} \dfrac{3\cos x - x\sin x}{\cos x} \right|$

$\left| \dfrac{3\cos 0 - 0 \cdot \sin 0}{\cos 0} \right| = \left| \dfrac{3 \cdot 1 - 0 \cdot 0}{1} \right| = \left| \dfrac{3 - 0}{1} \right| = \left| \dfrac{3}{1} \right| = \boxed{3}$

Example 3.6-3: Evaluate the limit of the following functions:

a. $\lim_{x \to 0} \dfrac{e^x - e^{-x} - 4x}{\sin 2x - 3x}$

b. $\lim_{x \to 0} \dfrac{\sin x}{\sqrt{x}}$

c. $\lim_{x \to 0} \dfrac{e^x - 1}{x^3}$

d. $\lim_{x \to 0} \dfrac{x + \sin 3x}{x - \sin 3x}$

e. $\lim_{x \to 1} \dfrac{x^3 - 2x^2 + 3x - 2}{x^3 - 2x + 1}$

f. $\lim_{x \to +\infty} \dfrac{\ln x}{2x}$

g. $\lim_{x \to +\infty} \dfrac{3\ln x}{2\sqrt{x}}$

h. $\lim_{x \to 0} \dfrac{\sqrt{2}xe^x}{1 - e^x}$

i. $\lim_{x \to \infty} \dfrac{\sqrt{5 + x^2}}{x^2}$

Solutions:

a. $\left| \lim_{x \to 0} \dfrac{e^x - e^{-x} - 4x}{\sin 2x - 3x} \right| = \left| \lim_{x \to 0} \dfrac{e^x - \frac{1}{e^x} - 4x}{\sin 2x - 3x} \right| = \left| \dfrac{e^0 - \frac{1}{e^0} - 4 \cdot 0}{\sin 2 \cdot 0 - 3 \cdot 0} \right| = \left| \dfrac{1 - \frac{1}{1} - 0}{0 - 0} \right| = \left| \dfrac{1-1}{0} \right| = \left| \dfrac{0}{0} \right|$ Apply the

L'Hopital's rule $\left| \lim_{x \to 0} \dfrac{e^x - e^{-x} - 4x}{\sin 2x - 3x} \right| = \left| \lim_{x \to 0} \dfrac{\frac{d}{dx}\left(e^x - e^{-x} - 4x\right)}{\frac{d}{dx}\left(\sin 2x - 3x\right)} \right| = \left| \lim_{x \to 0} \dfrac{\frac{d}{dx}e^x - \frac{d}{dx}e^{-x} - \frac{d}{dx}4x}{\frac{d}{dx}\sin 2x - \frac{d}{dx}3x} \right|$

$= \left| \lim_{x \to 0} \dfrac{e^x - e^{-x} \cdot \frac{d}{dx}(-x) - 4}{\cos 2x \cdot \frac{d}{dx}2x - 3} \right| = \left| \lim_{x \to 0} \dfrac{e^x - \left(e^{-x} \cdot -1\right) - 4}{\cos 2x \cdot 2 - 3} \right| = \left| \lim_{x \to 0} \dfrac{e^x + \frac{1}{e^x} - 4}{2\cos 2x - 3} \right| = \left| \dfrac{e^0 + \frac{1}{e^0} - 4}{2\cos(2 \cdot 0) - 3} \right|$

$= \left| \dfrac{1 + 1 - 4}{2\cos 0 - 3} \right| = \left| \dfrac{1 + 1 - 4}{(2 \cdot 1) - 3} \right| = \left| \dfrac{-2}{-1} \right| = \boxed{2}$

b. $\left| \lim_{x \to 0} \dfrac{\sin x}{\sqrt{x}} \right| = \left| \dfrac{\sin 0}{\sqrt{0}} \right| = \left| \dfrac{0}{0} \right|$ Apply the L'Hopital's rule $\left| \lim_{x \to 0} \dfrac{\sin x}{\sqrt{x}} \right| = \left| \lim_{x \to 0} \dfrac{\frac{d}{dx}\sin x}{\frac{d}{dx}\sqrt{x}} \right|$

$= \left| \lim_{x \to 0} \dfrac{\cos x}{\frac{1}{2\sqrt{x}}} \right| = \left| \lim_{x \to 0} \dfrac{\frac{\cos x}{1}}{\frac{1}{2\sqrt{x}}} \right| = \left| \lim_{x \to 0} \dfrac{\cos x \cdot 2\sqrt{x}}{1 \cdot 1} \right| = \left| \lim_{x \to 0} \dfrac{2\sqrt{x}\cos x}{1} \right| = \left| \lim_{x \to 0} 2\sqrt{x}\cos x \right|$

$= \left| 2 \cdot \sqrt{0} \cdot \cos 0 \right| = \left| 0 \cdot \cos 0 \right| = \left| 0 \cdot 1 \right| = \boxed{0}$

c. $\left| \lim_{x \to 0} \dfrac{e^x - 1}{x^3} \right| = \left| \dfrac{e^0 - 1}{0^3} \right| = \left| \dfrac{1-1}{0} \right| = \left| \dfrac{0}{0} \right|$ Apply the L'Hopital's rule $\left| \lim_{x \to 0} \dfrac{e^x - 1}{x^3} \right| = \left| \lim_{x \to 0} \dfrac{\frac{d}{dx}\left(e^x - 1\right)}{\frac{d}{dx}x^3} \right|$

$= \left| \lim_{x \to 0} \dfrac{\frac{d}{dx}e^x - \frac{d}{dx}1}{3x^2} \right| = \left| \lim_{x \to 0} \dfrac{e^x - 0}{3x^2} \right| = \left| \dfrac{e^0}{3 \cdot 0^2} \right| = \left| \dfrac{1}{0} \right| = \boxed{\infty}$

d. $\left| \lim_{x \to 0} \dfrac{x + \sin 3x}{x - \sin 3x} \right| = \left| \dfrac{0 + \sin(3 \cdot 0)}{0 - \sin(3 \cdot 0)} \right| = \left| \dfrac{0+0}{0-0} \right| = \left| \dfrac{0}{0} \right|$ Apply the L'Hopital's rule $\left| \lim_{x \to 0} \dfrac{x + \sin 3x}{x - \sin 3x} \right|$

$= \left| \lim_{x \to 0} \dfrac{\frac{d}{dx}(x + \sin 3x)}{\frac{d}{dx}(x - \sin 3x)} \right| = \left| \lim_{x \to 0} \dfrac{\frac{d}{dx}x + \frac{d}{dx}\sin 3x}{\frac{d}{dx}x - \frac{d}{dx}\sin 3x} \right| = \left| \lim_{x \to 0} \dfrac{1 + \cos 3x \cdot \frac{d}{dx}3x}{1 - \cos 3x \cdot \frac{d}{dx}3x} \right| = \left| \lim_{x \to 0} \dfrac{1 + \cos 3x \cdot 3}{1 - \cos 3x \cdot 3} \right|$

$= \left| \lim_{x \to 0} \dfrac{1 + 3\cos 3x}{1 - 3\cos 3x} \right| = \left| \dfrac{1 + 3\cos(3 \cdot 0)}{1 - 3\cos(3 \cdot 0)} \right| = \left| \dfrac{1 + 3\cos 0}{1 - 3\cos 0} \right| = \left| \dfrac{1 + (3 \cdot 1)}{1 - (3 \cdot 1)} \right| = \left| \dfrac{1+3}{1-3} \right| = \left| \dfrac{4}{-2} \right| = \boxed{-2}$

e. $\lim_{x \to 1} \dfrac{x^3 - 2x^2 + 3x - 2}{x^3 - 2x + 1} = \dfrac{1^3 - 2 \cdot 1^2 + 3 \cdot 1 - 2}{1^3 - 2 \cdot 1 + 1} = \dfrac{1 - 2 + 3 - 2}{1 - 2 + 1} = \dfrac{4 - 4}{2 - 2} = \dfrac{0}{0}$ Apply the L'Hopital's rule

$\lim_{x \to 1} \dfrac{x^3 - 2x^2 + 3x - 2}{x^3 - 2x + 1} = \lim_{x \to 1} \dfrac{\frac{d}{dx}\left(x^3 - 2x^2 + 3x - 2\right)}{\frac{d}{dx}\left(x^3 - 2x + 1\right)} = \lim_{x \to 1} \dfrac{\frac{d}{dx}x^3 - 2\frac{d}{dx}x^2 + 3\frac{d}{dx}x - \frac{d}{dx}2}{\frac{d}{dx}x^3 - 2\frac{d}{dx}x + \frac{d}{dx}1}$

$= \lim_{x \to 1} \dfrac{3x^2 - 2 \cdot 2x + 3 \cdot 1 - 0}{3x^2 - 2 \cdot 1 + 0} = \lim_{x \to 1} \dfrac{3x^2 - 4x + 3}{3x^2 - 2} = \dfrac{3 \cdot 1^2 - 4 \cdot 1 + 3}{3 \cdot 1^2 - 2} = \dfrac{3 - 4 + 3}{3 - 2} = \dfrac{2}{1} = \boxed{2}$

f. $\lim_{x \to +\infty} \dfrac{\ln x}{2x} = \dfrac{\ln \infty}{2 \cdot \infty} = \dfrac{\infty}{\infty}$ Apply the L'Hopital's rule $\lim_{x \to +\infty} \dfrac{\ln x}{2x} = \lim_{x \to +\infty} \dfrac{\frac{d}{dx}\ln x}{\frac{d}{dx}2x}$

$= \lim_{x \to +\infty} \dfrac{\frac{1}{x}}{2} = \lim_{x \to +\infty} \dfrac{1}{2x} = \dfrac{1}{2 \cdot \infty} = \dfrac{1}{\infty} = \boxed{0}$

g. $\lim_{x \to +\infty} \dfrac{3 \ln x}{2\sqrt{x}} = \dfrac{3 \ln \infty}{2 \cdot \sqrt{\infty}} = \dfrac{\infty}{\infty}$ Apply the L'Hopital's rule $\lim_{x \to +\infty} \dfrac{3 \ln x}{2\sqrt{x}} = \lim_{x \to +\infty} \dfrac{\frac{d}{dx}3 \ln x}{\frac{d}{dx}2\sqrt{x}}$

$= \lim_{x \to +\infty} \dfrac{3 \cdot \frac{1}{x}}{2 \cdot \frac{1}{2\sqrt{x}}} = \lim_{x \to +\infty} \dfrac{\frac{3}{x}}{\frac{1}{\sqrt{x}}} = \lim_{x \to +\infty} \dfrac{3\sqrt{x}}{x} = \dfrac{3 \cdot \sqrt{\infty}}{\infty} = \dfrac{\infty}{\infty}$ Apply the L'Hopital's rule

again $\lim_{x \to +\infty} \dfrac{3\sqrt{x}}{x} = \lim_{x \to +\infty} \dfrac{\frac{d}{dx}3\sqrt{x}}{\frac{d}{dx}x} = \lim_{x \to +\infty} \dfrac{\frac{3}{2\sqrt{x}}}{1} = \lim_{x \to +\infty} \dfrac{3}{2\sqrt{x}} = \dfrac{3}{2 \cdot \sqrt{\infty}} = \dfrac{3}{\infty} = \boxed{0}$

h. $\lim_{x \to 0} \dfrac{\sqrt{2}xe^x}{1 - e^x} = \dfrac{\sqrt{2} \cdot 0 \cdot e^0}{1 - e^0} = \dfrac{\sqrt{2} \cdot 0 \cdot 1}{1 - 1} = \dfrac{0}{0}$ Apply the L'Hopital's rule $\lim_{x \to 0} \dfrac{\sqrt{2}xe^x}{1 - e^x}$

$= \lim_{x \to 0} \dfrac{\frac{d}{dx}\sqrt{2}\,xe^x}{\frac{d}{dx}\left(1 - e^x\right)} = \lim_{x \to 0} \dfrac{\sqrt{2}\left(e^x \cdot \frac{d}{dx}x + x \cdot \frac{d}{dx}e^x\right)}{\frac{d}{dx}1 - \frac{d}{dx}e^x} = \lim_{x \to 0} \dfrac{\sqrt{2}\left(e^x \cdot 1 + x \cdot e^x\right)}{0 - e^x}$

$\lim_{x \to 0} \dfrac{\sqrt{2}\left(e^x + xe^x\right)}{-e^x} = \dfrac{\sqrt{2}\left(e^0 + 0 \cdot e^0\right)}{-e^0} = \dfrac{\sqrt{2}\left(1 + 0\right)}{-1} = \dfrac{\sqrt{2}}{-1} = \boxed{-\sqrt{2}}$

i. $\lim_{x \to \infty} \sqrt{\dfrac{5 + x^2}{x^2}} = \lim_{x \to \infty} \sqrt{\dfrac{5}{x^2} + 1} = \sqrt{\dfrac{5}{\infty^2} + 1} = \sqrt{\dfrac{5}{\infty} + 1} = \sqrt{0 + 1} = \sqrt{1} = \boxed{1}$

Example 3.6-4: Evaluate the limit of the following functions:

a. $\lim_{x \to 10} \dfrac{e^x - e^{10}}{x - 10}$

b. $\lim_{x \to 5} \dfrac{x^3 - 25x}{x^3 - 125}$

c. $\lim_{x \to \infty} \dfrac{\ln x}{\sqrt{x}}$

d. $\lim_{x \to \infty} \dfrac{x^4 + x^3 + 5}{e^x + 1}$

e. $\lim_{x \to \infty} \dfrac{3x^2 - 1}{e^{2x}}$

f. $\lim_{x \to \infty} \dfrac{2x^2}{5e^x + 2x}$

g. $\lim_{x \to \infty} \dfrac{e^x}{3e^x + 5x}$

h. $\lim_{x \to +\infty} \dfrac{7x + 5\ln x}{x + 2\ln x}$

i. $\lim_{x \to +\infty} \dfrac{2x^2 - \ln x}{3x^2 + 3\ln x}$

j. $\lim_{x \to 0} \dfrac{e^{5x} - e^{-5x}}{5\sin x}$

k. $\lim_{x \to 0} \dfrac{e^x - e^{-x}}{x}$

l. $\lim_{x \to 0} \dfrac{\cos x - 1}{\sin x}$

Solutions:

a. $\boxed{\lim_{x \to 10} \dfrac{e^x - e^{10}}{x - 10}} = \boxed{\dfrac{e^{10} - e^{10}}{10 - 10}} = \boxed{\dfrac{0}{0}}$ Apply the L'Hopital's rule $\boxed{\lim_{x \to 10} \dfrac{e^x - e^{10}}{x - 10}}$

$= \boxed{\lim_{x \to 10} \dfrac{\frac{d}{dx}\left(e^x - e^{10}\right)}{\frac{d}{dx}(x - 10)}} = \boxed{\lim_{x \to 10} \dfrac{\frac{d}{dx}e^x - \frac{d}{dx}e^{10}}{\frac{d}{dx}x - \frac{d}{dx}10}} = \boxed{\lim_{x \to 10} \dfrac{e^x - 0}{1 - 0}} = \boxed{\lim_{x \to 10} e^x} = \boxed{e^{10}}$

b. $\boxed{\lim_{x \to 5} \dfrac{x^3 - 25x}{x^3 - 125}} = \boxed{\dfrac{5^3 - 25 \cdot 5}{5^3 - 125}} = \boxed{\dfrac{125 - 125}{125 - 125}} = \boxed{\dfrac{0}{0}}$ Apply the L'Hopital's rule $\boxed{\lim_{x \to 5} \dfrac{x^3 - 25x}{x^3 - 125}}$

$= \boxed{\lim_{x \to 5} \dfrac{\frac{d}{dx}\left(x^3 - 25x\right)}{\frac{d}{dx}\left(x^3 - 125\right)}} = \boxed{\lim_{x \to 5} \dfrac{\frac{d}{dx}x^3 - \frac{d}{dx}25x}{\frac{d}{dx}x^3 - \frac{d}{dx}125}} = \boxed{\lim_{x \to 5} \dfrac{3x^2 - 25}{3x^2 - 0}} = \boxed{\dfrac{3 \cdot 5^2 - 25}{3 \cdot 5^2}} = \boxed{\dfrac{75 - 25}{75}} = \boxed{\dfrac{50}{75}} = \boxed{\dfrac{2}{3}}$

c. $\boxed{\lim_{x \to \infty} \dfrac{\ln x}{\sqrt{x}}} = \boxed{\dfrac{\ln \infty}{\sqrt{\infty}}} = \boxed{\dfrac{\infty}{\infty}}$ Apply the L'Hopital's rule $\boxed{\lim_{x \to \infty} \dfrac{\ln x}{\sqrt{x}}} = \boxed{\lim_{x \to \infty} \dfrac{\frac{d}{dx}\ln x}{\frac{d}{dx}\sqrt{x}}}$

$= \boxed{\lim_{x \to \infty} \dfrac{\frac{1}{x}}{\frac{1}{2\sqrt{x}}}} = \boxed{\lim_{x \to \infty} \dfrac{2\sqrt{x}}{x}} = \boxed{\lim_{x \to \infty} \dfrac{2}{\sqrt{x}}} = \boxed{\dfrac{2}{\sqrt{\infty}}} = \boxed{\dfrac{2}{\infty}} = \boxed{0}$

d. $\boxed{\lim_{x \to \infty} \dfrac{x^4 + x^3 + 5}{e^x + 1}} = \boxed{\dfrac{\infty^4 + \infty^3 + 5}{e^\infty + 1}} = \boxed{\dfrac{\infty}{\infty}}$ Apply the L'Hopital's rule $\boxed{\lim_{x \to \infty} \dfrac{x^4 + x^3 + 5}{e^x + 1}}$

$= \boxed{\lim_{x \to \infty} \dfrac{\frac{d}{dx}\left(x^4 + x^3 + 5\right)}{\frac{d}{dx}\left(e^x + 1\right)}} = \boxed{\lim_{x \to \infty} \dfrac{\frac{d}{dx}x^4 + \frac{d}{dx}x^3 + \frac{d}{dx}5}{\frac{d}{dx}e^x + \frac{d}{dx}1}} = \boxed{\lim_{x \to \infty} \dfrac{4x^3 + 3x^2 + 0}{e^x + 0}} = \boxed{\lim_{x \to \infty} \dfrac{4x^3 + 3x^2}{e^x}}$

$= \left| \dfrac{4 \cdot \infty^3 + 3 \cdot \infty^2}{e^\infty} \right| = \left| \dfrac{\infty}{\infty} \right|$ Apply the L'Hopital's rule again to the expression $\left| \lim_{x \to \infty} \dfrac{4x^3 + 3x^2}{e^x} \right|$

$= \left| \lim_{x \to \infty} \dfrac{\frac{d}{dx}\left(4x^3 + 3x^2\right)}{\frac{d}{dx} e^x} \right| = \left| \lim_{x \to \infty} \dfrac{\frac{d}{dx} 4x^3 + \frac{d}{dx} 3x^2}{\frac{d}{dx} e^x} \right| = \left| \lim_{x \to \infty} \dfrac{12x^2 + 6x}{e^x} \right| = \left| \dfrac{12 \cdot \infty^2 + 6 \cdot \infty}{e^\infty} \right| = \left| \dfrac{\infty}{\infty} \right|$

Apply the L'Hopital's rule again to the expression $\left| \lim_{x \to \infty} \dfrac{12x^2 + 6x}{e^x} \right| = \left| \lim_{x \to \infty} \dfrac{\frac{d}{dx}\left(12x^2 + 6x\right)}{\frac{d}{dx} e^x} \right|$

$= \left| \lim_{x \to \infty} \dfrac{\frac{d}{dx} 12x^2 + \frac{d}{dx} 6x}{\frac{d}{dx} e^x} \right| = \left| \lim_{x \to \infty} \dfrac{24x + 6}{e^x} \right| = \left| \dfrac{24 \cdot \infty + 6}{e^\infty} \right| = \left| \dfrac{\infty}{\infty} \right|$ Apply the L'Hopital's rule again

to the expression $\left| \lim_{x \to \infty} \dfrac{24x + 6}{e^x} \right| = \left| \lim_{x \to \infty} \dfrac{\frac{d}{dx}(24x + 6)}{\frac{d}{dx} e^x} \right| = \left| \lim_{x \to \infty} \dfrac{\frac{d}{dx} 24x + \frac{d}{dx} 6}{\frac{d}{dx} e^x} \right| = \left| \lim_{x \to \infty} \dfrac{24 + 0}{e^x} \right|$

$= \left| \lim_{x \to \infty} \dfrac{24}{e^x} \right| = \left| \dfrac{24}{e^\infty} \right| = \left| \dfrac{24}{\infty} \right| = \boxed{0}$

e. $\left| \lim_{x \to \infty} \dfrac{3x^2 - 1}{e^{2x}} \right| = \left| \dfrac{3 \cdot \infty^2 - 1}{e^{2 \cdot \infty}} \right| = \left| \dfrac{\infty}{\infty} \right|$ Apply the L'Hopital's rule $\left| \lim_{x \to \infty} \dfrac{3x^2 - 1}{e^{2x}} \right| = \left| \lim_{x \to \infty} \dfrac{\frac{d}{dx}\left(3x^2 - 1\right)}{\frac{d}{dx} e^{2x}} \right|$

$= \left| \lim_{x \to \infty} \dfrac{\frac{d}{dx} 3x^2 - \frac{d}{dx} 1}{\frac{d}{dx} e^{2x}} \right| = \left| \lim_{x \to \infty} \dfrac{6x - 0}{2e^{2x}} \right| = \left| \lim_{x \to \infty} \dfrac{6x}{2e^{2x}} \right| = \left| \dfrac{6 \cdot \infty}{2 \cdot e^{2 \cdot \infty}} \right| = \left| \dfrac{\infty}{\infty} \right|$ Apply the L'Hopital's

rule again to the expression $\left| \lim_{x \to \infty} \dfrac{6x}{2e^{2x}} \right| = \left| \lim_{x \to \infty} \dfrac{\frac{d}{dx} 6x}{\frac{d}{dx} 2e^{2x}} \right| = \left| \lim_{x \to \infty} \dfrac{6}{4e^{2x}} \right| = \left| \dfrac{6}{4 \cdot e^{2 \cdot \infty}} \right| = \left| \dfrac{6}{\infty} \right| = \boxed{0}$

f. $\left| \lim_{x \to \infty} \dfrac{2x^2}{5e^x + 2x} \right| = \left| \dfrac{2 \cdot \infty^2}{5 \cdot e^\infty + 2 \cdot \infty} \right| = \left| \dfrac{\infty}{\infty} \right|$ Apply the L'Hopital's rule $\left| \lim_{x \to \infty} \dfrac{2x^2}{5e^x + 2x} \right|$

$= \left| \lim_{x \to \infty} \dfrac{\frac{d}{dx} 2x^2}{\frac{d}{dx}\left(5e^x + 2x\right)} \right| = \left| \lim_{x \to \infty} \dfrac{\frac{d}{dx} 2x^2}{\frac{d}{dx} 5e^x + \frac{d}{dx} 2x} \right| = \left| \lim_{x \to \infty} \dfrac{4x}{5e^x + 2} \right| = \left| \dfrac{4 \cdot \infty}{5 \cdot e^\infty + 2} \right| = \left| \dfrac{\infty}{\infty} \right|$ Apply the

L'Hopital's rule again to the expression $\left| \lim_{x \to \infty} \dfrac{4x}{5e^x + 2} \right| = \left| \lim_{x \to \infty} \dfrac{\frac{d}{dx} 4x}{\frac{d}{dx}\left(5e^x + 2\right)} \right| = \left| \lim_{x \to \infty} \dfrac{\frac{d}{dx} 4x}{\frac{d}{dx} 5e^x + \frac{d}{dx} 2} \right|$

$$= \boxed{\lim_{x \to \infty} \frac{4}{5e^x + 0}} = \boxed{\lim_{x \to \infty} \frac{4}{5e^x}} = \boxed{\frac{4}{5 \cdot e^\infty}} = \boxed{\frac{4}{\infty}} = \boxed{0}$$

g. $\boxed{\lim_{x \to \infty} \frac{e^x}{3e^x + 5x}} = \boxed{\frac{e^\infty}{3 \cdot e^\infty + 5 \cdot \infty}} = \boxed{\frac{\infty}{\infty}}$ Apply the L'Hopital's rule $\boxed{\lim_{x \to \infty} \frac{e^x}{3e^x + 5x}}$

$$= \boxed{\lim_{x \to \infty} \frac{\frac{d}{dx} e^x}{\frac{d}{dx}\left(3e^x + 5x\right)}} = \boxed{\lim_{x \to \infty} \frac{\frac{d}{dx} e^x}{\frac{d}{dx} 3e^x + \frac{d}{dx} 5x}} = \boxed{\lim_{x \to \infty} \frac{e^x}{3e^x + 5}} = \boxed{\frac{e^\infty}{3 \cdot e^\infty + 5}} = \boxed{\frac{\infty}{\infty}}$$ Apply the

L'Hopital's rule again to the expression $\boxed{\lim_{x \to \infty} \frac{e^x}{3e^x + 5}} = \boxed{\lim_{x \to \infty} \frac{\frac{d}{dx} e^x}{\frac{d}{dx}\left(3e^x + 5\right)}} = \boxed{\lim_{x \to \infty} \frac{\frac{d}{dx} e^x}{\frac{d}{dx} 3e^x + \frac{d}{dx} 5}}$

$$= \boxed{\lim_{x \to \infty} \frac{e^x}{3e^x + 0}} = \boxed{\lim_{x \to \infty} \frac{e^x}{3e^x}} = \boxed{\lim_{x \to \infty} \frac{1}{3}} = \boxed{\frac{1}{3}}$$

h. $\boxed{\lim_{x \to +\infty} \frac{7x + 5\ln x}{x + 2\ln x}} = \boxed{\lim_{x \to +\infty} \frac{7 \cdot \infty + 5 \cdot \ln \infty}{\infty + 2 \cdot \ln \infty}} = \boxed{\frac{\infty}{\infty}}$ Apply the L'Hopital's rule $\boxed{\lim_{x \to +\infty} \frac{7x + 5\ln x}{x + 2\ln x}}$

$$= \boxed{\lim_{x \to +\infty} \frac{\frac{d}{dx}\left(7x + 5\ln x\right)}{\frac{d}{dx}\left(x + 2\ln x\right)}} = \boxed{\lim_{x \to +\infty} \frac{\frac{d}{dx} 7x + \frac{d}{dx} 5\ln x}{\frac{d}{dx} x + \frac{d}{dx} 2\ln x}} = \boxed{\lim_{x \to +\infty} \frac{7 + \frac{5}{x}}{1 + \frac{2}{x}}} = \boxed{\frac{7 + \frac{5}{\infty}}{1 + \frac{2}{\infty}}} = \boxed{\frac{7 + 0}{1 + 0}} = \boxed{\frac{7}{1}} = \boxed{7}$$

i. $\boxed{\lim_{x \to +\infty} \frac{2x^2 - \ln x}{3x^2 + 3\ln x}} = \boxed{\frac{2 \cdot \infty^2 - \ln \infty}{3 \cdot \infty^2 + 3\ln \infty}} = \boxed{\frac{\infty}{\infty}}$ Apply the L'Hopital's rule $\boxed{\lim_{x \to +\infty} \frac{2x^2 - \ln x}{3x^2 + 3\ln x}}$

$$= \boxed{\lim_{x \to +\infty} \frac{\frac{d}{dx}\left(2x^2 - \ln x\right)}{\frac{d}{dx}\left(3x^2 + 3\ln x\right)}} = \boxed{\lim_{x \to +\infty} \frac{\frac{d}{dx} 2x^2 - \frac{d}{dx}\ln x}{\frac{d}{dx} 3x^2 + \frac{d}{dx} 3\ln x}} = \boxed{\lim_{x \to +\infty} \frac{4x - \frac{1}{x}}{6x + \frac{3}{x}}} = \boxed{\frac{4 \cdot \infty - \frac{1}{\infty}}{6 \cdot \infty + \frac{3}{\infty}}} = \boxed{\frac{\infty - 0}{\infty + 0}}$$

$$= \boxed{\frac{\infty}{\infty}}$$ Apply the L'Hopital's rule again to the expression $\boxed{\lim_{x \to +\infty} \frac{4x - \frac{1}{x}}{6x + \frac{3}{x}}} = \boxed{\lim_{x \to +\infty} \frac{\frac{d}{dx}\left(4x - \frac{1}{x}\right)}{\frac{d}{dx}\left(6x + \frac{3}{x}\right)}}$

$$= \boxed{\lim_{x \to +\infty} \frac{\frac{d}{dx} 4x - \frac{d}{dx} \frac{1}{x}}{\frac{d}{dx} 6x + \frac{d}{dx} \frac{3}{x}}} = \boxed{\lim_{x \to +\infty} \frac{4 + \frac{1}{x^2}}{6 - \frac{3}{x^2}}} = \boxed{\frac{4 + \frac{1}{\infty^2}}{6 - \frac{3}{\infty^2}}} = \boxed{\frac{4 + 0}{6 - 0}} = \boxed{\frac{4}{6}} = \boxed{\frac{2}{3}}$$

j. $\boxed{\lim_{x \to 0} \frac{e^{5x} - e^{-5x}}{5\sin x}} = \boxed{\lim_{x \to 0} \frac{e^{5x} - \frac{1}{e^{5x}}}{5\sin x}} = \boxed{\frac{e^{5 \cdot 0} - \frac{1}{e^{5 \cdot 0}}}{5 \cdot \sin 0}} = \boxed{\frac{e^0 - \frac{1}{e^0}}{5 \cdot 0}} = \boxed{\frac{1 - \frac{1}{1}}{0}} = \boxed{\frac{1 - 1}{0}} = \boxed{\frac{0}{0}}$ Apply the

L'Hopital's rule $\boxed{\lim_{x \to 0} \frac{e^{5x} - e^{-5x}}{5\sin x}} = \boxed{\lim_{x \to 0} \frac{\frac{d}{dx}\left(e^{5x} - e^{-5x}\right)}{\frac{d}{dx} 5\sin x}} = \boxed{\lim_{x \to 0} \frac{\frac{d}{dx} e^{5x} - \frac{d}{dx} e^{-5x}}{\frac{d}{dx} 5\sin x}}$

$$= \boxed{\lim_{x\to 0} \frac{5e^{5x}+5e^{-5x}}{5\cos x}} = \boxed{\lim_{x\to 0} \frac{5e^{5x}+\frac{5}{e^{5x}}}{5\cos x}} = \boxed{\frac{5\cdot e^{5\cdot 0}+\frac{5}{e^{5\cdot 0}}}{5\cdot\cos 0}} = \boxed{\frac{5\cdot e^{0}+\frac{5}{e^{0}}}{5\cdot 1}} = \boxed{\frac{5\cdot 1+\frac{5}{1}}{5}} = \boxed{\frac{5+5}{5}} = \boxed{\frac{10}{5}} = \boxed{2}$$

k. $\boxed{\lim_{x\to 0} \frac{e^{x}-e^{-x}}{x}} = \boxed{\lim_{x\to 0} \frac{e^{x}-\frac{1}{e^{x}}}{x}} = \boxed{\frac{e^{0}-\frac{1}{e^{0}}}{0}} = \boxed{\frac{1-\frac{1}{1}}{0}} = \boxed{\frac{1-1}{0}} = \boxed{\frac{0}{0}}$ Apply the L'Hopital's rule

$$\boxed{\lim_{x\to 0} \frac{e^{x}-e^{-x}}{x}} = \boxed{\lim_{x\to 0} \frac{\frac{d}{dx}\left(e^{x}-e^{-x}\right)}{\frac{d}{dx}x}} = \boxed{\lim_{x\to 0} \frac{\frac{d}{dx}e^{x}-\frac{d}{dx}e^{-x}}{\frac{d}{dx}x}} = \boxed{\lim_{x\to 0} \frac{e^{x}+e^{-x}}{1}}$$

$$= \boxed{\lim_{x\to 0} \frac{e^{x}+\frac{1}{e^{x}}}{1}} = \boxed{\lim_{x\to 0}\left(e^{x}+\frac{1}{e^{x}}\right)} = \boxed{e^{0}+\frac{1}{e^{0}}} = \boxed{1+\frac{1}{1}} = \boxed{1+1} = \boxed{2}$$

l. $\boxed{\lim_{x\to 0} \frac{\cos x-1}{\sin x}} = \boxed{\frac{\cos 0-1}{\sin 0}} = \boxed{\frac{1-1}{0}} = \boxed{\frac{0}{0}}$ Apply the L'Hopital's rule $\boxed{\lim_{x\to 0} \frac{\cos x-1}{\sin x}}$

$$= \boxed{\lim_{x\to 0} \frac{\frac{d}{dx}\left(\cos x-1\right)}{\frac{d}{dx}\sin x}} = \boxed{\lim_{x\to 0} \frac{\frac{d}{dx}\cos x-\frac{d}{dx}1}{\frac{d}{dx}\sin x}} = \boxed{\lim_{x\to 0} \frac{-\sin x-0}{\cos x}} = \boxed{\frac{-\sin 0}{\cos 0}} = \boxed{\frac{0}{1}} = \boxed{0}$$

Example 3.6-5: Evaluate the limit of the following functions:

a. $\lim_{x\to 3} \frac{x^{3}-27}{x^{2}-9}$ b. $\lim_{x\to 2} \frac{x^{4}-16}{2x-4}$ c. $\lim_{x\to\infty} \frac{6x^{2}-5x}{7x^{2}+1}$

d. $\lim_{t\to 0} \frac{\sin t^{2}}{t}$ e. $\lim_{u\to 0} \frac{\sin 10u}{u}$ f. $\lim_{t\to 0} \frac{t-\sin t}{t^{2}}$

g. $\lim_{t\to 0} \frac{t-\sin t}{t^{3}}$ h. $\lim_{x\to +\infty} \frac{\sqrt{1+x}-1}{x^{2}}$ i. $\lim_{x\to 0} \frac{\sin x}{x^{3}}$

j. $\lim_{x\to 0} \frac{3-3\cos x}{x+x^{2}}$ k. $\lim_{x\to\infty} \frac{2+x^{2}}{x^{3}}$ l. $\lim_{x\to 0} \frac{\cos x-1}{\cos 3x-1}$

Solutions:

a. $\boxed{\lim_{x\to 3} \frac{x^{3}-27}{x^{2}-9}} = \boxed{\frac{3^{3}-27}{3^{2}-9}} = \boxed{\frac{27-27}{9-9}} = \boxed{\frac{0}{0}}$ Apply the L'Hopital's rule $\boxed{\lim_{x\to 3} \frac{x^{3}-27}{x^{2}-9}}$

$$= \boxed{\lim_{x\to 3} \frac{\frac{d}{dx}\left(x^{3}-27\right)}{\frac{d}{dx}\left(x^{2}-9\right)}} = \boxed{\lim_{x\to 3} \frac{\frac{d}{dx}x^{3}-\frac{d}{dx}27}{\frac{d}{dx}x^{2}-\frac{d}{dx}9}} = \boxed{\lim_{x\to 3} \frac{3x^{2}-0}{2x-0}} = \boxed{\lim_{x\to 3} \frac{3x^{2}}{2x}} = \boxed{\frac{3\cdot 3^{2}}{2\cdot 3}} = \boxed{\frac{27}{6}} = \boxed{\frac{9}{2}}$$

b. $\boxed{\lim_{x\to 2} \frac{x^{4}-16}{2x-4}} = \boxed{\frac{2^{4}-16}{2\cdot 2-4}} = \boxed{\frac{16-16}{4-4}} = \boxed{\frac{0}{0}}$ Apply the L'Hopital's rule $\boxed{\lim_{x\to 2} \frac{x^{4}-16}{2x-4}}$

$$= \left| \lim_{x \to 2} \frac{\frac{d}{dx}\left(x^4 - 16\right)}{\frac{d}{dx}\left(2x - 4\right)} \right| = \left| \lim_{x \to 2} \frac{\frac{d}{dx}x^4 - \frac{d}{dx}16}{\frac{d}{dx}2x - \frac{d}{dx}4} \right| = \left| \lim_{x \to 2} \frac{4x^3 - 0}{2 - 0} \right| = \left| \lim_{x \to 2} \frac{4x^3}{2} \right| = \left| \frac{4 \cdot 2^3}{2} \right| = \left| \frac{32}{2} \right| = \boxed{16}$$

c. $\left| \lim_{x \to \infty} \frac{6x^2 - 5x}{7x^2 + 1} \right| = \left| \frac{6 \cdot \infty^2 - 5 \cdot \infty}{7 \cdot \infty^2 + 1} \right| = \left| \frac{6 \cdot \infty^2 - 5 \cdot \infty}{7 \cdot \infty^2 + 1} \right| = \left| \frac{\infty}{\infty} \right|$ Apply the L'Hopital's rule $\left| \lim_{x \to \infty} \frac{6x^2 - 5x}{7x^2 + 1} \right|$

$$= \left| \lim_{x \to \infty} \frac{\frac{d}{dx}\left(6x^2 - 5x\right)}{\frac{d}{dx}\left(7x^2 + 1\right)} \right| = \left| \lim_{x \to \infty} \frac{\frac{d}{dx}6x^2 - \frac{d}{dx}5x}{\frac{d}{dx}7x^2 + \frac{d}{dx}1} \right| = \left| \lim_{x \to \infty} \frac{12x - 5}{14x + 0} \right| = \left| \lim_{x \to \infty} \frac{12x - 5}{14x} \right| = \left| \frac{12 \cdot \infty - 5}{14 \cdot \infty} \right|$$

$$= \left| \frac{\infty}{\infty} \right| \text{ Apply the L'Hopital's rule again to the expression } \left| \lim_{x \to \infty} \frac{12x - 5}{14x} \right| = \left| \lim_{x \to \infty} \frac{\frac{d}{dx}\left(12x - 5\right)}{\frac{d}{dx}14x} \right|$$

$$= \left| \lim_{x \to \infty} \frac{\frac{d}{dx}12x - \frac{d}{dx}5}{\frac{d}{dx}14x} \right| = \left| \lim_{x \to \infty} \frac{12 - 0}{14} \right| = \left| \frac{12}{14} \right| = \boxed{\frac{6}{7}}$$

d. $\left| \lim_{t \to 0} \frac{\sin t^2}{t} \right| = \left| \frac{\sin 0^2}{0} \right| = \left| \frac{0}{0} \right|$ Apply the L'Hopital's rule $\left| \lim_{t \to 0} \frac{\sin t^2}{t} \right| = \left| \lim_{t \to 0} \frac{\frac{d}{dt}\sin t^2}{\frac{d}{dt}t} \right|$

$$= \left| \lim_{t \to 0} \frac{\cos t^2 \cdot \frac{d}{dt}t^2}{1} \right| = \left| \lim_{t \to 0} \cos t^2 \cdot 2t \right| = \left| \lim_{t \to 0} 2t \cos t^2 \right| = \left| 2 \cdot 0 \cdot \cos 0^2 \right| = \left| 2 \cdot 0 \cdot 1 \right| = \boxed{0}$$

e. $\left| \lim_{u \to 0} \frac{\sin 10u}{u} \right| = \left| \frac{\sin 10 \cdot 0}{0} \right| = \left| \frac{\sin 0}{0} \right| = \left| \frac{0}{0} \right|$ Apply the L'Hopital's rule $\left| \lim_{u \to 0} \frac{\sin 10u}{u} \right|$

$$= \left| \lim_{u \to 0} \frac{\frac{d}{du}\sin 10u}{\frac{d}{du}u} \right| = \left| \lim_{u \to 0} \frac{\cos 10u \cdot \frac{d}{du}10u}{1} \right| = \left| \lim_{u \to 0} \frac{\cos 10u \cdot 10}{1} \right| = \left| \lim_{u \to 0} 10\cos 10u \right| = \left| 10\cos 10 \cdot 0 \right|$$

$$= \left| 10\cos 0 \right| = \left| 10 \cdot 1 \right| = \boxed{10}$$

f. $\left| \lim_{t \to 0} \frac{t - \sin t}{t^2} \right| = \left| \frac{0 - \sin 0}{0^2} \right| = \left| \frac{0}{0} \right|$ Apply the L'Hopital's rule $\left| \lim_{t \to 0} \frac{t - \sin t}{t^2} \right| = \left| \lim_{t \to 0} \frac{\frac{d}{dt}\left(t - \sin t\right)}{\frac{d}{dt}t^2} \right|$

$$= \left| \lim_{t \to 0} \frac{1 - \cos t}{2t} \right| = \left| \frac{1 - \cos 0}{2 \cdot 0} \right| = \left| \frac{1 - 1}{0} \right| = \left| \frac{0}{0} \right| \text{ Apply the L'Hopital's rule again to the expression}$$

$$\left| \lim_{t \to 0} \frac{1 - \cos t}{2t} \right| = \left| \lim_{t \to 0} \frac{\frac{d}{dt}\left(1 - \cos t\right)}{\frac{d}{dt}2t} \right| = \left| \lim_{t \to 0} \frac{\frac{d}{dt}1 - \frac{d}{dt}\cos t}{\frac{d}{dt}2t} \right| = \left| \lim_{t \to 0} \frac{0 + \sin t}{2} \right| = \left| \frac{\sin 0}{2} \right| = \left| \frac{0}{2} \right| = \boxed{0}$$

g. $\lim_{t \to 0} \dfrac{t - \sin t}{t^3} = \dfrac{0 - \sin 0}{0^3} = \dfrac{0 - 0}{0} = \dfrac{0}{0}$ Apply the L'Hopital's rule $\lim_{t \to 0} \dfrac{t - \sin t}{t^3}$

$= \lim_{t \to 0} \dfrac{\frac{d}{dt}(t - \sin t)}{\frac{d}{dt} t^3} = \lim_{t \to 0} \dfrac{\frac{d}{dt} t - \frac{d}{dt} \sin t}{3t^2} = \lim_{t \to 0} \dfrac{1 - \cos t}{3t^2} = \dfrac{1 - \cos 0}{3 \cdot 0^2} = \dfrac{1 - 1}{0} = \dfrac{0}{0}$ Apply the

rule again to the expression $\lim_{t \to 0} \dfrac{1 - \cos t}{3t^2} = \lim_{t \to 0} \dfrac{\frac{d}{dt}(1 - \cos t)}{\frac{d}{dt} 3t^2} = \lim_{t \to 0} \dfrac{\frac{d}{dt} 1 - \frac{d}{dt} \cos t}{\frac{d}{dt} 3t^2}$

$= \lim_{t \to 0} \dfrac{0 + \sin t}{6t} = \lim_{t \to 0} \dfrac{\sin t}{6t} = \dfrac{\sin 0}{6 \cdot 0} = \dfrac{0}{0}$ Apply the rule again to the expression

$\lim_{t \to 0} \dfrac{\sin t}{6t} = \lim_{t \to 0} \dfrac{\frac{d}{dt} \sin t}{\frac{d}{dt} 6t} = \lim_{t \to 0} \dfrac{\cos t}{6} = \dfrac{\cos 0}{6} = \boxed{\dfrac{1}{6}}$

h. $\lim_{x \to 0} \dfrac{\sqrt{1 + x} - 1}{x^2} = \dfrac{\sqrt{1 + 0} - 1}{0^2} = \dfrac{\sqrt{1} - 1}{0} = \dfrac{1 - 1}{0} = \dfrac{0}{0}$ Apply the L'Hopital's rule $\lim_{x \to 0} \dfrac{\sqrt{1 + x} - 1}{x^2}$

$= \lim_{x \to 0} \dfrac{\frac{d}{dx}(\sqrt{1 + x} - 1)}{\frac{d}{dx} x^2} = \dfrac{\frac{1}{2\sqrt{1+x}} - 0}{2x} = \lim_{x \to 0} \dfrac{\frac{1}{2\sqrt{1+x}}}{2x} = \dfrac{\frac{1}{2\sqrt{1+0}}}{2 \cdot 0} = \dfrac{\frac{1}{2}}{0} = \dfrac{\frac{1}{2}}{\frac{0}{1}} = \dfrac{1 \cdot 1}{2 \cdot 0} = \dfrac{1}{0} = \boxed{\infty}$

i. $\lim_{x \to 0} \dfrac{\sin x}{x^3} = \dfrac{\sin 0}{0^3} = \dfrac{0}{0}$ Apply the L'Hopital's rule $\lim_{x \to 0} \dfrac{\sin x}{x^3} = \lim_{x \to 0} \dfrac{\frac{d}{dx} \sin x}{\frac{d}{dx} x^3}$

$= \lim_{x \to 0} \dfrac{\cos x}{3x^2} = \dfrac{\cos 0}{3 \cdot 0^2} = \dfrac{1}{0} = \boxed{\infty}$

j. $\lim_{x \to 0} \dfrac{3 - 3\cos x}{x + x^2} = \dfrac{3 - 3\cos 0}{0 + 0^2} = \dfrac{3 - 3 \cdot 1}{0} = \dfrac{3 - 3}{0} = \dfrac{0}{0}$ Apply the L'Hopital's rule $\lim_{x \to 0} \dfrac{3 - 3\cos x}{x + x^2}$

$= \lim_{x \to 0} \dfrac{\frac{d}{dx}(3 - 3\cos x)}{\frac{d}{dx}(x + x^2)} = \lim_{x \to 0} \dfrac{\frac{d}{dx} 3 - \frac{d}{dx} 3\cos x}{\frac{d}{dx} x + \frac{d}{dx} x^2} = \lim_{x \to 0} \dfrac{0 + 3\sin x}{1 + 2x} = \dfrac{3\sin 0}{1 + 2 \cdot 0} = \dfrac{3 \cdot 0}{1} = \dfrac{0}{1} = \boxed{0}$

k. $\lim_{x \to \infty} \dfrac{2 + x^2}{x^3} = \dfrac{2 + \infty^2}{\infty^3} = \dfrac{\infty}{\infty}$ Apply the L'Hopital's rule $\lim_{x \to \infty} \dfrac{2 + x^2}{x^3} = \lim_{x \to \infty} \dfrac{\frac{d}{dx}(2 + x^2)}{\frac{d}{dx} x^3}$

$$= \boxed{\lim_{x\to\infty} \frac{\frac{d}{dx}2 + \frac{d}{dx}x^2}{\frac{d}{dx}x^3}} = \boxed{\lim_{x\to\infty} \frac{0+2x}{3x^2}} = \boxed{\lim_{x\to\infty} \frac{2x}{3x^2}} = \boxed{\lim_{x\to\infty} \frac{2}{3x}} = \boxed{\frac{2}{3\cdot\infty}} = \boxed{\frac{2}{\infty}} = \boxed{0}$$

Note: Think of the division of 2 by ∞ as the division of 2 by a very large number. In that case, the result would then be very close to zero.

l. $\boxed{\lim_{x\to 0} \frac{\cos x - 1}{\cos 3x - 1}} = \boxed{\frac{\cos 0 - 1}{\cos(3\cdot 0)-1}} = \boxed{\frac{\cos 0 - 1}{\cos 0 - 1}} = \boxed{\frac{1-1}{1-1}} = \boxed{\frac{0}{0}}$ Apply the L'Hopital's rule $\boxed{\lim_{x\to 0}\frac{\cos x-1}{\cos 3x-1}}$

$$= \boxed{\lim_{x\to 0}\frac{\frac{d}{dx}(\cos x - 1)}{\frac{d}{dx}(\cos 3x - 1)}} = \boxed{\lim_{x\to 0}\frac{\frac{d}{dx}\cos x - \frac{d}{dx}1}{\frac{d}{dx}\cos 3x - \frac{d}{dx}1}} = \boxed{\lim_{x\to 0}\frac{-\sin x - 0}{-3\sin 3x - 0}} = \boxed{\lim_{x\to 0}\frac{\sin x}{3\sin 3x}}$$

$$= \boxed{\frac{\sin 0}{3\sin(3\cdot 0)}} = \boxed{\frac{\sin 0}{3\sin 0}} = \boxed{\frac{0}{3\cdot 0}} = \boxed{\frac{0}{0}}$$ Apply the L'Hopital's rule again to the expression

$$\boxed{\lim_{x\to 0}\frac{\sin x}{3\sin 3x}} = \boxed{\lim_{x\to 0}\frac{\frac{d}{dx}\sin x}{\frac{d}{dx}3\sin 3x}} = \boxed{\lim_{x\to 0}\frac{\cos x}{9\cos 3x}} = \boxed{\frac{\cos 0}{9\cos(3\cdot 0)}} = \boxed{\frac{\cos 0}{9\cos 0}} = \boxed{\frac{1}{9\cdot 1}} = \boxed{\frac{1}{9}}$$

Solving indeterminate types of the form $0\cdot\infty$ and $\infty-\infty$:

Indeterminate forms of the type $0\cdot\infty$ and $\infty-\infty$ can sometimes be transformed to the indeterminate forms of the type $\frac{0}{0}$ or $\frac{\infty}{\infty}$ by changing the given expression algebraically. For example, the expression $\lim_{x\to\infty} x^2 e^{-2x}$ result in the indeterminate form of the type $\infty\cdot 0$ as the limit x approaches infinity , i.e., $\lim_{x\to\infty} x^2 e^{-2x} = \infty^2 \cdot e^{-2\cdot\infty} = \infty^2 \cdot \frac{1}{e^{2\cdot\infty}} = \infty\cdot\frac{1}{\infty} = \infty\cdot 0$. However, by rewriting the given expression in the form of $\lim_{x\to\infty}\frac{x^2}{e^{2x}}$ result in the indeterminate form of the type $\frac{\infty}{\infty}$ as the limit x approaches infinity. Thus, the L'Hopital's rule can be applied, i.e.,

$$\lim_{x\to\infty}\frac{x^2}{e^{2x}} = \lim_{x\to\infty}\frac{\frac{d}{dx}x^2}{\frac{d}{dx}e^{2x}} = \lim_{x\to\infty}\frac{2x}{e^{2x}\cdot\frac{d}{dx}2x} = \lim_{x\to\infty}\frac{2x}{e^{2x}\cdot 2} = \lim_{x\to\infty}\frac{x}{e^{2x}} = \frac{\infty}{e^{2\cdot\infty}} = \frac{\infty}{\infty}.$$

Applying the L'Hopital's rule again to the expression $\lim_{x\to\infty}\frac{x}{e^{2x}}$ we obtain: $\lim_{x\to\infty}\frac{x}{e^{2x}}$

$$= \lim_{x\to\infty}\frac{\frac{d}{dx}x}{\frac{d}{dx}e^{2x}} = \lim_{x\to\infty}\frac{1}{e^{2x}\cdot\frac{d}{dx}2x} = \lim_{x\to\infty}\frac{1}{e^{2x}\cdot 2} = \lim_{x\to\infty}\frac{1}{2e^{2x}} = \frac{1}{2e^{2\cdot\infty}} = \frac{1}{2\cdot\infty} = \frac{1}{\infty} = 0$$

The following are additional examples of expressions that result in the indeterminate forms of the type $0\cdot\infty$ and $\infty-\infty$ and can be changed to the indeterminate forms of the type $\frac{0}{0}$ or $\frac{\infty}{\infty}$:

Example 3.6-6: Evaluate the limit of the algebraic expression $\lim_{x \to 2} \left(\dfrac{4}{x^2 - 4} - \dfrac{1}{x - 2} \right)$.

Solution:

$$\boxed{\lim_{x \to 2} \left(\dfrac{4}{x^2 - 4} - \dfrac{1}{x - 2} \right)} = \boxed{\dfrac{4}{2^2 - 4} - \dfrac{1}{2 - 2}} = \boxed{\dfrac{4}{4 - 4} - \dfrac{1}{2 - 2}} = \boxed{\dfrac{4}{0} - \dfrac{1}{0}} = \boxed{\infty - \infty}$$ Since the limit leads to

the indeterminate form of the type $\infty - \infty$ we need to see if the given expression can be
rewritten in a different algebraic form. Let's change the algebraic expression to the form that
leads to the indeterminate form of the type $\dfrac{0}{0}$ by taking the common denominator of the
algebraic expression and simplifying the numerator and the denominator.

$$\boxed{\lim_{x \to 2} \left(\dfrac{4}{x^2 - 4} - \dfrac{1}{x - 2} \right)} = \boxed{\lim_{x \to 2} \dfrac{4(x-2) - (x^2 - 4)}{(x^2 - 4)(x - 2)}} = \boxed{\lim_{x \to 2} \dfrac{4x - 8 - x^2 + 4}{x^3 - 2x^2 - 4x + 8}} = \boxed{\lim_{x \to 2} \dfrac{-x^2 + 4x - 4}{x^3 - 2x^2 - 4x + 8}}$$

$$= \boxed{\dfrac{-2^2 + 4 \cdot 2 - 4}{2^3 - 2 \cdot 2^2 - 4 \cdot 2 + 8}} = \boxed{\dfrac{-4 + 8 - 4}{8 - 8 - 8 + 8}} = \boxed{\dfrac{0}{0}}$$ Apply the L'Hopital's rule to $\boxed{\lim_{x \to 2} \dfrac{-x^2 + 4x - 4}{x^3 - 2x^2 - 4x + 8}}$

$$= \boxed{\lim_{x \to 2} \dfrac{\frac{d}{dx}\left(-x^2 + 4x - 4\right)}{\frac{d}{dx}\left(x^3 - 2x^2 - 4x + 8\right)}} = \boxed{\lim_{x \to 2} \dfrac{-2x + 4}{3x^2 - 4x - 4}} = \boxed{\dfrac{-2 \cdot 2 + 4}{3 \cdot 2^2 - 4 \cdot 2 - 4}} = \boxed{\dfrac{-4 + 4}{12 - 8 - 4}} = \boxed{\dfrac{0}{0}}$$ Apply the

L'Hopital's rule again to $\boxed{\lim_{x \to 2} \dfrac{-2x + 4}{3x^2 - 4x - 4}} = \boxed{\lim_{x \to 2} \dfrac{\frac{d}{dx}(-2x + 4)}{\frac{d}{dx}\left(3x^2 - 4x - 4\right)}} = \boxed{\lim_{x \to 2} \dfrac{-2}{6x - 4}} = \boxed{\dfrac{-2}{6 \cdot 2 - 4}}$

$$= \boxed{\dfrac{-2}{12 - 4}} = \boxed{\dfrac{-2}{8}} = \boxed{-\dfrac{1}{4}}$$

Example 3.6-7: Evaluate the limit of the algebraic expression $\lim_{x \to \infty} x \sin \dfrac{1}{x}$.

Solution:

$$\boxed{\lim_{x \to \infty} x \sin \dfrac{1}{x}} = \boxed{\infty \cdot \sin \dfrac{1}{\infty}} = \boxed{\infty \cdot \sin 0} = \boxed{\infty \cdot 0}$$ Since the limit leads to the indeterminate form of

the type $\infty \cdot 0$ we need to see if the given expression can be rewritten in a different algebraic
form. Let's change the algebraic expression to the form that leads to the indeterminate form of
the type $\dfrac{0}{0}$ by letting $x = \dfrac{1}{t}$ and by allowing $t \to 0$.

$$\boxed{\lim_{x \to \infty} x \sin \dfrac{1}{x}} = \boxed{\lim_{t \to 0} \dfrac{1}{t} \sin \dfrac{1}{\frac{1}{t}}} = \boxed{\lim_{t \to 0} \dfrac{1}{t} \sin t} = \boxed{\lim_{t \to 0} \dfrac{\sin t}{t}} = \boxed{\dfrac{\sin 0}{0}} = \boxed{\dfrac{0}{0}}$$ Apply the

L'Hopital's rule to $\boxed{\lim_{t \to 0} \dfrac{\sin t}{t}} = \boxed{\lim_{t \to 0} \dfrac{\frac{d}{dt}\sin t}{\frac{d}{dt}t}} = \boxed{\lim_{t \to 0} \dfrac{\cos t}{1}} = \boxed{\lim_{t \to 0} \cos t} = \boxed{\cos 0} = \boxed{1}$

Solving indeterminate types of the form 1^{∞}, ∞^0, and 0^0:

Similarly, indeterminate forms of the type 1^{∞}, ∞^0, and 0^0 may first be transformed to the

indeterminate forms of the type $\dfrac{0}{0}$ or $\dfrac{\infty}{\infty}$ by multiplying both sides of the function

$\lim_{x \to a} y = \lim_{x \to a} f(x)$ by logarithm ($\ln$). For example, $\lim_{x \to 0}(\cos x)^{\frac{1}{x}}$ result in the

indeterminate form of the type 1^{∞} as the limit x approaches zero, i.e., $\lim_{x \to 0}(\cos x)^{\frac{1}{x}} = \cos 0^{\frac{1}{0}}$

$= \cos 0^{\frac{1}{0}} = 1^{\infty}$. However, if we let $y = (\cos x)^{\frac{1}{x}}$ and take the logarithm of both sides of the

equation the solution has an indeterminate limit of the type $\dfrac{0}{0}$, i.e., given $y = (\cos x)^{\frac{1}{x}}$, then

$\ln y = \ln (\cos x)^{\frac{1}{x}}$ and $\ln y = \lim_{x \to 0} \ln (\cos x)^{\frac{1}{x}} = \lim_{x \to 0} \dfrac{1}{x}\ln (\cos x)$; $\lim_{x \to 0} \dfrac{\ln (\cos x)}{x} = \dfrac{\ln (\cos 0)}{0}$

$= \dfrac{\ln 1}{0} = \dfrac{0}{0}$. Since the limit result in an indeterminate form of the type $\dfrac{0}{0}$ the L'Hopital's rule

can be applied to the expression $\lim_{x \to 0} \dfrac{\ln (\cos x)}{x} = \lim_{x \to 0} \dfrac{\frac{d}{dx}\ln (\cos x)}{\frac{d}{dx}x} = \lim_{x \to 0} \dfrac{\frac{1}{\cos x} \cdot \frac{d}{dx}\cos x}{1}$

$= \lim_{x \to 0} \dfrac{1}{\cos x} \cdot -\sin x = \lim_{x \to 0} -\dfrac{\sin x}{\cos x} = \lim_{x \to 0} -\tan x = -\tan 0 = 0$. Thus, $\ln y = 0$. To solve

for y we raise both sides of the equation by e and solve for y, i.e., $e^{\ln y} = e^0$; $y = e^0$; $y = 1$.

Therefore, indeterminate forms of the type 1^{∞}, ∞^0, and 0^0 may first be transformed to the

indeterminate forms of the type $\dfrac{0}{0}$ or $\dfrac{\infty}{\infty}$ by rewriting the function $\lim_{x \to a} y = \lim_{x \to a} f(x)$ in a

different form and solving the algebraic expression. The following are additional examples of

expressions that result in the indeterminate forms of the type 1^{∞}, ∞^0, and 0^0 which may be rewritten

in a different form and solved:

Example 3.6-8: Evaluate the limit of the algebraic expression $\lim_{x \to 0} x^{\sin x}$.

Solution:

$\boxed{\lim_{x \to 0} x^{\sin x}} = \boxed{0^{\sin 0}} = \boxed{0^0}$ Since the limit leads to the indeterminate form of the type 0^0 let

$y = x^{\sin x}$ then by taking the logarithm of both sides of the equation we obtain $\boxed{\lim_{x \to 0} \ln y}$

$= \boxed{\lim_{x \to 0} \ln x^{\sin x}} = \boxed{\lim_{x \to 0} \sin x \ln x} = \boxed{\lim_{x \to 0} \dfrac{\ln x}{\csc x}} = \boxed{\dfrac{\ln 0}{\csc 0}} = \boxed{\dfrac{\infty}{\infty}}$ Apply the L'Hopital's rule to

$$\boxed{\lim_{x\to 0}\frac{\ln x}{\csc x}}=\boxed{\lim_{x\to 0}\frac{\frac{d}{dx}\ln x}{\frac{d}{dx}\csc x}}=\boxed{\lim_{x\to 0}\frac{\frac{1}{x}}{-\csc x\cot x}}=\boxed{\lim_{x\to 0}\frac{\frac{1}{x}}{-\frac{1}{\sin x}\cdot\frac{\cos x}{\sin x}}}=\boxed{\lim_{x\to 0}\frac{\frac{1}{x}}{-\frac{\cos x}{\sin^2 x}}}$$

$$=\boxed{\lim_{x\to 0}-\frac{1\cdot\sin^2 x}{x\cdot\cos x}}=\boxed{\lim_{x\to 0}-\frac{\sin^2 x}{x\cos x}}=\boxed{-\frac{\sin^2 0}{0\cdot\cos 0}}=\boxed{-\frac{\sin 0}{0\cdot 1}}=\boxed{\frac{0}{0}}\text{ Apply the L'Hopital's rule}$$

again to the expression $\boxed{\lim_{x\to 0}-\frac{\sin^2 x}{x\cos x}}=\boxed{\lim_{x\to 0}-\frac{\frac{d}{dx}\sin^2 x}{\frac{d}{dx}x\cos x}}=\boxed{\lim_{x\to 0}-\frac{2\sin x\cdot\frac{d}{dx}\sin x}{\cos x\cdot\frac{d}{dx}x+x\cdot\frac{d}{dx}\cos x}}$

$$=\boxed{\lim_{x\to 0}-\frac{2\sin x\cdot\cos x}{\cos x\cdot 1+x\cdot-\sin x}}=\boxed{-\frac{2\cdot\sin 0\cdot\cos 0}{\cos 0-0\cdot\sin 0}}=\boxed{-\frac{2\cdot 0\cdot 1}{1-0\cdot 0}}=\boxed{-\frac{0}{1-0}}=\boxed{-\frac{0}{1}}=\boxed{0}$$

Thus, $\boxed{\ln y=0}$; $\boxed{e^{\ln y}=e^0}$; $\boxed{y=e^0}$; $\boxed{y=1}$

Example 3.6-9: Evaluate the limit of the algebraic expression $\lim_{x\to\infty}\left(\cos\frac{1}{x}\right)^x$.

Solution:

$\boxed{\lim_{x\to\infty}\left(\cos\frac{1}{x}\right)^x}=\boxed{\left(\cos\frac{1}{\infty}\right)^\infty}=\boxed{(\cos 0)^\infty}=\boxed{1^\infty}$ Since the limit leads to the indeterminate form of

the type 1^∞ let $y=\left(\cos\frac{1}{x}\right)^x$ then by taking the logarithm of both sides of the equation we

obtain $\boxed{\lim_{x\to\infty}\ln y}=\boxed{\lim_{x\to\infty}\ln\left(\cos\frac{1}{x}\right)^x}=\boxed{\lim_{x\to\infty}x\ln\left(\cos\frac{1}{x}\right)}=\boxed{\lim_{x\to\infty}\frac{\ln\left(\cos\frac{1}{x}\right)}{\frac{1}{x}}}=\boxed{\frac{\ln\left(\cos\frac{1}{\infty}\right)}{\frac{1}{\infty}}}$

$$=\boxed{\frac{\ln(\cos 0)}{0}}=\boxed{\frac{\ln 1}{0}}=\boxed{\frac{0}{0}}\text{ Apply the L'Hopital's rule to }\boxed{\lim_{x\to\infty}\frac{\ln\left(\cos\frac{1}{x}\right)}{\frac{1}{x}}}=\boxed{\lim_{x\to\infty}\frac{\frac{d}{dx}\ln\left(\cos\frac{1}{x}\right)}{\frac{d}{dx}\frac{1}{x}}}$$

$$=\boxed{\lim_{x\to\infty}\frac{\frac{1}{\cos(1/x)}\cdot\frac{d}{dx}\cos\frac{1}{x}}{-\frac{1}{x^2}}}=\boxed{\lim_{x\to\infty}\frac{\frac{1}{\cos(1/x)}\cdot-\sin\frac{1}{x}\cdot\frac{d}{dx}\frac{1}{x}}{-\frac{1}{x^2}}}=\boxed{\lim_{x\to\infty}\frac{\frac{1}{\cos(1/x)}\cdot-\sin\frac{1}{x}\cdot-\frac{1}{x^2}}{-\frac{1}{x^2}}}$$

$$=\boxed{\lim_{x\to\infty}\frac{\frac{1}{\cos(1/x)}\cdot-\sin\frac{1}{x}}{1}}=\boxed{\lim_{x\to\infty}\frac{1}{\cos(1/x)}\cdot-\sin\frac{1}{x}}=\boxed{\lim_{x\to\infty}-\frac{\sin(1/x)}{\cos(1/x)}}=\boxed{-\frac{\sin(1/\infty)}{\cos(1/\infty)}}=\boxed{-\frac{\sin 0}{\cos 0}}$$

$$=\boxed{-\frac{0}{1}}=\boxed{0}\text{ Thus, }\boxed{\ln y=0}\text{ ; }\boxed{e^{\ln y}=e^0}\text{ ; }\boxed{y=e^0}\text{ ; }\boxed{y=1}$$

Example 3.6-10: Evaluate the limit of the algebraic expression $\lim_{x \to \frac{\pi}{2}} (\tan x)^{\cos x}$.

Solution:

$$\boxed{\lim_{x \to \frac{\pi}{2}} (\tan x)^{\cos x}} = \boxed{\left(\tan \frac{\pi}{2} \right)^{\cos \frac{\pi}{2}}} = \boxed{\left(\frac{\sin \frac{\pi}{2}}{\cos \frac{\pi}{2}} \right)^{\cos \frac{\pi}{2}}} = \boxed{\left(\frac{1}{0} \right)^0} = \boxed{\infty^0}$$ Since the limit leads to the

indeterminate form of the type ∞^0 let $y = (\tan x)^{\cos x}$ then by taking the logarithm of both sides

of the equation we obtain $\boxed{\lim_{x \to \frac{\pi}{2}} \ln y} = \boxed{\lim_{x \to \frac{\pi}{2}} \ln (\tan x)^{\cos x}} = \boxed{\lim_{x \to \frac{\pi}{2}} \cos x \ln (\tan x)}$

$$= \boxed{\lim_{x \to \frac{\pi}{2}} \frac{\ln (\tan x)}{\sec x}} = \boxed{\frac{\ln \left(\tan \frac{\pi}{2} \right)}{\sec \frac{\pi}{2}}} = \boxed{\frac{\ln \infty}{\infty}} = \boxed{\frac{\infty}{\infty}}$$ Apply the L'Hopital's rule to $\boxed{\lim_{x \to \frac{\pi}{2}} \frac{\ln (\tan x)}{\sec x}}$

$$= \boxed{\lim_{x \to \frac{\pi}{2}} \frac{\frac{d}{dx} \ln (\tan x)}{\frac{d}{dx} \sec x}} = \boxed{\lim_{x \to \frac{\pi}{2}} \frac{\frac{1}{\tan x} \cdot \frac{d}{dx} \tan x}{\sec x \tan x}} = \boxed{\lim_{x \to \frac{\pi}{2}} \frac{\frac{1}{\tan x} \cdot \sec^2 x}{\sec x \tan x}} = \boxed{\lim_{x \to \frac{\pi}{2}} \frac{\sec^2 x}{\sec x \tan^2 x}}$$

$$= \boxed{\lim_{x \to \frac{\pi}{2}} \frac{\sec x}{\tan^2 x}} = \boxed{\lim_{x \to \frac{\pi}{2}} \frac{\frac{1}{\cos x}}{\frac{\sin^2 x}{\cos^2 x}}} = \boxed{\lim_{x \to \frac{\pi}{2}} \frac{\cos^2 x}{\cos x \sin^2 x}} = \boxed{\lim_{x \to \frac{\pi}{2}} \frac{\cos x}{\sin^2 x}} = \boxed{\frac{\cos \frac{\pi}{2}}{\sin^2 \frac{\pi}{2}}} = \boxed{\frac{0}{1}} = \boxed{0}$$

Thus, $\boxed{\ln y = 0}$; $\boxed{e^{\ln y} = e^0}$; $\boxed{y = e^0}$; $\boxed{y = 1}$

Section 3.6 Practice Problems - Evaluation of Indeterminate Forms Using L'Hopital's Rule

Evaluate the limit for the following functions by applying the L'Hopital's rule, if needed:

a. $\lim_{x \to \infty} \frac{\ln x}{x^2} =$

b. $\lim_{x \to 0} \frac{\sin x}{e^x - 1} =$

c. $\lim_{x \to 0} \frac{1 - \cos x}{x^3} =$

d. $\lim_{t \to 2} \frac{t - 2}{t^2 + 2t - 1} =$

e. $\lim_{x \to \frac{\pi}{2}} \frac{\cos x}{x - \frac{\pi}{2}} =$

f. $\lim_{t \to 0} \frac{t \cos t}{2 \sin t} =$

g. $\lim_{x \to 8} \frac{x - 8}{x^2 - 64} =$

h. $\lim_{x \to \pi} \frac{\sin x}{\pi - x} =$

i. $\lim_{x \to 0} \frac{\sin 8x}{3x} =$

j. $\lim_{t \to \infty} \frac{8t + 3}{4t - 2} =$

k. $\lim_{x \to 0} \frac{\cos x - 1}{x^2} =$

l. $\lim_{x \to \frac{\pi}{2}} \frac{1 - \sin x}{1 + \cos 2x} =$

Chapter 4
Integration (Part I)

Quick Reference to Chapter 4 Problems

$$\boxed{\int \left(x^6 + x^2 + 3\right)dx} = \; ; \quad \boxed{\int \left(\frac{1}{x^5} + \frac{1}{x^2} + 6\right)dx} = \; ; \quad \boxed{\int \left(\sqrt[3]{x} + \sqrt{x} + 5x\right)dx} =$$

$$\boxed{\int \frac{x}{\sqrt[3]{x^2 - 1}}\,dx} = \; ; \quad \boxed{\int \frac{4x^3 + 6x}{\sqrt{x^4 + 3x^2 + 5}}\,dx} = \; ; \quad \boxed{\int \frac{2x - 1}{\sqrt[4]{x^2 - x - 3}}\,dx} =$$

$$\boxed{\int \frac{\cos 2x + \sin 2x}{\sin 2x}\,dx} = \; ; \quad \boxed{\int e^{\sin 5x}\cos 5x\,dx} = \; ; \quad \boxed{\int e^{\tan 5x}\sec^2 5x\,dx} =$$

$$\boxed{\int \frac{dx}{\sqrt{16 - 9x^2}}} = \; ; \quad \boxed{\int \frac{x^2}{9 + x^6}\,dx} = \; ; \quad \boxed{\int \frac{dx}{9x^2 + 25}} =$$

$$\boxed{\int \left(e^x + 3\right)^2 e^x dx} = \; ; \quad \boxed{\int \frac{x + 6}{x + 5}\,dx} = \; ; \quad \boxed{\int \frac{1}{x^2 + 10x + 24}\,dx} =$$

Chapter 4 – Integration (Part I)

The objective of this chapter is to improve the student's ability to solve problems involving integration of various classes of functions. Integration using the basic integration formulas is addressed in Section 4.1. Integration of functions using the Substitution method is discussed in Section 4.2. How to find the integral of trigonometric functions is addressed in Section 4.3. Integration of expressions resulting in inverse trigonometric functions is addressed in Section 4.4. Finally, the steps in integrating expressions that result in exponential or logarithmic functions is discussed in Section 4.5. Each section is concluded by solving examples with practice problems to further enhance the student's ability.

4.1 Integration Using the Basic Integration Formulas

A function $F(x)$ is called a *solution* of the differential equation $\frac{dy}{dx} = f(x)$ over a defined interval I if the function $F(x)$ is differentiable at every point on that interval and if $\frac{d}{dx} F(x) = f(x)$ at every point on that interval as well. Note that the function $F(x)$ is referred to as the *antiderivative* of $f(x)$. The set of all antiderivatives of a function $f(x)$ is called the indefinite integral of f with respect to x. The indefinite integral is shown by the notation $\int f(x)\,dx$. The solution to the indefinite integral is given by $F(x)+c$ and is shown as $\int f(x)\,dx = F(x)+c$. The symbol $\int$ is referred to as the *integral sign*. The function $f(x)$ is called the *integrand* of the integral. The dx indicates that the variable of integration is x and c is the *constant of integration*. In the following examples we will solve problems using the following basic indefinite integration formulas:

$$\int a\,dx = ax+c \qquad a \neq 0$$

$$\int a\,f(x)\,dx = a\int f(x)\,dx \qquad a \neq 0$$

$$\int [f(x)+g(x)]\,dx = \int f(x)\,dx + \int g(x)\,dx$$

$$\int x^n\,dx = \frac{x^{n+1}}{n+1}+c \qquad n \neq -1$$

Let's integrate some integrals using the above basic integration formulas.

Example 4.1-1: Evaluate the following integrals.

a. $\int dx =$ 　　　　 b. $\int a\,dx =$ 　　　　 c. $\int 5\,dx =$

d. $\int 10a^2\,dx =$ 　　 e. $\int 3x\,dx =$ 　　 f.. $\int x^3\,dx =$

g. $\int (a+b)x^5\,dx =$ 　 h. $\int (x^3+x)\,dx =$ 　 i. $\int (x^6+x^2+3)\,dx =$

Solutions:

a. $\boxed{\int dx} = \boxed{\int x^0\,dx} = \boxed{\frac{1}{0+1}x^{0+1}+c} = \boxed{x+c}$

Check: Let $y = x + c$, then $y' = x^{1-1} + 0 = x^0 = 1$

b. $\boxed{\int a\,dx} = \boxed{a\int dx} = \boxed{a\int x^0 dx} = \boxed{\dfrac{a}{0+1}x^{0+1} + c} = \boxed{ax + c}$

Check: Let $y = ax + c$, then $y' = ax^{1-1} + 0 = ax^0 = a \cdot 1 = a$

c. $\boxed{\int 5\,dx} = \boxed{5\int dx} = \boxed{5\int x^0 dx} = \boxed{\dfrac{5}{0+1}x^{0+1} + c} = \boxed{5x + c}$

Check: Let $y = 5x + c$, then $y' = 5x^{1-1} + 0 = 5x^0 = 5 \cdot 1 = 5$

d. $\boxed{\int 10a^2 dx} = \boxed{10a^2\int dx} = \boxed{10a^2\int x^0 dx} = \boxed{\dfrac{10a^2}{0+1}x^{0+1} + c} = \boxed{10a^2 x + c}$

Check: Let $y = 10a^2 x + c$, then $y' = 10a^2 x^{1-1} + 0 = 10a^2 x^0 = 10a^2 \cdot 1 = 10a^2$

e. $\boxed{\int 3x\,dx} = \boxed{3\int x\,dx} = \boxed{3\int x^1 dx} = \boxed{\dfrac{3}{1+1}x^{1+1} + c} = \boxed{\dfrac{3}{2}x^2 + c}$

Check: Let $y = \dfrac{3}{2}x^2 + c$, then $y' = \dfrac{3}{2} \cdot 2x^{2-1} + 0 = 3x$

f. $\boxed{\int x^3 dx} = \boxed{\dfrac{1}{3+1}x^{3+1} + c} = \boxed{\dfrac{1}{4}x^4 + c}$

Check: Let $y = \dfrac{1}{4}x^4 + c$, then $y' = \dfrac{1}{4} \cdot 4x^{4-1} + 0 = x^3$

g. $\boxed{\int (a+b)x^5 dx} = \boxed{(a+b)\int x^5 dx} = \boxed{\dfrac{a+b}{5+1}x^{5+1} + c} = \boxed{\dfrac{a+b}{6}x^6 + c}$

or we can find the solution to the above integral in the following way – which is rather long:

$\boxed{\int (a+b)x^5 dx} = \boxed{\int \left(ax^5 + bx^5\right)dx} = \boxed{\int ax^5 dx + \int bx^5 dx} = \boxed{a\int x^5 dx + b\int x^5 dx} = \boxed{a\left(\dfrac{1}{5+1}x^{5+1} + c_1\right)}$

$\boxed{+b\left(\dfrac{1}{5+1}x^{5+1} + c_2\right)} = \boxed{\dfrac{a}{6}x^6 + ac_1 + \dfrac{b}{6}x^6 + bc_2} = \boxed{\left(\dfrac{a}{6} + \dfrac{b}{6}\right)x^6 + (ac_1 + bc_2)} = \boxed{\dfrac{a+b}{6}x^6 + c}$

Check: Let $y = \dfrac{a+b}{6}x^6 + c$, then $y' = \dfrac{a+b}{6} \cdot 6x^{6-1} + 0 = (a+b)x^5$

h. $\boxed{\int \left(x^3 + x\right)dx} = \boxed{\int x^3 dx + \int x\,dx} = \boxed{\dfrac{1}{3+1}x^{3+1} + \dfrac{1}{1+1}x^{1+1} + c} = \boxed{\dfrac{1}{4}x^4 + \dfrac{1}{2}x^2 + c}$

Check: Let $y = \dfrac{1}{4}x^4 + \dfrac{1}{2}x^2 + c$, then $y' = \dfrac{1}{4} \cdot 4x^{4-1} + \dfrac{1}{2} \cdot 2x^{2-1} + 0 = x^3 + x$

i. $\boxed{\int \left(x^6 + x^2 + 3\right)dx} = \boxed{\int x^6 dx + \int x^2 dx + \int 3dx} = \boxed{\dfrac{1}{6+1}x^{6+1} + \dfrac{1}{2+1}x^{2+1} + \dfrac{3}{0+1}x^{0+1} + c} = \boxed{\dfrac{1}{7}x^7 + \dfrac{1}{3}x^3 + 3x + c}$

Check: Let $y = \frac{1}{7}x^7 + \frac{1}{3}x^3 + 3x + c$, then $y' = \frac{1}{7} \cdot 7x^{7-1} + \frac{1}{3} \cdot 3x^{3-1} + 3x^{1-1} + 0 = x^6 + x^2 + 3$

Example 4.1-2: Evaluate the following integrals.

a. $\int \frac{dx}{x^3} =$

b. $\int \left(\frac{1}{x^5} + \frac{1}{x^2} + 6 \right) dx =$

c. $\int \sqrt{x^5}\, dx =$

d. $\int \sqrt[3]{x^2}\, dx =$

e. $\int \left(\sqrt[3]{x} + \sqrt{x} + 5x \right) dx =$

f. $\int \left(\sqrt[5]{x} + x^3 + 10 \right) dx =$

g. $\int \frac{1}{\sqrt{t}}\, dt =$

h. $\int \frac{1}{\sqrt[5]{x^2}}\, dx =$

i. $\int \left(\frac{1}{\sqrt[3]{z^2}} + z^2 \right) dz =$

Solutions:

a. $\boxed{\int \frac{dx}{x^3}} = \boxed{\int x^{-3}\, dx} = \boxed{\frac{1}{-3+1}x^{-3+1} + c} = \boxed{-\frac{1}{2}x^{-2} + c} = \boxed{-\frac{1}{2x^2} + c}$

Check: Let $y = -\frac{1}{2}x^{-2} + c$, then $y' = -\frac{1}{2} \cdot -2x^{-2-1} + 0 = x^{-3} = \frac{1}{x^3}$

b. $\boxed{\int \left(\frac{1}{x^5} + \frac{1}{x^2} + 6 \right) dx} = \boxed{\int \frac{1}{x^5}\, dx + \int \frac{1}{x^2}\, dx + \int 6\, dx} = \boxed{\int x^{-5}\, dx + \int x^{-2}\, dx + 6\int x^0\, dx} = \boxed{\frac{1}{-5+1}x^{-5+1} + \frac{1}{-2+1}x^{-2+1}}$

$\boxed{+\frac{6}{1+0}x^{0+1} + c} = \boxed{-\frac{1}{4}x^{-4} - x^{-1} + 6x + c} = \boxed{-\frac{1}{4x^4} - \frac{1}{x} + 6x + c}$

Check: Let $y = -\frac{1}{4}x^{-4} - x^{-1} + 6x + c$, then $y' = -\frac{1}{4} \cdot -4x^{-4-1} + x^{-1-1} + 6x^{1-1} + 0 = x^{-5} + x^{-2} + 6x^0$

$x^{-5} + x^{-2} + 6 = \frac{1}{x^5} + \frac{1}{x^2} + 6$

Exception: Note that we can not apply the same integration technique $\left(\int x^n\, dx = \frac{1}{n+1}x^{n+1} + c \right)$

in order to find the integral of $\int \frac{1}{x}\, dx$. This is because division by zero is undefined, i.e., $\int \frac{1}{x}\, dx$

$= \int x^{-1}\, dx = \frac{1}{-1+1}x^{-1+1} + c = \frac{1}{0}x^0 + c$. As we will see in Section 4.5, $\int \frac{1}{x}\, dx = \ln|x| + c$.

c. $\boxed{\int \sqrt{x^5}\, dx} = \boxed{\int x^{\frac{1}{5}}\, dx} = \boxed{\frac{1}{1+\frac{1}{5}}x^{\frac{1}{5}+1} + c} = \boxed{\frac{1}{\frac{5+1}{5}}x^{\frac{1+5}{5}} + c} = \boxed{\frac{1}{\frac{6}{5}}x^{\frac{6}{5}} + c} = \boxed{\frac{5}{6}x^{\frac{6}{5}} + c}$

Check: Let $y = \frac{5}{6}x^{\frac{6}{5}} + c$, then $y' = \frac{5}{6} \cdot \frac{6}{5}x^{\frac{6}{5}-1} + 0 = x^{\frac{6-5}{5}} = x^{\frac{1}{5}} = \sqrt{x^5}$

d. $\boxed{\int \sqrt[3]{x^2}\, dx} = \boxed{\int x^{\frac{2}{3}}\, dx} = \boxed{\frac{1}{1+\frac{2}{3}}x^{\frac{2}{3}+1} + c} = \boxed{\frac{1}{\frac{3+2}{3}}x^{\frac{2+3}{3}} + c} = \boxed{\frac{1}{\frac{5}{3}}x^{\frac{5}{3}} + c} = \boxed{\frac{3}{5}x^{\frac{5}{3}} + c}$

Check: Let $y = \frac{3}{5}x^{\frac{5}{3}} + c$, then $y' = \frac{3}{5} \cdot \frac{5}{3}x^{\frac{5}{3}-1} + 0 = x^{\frac{5-3}{3}} = x^{\frac{2}{3}} = \sqrt[3]{x^2}$

e. $\int\left(\sqrt[3]{x}+\sqrt{x}+5x\right)dx = \int\left(x^{\frac{1}{3}}+x^{\frac{1}{2}}+5x\right)dx = \int x^{\frac{1}{3}}dx+\int x^{\frac{1}{2}}dx+\int 5x\,dx = \dfrac{1}{1+\frac{1}{3}}x^{\frac{1}{3}+1}+\dfrac{1}{1+\frac{1}{2}}x^{\frac{1}{2}+1}$

$+\dfrac{5}{1+1}x^{1+1}+c = \dfrac{1}{\frac{3+1}{3}}x^{\frac{1+3}{3}}+\dfrac{1}{\frac{2+1}{2}}x^{\frac{1+2}{2}}+\dfrac{5}{2}x^2+c = \dfrac{1}{\frac{4}{3}}x^{\frac{4}{3}}+\dfrac{1}{\frac{3}{2}}x^{\frac{3}{2}}+\dfrac{5}{2}x^2+c = \dfrac{3}{4}x^{\frac{4}{3}}+\dfrac{2}{3}x^{\frac{3}{2}}+\dfrac{5}{2}x^2+c$

Check: Let $y=\dfrac{3}{4}x^{\frac{4}{3}}+\dfrac{2}{3}x^{\frac{3}{2}}+\dfrac{5}{2}x^2+c$, then $y'=\dfrac{3}{4}\cdot\dfrac{4}{3}x^{\frac{4}{3}-1}+\dfrac{2}{3}\cdot\dfrac{3}{2}x^{\frac{3}{2}-1}+\dfrac{5}{2}\cdot 2x^{2-1}+0 = x^{\frac{4-3}{3}}+x^{\frac{3-2}{2}}+5x$

$= x^{\frac{1}{3}}+x^{\frac{1}{2}}+5x = \sqrt[3]{x}+\sqrt{x}+5x$

f. $\int\left(\sqrt[5]{x}+x^3+10\right)dx = \int\left(x^{\frac{1}{5}}+x^3+10\right)dx = \int x^{\frac{1}{5}}dx+\int x^3dx+\int 10\,dx = \dfrac{1}{1+\frac{1}{5}}x^{\frac{1}{5}+1}+\dfrac{1}{1+3}x^{3+1}+10x+c$

$\dfrac{1}{\frac{5+1}{5}}x^{\frac{1+5}{5}}+\dfrac{1}{4}x^4+10x+c = \dfrac{5}{6}x^{\frac{6}{5}}+\dfrac{1}{4}x^4+10x+c$

Check: Let $y=\dfrac{5}{6}x^{\frac{6}{5}}+\dfrac{1}{4}x^4+10x+c$, then $y'=\dfrac{5}{6}\cdot\dfrac{6}{5}x^{\frac{6}{5}-1}+\dfrac{1}{4}\cdot 4x^{4-1}+10x^{1-1}+0 = x^{\frac{6-5}{5}}+x^3+10x^0$

$= x^{\frac{1}{5}}+x^3+10 = \sqrt[5]{x}+x^3+10$

g. $\int\dfrac{1}{\sqrt{t}}dt = \int\dfrac{1}{t^{\frac{1}{2}}}dt = \int t^{-\frac{1}{2}}dt = \dfrac{1}{1-\frac{1}{2}}t^{1-\frac{1}{2}}+c = \dfrac{1}{\frac{2-1}{2}}t^{\frac{2-1}{2}}+c = 2t^{\frac{1}{2}}+c = 2\sqrt{t}+c$

Check: Let $y=2t^{\frac{1}{2}}+c$, then $y'=2\cdot\dfrac{1}{2}t^{\frac{1}{2}-1}+0 = t^{\frac{1-2}{2}} = t^{-\frac{1}{2}} = \dfrac{1}{t^{\frac{1}{2}}} = \dfrac{1}{\sqrt{t}}$

h. $\int\dfrac{1}{\sqrt[5]{x^2}}dx = \int\dfrac{1}{x^{\frac{2}{5}}}dx = \int x^{-\frac{2}{5}}dx = \dfrac{1}{1-\frac{2}{5}}x^{1-\frac{2}{5}}+c = \dfrac{1}{\frac{5-2}{5}}x^{\frac{5-2}{5}}+c = \dfrac{5}{3}x^{\frac{3}{5}}+c = \dfrac{5}{3}\sqrt[5]{x^3}+c$

Check: Let $y=\dfrac{5}{3}x^{\frac{3}{5}}+c$, then $y'=\dfrac{5}{3}\cdot\dfrac{3}{5}x^{\frac{3}{5}-1}+0 = x^{\frac{3-5}{5}} = x^{-\frac{2}{5}} = \dfrac{1}{x^{\frac{2}{5}}} = \dfrac{1}{\sqrt[5]{x^2}}$

i. $\int\left(\dfrac{1}{\sqrt[3]{z^2}}+z^2\right)dz = \int\left(\dfrac{1}{z^{\frac{2}{3}}}+z^2\right)dz = \int\left(z^{-\frac{2}{3}}+z^2\right)dz = \int z^{-\frac{2}{3}}dz+\int z^2dz = \dfrac{1}{1-\frac{2}{3}}z^{1-\frac{2}{3}}+\dfrac{1}{2+1}z^{2+1}+c$

$= \dfrac{1}{\frac{3-2}{3}}z^{\frac{3-2}{3}}+\dfrac{1}{3}z^3+c = 3z^{\frac{1}{3}}+\dfrac{1}{3}z^3+c = 3\sqrt[3]{z}+\dfrac{1}{3}z^3+c$

Check: Let $y=3z^{\frac{1}{3}}+\dfrac{1}{3}z^3+c$, then $y'=3\cdot\dfrac{1}{3}z^{\frac{1}{3}-1}+\dfrac{1}{3}\cdot 3z^{3-1}+0 = z^{\frac{1-3}{3}}+z^2 = z^{-\frac{2}{3}}+z^2 = \dfrac{1}{\sqrt[3]{z^2}}+z^2$

Example 4.1-3: Evaluate the following integrals.

a. $\int \left(1+x^2\right)\sqrt{x}\,dx \;=$

b. $\int \left(x^2+1\right)(x-1)dx \;=$

c. $\int \left(5x-3\right)^2 dx \;=$

d. $\int x\left(2x+1\right)^2 dx \;=$

e. $\int \left(a+b\right)^3 dx \;=$

f. $\int \left(a+x\right)^3 dx \;=$

g. $\int \left(a-x\right)^3 dx \;=$

h. $\int \dfrac{y^4+4y^3-6y^2}{y^2}\,dy \;=$

i. $\int \dfrac{y^3+3y^2+5}{y^2}\,dy \;=$

Solutions:

a. $\boxed{\int \left(1+x^2\right)\sqrt{x}\,dx} = \boxed{\int \left(\sqrt{x}+x^2\sqrt{x}\right)dx} = \boxed{\int \sqrt{x}\,dx+\int x^2\sqrt{x}\,dx} = \boxed{\int x^{\frac{1}{2}}\,dx+\int x^2 x^{\frac{1}{2}}\,dx} = \boxed{\int x^{\frac{1}{2}}\,dx+\int x^{2+\frac{1}{2}}dx}$

$= \boxed{\int x^{\frac{1}{2}}\,dx+\int x^{\frac{5}{2}}dx} = \boxed{\dfrac{1}{1+\frac{1}{2}}x^{\frac{1}{2}+1}+\dfrac{1}{1+\frac{3}{2}}x^{\frac{5}{2}+1}+c} = \boxed{\dfrac{1}{\frac{2+1}{2}}x^{\frac{1+2}{2}}+\dfrac{1}{\frac{2+5}{2}}x^{\frac{5+2}{2}}+c} = \boxed{\dfrac{2}{3}x^{\frac{3}{2}}+\dfrac{2}{7}x^{\frac{7}{2}}+c}$

$= \boxed{\dfrac{2}{3}\sqrt{x^3}+\dfrac{2}{7}\sqrt{x^7}+c} = \boxed{\dfrac{2}{3}\sqrt{x^2\cdot x}+\dfrac{2}{7}\sqrt{x^2\cdot x^2\cdot x^2\cdot x}+c} = \boxed{\dfrac{2}{3}x\sqrt{x}+\dfrac{2}{7}x^3\sqrt{x}+c} = \boxed{\dfrac{2x\sqrt{x}}{3}+\dfrac{2x^3\sqrt{x}}{7}+c}$

Check: Let $y=\dfrac{2}{3}x^{\frac{3}{2}}+\dfrac{2}{7}x^{\frac{7}{2}}+c$, then $y'=\dfrac{2}{3}\cdot\dfrac{3}{2}x^{\frac{3}{2}-1}+\dfrac{2}{7}\cdot\dfrac{7}{2}x^{\frac{7}{2}-1}+0 = x^{\frac{3-2}{2}}+x^{\frac{7-2}{2}} = x^{\frac{1}{2}}+x^{\frac{5}{2}}$

b. $\boxed{\int \left(x^2+1\right)(x-1)dx} = \boxed{\int \left(x^3-x^2+x-1\right)dx} = \boxed{\int x^3 dx-\int x^2 dx+\int x\,dx-\int dx} = \boxed{\int x^3 dx-\int x^2 dx+\int x\,dx-\int x^0 dx}$

$= \boxed{\dfrac{1}{1+3}x^{3+1}-\dfrac{1}{1+2}x^{2+1}+\dfrac{1}{1+1}x^{1+1}-\dfrac{1}{1+0}x^{0+1}+c} = \boxed{\dfrac{1}{4}x^4-\dfrac{1}{3}x^3+\dfrac{1}{2}x^2-x+c}$

Check: Let $y=\dfrac{1}{4}x^4-\dfrac{1}{3}x^3+\dfrac{1}{2}x^2-x+c$, then $y'=\dfrac{1}{4}\cdot 4x^{4-1}-\dfrac{1}{3}\cdot 3x^{3-1}+\dfrac{1}{2}\cdot 2x^{2-1}-1+0 = x^3-x^2+x-1$

c. $\boxed{\int \left(5x-3\right)^2 dx} = \boxed{\int \left(25x^2-30x+9\right)dx} = \boxed{\int 25x^2 dx-\int 30x\,dx+\int 9\,dx} = \boxed{\dfrac{25}{2+1}x^{2+1}-\dfrac{30}{1+1}x^{1+1}+\dfrac{9}{0+1}x^{0+1}+c}$

$\boxed{\dfrac{25}{3}x^3-\dfrac{30}{2}x^2+9x+c} = \boxed{\dfrac{25}{3}x^3-15x^2+9x+c}$

Check: Let $y=\dfrac{25}{3}x^3-15x^2+9x+c$, then $y'=\dfrac{25}{3}\cdot 3x^{3-1}-15\cdot 2x^{2-1}+9+0 = 25x^2-30x+9$

d. $\boxed{\int x\left(2x+1\right)^2 dx} = \boxed{\int x\left(4x^2+4x+1\right)dx} = \boxed{\int \left(4x^3+4x^2+x\right)dx} = \boxed{\int 4x^3 dx+\int 4x^2 dx+\int x\,dx}$

$= \boxed{\dfrac{4}{3+1}x^{3+1}+\dfrac{4}{2+1}x^{2+1}+\dfrac{1}{1+1}x^{1+1}+c} = \boxed{\dfrac{4}{4}x^4+\dfrac{4}{3}x^3+\dfrac{1}{2}x^2+c} = \boxed{x^4+\dfrac{4}{3}x^3+\dfrac{1}{2}x^2+c}$

Check: Let $y = x^4 + \frac{4}{3}x^3 + \frac{1}{2}x^2 + c$, then $y' = 4x^{4-1} + \frac{4}{3} \cdot 3x^{3-1} + \frac{1}{2} \cdot 2x^{2-1} + 0 = 4x^3 + 4x^2 + x$

e. $\boxed{\int (a+b)^3 dx} = \boxed{(a+b)^3 \int dx} = \boxed{(a+b)^3 \int x^0 dx} = \boxed{(a+b)^3 \cdot \frac{1}{0+1}x^{0+1} + c} = \boxed{(a+b)^3 x + c}$

Check: Let $y = (a+b)^3 x + c$, then $y' = (a+b)^3 x^{1-1} + 0 = (a+b)^3 x^0 = (a+b)^3$

f. $\boxed{\int (a+x)^3 dx} = \boxed{\int \left(a^3 + x^3 + 3a^2 x + 3ax^2\right) dx} = \boxed{\int \left(x^3 + 3ax^2 + 3a^2 x + a^3\right) dx} = \boxed{\int x^3 dx + \int 3ax^2 dx + \int 3a^2 x\, dx}$

$\boxed{+ \int a^3 dx} = \boxed{\frac{1}{3+1}x^{3+1} + \frac{3a}{2+1}x^{2+1} + \frac{3a^2}{1+1}x^{1+1} + \frac{a^3}{0+1}x^{0+1} + c} = \boxed{\frac{1}{4}x^4 + ax^3 + \frac{3a^2}{2}x^2 + a^3 x + c}$

Check: Let $y = \frac{1}{4}x^4 + ax^3 + \frac{3a^2}{2}x^2 + a^3 x + c$, then $y' = \frac{1}{4} \cdot 4x^{4-1} + a \cdot 3x^{3-1} + \frac{3a^2}{2} \cdot 2x^{2-1} + a^3 x^{1-1} + 0$

$= x^3 + 3ax^2 + 3a^2 x + a^3 x^0 = x^3 + 3ax^2 + 3a^2 x + a^3 = a^3 + x^3 + 3a^2 x + 3ax^2 = (a+x)^3$

g. $\boxed{\int (a-x)^3 dx} = \boxed{\int \left(a^3 - x^3 + 3a^2 x - 3ax^2\right) dx} = \boxed{\int \left(-x^3 - 3ax^2 + 3a^2 x + a^3\right) dx} = \boxed{-\int x^3 dx - \int 3ax^2 dx + \int 3a^2 x\, dx}$

$\boxed{+ \int a^3 dx} = \boxed{-\frac{1}{3+1}x^{3+1} - \frac{3a}{2+1}x^{2+1} + \frac{3a^2}{1+1}x^{1+1} + \frac{a^3}{0+1}x^{0+1} + c} = \boxed{-\frac{1}{4}x^4 - ax^3 + \frac{3a^2}{2}x^2 + a^3 x + c}$

Check: Let $y = -\frac{1}{4}x^4 - ax^3 + \frac{3a^2}{2}x^2 + a^3 x + c$, then $y' = -\frac{1}{4} \cdot 4x^{4-1} - a \cdot 3x^{3-1} + \frac{3a^2}{2} \cdot 2x^{2-1} + a^3 x^{1-1} + 0$

$= -x^3 - 3ax^2 + 3a^2 x + a^3 x^0 = -x^3 - 3ax^2 + 3a^2 x + a^3 = a^3 - x^3 + 3a^2 x - 3ax^2 = (a-x)^3$

h. $\boxed{\int \frac{y^4 + 4y^3 - 6y^2}{y^2} dy} = \boxed{\int \left(\frac{y^4}{y^2} + \frac{4y^3}{y^2} - \frac{6y^2}{y^2}\right) dy} = \boxed{\int \left(y^2 + 4y - 6\right) dy} = \boxed{\int y^2 dy + \int 4y\, dy - \int 6\, dy}$

$= \boxed{\frac{1}{2+1}y^{2+1} + \frac{4}{1+1}y^{1+1} - \frac{6}{0+1}y^{0+1} + c} = \boxed{\frac{1}{3}y^3 + \frac{4}{2}y^2 - \frac{6}{1}y + c} = \boxed{\frac{1}{3}y^3 + 2y^2 - 6y + c}$

Check: Let $w = \frac{1}{3}y^3 + 2y^2 - 6y + c$, then $w' = \frac{1}{3} \cdot 3y^{3-1} + 2 \cdot 2y^{2-1} - 6y^{1-1} + 0 = y^2 + 4y - 6$

i. $\boxed{\int \frac{y^3 + 3y^2 + 5}{y^2} dy} = \boxed{\int \left(\frac{y^3}{y^2} + \frac{3y^2}{y^2} + \frac{5}{y^2}\right) dy} = \boxed{\int \left(y + 3 + 5y^{-2}\right) dy} = \boxed{\int y\, dy + \int 3\, dy + \int 5y^{-2} dy}$

$= \boxed{\frac{1}{1+1}y^{1+1} + \frac{3}{0+1}y^{0+1} + \frac{5}{-2+1}y^{-2+1} + c} = \boxed{\frac{1}{2}y^2 + 3y - 5y^{-1} + c} = \boxed{\frac{1}{2}y^2 + 3y - \frac{5}{y} + c}$

Check: Let $w = \frac{1}{2}y^2 + 3y - 5y^{-1} + c$, then $w' = \frac{1}{2} \cdot 2y^{2-1} + 3y^{1-1} - 5 \cdot -y^{-1-1} + 0 = y + 3 + 5y^{-2}$

Example 4.1-4: Evaluate the following integrals.

a. $\int \left(x^3 - 6\sqrt{x} \right) dx \;=\;$

b. $\int \sqrt{x-1}\; dx \;=\;$

c. $\int 30\sqrt{x+5}\; dx \;=\;$

d. $\int \left(x^2 - 5\sqrt{x} \right) dx \;=\;$

e. $\int \dfrac{2}{5}\sqrt{x+1}\; dx \;=\;$

f. $\int \sqrt{3x+2}\; dx \;=\;$

g. $\int (2x-1)^3\; dx \;=\;$

h. $\int (2x+1)^2\; dx \;=\;$

i. $\int 6x^2 \left(x^3 + 1 \right) dx \;=\;$

Solutions:

a. $\boxed{\int \left(x^3 - 6\sqrt{x} \right) dx} = \boxed{\int \left(x^3 - 6x^{\frac{1}{2}} \right) dx} = \boxed{\int x^3\, dx - \int 6 x^{\frac{1}{2}}\, dx} = \boxed{\dfrac{1}{3+1} x^{3+1} - \dfrac{6}{\frac{1}{2}+1} x^{\frac{1}{2}+1} + c} = \boxed{\dfrac{1}{4} x^4 - \dfrac{6}{\frac{1+2}{2}} x^{\frac{1+2}{2}} + c}$

$= \boxed{\dfrac{1}{4} x^4 - \dfrac{12}{3} x^{\frac{3}{2}} + c} = \boxed{\dfrac{1}{4} x^4 - 4x^{\frac{3}{2}} + c} = \boxed{\dfrac{1}{4} x^4 - 4\sqrt{x^3} + c} = \boxed{\dfrac{1}{4} x^4 - 4x\sqrt{x} + c}$

Check: Let $y = \dfrac{1}{4} x^4 - 4x^{\frac{3}{2}} + c$ then $y' = \dfrac{1}{4}\cdot 4\, x^{4-1} - 4\cdot\dfrac{3}{2} x^{\frac{3}{2}-1} + 0 = x^3 - 6x^{\frac{1}{2}} = x^3 - 6\sqrt{x}$

b. $\boxed{\int \sqrt{x-1}\; dx} = \boxed{\int (x-1)^{\frac{1}{2}}\, dx} = \boxed{\dfrac{1}{\frac{1}{2}+1} (x-1)^{\frac{1}{2}+1} + c} = \boxed{\dfrac{2}{3} (x-1)^{\frac{3}{2}} + c} = \boxed{\dfrac{2}{3}\sqrt{(x-1)^3} + c} = \boxed{\dfrac{2(x-1)}{3}\sqrt{x-1} + c}$

Check: Let $y = \dfrac{2}{3}(x-1)^{\frac{3}{2}} + c$ then $y' = \dfrac{2}{3}\cdot\dfrac{3}{2}(x-1)^{\frac{3}{2}-1} + 0 = (x-1)^{\frac{3-2}{2}} = (x-1)^{\frac{1}{2}} = \sqrt{x-1}$

c. $\boxed{\int 30\sqrt{x+5}\; dx} = \boxed{\int 30(x+5)^{\frac{1}{2}}\, dx} = \boxed{\dfrac{30}{\frac{1}{2}+1}(x+5)^{\frac{1}{2}+1} + c} = \boxed{\dfrac{30}{\frac{1+2}{2}}(x+5)^{\frac{1+2}{2}} + c} = \boxed{\dfrac{60}{3}(x+5)^{\frac{3}{2}} + c}$

$= \boxed{20\sqrt{(x+5)^3} + c} = \boxed{20(x+5)\sqrt{x+5} + c}$

Check: Let $y = \dfrac{60}{3}(x+5)^{\frac{3}{2}} + c$ then $y' = \dfrac{60}{3}\cdot\dfrac{3}{2}(x+5)^{\frac{3}{2}-1} + 0 = 30(x+5)^{\frac{3-2}{2}} = 30(x+5)^{\frac{1}{2}} = 30\sqrt{x+5}$

d. $\boxed{\int \left(x^2 - 5\sqrt{x} \right) dx} = \boxed{\int x^2\, dx - \int 5 x^{\frac{1}{2}}\, dx} = \boxed{\dfrac{1}{2+1} x^{2+1} - \dfrac{5}{\frac{1}{2}+1} x^{\frac{1}{2}+1} + c} = \boxed{\dfrac{1}{3} x^3 - \dfrac{5}{\frac{1+2}{2}} x^{\frac{1+2}{2}} + c} = \boxed{\dfrac{1}{3} x^3 - \dfrac{10}{3} x^{\frac{3}{2}} + c}$

$= \boxed{\dfrac{1}{3} x^3 - \dfrac{10}{3}\sqrt{x^3} + c} = \boxed{\dfrac{1}{3} x^3 - \dfrac{10}{3} x\sqrt{x} + c} = \boxed{\dfrac{1}{3} x\left(x^2 - 10\sqrt{x} \right) + c}$

Check: Let $y = \dfrac{1}{3} x^3 - \dfrac{10}{3} x^{\frac{3}{2}} + c$ then $y' = \dfrac{1}{3}\cdot 3x^{3-1} - \dfrac{10}{3}\cdot\dfrac{3}{2} x^{\frac{3}{2}-1} + 0 = \dfrac{3}{3} x^2 - \dfrac{30}{6} x^{\frac{3-2}{2}} = x^2 - 5x^{\frac{1}{2}}$

$= x^2 - 5\sqrt{x}$

e. $\boxed{\int \dfrac{2}{5}\sqrt{x+1}\; dx} = \boxed{\dfrac{2}{5}\int (x+1)^{\frac{1}{2}}\, dx} = \boxed{\dfrac{2}{5}\cdot\dfrac{1}{\frac{1}{2}+1}(x+1)^{\frac{1}{2}+1} + c} = \boxed{\dfrac{2}{5}\cdot\dfrac{1}{\frac{1+2}{2}}(x+1)^{\frac{1+2}{2}} + c} = \boxed{\dfrac{2}{5}\cdot\dfrac{2}{3}(x+1)^{\frac{3}{2}} + c}$

$$= \boxed{\frac{4}{15}(x+1)^{\frac{3}{2}}+c} = \boxed{\frac{4}{15}\sqrt{(x+1)^3}+c} = \boxed{\frac{4}{15}(x+1)\sqrt{x+1}+c}$$

Check: Let $y=\frac{4}{15}(x+1)^{\frac{3}{2}}+c$ then $y' = \frac{4}{15}\cdot\frac{3}{2}(x+1)^{\frac{3}{2}-1}+0 = \frac{12}{30}(x+1)^{\frac{3-2}{2}} = \frac{2}{5}(x+1)^{\frac{1}{2}} = \frac{2}{5}\sqrt{x+1}$

f. $\boxed{\int \sqrt{3x+2}\,dx} = \boxed{\int (3x+2)^{\frac{1}{2}}\,dx} = \boxed{\frac{1}{\frac{1}{2}+1}(3x+2)^{\frac{1}{2}+1}+c} = \boxed{\frac{1}{\frac{1+2}{2}}(3x+2)^{\frac{1+2}{2}}+c} = \boxed{\frac{2}{3}(3x+2)^{\frac{3}{2}}+c}$

$$= \boxed{\frac{2}{3}\sqrt{(3x+2)^3}+c} = \boxed{\frac{2}{3}(3x+2)\sqrt{3x+2}+c}$$

Check: Let $y=\frac{2}{3}(3x+2)^{\frac{3}{2}}+c$ then $y' = \frac{2}{3}\cdot\frac{3}{2}(3x+2)^{\frac{3}{2}-1}+0 = \frac{6}{6}(3x+2)^{\frac{3-2}{2}} = (3x+2)^{\frac{1}{2}} = \sqrt{3x+2}$

g. $\boxed{\int (2x-1)^3\,dx} = \boxed{\int \left[(2x)^3 -1^3 +3\cdot(2x)^2\cdot1-3\cdot2x\cdot1^2\right]dx} = \boxed{\int \left(8x^3 -1+12x^2 -6x\right)dx} = \boxed{8\int x^3\,dx+12\int x^2\,dx}$

$$\boxed{-6\int x\,dx-\int dx} = \boxed{\frac{8}{3+1}x^{3+1}+\frac{12}{2+1}x^{2+1}-\frac{6}{1+1}x^{1+1}-\frac{1}{0+1}x^{0+1}+c} = \boxed{2x^4+4x^3-3x^2-x+c}$$

Check: Let $y=2x^4+4x^3-3x^2-x+c$, then $y' = 2\cdot4x^{4-1}+4\cdot3x^{3-1}-3\cdot2x^{2-1}-x^{1-1}+0$

$$= 8x^3+12x^2-6x-1 = (2x-1)^3$$

h. $\boxed{\int (2x+1)^2\,dx} = \boxed{\int \left(4x^2+4x+1\right)dx} = \boxed{4\int x^2\,dx+4\int x\,dx+\int dx} = \boxed{\frac{4}{2+1}x^{2+1}+\frac{4}{1+1}x^{1+1}+\frac{1}{0+1}x^{0+1}+c}$

$$= \boxed{\frac{4}{3}x^3+\frac{4}{2}x^2+x+c} = \boxed{\frac{4}{3}x^3+2x^2+x+c}$$

Check: Let $y=\frac{4}{3}x^3+2x^2+x+c$ then $y' = \frac{4}{3}\cdot3x^{3-1}+2\cdot2x^{2-1}+1+0 = 4x^2+4x+1 = (2x+1)^2$

i. $\boxed{\int 6x^2\left(x^3+1\right)dx} = \boxed{\int \left(6x^5+6x^2\right)dx} = \boxed{6\int x^5\,dx+6\int x^2\,dx} = \boxed{\frac{6}{5+1}x^{5+1}+\frac{6}{2+1}x^{2+1}+c} = \boxed{\frac{6}{6}x^6+\frac{6}{3}x^3+c}$

$$= \boxed{x^6+\frac{2}{1}x^3+c} = \boxed{x^6+2x^3+c}$$

Check: Let $y=x^6+2x^3+c$, then $y' = 6x^{6-1}+2\cdot3x^{3-1}+0 = 6x^5+6x^2$

In the next section we will discuss a more systematic method of integration, referred to as the substitution method, in evaluating more difficult integrals.

1. Evaluate the following indefinite integrals:

   a. $\int -3\,dx =$

   b. $\int 5k\,dx =$

   c. $\int \frac{2}{3}x\,dx =$

   d. $\int x^5 dx =$

   e. $\int ax^7 dx =$

   f. $\int \left(x^4 + x^3\right) dx =$

   g. $\int \frac{dx}{x^4} =$

   h. $\int \left(\frac{1}{x^4} - \frac{1}{x^2}\right) dx =$

   i. $\int \sqrt[3]{x^6}\,dx =$

2. Evaluate the following indefinite integrals:

   a. $\int \left(\sqrt[3]{x^2} + x\right) dx =$

   b. $\int -\frac{1}{2\sqrt{t}}\,dt =$

   c. $\int \frac{1}{\sqrt[7]{x^2}}\,dx =$

   d. $\int (1+x)\sqrt{x}\,dx =$

   e. $\int \left(2x^2 + 1\right)(x-1)dx =$

   f. $\int x\,(3x-1)^2 dx =$

   g. $\int (2+x)^3 dx =$

   h. $\int (2-x)^3 dx =$

   i. $\int \frac{y^5 + 4y^2}{y^2}\,dy =$

4.2 Integration Using the Substitution Method

A method of solving integrals in which a change in variable can often change a difficult integral into a simpler form is referred to as the Substitution method. In this section we will learn how to use this method.

Given the integral $\int f[g(x)] \cdot g'(x) dx$, where $f(x)$ and $g'(x)$ are continuous functions, use the following steps in order to solve the integral by the Substitution method:

First - Let $u = g(x)$, then $\dfrac{du}{dx} = g'(x)$ which implies $du = g'(x) \cdot dx$ and $dx = \dfrac{du}{g'(x)}$.

Second - Replace $g(x)$ with u and dx with $\dfrac{du}{g'(x)}$ to obtain

$$\int f[g(x)] \cdot g'(x) dx = \int f[u] \cdot g'(x) \cdot \frac{du}{g'(x)} = \int f[u] \cdot du$$

Third - Solve the integral by integrating the function $f(u)$ with respect to u using integration formulas.

Fourth - Change the solution back to its original form by replacing u with $g(x)$.

Fifth - Check the answer by taking the derivative of the solution.

In the following examples we will solve integrals using the above substitution method.

Example 4.2-1: Evaluate the following indefinite integrals:

a. $\int 5x^4 (x^5 - 3)^2 dx =$ b. $\int x^2 (x^3 + 1)^2 dx =$ c. $\int 4x^3 (x^4 - 1)^3 dx =$

d. $\int 2x\sqrt{1-x^2}\, dx =$ e. $\int x\sqrt{x^2+3}\, dx =$ f. $\int \sqrt{10x+1}\, dx =$

g. $\int \sqrt[3]{5x-1}\, dx =$ h. $\int 6x\sqrt[5]{3x^2+2}\, dx =$ i. $\int \dfrac{1}{\sqrt{x+6}}\, dx =$

Solutions:

a. Given $\int 5x^4 (x^5 - 3)^2 dx$ let $u = x^5 - 3$, then $\dfrac{du}{dx} = \dfrac{d}{dx}(x^5 - 3) = 5x^4$ which implies that $du = 5x^4 dx$

and $dx = \dfrac{du}{5x^4}$. Substituting the equivalent values of $x^5 - 3$ and dx back into the integral we obtain

$$\boxed{\int 5x^4 (x^5 - 3)^2 dx} = \boxed{\int 5x^4 \cdot u^2 \cdot \frac{du}{5x^4}} = \boxed{\int u^2 du} = \boxed{\frac{1}{2+1} u^{2+1} + c} = \boxed{\frac{1}{3} u^3 + c} = \boxed{\frac{1}{3}(x^5 - 3)^3 + c}$$

Check: Let $y = \dfrac{1}{3}(x^5 - 3)^3 + c$, then $y' = \dfrac{1}{3} \cdot 3(x^5 - 3)^{3-1} \cdot 5x^4 + 0 = 5x^4 (x^5 - 3)^2$

b. Given $\int x^2 (x^3 + 1)^2 dx$ let $u = x^3 + 1$, then $\dfrac{du}{dx} = \dfrac{d}{dx}(x^3 + 1) = 3x^2$ which implies that $du = 3x^2 dx$

and $dx = \dfrac{du}{3x^2}$. Substituting the equivalent values of $x^3 + 1$ and dx back into the integral we obtain

$$\boxed{\int x^2\left(x^3+1\right)^2 dx}=\boxed{\int x^2\cdot u^2\cdot\dfrac{du}{3x^2}}=\boxed{\dfrac{1}{3}\int u^2 du}=\boxed{\dfrac{1}{3}\dfrac{1}{2+1}u^{2+1}+c}=\boxed{\dfrac{1}{9}u^3+c}=\boxed{\dfrac{1}{9}\left(x^3+1\right)^3+c}$$

Check: Let $y=\dfrac{1}{9}\left(x^3+1\right)^3+c$, then $y'=\dfrac{1}{9}\cdot 3\left(x^3+1\right)^{3-1}\cdot 3x^2+0=\dfrac{1}{3}\left(x^3+1\right)^2\cdot 3x^2=x^2\left(x^3+1\right)^2$

c. Given $\int 4x^3\left(x^4-1\right)^3 dx$ let $u=x^4-1$, then $\dfrac{du}{dx}=\dfrac{d}{dx}\left(x^4-1\right)=4x^3$ which implies that $du=4x^3 dx$

and $dx=\dfrac{du}{4x^3}$. Substituting the equivalent values of x^4-1 and dx back into the integral we obtain

$$\boxed{\int 4x^3\left(x^4-1\right)^3 dx}=\boxed{\int 4x^3\cdot u^3\cdot\dfrac{du}{4x^3}}=\boxed{\int u^3 du}=\boxed{\dfrac{1}{3+1}u^{3+1}+c}=\boxed{\dfrac{1}{4}u^4+c}=\boxed{\dfrac{1}{4}\left(x^4-1\right)^4+c}$$

Check: Let $y=\dfrac{1}{4}\left(x^4-1\right)^4+c$, then $y'=\dfrac{1}{4}\cdot 4\left(x^4-1\right)^{4-1}\cdot 4x^3+0=4x^3\left(x^4-1\right)^3$

d. Given $\int 2x\sqrt{1-x^2}\,dx$ let $u=1-x^2$, then $\dfrac{du}{dx}=\dfrac{d}{dx}\left(1-x^2\right)=-2x$ which implies that $du=-2x\,dx$

and $dx=\dfrac{du}{-2x}$. Substituting the equivalent values of $1-x^2$ and dx back into the integral we obtain

$$\boxed{\int 2x\sqrt{1-x^2}\,dx}=\boxed{\int 2x\cdot\sqrt{u}\cdot\dfrac{du}{-2x}}=\boxed{-\int u^{\frac{1}{2}}du}=\boxed{-\dfrac{1}{\frac{1}{2}+1}u^{\frac{1}{2}+1}+c}=\boxed{-\dfrac{2}{3}u^{\frac{3}{2}}+c}=\boxed{-\dfrac{2}{3}\left(1-x^2\right)^{\frac{3}{2}}+c}$$

Check: Let $y=-\dfrac{2}{3}\left(1-x^2\right)^{\frac{3}{2}}+c$, then $y'=-\dfrac{2}{3}\cdot\dfrac{3}{2}\left(1-x^2\right)^{\frac{3}{2}-1}\cdot-2x+0=2x\left(1-x^2\right)^{\frac{1}{2}}=2x\sqrt{1-x^2}$

e. Given $\int x\sqrt{x^2+3}\,dx$ let $u=x^2+3$, then $\dfrac{du}{dx}=\dfrac{d}{dx}\left(x^2+3\right)=2x$ which implies that $du=2x\,dx$

and $dx=\dfrac{du}{2x}$. Substituting the equivalent values of x^2+3 and dx back into the integral we obtain

$$\boxed{\int x\sqrt{x^2+3}\,dx}=\boxed{\int x\cdot\sqrt{u}\cdot\dfrac{du}{2x}}=\boxed{\dfrac{1}{2}\int u^{\frac{1}{2}}du}=\boxed{\dfrac{1}{2}\cdot\dfrac{1}{\frac{1}{2}+1}u^{\frac{1}{2}+1}+c}=\boxed{\dfrac{1}{2}\cdot\dfrac{2}{3}u^{\frac{3}{2}}+c}=\boxed{\dfrac{1}{3}u^{\frac{3}{2}}+c}=\boxed{\dfrac{1}{3}\left(x^2+3\right)^{\frac{3}{2}}+c}$$

Check: Let $y=\dfrac{1}{3}\left(x^2+3\right)^{\frac{3}{2}}+c$, then $y'=\dfrac{1}{3}\cdot\dfrac{3}{2}\left(x^2+3\right)^{\frac{3}{2}-1}\cdot 2x+0=x\left(x^2+3\right)^{\frac{1}{2}}=x\sqrt{x^2+3}$

f. Given $\int\sqrt{10x+1}\,dx$ let $u=10x+1$, then $\dfrac{du}{dx}=\dfrac{d}{dx}(10x+1)=10$ which implies that $du=10\,dx$ and

$dx=\dfrac{du}{10}$. Substituting the equivalent values of $10x+1$ and dx back into the integral we obtain

$$\boxed{\int\sqrt{10x+1}\,dx}=\boxed{\int\sqrt{u}\cdot\dfrac{du}{10}}=\boxed{\int\dfrac{u^{\frac{1}{2}}}{10}du}=\boxed{\dfrac{1}{10}\cdot\dfrac{1}{\frac{1}{2}+1}u^{\frac{1}{2}+1}+c}=\boxed{\dfrac{1}{10}\cdot\dfrac{2}{3}u^{\frac{3}{2}}+c}=\boxed{\dfrac{1}{15}u^{\frac{3}{2}}+c}=\boxed{\dfrac{1}{15}\left(10x+1\right)^{\frac{3}{2}}+c}$$

Check: Let $y = \dfrac{1}{15}(10x+1)^{\frac{3}{2}} + c$, then $y' = \dfrac{1}{15}\cdot\dfrac{3}{2}(10x+1)^{\frac{3}{2}-1}\cdot10+0 = (10x+1)^{\frac{1}{2}} = \sqrt{10x+1}$

g. Given $\int \sqrt[3]{5x-1}\,dx$ let $u = 5x-1$, then $\dfrac{du}{dx} = \dfrac{d}{dx}(5x-1) = 5$ which implies that $du = 5\,dx$ and

$dx = \dfrac{du}{5}$. Substituting the equivalent values of $5x-1$ and dx back into the integral we obtain

$$\boxed{\int\sqrt[3]{5x-1}\,dx} = \boxed{\int\sqrt[3]{u}\cdot\dfrac{du}{5}} = \boxed{\dfrac{1}{5}\int u^{\frac{1}{3}}\,du} = \boxed{\dfrac{1}{5}\cdot\dfrac{1}{\frac{1}{3}+1}u^{\frac{1}{3}+1}+c} = \boxed{\dfrac{1}{5}\cdot\dfrac{3}{4}u^{\frac{4}{3}}+c} = \boxed{\dfrac{3}{20}u^{\frac{4}{3}}+c} = \boxed{\dfrac{3}{20}(5x-1)^{\frac{4}{3}}+c}$$

Check: Let $y = \dfrac{3}{20}(5x-1)^{\frac{4}{3}} + c$, then $y' = \dfrac{3}{20}\cdot\dfrac{4}{3}(5x-1)^{\frac{4}{3}-1}\cdot5+0 = (5x-1)^{\frac{1}{3}} = \sqrt[3]{5x-1}$

h. Given $\int 6x\sqrt[5]{3x^2+2}\,dx$ let $u = 3x^2+2$ then $\dfrac{du}{dx} = \dfrac{d}{dx}\left(3x^2+2\right) = 6x$ which implies that $du = 6x\,dx$

and $dx = \dfrac{du}{6x}$. Substituting the equivalent values of $3x^2+2$ and dx back into the integral we obtain

$$\boxed{\int 6x\sqrt[5]{3x^2+2}\,dx} = \boxed{\int 6x\cdot\sqrt[5]{u}\cdot\dfrac{du}{6x}} = \boxed{\int u^{\frac{1}{5}}\,du} = \boxed{\dfrac{1}{\frac{1}{5}+1}u^{\frac{1}{5}+1}+c} = \boxed{\dfrac{5}{6}u^{\frac{6}{5}}+c} = \boxed{\dfrac{5}{6}\left(3x^2+2\right)^{\frac{6}{5}}+c}$$

Check: Let $y = \dfrac{5}{6}\left(3x^2+2\right)^{\frac{6}{5}} + c$, then $y' = \dfrac{5}{6}\cdot\dfrac{6}{5}\left(3x^2+2\right)^{\frac{6}{5}-1}\cdot6x+0 = 6x\left(3x^2+2\right)^{\frac{1}{5}} = 6x\sqrt[5]{3x^2+2}$

i. Given $\int\dfrac{1}{\sqrt{x+6}}\,dx$ let $u = x+6$, then $\dfrac{du}{dx} = \dfrac{d}{dx}(x+6) = 1$ which implies that $du = dx$. Substituting

the equivalent values of $x+6$ and dx back into the integral we obtain

$$\boxed{\int\dfrac{1}{\sqrt{x+6}}\,dx} = \boxed{\int\dfrac{1}{\sqrt{u}}\,du} = \boxed{\int u^{-\frac{1}{2}}\,du} = \boxed{\dfrac{1}{1-\frac{1}{2}}u^{1-\frac{1}{2}}+c} = \boxed{\dfrac{1}{\frac{2-1}{2}}u^{\frac{2-1}{2}}+c} = \boxed{2u^{\frac{1}{2}}+c} = \boxed{2(x+6)^{\frac{1}{2}}+c}$$

Check: Let $y = 2(x+6)^{\frac{1}{2}} + c$, then $y' = 2\cdot\dfrac{1}{2}(x+6)^{\frac{1}{2}-1}\cdot1+0 = (x+6)^{-\frac{1}{2}} = \dfrac{1}{(x+6)^{\frac{1}{2}}} = \dfrac{1}{\sqrt{x+6}}$

Example 4.2-2: Evaluate the following indefinite integrals:

a. $\int\dfrac{10}{\sqrt{x-5}}\,dx =$

b. $\int\dfrac{x}{\sqrt{x^2-3}}\,dx =$

c. $\int\dfrac{x}{\sqrt[3]{x^2-1}}\,dx =$

d. $\int\dfrac{x^2}{\sqrt{x^3+1}}\,dx =$

e. $\int\dfrac{5x^4}{\sqrt{x^5+3}}\,dx =$

f. $\int\dfrac{5x^4}{\sqrt[5]{x^5+3}}\,dx =$

g. $\int\dfrac{4x^3+6x}{\sqrt{x^4+3x^2+5}}\,dx =$

h. $\int\dfrac{3x^2+4x}{\sqrt[3]{x^3+2x^2-1}}\,dx =$

i. $\int\dfrac{2x-1}{\sqrt[4]{x^2-x-3}}\,dx =$

Solutions:

a. Given $\int \dfrac{10}{\sqrt{x-5}}\,dx$ let $u = x-5$, then $\dfrac{du}{dx} = \dfrac{d}{dx}(x-5) = 1$ which implies that $du = dx$. Substituting

the equivalent values of $x-5$ and dx back into the integral we obtain

$$\boxed{\int \dfrac{10}{\sqrt{x-5}}\,dx} = \boxed{\int \dfrac{10}{\sqrt{u}}\,du} = \boxed{10\int u^{-\frac{1}{2}}\,du} = \boxed{\dfrac{10}{1-\frac{1}{2}}u^{1-\frac{1}{2}} +c} = \boxed{\dfrac{10}{\frac{2-1}{2}}u^{\frac{2-1}{2}} +c} = \boxed{20u^{\frac{1}{2}} +c} = \boxed{20(x-5)^{\frac{1}{2}} +c}$$

Check: Let $y = 20(x-5)^{\frac{1}{2}} +c$, then $y' = 20\cdot\dfrac{1}{2}(x-5)^{\frac{1}{2}-1} +0 = 10(x-5)^{-\frac{1}{2}} = \dfrac{10}{(x-5)^{\frac{1}{2}}} = \dfrac{10}{\sqrt{x-5}}$

b. Given $\int \dfrac{x}{\sqrt{x^2-3}}\,dx$ let $u = x^2-3$, then $\dfrac{du}{dx} = \dfrac{d}{dx}(x^2-3) = 2x$ which implies that $du = 2x\,dx$ and

$dx = \dfrac{du}{2x}$. Substituting the equivalent values of x^2-3 and dx back into the integral we obtain

$$\boxed{\int \dfrac{x}{\sqrt{x^2-3}}\,dx} = \boxed{\int \dfrac{x}{\sqrt{u}}\cdot\dfrac{du}{2x}} = \boxed{\dfrac{1}{2}\int u^{-\frac{1}{2}}\,du} = \boxed{\dfrac{1}{2}\cdot\dfrac{1}{1-\frac{1}{2}}u^{1-\frac{1}{2}} +c} = \boxed{\dfrac{1}{2}\cdot\dfrac{1}{\frac{2-1}{2}}u^{\frac{2-1}{2}} +c} = \boxed{u^{\frac{1}{2}} +c} = \boxed{(x^2-3)^{\frac{1}{2}} +c}$$

Check: Let $y = (x^2-3)^{\frac{1}{2}} +c$, then $y' = \dfrac{1}{2}(x^2-3)^{\frac{1}{2}-1} +0 = \dfrac{1}{2}(x^2-3)^{-\frac{1}{2}}\cdot 2x = \dfrac{x}{(x^2-3)^{\frac{1}{2}}} = \dfrac{x}{\sqrt{x^2-3}}$

c. Given $\int \dfrac{x}{\sqrt[3]{x^2-1}}\,dx$ let $u = x^2-1$, then $\dfrac{du}{dx} = \dfrac{d}{dx}(x^2-1) = 2x$ which implies that $du = 2x\,dx$ and

$dx = \dfrac{du}{2x}$. Substituting the equivalent values of x^2-1 and dx back into the integral we obtain

$$\boxed{\int \dfrac{x}{\sqrt[3]{x^2-1}}\,dx} = \boxed{\int \dfrac{x}{\sqrt[3]{u}}\cdot\dfrac{du}{2x}} = \boxed{\dfrac{1}{2}\int u^{-\frac{1}{3}}\,du} = \boxed{\dfrac{1}{2}\cdot\dfrac{1}{1-\frac{1}{3}}u^{1-\frac{1}{3}} +c} = \boxed{\dfrac{1}{2}\cdot\dfrac{1}{\frac{3-1}{3}}u^{\frac{3-1}{3}} +c} = \boxed{\dfrac{3}{4}u^{\frac{2}{3}} +c} = \boxed{\dfrac{3}{4}(x^2-1)^{\frac{2}{3}} +c}$$

Check: Let $y = \dfrac{3}{4}(x^2-1)^{\frac{2}{3}} +c$, then $y' = \dfrac{3}{4}\cdot\dfrac{2}{3}(x^2-1)^{\frac{2}{3}-1}\cdot 2x +0 = \dfrac{1}{2}(x^2-1)^{\frac{2-3}{3}}\cdot 2x = x(x^2-1)^{-\frac{1}{3}}$

$$= \dfrac{x}{(x^2-1)^{\frac{1}{3}}} = \dfrac{x}{\sqrt[3]{x^2-1}}$$

d. Given $\int \dfrac{x^2}{\sqrt{x^3+1}}\,dx$ let $u = x^3+1$, then $\dfrac{du}{dx} = \dfrac{d}{dx}(x^3+1) = 3x^2$ which implies that $du = 3x^2\,dx$

and $dx = \dfrac{du}{3x^2}$. Substituting the equivalent values of x^3+1 and dx back into the integral we obtain

$$\boxed{\int \dfrac{x^2}{\sqrt{x^3+1}}\,dx} = \boxed{\int \dfrac{x^2}{\sqrt{u}}\cdot\dfrac{du}{3x^2}} = \boxed{\dfrac{1}{3}\int u^{-\frac{1}{2}}\,du} = \boxed{\dfrac{1}{3}\cdot\dfrac{1}{1-\frac{1}{2}}u^{1-\frac{1}{2}} +c} = \boxed{\dfrac{1}{3}\cdot\dfrac{1}{\frac{2-1}{2}}u^{\frac{2-1}{2}} +c} = \boxed{\dfrac{2}{3}u^{\frac{1}{2}} +c} = \boxed{\dfrac{2}{3}(x^3+1)^{\frac{1}{2}} +c}$$

Check: Let $y = \frac{2}{3}\left(x^3+1\right)^{\frac{1}{2}} + c$, then $y' = \frac{2}{3} \cdot \frac{1}{2}\left(x^3+1\right)^{\frac{1}{2}-1} \cdot 3x^2 + 0 = \frac{1}{3}\left(x^3+1\right)^{\frac{1-2}{2}} \cdot 3x^2 = x^2\left(x^3+1\right)^{-\frac{1}{2}}$

$$= \frac{x^2}{\left(x^3+1\right)^{\frac{1}{2}}} = \frac{x^2}{\sqrt{x^3+1}}$$

e. Given $\int \frac{5x^4}{\sqrt{x^5+3}}\,dx$ let $u = x^5+3$, then $\frac{du}{dx} = \frac{d}{dx}\left(x^5+3\right) = 5x^4$ which implies that $du = 5x^4\,dx$

and $dx = \frac{du}{5x^4}$. Substituting the equivalent values of x^5+3 and dx back into the integral we obtain

$$\boxed{\int \frac{5x^4}{\sqrt{x^5+3}}\,dx} = \boxed{\int \frac{5x^4}{\sqrt{u}} \cdot \frac{du}{5x^4}} = \boxed{\int u^{-\frac{1}{2}}\,du} = \boxed{\frac{1}{1-\frac{1}{2}}u^{1-\frac{1}{2}}+c} = \boxed{\frac{1}{\frac{2-1}{2}}u^{\frac{2-1}{2}}+c} = \boxed{2u^{\frac{1}{2}}+c} = \boxed{2\left(x^5+3\right)^{\frac{1}{2}}+c}$$

Check: Let $y = 2\left(x^5+3\right)^{\frac{1}{2}} + c$, then $y' = 2 \cdot \frac{1}{2}\left(x^5+3\right)^{\frac{1}{2}-1} \cdot 5x^4 + 0 = \left(x^5+3\right)^{\frac{1-2}{2}} \cdot 5x^4 = 5x^4\left(x^5+3\right)^{-\frac{1}{2}}$

$$= \frac{5x^4}{\left(x^5+3\right)^{\frac{1}{2}}} = \frac{5x^4}{\sqrt{x^5+3}}$$

f. Given $\int \frac{5x^4}{\sqrt[5]{x^5+3}}\,dx$ let $u = x^5+3$, then $\frac{du}{dx} = \frac{d}{dx}\left(x^5+3\right) = 5x^4$ which implies that $du = 5x^4\,dx$

and $dx = \frac{du}{5x^4}$. Substituting the equivalent values of x^5+3 and dx back into the integral we obtain

$$\boxed{\int \frac{5x^4}{\sqrt[5]{x^5+3}}\,dx} = \boxed{\int \frac{5x^4}{\sqrt[5]{u}} \cdot \frac{du}{5x^4}} = \boxed{\int u^{-\frac{1}{5}}\,du} = \boxed{\frac{1}{1-\frac{1}{5}}u^{1-\frac{1}{5}}+c} = \boxed{\frac{1}{\frac{5-1}{5}}u^{\frac{5-1}{5}}+c} = \boxed{\frac{5}{4}u^{\frac{4}{5}}+c} = \boxed{\frac{5}{4}\left(x^5+3\right)^{\frac{4}{5}}+c}$$

Check: Let $y = \frac{5}{4}\left(x^5+3\right)^{\frac{4}{5}} + c$, then $y' = \frac{5}{4} \cdot \frac{4}{5}\left(x^5+3\right)^{\frac{4}{5}-1} \cdot 5x^4 + 0 = \left(x^5+3\right)^{\frac{4-5}{5}} \cdot 5x^4 = 5x^4\left(x^5+3\right)^{-\frac{1}{5}}$

$$= \frac{5x^4}{\left(x^5+3\right)^{\frac{1}{5}}} = \frac{5x^4}{\sqrt[5]{x^5+3}}$$

g. Given $\int \frac{4x^3+6x}{\sqrt{x^4+3x^2+5}}\,dx$ let $u = x^4+3x^2+5$, then $\frac{du}{dx} = \frac{d}{dx}\left(x^4+3x^2+5\right) = 4x^3+6x$ which implies

that $du = \left(4x^3+6x\right)dx$ and $dx = \frac{du}{4x^3+6x}$. Substituting the equivalent values of x^4+3x^2+5 and dx

back into the integral we obtain $\boxed{\int \frac{4x^3+6x}{\sqrt{x^4+3x^2+5}}\,dx} = \boxed{\int \frac{4x^3+6x}{\sqrt{u}} \cdot \frac{du}{4x^3+6x}} = \boxed{\int \frac{du}{\sqrt{u}}} = \boxed{\int u^{-\frac{1}{2}}\,du}$

$$= \boxed{\frac{1}{1-\frac{1}{2}}u^{1-\frac{1}{2}}+c} = \boxed{\frac{1}{\frac{2-1}{2}}u^{\frac{2-1}{2}}+c} = \boxed{2u^{\frac{1}{2}}+c} = \boxed{2\left(x^4+3x^2+5\right)^{\frac{1}{2}}+c} = \boxed{2\sqrt{x^4+3x^2+5}+c}$$

Check: Let $y = 2\left(x^4 + 3x^2 + 5\right)^{\frac{1}{2}} + c$, then $y' = 2 \cdot \frac{1}{2}\left(x^4 + 3x^2 + 5\right)^{\frac{1}{2}-1}\left(4x^3 + 6x\right) + 0 = \left(x^4 + 3x^2 + 5\right)^{\frac{1-2}{2}}$

$$\times\left(4x^3 + 6x\right) = \left(x^4 + 3x^2 + 5\right)^{-\frac{1}{2}}\left(4x^3 + 6x\right) = \frac{4x^3 + 6x}{\left(x^4 + 3x^2 + 5\right)^{\frac{1}{2}}} = \frac{4x^3 + 6x}{\sqrt{x^4 + 3x^2 + 5}}$$

h. Given $\int \dfrac{3x^2 + 4x}{\sqrt[3]{x^3 + 2x^2 - 1}}\, dx$ let $u = x^3 + 2x^2 - 1$, then $\dfrac{du}{dx} = \dfrac{d}{dx}\left(x^3 + 2x^2 - 1\right) = 3x^2 + 4x$ which implies

that $du = \left(3x^2 + 4x\right)dx$ and $dx = \dfrac{du}{3x^2 + 4x}$. Substituting the equivalent values of $x^3 + 2x^2 - 1$ and dx

back into the integral we obtain $\boxed{\int \dfrac{3x^2 + 4x}{\sqrt[3]{x^3 + 2x^2 - 1}}\, dx} = \boxed{\int \dfrac{3x^2 + 4x}{\sqrt[3]{u}} \cdot \dfrac{du}{3x^2 + 4x}} = \boxed{\int \dfrac{du}{\sqrt[3]{u}}} = \boxed{\int u^{-\frac{1}{3}}\, du}$

$= \boxed{\dfrac{1}{1 - \frac{1}{3}} u^{1-\frac{1}{3}} + c} = \boxed{\dfrac{1}{\frac{3-1}{3}} u^{\frac{3-1}{3}} + c} = \boxed{\dfrac{3}{2} u^{\frac{2}{3}} + c} = \boxed{\dfrac{3}{2}\left(x^3 + 2x^2 - 1\right)^{\frac{2}{3}} + c} = \boxed{\dfrac{3}{2}\sqrt[3]{\left(x^3 + 2x^2 - 1\right)^2} + c}$

Check: Let $y = \dfrac{3}{2}\left(x^3 + 2x^2 - 1\right)^{\frac{2}{3}} + c$, then $y' = \dfrac{3}{2} \cdot \dfrac{2}{3}\left(x^3 + 2x^2 - 1\right)^{\frac{2}{3}-1}\left(3x^2 + 4x\right) + 0 = \left(x^3 + 2x^2 - 1\right)^{\frac{2-3}{3}}$

$$\times\left(3x^2 + 4x\right) = \left(x^3 + 2x^2 - 1\right)^{-\frac{1}{3}}\left(3x^2 + 4x\right) = \frac{3x^2 + 4x}{\left(x^3 + 2x^2 - 1\right)^{\frac{1}{3}}} = \frac{3x^2 + 4x}{\sqrt[3]{x^3 + 2x^2 - 1}}$$

i. Given $\int \dfrac{2x - 1}{\sqrt[4]{x^2 - x - 3}}\, dx$ let $u = x^2 - x - 3$, then $\dfrac{du}{dx} = \dfrac{d}{dx}\left(x^2 - x - 3\right) = 2x - 1$ which implies that

$du = (2x - 1)dx$ and $dx = \dfrac{du}{2x - 1}$. Substituting the equivalent values of $x^2 - x - 3$ and dx back into

the integral we obtain $\boxed{\int \dfrac{2x - 1}{\sqrt[4]{x^2 - x - 3}}\, dx} = \boxed{\int \dfrac{2x - 1}{\sqrt[4]{u}} \cdot \dfrac{du}{2x - 1}} = \boxed{\int \dfrac{du}{\sqrt[4]{u}}} = \boxed{\int u^{-\frac{1}{4}}\, du} = \boxed{\dfrac{1}{1 - \frac{1}{4}} u^{1-\frac{1}{4}} + c}$

$= \boxed{\dfrac{1}{\frac{4-1}{4}} u^{\frac{4-1}{4}} + c} = \boxed{\dfrac{4}{3} u^{\frac{3}{4}} + c} = \boxed{\dfrac{4}{3}\left(x^2 - x - 3\right)^{\frac{3}{4}} + c} = \boxed{\dfrac{4}{3}\sqrt[4]{\left(x^2 - x - 3\right)^3} + c}$

Check: Let $y = \dfrac{4}{3}\left(x^2 - x - 3\right)^{\frac{3}{4}} + c$, then $y' = \dfrac{4}{3} \cdot \dfrac{3}{4}\left(x^2 - x - 3\right)^{\frac{3}{4}-1}(2x - 1) + 0 = \left(x^2 - x - 3\right)^{\frac{3-4}{4}}(2x - 1)$

$$= \left(x^2 - x - 3\right)^{-\frac{1}{4}}(2x - 1) = \frac{2x - 1}{\left(x^2 - x - 3\right)^{\frac{1}{4}}} = \frac{2x - 1}{\sqrt[4]{x^2 - x - 3}}$$

Example 4.2-3: Evaluate the following indefinite integrals:

a. $\int \dfrac{x^2+2}{\sqrt[5]{x^3+6x+3}}\,dx =$

b. $\int \dfrac{5x^4-4x^3+1}{\sqrt{x^5-x^4+x}}\,dx =$

c. $\int \dfrac{x^4-2}{\sqrt{x^5-10x}}\,dx =$

d. $\int \dfrac{z}{\sqrt{2z^2-1}}\,dz =$

e. $\int \dfrac{x^3}{\sqrt{x^4-1}}\,dx =$

f. $\int \dfrac{dx}{(1+\sqrt{x})^2\,\sqrt{x}} =$

g. $\int \dfrac{x}{(4-x^2)^2}\,dx =$

h. $\int x^2(x^3-1)^{\frac{1}{5}}\,dx =$

i. $\int \dfrac{8x}{\sqrt{x^2-3}}\,dx =$

Solutions:

a. Given $\int \dfrac{x^2+2}{\sqrt[5]{x^3+6x+3}}\,dx$ let $u=x^3+6x+3$, then $\dfrac{du}{dx}=\dfrac{d}{dx}\left(x^3+6x+3\right)=3x^2+6$ which implies that

$du=\left(3x^2+6\right)dx$ and $dx=\dfrac{du}{3x^2+6}$. Substituting the equivalent values of x^3+6x+3 and dx back

into the integral we obtain $\boxed{\int \dfrac{x^2+2}{\sqrt[5]{x^3+6x+3}}\,dx} = \boxed{\int \dfrac{x^2+2}{\sqrt[5]{u}}\cdot\dfrac{du}{3(x^2+2)}} = \boxed{\dfrac{1}{3}\int\dfrac{du}{\sqrt[5]{u}}} = \boxed{\dfrac{1}{3}\int u^{-\frac{1}{5}}\,du}$

$= \boxed{\dfrac{1}{3}\cdot\dfrac{1}{1-\frac{1}{5}}u^{1-\frac{1}{5}}+c} = \boxed{\dfrac{1}{3}\cdot\dfrac{1}{\frac{5-1}{5}}u^{\frac{5-1}{5}}+c} = \boxed{\dfrac{5}{12}u^{\frac{4}{5}}+c} = \boxed{\dfrac{5}{12}\left(x^3+6x+3\right)^{\frac{4}{5}}+c} = \boxed{\dfrac{5}{12}\sqrt[5]{\left(x^3+6x+3\right)^4}+c}$

Check: Let $y=\dfrac{5}{12}\left(x^3+6x+3\right)^{\frac{4}{5}}+c$, then $y'=\dfrac{5}{12}\cdot\dfrac{4}{5}\left(x^3+6x+3\right)^{\frac{4}{5}-1}\cdot\left(3x^2+6\right)+0 = \dfrac{1}{3}\left(x^3+6x+3\right)^{\frac{4-5}{5}}$

$\times\left(3x^2+6\right) = \dfrac{1}{3}\left(x^3+6x+3\right)^{-\frac{1}{5}}\cdot 3\left(x^2+2\right) = \dfrac{x^2+2}{\left(x^3+6x+3\right)^{\frac{1}{5}}} = \dfrac{x^2+2}{\sqrt[5]{x^3+6x+3}}$

b. Given $\int \dfrac{5x^4-4x^3+1}{\sqrt{x^5-x^4+x}}\,dx$ let $u=x^5-x^4+x$, then $\dfrac{du}{dx}=\dfrac{d}{dx}\left(x^5-x^4+x\right)=5x^4-4x^3+1$ which implies

that $du=\left(5x^4-4x^3+1\right)dx$ and $dx=\dfrac{du}{5x^4-4x^3+1}$. Substituting the equivalent values of x^5-x^4+x

and dx back into the integral we obtain $\boxed{\int\dfrac{5x^4-4x^3+1}{\sqrt{x^5-x^4+x}}\,dx} = \boxed{\int\dfrac{5x^4-4x^3+1}{\sqrt{u}}\cdot\dfrac{du}{5x^4-4x^3+1}} = \boxed{\int\dfrac{du}{\sqrt{u}}}$

$\boxed{\int u^{-\frac{1}{2}}\,du} = \boxed{\dfrac{1}{1-\frac{1}{2}}u^{1-\frac{1}{2}}+c} = \boxed{\dfrac{1}{\frac{2-1}{2}}u^{\frac{2-1}{2}}+c} = \boxed{2u^{\frac{1}{2}}+c} = \boxed{2\left(x^5-x^4+x\right)^{\frac{1}{2}}+c} = \boxed{2\sqrt{x^5-x^4+x}+c}$

Check: Let $y=2\left(x^5-x^4+x\right)^{\frac{1}{2}}+c$, then $y' = 2\cdot\dfrac{1}{2}\left(x^5-x^4+x\right)^{\frac{1}{2}-1}\left(5x^4-4x^3+1\right)+0 = \left(x^5-x^4+x\right)^{-\frac{1}{2}}$

$$\times\left(5x^4-4x^3+1\right)=\dfrac{5x^4-4x^3+1}{\left(x^5-x^4+x\right)^{\frac{1}{2}}}=\dfrac{5x^4-4x^3+1}{\sqrt{x^5-x^4+x}}$$

c. Given $\displaystyle\int\dfrac{x^4-2}{\sqrt{x^5-10x}}\,dx$ let $u=x^5-10x$, then $\dfrac{du}{dx}=\dfrac{d}{dx}\left(x^5-10x\right)=5x^4-10$ which implies that

$du=\left(5x^4-10\right)dx$ and $dx=\dfrac{du}{5\left(x^4-2\right)}$. Substituting the equivalent values of x^5-10x and dx back

into the integral we obtain $\boxed{\displaystyle\int\dfrac{x^4-2}{\sqrt{x^5-10x}}\,dx}=\boxed{\displaystyle\int\dfrac{x^4-2}{\sqrt{u}}\cdot\dfrac{du}{5\left(x^4-2\right)}}=\boxed{\dfrac{1}{5}\displaystyle\int\dfrac{du}{\sqrt{u}}}=\boxed{\dfrac{1}{5}\displaystyle\int u^{-\frac{1}{2}}\,du}$

$=\boxed{\dfrac{1}{5}\cdot\dfrac{1}{1-\frac{1}{2}}u^{1-\frac{1}{2}}+c}=\boxed{\dfrac{1}{5}\cdot\dfrac{1}{\frac{2-1}{2}}u^{\frac{2-1}{2}}+c}=\boxed{\dfrac{2}{5}u^{\frac{1}{2}}+c}=\boxed{\dfrac{2}{5}\left(x^5-10x\right)^{\frac{1}{2}}+c}=\boxed{\dfrac{2}{5}\sqrt{x^5-10x}+c}$

Check: Let $y=\dfrac{2}{5}\left(x^5-10x\right)^{\frac{1}{2}}+c$, then $y'=\dfrac{2}{5}\cdot\dfrac{1}{2}\left(x^5-10x\right)^{\frac{1}{2}-1}\left(5x^4-10\right)+0=\dfrac{1}{5}\left(x^5-10x\right)^{-\frac{1}{2}}\left(5x^4-10\right)$

$=\dfrac{1}{5}\left(x^5-10x\right)^{-\frac{1}{2}}\cdot5\left(x^4-2\right)=\dfrac{x^4-2}{\left(x^5-10x\right)^{\frac{1}{2}}}=\dfrac{x^4-2}{\sqrt{x^5-10x}}$

d. Given $\displaystyle\int\dfrac{z}{\sqrt{2z^2-1}}\,dz$ let $u=2z^2-1$, then $\dfrac{du}{dz}=\dfrac{d}{dz}\left(2z^2-1\right)=4z$ which implies that $du=4z\,dz$

and $dz=\dfrac{du}{4z}$. Substituting the equivalent values of $2z^2-1$ and dx back into the integral we obtain

$\boxed{\displaystyle\int\dfrac{z}{\sqrt{2z^2-1}}\,dz}=\boxed{\displaystyle\int\dfrac{z}{\sqrt{u}}\cdot\dfrac{du}{4z}}=\boxed{\displaystyle\int\dfrac{u^{-\frac{1}{2}}}{4}\,du}=\boxed{\dfrac{1}{4}\cdot\dfrac{1}{1-\frac{1}{2}}u^{1-\frac{1}{2}}+c}=\boxed{\dfrac{1}{4}\cdot\dfrac{1}{\frac{2-1}{2}}u^{\frac{2-1}{2}}+c}=\boxed{\dfrac{1}{2}u^{\frac{1}{2}}+c}=\boxed{\dfrac{1}{2}\left(2z^2-1\right)^{\frac{1}{2}}+c}$

Check: Let $y=\dfrac{1}{2}\left(2z^2-1\right)^{\frac{1}{2}}+c$, then $y'=\dfrac{1}{2}\cdot\dfrac{1}{2}\left(2z^2-1\right)^{\frac{1}{2}-1}\cdot4z+0=\left(2z^2-1\right)^{-\frac{1}{2}}\cdot z=\dfrac{z}{\sqrt{2z^2-1}}$

e. Given $\displaystyle\int\dfrac{x^3}{\sqrt{x^4-1}}\,dx$ let $u=x^4-1$, then $\dfrac{du}{dx}=\dfrac{d}{dx}\left(x^4-1\right)=4x^3$ which implies that $du=4x^3dx$ and

$dx=\dfrac{du}{4x^3}$. Substituting the equivalent values of x^4-1 and dx back into the integral we obtain

$\boxed{\displaystyle\int\dfrac{x^3}{\sqrt{x^4-1}}\,dx}=\boxed{\displaystyle\int\dfrac{x^3}{\sqrt{u}}\cdot\dfrac{du}{4x^3}}=\boxed{\displaystyle\int\dfrac{u^{-\frac{1}{2}}}{4}\,du}=\boxed{\dfrac{1}{4}\cdot\dfrac{1}{1-\frac{1}{2}}u^{1-\frac{1}{2}}+c}=\boxed{\dfrac{1}{4}\cdot\dfrac{1}{\frac{2-1}{2}}u^{\frac{2-1}{2}}+c}=\boxed{\dfrac{1}{2}u^{\frac{1}{2}}+c}=\boxed{\dfrac{1}{2}\left(x^4-1\right)^{\frac{1}{2}}+c}$

Check: Let $y=\dfrac{1}{2}\left(x^4-1\right)^{\frac{1}{2}}+c$, then $y'=\dfrac{1}{2}\cdot\dfrac{1}{2}\left(x^4-1\right)^{\frac{1}{2}-1}\cdot4x^3+0=\left(x^4-1\right)^{-\frac{1}{2}}\cdot x^3=\dfrac{x^3}{\sqrt{x^4-1}}$

f. Given $\int \dfrac{dx}{\left(1+\sqrt{x}\right)^2 \sqrt{x}}$ let $u=1+\sqrt{x}$, then $\dfrac{du}{dx}=\dfrac{d}{dx}\left(1+\sqrt{x}\right)=\dfrac{1}{2\sqrt{x}}$ which implies that $du=\dfrac{dx}{2\sqrt{x}}$

and $dx=2\sqrt{x}\,du$. Substituting the equivalent values of $1+\sqrt{x}$ and dx back into the integral we obtain

$$\boxed{\int \dfrac{dx}{\left(1+\sqrt{x}\right)^2 \sqrt{x}}}=\boxed{\int \dfrac{2\sqrt{x}\,du}{u^2 \sqrt{x}}}=\boxed{\int \dfrac{2\,du}{u^2}}=\boxed{2\int u^{-2}du}=\boxed{\dfrac{2}{1-2}u^{1-2}+c}=\boxed{-2u^{-1}+c}=\boxed{-\dfrac{2}{u}+c}=\boxed{-\dfrac{2}{1+\sqrt{x}}+c}$$

Check: Let $y=-\dfrac{2}{1+\sqrt{x}}+c$, then $y'=-\dfrac{0-\frac{1}{2\sqrt{x}}\cdot 2}{\left(1+\sqrt{x}\right)^2}+0=\dfrac{\frac{1}{\sqrt{x}}}{\left(1+\sqrt{x}\right)^2}=\dfrac{1}{\sqrt{x}\left(1+\sqrt{x}\right)^2}$

g. Given $\int \dfrac{x}{\left(4-x^2\right)^2}dx$ let $u=4-x^2$, then $\dfrac{du}{dx}=\dfrac{d}{dx}\left(4-x^2\right)=-2x$ which implies that $du=-2xdx$

and $dx=\dfrac{du}{-2x}$. Substituting the equivalent values of $4-x^2$ and dx back into the integral we obtain

$$\boxed{\int \dfrac{x}{\left(4-x^2\right)^2}dx}=\boxed{\int \dfrac{x}{u^2}\cdot\dfrac{du}{-2x}}=\boxed{-\dfrac{1}{2}\int u^{-2}du}=\boxed{-\dfrac{1}{2}\cdot\dfrac{1}{-2+1}u^{-2+1}+c}=\boxed{\dfrac{1}{2}u^{-1}+c}=\boxed{\dfrac{1}{2u}+c}=\boxed{\dfrac{1}{2\left(4-x^2\right)}+c}$$

Check: Let $y=\dfrac{1}{2\left(4-x^2\right)}+c$, then $y'=\dfrac{0+4x}{4\left(4-x^2\right)^2}+0=\dfrac{4x}{4\left(4-x^2\right)^2}=\dfrac{x}{\left(4-x^2\right)^2}$

h. Given $\int x^2\left(x^3-1\right)^{\frac{1}{5}}dx$ let $u=x^3-1$, then $\dfrac{du}{dx}=\dfrac{d}{dx}\left(x^3-1\right)=3x^2$ which implies that $du=3x^2dx$

and $dx=\dfrac{du}{3x^2}$. Substituting the equivalent values of x^3-1 and dx back into the integral we obtain

$$\boxed{\int x^2\left(x^3-1\right)^{\frac{1}{5}}dx}=\boxed{\int x^2 u^{\frac{1}{5}}\cdot\dfrac{du}{3x^2}}=\boxed{\int \dfrac{u^{\frac{1}{5}}}{3}du}=\boxed{\dfrac{1}{3}\cdot\dfrac{1}{1+\frac{1}{5}}u^{1+\frac{1}{5}}+c}=\boxed{\dfrac{1}{3}\cdot\dfrac{1}{\frac{5+1}{5}}u^{\frac{5+1}{5}}+c}=\boxed{\dfrac{5}{18}u^{\frac{6}{5}}+c}=\boxed{\dfrac{5}{18}\left(x^3-1\right)^{\frac{6}{5}}+c}$$

Check: Let $y=\dfrac{5}{18}\left(x^3-1\right)^{\frac{6}{5}}+c$, then $y'=\dfrac{5}{18}\cdot\dfrac{6}{5}\left(x^3-1\right)^{\frac{6}{5}-1}\cdot 3x^2+0=\dfrac{1}{3}\left(x^3-1\right)^{\frac{6-5}{5}}\cdot 3x^2=x^2\left(x^3-1\right)^{\frac{1}{5}}$

i. Given $\int \dfrac{8x}{\sqrt{x^2-3}}dx$ let $u=x^2-3$, then $\dfrac{du}{dx}=\dfrac{d}{dx}\left(x^2-3\right)=2x$ which implies that $du=2xdx$ and

$dx=\dfrac{du}{2x}$. Substituting the equivalent values of x^2-3 and dx back into the integral we obtain

$$\boxed{\int \dfrac{8x}{\sqrt{x^2-3}}dx}=\boxed{\int \dfrac{8x}{\sqrt{u}}\cdot\dfrac{du}{2x}}=\boxed{4\int u^{-\frac{1}{2}}du}=\boxed{4\cdot\dfrac{1}{1-\frac{1}{2}}u^{1-\frac{1}{2}}+c}=\boxed{4\cdot\dfrac{1}{\frac{2-1}{2}}u^{\frac{2-1}{2}}+c}=\boxed{8u^{\frac{1}{2}}+c}=\boxed{8\left(x^2-3\right)^{\frac{1}{2}}+c}$$

Check: Let $y=8\left(x^2-3\right)^{\frac{1}{2}}+c$, then $y'=8\cdot\dfrac{1}{2}\left(x^2-3\right)^{\frac{1}{2}-1}\cdot 2x+0=4\left(x^2-3\right)^{-\frac{1}{2}}\cdot 2x=\dfrac{8x}{\sqrt{x^2-3}}$

Section 4.2 Practice Problems – Integration Using the Substitution Method

1. Evaluate the following indefinite integrals:

   a. $\int t^2 \left(1 + t^3\right) dt =$

   b. $\int 18x^2 \sqrt{6x^3 - 5}\, dx =$

   c. $\int \dfrac{5}{\sqrt{x+5}}\, dx =$

   d. $\int x \left(x^2 - 2\right)^{\frac{1}{5}} dx =$

   e. $\int \dfrac{3x}{\sqrt{x^2 + 3}}\, dx =$

   f. $\int \dfrac{t}{2} \sqrt{1 - t^2}\, dt =$

   g. $\int x^2 \left(1 - x^3\right)^2 dx =$

   h. $\int \dfrac{x^3}{\sqrt{x^4 + 3}}\, dx =$

   i. $\int x^8 \left(2x^9 + 1\right)^2 dx =$

2. Evaluate the following indefinite integrals:

   a. $\int 6x^2 \left(2x^3 - 1\right) dx =$

   b. $\int x \sqrt{1 + x^2}\, dx =$

   c. $\int \sqrt[5]{7x + 1}\, dx =$

   d. $\int x \sqrt[3]{3x^2 - 1}\, dx =$

   e. $\int \dfrac{x}{\sqrt{x^2 + 1}}\, dx =$

   f. $\int x \left(1 - x^2\right)^2 dx =$.

   g. $\int \dfrac{x^5}{\sqrt{x^6 + 3}}\, dx =$

   h. $\int \dfrac{3x}{\sqrt{x^2 - 1}}\, dx =$

   i. $\int \dfrac{5x^4 + 6x}{\sqrt{x^5 + 3x^2 + 1}}\, dx =$

4.3 Integration of Trigonometric Functions

In the following examples we will solve problems using the formulas below:

Table 4.3-1: Integration Formulas for Trigonometric Functions

1. $\int \sin x\,dx = -\cos x + c$	2. $\int \cos x\,dx = \sin x + c$	3. $\int \tan x\,dx = \ln\lvert \sec x \rvert + c$
4. $\int \cot x\,dx = \ln\lvert \sin x \rvert + c$	5. $\int \sec x\,dx = \ln\lvert \sec x + \tan x \rvert + c$	6. $\int \csc x\,dx = \ln\lvert \csc x - \cot x \rvert + c$
7. $\int \tan x \sec x\,dx = \sec x + c$	8. $\int \cot x \csc x\,dx = -\csc x + c$	9. $\int \sin^2 x\,dx = \dfrac{x}{2} - \dfrac{\sin 2x}{4} + c$
10. $\int \cos^2 x\,dx = \dfrac{x}{2} + \dfrac{\sin 2x}{4} + c$	11. $\int \tan^2 x\,dx = \tan x - x + c$	12. $\int \cot^2 x\,dx = -\cot x - x + c$
13. $\int \sec^2 x\,dx = \tan x + c$	14. $\int \csc^2 x\,dx = -\cot x + c$	

Additionally, the following formulas (identities) hold for the trigonometric functions:

1. Unit Circle Formulas

$$\sin^2 x + \cos^2 x = 1 \qquad\qquad \sec^2 x - \tan^2 x = 1 \qquad\qquad \csc^2 x - \cot^2 x = 1$$

2. Addition Formulas

$$\sin(x+y) = \sin x \cos y + \cos x \sin y \quad (1) \qquad\qquad \sin(x-y) = \sin x \cos y - \cos x \sin y \quad (2)$$

$$\cos(x+y) = \cos x \cos y - \sin x \sin y \quad (3) \qquad\qquad \cos(x-y) = \cos x \cos y + \sin x \sin y \quad (4)$$

$$\tan(x+y) = \frac{\tan x + \tan y}{1 - \tan x \tan y} \qquad\qquad \tan(x-y) = \frac{\tan x - \tan y}{1 + \tan x \tan y}$$

Note that from the identities (1) and (2) above we can obtain the formulas

$$\sin x \cos y = \frac{1}{2}\left[\sin(x-y) + \sin(x+y)\right] \qquad\qquad \cos x \sin y = \frac{1}{2}\left[\sin(x+y) - \cos(x-y)\right]$$

and from the identities (3) and (4) above we can obtain the formulas

$$\sin x \sin y = \frac{1}{2}\left[\cos(x-y) - \cos(x+y)\right] \qquad\qquad \cos x \cos y = \frac{1}{2}\left[\cos(x-y) + \cos(x+y)\right]$$

3. Half Angle Formulas

$$\sin^2 \frac{1}{2}x = \frac{1}{2}(1 - \cos x) \text{ or} \qquad 1 - \cos x = 2\sin^2 \frac{1}{2}x \quad \text{therefore} \qquad \sin^2 x = \frac{1}{2}(1 - \cos 2x)$$

$$\cos^2 \frac{1}{2}x = \frac{1}{2}(1 + \cos x) \text{ or} \qquad 1 + \cos x = 2\cos^2 \frac{1}{2}x \quad \text{therefore} \qquad \cos^2 x = \frac{1}{2}(1 + \cos 2x)$$

4. Double Angle Formulas

$$\sin 2x = 2 \sin x \cos x \text{ or} \qquad \sin x \cos x = \frac{1}{2}\sin 2x \qquad\qquad \cos 2x = \cos^2 x - \sin^2 x \qquad \text{and}$$

$$\sin^2 x \cos^2 x = \left(\tfrac{1}{2}\sin 2x\right)^2 = \tfrac{1}{4}\sin^2 2x = \tfrac{1}{8}(1-\cos 4x)$$

Also, the tangent, cotangent, secant, and cosecant functions are defined by

$$\tan x = \frac{\sin x}{\cos x} \qquad \cot x = \frac{1}{\tan x} = \frac{\cos x}{\sin x} \qquad \sec x = \frac{1}{\cos x} \qquad \csc x = \frac{1}{\sin x}$$

We should also note that sine, tangent, cotangent, and cosecant are odd functions. This implies that

$$\sin(-x) = -\sin x \qquad \tan(-x) = -\tan x \qquad \cot(-x) = -\cot x \qquad \csc(-x) = -\csc x$$

on the other hand, cosine and secant are even functions. This implies that

$$\cos(-x) = \cos x \qquad \sec(-x) = \sec x$$

Finally, we need to know how to differentiate the trigonometric functions (addressed in Chapter 3, Section 3.1) in order to check the answer to the given integrals below. The derivatives of trigonometric functions are repeated here and are as follows:

Table 4.3-2: Differentiation Formulas for Trigonometric Functions

$\dfrac{d}{dx}\sin u = \cos u \cdot \dfrac{du}{dx}$	$\dfrac{d}{dx}\cot u = -\csc^2 u \cdot \dfrac{du}{dx}$
$\dfrac{d}{dx}\cos u = -\sin u \cdot \dfrac{du}{dx}$	$\dfrac{d}{dx}\sec u = \sec u \tan u \cdot \dfrac{du}{dx}$
$\dfrac{d}{dx}\tan u = \sec^2 u \cdot \dfrac{du}{dx}$	$\dfrac{d}{dx}\csc u = -\csc u \cot u \cdot \dfrac{du}{dx}$

Let's integrate some trigonometric functions using the above integration formulas.

Example 4.3-1: Evaluate the following indefinite integrals:

a. $\displaystyle\int \sin 3x \, dx \;=$

b. $\displaystyle\int \sin \tfrac{1}{8} x \, dx \;=$

c. $\displaystyle\int \cos 5x \, dx \;=$

d. $\displaystyle\int \cos \tfrac{1}{4} x \, dx \;=$

e. $\displaystyle\int (\sin 4x + \cos 2x)\, dx \;=$

f. $\displaystyle\int \csc 5x \, dx \;=$

g. $\displaystyle\int \csc^2 5x \, dx \;=$

h. $\displaystyle\int \csc^2 \tfrac{1}{2} x \, dx \;=$

i. $\displaystyle\int x^2 \sec^2 x^3 \, dx \;=$

j. $\displaystyle\int x \sec^2 (x^2 + 1)\, dx \;=$

k. $\displaystyle\int x \csc^2 x^2 \, dx \;=$

l. $\displaystyle\int (2x+1)\csc^2 (x^2 + x)\, dx \;=$

Solutions:

a. Given $\displaystyle\int \sin 3x \, dx$ let $u = 3x$, then $\dfrac{du}{dx} = \dfrac{d}{dx}\,3x = 3$ which implies $dx = \dfrac{du}{3}$. Therefore,

$$\boxed{\int \sin 3x \, dx} = \boxed{\int \sin u \cdot \frac{du}{3}} = \boxed{\frac{1}{3}\int \sin u \, du} = \boxed{-\frac{1}{3}\cos u + c} = \boxed{-\frac{1}{3}\cos 3x + c}$$

Check: Let $y=-\frac{1}{3}\cos 3x+c$, then $y' = -\frac{1}{3}\cdot\frac{d}{dx}\cos 3x+\frac{d}{dx}c = -\frac{1}{3}\cdot-\sin 3x\cdot\frac{d}{dx}3x+0 = -\frac{1}{3}\cdot-\sin 3x\cdot 3$

$= \frac{3}{3}\cdot\sin 3x = \sin 3x$

b. Given $\int\sin\frac{x}{8}\,dx$ let $u=\frac{x}{8}$, then $\frac{du}{dx}=\frac{d}{dx}\frac{x}{8}=\frac{1}{8}$ which implies $dx=8\,du$. Therefore,

$$\boxed{\int\sin\frac{x}{8}\,dx}=\boxed{\int\sin u\cdot 8\,du}=\boxed{8\int\sin u\,du}=\boxed{-8\cos u+c}=\boxed{-8\cos\frac{x}{8}+c}$$

Check: Let $y=-8\cos\frac{x}{8}+c$, then $y' = -8\cdot\frac{d}{dx}\cos\frac{x}{8}+\frac{d}{dx}c = -8\cdot-\sin\frac{x}{8}\cdot\frac{d}{dx}\frac{x}{8}+0 = -8\cdot-\sin\frac{x}{8}\cdot\frac{1}{8}$

$= \frac{8}{8}\cdot\sin\frac{x}{8} = \sin\frac{x}{8}$

c. Given $\int\cos 5x\,dx$ let $u=5x$, then $\frac{du}{dx}=\frac{d}{dx}5x=5$ which implies $dx=\frac{du}{5}$. Therefore,

$$\boxed{\int\cos 5x\,dx}=\boxed{\int\cos u\cdot\frac{du}{5}}=\boxed{\frac{1}{5}\int\cos u\,du}=\boxed{\frac{1}{5}\sin u+c}=\boxed{\frac{1}{5}\sin 5x+c}$$

Check: Let $y=\frac{1}{5}\sin 5x+c$, then $y' = \frac{1}{5}\cdot\frac{d}{dx}\sin 5x+\frac{d}{dx}c = \frac{1}{5}\cdot\cos 5x\cdot\frac{d}{dx}5x+0 = \frac{1}{5}\cdot\cos 5x\cdot 5$

$= \frac{5}{5}\cdot\cos 5x = \cos 5x$

d. Given $\int\cos\frac{x}{4}\,dx$ let $u=\frac{x}{4}$, then $\frac{du}{dx}=\frac{d}{dx}\frac{x}{4}=\frac{1}{4}$ which implies $dx=4\,du$. Therefore,

$$\boxed{\int\cos\frac{x}{4}\,dx}=\boxed{\int\cos u\cdot 4\,du}=\boxed{4\int\cos u\,du}=\boxed{4\sin u+c}=\boxed{4\sin\frac{x}{4}+c}$$

Check: Let $y=4\sin\frac{x}{4}+c$, then $y' = 4\cdot\frac{d}{dx}\sin\frac{x}{4}+\frac{d}{dx}c = 4\cdot\cos\frac{x}{4}\cdot\frac{d}{dx}\frac{x}{4}+0 = 4\cdot\cos\frac{x}{4}\cdot\frac{1}{4}$

$= \frac{4}{4}\cdot\cos\frac{x}{4} = \cos\frac{x}{4}$

e. Given $\int(\sin 4x+\cos 2x)\,dx = \int\sin 4x\,dx+\int\cos 2x\,dx$ let:

a. $u=4x$, then $\frac{du}{dx}=\frac{d}{dx}4x$; $\frac{du}{dx}=4$; $du=4dx$; $dx=\frac{du}{4}$ and

b. $v=2x$, then $\frac{dv}{dx}=\frac{d}{dx}2x$; $\frac{dv}{dx}=2$; $dv=2dx$; $dx=\frac{dv}{2}$.

Therefore, $\boxed{\int\sin 4x\,dx+\int\cos 2x\,dx}=\boxed{\int\sin u\cdot\frac{du}{4}+\int\cos v\cdot\frac{dv}{2}}=\boxed{\frac{1}{4}\int\sin u\,du+\frac{1}{2}\int\cos v\,dv}=\boxed{-\frac{1}{4}\cos u+c_1}$

$\boxed{+\frac{1}{2}\sin v+c_2}=\boxed{-\frac{1}{4}\cos 4x+\frac{1}{2}\sin 2x+c_1+c_2}=\boxed{-\frac{1}{4}\cos 4x+\frac{1}{2}\sin 2x+c}$

Check: Let $y = -\dfrac{1}{4}\cos 4x + \dfrac{1}{2}\sin 2x + c$ then $y' = -\dfrac{1}{4}\cdot\dfrac{d}{dx}\cos 4x + \dfrac{1}{2}\cdot\dfrac{d}{dx}\sin 2x + \dfrac{d}{dx}c = \dfrac{1}{4}\cdot\sin 4x\cdot\dfrac{d}{dx}4x$

$+\dfrac{1}{2}\cdot\cos 2x\cdot\dfrac{d}{dx}2x + 0 = \dfrac{4}{4}\cdot\sin 4x + \dfrac{2}{2}\cdot\cos 2x = \sin 4x + \cos 2x$

f. Given $\displaystyle\int \csc 5x\,dx$ let $u = 5x$, then $\dfrac{du}{dx} = \dfrac{d}{dx}5x = 5$ which implies $du = 5dx$; $dx = \dfrac{du}{5}$. Therefore,

$$\boxed{\int \csc 5x\,dx} = \boxed{\int \csc u\cdot\dfrac{du}{5}} = \boxed{\dfrac{1}{5}\int \csc u\,du} = \boxed{\dfrac{1}{5}\ln|\csc u - \cot u| + c} = \boxed{\dfrac{1}{5}\ln|\csc 5x - \cot 5x| + c}$$

Check: Let $y = \dfrac{1}{5}\ln|\csc 5x - \cot 5x| + c$, then $y' = \dfrac{1}{5}\cdot\dfrac{d}{dx}\ln|\csc 5x - \cot 5x| + \dfrac{d}{dx}c = \dfrac{1}{5}\cdot\dfrac{1}{\csc 5x - \cot 5x}$

$\times\dfrac{d}{dx}\left(\csc 5x - \cot 5x\right) + \dfrac{d}{dx}c = \dfrac{1}{5}\cdot\dfrac{1}{\csc 5x - \cot 5x}\cdot\left(-\csc 5x\cdot\cot 5x\cdot 5 + \csc^2 5x\cdot 5\right) + 0$

$= \dfrac{1}{5}\cdot\dfrac{5\csc 5x\left(\csc 5x - \cot 5x\right)}{\csc 5x - \cot 5x} = \dfrac{5\csc 5x}{5} = \csc 5x$

g. Given $\displaystyle\int \csc^2 5x\,dx$ let $u = 5x$, then $\dfrac{du}{dx} = \dfrac{d}{dx}5x$; $\dfrac{du}{dx} = 5$; $du = 5dx$; $dx = \dfrac{du}{5}$. Therefore,

$$\boxed{\int \csc^2 5x\,dx} = \boxed{\int \csc^2 u\cdot\dfrac{du}{5}} = \boxed{\dfrac{1}{5}\int \csc^2 u\,du} = \boxed{-\dfrac{1}{5}\cot u + c} = \boxed{-\dfrac{1}{5}\cot 5x + c}$$

Check: Let $y = -\dfrac{1}{5}\cot 5x + c$, then $y' = -\dfrac{1}{5}\cdot\dfrac{d}{dx}\cot 5x + \dfrac{d}{dx}c = -\dfrac{1}{5}\cdot-\csc^2 5x\cdot\dfrac{d}{dx}5x + 0 = \dfrac{1}{5}\cdot\csc^2 5x\cdot 5$

$= \dfrac{5}{5}\cdot\csc^2 5x = \csc^2 5x$

h. Given $\displaystyle\int \csc^2 \dfrac{1}{2}x\,dx$ let $u = \dfrac{1}{2}x$, then $\dfrac{du}{dx} = \dfrac{d}{dx}\dfrac{x}{2}$; $\dfrac{du}{dx} = \dfrac{1}{2}$; $2du = dx$; $dx = 2du$. Therefore,

$$\boxed{\int \csc^2 \dfrac{1}{2}x\,dx} = \boxed{\int \csc^2 u\cdot 2du} = \boxed{2\int \csc^2 u\,du} = \boxed{-2\cot u + c} = \boxed{-2\cot \dfrac{1}{2}x + c}$$

Check: Let $y = -2\cot\dfrac{x}{2} + c$, then $y' = -2\cdot\dfrac{d}{dx}\cot\dfrac{x}{2} + \dfrac{d}{dx}c = -2\cdot-\csc^2\dfrac{x}{2}\cdot\dfrac{d}{dx}\dfrac{x}{2} + 0 = 2\csc^2\dfrac{x}{2}\cdot\dfrac{1}{2}$

$= \dfrac{2}{2}\csc^2\dfrac{x}{2} = \csc^2\dfrac{x}{2} = \csc^2\dfrac{1}{2}x$

i. Given $\displaystyle\int x^2 \sec^2 x^3\,dx$ let $u = x^3$, then $\dfrac{du}{dx} = \dfrac{d}{dx}x^3$; $\dfrac{du}{dx} = 3x^2$; $du = 3x^2 dx$; $dx = \dfrac{du}{3x^2}$. Therefore,

$$\boxed{\int x^2 \sec^2 x^3\,dx} = \boxed{\int x^2 \sec^2 u\cdot\dfrac{du}{3x^2}} = \boxed{\dfrac{1}{3}\int \sec^2 u\,du} = \boxed{\dfrac{1}{3}\tan u + c} = \boxed{\dfrac{1}{3}\tan x^3 + c}$$

Check: Let $y = \dfrac{1}{3}\tan x^3 + c$, then $y' = \dfrac{1}{3}\cdot\dfrac{d}{dx}\tan x^3 + \dfrac{d}{dx}c = \dfrac{1}{3}\cdot\sec^2 x^3\cdot\dfrac{d}{dx}x^3 + 0 = \dfrac{1}{3}\cdot\sec^2 x^3\cdot 3x^2$

$= \dfrac{3x^2}{3}\cdot\sec^2 x^3 = x^2 \sec^2 x^3$

j. Given $\int x \sec^2(x^2+1)dx$ let $u=x^2+1$, then $\frac{du}{dx}=\frac{d}{dx}(x^2+1)$; $\frac{du}{dx}=2x$; $du=2x\,dx$; $dx=\frac{du}{2x}$. Therefore,

$$\boxed{\int x\sec^2(x^2+1)dx}=\boxed{\int x\sec^2 u\cdot\frac{du}{2x}}=\boxed{\frac{1}{2}\int\sec^2 u\,du}=\boxed{\frac{1}{2}\tan u+c}=\boxed{\frac{1}{2}\tan(x^2+1)+c}$$

Check: Let $y=\frac{1}{2}\tan(x^2+1)+c$, then $y'=\frac{1}{2}\cdot\frac{d}{dx}\tan(x^2+1)+\frac{d}{dx}c=\frac{1}{2}\cdot\sec^2(x^2+1)\cdot\frac{d}{dx}(x^2+1)+0$

$=\frac{1}{2}\cdot\sec^2(x^2+1)\cdot 2x=\frac{2x}{2}\cdot\sec^2(x^2+1)=x\sec^2(x^2+1)$

k. Given $\int x\csc^2 x^2\,dx$ let $u=x^2$, then $\frac{du}{dx}=\frac{d}{dx}x^2$; $\frac{du}{dx}=2x$; $du=2x\,dx$; $dx=\frac{du}{2x}$. Therefore,

$$\boxed{\int x\csc^2 x^2\,dx}=\boxed{\int x\csc^2 u\cdot\frac{du}{2x}}=\boxed{\frac{1}{2}\int\csc^2 u\,du}=\boxed{-\frac{1}{2}\cot u+c}=\boxed{-\frac{1}{2}\cot x^2+c}$$

Check: Let $y=-\frac{1}{2}\cot x^2+c$, then $y'=-\frac{1}{2}\cdot\frac{d}{dx}\cot x^2+\frac{d}{dx}c=-\frac{1}{2}\cdot-\csc^2 x^2\cdot\frac{d}{dx}x^2+0=\frac{1}{2}\cdot\csc^2 x^2\cdot 2x$

$=\frac{2x}{2}\cdot\csc^2 x^2=x\csc^2 x^2$

l. Given $\int(2x+1)\csc^2(x^2+x)dx$ let $u=x^2+x$, then $\frac{du}{dx}=\frac{d}{dx}(x^2+x)$; $\frac{du}{dx}=2x+1$; $du=(2x+1)dx$;

$dx=\frac{du}{2x+1}$. Therefore, $\boxed{\int(2x+1)\csc^2(x^2+x)dx}=\boxed{\int(2x+1)\csc^2 u\cdot\frac{du}{(2x+1)}}=\boxed{\int\csc^2 u\,du}$

$=\boxed{-\cot u+c}=\boxed{-\cot(x^2+x)+c}$

Check: Let $y=-\cot(x^2+x)+c$, then $y'=-\frac{d}{dx}\cot(x^2+x)+\frac{d}{dx}c=\csc^2(x^2+x)\cdot\frac{d}{dx}(x^2+x)+0$

$=\csc^2(x^2+x)\cdot(2x+1)=(2x+1)\csc^2(x^2+x)$

Example 4.3-2: Evaluate the following indefinite integrals:

a. $\int\sec^2 3x\,dx=$ b. $\int\sec 3x\,dx=$ c. $\int x\csc x^2\,dx=$

d. $\int\sin^5(x+1)\cos(x+1)dx=$ e. $\int\cos^5 x\sin x\,dx=$ f. $\int\cos^5 3x\sin 3x\,dx=$

g. $\int\cos^4\frac{x}{3}\sin\frac{x}{3}dx=$ h. $\int x^2\cos x^3 dx=$ i. $\int e^x\sec e^x\,dx=$

Solutions:

a. Given $\int\sec^2 3x\,dx$ let $u=3x$, then $\frac{du}{dx}=\frac{d}{dx}3x$; $\frac{du}{dx}=3$; $du=3\,dx$; $dx=\frac{du}{3}$. Therefore,

$$\boxed{\int\sec^2 3x\,dx}=\boxed{\int\sec^2 u\cdot\frac{du}{3}}=\boxed{\frac{1}{3}\int\sec^2 u\,du}=\boxed{\frac{1}{3}\tan u+c}=\boxed{\frac{1}{3}\tan 3x+c}$$

Check: Let $y = \dfrac{1}{3}\tan 3x + c$, then $y' = \dfrac{d}{dx}\tan 3x + \dfrac{d}{dx}c = \dfrac{1}{3}\sec^2 3x \cdot 3 + 0 = \dfrac{3}{3}\sec^2 3x = \sec^2 3x$

b. Given $\displaystyle\int \sec 3x\, dx$ let $u = 3x$, then $\dfrac{du}{dx} = \dfrac{d}{dx}3x$; $\dfrac{du}{dx} = 3$; $du = 3\, dx$; $dx = \dfrac{du}{3}$. Therefore,

$$\boxed{\int \sec 3x\, dx} = \boxed{\int \sec u \cdot \frac{du}{3}} = \boxed{\frac{1}{3}\int \sec u\, du} = \boxed{\frac{1}{3}\ln\left|\sec u + \tan u\right| + c} = \boxed{\frac{1}{3}\ln\left|\sec 3x + \tan 3x\right| + c}$$

Check: Let $y = \dfrac{1}{3}\ln\left|\sec 3x + \tan 3x\right| + c$, then $y' = \dfrac{1}{3}\cdot\dfrac{1}{\sec 3x + \tan 3x}\cdot 3\sec 3x \tan 3x + 3\sec^2 3x + 0$

$$= \frac{1}{3}\cdot\frac{3\sec 3x\,(\sec 3x + \tan 3x)}{\sec 3x + \tan 3x} = \frac{3}{3}\sec 3x = \sec 3x$$

c. Given $\displaystyle\int x\csc x^2\, dx$ let $u = x^2$, then $\dfrac{du}{dx} = \dfrac{d}{dx}x^2$; $\dfrac{du}{dx} = 2x$; $du = 2x\, dx$; $dx = \dfrac{du}{2x}$. Therefore,

$$\boxed{\int x\csc x^2\, dx} = \boxed{\int x\csc u \cdot \frac{du}{2x}} = \boxed{\frac{1}{2}\int \csc u\, du} = \boxed{\frac{1}{2}\ln\left|\csc u - \cot u\right| + c} = \boxed{\frac{1}{2}\ln\left|\csc x^2 - \cot x^2\right| + c}$$

Check: Let $y = \dfrac{1}{2}\ln\left|\csc x^2 - \cot x^2\right| + c$, then $y' = \dfrac{1}{2}\cdot\dfrac{1}{\csc x^2 - \cot x^2}\cdot -2x\csc x^2 \cot x^2 + 2x\csc^2 x^2 + 0$

$$= \frac{1}{2}\cdot\frac{2x\csc x^2\left(\csc x^2 - \cot x^2\right)}{\csc x^2 - \cot x^2} = \frac{2}{2}x\csc x^2 = x\csc x^2$$

d. Given $\displaystyle\int \sin^5(x+1)\cos(x+1)\, dx$ let $u = \sin(x+1)$, then $\dfrac{du}{dx} = \dfrac{d}{dx}\sin(x+1)$; $\dfrac{du}{dx} = \cos(x+1)$;

$du = \cos(x+1)\cdot dx$; $dx = \dfrac{du}{\cos(x+1)}$. Therefore,

$$\boxed{\int \sin^5(x+1)\cos(x+1)\, dx} = \boxed{\int u^5 \cos(x+1)\cdot\frac{du}{\cos(x+1)}} = \boxed{\int u^5\, du} = \boxed{\frac{1}{6}u^6 + c} = \boxed{\frac{1}{6}\sin^6(x+1) + c}$$

Check: Let $y = \dfrac{1}{6}\sin^6(x+1) + c$, then $y' = \dfrac{1}{6}\cdot 6\sin^5(x+1)\cdot\cos(x+1) + 0 = \sin^5(x+1)\cos(x+1)$

e. Given $\displaystyle\int \cos^5 x \sin x\, dx$ let $u = \cos x$, then $\dfrac{du}{dx} = \dfrac{d}{dx}\cos x$; $\dfrac{du}{dx} = -\sin x$; $dx = \dfrac{du}{-\sin x}$. Therefore,

$$\boxed{\int \cos^5 x \sin x\, dx} = \boxed{\int u^5 \sin x \cdot\frac{du}{-\sin x}} = \boxed{-\int u^5\, du} = \boxed{-\frac{1}{6}u^6 + c} = \boxed{-\frac{1}{6}\cos^6 x + c}$$

Check: Let $y = -\dfrac{1}{6}\cos^6 x + c$, then $y' = -\dfrac{1}{6}\cdot 6\cos^5 x \cdot -\sin x + 0 = \dfrac{6}{6}\cos^5 x \sin x = \cos^5 x \sin x$

f. Given $\displaystyle\int \cos^5 3x \sin 3x\, dx$ let $u = \cos 3x$, then $\dfrac{du}{dx} = \dfrac{d}{dx}\cos 3x$; $\dfrac{du}{dx} = -3\sin 3x$; $dx = \dfrac{du}{-3\sin 3x}$. Thus,

$$\boxed{\int \cos^5 3x \sin 3x \, dx} = \boxed{\int u^5 \sin 3x \cdot \frac{du}{-3 \sin 3x}} = \boxed{-\frac{1}{3}\int u^5 du} = \boxed{-\frac{1}{3}\cdot\frac{1}{6}u^6 + c} = \boxed{-\frac{1}{18}\cos^6 3x + c}$$

Check: Let $y = -\frac{1}{18}\cos^6 3x + c$, then $y' = -\frac{1}{18}\cdot 6\cos^5 3x\cdot-3\sin 3x + 0 = \frac{18}{18}\cos^5 3x \sin 3x = \cos^5 3x \sin 3x$

g. Given $\int \cos^4 \frac{x}{3}\sin\frac{x}{3}\,dx$ let $u = \cos\frac{x}{3}$, then $\frac{du}{dx} = \frac{d}{dx}\cos\frac{x}{3}$; $\frac{du}{dx} = -\frac{1}{3}\sin\frac{x}{3}$; $dx = -\frac{3du}{\sin\frac{x}{3}}$. Therefore,

$$\boxed{\int \cos^4 \frac{x}{3}\sin\frac{x}{3}\,dx} = \boxed{\int u^4 \sin\frac{x}{3}\cdot-\frac{3\,du}{\sin\frac{x}{3}}} = \boxed{-3\int u^4 du} = \boxed{-\frac{3}{5}u^5 + c} = \boxed{-\frac{3}{5}\cos^5 \frac{x}{3} + c}$$

Check: Let $y = -\frac{3}{5}\cos^5 \frac{x}{3} + c$, then $y' = -\frac{3}{5}\cdot 5\cos^4 \frac{x}{3}\cdot-\sin\frac{x}{3}\cdot\frac{1}{3} + 0 = \frac{15}{15}\cos^4\frac{x}{3}\sin\frac{x}{3} = \cos^4\frac{x}{3}\sin\frac{x}{3}$

h. Given $\int x^2 \cos x^3 dx$ let $u = x^3$, then $\frac{du}{dx} = \frac{d}{dx}x^3$; $\frac{du}{dx} = 3x^2$; $du = 3x^2 \cdot dx$; $dx = \frac{du}{3x^2}$. Therefore,

$$\boxed{\int x^2 \cos x^3 dx} = \boxed{\int x^2 \cos u \cdot\frac{du}{3x^2}} = \boxed{\frac{1}{3}\int \cos u \, du} = \boxed{\frac{1}{3}\cdot\sin u + c} = \boxed{\frac{1}{3}\sin x^3 + c}$$

Check: Let $y = \frac{1}{3}\sin x^3 + c$, then $y' = \frac{1}{3}\cdot\cos x^3 \cdot 3x^2 + c = \frac{3}{3}x^2 \cos x^3 + 0 = x^2 \cos x^3$

i. Given $\int e^x \sec e^x \, dx$ let $u = e^x$, then $\frac{du}{dx} = \frac{d}{dx}e^x$; $\frac{du}{dx} = e^x$; $du = e^x \cdot dx$; $dx = \frac{du}{e^x}$. Therefore,

$$\boxed{\int e^x \sec e^x \, dx} = \boxed{\int e^x \sec u \cdot\frac{du}{e^x}} = \boxed{\int \sec u \, du} = \boxed{\ln|\sec u + \tan u| + c} = \boxed{\ln|\sec e^x + \tan e^x| + c}$$

Check: Let $y = \ln|\sec e^x + \tan e^x| + c$, then $y' = \frac{1}{\sec e^x + \tan e^x}\cdot e^x \sec e^x \tan e^x + e^x \sec^2 e^x + 0$

$$= \frac{e^x \sec e^x \left(\sec e^x + \tan e^x\right)}{\sec e^x + \tan e^x} = e^x \sec e^x$$

Example 4.3-3: Evaluate the following indefinite integrals:

a. $\int \tan^7 x \sec^2 x \, dx =$ b. $\int \tan^4(x-1)\sec^2(x-1)\,dx =$ c. $\int \cot^3 x \csc^2 x \, dx =$

d. $\int \cot^5 2x \csc^2 2x \, dx =$ e. $\int \tan 8x \, dx =$ f. $\int \tan\frac{x}{4}\,dx =$

g. $\int \sec 9x \tan 9x \, dx =$ h. $\int \sec\frac{x}{2}\tan\frac{x}{2}\,dx =$ i. $\int 5 \sec 3x \tan 3x \, dx =$

j. $\int x^2 \cot x^3 dx =$ k. $\int x \cot 3x^2 dx =$ l. $\int \sec\sqrt{x}\,\frac{dx}{\sqrt{x}} =$

Solutions:

a. $\int \tan^7 x \sec^2 x \, dx$ let $u = \tan x$, then $\dfrac{du}{dx} = \dfrac{d}{dx} \tan x$; $\dfrac{du}{dx} = \sec^2 x$; $du = \sec^2 x \, dx$; $dx = \dfrac{du}{\sec^2 x}$. Thus,

$$\boxed{\int \tan^7 x \sec^2 x \, dx} = \boxed{\int u^7 \cdot \sec^2 x \cdot \dfrac{du}{\sec^2 x}} = \boxed{\int u^7 \, du} = \boxed{\dfrac{1}{7+1} u^{7+1} + c} = \boxed{\dfrac{1}{8} u^8 + c} = \boxed{\dfrac{1}{8} \tan^8 x + c}$$

Check: Let $y = \dfrac{1}{8} \tan^8 x + c$, then $y' = \dfrac{1}{8} \cdot 8 (\tan x)^{8-1} \cdot \sec^2 x + 0 = (\tan x)^7 \sec^2 x = \tan^7 x \sec^2 x$

b. $\int \tan^4 (x-1) \sec^2 (x-1) \, dx$ let $u = \tan(x-1)$, then $\dfrac{du}{dx} = \dfrac{d}{dx} \tan(x-1)$; $\dfrac{du}{dx} = \sec^2(x-1) \text{c}$; $du = \sec^2(x-1) \, dx$

; $dx = \dfrac{du}{\sec^2 (x-1)}$. Therefore, $\boxed{\int \tan^4 (x-1) \sec^2 (x-1) dx} = \boxed{\int u^4 \cdot \sec^2 (x-1) \cdot \dfrac{du}{\sec^2 (x-1)}} = \boxed{\int u^4 \, du}$

$= \boxed{\dfrac{1}{4+1} u^{4+1} + c} = \boxed{\dfrac{1}{5} u^5 + c} = \boxed{\dfrac{1}{5} \tan^5 (x-1) + c}$

Check: Let $y = \dfrac{1}{5} \tan^5 (x-1) + c$, then $y' = \dfrac{1}{5} \cdot 5 [\tan(x-1)]^{5-1} \cdot \sec^2 (x-1) + 0 = \tan^4 (x-1) \sec^2 (x-1)$

c. Given $\int \cot^3 x \csc^2 x \, dx$ let $u = \cot x$, then $\dfrac{du}{dx} = \dfrac{d}{dx} \cot x$; $\dfrac{du}{dx} = -\csc^2 x \text{c}$; $du = -\csc^2 x \, dx$

; $dx = -\dfrac{du}{\csc^2 x}$. Therefore, $\boxed{\int \cot^3 x \csc^2 x \, dx} = \boxed{\int u^3 \cdot \csc^2 x \cdot \dfrac{-du}{\csc^2 x}} = \boxed{-\int u^3 \, du} = \boxed{\dfrac{-1}{3+1} u^{3+1} + c}$

$= \boxed{-\dfrac{1}{4} u^4 + c} = \boxed{-\dfrac{1}{4} \cot^4 x + c}$

Check: Let $y = -\dfrac{1}{4} \cot^4 x + c$, then $y' = -\dfrac{1}{4} \cdot 4 (\cot x)^{4-1} \cdot -\csc^2 x + 0 = \cot^3 x \csc^2 x$

d. Given $\int \cot^5 2x \csc^2 2x \, dx$ let $u = \cot 2x$, then $\dfrac{du}{dx} = \dfrac{d}{dx} \cot 2x$; $\dfrac{du}{dx} = -2\csc^2 2x$; $du = -2\csc^2 2x \, dx$

; $dx = -\dfrac{du}{2\csc^2 2x}$. Therefore, $\boxed{\int \cot^5 2x \csc^2 2x \, dx} = \boxed{\int u^5 \cdot \csc^2 2x \cdot \dfrac{-du}{2\csc^2 2x}} = \boxed{-\dfrac{1}{2} \int u^5 \, du}$

$= \boxed{-\dfrac{1}{2} \cdot \dfrac{1}{5+1} u^{5+1} + c} = \boxed{-\dfrac{1}{12} u^6 + c} = \boxed{-\dfrac{1}{12} \cot^6 2x + c}$

Check: Let $y = -\dfrac{1}{12} \cot^6 2x + c$, then $y' = -\dfrac{1}{12} \cdot 6 (\cot 2x)^{6-1} \cdot 2 \cdot -\csc^2 2x + 0 = \cot^5 2x \csc^2 2x$

e. Given $\int \tan 8x \, dx$ let $u = 8x$, then $\dfrac{du}{dx} = \dfrac{d}{dx} 8x$; $\dfrac{du}{dx} = 8$; $du = 8 \, dx$; $dx = \dfrac{du}{8}$. Therefore,

$$\boxed{\int \tan 8x \, dx} = \boxed{\int \tan u \cdot \frac{du}{8}} = \boxed{\frac{1}{8} \int \tan u \, du} = \boxed{\frac{1}{8} \ln|\sec u| + c} = \boxed{\frac{1}{8} \ln|\sec 8x| + c}$$

Check: Let $y = \frac{1}{8} \ln|\sec 8x| + c$, then $y' = \frac{1}{8} \cdot \frac{1}{\sec 8x} \cdot \sec 8x \tan 8x \cdot 8 + 0 = \frac{8}{8} \tan 8x = \tan 8x$

f. Given $\int \tan \frac{x}{4} \, dx$ let $u = \frac{x}{4}$, then $\frac{du}{dx} = \frac{d}{dx} \frac{x}{4}$; $\frac{du}{dx} = \frac{1}{4}$; $4du = dx$; $dx = 4du$. Therefore,

$$\boxed{\int \tan \frac{x}{4} \, dx} = \boxed{\int \tan u \cdot 4du} = \boxed{4 \int \tan u \, du} = \boxed{4 \ln|\sec u| + c} = \boxed{4 \ln\left|\sec \frac{x}{4}\right| + c}$$

Check: Let $y = 4 \ln\left|\sec \frac{x}{4}\right| + c$, then $y' = 4 \cdot \frac{1}{\sec \frac{x}{4}} \cdot \sec \frac{x}{4} \tan \frac{x}{4} \cdot \frac{1}{4} + 0 = \frac{4}{4} \tan \frac{x}{4} = \tan \frac{x}{4}$

g. Given $\int \sec 9x \tan 9x \, dx$ let $u = 9x$, then $\frac{du}{dx} = \frac{d}{dx} 9x$; $\frac{du}{dx} = 9$; $du = 9dx$; $dx = \frac{du}{9}$. Therefore,

$$\boxed{\int \sec 9x \tan 9x \, dx} = \boxed{\int \sec u \cdot \tan u \cdot \frac{du}{9}} = \boxed{\frac{1}{9} \int \sec u \tan u \, du} = \boxed{\frac{1}{9} \sec u + c} = \boxed{\frac{1}{9} \sec 9x + c}$$

Check: Let $y = \frac{1}{9} \sec 9x + c$ then $y' = \frac{1}{9} \cdot \sec 9x \tan 9x \cdot 9 + 0 = \frac{9}{9} \cdot \sec 9x \tan 9x = \sec 9x \tan 9x$

h. Given $\int \sec \frac{x}{2} \tan \frac{x}{2} \, dx$ let $u = \frac{x}{2}$, then $\frac{du}{dx} = \frac{d}{dx} \frac{x}{2}$; $\frac{du}{dx} = \frac{1}{2}$; $2du = dx$; $dx = 2du$. Therefore,

$$\boxed{\int \sec \frac{x}{2} \tan \frac{x}{2} \, dx} = \boxed{\int \sec u \cdot \tan u \cdot 2du} = \boxed{2 \int \sec u \tan u \, du} = \boxed{2 \sec u + c} = \boxed{2 \sec \frac{x}{2} + c}$$

Check: Let $y = 2 \sec \frac{x}{2} + c$, then $y' = 2 \cdot \sec \frac{x}{2} \tan \frac{x}{2} \cdot \frac{1}{2} + 0 = \frac{2}{2} \cdot \sec \frac{x}{2} \tan \frac{x}{2} = \sec \frac{x}{2} \tan \frac{x}{2}$

i. Given $\int 5 \sec 3x \tan 3x \, dx$ let $u = 3x$, then $\frac{du}{dx} = \frac{d}{dx} 3x$; $\frac{du}{dx} = 3$; $du = 3dx$; $dx = \frac{du}{3}$. Therefore,

$$\boxed{\int 5 \sec 3x \tan 3x \, dx} = \boxed{5 \int \sec u \cdot \tan u \cdot \frac{du}{3}} = \boxed{\frac{5}{3} \int \sec u \tan u \, du} = \boxed{\frac{5}{3} \sec u + c} = \boxed{\frac{5}{3} \sec 3x + c}$$

Check: Let $y = \frac{5}{3} \sec 3x + c$, then $y' = \frac{5}{3} \cdot \sec 3x \tan 3x \cdot 3 + 0 = \frac{15}{3} \cdot \sec 3x \tan 3x = 5 \sec 3x \tan 3x$

j. Given $\int x^2 \cot x^3 \, dx$ let $u = x^3$, then $\frac{du}{dx} = \frac{d}{dx} x^3$; $\frac{du}{dx} = 3x^2$; $du = 3x^2 dx$; $dx = \frac{du}{3x^2}$. Therefore,

$$\boxed{\int x^2 \cot x^3 \, dx} = \boxed{\int x^2 \cdot \cot u \cdot \frac{du}{3x^2}} = \boxed{\frac{1}{3} \int \cot u \cdot du} = \boxed{\frac{1}{3} \ln|\sin u| + c} = \boxed{\frac{1}{3} \ln|\sin x^3| + c}$$

Check: Let $y = \frac{1}{3} \ln|\sin x^3| + c$, then $y' = \frac{1}{3} \cdot \frac{1}{\sin x^3} \cdot \cos x^3 \cdot 3x^2 + 0 = \frac{3x^2}{3} \cdot \frac{\cos x^3}{\sin x^3} = x^2 \cot x^3$

k. Given $\int x \cot 3x^2 dx$ let $u = 3x^2$, then $\dfrac{du}{dx} = \dfrac{d}{dx} 3x^2$; $\dfrac{du}{dx} = 6x$; $du = 6x\, dx$; $dx = \dfrac{du}{6x}$. Therefore,

$$\boxed{\int x \cot 3x^2 dx} = \boxed{\int x \cdot \cot u \cdot \frac{du}{6x}} = \boxed{\frac{1}{6}\int \cot u \cdot du} = \boxed{\frac{1}{6}\ln|\sin u| + c} = \boxed{\frac{1}{6}\ln\left|\sin 3x^2\right| + c}$$

Check: Let $y = \dfrac{1}{6}\ln\left|\sin 3x^2\right| + c$, then $y' = \dfrac{1}{6} \cdot \dfrac{1}{\sin 3x^2} \cdot \cos 3x^2 \cdot 6x + 0 = \dfrac{6x}{6} \cdot \dfrac{\cos 3x^2}{\sin 3x^2} = x \cot 3x^2$

l. Given $\int \sec \sqrt{x} \, \dfrac{dx}{\sqrt{x}}$ let $u = x^{\frac{1}{2}}$, then $\dfrac{du}{dx} = \dfrac{d}{dx} x^{\frac{1}{2}}$; $\dfrac{du}{dx} = \dfrac{1}{2} x^{-\frac{1}{2}} = \dfrac{1}{2\sqrt{x}}$; $dx = 2\sqrt{x}\, du$. Therefore,

$$\boxed{\int \sec \sqrt{x} \, \frac{dx}{\sqrt{x}}} = \boxed{\int \sec u \cdot \frac{2\sqrt{x}\, du}{\sqrt{x}}} = \boxed{2\int \sec u \cdot du} = \boxed{2\ln|\sec u + \tan u| + c} = \boxed{2\ln\left|\sec \sqrt{x} + \tan \sqrt{x}\right| + c}$$

Check: Let $y = 2\ln\left|\sec \sqrt{x} + \tan \sqrt{x}\right| + c$, then $y' = 2 \cdot \dfrac{1}{\sec \sqrt{x} + \tan \sqrt{x}} \cdot \dfrac{\sec \sqrt{x} \tan \sqrt{x} + \sec^2 \sqrt{x}}{2\sqrt{x}} + 0$

$$= 2 \cdot \frac{1}{\sec \sqrt{x} + \tan \sqrt{x}} \cdot \frac{\sec \sqrt{x}\left(\tan \sqrt{x} + \sec \sqrt{x}\right)}{2\sqrt{x}} = \frac{2\sec \sqrt{x}}{2\sqrt{x}} = \frac{\sec \sqrt{x}}{\sqrt{x}}$$

Example 4.3-4: Evaluate the following indefinite integrals:

a. $\int \sec \sqrt[3]{x^2} \, \dfrac{dx}{\sqrt[3]{x}} =$

b. $\int \csc \sqrt[5]{x^3} \, \dfrac{dx}{\sqrt[5]{x^2}} =$

c. $\int 5x^2 \csc x^3 dx =$

d. $\int \dfrac{1}{\cos 3x} dx =$

e. $\int \dfrac{\cos 2x + \sin 2x}{\sin 2x} dx =$

f. $\int \left(\dfrac{1 + \cos x}{\sin x}\right) dx =$

g. $\int 5(\sin 3x \csc 3x) dx =$

h. $\int a\left(\cos 2t \sec 2t + t^2\right) dt =$

i. $\int e^{\sin 5x} \cos 5x \, dx =$

j. $\int e^{3\cos \frac{1}{2} x} \sin \dfrac{1}{2} x \, dx =$

k. $\int e^{\tan 5x} \sec^2 5x \, dx =$

l. $\int e^{\frac{1}{2}\cot 3x} \csc^2 3x \, dx =$

Solutions:

a. Given $\int \sec \sqrt[3]{x^2} \, \dfrac{dx}{\sqrt[3]{x}}$ let $u = x^{\frac{2}{3}}$, then $\dfrac{du}{dx} = \dfrac{d}{dx} x^{\frac{2}{3}}$; $\dfrac{du}{dx} = \dfrac{2}{3} x^{-\frac{1}{3}} = \dfrac{2}{3\sqrt[3]{x}}$; $dx = \dfrac{3}{2}\sqrt[3]{x}\, du$. Therefore,

$$\boxed{\int \sec \sqrt[3]{x^2} \, \frac{dx}{\sqrt[3]{x}}} = \boxed{\int \sec u \cdot \frac{3}{2} \cdot \frac{\sqrt[3]{x}\, du}{\sqrt[3]{x}}} = \boxed{\frac{3}{2}\int \sec u \cdot du} = \boxed{\frac{3}{2}\ln|\sec u + \tan u| + c} = \boxed{\frac{3}{2}\ln\left|\sec \sqrt[3]{x^2} + \tan \sqrt[3]{x^2}\right| + c}$$

Check: Let $y = \dfrac{3}{2}\ln\left|\sec \sqrt[3]{x^2} + \tan \sqrt[3]{x^2}\right| + c$, then $y' = \dfrac{3}{2} \cdot \dfrac{1}{\sec \sqrt[3]{x^2} + \tan \sqrt[3]{x^2}} \cdot \dfrac{2\sec \sqrt[3]{x^2} \tan \sqrt[3]{x^2} + 2\sec^2 \sqrt[3]{x^2}}{3\sqrt[3]{x}}$

$$= \frac{3}{2} \cdot \frac{1}{\sec \sqrt[3]{x^2} + \tan \sqrt[3]{x^2}} \cdot \frac{2}{3} \cdot \frac{\sec \sqrt[3]{x^2}\left(\sec \sqrt[3]{x^2} + \tan \sqrt[3]{x^2}\right)}{\sqrt[3]{x}} = \frac{6}{6} \cdot \frac{\sec \sqrt[3]{x^2}}{\sqrt[3]{x}} = \frac{\sec \sqrt[3]{x^2}}{\sqrt[3]{x}}$$

b. Given $\int \csc \sqrt[5]{x^3} \; \dfrac{dx}{\sqrt[5]{x^2}}$ let $u = x^{\frac{3}{5}}$, then $\dfrac{du}{dx} = \dfrac{d}{dx} x^{\frac{3}{5}}$; $\dfrac{du}{dx} = \dfrac{3}{5} x^{-\frac{2}{5}} = \dfrac{3}{5\sqrt[5]{x^2}}$; $dx = \dfrac{5}{3}\sqrt[5]{x^2} \; du$. Thus,

$$\boxed{\int \csc \sqrt[5]{x^3}\; \frac{dx}{\sqrt[5]{x^2}}} = \boxed{\int \csc u \cdot \frac{5}{3} \cdot \frac{\sqrt[5]{x^2}\,du}{\sqrt[5]{x^2}}} = \boxed{\frac{5}{3}\int \csc u \cdot du} = \boxed{\frac{5}{3}\ln\left|\csc u - \cot u\right| + c} = \boxed{\frac{5}{3}\ln\left|\csc \sqrt[5]{x^3} - \cot \sqrt[5]{x^3}\right| + c}$$

Check: Let $y = \dfrac{5}{3}\ln\left|\csc \sqrt[5]{x^3} - \cot \sqrt[5]{x^3}\right| + c$, then $y' = \dfrac{5}{3} \cdot \dfrac{1}{\csc \sqrt[5]{x^3} - \cot \sqrt[5]{x^3}} \cdot \dfrac{-3\csc\sqrt[5]{x^3}\cot\sqrt[5]{x^3} + 3\csc^2\sqrt[5]{x^3}}{5\sqrt[5]{x^2}}$

$$= \frac{5}{3} \cdot \frac{1}{\csc\sqrt[5]{x^3} - \cot\sqrt[5]{x^3}} \cdot \frac{3}{5} \cdot \frac{\csc\sqrt[5]{x^3}\left(\csc\sqrt[5]{x^3} - \cot\sqrt[5]{x^3}\right)}{\sqrt[5]{x^2}} = \frac{15}{15}\cdot\frac{\csc\sqrt[5]{x^3}}{\sqrt[5]{x^2}} = \frac{\csc\sqrt[5]{x^3}}{\sqrt[5]{x^2}}$$

c. Given $\int 5x^2 \csc x^3 \, dx$ let $u = x^3$, then $\dfrac{du}{dx} = \dfrac{d}{dx} x^3$; $\dfrac{du}{dx} = 3x^2$; $du = 3x^2 dx$; $dx = \dfrac{du}{3x^2}$. Therefore,

$$\boxed{\int 5x^2 \csc x^3\, dx} = \boxed{5\int x^2 \cdot \csc u \cdot \frac{du}{3x^2}} = \boxed{\frac{5}{3}\int \csc u \cdot du} = \boxed{\frac{5}{3}\ln\left|\csc u - \cot u\right| + c} = \boxed{\frac{5}{3}\ln\left|\csc x^3 - \cot x^3\right| + c}$$

Check: Let $y = \dfrac{5}{3}\ln\left|\csc x^3 - \cot x^3\right| + c$, then $y' = \dfrac{5}{3} \cdot \dfrac{-\csc x^3 \cot x^3 \cdot 3x^2 + \csc^2 x^3 \cdot 3x^2}{\csc x^3 - \cot x^3} + 0$

$$= \frac{5}{3} \cdot \frac{3x^2 \csc x^3\left(\csc x^3 - \cot x^3\right)}{\csc x^3 - \cot x^3} = \frac{15x^2}{3}\cdot \csc x^3 = 5x^2 \csc x^3$$

d. Given $\int \dfrac{1}{\cos 3x}\, dx = \int \sec 3x\, dx$ let $u = 3x$, then $\dfrac{du}{dx} = \dfrac{d}{dx} 3x$; $\dfrac{du}{dx} = 3$; $du = 3dx$; $dx = \dfrac{du}{3}$. Thus,

$$\boxed{\int \frac{1}{\cos 3x}\, dx} = \boxed{\int \sec 3x\, dx} = \boxed{\int \sec u \cdot \frac{du}{3}} = \boxed{\frac{1}{3}\int \sec u\, du} = \boxed{\frac{1}{3}\ln\left|\sec u + \tan u\right| + c} = \boxed{\frac{1}{3}\ln\left|\sec 3x + \tan 3x\right| + c}$$

Check: Let $y = \dfrac{1}{3}\ln\left|\sec 3x + \tan 3x\right| + c$, then $y' = \dfrac{1}{3} \cdot \dfrac{\sec 3x \tan 3x \cdot 3 + \sec^2 3x \cdot 3}{\sec 3x + \tan 3x} + 0$

$$= \frac{1}{3} \cdot \frac{3\sec 3x\left(\sec 3x + \tan 3x\right)}{\sec 3x + \tan 3x} = \frac{3}{3}\cdot \sec 3x = \sec 3x = \frac{1}{\cos 3x}$$

e. Given $\int \dfrac{\cos 2x + \sin 2x}{\sin 2x}\, dx = \int \left(\dfrac{\cos 2x}{\sin 2x} + 1\right) dx = \int (\cot 2x + 1)\, dx = \int \cot 2x\, dx + \int dx$ let $u = 2x$, then

$\dfrac{du}{dx} = \dfrac{d}{dx} 2x$; $\dfrac{du}{dx} = 2$; $du = 2dx$; $dx = \dfrac{du}{2}$. Therefore,

$$\boxed{\int \frac{\cos 2x + \sin 2x}{\sin 2x}\, dx} = \boxed{\int \cot 2x\, dx + \int dx} = \boxed{\frac{1}{2}\int \cot u\, du + x} = \boxed{\frac{1}{2}\ln\left|\sin u\right| + x + c} = \boxed{\frac{1}{2}\ln\left|\sin 2x\right| + x + c}$$

Check: Let $y = \frac{1}{2}\ln|\sin 2x| + x + c$, then $y' = \frac{1}{2}\cdot\frac{\cos 2x \cdot 2}{\sin 2x} + 1 + 0 = \frac{2}{2}\cdot\frac{\cos 2x}{\sin 2x} + 1 = \cot 2x + 1$

f. $\boxed{\int\left(\frac{1+\cos x}{\sin x}\right)dx} = \boxed{\int\left(\frac{1}{\sin x} + \frac{\cos x}{\sin x}\right)dx} = \boxed{\int \csc x\, dx + \int \cot x\, dx} = \boxed{\ln|\csc x - \cot x| + \ln|\sin x| + c}$

Check: Let $y = \ln|\csc x - \cot x| + \ln|\sin x| + c$, then $y' = \frac{-\csc x\cot x + \csc^2 x}{\csc x - \cot x} + \frac{\cos x}{\sin x} + 0$

$= \frac{\csc x(\csc x - \cot x)}{\csc x - \cot x} + \frac{\cos x}{\sin x} = \csc x + \frac{\cos x}{\sin x} = \frac{1}{\sin x} + \frac{\cos x}{\sin x} = \frac{1+\cos x}{\sin x}$

g. $\boxed{\int 5(\sin 3x \csc 3x)\,dx} = \boxed{5\int \sin 3x \cdot \frac{1}{\sin 3x}\,dx} = \boxed{5\int \frac{\sin 3x}{\sin 3x}\,dx} = \boxed{5\int dx} = \boxed{5x + c}$

Check: Let $y = 5x + c$, then $y' = 5\cdot x^{1-1} + 0 = 5\cdot x^0 = 5$

h. $\boxed{\int a\left(\cos 2t \sec 2t + t^2\right)dt} = \boxed{a\int \cos 2t \cdot \frac{1}{\cos 2t}\,dt + a\int t^2\,dt} = \boxed{a\int dt + a\int t^2\,dt} = \boxed{at + \frac{a}{3}t^3 + c}$

Check: Let $y = at + \frac{a}{3}t^3 + c$, then $y' = at^{1-1} + \frac{a}{3}\cdot 3t^2 + 0 = a + at^2$

i. Given $\int e^{\sin 5x}\cos 5x\, dx$ let $u = \sin 5x$, then $\frac{du}{dx} = \frac{d}{dx}\sin 5x$; $\frac{du}{dx} = 5\cos 5x$; $dx = \frac{du}{5\cos 5x}$. Thus,

$\boxed{\int e^{\sin 5x}\cos 5x\, dx} = \boxed{\int e^u \cos 5x \cdot \frac{du}{5\cos 5x}} = \boxed{\int \frac{e^u}{5}\,du} = \boxed{\frac{1}{5}\int e^u\,du} = \boxed{\frac{1}{5}e^u + c} = \boxed{\frac{1}{5}e^{\sin 5x} + c}$

Check: Let $y = \frac{1}{5}e^{\sin 5x} + c$, then $y' = \frac{1}{5}\cdot e^{\sin 5x}\cdot \cos 5x \cdot 5 + 0 = \frac{5}{5}\cdot e^{\sin 5x}\cos 5x = e^{\sin 5x}\cos 5x$

j. Given $\int e^{3\cos\frac{1}{2}x}\sin\frac{1}{2}x\, dx$ let $u = 3\cos\frac{1}{2}x$, then $\frac{du}{dx} = \frac{d}{dx}3\cos\frac{1}{2}x$; $\frac{du}{dx} = -\frac{3}{2}\sin\frac{1}{2}x$; $du = -\frac{3}{2}\sin\frac{1}{2}x\, dx$

; $dx = -\frac{2du}{3\sin\frac{1}{2}x}$. Therefore,

$\boxed{\int e^{3\cos\frac{1}{2}x}\sin\frac{1}{2}x\, dx} = \boxed{\int e^u \sin\frac{1}{2}x \cdot \frac{-2du}{3\sin\frac{1}{2}x}} = \boxed{-\frac{2}{3}\int e^u\,du} = \boxed{-\frac{2}{3}e^u + c} = \boxed{-\frac{2}{3}e^{3\cos\frac{1}{2}x} + c}$

Check: Let $y = -\frac{2}{3}e^{3\cos\frac{1}{2}x} + c$, then $y' = -\frac{2}{3}e^{3\cos\frac{1}{2}x}\cdot -3\sin\frac{1}{2}x\cdot\frac{1}{2} + 0 = e^{3\cos\frac{1}{2}x}\sin\frac{1}{2}x$

k. Given $\int e^{\tan 5x}\sec^2 5x\, dx$ let $u = \tan 5x$, then $\frac{du}{dx} = \frac{d}{dx}\tan 5x$; $\frac{du}{dx} = 5\sec^2 5x$; $du = 5\sec^2 5x\, dx$

; $dx = \frac{du}{5\sec^2 5x}$. Therefore,

$$\boxed{\int e^{\tan 5x}\sec^2 5x\,dx} = \boxed{\int e^u \sec^2 5x \cdot \frac{du}{5\sec^2 5x}} = \boxed{\frac{1}{5}\int e^u\,du} = \boxed{\frac{1}{5}e^u + c} = \boxed{\frac{1}{5}e^{\tan 5x} + c}$$

Check: Let $y = \dfrac{1}{5}e^{\tan 5x} + c$, then $y' = \dfrac{1}{5}e^{\tan 5x}\cdot\sec^2 5x\cdot 5 + 0 = e^{\tan 5x}\sec^2 5x$

l. Given $\int e^{\cot x}\csc^2 x\,dx$ let $u = \dfrac{1}{2}\cot 3x$, then $\dfrac{du}{dx} = \dfrac{d}{dx}\dfrac{1}{2}\cot 3x$; $\dfrac{du}{dx} = -\dfrac{3}{2}\csc^2 3x$; $du = -\dfrac{3}{2}\csc^2 3x\,dx$

; $dx = -\dfrac{2du}{3\csc^2 3x}$. Therefore,

$$\boxed{\int e^{\frac{1}{2}\cot 3x}\csc^2 3x\,dx} = \boxed{\int e^u \csc^2 3x\cdot\frac{-2du}{3\csc^2 3x}} = \boxed{-\frac{2}{3}\int e^u\,du} = \boxed{-\frac{2}{3}e^u + c} = \boxed{-\frac{2}{3}e^{\frac{1}{2}\cot 3x} + c}$$

Check: Let $y = -\dfrac{2}{3}e^{\frac{1}{2}\cot 3x} + c$, then $y' = -\dfrac{2}{3}e^{\frac{1}{2}\cot 3x}\cdot -\dfrac{1}{2}\csc^2 3x\cdot 3 + 0 = e^{\frac{1}{2}\cot 3x}\csc^2 3x$

- To integrate even powers of $\sin x$ and $\cos x$ use the following identities:

$$\sin^2 x + \cos^2 x = 1 \qquad\qquad \sin^2 x = \frac{1}{2}(1 - \cos 2x) \qquad\qquad \cos^2 x = \frac{1}{2}(1 + \cos 2x)$$

  To integrate odd powers of $\sin x$ and $\cos x$ use the following equalities:

$$\int \sin^{2n+1} x\,dx = \int \sin^{2n} x \sin x\,dx = \int\left(\sin^2 x\right)^n \sin x\,dx = \int\left(1 - \cos^2 x\right)^n \sin x\,dx \qquad (\text{ let } u = \cos x)$$

$$\int \cos^{2n+1} x\,dx = \int \cos^{2n} x \cos x\,dx = \int\left(\cos^2 x\right)^n \cos x\,dx = \int\left(1 - \sin^2 x\right)^n \cos x\,dx \qquad (\text{ let } u = \sin x)$$

- To integrate products of $\sin mx$, $\sin nx$, $\cos mx$, and $\cos nx$ (where m and n are integers) use the trigonometric identities below:

$$\int \sin mx \sin nx\,dx = \int \frac{1}{2}\left[\cos(m-n)x - \cos(m+n)x\right]dx$$

$$\int \cos mx \cos nx\,dx = \int \frac{1}{2}\left[\cos(m-n)x + \cos(m+n)x\right]dx$$

$$\int \sin mx \cos nx\,dx = \int \frac{1}{2}\left[\sin(m-n)x + \sin(m+n)x\right]dx$$

- To integrate $\tan^n x$, set

$$\tan^n x = \tan^{n-2} x \tan^2 x = \tan^{n-2} x\left(\sec^2 x - 1\right) = \tan^{n-2} x\sec^2 x - \tan^{n-2} x$$

- To integrate $\cot^n x$, set

$$\cot^n x = \cot^{n-2} x \cot^2 x = \cot^{n-2} x\left(\csc^2 x - 1\right) = \cot^{n-2} x\csc^2 x - \cot^{n-2} x$$

- To integrate $\sec^n x$

  For even powers, set $\sec^n x = \sec^{n-2} x\sec^2 x = \left(\tan^2 x + 1\right)^{\frac{n-2}{2}}\sec^2 x$

  For odd powers change the integrand to a product of even and odd functions, i.e., write

$\int \sec^3 x \, dx$ as $\int \sec^2 x \sec x \, dx$ (see Example 4.3-6, problem letter h).

- To integrate $\csc^n x$

  For even powers, set $\qquad\qquad\qquad \csc^n x = \csc^{n-2} x \csc^2 x = \left(\cot^2 x + 1 \right)^{\frac{n-2}{2}} \csc^2 x$

  For odd powers change the integrand to a product of even and odd functions, i.e., write

  $\int \csc^3 x \, dx$ as $\int \csc^2 x \csc x \, dx$ (see Example 4.3-6, problem letter i).

In the following examples we use the above general rules in order to solve integral of products and powers of trigonometric functions:

Example 4.3-5: Evaluate the following indefinite integrals:

a. $\int \sin 5x \cos 7x \, dx =$ 　　　　b. $\int \sin x \cos x \, dx =$ 　　　　c. $\int \cos 3x \cos 2x \, dx =$

d. $\int \sin 3x \sin 5x \, dx =$ 　　　　e. $\int \cos 3x \cos 5x \, dx =$ 　　　　f. $\int \sin^5 x \, dx =$

g. $\int \sin^3 x \, dx =$ 　　　　h. $\int \cos^5 x \, dx =$ 　　　　i. $\int \tan^4 x \, dx =$

j. $\int \sin^7 x \, dx =$ 　　　　k. $\int \sec^4 x \, dx =$ 　　　　l. $\int \cos^3 x \, dx =$

Solutions:

a. $\boxed{\int \sin 5x \cos 7x \, dx} = \boxed{\int \frac{1}{2} \left[\sin(5-7)x + \sin(5+7)x \right] dx} = \boxed{\int \frac{1}{2} \left[\sin(-2x) + \sin(12x) \right] dx} = \boxed{\frac{1}{2} \int \left(\sin 12x - \sin 2x \right) dx}$

$= \boxed{\frac{1}{2} \int \sin 12x \, dx - \frac{1}{2} \int \sin 2x \, dx} = \boxed{\frac{1}{2} \cdot -\frac{1}{12} \cos 12x - \frac{1}{2} \cdot -\frac{1}{2} \cos 2x + c} = \boxed{-\frac{1}{24} \cos 12x + \frac{1}{4} \cos 2x + c}$

Check: Let $y = -\frac{1}{24} \cos 12x + \frac{1}{4} \cos 2x + c$, then $y' = -\frac{1}{24} \cdot -12 \sin 12x + \frac{1}{4} \cdot -2 \sin 2x + 0 = \frac{12}{24} \sin 12x - \frac{2}{4} \sin 2x$

$= \frac{1}{2} \sin 12x - \frac{1}{2} \sin 2x = \frac{1}{2} \sin 12x + \frac{1}{2} \sin(-2x) = \frac{1}{2} \sin(5+7)x + \frac{1}{2} \sin(5-7)x = \sin 5x \cos 7x$

b. $\boxed{\int \sin x \cos x \, dx} = \boxed{\int \frac{1}{2} \left[\sin(1-1)x + \sin(1+1)x \right] dx} = \boxed{\int \frac{1}{2} \left[\sin(0x) + \sin(2x) \right] dx} = \boxed{\frac{1}{2} \int \left(0 + \sin 2x \right) dx}$

$= \boxed{\frac{1}{2} \int \sin 2x \, dx} = \boxed{\frac{1}{2} \cdot -\frac{1}{2} \cos 2x + c} = \boxed{-\frac{1}{4} \cos 2x + c}$

Check: Let $y = -\frac{1}{4} \cos 2x + c$, then $y' = -\frac{1}{4} \cdot -2 \sin 2x + 0 = \frac{1}{2} \sin 2x = \frac{1}{2} \cdot 2 \sin x \cos x = \sin x \cos x$

c. $\boxed{\int \cos 3x \cos 2x \, dx} = \boxed{\int \frac{1}{2} \left[\cos(3-2)x + \cos(3+2)x \right] dx} = \boxed{\int \frac{1}{2} \left(\cos x + \cos 5x \right) dx} = \boxed{\frac{1}{2} \int \cos x \, dx + \frac{1}{2} \int \cos 5x \, dx}$

$= \boxed{\frac{1}{2} \cdot \sin x + \frac{1}{2} \cdot \frac{1}{5} \sin 5x + c} = \boxed{\frac{1}{2} \sin x + \frac{1}{10} \sin 5x + c}$

Check: Let $y = \dfrac{1}{2}\sin x + \dfrac{1}{10}\sin 5x + c$, then $y' = \dfrac{1}{2}\cdot\cos x + \dfrac{1}{10}\cdot 5\cos 5x + 0 = \dfrac{1}{2}\cos x + \dfrac{5}{10}\cos 5x$

$$= \dfrac{1}{2}\cos x + \dfrac{1}{2}\cos 5x = \dfrac{1}{2}\cos(3-2)x + \dfrac{1}{2}\cos(3+2)x = \cos 3x\cos 2x$$

d. $\boxed{\displaystyle\int \sin 3x\sin 5x\,dx} = \boxed{\displaystyle\int \dfrac{1}{2}\left[\cos(3-5)x - \cos(3+5)x\right]dx} = \boxed{\displaystyle\int \dfrac{1}{2}\left[\cos(-2x) - \cos(8x)\right]dx} = \boxed{\dfrac{1}{2}\displaystyle\int(\cos 2x - \cos 8x)\,dx}$

$= \boxed{\dfrac{1}{2}\displaystyle\int \cos 2x\,dx - \dfrac{1}{2}\displaystyle\int \cos 8x\,dx} = \boxed{\dfrac{1}{2}\cdot\dfrac{1}{2}\sin 2x - \dfrac{1}{2}\cdot\dfrac{1}{8}\sin 8x + c} = \boxed{\dfrac{1}{4}\sin 2x - \dfrac{1}{16}\sin 8x + c}$

Check: Let $y = \dfrac{1}{4}\sin 2x - \dfrac{1}{16}\sin 8x + c$, then $y' = \dfrac{1}{4}\cdot\cos 2x\cdot 2 - \dfrac{1}{16}\cdot\cos 8x\cdot 8 + 0 = \dfrac{2}{4}\cos 2x - \dfrac{8}{16}\cos 8x$

$$= \dfrac{1}{2}\cos 2x - \dfrac{1}{2}\cos 8x = \dfrac{1}{2}\left[\cos(-2x) - \cos(8x)\right] = \dfrac{1}{2}\left[\cos(3-5)x - \cos(3+5)x\right] = \sin 3x\sin 5x$$

e. $\boxed{\displaystyle\int \cos 3x\cos 5x\,dx} = \boxed{\displaystyle\int \dfrac{1}{2}\left[\cos(3-5)x + \cos(3+5)x\right]dx} = \boxed{\displaystyle\int \dfrac{1}{2}\left[\cos(-2x) + \cos(8x)\right]dx} = \boxed{\displaystyle\int \dfrac{1}{2}(\cos 2x + \cos 8x)\,dx}$

$= \boxed{\dfrac{1}{2}\displaystyle\int \cos 2x\,dx + \dfrac{1}{2}\displaystyle\int \cos 8x\,dx} = \boxed{\dfrac{1}{2}\cdot\dfrac{1}{2}\sin 2x + \dfrac{1}{2}\cdot\dfrac{1}{8}\sin 8x + c} = \boxed{\dfrac{1}{4}\sin 2x + \dfrac{1}{16}\sin 8x + c}$

Check: Let $y = \dfrac{1}{4}\sin 2x + \dfrac{1}{16}\sin 8x + c$, then $y' = \dfrac{1}{4}\cdot\cos 2x\cdot 2 + \dfrac{1}{16}\cdot\cos 8x\cdot 8 + 0 = \dfrac{2}{4}\cos 2x + \dfrac{8}{16}\cos 8x$

$$= \dfrac{1}{2}\cos 2x + \dfrac{1}{2}\cos 8x = \dfrac{1}{2}\left[\cos(-2x) + \cos(8x)\right] = \dfrac{1}{2}\left[\cos(3-5)x + \cos(3+5)x\right] = \cos 3x\cos 5x$$

f. $\displaystyle\int \sin^5 x\,dx = \int \sin^4 x\sin x\,dx = \int\left(\sin^2 x\right)^2 \sin x\,dx = \int\left(1-\cos^2 x\right)^2 \sin x\,dx$. Let $u = \cos x$, then

$\dfrac{du}{dx} = \dfrac{d}{dx}\cos x$; $\dfrac{du}{dx} = -\sin x$; $du = -\sin x\,dx$; $dx = -\dfrac{du}{\sin x}$. Therefore,

$\boxed{\displaystyle\int \sin^5 x\,dx} = \boxed{\displaystyle\int\left(1-\cos^2 x\right)^2 \sin x\,dx} = \boxed{\displaystyle\int\left(1-u^2\right)^2 \sin x\,dx} = \boxed{\displaystyle\int\left(u^4 - 2u^2 + 1\right)\sin x\cdot\dfrac{du}{-\sin x}}$

$= \boxed{-\displaystyle\int\left(u^4 - 2u^2 + 1\right)du} = \boxed{-\displaystyle\int u^4\,du + 2\displaystyle\int u^2\,du - \displaystyle\int du} = \boxed{-\dfrac{1}{5}u^5 + \dfrac{2}{3}u^3 - u + c} = \boxed{-\dfrac{1}{5}\cos^5 x + \dfrac{2}{3}\cos^3 x - \cos x + c}$

Check: Let $y = -\dfrac{1}{5}\cos^5 x + \dfrac{2}{3}\cos^3 x - \cos x + c$, then $y' = -\dfrac{1}{5}\cdot 5\cos^4 x\cdot -\sin x + \dfrac{2}{3}\cdot 3\cos^2 x\cdot -\sin x + \sin x + 0$

$$= \sin x\cos^4 x - 2\sin x\cos^2 x + \sin x = \sin x\left(\cos^4 x - 2\cos^2 x + 1\right) = \sin x\left(1-\cos^2 x\right)^2$$

$$= \sin x\left(\sin^2 x\right)^2 = \sin x\sin^4 x = \sin^5 x$$

g. $\displaystyle\int \sin^3 x\,dx = \int \sin^2 x\sin x\,dx = \int\left(1-\cos^2 x\right)\sin x\,dx$. Let $u = \cos x$, then $\dfrac{du}{dx} = \dfrac{d}{dx}\cos x$; $\dfrac{du}{dx} = -\sin x$

$; \ du = -\sin x \, dx \ ; \ dx = -\dfrac{du}{\sin x}$. Therefore,

$$\boxed{\int \sin^3 x \, dx} = \boxed{\int \left(1-\cos^2 x\right)\sin x \, dx} = \boxed{\int \left(1-u^2\right)\sin x \cdot -\dfrac{du}{\sin x}} = \boxed{-\int \left(1-u^2\right)du} = \boxed{\int u^2 du - \int du}$$

$$= \boxed{\dfrac{1}{3}u^3 - u + c} = \boxed{\dfrac{1}{3}\cos^3 x - \cos x + c}$$

Check: Let $y = \dfrac{1}{3}\cos^3 x - \cos x + c$, then $y' = \dfrac{1}{3} \cdot 3\cos^2 x \cdot -\sin x + \sin x + 0 \ = \ -\sin x \cos^2 x + \sin x$

$$= \ \sin x \left(1 - \cos^2 x\right) = \ \sin x \sin^2 x \ = \ \sin^3 x$$

h. $\int \cos^5 x \, dx \ = \ \int \cos^4 x \cos x \, dx \ = \ \int \left(\cos^2 x\right)^2 \cos x \, dx \ = \ \int \left(1 - \sin^2 x\right)^2 \cos x \, dx$. Let $u = \sin x$, then

$\dfrac{du}{dx} = \dfrac{d}{dx}\sin x \ ; \ \dfrac{du}{dx} = \cos x \ ; \ du = \cos x \, dx \ ; \ dx = \dfrac{du}{\cos x}$. Therefore,

$$\boxed{\int \cos^5 x \, dx} = \boxed{\int \left(1-\sin^2 x\right)^2 \cos x \, dx} = \boxed{\int \left(1-u^2\right)^2 \cos x \cdot \dfrac{du}{\cos x}} = \boxed{\int \left(1-u^2\right)^2 du} = \boxed{\int \left(u^4 - 2u^2 + 1\right)du}$$

$$= \boxed{\int u^4 du - 2\int u^2 du + \int du} = \boxed{\dfrac{1}{5}u^5 - \dfrac{2}{3}u^3 + u + c} = \boxed{\dfrac{1}{5}\sin^5 x - \dfrac{2}{3}\sin^3 x + \sin x + c}$$

Check: Let $y = \dfrac{1}{5}\sin^5 x - \dfrac{2}{3}\sin^3 x + \sin x + c$, then $y' = \dfrac{1}{5} \cdot 5\sin^4 x \cdot \cos x - \dfrac{2}{3} \cdot 3\sin^2 x \cdot \cos x + \cos x + 0$

$$= \ \cos x \sin^4 x - 2\cos x \sin^2 x + \cos x \ = \ \cos x \left(\sin^4 x - 2\sin^2 x + 1\right) = \ \cos x \left(1 - \sin^2 x\right)^2$$

$$= \ \cos x \left(\cos^2 x\right)^2 \ = \ \cos x \cos^4 x \ = \ \cos^5 x$$

i. $\int \tan^4 x \, dx \ = \ \int \tan^2 x \tan^2 x \, dx \ = \ \int \tan^2 x \left(\sec^2 x - 1\right)dx \ = \ \int \tan^2 x \sec^2 x \, dx - \int \tan^2 x \, dx$

$$= \ \int \tan^2 x \sec^2 x \, dx - \int \left(\sec^2 x - 1\right)dx \ = \ \int \tan^2 x \sec^2 x \, dx - \int \sec^2 x \, dx + \int dx$$. To solve the first

integral let $u = \tan x$, then $\dfrac{du}{dx} = \dfrac{d}{dx}\tan x \ ; \ \dfrac{du}{dx} = \sec^2 x \ ; \ du = \sec^2 x \, dx \ ; \ dx = \dfrac{du}{\sec^2 x}$. Therefore,

$\int \tan^2 x \sec^2 x \, dx \ = \ \int u^2 \sec^2 x \cdot \dfrac{du}{\sec^2 x} \ = \ \int u^2 du \ = \ \dfrac{1}{3}u^3 \ = \ \dfrac{1}{3}\tan^3 x$. Grouping the terms we find

$$\boxed{\int \tan^4 x \, dx} = \boxed{\int \tan^2 x \sec^2 x \, dx - \int \sec^2 x \, dx + \int dx} = \boxed{\dfrac{1}{3}\tan^3 x - \int \sec^2 x \, dx + \int dx} = \boxed{\dfrac{1}{3}\tan^3 x - \tan x + x + c}$$

Check: Let $y = \dfrac{1}{3}\tan^3 x - \tan x + x + c$, then $y' = \dfrac{3}{3}\tan^2 x \cdot \sec^2 x - \sec^2 x + 1 + 0 \ = \ \tan^2 x \cdot \sec^2 x - \sec^2 x + 1$

$$= \ \sec^2 x \left(\tan^2 x - 1\right) + 1 \ = \ \left(\tan^2 x + 1\right)\left(\tan^2 x - 1\right) + 1 \ = \ \tan^4 x - \tan^2 x + \tan^2 x - 1 + 1 \ = \ \tan^4 x$$

j. $\int \sin^7 x \, dx = \int \sin^6 x \sin x \, dx = \int \left(\sin^2 x\right)^3 \sin x \, dx = \int \left(1-\cos^2 x\right)^3 \sin x \, dx$. Let $u = \cos x$, then

$\dfrac{du}{dx} = \dfrac{d}{dx}\cos x$; $\dfrac{du}{dx} = -\sin x$; $du = -\sin x \, dx$; $dx = -\dfrac{du}{\sin x}$. Therefore,

$$\boxed{\int \sin^7 x \, dx} = \boxed{\int \left(1-\cos^2 x\right)^3 \sin x \, dx} = \boxed{\int \left(1-u^2\right)^3 \sin x \, dx} = \boxed{\int \left(1-3u^2+3u^4-u^6\right)\sin x \cdot \dfrac{du}{-\sin x}}$$

$$= \boxed{-\int \left(1-3u^2+3u^4-u^6\right)du} = \boxed{\int u^6 du - 3\int u^4 du + 3\int u^2 du - \int du} = \boxed{\dfrac{1}{7}u^7 - 3\cdot\dfrac{1}{5}u^5 + 3\cdot\dfrac{1}{3}u^3 - u + c}$$

$$= \boxed{\dfrac{1}{7}\cos^7 x - \dfrac{3}{5}\cos^5 x + \cos^3 x - \cos x + c}$$

Check: Let $y = \dfrac{1}{7}\cos^7 x - \dfrac{3}{5}\cos^5 x + \cos^3 x - \cos x + c$, then $y' = \dfrac{1}{7}\cdot 7\cos^6 x \cdot -\sin x - \dfrac{3}{5}\cdot 5\cos^4 x \cdot -\sin x$

$\qquad + 3\cos^2 x \cdot -\sin x + \sin x + 0 = -\sin x \cos^6 x + 3\sin x \cos^4 x - 3\sin x \cos^2 x + \sin x$

$\qquad = \sin x\left(-\cos^6 x + 3\cos^4 x - 3\cos^2 x + 1\right) = \sin x\left(1-\cos^2 x\right)^3 = \sin x\left(\sin^2 x\right)^3 = \sin x \sin^6 x = \sin^7 x$

k. $\int \sec^4 x \, dx = \int \sec^2 x \sec^2 x \, dx = \int \sec^2 x\left(1+\tan^2 x\right)dx = \int \sec^2 x \, dx + \int \tan^2 x \sec^2 x \, dx$

$= \int \left(1+\tan^2 x\right)dx + \int \tan^2 x \sec^2 x \, dx = \int dx + \int \tan^2 x \, dx + \int \tan^2 x \sec^2 x \, dx$. To solve the third integral

let $u = \tan x$, then $\dfrac{du}{dx} = \dfrac{d}{dx}\tan x$; $\dfrac{du}{dx} = \sec^2 x$; $du = \sec^2 x \, dx$; $dx = \dfrac{du}{\sec^2 x}$. Therefore,

$\int \tan^2 x \sec^2 x \, dx = \int u^2 \sec^2 x \cdot \dfrac{du}{\sec^2 x} = \int u^2 du = \dfrac{1}{3}u^3 = \dfrac{1}{3}\tan^3 x$. Grouping the terms we find

$$\boxed{\int \sec^4 x \, dx} = \boxed{\int dx + \int \tan^2 x \, dx + \int \tan^2 x \sec^2 x \, dx} = \boxed{\int dx + \int \tan^2 x \, dx + \dfrac{1}{3}\tan^3 x} = \boxed{x + \tan x - x + \dfrac{1}{3}\tan^3 x + c}$$

$$\boxed{\dfrac{1}{3}\tan^3 x + \tan x + (x-x) + c} = \boxed{\dfrac{1}{3}\tan^3 x + \tan x + c}$$

Check: Let $y = \dfrac{1}{3}\tan^3 x + \tan x + c$, then $y' = \dfrac{3}{3}\tan^2 x \cdot \sec^2 x + \sec^2 x + 0 = \tan^2 x \cdot \sec^2 x + \sec^2 x$

$\qquad = \sec^2 x\left(\tan^2 x + 1\right) = \sec^2 x \sec^2 x = \sec^4 x$

l. $\int \cos^3 x \, dx = \int \left(\cos^2 x\right)\cos x \, dx = \int \left(1-\sin^2 x\right)\cos x \, dx$. Let $u = \sin x$, then $\dfrac{du}{dx} = \dfrac{d}{dx}\sin x$; $\dfrac{du}{dx} = \cos x$

; $du = \cos x \, dx$; $dx = \dfrac{du}{\cos x}$. Therefore,

$$\boxed{\int \cos^3 x \, dx} = \boxed{\int \left(1-\sin^2 x\right)\cos x \, dx} = \boxed{\int \left(1-u^2\right)\cos x \cdot \dfrac{du}{\cos x}} = \boxed{\int \left(1-u^2\right)du} = \boxed{-\int u^2 du + \int du}$$

$$= \boxed{-\frac{1}{3}u^3 + u + c} = \boxed{-\frac{1}{3}\sin^3 x + \sin x + c}$$

Check: Let $y = -\frac{1}{3}\sin^3 x + \sin x + c$, then $y' = -\frac{1}{3}\cdot 3\sin^2 x\cdot\cos x + \cos x + 0 = -\cos x\sin^2 x + \cos x$

$$= \cos x\left(1 - \sin^2 x\right) = \cos x\left(\cos^2 x\right) = \cos^3 x$$

Example 4.3-6: Evaluate the following indefinite integrals:

a. $\int \cos^4 x\, dx =$ b. $\int \cos^2 5x\, dx =$ c. $\int \sin^4 x\, dx =$

d. $\int \tan^3 x\, dx =$ e. $\int \cot^4 x\, dx =$ f. $\int \tan^6 x\, dx =$

g. $\int \cot^3 x\, dx =$ h. $\int \sec^3 x\, dx =$ i. $\int \csc^3 x\, dx =$

j. $\int \tan^5 x\, dx =$ k. $\int \cot^5 x\, dx =$ l. $\int \cot^6 x\, dx =$

Solutions:

a. $\boxed{\int \cos^4 x\, dx} = \boxed{\int\left(\cos^2 x\right)^2 dx} = \boxed{\int\left[\frac{1}{2}(1+\cos 2x)\right]^2 dx} = \boxed{\frac{1}{4}\int(1+\cos 2x)^2 dx} = \boxed{\frac{1}{4}\int\left(1+\cos^2 2x + 2\cos 2x\right)dx}$

$$= \boxed{\frac{1}{4}\int dx + \frac{1}{4}\int\cos^2 2x\, dx + \frac{2}{4}\int\cos 2x\, dx} = \boxed{\frac{x}{4} + \frac{1}{4}\int\frac{1}{2}(1+\cos 4x)\, dx + \frac{1}{2}\cdot\frac{1}{2}\sin 2x} = \boxed{\frac{x}{4} + \frac{1}{4}\cdot\frac{1}{2}\int dx}$$

$$\boxed{+\frac{1}{8}\int\cos 4x\, dx + \frac{1}{4}\sin 2x} = \boxed{\frac{x}{4} + \frac{x}{8} + \frac{1}{8}\cdot\frac{1}{4}\sin 4x + \frac{1}{4}\sin 2x + c} = \boxed{\frac{3}{8}x + \frac{1}{32}\sin 4x + \frac{1}{4}\sin 2x + c}$$

Check: Let $y = \frac{3x}{8} + \frac{1}{32}\sin 4x + \frac{1}{4}\sin 2x + c$, then $y' = \frac{3}{8} + \frac{4\cos 4x}{32} + \frac{2\cos 2x}{4} + 0 = \frac{3}{8} + \frac{1}{8}\cos 4x + \frac{2}{4}\cos 2x$

$$= \left(\frac{1}{4} + \frac{1}{8}\right) + \frac{1}{8}\cos 4x + \frac{2}{4}\cos 2x = \frac{1}{4} + \frac{1}{8}(1+\cos 4x) + \frac{2}{4}\cos 2x = \frac{1}{4} + \frac{1}{4}\cdot\frac{1}{2}(1+\cos 4x) + \frac{2}{4}\cos 2x$$

$$= \frac{1}{4} + \frac{1}{4}\cdot\cos^2 2x + \frac{2}{4}\cos 2x = \frac{1}{4}\left(1+\cos^2 2x + 2\cos 2x\right) = \frac{1}{4}(1+\cos 2x)^2 = \left[\frac{1}{2}(1+\cos 2x)\right]^2$$

$$= \left(\cos^2 x\right)^2 = \cos^4 x$$

b. $\boxed{\int\cos^2 5x\, dx} = \boxed{\int\left(1-\sin^2 5x\right)dx} = \boxed{\int dx - \int\sin^2 5x\, dx} = \boxed{\int dx - \int\frac{1}{2}(1-\cos 10x)dx} = \boxed{x - \frac{1}{2}\int dx + \frac{1}{2}\int\cos 10x\, dx}$

$$= \boxed{x - \frac{x}{2} + \frac{1}{2}\cdot\frac{1}{10}\sin 10x + c} = \boxed{x\left(1-\frac{1}{2}\right) + \frac{1}{20}\sin 10x + c} = \boxed{\frac{x}{2} + \frac{\sin 10x}{20} + c}$$

Check: Let $y = \frac{x}{2} + \frac{\sin 10x}{20} + c$, then $y' = \frac{1}{2} + \frac{1}{20}\cdot\cos 10x\cdot 10 + 0 = \frac{1}{2} + \frac{10}{20}\cdot\cos 10x = \frac{1}{2} + \frac{1}{2}\cos 10x$

$$= \left(1 - \frac{1}{2}\right) + \frac{1}{2}\cos 10x \; = \; 1 - \left(\frac{1}{2} - \frac{1}{2}\cos 10x\right) \; = \; 1 - \frac{1}{2}(1 - \cos 10x) \; = \; 1 - \sin^2 5x \; = \; \cos^2 5x$$

c. $\boxed{\int \sin^4 x \, dx} = \boxed{\int \left(\sin^2 x\right)^2 dx} = \boxed{\int \left[\frac{1}{2}(1 - \cos 2x)\right]^2 dx} = \boxed{\frac{1}{4}\int (1 - \cos 2x)^2 dx} = \boxed{\frac{1}{4}\int \left(1 + \cos^2 2x - 2\cos 2x\right)dx}$

$= \boxed{\frac{1}{4}\int dx + \frac{1}{4}\int \cos^2 2x \, dx - \frac{2}{4}\int \cos 2x \, dx} = \boxed{\frac{x}{4} + \frac{1}{4}\int \frac{1}{2}(1 + \cos 4x)\, dx - \frac{1}{2}\cdot\frac{1}{2}\sin 2x} = \boxed{\frac{x}{4} + \frac{1}{4}\cdot\frac{1}{2}\int dx}$

$\boxed{+\frac{1}{8}\int \cos 4x \, dx - \frac{1}{4}\sin 2x} = \boxed{\frac{x}{4} + \frac{x}{8} + \frac{1}{8}\cdot\frac{1}{4}\sin 4x - \frac{1}{4}\sin 2x + c} = \boxed{\frac{3}{8}x + \frac{1}{32}\sin 4x - \frac{1}{4}\sin 2x + c}$

Check: Let $y = \frac{3x}{8} + \frac{1}{32}\sin 4x - \frac{1}{4}\sin 2x + c$, then $y' = \frac{3}{8} + \frac{4\cos 4x}{32} - \frac{2\cos 2x}{4} + 0 = \frac{3}{8} + \frac{1}{8}\cos 4x - \frac{2}{4}\cos 2x$

$= \left(\frac{1}{4} + \frac{1}{8}\right) + \frac{1}{8}\cos 4x - \frac{2}{4}\cos 2x = \frac{1}{4} + \frac{1}{8}(1 + \cos 4x) - \frac{2}{4}\cos 2x = \frac{1}{4} + \frac{1}{4}\cdot\frac{1}{2}(1 + \cos 4x) - \frac{2}{4}\cos 2x$

$= \frac{1}{4} + \frac{1}{4}\cdot\cos^2 2x - \frac{2}{4}\cos 2x = \frac{1}{4}\left(1 + \cos^2 2x - 2\cos 2x\right) = \frac{1}{4}(1 - \cos 2x)^2 = \left[\frac{1}{2}(1 - \cos 2x)\right]^2$

$= \left(\sin^2 x\right)^2 = \sin^4 x$

d. $\int \tan^3 x \, dx = \int \tan^2 x \tan x \, dx = \int \left(\sec^2 x - 1\right)\tan x \, dx = \int \sec^2 x \tan x \, dx - \int \tan x \, dx$. To solve the first

integral let $u = \tan x$, then $\frac{du}{dx} = \frac{d}{dx}\tan x$; $\frac{du}{dx} = \sec^2 x$; $du = \sec^2 dx$; $dx = \frac{du}{\sec^2 x}$. Thus,

$\int \sec^2 x \tan x \, dx = \int \sec^2 x \cdot u \cdot \frac{du}{\sec^2 x} = \int u \, du = \frac{1}{2}u^2 = \frac{1}{2}\tan^2 x$. Combining the term

$\boxed{\int \tan^3 x \, dx} = \boxed{\int \tan^2 x \tan x \, dx} = \boxed{\int \left(\sec^2 x - 1\right)\tan x \, dx} = \boxed{\int \left(\sec^2 x \tan x - \tan x\right)dx} = \boxed{\int \sec^2 x \tan x \, dx}$

$\boxed{-\int \tan x \, dx} = \boxed{\frac{1}{2}\tan^2 x - \int \tan x \, dx} = \boxed{\frac{1}{2}\tan^2 x - \ln|\sec x| + c}$

Check: Let $y = \frac{1}{2}\tan^2 x - \ln|\sec x| + c$, then $y' = -\frac{1}{2}\cdot 2\tan x \cdot \sec^2 x - \frac{1}{\sec x}\cdot\sec x \tan x + 0$

$= \tan x \sec^2 x - \frac{\sec x \tan x}{\sec x} = \tan x \sec^2 x - \tan x = \tan x \left(\sec^2 x - 1\right) = \tan x \tan^2 x = \tan^3 x$

e. $\int \cot^4 x \, dx = \int \cot^2 x \cot^2 x \, dx = \int \cot^2 x \left(\csc^2 x - 1\right)dx = \int \cot^2 x \csc^2 x \, dx - \int \cot^2 x \, dx$

$= \int \cot^2 x \csc^2 x \, dx - \int \left(\csc^2 x - 1\right)dx = \int \cot^2 x \csc^2 x \, dx - \int \csc^2 x \, dx + \int dx$. To solve the first

integral let $u = \cot x$, then $\frac{du}{dx} = \frac{d}{dx}\cot x$; $\frac{du}{dx} = -\csc^2 x$; $du = -\csc^2 x \, dx$; $dx = -\frac{du}{\csc^2 x}$. Therefore,

$\int \cot^2 x \csc^2 x\, dx = \int u^2 \csc^2 x \cdot -\dfrac{du}{\csc^2 x} = -\int u^2 du = -\dfrac{1}{3}u^3 = -\dfrac{1}{3}\cot^3 x$. Grouping the terms we find

$$\boxed{\int \cot^4 x\, dx} = \boxed{\int \cot^2 x \csc^2 x\, dx - \int \csc^2 x\, dx + \int dx} = \boxed{-\dfrac{1}{3}\cot^3 x - \int \csc^2 x\, dx + \int dx} = \boxed{-\dfrac{1}{3}\cot^3 x + \cot x + x + c}$$

Check: Let $y = -\dfrac{1}{3}\cot^3 x + \cot x + x + c$, then $y' = -\dfrac{3}{3}\cot^2 x \cdot -\csc^2 x - \csc^2 x + 1 = \cot^2 x \csc^2 x - \csc^2 x + 1$

$$= \csc^2 x\left(\cot^2 x - 1\right) + 1 = \left(\cot^2 x + 1\right)\left(\cot^2 x - 1\right) + 1 = \cot^4 x - \cot^2 x + \cot^2 x - 1 + 1 = \cot^4 x$$

f. $\int \tan^6 x\, dx = \int \tan^4 x \tan^2 x\, dx = \int \tan^4 x \left(\sec^2 x - 1\right) dx = \int \left(\tan^4 x \sec^2 x - \tan^4 x\right) dx = \int \tan^4 x \sec^2 x\, dx$

$- \int \tan^4 x\, dx$. In example 4.3-5, problem letter i, we found that $\int \tan^4 x\, dx = \dfrac{1}{3}\tan^3 x - \tan x + x + c$. Thus,

$$\boxed{\int \tan^6 x\, dx} = \boxed{\int \tan^4 x \tan^2 x\, dx} = \boxed{\int \tan^4 x \left(\sec^2 x - 1\right) dx} = \boxed{\int \tan^4 x \sec^2 x\, dx - \int \tan^4 x\, dx}$$

$$= \boxed{\int \tan^4 x \sec^2 x\, dx - \left(\dfrac{1}{3}\tan^3 x - \tan x + x + c\right)} = \boxed{\dfrac{1}{5}\tan^5 x - \dfrac{1}{3}\tan^3 x + \tan x - x + c}$$

Check: Let $y = \dfrac{1}{5}\tan^5 x - \dfrac{1}{3}\tan^3 x + \tan x - x + c$, then $y' = \dfrac{1}{5} \cdot 5 \tan^4 x \cdot \sec^2 x - \dfrac{3}{3}\tan^2 x \cdot \sec^2 x + \sec^2 x - 1 + 0$

$$= \tan^4 x \cdot \sec^2 x - \tan^2 x \cdot \sec^2 x + \sec^2 x - 1 = \sec^2 x\left(\tan^4 x - \tan^2 x + 1\right) - 1$$

$$= \left(1 + \tan^2 x\right)\left(\tan^4 x - \tan^2 x + 1\right) - 1 = \tan^4 x - \tan^2 x + 1 + \tan^6 x - \tan^4 x + \tan^2 x - 1 = \tan^6 x$$

g. $\int \cot^3 x\, dx = \int \cot^2 x \cot x\, dx = \int \left(\csc^2 x - 1\right)\cot x\, dx = \int \csc^2 x \cot x\, dx - \int \cot x\, dx$. To solve the first

integral let $u = \cot x$, then $\dfrac{du}{dx} = \dfrac{d}{dx}\cot x$; $\dfrac{du}{dx} = -\csc^2 x$; $du = -\csc^2 dx$; $dx = -\dfrac{du}{\csc^2 x}$. Thus,

$\int \csc^2 x \cot x\, dx = \int \csc^2 x \cdot u \cdot -\dfrac{du}{\csc^2 x} = -\int u\, du = -\dfrac{1}{2}u^2 = -\dfrac{1}{2}\cot^2 x$. Combining the term

$$\boxed{\int \cot^3 x\, dx} = \boxed{\int \cot^2 x \cot x\, dx} = \boxed{\int \left(\csc^2 x - 1\right)\cot x\, dx} = \boxed{\int \left(\csc^2 x \cdot \cot x - \cot x\right) dx} = \boxed{\int \csc^2 x \cot x\, dx}$$

$$\boxed{- \int \cot x\, dx} = \boxed{-\dfrac{1}{2}\cot^2 x - \int \cot x\, dx} = \boxed{-\dfrac{1}{2}\cot^2 x - \ln|\sin x| + c}$$

Check: Let $y = -\dfrac{1}{2}\cot^2 x - \ln|\sin x| + c$, then $y' = -\dfrac{1}{2} \cdot 2\cot x \cdot -\csc^2 x - \dfrac{1}{\sin x} \cdot \cos x + 0$

$$= \cot x \csc^2 x - \dfrac{\cos x}{\sin x} = \cot x \csc^2 x - \cot x = \cot x\left(\csc^2 x - 1\right) = \cot x \cot^2 x = \cot^3 x$$

h. $\int \sec^3 x \, dx = \int \sec^2 x \sec x \, dx = \int \left(\tan^2 x + 1 \right) \sec x \, dx = \int \tan^2 x \sec x \, dx + \int \sec x \, dx = \int \tan x \cdot \tan x \sec x \, dx$

$+ \int \sec x \, dx$. To solve the first integral let $u = \tan x$ and $dv = \tan x \sec x \, dx$, then $du = \sec^2 x \, dx$ and

$\int dv = \int \tan x \sec x \, dx$ which implies $v = \sec x$. Using the integration by parts formula $\int u \, dv = uv - \int v \, du$

we obtain $\int \tan^2 x \sec x \, dx = \int \tan x \cdot \tan x \sec x \, dx = \tan x \cdot \sec x - \int \sec x \cdot \sec^2 x \, dx = \tan x \sec x - \int \sec^3 x \, dx$

Combining the terms we have $\int \sec^3 x \, dx = \int \tan x \cdot \tan x \sec x \, dx + \int \sec x \, dx = \tan x \sec x - \int \sec^3 x \, dx$.

$+ \int \sec x \, dx$. Moving the $\int \sec^3 x \, dx$ term from the right hand side of the equation to the left hand

side we obtain $\int \sec^3 x \, dx + \int \sec^3 x \, dx = 2 \int \sec^3 x \, dx = \tan x \sec x + \int \sec x \, dx$. Therefore,

$$\boxed{\int \sec^3 x \, dx} = \boxed{\frac{1}{2} \left(\tan x \sec x + \int \sec x \, dx \right)} = \boxed{\frac{1}{2} \tan x \sec x + \frac{1}{2} \ln \left| \sec x + \tan x \right| + c}$$

Check: Let $y = \frac{1}{2} \tan x \sec x + \frac{1}{2} \ln \left| \sec x + \tan x \right| + c$, then $y' = \dfrac{\sec^2 x \cdot \sec x + \sec x \tan x \cdot \tan x}{2}$

$+ \dfrac{\sec x \tan x + \sec^2 x}{2(\sec x + \tan x)} + 0 = \dfrac{\sec^3 x + \sec x \tan^2 x}{2} + \dfrac{\sec x (\sec x + \tan x)}{2(\sec x + \tan x)} = \dfrac{\sec^3 x + \sec x \tan^2 x}{2} + \dfrac{\sec x}{2}$

$= \dfrac{\sec^3 x + \sec x \tan^2 x + \sec x}{2} = \dfrac{\sec^3 x + \sec x \left(\tan^2 x + 1 \right)}{2} = \dfrac{\sec^3 x + \sec x \sec^2 x}{2} = \dfrac{\sec^3 x + \sec^3 x}{2}$

$= \dfrac{2 \sec^3 x}{2} = \sec^3 x$

i. $\int \csc^3 x \, dx = \int \csc^2 x \csc x \, dx = \int \left(1 + \cot^2 x \right) \csc x \, dx = \int \cot^2 x \csc x \, dx + \int \csc x \, dx = \int \cot x \cdot \cot x \csc x \, dx$

$+ \int \csc x \, dx$. To solve the first integral let $u = \cot x$ and $dv = \cot x \csc x \, dx$, then $du = -\csc^2 x \, dx$ and

$\int dv = \int \cot x \csc x \, dx$ which implies $v = -\csc x$. Using the integration by parts formula $\int u \, dv = uv - \int v \, du$

we obtain $\int \cot^2 x \csc x \, dx = \int \cot x \cdot \cot x \csc x \, dx = \cot x \cdot -\csc x - \int \csc x \cdot \csc^2 x \, dx = -\cot x \csc x - \int \csc^3 x \, dx$

Combining the terms we have $\int \csc^3 x \, dx = \int \cot x \cdot \cot x \csc x \, dx + \int \csc x \, dx = -\cot x \csc x - \int \csc^3 x \, dx$.

$+ \int \csc x \, dx$. Moving the $\int \csc^3 x \, dx$ term from the right hand side of the equation to the left hand

side we obtain $\int \csc^3 x \, dx + \int \csc^3 x \, dx = 2 \int \csc^3 x \, dx = -\cot x \csc x + \int \csc x \, dx$. Therefore,

$$\boxed{\int \csc^3 x \, dx} = \boxed{\frac{1}{2} \left(-\cot x \csc x + \int \csc x \, dx \right)} = \boxed{-\frac{1}{2} \cot x \csc x + \frac{1}{2} \ln \left| \csc x - \cot x \right| + c}$$

Check: Let $y=-\dfrac{1}{2}\cot x\csc x+\dfrac{1}{2}\ln|\csc x-\cot x|+c$, then $y'=\dfrac{-\left(-\csc^2 x\cdot\csc x-\csc x\cot x\cdot\cot x\right)}{2}$

$$+\dfrac{-\csc x\cot x+\csc^2 x}{2(\csc x-\cot x)}+0 = \dfrac{\csc^3 x+\csc x\cot^2 x}{2}+\dfrac{\csc x(\csc x-\cot x)}{2(\csc x-\cot x)} = \dfrac{\csc^3 x+\csc x\cot^2 x}{2}+\dfrac{\csc x}{2}$$

$$=\dfrac{\csc^3 x+\csc x\cot^2 x+\csc x}{2} = \dfrac{\csc^3 x+\csc x\left(\cot^2 x+1\right)}{2} = \dfrac{\csc^3 x+\csc x\csc^2 x}{2} = \dfrac{\sec^3 x+\sec^3 x}{2}$$

$$=\dfrac{2\csc^3 x}{2} = \csc^3 x$$

j. $\displaystyle\int\tan^5 x\,dx = \int\tan^3 x\tan^2 x\,dx = \int\tan^3 x\left(\sec^2 x-1\right)dx = \int\tan^3 x\sec^2 x\,dx-\int\tan^3 x\,dx$. To solve the

first integral let $u=\tan x$, then $\dfrac{du}{dx}=\dfrac{d}{dx}\tan x$; $\dfrac{du}{dx}=\sec^2 x$; $du=\sec^2 x\,dx$; $dx=\dfrac{du}{\sec^2 x}$. Therefore,

$\displaystyle\int\tan^3 x\sec^2 x\,dx = \int u^3\sec^2 x\cdot\dfrac{du}{\sec^2 x} = \int u^3\,du = \dfrac{1}{4}u^4+c = \dfrac{1}{4}\tan^4 x+c$. Also, in Example 4.3-6,

problem letter d, we found that $\displaystyle\int\tan^3 x\,dx = \dfrac{1}{2}\tan^2 x-\ln|\sec x|+c$. Grouping the terms together

$$\boxed{\int\tan^5 x\,dx} = \boxed{\int\tan^3 x\tan^2 x\,dx} = \boxed{\int\tan^3 x\left(\sec^2 x-1\right)dx} = \boxed{\int\tan^3 x\sec^2 x\,dx-\int\tan^3 x\,dx}$$

$$=\boxed{\dfrac{1}{4}\tan^4 x-\int\tan^3 x\,dx} = \boxed{\dfrac{1}{4}\tan^4 x-\left(\dfrac{1}{2}\tan^2 x-\ln|\sec x|+c\right)} = \boxed{\dfrac{1}{4}\tan^4 x-\dfrac{1}{2}\tan^2 x+\ln|\sec x|+c}$$

Check: Let $y=\dfrac{1}{4}\tan^4 x-\dfrac{1}{2}\tan^2 x+\ln|\sec x|+c$, then $y'=\dfrac{1}{4}\cdot 4\tan^3 x\cdot\sec^2 x-\dfrac{2}{2}\tan x\cdot\sec^2 x+\dfrac{\sec x\tan x}{\sec x}+0$

$$= \tan^3 x\sec^2 x-\tan x\sec^2 x+\tan x = \sec^2 x\left(\tan^3 x-\tan x\right)+\tan x = \left(1+\tan^2 x\right)\left(\tan^3 x-\tan x\right)+\tan x$$

$$= \tan^3 x-\tan x+\tan^5 x-\tan^3 +\tan x = \tan^5 x$$

k. $\displaystyle\int\cot^5 x\,dx = \int\cot^3 x\cot^2 x\,dx = \int\cot^3 x\left(\csc^2 x-1\right)dx = \int\cot^3 x\csc^2 x\,dx-\int\cot^3 x\,dx$. To solve the

first integral let $u=\cot x$, then $\dfrac{du}{dx}=\dfrac{d}{dx}\cot x$; $\dfrac{du}{dx}=-\csc^2 x$; $du=-\csc^2 x\,dx$; $dx=-\dfrac{du}{\csc^2 x}$. Thus,

$\displaystyle\int\cot^3 x\csc^2 x\,dx = \int u^3\csc^2 x\cdot-\dfrac{du}{\csc^2 x} = -\int u^3\,du = -\dfrac{1}{4}u^4+c = -\dfrac{1}{4}\cot^4 x+c$. Also, in example 4.3-6,

problem letter g, we found that $\displaystyle\int\cot^3 x\,dx = -\dfrac{1}{2}\cot^2 x-\ln|\sin x|+c$. Grouping the terms together

$$\boxed{\int\cot^5 x\,dx} = \boxed{\int\cot^3 x\cot^2 x\,dx} = \boxed{\int\cot^3 x\left(\csc^2 x-1\right)dx} = \boxed{\int\cot^3 x\csc^2 x\,dx-\int\cot^3 x\,dx}$$

$$=\boxed{-\dfrac{1}{4}\cot^4 x-\int\cot^3 x\,dx} = \boxed{-\dfrac{1}{4}\cot^4 x-\left(-\dfrac{1}{2}\cot^2 x-\ln|\sin x|+c\right)} = \boxed{-\dfrac{1}{4}\cot^4 x+\dfrac{1}{2}\cot^2 x+\ln|\sin x|+c}$$

Check: Let $y = -\frac{1}{4}\cot^4 x + \frac{1}{2}\cot^2 x + \ln|\sin x| + c$, then $y' = \frac{-4\cdot\cot^3 x\cdot -\csc^2 x}{4} + \frac{2\cdot\cot x\cdot -\csc^2 x}{2} + \frac{\cos x}{\sin x} + 0$

$= \cot^3 x\csc^2 x - \cot x\csc^2 x + \cot x = \csc^2 x\left(\cot^3 x - \cot x\right) + \cot x = \left(1 + \cot^2 x\right)\left(\cot^3 x - \cot x\right) + \cot x$

$= \cot^3 x - \cot x + \cot^5 x - \cot^3 x + \cot x = \cot^5 x$

l. $\int \cot^6 x\, dx = \int \cot^4 x\cot^2 x\, dx = \int \cot^4 x\left(\csc^2 x - 1\right)dx = \int\left(\cot^4 x\csc^2 x - \cot^4 x\right)dx = \int \cot^4 x\csc^2 x\, dx$

$-\int \cot^4 x\, dx$. In example 4.3-6, problem letter e, we found that $\int \cot^4 x\, dx = -\frac{1}{3}\cot^3 x + \cot x + x + c$.

Thus, $\boxed{\int \cot^6 x\, dx} = \boxed{\int \cot^4 x\cot^2 x\, dx} = \boxed{\int \cot^4 x\left(\csc^2 x - 1\right)dx} = \boxed{\int \cot^4 x\csc^2 x\, dx - \int \cot^4 x\, dx}$

$= \boxed{\int \cot^4 x\csc^2 x\, dx - \left(-\frac{1}{3}\cot^3 x + \cot x + x + c\right)} = \boxed{-\frac{1}{5}\cot^5 x + \frac{1}{3}\cot^3 x - \cot x - x + c}$

Check: Let $y = -\frac{1}{5}\cot^5 x + \frac{1}{3}\cot^3 x - \cot x - x + c$, then $y' = -\frac{5\cdot\cot^4 x\cdot -\csc^2 x}{5} + \frac{3\cdot\cot^2 x\cdot -\csc^2 x}{3} + \csc^2 x - 1$

$= \cot^4 x\cdot\csc^2 x - \cot^2 x\cdot\csc^2 x + \csc^2 x - 1 = \csc^2 x\left(\cot^4 x - \cot^2 x + 1\right) - 1$

$= \left(1 + \cot^2 x\right)\left(\cot^4 x - \cot^2 x + 1\right) - 1 = \cot^4 x - \cot^2 x + 1 + \cot^6 x - \cot^4 x + \cot^2 x - 1 = \cot^6 x$

Example 4.3-7: Evaluate the following indefinite integrals:

a. $\int \sin^2 x\cos^2 x\, dx =$ b. $\int \sin^2 x\cos^5 x\, dx =$ c. $\int \sin^4 x\cos^2 x\, dx =$

d. $\int \sin^3 x\cos^2 x\, dx =$ e. $\int \tan^2 x\sec x\, dx =$ f. $\int \tan^5 x\sec^3 x\, dx =$

g. $\int \tan^5 x\sec^4 x\, dx =$ h. $\int \tan^3 x\sec^4 x\, dx =$ i. $\int \tan^3 x\sec^3 x\, dx =$

j. $\int \cot^2 x\csc^2 x\, dx =$ k. $\int \cot^3 x\csc^3 x\, dx =$ l. $\int \cot^3 x\csc^4 x\, dx =$

Solutions:

a. $\boxed{\int \sin^2 x\cos^2 x\, dx} = \boxed{\frac{1}{4}\int \sin^2 2x\, dx} = \boxed{\frac{1}{4}\int \frac{1}{2}\left(1 - \cos 4x\right)dx} = \boxed{\frac{1}{8}\int\left(1 - \cos 4x\right)dx} = \boxed{\frac{1}{8}\int dx - \frac{1}{8}\int \cos 4x\, dx}$

$= \boxed{\frac{x}{8} - \frac{1}{8}\cdot\frac{1}{4}\sin 4x + c} = \boxed{\frac{1}{8}x - \frac{1}{32}\sin 4x + c}$

Check: Let $y = \frac{1}{8}x - \frac{1}{32}\sin 4x + c$, then $y' = \frac{1}{8} - \frac{4\cos 4x}{32} + 0 = \frac{1}{8} - \frac{1}{8}\cos 4x = \frac{1}{8}\left(1 - \cos 4x\right)$

$= \frac{1}{4}\cdot\frac{1}{2}\left(1 - \cos 4x\right) = \frac{1}{4}\cdot\sin^2 2x = \sin^2 x\cos^2 x$

b. $\boxed{\int \sin^2 x\cos^5 x\, dx} = \boxed{\int \sin^2 x\cos^4 x\cos x\, dx} = \boxed{\int \sin^2 x\left(\cos^2 x\right)^2\cos x\, dx} = \boxed{\int \sin^2 x\left(1 - \sin^2 x\right)^2\cos x\, dx}$

$$= \boxed{\int \sin^2 x \left(1+\sin^4 x - 2\sin^2 x\right)\cos x\, dx} \quad = \boxed{\int \left(\sin^2 x \cos x + \sin^6 x \cos x - 2\sin^4 x \cos x\right)dx}$$

$$= \boxed{\int \sin^2 x \cos x\, dx + \int \sin^6 x \cos x\, dx - 2\int \sin^4 x \cos x\, dx} = \boxed{\frac{1}{3}\sin^3 x + \frac{1}{7}\sin^7 x - \frac{2}{5}\sin^5 x + c}$$

Check: Let $y = \dfrac{1}{3}\sin^3 x + \dfrac{1}{7}\sin^7 x - \dfrac{2}{5}\sin^5 x + c$, then $y' = \dfrac{3\sin^2 x \cos x}{3} + \dfrac{7\sin^6 x \cos x}{7} - \dfrac{10\sin^4 x \cos x}{5} + 0$

$$= \sin^2 x \cos x + \sin^6 x \cos x - 2\sin^4 x \cos x = \sin^2 x \cos x\left(1+\sin^4 x - 2\sin^2 x\right)$$

$$= \sin^2 x \cos x\left(1+\sin^2 x\right)^2 = \sin^2 x \cos x\left(\cos^2 x\right)^2 = \sin^2 x \cos x \cos^4 x = \sin^2 x \cos^5 x$$

c. $\boxed{\int \sin^4 x \cos^2 x\, dx} = \boxed{\int \left(\sin x \cos x\right)^2 \sin^2 x\, dx} = \boxed{\int \left(\tfrac{1}{2}\sin 2x\right)^2 \cdot \tfrac{1}{2}\left(1-\cos 2x\right) dx} = \boxed{\frac{1}{8}\int \sin^2 2x\, dx}$

$$\boxed{-\frac{1}{8}\int \sin^2 2x \cos 2x\, dx} = \boxed{\frac{1}{8}\int \frac{1}{2}\left(1-\cos 4x\right)dx - \frac{1}{8}\int \sin^2 2x \cos 2x\, dx} = \boxed{\frac{1}{16}\int dx - \frac{1}{16}\int \cos 4x\, dx - \frac{1}{8}\cdot\frac{1}{6}\sin^3 2x}$$

$$\boxed{\frac{1}{16}x - \frac{1}{16}\cdot\frac{1}{4}\sin 4x - \frac{1}{48}\sin^3 2x + c} = \boxed{\frac{1}{16}x - \frac{1}{64}\sin 4x - \frac{1}{48}\sin^3 2x + c}$$

Check: Let $y = \dfrac{1}{16}x - \dfrac{1}{64}\sin 4x - \dfrac{1}{48}\sin^3 2x + c$, then $y' = \dfrac{1}{16} - \dfrac{4\cos 4x}{64} - \dfrac{6\sin^2 2x \cos 2x}{48} + 0$

$$= \frac{1}{16}\left(1-\cos 4x\right) - \frac{1}{8}\sin^2 2x \cos 2x = \frac{1}{8}\cdot\frac{1}{2}\left(1-\cos 4x\right) - \frac{1}{8}\sin^2 2x \cos 2x = \frac{1}{8}\cdot\sin^2 2x - \frac{1}{8}\sin^2 2x \cos 2x$$

$$= \left(\frac{1}{2}\sin 2x\right)^2 \cdot \frac{1}{2}\left(1-\cos 2x\right) = \left(\sin x \cos x\right)^2 \sin^2 x = \sin^2 x \cos^2 x \cdot \sin^2 x = \sin^4 x \cos^2 x$$

d. $\int \sin^3 x \cos^2 x\, dx = \int \sin^3 x \left(1-\sin^2 x\right)dx = \int \left(\sin^3 x - \sin^5 x\right)dx = \int \sin^3 x\, dx - \int \sin^5 x\, dx$. In Example,

4.3-5, problem letters f and g, we found that $\int \sin^5 x\, dx = -\dfrac{1}{5}\cos^5 x + \dfrac{2}{3}\cos^3 x - \cos x + c$

and $\int \sin^3 x\, dx = \dfrac{1}{3}\cos^3 x - \cos x + c$. Therefore,

$$\boxed{\int \sin^3 x \cos^2 x\, dx} = \boxed{\int \sin^3 x\, dx - \int \sin^5 x\, dx} = \boxed{\frac{1}{3}\cos^3 x - \cos x + c - \left(-\frac{1}{5}\cos^5 x + \frac{2}{3}\cos^3 x - \cos x + c\right)}$$

$$= \boxed{\frac{1}{5}\cos^5 x + \frac{1}{3}\cos^3 x - \frac{2}{3}\cos^3 x - \cos x + \cos x + c} = \boxed{\frac{1}{5}\cos^5 x - \frac{1}{3}\cos^3 x + c}$$

Check: Let $y = \dfrac{1}{5}\cos^5 x - \dfrac{1}{3}\cos^3 x + c$, then $y' = \dfrac{1}{5}\cdot 5\cos^4 x \cdot - \sin x - \dfrac{3}{3}\cdot\cos^2 x \cdot -\sin x + 0$

$$= -\sin x \cos^4 x + \sin x \cos^2 x = \sin x \cos^2 x\left(1-\cos^2 x\right) = \sin x \cos^2 x \sin^2 x = \sin^3 x \cos^2 x$$

e. $\int \tan^2 x \sec x\,dx = \int \left(\sec^2 x - 1\right)\sec x\,dx = \int \sec^3 x\,dx - \int \sec x\,dx$. In Example 4.3-6, problem letter

h, we found that $\int \sec^3 x\,dx = \dfrac{1}{2}\tan x \sec x + \dfrac{1}{2}\ln\left|\sec x + \tan x\right| + c$. Therefore,

$$\boxed{\int \tan^2 x \sec x\,dx} = \boxed{\int \sec^3 x\,dx - \int \sec x\,dx} = \boxed{\dfrac{1}{2}\tan x \sec x + \dfrac{1}{2}\ln\left|\sec x + \tan x\right| - \ln\left|\sec x + \tan x\right| + c}$$

$$\boxed{\dfrac{1}{2}\tan x \sec x - \dfrac{1}{2}\ln\left|\sec x + \tan x\right| + c}$$

Check: Let $y = \dfrac{1}{2}\tan x \sec x - \dfrac{1}{2}\ln\left|\sec x + \tan x\right| + c$, then $y' = \dfrac{1}{2}\left(\sec^3 x + \sec x \tan^2 x\right) - \dfrac{\sec x \tan x + \sec^2 x}{2\left(\sec x + \tan x\right)}$

$= \dfrac{1}{2}\left(\sec^3 x + \sec x \tan^2 x\right) - \dfrac{\sec x\left(\sec x + \tan x\right)}{2\left(\sec x + \tan x\right)} = \dfrac{1}{2}\sec^3 x + \dfrac{1}{2}\sec x \tan^2 x - \dfrac{\sec x}{2}$

$= \dfrac{1}{2}\sec x\left(\sec^2 x + \tan^2 x - 1\right) = \dfrac{1}{2}\sec x\left(\tan^2 x + \tan^2 x\right) = \dfrac{1}{2}\sec x \cdot 2\tan^2 x = \tan^2 x \sec x$

f. $\boxed{\int \tan^5 x \sec^3 x\,dx} = \boxed{\int \tan^4 x \sec^2 x \sec x \tan x\,dx} = \boxed{\int \left(\sec^2 x - 1\right)^2 \sec^2 x \sec x \tan x\,dx}$

$= \boxed{\int \left(\sec^4 x + 1 - 2\sec^2 x\right)\sec^2 x \sec x \tan x\,dx} = \boxed{\int \left(\sec^6 x + \sec^2 x - 2\sec^4 x\right)\sec x \tan x\,dx}$

$= \boxed{\dfrac{1}{7}\sec^7 x + \dfrac{1}{3}\sec^3 x - \dfrac{2}{5}\sec^5 x + c} = \boxed{\dfrac{1}{7}\sec^7 x - \dfrac{2}{5}\sec^5 x + \dfrac{1}{3}\sec^3 x + c}$

Check: Let $y = \dfrac{1}{7}\sec^7 x - \dfrac{2}{5}\sec^5 x + \dfrac{1}{3}\sec^3 x + c$, then $y' = \dfrac{7}{7}\sec^6 x \cdot \sec x \tan x - \dfrac{10}{5}\sec^4 x \cdot \sec x \tan x$

$+ \dfrac{3}{3}\sec^2 x \cdot \sec x \tan x + 0 = \sec^6 x \cdot \sec x \tan x - 2\sec^4 x \cdot \sec x \tan x + \sec^2 x \cdot \sec x \tan x$

$= \left(\sec^6 x + \sec^2 x - 2\sec^4 x\right)\sec x \tan x = \left(\sec^4 x + 1 - 2\sec^2 x\right)\sec^2 x \sec x \tan x$

$= \left(\sec^2 x - 1\right)^2 \sec^2 x \sec x \tan x = \tan^4 x \sec^2 x \sec x \tan x = \tan^5 x \sec^3 x$

g. $\boxed{\int \tan^5 x \sec^4 x\,dx} = \boxed{\int \tan^5 x \sec^2 x \sec^2 x\,dx} = \boxed{\int \tan^5 x\left(\tan^2 x + 1\right)\sec^2 x\,dx} = \boxed{\int \tan^7 x \sec^2 x\,dx}$

$\boxed{+ \int \tan^5 x \sec^2 x\,dx} = \boxed{\dfrac{1}{8}\tan^8 x + \dfrac{1}{6}\tan^6 x + c}$

Check: Let $y = \dfrac{1}{8}\tan^8 x + \dfrac{1}{6}\tan^6 x + c$, then $y' = \dfrac{1}{8}\cdot 8\tan^7 x \cdot \sec^2 x + \dfrac{1}{6}\cdot 6\tan^5 x \cdot \sec^2 x + 0$

$= \tan^7 x \sec^2 x + \tan^5 x \sec^2 x = \tan^5 x \sec^2 x\left(\tan^2 x + 1\right) = \tan^5 x \sec^2 x \sec^2 x = \tan^5 x \sec^4 x$

h. $\boxed{\int \tan^3 x \sec^4 x\,dx} = \boxed{\int \tan^3 x \sec^2 x \sec^2 x\,dx} = \boxed{\int \tan^3 x\left(\tan^2 x + 1\right)\sec^2 x\,dx} = \boxed{\int \tan^5 x \sec^2 x\,dx}$

$$\boxed{+\int \tan^3 x \sec^2 x\, dx} = \boxed{\frac{1}{6}\tan^6 x + \frac{1}{4}\tan^4 x + c}$$

Check: Let $y = \frac{1}{6}\tan^6 x + \frac{1}{4}\tan^4 x + c$, then $y' = \frac{1}{6}\cdot 6\tan^5 x \cdot \sec^2 x + \frac{1}{4}\cdot 4\tan^3 x \cdot \sec^2 x + 0$

$$= \tan^5 x \sec^2 x + \tan^3 x \sec^2 x = \tan^3 x \sec^2 x \left(\tan^2 x + 1\right) = \tan^3 x \sec^2 x \sec^2 x = \tan^3 x \sec^4 x$$

i. $\int \tan^3 x \sec^3 x\, dx = \int \tan^2 x \sec^2 x \cdot \tan x \sec x\, dx = \int \left(\sec^2 x - 1\right)\sec^2 x \cdot \tan x \sec x\, dx = \int \sec^4 x \cdot \tan x \sec x\, dx$

$\int \sec^2 x \cdot \tan x \sec x\, dx$. To solve the first and the second integral let $u = \sec x$, then $\dfrac{du}{dx} = \dfrac{d}{dx}\sec x$

; $\dfrac{du}{dx} = \sec x \tan x$; $du = \sec x \tan x\, dx$; $dx = \dfrac{du}{\sec x \tan x}$. Therefore,

$$\boxed{\int \tan^3 x \sec^3 x\, dx} = \boxed{\int \tan^2 x \sec^2 x \cdot \tan x \sec x\, dx} = \boxed{\int \sec^4 x \cdot \tan x \sec x\, dx - \int \sec^2 x \cdot \tan x \sec x\, dx}$$

$$= \boxed{\int u^4 \cdot \tan x \sec x \cdot \frac{du}{\tan x \sec x} - \int u^2 \cdot \tan x \sec x \cdot \frac{du}{\tan x \sec x}} = \boxed{\int u^4 du - \int u^2 du} = \boxed{\frac{1}{5}u^5 - \frac{1}{3}u^3 + c}$$

$$= \boxed{\frac{1}{5}\sec^5 x - \frac{1}{3}\sec^3 x + c}$$

Check: Let $y = \frac{1}{5}\sec^5 x - \frac{1}{3}\sec^3 x + c$, then $y' = \frac{5}{5}\sec^4 x \cdot \sec x \tan x - \frac{3}{3}\sec^2 x \cdot \sec x \tan x + 0$

$$= \sec^4 x \cdot \sec x \tan x - \sec^2 x \cdot \sec x \tan x = \left(\sec^4 x - \sec^2 x\right)\sec x \tan x = \left(\sec^2 x - 1\right)\sec^2 x \sec x \tan x$$

$$= \tan^2 x \sec^2 x \sec x \tan x = \tan^3 x \sec^3 x$$

j. Given $\int \cot^2 x \csc^2 x\, dx$ let $u = \cot x$, then $\dfrac{du}{dx} = \dfrac{d}{dx}\cot x$; $\dfrac{du}{dx} = -\csc^2 x$; $du = -\csc^2 x\, dx$; $dx = \dfrac{-du}{\csc^2 x}$.

Therefore, $\boxed{\int \cot^2 x \csc^2 x\, dx} = \boxed{\int u^2 \csc^2 x \cdot \frac{du}{-\csc^2 x}} = \boxed{-\int u^2 du} = \boxed{-\frac{1}{3}u^3 + c} = \boxed{-\frac{1}{3}\cot^3 x + c}$

Check: Let $y = -\frac{1}{3}\cot^3 x + c$, then $y' = -\frac{1}{3}\cdot 3\cot^2 x \cdot -\csc^2 x + 0 = \frac{3}{3}\cdot \cot^2 x \csc^2 x = \cot^2 x \csc^2 x$

k. $\boxed{\int \cot^3 x \csc^3 x\, dx} = \boxed{\int \cot x \cot^2 x \csc^3 x\, dx} = \boxed{\int \cot x \left(\csc^2 x - 1\right)\csc^2 x\, dx} = \boxed{\int \csc^4 x \cot x\, dx}$

$$\boxed{-\int \csc^2 x \cot x\, dx} = \boxed{-\frac{1}{5}\csc^5 x + \frac{1}{3}\csc^3 x + c}$$

Check: Let $y = -\frac{1}{5}\csc^5 x + \frac{1}{3}\csc^3 x + c$, then $y' = -\frac{1}{5}\cdot 5\csc^4 x \cdot -\csc x \cot x + \frac{1}{3}\cdot 3\csc^2 x \cdot -\csc x \cot x + 0$

$$= \csc^5 x \cdot \cot x - \csc^3 x \cdot \cot x = \csc^3 x \cot x \left(\csc^2 x - 1\right) = \csc^3 x \cot x \cot^2 x = \cot^3 x \csc^3 x$$

1. $\boxed{\int \cot^3 x \csc^4 x\, dx} = \boxed{\int \cot^3 x \csc^2 x \csc^2 x\, dx} = \boxed{\int \cot^3 x \left(\cot^2 x + 1\right)\csc^2 x\, dx} = \boxed{\int \cot^5 x \csc^2 x\, dx}$

$\boxed{+ \int \cot^3 x \csc^2 x\, dx} = \boxed{-\dfrac{1}{6}\cot^6 x - \dfrac{1}{4}\cot^4 x + c}$

Check: Let $y = -\dfrac{1}{6}\cot^6 x - \dfrac{1}{4}\cot^4 x + c$, then $y' = -\dfrac{1}{6}\cdot 6\cot^5 x \cdot -\csc^2 x - \dfrac{1}{4}\cdot 4\cot^3 x \cdot -\csc^2 x + 0$

$= \cot^5 x \csc^2 x + \cot^3 x \csc^2 x = \cot^3 x \csc^2 x\left(\cot^2 x + 1\right) = \cot^3 x \csc^2 x \csc^2 x = \cot^3 x \csc^4 x$

Section 4.3 Practice Problems – Integration of Trigonometric Functions

1. Evaluate the following integrals:

   a. $\int \sin \dfrac{3x}{5}\, dx =$

   b. $\int 3\cos 2x\, dx =$

   c. $\int \left(\sin 5x - \cos 7x\right) dx =$

   d. $\int 2\csc^2 3x\, dx =$

   e. $\int x\sec^2 x^2\, dx =$

   f. $\int 8\sec 5x\, dx =$

   g. $\int \sin^3 x \cos x\, dx =$

   h. $\int \cos^3 x \sin x\, dx =$

   i. $\int \dfrac{x}{2}\cos x^2 dx =$

2. Evaluate the following integrals:

   a. $\int e^{3x} \sec e^{3x}\, dx =$

   b. $\int \tan^9 x \sec^2 x\, dx =$

   c. $\int \cot^5 x \csc^2 x\, dx =$

   d. $\int \sec 2x \tan 2x\, dx =$

   e. $\int x \cot x^2 dx =$

   f. $\int \dfrac{1}{3}x^2 \csc x^3 dx =$

   g. $\int \left(\sin 5x \csc 5x\right) dx =$

   h. $\int \left(\cos 5t \sec 5t + 3t^2 + t\right) dt =$

   i. $\int e^{\cot x} \csc^2 x\, dx =$

3. Evaluate the following integrals:

   a. $\int \sin 3x \cos 5x\, dx =$

   b. $\int \cos 6x \cos 4x\, dx =$

   c. $\int \sin^5 3x\, dx =$

   d. $\int \tan^4 2x\, dx =$

   e. $\int \cos^2 ax\, dx =$

   f. $\int \tan^3 5x\, dx =$

   g. $\int \cot^4 3x\, dx =$

   h. $\int \cot^3 2x\, dx =$

   i. $\int \cot^6 3x\, dx =$

4.4   Integration of Expressions Resulting in Inverse Trigonometric Functions

In the following examples we will solve problems using the following formulas:

$$\int \frac{1}{\sqrt{a^2-x^2}}\,dx = \frac{1}{a} arc\sin\frac{x}{a}+c = \frac{1}{a}\sin^{-1}\frac{x}{a}+c$$

$$\int \frac{1}{a^2+x^2}\,dx = \frac{1}{a} arc\tan\frac{x}{a}+c = \frac{1}{a}\tan^{-1}\frac{x}{a}+c$$

$$\int \frac{1}{x\sqrt{x^2-a^2}}\,dx = \frac{1}{a} arc\sec\frac{x}{a}+c = \frac{1}{a}\sec^{-1}\frac{x}{a}+c$$

Let's integrate some algebraic expressions using the above integration formulas.

Example 4.4-1: Evaluate the following indefinite integrals:

a. $\displaystyle\int \frac{dx}{\sqrt{16-9x^2}} =$

b. $\displaystyle\int \frac{dx}{\sqrt{36-x^2}} =$

c. $\displaystyle\int \frac{x^3\,dx}{\sqrt{1-x^8}} =$

d. $\displaystyle\int \frac{x^2\,dx}{\sqrt{16-x^6}} =$

e. $\displaystyle\int \frac{x^2}{9+x^6}\,dx =$

f. $\displaystyle\int \frac{x}{x^4+5}\,dx =$

g. $\displaystyle\int \frac{dx}{9x^2+25} =$

h. $\displaystyle\int \frac{x^3\,dx}{49+4x^8} =$

i. $\displaystyle\int \frac{x^2\,dx}{5+9x^6} =$

Solutions:

a. **First** - Write the given integral in its standard form $\displaystyle\int \frac{dx}{\sqrt{a^2-x^2}} = arc\sin\frac{x}{a}+c$, i.e., $\displaystyle\int \frac{dx}{\sqrt{16-9x^2}}$

$$= \int \frac{dx}{\sqrt{4^2-(3x)^2}}$$

Second - Use substitution method by letting $u=3x$, then $\dfrac{du}{dx}=\dfrac{d}{dx}3x=3x$ which implies

$du=3x\,dx$; $dx=\dfrac{du}{3}$. Therefore, $\displaystyle\int \frac{dx}{\sqrt{4^2-(3x)^2}} = \int \frac{1}{\sqrt{4^2-u^2}}\cdot\frac{du}{3} = \frac{1}{3}\int \frac{du}{\sqrt{4^2-u^2}} = \frac{1}{3}arc\sin\frac{u}{4}$

Third - Write the answer in terms of the original variable, i.e., x . $\dfrac{1}{3}arc\sin\dfrac{u}{4}+c = \dfrac{1}{3}\boldsymbol{arc\sin}\dfrac{3x}{4}+c$

Fourth - Check the answer by differentiating the solution, i.e., let $y=\dfrac{1}{3}arc\sin\dfrac{3x}{4}+c$ then

$$y' = \frac{1}{3}\frac{1}{\sqrt{1-\left(\frac{3x}{4}\right)^2}}\cdot\frac{d}{dx}\frac{3x}{4}+0 = \frac{1}{3}\frac{1}{\sqrt{1-\frac{9x^2}{16}}}\cdot\frac{3}{4} = \frac{3}{12}\frac{1}{\sqrt{\frac{16-9x^2}{16}}} = \frac{1}{4}\frac{4}{\sqrt{16-9x^2}} = \frac{1}{\sqrt{16-9x^2}}$$

b. **First** - Write the given integral in its standard form $\displaystyle\int \frac{dx}{\sqrt{a^2-x^2}} = arc\sin\frac{x}{a}+c$, i.e., $\displaystyle\int \frac{dx}{\sqrt{36-x^2}}$

$$= \int \frac{dx}{\sqrt{6^2-x^2}}$$

Second - Use substitution method by letting $u = x$, then $\dfrac{du}{dx} = \dfrac{d}{dx}x = 1$ which implies $du = dx$.

Therefore, $\displaystyle\int \dfrac{dx}{\sqrt{36 - x^2}} = \int \dfrac{du}{\sqrt{6^2 - u^2}} = arc\sin\dfrac{u}{6} + c$

Third - Write the answer in terms of the original variable, i.e., x. $arc\sin\dfrac{u}{6} + c = \boldsymbol{arc\sin\dfrac{x}{6} + c}$

Fourth - Check the answer by differentiating the solution, i.e., let $y = arc\sin\dfrac{x}{6} + c$ then

$y' = \dfrac{1}{\sqrt{1 - \left(\frac{x}{6}\right)^2}} \cdot \dfrac{d}{dx}\dfrac{x}{6} + 0 = \dfrac{1}{\sqrt{1 - \frac{x^2}{36}}} \cdot \dfrac{1}{6} = \dfrac{1}{6}\dfrac{1}{\sqrt{\frac{36-x^2}{36}}} = \dfrac{1}{6}\dfrac{6}{\sqrt{36 - x^2}} = \dfrac{1}{\sqrt{36 - x^2}}$

c. **First** - Write the given integral in its standard form $\displaystyle\int \dfrac{dx}{\sqrt{a^2 - x^2}} = arc\sin\dfrac{x}{a} + c$, i.e., $\displaystyle\int \dfrac{x^3 dx}{\sqrt{1 - x^8}}$

$= \displaystyle\int \dfrac{x^3 dx}{\sqrt{1^2 - \left(x^4\right)^2}}$

Second - Use substitution method by letting $u = x^4$, then $\dfrac{du}{dx} = \dfrac{d}{dx}x^4 = 4x^3$ which implies

$du = 4x^3 dx$; $dx = \dfrac{du}{4x^3}$. Therefore, $\displaystyle\int \dfrac{x^3 dx}{\sqrt{1^2 - \left(x^4\right)^2}} = \int \dfrac{x^3}{\sqrt{1^2 - u^2}} \cdot \dfrac{du}{4x^3} = \dfrac{1}{4}\int \dfrac{du}{\sqrt{1^2 - u^2}} = \dfrac{1}{4}arc\sin\dfrac{u}{1} + c$

Third - Write the answer in terms of the original variable, i.e., x. $\dfrac{1}{4}arc\sin\dfrac{u}{1} + c = \boldsymbol{\dfrac{1}{4}arc\sin x^4 + c}$

Fourth - Check the answer by differentiating the solution, i.e., let $y = \dfrac{1}{4}arc\sin x^4 + c$ then

$y' = \dfrac{1}{4}\dfrac{1}{\sqrt{1 - \left(x^4\right)^2}} \cdot \dfrac{d}{dx}x^4 + 0 = \dfrac{1}{4}\dfrac{1}{\sqrt{1 - x^8}} \cdot 4x^3 = \dfrac{4}{4}\dfrac{x^3}{\sqrt{1 - x^8}} = \dfrac{x^3}{\sqrt{1 - x^8}}$

d. **First** - Write the given integral in its standard form $\displaystyle\int \dfrac{dx}{\sqrt{a^2 - x^2}} = arc\sin\dfrac{x}{a} + c$, i.e., $\displaystyle\int \dfrac{x^2 dx}{\sqrt{16 - x^6}}$

$= \displaystyle\int \dfrac{x^2 dx}{\sqrt{4^2 - \left(x^3\right)^2}}$

Second - Use substitution method by letting $u = x^3$, then $\dfrac{du}{dx} = \dfrac{d}{dx}x^3 = 3x^2$ which implies

$du = 3x^2 dx$; $dx = \dfrac{du}{3x^2}$. Therefore, $\displaystyle\int \dfrac{x^2 dx}{\sqrt{4^2 - \left(x^3\right)^2}} = \int \dfrac{x^2}{\sqrt{4^2 - u^2}} \cdot \dfrac{du}{3x^2} = \dfrac{1}{3}\int \dfrac{du}{\sqrt{4^2 - u^2}} = \dfrac{1}{3}arc\sin\dfrac{u}{4} + c$

Third - Write the answer in terms of the original variable, i.e., x. $\dfrac{1}{3}arc\sin\dfrac{u}{4} + c = \boldsymbol{\dfrac{1}{3}arc\sin\dfrac{x^3}{4} + c}$

Fourth - Check the answer by differentiating the solution, i.e., let $y = \dfrac{1}{3}arc\sin\dfrac{x^3}{4} + c$ then

$y' = \dfrac{1}{3}\dfrac{1}{\sqrt{1 - \left(\frac{x^3}{4}\right)^2}} \cdot \dfrac{d}{dx}\dfrac{x^3}{4} + 0 = \dfrac{1}{3}\dfrac{1}{\sqrt{1 - \frac{x^6}{16}}} \cdot \dfrac{3x^2}{4} = \dfrac{1}{4}\dfrac{x^2}{\sqrt{\frac{16-x^6}{16}}} = \dfrac{1}{4}\dfrac{4x^2}{\sqrt{16 - x^6}} = \dfrac{x^2}{\sqrt{16 - x^6}}$

e. **First** - Write the given integral in its standard form $\int \frac{dx}{a^2+x^2} = \frac{1}{a} arc\tan\frac{x}{a}+c$, i.e., $\int \frac{x^2}{9+x^6}dx$

$= \int \frac{dx}{3^2+\left(x^3\right)^2}$

Second - Use substitution method by letting $u=x^3$, then $\frac{du}{dx}=\frac{d}{dx}x^3=3x^2$ which implies

$du=3x^2 dx$; $dx=\frac{du}{3x^2}$. Therefore, $\int \frac{dx}{3^2+\left(x^3\right)^2} = \int \frac{1}{3^2+u^2}\cdot\frac{du}{3x^2} = \frac{1}{3}\int \frac{du}{3^2+u^2} = \frac{1}{9}arc\tan\frac{u}{3}+c$

Third - Write the answer in terms of the original variable, i.e., x. $\frac{1}{9}arc\tan\frac{u}{3}+c = \frac{1}{9}arc\tan\frac{x^3}{3}+c$

Fourth - Check the answer by differentiating the solution, i.e., let $y=\frac{1}{9}arc\tan\frac{x^3}{3}+c$ then

$y' = \frac{1}{9}\frac{1}{1+\left(\frac{x^3}{3}\right)^2}\cdot\frac{d}{dx}\frac{x^3}{3}+0 = \frac{1}{9}\frac{1}{1+\frac{x^6}{9}}\cdot\frac{3x^2}{3} = \frac{1}{9}\frac{x^2}{\frac{9+x^6}{9}} = \frac{1}{9}\frac{9x^2}{9+x^6} = \frac{x^2}{9+x^6}$

f. **First** - Write the given integral in its standard form $\int \frac{dx}{a^2+x^2} = \frac{1}{a} arc\tan\frac{x}{a}+c$, i.e., $\int \frac{x}{x^4+5}dx$

$= \int \frac{x}{5+x^4}dx = \int \frac{x\,dx}{\left(\sqrt{5}\right)^2+\left(x^2\right)^2}$

Second - Use substitution method by letting $u=x^2$, then $\frac{du}{dx}=\frac{d}{dx}x^2=2x$ which implies $du=2x\,dx$

; $dx=\frac{du}{2x}$. Therefore, $\int \frac{x\,dx}{\left(\sqrt{5}\right)^2+\left(x^2\right)^2} = \int \frac{x}{\left(\sqrt{5}\right)^2+u^2}\cdot\frac{du}{2x} = \frac{1}{2}\int \frac{du}{\left(\sqrt{5}\right)^2+u^2} = \frac{1}{2\sqrt{5}}arc\tan\frac{u}{\sqrt{5}}+c$

Third - Write the answer in terms of the original variable, i.e., x. $\frac{1}{2\sqrt{5}}arc\tan\frac{u}{\sqrt{5}}+c = \frac{1}{2\sqrt{5}}arc\tan\frac{x^2}{\sqrt{5}}+c$

Fourth - Check the answer by differentiating the solution, i.e., let $y=\frac{1}{2\sqrt{5}}arc\tan\frac{x^2}{\sqrt{5}}+c$ then

$y' = \frac{1}{2\sqrt{5}}\frac{1}{1+\left(\frac{x^2}{\sqrt{5}}\right)^2}\cdot\frac{d}{dx}\frac{x^2}{\sqrt{5}}+0 = \frac{1}{2\sqrt{5}}\frac{1}{1+\frac{x^4}{5}}\cdot\frac{2x}{\sqrt{5}} = \frac{1}{10}\frac{2x}{\frac{5+x^4}{5}} = \frac{1}{5}\frac{5x}{5+x^4} = \frac{x}{5+x^4}$

g. **First** - Write the given integral in its standard form $\int \frac{dx}{a^2+x^2} = \frac{1}{a} arc\tan\frac{x}{a}+c$, i.e., $\int \frac{dx}{9x^2+25}$

$= \int \frac{dx}{25+9x^2} = \int \frac{dx}{9\left(\frac{25}{9}+x^2\right)} = \frac{1}{9}\int \frac{dx}{\frac{25}{9}+x^2} = \frac{1}{9}\int \frac{dx}{\left(\frac{5}{3}\right)^2+x^2}$

Second - Use substitution method by letting $u=x$, then $\frac{du}{dx}=\frac{d}{dx}x=1$ which implies $du=dx$.

Therefore, $\frac{1}{9}\int \frac{dx}{\left(\frac{5}{3}\right)^2+u^2} = \frac{1}{9}\cdot\frac{1}{\frac{5}{3}}arc\tan\frac{u}{\frac{5}{3}}+c = \frac{1}{9}\cdot\frac{3}{5}arc\tan\frac{3u}{5}+c = \frac{1}{15}arc\tan\frac{3u}{5}+c$

Third - Write the answer in terms of the original variable, i.e., x. $\frac{1}{15}arc\tan\frac{3u}{5}+c = \frac{1}{15}arc\tan\frac{3x}{5}+c$

Fourth - Check the answer by differentiating the solution, i.e., let $y = \dfrac{1}{15} arc \tan \dfrac{3x}{5} + c$ then

$$y' = \dfrac{1}{15}\dfrac{1}{1+\left(\frac{3x}{5}\right)^2}\cdot\dfrac{d}{dx}\dfrac{3x}{5}+0 = \dfrac{1}{15}\dfrac{1}{1+\frac{9x^2}{25}}\cdot\dfrac{3}{5} = \dfrac{3}{75}\dfrac{1}{\frac{25+9x^2}{25}} = \dfrac{3}{75}\dfrac{25}{25+9x^2} = \dfrac{1}{75}\dfrac{75}{25+9x^2} = \dfrac{1}{25+9x^2}$$

Note that another way of solving this class of problems is by rewriting the integral in the following way:

$$\int \dfrac{dx}{9x^2+25} = \int \dfrac{dx}{25+9x^2} = \int \dfrac{dx}{5^2+(3x)^2}.$$ Now, let $u = 3x$, then $\dfrac{du}{dx} = \dfrac{d}{dx}3x = 3$ which implies

$du = 3dx$; $dx = \dfrac{du}{3}$. Therefore, $\int \dfrac{dx}{5^2+(3x)^2} = \int \dfrac{1}{5^2+u^2}\cdot\dfrac{du}{3} = \dfrac{1}{3}\int \dfrac{du}{5^2+u^2} = \dfrac{1}{3}\cdot\dfrac{1}{5} arc \tan \dfrac{u}{5} + c$

$= \dfrac{1}{15} arc \tan \dfrac{u}{5} + c = \dfrac{\mathbf{1}}{\mathbf{15}} \mathbf{arc} \tan \dfrac{\mathbf{3x}}{\mathbf{5}} + c$

h. **First** - Write the given integral in its standard form $\int \dfrac{dx}{a^2+x^2} = \dfrac{1}{a} arc \tan \dfrac{x}{a} + c$, i.e., $\int \dfrac{x^3 dx}{49+4x^8}$

$= \int \dfrac{x^3 dx}{4\left(\frac{49}{4}+x^8\right)} = \dfrac{1}{4}\int \dfrac{x^3 dx}{\frac{49}{4}+x^8} = \dfrac{1}{4}\int \dfrac{x^3 dx}{\left(\frac{7}{2}\right)^2+\left(x^4\right)^2}$

Second - Use substitution method by letting $u = x^4$, then $\dfrac{du}{dx} = \dfrac{d}{dx}x^4 = 4x^3$ which implies

$du = 4x^3 dx$; $dx = \dfrac{du}{4x^3}$. Therefore, $\dfrac{1}{4}\int \dfrac{x^3 dx}{\left(\frac{7}{2}\right)^2+\left(x^4\right)^2} = \dfrac{1}{4}\int \dfrac{x^3}{\left(\frac{7}{2}\right)^2+u^2}\cdot\dfrac{du}{4x^3} = \dfrac{1}{16}\int \dfrac{du}{\left(\frac{7}{2}\right)^2+u^2}$

$= \dfrac{1}{16}\cdot\dfrac{1}{\frac{7}{2}} arc \tan \dfrac{u}{\frac{7}{2}} + c = \dfrac{1}{16}\cdot\dfrac{2}{7} arc \tan \dfrac{2u}{7} + c = \dfrac{1}{56} arc \tan \dfrac{2u}{7} + c$

Third - Write the answer in terms of the original variable, i.e., x. $\dfrac{1}{56} arc \tan \dfrac{2u}{7} + c = \dfrac{\mathbf{1}}{\mathbf{56}} \mathbf{arc} \tan \dfrac{\mathbf{2x^4}}{\mathbf{7}} + c$

Fourth - Check the answer by differentiating the solution, i.e., let $y = \dfrac{1}{56} arc \tan \dfrac{2x^4}{7} + c$ then

$$y' = \dfrac{1}{56}\dfrac{1}{1+\left(\frac{2x^4}{7}\right)^2}\cdot\dfrac{d}{dx}\dfrac{2x^4}{7}+0 = \dfrac{1}{56}\dfrac{1}{1+\frac{4x^8}{49}}\cdot\dfrac{8x^3}{7} = \dfrac{1}{49}\dfrac{x^3}{\frac{49+4x^8}{49}} = \dfrac{1}{49}\dfrac{49x^3}{49+4x^8} = \dfrac{x^3}{49+4x^8}$$

or, the alternative approach would be to rearrange the integral in the following way:

$$\int \dfrac{x^3 dx}{49+4x^8} = \int \dfrac{x^3 dx}{7^2+\left(2x^4\right)^2}.$$ Now, let $u = 2x^4$, then $\dfrac{du}{dx} = \dfrac{d}{dx}2x^4 = 8x^3$ which implies $du = 8x^3 dx$

; $dx = \dfrac{du}{8x^3}$. Therefore, $\int \dfrac{x^3 dx}{7^2+\left(2x^4\right)^2} = \int \dfrac{x^3}{7^2+u^2}\cdot\dfrac{du}{8x^3} = \dfrac{1}{8}\int \dfrac{du}{7^2+u^2} = \dfrac{1}{8}\cdot\dfrac{1}{7} arc \tan \dfrac{u}{7} + c$

$= \dfrac{1}{56} arc \tan \dfrac{u}{7} + c = \dfrac{\mathbf{1}}{\mathbf{56}} \mathbf{arc} \tan \dfrac{\mathbf{2x^4}}{\mathbf{7}} + c$

i. **First** - Write the given integral in its standard form $\int \dfrac{dx}{a^2+x^2} = \dfrac{1}{a} arc \tan \dfrac{x}{a} + c$, i.e., $\int \dfrac{x^2 dx}{5+9x^6}$

$$= \int \frac{x^2 \, dx}{9\left(\frac{5}{9} + x^6\right)} = \frac{1}{9} \int \frac{x^2 \, dx}{\frac{5}{9} + x^6} = \frac{1}{9} \int \frac{x^2 \, dx}{\left(\frac{\sqrt{5}}{3}\right)^2 + \left(x^3\right)^2}$$

Second - Use substitution method by letting $u = x^3$, then $\dfrac{du}{dx} = \dfrac{d}{dx} x^3 = 3x^2$ which implies

$du = 3x^2 \, dx$; $dx = \dfrac{du}{3x^2}$. Therefore, $\dfrac{1}{9} \int \dfrac{x^2 \, dx}{\left(\frac{\sqrt{5}}{3}\right)^2 + \left(x^3\right)^2} = \dfrac{1}{9} \int \dfrac{x^2}{\left(\frac{\sqrt{5}}{3}\right)^2 + u^2} \cdot \dfrac{du}{3x^2} = \dfrac{1}{27} \int \dfrac{du}{\left(\frac{\sqrt{5}}{3}\right)^2 + u^2}$

$$= \frac{1}{27} \cdot \frac{1}{\frac{\sqrt{5}}{3}} \, arc \tan \frac{u}{\frac{\sqrt{5}}{3}} + c = \frac{1}{27} \cdot \frac{3}{\sqrt{5}} \, arc \tan \frac{3u}{\sqrt{5}} + c = \frac{1}{9\sqrt{5}} \, arc \tan \frac{3u}{\sqrt{5}} + c$$

Third - Write the answer in terms of the x variable, i.e., $\dfrac{1}{9\sqrt{5}} \, arc \tan \dfrac{3u}{\sqrt{5}} + c = \dfrac{1}{9\sqrt{5}} \, \boldsymbol{arc \tan} \dfrac{3x^3}{\sqrt{5}} + c$

Fourth - Check the answer by differentiating the solution, i.e., let $y = \dfrac{1}{9\sqrt{5}} \, arc \tan \dfrac{3x^3}{\sqrt{5}} + c$ then

$$y' = \frac{1}{9\sqrt{5}} \frac{1}{1 + \left(\frac{3x^3}{\sqrt{5}}\right)^2} \cdot \frac{d}{dx} \frac{3x^3}{\sqrt{5}} + 0 = \frac{1}{9\sqrt{5}} \frac{1}{1 + \frac{9x^6}{5}} \frac{9x^2}{\sqrt{5}} = \frac{1}{5} \frac{x^2}{\frac{5 + 9x^6}{5}} = \frac{1}{5} \frac{5x^2}{5 + 9x^6} = \frac{x^2}{5 + 9x^6}$$

or, the alternative approach would be to rearrange the integral in the following way:

$\int \dfrac{x^2 \, dx}{5 + 9x^6} = \int \dfrac{x^2 \, dx}{\left(\sqrt{5}\right)^2 + \left(3x^3\right)^2}$. Now, let $u = 3x^3$, then $\dfrac{du}{dx} = \dfrac{d}{dx} 3x^3 = 9x^2$ which implies $du = 9x^2 \, dx$

; $dx = \dfrac{du}{9x^2}$. Therefore, $\int \dfrac{x^2 \, dx}{\left(\sqrt{5}\right)^2 + \left(3x^3\right)^2} = \int \dfrac{x^2}{\left(\sqrt{5}\right)^2 + u^2} \cdot \dfrac{du}{9x^2} = \dfrac{1}{9} \int \dfrac{du}{\left(\sqrt{5}\right)^2 + u^2} = \dfrac{1}{9} \cdot \dfrac{1}{\sqrt{5}} \, arc \tan \dfrac{u}{\sqrt{5}} + c$

$$= \frac{1}{9\sqrt{5}} \, arc \tan \frac{u}{\sqrt{5}} + c = \frac{1}{9\sqrt{5}} \, \boldsymbol{arc \tan} \frac{3x^3}{\sqrt{5}} + c$$

Example 4.4-2: Evaluate the following indefinite integrals:

a. $\displaystyle\int \frac{dx}{x\sqrt{x^4 - 4}} =$

b. $\displaystyle\int \frac{dx}{x\sqrt{25\,x^2 - 9}} =$

c. $\displaystyle\int \frac{dx}{x\sqrt{x^2 - 25}} =$

d. $\displaystyle\int \frac{dx}{x\sqrt{6x^2 - 9}} =$

e. $\displaystyle\int \frac{dx}{x\sqrt{16x^4 - 25}} =$

f. $\displaystyle\int \frac{e^x}{e^{2x} + 4} \, dx =$

g. $\displaystyle\int \frac{e^x}{9\,e^{2x} + 16} \, dx =$

h. $\displaystyle\int \frac{dx}{\sqrt{25 - (x+4)^2}} =$

i. $\displaystyle\int \frac{dx}{\sqrt{3 - (x-2)^2}} =$

Solutions:

a. **First** - Write the given integral in its standard form $\displaystyle\int \dfrac{dx}{x\sqrt{x^2 - a^2}} = \dfrac{1}{a} \, arc \sec \dfrac{x}{a} + c$, i.e.,

$$\int \frac{dx}{x\sqrt{x^4 - 4}} = \int \frac{dx}{x\sqrt{\left(x^2\right)^2 - 2^2}}$$

Second - Use substitution method by letting $u = x^2$, then $\dfrac{du}{dx} = \dfrac{d}{dx}x^2 = 2x$ which implies

$du = 2x\,dx$; $dx = \dfrac{du}{2x}$. Therefore, $\displaystyle\int \dfrac{dx}{x\sqrt{\left(x^2\right)^2 - 2^2}} = \int \dfrac{1}{x\sqrt{u^2 - 2^2}} \cdot \dfrac{du}{2x} = \dfrac{1}{2}\int \dfrac{du}{x^2\sqrt{u^2 - 2^2}}$

$= \dfrac{1}{2}\int \dfrac{du}{u\sqrt{u^2 - 2^2}} = \dfrac{1}{2}\cdot\dfrac{1}{2}\,arc\sec\dfrac{u}{2} + c = \dfrac{1}{4}\,arc\sec\dfrac{u}{2} + c$

Third - Write the answer in terms of the original variable, i.e., x. $\dfrac{1}{4}\,arc\sec\dfrac{u}{2} + c = \dfrac{1}{4}\,arc\sec\dfrac{x^2}{2} + c$

Fourth - Check the answer by differentiating the solution, i.e., let $y = \dfrac{1}{4}\,arc\sec\dfrac{x^2}{2} + c$ then

$y' = \dfrac{1}{4}\dfrac{1}{\dfrac{x^2}{2}\sqrt{\left(\dfrac{x^2}{2}\right)^2 - 1}}\cdot\dfrac{d}{dx}\dfrac{x^2}{2} + 0 = \dfrac{1}{4}\dfrac{1}{\dfrac{x^2}{2}\sqrt{\dfrac{x^4}{4} - 1}}\cdot\dfrac{2x}{2} = \dfrac{1}{4}\dfrac{x}{\dfrac{x^2}{2}\sqrt{\dfrac{x^4 - 4}{4}}} = \dfrac{1}{4}\dfrac{4x}{x^2\sqrt{x^4 - 4}} = \dfrac{1}{x\sqrt{x^4 - 4}}$

b. **First** - Write the given integral in its standard form $\displaystyle\int \dfrac{dx}{x\sqrt{x^2 - a^2}} = \dfrac{1}{a}\,arc\sec\dfrac{x}{a} + c$, i.e.,

$\displaystyle\int \dfrac{dx}{x\sqrt{25x^2 - 9}} = \int \dfrac{dx}{\sqrt{25}x\sqrt{x^2 - \dfrac{9}{25}}} = \int \dfrac{dx}{5x\sqrt{x^2 - \dfrac{9}{25}}} = \dfrac{1}{5}\int \dfrac{dx}{x\sqrt{x^2 - \left(\dfrac{3}{5}\right)^2}}$

Second - Use substitution method by letting $u = x$, then $\dfrac{du}{dx} = \dfrac{d}{dx}x = 1$ which implies $du = dx$.

Therefore, $\dfrac{1}{5}\displaystyle\int \dfrac{dx}{x\sqrt{x^2 - \left(\dfrac{3}{5}\right)^2}} = \dfrac{1}{5}\int \dfrac{du}{u\sqrt{u^2 - \left(\dfrac{3}{5}\right)^2}} = \dfrac{1}{5}\cdot\dfrac{1}{\dfrac{3}{5}}\,arc\sec\dfrac{u}{\dfrac{3}{5}} + c = \dfrac{1}{3}\,arc\sec\dfrac{5u}{3} + c$

Third - Write the answer in terms of the original variable, i.e., x. $\dfrac{1}{3}\,arc\sec\dfrac{5u}{3} + c = \dfrac{1}{3}\,\textbf{\textit{arc\,sec}}\dfrac{5x}{3} + c$

Fourth - Check the answer by differentiating the solution, i.e., let $y = \dfrac{1}{3}\,arc\sec\dfrac{5x}{3} + c$ then

$y' = \dfrac{1}{3}\dfrac{1}{\dfrac{5x}{3}\sqrt{\left(\dfrac{5x}{3}\right)^2 - 1}}\cdot\dfrac{d}{dx}\dfrac{5x}{3} + 0 = \dfrac{1}{3}\dfrac{1}{\dfrac{5x}{3}\sqrt{\dfrac{25x^2}{9} - 1}}\cdot\dfrac{5}{3} = \dfrac{5}{9}\dfrac{1}{\dfrac{5x}{3}\sqrt{\dfrac{25x^2 - 9}{9}}} = \dfrac{5}{9}\dfrac{9}{5x\sqrt{25x^2 - 9}} = \dfrac{1}{x\sqrt{25x^2 - 9}}$

or, the alternative approach would be to rearrange the integral in the following way:

$\displaystyle\int \dfrac{dx}{x\sqrt{25x^2 - 9}} = \int \dfrac{dx}{x\sqrt{(5x)^2 - 3^2}}$. Now, let $u = 5x$, then $\dfrac{du}{dx} = \dfrac{d}{dx}5x = 5$ which implies $du = 5dx$

; $dx = \dfrac{du}{5}$. Therefore, $\displaystyle\int \dfrac{dx}{x\sqrt{(5x)^2 - 3^2}} = \int \dfrac{1}{\dfrac{u}{5}\sqrt{u^2 - 3^2}}\cdot\dfrac{du}{5} = \dfrac{1}{5}\int \dfrac{5\,du}{u\sqrt{u^2 - 3^2}} = \int \dfrac{du}{u\sqrt{u^2 - 3^2}}$

$= \dfrac{1}{3}\,arc\sec\dfrac{u}{3} + c = \dfrac{1}{3}\,\textbf{\textit{arc\,sec}}\dfrac{5x}{3} + c$

c. **First** - Write the given integral in its standard form $\displaystyle\int \dfrac{dx}{x\sqrt{x^2 - a^2}} = \dfrac{1}{a}\,arc\sec\dfrac{x}{a} + c$, i.e.,

$\displaystyle\int \dfrac{dx}{x\sqrt{x^2 - 25}} = \int \dfrac{dx}{x\sqrt{x^2 - 5^2}}$

Second - Use substitution method by letting $u = x$, then $\dfrac{du}{dx} = \dfrac{d}{dx} x = 1$ which implies $du = dx$.

Therefore, $\displaystyle\int \dfrac{dx}{x\sqrt{x^2 - 5^2}} = \int \dfrac{du}{u\sqrt{u^2 - 5^2}} = \dfrac{1}{5}\,arc\sec\dfrac{u}{5} + c$

Third - Write the answer in terms of the original variable, i.e., x. $\dfrac{1}{5}\,arc\sec\dfrac{u}{5} + c = \dfrac{1}{5}\,\textbf{arc sec}\,\dfrac{x}{5} + c$

Fourth - Check the answer by differentiating the solution, i.e., let $y = \dfrac{1}{5}\,arc\sec\dfrac{x}{5} + c$ then

$$y' = \dfrac{1}{5}\dfrac{1}{\frac{x}{5}\sqrt{\left(\frac{x}{5}\right)^2 - 1}}\cdot\dfrac{d}{dx}\dfrac{x}{5} + 0 = \dfrac{1}{5}\dfrac{1}{\frac{x}{5}\sqrt{\frac{x^2}{25} - 1}}\cdot\dfrac{1}{5} = \dfrac{1}{5}\dfrac{1}{\frac{x}{5}\sqrt{\frac{x^2 - 25}{25}}} = \dfrac{1}{5}\dfrac{5}{x\sqrt{x^2 - 25}} = \dfrac{1}{x\sqrt{x^2 - 25}}$$

d. **First** - Write the given integral in its standard form $\displaystyle\int \dfrac{dx}{x\sqrt{x^2 - a^2}} = \dfrac{1}{a}\,arc\sec\dfrac{x}{a} + c$, i.e.,

$$\int \dfrac{dx}{x\sqrt{6x^2 - 9}} = \int \dfrac{dx}{\sqrt{6}x\sqrt{x^2 - \frac{9}{6}}} = \int \dfrac{dx}{\sqrt{6}x\sqrt{x^2 - \left(\frac{3}{\sqrt{6}}\right)^2}} = \dfrac{1}{\sqrt{6}}\int \dfrac{dx}{x\sqrt{x^2 - \left(\frac{3}{\sqrt{6}}\right)^2}}$$

Second - Use substitution method by letting $u = x$, then $\dfrac{du}{dx} = \dfrac{d}{dx} x = 1$ which implies $du = dx$.

Therefore, $\dfrac{1}{\sqrt{6}}\displaystyle\int \dfrac{dx}{x\sqrt{x^2 - \left(\frac{3}{\sqrt{6}}\right)^2}} = \dfrac{1}{\sqrt{6}}\int \dfrac{du}{u\sqrt{u^2 - \left(\frac{3}{\sqrt{6}}\right)^2}} = \dfrac{1}{\sqrt{6}}\cdot\dfrac{1}{\frac{3}{\sqrt{6}}}\,arc\sec\dfrac{u}{\frac{3}{\sqrt{6}}} + c = \dfrac{1}{3}\,arc\sec\dfrac{\sqrt{6}u}{3} + c$

Third - Write the answer in terms of the original variable, i.e., x. $\dfrac{1}{3}\,arc\sec\dfrac{\sqrt{6}u}{3} + c = \dfrac{1}{3}\,\textbf{arc sec}\,\dfrac{\sqrt{6}x}{3} + c$

Fourth - Check the answer by differentiating the solution, i.e., let $y = \dfrac{1}{3}\,arc\sec\dfrac{\sqrt{6}x}{3} + c$ then y'

$$= \dfrac{1}{3}\dfrac{1}{\frac{\sqrt{6}x}{3}\sqrt{\left(\frac{\sqrt{6}x}{3}\right)^2 - 1}}\cdot\dfrac{d}{dx}\dfrac{\sqrt{6}x}{3} + 0 = \dfrac{1}{3}\dfrac{1}{\frac{\sqrt{6}x}{3}\sqrt{\frac{6x^2}{9} - 1}}\cdot\dfrac{\sqrt{6}}{3} = \dfrac{\sqrt{6}}{9}\dfrac{1}{\frac{\sqrt{6}x}{3}\sqrt{\frac{6x^2 - 9}{9}}} = \dfrac{\sqrt{6}}{9}\dfrac{9}{\sqrt{6}x\sqrt{6x^2 - 9}} = \dfrac{1}{x\sqrt{6x^2 - 9}}$$

or, the alternative approach would be to rearrange the integral in the following way:

$$\int \dfrac{dx}{x\sqrt{6x^2 - 9}} = \int \dfrac{dx}{x\sqrt{\left(\sqrt{6}x\right)^2 - 3^2}}.\ \text{Now, let } u = \sqrt{6}x, \text{ then } \dfrac{du}{dx} = \dfrac{d}{dx}\sqrt{6}x = \sqrt{6} \text{ which implies } du = \sqrt{6}\,dx$$

$; dx = \dfrac{du}{\sqrt{6}}$. Therefore, $\displaystyle\int \dfrac{dx}{x\sqrt{\left(\sqrt{6}x\right)^2 - 3^2}} = \int \dfrac{1}{\frac{u}{\sqrt{6}}\sqrt{u^2 - 3^2}}\cdot\dfrac{du}{\sqrt{6}} = \dfrac{1}{\sqrt{6}}\int \dfrac{\sqrt{6}\,du}{u\sqrt{u^2 - 3^2}} = \int \dfrac{du}{u\sqrt{u^2 - 3^2}}$

$= \dfrac{1}{3}\,arc\sec\dfrac{u}{3} + c = \dfrac{1}{3}\,\textbf{arc sec}\,\dfrac{\sqrt{6}x}{3} + c$

e. **First** - Write the given integral in its standard form $\displaystyle\int \dfrac{dx}{x\sqrt{x^2 - a^2}} = \dfrac{1}{a}\,arc\sec\dfrac{x}{a} + c$, i.e.,

$$\int \dfrac{dx}{x\sqrt{16x^4 - 25}} = \int \dfrac{dx}{\sqrt{16}x\sqrt{x^4 - \frac{25}{16}}} = \int \dfrac{dx}{4x\sqrt{\left(x^2\right)^2 - \left(\frac{5}{4}\right)^2}} = \dfrac{1}{4}\int \dfrac{dx}{x\sqrt{\left(x^2\right)^2 - \left(\frac{5}{4}\right)^2}}$$

Second - Use substitution method by letting $u = x^2$, then $\dfrac{du}{dx} = \dfrac{d}{dx}x^2 = 2x$ which implies

$du = 2x\,dx$; $dx = \dfrac{du}{2x}$. Thus, $\dfrac{1}{4}\displaystyle\int \dfrac{dx}{x\sqrt{\left(x^2\right)^2 - \left(\frac{5}{4}\right)^2}} = \dfrac{1}{4}\displaystyle\int \dfrac{1}{x\sqrt{u^2 - \left(\frac{5}{4}\right)^2}}\cdot \dfrac{du}{2x} = \dfrac{1}{8}\displaystyle\int \dfrac{du}{x^2\sqrt{u^2 - \left(\frac{5}{4}\right)^2}}$

$= \dfrac{1}{8}\displaystyle\int \dfrac{du}{u\sqrt{u^2 - \left(\frac{5}{4}\right)^2}} = \dfrac{1}{8}\cdot\dfrac{1}{\frac{5}{4}}\,arc\,sec\,\dfrac{u}{\frac{5}{4}} + c = \dfrac{1}{10}\,arc\,sec\,\dfrac{4u}{5} + c$

Third - Write the answer in terms of the original variable, i.e., x. $\dfrac{1}{10}\,arc\,sec\,\dfrac{4u}{5} + c = \dfrac{1}{10}\,\boldsymbol{arc\,sec}\,\dfrac{4x^2}{5} + c$

Fourth - Check the answer by differentiating the solution, i.e., let $y = \dfrac{1}{10}\,arc\,sec\,\dfrac{4x^2}{5} + c$ then

$y' = \dfrac{1}{10}\dfrac{1}{\frac{4x^2}{5}\sqrt{\left(\frac{4x^2}{5}\right)^2 - 1}}\cdot\dfrac{d}{dx}\dfrac{4x^2}{5} + 0 = \dfrac{1}{10}\dfrac{1}{\frac{4x^2}{5}\sqrt{\frac{16x^4}{25} - 1}}\cdot\dfrac{8x}{5} = \dfrac{8}{50}\dfrac{x}{\frac{4x^2}{5}\sqrt{\frac{16x^4 - 25}{25}}} = \dfrac{1}{x\sqrt{16x^4 - 25}}$

or, the alternative approach would be to rearrange the integral in the following way:

$\displaystyle\int \dfrac{dx}{x\sqrt{16x^4 - 25}} = \displaystyle\int \dfrac{dx}{x\sqrt{\left(4x^2\right)^2 - 5^2}}$. Now, let $u = 4x^2$, then $\dfrac{du}{dx} = \dfrac{d}{dx}4x^2 = 8x$ which implies

$du = 8x\,dx$; $dx = \dfrac{du}{8x}$. Therefore, $\displaystyle\int \dfrac{dx}{x\sqrt{\left(4x^2\right)^2 - 5^2}} = \displaystyle\int \dfrac{1}{x\sqrt{u^2 - 5^2}}\cdot\dfrac{du}{8x} = \dfrac{1}{8}\displaystyle\int \dfrac{du}{x^2\sqrt{u^2 - 5^2}}$

$= \dfrac{1}{8}\displaystyle\int \dfrac{du}{\frac{u}{4}\sqrt{u^2 - 5^2}} = \dfrac{1}{8}\displaystyle\int \dfrac{4\,du}{u\sqrt{u^2 - 5^2}} = \dfrac{1}{2}\displaystyle\int \dfrac{du}{u\sqrt{u^2 - 5^2}} = \dfrac{1}{2}\cdot\dfrac{1}{5}\,arc\,sec\,\dfrac{u}{5} + c = \dfrac{1}{10}\,\boldsymbol{arc\,sec}\,\dfrac{4x^2}{5} + c$

f. **First** - Write the given integral in its standard form $\displaystyle\int \dfrac{dx}{a^2 + x^2} = \dfrac{1}{a}\,arc\,tan\,\dfrac{x}{a} + c$, i.e., $\displaystyle\int \dfrac{e^x}{e^{2x} + 4}\,dx$

$= \displaystyle\int \dfrac{e^x}{4 + e^{2x}}\,dx = \displaystyle\int \dfrac{e^x}{2^2 + \left(e^x\right)^2}\,dx$

Second - Use substitution method by letting $u = e^x$, then $\dfrac{du}{dx} = \dfrac{d}{dx}e^x = e^x$ which implies

$du = e^x dx$; $dx = \dfrac{du}{e^x}$. Therefore, $\displaystyle\int \dfrac{e^x}{2^2 + \left(e^x\right)^2}\,dx = \displaystyle\int \dfrac{e^x}{2^2 + u^2}\cdot\dfrac{du}{e^x} = \displaystyle\int \dfrac{du}{2^2 + u^2} = \dfrac{1}{2}\,arc\,tan\,\dfrac{u}{2} + c$

Third - Write the answer in terms of the x variable, i.e., $\dfrac{1}{2}\,arc\,tan\,\dfrac{u}{2} + c = \dfrac{1}{2}\,\boldsymbol{arc\,tan}\,\dfrac{e^x}{2} + c$

Fourth - Check the answer by differentiating the solution, i.e., let $y = \dfrac{1}{2}\,arc\,tan\,\dfrac{e^x}{2} + c$ then

$y' = \dfrac{1}{2}\dfrac{1}{1 + \left(\frac{e^x}{2}\right)^2}\cdot\dfrac{d}{dx}\dfrac{e^x}{2} + 0 = \dfrac{1}{2}\dfrac{1}{1 + \frac{e^{2x}}{4}}\cdot\dfrac{e^x}{2} = \dfrac{1}{4}\dfrac{e^x}{\frac{4 + e^{2x}}{4}} = \dfrac{1}{4}\dfrac{4e^x}{4 + e^{2x}} = \dfrac{e^x}{4 + e^{2x}} = \dfrac{e^x}{e^{2x} + 4}$

g. **First** - Write the given integral in its standard form $\displaystyle\int \dfrac{dx}{a^2 + x^2} = \dfrac{1}{a}\,arc\,tan\,\dfrac{x}{a} + c$, i.e., $\displaystyle\int \dfrac{e^x}{9\,e^{2x} + 16}\,dx$

$$= \int \frac{e^x}{16+9\,e^{2x}}\,dx = \int \frac{e^x\,dx}{9\left(\frac{16}{9}+e^{2x}\right)} = \frac{1}{9}\int \frac{e^x\,dx}{\frac{16}{9}+e^{2x}} = \frac{1}{9}\int \frac{e^x\,dx}{\left(\frac{4}{3}\right)^2+\left(e^x\right)^2}$$

Second - Use substitution method by letting $u=e^x$, then $\dfrac{du}{dx}=\dfrac{d}{dx}e^x=e^x$ which implies

$du=e^x\,dx$; $dx=\dfrac{du}{e^x}$. Therefore, $\dfrac{1}{9}\displaystyle\int \dfrac{e^x\,dx}{\left(\frac{4}{3}\right)^2+\left(e^x\right)^2} = \dfrac{1}{9}\displaystyle\int \dfrac{e^x}{\left(\frac{4}{3}\right)^2+u^2}\cdot\dfrac{du}{e^x} = \dfrac{1}{9}\displaystyle\int \dfrac{du}{\left(\frac{4}{3}\right)^2+u^2}$

$$= \frac{1}{9}\cdot\frac{1}{\frac{4}{3}}\,arc\tan\frac{u}{\frac{4}{3}}+c = \frac{1}{9}\cdot\frac{3}{4}\,arc\tan\frac{3u}{4}+c = \frac{1}{12}\,arc\tan\frac{3u}{4}+c$$

Third - Write the answer in terms of the original variable, i.e., x . $\dfrac{1}{12}\,arc\tan\dfrac{3u}{4}+c = \dfrac{1}{12}\,\boldsymbol{arc\tan}\,\dfrac{3e^x}{4}+c$

Fourth - Check the answer by differentiating the solution, i.e., let $y=\dfrac{1}{12}\,arc\tan\dfrac{3e^x}{4}+c$ then

$$y' = \frac{1}{12}\frac{1}{1+\left(\frac{3e^x}{4}\right)^2}\cdot\frac{d}{dx}\frac{3e^x}{4}+0 = \frac{1}{12}\frac{1}{1+\frac{9e^{2x}}{16}}\cdot\frac{3e^x}{4} = \frac{1}{16}\frac{e^x}{\frac{16+9e^{2x}}{16}} = \frac{1}{16}\frac{16e^x}{16+9e^{2x}} = \frac{e^x}{16+9e^{2x}} = \frac{e^x}{9e^{2x}+16}$$

or, the alternative approach would be to rearrange the integral in the following way:

$$\int \frac{e^x}{16+9\,e^{2x}}\,dx = \int \frac{e^x\,dx}{4^2+\left(3e^x\right)^2}\;.$$ Now, let $u=3e^x$, then $\dfrac{du}{dx}=\dfrac{d}{dx}3e^x=3e^x$ which implies $du=3e^x\,dx$

; $dx=\dfrac{du}{3e^x}$. Therefore, $\displaystyle\int \dfrac{e^x\,dx}{4^2+\left(3e^x\right)^2} = \int \dfrac{e^x}{4^2+u^2}\cdot\dfrac{du}{3e^x} = \dfrac{1}{3}\int \dfrac{du}{4^2+u^2} = \dfrac{1}{3}\cdot\dfrac{1}{4}\,arc\tan\dfrac{u}{4}+c$

$$= \frac{1}{12}\,arc\tan\frac{u}{4}+c = \frac{1}{12}\,\boldsymbol{arc\tan}\,\frac{3e^x}{4}+c$$

h. Write the given integral $\displaystyle\int \dfrac{dx}{\sqrt{25-(x+4)^2}}$ in its standard form $\displaystyle\int \dfrac{dx}{\sqrt{a^2-x^2}} = arc\sin\dfrac{x}{a}+c$ by letting

$u=x+4$. Therefore, $\displaystyle\int \dfrac{dx}{\sqrt{25-(x+4)^2}} = \int \dfrac{dx}{\sqrt{25-u^2}} = \int \dfrac{dx}{\sqrt{5^2-u^2}} = arc\sin\dfrac{u}{5}+c = \boldsymbol{arc\sin}\,\dfrac{x+4}{5}+c$

Check: Let $y=arc\sin\dfrac{x+4}{5}+c$ then $y' = \dfrac{1}{\sqrt{1-\left(\frac{x+4}{5}\right)^2}}\cdot\dfrac{d}{dx}\dfrac{x+4}{5}+0 = \dfrac{1}{\sqrt{1-\frac{(x+4)^2}{25}}}\cdot\dfrac{1}{5} = \dfrac{1}{5}\dfrac{1}{\sqrt{\frac{25-(x+4)^2}{25}}}$

$$= \frac{1}{5}\frac{5}{\sqrt{25-(x+4)^2}} = \frac{1}{\sqrt{25-(x+4)^2}}$$

i. Write the given integral $\displaystyle\int \dfrac{dx}{\sqrt{3-(x-2)^2}}$ in its standard form $\displaystyle\int \dfrac{dx}{\sqrt{a^2-x^2}} = arc\sin\dfrac{x}{a}+c$ by letting

$u=x-2$. Therefore, $\displaystyle\int \dfrac{dx}{\sqrt{3-(x-2)^2}} = \int \dfrac{dx}{\sqrt{\left(\sqrt{3}\right)^2-u^2}} = arc\sin\dfrac{u}{\sqrt{3}}+c = \boldsymbol{arc\sin}\,\dfrac{x-2}{\sqrt{3}}+c$

Check: Let $y=arc\sin\dfrac{x-2}{\sqrt{3}}+c$ then $y' = \dfrac{1}{\sqrt{1-\left(\frac{x-2}{\sqrt{3}}\right)^2}}\cdot\dfrac{d}{dx}\dfrac{x-2}{\sqrt{3}}+0 = \dfrac{1}{\sqrt{1-\frac{(x-2)^2}{3}}}\cdot\dfrac{1}{\sqrt{3}} = \dfrac{1}{\sqrt{3}}\dfrac{1}{\sqrt{\frac{3-(x-2)^2}{3}}}$

. $$ = \frac{1}{\sqrt{3}}\frac{\sqrt{3}}{\sqrt{3-(x-2)^2}} = \frac{1}{\sqrt{3-(x-2)^2}} $$

Example 4.4-3: Evaluate the following indefinite integrals:

a. $\displaystyle\int \frac{dx}{9+(x+3)^2} =$ b. $\displaystyle\int \frac{x\,dx}{49+(x^2+9)^2} =$ c. $\displaystyle\int \frac{(x-2)^2}{9+(x-2)^6}\,dx =$

d. $\displaystyle\int \frac{(x+1)^3}{4+(x+1)^8}\,dx =$ e. $\displaystyle\int \frac{dx}{x\sqrt{x^2-36}} =$ f. $\displaystyle\int \frac{dx}{(x-3)\sqrt{(x-3)^2-49}} =$

g. $\displaystyle\int \frac{dx}{x^2-8x+17} =$ h. $\displaystyle\int \frac{dy}{y^2+20y+120} =$ i. $\displaystyle\int \frac{dt}{t^2+6t+13} =$

Solutions:

a. **First** - Write the given integral in its standard form $\displaystyle\int \frac{dx}{a^2+x^2} = \frac{1}{a}\,arc\tan\frac{x}{a}+c$, i.e., $\displaystyle\int \frac{dx}{9+(x+3)^2}$

Second - Use substitution method by letting $u=x+3$, then $\dfrac{du}{dx}=\dfrac{d}{dx}(x+3)=1$ which implies

$du=dx$. Therefore, $\displaystyle\int \frac{dx}{9+(x+3)^2} = \int \frac{du}{3^2+u^2} = \frac{1}{3}\,arc\tan\frac{u}{3}+c$

Third - Write the answer in terms of the x variable, i.e., $\dfrac{1}{3}\,arc\tan\dfrac{u}{3}+c = \dfrac{1}{3}\,\boldsymbol{arc\tan}\dfrac{x+3}{3}+c$

Fourth - Check the answer by differentiating the solution, i.e., let $y=\dfrac{1}{3}\,arc\tan\dfrac{x+3}{3}+c$ then

$$ y' = \frac{1}{3}\frac{1}{1+\left(\frac{x+3}{3}\right)^2}\cdot\frac{d}{dx}\left(\frac{x+3}{3}\right)+0 = \frac{1}{3}\frac{1}{1+\frac{(x+3)^2}{9}}\cdot\frac{1}{3} = \frac{1}{9}\frac{9}{9+(x+3)^2} = \frac{1}{9+(x+3)^2} $$

b. **First** - Write the given integral in its standard form $\displaystyle\int \frac{dx}{a^2+x^2} = \frac{1}{a}\,arc\tan\frac{x}{a}+c$, i.e., $\displaystyle\int \frac{x\,dx}{49+\left(x^2+9\right)^2}$

$$ = \int \frac{x\,dx}{7^2+\left(x^2+9\right)^2} $$

Second - Use substitution method by letting $u=x^2+9$, then $\dfrac{du}{dx}=\dfrac{d}{dx}\left(x^2+9\right)=2x$ which implies

$du=2x\,dx$; $dx=\dfrac{du}{2x}$. Therefore, $\displaystyle\int \frac{x\,dx}{7^2+\left(x^2+9\right)^2} = \int \frac{x}{7^2+u^2}\cdot\frac{du}{2x} = \frac{1}{2}\int \frac{du}{7^2+u^2} = \frac{1}{2}\cdot\frac{1}{7}\,arc\tan\frac{u}{7}+c$

Third - Write the answer in terms of the x variable, i.e., $\dfrac{1}{14}\,arc\tan\dfrac{u}{7}+c = \dfrac{1}{14}\,\boldsymbol{arc\tan}\dfrac{x^2+9}{7}+c$

Fourth - Check the answer by differentiating the solution, i.e., let $y=\dfrac{1}{14}\,arc\tan\dfrac{x^2+9}{7}+c$ then

$$y' = \frac{1}{14}\frac{1}{1+\left(\frac{x^2+9}{7}\right)^2}\cdot\frac{d}{dx}\left(\frac{x^2+9}{7}\right)+0 = \frac{1}{14}\frac{1}{1+\frac{(x^2+9)^2}{49}}\cdot\frac{2x}{7} = \frac{1}{49}\frac{49x}{49+(x^2+9)^2} = \frac{x}{49+(x^2+9)^2}$$

c. **First** - Write the given integral in its standard form $\int\frac{dx}{a^2+x^2} = \frac{1}{a}arc\tan\frac{x}{a}+c$, i.e., $\int\frac{(x-2)^2}{9+(x-2)^6}dx$

$$= \int\frac{(x-2)^2}{9+\left[(x-2)^3\right]^2}dx = \int\frac{(x-2)^2}{3^2+\left[(x-2)^3\right]^2}dx$$

Second - Use substitution method by letting $u=(x-2)^3$, then $\frac{du}{dx} = \frac{d}{dx}(x-2)^3 = 3(x-2)^2$ which

implies $du = 3(x-2)^2\,dx$; $dx = \frac{du}{3(x-2)^2}$. Thus, $\int\frac{(x-2)^2}{3^2+\left[(x-2)^3\right]^2}dx = \int\frac{(x-2)^2}{3^2+u^2}\cdot\frac{du}{3(x-2)^2}$

$$= \frac{1}{3}\int\frac{du}{3^2+u^2} = \frac{1}{3}\cdot\frac{1}{3}arc\tan\frac{u}{3}+c = \frac{1}{9}arc\tan\frac{u}{3}+c$$

Third - Write the answer in terms of the x variable, i.e., $\frac{1}{9}arc\tan\frac{u}{3}+c = \frac{1}{9}arc\tan\frac{(x-2)^3}{3}+c$

Fourth - Check the answer by differentiating the solution, i.e., let $y=\frac{1}{9}arc\tan\frac{(x-2)^3}{3}+c$ then

$$y' = \frac{1}{9}\frac{1}{1+\left(\frac{(x-2)^3}{3}\right)^2}\cdot\frac{d}{dx}\left(\frac{(x-2)^3}{3}\right)+0 = \frac{1}{9}\frac{1}{1+\frac{(x-2)^6}{9}}\cdot\frac{3(x-2)^2}{3} = \frac{1}{27}\frac{27(x-2)^2}{9+(x-2)^6} = \frac{(x-2)^2}{9+(x-2)^6}$$

d. **First** - Write the given integral in its standard form $\int\frac{dx}{a^2+x^2} = \frac{1}{a}arc\tan\frac{x}{a}+c$, i.e., $\int\frac{(x+1)^3}{4+(x+1)^8}dx$

$$= \int\frac{(x+1)^3}{4+\left[(x+1)^4\right]^2}dx = \int\frac{(x+1)^3}{2^2+\left[(x+1)^4\right]^2}dx$$

Second - Use substitution method by letting $u=(x+1)^4$, then $\frac{du}{dx} = \frac{d}{dx}(x+1)^4 = 4(x+1)^3$ which

implies $du = 4(x+1)^3\,dx$; $dx = \frac{du}{4(x+1)^3}$. Thus, $\int\frac{(x+1)^3}{2^2+\left[(x+1)^4\right]^2}dx = \int\frac{(x+1)^3}{2^2+u^2}\cdot\frac{du}{4(x+1)^3}$

$$= \frac{1}{4}\int\frac{du}{2^2+u^2} = \frac{1}{4}\cdot\frac{1}{2}arc\tan\frac{u}{2}+c = \frac{1}{8}arc\tan\frac{u}{2}+c$$

Third - Write the answer in terms of the x variable, i.e., $\frac{1}{8}arc\tan\frac{u}{2}+c = \frac{1}{8}arc\tan\frac{(x+1)^4}{2}+c$

Fourth - Check the answer by differentiating the solution, i.e., let $y=\frac{1}{8}arc\tan\frac{(x+1)^4}{2}+c$ then

$$y' = \frac{1}{8}\frac{1}{1+\left(\frac{(x+1)^4}{2}\right)^2}\cdot\frac{d}{dx}\left(\frac{(x+1)^4}{2}\right)+0 = \frac{1}{8}\frac{1}{1+\frac{(x+1)^8}{4}}\cdot\frac{4(x+1)^3}{2} = \frac{1}{16}\frac{16(x+1)^3}{4+(x+1)^8} = \frac{(x+1)^3}{4+(x+1)^8}$$

e. **First** - Write the integral in its standard form $\int \frac{1}{x\sqrt{x^2-a^2}}\,dx = \frac{1}{a}\,arc\,sec\,\frac{x}{a}+c$, i.e., $\int \frac{dx}{x\sqrt{x^2-36}}$

$= \int \frac{dx}{x\sqrt{x^2-6^2}}$

Second - Use substitution method by letting $u=x$, then $\frac{du}{dx}=\frac{d}{dx}x=1$ which implies $du=dx$.

Therefore, $\int \frac{dx}{x\sqrt{x^2-6^2}} = \int \frac{du}{u\sqrt{u^2-6^2}} = \frac{1}{6}arc\,sec\,\frac{u}{6}+c$

Third - Write the answer in terms of the original variable, i.e., x . $\frac{1}{6}arc\,sec\,\frac{u}{6}+c = \frac{1}{6}\boldsymbol{arc\,sec}\,\frac{\boldsymbol{x}}{6}+c$

Fourth - Check the answer by differentiating the solution, i.e., let $y=\frac{1}{6}arc\,sec\,\frac{x}{6}+c$ then

$y' = \frac{1}{6}\frac{1}{\frac{x}{6}\sqrt{\left(\frac{x}{6}\right)^2-1}}\cdot\frac{d}{dx}\frac{x}{6}+0 = \frac{1}{6}\frac{1}{\frac{x}{6}\sqrt{\frac{x^2}{36}-1}}\cdot\frac{1}{6} = \frac{1}{36}\frac{6}{x\sqrt{\frac{x^2-36}{36}}} = \frac{1}{36}\frac{36}{x\sqrt{x^2-36}} = \frac{1}{x\sqrt{x^2-36}}$

f. **First** - Write the given integral in its standard form $\int \frac{1}{x\sqrt{x^2-a^2}}\,dx = \frac{1}{a}\,arc\,sec\,\frac{x}{a}+c$, i.e.,

$\int \frac{dx}{(x-3)\sqrt{(x-3)^2-49}} = \int \frac{dx}{(x-3)\sqrt{(x-3)^2-7^2}}$

Second - Use substitution method by letting $u=x-3$, then $\frac{du}{dx}=\frac{d}{dx}x-3=1$ which implies $du=dx$.

Therefore, $\int \frac{dx}{(x-3)\sqrt{(x-3)^2-7^2}} = \int \frac{du}{u\sqrt{u^2-7^2}} = \frac{1}{7}arc\,sec\,\frac{u}{7}+c$

Third - Write the answer in terms of the original variable, i.e., x . $\frac{1}{7}arc\,sec\,\frac{u}{7}+c = \frac{1}{7}\boldsymbol{arc\,sec}\,\frac{\boldsymbol{x-3}}{7}+c$

Fourth - Check the answer by differentiating the solution, i.e., let $y=\frac{1}{7}arc\,sec\,\frac{x-3}{7}+c$ then y'

$= \frac{1}{7}\frac{1}{\frac{x-3}{7}\sqrt{\left(\frac{x-3}{7}\right)^2-1}}\cdot\frac{d}{dx}\frac{x-3}{7}+0 = \frac{1}{7}\frac{1}{\frac{x-3}{7}\sqrt{\frac{(x-3)^2}{49}-1}}\cdot\frac{1}{7} = \frac{1}{49}\frac{7}{x-3\sqrt{\frac{(x-3)^2-49}{49}}} = \frac{1}{49}\frac{49}{x-3\sqrt{(x-3)^2-49}}$

$= \frac{1}{x-3\sqrt{(x-3)^2-49}}$

g. **First** - Write the integral in its standard form $\int \frac{dx}{a^2+x^2} = \frac{1}{a}\,arc\,tan\,\frac{x}{a}+c$, i.e., $\int \frac{dx}{x^2-8x+17}$

$= \int \frac{dx}{\left(x^2-8x+16\right)+1} = \int \frac{dx}{(x-4)^2+1} = \int \frac{dx}{1+(x-4)^2}$

Second - Use substitution method by letting $u=x-4$, then $\frac{du}{dx}=\frac{d}{dx}(x-4)=1$ which implies

$du=dx$. Therefore, $\int \frac{dx}{1+(x-4)^2} = \int \frac{du}{1+u^2} = arc\,tan\,u+c$

Third - Write the answer in terms of the x variable, i.e., $arc\tan u+c = $ **$arc\tan(x-4)+c$**

Fourth - Check the answer by differentiating the solution, i.e., let $y=arc\tan(x-4)+c$ then

$$y' = \frac{1}{1+(x-4)^2}\cdot\frac{d}{dx}(x-4)+0 = \frac{1}{1+(x-4)^2}\cdot 1 = \frac{1}{1+(x-4)^2}$$

h. **First** - Write the integral in its standard form $\int\frac{dx}{a^2+x^2} = \frac{1}{a}arc\tan\frac{x}{a}+c$, i.e., $\int\frac{dy}{y^2+20y+120}$

$$= \int\frac{dy}{\left(y^2+20y+100\right)+20} = \int\frac{dy}{(y+10)^2+20} = \int\frac{dy}{20+(y+10)^2} = \int\frac{dy}{\left(\sqrt{20}\right)^2+(y+10)^2}$$

Second - Use substitution method by letting $u=y+10$, then $\frac{du}{dy}=\frac{d}{dy}(y+10)=1$ which implies

$du=dy$. Therefore, $\int\frac{dy}{\left(\sqrt{20}\right)^2+(y+10)^2} = \int\frac{du}{\left(\sqrt{20}\right)^2+u^2} = \frac{1}{\sqrt{20}}arc\tan\frac{u}{\sqrt{20}}+c$

Third - Write the answer in terms y, i.e., $\frac{1}{\sqrt{20}}arc\tan\frac{u}{\sqrt{20}}+c = \frac{1}{\sqrt{20}}$**$arc\tan\frac{y+10}{\sqrt{20}}+c$**

Fourth - Check the answer by differentiating the solution, i.e., let $w=\frac{1}{\sqrt{20}}arc\tan\frac{y+10}{\sqrt{20}}+c$ then

$$w' = \frac{1}{\sqrt{20}}\frac{1}{1+\left(\frac{y+10}{\sqrt{20}}\right)^2}\cdot\frac{d}{dy}\left(\frac{y+10}{\sqrt{20}}\right)+0 = \frac{1}{\sqrt{20}}\frac{1}{1+\frac{(y+10)^2}{20}}\cdot\frac{1}{\sqrt{20}} = \frac{1}{20}\frac{20}{20+(y+10)^2} = \frac{1}{20+(y+10)^2}$$

i. **First** - Write the integral in its standard form $\int\frac{dx}{a^2+x^2} = \frac{1}{a}arc\tan\frac{x}{a}+c$, i.e., $\int\frac{dt}{t^2+6t+13}$

$$= \int\frac{dt}{\left(t^2+6t+9\right)+4} = \int\frac{dt}{(t+3)^2+4} = \int\frac{dt}{4+(t+3)^2} = \int\frac{dt}{2^2+(t+3)^2}$$

Second - Use substitution method by letting $u=t+3$, then $\frac{du}{dt}=\frac{d}{dt}(t+3)=1$ which implies

$du=dt$. Therefore, $\int\frac{dt}{2^2+(t+3)^2} = \int\frac{du}{2^2+u^2} = \frac{1}{2}arc\tan\frac{u}{2}+c$

Third - Write the answer in terms of the variable t, i.e., $\frac{1}{2}arc\tan\frac{u}{2}+c = \frac{1}{2}$**$arc\tan\frac{t+3}{2}+c$**

Fourth - Check the answer by differentiating the solution, i.e., let $w=\frac{1}{2}arc\tan\frac{t+3}{2}+c$ then

$$w' = \frac{1}{2}\frac{1}{1+\left(\frac{t+3}{2}\right)^2}\cdot\frac{d}{dy}\left(\frac{t+3}{2}\right)+0 = \frac{1}{2}\frac{1}{1+\frac{(t+3)^2}{4}}\cdot\frac{1}{2} = \frac{1}{4}\frac{4}{4+(t+3)^2} = \frac{1}{4+(t+3)^2}$$

1. Evaluate the following indefinite integrals:

a. $\int \dfrac{dx}{\sqrt{25-9x^2}} =$

b. $\int \dfrac{dx}{\sqrt{4-x^2}} =$

c. $\int \dfrac{x^2\,dx}{\sqrt{25-x^6}} =$

d. $\int \dfrac{dx}{9x^2+16} =$

e. $\int \dfrac{x^2\,dx}{7+9x^6} =$

f. $\int \dfrac{dx}{x\sqrt{x^4-25}} =$

2. Evaluate the following indefinite integrals:

a. $\int \dfrac{dx}{x\sqrt{x^2-16}} =$

b. $\int \dfrac{dx}{x\sqrt{7x^2-4}} =$

c. $\int \dfrac{e^x}{4\,e^{2x}+9}\,dx =$

d. $\int \dfrac{dx}{\sqrt{25-(x-3)^2}} =$

e. $\int \dfrac{dx}{25+(x+4)^2} =$

f. $\int \dfrac{dx}{x^2-10x+26} =$

4.5   Integration of Expressions Resulting in Exponential or Logarithmic Functions

In the following examples we will solve problems using the following formulas:

$$\int \frac{1}{x}\,dx = \ln|x|+c$$

$$\int \ln x\,dx = x\ln|x|-x+c$$

$$\int a^x\,dx = \frac{a^x}{\ln a}+c \qquad a\rangle 0 \text{ and } a\neq 1$$

$$\int e^x\,dx = e^x+c$$

Let's integrate some exponential and algebraic expressions using the above integration formulas.

Example 4.5-1: Evaluate the following indefinite integrals:

a. $\int \dfrac{dx}{x+5} =$

b. $\int \dfrac{dx}{5x+1} =$

c. $\int \dfrac{x}{x^2-3}\,dx =$

d. $\int \dfrac{x^2}{2x^3+1}\,dx =$

e. $\int \dfrac{4x^3}{x^4-3}\,dx =$

f. $\int \left(\dfrac{5}{x+1}+\dfrac{2}{x-1}\right)dx =$

g. $\int \left(x^3-\dfrac{2x}{x^2+1}\right)dx =$

h. $\int xe^{x^2}\,dx =$

i. $\int 2x^2 e^{x^3}\,dx =$

Solutions:

a. Given $\int \dfrac{dx}{x+5}$ let $u=x+5$, then $\dfrac{du}{dx}=\dfrac{d}{dx}(x+5)$; $\dfrac{du}{dx}=1$; $dx=du$. Therefore,

$$\boxed{\int \frac{dx}{x+5}} = \boxed{\int \frac{1}{u}\cdot du} = \boxed{\ln|u|+c} = \boxed{\ln|x+5|+c}$$

Check: Let $y=\ln|x+5|+c$, then $y' = \dfrac{1}{x+5}+0 = \dfrac{1}{x+5}$

b. Given $\int \dfrac{dx}{5x+1}$ let $u=5x+1$, then $\dfrac{du}{dx}=\dfrac{d}{dx}(5x+1)$; $\dfrac{du}{dx}=5$; $du=5\,dx$; $dx=\dfrac{du}{5}$. Therefore,

$$\boxed{\int \frac{dx}{5x+1}} = \boxed{\int \frac{1}{u}\cdot \frac{du}{5}} = \boxed{\frac{1}{5}\int \frac{1}{u}\,du} = \boxed{\frac{1}{5}\ln|u|+c} = \boxed{\frac{1}{5}\ln|5x+1|+c}$$

Check: Let $y=\dfrac{1}{5}\ln|5x+1|+c$, then $y' = \dfrac{1}{5}\cdot\dfrac{1}{5x+1}\cdot 5+0 = \dfrac{5}{5}\cdot\dfrac{1}{5x+1} = \dfrac{1}{5x+1}$

c. Given $\int \dfrac{x}{x^2-3}\,dx$ let $u=x^2-3$, then $\dfrac{du}{dx}=\dfrac{d}{dx}(x^2-3)$; $\dfrac{du}{dx}=2x$; $du=2x\,dx$; $dx=\dfrac{du}{2x}$. Thus,

$$\boxed{\int \frac{x}{x^2-3}\,dx} = \boxed{\int \frac{x}{u}\cdot\frac{du}{2x}} = \boxed{\frac{1}{2}\int \frac{1}{u}\,du} = \boxed{\frac{1}{2}\ln|u|+c} = \boxed{\frac{1}{2}\ln|x^2-3|+c}$$

Check: Let $y = \frac{1}{2}\ln\left|x^2 - 3\right| + c$, then $y' = \frac{1}{2} \cdot \frac{1}{x^2 - 3} \cdot 2x + 0 = \frac{2}{2} \cdot \frac{x}{x^2 - 3} = \frac{x}{x^2 - 3}$

d. Given $\int \frac{x^2}{2x^3 + 1} dx$ let $u = 2x^3 + 1$, then $\frac{du}{dx} = \frac{d}{dx}\left(2x^3 + 1\right)$; $\frac{du}{dx} = 6x^2$; $du = 6x^2 dx$; $dx = \frac{du}{6x^2}$. Thus,

$$\boxed{\int \frac{x^2}{2x^3 + 1} dx} = \boxed{\int \frac{x^2}{u} \cdot \frac{du}{6x^2}} = \boxed{\frac{1}{6}\int \frac{1}{u} du} = \boxed{\frac{1}{6}\ln|u| + c} = \boxed{\frac{1}{6}\ln\left|2x^3 + 1\right| + c}$$

Check: Let $y = \frac{1}{6}\ln\left|2x^3 + 1\right| + c$, then $y' = \frac{1}{6} \cdot \frac{1}{2x^3 + 1} \cdot 6x^2 + 0 = \frac{6}{6} \cdot \frac{x^2}{2x^3 + 1} = \frac{x^2}{2x^3 + 1}$

e. Given $\int \frac{4x^3}{x^4 - 3} dx$ let $u = x^4 - 3$, then $\frac{du}{dx} = \frac{d}{dx}\left(x^4 - 3\right)$; $\frac{du}{dx} = x^4 - 3$; $du = 4x^3 dx$; $dx = \frac{du}{4x^3}$. Thus,

$$\boxed{\int \frac{4x^3}{x^4 - 3} dx} = \boxed{\int \frac{4x^3}{u} \cdot \frac{du}{4x^3}} = \boxed{\int \frac{1}{u} du} = \boxed{\ln|u| + c} = \boxed{\ln\left|x^4 - 3\right| + c}$$

Check: Let $y = \ln\left|x^4 - 3\right| + c$, then $y' = \frac{1}{x^4 - 3} \cdot 4x^3 + 0 = \frac{4x^3}{x^4 - 3}$

f. Given $\int \left(\frac{5}{x+1} + \frac{2}{x-1}\right) dx = \int \frac{5}{x+1} dx + \int \frac{2}{x-1} dx$ let $u_1 = x+1$ and $u_2 = x-1$ respectively. Therefore,

$$\boxed{\int \frac{5}{x+1} dx + \int \frac{2}{x-1} dx} = \boxed{5\int \frac{1}{u_1} du_1 + 2\int \frac{1}{u_2} du_2} = \boxed{5\ln|u_1| + 2\ln|u_2| + c} = \boxed{5\ln|x+1| + 2\ln|x-1| + c}$$

Check: Let $y = 5\ln|x+1| + 2\ln|x-1| + c$, then $y' = \frac{5}{x+1} + \frac{2}{x-1} + 0 = \frac{5}{x+1} + \frac{2}{x-1}$

g. Given $\int \left(x^3 - \frac{2x}{x^2 + 1}\right) dx = \int x^3 dx - \int \frac{2x}{x^2 + 1} dx$ let $u = x^2 + 1$, then $\frac{du}{dx} = 2x$; $dx = \frac{du}{2x}$. Therefore,

$$\boxed{\int x^3 dx - \int \frac{2x}{x^2 + 1} dx} = \boxed{\frac{1}{4}x^4 - \int \frac{2x}{u} \cdot \frac{du}{2x}} = \boxed{\frac{1}{4}x^4 - \int \frac{1}{u} du} = \boxed{\frac{1}{4}x^4 - \ln|u| + c} = \boxed{\frac{1}{4}x^4 - \ln\left|x^2 + 1\right| + c}$$

Check: Let $y = \frac{1}{4}x^4 - \ln\left|x^2 + 1\right| + c$, then $y' = \frac{1}{4} \cdot 4x^3 - \frac{1}{x^2 + 1} \cdot 2x + 0 = x^3 - \frac{2x}{x^2 + 1}$

h. Given $\int xe^{x^2} dx$ let $u = x^2$, then $\frac{du}{dx} = \frac{d}{dx}x^2$; $\frac{du}{dx} = 2x$; $du = 2x \, dx$; $dx = \frac{du}{2x}$. Therefore,

$$\boxed{\int xe^{x^2} dx} = \boxed{\int xe^u \cdot \frac{du}{2x}} = \boxed{\frac{1}{2}\int e^u du} = \boxed{\frac{1}{2}e^u + c} = \boxed{\frac{1}{2}e^{x^2} + c}$$

Check: Let $y = \frac{1}{2}e^{x^2} + c$, then $y' = \frac{1}{2} \cdot e^{x^2} \cdot 2x + 0 = \frac{2}{2} \cdot xe^{x^2} = xe^{x^2}$

i. Given $\int 2x^2 e^{x^3}\,dx$ let $u=x^3$, then $\frac{du}{dx}=\frac{d}{dx}x^3$; $\frac{du}{dx}=3x^2$; $du=3x^2\,dx$; $dx=\frac{du}{3x^2}$. Therefore,

$$\int 2x^2 e^{x^3}\,dx = \int 2x^2 e^u \cdot \frac{du}{3x^2} = \frac{2}{3}\int e^u\,du = \frac{2}{3}e^u +c = \frac{2}{3}e^{x^3}+c$$

Check: Let $y=\frac{2}{3}e^{x^3}+c$, then $y'=\frac{2}{3}\cdot e^{x^3}\cdot 3x^2+0 = \frac{6}{3}\cdot x^2 e^{x^3} = 2x^2 e^{x^3}$

Example 4.5-2: Evaluate the following indefinite integrals:

a. $\int e^y \sqrt{5+3e^y}\,dy =$ b. $\int \frac{e^x}{1-e^x}\,dx =$ c. $\int \frac{2e^{3x}}{1+e^{3x}}\,dx =$

d. $\int e^{-x}\,dx =$ e. $\int e^{-3x}\,dx =$ f. $\int \frac{e^{-5x+2}}{2}\,dx =$

g. $\int \left(x^2 + e^x - e^{-3x}\right)dx =$ h. $\int 1+\frac{e^{\frac{1}{x}}}{x^2}\,dx =$ i. $\int \frac{e^{\frac{1}{x^2}}}{x^3}\,dx =$

Solutions:

a. Given $\int e^y \sqrt{5+3e^y}\,dy$ let $u=5+3e^y$, then $\frac{du}{dy}=\frac{d}{dy}\left(5+3e^y\right)$; $\frac{du}{dy}=3e^y$; $dy=\frac{du}{3e^y}$. Therefore,

$$\int e^y \sqrt{5+3e^y}\,dy = \int e^y \cdot \sqrt{u}\cdot \frac{du}{3e^y} = \frac{1}{3}\int u^{\frac{1}{2}}\,du = \frac{1}{3}\cdot\frac{1}{1+\frac{1}{2}}u^{1+\frac{1}{2}} = \frac{1}{3}\cdot\frac{2}{3}u^{\frac{3}{2}} = \frac{2}{9}u^{\frac{3}{2}} = \frac{2}{9}\left(5+3e^y\right)^{\frac{3}{2}}$$

Check: Let $y=\frac{2}{9}\left(5+3e^y\right)^{\frac{3}{2}}$, then $y'=\frac{2}{9}\cdot\frac{3}{2}\left(5+3e^y\right)^{\frac{3}{2}-1}\cdot 3e^y = \frac{3}{3}e^y\left(5+3e^y\right)^{\frac{1}{2}} = e^y\sqrt{5+3e^y}$

b. Given $\int \frac{e^x}{1-e^x}\,dx$ let $u=1-e^x$, then $\frac{du}{dx}=\frac{d}{dx}\left(1-e^x\right)$; $\frac{du}{dx}=-e^x$; $du=-e^x\,dx$; $dx=-\frac{du}{e^x}$. Thus,

$$\int \frac{e^x}{1-e^x}\,dx = \int \frac{e^x}{1-e^x}\cdot -\frac{du}{e^x} = -\int \frac{1}{u}\,du = -\ln|u|+c = -\ln\left|1-e^x\right|+c$$

Check: Let $y=-\ln\left|1-e^x\right|+c$, then $y'=-\frac{1}{1-e^x}\cdot -e^x+0 = \frac{e^x}{1-e^x}$

c. Given $\int \frac{2e^{3x}}{1+e^{3x}}\,dx$ let $u=1+e^{3x}$, then $\frac{du}{dx}=\frac{d}{dx}\left(1+e^{3x}\right)$; $\frac{du}{dx}=3e^{3x}$; $du=3e^{3x}\,dx$; $dx=\frac{du}{3e^{3x}}$. Thus,

$$\int \frac{2e^{3x}}{1+e^{3x}}\,dx = \int \frac{2e^{3x}}{u}\cdot \frac{du}{3e^{3x}} = \frac{2}{3}\int \frac{1}{u}\,du = \frac{2}{3}\ln|u|+c = \frac{2}{3}\ln\left|1+e^{3x}\right|+c$$

Check: Let $y=\frac{2}{3}\ln\left|1+e^{3x}\right|+c$, then $y'=\frac{2}{3}\frac{1}{1+e^{3x}}\cdot 3e^{3x}+0 = \frac{6}{3}\cdot\frac{e^{3x}}{1+e^{3x}} = \frac{2e^{3x}}{1+e^{3x}}$

d. Given $\int e^{-x} dx$ let $u = -x$, then $\dfrac{du}{dx} = \dfrac{d}{dx}(-x)$; $\dfrac{du}{dx} = -1$; $du = -dx$; $dx = -du$. Therefore,

$$\boxed{\int e^{-x} dx} = \boxed{\int e^u \cdot -du} = \boxed{-\int e^u du} = \boxed{-e^u + c} = \boxed{-e^{-x} + c}$$

Check: Let $y = -e^{-x} + c$, then $y' = -e^{-x} \cdot -1 + 0 = e^{-x}$

e. Given $\int e^{-3x} dx$ let $u = -3x$, then $\dfrac{du}{dx} = \dfrac{d}{dx}(-3x)$; $\dfrac{du}{dx} = -3$; $du = -3dx$; $dx = -\dfrac{du}{3}$. Therefore,

$$\boxed{\int e^{-3x} dx} = \boxed{\int e^u \cdot -\dfrac{du}{3}} = \boxed{-\dfrac{1}{3}\int e^u du} = \boxed{-\dfrac{1}{3}e^u + c} = \boxed{-\dfrac{1}{3}e^{-3x} + c}$$

Check: Let $y = -\dfrac{1}{3}e^{-3x} + c$, then $y' = -\dfrac{1}{3}e^{-3x} \cdot -3 + 0 = \dfrac{3}{3}e^{-3x} = e^{-3x}$

f. Given $\int \dfrac{e^{-5x+2}}{2} dx$ let $u = -5x+2$, then $\dfrac{du}{dx} = \dfrac{d}{dx}(-5x+2)$; $\dfrac{du}{dx} = -5$; $du = -5dx$; $dx = -\dfrac{du}{5}$. Thus,

$$\boxed{\int \dfrac{e^{-5x+2}}{2} dx} = \boxed{\int \dfrac{e^u}{2} \cdot -\dfrac{du}{5}} = \boxed{-\dfrac{1}{10}\int e^u du} = \boxed{-\dfrac{1}{10}e^u + c} = \boxed{-\dfrac{1}{10}e^{-5x+2} + c}$$

Check: Let $y = -\dfrac{1}{10}e^{-5x+2} + c$, then $y' = -\dfrac{1}{10}e^{-5x+2} \cdot -5 + 0 = \dfrac{5}{10}e^{-5x+2} = \dfrac{1}{2}e^{-5x+2}$

g. $\boxed{\int \left(x^2 + e^x - e^{-3x}\right) dx} = \boxed{\int x^2 dx + \int e^x dx - \int e^{-3x} dx} = \boxed{\dfrac{1}{3}x^3 + e^x + \dfrac{1}{3}e^{-3x} + c}$

Check: Let $y = \dfrac{1}{3}x^3 + e^x + \dfrac{1}{3}e^{-3x} + c$ then $y' = \dfrac{1}{3} \cdot 3x^2 + e^x + \dfrac{1}{3} \cdot e^{-3x} \cdot -3 + 0 = x^2 + e^x - e^{-3x}$

h. $\int 1 + \dfrac{e^{\frac{1}{x}}}{x^2} dx = \int dx + \int \dfrac{e^{\frac{1}{x}}}{x^2} dx$ let $u = \dfrac{1}{x}$, then $\dfrac{du}{dx} = \dfrac{d}{dx}\dfrac{1}{x}$; $\dfrac{du}{dx} = -\dfrac{1}{x^2}$; $x^2 du = -dx$; $dx = -x^2 du$. Thus,

$$\boxed{\int 1 + \dfrac{e^{\frac{1}{x}}}{x^2} dx} = \boxed{x + \int \dfrac{e^u}{x^2} \cdot -x^2 du} = \boxed{x - \int e^u du} = \boxed{x - e^u + c} = \boxed{x - e^{\frac{1}{x}} + c}$$

Check: Let $y = x - e^{\frac{1}{x}} + c$, then $y' = 1 - e^{\frac{1}{x}} \cdot -\dfrac{1}{x^2} + 0 = 1 + \dfrac{e^{\frac{1}{x}}}{x^2}$

i. $\int \dfrac{e^{\frac{1}{x^2}}}{x^3} dx$ let $u = \dfrac{1}{x^2}$, then $\dfrac{du}{dx} = \dfrac{d}{dx}\dfrac{1}{x^2}$; $\dfrac{du}{dx} = -\dfrac{2}{x^3}$; $x^3 du = -2dx$; $dx = -\dfrac{x^3}{2} du$. Therefore,

$$\boxed{\int \dfrac{e^{\frac{1}{x^2}}}{x^3} dx} = \boxed{\int \dfrac{e^u}{x^3} \cdot -\dfrac{x^3}{2} du} = \boxed{-\dfrac{1}{2}\int e^u du} = \boxed{-\dfrac{1}{2}e^u + c} = \boxed{-\dfrac{1}{2}e^{\frac{1}{x^2}} + c}$$

Check: Let $y = -\dfrac{1}{2}e^{\frac{1}{x^2}} + c$, then $y' = -\dfrac{1}{2} \cdot e^{\frac{1}{x^2}} \cdot -\dfrac{2}{x^3} + 0 = \dfrac{2}{2} \cdot e^{\frac{1}{x^2}} \cdot \dfrac{1}{x^3} = \dfrac{e^{\frac{1}{x^2}}}{x^3}$

Example 4.5-3: Evaluate the following indefinite integrals:

a. $\int \dfrac{e^{x^{\frac{2}{3}}}}{3x^4}\,dx \;=\;$

b. $\int \dfrac{e^{\frac{2}{x}+5}}{x^2}\,dx \;=\;$

c. $\int \dfrac{e^{-\frac{1}{x^2}}}{x^3}\,dx \;=\;$

d. $\int \left(e^x - 1\right)e^x\,dx \;=\;$

e. $\int \left(e^x + 3\right)^2 e^x\,dx \;=\;$

f. $\int \left(e^{-x} - 1\right)^3 e^{-x}\,dx \;=\;$

g. $\int \dfrac{x^3}{1+4x^4}\,dx \;=\;$

h. $\int \dfrac{x^2}{1-x^3}\,dx \;=\;$

i. $\int \dfrac{10x^4}{1-3x^5}\,dx \;=\;$

Solutions:

a. Given $\int \dfrac{e^{x^{\frac{2}{3}}}}{3x^4}\,dx$ let $u = \dfrac{2}{x^3}$, then $\dfrac{du}{dx} = \dfrac{d}{dx}\dfrac{2}{x^3}$; $\dfrac{du}{dx} = -\dfrac{6}{x^4}$; $x^4\,du = -6\,dx$; $dx = -\dfrac{x^4}{6}\,du$. Therefore,

$$\boxed{\int \dfrac{e^{x^{\frac{2}{3}}}}{3x^4}\,dx} = \boxed{\int \dfrac{e^u}{3x^4}\cdot -\dfrac{x^4}{6}\,du} = \boxed{-\dfrac{1}{18}\int e^u\,du} = \boxed{-\dfrac{1}{18}e^u + c} = \boxed{-\dfrac{1}{18}e^{x^{\frac{2}{3}}} + c}$$

Check: Let $y = -\dfrac{1}{18}e^{x^{\frac{2}{3}}} + c$, then $y' = -\dfrac{1}{18}\cdot e^{x^{\frac{2}{3}}}\cdot -\dfrac{6}{x^4} + 0 = \dfrac{6}{18}\cdot e^{x^{\frac{2}{3}}}\cdot\dfrac{1}{x^4} = \dfrac{e^{x^{\frac{2}{3}}}}{3x^4}$

b. Given $\int \dfrac{e^{\frac{2}{x}+5}}{x^2}\,dx$ let $u = \dfrac{2}{x}+5$, then $\dfrac{du}{dx} = \dfrac{d}{dx}\left(\dfrac{2}{x}+5\right)$; $\dfrac{du}{dx} = -\dfrac{2}{x^2}$; $x^2\,du = -2\,dx$; $dx = -\dfrac{x^2}{2}\,du$. Thus,

$$\boxed{\int \dfrac{e^{\frac{2}{x}+5}}{x^2}\,dx} = \boxed{\int \dfrac{e^u}{x^2}\cdot -\dfrac{x^2}{2}\,du} = \boxed{-\dfrac{1}{2}\int e^u\,du} = \boxed{-\dfrac{1}{2}e^u + c} = \boxed{-\dfrac{1}{2}e^{\frac{2}{x}+5} + c}$$

Check: Let $y = -\dfrac{1}{2}e^{\frac{2}{x}+5} + c$, then $y' = -\dfrac{1}{2}\cdot e^{\frac{2}{x}+5}\cdot -\dfrac{2}{x^2} + 0 = \dfrac{2}{2}\cdot e^{\frac{2}{x}+5}\cdot\dfrac{1}{x^2} = \dfrac{e^{\frac{2}{x}+5}}{x^2}$

c. Given $\int \dfrac{e^{-\frac{1}{x^2}}}{x^3}\,dx$ let $u = -\dfrac{1}{x^2}$, then $\dfrac{du}{dx} = -\dfrac{d}{dx}x^{-2}$; $\dfrac{du}{dx} = 2x^{-3}$; $du = 2x^{-3}\,dx$; $dx = \dfrac{x^3}{2}\,du$. Thus,

$$\boxed{\int \dfrac{e^{-\frac{1}{x^2}}}{x^3}\,dx} = \boxed{\int \dfrac{e^u}{x^3}\cdot\dfrac{x^3}{2}\,du} = \boxed{\dfrac{1}{2}\int e^u\,du} = \boxed{\dfrac{1}{2}e^u + c} = \boxed{\dfrac{1}{2}e^{-\frac{1}{x^2}} + c}$$

Check: Let $y = \dfrac{1}{2}e^{-\frac{1}{x^2}} + c$, then $y' = \dfrac{1}{2}e^{-\frac{1}{x^2}}\cdot 2x^{-3} + 0 = \dfrac{2}{2}\cdot e^{-\frac{1}{x^2}}\cdot\dfrac{1}{x^3} = \dfrac{e^{-\frac{1}{x^2}}}{x^3}$

d. Given $\int \left(e^x - 1\right)e^x\,dx$ let $u = e^x - 1$, then $\dfrac{du}{dx} = \dfrac{d}{dx}\left(e^x - 1\right)$; $\dfrac{du}{dx} = e^x$; $du = e^x\,dx$; $dx = \dfrac{du}{e^x}$. Thus,

$$\boxed{\int \left(e^x - 1\right)e^x\,dx} = \boxed{\int u\cdot e^x\cdot\dfrac{du}{e^x}} = \boxed{\int u\,du} = \boxed{\dfrac{1}{2}u^2 + c} = \boxed{\dfrac{1}{2}\left(e^x - 1\right)^2 + c}$$

Check: Let $y = \frac{1}{2}\left(e^x - 1\right)^2 + c$, then $y' = \frac{1}{2} \cdot 2\left(e^x - 1\right) \cdot e^x + 0 = \frac{2}{2}\left(e^x - 1\right)e^x = \left(e^x - 1\right)e^x$

e. Given $\int \left(e^x + 3\right)^2 e^x dx$ let $u = e^x + 3$, then $\frac{du}{dx} = \frac{d}{dx}\left(e^x + 3\right)$; $\frac{du}{dx} = e^x$; $du = e^x dx$; $dx = \frac{du}{e^x}$. Thus,

$$\boxed{\int \left(e^x + 3\right)^2 e^x dx} = \boxed{\int u^2 \cdot e^x \cdot \frac{du}{e^x}} = \boxed{\int u^2 du} = \boxed{\frac{1}{3}u^3 + c} = \boxed{\frac{1}{3}\left(e^x + 3\right)^3 + c}$$

Check: Let $y = \frac{1}{3}\left(e^x + 3\right)^3 + c$, then $y' = \frac{1}{3} \cdot 3\left(e^x + 3\right)^2 \cdot e^x + 0 = \frac{3}{3}\left(e^x + 3\right)^2 e^x = \left(e^x + 3\right)^2 e^x$

f. Given $\int \left(e^{-x} - 1\right)^3 e^{-x} dx$ let $u = e^{-x} - 1$, then $\frac{du}{dx} = \frac{d}{dx}\left(e^{-x} - 1\right)$; $\frac{du}{dx} = -e^{-x}$; $dx = -\frac{du}{e^{-x}}$. Therefore,

$$\boxed{\int \left(e^{-x} - 1\right)^3 e^{-x} dx} = \boxed{\int u^3 \cdot e^{-x} \cdot -\frac{du}{e^{-x}}} = \boxed{-\int u^3 du} = \boxed{-\frac{1}{4}u^4 + c} = \boxed{-\frac{1}{4}\left(e^{-x} - 1\right)^4 + c}$$

Check: Let $y = -\frac{1}{4}\left(e^{-x} - 1\right)^4 + c$, then $y' = -\frac{1}{4} \cdot 4\left(e^{-x} - 1\right)^3 \cdot -e^{-x} + 0 = \frac{4}{4}\left(e^{-x} - 1\right)^3 e^{-x} = \left(e^{-x} - 1\right)^3 e^{-x}$

g. Given $\int \frac{x^3}{1 + 4x^4} dx$ let $u = 1 + 4x^4$, then $\frac{du}{dx} = \frac{d}{dx}\left(1 + 4x^4\right)$; $\frac{du}{dx} = 16x^3$; $dx = \frac{du}{16x^3}$. Therefore,

$$\boxed{\int \frac{x^3}{1 + 4x^4} dx} = \boxed{\int \frac{x^3}{u} \cdot \frac{du}{16x^3}} = \boxed{\frac{1}{16}\int \frac{1}{u} du} = \boxed{\frac{1}{16}\ln|u| + c} = \boxed{\frac{1}{16}\ln\left|1 + 4x^4\right| + c}$$

Check: Let $y = \frac{1}{16}\ln\left|1 + 4x^4\right| + c$, then $y' = \frac{1}{16} \cdot \frac{1}{1 + 4x^4} \cdot 16x^3 + 0 = \frac{16}{16} \cdot \frac{x^3}{1 + 4x^4} = \frac{x^3}{1 + 4x^4}$

h. Given $\int \frac{x^2}{1 - x^3} dx$ let $u = 1 - x^3$, then $\frac{du}{dx} = \frac{d}{dx}\left(1 - x^3\right)$; $\frac{du}{dx} = -3x^2$; $dx = -\frac{du}{3x^2}$. Therefore,

$$\boxed{\int \frac{x^2}{1 - x^3} dx} = \boxed{\int \frac{x^2}{u} \cdot -\frac{du}{3x^2}} = \boxed{-\frac{1}{3}\int \frac{1}{u} du} = \boxed{-\frac{1}{3}\ln|u| + c} = \boxed{-\frac{1}{3}\ln\left|1 - x^3\right| + c}$$

Check: Let $y = -\frac{1}{3}\ln\left|1 - x^3\right| + c$, then $y' = -\frac{1}{3} \cdot \frac{1}{1 - x^3} \cdot 3x^2 + 0 = \frac{3}{3} \cdot \frac{x^2}{1 - x^3} = \frac{x^2}{1 - x^3}$

i. Given $\int \frac{10x^4}{1 - 3x^5} dx$ let $u = 1 - 3x^5$, then $\frac{du}{dx} = \frac{d}{dx}\left(1 - 3x^5\right)$; $\frac{du}{dx} = -15x^4$; $dx = -\frac{du}{15x^4}$. Therefore,

$$\boxed{\int \frac{10x^4}{1 - 3x^5} dx} = \boxed{\int \frac{10x^4}{u} \cdot -\frac{du}{15x^4}} = \boxed{-\frac{10}{15}\int \frac{1}{u} du} = \boxed{-\frac{2}{3}\ln|u| + c} = \boxed{-\frac{2}{3}\ln\left|1 - 3x^5\right| + c}$$

Check: Let $y = -\frac{2}{3}\ln\left|1 - 3x^5\right| + c$, then $y' = -\frac{2}{3} \cdot \frac{1}{1 - 3x^5} \cdot -15x^4 + 0 = \frac{30}{3} \cdot \frac{x^4}{1 - 3x^5} = \frac{10x^4}{1 - 3x^5}$

Example 4.5-4: Evaluate the following indefinite integrals:

a. $\int \dfrac{x+3}{x+1}\,dx \; =$ b. $\int \dfrac{x+6}{x+5}\,dx \; =$ c. $\int \dfrac{x+8}{x+4}\,dx \; =$

d. $\int a^{3x}\,dx \; =$ e. $\int a^{x^2} x\,dx \; =$ f. $\int a^{x^2+5} x\,dx \; =$

g. $\int a^{x^3} x^2\,dx \; =$ h. $\int \dfrac{a^{x^4} x^3\,dx}{2} \; =$ i. $\int 3a^{x^3+1} x^2\,dx \; =$

Solutions:

a. $\boxed{\int \dfrac{x+3}{x+1}\,dx} = \boxed{\int \dfrac{(x+1)+2}{x+1}\,dx} = \boxed{\int \dfrac{x+1}{x+1}\,dx + \int \dfrac{2}{x+1}\,dx} = \boxed{\int dx + \int \dfrac{2}{x+1}\,dx} = \boxed{x + 2\ln|x+1| + c}$

    Check: Let $y = x + 2\ln|x+1| + c$, then $y' = 1 + \dfrac{2}{x+1}\cdot 1 + 0 = \dfrac{x+1+2}{x+1} = \dfrac{x+3}{x+1}$

b. $\boxed{\int \dfrac{x+6}{x+5}\,dx} = \boxed{\int \dfrac{(x+5)+1}{x+5}\,dx} = \boxed{\int \dfrac{x+5}{x+5}\,dx + \int \dfrac{1}{x+5}\,dx} = \boxed{\int dx + \int \dfrac{1}{x+5}\,dx} = \boxed{x + \ln|x+5| + c}$

    Check: Let $y = x + \ln|x+5| + c$, then $y' = 1 + \dfrac{1}{x+5}\cdot 1 + 0 = \dfrac{x+5+1}{x+5} = \dfrac{x+6}{x+5}$

c. $\boxed{\int \dfrac{x+8}{x+4}\,dx} = \boxed{\int \dfrac{(x+4)+4}{x+4}\,dx} = \boxed{\int \dfrac{x+4}{x+4}\,dx + \int \dfrac{4}{x+4}\,dx} = \boxed{\int dx + \int \dfrac{4}{x+4}\,dx} = \boxed{x + 4\ln|x+4| + c}$

    Check: Let $y = x + 4\ln|x+4| + c$, then $y' = 1 + \dfrac{4}{x+4}\cdot 1 + 0 = \dfrac{x+4+4}{x+4} = \dfrac{x+8}{x+4}$

d. Given $\int a^{3x}\,dx$ let $u = 3x$, then $\dfrac{du}{dx} = \dfrac{d}{dx} 3x$; $\dfrac{du}{dx} = 3$; $du = 3dx$; $dx = \dfrac{du}{3}$. Therefore,

    $\boxed{\int a^{3x}\,dx} = \boxed{\int a^u \cdot \dfrac{du}{3}} = \boxed{\dfrac{1}{3}\int a^u\,du} = \boxed{\dfrac{1}{3}\dfrac{a^u}{\ln a} + c} = \boxed{\dfrac{1}{3}\dfrac{a^{3x}}{\ln a} + c}$

    Check: Let $y = \dfrac{1}{3}\dfrac{a^{3x}}{\ln a} + c$, then $y' = \dfrac{1}{3\ln a}\cdot a^{3x}\ln a \cdot 3 + 0 = \dfrac{3\ln a}{3\ln a}\cdot a^{3x} = a^{3x}$

Reminder: In Section 3.3 "Differentiation of Exponential and Logarithmic Functions" we learned that the derivative of $a^u = e^{u\ln a}$ is equal to the following:

$\dfrac{d}{dx}a^u = \dfrac{d}{dx}e^{u\ln a} = e^{u\ln a}\cdot\dfrac{d}{dx}u\ln a = e^{\ln a^u}\cdot\ln a\cdot\dfrac{du}{dx} = a^u\ln a\dfrac{du}{dx}$. Therefore, $\dfrac{d}{dx}a^u = a^u\ln a\dfrac{du}{dx}$

e. Given $\int a^{x^2} x\,dx$ let $u = x^2$, then $\dfrac{du}{dx} = \dfrac{d}{dx}x^2$; $\dfrac{du}{dx} = 2x$; $du = 2x\,dx$; $dx = \dfrac{du}{2x}$. Therefore,

    $\boxed{\int a^{x^2} x\,dx} = \boxed{\int a^u \cdot x \cdot \dfrac{du}{2x}} = \boxed{\dfrac{1}{2}\int a^u\,du} = \boxed{\dfrac{1}{2}\dfrac{a^u}{\ln a} + c} = \boxed{\dfrac{1}{2}\dfrac{a^{x^2}}{\ln a} + c}$

Check: Let $y = \dfrac{1}{2}\dfrac{a^{x^2}}{\ln a} + c$, then $y' = \dfrac{1}{2\ln a} \cdot a^{x^2} \ln a \cdot 2x + 0 = \dfrac{2\ln a}{2\ln a} \cdot a^{x^2} \cdot x = a^{x^2} x$

f. Given $\displaystyle\int a^{x^2+5} x \, dx$ let $u = x^2 + 5$, then $\dfrac{du}{dx} = \dfrac{d}{dx}\left(x^2 + 5\right)$; $\dfrac{du}{dx} = 2x$; $du = 2x \, dx$; $dx = \dfrac{du}{2x}$. Therefore,

$$\boxed{\int a^{x^2+5} x \, dx} = \boxed{\int a^u \cdot x \cdot \dfrac{du}{2x}} = \boxed{\dfrac{1}{2}\int a^u \, du} = \boxed{\dfrac{1}{2}\dfrac{a^u}{\ln a} + c} = \boxed{\dfrac{1}{2}\dfrac{a^{x^2+5}}{\ln a} + c}$$

Check: Let $y = \dfrac{1}{2}\dfrac{a^{x^2+5}}{\ln a} + c$, then $y' = \dfrac{1}{2\ln a} \cdot a^{x^2+5} \ln a \cdot 2x + 0 = \dfrac{2\ln a}{2\ln a} \cdot a^{x^2+5} \cdot x = a^{x^2+5} x$

g. Given $\displaystyle\int a^{x^3} x^2 \, dx$ let $u = x^3$, then $\dfrac{du}{dx} = \dfrac{d}{dx} x^3$; $\dfrac{du}{dx} = 3x^2$; $du = 3x^2 \, dx$; $dx = \dfrac{du}{3x^2}$. Therefore,

$$\boxed{\int a^{x^3} x^2 \, dx} = \boxed{\int a^u \cdot x^2 \cdot \dfrac{du}{3x^2}} = \boxed{\dfrac{1}{3}\int a^u \, du} = \boxed{\dfrac{1}{3}\dfrac{a^u}{\ln a} + c} = \boxed{\dfrac{1}{3}\dfrac{a^{x^3}}{\ln a} + c}$$

Check: Let $y = \dfrac{1}{3}\dfrac{a^{x^3}}{\ln a} + c$, then $y' = \dfrac{1}{3\ln a} \cdot a^{x^3} \ln a \cdot 3x^2 + 0 = \dfrac{3\ln a}{3\ln a} \cdot a^{x^3} \cdot x^2 = a^{x^3} x^2$

h. Given $\displaystyle\int \dfrac{a^{x^4} x^3 \, dx}{2}$ let $u = x^4$, then $\dfrac{du}{dx} = \dfrac{d}{dx} x^4$; $\dfrac{du}{dx} = 4x^3$; $du = 4x^3 \, dx$; $dx = \dfrac{du}{4x^3}$. Therefore,

$$\boxed{\int \dfrac{a^{x^4} x^3 \, dx}{2}} = \boxed{\int a^u \cdot x^3 \cdot \dfrac{du}{8x^3}} = \boxed{\dfrac{1}{8}\int a^u \, du} = \boxed{\dfrac{1}{8}\dfrac{a^u}{\ln a} + c} = \boxed{\dfrac{1}{8}\dfrac{a^{x^4}}{\ln a} + c}$$

Check: Let $y = \dfrac{1}{8}\dfrac{a^{x^4}}{\ln a} + c$, then $y' = \dfrac{1}{8\ln a} \cdot a^{x^4} \ln a \cdot 4x^3 + 0 = \dfrac{4\ln a}{8\ln a} \cdot a^{x^4} \cdot x^3 = \dfrac{1}{2} a^{x^4} x^3$

i. Given $\displaystyle\int 3a^{x^3+1} x^2 \, dx$ let $u = x^3 + 1$, then $\dfrac{du}{dx} = \dfrac{d}{dx}\left(x^3 + 1\right)$; $\dfrac{du}{dx} = 3x^2$; $du = 3x^2 \, dx$; $dx = \dfrac{du}{3x^2}$. Thus,

$$\boxed{\int 3a^{x^3+1} x^2 \, dx} = \boxed{3\int a^u \cdot x^2 \cdot \dfrac{du}{3x^2}} = \boxed{\dfrac{3}{3}\int a^u \, du} = \boxed{\dfrac{a^u}{\ln a} + c} = \boxed{\dfrac{a^{x^3+1}}{\ln a} + c}$$

Check: Let $y = \dfrac{a^{x^3+1}}{\ln a} + c$, then $y' = \dfrac{1}{\ln a} \cdot a^{x^3+1} \ln a \cdot 3x^2 + 0 = \dfrac{\ln a}{\ln a} \cdot a^{x^3+1} \cdot 3x^2 = 3a^{x^3+1} x^2$

Example 4.5-5: Evaluate the following indefinite integrals:

a. $\displaystyle\int a^{x^2} 2x \, dx =$

b. $\displaystyle\int \left(a^{x+1} + e^{x+1}\right) dx =$

c. $\displaystyle\int \left(a^{2x+3} + 3\right) dx =$

d. $\displaystyle\int 5a^{5x+3} \, dx =$

e. $\displaystyle\int \left(e^x + 1\right)^5 e^x dx =$

f. $\displaystyle\int \dfrac{x+2}{x+1} \, dx =$

Solutions:

a. Given $\int a^{x^2} 2x\, dx$ let $u = x^2$, then $\dfrac{du}{dx} = \dfrac{d}{dx}x^2$; $\dfrac{du}{dx} = 2x$; $du = 2x\, dx$; $dx = \dfrac{du}{2x}$. Thus,

$$\boxed{\int a^{x^2} 2x\, dx} = \boxed{\int a^u \cdot 2x \cdot \dfrac{du}{2x}} = \boxed{\int a^u\, du} = \boxed{\dfrac{a^u}{\ln a} + c} = \boxed{\dfrac{a^{x^2}}{\ln a} + c}$$

Check: Let $y = \dfrac{a^{x^2}}{\ln a} + c$, then $y' = \dfrac{1}{\ln a} \cdot a^{x^2} \ln a \cdot 2x + 0 = \dfrac{\ln a}{\ln a} \cdot a^{x^2} \cdot 2x = a^{x^2} 2x$

b. Given $\int \left(a^{x+1} + e^{x+1}\right) dx = \int a^{x+1} dx + \int e^{x+1} dx$ let $u = x+1$, then $\dfrac{du}{dx} = \dfrac{d}{dx}(x+1)$; $\dfrac{du}{dx} = 1$; $dx = du$. Thus,

$$\boxed{\int a^{x+1} dx + \int e^{x+1} dx} = \boxed{\int a^u\, du + \int e^u\, du} = \boxed{\dfrac{a^u}{\ln a} + e^u + c} = \boxed{\dfrac{a^{x+1}}{\ln a} + e^{x+1} + c}$$

Check: Let $y = \dfrac{a^{x+1}}{\ln a} + e^{x+1} + c$, then $y' = \dfrac{1}{\ln a} \cdot a^{x+1} \ln a \cdot 1 + e^{x+1} \cdot 1 + 0 = \dfrac{\ln a}{\ln a} \cdot a^{x+1} + e^{x+1} = a^{x+1} + e^{x+1}$

c. Given $\int \left(a^{2x+3} + 3\right) dx = \int a^{2x+3} dx + \int 3\, dx$ let $u = 2x+3$, then $\dfrac{du}{dx} = \dfrac{d}{dx}(2x+3)$; $\dfrac{du}{dx} = 2$; $dx = \dfrac{du}{2}$. Thus,

$$\boxed{\int a^{2x+3} dx + \int 3\, dx} = \boxed{\int a^u \cdot \dfrac{du}{2} + 3x} = \boxed{\dfrac{1}{2}\int a^u\, du + 3x} = \boxed{\dfrac{1}{2}\dfrac{a^u}{\ln a} + 3x + c} = \boxed{\dfrac{a^{2x+3}}{2\ln a} + 3x + c}$$

Check: Let $y = \dfrac{a^{2x+3}}{2\ln a} + 3x + c$, then $y' = \dfrac{1}{2\ln a} \cdot a^{2x+3} \ln a \cdot 2 + 3 + 0 = \dfrac{2\ln a}{2\ln a} \cdot a^{2x+3} + 3 = a^{2x+3} + 3$

d. Given $\int 5a^{5x+3}\, dx$ let $u = 5x+3$, then $\dfrac{du}{dx} = \dfrac{d}{dx}(5x+3)$; $\dfrac{du}{dx} = 5$; $du = 5\, dx$; $dx = \dfrac{du}{5}$. Thus,

$$\boxed{\int 5a^{5x+3}\, dx} = \boxed{5\int a^u \cdot \dfrac{du}{5} + 3x} = \boxed{\dfrac{5}{5}\int a^u\, du} = \boxed{\int a^u\, du} = \boxed{\dfrac{a^u}{\ln a} + c} = \boxed{\dfrac{a^{5x+3}}{\ln a} + c}$$

Check: Let $y = \dfrac{a^{5x+3}}{\ln a} + c$, then $y' = \dfrac{1}{\ln a} \cdot a^{5x+3} \ln a \cdot 5 + 0 = \dfrac{5\ln a}{\ln a} \cdot a^{5x+3} = 5a^{5x+3}$

e. Given $\int \left(e^x + 1\right)^5 e^x dx$ let $u = e^x + 1$, then $\dfrac{du}{dx} = \dfrac{d}{dx}\left(e^x + 1\right)$; $\dfrac{du}{dx} = e^x$; $du = e^x dx$; $dx = \dfrac{du}{e^x}$. Therefore,

$$\boxed{\int \left(e^x + 1\right)^5 e^x dx} = \boxed{\int u^5 \cdot e^x \cdot \dfrac{du}{e^x}} = \boxed{\int u^5\, du} = \boxed{\dfrac{1}{6}u^6 + c} = \boxed{\dfrac{1}{6}\left(e^x + 1\right)^6 + c}$$

Check: Let $y = \frac{1}{6}\left(e^x + 1\right)^6 + c$, then $y' = \frac{1}{6} \cdot 6 \left(e^x + 1\right)^5 \cdot e^x + 0 = \frac{6}{6}\left(e^x + 1\right)^5 e^x = \left(e^x + 1\right)^5 e^x$

f. $\boxed{\int \frac{x+2}{x+1}\,dx} = \boxed{\int \frac{(x+1)+1}{x+1}\,dx} = \boxed{\int \frac{x+1}{x+1}\,dx + \int \frac{1}{x+1}\,dx} = \boxed{\int dx + \int \frac{1}{x+1}\,dx} = \boxed{x + \ln|x+1| + c}$

Check: Let $y = x + \ln|x+1| + c$, then $y' = 1 + \frac{1}{x+1} \cdot 1 + 0 = \frac{x+1+1}{x+1} = \frac{x+2}{x+1}$

In the following examples we will solve problems using the following two formulas:

$$\int \frac{1}{x^2 - a^2}\,dx = \frac{1}{2a} \ln \left| \frac{x-a}{x+a} \right| + c$$

$$\int \frac{1}{a^2 - x^2}\,dx = \frac{1}{2a} \ln \left| \frac{a+x}{a-x} \right| + c$$

Example 4.5-6: Evaluate the following indefinite integrals:

a. $\int \frac{1}{x^2 - 4}\,dx =$

b. $\int \frac{1}{36x^2 - 25}\,dx =$

c. $\int \frac{1}{(x-1)^2 - 16}\,dx =$

d. $\int \frac{1}{x^2 + 10x + 24}\,dx =$

e. $\int \frac{1}{x^2 - 4x + 3}\,dx =$

f. $\int \frac{1}{x^2 - 4x + 1}\,dx =$

g. $\int \frac{1}{x^2 + 18x + 75}\,dx =$

h. $\int \frac{1}{16 - 9x^2}\,dx =$

i. $\int \frac{1}{4 - (x-1)^2}\,dx =$

j. $\int \frac{1}{16 - (t+3)^2}\,dt =$

k. $\int \frac{2x+3}{x^2 + 3x}\,dx + \int \frac{1}{36 - x^2}\,dx =$

l. $\int \frac{dx}{x^2 - 5} + \int \frac{3\,dx}{4 - x^2} =$

Solutions:

a. $\boxed{\int \frac{1}{x^2 - 4}\,dx} = \boxed{\int \frac{1}{x^2 - 2^2}\,dx} = \boxed{\frac{1}{2 \cdot 2} \ln \left| \frac{x-2}{x+2} \right| + c} = \boxed{\frac{1}{4} \ln \left| \frac{x-2}{x+2} \right| + c}$

Check: Let $y = \frac{1}{4} \ln \left| \frac{x-2}{x+2} \right| + c$ then $y' = \frac{1}{4} \cdot \frac{1}{\frac{x-2}{x+2}} \cdot \frac{1 \cdot (x+2) - 1 \cdot (x-2)}{(x+2)^2} + 0 = \frac{1}{4} \cdot \frac{x+2}{x-2} \cdot \frac{x+2-x+2}{(x+2)^2}$

$= \frac{1}{4} \cdot \frac{1}{x-2} \cdot \frac{4}{x+2} = \frac{1}{(x-2)(x+2)} = \frac{1}{x^2 + 2x - 2x - 4} = \frac{1}{x^2 - 4}$

b. $\boxed{\int \frac{1}{36x^2 - 25}\,dx} = \boxed{\int \frac{1}{36\left(x^2 - \frac{25}{36}\right)}\,dx} = \boxed{\frac{1}{36} \int \frac{1}{x^2 - \left(\frac{5}{6}\right)^2}\,dx} = \boxed{\frac{1}{36} \cdot \frac{1}{2 \cdot \frac{5}{6}} \ln \left| \frac{x - \frac{5}{6}}{x + \frac{5}{6}} \right| + c} = \boxed{\frac{1}{60} \ln \left| \frac{6x-5}{6x+5} \right| + c}$

Check: Let $y = \frac{1}{60} \ln \left| \frac{6x-5}{6x+5} \right| + c$ then $y' = \frac{1}{60} \cdot \frac{1}{\frac{6x-5}{6x+5}} \cdot \frac{6 \cdot (6x+5) - 6 \cdot (6x-5)}{(6x+5)^2} + 0 = \frac{1}{60} \cdot \frac{6x+5}{6x-5}$

$$\times \frac{36x+30-36x+30}{(6x+5)^2} = \frac{1}{60} \cdot \frac{1}{6x-5} \cdot \frac{60}{6x+5} = \frac{1}{(6x-5)(6x+5)} = \frac{1}{36x^2+30x-30x-25} = \frac{1}{36x^2-25}$$

c. $\boxed{\int \frac{1}{(x-1)^2-16}\,dx} = \boxed{\int \frac{1}{(x-1)^2-4^2}\,dx} = \boxed{\frac{1}{2\cdot 4}\ln\left|\frac{(x-1)-4}{(x-1)+4}\right|+c} = \boxed{\frac{1}{8}\ln\left|\frac{x-5}{x+3}\right|+c}$

Check: Let $y = \frac{1}{8}\ln\left|\frac{x-5}{x+3}\right|+c$ then $y' = \frac{1}{8} \cdot \frac{1}{\frac{x-5}{x+3}} \cdot \frac{1\cdot(x+3)-1\cdot(x-5)}{(x+3)^2}+0 = \frac{1}{8} \cdot \frac{x+3}{x-5} \cdot \frac{x+3-x+5}{(x+3)^2}$

$$= \frac{1}{8} \cdot \frac{1}{x-5} \cdot \frac{8}{x+3} = \frac{1}{(x-5)(x+3)} = \frac{1}{x^2+3x-5x-15} = \frac{1}{x^2-2x-15} = \frac{1}{(x-1)^2-16}$$

d. $\boxed{\int \frac{1}{x^2+10x+24}\,dx} = \boxed{\int \frac{1}{(x^2+10x+25)-1}\,dx} = \boxed{\int \frac{1}{(x+5)^2-1}\,dx} = \boxed{\frac{1}{2\cdot 1}\ln\left|\frac{(x+5)-1}{(x+5)+1}\right|+c} = \boxed{\frac{1}{2}\ln\left|\frac{x+4}{x+6}\right|+c}$

Check: Let $y = \frac{1}{2}\ln\left|\frac{x+4}{x+6}\right|+c$ then $y' = \frac{1}{2} \cdot \frac{1}{\frac{x+4}{x+6}} \cdot \frac{1\cdot(x+6)-1\cdot(x+4)}{(x+6)^2}+0 = \frac{1}{2} \cdot \frac{x+6}{x+4} \cdot \frac{x+6-x-4}{(x+6)^2}$

$$= \frac{1}{2} \cdot \frac{1}{x+4} \cdot \frac{2}{x+6} = \frac{1}{(x+4)(x+6)} = \frac{1}{x^2+6x+4x+24} = \frac{1}{x^2+10x+24}$$

e. $\boxed{\int \frac{1}{x^2-4x+3}\,dx} = \boxed{\int \frac{1}{(x^2-4x+4)-1}\,dx} = \boxed{\int \frac{1}{(x-2)^2-1}\,dx} = \boxed{\frac{1}{2\cdot 1}\ln\left|\frac{(x-2)-1}{(x-2)+1}\right|+c} = \boxed{\frac{1}{2}\ln\left|\frac{x-3}{x-1}\right|+c}$

Check: Let $y = \frac{1}{2}\ln\left|\frac{x-3}{x-1}\right|+c$ then $y' = \frac{1}{2} \cdot \frac{1}{\frac{x-3}{x-1}} \cdot \frac{1\cdot(x-1)-1\cdot(x-3)}{(x-1)^2}+0 = \frac{1}{2} \cdot \frac{x-1}{x-3} \cdot \frac{x-1-x+3}{(x-1)^2}$

$$= \frac{1}{2} \cdot \frac{1}{x-3} \cdot \frac{2}{x-1} = \frac{1}{(x-3)(x-1)} = \frac{1}{x^2-x-3x+3} = \frac{1}{x^2-4x+3}$$

f. $\boxed{\int \frac{1}{x^2-4x+1}\,dx} = \boxed{\int \frac{1}{(x^2-4x+4)-3}\,dx} = \boxed{\int \frac{1}{(x-2)^2-(\sqrt{3})^2}\,dx} = \boxed{\frac{1}{2\sqrt{3}}\ln\left|\frac{(x-2)-\sqrt{3}}{(x-2)+\sqrt{3}}\right|+c}$

Check: Let $y = \frac{1}{2\sqrt{3}}\ln\left|\frac{x-2-\sqrt{3}}{x-2+\sqrt{3}}\right|+c$ then $y' = \frac{1}{3.46} \cdot \frac{1}{\frac{x-3.73}{x-0.27}} \cdot \frac{1\cdot(x-0.27)-1\cdot(x-3.73)}{(x-0.27)^2}+0 = \frac{1}{3.46}$

$$\times \frac{x-0.27}{x-3.73} \cdot \frac{x-0.27-x+3.73}{(x-0.27)^2} = \frac{1}{3.46} \cdot \frac{1}{x-3.73} \cdot \frac{3.46}{x-0.27} = \frac{3.46}{3.46(x-3.73)(x-0.27)}$$

$$= \frac{1}{x^2-0.27x-3.73x+1} = \frac{1}{x^2-4x+1}$$

g. $\boxed{\int \frac{1}{x^2+18x+75}\,dx} = \boxed{\int \frac{1}{(x^2+18x+81)-6}\,dx} = \boxed{\int \frac{1}{(x+9)^2-(\sqrt{6})^2}\,dx} = \boxed{\frac{1}{2\cdot\sqrt{6}}\ln\left|\frac{(x+9)-\sqrt{6}}{(x+9)+\sqrt{6}}\right|+c}$

$$= \boxed{\frac{1}{2\cdot 2.45}\ln\left|\frac{(x+9)-2.45}{(x+9)+2.45}\right|+c} = \boxed{\frac{1}{4.9}\ln\left|\frac{x+6.55}{x+11.45}\right|+c}$$

Check: Let $y = \dfrac{1}{4.9} \ln \left| \dfrac{x+6.55}{x+11.45} \right| + c$ then $y' = \dfrac{1}{4.9} \cdot \dfrac{1}{\frac{x+6.55}{x+11.45}} \cdot \dfrac{1 \cdot (x+11.45) - 1 \cdot (x+6.55)}{(x+11.45)^2} + 0 = \dfrac{1}{4.9} \cdot \dfrac{x+11.45}{x+6.55}$

$\times \dfrac{x+11.45-x-6.55}{(x+11.45)^2} = \dfrac{1}{4.9} \cdot \dfrac{1}{x+6.55} \cdot \dfrac{4.9}{x+11.45} = \dfrac{1}{(x+6.55)(x+11.45)} = \dfrac{1}{x^2+18x+75}$

h. $\boxed{\displaystyle\int \dfrac{1}{16-9x^2}\,dx} = \boxed{\displaystyle\int \dfrac{1}{9\left(\frac{16}{9}-x^2\right)}\,dx} = \boxed{\dfrac{1}{9}\displaystyle\int \dfrac{1}{\frac{16}{9}-x^2}\,dx} = \boxed{\dfrac{1}{9}\displaystyle\int \dfrac{1}{\left(\frac{4}{3}\right)^2-x^2}\,dx} = \boxed{\dfrac{1}{9}\cdot\dfrac{1}{2\cdot\frac{4}{3}}\ln\left|\dfrac{\frac{4}{3}+x}{\frac{4}{3}-x}\right|+c}$

$= \boxed{\dfrac{1}{9}\cdot\dfrac{3}{8}\ln\left|\dfrac{\frac{4+3x}{3}}{\frac{4-3x}{3}}\right|+c} = \boxed{\dfrac{3}{72}\ln\left|\dfrac{3(4+3x)}{3(4-3x)}\right|+c} = \boxed{\dfrac{1}{24}\ln\left|\dfrac{4+3x}{4-3x}\right|+c}$

Check: Let $y = \dfrac{1}{24}\ln\left|\dfrac{4+3x}{4-3x}\right|+c$ then $y' = \dfrac{1}{24}\cdot\dfrac{1}{\frac{4+3x}{4-3x}}\cdot\dfrac{3\cdot(4-3x)+3\cdot(4+3x)}{(4-3x)^2}+0 = \dfrac{1}{24}\cdot\dfrac{4-3x}{4+3x}$

$\times \dfrac{12-9x+12+9x}{(4-3x)^2} = \dfrac{1}{24}\cdot\dfrac{1}{4+3x}\cdot\dfrac{24}{4-3x} = \dfrac{1}{(4+3x)(4-3x)} = \dfrac{1}{16-12x+12x-9x^2} = \dfrac{1}{16-9x^2}$

i. $\boxed{\displaystyle\int \dfrac{1}{4-(x-1)^2}\,dx} = \boxed{\displaystyle\int \dfrac{1}{2^2-(x-1)^2}\,dx} = \boxed{\dfrac{1}{2\cdot2}\ln\left|\dfrac{2+(x-1)}{2-(x-1)}\right|+c} = \boxed{\dfrac{1}{4}\ln\left|\dfrac{2+x-1}{2-x+1}\right|+c} = \boxed{\dfrac{1}{4}\ln\left|\dfrac{1+x}{3-x}\right|+c}$

Check: Let $y = \dfrac{1}{4}\ln\left|\dfrac{1+x}{3-x}\right|+c$ then $y' = \dfrac{1}{4}\cdot\dfrac{1}{\frac{1+x}{3-x}}\cdot\dfrac{1\cdot(3-x)+1\cdot(1+x)}{(3-x)^2}+0 = \dfrac{1}{4}\cdot\dfrac{3-x}{1+x}\cdot\dfrac{3-x+1+x}{(3-x)^2}$

$= \dfrac{1}{4}\cdot\dfrac{1}{1+x}\cdot\dfrac{4}{3-x} = \dfrac{1}{(1+x)(3-x)} = \dfrac{1}{3-x+3x-x^2} = \dfrac{1}{3+2x-x^2} = \dfrac{1}{4-(x-1)^2}$

j. $\boxed{\displaystyle\int \dfrac{1}{16-(t+3)^2}\,dt} = \boxed{\displaystyle\int \dfrac{1}{4^2-(t+3)^2}\,dx} = \boxed{\dfrac{1}{2\cdot4}\ln\left|\dfrac{4+(t+3)}{4-(t+3)}\right|+c} = \boxed{\dfrac{1}{8}\ln\left|\dfrac{4+t+3}{4-t-3}\right|+c} = \boxed{\dfrac{1}{8}\ln\left|\dfrac{7+t}{1-t}\right|+c}$

Check: Let $y = \dfrac{1}{8}\ln\left|\dfrac{7+t}{1-t}\right|+c$ then $y' = \dfrac{1}{8}\cdot\dfrac{1}{\frac{7+t}{1-t}}\cdot\dfrac{1\cdot(1-t)+1\cdot(7+t)}{(1-t)^2}+0 = \dfrac{1}{8}\cdot\dfrac{1-t}{7+t}\cdot\dfrac{1-t+7+t}{(1-t)^2}$

$= \dfrac{1}{8}\cdot\dfrac{1}{7+t}\cdot\dfrac{8}{1-t} = \dfrac{1}{(7+t)(1-t)} = \dfrac{1}{7-7t+t-t^2} = \dfrac{1}{7-6t-t^2} = \dfrac{1}{16-(t+3)^2}$

k. $\boxed{\displaystyle\int \dfrac{2x+3}{x^2+3x}\,dx + \displaystyle\int \dfrac{1}{36-x^2}\,dx} = \boxed{\displaystyle\int \dfrac{2x+3}{u}\dfrac{du}{2x+3} + \displaystyle\int \dfrac{dx}{6^2-x^2}} = \boxed{\displaystyle\int \dfrac{du}{u} + \displaystyle\int \dfrac{dx}{6^2-x^2}} = \boxed{\displaystyle\int \dfrac{du}{u} + \dfrac{1}{2\cdot6}\ln\left|\dfrac{6+x}{6-x}\right|}$

$= \boxed{\ln|u| + \dfrac{1}{12}\ln\left|\dfrac{6+x}{6-x}\right|+c} = \boxed{\ln\left|x^2+3x\right| + \dfrac{1}{12}\ln\left|\dfrac{6+x}{6-x}\right|+c}$

Check: Let $y_1 = \ln\left|x^2+3x\right|+c$ then $y_1' = \dfrac{2x+3}{x^2+3x}+0 = \dfrac{2x+3}{x^2+3x}$. Let $y_2 = \dfrac{1}{12}\ln\left|\dfrac{6+x}{6-x}\right|+c$

then $y_2' = \dfrac{1}{12} \cdot \dfrac{1}{\frac{6+x}{6-x}} \cdot \dfrac{1\cdot(6+x)+1\cdot(6-x)}{(6-x)^2} + 0 = \dfrac{1}{12} \cdot \dfrac{6-x}{6+x} \cdot \dfrac{6+x+6-x}{(6-x)^2} = \dfrac{1}{12} \cdot \dfrac{1}{6+x} \cdot \dfrac{12}{6-x}$

$= \dfrac{1}{(6+x)(6-x)} = \dfrac{1}{36-6x+6x-x^2} = \dfrac{1}{36-x^2}$

l. $\boxed{\displaystyle\int \dfrac{dx}{x^2-5} + \int \dfrac{3\,dx}{4-x^2}} = \boxed{\displaystyle\int \dfrac{dx}{x^2-\left(\sqrt{5}\right)^2} + \int \dfrac{3\,dx}{2^2-x^2}} = \boxed{\dfrac{1}{2\sqrt{5}} \ln\left|\dfrac{x-\sqrt{5}}{x+\sqrt{5}}\right| + \dfrac{3}{4}\ln\left|\dfrac{2+x}{2-x}\right| + c}$

Check: Let $y_1 = \dfrac{1}{2\cdot\sqrt{5}} \ln\left|\dfrac{x-\sqrt{5}}{x+\sqrt{5}}\right| + c$ then $y_1' = \dfrac{1}{2\sqrt{5}} \cdot \dfrac{1}{\frac{x-\sqrt{5}}{x+\sqrt{5}}} \cdot \dfrac{1\cdot(x+\sqrt{5})-1\cdot(x-\sqrt{5})}{\left(x+\sqrt{5}\right)^2} + 0 = \dfrac{1}{2\sqrt{5}} \cdot \dfrac{x+\sqrt{5}}{x-\sqrt{5}}$

$\times \dfrac{x+\sqrt{5}-x+\sqrt{5}}{\left(x+\sqrt{5}\right)^2} = \dfrac{1}{2\sqrt{5}} \cdot \dfrac{1}{x-\sqrt{5}} \cdot \dfrac{2\sqrt{5}}{x+\sqrt{5}} = \dfrac{1}{\left(x-\sqrt{5}\right)\left(x+\sqrt{5}\right)} = \dfrac{1}{x^2+\sqrt{5}x-\sqrt{5}x-5} = \dfrac{1}{x^2-5}$.

Let $y_2 = \dfrac{3}{4} \ln\left|\dfrac{2+x}{2-x}\right| + c$ then $y_2' = \dfrac{3}{4} \cdot \dfrac{1}{\frac{2+x}{2-x}} \cdot \dfrac{1\cdot(2+x)+1\cdot(2-x)}{(2-x)^2} + 0 = \dfrac{3}{4} \cdot \dfrac{2-x}{2+x} \cdot \dfrac{2+x+2-x}{(2-x)^2}$

$= \dfrac{3}{4} \cdot \dfrac{1}{2+x} \cdot \dfrac{4}{2-x} = \dfrac{3}{(2+x)(2-x)} = \dfrac{3}{4-2x+2x-x^2} = \dfrac{3}{4-x^2}$

Section 4.5 Practice Problems – Integration of Expressions Resulting in Exponential or Logarithmic Functions

1. Evaluate the following indefinite integrals.

a. $\displaystyle\int \dfrac{dx}{2x+1} =$

b. $\displaystyle\int \dfrac{x}{x^2-a}\,dx =$

c. $\displaystyle\int \dfrac{x^3}{x^4-1}\,dx =$

d. $\displaystyle\int\left(\dfrac{1}{x+3}+\dfrac{3}{x-5}\right)dx =$

e. $\displaystyle\int xe^{3x^2}\,dx =$

f. $\displaystyle\int \dfrac{e^{5x}}{1-e^{5x}}\,dx =$

g. $\displaystyle\int 3e^{-ax}\,dx =$

h. $\displaystyle\int\left(x^3+e^{2x}-e^{-5x}\right)dx =$

i. $\displaystyle\int \dfrac{e^{\frac{1}{x^3}}}{5x^4}\,dx =$

2. Evaluate the following indefinite integrals.

a. $\displaystyle\int\left(3e^{5x}+5\right)e^{5x}\,dx =$

b. $\displaystyle\int\left(e^x-1\right)^5 e^x\,dx =$

c. $\displaystyle\int \dfrac{x^2}{1+x^3}\,dx =$

d. $\displaystyle\int \dfrac{x^4}{1+x^5}\,dx =$

e. $\displaystyle\int \dfrac{x+7}{x+6}\,dx =$

f. $\displaystyle\int \dfrac{x+9}{x+5}\,dx =$

g. $\displaystyle\int a^{x^2+k}x\,dx =$

h. $\displaystyle\int \dfrac{2}{3}a^{x^3+5}x^2\,dx =$

i. $\displaystyle\int\left(e^{2x}+3\right)^3 e^{2x}\,dx =$

j. $\displaystyle\int \dfrac{1}{(x+1)^2-25}\,dx =$

k. $\displaystyle\int \dfrac{1}{x^2+6x+8}\,dx =$

l. $\displaystyle\int \dfrac{1}{9-(x-1)^2}\,dx =$

Chapter 5
Integration (Part II)

Quick Reference to Chapter 5 Problems

$$\boxed{\int e^{2x}\cos 2x\, dx} = ; \quad \boxed{\int e^{x}\sin x\, dx} = ; \quad \boxed{\int x\cos 3x\, dx} =$$

$$\boxed{\int \frac{x^2\, dx}{\sqrt{36-x^2}}} = ; \quad \boxed{\int \frac{dx}{\left(9+x^2\right)^2}} = ; \quad \boxed{\int \frac{dx}{x^4\sqrt{x^2-1}}} =$$

Case I - The Denominator Has Distinct Linear Factors *320*

$$\boxed{\int \frac{x+1}{x(x-2)(x+3)}\, dx} = ; \quad \boxed{\int \frac{1}{(x+1)(x+2)}\, dx} = ; \quad \boxed{\int \frac{x^2+1}{x(x-1)(x+1)}\, dx} =$$

Case II - The Denominator Has Repeated Linear Factors *327*

$$\boxed{\int \frac{x+3}{x(x-1)^2}\, dx} = ; \quad \boxed{\int \frac{1}{x^2(x-1)}\, dx} = ; \quad \boxed{\int \frac{5}{x(x-1)^2}\, dx} =$$

Case III - The Denominator Has Distinct Quadratic Factors *334*

$$\boxed{\int \frac{x^2-x+3}{x(x^2+1)}\, dx} = ; \quad \boxed{\int \frac{1}{x(x^2+25)}\, dx} = ; \quad \boxed{\int \frac{1}{x^2(x^2+16)}\, dx} =$$

Case IV - The Denominator Has Repeated Quadratic Factors *344*

$$\boxed{\int \frac{x^2}{\left(x^2+1\right)^2}\, dx} = ; \quad \boxed{\int \frac{x^2+1}{\left(x^2+4\right)^2}\, dx} = ; \quad \boxed{\int \frac{x^3}{\left(x^2+2\right)^2}\, dx} =$$

$$\boxed{\int \cosh \frac{1}{5}x\, dx} = ; \quad \boxed{\int (\sinh 4x + \cosh 2x)\, dx} = ; \quad \boxed{\int x^2\, \csc h^2 x^3\, dx} =$$

Chapter 5 – Integration (Part II)

The objective of this chapter is to improve the student's ability to solve additional problems involving integration. A method used to integrate functions, known as Integration by Parts, is addressed in Section 5.1. Integration of functions using the Trigonometric Substitution method is discussed in Section 5.2. Integration of functions using the Partial Fractions technique is addressed in Section 5.3. Four different cases, depending on the denominator having distinct linear factors, repeated linear factors, distinct quadratic factors, or repeated quadratic factors, are addressed in this section. Finally, integration of hyperbolic functions is discussed in Section 5.4. Each section is concluded by solving examples with practice problems to further enhance the student's ability.

5.1   Integration by Parts

Integration by parts is a technique for replacing hard to integrate integrals by ones that are easier to integrate. This technique applies mainly to integrals that are in the form of $\int f(x)g(x)\,dx$ where in most cases $f(x)$ can be differentiated several times to become zero and $g(x)$ can be integrated several times without difficulty. For example, given the integral $\int x^2 e^{3x}\,dx$ the function $f(x)=x^2$ can be differentiated three times to become zero and the function $g(x)=e^{3x}$ can be integrated several times easily. On the other hand, integrals such as $\int e^{2x}\cos 2x\,dx$ and $\int e^{-3x}\sin 3x\,dx$ do not fall under the category described above. In this section we will learn how to apply Integration by Parts method in solving various integrals. The formula for integration by parts comes from the product rule, i.e.,

$$\frac{d}{dx}(uv)=u\frac{dv}{dx}+v\frac{du}{dx}$$

where u and v are differentiable functions of x, multiplying both sides of the equation by dx we obtain

$$d(uv)=u\,dv+v\,du$$

rearranging the terms we then have

$$u\,dv=d(uv)-v\,du$$

integrating both sides of the equation we obtain

$$\int u\,dv=uv-\int v\,du$$

The above formula is referred to as the Integration by Parts Formula. Note that in using the above equality we must first select dv such that it is easily integrable and second ensure that $\int u\,dv$ is easier to evaluate than $\int u\,dv$. In the following examples we will solve problems using the Integration by Parts method.

Example 5.1-1: Evaluate the following indefinite integrals:

a. $\int x\,e^x dx =$

b. $\int x^2 e^{3x}\,dx =$

c. $\int x\,e^{-2x}\,dx =$

d. $\int \ln x\,dx =$

e. $\int x\ln x\,dx =$

f. $\int x^3 \ln x\,dx =$

g. $\int e^{2x} \cos 2x \, dx =$ h. $\int e^x \sin x \, dx =$ i. $\int x \cos 3x \, dx =$

Solutions:

a. Given $\int xe^x \, dx$ let $u = x$ and $dv = e^x dx$ then $du = dx$ and $\int dv = \int e^x dx$ which implies $v = e^x$.

Using the integration by parts formula $\int u \, dv = uv - \int v \, du$ we obtain

$$\boxed{\int xe^x \, dx} = \boxed{xe^x - \int e^x dx} = \boxed{xe^x - e^x + c} = \boxed{e^x(x-1) + c}$$

Check: Let $y = e^x(x-1) + c$, then $y' = e^x \cdot (x-1) + 1 \cdot e^x + 0 = xe^x - e^x + e^x = xe^x - e^x + e^x = \boldsymbol{xe^x}$

b. Given $\int x^2 e^{3x} \, dx$ let $u = x^2$ and $dv = e^{3x} dx$ then $du = 2x \, dx$ and $\int dv = \int e^{3x} dx$ which implies $v = \frac{1}{3} e^{3x}$.

Using the integration by parts formula $\int u \, dv = uv - \int v \, du$ we obtain

$$\int x^2 e^{3x} \, dx = x^2 \cdot \frac{1}{3} e^{3x} - \frac{1}{3} \int e^{3x} \cdot 2x \, dx = \frac{x^2 e^{3x}}{3} - \frac{2}{3} \int x e^{3x} \, dx \qquad (1)$$

To integrate $\int x e^{3x} \, dx$ let $u = x$ and $dv = e^{3x} dx$ then $du = dx$ and $\int dv = \int e^{3x} dx$ which implies

$v = \frac{1}{3} e^{3x}$. Therefore, $\int x e^{3x} \, dx = x \cdot \frac{1}{3} e^{3x} - \int \frac{1}{3} e^{3x} dx = \frac{xe^{3x}}{3} - \frac{e^{3x}}{9} + c \qquad (2)$

Combining equations (1) and (2) together we obtain:

$$\boxed{\int x^2 e^{3x} \, dx} = \boxed{\frac{x^2 e^{3x}}{3} - \frac{2}{3} \int x e^{3x} \, dx} = \boxed{\frac{x^2 e^{3x}}{3} - \frac{2}{3} \left(\frac{xe^{3x}}{3} - \frac{e^{3x}}{9} + c \right)} = \boxed{\frac{1}{3} x^2 e^{3x} - \frac{2}{9} xe^{3x} + \frac{2}{27} e^{3x} + c}$$

Check: Let $y = \frac{1}{3} x^2 e^{3x} - \frac{2}{9} xe^{3x} + \frac{2}{27} e^{3x} + c$, then $y' = \frac{1}{3} \left(2x \cdot e^{3x} + 3e^{3x} \cdot x^2 \right) - \frac{2}{9} \left(1 \cdot e^{3x} + 3e^{3x} \cdot x \right)$

$+ \frac{2}{27} \cdot 3e^{3x} + 0 = \frac{2}{3} xe^{3x} + x^2 e^{3x} - \frac{2}{9} e^{3x} - \frac{2}{3} xe^{3x} + \frac{2}{9} e^{3x} = x^2 e^{3x}$

c. Given $\int x e^{-2x} \, dx$ let $u = x$ and $dv = e^{-2x} dx$ then $du = dx$ and $\int dv = \int e^{-2x} dx$ which implies

$v = -\frac{1}{2} e^{-2x}$. Using the integration by parts formula $\int u \, dv = uv - \int v \, du$ we obtain

$$\boxed{\int x e^{-2x} \, dx} = \boxed{x \cdot -\frac{1}{2} e^{-2x} + \frac{1}{2} \int e^{-2x} dx} = \boxed{-\frac{1}{2} x e^{-2x} + \frac{1}{2} \int e^{-2x} dx} = \boxed{-\frac{1}{2} x e^{-2x} - \frac{1}{4} e^{-2x} + c}$$

Check: Let $y = -\frac{1}{2} x e^{-2x} - \frac{1}{4} e^{-2x} + c$, then $y' = -\frac{1}{2} \left(1 \cdot e^{-2x} - 2e^{-2x} \cdot x \right) - \frac{1}{4} \cdot -2e^{-2x} + 0 = -\frac{1}{2} e^{-2x}$

$xe^{-2x} + \frac{1}{2} e^{-2x} = xe^{-2x}$

d. Given $\int \ln x \, dx$ let $u = \ln x$ and $dv = dx$ then $du = \dfrac{1}{x} dx$ and $\int dv = \int dx$ which implies $v = x$.

Using the integration by parts formula $\int u \, dv = u \, v - \int v \, du$ we obtain

$$\boxed{\int \ln x \, dx} = \boxed{\ln x \cdot x - \int x \cdot \frac{1}{x} dx} = \boxed{x \ln x - \int dx} = \boxed{x \ln x - x + c}$$

Check: Let $y = x \ln x - x + c$, then $y' = \left(1 \cdot \ln x + \dfrac{1}{x} \cdot x \right) - 1 + 0 = \ln x + 1 - 1 = \ln x$

e. Given $\int x \ln x \, dx$ let $u = \ln x$ and $dv = x \, dx$ then $du = \dfrac{1}{x} dx$ and $\int dv = \int x \, dx$ which implies $v = \dfrac{1}{2} x^2$.

Using the integration by parts formula $\int u \, dv = u \, v - \int v \, du$ we obtain

$$\boxed{\int x \ln x \, dx} = \boxed{\ln x \cdot \frac{1}{2} x^2 - \int \frac{1}{2} x^2 \cdot \frac{1}{x} dx} = \boxed{\frac{1}{2} x^2 \ln x - \frac{1}{2} \int x \, dx} = \boxed{\frac{1}{2} x^2 \ln x - \frac{1}{4} x^2 + c}$$

Check: Let $y = \dfrac{1}{2} x^2 \ln x - \dfrac{1}{4} x^2 + c$, then $y' = \dfrac{1}{2} \left(2x \cdot \ln x + \dfrac{1}{x} \cdot x^2 \right) - \dfrac{1}{4} \cdot 2x + 0 = x \ln x + \dfrac{1}{2} x - \dfrac{1}{2} x = x \ln x$

f. Given $\int x^3 \ln x \, dx$ let $u = \ln x$ and $dv = x^3 dx$ then $du = \dfrac{1}{x} dx$ and $\int dv = \int x^3 dx$ which implies $v = \dfrac{1}{4} x^4$.

Using the integration by parts formula $\int u \, dv = u \, v - \int v \, du$ we obtain

$$\boxed{\int x \ln x \, dx} = \boxed{\ln x \cdot \frac{1}{4} x^4 - \int \frac{1}{4} x^4 \cdot \frac{1}{x} dx} = \boxed{\frac{1}{4} x^4 \ln x - \frac{1}{4} \int x^3 dx} = \boxed{\frac{1}{4} x^4 \ln x - \frac{1}{16} x^4 + c}$$

Check: Let $y = \dfrac{1}{4} x^4 \ln x - \dfrac{1}{16} x^4 + c$, then $y' = \dfrac{1}{4} \left(4x^3 \cdot \ln x + \dfrac{1}{x} \cdot x^4 \right) - \dfrac{1}{16} \cdot 4x^3 + 0 = x^3 \ln x + \dfrac{1}{4} x^3$

$- \dfrac{1}{4} x^3 = x^3 \ln x$

g. Given $\int e^{2x} \cos 2x \, dx$ let $u = e^{2x}$ and $dv = \cos 2x \, dx$ then $du = 2e^{2x} dx$ and $\int dv = \int \cos 2x \, dx$ which

implies $v = \dfrac{1}{2} \sin 2x$. Using the integration by parts formula $\int u \, dv = u \, v - \int v \, du$ we

obtain $\int e^{2x} \cos 2x \, dx = e^{2x} \cdot \dfrac{1}{2} \sin 2x - \dfrac{1}{2} \int \sin 2x \cdot 2e^{2x} dx = \dfrac{e^{2x} \sin 2x}{2} - \int e^{2x} \sin 2x \, dx \qquad (1)$

To integrate $\int e^{2x} \sin 2x \, dx$ let $u = e^{2x}$ and $dv = \sin 2x \, dx$ then $du = 2e^{2x} dx$ and $\int dv = \int \sin 2x \, dx$ which

implies $v = -\dfrac{1}{2} \cos 2x$. Thus, $\int e^{2x} \sin 2x \, dx = e^{2x} \cdot -\dfrac{1}{2} \cos 2x + \dfrac{1}{2} \int \cos 2x \cdot 2e^{2x} dx = -\dfrac{e^{2x} \cos 2x}{2}$

$+ \int e^{2x} \cos 2x \, dx \qquad (2)$. Combining equations (1) and (2) together we obtain:

$$\int e^{2x}\cos 2x\,dx = \frac{e^{2x}\sin 2x}{2}-\int e^{2x}\sin 2x\,dx = \frac{e^{2x}\sin 2x}{2}-\left(-\frac{e^{2x}\cos 2x}{2}+\int e^{2x}\cos 2x\,dx\right)=\frac{e^{2x}\sin 2x}{2}$$

$+\dfrac{e^{2x}\cos 2x}{2}-\int e^{2x}\cos 2x\,dx$. Taking the $\int e^{2x}\cos 2x\,dx$ from the right hand side of the equation

to the left hand side we obtain $\int e^{2x}\cos 2x\,dx+\int e^{2x}\cos 2x\,dx = \dfrac{e^{2x}\sin 2x}{2}+\dfrac{e^{2x}\cos 2x}{2}$ which implies

$$\boxed{2\int e^{2x}\cos 2x\,dx}=\boxed{\frac{e^{2x}\sin 2x}{2}+\frac{e^{2x}\cos 2x}{2}}\text{ and }\boxed{\int e^{2x}\cos 2x\,dx}=\boxed{\frac{1}{4}\left(e^{2x}\sin 2x+e^{2x}\cos 2x\right)+c}$$

Check: Let $y=\dfrac{1}{4}\left(e^{2x}\sin 2x+e^{2x}\cos 2x\right)+c$, then $y'=\dfrac{2}{4}e^{2x}\cdot\sin 2x+\dfrac{2}{4}\cos 2x\cdot e^{2x}+\dfrac{2}{4}e^{2x}\cdot\cos 2x$

$-\dfrac{2}{4}\sin 2x\cdot e^{2x}+0 = \dfrac{2}{4}e^{2x}\sin 2x+\dfrac{4}{4}e^{2x}\cos 2x-\dfrac{2}{4}e^{2x}\sin 2x = \dfrac{4}{4}e^{2x}\cos 2x = e^{2x}\cos 2x$

h. Given $\int e^x\sin x\,dx$ let $u=e^x$ and $dv=\sin x\,dx$ then $du=e^x dx$ and $\int dv=\int\sin x\,dx$ which implies

$v=-\cos x$. Using the integration by parts formula $\int u\,dv=u\,v-\int v\,du$ we obtain

$$\int e^x\sin x\,dx = e^x\cdot-\cos x-\int-\cos x\cdot e^x dx = -e^x\cos x+\int e^x\cos x\,dx \qquad (1)$$

To integrate $\int e^x\cos x\,dx$ let $u=e^x$ and $dv=\cos x\,dx$ then $du=e^x dx$ and $\int dv=\int\cos x\,dx$ which

implies $v=\sin x$. Thus, $\int e^x\cos x\,dx = e^x\cdot\sin x-\int\sin x\cdot e^x dx = e^x\sin x-\int e^x\sin x\,dx \qquad (2)$

Combining equations (1) and (2) together we obtain:

$$\int e^x\sin x\,dx = -e^x\cos x+\int e^x\cos x\,dx = -e^x\cos x+e^x\sin x-\int e^x\sin x\,dx$$

Taking the $\int e^x\sin x\,dx$ from the right hand side of the equation to the left hand side we obtain

$$\boxed{\int e^x\sin x\,dx+\int e^x\sin x\,dx}=\boxed{-e^x\cos x+e^x\sin x}\text{ which implies }\boxed{2\int e^x\sin x\,dx}=\boxed{-e^x\cos x+e^x\sin x}$$

and $\boxed{\int e^x\sin x\,dx}=\boxed{-\dfrac{1}{2}e^x\cos x+\dfrac{1}{2}e^x\sin x+c}$

Check: Let $y=-\dfrac{1}{2}e^x\cos x+\dfrac{1}{2}e^x\sin x+c$, then $y'=-\dfrac{1}{2}e^x\cdot\cos x+\dfrac{1}{2}\sin x\cdot e^x+\dfrac{1}{2}e^x\cdot\sin x+\dfrac{1}{2}\cos x\cdot e^x$

$$=-\dfrac{1}{2}e^x\cos x+\dfrac{1}{2}e^x\sin x+\dfrac{1}{2}e^x\sin x+\dfrac{1}{2}e^x\cos x = \dfrac{1}{2}e^x\sin x+\dfrac{1}{2}e^x\sin x = e^x\sin x$$

i. Given $\int x\cos 3x\,dx$ let $u=x$ and $dv=\cos 3x\,dx$ then $du=dx$ and $\int dv=\int\cos 3x\,dx$ which implies

$v = \dfrac{1}{3}\sin 3x$. Using the integration by parts formula $\int u\,dv = u\,v - \int v\,du$ we obtain

$$\boxed{\int x\cos 3x\,dx} = \boxed{x \cdot \dfrac{1}{3}\sin 3x - \dfrac{1}{3}\int \sin 3x\,dx} = \boxed{\dfrac{1}{3}x\,\sin 3x + \dfrac{1}{9}\cos 3x + c}$$

Check: Let $y = \dfrac{1}{3}x\sin 3x + \dfrac{1}{9}\cos 3x + c$, then $y' = \dfrac{1}{3}\left(1\cdot \sin 3x + \cos 3x \cdot 3 \cdot x\right) - \dfrac{1}{9}\cdot \sin 3x \cdot 3 + 0 = \dfrac{1}{3}\sin 3x$

$+\dfrac{3}{3}x\cos 3x - \dfrac{1}{3}\sin 3x = x\cos 3x$

Example 5.1-2: Evaluate the following indefinite integrals:

a. $\displaystyle\int x\sec^2 5x\,dx =$ 　　　　　　b. $\displaystyle\int x\sec^2(x+1)\,dx =$ 　　　　　　c. $\displaystyle\int \dfrac{x}{3}\sin 2x\,dx =$

d. $\displaystyle\int arc\sin 6x\,dx =$ 　　　　　　e. $\displaystyle\int \dfrac{1}{5}arc\sin y\,dy =$ 　　　　　　f. $\displaystyle\int arc\cos x\,dx =$

g. $\displaystyle\int arc\cos \dfrac{x}{2}\,dx =$ 　　　　　　h. $\displaystyle\int arc\tan 10x\,dx =$ 　　　　　　i. $\displaystyle\int \dfrac{x\,e^x}{(1+x)^2}\,dx =$

Solutions:

a. Given $\displaystyle\int x\sec^2 5x\,dx$ let $u = x$ and $dv = \sec^2 5x\,dx$ then $du = dx$ and $\int dv = \int \sec^2 5x\,dx$ which implies

$v = \dfrac{1}{5}\tan 5x$. Using the integration by parts formula $\int u\,dv = u\,v - \int v\,du$ we obtain

$$\boxed{\int x\sec^2 5x\,dx} = \boxed{x \cdot \dfrac{1}{5}\tan 5x - \dfrac{1}{5}\int \tan 5x\,dx} = \boxed{\dfrac{1}{5}x\,\tan 5x - \dfrac{1}{25}\ln\left|\sec 5x\right| + c}$$

Check: Let $y = \dfrac{1}{5}x\tan 5x - \dfrac{1}{25}\ln\left|\sec 5x\right| + c$, then $y' = \dfrac{1}{5}\left(1\cdot \tan 5x + \sec^2 5x \cdot 5 \cdot x\right) - \dfrac{\sec 5x\tan 5x}{25\sec 5x}\cdot 5 + 0$

$= \dfrac{\tan 5x}{5} + \dfrac{5x\sec^2 5x}{5} - \dfrac{5\sec 5x\tan 5x}{25\sec 5x} = \dfrac{\tan 5x}{5} + x\sec^2 5x - \dfrac{\tan 5x}{5} = x\sec^2 5x$

b. Given $\displaystyle\int x\sec^2(x+1)\,dx$ let $u = x$ and $dv = \sec^2(x+1)\,dx$ then $du = dx$ and $\int dv = \int \sec^2(x+1)\,dx$

which implies $v = \tan(x+1)$. Using the integration by parts formula $\int u\,dv = u\,v - \int v\,du$ we obtain

$$\boxed{\int x\sec^2(x+1)\,dx} = \boxed{x \cdot \tan(x+1) - \int \tan(x+1)\,dx} = \boxed{x\,\tan(x+1) - \ln\left|\sec(x+1)\right| + c}$$

Check: Let $y = x\tan(x+1) - \ln\left|\sec(x+1)\right| + c$, then $y' = 1\cdot \tan(x+1) + \sec^2(x+1)\cdot x - \dfrac{\sec(x+1)\tan(x+1)}{\sec(x+1)} + 0$

$= \tan(x+1) + x\sec^2(x+1) - \tan(x+1) = x\sec^2(x+1)$

c. Given $\displaystyle\int \dfrac{x}{3}\sin 2x\,dx$ let $u = \dfrac{x}{3}$ and $dv = \sin 2x\,dx$ then $du = \dfrac{dx}{3}$ and $\int dv = \int \sin 2x\,dx$ which implies

$v = -\dfrac{1}{2}\cos 2x$. Using the integration by parts formula $\int u\,dv = u\,v - \int v\,du$ we obtain

$$\boxed{\int \dfrac{x}{3}\sin 2x\,dx} = \boxed{\dfrac{x}{3}\cdot -\dfrac{1}{2}\cos 2x + \dfrac{1}{2}\int \cos 2x\,\dfrac{dx}{3}} = \boxed{-\dfrac{1}{6}\,x\cos 2x + \dfrac{1}{12}\sin 2x + c}$$

Check: Let $y = -\dfrac{1}{6}x\cos 2x + \dfrac{1}{12}\sin 2x + c$, then $y' = -\dfrac{1}{6}\left(1\cdot\cos 2x - \sin 2x\cdot 2\cdot x\right) + \dfrac{1}{12}\cos 2x\cdot 2 + 0$

$$= -\dfrac{1}{6}\cos 2x + \dfrac{1}{3}x\sin 2x + \dfrac{1}{6}\cos 2x = \dfrac{x}{3}\sin 2x$$

d. Given $\int arc\sin 6x\,dx$ let $u = arc\sin 6x$ and $dv = dx$ then $du = \dfrac{6\,dx}{\sqrt{1-(6x)^2}}$ and $\int dv = \int dx$ which

implies $v = x$. Using the integration by parts formula $\int u\,dv = u\,v - \int v\,du$ we obtain

$$\int arc\sin 6x\,dx = arc\sin 6x\cdot x - \int x\cdot\dfrac{6\,dx}{\sqrt{1-36x^2}} = x\,arc\sin 6x - 6\int \dfrac{x\,dx}{\sqrt{1-36x^2}} \qquad (1)$$

To integrate $\int \dfrac{x\,dx}{\sqrt{1-36x^2}}$ use the substitution method by letting $w = 1-36x^2$ then $\dfrac{dw}{dx} = -72x$ and

$dx = -\dfrac{dw}{72x}$ Therefore, $\int \dfrac{x\,dx}{\sqrt{1-36x^2}} = \int \dfrac{x}{\sqrt{w}}\,\dfrac{dw}{-72x} = -\dfrac{1}{72}\int \dfrac{dw}{\sqrt{w}} = -\dfrac{1}{72}\int w^{-\frac{1}{2}}\,dw = -\dfrac{1}{72}\cdot\dfrac{1}{1-\frac{1}{2}}w^{1-\frac{1}{2}}$

$$= -\dfrac{1}{72}\cdot\dfrac{1}{\frac{2-1}{2}}w^{\frac{2-1}{2}} = -\dfrac{1}{72}\cdot\dfrac{2}{1}w^{\frac{1}{2}} = -\dfrac{1}{36}w^{\frac{1}{2}} = -\dfrac{1}{36}\left(1-36x^2\right)^{\frac{1}{2}} \qquad (2)$$

Combining equations (1) and (2) together we obtain

$$\boxed{\int arc\sin 6x\,dx} = \boxed{x\,arc\sin 6x - 6\int \dfrac{x\,dx}{\sqrt{1-36x^2}}} = \boxed{x\,arc\sin 6x + \dfrac{6}{36}\left(1-36x^2\right)^{\frac{1}{2}} + c} = \boxed{x\,arc\sin 6x + \dfrac{1}{6}\left(1-36x^2\right)^{\frac{1}{2}} + c}$$

Check: Let $y = x\,arc\sin 6x + \dfrac{1}{6}\left(1-36x^2\right)^{\frac{1}{2}} + c$, then $y' = arc\sin 6x + \dfrac{6x}{\sqrt{1-36x^2}} - \dfrac{1}{12}\dfrac{72x}{\sqrt{1-36x^2}} + 0$

$$= arc\sin 6x + \dfrac{6x}{\sqrt{1-36x^2}} - \dfrac{6x}{\sqrt{1-36x^2}} = arc\sin 6x$$

e. Given $\int \dfrac{1}{5}arc\sin y\,dy$ let $u = arc\sin y$ and $dv = dy$ then $du = \dfrac{dy}{\sqrt{1-y^2}}$ and $\int dv = \int dy$ which implies

$v = y$. Using the integration by parts formula $\int u\,dv = u\,v - \int v\,du$ we obtain

$$\int \dfrac{1}{5}arc\sin y\,dy = \dfrac{1}{5}arc\sin y\cdot y - \dfrac{1}{5}\int y\cdot\dfrac{dy}{\sqrt{1-y^2}} = \dfrac{1}{5}y\,arc\sin y - \dfrac{1}{5}\int \dfrac{y\,dy}{\sqrt{1-y^2}} \qquad (1)$$

To integrate $\int \frac{y\,dy}{\sqrt{1-y^2}}$ use the substitution method by letting $w = 1 - y^2$ then $\frac{dw}{dy} = -2y$ and

$dy = -\frac{dw}{2y}$ Therefore, $\int \frac{y\,dy}{\sqrt{1-y^2}} = \int \frac{y}{\sqrt{w}}\frac{dw}{-2y} = -\frac{1}{2}\int \frac{dw}{\sqrt{w}} = -\frac{1}{2}\int w^{-\frac{1}{2}}dw = -\frac{1}{2}\cdot\frac{1}{1-\frac{1}{2}}w^{1-\frac{1}{2}}$

$= -\frac{1}{2}\cdot\frac{1}{\frac{2-1}{2}}w^{\frac{2-1}{2}} = -\frac{1}{2}\cdot\frac{2}{1}w^{\frac{1}{2}} = -w^{\frac{1}{2}} = -\left(1-y^2\right)^{\frac{1}{2}}$ (2)

Combining equations (1) and (2) together we obtain:

$$\boxed{\int \frac{1}{5}arc\sin y\,dy} = \boxed{\frac{1}{5}y\,arc\sin y - \frac{1}{5}\int \frac{y\,dy}{\sqrt{1-y^2}}} = \boxed{\frac{1}{5}y\,arc\sin y + \frac{1}{5}\left(1-y^2\right)^{\frac{1}{2}} + c}$$

Check: Let $w = \frac{1}{5}y\,arc\sin y + \frac{1}{5}\left(1-y^2\right)^{\frac{1}{2}} + c$, then $w' = \frac{1}{5}arc\sin y + \frac{1}{5}\frac{y}{\sqrt{1-y^2}} - \frac{1}{5}\cdot\frac{1}{2}\cdot\frac{2y}{\sqrt{1-y^2}} + 0$

$= \frac{1}{5}arc\sin y + \frac{1}{5}\frac{y}{\sqrt{1-y^2}} - \frac{1}{5}\frac{y}{\sqrt{1-y^2}} = \frac{1}{5}arc\sin y$

f. Given $\int arc\cos x\,dx$ let $u = arc\cos x$ and $dv = dx$ then $du = \frac{-dx}{\sqrt{1-x^2}}$ and $\int dv = \int dx$ which implies

$v = x$. Using the integration by parts formula $\int u\,dv = u\,v - \int v\,du$ we obtain

$\int arc\cos x\,dx = arc\cos x\cdot x - \int x\cdot\frac{-dx}{\sqrt{1-x^2}} = x\,arc\cos x + \int \frac{x\,dx}{\sqrt{1-x^2}}$ (1)

To integrate $\int \frac{x\,dx}{\sqrt{1-x^2}}$ use the substitution method by letting $w = 1 - x^2$ then $\frac{dw}{dx} = -2x$ and

$dx = -\frac{dw}{2x}$ Therefore, $\int \frac{x\,dx}{\sqrt{1-x^2}} = \int \frac{x}{\sqrt{w}}\frac{dw}{-2x} = -\frac{1}{2}\int \frac{dw}{\sqrt{w}} = -\frac{1}{2}\int w^{-\frac{1}{2}}dw = -\frac{1}{2}\cdot\frac{1}{1-\frac{1}{2}}w^{1-\frac{1}{2}}$

$= -\frac{1}{2}\cdot\frac{1}{\frac{2-1}{2}}w^{\frac{2-1}{2}} = -\frac{1}{2}\cdot\frac{2}{1}w^{\frac{1}{2}} = -w^{\frac{1}{2}} = -\left(1-x^2\right)^{\frac{1}{2}}$ (2)

Combining equations (1) and (2) together we obtain:

$$\boxed{\int arc\cos x\,dx} = \boxed{x\,arc\cos x + \int \frac{x\,dx}{\sqrt{1-x^2}}} = \boxed{x\,arc\cos x - \left(1-x^2\right)^{\frac{1}{2}} + c}$$

Check: Let $y = x\,arc\cos x - \left(1-x^2\right)^{\frac{1}{2}} + c$, then $y' = arc\cos x - \frac{x}{\sqrt{1-x^2}} - \frac{1}{2}\frac{-2x}{\sqrt{1-x^2}} + 0 = arc\cos x$

$-\frac{x}{\sqrt{1-x^2}} + \frac{x}{\sqrt{1-x^2}} = arc\cos x$

g. Given $\int arc\cos\frac{x}{2}\,dx$ let $u=arc\cos\frac{x}{2}$ and $dv=dx$ then $du=\dfrac{-dx}{2\sqrt{1-\left(\frac{x}{2}\right)^2}}$ and $\int dv=\int dx$ which

implies $v=x$. Using the integration by parts formula $\int u\,dv=u\,v-\int v\,du$ we obtain

$$\int arc\cos\frac{x}{2}\,dx = arc\cos\frac{x}{2}\cdot x-\int x\cdot\frac{-dx}{2\sqrt{1-\left(\frac{x}{2}\right)^2}} = x\,arc\cos\frac{x}{2}+\frac{1}{2}\int\frac{x\,dx}{\sqrt{1-\left(\frac{x}{2}\right)^2}} \qquad (1)$$

To integrate $\int\dfrac{x\,dx}{\sqrt{1-\left(\frac{x}{2}\right)^2}}$ use the substitution method by letting $w=1-\left(\frac{x}{2}\right)^2$ then $\dfrac{dw}{dx}=-\dfrac{2x}{2}\cdot\dfrac{1}{2}=-\dfrac{x}{2}$

and $dx=-\dfrac{2}{x}dw$ Therefore, $\int\dfrac{x\,dx}{\sqrt{1-\left(\frac{x}{2}\right)^2}} = \int\dfrac{x}{\sqrt{w}}\dfrac{-2dw}{x} = -2\int\dfrac{dw}{\sqrt{w}} = -2\int w^{-\frac{1}{2}}dw = -2\cdot\dfrac{1}{1-\frac{1}{2}}w^{1-\frac{1}{2}}$

$$= -2\cdot\dfrac{1}{\frac{2-1}{2}}w^{\frac{2-1}{2}} = -2\cdot\dfrac{2}{1}w^{\frac{1}{2}} = -4w^{\frac{1}{2}} = -4\left[1-\left(\frac{x}{2}\right)^2\right]^{\frac{1}{2}} \qquad (2)$$

Combining equations (1) and (2) together we obtain:

$$\boxed{\int arc\cos\frac{x}{2}\,dx} = \boxed{x\,arc\cos\frac{x}{2}+\frac{1}{2}\int\frac{x\,dx}{\sqrt{1-\left(\frac{x}{2}\right)^2}}} = \boxed{x\,arc\cos\frac{x}{2}-\frac{4}{2}\left[1-\left(\frac{x}{2}\right)^2\right]^{\frac{1}{2}}+c} = \boxed{x\,arc\cos\frac{x}{2}-2\left[1-\left(\frac{x}{2}\right)^2\right]^{\frac{1}{2}}+c}$$

Check: Let $y=x\,arc\cos\frac{x}{2}-2\left[1-\left(\frac{x}{2}\right)^2\right]^{\frac{1}{2}}+c$, then $y' = arc\cos\frac{x}{2}-\dfrac{x}{2\sqrt{1-\left(\frac{x}{2}\right)^2}}-\dfrac{x}{\sqrt{1-\left(\frac{x}{2}\right)^2}}\cdot-\dfrac{1}{2}+0$

$$= arc\cos\frac{x}{2}-\dfrac{x}{2\sqrt{1-\left(\frac{x}{2}\right)^2}}+\dfrac{x}{2\sqrt{1-\left(\frac{x}{2}\right)^2}} = arc\cos\frac{x}{2}$$

h. Given $\int arc\tan 10x\,dx$ let $u=arc\tan 10x$ and $dv=dx$ then $du=\dfrac{10\,dx}{1+(10x)^2}$ and $\int dv=\int dx$ which

implies $v=x$. Using the integration by parts formula $\int u\,dv=u\,v-\int v\,du$ we obtain

$$\int arc\tan 10x\,dx = arc\tan 10x\cdot x-\int x\cdot\frac{10\,dx}{1+(10x)^2} = x\,arc\tan 10x-10\int\frac{x\,dx}{1+(10x)^2} \qquad (1)$$

To integrate $\int\dfrac{x\,dx}{1+(10x)^2}$ use the substitution method by letting $w=1+(10x)^2$ then $\dfrac{dw}{dx}=200x$

And $dx=\dfrac{dw}{200x}$ Thus, $\int\dfrac{x\,dx}{1+(10x)^2} = \int\dfrac{x}{w}\dfrac{dw}{200x} = \dfrac{1}{200}\int\dfrac{dw}{w} = \dfrac{1}{200}\ln|w| = \dfrac{1}{200}\ln\left|1+(10x)^2\right|$ (2)

Combining equations (1) and (2) together we obtain:

$$\boxed{\int arc\tan 10x\, dx} = \boxed{x\, arc\tan 10x - 10\int \frac{x\, dx}{1+(10x)^2}} = \boxed{x\, arc\tan 10x - \frac{1}{20}\ln\left|1+(10x)^2\right| + c}$$

Check: Let $y = x\, arc\tan 10x - \frac{1}{20}\ln\left|1+(10x)^2\right| + c$, then $y' = arc\tan 10x + \frac{x}{1+(10x)^2} - \frac{1}{20}\frac{20x}{1+(10x)^2} + 0$

$$arc\tan 10x + \frac{x}{1+(10x)^2} - \frac{x}{1+(10x)^2} = arc\tan 10x$$

i. Given $\int \frac{x\,e^x}{(1+x)^2}\, dx$ let $u = xe^x$ and $dv = (1+x)^{-2}\, dx$ then $du = e^x(1+x)dx$ and $\int dv = \int (1+x)^{-2}\, dx$

which implies $v = -\frac{1}{1+x}$. Using the integration by parts formula $\int u\, dv = u\,v - \int v\, du$ we obtain

$$\boxed{\int \frac{x\,e^x}{(1+x)^2}\, dx} = \boxed{xe^x \cdot \frac{-1}{1+x} - \int \frac{-1}{(1+x)} \cdot e^x(1+x)dx} = \boxed{\frac{-xe^x}{1+x} + \int e^x dx} = \boxed{\frac{-xe^x}{1+x} + e^x + c} = \boxed{\frac{-xe^x + e^x(1+x)}{1+x} + c}$$

$$= \boxed{\frac{-xe^x + e^x + xe^x}{1+x} + c} = \boxed{\frac{e^x}{1+x} + c}$$

Check: Let $y = \frac{e^x}{1+x} + c$, then $y' = \frac{e^x(1+x) - e^x}{(1+x)^2} + 0 = \frac{e^x + xe^x - e^x}{(1+x)^2} = \frac{xe^x}{(1+x)^2}$

Example 5.1-3: Evaluate the following indefinite integrals:

a. $\int \sin^3 x\, dx =$
                          b. $\int \sin^2 x\, dx =$
                      c. $\int \arctan x\, dx =$

d. $\int \sin(\ln x)\, dx =$
                  e. $\int x^2 e^x dx =$
                     f. $\int x^3 \sin x\, dx =$

g. $\int x^2 \cos 3x\, dx =$
                 h. $\int e^{-x} \cos x\, dx =$
                  i. $\int e^{-3x} \sin 3x\, dx =$

Solutions:

a. Given $\int \sin^3 x\, dx = \int \sin^2 x \cdot \sin x\, dx$ let $u = \sin^2 x$ and $dv = \sin x\, dx$ then $du = 2\sin x \cos x\, dx$ and

$\int dv = \int \sin x\, dx$ which implies $v = -\cos x$. Using the integration by parts formula $\int u\, dv = u\,v - \int v\, du$

we obtain $\int \sin^3 x\, dx = \sin^2 x \cdot -\cos x + \int \cos x \cdot 2\sin x \cos x\, dx = -\sin^2 x \cos x + 2\int \cos^2 x \sin x\, dx$ (1)

To integrate $\int \cos^2 x \sin x\, dx$ use the integration by parts method again, i.e., let $u = \cos^2 x$ and

$dv = \sin x\, dx$ then $du = -2\sin x \cos x\, dx$ and $\int dv = \int \sin x\, dx$ which implies $v = -\cos x$. Therefore,

$\int \cos^2 x \sin x\, dx = \cos^2 x \cdot -\cos x - \int \cos x \cdot 2\sin x \cos x\, dx = -\cos^3 x - 2\int \cos^2 x \sin x\, dx$. Taking the

integral $-2\int \cos^2 x \sin x\, dx$ from the right hand side of the equation to the left side we obtain

$\int \cos^2 x \sin x \, dx + 2\int \cos^2 x \sin x \, dx \;=\; -\cos^3 x$. Therefore, $\int \cos^2 x \sin x \, dx \;=\; -\dfrac{1}{3}\cos^3 x$ \qquad (2)

Combining equations (1) and (2) together we have

$$\boxed{\int \sin^3 x \, dx} = \boxed{-\sin^2 x \cos x + 2\int \cos^2 x \sin x \, dx} = \boxed{-\sin^2 x \cos x + 2\cdot -\frac{1}{3}\cos^3 x + c} = \boxed{-\sin^2 x \cos x - \frac{2}{3}\cos^3 x + c}$$

$$\boxed{-\left(1-\cos^2 x\right)\cos x - \frac{2}{3}\cos^3 x + c} = \boxed{-\cos x + \left(\cos^3 x - \frac{2}{3}\cos^3 x\right) + c} = \boxed{\frac{1}{3}\cos^3 x - \cos x + c}$$

Note that another method of solving the above problem (as was shown in Section 4.3) is in the following way:

$$\int \sin^3 x \, dx = \int \sin^2 x \cdot \sin x \, dx = \int \left(1-\cos^2 x\right)\cdot \sin x \, dx \text{ let } u = \cos x, \text{ then } \frac{du}{dx} = -\sin x \text{ and } dx = -\frac{du}{\sin x}.$$

Therefore, $\boxed{\int \sin^3 x \, dx} = \boxed{\int \sin^2 x \cdot \sin x \, dx} = \boxed{\int \left(1-\cos^2 x\right)\cdot \sin x \, dx} = \boxed{\int \left(1-u^2\right)\cdot \sin x \cdot -\frac{du}{\sin x}}$

$= \boxed{-\int \left(1-u^2\right)du} = \boxed{\int u^2 - 1 \, du} = \boxed{\frac{1}{3}u^3 - u + c} = \boxed{\frac{1}{3}\cos^3 x - \cos x + c}$

Check: Let $y = \dfrac{1}{3}\cos^3 x - \cos x + c$, then $y' = \dfrac{1}{3}\cdot 3\cos^2 x \cdot -\sin x + \sin x + 0 \;=\; -\cos^2 x \sin x + \sin x$

$\qquad = \sin x \left(1-\cos^2 x\right) = \sin x \sin^2 x = \sin^3 x$

b. Given $\int \sin^2 x \, dx = \int \sin x \cdot \sin x \, dx$ let $u = \sin x$ and $dv = \sin x \, dx$ then $du = \cos x \, dx$ and $\int dv = \int \sin x \, dx$

which implies $v = -\cos x$. Using the integration by parts formula $\int u \, dv = u v - \int v \, du$ we obtain

$\int \sin^2 x \, dx \;=\; \sin x \cdot -\cos x + \int \cos x \cdot \cos x \, dx \;=\; -\sin x \cos x + \int \cos^2 x \, dx \;=\; -\sin x \cos x + \int \left(1-\sin^2 x\right)dx$

$= -\sin x \cos x + x - \int \sin^2 x \, dx$. Taking the integral $\int \sin^2 x \, dx$ from the right hand side of the

equation to the left side we have $\int \sin^2 x \, dx + \int \sin^2 x \, dx \;=\; -\sin x \cos x + x$. Therefore,

$$\boxed{2\int \sin^2 x \, dx} = \boxed{-\sin x \cos x + x} \text{ and } \boxed{\int \sin^2 x \, dx} = \boxed{-\frac{1}{2}\sin x \cos x + \frac{x}{2} + c} = \boxed{-\frac{1}{4}\sin 2x + \frac{x}{2} + c}$$

or, we can solve the given integral in the following way:

$$\boxed{\int \sin^2 x \, dx} = \boxed{\int \frac{1}{2}\left(1-\cos 2x\right)dx} = \boxed{\frac{1}{2}\int dx - \frac{1}{2}\int \cos 2x \, dx} = \boxed{\frac{x}{2} - \frac{1}{2}\cdot \frac{\sin 2x}{2} + c} = \boxed{-\frac{1}{4}\sin 2x + \frac{x}{2} + c}$$

Check: Let $y = -\dfrac{1}{4}\sin 2x + \dfrac{x}{2} + c$, then $y' \;=\; -\dfrac{1}{4}\cos 2x \cdot 2 + \dfrac{1}{2} + 0 \;=\; -\dfrac{1}{2}\cos 2x + \dfrac{1}{2} \;=\; \dfrac{1}{2}\left(1-\cos 2x\right) = \sin^2 x$ or,

Let $y = -\frac{1}{2}\sin x \cos x + \frac{x}{2} + c$, then $y' = -\frac{1}{2}\cos x \cos x + \frac{1}{2}\sin x \sin x + \frac{1}{2} + 0 = -\frac{1}{2}\cos^2 x + \frac{1}{2}\sin^2 x + \frac{1}{2}$

$$= -\frac{1}{2}\left(1 - \sin^2 x\right) + \frac{1}{2}\sin^2 x + \frac{1}{2} = -\frac{1}{2} + \frac{1}{2}\sin^2 x + \frac{1}{2}\sin^2 x + \frac{1}{2} = \frac{1}{2}\sin^2 x + \frac{1}{2}\sin^2 x = \sin^2 x$$

c. Given $\int \arctan x\, dx$ let $u = arc\tan x$ and $dv = dx$ then $du = \dfrac{dx}{1+x^2}$ and $\int dv = \int dx$ which implies

$v = x$. Substituting the integral with its equivalent value $\int u\, dv = uv - \int v\, du$ we obtain

$$\int x \arctan x\, dx = arc\tan x \cdot x - \int x \cdot \frac{dx}{1+x^2} = x\, arc\tan x - \int \frac{x\, dx}{1+x^2} \tag{1}$$

To integrate $\int \dfrac{x\, dx}{1+x^2}$ use the substitution method by letting $w = 1 + x^2$ then $\dfrac{dw}{dx} = 2x$ and $dx = \dfrac{dw}{2x}$

Therefore, $\int \dfrac{x\, dx}{1+x^2} = \int \dfrac{x}{w}\dfrac{dw}{2x} = \dfrac{1}{2}\int \dfrac{dw}{w} = \dfrac{1}{2}\ln|w| = \dfrac{1}{2}\ln\left|1+x^2\right|$ $\tag{2}$

Combining equations (1) and (2) together we obtain:

$$\boxed{\int x \arctan x\, dx} = \boxed{x\, arc\tan x - \int \frac{x\, dx}{1+x^2}} = \boxed{x\, arc\tan x - \frac{1}{2}\ln\left|1+x^2\right| + c}$$

Check: Let $y = x\, arc\tan x - \dfrac{1}{2}\ln\left|1+x^2\right| + c$, then $y' = arc\tan x + \dfrac{x}{1+x^2} - \dfrac{1}{2}\dfrac{2x}{1+x^2} + 0 = arc\tan x$

$$+ \frac{x}{1+x^2} - \frac{x}{1+x^2} = arc\tan x$$

d. Given $\int \sin(\ln x)\, dx$ let $u = \sin(\ln x)$ and $dv = dx$ then $du = \dfrac{\cos(\ln x)}{x}\, dx$ and $\int dv = \int dx$ which

implies $v = x$. Using the integration by parts formula $\int u\, dv = uv - \int v\, du$ we obtain

$$\int \sin(\ln x)\, dx = \sin(\ln x) \cdot x - \int x \cdot \frac{\cos(\ln x)}{x}\, dx = x \sin(\ln x) - \int \cos(\ln x)\, dx \tag{1}$$

To integrate $\int \cos(\ln x)\, dx$ use the integration by parts formula again, i.e., let $u = \cos(\ln x)$ and

$dv = dx$ then $du = \dfrac{-\sin(\ln x)}{x}\, dx$ and $\int dv = \int dx$ which implies $v = x$. Therefore, $\int \cos(\ln x)\, dx$

$$= \cos(\ln x) \cdot x + \int x \cdot \frac{\sin(\ln x)}{x}\, dx = x \cos(\ln x) + \int \sin(\ln x)\, dx \tag{2}$$

Combining equations (1) and (2) together we have

$$\int \sin(\ln x)\, dx = x \sin(\ln x) - \int \cos(\ln x)\, dx = x \sin(\ln x) - x \cos(\ln x) - \int \sin(\ln x)\, dx$$

Taking the integral $-\int \sin(\ln x)\, dx$ from the right hand side of the equation to the left hand side we obtain $\int \sin(\ln x)\,dx + \int \sin(\ln x)\,dx = x\sin(\ln x) - x\cos(\ln x) + c$ Therefore,

$$\boxed{2\int \sin(\ln x)\,dx} = \boxed{x\sin(\ln x) - x\cos(\ln x) + c}\ \text{ and thus }\ \boxed{\int \sin(\ln x)\,dx} = \boxed{\frac{x}{2}\sin(\ln x) - \frac{x}{2}\cos(\ln x) + c}$$

Check: Let $y = \frac{x}{2}\sin(\ln x) - \frac{x}{2}\cos(\ln x) + c$, then $y' = \frac{\sin(\ln x)}{2} + \frac{x\cos(\ln x)}{2x} - \frac{\cos(\ln x)}{2} + \frac{x\sin(\ln x)}{2x} + 0$

$$= \frac{\sin(\ln x)}{2} + \frac{\cos(\ln x)}{2} - \frac{\cos(\ln x)}{2} + \frac{\sin(\ln x)}{2} = \frac{\sin(\ln x)}{2} + \frac{\sin(\ln x)}{2} = \sin(\ln x)$$

e. Given $\int x^2 e^x dx$ let $u = x^2$ and $dv = e^x dx$ then $du = 2x\,dx$ and $\int dv = \int e^x dx$ which implies $v = e^x$.

Using the integration by parts formula $\int u\,dv = uv - \int v\,du$ we obtain

$$\int x^2 e^x dx = x^2 \cdot e^x - \int e^x \cdot 2x\,dx = x^2 e^x - 2\int x e^x\,dx \tag{1}$$

To integrate $\int xe^x dx$ use the integration by parts formula again, i.e., let $u = x$ and $dv = e^x dx$

Then $du = dx$ and $\int dv = \int dx$ which implies $v = e^x$. Therefore,

$$\int xe^x dx = x \cdot e^x - \int e^x \cdot dx = xe^x - \int e^x dx = xe^x - e^x \tag{2}$$

Combining equations (1) and (2) together we have

$$\boxed{\int x^2 e^x dx} = \boxed{x^2 e^x - 2\int x e^x\,dx} = \boxed{x^2 e^x - 2\left(xe^x - e^x\right) + c} = \boxed{x^2 e^x - 2xe^x + 2e^x + c}$$

Check: Let $y = x^2 e^x - 2xe^x + 2e^x + c$, then $y' = 2xe^x + x^2 e^x - 2e^x - 2xe^x + 2e^x + 0 = x^2 e^x$

f. Given $\int x^3 \sin x\,dx$ let $u = x^3$ and $dv = \sin x\,dx$ then $du = 3x^2 dx$ and $\int dv = \int \sin x\,dx$ which implies

$v = -\cos x$. Using the integration by parts formula $\int u\,dv = uv - \int v\,du$ we obtain

$$\int x^3 \sin x\,dx = x^3 \cdot -\cos x + \int \cos x \cdot 3x^2 dx = -x^3 \cos x + 3\int x^2 \cos x\,dx \tag{1}$$

To integrate $\int x^2 \cos x\,dx$ use the integration by parts formula again, i.e., let $u = x^2$ and $dv = \cos x\,dx$

then $du = 2x\,dx$ and $\int dv = \int \cos x\,dx$ which implies $v = \sin x$. Using the integration by parts

formula $\int u\,dv = uv - \int v\,du$ we obtain

$$\int x^2 \cos x\,dx = x^2 \cdot \sin x - \int \sin x \cdot 2x\,dx = x^2 \sin x - 2\int x \sin x\,dx \tag{2}$$

To integrate $\int x\sin x\,dx$ use the integration by parts formula again, i.e., let $u = x$ and $dv = \sin x\,dx$

then $du = dx$ and $\int dv = \int \sin x\, dx$ which implies $v = -\cos x$. Using the integration by parts

formula $\int u\, dv = u\, v - \int v\, du$ we obtain

$$\int x \sin x\, dx = x \cdot -\cos x + \int \cos x \cdot dx = -x \cos x + \sin x \qquad (3)$$

Combining equations (1), (2) and (3) together we have

$$\boxed{\int x^3 \sin x\, dx} = \boxed{-x^3 \cos x + 3\int x^2 \cos x\, dx} = \boxed{-x^3 \cos x + 3x^2 \sin x - 6\int x \sin x\, dx} = \boxed{-x^3 \cos x + 3x^2 \sin x}$$

$$\boxed{+6x \cos x - 6\sin x + c}$$

Check: Let $y = -x^3 \cos x + 3x^2 \sin x + 6x \cos x - 6\sin x + c$, then $y' = \left(-3x^2 \cos x + x^3 \sin x\right)$

$$+\left(6x \sin x + 3x^2 \cos x\right) + \left(6\cos x - 6x \sin x\right) - 6\cos x + 0 = x^3 \sin x$$

g. Given $\int x^2 \cos 3x\, dx$ let $u = x^2$ and $dv = \cos 3x\, dx$ then $du = 2x\, dx$ and $\int dv = \int \cos 3x\, dx$ which

implies $v = \dfrac{\sin 3x}{3}$. Using the integration by parts formula $\int u\, dv = u\, v - \int v\, du$ we obtain

$$\int x^2 \cos 3x\, dx = x^2 \cdot \frac{\sin 3x}{3} - \int \frac{\sin 3x}{3} \cdot 2x\, dx = \frac{1}{3} x^2 \sin 3x - \frac{2}{3}\int x \sin 3x\, dx \qquad (1)$$

To integrate $\int x \sin 3x\, dx$ use the integration by parts formula again, i.e., let $u = x$ and $dv = \sin 3x\, dx$

then $du = dx$ and $\int dv = \int \sin 3x\, dx$ which implies $v = \dfrac{-\cos 3x}{3}$. Using the integration by parts

formula $\int u\, dv = u\, v - \int v\, du$ we obtain

$$\int x \sin 3x\, dx = x \cdot \frac{-\cos 3x}{3} + \int \frac{\cos 3x}{3} \cdot dx = -\frac{1}{3} x \cos 3x + \frac{1}{3}\int \cos 3x\, dx \qquad (2)$$

Combining equations (1) and (2) together we have

$$\boxed{\int x^2 \cos 3x\, dx} = \boxed{\frac{1}{3} x^2 \sin 3x - \frac{2}{3}\int x \sin 3x\, dx} = \boxed{\frac{1}{3} x^2 \sin 3x - \frac{2}{3} \cdot -\frac{1}{3} x \cos 3x - \frac{2}{3} \cdot \frac{1}{3}\int \cos 3x\, dx}$$

$$= \boxed{\frac{1}{3} x^2 \sin 3x + \frac{2}{9} x \cos 3x - \frac{2}{27} \sin 3x + c}$$

Check: Let $y = \dfrac{1}{3} x^2 \sin 3x + \dfrac{2}{9} x \cos 3x - \dfrac{2}{27} \sin 3x + c$, then $y' = \dfrac{1}{3}\left(2x \sin 3x + 3x^2 \cos 3x\right)$

$$+\frac{2}{9}\left(\cos 3x - 3x \sin 3x\right) - \frac{6}{27}\cos 3x + c = \frac{2}{3} x \sin 3x + \frac{3}{3} x^2 \cos 3x + \frac{2}{9}\cos 3x - \frac{2}{3} x \sin 3x - \frac{2}{9}\cos 3x$$

$$= \frac{3}{3} x^2 \cos 3x = x^2 \cos 3x$$

h. Given $\int e^{-x}\cos x\,dx$ let $u=\cos x$ and $dv=e^{-x}dx$ then $du=-\sin x\,dx$ and $\int dv=\int e^{-x}dx$ which

implies $v=-e^{-x}$. Using the integration by parts formula $\int u\,dv=uv-\int v\,du$ we obtain

$$\int e^{-x}\cos x\,dx = \cos x\cdot-e^{-x}-\int e^{-x}\cdot\sin x\,dx = -e^{-x}\cos x-\int e^{-x}\sin x\,dx \qquad (1)$$

To integrate $\int e^{-x}\sin x\,dx$ use the integration by parts formula again, i.e., let $u=\sin x$ and

$dv=e^{-x}dx$ then $du=\cos x\,dx$ and $\int dv=\int e^{-x}dx$ which implies $v=-e^{-x}$. Therefore,

$$\int e^{-x}\sin x\,dx = \sin x\cdot-e^{-x}+\int e^{-x}\cdot\cos x\,dx = -e^{-x}\sin x+\int e^{-x}\cos x\,dx \qquad (2)$$

Combining equations (1) and (2) together we have

$$\int e^{-x}\cos x\,dx = -e^{-x}\cos x-\int e^{-x}\sin x\,dx = -e^{-x}\cos x+e^{-x}\sin x-\int e^{-x}\cos x\,dx$$

Taking the integral $\int e^{-x}\cos x\,dx$ from the right hand side of the equation to the left hand side

we obtain $\int e^{-x}\cos x\,dx+\int e^{-x}\cos x\,dx = -e^{-x}\cos x+e^{-x}\sin x$ therefore

$$\boxed{2\int e^{-x}\cos x\,dx} = \boxed{-e^{-x}\cos x+\sin x\,e^{-x}} \text{ and thus } \boxed{\int e^{-x}\cos x\,dx} = \boxed{-\frac{1}{2}e^{-x}\cos x+\frac{1}{2}e^{-x}\sin x+c}$$

Check: Let $y=\dfrac{-e^{-x}\cos x}{2}+\dfrac{e^{-x}\sin x}{2}+c$, then $y' = -\frac{1}{2}\left(-e^{-x}\cos x-e^{-x}\sin x\right)+\frac{1}{2}\left(-e^{-x}\sin x+e^{-x}\cos x\right)+0$

$$= \frac{1}{2}e^{-x}\cos x+\frac{1}{2}e^{-x}\sin x-\frac{1}{2}e^{-x}\sin x+\frac{1}{2}e^{-x}\cos x = \frac{1}{2}e^{-x}\cos x+\frac{1}{2}e^{-x}\cos x = e^{-x}\cos x$$

i. Given $\int e^{-3x}\sin 3x\,dx$ let $u=\sin 3x$ and $dv=e^{-3x}dx$ then $du=3\cos 3x\,dx$ and $\int dv=\int e^{-3x}dx$ which

implies $v=-\frac{1}{3}e^{-3x}$. Using the integration by parts formula $\int u\,dv=uv-\int v\,du$ we obtain

$$\int e^{-3x}\sin 3x\,dx = \sin 3x\cdot-\frac{1}{3}e^{-3x}+\frac{1}{3}\int e^{-3x}\cdot3\cos 3x\,dx = -\frac{1}{3}e^{-3x}\sin 3x+\int e^{-3x}\cos 3x\,dx \qquad (1)$$

To integrate $\int e^{-3x}\cos 3x\,dx$ use the integration by parts formula again, i.e., let $u=\cos 3x$ and

$dv=e^{-3x}dx$ then $du=-3\sin 3x$ and $\int dv=\int e^{-3x}dx$ which implies $v=-\frac{1}{3}e^{-3x}$. Therefore,

$$\int e^{-3x}\cos 3x\,dx = \cos 3x\cdot-\frac{1}{3}e^{-3x}-\frac{1}{3}\int e^{-3x}\cdot3\sin 3x\,dx = -\frac{1}{3}e^{-3x}\cos 3x-\int e^{-3x}\sin 3x\,dx \qquad (2)$$

Combining equations (1) and (2) together we have

$$\int e^{-3x}\sin 3x\,dx = -\frac{1}{3}e^{-3x}\sin 3x+\int e^{-3x}\cos 3x\,dx = -\frac{1}{3}e^{-3x}\sin 3x-\frac{1}{3}e^{-3x}\cos 3x-\int e^{-3x}\sin 3x\,dx$$

Taking the integral $\int e^{-3x}\sin 3x\,dx$ from the right hand side of the equation to the left hand side we obtain $\int e^{-3x} \sin 3x\, dx + \int e^{-3x} \sin 3x\, dx = -\dfrac{1}{3} e^{-3x} \sin 3x - \dfrac{1}{3} e^{-3x} \cos 3x$ therefore

$$\boxed{2\int e^{-3x} \sin 3x\, dx} = \boxed{-\dfrac{1}{3} e^{-3x} \sin 3x - \dfrac{1}{3} e^{-3x} \cos 3x} \text{ and thus } \boxed{\int e^{-3x} \sin 3x\, dx} = \boxed{-\dfrac{1}{6} e^{-3x} \sin 3x - \dfrac{1}{6} e^{-3x} \cos 3x}$$

Check: Let $y = -\dfrac{1}{6} e^{-3x} \sin 3x - \dfrac{1}{6} e^{-3x} \cos 3x$, then $y' = -\dfrac{1}{6}\left(-3e^{-3x} \sin 3x + 3e^{-3x} \cos 3x\right)$

$-\dfrac{1}{6}\left(-3e^{-3x} \cos 3x - 3e^{-3x} \sin 3x\right) + 0 = \dfrac{1}{2} e^{-3x} \sin 3x - \dfrac{1}{2} e^{-3x} \cos 3x + \dfrac{1}{2} e^{-3x} \cos 3x + \dfrac{1}{2} e^{-3x} \sin 3x$

$= \dfrac{1}{2} e^{-3x} \sin 3x + \dfrac{1}{2} e^{-3x} \sin 3x = e^{-3x} \sin 3x$

Example 5.1-4: Evaluate the following indefinite integrals:

a. $\int \dfrac{x}{3}(x+1)^4\, dx =$

b. $\int (x-3)(3x-1)^3\, dx =$

c. $\int (x+1)\csc^2 x\, dx =$

d. $\int x\sec^2 x\, dx =$

e. $\int x\sqrt{x-5}\, dx =$

f. $\int \ln\left(x^2+1\right) dx =$

g. $\int \dfrac{1}{5} e^x \sin 3x\, dx =$

h. $\int \cos^{-1} 3x\, dx =$

i. $\int \tan^{-1} 5x\, dx =$

Solutions:

a. Given $\int \dfrac{x}{3}(x+1)^4\, dx = \dfrac{1}{3}\int x(x+1)^4\, dx$ let $u = x$ and $dv = (x+1)^4\, dx$ then $du = dx$ and $\int dv = \int (x+1)^4\, dx$

which implies $v = \dfrac{1}{5}(x+1)^5$. Using the integration by parts formula $\int u\, dv = u\, v - \int v\, du$ we obtain

$$\boxed{\dfrac{1}{3}\int x(x+1)^4\, dx} = \boxed{\dfrac{x}{3}\cdot\dfrac{(x+1)^5}{5} - \dfrac{1}{3\cdot 5}\int (x+1)^5 \cdot dx} = \boxed{\dfrac{x(x+1)^5}{15} - \dfrac{1}{15}\cdot\dfrac{1}{6}(x+1)^{5+1} + c} = \boxed{\dfrac{x(x+1)^5}{15} - \dfrac{1}{90}(x+1)^6 + c}$$

Check: Let $y = \dfrac{x(x+1)^5}{15} - \dfrac{1}{90}(x+1)^6 + c$, then $y' = \dfrac{1}{15}\left[(x+1)^5 + 5x(x+1)^4\right] - \dfrac{6}{90}(x+1)^5 + 0 = \dfrac{(x+1)^5}{15}$

$+\dfrac{5x}{15}(x+1)^4 - \dfrac{(x+1)^5}{15} = \dfrac{5x}{15}(x+1)^4 = \dfrac{x}{3}(x+1)^4$

b. Given $\int (x-3)(3x-1)^3\, dx$ let $u = x-3$ and $dv = (3x-1)^3\, dx$ then $du = dx$ and $\int dv = \int (3x-1)^3\, dx$

which implies $v = \dfrac{1}{12}(3x-1)^4$. Using the integration by parts formula $\int u\, dv = u\, v - \int v\, du$ we obtain

$$\boxed{\int (x-3)(3x-1)^3\, dx} = \boxed{(x-3)\cdot\dfrac{(3x-1)^4}{12} - \dfrac{1}{12}\int (3x-1)^4\, dx} = \boxed{\dfrac{(x-3)(3x-1)^4}{12} - \dfrac{1}{12}\cdot\dfrac{1}{15}(3x-1)^{4+1} + c}$$

$$= \boxed{\dfrac{(x-3)(3x-1)^4}{12} - \dfrac{1}{180}(3x-1)^5 + c}$$

Check: Let $y = \dfrac{(x-3)(3x-1)^4}{12} - \dfrac{1}{180}(3x-1)^5 + c$, then $y' = \dfrac{1}{12}\Big[(3x-1)^4 + 12(x-3)(x+1)^3\Big] - \dfrac{15}{180}(3x-1)^4 + 0$

$$= \dfrac{1}{12}(3x-1)^4 + \dfrac{12(x-3)(x+1)^3}{12} - \dfrac{1}{12}(3x-1)^4 = (x-3)(x+1)^3$$

c. Given $\int (x+1)\csc^2 x\, dx$ let $u = x+1$ and $dv = \csc^2 x\, dx$ then $du = dx$ and $\int dv = \int \csc^2 x\, dx$ which

implies $v = -\cot x$. Using the integration by parts formula $\int u\, dv = u\,v - \int v\, du$ we obtain

$$\boxed{\int (x+1)\csc^2 x\, dx} = \boxed{(x+1)\cdot -\cot x + \int \cot x\, dx} = \boxed{-(x+1)\cot x + \ln|\sin x| + c}$$

Check: Let $y = -(x+1)\cot x + \ln|\sin x| + c$, then $y' = -\Big[\cot x - (x+1)\csc^2 x\Big] + \dfrac{\cos x}{\sin x} + 0 = -\cot x$

$$+ (x+1)\csc^2 x + \cot x = (x+1)\csc^2 x$$

d. Given $\int x\sec^2 x\, dx$ let $u = x$ and $dv = \sec^2 x\, dx$ then $du = dx$ and $\int dv = \int \sec^2 x\, dx$ which implies

$v = \tan x$. Using the integration by parts formula $\int u\, dv = u\,v - \int v\, du$ we obtain

$$\boxed{\int x\sec^2 x\, dx} = \boxed{x\cdot \tan x - \int \tan x\, dx} = \boxed{x\tan x - \ln|\sec x| + c}$$

Check: Let $y = x\tan x - \ln|\sec x| + c$, then $y' = \tan x + x\sec^2 x - \dfrac{\sec x \tan x}{\sec x} + 0 = \tan x + x\sec^2 x - \tan x$

$$= x\sec^2 x$$

e. Given $\int x\sqrt{x-5}\, dx$ let $u = x$ and $dv = \sqrt{x-5}\, dx$ then $du = dx$ and $\int dv = \int \sqrt{x-5}\, dx$ which implies

$v = \dfrac{2}{3}(x-5)^{\frac{3}{2}}$. Using the integration by parts formula $\int u\, dv = u\,v - \int v\, du$ we obtain

$$\boxed{\int x\sqrt{x-5}\, dx} = \boxed{x\cdot \dfrac{2}{3}(x-5)^{\frac{3}{2}} - \int \dfrac{2}{3}(x-5)^{\frac{3}{2}}\, dx} = \boxed{\dfrac{2}{3}x(x-5)^{\frac{3}{2}} - \dfrac{2}{3}\cdot\dfrac{1}{1+\frac{3}{2}}(x-5)^{\frac{3}{2}+1} + c} = \boxed{\dfrac{2}{3}x(x-5)^{\frac{3}{2}}}$$

$$\boxed{-\dfrac{2}{3}\cdot\dfrac{2}{5}(x-5)^{\frac{5}{2}} + c} = \boxed{\dfrac{2}{3}x(x-5)^{\frac{3}{2}} - \dfrac{4}{15}(x-5)^{\frac{5}{2}} + c}$$

Check: Let $y = \dfrac{2}{3}x(x-5)^{\frac{3}{2}} - \dfrac{4}{15}(x-5)^{\frac{5}{2}} + c$, then $y' = \dfrac{2}{3}(x-5)^{\frac{3}{2}} + \dfrac{2}{3}\cdot\dfrac{3}{2}x(x-5)^{\frac{1}{2}} - \dfrac{4}{15}\cdot\dfrac{5}{2}(x-5)^{\frac{3}{2}} + 0$

$$= \dfrac{2}{3}(x-5)^{\frac{3}{2}} + x(x-5)^{\frac{1}{2}} - \dfrac{2}{3}(x-5)^{\frac{3}{2}} = x(x-5)^{\frac{1}{2}} = x\sqrt{x-5}$$

f. Given $\int \ln\left(x^2+1\right)dx$ let $u = \ln\left(x^2+1\right)$ and $dv = dx$ then $du = \dfrac{2x}{x^2+1}\, dx$ and $\int dv = \int dx$ which implies

$v = x$. Using the integration by parts formula $\int u\, dv = u\,v - \int v\, du$ we obtain

$$\boxed{\int \ln\left(x^2+1\right)dx} = \boxed{\ln\left(x^2+1\right)\cdot x - \int x\cdot\frac{2x}{x^2+1}dx} = \boxed{x\ln\left(x^2+1\right)-2\int\frac{x^2}{x^2+1}dx} = \boxed{x\ln\left(x^2+1\right)-2\int\left(1-\frac{1}{x^2+1}\right)dx}$$

$$= \boxed{x\ln\left(x^2+1\right)-2\int dx+2\int\frac{1}{x^2+1}dx} = \boxed{x\ln\left(x^2+1\right)-2x+2\tan^{-1}x+c}$$

Check: Let $y = x\ln\left(x^2+1\right)-2x+2\tan^{-1}x+c$, then $y' = 1\cdot\ln\left(x^2+1\right)+\frac{2x}{x^2+1}\cdot x-2+\frac{2}{1+x^2}+0$

$$= \ln\left(x^2+1\right)+\frac{2x^2}{x^2+1}-2+\frac{2}{x^2+1} = \ln\left(x^2+1\right)+\frac{2x^2-2x^2-2+2}{x^2+1} = \ln\left(x^2+1\right)+\frac{0}{x^2+1} = \ln\left(x^2+1\right)$$

g. Given $\frac{1}{5}\int e^x\sin 3x\,dx$ let $u=e^x$ and $dv=\sin 3x\,dx$ then $du=e^x\,dx$ and $\int dv=\int\sin 3x\,dx$ which

implies $v=-\frac{1}{3}\cos 3x$. Using the integration by parts formula $\int u\,dv=uv-\int v\,du$ we obtain

$$\frac{1}{5}\int e^x\sin 3x\,dx = \frac{1}{5}e^x\cdot-\frac{1}{3}\cos 3x+\frac{1}{15}\int\cos 3x\cdot e^x\,dx = -\frac{1}{15}e^x\cos 3x+\frac{1}{15}\int e^x\cos 3x\,dx \qquad (1)$$

To integrate $\int e^x\cos 3x\,dx$ use the integration by parts method again, i.e., let $u=e^x$ and

$dv=\cos 3x\,dx$ then $du=e^x\,dx$ and $\int dv=\int\cos 3x\,dx$ which implies $v=\frac{1}{3}\sin 3x$. Therefore,

$$\int e^x\cos 3x\,dx = e^x\cdot\frac{1}{3}\sin 3x-\frac{1}{3}\int\sin 3x\cdot e^x\,dx = \frac{1}{3}e^x\sin 3x-\frac{1}{3}\int e^x\sin 3x\,dx \qquad (2)$$

Combining equations (1) and (2) together we have

$$\frac{1}{5}\int e^x\sin 3x\,dx = -\frac{1}{15}e^x\cos 3x+\frac{1}{15}\int e^x\cos 3x\,dx = -\frac{1}{15}e^x\cos 3x+\frac{1}{45}e^x\sin 3x-\frac{1}{45}\int e^x\sin 3x\,dx$$

Taking the integral $-\frac{1}{45}\int e^x\sin 3x\,dx$ from the right hand side of the equation to the left side we

obtain $\frac{1}{5}\int e^x\sin 3x\,dx+\frac{1}{45}\int e^x\sin 3x\,dx = -\frac{1}{15}e^x\cos 3x+\frac{1}{45}e^x\sin 3x+c$. Therefore, $\boxed{\frac{2}{9}\int e^x\sin 3x\,dx}$

$$= \boxed{-\frac{1}{15}e^x\cos 3x+\frac{1}{45}e^x\sin 3x+c}\text{ which implies }\boxed{\int e^x\sin 3x\,dx} = \boxed{-\frac{3}{10}e^x\cos 3x+\frac{1}{10}e^x\sin 3x+c}$$

Check: Let $y=-\frac{3}{10}e^x\cos 3x+\frac{1}{10}e^x\sin 3x+c$, then $y' = -\frac{3}{10}e^x\cdot\cos 3x+\frac{3}{10}\sin 3x\cdot 3\cdot e^x+\frac{1}{10}e^x\cdot\sin 3x$

$$+\frac{1}{10}\cos 3x\cdot 3\cdot e^x+0 = -\frac{3}{10}e^x\cos 3x+\frac{9}{10}e^x\sin 3x+\frac{1}{10}e^x\sin 3x+\frac{3}{10}e^x\cos 3x = \frac{9}{10}e^x\sin 3x$$

$$+\frac{1}{10}e^x\sin 3x = \frac{9+1}{10}e^x\sin 3x = \frac{10}{10}e^x\sin 3x = e^x\sin 3x$$

h. Given $\int\cos^{-1}3x\,dx$ let $u=\cos^{-1}3x$ and $dv=dx$ then $du=-\frac{3}{\sqrt{1-9x^2}}dx$ and $\int dv=\int dx$ which implies $v = x$. Using the integration by parts formula $\int u\,dv = u\,v - \int v\,du$ we obtain

$$\int \cos^{-1}3x\,dx = \cos^{-1}3x \cdot x + \int x \cdot \frac{3\,dx}{\sqrt{1-9x^2}} = x\cos^{-1}3x + \int \frac{3x}{\sqrt{1-9x^2}}\,dx$$

To integrate $\int \frac{3x}{\sqrt{1-9x^2}}\,dx$ use the substitution method by letting $w = 1-9x^2$ then $\frac{dw}{dx} = -18x$ which

implies $dx = -\frac{dw}{18x}$. Thus, $\int \frac{3x}{\sqrt{1-9x^2}}\,dx = \int \frac{3x}{\sqrt{w}} \cdot -\frac{dw}{18x} = -\frac{1}{6}\int \frac{dw}{\sqrt{w}} = -\frac{1}{6}\int \frac{1}{w^{\frac{1}{2}}}\,dw = -\frac{1}{6}\int w^{-\frac{1}{2}}\,dw$

$= -\frac{1}{6} \cdot 2w^{\frac{1}{2}} = -\frac{\sqrt{1-9x^2}}{3}$ and $\boxed{\int \cos^{-1}3x\,dx} = \boxed{x\cos^{-1}3x + \int \frac{3x}{\sqrt{1-9x^2}}\,dx} = \boxed{x\cos^{-1}3x - \frac{\sqrt{1-9x^2}}{3} + c}$

Check: Let $y = x\cos^{-1}3x - \frac{\sqrt{1-9x^2}}{3} + c$, then $y' = \cos^{-1}3x - \frac{3x}{\sqrt{1-9x^2}} + \frac{18x}{6\sqrt{1-9x^2}} + 0 = \cos^{-1}3x$

i. Given $\int \tan^{-1}5x\,dx$ let $u = \tan^{-1}5x$ and $dv = dx$ then $du = \frac{5}{1+25x^2}\,dx$ and $\int dv = \int dx$ which

implies $v = x$. Using the integration by parts formula $\int u\,dv = u\,v - \int v\,du$ we obtain

$$\int \tan^{-1}5x\,dx = \tan^{-1}5x \cdot x - \int x \cdot \frac{5\,dx}{1+25x^2} = x\tan^{-1}5x - \int \frac{5x}{1+25x^2}\,dx$$

To integrate $\int \frac{5x}{1+25x^2}\,dx$ use the substitution method by letting $w = 1+25x^2$ then $\frac{dw}{dx} = 50x$ which

implies $dx = \frac{dw}{50x}$. Thus, $\int \frac{5x}{1+25x^2}\,dx = \int \frac{5x}{w} \cdot \frac{dw}{50x} = \frac{1}{10}\int \frac{dw}{w} = \frac{1}{10}\ln|w| = \frac{1}{10}\ln\left|1+25x^2\right|$ and

$\boxed{\int \tan^{-1}5x\,dx} = \boxed{x\tan^{-1}5x - \int \frac{5x}{1+25x^2}\,dx} = \boxed{x\tan^{-1}5x - \frac{1}{10}\ln\left|1+25x^2\right| + c}$

Check: Let $y = x\tan^{-1}5x - \frac{1}{10}\ln\left|1+25x^2\right| + c$, then $y' = \tan^{-1}5x + \frac{5x}{1+25x^2} - \frac{1}{10}\frac{50x}{1+25x^2} + 0 = \tan^{-1}5x$

Example 5.1-5: Evaluate the following indefinite integrals:

a. $\int \sinh^{-1}5x\,dx =$ b. $\int x\tan^{-1}x\,dx =$ c. $\int \sin x\sin 7x\,dx =$

d. $\int \cos 5x\cos 7x\,dx =$ e. $\int \frac{x}{5}e^{-x}\,dx =$ f. $\int x\sinh 3x\,dx =$

Solutions:

a. Given $\int \sinh^{-1}5x\,dx$ let $u = \sinh^{-1}5x$ and $dv = dx$ then $du = \frac{5}{\sqrt{1+25x^2}}\,dx$ and $\int dv = \int x\,dx$ which

implies $v = x$. Using the integration by parts formula $\int u\,dv = u\,v - \int v\,du$ we obtain

$$\int \sinh^{-1} 5x \, dx \; = \; \sinh^{-1} 5x \cdot x - \int x \cdot \frac{5 \, dx}{\sqrt{1+25x^2}} \; = \; x \sinh^{-1} 5x - \int \frac{5x \, dx}{\sqrt{1+25x^2}} \qquad (1)$$

To get the integral of $\int \dfrac{5x \, dx}{\sqrt{1+25x^2}}$ use the substitution method by letting $w = 1 + 25x^2$ then

$dw = 50x \, dx$ which implies $dx = \dfrac{dw}{50x}$. Therefore, $\int \dfrac{5x \, dx}{\sqrt{1+25x^2}} = \int \dfrac{5x}{\sqrt{w}} \cdot \dfrac{dw}{50x} = \dfrac{5}{50} \int \dfrac{1}{\sqrt{w}} \, dw$

$$= \frac{1}{10} \int \frac{1}{w^{\frac{1}{2}}} \, dw \; = \; \frac{1}{10} \int w^{-\frac{1}{2}} \, dw \; = \; \frac{1}{10} \cdot \frac{1}{1-\frac{1}{2}} w^{1-\frac{1}{2}} \; = \; \frac{2}{10} w^{\frac{1}{2}} \; = \; \frac{1}{5} w^{\frac{1}{2}} \; = \; \frac{1}{5} \sqrt{1+25x^2} \qquad (2)$$

Combining equations (1) and (2) together we have

$$\boxed{\int \sinh^{-1} 5x \, dx} = \boxed{x \sinh^{-1} 5x - \int \frac{5x \, dx}{\sqrt{1+25x^2}}} = \boxed{x \sinh^{-1} 5x - \frac{1}{5}\left(1+25x^2\right)^{\frac{1}{2}} + c}$$

Check: Let $y = x \sinh^{-1} 5x - \dfrac{1}{5}\left(1+25x^2\right)^{\frac{1}{2}} + c$, then $y' = \sinh^{-1} 5x + \dfrac{5x}{\sqrt{1+25x^2}} - \dfrac{50x}{10\sqrt{1+25x^2}} + 0$

$$\sinh^{-1} 5x + \frac{5x}{\sqrt{1+25x^2}} - \frac{5x}{\sqrt{1+25x^2}} \; = \; \sinh^{-1} 5x$$

b. Given $\int x \tan^{-1} x \, dx$ let $u = \tan^{-1} x$ and $dv = x \, dx$ then $du = \dfrac{1}{1+x^2} \, dx$ and $\int dv = \int x \, dx$ which

implies $v = \dfrac{1}{2} x^2$. Using the integration by parts formula $\int u \, dv = u \, v - \int v \, du$ we obtain

$$\boxed{\int x \tan^{-1} x \, dx} = \boxed{\tan^{-1} x \cdot \frac{x^2}{2} - \frac{1}{2} \int x^2 \cdot \frac{dx}{1+x^2}} = \boxed{\frac{1}{2} x^2 \tan^{-1} x - \frac{1}{2} \int \frac{x^2}{1+x^2} dx} = \boxed{\frac{1}{2} x^2 \tan^{-1} x - \frac{1}{2} \int \left(1 - \frac{1}{1+x^2}\right) dx}$$

$$= \boxed{\frac{1}{2} x^2 \tan^{-1} x - \frac{1}{2} \int dx + \frac{1}{2} \int \frac{1}{1+x^2} dx} = \boxed{\frac{1}{2} x^2 \tan^{-1} x - \frac{1}{2} x + \frac{1}{2} \tan^{-1} x + c}$$

Check: Let $y = \dfrac{1}{2} x^2 \tan^{-1} x - \dfrac{1}{2} x + \dfrac{1}{2} \tan^{-1} x + c$, then $y' = \dfrac{1}{2} \cdot 2x \cdot \tan^{-1} x + \dfrac{1}{2} \dfrac{x^2}{1+x^2} - \dfrac{1}{2} + \dfrac{1}{2} \cdot \dfrac{1}{1+x^2} + 0$

$$x \tan^{-1} x + \frac{1}{2} \cdot \frac{x^2}{1+x^2} + \frac{1}{2} \cdot \frac{1}{1+x^2} - \frac{1}{2} \; = \; x \tan^{-1} x + \frac{1}{2} \frac{x^2+1}{1+x^2} - \frac{1}{2} \; = \; x \tan^{-1} x + \frac{1}{2} - \frac{1}{2} \; = \; x \tan^{-1} x$$

c. Given $\int \sin x \sin 7x \, dx$ let $u = \sin x$ and $dv = \sin 7x \, dx$ then $du = \cos x \, dx$ and $\int dv = \int \sin 7x \, dx$ which

implies $v = -\dfrac{1}{7} \cos 7x$. Using the integration by parts formula $\int u \, dv = u \, v - \int v \, du$ we obtain

$$\int \sin x \sin 7x \, dx \; = \; \sin x \cdot -\frac{1}{7} \cos 7x + \frac{1}{7} \int \cos 7x \cdot \cos x \, dx \; = \; -\frac{1}{7} \sin x \cos 7x + \frac{1}{7} \int \cos x \cdot \cos 7x \, dx \qquad (1)$$

To integrate $\int \cos x \cdot \cos 7x \, dx$ use the integration by parts method again, i.e., let $u = \cos x$ and

$dv = \cos 7x \, dx$ then $du = -\sin x \, dx$ and $\int dv = \int \cos 7x \, dx$ which implies $v = \frac{1}{7}\sin 7x$. Therefore,

$$\int \cos x \cdot \cos 7x \, dx = \cos x \cdot \frac{1}{7}\sin 7x + \frac{1}{7}\int \sin 7x \cdot \sin x \, dx = \frac{1}{7}\cos x \sin 7x + \frac{1}{7}\int \sin x \sin 7x \, dx \qquad (2)$$

Combining equations (1) and (2) together we have

$$\int \sin x \sin 7x \, dx = -\frac{1}{7}\sin x \cos 7x + \frac{1}{7}\int \cos x \cdot \cos 7x \, dx = -\frac{1}{7}\sin x \cos 7x + \frac{1}{49}\cos x \sin 7x + \frac{1}{49}\int \sin x \sin 7x \, dx$$

Taking the integral $\frac{1}{49}\int \sin x \sin 7x \, dx$ to the left hand side and simplifying we have

$$\boxed{\int \sin x \sin 7x \, dx} = \boxed{-\frac{1}{7}\cdot\frac{49}{48}\sin x \cos 7x + \frac{1}{49}\cdot\frac{49}{48}\cos x \sin 7x + c} = \boxed{-\frac{7}{48}\sin x \cos 7x + \frac{1}{48}\cos x \sin 7x + c}$$

Check: Let $y = -\frac{7}{48}\sin x \cos 7x + \frac{1}{48}\cos x \sin 7x + c$, then $y' = -\frac{7}{48}\cos x \cdot \cos 7x + \frac{49}{48}\sin 7x \cdot \sin x$

$$-\frac{1}{48}\sin x \cdot \sin 7x + \frac{7}{48}\cos 7x \cdot \cos x + 0 = \frac{49}{48}\sin 7x \cdot \sin x - \frac{1}{48}\sin x \cdot \sin 7x = \frac{49-1}{48}\sin 7x \cdot \sin x$$

$$= \frac{48}{48}\sin 7x \cdot \sin x = \sin 7x \sin x$$

d. Given $\int \cos 5x \cos 7x \, dx$ let $u = \cos 5x$ and $dv = \cos 7x \, dx$ then $du = -5\sin 5x \, dx$ and $\int dv = \int \cos 7x \, dx$

which implies $v = \frac{1}{7}\sin 7x$. Using the integration by parts formula $\int u \, dv = u v - \int v \, du$ we obtain

$$\int \cos 5x \cos 7x \, dx = \cos 5x \cdot \frac{1}{7}\sin 7x + \frac{5}{7}\int \sin 7x \cdot \sin 5x \, dx = \frac{1}{7}\cos 5x \sin 7x + \frac{5}{7}\int \sin 5x \cdot \sin 7x \, dx \qquad (1)$$

To integrate $\int \sin 5x \cdot \sin 7x \, dx$ use the integration by parts method again, i.e., let $u = \sin 5x$ and

$dv = \sin 7x \, dx$ then $du = 5\cos 5x \, dx$ and $\int dv = \int \sin 7x \, dx$ which implies $v = -\frac{1}{7}\cos 7x$. Therefore,

$$\int \sin 5x \cdot \sin 7x \, dx = \sin 5x \cdot -\frac{1}{7}\cos 7x + \frac{5}{7}\int \cos 7x \cdot \cos 5x \, dx = -\frac{1}{7}\sin 5x \cos 7x + \frac{5}{7}\int \cos 5x \cos 7x \, dx \quad (2)$$

Combining equations (1) and (2) together we have

$$\int \cos 5x \cos 7x \, dx = \frac{1}{7}\cos 5x \sin 7x + \frac{5}{7}\int \sin 5x \cdot \sin 7x \, dx = \frac{1}{7}\cos 5x \sin 7x - \frac{5}{49}\sin 5x \cos 7x + \frac{25}{49}\int \cos 5x \cos 7x \, dx$$

Taking the integral $\frac{25}{49}\int \cos 5x \cos 7x \, dx$ to the left hand side and simplifying we have

$$\boxed{\int \cos 5x \cos 7x \, dx} = \boxed{\frac{1}{7}\cdot\frac{49}{24}\cos 5x \sin 7x - \frac{5}{49}\cdot\frac{49}{24}\sin 5x \cos 7x + c} = \boxed{\frac{7}{24}\cos 5x \sin 7x - \frac{5}{24}\sin 5x \cos 7x + c}$$

Check: Let $y = \frac{7}{24}\cos 5x \sin 7x - \frac{5}{24}\sin 5x \cos 7x + c$, then $y' = -\frac{35}{24}\sin 5x \cdot \sin 7x + \frac{49}{24}\cos 7x \cdot \cos 5x$

$$-\frac{25}{24}\cos 5x \cdot \cos 7x + \frac{35}{24}\sin 5x \cdot \sin 7x + 0 \; = \; \frac{49}{24}\cos 5x \cos 7x - \frac{25}{24}\cos 5x \cos 7x \; = \; \frac{49-25}{24}\cos 5x \cos 7x$$

$$= \frac{24}{24}\cos 5x \cos 7x \; = \; \cos 5x \cos 7x$$

e. Given $\int \frac{x}{5}e^{-x}\,dx$ let $u = \frac{x}{5}$ and $dv = e^{-x}\,dx$ then $du = \frac{1}{5}dx$ and $\int dv = \int e^{-x}\,dx$ which implies

$v = -e^{-x}$. Using the integration by parts formula $\int u\,dv = u\,v - \int v\,du$ we obtain

$$\boxed{\int \frac{x}{5}e^{-x}\,dx} = \boxed{\frac{x}{5}\cdot -e^{-x} + \int e^{-x}\cdot\frac{dx}{5}} = \boxed{-\frac{xe^{-x}}{5} - \frac{e^{-x}}{5} + c} = \boxed{-\frac{e^{-x}}{5}(x+1)+c}$$

Check: Let $y = -\dfrac{xe^{-x}}{5} - \dfrac{e^{-x}}{5} + c$, then $y' \; = \; -\dfrac{e^{-x}}{5} + \dfrac{xe^{-x}}{5} + \dfrac{e^{-x}}{5} + 0 \; = \; \dfrac{xe^{-x}}{5}$

f. Given $\int x\sinh 3x\,dx$ let $u = x$ and $dv = \sinh 3x\,dx$ then $du = dx$ and $\int dv = \int \sinh 3x\,dx$ which

implies $v = \frac{1}{3}\cosh 3x\,dx$. Using the integration by parts formula $\int u\,dv = u\,v - \int v\,du$ we obtain

$$\boxed{\int x\sinh 3x\,dx} = \boxed{x\cdot\frac{1}{3}\cosh 3x - \int\frac{1}{3}\cosh 3x\cdot dx} = \boxed{\frac{1}{3}x\cosh 3x - \frac{1}{3}\int\cosh 3x\cdot dx} = \boxed{\frac{1}{3}x\cosh 3x - \frac{1}{9}\sinh 3x + c}$$

Check: Let $y = \frac{1}{3}x\cosh 3x - \frac{1}{9}\sinh 3x + c$, then $y' \; = \; \frac{1}{3}\cosh 3x + \frac{1}{3}\cdot 3x\sinh 3x - \frac{1}{9}\cdot 3\cosh 3x + 0 \; = \; \frac{1}{3}\cosh 3x$

$$+ x\sinh 3x - \frac{1}{3}\cosh 3x \; = \; x\sinh 3x$$

Section 5.1 Practice Problems – Integration by Parts

1. Evaluate the following integrals using the integration by parts method.

a. $\int xe^{4x}\,dx \; =$

b. $\int \frac{x}{2}\cos x\,dx \; =$

c. $\int (5-x)e^{5x}\,dx \; =$

d. $\int x\sin 5x\,dx \; =$

e. $\int x\sqrt{3-x}\,dx \; =$

f. $\int x^3 e^{3x}\,dx \; =$

g. $\int \cos(\ln x)\,dx \; =$

h. $\int \frac{x}{3}\tan^{-1}x\,dx \; =$

i. $\int \ln x^5\,dx \; =$

j. $\int x\,e^{-ax}\,dx \; =$

k. $\int e^x\sin 3x\,dx \; =$

l. $\int e^x\cos 5x\,dx \; =$

2. Evaluate the following integrals using the integration by parts method.

a. $\int x\sec^2 x\,dx \; =$

b. $\int \arcsin 3y\,dy \; =$

c. $\int \arctan x\,dx \; =$

d. $\int \sin^3 5x\,dx \; =$

e. $\int x^2\cos x\,dx \; =$

f. $\int e^{-2x}\cos 3x\,dx \; =$

g. $\int x(5x-1)^3\,dx \; =$

h. $\int x\csc^2 x\,dx \; =$

i. $\int \frac{2}{3}\cos^{-1}5x\,dx \; =$

j. $\int \sinh^{-1}x\,dx \; =$

k. $\int x\sec^2 10x\,dx \; =$

l. $\int \frac{x}{5}\sinh 7x\,dx \; =$

5.2   Integration Using Trigonometric Substitution

Many integration problems involve radical expressions of the form

$$\sqrt{a^2 - b^2 x^2} \qquad\qquad \sqrt{a^2 + b^2 x^2} \qquad\qquad \sqrt{b^2 x^2 - a^2}$$

In such instances we can use a trigonometric substitution by letting

$$x = \frac{a}{b}\sin t \qquad\qquad x = \frac{a}{b}\tan t \qquad\qquad x = \frac{a}{b}\sec t \qquad \text{respectively to obtain}$$

$$\sqrt{a^2 - b^2 x^2} = \sqrt{a^2 - b^2 \cdot \frac{a^2}{b^2}\sin^2 t} = \sqrt{a^2 - a^2 \sin^2 t} = a\sqrt{\left(1 - \sin^2 t\right)} = a\sqrt{\cos^2 t} = a\cos t$$

$$\sqrt{a^2 + b^2 x^2} = \sqrt{a^2 + b^2 \cdot \frac{a^2}{b^2}\tan^2 t} = \sqrt{a^2 + a^2 \tan^2 t} = a\sqrt{\left(1 + \tan^2 t\right)} = a\sqrt{\sec^2 t} = a\sec t$$

$$\sqrt{b^2 x^2 - a^2} = \sqrt{b^2 \cdot \frac{a^2}{b^2}\sec^2 t - a^2} = \sqrt{a^2 \sec^2 t - a^2} = a\sqrt{\left(\sec^2 t - 1\right)} = a\sqrt{\tan^2 t} = a\tan t$$

Notice that using trigonometric substitution result in elimination of the radical expression. This in effect reduces the difficulty of solving integrals with radical expressions.

Reminder 1:

Given $x = \frac{a}{b}\sin t$, then $t = \sin^{-1}\frac{bx}{a}$ for $-1 \le x \le 1$ and $-\frac{\pi}{2} \le t \le \frac{\pi}{2}$

Given $x = \frac{a}{b}\tan t$, then $t = \tan^{-1}\frac{bx}{a}$ for all x and $-\frac{\pi}{2}\langle t \langle \frac{\pi}{2}$

Given $x = \frac{a}{b}\sec t$, then $t = \sec^{-1}\frac{bx}{a}$ for $x \ge 1$ or $x \le -1$ and $0 \le t \langle \frac{\pi}{2}$ or $\pi \le t \langle \frac{3\pi}{2}$

Reminder 2:

In solving this class of integrals the integrand in the original variable may be obtained by the use of a right triangle. For example, in a right triangle

- $\sin t = \frac{opposite}{hypotenuse} = \frac{x}{a}$. Therefore, using the Pythagorean theorem, the adjacent side (w) is equal to

$$a = \sqrt{x^2 + w^2} \;\; ; \;\; a^2 = x^2 + w^2 \;\; ; \;\; w^2 = a^2 - x^2 \;\; ; \;\; w = \sqrt{a^2 - x^2}$$

- $\cos t = \frac{adjacent}{hypotenuse} = \frac{x}{a}$. Therefore, the opposite side (w) is equal to

$$a = \sqrt{x^2 + w^2} \;\; ; \;\; a^2 = x^2 + w^2 \;\; ; \;\; w^2 = a^2 - x^2 \;\; ; \;\; w = \sqrt{a^2 - x^2}$$

- $\tan t = \frac{opposite}{adjacent} = \frac{x}{a}$. Therefore, the hypotenuse (w) is equal to $w = \sqrt{a^2 + x^2}$

- $\cot t = \frac{adjacent}{opposite} = \frac{x}{a}$. Therefore, the hypotenuse (w) is equal to $w = \sqrt{a^2 + x^2}$

- $\sec t = \dfrac{hypotenuse}{adjacent} = \dfrac{x}{a}$. Therefore, the opposite side (w) is equal to

$x = \sqrt{a^2 + w^2}$; $x^2 = a^2 + w^2$; $w^2 = x^2 - a^2$; $w = \sqrt{x^2 - a^2}$

- $\csc t = \dfrac{hypotenuse}{opposite} = \dfrac{x}{a}$. Therefore, the adjacent side (w) is equal to

$x = \sqrt{a^2 + w^2}$; $x^2 = a^2 + w^2$; $w^2 = x^2 - a^2$; $w = \sqrt{x^2 - a^2}$

Let's integrate some integrals using the above trigonometric substitution method:

Example 5.2-1: Use trigonometric substitution to evaluate the following indefinite integrals:

a. $\displaystyle\int \dfrac{dx}{x^2\sqrt{4-x^2}} =$

b. $\displaystyle\int \dfrac{x^2\,dx}{\sqrt{25-x^2}} =$

c. $\displaystyle\int \dfrac{\sqrt{1+x^2}}{x^2}\,dx =$

d. $\displaystyle\int \dfrac{dx}{\left(9+x^2\right)^2} =$

e. $\displaystyle\int \dfrac{x^2}{\sqrt{x^2-1}}\,dx =$

f. $\displaystyle\int \dfrac{dx}{x^4\sqrt{x^2-1}} =$

Solutions:

a. Given $\displaystyle\int \dfrac{dx}{x^2\sqrt{4-x^2}}$ let $x = 2\sin t$, then $dx = 2\cos t\,dt$ and $\sqrt{4-x^2} = \sqrt{4-4\sin^2 t} = \sqrt{4\left(1-\sin^2 t\right)}$

$= \sqrt{4\cos^2 t} = 2\cos t$. Substituting these values back into the original integral we obtain:

$$\int \dfrac{dx}{x^2\sqrt{4-x^2}} = \int \dfrac{2\cos t\,dt}{(2\sin t)^2 \cdot 2\cos t} = \int \dfrac{2\cos t}{4\sin^2 t \cdot 2\cos t}\,dt = \int \dfrac{1}{4\sin^2 t}\,dt = \dfrac{1}{4}\int \csc^2 t\,dt = -\dfrac{1}{4}\cot t + c$$

$$= -\dfrac{1}{4}\dfrac{\cos t}{\sin t} + c = -\dfrac{1}{4}\dfrac{\frac{\sqrt{4-x^2}}{2}}{\frac{x}{2}} + c = -\dfrac{1}{4}\left(\dfrac{\sqrt{4-x^2}\cdot 2}{2\cdot x}\right) + c = -\dfrac{1}{4}\left(\dfrac{\sqrt{4-x^2}}{x}\right) + c$$

Check: To check the answer we start with the solution and find its derivative. The derivative should match with the integrand, i.e., the algebraic expression inside the integral. Note that not all the steps in finding the derivative is given. At this level, it is expected that students are able to work through the details that are not shown (review differentiation techniques described in Chapters 2 and 3).

Let $y = -\dfrac{1}{4}\dfrac{\sqrt{4-x^2}}{x} + c$, then $y' = -\dfrac{1}{4}\dfrac{\frac{-2x}{2\sqrt{4-x^2}}\cdot x - 1\cdot\sqrt{4-x^2}}{x^2} + 0 = -\dfrac{1}{4}\dfrac{\frac{-2x^2}{2\sqrt{4-x^2}} - \frac{\sqrt{4-x^2}}{1}}{x^2}$

$= -\dfrac{1}{4}\dfrac{\frac{-x^2-\sqrt{4-x^2}\cdot\sqrt{4-x^2}}{\sqrt{4-x^2}}}{x^2} = -\dfrac{1}{4}\dfrac{\frac{-x^2-\left(4-x^2\right)}{\sqrt{4-x^2}}}{x^2} = -\dfrac{1}{4}\dfrac{-x^2-4+x^2}{x^2\sqrt{4-x^2}} = -\dfrac{1}{4}\dfrac{-4}{x^2\sqrt{4-x^2}} = \dfrac{1}{x^2\sqrt{4-x^2}}$

b. Given $\int \dfrac{x^2 dx}{\sqrt{25-x^2}}$ let $x=5\sin t$, then $dx=5\cos t\, dt$ and $\sqrt{25-x^2}=\sqrt{25-25\sin^2 t}=\sqrt{25\left(1-\sin^2 t\right)}$

$=\sqrt{25\cos^2 t}=5\cos t$. Substituting these values back into the original integral we obtain:

$$\boxed{\int \dfrac{x^2 dx}{\sqrt{25-x^2}}}=\boxed{\int \dfrac{25\sin^2 t\cdot 5\cos t\, dt}{5\cos t}}=\boxed{\int \dfrac{25\sin^2 t}{1}\,dt}=\boxed{25\int \sin^2 t\, dt}=\boxed{25\int \dfrac{1-\cos 2t}{2}\,dt}$$

$$=\boxed{\dfrac{25}{2}\int \left(1-\cos 2t\right)dt}=\boxed{\dfrac{25}{2}\left(t-\dfrac{1}{2}\sin 2t\right)+c}=\boxed{\dfrac{25}{2}t-\dfrac{25}{2}\sin t\cos t+c}=\boxed{\dfrac{25}{2}\left(\sin^{-1}\dfrac{x}{5}\right)-\dfrac{25}{2}\left(\dfrac{x}{5}\cdot\dfrac{\sqrt{25-x^2}}{5}\right)+c}$$

$$=\boxed{\dfrac{25}{2}\left(\sin^{-1}\dfrac{x}{5}\right)-\dfrac{25}{2}\left(\dfrac{x}{25}\sqrt{25-x^2}\right)+c}=\boxed{\dfrac{25}{2}\left(\sin^{-1}\dfrac{x}{5}\right)-\dfrac{x}{2}\sqrt{25-x^2}+c}$$

Check: Let $y=\dfrac{25}{2}\sin^{-1}\dfrac{x}{5}-\dfrac{x}{2}\sqrt{25-x^2}+c$, then $y'=\dfrac{25}{2}\dfrac{1}{\sqrt{1-\frac{x^2}{25}}}\cdot\dfrac{1}{5}-\left(\dfrac{1}{2}\sqrt{25-x^2}+\dfrac{-2x}{2\sqrt{25-x^2}}\cdot\dfrac{x}{2}\right)$

$$=\dfrac{25}{2}\dfrac{5}{\sqrt{25-x^2}}\cdot\dfrac{1}{5}-\left(\dfrac{\sqrt{25-x^2}}{2}-\dfrac{x^2}{2\sqrt{25-x^2}}\right)=\dfrac{25}{2\sqrt{25-x^2}}-\left(\dfrac{2\left(25-x^2\right)-2x^2}{4\sqrt{25-x^2}}\right)=\dfrac{25}{2\sqrt{25-x^2}}$$

$$-\left(\dfrac{50-4x^2}{4\sqrt{25-x^2}}\right)=\dfrac{25}{2\sqrt{25-x^2}}-\dfrac{25-2x^2}{2\sqrt{25-x^2}}=\dfrac{25-25+2x^2}{2\sqrt{25-x^2}}=\dfrac{2x^2}{2\sqrt{25-x^2}}=\dfrac{x^2}{\sqrt{25-x^2}}$$

c. Given $\int \dfrac{\sqrt{1+x^2}}{x^2}\,dx$ let $x=\tan t$, then $dx=\sec^2 t\, dt$ and $\sqrt{1+x^2}=\sqrt{1+\tan^2 t}=\sqrt{\sec^2 t}=\sec t$

Substituting these values back into the original integral we obtain $\boxed{\int \dfrac{\sqrt{1+x^2}}{x^2}\,dx}=\boxed{\int \dfrac{\sec t\cdot\sec^2 t\, dt}{\tan^2 t}}$

$$=\boxed{\int \dfrac{\sec t\cdot\left(1+\tan^2 t\right)dt}{\tan^2 t}}=\boxed{\int \dfrac{\sec t+\sec t\tan^2 t}{\tan^2 t}\,dt}=\boxed{\int \dfrac{\sec t\tan^2 t}{\tan^2 t}\,dt+\int \dfrac{\sec t}{\tan^2 t}\,dt}=\boxed{\int \sec t\, dt+\int \dfrac{\sec t}{\tan^2 t}\,dt}$$

$$=\boxed{\ln\left|\sec t+\tan t\right|+\int \dfrac{\frac{1}{\cos t}}{\frac{\sin^2 t}{\cos^2 t}}\,dt}=\boxed{\ln\left|\sec t+\tan t\right|+\int \dfrac{\cos^2 t}{\cos t\sin^2 t}\,dt}=\boxed{\ln\left|\sec t+\tan t\right|+\int \dfrac{\cos t}{\sin^2 t}\,dt}$$

$$=\boxed{\ln\left|\sec t+\tan t\right|+\int \sin^{-2} t\cos t\, dt}=\boxed{\ln\left|\sec t+\tan t\right|+\sin^{-1}t+c}=\boxed{\ln\left|\sec t+\tan t\right|+\dfrac{1}{\sin t}+c}$$

$$=\boxed{\ln\left|\sec t+\tan t\right|-\csc t+c}=\boxed{\ln\left|\sqrt{1+x^2}+x\right|-\dfrac{\sqrt{1+x^2}}{x}+c}$$

Check: Let $y = \ln\left|\sqrt{1+x^2}+x\right| - \dfrac{\sqrt{1+x^2}}{x}+c$, then $y' = \dfrac{1}{\sqrt{1+x^2}+x} \cdot \left(\dfrac{2x}{2\sqrt{1+x^2}}+1\right) - \dfrac{\frac{x^2}{2\sqrt{1+x^2}}\cdot x-\sqrt{1+x^2}}{x^2}$

$= \dfrac{1}{\sqrt{1+x^2}+x} \cdot \left(\dfrac{2x+2\sqrt{1+x^2}}{2\sqrt{1+x^2}}\right) - \left(\dfrac{x^2-\left(1+x^2\right)}{x^2\sqrt{1+x^2}}\right) = \dfrac{1}{\sqrt{1+x^2}+x} \cdot \dfrac{2\left(x+\sqrt{1+x^2}\right)}{2\sqrt{1+x^2}} - \left(\dfrac{x^2-1-x^2}{x^2\sqrt{1+x^2}}\right)$

$= \dfrac{1}{\sqrt{1+x^2}} + \dfrac{1}{x^2\sqrt{1+x^2}} = \dfrac{x^2\sqrt{1+x^2}+\sqrt{1+x^2}}{x^2\left(1+x^2\right)} = \dfrac{\sqrt{1+x^2}\left(1+x^2\right)}{x^2\left(1+x^2\right)} = \dfrac{\sqrt{1+x^2}}{x^2}$

d. Given $\displaystyle\int \dfrac{dx}{\left(9+x^2\right)^2}$ let $x = 3\tan t$, then $dx = 3\sec^2 t\, dt$ and $9+x^2 = 9+\left(3\tan t\right)^2 = 9+9\tan^2 t$

$= 9\left(1+\tan^2 t\right) = 9\sec^2 t$ Substituting these values back into the original integral we obtain

$\boxed{\displaystyle\int \dfrac{dx}{\left(9+x^2\right)^2}} = \boxed{\displaystyle\int \dfrac{3\sec^2 t\, dt}{\left(9\sec^2 t\right)^2}} = \boxed{\displaystyle\int \dfrac{3\sec^2 t\, dt}{81\sec^2 t \sec^2 t}} = \boxed{\dfrac{1}{27}\displaystyle\int \dfrac{dt}{\sec^2 t}} = \boxed{\dfrac{1}{27}\displaystyle\int \cos^2 t\, dt} = \boxed{\dfrac{1}{27}\cdot\dfrac{1}{2}\displaystyle\int \left(1+\cos 2t\right)dt}$

$= \boxed{\dfrac{1}{54}\left(t+\dfrac{1}{2}\sin 2t\right)+c} = \boxed{\dfrac{1}{54}\left(t+\sin t\cos t\right)+c} = \boxed{\dfrac{1}{54}\left(\tan^{-1}\dfrac{x}{3}+\dfrac{x}{\sqrt{9+x^2}}\cdot\dfrac{3}{\sqrt{9+x^2}}\right)+c} = \boxed{\dfrac{1}{54}\left(\tan^{-1}\dfrac{x}{3}+\dfrac{3x}{9+x^2}\right)+c}$

Check: Let $\dfrac{1}{54}\left(\tan^{-1}\dfrac{x}{3}+\dfrac{3x}{9+x^2}\right)+c$ then $y' = \dfrac{1}{54}\cdot\dfrac{1}{3\left(1+\frac{x^2}{9}\right)} + \dfrac{1}{54}\cdot\dfrac{3\left(9+x^2\right)-2x\cdot 3x}{\left(9+x^2\right)^2} = \dfrac{1}{54}\cdot\dfrac{9}{3\left(9+x^2\right)}$

$+\dfrac{1}{54}\cdot\dfrac{27+3x^2-6x^2}{\left(9+x^2\right)^2} = \dfrac{1}{54}\cdot\dfrac{3}{\left(9+x^2\right)} + \dfrac{1}{54}\cdot\dfrac{27-3x^2}{\left(9+x^2\right)^2} = \dfrac{1}{18\left(9+x^2\right)} + \dfrac{9-x^2}{18\left(9+x^2\right)^2} = \dfrac{9+x^2+9-x^2}{18\left(9+x^2\right)^2}$

$= \dfrac{18}{18\left(9+x^2\right)^2} = \dfrac{1}{\left(9+x^2\right)^2}$

e. Given $\displaystyle\int \dfrac{x^2}{\sqrt{x^2-1}}\,dx$ let $x = \sec t$, then $dx = \sec t\tan t$ and $\sqrt{x^2-1} = \sqrt{\sec^2 t-1} = \sqrt{\tan^2 t} = \tan t$

Substituting these values back into the original integral we obtain $\displaystyle\int \dfrac{x^2}{\sqrt{x^2-1}}\,dx = \displaystyle\int \dfrac{\sec^2 t}{\tan t}\sec t\tan t\, dt$

$= \displaystyle\int \sec^3 t\, dt = \displaystyle\int \sec^2 t\sec t\, dt$. Let $u = \sec t$ and $dv = \sec^2 t$, then $du = \sec t\tan t$ and $v = \tan t$.

Using the substitution formula $uv - \displaystyle\int v\, du$ the integral $\displaystyle\int \sec^2 t\sec t\, dt$ can be rewritten as

$\displaystyle\int \sec^3 t\, dt = \sec t\tan t - \displaystyle\int \sec t\tan^2 t\ dt = \sec t\tan t - \displaystyle\int \sec t\left(\sec^2 t-1\right)dt = \sec t\tan t - \displaystyle\int \left(\sec^3 t-\sec t\right)dt$

$$= \sec t \tan t - \int \sec^3 t \; dt + \int \sec t \; dt$$

Note that $\int \sec^3 t \; dt = \sec t \tan t - \int \sec^3 t \; dt + \int \sec t \; dt$ therefore by moving $-\int \sec^3 t \; dt$ to the left

hand side of the equality we obtain

$$\boxed{\int \sec^3 t \; dt + \int \sec^3 t \; dt} = \boxed{\sec t \tan t + \int \sec t \; dt} \text{ thus } \boxed{2\int \sec^3 t \; dt} = \boxed{\sec t \tan t + \int \sec t \; dt} \text{ and } \boxed{\int \sec^3 t \; dt}$$

$$\boxed{\frac{1}{2}\left(\sec t \tan t + \int \sec t \; dt\right)} = \boxed{\frac{1}{2}\left(\sec t \tan t + \ln|\sec t + \tan t|\right) + c} = \boxed{\frac{1}{2}\left(x\sqrt{x^2-1} + \ln\left|x+\sqrt{x^2-1}\right|\right) + c}$$

Check: Let $y = \frac{1}{2}\left(x\sqrt{x^2-1} + \ln\left|x+\sqrt{x^2-1}\right|\right) + c$, then $y' = \frac{1}{2}\left(\sqrt{x^2-1} + \frac{2x^2}{2\sqrt{x^2-1}}\right) + \frac{1}{2}\frac{1}{x+\sqrt{x^2-1}}$

$$\times\left(1 + \frac{2x}{2\sqrt{x^2-1}}\right) = \frac{1}{2}\frac{2(x^2-1)+2x^2}{2\sqrt{x^2-1}} + \frac{1}{2}\frac{1}{x+\sqrt{x^2-1}}\frac{2\left(x+\sqrt{x^2-1}\right)}{2\sqrt{x^2-1}} = \frac{1}{2}\left(\frac{4x^2-2}{2\sqrt{x^2-1}} + \frac{1}{\sqrt{x^2-1}}\right)$$

$$= \frac{1}{2}\frac{4x^2-2+2}{2\sqrt{x^2-1}} = \frac{4x^2}{4\sqrt{x^2-1}} = \frac{x^2}{\sqrt{x^2-1}}$$

f. Given $\int \dfrac{dx}{x^4\sqrt{x^2-1}}$ let $x = \sec t$, then $dx = \sec t \tan t \; dt$ and $\sqrt{x^2-1} = \sqrt{\sec^2 t -1} = \sqrt{\tan^2 t} = \tan t$

Substituting these values back into the original integral we obtain

$$\boxed{\int \frac{dx}{x^4\sqrt{x^2-1}}} = \boxed{\int \frac{\sec t \tan t \; dt}{\sec^4 t \tan t}} = \boxed{\int \frac{dt}{\sec^3 t}} = \boxed{\int \cos^3 t \; dt} = \boxed{\int \cos^2 t \cos t \; dt} = \boxed{\int \left(1 - \sin^2 t\right) \cos t \; dt}$$

$$= \boxed{\int \left(\cos t - \sin^2 t \cos t\right) dt} = \boxed{\int \cos t \; dt - \int \sin^2 t \cos t \; dt} = \boxed{\sin t - \frac{1}{3}\sin^3 t + c} = \boxed{\frac{\sqrt{x^2-1}}{x} - \frac{1}{3}\left(\frac{\sqrt{x^2-1}}{x}\right)^3 + c}$$

Check: Let $y = \dfrac{\sqrt{x^2-1}}{x} - \dfrac{1}{3}\left(\dfrac{\sqrt{x^2-1}}{x}\right)^3 + c$, then $y' = \dfrac{\frac{2x^2}{2\sqrt{x^2-1}} - \sqrt{x^2-1}}{x^2} - \dfrac{3}{3}\left(\dfrac{\sqrt{x^2-1}}{x}\right)^2 \cdot \dfrac{\frac{2x^2}{2\sqrt{x^2-1}} - \sqrt{x^2-1}}{x^2}$

$$= \frac{\frac{x^2-x^2+1}{\sqrt{x^2-1}}}{x^2} - \left(\frac{x^2-1}{x^2}\right)\cdot\frac{\frac{x^2-x^2+1}{\sqrt{x^2-1}}}{x^2} = \frac{1}{x^2\sqrt{x^2-1}} - \left(\frac{x^2-1}{x^2}\right)\cdot\frac{1}{x^2\sqrt{x^2-1}} = \frac{1}{x^2\sqrt{x^2-1}} - \frac{x^2-1}{x^4\sqrt{x^2-1}}$$

$$= \frac{x^2-x^2+1}{x^4\sqrt{x^2-1}} = \frac{1}{x^4\sqrt{x^2-1}}$$

Example 5.2-2: Use trigonometric substitution to evaluate the following indefinite integrals:

a. $\int \dfrac{dx}{\left(1-x^2\right)^{\frac{3}{2}}} =$

b. $\int x^2 \sqrt{4-x^2}\, dx =$

c. $\int \dfrac{dx}{\left(4+9x^2\right)^{\frac{3}{2}}} =$

d. $\int \dfrac{x^2}{\sqrt{4+x^2}}\, dx =$

e. $\int \sqrt{a^2-x^2}\, dx =$

f. $\int \sqrt{x^2-a^2}\, dx =$

g. $\int \dfrac{x^2}{\sqrt{9-x^2}}\, dx =$

h. $\int \sqrt{9-x^2}\, dx =$

i. $\int \dfrac{1}{x^2\sqrt{4+x^2}}\, dx =$

Solutions:

a. Given $\int \dfrac{dx}{\left(1-x^2\right)^{\frac{3}{2}}}$ let $x = \sin t$, then $dx = \cos t\, dt$ and $1-x^2 = 1-\sin^2 t = \cos^2 t$. Therefore,

$$\boxed{\int \dfrac{dx}{\left(1-x^2\right)^{\frac{3}{2}}}} = \boxed{\int \dfrac{\cos t\, dt}{\left(\cos^2 t\right)^{\frac{3}{2}}}} = \boxed{\int \dfrac{\cos t}{\cos^3 t}\, dt} = \boxed{\int \dfrac{1}{\cos^2 t}\, dt} = \boxed{\int \sec^2 t\, dt} = \boxed{\tan t + c} = \boxed{\dfrac{\sin t}{\cos t} + c} = \boxed{\dfrac{x}{\sqrt{1-x^2}} + c}$$

Check: Let $y = \dfrac{x}{\sqrt{1-x^2}} + c$, then $y' = \dfrac{1 \cdot \sqrt{1-x^2} - \dfrac{-2x}{2\sqrt{1-x^2}} \cdot x}{1-x^2} = \dfrac{\dfrac{2-2x^2+2x^2}{2\sqrt{1-x^2}}}{1-x^2} = \dfrac{2}{2\left(1-x^2\right)^{\frac{3}{2}}} = \dfrac{1}{\left(1-x^2\right)^{\frac{3}{2}}}$

b. Given $\int x^2 \sqrt{4-x^2}\, dx$ let $x = 2\sin t$, then $dx = 2\cos t\, dt$ and $\sqrt{4-x^2} = \sqrt{4-(2\sin t)^2} = \sqrt{4-4\sin^2 t}$

$\sqrt{4\left(1-\sin^2 t\right)} = \sqrt{4\cos^2 t} = 2\cos t$. Therefore,

$$\boxed{\int x^2 \sqrt{4-x^2}\, dx} = \boxed{\int 4\sin^2 t \cdot 2\cos t \cdot 2\cos t\, dt} = \boxed{16\int \sin^2 t \cos^2 t\, dt} = \boxed{16\int \dfrac{1}{4}(1-\cos 2t)(1+\cos 2t)\, dt}$$

$$= \boxed{4\int\left(1-\cos^2 2t\right)dt} = \boxed{\int 4\, dt - 4\int \cos^2 2t\, dt} = \boxed{\int 4\, dt - 4\int \dfrac{1}{2}(1+\cos 4t)\, dt} = \boxed{\int 4\, dt - \int 2\, dt - 2\int \cos 4t\, dt}$$

$$= \boxed{4t - 2t - \dfrac{1}{2}\sin 4t + c} = \boxed{4t - 2t - \dfrac{4}{2}\sin 2t \cos 2t + c} = \boxed{2t - 2(\sin t \cos t)\left(\cos^2 t - \sin^2 t\right) + c}$$

$$= \boxed{2\sin^{-1}\dfrac{x}{2} - 2\left(\dfrac{x}{2} \cdot \dfrac{\sqrt{4-x^2}}{2}\right)\left(\dfrac{4-x^2}{4} - \dfrac{x^2}{4}\right) + c} = \boxed{2\sin^{-1}\dfrac{x}{2} - \sqrt{4-x^2}\left(\dfrac{2x-x^3}{4}\right) + c}$$

Check: Let $y = 2\sin^{-1}\dfrac{x}{2} - \sqrt{4-x^2}\left(\dfrac{2x-x^3}{4}\right) + c$, then $y' = \dfrac{2}{2\sqrt{1-\frac{x^2}{4}}} - \dfrac{-2x\left(2x-x^3\right)}{8\sqrt{4-x^2}} + \left(2-3x^2\right)\left(4-x^2\right)^{\frac{1}{2}}$

$$= \dfrac{2}{\sqrt{4-x^2}} - \dfrac{-2x^2+x^4+\left(2-3x^2\right)\left(4-x^2\right)}{4\sqrt{4-x^2}} = \dfrac{2}{\sqrt{4-x^2}} - \dfrac{-2x^2+x^4+8-14x^2+3x^4}{4\sqrt{4-x^2}}$$

$$= \frac{8-4x^4+16x^2-8}{4\sqrt{4-x^2}} = \frac{-4x^4+16x^2}{4\sqrt{4-x^2}} = \frac{-4x^4+16x^2}{4\sqrt{4-x^2}} \cdot \frac{\sqrt{4-x^2}}{\sqrt{4-x^2}} = \frac{4(4-x^2)x^2\sqrt{4-x^2}}{4(4-x^2)} = x^2\sqrt{4-x^2}$$

c. Given $\displaystyle\int \frac{dx}{\left(4+9x^2\right)^{\frac{3}{2}}} = \int \frac{dx}{\left(\frac{4}{9}+x^2\right)^{\frac{3}{2}}}$ let $x=\frac{2}{3}\tan t$, then $dx=\frac{2}{3}\sec^2 t\, dt$ and $4+9x^2 = 4+9\cdot\left(\frac{2}{3}\tan t\right)^2$

$$= 4+9\cdot\frac{4}{9}\tan^2 t = 4+4\tan^2 t = 4\left(1+\tan^2 t\right) = 4\sec^2 t .\text{ Therefore,}$$

$$\boxed{\int \frac{dx}{\left(4+9x^2\right)^{\frac{3}{2}}}} = \boxed{\int \frac{\frac{2}{3}\sec^2 t}{\left(4\sec^2 t\right)^{\frac{3}{2}}}dt} = \boxed{\int \frac{\frac{2}{3}\sec^2 t}{4^{\frac{3}{2}}\sec^{2\times\frac{3}{2}} t}dt} = \boxed{\int \frac{\frac{2}{3}\sec^2 t}{\sqrt[2]{4^3}\ \sec^3 t}dt} = \boxed{\frac{2}{3}\int \frac{dt}{\sqrt[2]{64}\ \sec t}} = \boxed{\frac{2}{3}\int \frac{dt}{8\sec t}}$$

$$= \boxed{\frac{1}{12}\int \frac{dt}{\sec t}} = \boxed{\frac{1}{12}\int \cos t\, dt} = \boxed{\frac{1}{12}\sin t+c} = \boxed{\frac{1}{12}\frac{3x}{\sqrt{4+9x^2}}+c} = \boxed{\frac{x}{4\sqrt{4+9x^2}}+c}$$

Check: Let $y=\dfrac{x}{4\sqrt{4+9x^2}}+c$, then $y' = \dfrac{1\cdot 4\sqrt{4+9x^2}-4\cdot\dfrac{18x}{2\sqrt{4+9x^2}}\cdot x}{16\left(4+9x^2\right)} = \dfrac{\dfrac{4\left(4+9x^2\right)-36x^2}{\sqrt{4+9x^2}}}{16\left(4+9x^2\right)} = \dfrac{1}{\left(4+9x^2\right)^{\frac{3}{2}}}$

d. Given $\displaystyle\int \frac{x^2 dx}{\sqrt{4+x^2}}$ let $x=2\tan t$, then $dx=2\sec^2 t\, dt$ and $\sqrt{4+x^2} = \sqrt{4+(2\tan t)^2} = \sqrt{4+4\tan^2 t}$

$$= \sqrt{4\left(1+\tan^2 t\right)} = \sqrt{4\sec^2 t} = 2\sec t .\text{ Therefore,}$$

$$\int \frac{x^2 dx}{\sqrt{4+x^2}} = \int \frac{4\tan^2 t\cdot 2\sec^2 t\, dt}{2\sec t} = 4\int \tan^2 t \sec t\, dt = 4\int\left(\sec^2 t-1\right)\sec t\, dt = 4\int\left(\sec^3 t-\sec t\right)dt$$

$= 4\int \sec^3 t\, dt-4\int \sec t\, dt$. To solve $\int \sec^3 t\, dt = \int \sec^2 t\cdot\sec t\, dt$ use substitution method by

letting $u=\sec t$ and $dv=\sec^2 t$ then $du=\sec t \tan t\, dt$ and $v=\tan t$. Using the substitution

formula $uv-\int v\, du$ we obtain $\int \sec^2 t\cdot\sec t\, dt = \sec t \tan t-\int \tan t\cdot\sec t \tan t\, dt$

$= \sec t \tan t-\int \sec t \tan^2 t\, dt = \sec t \tan t-\int \sec t\left(\sec^2 t-1\right)dt = \sec t \tan t-\int\left(\sec^3 t-\sec t\right)dt$

$= \sec t \tan t-\int \sec^3 t\, dt+\int \sec t\, dt$. Again at this point we know that $\int \sec^3 t\, dt$

$= \sec t \tan t-\int \sec^3 t\, dt+\int \sec t\, dt$ bringing $-\int \sec^3 t\, dt$ into the left hand side of the equation we

obtain $\int \sec^3 t\, dt+\int \sec^3 t\, dt = \sec t \tan t+\int \sec t\, dt$. Therefore $2\int \sec^3 t\, dt = \sec t \tan t+\int \sec t\, dt$

thus $\int \sec^3 t\, dt = \frac{1}{2}\left[\sec t \tan t+\int \sec t\, dt\right] = \frac{1}{2}\left[\sec t \tan t+\ln\left|\sec t+\tan t\right|\right]$. Now substituting this

value into $4\int \sec^3 t\, dt-4\int \sec t\, dt$ we have

$$4\int \sec^3 t \, dt - 4\int \sec t \, dt = \boxed{4 \cdot \frac{1}{2}\left[\sec t \tan t + \ln|\sec t + \tan t|\right] - 4\int \sec t \, dt} = \boxed{2\left[\sec t \tan t + \ln|\sec t + \tan t|\right]}$$

$$\boxed{-4\ln|\sec t + \tan t| + c} = \boxed{2\sec t \tan t + 2\ln|\sec t + \tan t| - 4\ln|\sec t + \tan t| + c} = \boxed{2\sec t \tan t}$$

$$\boxed{-2\ln|\sec t + \tan t| + c} = \boxed{2 \cdot \frac{\sqrt{4+x^2}}{2} \cdot \frac{x}{2} - 2\ln\left|\frac{\sqrt{4+x^2}}{2} + \frac{x}{2}\right| + c} = \boxed{\frac{x\sqrt{4+x^2}}{2} - 2\ln\left|\frac{x+\sqrt{4+x^2}}{2}\right| + c}$$

Check: Let $y = \dfrac{x\sqrt{4+x^2}}{2} - 2\ln\left|\dfrac{x+\sqrt{4+x^2}}{2}\right| + c$, then $y' = \dfrac{1}{2}\left(\sqrt{4+x^2} + \dfrac{2x^2}{2\sqrt{4+x^2}}\right)$

$$-2\left[\frac{2}{x+\sqrt{4+x^2}} \cdot \frac{1}{2} \cdot \left(1 + \frac{2x}{2\sqrt{4+x^2}}\right)\right] + 0 = \frac{1}{2}\left(\frac{4+2x^2}{\sqrt{4+x^2}}\right) - \frac{2\sqrt{4+x^2}+2x}{\left(x+\sqrt{4+x^2}\right)\sqrt{4+x^2}}$$

$$= \frac{\left(2+x^2\right)\left(x+\sqrt{4+x^2}\right) - 2\sqrt{4+x^2} - 2x}{\left(x+\sqrt{4+x^2}\right)\sqrt{4+x^2}} = \frac{\left(2+x^2\right)\left(x+\sqrt{4+x^2}\right) - 2\sqrt{4+x^2} - 2x}{\left(x+\sqrt{4+x^2}\right)\sqrt{4+x^2}} \cdot \frac{x-\sqrt{4+x^2}}{x-\sqrt{4+x^2}}$$

$$= \frac{\left(2+x^2\right)\left(x+\sqrt{4+x^2}\right) - 2\sqrt{4+x^2} - 2x}{\left(x+\sqrt{4+x^2}\right)\sqrt{4+x^2}} \cdot \frac{x-\sqrt{4+x^2}}{x-\sqrt{4+x^2}} = \frac{x^4 - x^3\sqrt{4+x^2} - x^3\sqrt{4+x^2} - x^2\left(4+x^2\right)}{-4\sqrt{4+x^2}}$$

$$= \frac{x^4 - 4x^2 - x^4}{-4\sqrt{4+x^2}} = \frac{-4x^2}{-4\sqrt{4+x^2}} = \frac{x^2}{\sqrt{4+x^2}}$$

e. Given $\int \sqrt{a^2 - x^2} \, dx$ let $x = a\sin t$, then $dx = a\cos t \, dt$ and $\sqrt{a^2 - x^2} = \sqrt{a^2 - a^2\sin^2 t}$

$$= \sqrt{a^2\left(1 - \sin^2 t\right)} = \sqrt{a^2\cos^2 t} = a\cos t . \text{ Therefore,}$$

$$\boxed{\int \sqrt{a^2 - x^2} \, dx} = \boxed{\int a\cos t \cdot a\cos t \, dt} = \boxed{\int a^2\cos^2 t \, dt} = \boxed{\frac{a^2}{2}\int (1+\cos 2t)\, dt} = \boxed{\frac{a^2}{2}\left(t + \frac{1}{2}\sin 2t\right) + c}$$

$$= \boxed{\frac{a^2}{2}(t + \sin t \cos t) + c} = \boxed{\frac{a^2}{2}\left(\sin^{-1}\frac{x}{a} + \frac{x}{a} \cdot \frac{\sqrt{a^2-x^2}}{a}\right) + c} = \boxed{\frac{a^2}{2}\sin^{-1}\frac{x}{a} + \frac{x\sqrt{a^2-x^2}}{2} + c}$$

Check: Let $y = \dfrac{a^2}{2}\sin^{-1}\dfrac{x}{a} + \dfrac{x\sqrt{a^2-x^2}}{2} + c$, then $y' = \dfrac{a^2}{2} \dfrac{1}{\sqrt{1-\left(\frac{x}{a}\right)^2}} \cdot \dfrac{1}{a} + \dfrac{1}{2}\sqrt{a^2-x^2} + \dfrac{-2x^2}{2\sqrt{a^2-x^2}}$

$$= \frac{a^2}{2} \cdot \frac{a}{a\sqrt{a^2-x^2}} + \frac{a^2-x^2-x^2}{2\sqrt{a^2-x^2}} = \frac{a^2}{2\sqrt{a^2-x^2}} + \frac{a^2-2x^2}{2\sqrt{a^2-x^2}} = \frac{2a^2-2x^2}{2\sqrt{a^2-x^2}} = \frac{a^2-x^2}{\sqrt{a^2-x^2}}$$

$$= \frac{a^2-x^2}{\sqrt{a^2-x^2}} \cdot \frac{\sqrt{a^2-x^2}}{\sqrt{a^2-x^2}} = \frac{\left(a^2-x^2\right)\sqrt{a^2-x^2}}{a^2-x^2} = \sqrt{a^2-x^2}$$

f. Given $\int \sqrt{x^2 - a^2}\, dx$ let $x = a \sec t$, then $dx = a \sec t \tan t\, dt$ and $\sqrt{x^2 - a^2} = \sqrt{a^2 \sec^2 t - a^2}$

$= \sqrt{a^2 \tan^2 t} = a \tan t$. Therefore,

$$\boxed{\int \sqrt{x^2 - a^2}\, dx} = \boxed{\int a \tan t \cdot a \sec t \tan t\, dt} = \boxed{\int a^2 \sec t \tan^2 t\, dt} = \boxed{a^2 \int \sec t \left(\sec^2 t - 1\right) dt} = \boxed{a^2 \int \sec^3 t\, dt}$$

$$\boxed{-a^2 \int \sec t\, dt} = \boxed{\frac{a^2}{2}\left(\tan t \sec t + \ln|\sec t + \tan t|\right) - a^2 \ln|\sec t + \tan t| + c} = \boxed{\frac{a^2}{2}\left(\tan t \sec t - \ln|\sec t + \tan t|\right) + c}$$

$$= \boxed{\frac{a^2}{2}\left(\frac{x}{a} \cdot \frac{\sqrt{x^2-a^2}}{a} - \ln\left|\frac{x}{a} + \frac{\sqrt{x^2-a^2}}{a}\right|\right) + c} = \boxed{\frac{x}{2}\sqrt{x^2-a^2} - \frac{a^2}{2}\ln\left|\frac{x+\sqrt{x^2-a^2}}{a}\right| + c}$$

$$= \boxed{\frac{x}{2}\sqrt{x^2-a^2} - \frac{a^2}{2}\ln\left|x+\sqrt{x^2-a^2}\right| + \frac{a^2}{2}\ln a + c} = \boxed{\frac{x}{2}\sqrt{x^2-a^2} - \frac{a^2}{2}\ln\left|x+\sqrt{x^2-a^2}\right| + c}$$

Check: Let $y = \frac{x}{2}\sqrt{x^2-a^2} - \frac{a^2}{2}\ln\left|x+\sqrt{x^2-a^2}\right| + c$, then $y' = \frac{1}{2}\left(\sqrt{x^2-a^2} + \frac{2x^2}{2\sqrt{x^2-a^2}}\right)$

$$-\frac{a^2}{2}\cdot\frac{1}{x+\sqrt{x^2-a^2}}\cdot\left(1+\frac{2x}{2\sqrt{x^2-a^2}}\right) + 0 = \frac{x^2-a^2+x^2}{2\sqrt{x^2-a^2}} - \frac{a^2}{2}\cdot\frac{1}{x+\sqrt{x^2-a^2}}\cdot\frac{x+\sqrt{x^2-a^2}}{\sqrt{x^2-a^2}}$$

$$= \frac{x^2-a^2+x^2}{2\sqrt{x^2-a^2}} - \frac{a^2}{2\sqrt{x^2-a^2}} = \frac{2x^2-2a^2}{2\sqrt{x^2-a^2}} = \frac{x^2-a^2}{\sqrt{x^2-a^2}} = \frac{x^2-a^2}{\sqrt{x^2-a^2}}\cdot\frac{\sqrt{x^2-a^2}}{\sqrt{x^2-a^2}} = \sqrt{x^2-a^2}$$

g. Given $\int \frac{x^2}{\sqrt{9-x^2}}\, dx$ let $x = 3\sin t$, then $dx = 3\cos t\, dt$ and $\sqrt{9-x^2} = \sqrt{9-9\sin^2 t} = \sqrt{9\left(1-\sin^2 t\right)}$

$= \sqrt{9\cos^2 t} = 3\cos t$. Substituting these values back into the original integral we obtain:

$$\boxed{\int \frac{x^2\, dx}{\sqrt{9-x^2}}} = \boxed{\int \frac{9\sin^2 t \cdot 3\cos t\, dt}{3\cos t}} = \boxed{\int \frac{9\sin^2 t}{1}\, dt} = \boxed{9\int \sin^2 t\, dt} = \boxed{9\int \frac{1-\cos 2t}{2}\, dt}$$

$$= \boxed{\frac{9}{2}\int\left(1-\cos 2t\right) dt} = \boxed{\frac{9}{2}\left(t-\frac{1}{2}\sin 2t\right) + c} = \boxed{\frac{9}{2}t - \frac{9}{2}\sin t\cos t + c} = \boxed{\frac{9}{2}\left(\sin^{-1}\frac{x}{3}\right) - \frac{9}{2}\left(\frac{x}{3}\cdot\frac{\sqrt{9-x^2}}{3}\right) + c}$$

$$= \boxed{\frac{9}{2}\left(\sin^{-1}\frac{x}{3}\right) - \frac{9}{2}\left(\frac{x}{9}\sqrt{9-x^2}\right) + c} = \boxed{\frac{9}{2}\left(\sin^{-1}\frac{x}{3}\right) - \frac{x}{2}\sqrt{9-x^2} + c}$$

Check: Let $y = \frac{9}{2}\sin^{-1}\frac{x}{3} - \frac{x}{2}\sqrt{9-x^2} + c$, then $y' = \frac{9}{2}\frac{1}{\sqrt{1-\frac{x^2}{9}}}\cdot\frac{1}{3} - \left(\frac{1}{2}\sqrt{9-x^2} + \frac{-2x}{2\sqrt{9-x^2}}\cdot\frac{x}{2}\right)$

$$= \frac{9}{2} \frac{3}{\sqrt{9-x^2}} \cdot \frac{1}{3} - \left(\frac{\sqrt{9-x^2}}{2} - \frac{x^2}{2\sqrt{9-x^2}} \right) = \frac{9}{2\sqrt{9-x^2}} - \left(\frac{2\left(9-x^2\right)-2x^2}{4\sqrt{9-x^2}} \right) = \frac{9}{2\sqrt{9-x^2}}$$

$$- \left(\frac{18-4x^2}{4\sqrt{9-x^2}} \right) = \frac{9}{2\sqrt{9-x^2}} - \frac{9-2x^2}{2\sqrt{9-x^2}} = \frac{9-9+2x^2}{2\sqrt{9-x^2}} = \frac{2x^2}{2\sqrt{9-x^2}} = \frac{x^2}{\sqrt{9-x^2}}$$

h. Given $\int \sqrt{9-x^2}\, dx$ let $x = 3\sin t$, then $dx = 3\cos t\, dt$ and $\sqrt{9-x^2} = \sqrt{9-9\sin^2 t} = \sqrt{9\left(1-\sin^2 t\right)}$

$= \sqrt{9\cos^2 t} = 9\cos t$. Therefore,

$$\boxed{\int \sqrt{9-x^2}\, dx} = \boxed{\int 3\cos t \cdot 3\cos t\, dt} = \boxed{\int 9\cos^2 t\, dt} = \boxed{\frac{9}{2}\int (1+\cos 2t)\, dt} = \boxed{\frac{9}{2}\left(t + \frac{1}{2}\sin 2t \right) + c}$$

$$= \boxed{\frac{9}{2}\left(t + \sin t \cos t \right) + c} = \boxed{\frac{9}{2}\left(\sin^{-1}\frac{x}{3} + \frac{x}{3} \cdot \frac{\sqrt{9-x^2}}{3} \right) + c} = \boxed{\frac{9}{2}\sin^{-1}\frac{x}{3} + \frac{x\sqrt{9-x^2}}{2} + c}$$

Check: Let $y = \frac{9}{2}\sin^{-1}\frac{x}{3} + \frac{x\sqrt{9-x^2}}{2} + c$, then $y' = \frac{9}{2}\frac{1}{\sqrt{1-\left(\frac{x}{3}\right)^2}} \cdot \frac{1}{3} + \frac{1}{2}\sqrt{9-x^2} + \frac{-2x^2}{2\sqrt{9-x^2}}$

$$= \frac{9}{2} \cdot \frac{3}{3\sqrt{9-x^2}} + \frac{9-x^2-x^2}{2\sqrt{9-x^2}} = \frac{9}{2\sqrt{9-x^2}} + \frac{9-2x^2}{2\sqrt{9-x^2}} = \frac{9+9-2x^2}{2\sqrt{9-x^2}} = \frac{18-2x^2}{2\sqrt{9-x^2}} = \frac{2\left(9-x^2\right)}{2\sqrt{9-x^2}}$$

$$= \frac{9-x^2}{2\sqrt{9-x^2}} = \frac{9-x^2}{\sqrt{9-x^2}} \cdot \frac{\sqrt{9-x^2}}{\sqrt{9-x^2}} = \frac{\left(9-x^2\right)\sqrt{9-x^2}}{9-x^2} = \sqrt{9-x^2}$$

i. Given $\int \frac{1}{x^2\sqrt{4+x^2}}\, dx$ let $x = 2\tan t$, then $dx = 2\sec^2 t\, dt$ and $\sqrt{4+x^2} = \sqrt{4+4\tan^2 t} = 2\sqrt{1+\tan^2 t}$

$= 2\sqrt{\sec^2 t} = 2\sec t$. Therefore,

$$\boxed{\int \frac{dx}{x^2\sqrt{4+x^2}}\, dx} = \boxed{\int \frac{2\sec^2 t}{4\tan^2 t \cdot \left(2\sec t\right)}\, dt} = \boxed{\frac{1}{4}\int \frac{\sec dt}{\tan^2 t}\, dt} = \boxed{\frac{1}{4}\int \frac{1}{\cos t} \cdot \frac{\cos^2 t}{\sin^2 t}\, dt} = \boxed{\frac{1}{4}\int \frac{\cos t}{\sin^2 t}\, dt} \text{ let } u = \sin t$$

$$= \boxed{\frac{1}{4}\int \frac{\cos t}{u^2} \cdot \frac{du}{\cos t}} = \boxed{\frac{1}{4}\int \frac{1}{u^2}\, du} = \boxed{-\frac{1}{4}u^{-1} + c} = \boxed{-\frac{1}{4u} + c} = \boxed{-\frac{1}{4\sin t} + c} = \boxed{-\frac{1}{4}\frac{\sqrt{4+x^2}}{x} + c}$$

Check: Let $y = -\frac{1}{4}\frac{\sqrt{4+x^2}}{x} + c$, then $y' = -\frac{1}{4}\frac{\frac{2x}{2\sqrt{4+x^2}} \cdot x - 1 \cdot \sqrt{4+x^2}}{x^2} = -\frac{1}{4}\frac{\frac{2x^2}{2\sqrt{4+x^2}} - \sqrt{4+x^2}}{x^2}$

$$= -\frac{1}{4}\frac{\frac{x^2 - \left(\sqrt{4+x^2}\right)\left(\sqrt{4+x^2}\right)}{\sqrt{4+x^2}}}{x^2} = -\frac{1}{4}\frac{\frac{x^2-4-x^2}{\sqrt{4+x^2}}}{4x^2} = -\frac{-4}{4x^2\sqrt{4+x^2}} = \frac{1}{x^2\sqrt{4+x^2}}$$

The following are additional standard forms of integration that have already been derived. Trigonometric substitution can be used in most of these cases in order to confirm the result.

Table 5.2-1: Integration Formulas

1. $\int \dfrac{dx}{a^2+x^2} = \dfrac{1}{a}\tan^{-1}\dfrac{x}{a}+c$	2. $\int \dfrac{dx}{\left(a^2+x^2\right)^2} = \dfrac{x}{2a^2\left(a^2+x^2\right)}+\dfrac{1}{2a^3}\tan^{-1}\dfrac{x}{a}+c$
3. $\int \dfrac{dx}{a^2-x^2} = \dfrac{1}{2a}\ln\left|\dfrac{x+a}{x-a}\right|+c = \dfrac{1}{2a}\ln\left|\dfrac{a+x}{a-x}\right|+c$	4. $\int \dfrac{dx}{\left(a^2-x^2\right)^2} = \dfrac{x}{2a^2\left(a^2-x^2\right)}+\dfrac{1}{2a^2}\int\dfrac{dx}{a^2-x^2}$
5. $\int \dfrac{dx}{\sqrt{a^2+x^2}} = \sinh^{-1}\dfrac{x}{a}+c = \ln\left|x+\sqrt{a^2+x^2}\right|+c$	6. $\int \sqrt{a^2+x^2}\,dx = \dfrac{x}{2}\sqrt{a^2+x^2}+\dfrac{a^2}{2}\sinh^{-1}\dfrac{x}{a}+c$
7. $\int x^2\sqrt{a^2+x^2}\,dx = \dfrac{x\left(a^2+2x^2\right)\sqrt{a^2+x^2}}{8}-\dfrac{a^4}{8}\sinh^{-1}\dfrac{x}{a}+c$	8. $\int \dfrac{\sqrt{a^2+x^2}}{x}\,dx = \sqrt{a^2+x^2}-a\sinh^{-1}\left|\dfrac{a}{x}\right|+c$
9. $\int \dfrac{\sqrt{a^2+x^2}}{x}\,dx = \sinh^{-1}\dfrac{x}{a}-\dfrac{\sqrt{a^2+x^2}}{x}+c$	10. $\int \dfrac{x^2}{\sqrt{a^2+x^2}}\,dx = -\dfrac{a^2}{2}\sinh^{-1}\dfrac{x}{a}+\dfrac{x\sqrt{a^2+x^2}}{2}+c$
11. $\int \dfrac{dx}{x\sqrt{a^2+x^2}} = -\dfrac{1}{a}\ln\left|\dfrac{a+\sqrt{a^2+x^2}}{x}\right|+c$	12. $\int \dfrac{dx}{x^2\sqrt{a^2+x^2}} = -\dfrac{\sqrt{a^2+x^2}}{a^2x}+c$
13. $\int \dfrac{dx}{\sqrt{a^2-x^2}} = \sin^{-1}\dfrac{x}{a}+c$	14. $\int \sqrt{a^2-x^2}\,dx = \dfrac{x}{2}\sqrt{a^2-x^2}+\dfrac{a^2}{2}\sin^{-1}\dfrac{x}{a}+c$
15. $\int x^2\sqrt{a^2-x^2}\,dx = \dfrac{a^4}{8}\sin^{-1}\dfrac{x}{a}-\dfrac{1}{8}x\sqrt{a^2-x^2}\left(a^2-2x^2\right)+c$	16. $\int \dfrac{\sqrt{a^2-x^2}}{x}\,dx = \sqrt{a^2-x^2}-a\ln\left|\dfrac{a+\sqrt{a^2-x^2}}{x}\right|+c$
17. $\int \dfrac{\sqrt{a^2-x^2}}{x^2}\,dx = -\sin^{-1}\dfrac{x}{a}-\dfrac{\sqrt{a^2-x^2}}{x}+c$	18. $\int \dfrac{x^2}{\sqrt{a^2-x^2}}\,dx = \dfrac{a^2}{2}\sin^{-1}\dfrac{x}{a}-\dfrac{1}{2}x\sqrt{a^2-x^2}+c$
19. $\int \dfrac{dx}{x\sqrt{a^2-x^2}} = -\dfrac{1}{a}\ln\left|\dfrac{a+\sqrt{a^2-x^2}}{x}\right|+c$	20. $\int \dfrac{dx}{x^2\sqrt{a^2-x^2}} = \dfrac{\sqrt{a^2-x^2}}{a^2x}+c$
21. $\int \dfrac{dx}{\sqrt{x^2-a^2}} = \cosh^{-1}\dfrac{x}{a}+c = \ln\left|x+\sqrt{x^2-a^2}\right|+c$	22. $\int \sqrt{x^2-a^2}\,dx = \dfrac{x}{2}\sqrt{x^2-a^2}-\dfrac{a^2}{2}\cosh^{-1}\dfrac{x}{a}+c$
23. $\int x^2\sqrt{x^2-a^2}\,dx = \dfrac{x}{8}\left(2x^2-a^2\right)\sqrt{x^2-a^2}-\dfrac{a^4}{8}\cosh^{-1}\dfrac{x}{a}+c$	24. $\int \dfrac{\sqrt{x^2-a^2}}{x}\,dx = \sqrt{x^2-a^2}-a\sec^{-1}\left|\dfrac{x}{a}\right|+c$
25. $\int \dfrac{\sqrt{x^2-a^2}}{x^2}\,dx = \cosh^{-1}\dfrac{x}{a}-\dfrac{\sqrt{x^2-a^2}}{x}+c$	26. $\int \dfrac{x^2}{\sqrt{x^2-a^2}}\,dx = \dfrac{a^2}{2}\cosh^{-1}\dfrac{x}{a}+\dfrac{x}{2}\sqrt{x^2-a^2}+c$
27. $\int \dfrac{dx}{x\sqrt{x^2-a^2}} = \dfrac{1}{a}\sec^{-1}\left|\dfrac{x}{a}\right|+c = \dfrac{1}{a}\cos^{-1}\left|\dfrac{a}{x}\right|+c$	28. $\int \dfrac{dx}{x^2\sqrt{x^2-a^2}} = \dfrac{\sqrt{x^2-a^2}}{a^2x}+c$
29. $\int \dfrac{dx}{\sqrt{2ax-x^2}} = \sin^{-1}\left(\dfrac{x-a}{a}\right)+c$	30. $\int \sqrt{a^2+x^2}\,dx = \dfrac{x}{2}\sqrt{a^2+x^2}+\dfrac{a^2}{2}\ln\left(x+\sqrt{x^2+a^2}\right)+c$

Evaluate the following indefinite integrals:

a. $\displaystyle \int \frac{dx}{x^2\sqrt{16-x^2}} =$

b. $\displaystyle \int \frac{x^2}{\sqrt{9-x^2}}\, dx =$

c. $\displaystyle \int \frac{dx}{x\sqrt{9+4x^2}} =$

d. $\displaystyle \int \frac{1}{\left(49+x^2\right)^2}\, dx =$

e. $\displaystyle \int \left(\frac{x^2}{\sqrt{x^2-1}} + 5x \right) dx =$

f. $\displaystyle \int \sqrt{x^2-25}\, dx =$

g. $\displaystyle \int \sqrt{36-x^2}\, dx =$

h. $\displaystyle \int \frac{dx}{\left(9+36x^2\right)^{\frac{3}{2}}} =$

i. $\displaystyle \int \frac{\sqrt{9-4x^2}}{x}\, dx =$

5.3   Integration by Partial Fractions

The primary objective of this section is to show that rational functions can be integrated by breaking them into simpler parts. A function $F(x) = \dfrac{f(x)}{g(x)}$, where $f(x)$ and $g(x)$ are polynomials, is referred to as a rational function. Depending on the degree of the $f(x)$, the function $F(x)$ is either a proper or an improper rational fraction.

- A proper rational fraction is a fraction where the degree of the numerator $f(x)$ is less than the degree of the denominator $g(x)$. For example, $\dfrac{x+1}{x^2-x-6}$, $\dfrac{1}{x^3-1}$, $\dfrac{1}{x^2+2x-3}$, and $\dfrac{x}{x^2+1}$ are proper rational fractions.

- An improper rational fraction is a fraction where the degree of the numerator $f(x)$ is at least as large as the degree of the denominator $g(x)$. In such cases, long division is used in order to reduce the fraction to the sum of a polynomial and a proper rational fraction. For example, $\dfrac{x^3+2}{x^2-x-6}$, $\dfrac{x^3}{x^2+1}$, and $\dfrac{x^4-x^3-x-1}{x^3-x^2}$ are improper rational fractions. (See Section 6.3 of *Mastering Algebra – An Introduction* for solved problems on long division.)

A fraction, depending on its classification of the denominator, can be represented in four different cases. These cases are as follows:

CASE I - The Denominator Has Distinct Linear Factors

In this case the linear factors of the form $ax+b$ appear only once in the denominator. To solve this class of rational fractions we equate each proper rational fraction with a single fraction of the form $\dfrac{A}{ax+b}$, $\dfrac{B}{cx+d}$, $\dfrac{C}{ex+f}$, etc. The following examples show the steps as to how this class of integrals are solved.

Example 5.3-1: Evaluate the integral $\displaystyle\int \frac{x+1}{x^3+x^2-6x}\,dx$.

First - Check to see if the integrand is a proper or an improper rational fraction. If the integrand is an improper rational fraction use synthetic division (long division) to reduce the rational fraction to the sum of a polynomial and a proper rational fraction.

Second - Factor the denominator x^3+x^2-6x into $x(x-2)(x+3)$.

Third - Write the linear factors in partial fraction form. Since each linear factor in the denominator is occurring only once, the integrand can be represented in the following way:

$$\frac{x+1}{x^3+x^2-6x} = \frac{x+1}{x(x-2)(x+3)} = \frac{A}{x} + \frac{B}{x-2} + \frac{C}{x+3}$$

Fourth - Solve for the constants A, B, and C by equating coefficients of the like powers.

$$\frac{x+1}{x^3+x^2-6x} = \frac{A(x-2)(x+3)+Bx(x+3)+Cx(x-2)}{x(x-2)(x+3)}$$

$$x+1 = A\left(x^2 + 3x - 2x - 6\right) + B\left(x^2 + 3x\right) + C\left(x^2 - 2x\right) = Ax^2 + Ax - 6A + Bx^2 + 3Bx + Cx^2 - 2Cx$$

$$x+1 = (A + B + C)x^2 + (A + 3B - 2C)x - 6A \quad \text{therefore,}$$

$$A + B + C = 0 \qquad\qquad A + 3B - 2C = 1 \qquad\qquad -6A = 1$$

which result in having $A = -\dfrac{1}{6}$, $B = \dfrac{3}{10}$, and $C = -\dfrac{2}{15}$

Fifth - Rewrite the integral in its equivalent partial fraction form by substituting the constants with their specific values.

$$\int \frac{x+1}{x^3 + x^2 - 6x}\,dx = \int \frac{A}{x}\,dx + \int \frac{B}{x-2}\,dx + \int \frac{C}{x+3}\,dx = -\frac{1}{6}\int\frac{1}{x}\,dx + \frac{3}{10}\int\frac{1}{x-2}\,dx - \frac{2}{15}\int\frac{1}{x+3}\,dx$$

Sixth - Integrate each integral individually using integration methods learned in previous sections.

$$-\frac{1}{6}\int\frac{1}{x}\,dx + \frac{3}{10}\int\frac{1}{x-2}\,dx - \frac{2}{15}\int\frac{1}{x+3}\,dx = -\frac{1}{6}\ln|x| + \frac{3}{10}\ln|x-2| - \frac{2}{15}\ln|x+3| + c$$

Seventh - Check the answer by differentiating the solution. The result should match the integrand.

Let $y = -\dfrac{1}{6}\ln|x| + \dfrac{3}{10}\ln|x-2| - \dfrac{2}{15}\ln|x+3| + c$, then $y' = -\dfrac{1}{6}\cdot\dfrac{1}{x}\cdot 1 + \dfrac{3}{10}\cdot\dfrac{1}{x-2}\cdot 1 - \dfrac{2}{15}\cdot\dfrac{1}{x+3}\cdot 1 + 0$

$$= \frac{-150(x-2)(x+3) + 270x(x+3) - 120x(x-2)}{900x(x-2)(x+3)} = \frac{-150\left(x^2 + x - 6\right) + 270\left(x^2 + 3x\right) - 120\left(x^2 - 2x\right)}{900x\left(x^3 + x^2 - 6x\right)}$$

$$= \frac{-150x^2 - 150x + 900 + 270x^2 + 810x - 120x^2 + 240x}{900\left(x^3 + x^2 - 6x\right)} = \frac{(-150 + 270 - 120)x^2 + (-150 + 810 + 240)x + 900}{900\left(x^3 + x^2 - 6x\right)}$$

$$= \frac{900x + 900}{900\left(x^3 + x^2 - 6x\right)} = \frac{900(x+1)}{900\left(x^3 + x^2 - 6x\right)} = \frac{x+1}{x^3 + x^2 - 6x}$$

Example 5.3-2: Evaluate the integral $\displaystyle\int \frac{dx}{x^2 + 3x + 2}$.

First - Check to see if the integrand is a proper or an improper rational fraction. If the integrand is an improper rational fraction use synthetic division (long division) to reduce the rational fraction to the sum of a polynomial and a proper rational fraction.

Second - Factor the denominator $x^2 + 3x + 2$ into $(x+1)(x+2)$.

Third - Write the linear factors in partial fraction form. Since each linear factor in the denominator is occurring only once, the integrand can be represented in the following way:

$$\frac{1}{x^2 + 3x + 2} = \frac{1}{(x+1)(x+2)} = \frac{A}{x+1} + \frac{B}{x+2}$$

Fourth - Solve for the constants A and B by equating coefficients of the like powers.

$$\frac{1}{x^2 + 3x + 2} = \frac{A(x+2) + B(x+1)}{(x+1)(x+2)}$$

$$1 = A(x+2)+B(x+1) = Ax+2A+Bx+B$$

$$1 = (A+B)x+(2A+B) \text{ therefore,}$$

$$A+B=0 \qquad\qquad 2A+B=1$$

which result in having $A=1$ and $B=-1$

Fifth - Rewrite the integral in its equivalent partial fraction form by substituting the constants with their specific values.

$$\int \frac{1}{x^2+3x+2}dx = \int \frac{A}{x+1}dx + \int \frac{B}{x+2}dx = \int \frac{1}{x+1}dx - \int \frac{1}{x+2}dx$$

Sixth - Integrate each integral individually using integration methods learned in previous sections.

$$\int \frac{1}{x+1}dx - \int \frac{1}{x+2}dx = \ln|x+1|-\ln|x+2|+c$$

Seventh - Check the answer by differentiating the solution. The result should match the integrand.

Let $y=\ln|x+1|-\ln|x+2|+c$, then $y' = \frac{1}{x+1}\cdot 1 - \frac{1}{x+2}\cdot 1 + 0 = \frac{(x+2)-(x+1)}{(x+1)(x+2)} = \frac{1}{x^2+3x+2}$

Example 5.3-3: Evaluate the integral $\int \frac{x\,dx}{x^2-5x+6}$.

First - Check to see if the integrand is a proper or an improper rational fraction. If the integrand is an improper rational fraction use synthetic division (long division) to reduce the rational fraction to the sum of a polynomial and a proper rational fraction.

Second - Factor the denominator x^2-5x+6 into $(x-2)(x-3)$.

Third - Write the linear factors in partial fraction form. Since each linear factor in the denominator is occurring only once, the integrand can be represented in the following way:

$$\frac{x}{x^2-5x+6} = \frac{x}{(x-2)(x-3)} = \frac{A}{x-2}+\frac{B}{x-3}$$

Fourth - Solve for the constants A and B by equating coefficients of the like powers.

$$\frac{x}{x^2-5x+6} = \frac{A(x-3)+B(x-2)}{(x-2)(x-3)}$$

$$x = A(x-3)+B(x-2) = Ax-3A+Bx-2B$$

$$x = (A+B)x-(3A+2B) \text{ therefore,}$$

$$A+B=1 \qquad\qquad 3A+2B=0$$

which result in having $A=-2$ and $B=3$

Fifth - Rewrite the integral in its equivalent partial fraction form by substituting the constants with their specific values.

$$\int \frac{x\,dx}{x^2-5x+6} = \int \frac{A}{x-2}dx + \int \frac{B}{x-3}dx = -\int \frac{2}{x-2}dx + \int \frac{3}{x-3}dx$$

Sixth - Integrate each integral individually using integration methods learned in previous sections.

$$-\int \frac{2}{x-2}dx + \int \frac{3}{x-3}dx = -2\ln|x-2| + 3\ln|x-3| + c$$

Seventh - Check the answer by differentiating the solution. The result should match the integrand.

Let $y = -2\ln|x-2| + 3\ln|x-3| + c$, then $y' = -2\cdot\frac{1}{x-2}\cdot 1 + 3\cdot\frac{1}{x-3}\cdot 1 + 0 = \frac{-2(x-3)+3(x-2)}{(x-2)(x-3)}$

$$= \frac{-2x+6+3x-6}{x^2-5x+6} = \frac{x}{x^2-5x+6}$$

Example 5.3-4: Evaluate the integral $\int \frac{x^2+1}{x^3-x}dx$.

First - Check to see if the integrand is a proper or an improper rational fraction. If the integrand is an improper rational fraction use synthetic division (long division) to reduce the rational fraction to the sum of a polynomial and a proper rational fraction.

Second - Factor the denominator x^3-x into $x(x^2-1) = x(x-1)(x+1)$.

Third - Write the linear factors in partial fraction form. Since each linear factor in the denominator is occurring only once, the integrand can be represented in the following way:

$$\frac{x^2+1}{x^3-x} = \frac{x^2+1}{x(x-1)(x+1)} = \frac{A}{x} + \frac{B}{x-1} + \frac{C}{x+1}$$

Fourth - Solve for the constants A, B, and C by equating coefficients of the like powers.

$$\frac{x^2+1}{x^3-x} = \frac{A(x-1)(x+1)+Bx(x+1)+Cx(x-1)}{x(x-1)(x+1)}$$

$$x^2+1 = A(x^2+x-x-1)+B(x^2+x)+C(x^2-x) = Ax^2-A+Bx^2+Bx+Cx^2-Cx$$

$$x^2+1 = (A+B+C)x^2 + (B-C)x - A \text{ therefore,}$$

$$A+B+C=1 \qquad\qquad B-C=0 \qquad\qquad -A=1$$

which result in having $A=-1$, $B=1$, and $C=1$

Fifth - Rewrite the integral in its equivalent partial fraction form by substituting the constants with their specific values.

$$\int \frac{x^2+1}{x^3-x}dx = \int \frac{A}{x}dx + \int \frac{B}{x-1}dx + \int \frac{C}{x+1}dx = -\int \frac{1}{x}dx + \int \frac{1}{x-1}dx + \int \frac{1}{x+1}dx$$

Sixth - Integrate each integral individually using integration methods learned in previous sections.

$$-\int \frac{1}{x}dx + \int \frac{1}{x-1}dx + \int \frac{1}{x+1}dx = -\ln|x| + \ln|x-1| + \ln|x+1| + c$$

Seventh - Check the answer by differentiating the solution. The result should match the integrand.

Let $y = -\ln|x| + \ln|x-1| + \ln|x+1| + c$, then $y' = -\dfrac{1}{x} + \dfrac{1}{x-1} + \dfrac{1}{x+1} + 0 = \dfrac{-(x-1)(x+1) + x(x+1) + x(x-1)}{x(x-1)(x+1)}$

$$= \dfrac{-x^2 + 1 + x^2 + x + x^2 - x}{x(x^2 - 1)} = \dfrac{x^2 + 1}{x^3 - x}$$

Example 5.3-5: Evaluate the integral $\displaystyle\int \dfrac{x-3}{x(x^2 + x - 2)}\, dx$.

First - Check to see if the integrand is a proper or an improper rational fraction. If the integrand is an improper rational fraction use synthetic division (long division) to reduce the rational fraction to the sum of a polynomial and a proper rational fraction.

Second - Factor the denominator $x^2 + x - 2$ into $(x+2)(x-1)$.

Third - Write the linear factors in partial fraction form. Since each linear factor in the denominator is occurring only once, the integrand can be represented in the following way:

$$\dfrac{x-3}{x(x+2)(x-1)} = \dfrac{A}{x} + \dfrac{B}{x+2} + \dfrac{C}{x-1}$$

Fourth - Solve for the constants A, B, and C by equating coefficients of the like powers.

$$\dfrac{x-3}{x(x+2)(x-1)} = \dfrac{A(x+2)(x-1) + Bx(x-1) + Cx(x+2)}{x(x+2)(x-1)}$$

$$x-3 = A(x^2 + 2x - x - 2) + B(x^2 - x) + C(x^2 + 2x) = Ax^2 + Ax - 2A + Bx^2 - Bx + Cx^2 + 2Cx$$

$$x-3 = (A+B+C)x^2 + (A-B+2C)x - 2A \quad \text{therefore,}$$

$$A+B+C = 0 \qquad\qquad A-B+2C = 1 \qquad\qquad -2A = -3$$

which result in having $A = \dfrac{3}{2}$, $B = -\dfrac{5}{6}$, and $C = -\dfrac{2}{3}$

Fifth - Rewrite the integral in its equivalent partial fraction form by substituting the constants with their specific values.

$$\int \dfrac{x-3}{x(x^2 + x - 2)}\, dx = \int \dfrac{A}{x}\, dx + \int \dfrac{B}{x+2}\, dx + \int \dfrac{C}{x-1}\, dx = \dfrac{3}{2}\int \dfrac{1}{x}\, dx - \dfrac{5}{6}\int \dfrac{1}{x+2}\, dx - \dfrac{2}{3}\int \dfrac{1}{x-1}\, dx$$

Sixth - Integrate each integral individually using integration methods learned in previous sections.

$$\dfrac{3}{2}\int \dfrac{1}{x}\, dx - \dfrac{5}{6}\int \dfrac{1}{x+2}\, dx - \dfrac{2}{3}\int \dfrac{1}{x-1}\, dx = \dfrac{3}{2}\ln|x| - \dfrac{5}{6}\ln|x+2| - \dfrac{2}{3}\ln|x-1| + c$$

Seventh - Check the answer by differentiating the solution. The result should match the integrand.

Let $y = \dfrac{3}{2}\ln|x| - \dfrac{5}{6}\ln|x+2| - \dfrac{2}{3}\ln|x-1| + c$, then $y' = \dfrac{3}{2}\cdot\dfrac{1}{x}\cdot 1 - \dfrac{5}{6}\cdot\dfrac{1}{x+2}\cdot 1 - \dfrac{2}{3}\cdot\dfrac{1}{x-1}\cdot 1 + 0$

$$= \dfrac{9(x+2)(x-1) - 5x(x-1) - 4x(x+2)}{6x(x+2)(x-1)} = \dfrac{9(x^2 + x - 2) - 5(x^2 - x) - 4(x^2 + 2x)}{6x(x^2 + x - 2)} = \dfrac{(9-5-4)x^2 + (9+5-8)x - 18}{6x(x^2 + x - 2)}$$

$$= \frac{6x-18}{6x\left(x^2+x-2\right)} = \frac{6(x-3)}{6x\left(x^2+x-2\right)} = \frac{x-3}{x\left(x^2+x-2\right)}$$

Example 5.3-6: Evaluate the integral $\int \frac{x^3+2}{x^2-x-6}\,dx$.

First - Check to see if the integrand is a proper or an improper rational fraction. If the integrand is an improper rational fraction use synthetic division (long division) to reduce the rational fraction to the sum of a polynomial and a proper rational fraction. In this case the integrand is an improper rational fraction, i.e., the degree of the numerator is greater than the degree of the denominator. Applying the long division method we obtain

$$\frac{x^3+2}{x^2-x-6} = (x+1)+\frac{7x+8}{x^2-x-6}$$

To integrate the second term we proceed with the following steps:

Second - Factor the denominator x^2-x-6 into $(x-3)(x+2)$.

Third - Write the linear factors in partial fraction form. Since each linear factor in the denominator is occurring only once, the integrand can be represented in the following way:

$$\frac{7x+8}{x^2-x-6} = \frac{7x+8}{(x-3)(x+2)} = \frac{A}{x-3}+\frac{B}{x+2}$$

Fourth - Solve for the constants A and B by equating coefficients of the like powers.

$$\frac{7x+8}{x^2-x-6} = \frac{A(x+2)+B(x-3)}{(x-3)(x+2)}$$

$$7x+8 = Ax+2A+Bx-3B$$

$$7x+8 = (A+B)x+(2A-3B) \quad \text{therefore,}$$

$$A+B=7 \qquad\qquad\qquad 2A-3B=8$$

which result in having $A=\dfrac{29}{5}$ and $B=\dfrac{6}{5}$.

Fifth - Rewrite the integral in its equivalent partial fraction form by substituting the constants with their specific values.

$$\int\frac{x^3+2}{x^2-x-6}\,dx = \int(x+1)dx+\int\frac{7x+8}{x^2-x-6}\,dx = \int(x+1)dx+\int\frac{7x+8}{(x-3)(x+2)}\,dx = \frac{1}{2}(x+1)^2 +\int\frac{A}{x-3}\,dx+\int\frac{B}{x+2}\,dx$$

$$= \frac{1}{2}(x+1)^2 +\frac{29}{5}\int\frac{1}{x-3}\,dx+\frac{6}{5}\int\frac{1}{x+2}\,dx$$

Sixth - Integrate each integral individually using integration methods learned in previous sections.

$$\frac{1}{2}(x+1)^2 +\frac{29}{5}\int\frac{1}{x-3}\,dx+\frac{6}{5}\int\frac{1}{x+2}\,dx = \frac{1}{2}(x+1)^2 +\frac{29}{5}\ln|x-3|+\frac{6}{5}\ln|x+2|+c$$

Seventh - Check the answer by differentiating the solution. The result should match the integrand.

Let $y = \frac{1}{2}(x+1)^2 + \frac{29}{5}\ln|x-3| + \frac{6}{5}\ln|x+2| + c$, then $y' = (x+1) + \frac{29}{5} \cdot \frac{1}{x-3} \cdot 1 + \frac{6}{5} \cdot \frac{1}{x+2} \cdot 1 + 0$

$= (x+1) + \frac{29}{5} \cdot \frac{1}{x-3} + \frac{6}{5} \cdot \frac{1}{x+2} = \frac{5(x+1)(x-3)(x+2) + 29(x+2) + 6(x-3)}{5(x-3)(x+2)} = \frac{5x^3 - 35x - 30 + 29x + 58 + 6x - 18}{5(x-3)(x+2)}$

$= \frac{5x^3 + 10}{5(x-3)(x+2)} = \frac{5(x^3+2)}{5(x^2-x-6)} = \frac{x^3+2}{x^2-x-6}$

Example 5.3-7: Evaluate the integral $\int \frac{1}{49-x^2} dx$.

First - Check to see if the integrand is a proper or an improper rational fraction. If the integrand is an improper rational fraction use synthetic division (long division) to reduce the rational fraction to the sum of a polynomial and a proper rational fraction.

Second - Factor the denominator $49 - x^2$ into $(7-x)(7+x)$.

Third - Write the linear factors in partial fraction form. Since each linear factor in the denominator is occurring only once, the integrand can be represented in the following way:

$$\frac{1}{49-x^2} = \frac{1}{(7-x)(7+x)} = \frac{A}{7-x} + \frac{B}{7+x}$$

Fourth - Solve for the constants A and B by equating coefficients of the like powers.

$$\frac{1}{49-x^2} = \frac{A(7+x) + B(7-x)}{(7-x)(7+x)}$$

$$1 = A(7+x) + B(7-x) = 7A + Ax + 7B - Bx$$

$$1 = (A-B)x + (7A+7B) \quad \text{therefore,}$$

$$7A + 7B = 1 \qquad\qquad\qquad\qquad A - B = 0$$

which result in having $A = \frac{1}{14}$, and $B = \frac{1}{14}$

Fifth - Rewrite the integral in its equivalent partial fraction form by substituting the constants with their specific values.

$$\int \frac{1}{49-x^2} dx = \int \frac{A}{7-x} dx + \int \frac{B}{7+x} dx = \frac{1}{14} \int \frac{1}{7-x} dx + \frac{1}{14} \int \frac{1}{7+x} dx$$

Sixth - Integrate each integral individually using integration methods learned in previous sections.

$$\frac{1}{14} \int \frac{1}{7-x} dx + \frac{1}{14} \int \frac{1}{7+x} dx = \frac{1}{14}\ln|7-x| + \frac{1}{14}\ln|7+x| + c$$

Seventh - Check the answer by differentiating the solution. The result should match the

integrand. Let $y = \frac{1}{14}\ln|7-x| + \frac{1}{14}\ln|7+x| + c$, then $y' = \frac{1}{14} \cdot \frac{1}{7-x} + \frac{1}{14} \cdot \frac{1}{7+x} + 0 = \frac{7+x+7-x}{14(7-x)(7+x)}$

$= \frac{7+7}{14(49+7x-7x-x^2)} = \frac{14}{14(49-x^2)} = \frac{1}{49-x^2}$

CASE II - The Denominator Has Repeated Linear Factors

In this case each linear factor of the form $ax+b$ appears n times in the denominator. To solve this class of rational fractions we equate each proper rational fraction, that appears n times in the denominator, with a sum of n partial fractions of the form $\dfrac{M_1}{ax+b} + \dfrac{M_2}{(ax+b)^2} + ... + \dfrac{M_n}{(ax+b)^n}$. The following examples show the steps as to how this class of integrals are solved.

Example 5.3-8: Evaluate the integral $\int \dfrac{x+3}{x^3-2x^2+x}\,dx$.

First - Check to see if the integrand is a proper or an improper rational fraction. If the integrand is an improper rational fraction use synthetic division (long division) to reduce the rational fraction to the sum of a polynomial and a proper rational fraction.

Second - Factor the denominator x^3-2x^2+x into $x\left(x^2-2x+1\right) = x(x-1)^2$.

Third - Write the linear factors in partial fraction form. Since one of the factors in the denominator is repeated, the integrand can be represented in the following way:

$$\frac{x+3}{x^3-2x^2+x} = \frac{x+3}{x\left(x^2-2x+1\right)} = \frac{x+3}{x(x-1)^2} = \frac{A}{x} + \frac{B}{x-1} + \frac{C}{(x-1)^2}$$

Fourth - Solve for the constants A, B, and C by equating coefficients of the like powers.

$$\frac{x+3}{x^3-2x^2+x} = \frac{A(x-1)^2 + Bx(x-1) + Cx}{x(x-1)(x-1)^2}$$

$$x+3 = A\left(x^2-2x+1\right) + B\left(x^2-x\right) + Cx = Ax^2 - 2Ax + A + Bx^2 - Bx + Cx$$

$$x+3 = (A+B)x^2 + (-2A-B+C)x + A \quad \text{therefore,}$$

$$A+B=0 \qquad\qquad -2A-B+C=1 \qquad\qquad A=3$$

which result in having $A=3$, $B=-3$, and $C=4$

Fifth - Rewrite the integral in its equivalent partial fraction form by substituting the constants with their specific values.

$$\int \frac{x+3}{x^3-2x^2+x}\,dx = \int \frac{A}{x}\,dx + \int \frac{B}{x-1}\,dx + \int \frac{C}{(x-1)^2}\,dx = 3\int \frac{1}{x}\,dx - 3\int \frac{1}{x-1}\,dx + 4\int \frac{1}{(x-1)^2}\,dx$$

Sixth - Integrate each integral individually using integration methods learned in previous sections.

$$3\int \frac{1}{x}\,dx - 3\int \frac{1}{x-1}\,dx + 4\int \frac{1}{(x-1)^2}\,dx = 3\ln|x| - 3\ln|x-1| - \frac{4}{x-1} + c$$

Seventh - Check the answer by differentiating the solution. The result should match the integrand.

Let $y = 3\ln|x| - 3\ln|x-1| - \dfrac{4}{x-1} + c$, then $y' = 3\cdot\dfrac{1}{x} - 3\cdot\dfrac{1}{x-1} + \dfrac{4}{(x-1)^2} + 0 = \dfrac{3(x-1)^2 - 3x(x-1) + 4x}{x(x-1)^2}$

$$= \frac{3x^2 - 6x + 3 - 3x^2 + 3x + 4x}{x^3 - 2x^2 + x} = \frac{x+3}{x^3 - 2x^2 + x}$$

Example 5.3-9: Evaluate the integral $\int \frac{dx}{x^3 - x^2}$.

First - Check to see if the integrand is a proper or an improper rational fraction. If the integrand is an improper rational fraction use synthetic division (long division) to reduce the rational fraction to the sum of a polynomial and a proper rational fraction.

Second - Factor the denominator $x^3 - x^2$ into $x^2(x-1)$.

Third - Write the linear factors in partial fraction form. Since one of the factors in the denominator is repeated, the integrand can be represented in the following way:

$$\frac{1}{x^3 - x^2} = \frac{1}{x^2(x-1)} = \frac{A}{x} + \frac{B}{x^2} + \frac{C}{x-1}$$

Fourth - Solve for the constants A, B, and C by equating coefficients of the like powers.

$$\frac{1}{x^3 - x^2} = \frac{Ax(x-1) + B(x-1) + Cx^2}{x^2(x-1)}$$

$$1 = A\left(x^2 - x\right) + B(x-1) + Cx^2 = Ax^2 - Ax + Bx - B + Cx^2$$

$$1 = (A+C)x^2 + (-A+B)x - B \quad \text{therefore,}$$

$$A + C = 0 \qquad\qquad -A + B = 0 \qquad\qquad -B = 1$$

which result in having $A = -1$, $B = -1$, and $C = 1$

Fifth - Rewrite the integral in its equivalent partial fraction form by substituting the constants with their specific values.

$$\int \frac{dx}{x^3 - x^2} = \int \frac{A}{x} dx + \int \frac{B}{x^2} dx + \int \frac{C}{x-1} dx = -\int \frac{1}{x} dx - \int \frac{1}{x^2} dx + \int \frac{1}{x-1} dx$$

Sixth - Integrate each integral individually using integration methods learned in previous sections.

$$-\int \frac{1}{x} dx - \int \frac{1}{x^2} dx + \int \frac{1}{x-1} dx = -\ln|x| + \frac{1}{x} + \ln|x-1| + c = \frac{1}{x} + \ln|x-1| - \ln|x| + c$$

Seventh - Check the answer by differentiating the solution. The result should match the integrand.

Let $y = \frac{1}{x} + \ln|x-1| - \ln|x| + c$, then $y' = -\frac{1}{x^2} + \frac{1}{x-1} - \frac{1}{x} + 0 = \frac{-(x-1) + x^2 - x(x-1)}{x^2(x-1)}$

$$= \frac{-x + 1 + x^2 - x^2 + x}{x^3 - x^2} = \frac{1}{x^3 - x^2}$$

Example 5.3-10: Evaluate the integral $\int \frac{5dx}{x^3 - 2x^2 + x}$.

First - Check to see if the integrand is a proper or an improper rational fraction. If the integrand is an improper rational fraction use synthetic division (long division) to reduce the rational fraction to the sum of a polynomial and a proper rational fraction.

Second – Factor the denominator $x^3 - 2x^2 + x$ into $x(x-1)^2$.

Third - Write the linear factors in partial fraction form. Since one of the factors in the denominator is repeated, the integrand can be represented in the following way:

$$\frac{5}{x(x-1)^2} = \frac{A}{x} + \frac{B}{x-1} + \frac{C}{(x-1)^2}$$

Fourth - Solve for the constants A, B, and C by equating coefficients of the like powers.

$$\frac{5}{x^3 - 2x^2 + x} = \frac{A(x-1)^2 + Bx(x-1) + Cx}{x(x-1)^2}$$

$$5 = A(x^2 - 2x + 1) + B(x^2 - x) + Cx = Ax^2 - 2Ax + A + Bx^2 - Bx + Cx$$

$$5 = (A+B)x^2 + (-2A - B + C)x + A \text{ therefore,}$$

$$A + B = 0 \qquad\qquad -2A - B + C = 0 \qquad\qquad A = 5$$

which result in having $A = 5$, $B = -5$, and $C = 5$

Fifth - Rewrite the integral in its equivalent partial fraction form by substituting the constants with their specific values.

$$\int \frac{5 dx}{x^3 - 2x^2 + x} = \int \frac{A}{x} dx + \int \frac{B}{x-1} dx + \int \frac{C}{(x-1)^2} dx = 5\int \frac{1}{x} dx - 5\int \frac{1}{x-1} dx + 5\int \frac{1}{(x-1)^2} dx$$

Sixth - Integrate each integral individually using integration methods learned in previous sections.

$$5\int \frac{1}{x} dx - 5\int \frac{1}{x-1} dx + 5\int \frac{1}{(x-1)^2} dx = 5\ln|x| - 5\ln|x-1| - \frac{5}{x-1} + c$$

Seventh - Check the answer by differentiating the solution. The result should match the integrand.

Let $y = 5\ln|x| - 5\ln|x-1| - \frac{5}{x-1} + c$, then $y' = 5 \cdot \frac{1}{x} \cdot 1 - 5 \cdot \frac{1}{x-1} \cdot 1 + 5 \cdot \frac{1}{(x-1)^2} \cdot 1 + 0 = \frac{5}{x} - \frac{5}{x-1} + \frac{5}{(x-1)^2}$

$$= \frac{5(x-1)^2 - 5x(x-1) + 5x}{x(x-1)^2} = \frac{5(x^2 - 2x + 1) - 5x^2 + 5x + 5x}{x^3 - 2x^2 + x} = \frac{5x^2 - 10x + 5 - 5x^2 + 5x + 5x}{x^3 - 2x^2 + x} = \frac{5}{x^3 - 2x^2 + x}$$

Example 5.3-11: Evaluate the integral $\int \frac{x+6}{(x+2)(x-3)^2} dx$.

First - Check to see if the integrand is a proper or an improper rational fraction. If the integrand is an improper rational fraction use synthetic division (long division) to reduce the rational fraction to the sum of a polynomial and a proper rational fraction.

Second – Factor the denominator. However, the denominator is already in its reduced form of $(x+2)(x-3)^2$.

Third - Write the linear factors in partial fraction form. Since one of the factors in the denominator is repeated, the integrand can be represented in the following way:

$$\frac{x+6}{(x+2)(x-3)^2} = \frac{A}{x+2}+\frac{B}{x-3}+\frac{C}{(x-3)^2}$$

Fourth - Solve for the constants A, B, and C by equating coefficients of the like powers.

$$\frac{x+6}{(x+2)(x-3)^2} = \frac{A(x-3)^2 +B(x+2)(x-3)+C(x+2)}{(x+2)(x-3)^2}$$

$$x+6 = A(x^2 -6x+9)+B(x^2 -x-6)+C(x+2) = Ax^2 -6Ax+9A+Bx^2 -Bx-6B+Cx+2C$$

$$x+6 = (A+B)x^2 +(-6A-B+C)x+(9A-6B+2C) \text{ therefore,}$$

$$A+B=0 \qquad\qquad -6A-B+C=1 \qquad\qquad 9A-6B+2C=6$$

which result in having $A=\dfrac{4}{25}$, $B=-\dfrac{4}{25}$, and $C=\dfrac{9}{5}$

Fifth - Rewrite the integral in its equivalent partial fraction form by substituting the constants with their specific values.

$$\int\frac{x+6}{(x+2)(x-3)^2}dx = \int\frac{A}{x+2}dx+\int\frac{B}{x-3}dx+\int\frac{C}{(x-3)^2}dx = \frac{4}{25}\int\frac{1}{x+2}dx-\frac{4}{25}\int\frac{1}{x-3}dx+\frac{9}{5}\int\frac{1}{(x-3)^2}dx$$

Sixth - Integrate each integral individually using integration methods learned in previous sections.

$$\frac{4}{25}\int\frac{1}{x+2}dx-\frac{4}{25}\int\frac{1}{x-3}dx+\frac{9}{5}\int\frac{1}{(x-3)^2}dx = \frac{4}{25}\ln|x+2|-\frac{4}{25}\ln|x-3|-\frac{9}{5}\frac{1}{(x-3)}+c$$

Seventh - Check the answer by differentiating the solution. The result should match the integrand.

Let $y=\dfrac{4}{25}\ln|x+2|-\dfrac{4}{25}\ln|x-3|-\dfrac{9}{5}\dfrac{1}{(x-3)}+c$, then $y' = \dfrac{4}{25}\cdot\dfrac{1}{x+2}-\dfrac{4}{25}\cdot\dfrac{1}{x-3}+\dfrac{9}{5}\dfrac{1}{(x-3)^2}+0$

$$= \frac{4(x-3)^2 -4(x+2)(x-3)+45(x+2)}{25(x+2)(x-3)^2} = \frac{4(x^2 -6x+9)-4(x^2 -x-6)+45(x+2)}{25(x+2)(x-3)^2} = \frac{4x^2 -24x+36-4x^2}{25(x+2)(x-3)^2}$$

$$+\frac{4x+24+45x+90}{25(x+2)(x-3)^2} = \frac{25x+150}{25(x+2)(x-3)^2} = \frac{25(x+6)}{25(x+2)(x-3)^2} = \frac{x+6}{(x+2)(x-3)^2}$$

Example 5.3-12: Evaluate the integral $\int\dfrac{x+5}{x^3 +4x^2 +4x}dx$.

First - Check to see if the integrand is a proper or an improper rational fraction. If the integrand is an improper rational fraction use synthetic division (long division) to reduce the rational fraction to the sum of a polynomial and a proper rational fraction.

Second - Factor the denominator $x^3 + 4x^2 + 4x$ into $x(x^2 + 4x + 4) = x(x+2)^2$.

Third - Write the linear factors in partial fraction form. Since one of the factors in the denominator is repeated, the integrand can be represented in the following way:

$$\frac{x+5}{x^3+4x^2+4x} = \frac{x+5}{x(x^2+4x+4)} = \frac{x+5}{x(x+2)^2} = \frac{A}{x} + \frac{B}{x+2} + \frac{C}{(x+2)^2}$$

Fourth - Solve for the constants A, B, and C by equating coefficients of the like powers.

$$\frac{x+5}{x^3+4x^2+4x} = \frac{A(x+2)^2 + Bx(x+2) + Cx}{x(x+2)(x+2)^2}$$

$$x+5 = A(x^2+4x+4) + B(x^2+2x) + Cx = Ax^2 + 4Ax + 4A + Bx^2 + 2Bx + Cx$$

$$x+5 = (A+B)x^2 + (4A+2B+C)x + 4A \quad \text{therefore,}$$

$$A+B=0 \qquad\qquad\qquad 4A+2B+C=1 \qquad\qquad\qquad 4A=5$$

which result in having $A = \frac{5}{4}$, $B = -\frac{5}{4}$, and $C = -\frac{3}{2}$

Fifth - Rewrite the integral in its equivalent partial fraction form by substituting the constants with their specific values.

$$\int \frac{x+5}{x^3+4x^2+4x}\,dx = \int \frac{A}{x}\,dx + \int \frac{B}{x+2}\,dx + \int \frac{C}{(x+2)^2}\,dx = \frac{5}{4}\int \frac{1}{x}\,dx - \frac{5}{4}\int \frac{1}{x+2}\,dx - \frac{3}{2}\int \frac{1}{(x+2)^2}\,dx$$

Sixth - Integrate each integral individually using integration methods learned in previous sections.

$$\frac{5}{4}\int \frac{1}{x}\,dx - \frac{5}{4}\int \frac{1}{x+2}\,dx - \frac{3}{2}\int \frac{1}{(x+2)^2}\,dx = \frac{5}{4}\ln|x| - \frac{5}{4}\ln|x+2| + \frac{3}{2}\cdot\frac{1}{x+2} + c$$

Seventh - Check the answer by differentiating the solution. The result should match the integrand.

Let $y = \frac{5}{4}\ln|x| - \frac{5}{4}\ln|x+2| + \frac{3}{2}\cdot\frac{1}{x+2} + c$, then $y' = \frac{5}{4}\cdot\frac{1}{x} - \frac{5}{4}\cdot\frac{1}{x+2} - \frac{3}{2(x+2)^2} + 0 = \frac{5(x+2)^2 - 5x(x+2) - 6x}{4x(x+2)^2}$

$$= \frac{5x^2+20x+20-5x^2-10x-6x}{4(x^3+4x^2+4x)} = \frac{4x+20}{4(x^3+4x^2+4x)} = \frac{4(x+5)}{4(x^3+4x^2+4x)} = \frac{x+5}{x^3+4x^2+4x}$$

Example 5.3-13: Evaluate the integral $\int \frac{1}{x^5+2x^4+x^3}\,dx$.

First - Check to see if the integrand is a proper or an improper rational fraction. If the integrand is an improper rational fraction use synthetic division (long division) to reduce the rational fraction to the sum of a polynomial and a proper rational fraction.

Second - Factor the denominator $x^5 + 2x^4 + x^3$ into $x^3(x^2 + 2x + 1) = x^3(x+1)^2$.

Third - Write the linear factors in partial fraction form. Since both factors in the denominator are repeated, the integrand can be represented in the following way:

$$\frac{1}{x^5+2x^4+x^3}=\frac{1}{x^3\left(x^2+2x+1\right)}=\frac{1}{x^3(x+1)^2}=\frac{A}{x}+\frac{B}{x^2}+\frac{C}{x^3}+\frac{D}{x+1}+\frac{E}{(x+1)^2}$$

Fourth - Solve for the constants A, B, C, D, and E by equating coefficients of the like powers.

$$\frac{1}{x^5+2x^4+x^3}=\frac{Ax^2(x+1)^2+Bx(x+1)^2+C(x+1)^2+Dx^3(x+1)+Ex^3}{x^3(x+1)^2}$$

$$1=Ax^2\left(x^2+2x+1\right)+Bx\left(x^2+2x+1\right)+C\left(x^2+2x+1\right)+Dx^3(x+1)+Ex^3$$

$$1=Ax^4+2Ax^3+Ax^2+Bx^3+2Bx^2+Bx+Cx^2+2Cx+C+Dx^4+Dx^3+Ex^3$$

$$1=(A+D)x^4+(2A+B+D+E)x^3+(A+2B+C)x^2+(B+2C)x+C \text{ therefore,}$$

$$A+D=0 \qquad 2A+B+D+E=0 \qquad A+2B+C=0 \qquad B+2C=0 \qquad C=1$$

which result in having $A=3$, $B=-2$, $C=1$, $D=-3$, and $E=-1$

Fifth - Rewrite the integral in its equivalent partial fraction form by substituting the constants with their specific values.

$$\int\frac{1}{x^5+2x^4+x^3}dx=3\int\frac{1}{x}dx-2\int\frac{1}{x^2}dx+\int\frac{1}{x^3}dx-3\int\frac{1}{x+1}dx-\int\frac{1}{(x+1)^2}dx$$

Sixth - Integrate each integral individually using integration methods learned in previous sections.

$$3\int\frac{1}{x}dx-2\int\frac{1}{x^2}dx+\int\frac{1}{x^3}dx-3\int\frac{1}{x+1}dx-\int\frac{1}{(x+1)^2}dx=3\ln|x|+\frac{2}{x}-\frac{1}{2x^2}-3\ln|x+1|+\frac{1}{x+1}+c$$

Seventh - Check the answer by differentiating the solution. The result should match the integrand.

Let $y=3\ln|x|+\frac{2}{x}-\frac{1}{2x^2}-3\ln|x+1|+\frac{1}{x+1}+c$, then $y'=3\cdot\frac{1}{x}-\frac{2}{x^2}+\frac{1}{x^3}-3\cdot\frac{1}{x+1}-\frac{1}{(x+1)^2}+0$

$$=\frac{3x^2(x+1)^2-2x(x+1)^2+(x+1)^2-3x^3(x+1)-x^3}{x^3(x+1)^2}=\frac{3x^4+3x^2+6x^3-2x^3-2x-4x^2+x^2+1+2x-3x^4}{x^3\left(x^2+2x+1\right)}$$

$$+\frac{-3x^3-x^3}{x^3\left(x^2+2x+1\right)}=\frac{1}{x^5+2x^4+x^3}$$

Example 5.3-14: Evaluate the integral $\int\frac{1}{x^4-6x^3+9x^2}dx$.

First - Check to see if the integrand is a proper or an improper rational fraction. If the integrand is an improper rational fraction use synthetic division (long division) to reduce the rational fraction to the sum of a polynomial and a proper rational fraction.

Second - Factor the denominator $x^4-6x^3+9x^2$ into $x^2\left(x^2-6x+9\right)=x^2(x-3)^2$.

Third - Write the linear factors in partial fraction form. Since one of the factors in the denominator is repeated, the integrand can be represented in the following way:

$$\frac{1}{x^4 - 6x^3 + 9x^2} = \frac{1}{x^2\left(x^2 - 6x + 9\right)} = \frac{1}{x^2(x-3)^2} = \frac{A}{x} + \frac{B}{x^2} + \frac{C}{x-3} + \frac{D}{(x-3)^2}$$

Fourth - Solve for the constants A, B, and C by equating coefficients of the like powers.

$$\frac{1}{x^4 - 6x^3 + 9x^2} = \frac{Ax(x-3)^2 + B(x-3)^2 + Cx^2(x-3) + Dx^2}{x^2(x-3)^2}$$

$$1 = Ax(x-3)^2 + B(x-3)^2 + Cx^2(x-3) + Dx^2 = Ax\left(x^2 + 9 - 6x\right) + B\left(x^2 + 9 - 6x\right) + Cx^2(x-3) + Dx^2$$

$$1 = Ax^3 + 9Ax - 6Ax^2 + Bx^2 + 9B - 6Bx + Cx^3 - 3Cx^2 + Dx^2$$

$$1 = (A+C)x^3 + (-6A + B - 3C + D)x^2 + (9A - 6B)x + 9B \quad \text{therefore,}$$

$$A + C = 0 \qquad\qquad -6A + B - 3C + D = 0 \qquad\qquad 9A - 6B = 0 \qquad\qquad 9B = 1$$

which result in having $A = \dfrac{6}{81}$, $B = \dfrac{1}{9}$, $C = -\dfrac{6}{81}$, and $D = \dfrac{1}{9}$

Fifth - Rewrite the integral in its equivalent partial fraction form by substituting the constants with their specific values.

$$\int \frac{dx}{x^4 - 6x^3 + 9x^2} = \int \frac{A}{x}dx + \int \frac{B}{x^2}dx + \int \frac{C}{x-3}dx + \int \frac{D}{(x-3)^2}dx = \frac{6}{81}\int \frac{dx}{x} + \frac{1}{9}\int \frac{dx}{x^2} - \frac{6}{81}\int \frac{dx}{x-3} + \frac{1}{9}\int \frac{dx}{(x-3)^2}$$

Sixth - Integrate each integral individually using integration methods learned in previous sections.

$$\frac{6}{81}\int \frac{dx}{x} + \frac{1}{9}\int \frac{dx}{x^2} - \frac{6}{81}\int \frac{dx}{x-3} + \frac{1}{9}\int \frac{dx}{(x-3)^2} = \frac{6}{81}\ln|x| - \frac{1}{9x} - \frac{6}{81}\ln|x-3| - \frac{1}{9(x-3)} + c$$

Seventh - Check the answer by differentiating the solution. The result should match the integrand.

Let $y = \dfrac{6}{81}\ln|x| - \dfrac{1}{9x} - \dfrac{6}{81}\ln|x-3| - \dfrac{1}{9(x-3)} + c$, then $y' = \dfrac{6}{81} \cdot \dfrac{1}{x} + \dfrac{1}{9x^2} - \dfrac{6}{81} \cdot \dfrac{1}{x-3} + \dfrac{1}{9(x-3)^2} + 0$

$$= \frac{6x(x-3)^2 + 9(x-3)^2 - 6x^2(x-3) + 9x^2}{81x^2(x-3)^2} = \frac{6x\left(x^2 - 6x + 9\right) + 9\left(x^2 - 6x + 9\right) - 6x^3 + 18x^2 + 9x^2}{81x^2\left(x^2 - 6x + 9\right)}$$

$$= \frac{6x^3 - 36x^2 + 54x + 9x^2 - 54x + 81 - 6x^3 + 18x^2 + 9x^2}{81x^2\left(x^2 - 6x + 9\right)} = \frac{81}{81\left(x^4 - 6x^3 + 9x^2\right)} = \frac{1}{x^4 - 6x^3 + 9x^2}$$

CASE III - The Denominator Has Distinct Quadratic Factors

In this case the quadratic factors of the form ax^2+bx+c appear only once in the denominator and are irreducible. To solve this class of rational fractions we equate each proper rational fraction with a single fraction of the form $\dfrac{Ax+B}{ax^2+bx+c}$, $\dfrac{Cx+D}{cx^2+dx+e}$, $\dfrac{Ex+F}{ex^2+fx+g}$, etc. The following examples show the steps as to how this class of integrals are solved.

Example 5.3-15: Evaluate the integral $\int \dfrac{x^2-x+3}{x^3+x}dx$.

First - Check to see if the integrand is a proper or an improper rational fraction. If the integrand is an improper rational fraction use synthetic division (long division) to reduce the rational fraction to the sum of a polynomial and a proper rational fraction.

Second - Factor the denominator x^3+x into $x(x^2+1)$.

Third - Write the factors in partial fraction form. Since one of the factors in the denominator is in quadratic form, the integrand can be represented in the following way:

$$\frac{x^2-x+3}{x^3+x} = \frac{x^2-x+3}{x(x^2+1)} = \frac{A}{x}+\frac{Bx+C}{x^2+1}$$

Fourth - Solve for the constants A, B, and C by equating coefficients of the like powers.

$$\frac{x^2-x+3}{x^3+x} = \frac{A(x^2+1)+(Bx+C)x}{x(x^2+1)}$$

$$x^2-x+3 = A(x^2+1)+(Bx+C)x = Ax^2+A+Bx^2+Cx$$

$$x^2-x+3 = (A+B)x^2+Cx+A \quad \text{therefore,}$$

$$A+B=1 \qquad\qquad C=-1 \qquad\qquad A=3$$

which result in having $A=3$, $B=-2$, and $C=-1$

Fifth - Rewrite the integral in its equivalent partial fraction form by substituting the constants with their specific values.

$$\int\frac{x^2-x+3}{x^3+x}dx = \int\frac{A}{x}dx+\int\frac{Bx+C}{x^2+1}dx = 3\int\frac{1}{x}dx+\int\frac{-2x-1}{x^2+1}dx = 3\int\frac{1}{x}dx-\int\frac{2x}{x^2+1}dx-\int\frac{1}{x^2+1}dx$$

Sixth - Integrate each integral individually using integration methods learned in previous sections. To solve the second integral let $u=x^2+1$.

$$3\int\frac{1}{x}dx-\int\frac{2x}{x^2+1}dx-\int\frac{1}{x^2+1}dx = 3\int\frac{1}{x}dx-\int\frac{2x}{u}\cdot\frac{du}{2x}-\int\frac{1}{x^2+1}dx = 3\int\frac{1}{x}dx-\int\frac{1}{u}du-\int\frac{1}{x^2+1}dx$$

$$= 3\ln|x|-\ln|u|-\tan^{-1}x+c = 3\ln|x|-\ln|x^2+1|-\tan^{-1}x+c$$

Seventh - Check the answer by differentiating the solution. The result should match the integrand.

Let $y = 3\ln|x| - \ln|x^2+1| - \tan^{-1}x + c$, then $y' = 3 \cdot \dfrac{1}{x} + \dfrac{-1}{x^2+1} \cdot 2x - \dfrac{1}{1+x^2} + 0 = \dfrac{3}{x} - \dfrac{2x}{x^2+1} - \dfrac{1}{x^2+1}$

$= \dfrac{3(x^2+1) - 2x^2 - x}{x(x^2+1)} = \dfrac{3x^2 + 3 - 2x^2 - x}{x^3+x} = \dfrac{x^2 - x + 3}{x^3+x}$

Example 5.3-16: Evaluate the integral $\displaystyle\int \dfrac{1}{x^3+25x}\,dx$.

First - Check to see if the integrand is a proper or an improper rational fraction. If the integrand is an improper rational fraction use synthetic division (long division) to reduce the rational fraction to the sum of a polynomial and a proper rational fraction.

Second - Factor the denominator x^3+25x into $x(x^2+25)$.

Third - Write the factors in partial fraction form. Since one of the factors in the denominator is in quadratic form, the integrand can be represented in the following way:

$$\dfrac{1}{x^3+25x} = \dfrac{1}{x(x^2+25)} = \dfrac{A}{x} + \dfrac{Bx+C}{x^2+25}$$

Fourth - Solve for the constants A , B , and C by equating coefficients of the like powers.

$$\dfrac{1}{x^3+25x} = \dfrac{A(x^2+25)+(Bx+C)x}{x(x^2+25)}$$

$$1 = A(x^2+25)+(Bx+C)x = Ax^2+25A+Bx^2+Cx$$

$$1 = (A+B)x^2+Cx+25A \quad \text{therefore,}$$

$$25A = 1 \qquad\qquad\qquad C = 0 \qquad\qquad\qquad A+B = 0$$

which result in having $A = \dfrac{1}{25}$, $B = -\dfrac{1}{25}$, and $C = 0$

Fifth - Rewrite the integral in its equivalent partial fraction form by substituting the constants with their specific values.

$$\int \dfrac{1}{x^3+25x}\,dx = \int \dfrac{A}{x}\,dx + \int \dfrac{Bx+C}{x^2+25}\,dx = \dfrac{1}{25}\int \dfrac{1}{x}\,dx - \dfrac{1}{25}\int \dfrac{x}{x^2+25}\,dx$$

Sixth - Integrate each integral individually using integration methods learned in previous sections. To solve the second integral let $u = x^2+25$.

$$\dfrac{1}{25}\int \dfrac{1}{x}\,dx - \dfrac{1}{25}\int \dfrac{x}{x^2+25}\,dx = \dfrac{1}{25}\int \dfrac{1}{x}\,dx - \dfrac{1}{25}\int \dfrac{x}{u}\cdot\dfrac{du}{2x} = \dfrac{1}{25}\int \dfrac{1}{x}\,dx - \dfrac{1}{50}\int \dfrac{1}{u}\,du = \dfrac{1}{25}\ln|x| - \dfrac{1}{50}\ln|u| + c$$

$$= \dfrac{1}{25}\ln|x| - \dfrac{1}{50}\ln\left|x^2+25\right| + c$$

Seventh - Check the answer by differentiating the solution. The result should match the integrand.

Let $y = \dfrac{1}{25}\ln|x| - \dfrac{1}{50}\ln\left|x^2+25\right| + c$, then $y' = \dfrac{1}{25}\cdot\dfrac{1}{x} - \dfrac{1}{50}\cdot\dfrac{1}{x^2+25}\cdot 2x + 0 = \dfrac{1}{25x} - \dfrac{x}{25\left(x^2+25\right)}$

$= \dfrac{\left(x^2+25\right)-x^2}{25x\left(x^2+25\right)} = \dfrac{x^2+25-x^2}{25\left(x^3+25x\right)} = \dfrac{25}{25\left(x^3+25x\right)} = \dfrac{1}{x^3+25x}$

Example 5.3-17: Evaluate the integral $\int \dfrac{1}{x^4+16x^2}\,dx$.

First - Check to see if the integrand is a proper or an improper rational fraction. If the integrand is an improper rational fraction use synthetic division (long division) to reduce the rational fraction to the sum of a polynomial and a proper rational fraction.

Second - Factor the denominator x^4+16x^2 into $x^2\left(x^2+16\right)$.

Third - Write the factors in partial fraction form. Since the factors in the denominator are in quadratic form, the integrand can be represented in the following way:

$$\dfrac{1}{x^4+16x^2} = \dfrac{1}{x^2\left(x^2+16\right)} = \dfrac{Ax+B}{x^2} + \dfrac{Cx+D}{x^2+16}$$

Fourth - Solve for the constants A, B, C, and D by equating coefficients of the like powers.

$$\dfrac{1}{x^4+16x^2} = \dfrac{(Ax+B)\left(x^2+16\right)+x^2(Cx+D)}{x^2\left(x^2+16\right)}$$

$$1 = (Ax+B)\left(x^2+16\right)+x^2(Cx+D) = Ax^3+16Ax+Bx^2+16B+Cx^3+Dx^2$$

$$1 = (A+C)x^3+(B+D)x^2+16Ax+16B \quad \text{therefore,}$$

$$A+C=0 \qquad\qquad B+D=0 \qquad\qquad 16A=0 \qquad\qquad 16B=1$$

which result in having $A=0$, $B=\dfrac{1}{16}$, $C=0$, and $D=-\dfrac{1}{16}$

Fifth - Rewrite the integral in its equivalent partial fraction form by substituting the constants with their specific values.

$$\int\dfrac{1}{x^4+16x^2}\,dx = \int\dfrac{Ax+B}{x^2}\,dx+\int\dfrac{Cx+D}{x^2+16}\,dx = \int\dfrac{B}{x^2}\,dx+\int\dfrac{D}{x^2+16}\,dx = \dfrac{1}{16}\int\dfrac{1}{x^2}\,dx-\dfrac{1}{16}\int\dfrac{1}{x^2+16}\,dx$$

Sixth - Integrate each integral individually using integration methods learned in previous sections.

$$\dfrac{1}{16}\int\dfrac{1}{x^2}\,dx-\dfrac{1}{16}\int\dfrac{1}{x^2+16}\,dx = \dfrac{1}{16}\int\dfrac{1}{x^2}\,dx-\dfrac{1}{16}\int\dfrac{1}{16\left(x^2+\frac{1}{16}\right)}\,dx = \dfrac{1}{16}\int\dfrac{1}{x^2}\,dx-\dfrac{1}{256}\int\dfrac{1}{\left(x^2+\frac{1}{16}\right)}\,dx$$

$$= \dfrac{1}{16}\cdot-\dfrac{1}{x}-\dfrac{1}{256}\cdot 4\tan^{-1}\dfrac{x}{4}+c = -\dfrac{1}{16x}-\dfrac{1}{64}\tan^{-1}\dfrac{x}{4}+c$$

Seventh - Check the answer by differentiating the solution. The result should match the integrand.

Let $y=-\dfrac{1}{16x}-\dfrac{1}{64}\tan^{-1}\dfrac{x}{4}+c$, then $y'=\dfrac{1}{16x^2}-\dfrac{1}{64}\cdot\dfrac{1}{1+\frac{x^2}{16}}\cdot\dfrac{1}{4}+0=\dfrac{1}{16x^2}-\dfrac{1}{256}\cdot\dfrac{16}{16+x^2}$

$=\dfrac{1}{16x^2}-\dfrac{1}{16(16+x^2)}=\dfrac{(16+x^2)-x^2}{16x^2(16+x^2)}=\dfrac{16+x^2-x^2}{16x^2(16+x^2)}=\dfrac{16}{16x^2(16+x^2)}=\dfrac{1}{x^4+16x^2}$

Example 5.3-18: Evaluate the integral $\int\dfrac{1}{x^3-8}\,dx$.

First - Check to see if the integrand is a proper or an improper rational fraction. If the integrand is an improper rational fraction use synthetic division (long division) to reduce the rational fraction to the sum of a polynomial and a proper rational fraction.

Second - Factor the denominator x^3-8 into $(x-2)(x^2+2x+4)$.

Third - Write the factors in partial fraction form. Since one of the factors in the denominator is in quadratic form, the integrand can be represented in the following way:

$$\dfrac{1}{x^3-8}=\dfrac{1}{(x-2)(x^2+2x+4)}=\dfrac{A}{x-2}+\dfrac{Bx+C}{x^2+2x+4}$$

Fourth - Solve for the constants A, B, and C by equating coefficients of the like powers.

$$\dfrac{1}{x^3-8}=\dfrac{A(x^2+2x+4)+(Bx+C)(x-2)}{(x-2)(x^2+2x+4)}$$

$$1=A(x^2+2x+4)+(Bx+C)(x-2)=Ax^2+2Ax+4A+Bx^2-2Bx+Cx-2C$$

$$1=(A+B)x^2+(2A-2B+C)x+(4A-2C)\ \ \text{therefore,}$$

$$A+B=0 \qquad\qquad 2A-2B+C=0 \qquad\qquad 4A-2C=1$$

which result in having $A=\dfrac{1}{12}$, $B=-\dfrac{1}{12}$, and $C=-\dfrac{1}{3}$

Fifth - Rewrite the integral in its equivalent partial fraction form by substituting the constants with their specific values.

$$\int\dfrac{1}{x^3-8}\,dx=\int\dfrac{A}{x-2}\,dx+\int\dfrac{Bx+C}{x^2+2x+4}\,dx=\dfrac{1}{12}\int\dfrac{1}{x-2}\,dx+\int\dfrac{-\frac{1}{12}x-\frac{1}{3}}{x^2+2x+4}\,dx=\dfrac{1}{12}\int\dfrac{1}{x-2}\,dx-\dfrac{1}{12}\int\dfrac{x+4}{x^2+2x+4}\,dx$$

Sixth - Integrate each integral individually using integration methods learned in previous sections. To solve the second integral let $u=x^2+2x+4$, then $\dfrac{du}{dx}=2x+2$ and $dx=\dfrac{du}{2x+2}$.

Also, $x+4$ can be rewritten as $x+4=(x+1)+3=\dfrac{1}{2}(2x+2)+3$. Therefore,

$$\frac{1}{12}\int\frac{1}{x-2}dx-\frac{1}{12}\int\frac{x+4}{x^2+2x+4}dx \;=\; \frac{1}{12}\int\frac{1}{x-2}dx-\frac{1}{12}\int\frac{(x+1)+3}{x^2+2x+4}dx \;=\; \frac{1}{12}\int\frac{1}{x-2}dx-\frac{1}{12}\int\frac{\frac{1}{2}(2x+2)+3}{x^2+2x+4}dx$$

$$=\;\frac{1}{12}\int\frac{1}{x-2}dx-\frac{1}{24}\int\frac{2x+2}{x^2+2x+4}dx-\frac{1}{12}\int\frac{3}{x^2+2x+4}dx \;=\; \frac{1}{12}\int\frac{1}{x-2}dx-\frac{1}{24}\int\frac{2x+2}{u}\cdot\frac{du}{2x+2}-\frac{3}{12}\int\frac{1}{(x+1)^2+3}dx$$

$$=\;\frac{1}{12}\int\frac{1}{x-2}dx-\frac{1}{24}\int\frac{1}{u}du-\frac{1}{4}\int\frac{1}{(x+1)^2+3}dx \;=\; \frac{1}{12}\ln|x-2|-\frac{1}{24}\ln|u|-\frac{1}{4\sqrt{3}}\tan^{-1}\frac{x+1}{\sqrt{3}}+c$$

$$=\;\frac{1}{12}\ln|x-2|-\frac{1}{24}\ln\left|x^2+2x+4\right|-\frac{\sqrt{3}}{4\cdot3}\tan^{-1}\frac{x+1}{\sqrt{3}}+c \;=\; \frac{1}{12}\ln|x-2|-\frac{1}{24}\ln\left|x^2+2x+4\right|-\frac{\sqrt{3}}{12}\tan^{-1}\frac{x+1}{\sqrt{3}}+c$$

Seventh - Check the answer by differentiating the solution. The result should match the integrand.

Let $y=\dfrac{1}{12}\ln|x-2|-\dfrac{1}{24}\ln\left|x^2+2x+4\right|-\dfrac{\sqrt{3}}{12}\tan^{-1}\dfrac{x+1}{\sqrt{3}}+c$, then

$$y'=\frac{1}{12}\cdot\frac{1}{x-2}-\frac{1}{24}\cdot\frac{1}{x^2+2x+4}\cdot(2x+2)-\frac{\sqrt{3}}{12}\cdot\frac{1}{1+\left(\frac{x+1}{\sqrt{3}}\right)^2}\cdot\frac{(1\cdot\sqrt{3})-0\cdot(x+1)}{(\sqrt{3})^2}+0 \;=\; \frac{1}{12(x-2)}-\frac{x+1}{12(x^2+2x+4)}$$

$$-\frac{\sqrt{3}}{12}\cdot\frac{3}{3+(x+1)^2}\cdot\frac{\sqrt{3}}{3} \;=\; \frac{1}{12(x-2)}-\frac{x+1}{12(x^2+2x+4)}-\frac{3}{12(x^2+2x+4)} \;=\; \frac{1}{12(x-2)}+\frac{-x-1-3}{12(x^2+2x+4)}$$

$$=\;\frac{1}{12(x-2)}+\frac{-x-4}{12(x^2+2x+4)} \;=\; \frac{(x^2+2x+4)+(-x-4)(x-2)}{12(x-2)(x^2+2x+4)} \;=\; \frac{x^2+2x+4-x^2+2x-4x+8}{12(x-2)(x^2+2x+4)}$$

$$=\;\frac{4+8}{12(x-2)(x^2+2x+4)} \;=\; \frac{12}{12(x-2)(x^2+2x+4)} \;=\; \frac{1}{x^3+2x^2+4x-2x^2-4x-8} \;=\; \frac{1}{x^3-8}$$

Example 5.3-19: Evaluate the integral $\int\dfrac{1}{x^3+8}dx$.

First - Check to see if the integrand is a proper or an improper rational fraction. If the integrand is an improper rational fraction use synthetic division (long division) to reduce the rational fraction to the sum of a polynomial and a proper rational fraction.

Second - Factor the denominator x^3+8 into $(x+2)(x^2-2x+4)$.

Third - Write the factors in partial fraction form. Since one of the factors in the denominator is in quadratic form, the integrand can be represented in the following way:

$$\frac{1}{x^3+8} \;=\; \frac{1}{(x+2)(x^2-2x+4)} \;=\; \frac{A}{x+2}+\frac{Bx+C}{x^2-2x+4}$$

Fourth - Solve for the constants A, B, and C by equating coefficients of the like powers.

$$\frac{1}{x^3+8} \;=\; \frac{A(x^2-2x+4)+(Bx+C)(x+2)}{(x+2)(x^2-2x+4)}$$

$$1 = A\left(x^2 - 2x + 4\right) + (Bx + C)(x + 2) = Ax^2 - 2Ax + 4A + Bx^2 + 2Bx + Cx + 2C$$

$$1 = (A + B)x^2 + (-2A + 2B + C)x + (4A + 2C) \quad \text{therefore,}$$

$$A + B = 0 \qquad\qquad -2A + 2B + C = 0 \qquad\qquad 4A + 2C = 1$$

which result in having $A = \dfrac{1}{12}$, $B = -\dfrac{1}{12}$, and $C = \dfrac{1}{3}$

Fifth - Rewrite the integral in its equivalent partial fraction form by substituting the constants with their specific values.

$$\int \frac{1}{x^3 + 8}\, dx = \int \frac{A}{x+2}\, dx + \int \frac{Bx + C}{x^2 - 2x + 4}\, dx = \frac{1}{12}\int \frac{1}{x+2}\, dx + \int \frac{-\frac{1}{12}x + \frac{1}{3}}{x^2 - 2x + 4}\, dx = \frac{1}{12}\int \frac{1}{x+2}\, dx - \frac{1}{12}\int \frac{x-4}{x^2 - 2x + 4}\, dx$$

Sixth - Integrate each integral individually using integration methods learned in previous

sections. To solve the second integral let $u = x^2 - 2x + 4$, then $\dfrac{du}{dx} = 2x - 2$ and $dx = \dfrac{du}{2x - 2}$.

Also, $x - 4$ can be rewritten as $x - 4 = (x - 1) - 3 = \dfrac{1}{2}(2x - 2) - 3$. Therefore,

$$\frac{1}{12}\int \frac{1}{x+2}\, dx - \frac{1}{12}\int \frac{x-4}{x^2 - 2x + 4}\, dx = \frac{1}{12}\int \frac{1}{x+2}\, dx - \frac{1}{12}\int \frac{(x-1)-3}{x^2 - 2x + 4}\, dx = \frac{1}{12}\int \frac{1}{x+2}\, dx - \frac{1}{12}\int \frac{\frac{1}{2}(2x-2)-3}{x^2 - 2x + 4}\, dx$$

$$= \frac{1}{12}\int \frac{1}{x+2}\, dx - \frac{1}{24}\int \frac{2x-2}{x^2 - 2x + 4}\, dx + \frac{1}{12}\int \frac{3}{x^2 - 2x + 4}\, dx = \frac{1}{12}\int \frac{1}{x+2}\, dx - \frac{1}{24}\int \frac{2x-2}{u}\cdot\frac{du}{2x+2} + \frac{3}{12}\int \frac{1}{(x-1)^2 + 3}\, dx$$

$$= \frac{1}{12}\int \frac{1}{x+2}\, dx - \frac{1}{24}\int \frac{1}{u}\, du + \frac{1}{4}\int \frac{1}{(x-1)^2 + 3}\, dx = \frac{1}{12}\ln|x+2| - \frac{1}{24}\ln|u| + \frac{1}{4\sqrt{3}}\tan^{-1}\frac{x-1}{\sqrt{3}} + c$$

$$= \frac{1}{12}\ln|x+2| - \frac{1}{24}\ln\left|x^2 - 2x + 4\right| + \frac{\sqrt{3}}{4\cdot 3}\tan^{-1}\frac{x-1}{\sqrt{3}} + c = \frac{1}{12}\ln|x+2| - \frac{1}{24}\ln\left|x^2 - 2x + 4\right| + \frac{\sqrt{3}}{12}\tan^{-1}\frac{x-1}{\sqrt{3}} + c$$

Seventh - Check the answer by differentiating the solution. The result should match the integrand.

Let $y = \dfrac{1}{12}\ln|x+2| - \dfrac{1}{24}\ln\left|x^2 - 2x + 4\right| + \dfrac{\sqrt{3}}{12}\tan^{-1}\dfrac{x-1}{\sqrt{3}} + c$, then

$$y' = \frac{1}{12}\cdot\frac{1}{x+2} - \frac{1}{24}\cdot\frac{1}{x^2 - 2x + 4}\cdot(2x-2) + \frac{\sqrt{3}}{12}\cdot\frac{1}{1+\left(\frac{x-1}{\sqrt{3}}\right)^2}\cdot\frac{(1\cdot\sqrt{3})-0\cdot(x-1)}{(\sqrt{3})^2} + 0 = \frac{1}{12(x+2)} - \frac{x-1}{12\left(x^2 - 2x + 4\right)}$$

$$+ \frac{\sqrt{3}}{12}\cdot\frac{3}{3 + (x-1)^2}\cdot\frac{\sqrt{3}}{3} = \frac{1}{12(x+2)} - \frac{x-1}{12\left(x^2 - 2x + 4\right)} + \frac{3}{12\left(x^2 - 2x + 4\right)} = \frac{1}{12(x+2)} + \frac{-x+1+3}{12\left(x^2 - 2x + 4\right)}$$

$$= \frac{1}{12(x+2)} + \frac{-x+4}{12\left(x^2 - 2x + 4\right)} = \frac{\left(x^2 - 2x + 4\right) + (-x+4)(x+2)}{12(x+2)\left(x^2 - 2x + 4\right)} = \frac{x^2 - 2x + 4 - x^2 - 2x + 4x + 8}{12(x+2)\left(x^2 - 2x + 4\right)}$$

$$= \frac{4+8}{12(x+2)(x^2-2x+4)} = \frac{12}{12(x+2)(x^2-2x+4)} = \frac{1}{x^3-2x^2+4x+2x^2-4x+8} = \frac{1}{x^3+8}$$

Example 5.3-20: Evaluate the integral $\int \frac{x^2}{16-x^4}dx$.

First - Check to see if the integrand is a proper or an improper rational fraction. If the integrand is an improper rational fraction use synthetic division (long division) to reduce the rational fraction to the sum of a polynomial and a proper rational fraction.

Second - Factor the denominator $16-x^4$ into $(4-x^2)(4+x^2) = (2-x)(2+x)(4+x^2)$.

Third - Write the factors in partial fraction form. Since one of the factors in the denominator is in quadratic form, the integrand can be represented in the following way:

$$\frac{x^2}{16-x^4} = \frac{x^2}{(2-x)(2+x)(4+x^2)} = \frac{A}{2-x} + \frac{B}{2+x} + \frac{Cx+D}{4+x^2}$$

Fourth - Solve for the constants A, B, C, and D by equating coefficients of the like powers.

$$\frac{x^2}{16-x^4} = \frac{A(2+x)(4-x^2) + B(2-x)(4-x^2) + (2-x)(2+x)(Cx+D)}{(2-x)(2+x)(4+x^2)}$$

$$x^2 = A(2+x)(4+x^2) + B(2-x)(4+x^2) + (2-x)(2+x)(Cx+D)$$

$$x^2 = 8A + 2Ax^2 + 4Ax + Ax^3 + 8B + 2Bx^2 - 4Bx - Bx^3 + 4Cx + 4D - Cx^3 - Dx^2$$

$$x^2 = (A-B-C)x^3 + (2A+2B-D)x^2 + (4A-4B+4C)x + (8A+8B+4D) \quad \text{therefore,}$$

$$A-B-C=0 \qquad 2A+2B-D=1 \qquad 4A-4B+4C=0 \qquad 8A+8B+4D=0$$

which result in having $A=\frac{1}{8}$, $B=\frac{1}{8}$, $C=0$, and $D=-\frac{1}{2}$

Fifth - Rewrite the integral in its equivalent partial fraction form by substituting the constants with their specific values.

$$\int \frac{x^2}{16-x^4}dx = \int \frac{A}{2-x}dx + \int \frac{B}{2+x}dx + \int \frac{Cx+D}{4+x^2}dx = \frac{1}{8}\int \frac{1}{2-x}dx + \frac{1}{8}\int \frac{1}{2+x}dx - \frac{1}{2}\int \frac{1}{4+x^2}dx$$

Sixth - Integrate each integral individually using integration methods learned in previous sections.

$$\frac{1}{8}\int \frac{1}{2-x}dx + \frac{1}{8}\int \frac{1}{2+x}dx - \frac{1}{2}\int \frac{1}{4+x^2}dx = \frac{1}{8}\int \frac{1}{2-x}dx + \frac{1}{8}\int \frac{1}{2+x}dx - \frac{1}{2}\int \frac{1}{2^2+x^2}dx$$

$$= \frac{1}{8}\ln|2-x| + \frac{1}{8}\ln|2+x| - \frac{1}{2}\cdot\frac{1}{2}\tan^{-1}\frac{x}{2} + c = \frac{1}{8}\ln|2-x| + \frac{1}{8}\ln|2+x| - \frac{1}{4}\tan^{-1}\frac{x}{2} + c$$

Seventh - Check the answer by differentiating the solution. The result should match the integrand.

Let $y = \frac{1}{8}\ln|2-x| + \frac{1}{8}\ln|2+x| - \frac{1}{4}\tan^{-1}\frac{x}{2} + c$, then $y' = \frac{1}{8}\cdot\frac{1}{2-x} + \frac{1}{8}\cdot\frac{1}{2+x} - \frac{1}{4}\cdot\frac{1}{1+\frac{x^2}{4}}\cdot\frac{1}{2} + 0$

$= \frac{1}{8(2-x)} + \frac{1}{8(2+x)} - \frac{4}{8(4+x^2)} = \frac{(2+x)(4+x^2)+(2-x)(4+x^2)-4(2-x)(2+x)}{8(2-x)(2+x)(4+x^2)} = \frac{8+2x^2+4x+x^3+8}{8(4-x^2)(4+x^2)}$

$+ \frac{2x^2-4x-x^3-16+4x^2}{8(4-x^2)(4+x^2)} = \frac{(2x^2+2x^2+4x^2)+(8+8-16)}{8(16-x^4)} = \frac{8x^2}{8(16-x^4)} = \frac{x^2}{16-x^4}$

Example 5.3-21: Evaluate the integral $\int \frac{5}{x^4-1}\,dx$.

First - Check to see if the integrand is a proper or an improper rational fraction. If the integrand is an improper rational fraction use synthetic division (long division) to reduce the rational fraction to the sum of a polynomial and a proper rational fraction.

Second - Factor the denominator x^4-1 into $(x^2-1)(x^2+1) = (x-1)(x+1)(x^2+1)$.

Third - Write the factors in partial fraction form. Since one of the factors in the denominator is in quadratic form, the integrand can be represented in the following way:

$$\frac{5}{x^4-1} = \frac{5}{(x-1)(x+1)(x^2+1)} = \frac{A}{x-1} + \frac{B}{x+1} + \frac{Cx+D}{x^2+1}$$

Fourth - Solve for the constants A, B, C, and D by equating coefficients of the like powers.

$$\frac{5}{x^4-1} = \frac{A(x+1)(x^2+1)+B(x-1)(x^2+1)+(x-1)(x+1)(Cx+D)}{(x-1)(x+1)(x^2+1)}$$

$$5 = A(x+1)(x^2+1)+B(x-1)(x^2+1)+(x-1)(x+1)(Cx+D)$$

$$5 = Ax^3+Ax^2+Ax+A+Bx^3-Bx^2+Bx-B+Cx^3+Dx^2-Cx-D$$

$$5 = (A+B+C)x^3+(A-B+D)x^2+(A+B-C)x+(A-B-D) \text{ therefore,}$$

$$A+B+C=0 \qquad A-B+D=0 \qquad A+B-C=0 \qquad A-B-D=5$$

which result in having $A=\frac{5}{4}$, $B=-\frac{5}{4}$, $C=0$, and $D=-\frac{5}{2}$

Fifth - Rewrite the integral in its equivalent partial fraction form by substituting the constants with their specific values.

$$\int \frac{5}{x^4-1}\,dx = \int \frac{A}{x-1}\,dx + \int \frac{B}{x+1}\,dx + \int \frac{Cx+D}{x^2+1}\,dx = \frac{5}{4}\int \frac{1}{x-1}\,dx - \frac{5}{4}\int \frac{1}{x+1}\,dx - \frac{5}{2}\int \frac{1}{x^2+1}\,dx$$

Sixth - Integrate each integral individually using integration methods learned in previous sections.

$$\frac{5}{4}\int \frac{1}{x-1}dx - \frac{5}{4}\int \frac{1}{x+1}dx - \frac{5}{2}\int \frac{1}{x^2+1}dx = \frac{5}{4}\ln|x-1| - \frac{5}{4}\ln|x+1| - \frac{5}{2}\tan^{-1}x + c$$

Seventh - Check the answer by differentiating the solution. The result should match the integrand.

Let $y = \frac{5}{4}\ln|x-1| - \frac{5}{4}\ln|x+1| - \frac{5}{2}\tan^{-1}x + c$, then $y' = \frac{5}{4}\cdot\frac{1}{x-1} - \frac{5}{4}\cdot\frac{1}{x+1} - \frac{5}{2}\cdot\frac{1}{x^2+1}\cdot 1 + 0$

$$= \frac{5}{4(x-1)} - \frac{5}{4(x+1)} - \frac{5}{2(x^2+1)} = \frac{5(x+1)(x^2+1) - 5(x-1)(x^2+1) - 10(x-1)(x+1)}{4(x-1)(x+1)(x^2+1)} = \frac{5x^3 + 5x^2 + 5x + 5 - 5x^3}{4(x^2-1)(4+x^2)}$$

$$\frac{+5x^2 - 5x + 5 - 10x^2 + 10}{4(x^2-1)(4+x^2)} = \frac{20}{4(x^4-1)} = \frac{5}{x^4-1}$$

Example 5.3-22: Evaluate the integral $\int \frac{1}{x^3-64}dx$.

First - Check to see if the integrand is a proper or an improper rational fraction. If the integrand is an improper rational fraction use synthetic division (long division) to reduce the rational fraction to the sum of a polynomial and a proper rational fraction.

Second - Factor the denominator $x^3 - 64$ into $(x-4)(x^2+4x+16)$.

Third - Write the factors in partial fraction form. Since one of the factors in the denominator is in quadratic form, the integrand can be represented in the following way:

$$\frac{1}{x^3-64} = \frac{1}{(x-4)(x^2+4x+16)} = \frac{A}{x-4} + \frac{Bx+C}{x^2+4x+16}$$

Fourth - Solve for the constants A, B, and C by equating coefficients of the like powers.

$$\frac{1}{x^3-64} = \frac{A(x^2+4x+16) + (Bx+C)(x-4)}{(x-4)(x^2+4x+16)}$$

$$1 = A(x^2+4x+16) + (Bx+C)(x-4) = Ax^2 + 4Ax + 16A + Bx^2 - 4Bx + Cx - 4C$$

$$1 = (A+B)x^2 + (4A-4B+C)x + (16A-4C) \text{ therefore,}$$

$$A+B=0 \qquad\qquad 4A-4B+C=0 \qquad\qquad 16A-4C=1$$

which result in having $A = \frac{1}{48}$, $B = -\frac{1}{48}$, and $C = -\frac{1}{6}$

Fifth - Rewrite the integral in its equivalent partial fraction form by substituting the constants with their specific values.

$$\int \frac{1}{x^3-64}dx = \int \frac{A}{x-4}dx + \int \frac{Bx+C}{x^2+4x+16}dx = \frac{1}{48}\int \frac{1}{x-4}dx + \int \frac{-\frac{1}{48}x - \frac{1}{6}}{x^2+4x+16}dx = \frac{1}{48}\int \frac{1}{x-4}dx - \frac{1}{48}\int \frac{x+8}{x^2+4x+16}dx$$

Sixth - Integrate each integral individually using integration methods learned in previous sections. To solve the second integral let $u = x^2 + 4x + 16$, then $\dfrac{du}{dx} = 2x + 4$ and $dx = \dfrac{du}{2x+4}$.

Also, $x+8$ can be rewritten as $x+8 = (x+2)+6 = \dfrac{1}{2}(2x+4)+6$. Therefore,

$$\frac{1}{48}\int \frac{1}{x-4}\,dx - \frac{1}{48}\int \frac{x+8}{x^2+4x+16}\,dx = \frac{1}{48}\int \frac{1}{x-4}\,dx - \frac{1}{48}\int \frac{(x+2)+6}{x^2+4x+16}\,dx = \frac{1}{48}\int \frac{1}{x-4}\,dx - \frac{1}{48}\int \frac{\frac{1}{2}(2x+4)+6}{x^2+4x+16}\,dx$$

$$= \frac{1}{48}\int \frac{1}{x-4}\,dx - \frac{1}{96}\int \frac{2x+4}{x^2+4x+16}\,dx - \frac{1}{48}\int \frac{6}{x^2+4x+16}\,dx = \frac{1}{48}\int \frac{1}{x-4}\,dx - \frac{1}{96}\int \frac{2x+4}{u}\cdot\frac{du}{2x+4} - \frac{6}{48}\int \frac{1}{(x+2)^2+12}\,dx$$

$$= \frac{1}{48}\int \frac{1}{x-4}\,dx - \frac{1}{96}\int \frac{1}{u}\,du - \frac{1}{8}\int \frac{1}{(x+2)^2+12}\,dx = \frac{1}{48}\ln\left|x-4\right| - \frac{1}{96}\ln\left|u\right| - \frac{1}{8\sqrt{12}}\tan^{-1}\frac{x+2}{\sqrt{12}} + c$$

$$= \frac{1}{48}\ln\left|x-4\right| - \frac{1}{96}\ln\left|x^2+4x+16\right| - \frac{\sqrt{12}}{8\cdot 12}\tan^{-1}\frac{x+2}{\sqrt{12}} + c = \boxed{\frac{1}{48}\ln\left|x-4\right| - \frac{1}{96}\ln\left|x^2+4x+16\right| - \frac{\sqrt{12}}{96}\tan^{-1}\frac{x+2}{\sqrt{12}} + c}$$

Seventh - Check the answer by differentiating the solution. The result should match the integrand.

Let $y = \dfrac{1}{48}\ln\left|x-4\right| - \dfrac{1}{96}\ln\left|x^2+4x+16\right| - \dfrac{\sqrt{12}}{96}\tan^{-1}\dfrac{x+2}{\sqrt{12}} + c$, then

$$y' = \frac{1}{48}\cdot\frac{1}{x-4} - \frac{1}{96}\cdot\frac{1}{x^2+4x+16}\cdot(2x+4) - \frac{\sqrt{12}}{96}\cdot\frac{1}{1+\left(\frac{x+2}{\sqrt{12}}\right)^2}\cdot\frac{\left(1\cdot\sqrt{12}\right)-0\cdot(x+2)}{\left(\sqrt{12}\right)^2} + 0 = \frac{1}{48(x-4)} - \frac{x+2}{48\left(x^2+4x+16\right)}$$

$$- \frac{\sqrt{12}}{96}\cdot\frac{12}{12+(x+2)^2}\cdot\frac{\sqrt{12}}{12} = \frac{1}{48(x-4)} - \frac{x+2}{48\left(x^2+4x+16\right)} - \frac{6}{48\left(x^2+4x+16\right)} = \frac{1}{48(x-4)} + \frac{-x-2-6}{48\left(x^2+4x+16\right)}$$

$$= \frac{1}{48(x-4)} + \frac{-x-8}{48\left(x^2+4x+16\right)} = \frac{\left(x^2+4x+16\right)+(-x-8)(x-4)}{48(x-4)\left(x^2+4x+16\right)} = \frac{x^2+4x+16-x^2+4x-8x+32}{48(x-4)\left(x^2+4x+16\right)}$$

$$= \frac{16+32}{48(x-4)\left(x^2+4x+16\right)} = \frac{48}{48(x-4)\left(x^2+4x+16\right)} = \frac{1}{x^3+4x^2+16x-4x^2-16x-64} = \frac{1}{x^3-64}$$

CASE IV - The Denominator Has Repeated Quadratic Factors

In this case each irreducible quadratic factor of the form ax^2+bx+c appears n times in the denominator. To solve this class of rational fractions we equate each proper rational fraction, that appears n times in the denominator, with a sum of n partial fractions of the form $\dfrac{M_1 x+N_1}{ax^2+bx+c}+\dfrac{M_2 x+N_2}{\left(ax^2+bx+c\right)^2}+...+\dfrac{M_n x+N_n}{\left(ax^2+bx+c\right)^n}$. The following examples show the steps as to how this class of integrals are solved.

Example 5.3-23: Evaluate the integral $\displaystyle\int\dfrac{x^2}{x^4+2x^2+1}dx$.

First - Check to see if the integrand is a proper or an improper rational fraction. If the integrand is an improper rational fraction use synthetic division (long division) to reduce the rational fraction to the sum of a polynomial and a proper rational fraction.

Second - Factor the denominator x^4+2x^2+1 into $\left(x^2+1\right)^2$.

Third - Write the factors in partial fraction form. Since the quadratic form in the denominator is repeated, the integrand can be represented in the following way:

$$\frac{x^2}{x^4+2x^2+1}=\frac{x^2}{\left(x^2+1\right)^2}=\frac{Ax+B}{x^2+1}+\frac{Cx+D}{\left(x^2+1\right)^2}$$

Fourth - Solve for the constants A, B, C, and D by equating coefficients of the like powers.

$$\frac{x^2}{x^4+2x^2+1}=\frac{(Ax+B)\left(x^2+1\right)+Cx+D}{\left(x^2+1\right)^2}$$

$$x^2=(Ax+B)\left(x^2+1\right)+Cx+D=Ax^3+Ax+Bx^2+B+Cx+D$$

$$x^2=Ax^3+Bx^2+(A+C)x+(B+D)\ \ \text{therefore,}$$

$$A=0\qquad\qquad B=1\qquad\qquad A+C=0\qquad\qquad B+D=0$$

which result in having $A=0$, $B=1$, $C=0$, and $D=-1$

Fifth - Rewrite the integral in its equivalent partial fraction form by substituting the constants with their specific values.

$$\int\frac{x^2}{x^4+2x^2+1}dx=\int\frac{Ax+B}{x^2+1}dx+\int\frac{Cx+D}{\left(x^2+1\right)^2}dx=\int\frac{0+1}{x^2+1}dx+\int\frac{0-1}{\left(x^2+1\right)^2}dx=\int\frac{1}{x^2+1}dx-\int\frac{1}{\left(x^2+1\right)^2}dx$$

Sixth - Integrate each integral individually using integration methods learned in previous sections. To solve the second integral let $x=\tan t$, then $\dfrac{dx}{dt}=\sec^2 t$ which implies $dx=\sec^2 t\,dt$.

$$\int\frac{1}{x^2+1}dx-\int\frac{1}{\left(x^2+1\right)^2}dx=arc\tan x-\int\frac{\sec^2 t}{\left(\tan^2 t+1\right)^2}dt=arc\tan x-\int\frac{\sec^2 t}{\left(\sec^2 t\right)^2}dt=arc\tan x-\int\frac{\sec^2 t}{\sec^4 t}dt$$

$$= arc\tan x - \int \frac{1}{\sec^2 t}\, dt \;=\; arc\tan x - \int \cos^2 t\, dt \;=\; \tan^{-1} x - \frac{1}{2}\int (1+\cos 2t)\, dt \;=\; \tan^{-1} x - \frac{1}{2}\left(t + \frac{1}{2}\sin 2t\right) + c$$

$$= \tan^{-1} x - \frac{1}{2}(t + \sin t \cos t) + c \;=\; \tan^{-1} x - \frac{1}{2}\tan^{-1} x - \frac{1}{2}\cdot\frac{x}{\sqrt{1+x^2}}\cdot\frac{1}{\sqrt{1+x^2}} + c \;=\; \frac{1}{2}\tan^{-1} x - \frac{x}{2\left(x^2 + 1\right)} + c$$

Or, we could use the already derived integration formulas by using Tables 5.2-1 and 5.4-3.

Note – Since the objective of this section is to teach the process for solving integrals using the Partial Fractions method, in the remaining example problems, we will use the already derived integration formulas summarized primarily in Tables 5.2-1 and 5.4-3 in order to solve this class of problems.

Seventh - Check the answer by differentiating the solution. The result should match the integrand.

Let $y = \dfrac{1}{2}\tan^{-1} x - \dfrac{x}{2\left(x^2+1\right)} + c$, then $y' = \dfrac{1}{2}\cdot\dfrac{1}{x^2+1} - \dfrac{1}{2}\cdot\dfrac{1\cdot\left(x^2+1\right) - 2x\cdot x}{\left(x^2+1\right)^2} + 0 = \dfrac{1}{2\left(x^2+1\right)} - \dfrac{x^2 - 2x^2 + 1}{2\left(x^2+1\right)^2}$

$$= \frac{1}{2\left(x^2+1\right)} - \frac{-x^2+1}{2\left(x^2+1\right)^2} = \frac{x^2+1+x^2-1}{2\left(x^2+1\right)^2} = \frac{2x^2}{2\left(x^2+1\right)^2} = \frac{x^2}{\left(x^2+1\right)^2} = \frac{x^2}{x^4+2x^2+1}$$

Example 5.3-24: Evaluate the integral $\displaystyle\int \frac{x^2+1}{x^4+8x^2+16}\, dx$.

First - Check to see if the integrand is a proper or an improper rational fraction. If the integrand is an improper rational fraction use synthetic division (long division) to reduce the rational fraction to the sum of a polynomial and a proper rational fraction.

Second - Factor the denominator $x^4 + 8x^2 + 16$ into $\left(x^2+4\right)^2$.

Third - Write the factors in partial fraction form. Since the quadratic form in the denominator is repeated, the integrand can be represented in the following way:

$$\frac{x^2+1}{x^4+8x^2+16} = \frac{x^2+1}{\left(x^2+4\right)^2} = \frac{Ax+B}{x^2+4} + \frac{Cx+D}{\left(x^2+4\right)^2}$$

Fourth - Solve for the constants A , B , C , and D by equating coefficients of the like powers.

$$\frac{x^2+1}{x^4+8x^2+16} = \frac{(Ax+B)\left(x^2+4\right)+Cx+D}{\left(x^2+4\right)^2}$$

$$x^2+1 = (Ax+B)\left(x^2+4\right)+Cx+D = Ax^3 + 4Ax + Bx^2 + 4B + Cx + D$$

$$x^2+1 = Ax^3 + Bx^2 + (4A+C)x + (4B+D) \quad \text{therefore,}$$

$A=0$	$B=1$	$4A+C=0$	$4B+D=1$

which result in having $A=0$, $B=1$, $C=0$, and $D=-3$

Fifth - Rewrite the integral in its equivalent partial fraction form by substituting the constants with their specific values.

$$\int \frac{x^2+1}{x^4+8x^2+16}dx = \int \frac{Ax+B}{x^2+4}dx + \int \frac{Cx+D}{\left(x^2+4\right)^2}dx = \int \frac{0+1}{x^2+4}dx + \int \frac{0-3}{\left(x^2+4\right)^2}dx = \int \frac{1}{x^2+4}dx - \int \frac{3}{\left(x^2+4\right)^2}dx$$

Sixth - Integrate each integral individually by using Tables 5.2-1 and 5.4-3.

$$\int \frac{1}{x^2+4}dx - \int \frac{3}{\left(x^2+4\right)^2}dx = \frac{1}{2}\tan^{-1}\frac{x}{2} - \frac{3}{16}\tan^{-1}\frac{x}{2} - \frac{3x}{8\left(x^2+4\right)} + c = \frac{5}{16}\tan^{-1}\frac{x}{2} - \frac{3x}{8\left(x^2+4\right)} + c$$

Seventh - Check the answer by differentiating the solution. The result should match the integrand.

Let $y = \frac{5}{16}\tan^{-1}\frac{x}{2} - \frac{3x}{8\left(x^2+4\right)} + c$, then $y' = \frac{5}{16}\cdot\frac{1}{\left(\frac{x}{2}\right)^2+1}\cdot\frac{1}{2} - \frac{3}{8}\cdot\frac{1\cdot\left(x^2+4\right)-2x\cdot x}{\left(x^2+4\right)^2} + 0$

$$= \frac{5}{8\left(x^2+4\right)} - \frac{3}{8}\cdot\frac{x^2-2x^2+4}{\left(x^2+4\right)^2} = \frac{5}{8\left(x^2+4\right)} - \frac{3}{8}\cdot\frac{-x^2+4}{\left(x^2+4\right)^2} = \frac{5x^2+20+3x^2-12}{8\left(x^2+4\right)^2} = \frac{x^2+1}{\left(x^2+4\right)^2} = \frac{x^2+1}{x^4+8x^2+16}$$

Example 5.3-25: Evaluate the integral $\int \frac{1}{x^4+10x^2+25}dx$.

First - Check to see if the integrand is a proper or an improper rational fraction. If the integrand is a rational fraction use synthetic division (long division) to reduce the rational fraction to the sum of a polynomial and a proper rational fraction.

Second - Factor the denominator x^4+10x^2+25 into $\left(x^2+5\right)^2$.

Third - Write the factors in partial fraction form. Since the quadratic form in the denominator is repeated, the integrand can be represented in the following way:

$$\frac{1}{x^4+10x^2+25} = \frac{1}{\left(x^2+5\right)^2} = \frac{Ax+B}{x^2+5} + \frac{Cx+D}{\left(x^2+5\right)^2}$$

Fourth - Solve for the constants A, B, C, and D by equating coefficients of the like powers.

$$\frac{1}{x^4+10x^2+25} = \frac{(Ax+B)\left(x^2+5\right)+Cx+D}{\left(x^2+5\right)^2}$$

$$1 = (Ax+B)\left(x^2+5\right)+Cx+D = Ax^3+5Ax+Bx^2+5B+Cx+D$$

$$1 = Ax^3+Bx^2+(5A+C)x+(5B+D) \text{ therefore,}$$

$$A=0 \qquad\qquad B=0 \qquad\qquad 5A+C=0 \qquad\qquad 5B+D=1$$

which result in having $A=0$, $B=0$, $C=0$, and $D=1$

Fifth - Rewrite the integral in its equivalent partial fraction form by substituting the constants with their specific values.

$$\int \frac{1}{x^4+10x^2+25}dx = \int \frac{Ax+B}{x^2+5}dx + \int \frac{Cx+D}{(x^2+5)^2}dx = \int \frac{0+0}{x^2+5}dx + \int \frac{0+1}{(x^2+5)^2}dx = \int \frac{1}{(x^2+5)^2}dx$$

Sixth - Integrate each integral individually by using Tables 5.2-1 and 5.4-3.

$$\int \frac{1}{(x^2+5)^2}dx = \frac{1}{10\sqrt{5}}\tan^{-1}\frac{x}{\sqrt{5}} + \frac{x}{10(x^2+5)} + c$$

Seventh - Check the answer by differentiating the solution. The result should match the integrand.

Let $y = \frac{1}{10\sqrt{5}}\tan^{-1}\frac{x}{\sqrt{5}} + \frac{x}{10(x^2+5)} + c$, then $y' = \frac{1}{10\sqrt{5}}\cdot\frac{1}{\left(\frac{x}{\sqrt{5}}\right)^2+1}\cdot\frac{1}{\sqrt{5}} + \frac{1\cdot(x^2+5)-2x\cdot x}{10(x^2+5)^2} + 0$

$$= \frac{5}{50(x^2+5)} + \frac{1}{10}\cdot\frac{x^2-2x^2+5}{(x^2+5)^2} = \frac{1}{10(x^2+5)} + \frac{-x^2+5}{10(x^2+5)^2} = \frac{x^2+5-x^2+5}{10(x^2+5)^2} = \frac{1}{(x^2+5)^2} = \frac{1}{x^4+10x^2+25}$$

Example 5.3-26: Evaluate the integral $\int \frac{x^3}{x^4+4x^2+4}dx$.

First - Check to see if the integrand is a proper or an improper rational fraction. If the integrand is a rational fraction use synthetic division (long division) to reduce the rational fraction to the sum of a polynomial and a proper rational fraction.

Second - Factor the denominator x^4+4x^2+4 into $(x^2+2)^2$.

Third - Write the factors in partial fraction form. Since the quadratic form in the denominator is repeated, the integrand can be represented in the following way:

$$\frac{x^3}{x^4+4x^2+4} = \frac{x^3}{(x^2+2)^2} = \frac{Ax+B}{x^2+2} + \frac{Cx+D}{(x^2+2)^2}$$

Fourth - Solve for the constants A, B, C, and D by equating coefficients of the like powers.

$$\frac{x^3}{x^4+4x^2+4} = \frac{(Ax+B)(x^2+2)+Cx+D}{(x^2+2)^2}$$

$$x^3 = (Ax+B)(x^2+2)+Cx+D = Ax^3+2Ax+Bx^2+2B+Cx+D$$

$$x^3 = Ax^3+Bx^2+(2A+C)x+(2B+D) \text{ therefore,}$$

$$A=1 \qquad\qquad B=0 \qquad\qquad 2A+C=0 \qquad\qquad 2B+D=0$$

which result in having $A=1$, $B=0$, $C=-2$, and $D=0$

Fifth - Rewrite the integral in its equivalent partial fraction form by substituting the constants with their specific values.

$$\int \frac{x^3}{x^4+4x^2+4}dx = \int \frac{Ax+B}{x^2+2}dx + \int \frac{Cx+D}{(x^2+2)^2}dx = \int \frac{x+0}{x^2+2}dx + \int \frac{-2x+0}{(x^2+2)^2}dx = \int \frac{x}{x^2+2}dx - \int \frac{2x}{(x^2+2)^2}dx$$

Sixth - Integrate each integral individually by using Tables 5.2-1 and 5.4-3.

$$\int \frac{x}{x^2+2}\,dx - \int \frac{2x}{\left(x^2+2\right)^2}\,dx = \int \frac{x}{u}\cdot\frac{1}{2x}\,du - 2\int \frac{x}{u^2}\cdot\frac{1}{2x}\,du = \frac{1}{2}\int \frac{1}{u}\,du - \frac{2}{2}\int \frac{1}{u^2}\,du = \frac{1}{2}\int \frac{1}{u}\,du - \int u^{-2}\,du$$

$$= \frac{1}{2}\ln|u| - \frac{1}{-2+1}u^{-2+1} + c = \frac{1}{2}\ln|u| + u^{-1} + c = \frac{1}{2}\ln|u| + \frac{1}{u} + c = \frac{1}{2}\ln\left|x^2+2\right| + \frac{1}{x^2+2} + c$$

Seventh - Check the answer by differentiating the solution. The result should match the integrand.

Let $y = \frac{1}{2}\ln\left|x^2+2\right| + \frac{1}{x^2+2} + c$, then $y' = \frac{1}{2}\cdot\frac{2x}{x^2+2} - \frac{2x}{\left(x^2+2\right)^2} + 0 = \frac{x}{x^2+2} - \frac{2x}{\left(x^2+2\right)^2} = \frac{x\left(x^2+2\right)-2x}{\left(x^2+2\right)^2}$

$$= \frac{x^3+2x-2x}{\left(x^2+2\right)^2} = \frac{x^3}{\left(x^2+2\right)^2} = \frac{x^3}{x^4+4x^2+4}$$

Example 5.3-27: Evaluate the integral $\int \frac{2x^2+x+7}{x^4+8x^2+16}\,dx$.

First - Check to see if the integrand is a proper or an improper rational fraction. If the integrand is a rational fraction use synthetic division (long division) to reduce the rational fraction to the sum of a polynomial and a proper rational fraction.

Second - Factor the denominator x^4+8x^2+16 into $\left(x^2+4\right)^2$.

Third - Write the factors in partial fraction form. Since the quadratic form in the denominator is repeated, the integrand can be represented in the following way:

$$\frac{2x^2+x+7}{x^4+8x^2+16} = \frac{2x^2+x+7}{\left(x^2+4\right)^2} = \frac{Ax+B}{x^2+4} + \frac{Cx+D}{\left(x^2+4\right)^2}$$

Fourth - Solve for the constants A, B, C, and D by equating coefficients of the like powers.

$$\frac{2x^2+x+7}{x^4+8x^2+16} = \frac{(Ax+B)\left(x^2+4\right)+Cx+D}{\left(x^2+4\right)^2}$$

$$2x^2+x+7 = (Ax+B)\left(x^2+4\right)+Cx+D = Ax^3+4Ax+Bx^2+4B+Cx+D$$

$$2x^2+x+7 = Ax^3+Bx^2+(4A+C)x+(4B+D) \quad \text{therefore,}$$

$$A=0 \qquad\qquad B=2 \qquad\qquad 4A+C=1 \qquad\qquad 4B+D=7$$

which result in having $A=0$, $B=2$, $C=1$, and $D=-1$

Fifth - Rewrite the integral in its equivalent partial fraction form by substituting the constants with their specific values.

$$\int \frac{2x^2+x+7}{x^4+8x^2+16}\,dx = \int \frac{Ax+B}{x^2+4}\,dx + \int \frac{Cx+D}{\left(x^2+4\right)^2}\,dx = \int \frac{0+2}{x^2+4}\,dx + \int \frac{x-1}{\left(x^2+4\right)^2}\,dx = \int \frac{2}{x^2+4}\,dx + \int \frac{x-1}{\left(x^2+4\right)^2}\,dx$$

$$= \int \frac{2}{x^2+4}\,dx + \int \frac{x}{\left(x^2+4\right)^2}\,dx - \int \frac{1}{\left(x^2+4\right)^2}\,dx$$

Sixth - Integrate each integral individually by using Tables 5.2-1 and 5.4-3.

$$\int \frac{2}{x^2+4}\,dx + \int \frac{x}{\left(x^2+4\right)^2}\,dx - \int \frac{1}{\left(x^2+4\right)^2}\,dx = 2\cdot\frac{1}{2}\tan^{-1}\frac{x}{2} + \int \frac{x}{u^2}\cdot\frac{1}{2x}\,du - \frac{1}{16}\tan^{-1}\frac{x}{2} - \frac{x}{8\left(x^2+4\right)} + c$$

$$= 2\cdot\frac{1}{2}\tan^{-1}\frac{x}{2} + \frac{1}{2}\cdot\frac{1}{-2+1}u^{-2+1} - \frac{1}{16}\tan^{-1}\frac{x}{2} - \frac{x}{8\left(x^2+4\right)} + c = \tan^{-1}\frac{x}{2} - \frac{1}{2u} - \frac{1}{16}\tan^{-1}\frac{x}{2} - \frac{x}{8\left(x^2+4\right)} + c$$

$$= \tan^{-1}\frac{x}{2} - \frac{1}{2\left(x^2+4\right)} - \frac{1}{16}\tan^{-1}\frac{x}{2} - \frac{x}{8\left(x^2+4\right)} + c = \frac{15}{16}\tan^{-1}\frac{x}{2} + \frac{-4-x}{8\left(x^2+4\right)} + c = \frac{15}{16}\tan^{-1}\frac{x}{2} - \frac{x+4}{8\left(x^2+4\right)} + c$$

Seventh - Check the answer by differentiating the solution. The result should match the integrand.

Let $y = \dfrac{15}{16}\tan^{-1}\dfrac{x}{2} - \dfrac{x+4}{8\left(x^2+4\right)} + c$, then $y' = \dfrac{15}{16}\cdot\dfrac{1}{\left(\frac{x}{2}\right)^2+1}\cdot\dfrac{1}{2} - \dfrac{1\cdot\left(x^2+4\right)-2x\cdot(x+4)}{8\left(x^2+4\right)^2} + 0$

$$= \frac{15}{8\left(x^2+4\right)} - \frac{1}{8}\cdot\frac{x^2-2x^2-8x+4}{\left(x^2+4\right)^2} = \frac{15}{8\left(x^2+4\right)} + \frac{x^2+8x-4}{8\left(x^2+4\right)^2} = \frac{15x^2+60+x^2+8x-4}{8\left(x^2+4\right)^2} = \frac{16x^2+8x+56}{8\left(x^2+4\right)^2}$$

$$= \frac{8\left(2x^2+x+7\right)}{8\left(x^2+4\right)^2} = \frac{2x^2+x+7}{\left(x^2+4\right)^2} = \frac{2x^2+x+7}{x^4+8x+16}$$

Section 5.3 Practice Problems – Integration by Partial Fractions

Evaluate the following integrals:

a. $\displaystyle\int \frac{dx}{x^2+5x+6} =$

b. $\displaystyle\int \frac{x^2+1}{x^3-4x}\,dx =$

c. $\displaystyle\int \frac{1}{36-x^2}\,dx =$

d. $\displaystyle\int \frac{x+5}{x^3+2x^2+x}\,dx =$

e. $\displaystyle\int \frac{1}{x^3-2x^2+x}\,dx =$

f. $\displaystyle\int \frac{x^2+3}{x^2-1}\,dx =$

g. $\displaystyle\int \frac{1}{x^3-1}\,dx =$

h. $\displaystyle\int \frac{1}{x^4-1}\,dx =$

i. $\displaystyle\int \frac{1}{x^3+64}\,dx =$

5.4 Integration of Hyperbolic Functions

In the following examples we will solve problems using the formulas below:

Table 5.4-1: Integration Formulas for Hyperbolic Functions

1. $\int \sinh x \, dx = \cosh x + c$	2. $\int \cosh x \, dx = \sinh x + c$	3. $\int \tanh x \, dx = \ln \cosh x + c$
4. $\int \coth x \, dx = \ln\|\sinh x\| + c$	5. $\int \sec h \, x \, dx = \sin^{-1}(\tanh x) + c$	6. $\int \csc h \, x \, dx = \ln\left\|\tanh \frac{x}{2}\right\| + c$
7. $\int \tanh x \sec h \, x \, dx = -\sec h \, x + c$	8. $\int \coth x \csc h \, x \, dx = -\csc h \, x + c$	9. $\int \sinh^2 x \, dx = \frac{\sinh 2x}{4} - \frac{x}{2} + c$
10. $\int \cosh^2 x \, dx = \frac{\sinh 2x}{4} + \frac{x}{2} + c$	11. $\int \tanh^2 x \, dx = x - \tanh x + c$	12. $\int \coth^2 x \, dx = x - \coth x + c$
13. $\int \sec h^2 x \, dx = \tanh x + c$	14. $\int \csc h^2 x \, dx = -\coth x + c$	

Additionally, the following formulas, similar to the trigonometric functions, hold for the hyperbolic functions:

1. **Unit Formulas**

$$\cosh^2 x - \sinh^2 x = 1 \qquad \tanh h^2 x + \sec h^2 x = 1 \qquad \coth h^2 x - \csc h^2 x = 1$$

2. **Addition Formulas**

$$\sinh(x \pm y) = \sinh x \cosh y \pm \cosh x \sinh y \qquad \cosh(x \pm y) = \cosh x \cosh y \pm \sinh h \, x \sinh y$$

$$\tanh(x \pm y) = \frac{\tanh x \pm \tanh y}{1 \pm \tanh x \tanh y} \qquad \coth(x \pm y) = \frac{\coth x \coth y \pm 1}{\coth y \pm \coth x}$$

3. **Half Angle Formulas**

$$\sinh \frac{1}{2} x = \sqrt{\frac{\cosh x - 1}{2}} \quad x \rangle 0 \qquad\qquad \sinh \frac{1}{2} x = -\sqrt{\frac{\cosh x - 1}{2}} \quad x \langle 0$$

$$\cosh \frac{1}{2} x = \sqrt{\frac{\cosh x + 1}{2}}$$

$$\tanh \frac{1}{2} x = \sqrt{\frac{\cosh x - 1}{\cosh x + 1}} \quad x \rangle 0 \qquad\qquad \sinh \frac{1}{2} x = -\sqrt{\frac{\cosh x - 1}{\cosh x + 1}} \quad x \langle 0 \qquad \text{or,}$$

$$\tanh \frac{1}{2} x = \frac{\sinh x}{\cosh x + 1} = \frac{\cosh x - 1}{\sinh x}$$

4. **Double Angle Formulas**

$$\sinh 2x = 2 \sinh x \cosh x \qquad\qquad \cosh 2x = \cosh^2 x + \sinh^2 x = 2\cosh^2 x - 1 = 1 + 2\sinh^2 x$$

$$\tanh 2x = \frac{2 \tanh x}{1 + \tanh^2 x}$$

Also, the hyperbolic functions are defined by

$$\sinh x = \frac{1}{2}\left(e^x - e^{-x}\right) \qquad\qquad\qquad \cosh x = \frac{1}{2}\left(e^x + e^{-x}\right)$$

$$\tanh x = \frac{\sinh x}{\cosh x} = \frac{\frac{1}{2}\left(e^x - e^{-x}\right)}{\frac{1}{2}\left(e^x + e^{-x}\right)} = \frac{e^x - e^{-x}}{e^x + e^{-x}} \qquad\qquad \coth x = \frac{\cosh x}{\sinh x} = \frac{\frac{1}{2}\left(e^x + e^{-x}\right)}{\frac{1}{2}\left(e^x - e^{-x}\right)} = \frac{e^x + e^{-x}}{e^x - e^{-x}}$$

$$\sec h\, x = \frac{1}{\cosh x} = \frac{1}{\frac{1}{2}\left(e^x + e^{-x}\right)} = \frac{2}{e^x + e^{-x}} \qquad\qquad \csc h\, x = \frac{1}{\sinh x} = \frac{1}{\frac{1}{2}\left(e^x - e^{-x}\right)} = \frac{2}{e^x - e^{-x}}$$

Also note that the negative argument of the hyperbolic functions is equal to the following:

$$\sinh\left(-x\right) = -\sinh x \qquad\qquad \tanh\left(-x\right) = -\tanh x \qquad\qquad \coth\left(-x\right) = -\coth x$$

$$\csc h\left(-x\right) = -\csc h\, x \qquad\qquad \cosh\left(-x\right) = \cosh x \qquad\qquad \sec h\left(-x\right) = \sec h\, x$$

Finally, we need to know how to differentiate the hyperbolic functions (addressed in Chapter 3, Section 3.4) in order to check the answer to the given integrals below. The derivatives of hyperbolic functions are repeated here and are as follows:

Table 5.4-2: Differentiation Formulas for Hyperbolic Functions

$\dfrac{d}{dx}\sinh u = \cosh u \cdot \dfrac{du}{dx}$	$\dfrac{d}{dx}\coth u = -\csc h^2 u \cdot \dfrac{du}{dx}$
$\dfrac{d}{dx}\cosh u = \sinh u \cdot \dfrac{du}{dx}$	$\dfrac{d}{dx}\sec h\, u = -\sec h\, u \tanh u \cdot \dfrac{du}{dx}$
$\dfrac{d}{dx}\tanh u = \sec h^2 u \cdot \dfrac{du}{dx}$	$\dfrac{d}{dx}\csc h\, u = -\csc h\, u \coth u \cdot \dfrac{du}{dx}$

Let's integrate some hyperbolic functions using the above integration formulas.

Example 5.4-1: Evaluate the following integrals:

a. $\displaystyle\int \sinh 5x\, dx =$

b. $\displaystyle\int \sinh \frac{1}{6}x\, dx =$

c. $\displaystyle\int \cosh 7x\, dx =$

d. $\displaystyle\int \cosh \frac{1}{5}x\, dx =$

e. $\displaystyle\int \left(\sinh 4x + \cosh 2x\right) dx =$

f. $\displaystyle\int \csc h\, 8x\, dx =$

g. $\displaystyle\int \csc h^2 5x\, dx =$

h. $\displaystyle\int \csc h^2 \frac{1}{4}x\, dx =$

i. $\displaystyle\int x^2 \sec h^2 x^3\, dx =$

j. $\displaystyle\int x \sec h^2 \left(x^2 + 5\right) dx =$

k. $\displaystyle\int x^2 \csc h^2 x^3\, dx =$

l. $\displaystyle\int 2x \csc h^2 x^2\, dx =$

Solutions:

a. Given $\displaystyle\int \sinh 5x\, dx$ let $u = 5x$, then $\dfrac{du}{dx} = \dfrac{d}{dx}5x = 5$ which implies $dx = \dfrac{du}{5}$. Therefore,

$$\boxed{\int \sinh 5x\, dx} = \boxed{\int \sinh u \cdot \frac{du}{5}} = \boxed{\frac{1}{5}\int \sinh u\, du} = \boxed{\frac{1}{5}\cosh u + c} = \boxed{\mathbf{\frac{1}{5}\cosh 5x + c}}$$

Check: Let $y = \frac{1}{5}\cosh 5x + c$, then $y' = \frac{1}{5}\cdot\frac{d}{dx}\sinh 5x + \frac{d}{dx}c = \frac{1}{5}\cdot\sinh 5x \cdot\frac{d}{dx}5x + 0 = \frac{1}{5}\cdot\sinh 5x\cdot 5$

$$= \frac{5}{5}\cdot\sinh 5x = \sinh 5x$$

b. Given $\int \sinh\frac{x}{6}\,dx$ let $u = \frac{x}{6}$, then $\frac{du}{dx} = \frac{d}{dx}\frac{x}{6} = \frac{1}{6}$ which implies $dx = 6\,du$. Therefore,

$$\boxed{\int \sinh\frac{x}{6}\,dx} = \boxed{\int \sinh u \cdot 6\,du} = \boxed{6\int \sinh u\, du} = \boxed{6\cosh u + c} = \boxed{\mathbf{6\cosh\frac{x}{6} + c}}$$

Check: Let $y = 6\cosh\frac{x}{6} + c$, then $y' = 6\cdot\frac{d}{dx}\cosh\frac{x}{6} + \frac{d}{dx}c = 6\cdot\sinh\frac{x}{6}\cdot\frac{d}{dx}\frac{x}{6} + 0 = 6\cdot\sinh\frac{x}{6}\cdot\frac{1}{6}$

$$= \frac{6}{6}\cdot\sinh\frac{x}{6} = \sinh\frac{x}{6}$$

c. Given $\int \cosh 7x\, dx$ let $u = 7x$, then $\frac{du}{dx} = \frac{d}{dx}7x = 7$ which implies $dx = \frac{du}{7}$. Therefore,

$$\boxed{\int \cosh 7x\, dx} = \boxed{\int \cosh u \cdot\frac{du}{7}} = \boxed{\frac{1}{7}\int \cosh u\, du} = \boxed{\frac{1}{7}\sinh u + c} = \boxed{\mathbf{\frac{1}{7}\sinh 7x + c}}$$

Check: Let $y = \frac{1}{7}\sinh 7x + c$, then $y' = \frac{1}{7}\cdot\frac{d}{dx}\sinh 7x + \frac{d}{dx}c = \frac{1}{7}\cdot\cosh 7x\cdot\frac{d}{dx}7x + 0 = \frac{1}{7}\cdot\cosh 7x\cdot 7$

$$= \frac{7}{7}\cdot\cosh 7x = \cosh 7x$$

d. Given $\int \cosh\frac{x}{5}\,dx$ let $u = \frac{x}{5}$, then $\frac{du}{dx} = \frac{d}{dx}\frac{x}{5} = \frac{1}{5}$ which implies $dx = 5\,du$. Therefore,

$$\boxed{\int \cosh\frac{x}{5}\,dx} = \boxed{\int \cosh u \cdot 5\,du} = \boxed{5\int \cosh u\, du} = \boxed{5\sinh u + c} = \boxed{\mathbf{5\sinh\frac{x}{5} + c}}$$

Check: Let $y = 5\sinh\frac{x}{5} + c$, then $y' = 5\cdot\frac{d}{dx}\sinh\frac{x}{5} + \frac{d}{dx}c = 5\cdot\cosh\frac{x}{5}\cdot\frac{d}{dx}\frac{x}{5} + 0 = 5\cdot\cosh\frac{x}{5}\cdot\frac{1}{5}$

$$= \frac{5}{5}\cdot\cosh\frac{x}{5} = \cosh\frac{x}{5}$$

e. Given $\int (\sinh 4x + \cosh 2x)\,dx = \int \sinh 4x\, dx + \int \cosh 2x\, dx$ let:

a. $u = 4x$, then $\frac{du}{dx} = \frac{d}{dx}4x$; $\frac{du}{dx} = 4$; $du = 4dx$; $dx = \frac{du}{4}$ and

b. $v = 2x$, then $\frac{dv}{dx} = \frac{d}{dx}2x$; $\frac{dv}{dx} = 2$; $dv = 2dx$; $dx = \frac{dv}{2}$.

Therefore, $\boxed{\int \sinh 4x\, dx + \int \cosh 2x\, dx} = \boxed{\int \sinh u \cdot\frac{du}{4} + \int \cosh v \cdot\frac{dv}{2}} = \boxed{\frac{1}{4}\int \sinh u\, du + \frac{1}{2}\int \cosh v\, dv}$

$$= \boxed{\frac{1}{4}\cosh u + c_1 + \frac{1}{2}\sinh v + c_2} = \boxed{\frac{1}{4}\cosh 4x + \frac{1}{2}\sinh 2x + c_1 + c_2} = \boxed{\frac{1}{4}\cosh 4x + \frac{1}{2}\sinh 2x + c}$$

Check: Let $y = \frac{1}{4}\cosh 4x + \frac{1}{2}\sinh 2x + c$ then $y' = \frac{1}{4}\cdot\frac{d}{dx}\cosh 4x + \frac{1}{2}\cdot\frac{d}{dx}\sinh 2x + \frac{d}{dx}c = \frac{1}{4}\cdot\sinh 4x\cdot\frac{d}{dx}4x$

$+ \frac{1}{2}\cdot\cosh 2x\cdot\frac{d}{dx}2x + 0 = \frac{4}{4}\cdot\sinh 4x + \frac{2}{2}\cdot\cosh 2x = \sinh 4x + \cosh 2x$

f. Given $\int \csc h\, 8x\, dx$ let $u = 8x$, then $\frac{du}{dx} = \frac{d}{dx}8x = 8$ which implies $du = 8dx$; $dx = \frac{du}{8}$. Therefore,

$$\boxed{\int \csc h\, 8x\, dx} = \boxed{\int \csc h\, u\cdot\frac{du}{8}} = \boxed{\frac{1}{8}\int \csc h\, u\, du} = \boxed{\frac{1}{8}\ln\left|\tanh\frac{u}{2}\right| + c} = \boxed{\frac{1}{8}\ln\left|\tanh\frac{8x}{2}\right| + c} = \boxed{\frac{1}{8}\ln\left|\tanh 4x\right| + c}$$

Check: Let $y = \frac{1}{8}\ln\left|\tanh 4x\right| + c$, then $y' = \frac{1}{8}\cdot\frac{d}{dx}\ln\left|\tanh 4x\right| + \frac{d}{dx}c = \frac{1}{8}\cdot\frac{1}{\tanh 4x}\cdot\frac{d}{dx}(\tanh 4x) + 0$

$$= \frac{1}{8}\cdot\frac{1}{\tanh 4x}\cdot\left(\sec h^2 4x\cdot 4\right) + 0 = \frac{1}{8}\cdot\frac{4\sec h^2 4x}{\tanh 4x} = \frac{1}{2}\cdot\frac{\sec h^2 4x}{\tanh 4x} = \frac{1}{2}\cdot\frac{1-\tanh^2 4x}{\tanh 4x} = \frac{1}{2}\cdot\frac{1-\frac{\sinh^2 4x}{\cosh^2 4x}}{\frac{\sinh 4x}{\cosh 4x}}$$

$$= \frac{1}{2}\cdot\frac{\frac{\cosh^2 4x - \sinh^2 4x}{\cosh^2 4x}}{\frac{\sinh 4x}{\cosh 4x}} = \frac{1}{2}\cdot\frac{\frac{1}{\cosh^2 4x}}{\frac{\sinh 4x}{\cosh 4x}} = \frac{1}{2}\cdot\frac{\cosh 4x}{\cosh^2 4x\cdot\sinh 4x} = \frac{1}{2\cosh 4x\cdot\sinh 4x} = \frac{1}{\sinh 2\cdot 4x}$$

$$= \frac{1}{\sinh 8x} = \csc h\, 8x$$

g. Given $\int \csc h^2 5x\, dx$ let $u = 5x$, then $\frac{du}{dx} = \frac{d}{dx}5x$; $\frac{du}{dx} = 5$; $du = 5dx$; $dx = \frac{du}{5}$. Therefore,

$$\boxed{\int \csc h^2 5x\, dx} = \boxed{\int \csc h^2 u\cdot\frac{du}{5}} = \boxed{\frac{1}{5}\int \csc h^2 u\, du} = \boxed{-\frac{1}{5}\coth u + c} = \boxed{-\frac{1}{5}\coth 5x + c}$$

Check: Let $y = -\frac{1}{5}\coth 5x + c$, then $y' = -\frac{1}{5}\cdot\frac{d}{dx}(\coth 5x)\cdot\frac{d}{dx}5x + \frac{d}{dx}c = -\frac{1}{5}\cdot\left(-\csc h^2 5x\right)\cdot 5 + 0$

$$= \frac{5}{5}\cdot\csc h^2 5x = \csc h^2 5x$$

h. Given $\int \csc h^2 \frac{1}{4}x\, dx$ let $u = \frac{1}{4}x$, then $\frac{du}{dx} = \frac{d}{dx}\frac{x}{4}$; $\frac{du}{dx} = \frac{1}{4}$; $4du = dx$; $dx = 4du$. Therefore,

$$\boxed{\int \csc h^2 \frac{1}{4}x\, dx} = \boxed{\int \csc h^2 u\cdot 4du} = \boxed{4\int \csc h^2 u\, du} = \boxed{-4\coth u + c} = \boxed{-4\coth \frac{1}{4}x + c}$$

Check: Let $y = -4\coth\frac{x}{4} + c$, then $y' = -4\cdot\frac{d}{dx}\coth\frac{x}{4} + \frac{d}{dx}c = -4\cdot-\csc h^2\frac{x}{4}\cdot\frac{d}{dx}\frac{x}{2} + 0 = 4\csc h^2\frac{x}{4}\cdot\frac{1}{4}$

$$= \frac{4}{4}\csc h^2\frac{x}{4} = \csc h^2\frac{x}{4} = \csc h^2\frac{1}{4}x$$

i. Given $\int x^2 \sec h^2 x^3\, dx$ let $u = x^3$, then $\frac{du}{dx} = \frac{d}{dx}x^3$; $\frac{du}{dx} = 3x^2$; $du = 3x^2 dx$; $dx = \frac{du}{3x^2}$. Therefore,

$$\boxed{\int x^2 \sec h^2 x^3 \, dx} = \boxed{\int x^2 \sec h^2 u \cdot \frac{du}{3x^2}} = \boxed{\frac{1}{3}\int \sec h^2 u \, du} = \boxed{\frac{1}{3}\tanh u + c} = \boxed{\frac{1}{3}\tanh x^3 + c}$$

Check: Let $y = \frac{1}{3}\tanh x^3 + c$, then $y' = \frac{1}{3}\cdot\frac{d}{dx}\tanh x^3 + \frac{d}{dx}c = \frac{1}{3}\cdot\sec h^2 x^3 \cdot \frac{d}{dx}x^3 + 0 = \frac{1}{3}\cdot\sec h^2 x^3 \cdot 3\, x^2$

$\qquad = \frac{3x^2}{3}\cdot\sec h^2 x^3 = x^2 \sec h^2 x^3$

j. Given $\int x \sec h^2 (x^2 + 5)dx$ let $u = x^2 + 5$, then $\frac{du}{dx} = \frac{d}{dx}(x^2 + 5);\ \frac{du}{dx} = 2x;\ du = 2x\, dx;\ dx = \frac{du}{2x}$. Therefore,

$$\boxed{\int x \sec h^2 (x^2 + 5)dx} = \boxed{\int x \sec h^2 u \cdot \frac{du}{2x}} = \boxed{\frac{1}{2}\int \sec h^2 u \, du} = \boxed{\frac{1}{2}\tanh u + c} = \boxed{\frac{1}{2}\tanh (x^2 + 5) + c}$$

Check: Let $y = \frac{1}{2}\tanh (x^2 + 5) + c$, then $y' = \frac{1}{2}\cdot\frac{d}{dx}\tanh (x^2 + 5) + \frac{d}{dx}c = \frac{1}{2}\cdot\sec h^2 (x^2 + 5)\cdot \frac{d}{dx}(x^2 + 5) + 0$

$\qquad = \frac{1}{2}\cdot\sec h^2 (x^2 + 5)\cdot 2x = \frac{2x}{2}\cdot\sec h^2 (x^2 + 5) = x \sec h^2 (x^2 + 5)$

k. Given $\int x^2 \csc h^2 x^3 dx$ let $u = x^3$, then $\frac{du}{dx} = \frac{d}{dx}x^3;\ \frac{du}{dx} = 3x^2;\ du = 3x^2 dx;\ dx = \frac{du}{3x^2}$. Therefore,

$$\boxed{\int x^2 \csc h^2 x^3 dx} = \boxed{\int x^2 \csc h^2 u \cdot \frac{du}{3x^2}} = \boxed{\frac{1}{3}\int \csc h^2 u \, du} = \boxed{-\frac{1}{3}\coth u + c} = \boxed{-\frac{1}{3}\coth x^3 + c}$$

Check: Let $y = -\frac{1}{3}\coth x^3 + c$, then $y' = -\frac{1}{3}\cdot\frac{d}{dx}\coth x^3 + \frac{d}{dx}c = -\frac{1}{3}\cdot-\csc h^2 x^3 \cdot \frac{d}{dx}x^3 + 0$

$\qquad = \frac{1}{3}\cdot\csc h^2 x^3 \cdot 3x^2 = \frac{3x^2}{3}\cdot\csc h^2 x^3 = x^2 \csc h^2 x^3$

l. Given $\int 2x \csc h^2 x^2 dx$ let $u = x^2$, then $\frac{du}{dx} = \frac{d}{dx}x^2;\ \frac{du}{dx} = 2x;\ du = 2x\, dx;\ dx = \frac{du}{2x}$. Therefore,

$$\boxed{\int 2x \csc h^2 x^2 dx} = \boxed{\int 2x \csc h^2 u \cdot \frac{du}{2x}} = \boxed{\int \csc h^2 u \, du} = \boxed{-\coth u + c} = \boxed{-\coth x^2 + c}$$

Check: Let $y = -\coth x^2 + c$, then $y' = -\frac{d}{dx}\coth x^2 + \frac{d}{dx}c = \csc h^2 x^2 \cdot \frac{d}{dx}x^2 + 0 = \csc h^2 x^2 \cdot 2x$

$\qquad = 2x \csc h^2 x^2$

Example 5.4-2: Evaluate the following indefinite integrals:

a. $\int \sec h^2 5x \, dx =$

   b. $\int x^2 \sinh x^3 dx =$

   c. $\int x \csc h\, x^2 dx =$

d. $\int \sinh^5 (x+1)\cosh (x+1) dx =$

   e. $\int \cosh^5 x \sinh x \, dx =$

   f. $\int \cosh^5 5x \sinh 5x \, dx =$

g. $\int \cosh^4 \frac{x}{2}\sinh \frac{x}{2} dx =$

   h. $\int x^2 \cosh x^3 dx =$

   i. $\int \frac{e^{2x}}{3}\sec h\, e^{2x} \, dx =$

j. $\int \sec h^2 (5x-1)dx =$

   k. $\int \csc h\, 7x \coth 7x \, dx =$

   l. $\int \tanh^2 10x \, dx =$

m. $\int \cosh^3 \frac{x}{5} dx =$

   n. $\int \sinh^3 x \, dx =$

   o. $\int \frac{x}{2}\sinh x \, dx =$

Solutions:

a. Given $\int \sec h^2 5x \, dx$ let $u = 5x$, then $\dfrac{du}{dx} = \dfrac{d}{dx} 5x$; $\dfrac{du}{dx} = 5$; $du = 5 \, dx$; $dx = \dfrac{du}{5}$. Therefore,

$$\boxed{\int \sec h^2 5x \, dx} = \boxed{\int \sec h^2 u \cdot \dfrac{du}{5}} = \boxed{\dfrac{1}{5} \int \sec h^2 u \, du} = \boxed{\dfrac{1}{5} \tanh u + c} = \boxed{\mathbf{\dfrac{1}{5} \tanh 5x + c}}$$

Check: Let $y = \dfrac{1}{5} \tanh 5x + c$, then $y' = \dfrac{1}{5} \cdot \dfrac{d}{dx} \tanh 5x + \dfrac{d}{dx} c = \dfrac{1}{5} \sec h^2 5x \cdot 5 + 0 = \dfrac{5}{5} \sec h^2 5x = \sec h^2 5x$

b. Given $\int x^2 \sinh x^3 dx$ let $u = x^3$, then $\dfrac{du}{dx} = \dfrac{d}{dx} x^3 = 3x^2$ which implies $du = 3x^2 dx$; $dx = \dfrac{du}{3x^2}$. Thus,

$$\boxed{\int x^2 \sinh x^3 dx} = \boxed{\int x^2 \sinh u \cdot \dfrac{du}{3x^2}} = \boxed{\dfrac{1}{3} \int \sinh u \, du} = \boxed{\dfrac{1}{3} \cosh u + c} = \boxed{\mathbf{\dfrac{1}{3} \cosh x^3 + c}}$$

Check: Let $y = \dfrac{1}{3} \cosh x^3 + c$, then $y' = \dfrac{1}{3} \cdot \dfrac{d}{dx} \cosh x^3 + \dfrac{d}{dx} c = \dfrac{1}{3} \cdot \sinh x^3 \cdot \dfrac{d}{dx} x^3 + 0 = \dfrac{1}{3} \cdot \sinh x^3 \cdot 3x^2$

$$= \dfrac{3x^2}{3} \cdot \sinh x^3 = x^2 \sinh x^3$$

c. Given $\int x \csc h \, x^2 dx$ let $u = x^2$, then $\dfrac{du}{dx} = \dfrac{d}{dx} x^2$; $\dfrac{du}{dx} = 2x$; $du = 2x \, dx$; $dx = \dfrac{du}{2x}$. Therefore,

$$\boxed{\int x \csc h \, x^2 dx} = \boxed{\int x \csc h \, u \cdot \dfrac{du}{2x}} = \boxed{\dfrac{1}{2} \int \csc h \, u \, du} = \boxed{\dfrac{1}{2} \ln \left| \tanh \dfrac{u}{2} \right| + c} = \boxed{\mathbf{\dfrac{1}{2} \ln \left| \tanh \dfrac{x^2}{2} \right| + c}}$$

Check: Let $y = \dfrac{1}{2} \ln \left| \tanh \dfrac{x^2}{2} \right| + c$, then $y' = \dfrac{1}{2} \cdot \dfrac{d}{dx} \ln \left| \tanh \dfrac{x^2}{2} \right| + \dfrac{d}{dx} c = \dfrac{1}{2} \cdot \dfrac{1}{\tanh \frac{x^2}{2}} \cdot \dfrac{d}{dx} \tanh \dfrac{x^2}{2} + 0$

$$= \dfrac{1}{2} \cdot \dfrac{1}{\tanh \frac{x^2}{2}} \cdot \sec h^2 \dfrac{x^2}{2} \cdot \dfrac{d}{dx} \dfrac{x^2}{2} + 0 = \dfrac{1}{2} \cdot \dfrac{\sec h^2 \frac{x^2}{2}}{\tanh \frac{x^2}{2}} \cdot \dfrac{2x}{2} = \dfrac{x}{2} \cdot \dfrac{1 - \tanh^2 \frac{x^2}{2}}{\tanh \frac{x^2}{2}} = \dfrac{x}{2} \cdot \dfrac{1 - \frac{\sinh^2 \frac{x^2}{2}}{\cosh^2 \frac{x^2}{2}}}{\frac{\sinh \frac{x^2}{2}}{\cosh \frac{x^2}{2}}}$$

$$= \dfrac{x}{2} \cdot \dfrac{\frac{\cosh^2 \frac{x^2}{2} - \sinh^2 \frac{x^2}{2}}{\cosh^2 \frac{x^2}{2}}}{\frac{\sinh \frac{x^2}{2}}{\cosh \frac{x^2}{2}}} = \dfrac{x}{2} \cdot \dfrac{\frac{1}{\cosh^2 \frac{x^2}{2}}}{\frac{\sinh \frac{x^2}{2}}{\cosh \frac{x^2}{2}}} = \dfrac{x}{2} \cdot \dfrac{\cosh \frac{x^2}{2}}{\cosh^2 \frac{x^2}{2} \cdot \sinh \frac{x^2}{2}} = \dfrac{x}{2 \cosh \frac{x^2}{2} \cdot \sinh \frac{x^2}{2}} = \dfrac{x}{\sinh 2 \cdot \frac{x^2}{2}}$$

$$= \dfrac{x}{\sinh x^2} = x \csc h \, x^2$$

d. Given $\int \sinh^5 (x+1) \cosh (x+1) \, dx$ let $u = \sinh (x+1)$, then $\dfrac{du}{dx} = \dfrac{d}{dx} \sinh (x+1)$; $\dfrac{du}{dx} = \cosh (x+1)$;

$du = \cosh (x+1) \cdot dx$; $dx = \dfrac{du}{\cosh (x+1)}$. Therefore,

$$\boxed{\int \sinh^5(x+1)\cosh(x+1)\,dx} = \boxed{\int u^5 \cosh(x+1)\cdot\frac{du}{\cosh(x+1)}} = \boxed{\int u^5\,du} = \boxed{\frac{1}{6}u^6 + c} = \boxed{\frac{1}{6}\sinh^6(x+1)+c}$$

Check: Let $y = \frac{1}{6}\sinh^6(x+1)+c$, then $y' = \frac{1}{6}\cdot 6\sinh^5(x+1)\cdot\cosh(x+1)+0 = \sinh^5(x+1)\cosh(x+1)$

e. Given $\int \cosh^5 x \sinh x\,dx$ let $u = \cosh x$, then $\frac{du}{dx} = \frac{d}{dx}\cosh x$; $\frac{du}{dx} = \sinh x$; $dx = \frac{du}{\sinh x}$. Therefore,

$$\boxed{\int \cosh^5 x \sinh x\,dx} = \boxed{\int u^5 \sinh x\cdot\frac{du}{\sinh x}} = \boxed{\int u^5\,du} = \boxed{\frac{1}{6}u^6+c} = \boxed{\frac{1}{6}\cosh^6 x+c}$$

Check: Let $y = \frac{1}{6}\cosh^6 x+c$, then $y' = \frac{1}{6}\cdot 6\cosh^5 x\cdot\sinh x+0 = \frac{6}{6}\cosh^5 x \sinh x = \cosh^5 x \sinh x$

f. Given $\int \cosh^5 5x \sinh 5x\,dx$ let $u = \cosh 5x$, then $\frac{du}{dx} = \frac{d}{dx}\cosh 5x$; $\frac{du}{dx} = 5\sinh 5x$; $dx = \frac{du}{5\sinh 5x}$. Thus,

$$\boxed{\int \cosh^5 5x \sin 5\,5x\,dx} = \boxed{\int u^5 \sinh 5x\cdot\frac{du}{5\sinh 5x}} = \boxed{\frac{1}{5}\int u^5\,du} = \boxed{\frac{1}{5}\cdot\frac{1}{6}u^6+c} = \boxed{\frac{1}{30}\cosh^6 5x+c}$$

Check: Let $y = \frac{1}{30}\cosh^6 5x+c$, then $y' = \frac{6\cosh^5 5x\cdot 5\sinh 5x}{30}+0 = \frac{30}{30}\cosh^5 5x \sinh 5x = \cosh^5 hx \sinh 5x$

g. Given $\int \cosh^4 \frac{x}{2}\sinh\frac{x}{2}\,dx$ let $u = \cosh\frac{x}{2}$, then $\frac{du}{dx} = \frac{d}{dx}\cosh\frac{x}{2}$; $\frac{du}{dx} = \frac{1}{2}\sinh\frac{x}{2}$; $dx = \frac{2\,du}{\sin\frac{x}{2}}$. Therefore,

$$\boxed{\int \cosh^4 \frac{x}{2}\sinh\frac{x}{2}\,dx} = \boxed{\int u^4 \sinh\frac{x}{2}\cdot\frac{2\,du}{\sinh\frac{x}{2}}} = \boxed{2\int u^4\,du} = \boxed{\frac{2}{5}u^5+c} = \boxed{\frac{2}{5}\cosh^5\frac{x}{2}+c}$$

Check: Let $y = \frac{2}{5}\cosh^5\frac{x}{2}+c$, then $y' = \frac{2}{5}\cdot 5\cosh^4\frac{x}{2}\cdot\sinh\frac{x}{2}\cdot\frac{1}{2}+0 = \frac{10}{10}\cosh^4\frac{x}{2}\sin\frac{x}{2} = \cosh^4\frac{x}{2}\sinh\frac{x}{2}$

h. Given $\int x^2 \cosh x^3\,dx$ let $u = x^3$, then $\frac{du}{dx} = \frac{d}{dx}x^3$; $\frac{du}{dx} = 3x^2$; $du = 3x^2\cdot dx$; $dx = \frac{du}{3x^2}$. Therefore,

$$\boxed{\int x^2 \cosh x^3\,dx} = \boxed{\int x^2 \cosh u\cdot\frac{du}{3x^2}} = \boxed{\frac{1}{3}\int \cosh u\,du} = \boxed{\frac{1}{3}\cdot\sinh u+c} = \boxed{\frac{1}{3}\sinh x^3+c}$$

Check: Let $y = \frac{1}{3}\sinh x^3+c$, then $y' = \frac{1}{3}\cdot\cosh x^3\cdot 3x^2+c = \frac{3}{3}x^2\cosh x^3+0 = x^2\cosh x^3$

i. Given $\int \frac{e^{2x}}{3}\operatorname{sec}h\,e^{2x}\,dx$ let $u = e^{2x}$, then $\frac{du}{dx} = \frac{d}{dx}e^{2x}$; $\frac{du}{dx} = 2e^{2x}$; $du = 2e^{2x}\cdot dx$; $dx = \frac{du}{2e^{2x}}$. Thus,

$$\boxed{\int \frac{e^{2x}}{3}\operatorname{sec}h\,e^{2x}\,dx} = \boxed{\frac{1}{3}\int e^{2x}\operatorname{sec}h\,u\cdot\frac{du}{2e^{2x}}} = \boxed{\frac{1}{6}\int \operatorname{sec}h\,u\,du} = \boxed{\frac{1}{6}\sin^{-1}(\tanh u)+c} = \boxed{\frac{1}{6}\sin^{-1}\left(\tanh e^{2x}\right)+c}$$

Check: Let $y = \frac{1}{6}\sin^{-1}\left(\tanh e^{2x}\right)+c$, then $y' = \frac{1}{6\sqrt{1-\tanh^2 e^{2x}}}\cdot\frac{d}{dx}\tanh e^{2x}+0 = \frac{\operatorname{sec}h^2 e^{2x}}{6\sqrt{\operatorname{sec}h^2 e^{2x}}}\cdot\frac{d}{dx}e^{2x}$

$$= \frac{\sec h^2 e^{2x}}{6 \sec h\, e^{2x}} \cdot 2 e^{2x} = \frac{2 e^{2x}}{6} \cdot \frac{\sec h^2 e^{2x}}{\sec h\, e^{2x}} = \frac{e^{2x}}{3} \cdot \sec h\, e^{2x}$$

j. Given $\int \sec h^2 (5x-1) dx$ let $u = 5x-1$, then $\frac{du}{dx} = \frac{d}{dx}(5x-1)$; $\frac{du}{dx} = 5$; $du = 5dx$; $dx = \frac{du}{5}$. Therefore,

$$\boxed{\int \sec h^2 (5x-1) dx} = \boxed{\int \sec h^2 u \cdot \frac{du}{5}} = \boxed{\frac{1}{5}\int \sec h^2 u\, du} = \boxed{\frac{1}{5}\tanh u + c} = \boxed{\mathbf{\frac{1}{5}\tanh (5x-1) + c}}$$

Check: Let $y = \frac{1}{5}\tanh (5x-1) + c$, then $y' = \frac{1}{5} \cdot \sec h^2 (5x-1) \cdot \frac{d}{dx}(5x-1) + 0 = \frac{1}{5} \cdot \sec h^2 (5x-1) \cdot 5$

$$\frac{5}{5} \cdot \sec h^2 (5x-1) = \sec h^2 (5x-1)$$

k. Given $\int \csc h\, 7x \coth 7x\, dx$ let $u = 7x$, then $\frac{du}{dx} = \frac{d}{dx} 7x$; $\frac{du}{dx} = 7$; $du = 7dx$; $dx = \frac{du}{7}$. Therefore,

$$\boxed{\int \csc h\, 7x \coth 7x\, dx} = \boxed{\int \csc h\, u \coth u\, \frac{du}{7}} = \boxed{\frac{1}{7}\int \csc h\, u \coth u\, du} = \boxed{-\frac{1}{7}\csc h\, u + c} = \boxed{\mathbf{-\frac{1}{7}\csc h\, 7x + c}}$$

Check: Let $y = -\frac{1}{7}\csc h\, 7x + c$, then $y' = -\frac{1}{7} \cdot -\csc h\, 7x \coth 7x \cdot \frac{d}{dx} 7x + 0 = \frac{1}{7} \cdot \csc h\, 7x \coth 7x \cdot 7$

$$\frac{7}{7} \cdot \csc h\, 7x \coth 7x = \csc h\, 7x \coth 7x$$

l. Given $\int \tanh^2 10x\, dx$ let $u = 10x$, then $\frac{du}{dx} = \frac{d}{dx} 10x$; $\frac{du}{dx} = 10$; $du = 10\, dx$; $dx = \frac{du}{10}$. Therefore,

$$\boxed{\int \tanh^2 10x\, dx} = \boxed{\int \tanh^2 u \cdot \frac{du}{10}} = \boxed{\frac{1}{10}\int \tanh^2 u\, du} = \boxed{\frac{1}{10}(u - \tanh u) + c} = \boxed{\mathbf{\frac{1}{10}(10x - \tanh 10x) + c}}$$

Check: Let $y = \frac{1}{10}(10x - \tanh 10x) + c$, then $y' = \frac{1}{10}\left(10 - \sec h^2 10x \cdot \frac{d}{dx} 10x\right) + 0 = \frac{1}{10}\left(10 - \sec h^2 10x \cdot 10\right)$

$$\frac{10\left(1 - \sec h^2 10x\right)}{10} = 1 - \sec h^2 10x = \tanh^2 10x$$

m. $\int \cosh^3 \frac{x}{5}\, dx = \int \cosh^2 \frac{x}{5} \cosh \frac{x}{5}\, dx = \int \left(1 + \sinh^2 \frac{x}{5}\right)\cosh \frac{x}{5}\, dx = \int \left(\cosh \frac{x}{5} + \sinh^2 \frac{x}{5} \cosh \frac{x}{5}\right) dx$

$= \int \cosh \frac{x}{5}\, dx + \int \sinh^2 \frac{x}{5} \cosh \frac{x}{5}\, dx$ let $u = \frac{x}{5}$, then $\frac{du}{dx} = \frac{d}{dx}\frac{x}{5}$; $\frac{du}{dx} = \frac{1}{5}$; $du = \frac{1}{5} dx$; $dx = 5\, du$. Therefore,

$$\boxed{\int \cosh \frac{x}{5}\, dx + \int \sinh^2 \frac{x}{5} \cosh \frac{x}{5}\, dx} = \boxed{\int \cosh u\, 5du + \int \sinh^2 u \cosh u\, 5du} = \boxed{5\sinh u + 5\int \sinh^2 u \cosh u\, du}$$

To solve the second integral let $w = \sinh u$, then $\frac{dw}{dx} = \frac{d}{dx}\sinh u$; $\frac{dw}{dx} = \cosh u$; $dx = \frac{dw}{\cosh u}$ thus,

$$= \boxed{5\sinh u + 5\int w^2 \cosh u \cdot \frac{dw}{\cosh u}} = \boxed{5\sinh u + 5\int w^2 dw} = \boxed{5\sinh u + \frac{5}{3} w^3 + c} = \boxed{\mathbf{\frac{5}{3}\sinh^3 \frac{x}{5} + 5\sinh \frac{x}{5} + c}}$$

Check: Let $y = 5\sinh\frac{x}{5} + \frac{5}{3}\sinh^3\frac{x}{5} + c$, then $y' = 5\cdot\cosh\frac{x}{5}\cdot\frac{1}{5} + \frac{5}{3}\cdot 3\sinh^2\frac{x}{5}\cdot\cosh\frac{x}{5}\cdot\frac{1}{5} + 0$

$$= \frac{5}{5}\cdot\cosh\frac{x}{5} + \frac{15}{15}\cdot\sinh^2\frac{x}{5}\cdot\cosh\frac{x}{5} = \cosh\frac{x}{5}\left(1 + \sinh^2\frac{x}{5}\right) = \cosh\frac{x}{5}\cdot\cosh^2\frac{x}{5} = \cosh^3\frac{x}{5}$$

n. $\int \sinh^3 x \, dx = \int \sinh^2 x \sinh x \, dx = \int \left(\cosh^2 x - 1\right)\sinh x \, dx = \int \left(\cosh^2 x \sinh x - \sinh x\right) dx$

$= \int \cosh^2 x \sinh x \, dx - \int \sinh x \, dx = \int \cosh^2 x \sinh x \, dx - \cosh x$. To solve the first integral let $u = \cosh x$,

then $\dfrac{du}{dx} = \dfrac{d}{dx}\cosh x$; $\dfrac{du}{dx} = \sinh x$; $dx = \dfrac{du}{\sinh x}$. Therefore,

$$\boxed{\int \cosh^2 x \sinh x \, dx - \cosh x} = \boxed{\int u^2 \sinh x \cdot \frac{du}{\sinh x} - \cosh x} = \boxed{\int u^2 du - \cosh x} = \boxed{\frac{1}{3}u^3 - \cosh x + c}$$

$$= \boxed{\frac{\cosh^3 x}{3} - \cosh x + c}$$

Check: Let $y = \frac{1}{3}\cosh^3 x - \cosh + c$, then $y' = \frac{1}{3}\cdot 3\cosh^2 x\cdot\sinh x - \sinh x + 0 = \cosh^2 x\cdot\sinh x - \sinh x$

$$= \sinh x\left(\cosh^2 x - 1\right) = \sinh x\cdot\sinh^2 x = \sinh^3 x$$

o. Given $\int \frac{x}{2}\sinh x \, dx = \frac{1}{2}\int x\sinh x \, dx$ let $u = x$ and $dv = \sinh x \, dx$ then $du = dx$ and $\int dv = \int \sinh x \, dx$

which implies $v = \cosh x$. Using the substitution by parts formula $\int u \, dv = u\,v - \int v \, du$ we obtain

$$\boxed{\frac{1}{2}\int x\sinh x \, dx} = \boxed{\frac{1}{2}x\cosh x - \frac{1}{2}\int \cosh x \, dx} = \boxed{\frac{1}{2}x\cosh x - \frac{1}{2}\sinh x + c}$$

Check: Let $y = \frac{1}{2}x\cosh x - \frac{1}{2}\sinh x + c$, then $y' = \frac{1}{2}\left(\cosh x + x\sinh x\right) - \frac{1}{2}\cosh x + 0 = \frac{\cosh x}{2} + \frac{x\sinh x}{2}$

$$-\frac{\cosh x}{2} = \frac{x\sinh x}{2} = \frac{1}{2}x\sinh x$$

Example 5.4-3: Evaluate the following indefinite integrals:

a. $\int \tanh^8 x \sec h^2 x \, dx =$

b. $\int \tanh^5 (x+3)\sec h^2 (x+3)\, dx =$

c. $\int \coth^3 x \csc h^2 x \, dx =$

d. $\int \coth^5 3x \csc h^2 3x \, dx =$

e. $\int \tanh 5x \, dx =$

f. $\int 2\tanh\frac{x}{3}\, dx =$

g. $\int \sec h \, 5x \tanh 5x \, dx =$

h. $\int \sec h\frac{x}{2}\tanh\frac{x}{2}\, dx =$

i. $\int \csc h^2 (1-2x)\, dx =$

j. $\int x^2 \coth x^3 dx =$

k. $\int x^2 \sec h \, 5x^3 \, dx =$

l. $\int \frac{\sec h \sqrt{x}}{\sqrt{x}}\, dx =$

Solutions:

a. $\int \tanh^8 x \sec h^2 x\, dx$ let $u = \tanh x$, then $\frac{du}{dx} = \frac{d}{dx}\tanh x$; $\frac{du}{dx} = \sec h^2 x$; $du = \sec h^2 x\, dx$; $dx = \frac{du}{\sec h^2 x}$. Thus,

$$\int \tanh^8 x \sec h^2 x\, dx = \int u^8 \cdot \sec h^2 x \cdot \frac{du}{\sec h^2 x} = \int u^8 du = \frac{1}{8+1}u^{8+1}+c = \frac{1}{9}u^9+c = \frac{1}{9}\tanh^9 x+c$$

Check: Let $y = \frac{1}{9}\tanh^9 x+c$ then $y' = \frac{1}{9}\cdot 9(\tanh x)^{9-1}\cdot \sec h^2 x+0 = (\tanh x)^8 \sec h^2 x = \tanh^8 x \sec h^2 x$

b. $\int \tanh^5(x+3)\sec h^2(x+3)\, dx$ let $u = \tanh(x+3)$, then $\frac{du}{dx} = \frac{d}{dx}\tanh(x+3)$; $\frac{du}{dx} = \sec h^2(x+3)c$;

$du = \sec h^2(x+3)dx$; $dx = \frac{du}{\sec h^2(x+3)}$. Thus, $\int \tanh^5(x+3)\sec h^2(x+3)\, dx = \int u^5 \cdot \sec^2(x+3)\cdot \frac{du}{\sec^2(x+3)}$

$= \int u^5 du = \frac{1}{5+1}u^{5+1}+c = \frac{1}{6}u^6+c = \frac{1}{6}\tanh^6(x+3)+c$

Check: Let $y = \frac{1}{6}\tanh^6(x+3)+c$ then $y' = \frac{1}{6}\cdot 6[\tanh(x+3)]^{6-1}\cdot \sec h^2(x+3)+0 = \tanh^5(x+3)\sec h^2(x+3)$

c. Given $\int \coth^3 x \csc h^2 x\, dx$ let $u = \coth x$, then $\frac{du}{dx} = \frac{d}{dx}\coth x$; $\frac{du}{dx} = -\csc h^2 x\,c$; $du = -\csc h^2 x\, dx$

; $dx = -\frac{du}{\csc h^2 x}$. Therefore, $\int \coth^3 x \csc h^2 x\, dx = \int u^3 \cdot \csc h^2 x \cdot \frac{-du}{\csc h^2 x} = -\int u^3 du = \frac{-1}{3+1}u^{3+1}+c$

$= -\frac{1}{4}u^4+c = -\frac{1}{4}\coth^4 x+c$

Check: Let $y = -\frac{1}{4}\coth^4 x+c$ then $y' = -\frac{1}{4}\cdot 4(\coth x)^{4-1}\cdot -\csc h^2 x+0 = \coth^3 x \csc h^2 x$

d. Given $\int \coth^5 3x \csc h^2 3x\, dx$ let $u = \coth 3x$, then $\frac{du}{dx} = \frac{d}{dx}\coth 3x$; $\frac{du}{dx} = -3\csc h^2 3x\,c$; $du = -3\csc h^2 3x\, dx$

; $dx = -\frac{du}{3\csc h^2 3x}$. Therefore, $\int \coth^5 3x \csc h^2 3x\, dx = \int u^5 \cdot \csc h^2 3x \cdot \frac{-du}{3\csc h^2 3x} = -\frac{1}{3}\int u^5 du$

$= -\frac{1}{3}\cdot\frac{1}{5+1}u^{5+1}+c = -\frac{1}{18}u^6+c = -\frac{1}{18}\coth^6 3x+c$

Check: Let $y = -\frac{1}{18}\coth^6 3x+c$ then $y' = -\frac{1}{18}\cdot 6(\coth 3x)^{6-1}\cdot 3\cdot -\csc^2 3x+0 = \coth^5 3x \csc h^2 3x$

e. Given $\int \tanh 5x\, dx$ let $u = 5x$, then $\frac{du}{dx} = \frac{d}{dx}5x$; $\frac{du}{dx} = 5$; $du = 5\, dx$; $dx = \frac{du}{5}$. Therefore,

$$\boxed{\int \tanh 5x\, dx} = \boxed{\int \tanh u \cdot \frac{du}{5}} = \boxed{\frac{1}{5}\int \tanh u\, du} = \boxed{\frac{1}{5}\ln\cosh u + c} = \boxed{\frac{1}{5}\ln\cosh 5x + c}$$

Check: Let $y = \frac{1}{5}\ln\cosh 5x + c$ then $y' = \frac{1}{5}\cdot\frac{1}{\cosh 5x}\cdot\sinh 5x \cdot 5 + 0 = \frac{5}{5}\cdot\frac{\sinh 5x}{\cosh 5x} = \tanh 5x$

f. Given $\int 2\tanh\frac{x}{3}\, dx$ let $u = \frac{x}{3}$, then $\frac{du}{dx} = \frac{d}{dx}\frac{x}{3}$; $\frac{du}{dx} = \frac{1}{3}$; $3du = dx$; $dx = 3du$. Therefore,

$$\boxed{\int 2\tanh\frac{x}{3}\, dx} = \boxed{2\int \tanh u \cdot 3du} = \boxed{6\int \tanh u\, du} = \boxed{6\ln\cosh u + c} = \boxed{6\ln\cosh\frac{x}{3} + c}$$

Check: Let $y = 6\ln\cosh\frac{x}{3} + c$, then $y' = 6\cdot\frac{1}{\cosh\frac{x}{3}}\cdot\sinh\frac{x}{3}\cdot\frac{1}{3} + 0 = \frac{6}{3}\tanh\frac{x}{3} = 2\tanh\frac{x}{3}$

g. Given $\int \sec h\, 5x\tanh 5x\, dx$ let $u = 5x$, then $\frac{du}{dx} = \frac{d}{dx}5x$; $\frac{du}{dx} = 5$; $du = 5dx$; $dx = \frac{du}{5}$. Therefore,

$$\boxed{\int \sec h\, 5x\tanh 5x\, dx} = \boxed{\int \sec h\, u \cdot \tanh u \cdot \frac{du}{5}} = \boxed{\frac{1}{5}\int \sec h\, u \tanh u\, du} = \boxed{-\frac{1}{5}\sec h\, u + c} = \boxed{-\frac{1}{5}\sec h\, 5x + c}$$

Check: Let $y = -\frac{1}{5}\sec h\, 5x + c$, then $y' = \frac{1}{5}\cdot\sec h\, 5x\tanh 5x\cdot 5 + 0 = \frac{5\sec h\, 5x\tanh 5x}{5} = \sec h\, 5x\tanh 5x$

h. Given $\int \sec h\frac{x}{2}\tanh\frac{x}{2}\, dx$ let $u = \frac{x}{2}$, then $\frac{du}{dx} = \frac{d}{dx}\frac{x}{2}$; $\frac{du}{dx} = \frac{1}{2}$; $2du = dx$; $dx = 2du$. Therefore,

$$\boxed{\int \sec h\frac{x}{2}\tanh\frac{x}{2}\, dx} = \boxed{\int \sec h\, u \cdot \tanh u \cdot 2du} = \boxed{2\int \sec h\, u \tanh u\, du} = \boxed{-2\sec h\, u + c} = \boxed{-2\sec h\frac{x}{2} + c}$$

Check: Let $y = -2\sec h\frac{x}{2} + c$ then $y' = -2\cdot -\sec h\frac{x}{2}\tanh\frac{x}{2}\cdot\frac{1}{2} + 0 = \frac{2}{2}\cdot\sec h\frac{x}{2}\tanh\frac{x}{2} = \sec h\frac{x}{2}\tanh\frac{x}{2}$

i. Given $\int \csc h^2(1-2x)\, dx$ let $u = 1-2x$, then $\frac{du}{dx} = \frac{d}{dx}1-2x$; $\frac{du}{dx} = -2$; $du = -2dx$; $dx = \frac{du}{-2}$. Thus,

$$\boxed{\int \csc h^2(1-2x)\, dx} = \boxed{\int \csc h^2 u \cdot -\frac{du}{2}} = \boxed{-\frac{1}{2}\int \csc h^2 u\, du} = \boxed{\frac{1}{2}\coth u + c} = \boxed{\frac{1}{2}\coth(1-2x) + c}$$

Check: Let $y = \frac{1}{2}\coth(1-2x) + c$ then $y' = \frac{1}{2}\cdot -\csc h^2(1-2x)\cdot -2 + 0 = \csc h^2(1-2x)$

j. Given $\int x^2\coth x^3\, dx$ let $u = x^3$, then $\frac{du}{dx} = \frac{d}{dx}x^3$; $\frac{du}{dx} = 3x^2$; $du = 3x^2 dx$; $dx = \frac{du}{3x^2}$. Therefore,

$$\boxed{\int x^2\coth x^3\, dx} = \boxed{\int x^2 \cdot \coth u \cdot \frac{du}{3x^2}} = \boxed{\frac{1}{3}\int \coth u \cdot du} = \boxed{\frac{1}{3}\ln|\sinh u| + c} = \boxed{\frac{1}{3}\ln|\sinh x^3| + c}$$

Check: Let $y = \frac{1}{3}\ln|\sinh x^3| + c$ then $y' = \frac{1}{3}\cdot\frac{1}{\sinh x^3}\cdot\cosh x^3\cdot 3x^2 + 0 = \frac{3x^2}{3}\cdot\frac{\cosh x^3}{\sinh x^3} = x^2\coth x^3$

k. Given $\int x^2 \sec h\, 5x^3\, dx$ let $u = 5x^3$, then $\dfrac{du}{dx} = \dfrac{d}{dx}5x^3$; $\dfrac{du}{dx} = 15x^2$; $du = 15x^2\, dx$; $dx = \dfrac{du}{15x^2}$. Thus,

$$\boxed{\int x^2 \sec h\, 5x^3\, dx} = \boxed{\int x^2 \cdot \sec h\, u \cdot \dfrac{du}{15x^2}} = \boxed{\dfrac{1}{15}\int \sec h\, u \cdot du} = \boxed{\dfrac{1}{15}\sin^{-1}(\tanh u) + c} = \boxed{\dfrac{1}{15}\sin^{-1}\left(\tanh 5x^3\right) + c}$$

Check: Let $y = \dfrac{1}{15}\sin^{-1}\left(\tanh 5x^3\right) + c$ then $y' = \dfrac{1}{15\sqrt{1 - \tanh^2 5x^3}} \cdot \dfrac{d}{dx}\tanh 5x^3 + 0 = \dfrac{\sec h^2 5x^3}{15\sqrt{\sec h^2 5x^3}} \cdot \dfrac{d}{dx}5x^3$

$$= \dfrac{\sec h^2 5x^3}{15\sec h\, 5x^3} \cdot 15x^2 = \dfrac{15x^2}{15} \cdot \dfrac{\sec h^2 5x^3}{\sec h\, 5x^3} = x^2 \sec h\, 5x^3$$

l. Given $\int \dfrac{\sec h\, \sqrt{x}}{\sqrt{x}}\, dx$ let $u = x^{\frac{1}{2}}$, then $\dfrac{du}{dx} = \dfrac{d}{dx}x^{\frac{1}{2}}$; $\dfrac{du}{dx} = \dfrac{1}{2}x^{-\frac{1}{2}} = \dfrac{1}{2\sqrt{x}}$; $dx = 2\sqrt{x}\, du$. Therefore,

$$\boxed{\int \dfrac{\sec h\, \sqrt{x}}{\sqrt{x}}\, dx} = \boxed{\int \dfrac{\sec h\, u}{\sqrt{x}} \cdot 2\sqrt{x}\, du} = \boxed{2\int \sec h\, u \cdot du} = \boxed{2\sin^{-1}(\tanh u) + c} = \boxed{2\sin^{-1}\left(\tanh \sqrt{x}\right) + c}$$

Check: Let $y = 2\sin^{-1}\left(\tanh \sqrt{x}\right) + c$ then $y' = \dfrac{2}{\sqrt{1 - \tanh^2 x^{\frac{1}{2}}}} \cdot \dfrac{d}{dx}\tanh x^{\frac{1}{2}} + 0 = \dfrac{2\sec h^2 x^{\frac{1}{2}}}{\sqrt{\sec h^2 x^{\frac{1}{2}}}} \cdot \dfrac{d}{dx}x^{\frac{1}{2}}$

$$= \dfrac{2\sec h^2 x^{\frac{1}{2}}}{\sec h\, x^{\frac{1}{2}}} \cdot \dfrac{1}{2x^{\frac{1}{2}}} = \dfrac{2}{2x^{\frac{1}{2}}} \cdot \dfrac{\sec h^2 x^{\frac{1}{2}}}{\sec h\, x^{\frac{1}{2}}} = \dfrac{\sec h\, x^{\frac{1}{2}}}{x^{\frac{1}{2}}} = \dfrac{\sec h\, \sqrt{x}}{\sqrt{x}}$$

Example 5.4-4: Evaluate the following indefinite integrals:

a. $\int \sec h\, \sqrt[5]{x^2}\, \dfrac{dx}{\sqrt[5]{x^3}} =$

b. $\int \dfrac{x\, dx}{\sinh x^2} =$

c. $\int x^2 \sinh x\, dx =$

d. $\int \dfrac{1}{\cosh 7x}\, dx =$

e. $\int \dfrac{\cosh 5x + \sinh 5x}{\sinh 5x}\, dx =$

f. $\int \left(\dfrac{1 + \cosh x}{\sinh x}\right) dx =$

g. $\int 2\sinh\dfrac{3x}{2}\csc h\,\dfrac{3x}{2}\, dx =$

h. $\int \left(\cosh x \sec h\, x + e^{3x}\right) dx =$

i. $\int e^{\sinh 8x}\cosh 8x\, dx =$

j. $\int e^{\cosh \frac{x}{3}}\sinh\dfrac{x}{3}\, dx =$

k. $\int e^{\tanh 5x}\sec h^2 5x\, dx =$

l. $\int e^{\frac{1}{3}\coth 7x}\csc h^2 7x\, dx =$

Solutions:

a. Given $\int \sec h\, \sqrt[5]{x^2}\, \dfrac{dx}{\sqrt[5]{x^3}}$ let $u = x^{\frac{2}{5}}$, then $\dfrac{du}{dx} = \dfrac{d}{dx}x^{\frac{2}{5}}$; $\dfrac{du}{dx} = \dfrac{2}{5}x^{-\frac{3}{5}} = \dfrac{2}{5\sqrt[5]{x^3}}$; $dx = \dfrac{5}{2}\sqrt[5]{x^3}\, du$. Therefore,

$$\boxed{\int \sec h\, \sqrt[5]{x^2}\, \dfrac{dx}{\sqrt[5]{x^3}}} = \boxed{\int \sec h\, u \cdot \dfrac{5\sqrt[5]{x^3}\, du}{2\sqrt[5]{x^3}}} = \boxed{\dfrac{5}{2}\int \sec h\, u \cdot du} = \boxed{\dfrac{5}{2}\sin^{-1}(\tanh u) + c} = \boxed{\dfrac{5}{2}\sin^{-1}\left(\tanh \sqrt[5]{x^2}\right) + c}$$

Check: Let $y = \dfrac{5}{2}\sin^{-1}\left(\tanh\sqrt[5]{x^2}\right)+c$, then $y' = \dfrac{5}{2\sqrt{1-\tanh^2 x^{\frac{2}{5}}}}\cdot\dfrac{d}{dx}\tanh x^{\frac{2}{5}}+0 = \dfrac{5\sec h^2 x^{\frac{2}{5}}}{2\sqrt{\sec h^2 x^{\frac{2}{5}}}}\cdot\dfrac{d}{dx}x^{\frac{2}{5}}$

$$= \dfrac{5\sec h^2 x^{\frac{2}{5}}}{2\sec h\, x^{\frac{2}{5}}}\cdot\dfrac{2}{5x^{\frac{3}{5}}} = \dfrac{10}{10x^{\frac{3}{5}}}\cdot\dfrac{\sec h^2 x^{\frac{2}{5}}}{\sec h\, x^{\frac{2}{5}}} = \dfrac{\sec h\, x^{\frac{2}{5}}}{x^{\frac{3}{5}}} = \dfrac{\sec h\,\sqrt[5]{x^2}}{\sqrt[5]{x^3}}$$

b. Given $\displaystyle\int\dfrac{x\,dx}{\sinh x^2} = \int x\csc h\,x^2\,dx$ let $u = x^2$, then $\dfrac{du}{dx}=\dfrac{d}{dx}x^2$; $\dfrac{du}{dx}=2x$; $du = 2x\,dx$; $dx=\dfrac{du}{2x}$. Thus,

$$\boxed{\int\dfrac{x\,dx}{\sinh x^2}} = \boxed{\int\dfrac{x}{\sinh u}\dfrac{du}{2x}} = \boxed{\dfrac{1}{2}\int\dfrac{du}{\sinh u}} = \boxed{\dfrac{1}{2}\int\csc h\,u\,du} = \boxed{\dfrac{1}{2}\ln\left|\tanh\dfrac{u}{2}\right|+c} = \boxed{\dfrac{1}{2}\ln\left|\tanh\dfrac{x^2}{2}\right|+c}$$

Check: Let $y = \dfrac{1}{2}\ln\left|\tanh\dfrac{x^2}{2}\right|+c$, then $y' = \dfrac{1}{2}\cdot\dfrac{1}{\tanh\frac{x^2}{2}}\cdot\sec h^2\dfrac{x^2}{2}\cdot\dfrac{2x}{2} = \dfrac{2x}{4}\cdot\dfrac{\sec h^2\frac{x^2}{2}}{\tanh\frac{x^2}{2}} = \dfrac{x}{2}\cdot\dfrac{1-\tanh^2\frac{x^2}{2}}{\tanh\frac{x^2}{2}}$

$$= \dfrac{x}{2}\cdot\dfrac{1-\dfrac{\sinh^2\frac{x^2}{2}}{\cosh^2\frac{x^2}{2}}}{\dfrac{\sinh\frac{x^2}{2}}{\cosh\frac{x^2}{2}}} = \dfrac{x}{2}\cdot\dfrac{\dfrac{\cosh^2\frac{x^2}{2}-\sinh^2\frac{x^2}{2}}{\cosh^2\frac{x^2}{2}}}{\dfrac{\sinh\frac{x^2}{2}}{\cosh\frac{x^2}{2}}} = \dfrac{x}{2}\cdot\dfrac{\dfrac{1}{\cosh^2\frac{x^2}{2}}}{\dfrac{\sinh\frac{x^2}{2}}{\cosh\frac{x^2}{2}}} = \dfrac{x}{2}\cdot\dfrac{\cosh\frac{x^2}{2}}{\cosh^2\frac{x^2}{2}\cdot\sinh\frac{x^2}{2}} = \dfrac{x}{2\cosh\frac{x^2}{2}\cdot\sinh\frac{x^2}{2}}$$

$$= \dfrac{x}{\sinh 2\cdot\frac{x^2}{2}} = \dfrac{x}{\sinh x^2}$$

c. Given $\displaystyle\int x^2\sinh x\,dx$ let $u = x^2$ and $dv = \sinh x\,dx$ then $du = 2x\,dx$ and $\int dv = \int\sinh x\,dx$ which implies

implies $v = \cosh x$. Using the integration by parts formula $\int u\,dv = uv-\int v\,du$ we obtain

$$\int x^2\sinh x\,dx = x^2\cdot\cosh x-\int\cosh x\cdot 2x\,dx = x^2\cosh x-2\int x\cosh x\,dx \qquad (1)$$

To integrate $\int x\cosh x\,dx$ let $u = x$ and $dv = \cosh x\,dx$ then $du = dx$ and $\int dv = \int\cosh x\,dx$ which

implies $v = \sinh x$. Using the integration by parts formula again we have

$$\int x\cosh x\,dx = x\cdot\sinh x-\int\sinh x\cdot dx = x\sinh x-\cosh x+c \qquad (2)$$

Combining equations (1) and (2) together we obtain

$$\boxed{\int x^2\sinh x\,dx} = \boxed{x^2\cosh x-2\int x\cosh x\,dx} = \boxed{x^2\cosh x-2\left(x\sinh x-\cosh x+c\right)} = \boxed{\left(x^2+2\right)\cosh x-2x\sinh x+c}$$

Check: Let $y = \left(x^2+2\right)\cosh x-2x\sinh x+c$, then $y' = 2x\cosh x+\left(x^2+2\right)\sinh h\,x-2\sinh x-2x\cosh x+0$

$$= \left(x^2+2\right)\sinh h\,x-2\sinh x = x^2\sinh h\,x+2\sinh x-2\sinh x = x^2\sinh h\,x$$

d. Given $\int \dfrac{1}{\cosh 7x}\,dx = \int \sec h\, 7x\, dx$ let $u = 7x$, then $\dfrac{du}{dx} = \dfrac{d}{dx}7x$; $\dfrac{du}{dx} = 7$; $du = 7dx$; $dx = \dfrac{du}{7}$. Thus,

$$\boxed{\int \dfrac{1}{\cosh 7x}\,dx} = \boxed{\int \sec h\, 7x\, dx} = \boxed{\int \sec h\, u\cdot\dfrac{du}{7}} = \boxed{\dfrac{1}{7}\int \sec h\, u\, du} = \boxed{\dfrac{1}{7}\sin^{-1}(\tanh u)+c} = \boxed{\dfrac{1}{7}\sin^{-1}(\tanh 7x)+c}$$

Check: Let $y = \dfrac{1}{7}\sin^{-1}(\tanh 7x)+c$, then $y' = \dfrac{1}{7\sqrt{1-\tanh^2 7x}}\cdot\dfrac{d}{dx}\tanh 7x + 0 = \dfrac{\sec h^2 7x}{7\sqrt{\sec h^2 7x}}\cdot\dfrac{d}{dx}7x$

$= \dfrac{\sec h^2 7x}{7\sec h\, 7x}\cdot 7 = \dfrac{7}{7}\cdot\dfrac{\sec h^2 7x}{\sec h\, 7x} = \sec h\, 7x = \dfrac{1}{\cosh 7x}$

e. Given $\int \dfrac{\cosh 5x + \sinh 5x}{\sinh 5x}\,dx = \int\left(\dfrac{\cosh hx}{\sinh 5x}+1\right)dx = \int(\coth 5x+1)\,dx = \int\coth 5x\, dx + \int dx$ let $u = 5x$, then

$\dfrac{du}{dx} = \dfrac{d}{dx}5x$; $\dfrac{du}{dx} = 5$; $du = 5dx$; $dx = \dfrac{du}{5}$. Therefore,

$$\boxed{\int \dfrac{\cosh 5x + \sinh 5x}{\sinh 5x}\,dx} = \boxed{\int\left(\dfrac{\cosh hx}{\sinh 5x}+1\right)dx} = \boxed{\int\coth 5x\, dx + \int dx} = \boxed{\int\coth u\cdot\dfrac{du}{5x}+x} = \boxed{\dfrac{1}{5}\int\coth u\, du + x}$$

$$= \boxed{\dfrac{1}{5}\ln|\sinh u| + x + c} = \boxed{\dfrac{1}{5}\ln|\sinh 5x| + x + c}$$

Check: Let $y = \dfrac{1}{5}\ln|\sinh 5x| + x + c$, then $y' = \dfrac{1}{5}\cdot\dfrac{\cosh 5x\cdot 5}{\sinh 5x}+1+0 = \dfrac{5}{5}\cdot\dfrac{\cosh 5x}{\sinh 5x}+1 = \coth 5x + 1$

f. $\boxed{\int\left(\dfrac{1+\cosh x}{\sinh x}\right)dx} = \boxed{\int\left(\dfrac{1}{\sinh x}+\dfrac{\cosh x}{\sinh x}\right)dx} = \boxed{\int\csc h\, x\, dx + \int\coth x\, dx} = \boxed{\ln\left|\tanh\dfrac{x}{2}\right| + \ln|\sinh x| + c}$

Check: Let $y = \ln\left|\tanh\dfrac{x}{2}\right| + \ln|\sinh x| + c$, then $y' = \dfrac{1}{\tanh\frac{x}{2}}\cdot\dfrac{d}{dx}\tanh\dfrac{x}{2} + \dfrac{1}{\sinh x}\cdot\dfrac{d}{dx}\sinh x + 0$

$= \dfrac{1}{\tanh\frac{x}{2}}\cdot\sec h^2\dfrac{x}{2}\cdot\dfrac{1}{2}+\dfrac{1}{\sinh x}\cdot\cosh x = \dfrac{\sec h^2\frac{x}{2}}{2\tanh\frac{x}{2}}+\dfrac{\cosh x}{\sinh x} = \dfrac{1}{2}\cdot\dfrac{1-\tanh^2\frac{x}{2}}{\tanh\frac{x}{2}}+\dfrac{\cosh x}{\sinh x} = \dfrac{1}{2}\cdot\dfrac{1-\frac{\sinh^2\frac{x}{2}}{\cosh^2\frac{x}{2}}}{\frac{\sinh\frac{x}{2}}{\cosh\frac{x}{2}}}$

$+\dfrac{\cosh x}{\sinh x} = \dfrac{1}{2}\cdot\dfrac{\frac{\cosh^2\frac{x}{2}-\sinh^2\frac{x}{2}}{\cosh^2\frac{x}{2}}}{\frac{\sinh\frac{x}{2}}{\cosh\frac{x}{2}}}+\dfrac{\cosh x}{\sinh x} = \dfrac{1}{2}\cdot\dfrac{\frac{1}{\cosh^2\frac{x}{2}}}{\frac{\sinh\frac{x}{2}}{\cosh\frac{x}{2}}}+\dfrac{\cosh x}{\sinh x} = \dfrac{1}{2}\cdot\dfrac{\cosh\frac{x}{2}}{\cosh^2\frac{x}{2}\cdot\sinh\frac{x}{2}}+\dfrac{\cosh x}{\sinh x}$

$= \dfrac{1}{2\cosh\frac{x}{2}\cdot\sinh\frac{x}{2}}+\dfrac{\cosh x}{\sinh x} = \dfrac{1}{\sinh 2\cdot\frac{x}{2}}+\dfrac{\cosh x}{\sinh x} = \dfrac{1}{\sinh x}+\dfrac{\cosh x}{\sinh x} = \dfrac{1+\cosh x}{\sinh x}$

g. $\boxed{\int 2\sinh\dfrac{3x}{2}\csc h\dfrac{3x}{2}\,dx} = \boxed{2\int\sinh\dfrac{3x}{2}\cdot\dfrac{1}{\sinh\frac{3x}{2}}\,dx} = \boxed{2\int\dfrac{\sinh\frac{3x}{2}}{\sinh\frac{3x}{2}}\,dx} = \boxed{2\int dx} = \boxed{2x+c}$

Check: Let $y = 2x + c$, then $y' = 2 \cdot x^{1-1} + 0 = 2 \cdot x^0 = 2$

h. $\boxed{\int \left(\cosh x \sec h\, x + e^{3x} \right) dx} = \boxed{\int \left(\cosh x \cdot \dfrac{1}{\cosh x} + e^{3x} \right) dx} = \boxed{\int \left(1 + e^{3x} \right) dx} = \boxed{\int dx + \int e^{3x} dx} = \boxed{x + \dfrac{1}{3} e^{3x} + c}$

Check: Let $y = x + \dfrac{1}{3} e^{3x} + c$, then $y' = x^{1-1} + \dfrac{1}{3} \cdot e^{3x} \cdot 3 + 0 = x^0 + \dfrac{3}{3} \cdot e^{3x} = 1 + e^{3x}$

i. Given $\int e^{\sinh 8x} \cosh 8x\, dx$ let $u = \sinh 8x$, then $\dfrac{du}{dx} = \dfrac{d}{dx} \sinh 8x$; $\dfrac{du}{dx} = 8 \cosh 8x$; $dx = \dfrac{du}{8 \cosh 8x}$. Thus,

$\boxed{\int e^{\sinh 8x} \cosh 8x\, dx} = \boxed{\int e^u \cosh 8x \cdot \dfrac{du}{8 \cosh 8x}} = \boxed{\int \dfrac{e^u}{8} du} = \boxed{\dfrac{1}{8} \int e^u du} = \boxed{\dfrac{1}{8} e^u + c} = \boxed{\dfrac{1}{8} e^{\sinh 8x} + c}$

Check: Let $y = \dfrac{1}{8} e^{\sinh 8x} + c$, then $y' = \dfrac{1}{8} \cdot e^{\sinh 8x} \cdot \cosh 8x \cdot 8 + 0 = \dfrac{8}{8} \cdot e^{\sinh 8x} \cosh 8x = e^{\sinh 8x} \cosh 8x$

j. Given $\int e^{\cosh \frac{x}{3}} \sinh \dfrac{x}{3}\, dx$ let $u = \cosh \dfrac{x}{3}$, then $\dfrac{du}{dx} = \dfrac{d}{dx} \cosh \dfrac{x}{3}$; $\dfrac{du}{dx} = \dfrac{1}{3} \sinh \dfrac{x}{3}$; $dx = \dfrac{3du}{\sin \frac{x}{3}}$. Therefore,

$\boxed{\int e^{\cosh \frac{x}{3}} \sinh \dfrac{x}{3}\, dx} = \boxed{\int e^u \sinh \dfrac{x}{3} \cdot \dfrac{3du}{\sinh \frac{x}{3}}} = \boxed{3 \int e^u du} = \boxed{3e^u + c} = \boxed{3e^{\cosh \frac{x}{3}} + c}$

Check: Let $y = 3e^{\cosh \frac{x}{3}} + c$, then $y' = 3e^{\cosh \frac{x}{3}} \cdot \sinh \dfrac{x}{3} \cdot \dfrac{1}{3} + 0 = \dfrac{3}{3} e^{\cosh \frac{x}{3}} \cdot \sinh \dfrac{x}{3} = e^{\cosh \frac{x}{3}} \sinh \dfrac{x}{3}$

k. Given $\int e^{\tanh 5x} \sec h^2 5x\, dx$ let $u = \tanh 5x$, then $\dfrac{du}{dx} = \dfrac{d}{dx} \tanh 5x$; $\dfrac{du}{dx} = 5 \sec h^2 5x$; $du = 5 \sec h^2 5x\, dx$

; $dx = \dfrac{du}{5 \sec h^2 5x}$. Therefore,

$\boxed{\int e^{\tanh 5x} \sec h^2 5x\, dx} = \boxed{\int e^u \sec h^2 5x \cdot \dfrac{du}{5 \sec h^2 5x}} = \boxed{\dfrac{1}{5} \int e^u du} = \boxed{\dfrac{1}{5} e^u + c} = \boxed{\dfrac{1}{5} e^{\tanh 5x} + c}$

Check: Let $y = \dfrac{1}{5} e^{\tanh 5x} + c$, then $y' = \dfrac{1}{5} e^{\tanh 5x} \cdot \sec h^2 5x \cdot 5 + 0 = \dfrac{5}{5} e^{\tanh 5x} \sec h^2 5x = e^{\tanh 5x} \sec h^2 5x$

l. Given $\int e^{\frac{1}{3} \coth 7x} \csc h^2 7x\, dx$ let $u = \dfrac{1}{3} \coth 7x$, then $\dfrac{du}{dx} = \dfrac{d}{dx} \dfrac{1}{3} \coth 7x$; $\dfrac{du}{dx} = -\dfrac{7}{3} \csc h^2 7x$

; $du = -\dfrac{7}{3} \csc h^2 7x\, dx$; $dx = -\dfrac{3du}{7 \csc h^2 7x}$. Therefore,

$\boxed{\int e^{\frac{1}{3} \coth 7x} \csc h^2 7x\, dx} = \boxed{\int e^u \csc h^2 7x \cdot \dfrac{-3du}{7 \csc h^2 7x}} = \boxed{-\dfrac{3}{7} \int e^u du} = \boxed{-\dfrac{3}{7} e^u + c} = \boxed{-\dfrac{3}{7} e^{\frac{1}{3} \coth 7x} + c}$

Check: Let $y = -\dfrac{3}{7} e^{\frac{1}{3} \coth 7x} + c$, then $y' = -\dfrac{3}{7} e^{\frac{1}{3} \coth 7x} \cdot -\dfrac{1}{3} \csc h^2 7x \cdot 7 + 0 = e^{\frac{1}{3} \coth 7x} \csc h^2 7x$

- To integrate even powers of $\sinh x$ and $\cosh x$ use the following identities:

$$\cosh^2 x - \sinh^2 x = 1 \qquad\qquad \sinh^2 x = \frac{1}{2}\left(\cosh 2x - 1\right) \qquad\qquad \cosh^2 x = \frac{1}{2}\left(\cosh 2x + 1\right)$$

To integrate odd powers of $\sin x$ and $\cos x$ use the following equalities:

$$\int \sinh^{2n+1} x\, dx = \int \sinh^{2n} x \sinh x\, dx = \int \left(\sinh^2 x\right)^n \sinh x\, dx = \int \left(\cosh^2 x - 1\right)^n \sinh x\, dx \qquad (\text{ let } u = \cosh x)$$

$$\int \cosh^{2n+1} x\, dx = \int \cosh^{2n} x \cosh x\, dx = \int \left(\cosh^2 x\right)^n \cosh x\, dx = \int \left(1 + \sinh^2 x\right)^n \cosh x\, dx \qquad (\text{ let } u = \sinh x)$$

- To integrate products of $\sinh x$, $\sinh y$, $\cosh x$, and $\cosh y$ use the identities below:

$$\sinh x \sinh y = \frac{1}{2}\left[\cosh\left(x+y\right) - \cosh\left(x-y\right)\right]$$

$$\cosh x \cosh y = \frac{1}{2}\left[\cosh\left(x+y\right) + \cosh\left(x-y\right)\right]$$

$$\sinh x \cosh y = \frac{1}{2}\left[\sinh\left(x+y\right) + \sinh\left(x-y\right)\right]$$

- To integrate $\tanh^n x$, set

$$\tanh^n x = \tanh^{n-2} x \tanh^2 x = \tanh^{n-2} x\left(1 - \sec h^2 x\right) = \tanh^{n-2} x - \tanh^{n-2} x \sec h^2 x$$

- To integrate $\coth^n x$, set

$$\coth^n x = \coth^{n-2} x \coth^2 x = \coth^{n-2} x\left(1 + \csc h^2 x\right) = \coth^{n-2} x + \coth^{n-2} x \csc h^2 x$$

- To integrate $\sec h^n x$

For even powers, set $$\sec h^n x = \sec h^{n-2} x \sec h^2 x = \left(1 - \tanh^2 x\right)^{\frac{n-2}{2}} \sec h^2 x$$

For odd powers change the integrand to a product of even and odd functions, i.e., write

$\int \sec h^3 x\, dx$ as $\int \sec h^2 x \sec h\, x\, dx$ (see Example 5.4-6, problem letter h).

- To integrate $\csc h^n x$

For even powers, set $$\csc h^n x = \csc h^{n-2} x \csc h^2 x = \left(\coth^2 x - 1\right)^{\frac{n-2}{2}} \csc h^2 x$$

For odd powers change the integrand to a product of even and odd functions, i.e., write

$\int \csc h^3 x\, dx$ as $\int \csc h^2 x \csc h\, x\, dx$ (see Example 5.4-6, problem letter i).

In the following examples we use the above general rules in order to solve integral of products and powers of hyperbolic functions:

Example 5.4-5: Evaluate the following indefinite integrals:

a. $\int \sinh 5x \cosh 7x\, dx =$

b. $\int \sinh x \cosh x\, dx =$

c. $\int \cosh 3x \cosh 2x\, dx =$

d. $\int \sinh 3x \sinh 5x\, dx =$

e. $\int \cosh 3x \cosh 5x\, dx =$

f. $\int \sinh^5 x\, dx =$

g. $\int \sinh^3 x\, dx =$

h. $\int \cosh^5 x\, dx =$

i. $\int \tanh^4 x\, dx =$

j. $\int \sinh^7 x\, dx =$

k. $\int \sec h^4 x\, dx =$

l. $\int \cosh^3 x\, dx =$

Solutions:

a. $\boxed{\int \sinh 5x \cosh 7x\, dx} = \boxed{\int \frac{1}{2}\left[\sinh\left(5+7\right)x + \sinh\left(5-7\right)x\right]dx} = \boxed{\int \frac{1}{2}\left[\sinh\left(12x\right) + \sinh\left(-2x\right)\right]dx}$

$= \boxed{\frac{1}{2}\int \left(\sinh 12x - \sinh 2x\right)dx} = \boxed{\frac{1}{2}\int \sinh 12x\, dx - \frac{1}{2}\int \sinh 2x\, dx} = \boxed{\frac{1}{2}\cdot\frac{1}{12}\cosh 12x - \frac{1}{2}\cdot\frac{1}{2}\cosh 2x + c}$

$= \boxed{\frac{1}{24}\cosh 12x - \frac{1}{4}\cosh 2x + c}$

Check: Let $y = \frac{1}{24}\cosh 12x - \frac{1}{4}\cosh 2x + c$, then $y' = \frac{1}{24}\cdot 12\sinh 12x - \frac{1}{4}\cdot 2\sinh 2x + 0 = \frac{12}{24}\sinh 12x - \frac{2}{4}\sinh 2x$

$= \frac{1}{2}\sinh 12x - \frac{1}{2}\sinh 2x = \frac{1}{2}\sinh 12x + \frac{1}{2}\sinh\left(-2x\right) = \frac{1}{2}\sinh\left(5+7\right)x + \frac{1}{2}\sinh\left(5-7\right)x = \sinh 5x \cosh 7x$

b. $\boxed{\int \sinh x \cosh x\, dx} = \boxed{\int \frac{1}{2}\left[\sinh\left(1+1\right)x + \sinh\left(1-1\right)x\right]dx} = \boxed{\int \frac{1}{2}\left[\sinh\left(2x\right) + \sinh\left(0x\right)\right]dx} = \boxed{\frac{1}{2}\int \left(\sinh 2x + 0\right)dx}$

$= \boxed{\frac{1}{2}\int \sinh 2x\, dx} = \boxed{\frac{1}{2}\cdot\frac{1}{2}\cosh 2x} = \boxed{\frac{1}{4}\cosh 2x + c}$

Check: Let $y = \frac{1}{4}\cosh 2x + c$, then $y' = \frac{1}{4}\cdot 2\sinh 2x + 0 = \frac{1}{2}\sinh 2x = \frac{1}{2}\cdot 2\sinh x \cosh x = \sinh x \cosh x$

c. $\boxed{\int \cosh 3x \cosh 2x\, dx} = \boxed{\int \frac{1}{2}\left[\cosh\left(3+2\right)x + \cosh\left(3-2\right)x\right]dx} = \boxed{\int \frac{1}{2}\left(\cosh 5x + \cosh x\right)dx} = \boxed{\frac{1}{2}\int \cosh 5x\, dx}$

$\boxed{+\frac{1}{2}\int \cosh x\, dx} = \boxed{\frac{1}{2}\cdot\frac{1}{5}\sinh 5x + \frac{1}{2}\sinh x + c} = \boxed{\frac{1}{10}\sinh 5x + \frac{1}{2}\sinh x + c}$

Check: Let $y = \frac{1}{10}\sinh 5x + \frac{1}{2}\sinh x + c$, then $y' = \frac{1}{10}\cdot 5\cosh 5x + \frac{1}{2}\cdot\cosh x + 0 = \frac{5}{10}\cosh 5x + \frac{1}{2}\cosh x$

$= \frac{1}{2}\cosh 5x + \frac{1}{2}\cosh x = \frac{1}{2}\cosh\left(3+2\right)x + \frac{1}{2}\cosh\left(3-2\right)x = \cosh 3x \cosh 2x$

d. $\boxed{\int \sinh 3x \sinh 5x\, dx} = \boxed{\int \frac{1}{2}\left[\cosh\left(3+5\right)x - \cosh\left(3-5\right)x\right]dx} = \boxed{\int \frac{1}{2}\left[\cosh\left(8x\right) - \cosh\left(-2x\right)\right]dx}$

$= \boxed{\frac{1}{2}\int \left(\cosh 8x - \cosh 2x\right)dx} = \boxed{\frac{1}{2}\int \cosh 8x\, dx - \frac{1}{2}\int \cosh 2x\, dx} = \boxed{\frac{1}{2}\cdot\frac{1}{8}\sinh 8x - \frac{1}{2}\cdot\frac{1}{2}\sinh 2x + c}$

$= \boxed{\frac{1}{16}\sinh 8x - \frac{1}{4}\sinh 2x + c}$

Check: Let $y = \frac{1}{16}\sinh 8x - \frac{1}{4}\sinh 2x + c$, then $y' = \frac{1}{16}\cdot\cosh 8x \cdot 8 - \frac{1}{4}\cdot\cosh 2x \cdot 2 + 0 = \frac{8}{16}\cosh 8x - \frac{2}{4}\cosh 2x$

$$= \frac{1}{2}\cosh 8x - \frac{1}{2}\cosh 2x = \frac{1}{2}\left[\cosh\left(8x\right) - \cosh\left(-2x\right)\right] = \frac{1}{2}\left[\cosh\left(3+5\right)x - \cosh\left(3-5\right)x\right] = \sinh 3x \sinh 5x$$

e. $\boxed{\int \cosh 3x \cosh 5x \, dx} = \boxed{\int \frac{1}{2}\left[\cosh\left(3+5\right)x + \cosh\left(3-5\right)x\right]dx} = \boxed{\int \frac{1}{2}\left[\cosh\left(8x\right) + \cosh\left(-2x\right)\right]dx}$

$= \boxed{\int \frac{1}{2}\left(\cosh 8x + \cosh 2x\right)dx} = \boxed{\frac{1}{2}\int\cosh 8x\, dx + \frac{1}{2}\int\cosh 2x\, dx} = \boxed{\frac{1}{2}\cdot\frac{1}{8}\sinh 8x + \frac{1}{2}\cdot\frac{1}{2}\sinh 2x + c}$

$= \boxed{\mathbf{\frac{1}{16}\sinh 8x + \frac{1}{4}\sinh 2x + c}}$

Check: Let $y = \frac{1}{16}\sinh 8x + \frac{1}{4}\sinh 2x + c$, then $y' = \frac{1}{16}\cdot\cosh 8x \cdot 8 + \frac{1}{4}\cdot\cosh 2x \cdot 2 + 0 = \frac{8}{16}\cosh 8x + \frac{2}{4}\cosh 2x$

$$= \frac{1}{2}\cosh 8x + \frac{1}{2}\cosh 2x = \frac{1}{2}\left[\cosh\left(8x\right) + \cosh\left(-2x\right)\right] = \frac{1}{2}\left[\cosh\left(3+5\right)x + \cosh\left(3-5\right)x\right] = \cosh 3x \cosh 5x$$

f. $\int \sinh^5 x \, dx = \int \sinh^4 x \sinh x \, dx = \int\left(\sinh^2 x\right)^2 \sinh x \, dx = \int\left(\cosh^2 x - 1\right)^2 \sinh x \, dx$. Let $u = \cosh x$, then

$\frac{du}{dx} = \frac{d}{dx}\cosh x$; $\frac{du}{dx} = \sinh x$; $du = \sinh x \, dx$; $dx = \frac{du}{\sinh x}$. Therefore,

$\boxed{\int \sinh^5 x \, dx} = \boxed{\int\left(\cosh^2 x - 1\right)^2 \sinh x \, dx} = \boxed{\int\left(u^2 - 1\right)^2 \sinh x \, dx} = \boxed{\int\left(u^4 - 2u^2 + 1\right)\sinh x \cdot \frac{du}{\sinh x}}$

$= \boxed{\int\left(u^4 - 2u^2 + 1\right)du} = \boxed{\int u^4 du - 2\int u^2 du + \int du} = \boxed{\frac{1}{5}u^5 - \frac{2}{3}u^3 + u + c} = \boxed{\mathbf{\frac{1}{5}\cosh^5 x - \frac{2}{3}\cosh^3 x + \cosh x + c}}$

Check: Let $y = \frac{1}{5}\cosh^5 x - \frac{2}{3}\cosh^3 x + \cosh x + c$, then $y' = \frac{1}{5}\cdot 5\cosh^4 x \cdot \sinh x - \frac{2}{3}\cdot 3\cosh^2 x \cdot \sinh x + \sinh x$

$$= \sinh x \cosh^4 x - 2\sinh x \cosh^2 x + \sinh x = \sinh x\left(\cosh^4 x - 2\cosh^2 x + 1\right) = \sinh x\left(\cosh^2 x - 1\right)^2$$

$$= \sinh x\left(\sinh^2 x\right)^2 = \sinh x \sinh^4 x = \sinh^5 x$$

g. $\int \sinh^3 x \, dx = \int \sinh^2 x \sinh x \, dx = \int\left(\cosh^2 x - 1\right)\sinh x \, dx$. Let $u = \cosh x$, then $\frac{du}{dx} = \frac{d}{dx}\cosh x$

; $\frac{du}{dx} = \sinh x$; $du = \sinh x \, dx$; $dx = \frac{du}{\sinh x}$. Therefore,

$\boxed{\int \sinh^3 x \, dx} = \boxed{\int\left(\cosh^2 x - 1\right)\sinh x \, dx} = \boxed{\int\left(u^2 - 1\right)\sinh x \cdot \frac{du}{\sinh x}} = \boxed{\int\left(u^2 - 1\right)du} = \boxed{\int u^2 du - \int du}$

$= \boxed{\frac{1}{3}u^3 - u + c} = \boxed{\mathbf{\frac{1}{3}\cosh^3 x - \cosh x + c}}$

Check: Let $y = \frac{1}{3}\cosh^3 x - \cosh x + c$, then $y' = \frac{1}{3} \cdot 3\cosh^2 x \cdot \sinh x - \sinh x + 0 = \sinh x \cosh^2 x - \sinh x$

$$= \sinh x \left(\cosh^2 x - 1 \right) = \sinh x \sinh^2 x = \sinh^3 x$$

h. $\int \cosh^5 x\, dx = \int \cosh^4 x \cosh x\, dx = \int \left(\cosh^2 x \right)^2 \cosh x\, dx = \int \left(1 + \sinh^2 x \right)^2 \cos x\, dx$. Let $u = \sinh x$,

then $\dfrac{du}{dx} = \dfrac{d}{dx}\sinh x$; $\dfrac{du}{dx} = \cosh x$; $du = \cosh x\, dx$; $dx = \dfrac{du}{\cosh x}$. Therefore,

$$\boxed{\int \cosh^5 x\, dx} = \boxed{\int \left(1 + \sinh^2 x\right)^2 \cosh x\, dx} = \boxed{\int \left(1 + u^2\right)^2 \cosh x \cdot \frac{du}{\cosh x}} = \boxed{\int \left(1 + u^2\right)^2 du} = \boxed{\int \left(u^4 + 2u^2 + 1\right)du}$$

$$= \boxed{\int u^4\, du + 2\int u^2\, du + \int du} = \boxed{\frac{1}{5}u^5 + \frac{2}{3}u^3 + u + c} = \boxed{\frac{1}{5}\sinh^5 x + \frac{2}{3}\sinh^3 x + \sinh x + c}$$

Check: Let $y = \frac{1}{5}\sinh^5 x + \frac{2}{3}\sinh^3 x + \sinh x + c$, then $y' = \frac{5}{5}\sinh^4 x \cdot \cosh x + \frac{6}{3}\sinh^2 x \cdot \cosh x + \cosh x + 0$

$$= \cosh x \sinh^4 x + 2\cosh x \sinh^2 x + \cosh x = \cosh x \left(\sinh^4 x + 2\sinh^2 x + 1 \right) = \cosh x \left(1 + \sinh^2 x \right)^2$$

$$= \cosh x \left(\cosh^2 x \right)^2 = \cosh x \cosh^4 x = \cosh^5 x$$

i. $\int \tanh^4 x\, dx = \int \tanh^2 x \tanh^2 x\, dx = \int \tanh^2 x \left(1 - \sec h^2 x\right) dx = \int \tanh^2 x\, dx - \int \tanh^2 x \sec h^2 x\, dx$

$$= \int \left(1 - \sec h^2 x\right) dx - \int \tanh^2 x \sec h^2 x\, dx = -\int \tanh^2 x \sec h^2 x\, dx - \int \sec h^2 x\, dx + \int dx. \text{ To solve the first}$$

integral let $u = \tanh x$, then $\dfrac{du}{dx} = \dfrac{d}{dx}\tanh x$; $\dfrac{du}{dx} = \sec h^2 x$; $du = \sec h^2 x\, dx$; $dx = \dfrac{du}{\sec h^2 x}$. Thus,

$$-\int \tanh^2 x \sec h^2 x\, dx = -\int u^2 \sec h^2 x \cdot \frac{du}{\sec h^2 x} = -\int u^2\, du = -\frac{1}{3}u^3 = -\frac{1}{3}\tanh^3 x. \text{ Therefore,}$$

$$\boxed{\int \tanh^4 x\, dx} = \boxed{\int \tanh^2 x \tanh^2 x\, dx} = \boxed{-\int \tanh^2 x \sec h^2 x\, dx - \int \sec h^2 x\, dx + \int dx} = \boxed{-\frac{1}{3}\tanh^3 x - \tanh x + x + c}$$

Check: Let $y = -\frac{1}{3}\tanh^3 x - \tanh x + x + c$, then $y' = -\tanh^2 x \sec h^2 x - \sec h^2 x + 1$

$$= \sec h^2 x \left(-\tanh^2 x - 1 \right) + 1 = \left(1 - \tanh^2 x \right)\left(-\tanh^2 x - 1 \right) + 1 = -\tanh^2 x - 1 + \tanh^4 x + \tanh^2 x + 1$$

$$= \left(-\tanh^2 x + \tanh^2 x \right) + \left(-1 + 1 \right) + \tanh^4 x = \tanh^4 x$$

j. $\int \sinh^7 x\, dx = \int \sinh^6 x \sinh x\, dx = \int \left(\sinh^2 x \right)^3 \sinh x\, dx = \int \left(\cosh^2 x - 1 \right)^3 \sinh x\, dx$. Let $u = \cosh x$, then

$\dfrac{du}{dx} = \dfrac{d}{dx}\cosh x$; $\dfrac{du}{dx} = \sinh x$; $du = \sinh x\, dx$; $dx = \dfrac{du}{\sinh x}$. Therefore,

$$\boxed{\int \sinh^7 x\, dx} = \boxed{\int \left(\cosh^2 x - 1 \right)^3 \sinh x\, dx} = \boxed{\int \left(u^2 - 1 \right)^3 \sinh x\, dx} = \boxed{\int \left(u^6 - 3u^4 + 3u^2 - 1 \right)\sinh x \cdot \frac{du}{\sinh x}}$$

$$= \boxed{\int \left(u^6 - 3u^4 + 3u^2 - 1\right)du} = \boxed{\int u^6 du - 3\int u^4 du + 3\int u^2 du - \int du} = \boxed{\frac{1}{7}u^7 - 3\cdot\frac{1}{5}u^5 + 3\cdot\frac{1}{3}u^3 - u + c}$$

$$= \boxed{\frac{1}{7}\cosh^7 x - \frac{3}{5}\cosh^5 x + \cosh^3 x - \cosh x + c}$$

Check: Let $y = \frac{1}{7}\cosh^7 x - \frac{3}{5}\cosh^5 x + \cosh^3 x - \cosh x + c$, then $y' = \frac{7}{7}\cosh^6 x\cdot\sinh x - \frac{15}{5}\cosh^4 x\cdot\sinh x$

$$+ 3\cosh^2 x\cdot\sinh x - \sinh x + 0 = \sinh x \cosh^6 x - 3\sinh x \cosh^4 x + 3\sinh x \cosh^2 x - \sinh x$$

$$= \sinh x\left(\cosh^6 x - 3\cosh^4 x + 3\cosh^2 x - 1\right) = \sinh x\left(\cosh^2 x - 1\right)^3 = \sinh x\left(\sinh^2 x\right)^3$$

$$= \sinh x \sinh^6 x = \sinh^7 x$$

k. $\int \sec h^4 x\, dx = \int \sec h^2 x \sec h^2 x\, dx = \int \sec h^2 x\left(1 - \tanh^2 x\right)dx = \int \sec h^2 x\, dx - \int \tanh^2 x \sec h^2 x\, dx$

$= \int\left(1 - \tanh^2 x\right)dx - \int \tanh^2 x \sec h^2 x\, dx = \int dx - \int \tanh^2 x\, dx - \int \tanh^2 x \sec h^2 x\, dx$. To solve the third

integral let $u = \tanh x$, then $\frac{du}{dx} = \frac{d}{dx}\tanh x$; $\frac{du}{dx} = \sec h^2 x$; $du = \sec h^2 x\, dx$; $dx = \frac{du}{\sec h^2 x}$. Therefore,

$-\int \tanh^2 x \sec h^2 x\, dx = -\int u^2 \sec h^2 x \cdot \frac{du}{\sec h^2 x} = -\int u^2 du = -\frac{1}{3}u^3 = -\frac{1}{3}\tanh^3 x$. Thus,

$$\boxed{\int \sec h^4 x\, dx} = \boxed{\int \sec h^2 x \sec h^2 x\, dx} = \boxed{\int dx - \int \tanh^2 x\, dx - \int \tanh^2 x \sec h^2 x\, dx} = \boxed{\int dx - \int \tanh^2 x\, dx - \frac{1}{3}\tanh^3 x}$$

$$= \boxed{x - (x - \tanh x) - \frac{1}{3}\tanh^3 x + c} = \boxed{-\frac{1}{3}\tanh^3 x + \tanh x + (x - x) + c} = \boxed{-\frac{1}{3}\tanh^3 x + \tanh x + c}$$

Check: Let $y = -\frac{1}{3}\tanh^3 x + \tanh x + c$, then $y' = -\frac{3}{3}\tanh^2 x\cdot\sec h^2 x + \sec h^2 x = -\tanh^2 x \sec h^2 x + \sec h^2 x$

$$= \sec h^2 x\left(1 - \tanh^2 x\right) = \sec h^2 x \sec h^2 x = \sec h^4 x$$

l. $\int \cosh^3 x\, dx = \int\left(\cosh^2 x\right)\cosh x\, dx = \int\left(1 + \sinh^2 x\right)\cosh x\, dx$. Let $u = \sinh x$, then $\frac{du}{dx} = \frac{d}{dx}\sinh x$

; $\frac{du}{dx} = \cosh x$; $du = \cosh x\, dx$; $dx = \frac{du}{\cosh x}$. Therefore,

$$\boxed{\int \cosh^3 x\, dx} = \boxed{\int\left(1 + \sinh^2 x\right)\cosh x\, dx} = \boxed{\int\left(1 + u^2\right)\cosh x\cdot\frac{du}{\cosh x}} = \boxed{\int\left(1 + u^2\right)du} = \boxed{\int u^2 du + \int du}$$

$$= \boxed{\frac{1}{3}u^3 + u + c} = \boxed{\frac{1}{3}\sinh^3 x + \sinh x + c}$$

Check: Let $y = \frac{1}{3}\sinh^3 x + \sinh x + c$, then $y' = \frac{1}{3}\cdot 3\sinh^2 x\cdot\cosh x + \cosh x + 0 = \cosh x \sinh^2 x + \cosh x$

$$= \cosh x\left(1 + \sinh^2 x\right) = \cosh x\left(\cosh^2 x\right) = \cosh^3 x$$

Example 5.4-6: Evaluate the following indefinite integrals:

a. $\int \cosh^4 x\, dx =$ b. $\int \cosh^2 5x\, dx =$ c. $\int \sinh^4 x\, dx =$

d. $\int \tanh^3 x\, dx =$ e. $\int \coth^4 x\, dx =$ f. $\int \tanh^6 x\, dx =$

g. $\int \coth^3 x\, dx =$ h. $\int \sec h^3 x\, dx =$ i. $\int \csc h^3 x\, dx =$

Solutions:

a. $\boxed{\int \cosh^4 x\, dx} = \boxed{\int \left(\cosh^2 x\right)^2 dx} = \boxed{\int \left[\frac{1}{2}(\cosh 2x +1)\right]^2 dx} = \boxed{\frac{1}{4}\int (\cosh 2x +1)^2 dx} = \boxed{\frac{1}{4}\int \left(1+\cosh^2 2x + 2\cosh 2x\right)dx}$

$= \boxed{\frac{1}{4}\int dx + \frac{1}{4}\int \cosh^2 2x\, dx + \frac{2}{4}\int \cosh 2x\, dx} = \boxed{\frac{x}{4} + \frac{1}{4}\int \frac{1}{2}(1+\cosh 4x)\, dx + \frac{1}{2}\cdot\frac{1}{2}\sinh 2x} = \boxed{\frac{x}{4} + \frac{1}{4}\cdot\frac{1}{2}\int dx}$

$\boxed{+ \frac{1}{8}\int \cosh 4x\, dx + \frac{1}{4}\sinh 2x} = \boxed{\frac{x}{4} + \frac{x}{8} + \frac{1}{8}\cdot\frac{1}{4}\sinh 4x + \frac{1}{4}\sinh 2x + c} = \boxed{\frac{3}{8}x + \frac{1}{32}\sinh 4x + \frac{1}{4}\sinh 2x + c}$

Check: Let $y = \frac{3x}{8} + \frac{1}{32}\sinh 4x + \frac{1}{4}\sinh 2x + c$, then $y' = \frac{3}{8} + \frac{4\cosh 4x}{32} + \frac{2\cosh 2x}{4} = \frac{3}{8} + \frac{1}{8}\cosh 4x + \frac{2}{4}\cosh 2x$

$= \left(\frac{1}{4}+\frac{1}{8}\right) + \frac{1}{8}\cosh 4x + \frac{2}{4}\cosh 2x = \frac{1}{4} + \frac{1}{8}(1+\cosh 4x) + \frac{2}{4}\cosh 2x = \frac{1}{4} + \frac{1}{4}\cdot\frac{1}{2}(1+\cosh 4x) + \frac{2}{4}\cosh 2x$

$= \frac{1}{4} + \frac{1}{4}\cdot\cosh^2 2x + \frac{2}{4}\cosh 2x = \frac{1}{4}\left(1+\cosh^2 2x + 2\cosh 2x\right) = \frac{1}{4}(\cosh 2x+1)^2 = \left[\frac{1}{2}(\cosh 2x+1)\right]^2$

$= \left(\cosh^2 x\right)^2 = \cosh^4 x$

b. $\boxed{\int \cosh^2 5x\, dx} = \boxed{\int \left(1+\sinh^2 5x\right)dx} = \boxed{\int dx + \int \sinh^2 5x\, dx} = \boxed{\int dx + \int \frac{1}{2}(\cosh 10x -1)\, dx} = \boxed{x + \frac{1}{2}\int \cosh 10x\, dx}$

$\boxed{-\frac{1}{2}\int dx} = \boxed{x + \frac{1}{2}\cdot\frac{1}{10}\sinh 10x - \frac{x}{2} + c} = \boxed{x\left(1-\frac{1}{2}\right) + \frac{1}{20}\sinh 10x + c} = \boxed{\frac{x}{2} + \frac{\sinh 10x}{20} + c}$

Check: Let $y = \frac{x}{2} + \frac{\sinh 10x}{20} + c$, then $y' = \frac{1}{2} + \frac{1}{20}\cdot\cosh 10x\cdot 10 + 0 = \frac{1}{2} + \frac{10}{20}\cdot\cosh 10x = \frac{1}{2} + \frac{1}{2}\cosh 10x$

$= \left(1-\frac{1}{2}\right) + \frac{1}{2}\cosh 10x = 1 + \left(\frac{1}{2}\cosh 10x - \frac{1}{2}\right) = 1 + \frac{1}{2}(\cosh 10x -1) = 1+\sinh^2 5x = \cosh^2 5x$

c. $\boxed{\int \sinh^4 x\, dx} = \boxed{\int \left(\sinh^2 x\right)^2 dx} = \boxed{\int \left[\frac{1}{2}(\cosh 2x -1)\right]^2 dx} = \boxed{\frac{1}{4}\int (\cosh 2x -1)^2 dx} = \boxed{\frac{1}{4}\int \left(\cosh^2 2x - 2\cosh 2x +1\right)dx}$

$= \boxed{\frac{1}{4}\int \cosh^2 2x\, dx - \frac{2}{4}\int \cosh 2x\, dx + \frac{1}{4}\int dx} = \boxed{\frac{1}{4}\int \frac{1}{2}(1+\cosh 4x)\, dx - \frac{1}{2}\cdot\frac{1}{2}\sinh 2x + \frac{x}{4}} = \boxed{\frac{1}{4}\cdot\frac{1}{2}\int dx}$

$$\boxed{+\frac{1}{8}\int \cosh 4x\,dx - \frac{1}{4}\sinh 2x + \frac{x}{4}} = \boxed{\left(\frac{x}{8}+\frac{x}{4}\right)+\frac{1}{8}\cdot\frac{1}{4}\sinh 4x - \frac{1}{4}\sinh 2x + c} = \boxed{\frac{3}{8}x + \frac{1}{32}\sinh 4x - \frac{1}{4}\sinh 2x + c}$$

Check:　Let $y = \dfrac{3x}{8} + \dfrac{1}{32}\sinh 4x - \dfrac{1}{4}\sinh 2x + c$, then $y' = \dfrac{3}{8} + \dfrac{4\cosh 4x}{32} - \dfrac{2\cosh 2x}{4} = \dfrac{3}{8} + \dfrac{1}{8}\cosh 4x - \dfrac{2}{4}\cosh 2x$

$$= \left(\frac{1}{4}+\frac{1}{8}\right) + \frac{1}{8}\cosh 4x - \frac{2}{4}\cosh 2x = \frac{1}{4} + \frac{1}{8}(1+\cosh 4x) - \frac{2}{4}\cosh 2x = \frac{1}{4} + \frac{1}{4}\cdot\frac{1}{2}(1+\cosh 4x) - \frac{2}{4}\cosh 2x$$

$$= \frac{1}{4} + \frac{1}{4}\cdot\cosh^2 2x - \frac{2}{4}\cosh 2x = \frac{1}{4}\left(1+\cosh^2 2x - 2\cosh 2x\right) = \frac{1}{4}(\cosh 2x - 1)^2 = \left[\frac{1}{2}(\cosh 2x - 1)\right]^2$$

$$= \left(\sinh^2 x\right)^2 = \sinh^4 x$$

d.　$\displaystyle\int \tanh^3 x\,dx = \int \tanh^2 x \tanh x\,dx = \int \left(1 - \operatorname{sech}^2 x\right)\tanh x\,dx = -\int \operatorname{sech}^2 x \tanh x\,dx + \int \tanh x\,dx$. To solve

the first integral let $u = \tanh x$, then $\dfrac{du}{dx} = \dfrac{d}{dx}\tanh x$; $\dfrac{du}{dx} = \operatorname{sech}^2 x$; $du = \operatorname{sech}^2 dx$; $dx = \dfrac{du}{\operatorname{sech}^2 x}$. Thus,

$$-\int \operatorname{sech}^2 x \tanh x\,dx = -\int \operatorname{sech}^2 x \cdot u \cdot \frac{du}{\operatorname{sech}^2 x} = -\int u\,du = -\frac{1}{2}u^2 = -\frac{1}{2}\tan^2 x.$$ Combining the term

$$\boxed{\int \tanh^3 x\,dx} = \boxed{\int \tanh^2 x \tanh x\,dx} = \boxed{\int \left(1-\operatorname{sech}^2 x\right)\tanh x\,dx} = \boxed{\int \left(-\operatorname{sech}^2 x \tanh x + \tanh x\right)dx}$$

$$= \boxed{-\int \operatorname{sech}^2 x \tanh x\,dx + \int \tanh x\,dx} = \boxed{-\frac{1}{2}\tanh^2 x + \int \tanh x\,dx} = \boxed{-\frac{1}{2}\tanh^2 x + \ln \cosh x + c}$$

Check:　Let $y = -\dfrac{1}{2}\tanh^2 x + \ln \cosh x + c$, then $y' = -\dfrac{1}{2}\cdot 2\tanh x \cdot \operatorname{sech}^2 x + \dfrac{1}{\cosh x}\cdot \sinh x + 0$

$$= -\tanh x \operatorname{sech}^2 x + \frac{\sinh x}{\cosh x} = -\tanh x \operatorname{sech}^2 x + \tanh x = \tanh x\left(1 - \operatorname{sech}^2 x\right) = \tanh x \tanh^2 x = \tanh^3 x$$

e.　$\displaystyle\int \coth^4 x\,dx = \int \coth^2 x \coth^2 x\,dx = \int \coth^2 x \left(\operatorname{csch}^2 x + 1\right)dx = \int \coth^2 x \operatorname{csch}^2 x\,dx + \int \coth^2 x\,dx$

$$= \int \coth^2 x \operatorname{csch}^2 x\,dx + \int \left(\operatorname{csch}^2 x + 1\right)dx = \int \coth^2 x \operatorname{csch}^2 x\,dx + \int \operatorname{csch}^2 x\,dx + \int dx.$$

To solve the first integral let $u = \coth x$, then $\dfrac{du}{dx} = \dfrac{d}{dx}\coth x$; $\dfrac{du}{dx} = -\operatorname{csch}^2 x$; $du = -\operatorname{csch}^2 x\,dx$

; $dx = -\dfrac{du}{\operatorname{csch}^2 x}$. Grouping the terms together we find $\displaystyle\int \coth^2 x \operatorname{csch}^2 x\,dx = \int u^2 \operatorname{csch}^2 x \cdot -\frac{du}{\operatorname{csch}^2 x}$

$$= -\int u^2\,du = -\frac{1}{3}u^3 = -\frac{1}{3}\coth^3 x.$$ Therefore,

$$\boxed{\int \coth^4 x\,dx} = \boxed{\int \coth^2 x \coth^2 x\,dx} = \boxed{\int \coth^2 x \left(\operatorname{csch}^2 x + 1\right)dx} = \boxed{\int \coth^2 x \operatorname{csch}^2 x\,dx + \int \operatorname{csch}^2 x\,dx + \int dx}$$

$$= \boxed{-\frac{1}{3}\coth^3 x + \int \csc h^2 x\, dx + \int dx} = \boxed{-\frac{1}{3}\coth^3 x - \coth x + x + c}$$

Check: Let $y = -\frac{1}{3}\coth^3 x - \coth x + x + c$, then $y' = -\frac{3}{3}\coth^2 x \cdot -\csc h^2 x + \csc h^2 x + 1 + 0$

$$= \coth^2 x \csc h^2 x + \csc h^2 x + 1 = \csc h^2 x \left(\coth^2 x + 1\right) + 1 = \left(\coth^2 x - 1\right)\left(\coth^2 x + 1\right) + 1$$

$$= \coth^4 x + \coth^2 x - \coth^2 x - 1 + 1 = \coth^4 x$$

f. $\int \tanh^6 x\, dx = \int \tanh^4 x \tanh^2 x\, dx = \int \tanh^4 x \left(1 - \sec h^2 x\right) dx = \int \left(\tanh^4 x - \tanh^4 x \sec h^2 x\right) dx =$

$\int \tanh^4 x\, dx - \int \tanh^4 x \sec h^2 x\, dx$. In example 5.4-5, problem letter i, we found that

$\int \tanh^4 x\, dx = -\frac{1}{3}\tanh^3 x - \tanh x + x + c$. Therefore,

$$\boxed{\int \tanh^6 x\, dx} = \boxed{\int \tanh^4 x \tanh^2 x\, dx} = \int \tanh^4 x \left(1 - \sec h^2 x\right) dx = \int \tanh^4 x\, dx - \int \tanh^4 x \sec h^2 x\, dx$$

$$= \boxed{\left(-\frac{1}{3}\tanh^3 x - \tanh x + x\right) - \int \tanh^4 x \sec h^2 x\, dx} = \boxed{-\frac{1}{5}\tanh^5 x - \frac{1}{3}\tanh^3 x - \tanh x + x + c}$$

Check: Let $y = -\frac{1}{5}\tanh^5 x - \frac{1}{3}\tanh^3 x - \tanh x + x + c$, then $y' = -\frac{5}{5}\tanh^4 x \cdot \sec h^2 x - \frac{3}{3}\tanh^2 x \cdot \sec h^2 x$

$$- \sec h^2 x + 1 + 0 = -\tanh^4 x \sec h^2 x - \tanh^2 x \sec h^2 x - \sec h^2 x + 1 = -\sec h^2 x \left(\tanh^4 x + \tanh^2 x + 1\right) + 1$$

$$= -\left(1 - \tanh^2 x\right)\left(\tanh^4 x + \tanh^2 x + 1\right) + 1 = -\left(\tanh^4 x + \tanh^2 x + 1 - \tanh^6 x - \tanh^4 x - \tanh^2 x\right) + 1$$

$$= -\tanh^4 x - \tanh^2 x - 1 + \tanh^6 x + \tanh^4 x + \tanh^2 x + 1 = \tanh^6 x$$

g. $\int \coth^3 x\, dx = \int \coth^2 x \coth x\, dx = \int \left(\csc h^2 x + 1\right)\coth x\, dx = \int \csc h^2 x \coth x\, dx + \int \coth x\, dx$. To solve the

first integral let $u = \coth x$, then $\frac{du}{dx} = \frac{d}{dx}\coth x$; $\frac{du}{dx} = -\csc h^2 x$; $du = -\csc h^2 dx$; $dx = -\frac{du}{\csc h^2 x}$. Thus,

$\int \csc h^2 x \coth x\, dx = \int \csc h^2 x \cdot u \cdot -\frac{du}{\csc h^2 x} = -\int u\, du = -\frac{1}{2}u^2 = -\frac{1}{2}\coth^2 x$. Combining the term

$$\boxed{\int \coth^3 x\, dx} = \boxed{\int \coth^2 x \coth x\, dx} = \boxed{\int \left(\csc h^2 x + 1\right)\coth x\, dx} = \boxed{\int \left(\csc h^2 x \cdot \coth x + \coth x\right) dx}$$

$$\boxed{\int \csc h^2 x \coth x\, dx + \int \coth x\, dx} = \boxed{-\frac{1}{2}\coth^2 x + \int \coth x\, dx} = \boxed{-\frac{1}{2}\coth^2 x + \ln|\sinh x| + c}$$

Check: Let $y = -\frac{1}{2}\coth^2 x + \ln|\sinh x| + c$, then $y' = -\frac{1}{2} \cdot 2\coth x \cdot -\csc h^2 x + \frac{1}{\sinh x} \cdot \cosh x + 0$

$$= \coth x \csc h^2 x + \frac{\cosh x}{\sinh x} = \coth x \csc h^2 x + \coth x = \coth x \left(\csc h^2 x + 1\right) = \coth x \coth^2 x = \coth^3 x$$

h. $\int \sec h^3 x\, dx \;=\; \int \sec h^2 x \sec h\, x\, dx \;=\; \int \left(1 - \tanh^2 x\right) \sec h\, x\, dx \;=\; -\int \tanh^2 x \sec h\, x\, dx + \int \sec h\, x\, dx$

$= -\int \tanh x \cdot \tanh x \sec h\, x\, dx + \int \sec h\, x\, dx$. To solve the first integral let $u = \tanh x$ and $dv = \tanh x \sec h\, x\, dx$,

then $du = \sec h^2 x\, dx$ and $\int dv = \int \tanh x \sec h\, x\, dx$ which implies $v = -\sec h\, x$. Using the integration by

parts formula $\int u\, dv = u v - \int v\, du$ we obtain

$-\int \tanh^2 x \sec h\, x\, dx \;=\; -\int \tanh x \cdot \tanh x \sec h\, x\, dx \;=\; \tanh x \sec h\, x - \int \sec h\, x \sec h^2 x\, dx \;=\; \tanh x \sec h\, x$

$-\int \sec h^3 x\, dx$. Combining the terms we have

$\int \sec h^3 x\, dx \;=\; \int \tanh x \cdot \tanh x \sec h\, x\, dx - \int \sec h\, x\, dx \;=\; \tanh x \sec h\, x - \int \sec h^3 x\, dx + \int \sec h\, x\, dx$. Moving the

$\int \sec h^3 x\, dx$ term from the right hand side of the equation to the left hand side we obtain

$\int \sec h^3 x\, dx + \int \sec h^3 x\, dx \;=\; 2\int \sec h^3 x\, dx \;=\; \tanh x \sec h\, x + \int \sec h\, x\, dx$. Therefore,

$$\boxed{\int \sec h^3 x\, dx} = \boxed{\frac{1}{2}\left(\tanh x \sec h\, x + \int \sec h\, x\, dx\right)} = \boxed{\frac{1}{2}\,\tanh x \sec h\, x + \frac{1}{2}\sin^{-1}\left(\tanh x\right) + c}$$

Check: Let $y = \dfrac{1}{2}\tanh x \sec h\, x + \dfrac{1}{2}\sin^{-1}(\tanh x) + c$, then $y' = \dfrac{\sec h^2 x \cdot \sec h\, x - \sec h\, x \tanh x \cdot \tanh x}{2}$

$+ \dfrac{\sec h^2 x}{2\sqrt{1 - \tanh^2 x}} + 0 = \dfrac{\sec h^3 x - \sec h\, x \tanh^2 x}{2} + \dfrac{\sec h^2 x}{2\sqrt{\sec h^2 x}} = \dfrac{\sec h^3 x - \sec h\, x \tanh^2 x}{2} + \dfrac{\sec h\, x}{2}$

$= \dfrac{\sec h^3 x - \sec h\, x \tanh^2 x + \sec h\, x}{2} = \dfrac{\sec h^3 x + \sec h\, x \left(1 - \tanh^2 x\right)}{2} = \dfrac{\sec h^3 x + \sec h\, x \sec h^2 x}{2}$

$= \dfrac{\sec h^3 x + \sec h^3 x}{2} = \dfrac{2\sec h^3 x}{2} = \sec h^3 x$

i. $\int \csc h^3 x\, dx \;=\; \int \csc h^2 x \csc h\, x\, dx \;=\; \int \left(\coth^2 x - 1\right) \csc h\, x\, dx \;=\; \int \coth^2 x \csc h\, x\, dx - \int \csc h\, x\, dx$

$= \int \coth x \cdot \coth x \csc h\, x\, dx - \int \csc h\, x\, dx$. To solve the first integral let $u = \coth x$ and $dv = \coth x \csc h\, x\, dx$,

then $du = -\csc h^2 x\, dx$ and $\int dv = \int \coth x \csc h\, x\, dx$ which implies $v = -\csc h\, x$. Using the integration by

parts formula $\int u\, dv = u v - \int v\, du$ we obtain

$\int \coth^2 x \csc h\, x\, dx \;=\; \int \coth x \cdot \coth x \csc h\, x\, dx \;=\; \coth x \cdot -\csc h\, x - \int \csc h\, x \cdot \csc h^2 x\, dx \;=\; -\coth x \csc h\, x$

$-\int \csc h^3 x\, dx$. Combining the terms we have

$\int \csc h^3 x\, dx \;=\; \int \coth x \cdot \coth x \csc h\, x\, dx - \int \csc h\, x\, dx \;=\; -\coth x \csc h\, x - \int \csc h^3 x\, dx - \int \csc h\, x\, dx$. Moving the $\int \csc h^3 x \, dx$ term from the right hand side of the equation to the left hand side we obtain

$\int \csc h^3 x \, dx + \int \csc h^3 x \, dx = 2\int \csc h^3 x \, dx = -\coth x \csc h \, x - \int \csc h \, x \, dx$. Therefore,

$$\boxed{\int \csc h^3 x \, dx} = \boxed{\frac{1}{2}\left(-\coth x \csc h \, x - \int \csc h \, x \, dx\right)} = \boxed{-\frac{1}{2}\coth x \csc h \, x - \frac{1}{2}\ln\left|\tanh \frac{x}{2}\right| + c}$$

Check: Let $y = -\frac{1}{2}\coth x \csc h \, x - \frac{1}{2}\ln\left|\tanh \frac{x}{2}\right| + c$, then $y' = \dfrac{-\left(-\csc h^2 x \cdot \csc h \, x - \csc h \, x \coth x \cdot \coth x\right)}{2}$

$-\dfrac{\sec h^2 \frac{x}{2}}{4\left(\tanh \frac{x}{2}\right)} + 0 = \dfrac{\csc h^3 x + \csc h \, x \coth^2 x}{2} - \dfrac{\sec h^2 \frac{x}{2}}{4\left(\tanh \frac{x}{2}\right)}$. The 2nd term is simplified as follows:

$$\dfrac{\sec h^2 \frac{x}{2}}{4\left(\tanh \frac{x}{2}\right)} = \frac{1}{4} \cdot \dfrac{1 - \tanh^2 \frac{x}{2}}{\tanh \frac{x}{2}} = \frac{1}{4} \cdot \dfrac{1 - \frac{\sinh^2 \frac{x}{2}}{\cosh^2 \frac{x}{2}}}{\frac{\sinh \frac{x}{2}}{\cosh \frac{x}{2}}} = \frac{1}{4} \cdot \dfrac{\frac{\cosh^2 \frac{x}{2} - \sinh^2 \frac{x}{2}}{\cosh^2 \frac{x}{2}}}{\frac{\sinh \frac{x}{2}}{\cosh \frac{x}{2}}} = \frac{1}{4} \cdot \dfrac{\frac{1}{\cosh^2 \frac{x}{2}}}{\frac{\sinh \frac{x}{2}}{\cosh \frac{x}{2}}} = \frac{1}{4} \cdot \dfrac{\cosh \frac{x}{2}}{\cosh^2 \frac{x}{2} \cdot \sinh \frac{x}{2}}$$

$$= \dfrac{1}{2 \cdot 2\cosh \frac{x}{2} \cdot \sinh \frac{x}{2}} = \dfrac{1}{2 \cdot \sinh 2 \cdot \frac{x}{2}} = \frac{1}{2}\dfrac{1}{\sinh x} = \dfrac{\csc h \, x}{2}$$. Therefore,

$$\dfrac{\csc h^3 x + \csc h \, x \coth^2 x}{2} - \dfrac{\sec h^2 \frac{x}{2}}{4\left(\tanh \frac{x}{2}\right)} = \dfrac{\csc h^3 x + \csc h \, x \coth^2 x}{2} - \dfrac{\csc h \, x}{2}$$

$$= \dfrac{\csc h^3 x + \csc h \, x \coth^2 x - \csc h \, x}{2} = \dfrac{\csc h^3 x + \csc h \, x \left(\coth^2 x - 1\right)}{2} = \dfrac{\csc h^3 x + \csc h \, x \csc h^2 x}{2}$$

$$= \dfrac{\sec h^3 x + \sec h^3 x}{2} = \dfrac{2\csc h^3 x}{2} = \csc h^3 x$$

Example 5.4-7: Evaluate the following indefinite integrals:

a. $\int \sinh^2 x \cosh^2 x \, dx =$ b. $\int \sinh^2 x \cosh^5 x \, dx =$ c. $\int \sinh^4 x \cosh^2 x \, dx =$

d. $\int \sinh^3 x \cosh^2 x \, dx =$ e. $\int \tanh^5 x \sec h^4 x \, dx =$ f. $\int \tanh^3 x \sec h^3 x \, dx =$

g. $\int \coth^2 x \csc h^2 x \, dx =$ h. $\int \coth^3 x \csc h^3 x \, dx =$ i. $\int \coth^3 x \csc h^4 x \, dx =$

Solutions:

a. $\boxed{\int \sinh^2 x \cosh^2 x \, dx} = \boxed{\frac{1}{4}\int \sinh^2 2x \, dx} = \boxed{\frac{1}{4}\int \frac{1}{2}\left(\cosh 4x - 1\right)dx} = \boxed{\frac{1}{8}\int \left(\cosh 4x - 1\right)dx} = \boxed{\frac{1}{8}\int \cosh 4x \, dx - \frac{1}{8}\int dx}$

$= \boxed{\frac{1}{8} \cdot \frac{1}{4}\sinh 4x - \frac{x}{8} + c} = \boxed{\frac{1}{32}\sinh 4x - \frac{1}{8}x + c}$

Check: Let $y = \frac{1}{32}\sinh 4x - \frac{1}{8}x + c$, then $y' = \dfrac{4\cosh 4x}{32} - \frac{1}{8} + 0 = \frac{1}{8}\cosh 4x - \frac{1}{8} = \frac{1}{8}\left(\cosh 4x - 1\right)$

$$= \frac{1}{4}\cdot\frac{1}{2}\left(\cosh 4x-1\right) = \frac{1}{4}\cdot\sinh^2 2x = \sinh^2 x\cosh^2 x$$

b. $\boxed{\int \sinh^2 x\cosh^5 x\,dx} = \boxed{\int \sinh^2 x\cosh^4 x\cosh x\,dx} = \boxed{\int \sinh^2 x\left(\cosh^2 x\right)^2\cosh x\,dx} = \boxed{\int \sinh^2 x\left(1+\sinh^2 x\right)^2\cos x\,dx}$

$= \boxed{\int \sinh^2 x\left(1+\sinh^4 x+2\sinh^2 x\right)\cosh x\,dx} = \boxed{\int\left(\sinh^2 x\cosh x+\sinh^6 x\cosh x+2\sinh^4 x\cosh x\right)dx}$

$= \boxed{\int \sinh^2 x\cosh x\,dx+\int \sinh^6 x\cosh x\,dx+2\int \sinh^4 x\cosh x\,dx} = \boxed{\frac{1}{3}\sinh^3 x+\frac{1}{7}\sinh^7 x+\frac{2}{5}\sinh^5 x+c}$

Check: Let $y=\frac{1}{3}\sinh^3 x+\frac{1}{7}\sinh^7 x+\frac{2}{5}\sinh^5 x+c$, then $y'=\dfrac{3\sinh^2 x\cosh x}{3}+\dfrac{7\sinh^6 x\cosh x}{7}$

$+\dfrac{10\sinh^4 x\cosh x}{5} = \sinh^2 x\cosh x+\sinh^6 x\cosh x+2\sinh^4 x\cosh x = \sinh^2 x\cosh x\left(1+\sinh^4 x+2\sinh^2 x\right)$

$= \sinh^2 x\cosh x\left(1+\sinh^2 x\right)^2 = \sinh^2 x\cosh x\left(\cosh^2 x\right)^2 = \sinh^2 x\cosh x\cosh^4 x = \sinh^2 x\cosh^5 x$

c. $\boxed{\int \sinh^4 x\cosh^2 x\,dx} = \boxed{\int\left(\sinh x\cosh x\right)^2\sinh^2 x\,dx} = \boxed{\int\left(\frac{1}{2}\sinh 2x\right)^2\cdot\frac{1}{2}\left(\cosh 2x-1\right)dx} = \boxed{-\frac{1}{8}\int \sinh^2 2x\,dx}$

$\boxed{+\frac{1}{8}\int \sinh^2 2x\cosh 2x\,dx} = \boxed{-\frac{1}{8}\int\frac{1}{2}\left(\cosh 4x-1\right)dx+\frac{1}{8}\int \sinh^2 2x\cosh 2x\,dx} = \boxed{\frac{1}{16}\int dx-\frac{1}{16}\int\cosh 4x\,dx}$

$= \boxed{+\frac{1}{8}\cdot\frac{1}{6}\sinh^3 2x} = \boxed{\frac{1}{16}x-\frac{1}{16}\cdot\frac{1}{4}\sinh 4x+\frac{1}{48}\sinh^3 2x+c} = \boxed{\frac{1}{16}x-\frac{1}{64}\sinh 4x+\frac{1}{48}\sinh^3 2x+c}$

Check: Let $y=\frac{1}{16}x-\frac{1}{64}\sinh 4x+\frac{1}{48}\sinh^3 2x+c$, then $y'=\frac{1}{16}-\dfrac{4\cosh 4x}{64}+\dfrac{6\sinh^2 2x\cosh 2x}{48}+0$

$= -\frac{1}{16}\left(\cosh 4x-1\right)+\frac{1}{8}\sinh^2 2x\cos 2x = -\frac{1}{8}\cdot\frac{1}{2}\left(\cosh 4x-1\right)+\frac{1}{8}\sinh^2 2x\cosh 2x$

$= -\frac{1}{8}\cdot\sinh^2 2x+\frac{1}{8}\sinh^2 2x\cosh 2x = \left(\frac{1}{2}\sinh 2x\right)^2\cdot\frac{1}{2}\left(\cosh 2x-1\right) = \left(\sinh x\cosh x\right)^2\sinh^2 x$

$= \sinh^2 x\cosh^2 x\cdot\sinh^2 x = \sinh^4 x\cosh^2 x$

d. $\int \sinh^3 x\cosh^2 x\,dx = \int \sinh^3 x\left(1+\sinh^2 x\right)dx = \int\left(\sinh^3 x+\sinh^5 x\right)dx = \int \sinh^3 x\,dx+\int \sinh^5 x\,dx$. In

Example, 5.4-5, problem letters f and g, we found $\int \sinh^5 x\,dx = \frac{1}{5}\cosh^5 x-\frac{2}{3}\cosh^3 x+\cosh x+c$ and

$\int \sinh^3 x\,dx = \frac{1}{3}\cosh^3 x-\cosh x+c$. Therefore,

$\boxed{\int \sinh^3 x\cosh^2 x\,dx} = \boxed{\int \sinh^3 x\,dx+\int \sinh^5 x\,dx} = \boxed{\frac{1}{3}\cosh^3 x-\cosh x+c+\left(\frac{1}{5}\cosh^5 x-\frac{2}{3}\cosh^3 x+\cosh x+c\right)}$

$$= \boxed{\frac{1}{5}\cosh^5 x + \frac{1}{3}\cosh^3 x - \frac{2}{3}\cosh^3 x - \cosh x + \cosh x + c} = \boxed{\frac{1}{5}\cosh^5 x - \frac{1}{3}\cosh^3 x + c}$$

Check: Let $y = \frac{1}{5}\cosh^5 x - \frac{1}{3}\cosh^3 x + c$, then $y' = \frac{1}{5}\cdot 5\cosh^4 x\cdot\sinh x - \frac{3}{3}\cdot\cosh^2 x\cdot\sinh x + 0$

$$= \sinh x\cosh^4 x - \sinh x\cosh^2 x = \sinh x\cosh^2 x\left(\cosh^2 x - 1\right) = \sinh x\cosh^2 x\sinh^2 x = \sinh^3 x\cosh^2 x$$

e. $\boxed{\int \tanh^5 x\sec h^4 x\, dx} = \boxed{\int \tanh^5 x\sec h^2 x\sec h^2 x\, dx} = \boxed{\int \tanh^5 x\left(1 - \tanh^2 x\right)\sec h^2 x\, dx} = \boxed{-\int \tanh^7 x\sec h^2 x\, dx}$

$\boxed{+\int \tanh^5 x\sec h^2 x\, dx} = \boxed{-\frac{1}{8}\tanh^8 x + \frac{1}{6}\tanh^6 x + c}$

Check: Let $y = -\frac{1}{8}\tanh^8 x + \frac{1}{6}\tanh^6 x + c$, then $y' = -\frac{1}{8}\cdot 8\tanh^7 x\cdot\sec h^2 x + \frac{1}{6}\cdot 6\tanh^5 x\cdot\sec h^2 x + 0$

$$= -\tanh^7 x\sec h^2 x + \tanh^5 x\sec h^2 x = \tanh^5 x\sec h^2 x\left(1 - \tanh^2 x\right) = \tanh^5 x\sec h^2 x\sec h^2 x$$

$$= \tanh^5 x\sec h^4 x$$

f. $\int \tanh^3 x\sec h^3 x\, dx = \int \tanh^2 x\sec h^2 x\cdot\tanh x\sec h x\, dx = \int\left(1 - \sec h^2 x\right)\sec h^2 x\cdot\tanh x\sec h x\, dx$

$$= -\int \sec h^4 x\cdot\tanh x\sec h x\, dx + \int \sec h^2 x\cdot\tanh x\sec h x\, dx. \text{ To solve the first and the second integral}$$

let $u = \sec h x$, then $\dfrac{du}{dx} = \dfrac{d}{dx}\sec h x$; $\dfrac{du}{dx} = -\sec h x\tanh x$; $dx = -\dfrac{du}{\sec h x\tanh x}$. Therefore,

$\boxed{\int \tanh^3 x\sec h^3 x\, dx} = \boxed{\int \tanh^2 x\sec h^2 x\cdot\tanh x\sec h x\, dx} = \boxed{-\int \sec h^4 x\cdot\tanh x\sec h x\, dx}$

$\boxed{+\int \sec h^2 x\cdot\tanh x\sec h x\, dx} = \boxed{-\int u^4\cdot-\frac{\tanh x\sec h x}{\tanh x\sec h x}\, du + \int u^2\cdot-\frac{\tanh x\sec h x}{\tanh x\sec h x}\, du} = \boxed{\int u^4\, du - \int u^2\, du}$

$$= \boxed{\frac{1}{5}u^5 - \frac{1}{3}u^3 + c} = \boxed{\frac{1}{5}\sec h^5 x - \frac{1}{3}\sec h^3 x + c}$$

Check: Let $y = \frac{1}{5}\sec h^5 x - \frac{1}{3}\sec h^3 x + c$, then $y' = \frac{5}{5}\sec h^4 x\cdot-\sec h x\tanh x - \frac{3}{3}\sec h^2 x\cdot-\sec h x\tanh x$

$$= -\sec h^4 x\cdot\sec h x\tanh x + \sec h^2 x\cdot\sec h x\tanh x = \left(-\sec h^4 x + \sec h^2 x\right)\sec h x\tanh x$$

$$= \left(1 - \sec h^2 x\right)\sec h^2 x\sec h x\tanh x = \tanh^2 x\sec h^2 x\sec h x\tanh x = \tanh^3 x\sec h^3 x$$

g. Given $\int \coth^2 x\csc h^2 x\, dx$ let $u = \coth x$, then $\dfrac{du}{dx} = \dfrac{d}{dx}\coth x$; $\dfrac{du}{dx} = -\csc h^2 x$; $dx = \dfrac{-du}{\csc h^2 x}$. Thus,

$\boxed{\int \coth^2 x\csc h^2 x\, dx} = \boxed{\int \frac{u^2\csc h^2 x}{-\csc h^2 x}\, du} = \boxed{-\int u^2\, du} = \boxed{-\frac{1}{3}u^3 + c} = \boxed{-\frac{1}{3}\coth^3 x + c}$

Check: Let $y=-\frac{1}{3}\coth^3 x+c$, then $y' = -\frac{1}{3}\cdot 3\coth^2 x\cdot-\csc h^2 x+0 = \frac{3}{3}\cdot\coth^2 x\csc h^2 x = \coth^2 x\csc h^2 x$

h. $\boxed{\int\coth^3 x\csc h^3 x\,dx} = \boxed{\int\coth x\coth^2 x\csc h^3 x\,dx} = \boxed{\int\coth x\left(1+\csc h^2 x\right)\csc h^2 x\,dx} = \boxed{\int\csc h^4 x\coth x\,dx}$

$\boxed{+\int\csc h^2 x\coth x\,dx} = \boxed{-\frac{1}{5}\csc h^5 x-\frac{1}{3}\csc h^3 x+c}$

Check: Let $y=-\frac{1}{5}\csc h^5 x-\frac{1}{3}\csc h^3 x+c$, then $y' = -\frac{5}{5}\csc h^4 x\cdot-\csc h x\coth x-\frac{3}{3}\csc h^2 x\cdot-\csc h x\coth x$

$= \csc h^5 x\coth x+\csc h^3 x\coth x = \csc h^3 x\coth x\left(\csc h^2 x+1\right) = \csc h^3 x\coth x\coth^2 x = \coth^3 x\csc h^3 x$

i. $\boxed{\int\coth^3 x\csc h^4 x\,dx} = \boxed{\int\coth^3 x\csc h^2 x\csc h^2 x\,dx} = \boxed{\int\coth^3 x\left(\coth^2 x-1\right)\csc h^2 x\,dx} = \boxed{\int\coth^5 x\csc h^2 x\,dx}$

$\boxed{-\int\coth^3 x\csc h^2 x\,dx} = \boxed{-\frac{1}{6}\coth^6 x+\frac{1}{4}\coth^4 x+c}$

Check: Let $y=-\frac{1}{6}\coth^6 x+\frac{1}{4}\coth^4 x+c$, then $y' = -\frac{1}{6}\cdot 6\coth^5 x\cdot-\csc h^2 x+\frac{1}{4}\cdot 4\coth^3 x\cdot-\csc h^2 x+0$

$= \coth^5 x\csc h^2 x-\coth^3 x\csc h^2 x = \coth^3 x\csc h^2 x\left(\coth^2 x-1\right) = \coth^3 x\csc h^2 x\csc h^2 x$

$= \coth^3 x\csc h^4 x$

Table 5.4-3 provides a summary of the basic integration formulas covered in this book.

Section 5.4 Practice Problems – Integration of Hyperbolic Functions

1. Evaluate the following integrals:

a. $\int\cosh 3x\,dx =$

b. $\int\left(\sinh 2x-e^{3x}\right)dx =$

c. $\int\csc h\,5x\,dx =$

d. $\int x^2\sec h^2 x^3\,dx =$

e. $\int\frac{2}{3}x^3\csc h^2\left(x^4+1\right)dx =$

f. $\int x^3\csc h\left(2x^4+5\right)dx =$

g. $\int\cosh^7(x+1)\sinh(x+1)\,dx =$

h. $\int\csc h\,(5x+3)\coth(5x+3)\,dx =$

i. $\int e^{x+1}\sec h\,e^{x+1}\,dx =$

2. Evaluate the following integrals:

a. $\int\tanh^5 x\sec h^2 x\,dx =$

b. $\int\coth^6(x+1)\csc h^2(x+1)\,dx =$

c. $\int e^{3x}\tanh e^{3x}\,dx =$

d. $\int x^3\sec h\left(x^4+1\right)dx =$

e. $\int\sec h\,(3x+2)\,dx =$

f. $\int e^{\cosh(3x+5)}\sinh(3x+5)\,dx =$

g. $\int\tanh^5 x\,dx =$

h. $\int\coth^5 x\,dx =$

i. $\int\coth^6 x\,dx =$

Table 5.4-3: Basic Integration Formulas

1. $\int a\,dx = ax+c \qquad\qquad a\neq 0$	2. $\int x^n dx = \dfrac{x^{n+1}}{n+1}+c \qquad\qquad n\neq -1$
3. $\int a\,f(x)\,dx = a\int f(x)\,dx \qquad a\neq 0$	4. $\int\left[f(x)+g(x)\right]dx = \int f(x)\,dx+\int g(x)\,dx$
5. $\int \sin x\,dx = -\cos x+c$	6. $\int \cos x\,dx = \sin x+c$
7. $\int \tan x\,dx = \ln\left|\sec x\right|+c$	8. $\int \cot x\,dx = \ln\left|\sin x\right|+c$
9. $\int \sec x\,dx = \ln\left|\sec x+\tan x\right|+c$	10. $\int \csc x\,dx = \ln\left|\csc x-\cot x\right|+c$
11. $\int \tan x\sec x\,dx = \sec x+c$	12. $\int \cot x\csc x\,dx = -\csc x+c$
13. $\int \sin^2 x\,dx = \dfrac{x}{2}-\dfrac{\sin 2x}{4}+c$	14. $\int \cos^2 x\,dx = \dfrac{x}{2}+\dfrac{\sin 2x}{4}+c$
15. $\int \tan^2 x\,dx = \tan x-x+c$	16. $\int \tan^2 x\,dx = \tan x-x+c$
17. $\int \cot^2 x\,dx = -\cot x-x+c$	18. $\int \sec^2 x\,dx = \tan x+c$
19. $\int \csc^2 x\,dx = -\cot x+c$	20. $\int \dfrac{1}{x}\,dx = \ln\left|x\right|+c$
21. $\int \ln x\,dx = x\ln\left|x\right|-x+c$	22. $\int a^x dx = \dfrac{a^x}{\ln a}+c \qquad a\rangle 0\ and\ a\neq 1$
23. $\int e^x dx = e^x +c$	24. $\int \dfrac{1}{\sqrt{a^2-x^2}}\,dx = \dfrac{1}{a}arc\sin\dfrac{x}{a}+c = \dfrac{1}{a}\sin^{-1}\dfrac{x}{a}+c$
25. $\int \dfrac{1}{a^2+x^2}\,dx = \dfrac{1}{a}arc\tan\dfrac{x}{a}+c = \dfrac{1}{a}\tan^{-1}\dfrac{x}{a}+c$	26. $\int \dfrac{1}{a^2-x^2}\,dx = \dfrac{1}{2a}\ln\left|\dfrac{a+x}{a-x}\right|+c$
27. $\int \dfrac{1}{x^2-a^2}\,dx = \dfrac{1}{2a}\ln\left|\dfrac{x-a}{x+a}\right|+c$	28. $\int \dfrac{1}{x\sqrt{x^2-a^2}}\,dx = \dfrac{1}{a}arc\sec\dfrac{x}{a}+c = \dfrac{1}{a}\sec^{-1}\dfrac{x}{a}+c$
29. $\int \sinh x\,dx = \cosh x+c$	30. $\int \cosh x\,dx = \sinh x+c$
31. $\int \tanh x\,dx = \ln\cosh x+c$	32. $\int \coth x\,dx = \ln\left|\sinh x\right|+c$
33. $\int \sec h\,x\,dx = \sin^{-1}(\tanh x)+c$	34. $\int \csc h\,x\,dx = \ln\left|\tanh\dfrac{x}{2}\right|+c$
35. $\int \tanh x\sec h\,x\,dx = -\sec h\,x+c$	36. $\int \coth x\csc h\,x\,dx = -\csc h\,x+c$
37. $\int \sinh^2 x\,dx = \dfrac{\sinh 2x}{4}-\dfrac{x}{2}+c$	38. $\int \cosh^2 x\,dx = \dfrac{\sinh 2x}{4}+\dfrac{x}{2}+c$
39. $\int \tanh^2 x\,dx = x-\tanh x+c$	40. $\int \coth^2 x\,dx = x-\coth x+c$
41. $\int \sec h^2 x\,dx = \tanh x+c$	42. $\int \csc h^2 x\,dx = -\coth x+c$

Appendix - Exercise Solutions
Chapter 1 Solutions:

1. List the first four and tenth terms of the given sequence.

a. Given $a_n = \dfrac{2n+1}{-2n}$, then
$a_1 = \dfrac{2\cdot1+1}{-2\cdot1} = -\dfrac{3}{2} = -1.5$ $a_2 = \dfrac{2\cdot2+1}{-2\cdot2} = -\dfrac{5}{4} = -1.25$

$a_3 = \dfrac{2\cdot3+1}{-2\cdot3} = -\dfrac{7}{6} = -1.17$ $a_4 = \dfrac{2\cdot4+1}{-2\cdot4} = -\dfrac{9}{8} = -1.13$ $a_{10} = \dfrac{2\cdot10+1}{-2\cdot10} = -\dfrac{21}{20} = -1.05$

b. Given $b_k = \dfrac{k(k+1)}{k^2}$, then
$b_1 = \dfrac{1\cdot(1+1)}{1^2} = \dfrac{1\cdot2}{1} = \dfrac{2}{1} = 2$ $b_2 = \dfrac{2\cdot(2+1)}{2^2} = \dfrac{2\cdot3}{4} = \dfrac{6}{4} = 1.5$

$b_3 = \dfrac{3\cdot(3+1)}{3^2} = \dfrac{3\cdot4}{9} = \dfrac{12}{9} = 1.33$ $b_4 = \dfrac{4\cdot(4+1)}{4^2} = \dfrac{4\cdot5}{16} = \dfrac{20}{16} = 1.25$ $b_{10} = \dfrac{10\cdot(10+1)}{10^2} = \dfrac{110}{100} = 1.1$

c. Given $d_n = 3-(-2)^n$, then
$d_1 = 3-(-2)^1 = 3-(-2) = 3+2 = 5$

$d_2 = 3-(-2)^2 = 3-(+4) = 3-4 = -1$ $d_3 = 3-(-2)^3 = 3-(-8) = 3+8 = 11$

$d_4 = 3-(-2)^4 = 3-(+16) = 3-16 = -13$ $d_{10} = 3-(-2)^{10} = 3-(+1024) = 3-1024 = -1021$

d. Given $k_n = \left(-\dfrac{1}{2}\right)^n \dfrac{(-1)^{n+1}}{n+2}$, then
$k_1 = \left(-\dfrac{1}{2}\right)^1 \dfrac{(-1)^{1+1}}{1+2} = -\dfrac{1}{2}\cdot\dfrac{(-1)^2}{3} = -\dfrac{1}{2}\cdot\dfrac{1}{3} = -\dfrac{1}{6}$

$k_2 = \left(-\dfrac{1}{2}\right)^2 \dfrac{(-1)^{2+1}}{2+2} = \dfrac{1}{4}\cdot\dfrac{(-1)^3}{4} = \dfrac{1}{4}\cdot-\dfrac{1}{4} = -\dfrac{1}{16}$ $k_3 = \left(-\dfrac{1}{2}\right)^3 \dfrac{(-1)^{3+1}}{3+2} = -\dfrac{1}{8}\cdot\dfrac{(-1)^4}{5} = -\dfrac{1}{8}\cdot\dfrac{1}{5} = -\dfrac{1}{40}$

$k_4 = \left(-\dfrac{1}{2}\right)^4 \dfrac{(-1)^{4+1}}{4+2} = \dfrac{1}{16}\cdot\dfrac{(-1)^5}{6} = \dfrac{1}{16}\cdot-\dfrac{1}{6} = -\dfrac{1}{96}$ $k_{10} = \left(-\dfrac{1}{2}\right)^{10} \dfrac{(-1)^{10+1}}{10+2} = \dfrac{1}{1024}\cdot\dfrac{(-1)^{11}}{12} = \dfrac{-1}{12{,}288}$

2. Write s_3, s_4, s_5, s_8, and s_{10} for the following sequences.

a. Given $s_n = \dfrac{n(n+1)}{2n^{-1}}$, then
$s_3 = \dfrac{3\cdot(3+1)}{2\cdot3^{-1}} = \dfrac{3\cdot4}{2\cdot3^{-1}} = \dfrac{3\cdot4\cdot3}{2} = \dfrac{36}{2} = 18$

$s_4 = \dfrac{4\cdot(4+1)}{2\cdot4^{-1}} = \dfrac{4\cdot5}{2\cdot4^{-1}} = \dfrac{4\cdot5\cdot4}{2} = \dfrac{80}{2} = 40$ $s_5 = \dfrac{5\cdot(5+1)}{2\cdot5^{-1}} = \dfrac{5\cdot6}{2\cdot5^{-1}} = \dfrac{5\cdot6\cdot5}{2} = \dfrac{150}{2} = 75$

$s_8 = \dfrac{8\cdot(8+1)}{2\cdot8^{-1}} = \dfrac{8\cdot9}{2\cdot8^{-1}} = \dfrac{8\cdot9\cdot8}{2} = \dfrac{576}{2} = 288$ $s_{10} = \dfrac{10\cdot(10+1)}{2\cdot10^{-1}} = \dfrac{10\cdot11}{2\cdot10^{-1}} = \dfrac{10\cdot11\cdot10}{2} = \dfrac{1100}{2} = 550$

b. Given $s_n = (-1)^{n+1}2^{n-2}$, then
$s_3 = (-1)^{3+1}\cdot2^{3-2} = (-1)^4\cdot2^1 = 1\cdot2 = 2$

$s_4 = (-1)^{4+1}\cdot2^{4-2} = (-1)^5\cdot2^2 = -1\cdot4 = -4$ $s_5 = (-1)^{5+1}\cdot2^{5-2} = (-1)^6\cdot2^3 = 1\cdot8 = 8$

$s_8 = (-1)^{8+1}\cdot2^{8-2} = (-1)^9\cdot2^6 = -1\cdot64 = -64$ $s_{10} = (-1)^{10+1}\cdot2^{10-2} = (-1)^{11}\cdot2^8 = -1\cdot256 = -256$

c. Given $s_n = \dfrac{(-2)^{n+1}(n-2)}{2n}$, then
$s_3 = \dfrac{(-2)^{3+1}(3-2)}{2\cdot3} = \dfrac{(-2)^4\cdot1}{6} = \dfrac{16}{6} = 2.67$

$s_4 = \dfrac{(-2)^{4+1}(4-2)}{2\cdot4} = \dfrac{(-2)^5\cdot2}{8} = \dfrac{-32\cdot2}{8} = -8$ $s_5 = \dfrac{(-2)^{5+1}(5-2)}{2\cdot5} = \dfrac{(-2)^6\cdot3}{10} = \dfrac{64\cdot3}{10} = 19.2$

$s_8 = \dfrac{(-2)^{8+1}(8-2)}{2\cdot8} = \dfrac{(-2)^9\cdot6}{16} = \dfrac{-512\cdot6}{16} = -192$ $s_{10} = \dfrac{(-2)^{10+1}(10-2)}{2\cdot10} = \dfrac{(-2)^{11}\cdot8}{20} = \dfrac{-2048\cdot8}{20} = -819.2$

3. Write the first five terms of the following sequences.

a. Given $a_n = (-1)^{n+1}(n+2)$, then $a_1 = (-1)^{1+1}(1+2) = (-1)^2 \cdot 3 = 1 \cdot 3 = \mathbf{3}$

$a_2 = (-1)^{2+1}(2+2) = (-1)^3 \cdot 4 = -1 \cdot 4 = \mathbf{-4}$ $a_3 = (-1)^{3+1}(3+2) = (-1)^4 \cdot 5 = 1 \cdot 5 = \mathbf{5}$

$a_4 = (-1)^{4+1}(4+2) = (-1)^5 \cdot 6 = -1 \cdot 6 = \mathbf{-6}$ $a_5 = (-1)^{5+1}(5+2) = (-1)^6 \cdot 7 = 1 \cdot 7 = \mathbf{7}$

b. Given $a_i = 3\left(\dfrac{1}{100}\right)^{i-2}$, then $a_1 = 3\left(\dfrac{1}{100}\right)^{1-2} = 3\left(\dfrac{1}{100}\right)^{-1} = 3 \cdot \dfrac{1}{\frac{1}{100}} = 3 \cdot \dfrac{100}{1} = \mathbf{300}$

$a_2 = 3\left(\dfrac{1}{100}\right)^{2-2} = 3\left(\dfrac{1}{100}\right)^{0} = 3 \cdot 1 = \mathbf{3}$ $a_3 = 3\left(\dfrac{1}{100}\right)^{3-2} = 3\left(\dfrac{1}{100}\right)^{1} = 3 \cdot \dfrac{1}{100^1} = \dfrac{\mathbf{3}}{\mathbf{100}}$

$a_4 = 3\left(\dfrac{1}{100}\right)^{4-2} = 3\left(\dfrac{1}{100}\right)^{2} = 3 \cdot \dfrac{1}{100^2} = \dfrac{\mathbf{3}}{\mathbf{10,000}}$ $a_5 = 3\left(\dfrac{1}{100}\right)^{5-2} = 3\left(\dfrac{1}{100}\right)^{3} = 3 \cdot \dfrac{1}{100^3} = \dfrac{\mathbf{3}}{\mathbf{1,000,000}}$

c. Given $c_i = 3\left(-\dfrac{1}{5}\right)^{i-1}$, then $c_1 = 3\left(-\dfrac{1}{5}\right)^{1-1} = 3\left(-\dfrac{1}{5}\right)^{0} = 3 \cdot 1 = \mathbf{3}$

$c_2 = 3\left(-\dfrac{1}{5}\right)^{2-1} = 3\left(-\dfrac{1}{5}\right)^{1} = 3 \cdot -\dfrac{1}{5} = -\dfrac{3}{5} = \mathbf{-0.6}$ $c_3 = 3\left(-\dfrac{1}{5}\right)^{3-1} = 3\left(-\dfrac{1}{5}\right)^{2} = 3 \cdot \dfrac{1}{5^2} = \dfrac{3}{25} = \mathbf{0.12}$

$c_4 = 3\left(-\dfrac{1}{5}\right)^{4-1} = 3\left(-\dfrac{1}{5}\right)^{3} = 3 \cdot -\dfrac{1}{5^3} = -\dfrac{3}{125} = \mathbf{-0.024}$ $c_5 = 3\left(-\dfrac{1}{5}\right)^{5-1} = 3\left(-\dfrac{1}{5}\right)^{4} = 3 \cdot \dfrac{1}{5^4} = \dfrac{3}{625} = \mathbf{0.0048}$

d. Given $a_n = (3n-5)^2$, then $a_1 = (3 \cdot 1 - 5)^2 = (3-5)^2 = (-2)^2 = \mathbf{4}$

$a_2 = (3 \cdot 2 - 5)^2 = (6-5)^2 = 1^2 = \mathbf{1}$ $a_3 = (3 \cdot 3 - 5)^2 = (9-5)^2 = 4^2 = \mathbf{16}$

$a_4 = (3 \cdot 4 - 5)^2 = (12-5)^2 = 7^2 = \mathbf{49}$ $a_5 = (3 \cdot 5 - 5)^2 = (15-5)^2 = 10^2 = \mathbf{100}$

e. Given $u_k = ar^{k-2} + 2$, then $u_1 = ar^{1-2} + 2 = ar^{-1} + 2 = \dfrac{\mathbf{a}}{\mathbf{r}} + \mathbf{2}$

$u_2 = ar^{2-2} + 2 = ar^0 + 2 = \mathbf{a+2}$ $u_3 = ar^{3-2} + 2 = ar^1 + 2 = \mathbf{ar+2}$

$u_4 = ar^{4-2} + 2 = \mathbf{ar^2 + 2}$ $u_5 = ar^{5-2} + 2 = \mathbf{ar^3 + 2}$

f. Given $b_k = -3\left(\dfrac{2}{3}\right)^{k-2}$, then $b_1 = -3 \cdot \left(\dfrac{2}{3}\right)^{1-2} = -3 \cdot \left(\dfrac{2}{3}\right)^{-1} = -3 \cdot \dfrac{1}{\frac{2}{3}} = -3 \cdot \dfrac{3}{2} = -\dfrac{\mathbf{9}}{\mathbf{2}}$

$b_2 = -3 \cdot \left(\dfrac{2}{3}\right)^{2-2} = -3 \cdot \left(\dfrac{2}{3}\right)^{0} = -3 \cdot 1 = \mathbf{-3}$ $b_3 = -3 \cdot \left(\dfrac{2}{3}\right)^{3-2} = -3 \cdot \dfrac{2}{3} = -\dfrac{3 \cdot 2}{3} = \mathbf{-2}$

$b_4 = -3 \cdot \left(\dfrac{2}{3}\right)^{4-2} = -3 \cdot \left(\dfrac{2}{3}\right)^{2} = -3 \cdot \dfrac{4}{9} = -\dfrac{\mathbf{4}}{\mathbf{3}}$ $b_5 = -3 \cdot \left(\dfrac{2}{3}\right)^{5-2} = -3 \cdot \left(\dfrac{2}{3}\right)^{3} = -3 \cdot \dfrac{8}{27} = -\dfrac{\mathbf{8}}{\mathbf{9}}$

g. Given $c_j = \dfrac{j}{j+1} + j$, then $c_1 = \dfrac{1}{1+1} + 1 = \dfrac{1}{2} + 1 = \dfrac{1+2}{2} = \dfrac{\mathbf{3}}{\mathbf{2}}$

$c_2 = \dfrac{2}{2+1} + 2 = \dfrac{2}{3} + 2 = \dfrac{2+6}{3} = \dfrac{\mathbf{8}}{\mathbf{3}}$ $c_3 = \dfrac{3}{3+1} + 3 = \dfrac{3}{4} + 3 = \dfrac{3+12}{4} = \dfrac{\mathbf{15}}{\mathbf{4}}$

$c_4 = \dfrac{4}{4+1} + 4 = \dfrac{4}{5} + 4 = \dfrac{4+20}{5} = \dfrac{\mathbf{24}}{\mathbf{5}}$ $c_5 = \dfrac{5}{5+1} + 5 = \dfrac{5}{6} + 5 = \dfrac{5+30}{6} = \dfrac{\mathbf{35}}{\mathbf{6}}$

h. Given $y_n = \left(1 - \dfrac{1}{n+2}\right)^{n+1}$, then $y_1 = \left(1 - \dfrac{1}{1+2}\right)^{1+1} = \left(1 - \dfrac{1}{3}\right)^{2} = \left(\dfrac{3-1}{3}\right)^{2} = \left(\dfrac{2}{3}\right)^{2} = \mathbf{0.67^2}$

$$y_2 = \left(1-\frac{1}{2+2}\right)^{2+1} = \left(1-\frac{1}{4}\right)^3 = \left(\frac{4-1}{4}\right)^3 = \left(\frac{3}{4}\right)^3 = \mathbf{0.75^3} \qquad y_3 = \left(1-\frac{1}{3+2}\right)^{3+1} = \left(1-\frac{1}{5}\right)^4 = \left(\frac{5-1}{5}\right)^4 = \left(\frac{4}{5}\right)^4 = \mathbf{0.8^4}$$

$$y_4 = \left(1-\frac{1}{4+2}\right)^{4+1} = \left(1-\frac{1}{6}\right)^5 = \left(\frac{6-1}{6}\right)^5 = \left(\frac{5}{6}\right)^5 = \mathbf{0.83^5} \qquad y_5 = \left(1-\frac{1}{5+2}\right)^{5+1} = \left(1-\frac{1}{7}\right)^6 = \left(\frac{7-1}{7}\right)^6 = \left(\frac{6}{7}\right)^6 = \mathbf{0.86^6}$$

i. Given $u_k = 1-(-1)^{k+1}$, then $u_1 = 1-(-1)^{1+1} = 1-(-1)^2 = 1-1 = \mathbf{0}$

$$u_2 = 1-(-1)^{2+1} = 1-(-1)^3 = 1-(-1) = 1+1 = \mathbf{2} \qquad u_3 = 1-(-1)^{3+1} = 1-(-1)^4 = 1-1 = \mathbf{0}$$

$$u_4 = 1-(-1)^{4+1} = 1-(-1)^5 = 1-(-1) = 1+1 = \mathbf{2} \qquad u_5 = 1-(-1)^{5+1} = 1-(-1)^6 = 1-1 = \mathbf{0}$$

j. Given $y_k = \dfrac{k}{2^{k-1}}$, then $y_1 = \dfrac{1}{2^{1-1}} = \dfrac{1}{2^0} = \dfrac{1}{1} = \mathbf{1}$

$$y_2 = \frac{2}{2^{2-1}} = \frac{2}{2^1} = \frac{2}{2} = \mathbf{1} \qquad y_3 = \frac{3}{2^{3-1}} = \frac{3}{2^2} = \frac{3}{4} = \mathbf{0.75}$$

$$y_4 = \frac{4}{2^{4-1}} = \frac{4}{2^3} = \frac{4}{8} = \mathbf{0.5} \qquad y_5 = \frac{5}{2^{5-1}} = \frac{5}{2^4} = \frac{5}{16} = \mathbf{0.313}$$

k. Given $y_n = 9^{\frac{1}{k}}(k-2)$, then $y_1 = 9^{\frac{1}{1}}(1-2) = 9\cdot-1 = \mathbf{-9}$

$$y_2 = 9^{\frac{1}{2}}(2-2) = 9^{\frac{1}{2}}\cdot 0 = \mathbf{0} \qquad y_3 = 9^{\frac{1}{3}}(3-2) = 9^{\frac{1}{3}}\cdot 1 = 9^{\frac{1}{3}} = \mathbf{\sqrt[3]{9}}$$

$$y_4 = 9^{\frac{1}{4}}(4-2) = 9^{\frac{1}{4}}\cdot 2 = \mathbf{2\sqrt[4]{9}} \qquad y_5 = 9^{\frac{1}{5}}(5-2) = 9^{\frac{1}{5}}\cdot 3 = \mathbf{3\sqrt[5]{9}}$$

l. Given $c_n = \dfrac{n^2-2}{n+1}$, then $c_1 = \dfrac{1^2-2}{1+1} = \dfrac{1-2}{2} = \mathbf{-\dfrac{1}{2}}$

$$c_2 = \frac{2^2-2}{2+1} = \frac{4-2}{3} = \mathbf{\frac{2}{3}} \qquad c_3 = \frac{3^2-2}{3+1} = \frac{9-2}{4} = \mathbf{\frac{7}{4}}$$

$$c_4 = \frac{4^2-2}{4+1} = \frac{16-2}{5} = \mathbf{\frac{14}{5}} \qquad c_5 = \frac{5^2-2}{5+1} = \frac{25-2}{6} = \mathbf{\frac{23}{6}}$$

4. Given $n!$ read as "n factorial" which is defined as $n! = n\,(n-1)(n-2)(n-3)\cdots 5\cdot 4\cdot 3\cdot 2\cdot 1$, find

a. $8! = 8\cdot 7\cdot 6\cdot 5\cdot 4\cdot 3\cdot 2\cdot 1 = \mathbf{40,320}$

b. Given $a_n = \dfrac{2n+1}{n!}$, then $a_1 = \dfrac{2\cdot1+1}{1!} = \dfrac{2+1}{1!} = \dfrac{3}{1} = \mathbf{3}$

$$a_2 = \frac{2\cdot2+1}{2!} = \frac{4+1}{2!} = \frac{5}{2\cdot1} = \frac{5}{2} = \mathbf{2.5} \qquad a_3 = \frac{2\cdot3+1}{3!} = \frac{6+1}{3!} = \frac{7}{3\cdot2\cdot1} = \frac{7}{6} = \mathbf{1.17}$$

$$a_4 = \frac{2\cdot4+1}{4!} = \frac{8+1}{4!} = \frac{9}{4\cdot3\cdot2\cdot1} = \frac{3}{8} = \mathbf{0.375}$$

c. Given $c_n = \dfrac{1+3^{n-1}}{(n!)^2}$, then $c_{10} = \dfrac{1+3^{10-1}}{(10!)^2} = \dfrac{1+3^9}{10!10!} = \mathbf{\dfrac{19,684}{10!10!}}$ and $c_{12} = \dfrac{1+3^{12-1}}{(12!)^2} = \dfrac{1+3^{11}}{12!12!} = \mathbf{\dfrac{177,148}{12!12!}}$

d. The first, fifth, tenth, and fifteenth terms of $y_n = \dfrac{n!\,(n-1)}{2+n!}$.

$$y_1 = \frac{1!\,(1-1)}{2+1!} = \frac{1!\cdot0}{2+1!} = \frac{0}{3} = \mathbf{0}$$

$$y_5 = \frac{5!\,(5-1)}{2+5!} = \frac{5!\cdot4}{2+5!} = \frac{(5\cdot4\cdot3\cdot2\cdot1)\cdot4}{2+(5\cdot4\cdot3\cdot2\cdot1)} = \frac{120\cdot4}{2+120} = \frac{480}{122} = \mathbf{3.934}$$

$$y_{10} = \frac{10!\,(10-1)}{2+10!} = \frac{10!\cdot9}{2+10!} = \frac{(10\cdot9\cdot8\cdot7\cdot6\cdot5\cdot4\cdot3\cdot2\cdot1)\cdot9}{2+(10\cdot9\cdot8\cdot7\cdot6\cdot5\cdot4\cdot3\cdot2\cdot1)} = \frac{3,628,800\cdot9}{2+3,628,800} = \frac{32,659,200}{3,628,802} = 8.9999 \approx \mathbf{9}$$

or, a quicker way of solving this problem is as follows: $y_{10} = \dfrac{10!\,(10-1)}{2+10!} = \dfrac{10!\cdot 9}{2+10!} \approx \dfrac{10!\cdot 9}{10!} = \dfrac{\cancel{10!}\cdot 9}{\cancel{10!}} = 9$

$y_{15} = \dfrac{15!\,(15-1)}{2+15!} = \dfrac{15!\cdot 14}{2+15!} = \dfrac{(15\cdot 14\cdot 13\cdot 12\cdot 11\cdot 10\cdot 9\cdot 8\cdot 7\cdot 6\cdot 5\cdot 4\cdot 3\cdot 2\cdot 1)\cdot 14}{2+(15\cdot 14\cdot 13\cdot 12\cdot 11\cdot 10\cdot 9\cdot 8\cdot 7\cdot 6\cdot 5\cdot 4\cdot 3\cdot 2\cdot 1)} \approx \mathbf{14}$ or

$y_{15} = \dfrac{15!\,(15-1)}{2+15!} = \dfrac{15!\cdot 14}{2+15!} \approx \dfrac{15!\cdot 14}{15!} = \dfrac{\cancel{15!}\cdot 14}{\cancel{15!}} = \mathbf{14}$

5. Write the first three terms of the following sequences.

a. Given $c_n = \dfrac{(2n-3)(n+1)}{(n-4)n}$, then

$c_1 = \dfrac{[(2\cdot 1)-3](1+1)}{(1-4)\cdot 1} = \dfrac{(2-3)\cdot 2}{-3\cdot 1} = \dfrac{-1\cdot 2}{-3} = \dfrac{2}{3}$

$c_2 = \dfrac{[(2\cdot 2)-3](2+1)}{(2-4)\cdot 2} = \dfrac{(4-3)\cdot 3}{-2\cdot 2} = \dfrac{1\cdot 3}{-4} = -\dfrac{3}{4}$

$c_3 = \dfrac{[(2\cdot 3)-3](3+1)}{(3-4)\cdot 3} = \dfrac{(6-3)\cdot 4}{-1\cdot 3} = \dfrac{3\cdot 4}{-3} = -\dfrac{12}{4} = \mathbf{-3}$

b. Given, $a_n = \left(\dfrac{1}{n-1}\right)\left(\dfrac{n-2}{2+n}\right)$, then

$a_1 = \left(\dfrac{1}{1-1}\right)\left(\dfrac{1-2}{2+1}\right) = \left(\dfrac{1}{0}\right)\left(\dfrac{-1}{3}\right)$ **which is undefined**

$a_2 = \left(\dfrac{1}{2-1}\right)\left(\dfrac{2-2}{2+2}\right) = \left(\dfrac{1}{1}\right)\left(\dfrac{0}{4}\right) = \mathbf{0}$

$a_3 = \left(\dfrac{1}{3-1}\right)\left(\dfrac{3-2}{2+3}\right) = \left(\dfrac{1}{2}\right)\left(\dfrac{1}{5}\right) = \dfrac{1}{10}$

c. Given $s_n = (-1)^{n+1}2^{n+1}$, then

$s_1 = (-1)^{1+1}\cdot 2^{1+1} = (-1)^2\cdot 2^2 = 1\cdot 4 = \mathbf{4}$

$s_2 = (-1)^{2+1}\cdot 2^{2+1} = (-1)^3\cdot 2^3 = -1\cdot 8 = \mathbf{-8}$

$s_3 = (-1)^{3+1}\cdot 2^{3+1} = (-1)^4\cdot 2^4 = 1\cdot 16 = \mathbf{16}$

d. Given $y_k = (-1)^{k+1}\dfrac{k(k-1)}{2}$, then

$y_1 = (-1)^{1+1}\cdot\dfrac{1\cdot(1-1)}{2} = (-1)^2\cdot\dfrac{1\cdot 0}{2} = \dfrac{0}{2} = \mathbf{0}$

$y_2 = (-1)^{2+1}\cdot\dfrac{2\cdot(2-1)}{2} = (-1)^3\cdot\dfrac{2\cdot 1}{2} = -\dfrac{2}{2} = \mathbf{1}$

$y_3 = (-1)^{3+1}\cdot\dfrac{3\cdot(3-1)}{2} = (-1)^4\cdot\dfrac{3\cdot 2}{2} = \dfrac{6}{2} = \mathbf{3}$

e. Given $b_n = n^2\left(\dfrac{n-1}{2+n}\right)$, then

$b1 = 1^2\cdot\left(\dfrac{1-1}{2+1}\right) = 1\cdot\left(\dfrac{0}{3}\right) = \mathbf{0}$

$b_2 = 2^2\cdot\left(\dfrac{2-1}{2+2}\right) = 4\cdot\left(\dfrac{1}{4}\right) = \dfrac{4}{4} = \mathbf{1}$

$b_3 = 3^2\cdot\left(\dfrac{3-1}{2+3}\right) = 9\cdot\left(\dfrac{2}{5}\right) = \dfrac{18}{5} = \mathbf{3.6}$

f. Given $x_a = (5-a)^{a+1}2^a$, then

$x_1 = (5-1)^{1+1}\cdot 2^1 = 4^2\cdot 2 = 16\cdot 2 = \mathbf{32}$

$x_2 = (5-2)^{2+1}\cdot 2^2 = 3^3\cdot 4 = 27\cdot 4 = \mathbf{108}$

$x_3 = (5-3)^{3+1}\cdot 2^3 = 2^4\cdot 8 = 16\cdot 8 = \mathbf{128}$

Section 1.2 Solutions - Series

1. Given $\displaystyle\sum_{i=1}^{n} a_i = 10$ and $\displaystyle\sum_{i=1}^{n} b_i = 25$, find

a. $\displaystyle\sum_{i=1}^{n}(2a_i + 4b_i) = \sum_{i=1}^{n}2a_i + \sum_{i=1}^{n}4b_i = 2\sum_{i=1}^{n}a_i + 4\sum_{i=1}^{n}b_i = (2\cdot 10)+(4\cdot 25) = 20+100 = \mathbf{120}$

b. $\displaystyle\sum_{i=1}^{n}(-a_i + b_i) = \sum_{i=1}^{n}-a_i + \sum_{i=1}^{n}b_i = -\sum_{i=1}^{n}a_i + \sum_{i=1}^{n}b_i = -10+25 = \mathbf{15}$

c. $\displaystyle\sum_{i=1}^{n}(3a_i + 5b_i) = \sum_{i=1}^{n}3a_i + \sum_{i=1}^{n}5b_i = 3\sum_{i=1}^{n}a_i + 5\sum_{i=1}^{n}b_i = (3\cdot 10)+(5\cdot 25) = 30+125 = \mathbf{155}$

d. $\displaystyle\sum_{i=1}^{n}\left(\dfrac{1}{2}a_i + \dfrac{1}{5}b_i\right) = \sum_{i=1}^{n}\dfrac{1}{2}a_i + \sum_{i=1}^{n}\dfrac{1}{5}b_i = \dfrac{1}{2}\sum_{i=1}^{n}a_i + \dfrac{1}{5}\sum_{i=1}^{n}b_i = \left(\dfrac{1}{2}\cdot 10\right)+\left(\dfrac{1}{5}\cdot 25\right) = 5+5 = \mathbf{10}$

2. Evaluate each of the following series.

a. $\sum_{k=1}^{5} 2+k = (2+1)+(2+2)+(2+3)+(2+4)+(2+5) = 3+4+5+6+7 = \mathbf{25}$

b. $\sum_{n=0}^{6}\frac{1}{(-2)^{n+1}} = \frac{1}{(-2)^{0+1}}+\frac{1}{(-2)^{1+1}}+\frac{1}{(-2)^{2+1}}+\frac{1}{(-2)^{3+1}}+\frac{1}{(-2)^{4+1}}+\frac{1}{(-2)^{5+1}}+\frac{1}{(-2)^{6+1}} = \frac{1}{(-2)^1}+\frac{1}{(-2)^2}+\frac{1}{(-2)^3}+\frac{1}{(-2)^4}$

$+\frac{1}{(-2)^5}+\frac{1}{(-2)^6}+\frac{1}{(-2)^7} = -\frac{1}{2}+\frac{1}{4}-\frac{1}{8}+\frac{1}{16}-\frac{1}{32}+\frac{1}{64}-\frac{1}{128} = -0.5+0.25-0.125+0.0625-0.031+0.016-0.008$

$= (-0.5-0.125-0.031-0.008)+(0.25+0.0625+0.016) = -0.664+0.328 = \mathbf{-0.336}$

c. $\sum_{n=0}^{4}(-1)^{n+1} = (-1)^{0+1}+(-1)^{1+1}+(-1)^{2+1}+(-1)^{3+1}+(-1)^{4+1} = (-1)^1+(-1)^2+(-1)^3+(-1)^4+(-1)^5 = -1+1-1+1-1 = \mathbf{-1}$

d. $\sum_{j=-3}^{3} j-3j^2 = [-3-3\cdot(-3)^2]+[-2-3\cdot(-2)^2]+[-1-3\cdot(-1)^2]+[0-3\cdot 0^2]+[1-3\cdot 1^2]+[2-3\cdot 2^2]+[3-3\cdot 3^2]$

$= [-3-(3\cdot 9)]+[-2-(3\cdot 4)]+[-1-(3\cdot 1)]+0+[1-(3\cdot 1)]+[2-(3\cdot 4)]+[3-(3\cdot 9)] = [-3-27]+[-2-12]+[-1-3]+[1-3]$

$+[2-12]+[3-27] = -30-14-4-2-10-24 = \mathbf{-84}$

e. $\sum_{a=3}^{5}(a+2)^a = (3+2)^3+(4+2)^4+(5+2)^5 = 5^3+6^4+7^5 = 125+1296+16807 = \mathbf{18{,}228}$

f. $\sum_{i=0}^{5}\frac{(-1)^{i+1}}{2^i} = \frac{(-1)^{0+1}}{2^0}+\frac{(-1)^{1+1}}{2^1}+\frac{(-1)^{2+1}}{2^2}+\frac{(-1)^{3+1}}{2^3}+\frac{(-1)^{4+1}}{2^4}+\frac{(-1)^{5+1}}{2^5} = \frac{(-1)^1}{1}+\frac{(-1)^2}{2}+\frac{(-1)^3}{4}+\frac{(-1)^4}{8}+\frac{(-1)^5}{16}+\frac{(-1)^6}{32}$

$= -1+\frac{1}{2}-\frac{1}{4}+\frac{1}{8}-\frac{1}{16}+\frac{1}{32} = -1+0.5-0.25+0.125-0.063+0.031 = -1.313+0.656 = \mathbf{-0.657}$

g. $\sum_{k=-2}^{3}(2k-3)^{k+2} = [(2\cdot -2)-3]^{-2+2}+[(2\cdot -1)-3]^{-1+2}+[(2\cdot 0)-3]^{0+2}+[(2\cdot 1)-3]^{1+2}+[(2\cdot 2)-3]^{2+2}+[(2\cdot 3)-3]^{3+2}$

$= [-4-3]^0+[-2-3]^1+[0-3]^2+[2-3]^3+[4-3]^4+[6-3]^5 = (-7)^0+(-5)^1+(-3)^2+(-1)^3+1^4+3^5 = \mathbf{248}$

h. $\sum_{m=1}^{5}\left(\frac{1}{m}-1\right)^2 = \left(\frac{1}{1}-1\right)^2+\left(\frac{1}{2}-1\right)^2+\left(\frac{1}{3}-1\right)^2+\left(\frac{1}{4}-1\right)^2+\left(\frac{1}{5}-1\right)^2 = (1-1)^2+(0.5-1)^2+(0.333-1)^2+(0.25-1)^2+(0.2-1)^2$

$= 0^2+(-0.5)^2+(0.667)^2+(-0.75)^2+(-0.8)^2 = 0.25+0.445+0.563+0.64 = \mathbf{1.898}$

3. Find the sum of the following series within the specified range.

a. $\sum_{i=-3}^{3}10^i = 10^{-3}+10^{-2}+10^{-1}+10^0+10^1+10^2+10^3 = 0.001+0.01+0.1+1+10+100+1000 = \mathbf{1111.111}$

b. $\sum_{n=0}^{6}\frac{n-1}{2^n} = \frac{0-1}{2^0}+\frac{1-1}{2^1}+\frac{2-1}{2^2}+\frac{3-1}{2^3}+\frac{4-1}{2^4}+\frac{5-1}{2^5}+\frac{6-1}{2^6} = -\frac{1}{1}+\frac{0}{2}+\frac{1}{4}+\frac{2}{8}+\frac{3}{16}+\frac{4}{32}+\frac{5}{64} = -1+0.25+0.25+0.1875$

$+0.125+0.08 = \mathbf{-0.108}$

c. $\sum_{a=0}^{4}\frac{1}{10^a} = \frac{1}{10^0}+\frac{1}{10^1}+\frac{1}{10^2}+\frac{1}{10^3}+\frac{1}{10^4} = \frac{1}{1}+\frac{1}{10}+\frac{1}{100}+\frac{1}{1000}+\frac{1}{10000} = 1+0.1+0.01+0.001+0.0001 = \mathbf{1.1111}$

d. $\sum_{n=1}^{5}\left(n^2-n\right) = (1^2-1)+(2^2-2)+(3^2-3)+(4^2-4)+(5^2-5) = 0+(4-2)+(9-3)+(16-4)+(25-5) = 2+6+12+20 = \mathbf{40}$

e. $\sum_{m=0}^{6}(-1)^{m+1} = (-1)^{0+1}+(-1)^{1+1}+(-1)^{2+1}+(-1)^{3+1}+(-1)^{4+1}+(-1)^{5+1}+(-1)^{6+1} = (-1)^1+(-1)^2+(-1)^3+(-1)^4+(-1)^5$

$+(-1)^6+(-1)^7 = -1+1-1+1-1+1-1 = \boldsymbol{-1}$

f. $\sum_{k=0}^{5}\dfrac{1+(-1)^k}{2^k} = \dfrac{1+(-1)^0}{2^0}+\dfrac{1+(-1)^1}{2^1}+\dfrac{1+(-1)^2}{2^2}+\dfrac{1+(-1)^3}{2^3}+\dfrac{1+(-1)^4}{2^4}+\dfrac{1+(-1)^5}{2^5} = \dfrac{1+1}{1}+\dfrac{1-1}{2}+\dfrac{1+1}{4}+\dfrac{1-1}{8}+\dfrac{1+1}{16}+\dfrac{1-1}{32}$

$= \dfrac{2}{1}+\dfrac{0}{2}+\dfrac{2}{4}+\dfrac{0}{8}+\dfrac{2}{16}+\dfrac{0}{32} = 2+0.5+0.125 = \boldsymbol{2.625}$

g. $\sum_{a=1}^{6}[5(a-1)+3] = [5(1-1)+3]+[5(2-1)+3]+[5(3-1)+3]+[5(4-1)+3]+[5(5-1)+3]+[5(6-1)+3] = [5\cdot0+3]+[5\cdot1+3]$

$+[5\cdot2+3]+[5\cdot3+3]+[5\cdot4+3]+[5\cdot5+3] = 3+8+13+18+23+28 = \boldsymbol{93}$

h. $\sum_{k=0}^{5}\left(-\dfrac{1}{3}\right)^{k-1} = \left(-\dfrac{1}{3}\right)^{0-1}+\left(-\dfrac{1}{3}\right)^{1-1}+\left(-\dfrac{1}{3}\right)^{2-1}+\left(-\dfrac{1}{3}\right)^{3-1}+\left(-\dfrac{1}{3}\right)^{4-1}+\left(-\dfrac{1}{3}\right)^{5-1} = \left(-\dfrac{1}{3}\right)^{-1}+\left(-\dfrac{1}{3}\right)^{0}+\left(-\dfrac{1}{3}\right)^{1}+\left(-\dfrac{1}{3}\right)^{2}$

$+\left(-\dfrac{1}{3}\right)^{3}+\left(-\dfrac{1}{3}\right)^{4} = -3+1-\dfrac{1}{3}+\dfrac{1}{9}-\dfrac{1}{27}+\dfrac{1}{81} = -3+1-0.33+0.11-0.04+0.01 = \boldsymbol{-2.25}$

i. $\sum_{j=1}^{5}\left(j-3j^2\right) = \left(1-3\cdot1^2\right)+\left(2-3\cdot2^2\right)+\left(3-3\cdot3^2\right)+\left(4-3\cdot4^2\right)+\left(5-3\cdot5^2\right) = (1-3)+(2-12)+(3-27)+(4-48)+(5-75)$

$= -2-10-24-44-72 = \boldsymbol{-152}$

j. $\sum_{n=1}^{4}\dfrac{n+1}{n}-\sum_{n=1}^{4}\dfrac{n^2}{n+1} = \left(\dfrac{1+1}{1}+\dfrac{2+1}{2}+\dfrac{3+1}{3}+\dfrac{4+1}{4}\right)-\left(\dfrac{1^2}{1+1}+\dfrac{2^2}{2+1}+\dfrac{3^2}{3+1}+\dfrac{4^2}{4+1}\right) = \left(\dfrac{2}{1}+\dfrac{3}{2}+\dfrac{4}{3}+\dfrac{5}{4}\right)-\left(\dfrac{1}{2}+\dfrac{4}{3}+\dfrac{9}{4}+\dfrac{16}{5}\right)$

$= (2+1.5+1.33+1.25)-(0.5+1.33+2.25+3.2) = 6.08-7.28 = \boldsymbol{-1.2}$

k. $\sum_{k=1}^{5}5k^{-1} = 5\cdot1^{-1}+5\cdot2^{-1}+5\cdot3^{-1}+5\cdot4^{-1}+5\cdot5^{-1} = \dfrac{5}{1}+\dfrac{5}{2}+\dfrac{5}{3}+\dfrac{5}{4}+\dfrac{5}{5} = 5+2.5+1.67+1.25+1 = \boldsymbol{11.42}$

l. $\sum_{i=1}^{4}(-0.1)^{2i-5} = (-0.1)^{2\cdot1-5}+(-0.1)^{2\cdot2-5}+(-0.1)^{2\cdot3-5}+(-0.1)^{2\cdot4-5} = (-0.1)^{-3}+(-0.1)^{-1}+(-0.1)^{1}+(-0.1)^{3}$

$= \dfrac{1}{(-0.1)^3}+\dfrac{1}{(-0.1)^1}-0.1-0.001 = \dfrac{1}{-0.001}+\dfrac{1}{-0.1}-0.1-0.001 = -1000-10-0.1-0.001 = \boldsymbol{-1010.101}$

4. Rewrite the following terms using the sigma notation.

a. $\dfrac{1}{2}+\dfrac{1}{3}+\dfrac{1}{4}+\dfrac{1}{5}+\dfrac{1}{6}+\dfrac{1}{7} = \sum_{n=0}^{5}\dfrac{1}{n+2}$

b. $\dfrac{1}{2}+\dfrac{2}{3}+\dfrac{3}{4}+\dfrac{4}{5}+\dfrac{5}{6}+\dfrac{6}{7} = \sum_{n=1}^{6}\dfrac{n}{n+1}$

c. $2+4+8+16+32+64 = \sum_{k=0}^{5}2^{k+1}$

d. $1+\dfrac{1}{2}+\dfrac{1}{3}+\dfrac{1}{4}+\dfrac{1}{5}+\dfrac{1}{6} = \sum_{k=1}^{6}\dfrac{1}{k}$

e. $0+\dfrac{1}{2}+\dfrac{2}{3}+\dfrac{3}{4}+\dfrac{4}{5}+\dfrac{5}{6} = \sum_{i=1}^{6}\dfrac{n-1}{n}$

f. $1-\dfrac{1}{2}+\dfrac{1}{3}-\dfrac{1}{4}+\dfrac{1}{5}-\dfrac{1}{6} = \sum_{n=1}^{6}\dfrac{(-1)^{n+1}}{n}$

Section 1.3 Solutions - Arithmetic Sequences and Arithmetic Series

1. Find the next seven terms of the following arithmetic sequences.

a. Substituting $s_1=3$ and $d=2$ into the general arithmetic sequence $s_n = s_1+(n-1)d$ for $n=2,3,4,5,6,\text{ and }7$ we obtain

$s_2 = s_1+(2-1)d = s_1+d = 3+2 = \boldsymbol{5}$

$s_3 = s_1+(3-1)d = s_1+2d = 3+(2\cdot2) = 3+4 = \boldsymbol{7}$

$$s_4 = s_1 + (4-1)d = s_1 + 3d = 3 + (3 \cdot 2) = 3 + 6 = \mathbf{9}$$
$$s_5 = s_1 + (5-1)d = s_1 + 4d = 3 + (4 \cdot 2) = 3 + 8 = \mathbf{11}$$
$$s_6 = s_1 + (6-1)d = s_1 + 5d = 3 + (5 \cdot 2) = 3 + 10 = \mathbf{13}$$
$$s_7 = s_1 + (7-1)d = s_1 + 6d = 3 + (6 \cdot 2) = 3 + 12 = \mathbf{15}$$

Thus, the first seven terms of the arithmetic sequence are $(3, 5, 7, 9, 11, 13, 15)$

b. Substituting $s_1 = -3$ and $d = 2$ into the general arithmetic sequence $s_n = s_1 + (n-1)d$ for $n = 2, 3, 4, 5, 6, and\ 7$ we obtain

$$s_2 = s_1 + (2-1)d = s_1 + d = -3 + 2 = \mathbf{-1}$$
$$s_3 = s_1 + (3-1)d = s_1 + 2d = -3 + (2 \cdot 2) = -3 + 4 = \mathbf{1}$$
$$s_4 = s_1 + (4-1)d = s_1 + 3d = -3 + (3 \cdot 2) = -3 + 6 = \mathbf{3}$$
$$s_5 = s_1 + (5-1)d = s_1 + 4d = -3 + (4 \cdot 2) = -3 + 8 = \mathbf{5}$$
$$s_6 = s_1 + (6-1)d = s_1 + 5d = -3 + (5 \cdot 2) = -3 + 10 = \mathbf{7}$$
$$s_7 = s_1 + (7-1)d = s_1 + 6d = -3 + (6 \cdot 2) = -3 + 12 = \mathbf{9}$$

Thus, the first seven terms of the arithmetic sequence are $(-3, -1, 1, 3, 5, 7, 9)$

c. Substituting $s_1 = 10$ and $d = 0.8$ into the general arithmetic sequence $s_n = s_1 + (n-1)d$ for $n = 2, 3, 4, 5, 6, and\ 7$ we obtain

$$s_2 = s_1 + (2-1)d = s_1 + d = 10 + 0.8 = \mathbf{10.8}$$
$$s_3 = s_1 + (3-1)d = s_1 + 2d = 10 + (2 \cdot 0.8) = 10 + 1.6 = \mathbf{11.6}$$
$$s_4 = s_1 + (4-1)d = s_1 + 3d = 10 + (3 \cdot 0.8) = 10 + 2.4 = \mathbf{12.4}$$
$$s_5 = s_1 + (5-1)d = s_1 + 4d = 10 + (4 \cdot 0.8) = 10 + 3.2 = \mathbf{13.2}$$
$$s_6 = s_1 + (6-1)d = s_1 + 5d = 10 + (5 \cdot 0.8) = 10 + 4 = \mathbf{14}$$
$$s_7 = s_1 + (7-1)d = s_1 + 6d = 10 + (6 \cdot 0.8) = 10 + 4.8 = \mathbf{14.8}$$

Thus, the first seven terms of the arithmetic sequence are $(10, 10.8, 11.6, 12.4, 13.2, 14, 14.8)$

2. Find the general term and the eighth term of the following arithmetic sequences.

a. Given $s_1 = 3$ and $d = 4$, the nth term of the arithmetic sequence is equal to $s_n = s_1 + (n-1)d = 3 + (n-1) \cdot 4 = 3 + 4n - 4$

$= \mathbf{4n-1}$. Substituting $n = 8$ into the general equation $s_n = 4n - 1$ we have $s_8 = 4 \cdot 8 - 1 = 32 - 1 = \mathbf{31}$

b. Given $s_1 = -3$ and $d = 5$, the nth term of the arithmetic sequence is equal to $s_n = s_1 + (n-1)d = -3 + (n-1) \cdot 5 = -3 + 5n - 5$

$= \mathbf{5n-8}$. Substituting $n = 8$ into the general equation $s_n = 5n - 8$ we have $s_8 = 5 \cdot 8 - 8 = 40 - 8 = \mathbf{32}$

c. Given $s_1 = 8$ and $d = -1.2$, the nth term of the arithmetic sequence is equal to $s_n = s_1 + (n-1)d = 8 + (n-1) \cdot -1.2$

$= 8 - 1.2n + 1.2 = \mathbf{-1.2n+9.2}$. Substituting $n = 8$ into the general equation $s_n = -1.2n + 9.2$ we have $s_8 = -1.2 \cdot 8 + 9.2$

$= -9.6 + 9.2 = \mathbf{-0.4}$

3. Find the next six terms in each of the following arithmetic sequences.

a. Given the arithmetic sequence $5, 8, \cdots$, $s_1 = 5$ and $d = 8 - 5 = 3$. Therefore, using the general arithmetic equation

$s_n = s_1 + (n-1)d$ or $s_{n+1} = s_n + d$ the next six terms are as follows:

$$s_3 = s_2 + d = 8 + 3 = \mathbf{11} \qquad\qquad s_4 = s_3 + d = 11 + 3 = \mathbf{14} \qquad\qquad s_5 = s_4 + d = 14 + 3 = \mathbf{17}$$

$$s_6 = s_5 + d = 17 + 3 = \mathbf{20} \qquad\qquad s_7 = s_6 + d = 20 + 3 = \mathbf{23} \qquad\qquad s_8 = s_7 + d = 23 + 3 = \mathbf{26}$$

Thus, the first eight terms of the arithmetic sequence are $(5, 8, 11, 14, 17, 20, 23, 26)$

b. Given the arithmetic sequence $x, x+4, \cdots$, $s_1 = x$ and $d = (x+4) - x = 4$. Therefore, using the general arithmetic equation

$s_n = s_1 + (n-1)d$ or $s_{n+1} = s_n + d$ the next six terms are as follows:

$$s_3 = s_2 + d = (x+4) + 4 = \mathbf{x+8} \qquad\qquad\qquad s_4 = s_3 + d = (x+8) + 4 = \mathbf{x+12}$$

$$s_5 = s_4 + d = (x+12) + 4 = x + 16$$

$$s_6 = s_5 + d = (x+16) + 4 = x + 20$$

$$s_7 = s_6 + d = (x+20) + 4 = x + 24$$

$$s_8 = s_7 + d = (x+24) + 4 = x + 28$$

Thus, the first eight terms of the arithmetic sequence are $(x, x+4, x+8, x+12, x+16, x+20, x+24, x+28)$

c. Given the arithmetic sequence $3x+1, 3x+4, \cdots$, $s_1 = 3x+1$ and $d = (3x+4) - (3x+1) = 3$. Therefore, using the general

arithmetic equation $s_n = s_1 + (n-1)d$ or $s_{n+1} = s_n + d$ the next six terms are as follows:

$$s_3 = s_2 + d = (3x+4) + 3 = 3x + 7$$

$$s_4 = s_3 + d = (3x+7) + 3 = 3x + 10$$

$$s_5 = s_4 + d = (3x+10) + 3 = 3x + 13$$

$$s_6 = s_5 + d = (3x+13) + 3 = 3x + 16$$

$$s_7 = s_6 + d = (3x+16) + 3 = 3x + 19$$

$$s_8 = s_7 + d = (3x+19) + 3 = 3x + 22$$

Thus, the first eight terms of the arithmetic sequence are $(3x+1, 3x+4, 3x+7, 3x+10, 3x+13, 3x+16, 3x+19, 3x+22)$

d. Given the arithmetic sequence $w, w-10, \cdots$, $s_1 = w$ and $d = (w-10) - w = -10$. Using the general arithmetic equation

$s_n = s_1 + (n-1)d$ or $s_{n+1} = s_n + d$ the next six terms are as follows:

$$s_3 = s_2 + d = (w-10) - 10 = w - 20$$

$$s_4 = s_3 + d = (w-20) - 10 = w - 30$$

$$s_5 = s_4 + d = (w-30) - 10 = w - 40$$

$$s_6 = s_5 + d = (w-40) - 10 = w - 50$$

$$s_7 = s_6 + d = (w-50) - 10 = w - 60$$

$$s_8 = s_7 + d = (w-60) - 10 = w - 70$$

Thus, the first eight terms of the arithmetic sequence are $(w, w-10, w-20, w-30, w-40, w-50, w-60, w-70)$

4. Find the sum of the following arithmetic series.

a. The first three terms of the given series are $\displaystyle\sum_{i=10}^{20} (2i-4) = (2 \cdot 10 - 4) + (2 \cdot 11 - 4) + (2 \cdot 12 - 4) + \cdots = 16 + 18 + 20 + \cdots$.

Therefore, $s_1 = 16$, $d = 18 - 16 = 2$, and $n = 11$. Substituting s_1, d, and n into the arithmetic series formula

$S_n = \dfrac{n}{2}[2s_1 + (n-1)d]$ we can obtain $S_{11} = \dfrac{11}{2}[2 \cdot 16 + (11-1) \cdot 2] = 5.5 \cdot (32 + 10 \cdot 2) = 5.5 \cdot (32 + 20) = 5.5 \cdot 52 = \mathbf{286}$

b. The first three terms of the given series are $\displaystyle\sum_{k=1}^{1000} k = 1 + 2 + 3 + \cdots$. Therefore, $s_1 = 1$, $d = 2 - 1 = 1$, and $n = 1000$.

Substituting s_1, d, and n into the arithmetic series formula $S_n = \dfrac{n}{2}[2s_1 + (n-1)d]$ we obtain

$$S_{1000} = \dfrac{1000}{2}[2 \cdot 1 + (1000 - 1) \cdot 1] = 500 \cdot [2 + 999] = 500 \cdot 1001 = \mathbf{500500}$$

c. The first three terms of the given series are $\displaystyle\sum_{k=1}^{100} (2k-3) = (2 \cdot 1 - 3) + (2 \cdot 2 - 3) + (2 \cdot 3 - 3) + \cdots = -1 + 1 + 3 + \cdots$.

Therefore, $s_1 = -1$, $d = 1 - (-1) = 2$, and $n = 100$. Substituting s_1, d, and n into the arithmetic series formula

$S_n = \dfrac{n}{2}[2s_1 + (n-1)d]$ we obtain $S_{100} = \dfrac{100}{2}[(2 \cdot -1) + (100 - 1) \cdot 2] = 50 \cdot (-2 + 99 \cdot 2) = 50 \cdot (-2 + 198) = 50 \cdot 196 = \mathbf{9800}$

d. The first three terms of the given series are $\displaystyle\sum_{i=1}^{15} 3i = (3 \cdot 1) + (3 \cdot 2) + (3 \cdot 3) + \cdots = 3 + 6 + 9 + \cdots$.

Therefore, $s_1 = 3$, $d = 6 - 3 = 3$, and $n = 15$. Substituting s_1, d, and n into the arithmetic series formula

$S_n = \dfrac{n}{2}[2s_1 + (n-1)d]$ we obtain $S_{15} = \dfrac{15}{2}[(2 \cdot 3) + (15 - 1) \cdot 3] = 12.5 \cdot (6 + 14 \cdot 3) = 12.5 \cdot 48 = \mathbf{600}$

e. The first three terms of the given series are $\displaystyle\sum_{i=1}^{10} (i+1) = (1+1) + (2+1) + (3+1) + \cdots = 2 + 3 + 4 + \cdots$.

Therefore, $s_1 = 2$, $d = 3 - 2 = 1$, and $n = 10$. Substituting s_1, d, and n into the arithmetic series formula $S_n = \frac{n}{2}[2s_1 + (n-1)d]$ we obtain $S_{10} = \frac{10}{2}[(2\cdot 2) + (10-1)\cdot 1] = 5\cdot(4+9) = 5\cdot 13 = \mathbf{65}$

f. The first three terms of the given series are $\sum_{k=5}^{15}(2k-1) = (2\cdot 5-1)+(2\cdot 6-1)+(2\cdot 7-1)+\cdots = 9+11+13+\cdots$.

Therefore, $s_1 = 9$, $d = 11 - 9 = 2$, and $n = 11$. Substituting s_1, d, and n into the arithmetic series formula $S_n = \frac{n}{2}[2s_1 + (n-1)d]$ we obtain $S_{11} = \frac{11}{2}[(2\cdot 9)+(11-1)\cdot 2] = 5.5\cdot(18+20) = 5.5\cdot 38 = \mathbf{209}$

g. The first three terms of the given series are $\sum_{i=4}^{10}(3i+4) = (3\cdot 4+4)+(3\cdot 5+4)+(3\cdot 6+4)+\cdots = 16+19+22+\cdots$.

Therefore, $s_1 = 16$, $d = 19 - 16 = 3$, and $n = 7$. Substituting s_1, d, and n into the arithmetic series formula $S_n = \frac{n}{2}[2s_1 + (n-1)d]$ we obtain $S_7 = \frac{7}{2}[(2\cdot 16)+(7-1)\cdot 3] = 3.5\cdot(32+18) = 3.5\cdot 50 = \mathbf{175}$

h. The first three terms of the given series are $\sum_{j=5}^{13}(3j+1) = (3\cdot 5+1)+(3\cdot 6+1)+(3\cdot 7+1)+\cdots = 16+19+22+\cdots$.

Therefore, $s_1 = 16$, $d = 19 - 16 = 3$, and $n = 9$. Substituting s_1, d, and n into the arithmetic series formula $S_n = \frac{n}{2}[2s_1 + (n-1)d]$ we obtain $S_9 = \frac{9}{2}[(2\cdot 16)+(9-1)\cdot 3] = 4.5\cdot(32+24) = 4.5\cdot 56 = \mathbf{252}$

i. The first three terms of the given series are $\sum_{k=7}^{18}(4k-3) = (4\cdot 7-3)+(4\cdot 8-3)+(4\cdot 9-3)+\cdots = 25+29+33+\cdots$.

Therefore, $s_1 = 25$, $d = 29 - 25 = 4$, and $n = 12$. Substituting s_1, d, and n into the arithmetic series formula $S_n = \frac{n}{2}[2s_1 + (n-1)d]$ we obtain $S_{12} = \frac{12}{2}[(2\cdot 25)+(12-1)\cdot 4] = 6\cdot(50+44) = 6\cdot 94 = \mathbf{564}$

5. The first term of an arithmetic sequence is 6 and the third term is 24. Find the tenth term.

Since $s_1 = 6$ and $s_3 = 24$ we use the general formula $s_n = s_1 + (n-1)d$ in order to solve for d. Therefore,

$s_3 = s_1 + (3-1)d$; $24 = 6+2d$; $24-6 = 2d$; $d = \frac{18}{2}$; $d = 9$. Then, $s_{10} = s_1 + (10-1)d = s_1 + 9\cdot d = 6+9\cdot 9 = \mathbf{87}$

6. Given the first term s_1 and d, find S_{50} for each of the following arithmetic sequences.

a. Given $s_1 = 2$ and $d = 5$, use the nth term for an arithmetic series $S_n = \frac{n}{2}[2s_1 + (n-1)d]$ to find S_{50}.

$S_{50} = \frac{50}{2}[(2\cdot 2)+(50-1)\cdot 5] = \frac{50}{2}(4+245) = 25\cdot 249 = \mathbf{6225}$

b. Given $s_1 = -5$ and $d = 6$, use the nth term for an arithmetic series $S_n = \frac{n}{2}[2s_1 + (n-1)d]$ to find S_{50}.

$S_{50} = \frac{50}{2}[(2\cdot -5)+(50-1)\cdot 6] = \frac{50}{2}(-10+294) = 25\cdot 284 = \mathbf{7100}$

c. Given $s_1 = 30$ and $d = 10$, use the nth term for an arithmetic series $S_n = \frac{n}{2}[2s_1 + (n-1)d]$ to find S_{50}.

$S_{50} = \frac{50}{2}[(2\cdot 30)+(50-1)\cdot 10] = \frac{50}{2}(60+490) = 25\cdot 550 = \mathbf{13750}$

7. Find the sum of the following sequences for the indicated values.

a. Given the sequence $-8, 6, \cdots$ the first term s_1 and the common difference d are equal to $s_1 = -8$ and $d = 6 - (-8) = 14$.

Thus, using the general arithmetic series $S_n = \frac{n}{2}[2s_1 + (n-1)d]$, S_{15} is equal to:

$$S_{15} = \frac{15}{2}\big[(2 \cdot -8) + (15-1) \cdot 14\big] = \frac{15}{2}(-16 + 196) = 7.5 \cdot 180 = \mathbf{1350}$$

b. Given the sequence $-20, 20, \cdots$ the first term s_1 and the common difference d are equal to $s_1 = -20$ and $d = 20 - (-20)$

$= 40$. Thus, using the general arithmetic series $S_n = \frac{n}{2}\big[2s_1 + (n-1)d\big]$, S_{100} is equal to:

$$S_{100} = \frac{100}{2}\big[(2 \cdot -20) + (100-1) \cdot 40\big] = 50(-40 + 3960) = 50 \cdot 3920 = \mathbf{196{,}000}$$

Section 1.4 Solutions - Geometric Sequences and Geometric Series

---

1. Find the next four terms of the following geometric sequences.

   a. Substituting $s_1 = 3$, $r = 0.5$ into $s_n = s_1 r^{n-1}$ we obtain

   $s_2 = 3 \cdot r^{2-1} = 3r = 3 \cdot 0.5 = \mathbf{1.5}$ $s_3 = 3 \cdot r^{3-1} = 3r^2 = 3 \cdot 0.5^2 = 3 \cdot 0.25 = \mathbf{0.75}$

   $s_4 = 3 \cdot r^{4-1} = 3r^3 = 3 \cdot 0.5^3 = 3 \cdot 0.125 = \mathbf{0.375}$ $s_5 = 3 \cdot r^{5-1} = 3r^4 = 3 \cdot 0.5^4 = 3 \cdot 0.0625 = \mathbf{0.1875}$

   Thus, the first five terms of the geometric sequence are $(3, 1.5, 0.75, 0.375, 0.1875)$

   b. Substituting $s_1 = -5$, $r = 2$ into $s_n = s_1 r^{n-1}$ we obtain

   $s_2 = -5 \cdot r^{2-1} = -5r = -5 \cdot 2 = \mathbf{-10}$ $s_3 = -5 \cdot r^{3-1} = -5r^2 = -5 \cdot 2^2 = -5 \cdot 4 = \mathbf{-20}$

   $s_4 = -5 \cdot r^{4-1} = -5r^3 = -5 \cdot 2^3 = -5 \cdot 8 = \mathbf{-40}$ $s_5 = -5 \cdot r^{5-1} = -5r^4 = -5 \cdot 2^4 = -5 \cdot 16 = \mathbf{-80}$

   Thus, the first five terms of the geometric sequence are $(-5, -10, -20, -40, -80)$

   c. Substituting $s_1 = 5$, $r = 0.75$ into $s_n = s_1 r^{n-1}$ we obtain

   $s_2 = 5 \cdot r^{2-1} = 5r = 5 \cdot 0.75 = \mathbf{3.75}$ $s_3 = 5 \cdot r^{3-1} = 5r^2 = 5 \cdot 0.75^2 = 5 \cdot 0.563 = \mathbf{2.81}$

   $s_4 = 5 \cdot r^{4-1} = 5r^3 = 5 \cdot 0.75^3 = 5 \cdot 0.42 = \mathbf{2.11}$ $s_5 = 5 \cdot r^{5-1} = 5r^4 = 5 \cdot 0.75^4 = 5 \cdot 0.316 = \mathbf{1.58}$

   Thus, the first five terms of the geometric sequence are $(5, 3.75, 2.81, 2.11, 1.58)$

2. Find the eighth and the general term of the following geometric sequences.

   a. Substituting $s_1 = 2$, $r = \sqrt{3}$ into $s_n = s_1 r^{n-1}$ the eighth and the nth term are equal to:

   $$s_8 = 2r^{8-1} = 2r^7 = 2 \cdot \left(\sqrt{3}\right)^7 = 2 \cdot 3^{\frac{7}{2}} = 2 \cdot 46.76 = \mathbf{93.53} \text{ and } s_n = 2 \cdot \left(\sqrt{3}\right)^{n-1} = 2 \cdot 3^{\frac{n-1}{2}}$$

   b. Substituting $s_1 = -4$, $r = 1.2$ into $s_n = s_1 r^{n-1}$ the eighth and the nth term are equal to:

   $$s_8 = -4r^{8-1} = -4r^7 = -4 \cdot (1.2)^7 = -4 \cdot 1.2^7 = -4 \cdot 3.583 = \mathbf{-14.33} \text{ and } s_n = -4 \cdot (1.2)^{n-1}$$

   c. Substituting $s_1 = 4$, $r = -2.5$ into $s_n = s_1 r^{n-1}$ the eighth and the nth term are equal to:

   $$s_8 = 4r^{8-1} = 4r^7 = 4 \cdot (-2.5)^7 = -4 \cdot 2.5^7 = -4 \cdot 610.35 = \mathbf{-2441.4} \text{ and } s_n = 4 \cdot (-2.5)^{n-1}$$

3. Find the next six terms and the nth term in each of the following geometric sequences.

   a. Given $1, \frac{1}{4}, \cdots$, then $s_1 = 1$ and $r = \frac{\frac{1}{4}}{1} = \frac{1}{4}$. Using the general geometric equation $s_n = s_1 r^{n-1}$ the next six terms are:

   $s_3 = 1 \cdot r^{3-1} = r^2 = \left(\frac{1}{4}\right)^2 = \frac{1}{4^2}$ $s_4 = 1 \cdot r^{4-1} = r^3 = \left(\frac{1}{4}\right)^3 = \frac{1}{4^3}$

   $s_5 = 1 \cdot r^{5-1} = r^4 = \left(\frac{1}{4}\right)^4 = \frac{1}{4^4}$ $s_6 = 1 \cdot r^{6-1} = r^5 = \left(\frac{1}{4}\right)^5 = \frac{1}{4^5}$

   $s_7 = 1 \cdot r^{7-1} = r^6 = \left(\frac{1}{4}\right)^6 = \frac{1}{4^6}$ $s_8 = 1 \cdot r^{8-1} = r^7 = \left(\frac{1}{4}\right)^7 = \frac{1}{4^7}$

Thus, the first eight terms of the geometric sequence are $\left(1, \dfrac{1}{4}, \dfrac{1}{4^2}, \dfrac{1}{4^3}, \dfrac{1}{4^4}, \dfrac{1}{4^5}, \dfrac{1}{4^6}, \dfrac{1}{4^7}\right)$ and the nth term is equal to

$$s_n = 1 \cdot \left(\frac{1}{4}\right)^{n-1} = \frac{1^{n-1}}{4^{n-1}} = \frac{1}{4^{n-1}}$$

b. Given $-\dfrac{1}{2}, \dfrac{1}{4}, \cdots$, then $s_1 = -\dfrac{1}{2}$ and $r = \dfrac{\frac{1}{4}}{-\frac{1}{2}} = -\dfrac{1 \cdot 2}{4 \cdot 1} = -\dfrac{2}{4} = -\dfrac{1}{2}$. Using the general geometric equation $s_n = s_1 r^{n-1}$

the next six terms are:

$$s_3 = -\frac{1}{2} \cdot r^{3-1} = -\frac{1}{2}r^2 = -\frac{1}{2} \cdot \left(-\frac{1}{2}\right)^2 = -\frac{1}{2^3}$$
$$s_4 = -\frac{1}{2} \cdot r^{4-1} = -\frac{1}{2}r^3 = -\frac{1}{2} \cdot \left(-\frac{1}{2}\right)^3 = \frac{1}{2^4}$$

$$s_5 = -\frac{1}{2} \cdot r^{5-1} = -\frac{1}{2}r^4 = -\frac{1}{2} \cdot \left(-\frac{1}{2}\right)^4 = -\frac{1}{2^5}$$
$$s_6 = -\frac{1}{2} \cdot r^{6-1} = r^5 = -\frac{1}{2} \cdot \left(-\frac{1}{2}\right)^5 = \frac{1}{2^6}$$

$$s_7 = -\frac{1}{2} \cdot r^{7-1} = -\frac{1}{2}r^6 = -\frac{1}{2} \cdot \left(-\frac{1}{2}\right)^6 = -\frac{1}{2^7}$$
$$s_8 = -\frac{1}{2} \cdot r^{8-1} = -\frac{1}{2}r^7 = -\frac{1}{2}\left(-\frac{1}{2}\right)^7 = \frac{1}{2^8}$$

Thus, the first eight terms of the geometric sequence are $\left(-\dfrac{1}{2}, \dfrac{1}{2^2}, -\dfrac{1}{2^3}, \dfrac{1}{2^4}, -\dfrac{1}{2^5}, \dfrac{1}{2^6}, -\dfrac{1}{2^7}, \dfrac{1}{2^8}\right)$ and the nth term is

equal to $s_n = -\dfrac{1}{2} \cdot \left(-\dfrac{1}{2}\right)^{n-1} = -\dfrac{1}{2} \cdot \dfrac{(-1)^{n-1}}{2^{n-1}} = -\dfrac{(-1)^{n-1}}{2 \cdot 2^{n-1}} = -\dfrac{(-1)^{n-1}}{2^{n-1+1}} = -\dfrac{(-1)^{n-1}}{2^n}$

c. Given $\dfrac{1}{3}p, -3p, \cdots$, then $s_1 = \dfrac{p}{3}$ and $r = \dfrac{-3p}{\frac{p}{3}} = \dfrac{\frac{-3p}{1}}{\frac{p}{3}} = \dfrac{-3p \cdot 3}{1 \cdot p} = \dfrac{-9p}{p} - 9$. Using the general geometric equation

$s_n = s_1 r^{n-1}$ the next six terms are:

$$s_3 = \frac{p}{3} \cdot r^{3-1} = \frac{p}{3} \cdot r^2 = \frac{p}{3} \cdot (-9)^2 = \frac{p}{3} \cdot 9^2 = 3^3 p$$
$$s_4 = \frac{p}{3} \cdot r^{4-1} = \frac{p}{3} \cdot r^3 = \frac{p}{3} \cdot (-9)^3 = -\frac{p}{3} \cdot 9^3 = -3^5 p$$

$$s_5 = \frac{p}{3} \cdot r^{5-1} = \frac{p}{3} \cdot r^4 = \frac{p}{3} \cdot (-9)^4 = \frac{p}{3} \cdot 9^4 = 3^7 p$$
$$s_6 = \frac{p}{3} \cdot r^{6-1} = \frac{p}{3} \cdot r^5 = \frac{p}{3} \cdot (-9)^5 = -\frac{p}{3} \cdot 9^5 = -3^9 p$$

$$s_7 = \frac{p}{3} \cdot r^{7-1} = \frac{p}{3} \cdot r^6 = \frac{p}{3} \cdot (-9)^6 = \frac{p}{3} \cdot 9^6 = 3^{11} p$$
$$s_8 = \frac{p}{3} \cdot r^{8-1} = r^7 = \frac{p}{3} \cdot (-9)^7 = -\frac{p}{3} \cdot 9^7 = -3^{13} p$$

Thus, the first eight terms of the geometric sequence are $\left(\dfrac{1}{3p}, -3p, 3^3 p, -3^5 p, 3^7 p, -3^9 p, 3^{11} p, -3^{13} p\right)$ and the nth

term is equal to $s_n = \dfrac{p}{3} \cdot (-9)^{n-1} = \dfrac{p}{3} \cdot (-1)^{n-1}\left(3^2\right)^{n-1} = \dfrac{p}{3} \cdot (-1)^{n-1} \cdot 3^{2n-2} = p(-1)^{n-1}3^{2n-2-1} = p(-1)^{n-1}3^{2n-3}$

4. Given the following terms of a geometric sequence, find the common ratio r .

a. Substitute $s_1 = 25$ and $s_4 = \dfrac{1}{5}$ into $s_n = s_1 r^{n-1}$ and solve for r , i.e., $s_4 = s_1 r^{4-1}$; $\dfrac{1}{5} = 25r^3$; $\dfrac{\frac{1}{5}}{25} = r^3$; $\dfrac{1}{125} = r^3$

; $\dfrac{1}{5^3} = r^3$; $\left(\dfrac{1}{5^3}\right)^{\frac{1}{3}} = \left(r^3\right)^{\frac{1}{3}}$; $\dfrac{1}{5^{3 \times \frac{1}{3}}} = r^{3 \times \frac{1}{3}}$; $\dfrac{1}{5} = r$; $r = \dfrac{1}{5}$

b. Substitute $s_1 = 4$ and $s_5 = \dfrac{1}{64}$ into $s_n = s_1 r^{n-1}$ and solve for r , i.e., $s_5 = s_1 r^{5-1}$; $\dfrac{1}{64} = 4r^{5-1}$; $\dfrac{\frac{1}{64}}{4} = r^4$; $\dfrac{\frac{1}{64}}{\frac{4}{1}} = r^4$

; $\dfrac{1}{64 \times 4} = r^4$; $\dfrac{1}{256} = r^4$; $\dfrac{1}{4^4} = r^4$; $\left(\dfrac{1}{4^4}\right)^{\frac{1}{4}} = \left(r^4\right)^{\frac{1}{4}}$; $\dfrac{1}{4^{4 \times \frac{1}{4}}} = r^{4 \times \frac{1}{4}}$; $\dfrac{1}{4} = r$; $r = \dfrac{1}{4}$

c. Substitute $s_1 = 3$ and $s_8 = 1$ into $s_n = s_1 r^{n-1}$ and solve for r , i.e., $s_8 = s_1 r^{8-1}$; $1 = 3r^7$; $\dfrac{1}{3} = r^7$; $\left(\dfrac{1}{3}\right)^{\frac{1}{7}} = \left(r^7\right)^{\frac{1}{7}}$

; $\dfrac{1}{3^{\frac{1}{7}}} = r^{7 \times \frac{1}{7}}$; $\dfrac{1}{3^{\frac{1}{7}}} = r$; $r = \dfrac{1}{\sqrt[7]{3}}$

5. Write the first five terms of the following geometric sequences.

a. Given $s_n = \left(-\dfrac{1}{3}\right)^{2n-1}$, then

$s_1 = \left(-\dfrac{1}{3}\right)^{2\cdot1-1} = \left(-\dfrac{1}{3}\right)^{2-1} = -\dfrac{1}{3}$

$s_2 = \left(-\dfrac{1}{3}\right)^{2\cdot2-1} = \left(-\dfrac{1}{3}\right)^{4-1} = \left(-\dfrac{1}{3}\right)^{3} = -\dfrac{1}{3^3}$

$s_3 = \left(-\dfrac{1}{3}\right)^{2\cdot3-1} = \left(-\dfrac{1}{3}\right)^{6-1} = \left(-\dfrac{1}{3}\right)^{5} = -\dfrac{1}{3^5}$

$s_4 = \left(-\dfrac{1}{3}\right)^{2\cdot4-1} = \left(-\dfrac{1}{3}\right)^{8-1} = \left(-\dfrac{1}{3}\right)^{7} = -\dfrac{1}{3^7}$

$s_5 = \left(-\dfrac{1}{3}\right)^{2\cdot5-1} = \left(-\dfrac{1}{3}\right)^{10-1} = \left(-\dfrac{1}{3}\right)^{9} = -\dfrac{1}{3^9}$

Thus, the first five terms of the geometric sequence are $\left(-\dfrac{1}{3}, -\dfrac{1}{3^3}, -\dfrac{1}{3^5}, -\dfrac{1}{3^7}, -\dfrac{1}{3^9}\right)$

b. Given $s_n = \left(\dfrac{1}{3}\right)^{2n+2}$, then

$s_1 = \left(\dfrac{1}{3}\right)^{2\cdot1+2} = \left(\dfrac{1}{3}\right)^{2+2} = \left(\dfrac{1}{3}\right)^{4} = \dfrac{1}{3^4}$

$s_2 = \left(\dfrac{1}{3}\right)^{2\cdot2+2} = \left(\dfrac{1}{3}\right)^{4+2} = \left(\dfrac{1}{3}\right)^{6} = \dfrac{1}{3^6}$

$s_3 = \left(\dfrac{1}{3}\right)^{2\cdot3+2} = \left(\dfrac{1}{3}\right)^{6+2} = \left(\dfrac{1}{3}\right)^{8} = \dfrac{1}{3^8}$

$s_4 = \left(\dfrac{1}{3}\right)^{2\cdot4+2} = \left(\dfrac{1}{3}\right)^{8+2} = \left(\dfrac{1}{3}\right)^{10} = \dfrac{1}{3^{10}}$

$s_5 = \left(\dfrac{1}{3}\right)^{2\cdot5+2} = \left(\dfrac{1}{3}\right)^{10+2} = \left(\dfrac{1}{3}\right)^{12} = \dfrac{1}{3^{12}}$

Thus, the first five terms of the geometric sequence are $\left(\dfrac{1}{3^4}, \dfrac{1}{3^6}, \dfrac{1}{3^8}, \dfrac{1}{3^{10}}, \dfrac{1}{3^{12}}\right)$

c. Given $s_n = \left(-\dfrac{1}{5}\right)^{2n-3}$, then

$s_1 = \left(-\dfrac{1}{5}\right)^{2\cdot1-3} = \left(-\dfrac{1}{5}\right)^{2-3} = \left(-\dfrac{1}{5}\right)^{-1} = -5$

$s_2 = \left(-\dfrac{1}{5}\right)^{2\cdot2-3} = \left(-\dfrac{1}{5}\right)^{4-3} = -\dfrac{1}{5}$

$s_3 = \left(-\dfrac{1}{5}\right)^{2\cdot3-3} = \left(-\dfrac{1}{5}\right)^{6-3} = \left(-\dfrac{1}{5}\right)^{3} = -\dfrac{1}{5^3}$

$s_4 = \left(-\dfrac{1}{5}\right)^{2\cdot4-3} = \left(-\dfrac{1}{5}\right)^{8-3} = \left(-\dfrac{1}{5}\right)^{5} = -\dfrac{1}{5^5}$

$s_5 = \left(-\dfrac{1}{5}\right)^{2\cdot5-3} = \left(-\dfrac{1}{5}\right)^{10-3} = \left(-\dfrac{1}{5}\right)^{7} = -\dfrac{1}{5^7}$

Thus, the first five terms of the geometric sequence are $\left(-5, -\dfrac{1}{5}, -\dfrac{1}{5^3}, -\dfrac{1}{5^5}, -\dfrac{1}{5^7}\right)$

d. Given $s_n = \left(-\dfrac{1}{2}\right)^{n}$, then

$s_1 = \left(-\dfrac{1}{2}\right)^{1} = -\dfrac{1}{2} = -0.5$

$s_2 = \left(-\dfrac{1}{2}\right)^{2} = \dfrac{1}{2^2} = \dfrac{1}{4} = 0.25$

$s_3 = \left(-\dfrac{1}{2}\right)^{3} = -\dfrac{1}{2^3} = -\dfrac{1}{8} = -0.125$

$s_4 = \left(-\dfrac{1}{2}\right)^{4} = \dfrac{1}{2^4} = \dfrac{1}{16} = 0.063$

$s_5 = \left(-\dfrac{1}{2}\right)^{5} = -\dfrac{1}{2^5} = -\dfrac{1}{32} = -0.031$

Thus, the first five terms of the geometric sequence are $\left(-0.5, 0.25, -0.125, 0.063, -0.031\right)$

6. Evaluate the sum of the following geometric series.

a. $\displaystyle\sum_{k=1}^{6} 3^{k-1} = 3^{1-1} + 3^{2-1} + 3^{3-1} + 3^{4-1} + 3^{5-1} + 3^{6-1} = 3^0 + 3^1 + 3^2 + 3^3 + 3^4 + 3^5 = 1 + 3 + 9 + 27 + 81 + 243 = \mathbf{364}$

or we can use the geometric series formula $S_n = \dfrac{s_1\left(1-r^n\right)}{1-r}$ where $s_1 = 1$, $r = \dfrac{3}{1} = 3$, and $n = 6$. Therefore,

$S_6 = \dfrac{1\cdot\left(1-3^6\right)}{1-3} = \dfrac{1-729}{-2} = \dfrac{728}{2} = \mathbf{364}$

b. $\displaystyle\sum_{k=3}^{10} (-2)^{k-3} = (-2)^{3-3} + (-2)^{4-3} + (-2)^{5-3} + (-2)^{6-3} + (-2)^{7-3} + (-2)^{8-3} + (-2)^{9-3} + (-2)^{10-3} = (-2)^0 + (-2)^1 + (-2)^2$

$+ (-2)^3 + (-2)^4 + (-2)^5 + (-2)^6 + (-2)^7 = 1 - 2 + 4 - 8 + 16 - 32 + 64 - 128 = \mathbf{-85}$

or we can use the geometric series formula $S_n = \dfrac{s_1\left(1-r^n\right)}{1-r}$ where $s_1 = 1$, $r = \dfrac{-2}{1} = -2$, and $n = 8$. Therefore,

$$S_8 = \frac{1 \cdot \left[1-(-2)^8\right]}{1-(-2)} = \frac{1-256}{1+2} = -\frac{255}{3} = \mathbf{-85}$$

c. $\displaystyle\sum_{j=4}^{8} 4\left(-\frac{1}{2}\right)^{j+1} = 4\sum_{j=4}^{8}\left(-\frac{1}{2}\right)^{j+1} = 4\left[\left(-\frac{1}{2}\right)^{4+1}+\left(-\frac{1}{2}\right)^{5+1}+\left(-\frac{1}{2}\right)^{6+1}+\left(-\frac{1}{2}\right)^{7+1}+\left(-\frac{1}{2}\right)^{8+1}\right] = 4\left[\left(-\frac{1}{2}\right)^5+\left(-\frac{1}{2}\right)^6\right]$

$+4\left[\left(-\frac{1}{2}\right)^7+\left(-\frac{1}{2}\right)^8+\left(-\frac{1}{2}\right)^9\right] = 4(-0.03+0.012-0.008+0.004-0.002) = 4(-0.024) = \mathbf{-0.096}$

or we can use the geometric series formula $S_n = \dfrac{s_1\left(1-r^n\right)}{1-r}$ where $s_1 = -0.12$, $r = \dfrac{0.012}{-0.03} = -0.4$, and $n = 5$. Therefore,

$$S_5 = \frac{-0.12 \cdot \left[1-(-0.4)^5\right]}{1-(-0.4)} = \frac{-0.12\cdot(1+0.0102)}{1+0.4} = -\frac{0.1212}{1.4} = \mathbf{-0.09}$$

d. $\displaystyle\sum_{m=1}^{4}(-2)^{m-3} = (-2)^{1-3}+(-2)^{2-3}+(-2)^{3-3}+(-2)^{4-3} = (-2)^{-2}+(-2)^{-1}+(-2)^0+(-2)^1 = \frac{1}{(-2)^2}+\frac{1}{-2}+1-2$

$= \dfrac{1}{4}-\dfrac{1}{2}+1-2 = 0.25-0.5-1 = \mathbf{-1.25}$

or we can use the geometric series formula $S_n = \dfrac{s_1\left(1-r^n\right)}{1-r}$ where $s_1 = 0.25$, $r = \dfrac{-0.5}{0.25} = -2$, and $n = 4$. Therefore,

$$S_4 = \frac{0.25 \cdot \left[1-(-2)^4\right]}{1-(-2)} = \frac{0.25\cdot(1-16)}{1+2} = -\frac{3.75}{3} = \mathbf{-1.25}$$

e. $\displaystyle\sum_{n=5}^{10}(-3)^{n-4} = (-3)^{5-4}+(-3)^{6-4}+(-3)^{7-4}+(-3)^{8-4}+(-3)^{9-4}+(-3)^{10-4} = (-3)^1+(-3)^2+(-3)^3+(-3)^4+(-3)^5+(-3)^6$

$= -3+9-27+81-243+729 = \mathbf{546}$

or we can use the geometric series formula $S_n = \dfrac{s_1\left(1-r^n\right)}{1-r}$ where $s_1 = -3$, $r = \dfrac{9}{-3} = -3$, and $n = 6$. Therefore,

$$S_6 = \frac{-3 \cdot \left[1-(-3)^6\right]}{1-(-3)} = \frac{-3\cdot(1-729)}{1+3} = \frac{2184}{4} = \mathbf{546}$$

f. $\displaystyle\sum_{k=1}^{5}(-3)^{k-1} = (-3)^{1-1}+(-3)^{2-1}+(-3)^{3-1}+(-3)^{4-1}+(-3)^{5-1} = 1+(-3)^1+(-3)^2+(-3)^3+(-3)^4 = 1-3+9-27+81 = \mathbf{61}$

or we can use the geometric series formula $S_n = \dfrac{s_1\left(1-r^n\right)}{1-r}$ where $s_1 = 1$, $r = \dfrac{-3}{1} = -3$, and $n = 5$. Therefore,

$$S_5 = \frac{1 \cdot \left[1-(-3)^5\right]}{1-(-3)} = \frac{1+243}{1+3} = \frac{244}{4} = \mathbf{61}$$

g. $\displaystyle\sum_{m=1}^{5} 4^m = 4^1+4^2+4^3+4^4+4^5 = 4+16+64+256+1024 = \mathbf{1364}$

or we can use the geometric series formula $S_n = \dfrac{s_1\left(1-r^n\right)}{1-r}$ where $s_1 = 4$, $r = \dfrac{16}{4} = 4$, and $n = 5$. Therefore,

$$S_5 = \frac{4 \cdot \left(1-4^5\right)}{1-4} = \frac{4\cdot(1-1024)}{-3} = \frac{4092}{3} = \mathbf{1364}$$

h. $\displaystyle\sum_{j=1}^{4} \frac{3^j}{27} = \frac{1}{27}\sum_{j=1}^{4} 3^j = \frac{1}{27}\left(3^1+3^2+3^3+3^4\right) = \frac{1}{27}(3+9+27+81) = \frac{120}{27} = \mathbf{4.44}$

or we can use the geometric series formula $S_n = \dfrac{s_1\left(1-r^n\right)}{1-r}$ where, $s_1 = \dfrac{3}{27} = 0.111$, $r = \dfrac{9}{3} = 3$, and $n = 4$. Therefore,

$$S_4 = \frac{0.111\cdot\left(1-3^4\right)}{1-3} = \frac{0.111\cdot\left(1-81\right)}{-2} = \frac{8.88}{2} = \mathbf{4.44}$$

i. $\displaystyle\sum_{k=3}^{6} 6\left(\frac{1}{2}\right)^{k+1} = 6\sum_{k=3}^{6} 0.5^{k+1} = 6\left[0.5^{3+1}+0.5^{4+1}+0.5^{5+1}+0.5^{6+1}\right] = 6\left[0.5^4+0.5^5+0.5^6+0.5^7\right] = 6\left[0.063+0.031\right]$

$+6\left[0.016+0.008\right] = 6\left(0.118\right) = 0.708 \approx \mathbf{0.7}$

or we can use the geometric series formula $S_n = \dfrac{s_1\left(1-r^n\right)}{1-r}$ where, $s_1 = 6\cdot0.5^4 = 0.375$, $r = \dfrac{0.031}{0.063} = 0.5$, and $n = 4$. Thus,

$$S_4 = \frac{0.375\cdot\left(1-0.5^4\right)}{1-0.5} = \frac{0.375\cdot\left(1-0.063\right)}{0.5} = \frac{0.351}{0.5} = 0.702 \approx \mathbf{0.7}$$

7. Given the first term s_1 and r, find S_8 for each of the following geometric sequences.

a. Given $s_1 = 3$ and $r = 3$ use the geometric series formula $S_n = \dfrac{s_1\left(1-r^n\right)}{1-r}$ to find S_8, i.e.,

$$S_8 = \frac{3\cdot\left(1-3^8\right)}{1-3} = \frac{3\cdot\left(1-6561\right)}{-2} = \frac{19680}{2} = \mathbf{9840}$$

b. Given $s_1 = -8$ and $r = 0.5$ use the geometric series formula $S_n = \dfrac{s_1\left(1-r^n\right)}{1-r}$ to find S_8, i.e.,

$$S_8 = \frac{-8\cdot\left(1-0.5^8\right)}{1-0.5} = \frac{-8\cdot\left(1-0.0039\right)}{0.5} = -\frac{7.968}{0.5} = \mathbf{-15.94}$$

c. Given $s_1 = 2$ and $r = -2.5$ use the geometric series formula $S_n = \dfrac{s_1\left(1-r^n\right)}{1-r}$ to find S_8, i.e.,

$$S_8 = \frac{2\cdot\left[1-\left(-2.5\right)^8\right]}{1-\left(-2.5\right)} = \frac{2\cdot\left[1-1525.88\right]}{1+2.5} = \frac{-2\cdot1524.88}{3.5} = -\frac{3049.76}{3.5} = \mathbf{-871.36}$$

8. Solve for x and y.

a. Given $\displaystyle\sum_{i=3}^{7}\left(ix+2\right) = 30$, then $\left(3x+2\right)+\left(4x+2\right)+\left(5x+2\right)+\left(6x+2\right)+\left(7x+2\right)=30$; $\left(3x+4x+5x+6x+7x\right)+10=30$

; $25x+10 = 30$; $25x = 30-10$; $x = \dfrac{20}{25}$; $x = \mathbf{0.8}$

b. Expanding $\displaystyle\sum_{i=1}^{4}\left(ix+y\right) = 20$ we obtain $\left(x+y\right)+\left(2x+y\right)+\left(3x+y\right)+\left(4x+y\right)=20$; $10x+4y = 20$.

Expanding $\displaystyle\sum_{i=2}^{6}\left(ix+y\right) = 10$ we obtain $\left(2x+y\right)+\left(3x+y\right)+\left(4x+y\right)+\left(5x+y\right)+\left(6x+y\right)=10$; $20x+5y = 10$. Using

substitution method we obtain $x = \mathbf{-2}$ and $y = \mathbf{10}$

Section 1.5 Solutions – Limits of Sequences and Series

1. State which of the following sequences are convergent.

To see if a sequence is *convergent* or *divergent* consider the nth term of the sequence and let it approach infinity.

a. $\lim_{n\to\infty}\dfrac{n+1}{2} \approx \lim_{n\to\infty}\dfrac{n}{2} = \dfrac{\infty}{2} = \infty$ 　　　　　　　　　　　　　　　**The sequence diverges**

b. $\lim_{n\to\infty}\dfrac{n^2-1}{n} \approx \lim_{n\to\infty}\dfrac{n^2}{n} = \lim_{n\to\infty}\dfrac{n^{2=1}}{\not{n}} = \lim_{n\to\infty}n = \infty$ 　　　**The sequence diverges**

c. $\lim_{n\to\infty}2^{n+1} \approx \lim_{n\to\infty}2^n = 2^\infty = \infty$ 　　　　　　　　　　　　　**The sequence diverges**

d. $\lim_{n\to\infty}\dfrac{1}{4^{n+1}} \approx \lim_{n\to\infty}\dfrac{1}{4^{n}} = \dfrac{1}{4^{\infty}} = \dfrac{1}{\infty} = 0$ **The sequence converges**

e. $\lim_{n\to\infty}\dfrac{n-1}{n^{2}} \approx \lim_{n\to\infty}\dfrac{n}{n^{2}} = \lim_{n\to\infty}\dfrac{\not{n}}{n^{2=1}} = \lim_{n\to\infty}\dfrac{1}{n} = \dfrac{1}{\infty} = 0$ **The sequence converges**

f. $\lim_{n\to\infty}\left(\dfrac{1}{5}\right)^{n+1} \approx \lim_{n\to\infty}\left(\dfrac{1}{5}\right)^{n} = \lim_{n\to\infty}0.2^{n} = 0.2^{\infty} = 0$ **The sequence converges**

2. State which of the following geometric sequences are convergent.

  a. The sequence $\dfrac{1}{4},\dfrac{1}{16},\dfrac{1}{64},\dfrac{1}{256},\cdots,\dfrac{1}{4^{n}},\cdots$ **converges to 0** since, for large values of n, the absolute value of the difference

  between $\dfrac{1}{4^{n}}$ and 0 is very small.

  b. The sequence $-5, 25, -125, 625, -3125, \cdots, (-5)^{n}, \cdots$ **diverges** since, as n increases, the nth term increases without bound.

  c. The sequence $2, -2, 2, -2, \cdots, 2(-1)^{n+1}, \cdots$ **diverges** since, as n increases, the nth term oscillates back and forth between
  $+2$ and -2.

  d. The sequence $1,\dfrac{1}{2},\dfrac{1}{4},\dfrac{1}{8},\cdots,\left(\dfrac{1}{2}\right)^{n-1},\cdots$ **converges to 0** since, for large values of n, the absolute value of the difference

  between $\left(\dfrac{1}{2}\right)^{n-1}$ and 0 is very small.

  e. The sequence $-9, 27, -81, 243, \cdots, (-1)^{n}3^{n+1}, \cdots$ **diverges** since, as n increases, the nth term oscillates back and forth from
  a large positive number to a large negative number.

  f. The sequence $1,\dfrac{1}{3},\dfrac{1}{9},\dfrac{1}{27},\dfrac{1}{81},\cdots,\left(\dfrac{1}{3}\right)^{n-1},\cdots$ **converges to 0** since, for large values of n, the absolute value of the

  difference between $\left(\dfrac{1}{3}\right)^{n-1}$ and 0 is very small.

Again note that an easier way of knowing if a sequence is convergent or divergent is by considering the nth term and letting it approach to infinity as shown in practice problems 1.5-1 and 1.5-3.

3. State whether or not the following sequences converges or diverges as $n\to\infty$. If it does converge, find the limit.

  a. $\lim_{n\to\infty}\dfrac{n^{2}}{n^{3}-4} \approx \lim_{n\to\infty}\dfrac{n^{2}}{n^{3}} = \lim_{n\to\infty}\dfrac{n^{2}}{n^{3=1}} = \lim_{n\to\infty}\dfrac{1}{n} = \dfrac{1}{\infty} = 0$ **converges to 0**

  b. $\lim_{n\to\infty}\dfrac{5n+1}{\sqrt{n^{2}+1}} \approx \lim_{n\to\infty}\dfrac{5n}{\sqrt{n^{2}}} = \lim_{n\to\infty}\dfrac{5n}{n} = \lim_{n\to\infty}\dfrac{5\not{n}}{\not{n}} = \lim_{n\to\infty}5 = 5$ **converges to 5**

  c. $\lim_{n\to\infty}\dfrac{25^{n}}{5^{n+1}} \approx \lim_{n\to\infty}\dfrac{5^{2n}}{5^{n}} = \lim_{n\to\infty}\dfrac{5^{2n}\cdot 5^{-n}}{1} = \lim_{n\to\infty}5^{2n-n} = \lim_{n\to\infty}5^{n} = 5^{\infty} = \infty$ **diverges**

  d. $\lim_{n\to\infty}\dfrac{5^{n}+25}{125^{n}} \approx \lim_{n\to\infty}\dfrac{5^{n}}{5^{3n}} = \lim_{n\to\infty}\dfrac{1}{5^{3n}\cdot 5^{-n}} = \lim_{n\to\infty}\dfrac{1}{5^{3n-n}} = \lim_{n\to\infty}\dfrac{1}{5^{2n}} = \dfrac{1}{5^{2\cdot\infty}} = \dfrac{1}{5^{\infty}} = \dfrac{1}{\infty} = 0$

  converges to 0

  e. $\lim_{n\to\infty}\dfrac{(n+2)^{2}}{n^{2}} \approx \lim_{n\to\infty}\dfrac{n^{2}}{n^{2}} = \lim_{n\to\infty}\dfrac{n^{2}}{n^{2}} = \lim_{n\to\infty}1 = 1$ **converges to 1**

  f. $\lim_{n\to\infty}\dfrac{2^{n}}{2^{n}+1} \approx \lim_{n\to\infty}\dfrac{2^{n}}{2^{n}} = \lim_{n\to\infty}\dfrac{2^{\not{n}}}{2^{\not{n}}} = \lim_{n\to\infty}1 = 1$ **converges to 1**

g. $\lim_{n\to\infty}\dfrac{\sqrt{n^2+2n}}{\sqrt{n^4+1}} \approx \lim_{n\to\infty}\dfrac{\sqrt{n^2}}{\sqrt{n^4}} = \lim_{n\to\infty}\dfrac{n}{n^2} = \lim_{n\to\infty}\dfrac{\not{n}}{n^{\not{2}}} = \lim_{n\to\infty}\dfrac{1}{n} = \dfrac{1}{\infty} = 0$ **converges to 0**

h. $\lim_{n\to\infty}\dfrac{5}{n^2+1} \approx \lim_{n\to\infty}\dfrac{5}{n^2} = \dfrac{5}{\infty^2} = \dfrac{5}{\infty} = \mathbf{0}$ **converges to 0**

i. $\lim_{n\to\infty}\dfrac{n+1}{n-1} \approx \lim_{n\to\infty}\dfrac{n}{n} = \lim_{n\to\infty}\dfrac{\not{n}}{\not{n}} = \lim_{n\to\infty}1 = \mathbf{1}$ **converges to 1**

j. $\lim_{n\to\infty}\dfrac{n}{n^3-1} \approx \lim_{n\to\infty}\dfrac{n}{n^3} = \lim_{n\to\infty}\dfrac{\not{n}}{n^{\not{3}=2}} = \lim_{n\to\infty}\dfrac{1}{n^2} = \dfrac{1}{\infty^2} = \dfrac{1}{\infty} = \mathbf{0}$ **converges to 0**

k. $\lim_{n\to\infty}10^{\frac{1}{n}} = 10^{\frac{1}{\infty}} = 10^0 = \mathbf{1}$ **converges to 1**

l. $\lim_{n\to\infty}\dfrac{(n-1)^2}{(1-n)(1+n)} \approx \lim_{n\to\infty}\dfrac{n^2}{-n\cdot n} = \lim_{n\to\infty}-\dfrac{n^2}{n^2} = \lim_{n\to\infty}-\dfrac{n^{\not{2}}}{n^{\not{2}}} = \lim_{n\to\infty}-1 = \mathbf{-1}$ **converges to −1**

m. $\lim_{n\to\infty}100^{-\frac{1}{n}} = \lim_{n\to\infty}\dfrac{1}{100^{\frac{1}{n}}} = \dfrac{1}{100^{\frac{1}{\infty}}} = \dfrac{1}{100^0} = \dfrac{1}{1} = \mathbf{1}$ **converges to 1**

n. $\lim_{n\to\infty}3^{\frac{3}{n}} = 3^{\frac{3}{\infty}} = 3^0 = \mathbf{1}$ **converges to 1**

o. $\lim_{n\to\infty}\dfrac{n+100}{n^3-10} \approx \lim_{n\to\infty}\dfrac{n}{n^3} = \lim_{n\to\infty}\dfrac{\not{n}}{n^{\not{3}=2}} = \lim_{n\to\infty}\dfrac{1}{n^2} = \dfrac{1}{\infty^2} = \dfrac{1}{\infty} = \mathbf{0}$ **converges to 0**

p. $\lim_{n\to\infty}\dfrac{100^{\frac{1}{n}}}{n^2+3} \approx \lim_{n\to\infty}\dfrac{100^{\frac{1}{n}}}{n^2} = \dfrac{100^{\frac{1}{\infty}}}{\infty^2} = \dfrac{100^0}{\infty} = \dfrac{1}{\infty} = \mathbf{0}$ **converges to 0**

q. $\lim_{n\to\infty}\dfrac{1}{n+1}-1 \approx \lim_{n\to\infty}\dfrac{1}{n}-1 = \dfrac{1}{\infty}-1 = 0-1 = \mathbf{-1}$ **converges to −1**

r. $\lim_{n\to\infty}(0.25)^{-n} = \lim_{n\to\infty}\dfrac{1}{0.25^n} = \dfrac{1}{0.25^{\infty}} = \dfrac{1}{0} = \boldsymbol{\infty}$ **diverges**

s. $\lim_{n\to\infty}\dfrac{\sqrt{n+1}}{\sqrt{n}+1} \approx \lim_{n\to\infty}\dfrac{\sqrt{n}}{\sqrt{n}} = \lim_{n\to\infty}\dfrac{\sqrt{\not{n}}}{\sqrt{\not{n}}} = \lim_{n\to\infty}1 = \mathbf{1}$ **converges to 1**

t. $\lim_{n\to\infty}\dfrac{\sqrt{n}}{\sqrt{n+1}}+2 \approx \lim_{n\to\infty}\dfrac{\sqrt{n}}{\sqrt{n}}+2 = \lim_{n\to\infty}\dfrac{\sqrt{\not{n}}}{\sqrt{\not{n}}}+2 = \lim_{n\to\infty}(1+2) = 1+2 = \mathbf{3}$ **converges to 3**

4. Find the sum of the following geometric series.

  a. Given $\sum_{j=0}^{\infty}3\left(\dfrac{1}{8}\right)^j$, then $s_1=3$ and $r=\dfrac{1}{8}$. Since $|r|\langle\,1$ we can use the equation $S_{\infty} = \sum_{n=0}^{\infty}s_1 r^n = \sum_{n=1}^{\infty}s_1 r^{n-1} = \dfrac{s_1}{1-r}$

  to obtain the sum, i.e., $\sum_{j=0}^{\infty}3\left(\dfrac{1}{8}\right)^j = \dfrac{3}{1-\dfrac{1}{8}} = \dfrac{3}{\dfrac{8-1}{8}} = \dfrac{3}{\dfrac{7}{8}} = \dfrac{\dfrac{3}{1}}{\dfrac{7}{8}} = \dfrac{3\times 8}{1\times 7} = \dfrac{\mathbf{24}}{\mathbf{7}}$

  b. Given $\sum_{j=0}^{\infty}3\left(-\dfrac{1}{4}\right)^j$, then $s_1=3$ and $r=-\dfrac{1}{4}$. Since $|r|\langle\,1$ we can use the equation $S_{\infty} = \sum_{n=0}^{\infty}s_1 r^n = \sum_{n=1}^{\infty}s_1 r^{n-1} = \dfrac{s_1}{1-r}$

  to obtain the sum, i.e., $\sum_{j=0}^{\infty}3\left(-\dfrac{1}{4}\right)^j = \dfrac{3}{1-\left(-\dfrac{1}{4}\right)} = \dfrac{3}{1+\dfrac{1}{4}} = \dfrac{3}{\dfrac{4+1}{4}} = \dfrac{3}{\dfrac{5}{4}} = \dfrac{\dfrac{3}{1}}{\dfrac{5}{4}} = \dfrac{3\times 4}{1\times 5} = \dfrac{\mathbf{12}}{\mathbf{5}}$

  c. Given $\sum_{k=1}^{\infty}3\left(\dfrac{3}{2}\right)^{k-1}$, then $s_1=3$ and $r=\dfrac{3}{2}$. Since $|r|\rangle 1$ the geometric series $\sum_{k=1}^{\infty}3\left(\dfrac{3}{2}\right)^{k-1}$ **has no finite sum.**

d. Given $\displaystyle\sum_{k=1}^{\infty}\frac{5}{100^{k+1}}=\sum_{k=1}^{\infty}\frac{5}{100^2\cdot100^{k-1}}=\sum_{k=1}^{\infty}\frac{5}{10000}\left(\frac{1}{100^{k-1}}\right)=\sum_{k=1}^{\infty}\frac{5}{10000}\left(\frac{1}{100}\right)^{k-1}$, then $s_1=\dfrac{5}{10000}$ and $r=\dfrac{1}{100}$.

Since $|r|\langle\,1$ we can use the equation $S_\infty=\displaystyle\sum_{n=0}^{\infty}s_1r^n=\sum_{n=1}^{\infty}s_1r^{n-1}=\frac{s_1}{1-r}$ to obtain the sum, i.e., $\displaystyle\sum_{k=1}^{\infty}\frac{5}{10000}\left(\frac{1}{100}\right)^{k-1}$

$=\dfrac{\dfrac{5}{10000}}{1-\dfrac{1}{100}}=\dfrac{\dfrac{5}{10000}}{\dfrac{100-1}{100}}=\dfrac{\dfrac{5}{10000}}{\dfrac{99}{100}}=\dfrac{5\times100}{10000\times99}=\dfrac{500}{990000}=\dfrac{5}{9900}=\mathbf{\dfrac{1}{1980}}$

e. Give $\displaystyle\sum_{j=0}^{\infty}\left(\frac{1}{3}\right)^j$, then $s_1=1$ and $r=\dfrac{1}{3}$. Since $|r|\langle\,1$ we can use the equation $S_\infty=\displaystyle\sum_{n=0}^{\infty}s_1r^n=\sum_{n=1}^{\infty}s_1r^{n-1}=\frac{s_1}{1-r}$

to obtain the sum, i.e., $\displaystyle\sum_{j=0}^{\infty}\left(\frac{1}{3}\right)^j=\dfrac{1}{1-\dfrac{1}{3}}=\dfrac{1}{\dfrac{3-1}{3}}=\dfrac{1}{\dfrac{2}{3}}=\dfrac{1}{\dfrac{2}{3}}=\dfrac{1\times3}{1\times2}=\mathbf{\dfrac{3}{2}}$

f. Given $\displaystyle\sum_{j=0}^{\infty}\left(-\frac{1}{5}\right)^j$, then $s_1=1$ and $r=-\dfrac{1}{5}$. Since $|r|\langle\,1$ we can use the equation $S_\infty=\displaystyle\sum_{n=0}^{\infty}s_1r^n=\sum_{n=1}^{\infty}s_1r^{n-1}=\frac{s_1}{1-r}$

to obtain the sum, i.e., $\displaystyle\sum_{j=0}^{\infty}\left(-\frac{1}{5}\right)^j=\dfrac{1}{1-\left(-\dfrac{1}{5}\right)}=\dfrac{1}{1+\dfrac{1}{5}}=\dfrac{1}{\dfrac{5+1}{5}}=\dfrac{1}{\dfrac{6}{5}}=\dfrac{1}{\dfrac{6}{5}}=\dfrac{1\times5}{1\times6}=\mathbf{\dfrac{5}{6}}$

5. Find the sum of the following infinite geometric series.

a. Given the series $5-1+\dfrac{1}{5}-\dfrac{1}{25}+\cdots$, $s_1=5$ and $r=-\dfrac{1}{5}$. Since $|r|=\left|-\dfrac{1}{5}\right|=\dfrac{1}{5}=0.2\,\langle\,1$ we can use the equation

$S_\infty=\dfrac{s_1}{1-r}$ to obtain the sum, i.e., $5-1+\dfrac{1}{5}-\dfrac{1}{25}+\cdots=\dfrac{5}{1-\left(-\dfrac{1}{5}\right)}=\dfrac{5}{1+\dfrac{1}{5}}=\dfrac{5}{\dfrac{5+1}{5}}=\dfrac{5}{\dfrac{6}{5}}=\dfrac{\dfrac{5}{1}}{\dfrac{6}{5}}=\dfrac{5\times5}{1\times6}=\mathbf{\dfrac{25}{6}}$

b. Given the series $-\dfrac{1}{2}+2-8+32+\cdots$ $s_1=-\dfrac{1}{2}$ and $r=\dfrac{2}{-\frac{1}{2}}=-4$. Since $|r|=|-4|=4$ is greater than one the geometric

series $-\dfrac{1}{2}+2-8+32+\cdots$ **has no finite solution.**

c. Given the series $1+\dfrac{1}{6}+\dfrac{1}{36}+\dfrac{1}{216}+\cdots$ $s_1=1$ and $r=\dfrac{\frac{1}{6}}{1}=\dfrac{1}{6}$. Since $|r|=\left|\dfrac{1}{6}\right|=\dfrac{1}{6}=0.17\,\langle\,1$ we can use the equation

$S_\infty=\dfrac{s_1}{1-r}$ to obtain the sum, i.e., $1+\dfrac{1}{6}+\dfrac{1}{36}+\dfrac{1}{216}+\cdots=\dfrac{1}{1-\dfrac{1}{6}}=\dfrac{1}{\dfrac{6-1}{6}}=\dfrac{1}{\dfrac{5}{6}}=\dfrac{\dfrac{1}{1}}{\dfrac{5}{6}}=\dfrac{1\times6}{1\times5}=\mathbf{\dfrac{6}{5}}$

d. Given the series $1+\dfrac{1}{10}+\dfrac{1}{100}+\dfrac{1}{1000}+\cdots$ $s_1=1$ and $r=\dfrac{\frac{1}{10}}{1}=\dfrac{1}{10}$. Since $|r|=\left|\dfrac{1}{10}\right|=\dfrac{1}{10}=0.1\,\langle\,1$ we can use the equation

$S_\infty=\dfrac{s_1}{1-r}$ to obtain the sum, i.e., $1+\dfrac{1}{10}+\dfrac{1}{100}+\dfrac{1}{1000}+\cdots=\dfrac{1}{1-\dfrac{1}{10}}=\dfrac{1}{\dfrac{10-1}{10}}=\dfrac{1}{\dfrac{9}{10}}=\dfrac{\dfrac{1}{1}}{\dfrac{9}{10}}=\dfrac{1\times10}{1\times9}=\mathbf{\dfrac{10}{9}}$

6. Write the following repeating decimals as the quotient of two positive integers.

a. Given $0.\overline{666666}\cdots=0.66+0.0066+0.000066$, which is a gemetric series, then $s_1=0.66$ and $r=\dfrac{0.0066}{0.66}=0.01$. Since th ratio r is less than one, we can use the infinite geometric series equation $s_\infty = \dfrac{s_1}{1-r}$ to obtain the sum of the infinite

series $0.66 + 0.0066 + 0.000066$, i.e., $s_\infty = \dfrac{s_1}{1-r} = \dfrac{0.66}{1-0.01} = \dfrac{0.66}{0.99} = \dfrac{66}{99} = \dfrac{22}{33}$. Thus, $\mathbf{0.666666\cdots = \dfrac{22}{33}}$

b. Given $3.027027027\,\overline{}\cdots$, consider the decimal portion of the number $3.027027027\,\overline{}\cdots$ and write it in its equivalent form of

$0.027027027\,\overline{}\cdots = 0.027 + 0.000027 + 0.000000027\cdots$. Since this is a geometric series, then $s_1 = 0.027$ and $r = \dfrac{0.000027}{0.027}$

$= 0.001$. Since the ratio r is less than one, we can use the infinite geometric series equation $s_\infty = \dfrac{s_1}{1-r}$ to obtain the sum

of the infinite series $0.027 + 0.000027 + 0.000000027\cdots$, i.e., $s_\infty = \dfrac{s_1}{1-r} = \dfrac{0.027}{1-0.001} = \dfrac{0.027}{0.999} = \dfrac{27}{999} = \dfrac{3}{111}$. Thus,

$\mathbf{3.027027027\overline{}\cdots = 3\dfrac{3}{111}}$

c. $0.111\overline{11}\cdots = 0.11 + 0.0011 + 0.000011$, which is a gemetric series, then $s_1 = 0.11$ and $r = \dfrac{0.0011}{0.11} = 0.01$. Since the

ratio r is less than one, we can use the infinite geometric series equation $s_\infty = \dfrac{s_1}{1-r}$ to obtain the sum of the infinite series

$0.11 + 0.0011 + 0.000011$, i.e., $s_\infty = \dfrac{s_1}{1-r} = \dfrac{0.11}{1-0.01} = \dfrac{0.11}{0.99} = \dfrac{11}{99} = \dfrac{1}{9}$. Thus, $\mathbf{0.111\overline{11}\cdots = \dfrac{1}{9}}$

Section 1.6 Solutions - The Factorial Notation

1. Expand and simplify the following factorial expressions.

a. $11! = 11\cdot10\cdot9\cdot8\cdot7\cdot6\cdot5\cdot4\cdot3\cdot2\cdot1 = \mathbf{39{,}916{,}800}$

b. $(10-3)! = 7! = 7\cdot6\cdot5\cdot4\cdot3\cdot2\cdot1 = \mathbf{5040}$

c. $\dfrac{12!}{5!} = \dfrac{12\cdot11\cdot10\cdot9\cdot8\cdot7\cdot6\cdot5!}{5!} = \dfrac{12\cdot11\cdot10\cdot9\cdot8\cdot7\cdot6\cdot\cancel{5}!}{\cancel{5}!} = \dfrac{12\cdot11\cdot10\cdot9\cdot8\cdot7\cdot6}{1} = \mathbf{3{,}991{,}680}$

d. $\dfrac{14!}{10!} = \dfrac{14\cdot13\cdot12\cdot11\cdot10!}{10!} = \dfrac{14\cdot13\cdot12\cdot11\cdot\cancel{10}!}{\cancel{10}!} = \dfrac{14\cdot13\cdot12\cdot11}{1} = \mathbf{24{,}024}$

e. $\dfrac{15!}{8!\,4!} = \dfrac{15\cdot14\cdot13\cdot12\cdot11\cdot10\cdot9\cdot8!}{8!\,4!} = \dfrac{15\cdot14\cdot13\cdot12\cdot11\cdot10\cdot9\cdot\cancel{8}!}{\cancel{8}!\cdot4\cdot3\cdot2\cdot1} = \dfrac{15\cdot14\cdot13\cdot\overset{3}{\cancel{12}}\cdot11\cdot\overset{5}{\cancel{10}}\cdot\overset{3}{\cancel{9}}}{4\cdot3\cdot2\cdot1} = \dfrac{15\cdot14\cdot13\cdot3\cdot11\cdot5\cdot3}{1} = \mathbf{1{,}351{,}350}$

f. $\dfrac{10!}{4!\,(10-2)!} = \dfrac{10!}{4!\,8!} = \dfrac{10\cdot9\cdot8!}{4!\,8!} = \dfrac{10\cdot9\cdot\cancel{8}!}{4!\,\cancel{8}!} = \dfrac{10\cdot9}{4\cdot3\cdot2\cdot1} = \dfrac{\overset{5}{\cancel{10}}\cdot\overset{3}{\cancel{9}}}{4\cdot3\cdot2\cdot1} = \dfrac{5\cdot3}{4} = \dfrac{\mathbf{15}}{\mathbf{4}}$

g. $\dfrac{12!\,6!}{14!} = \dfrac{12!\,6!}{14\cdot13\cdot12!} = \dfrac{\cancel{12}!\,6!}{14\cdot13\cdot\cancel{12}!} = \dfrac{6!}{14\cdot13} = \dfrac{6\cdot5\cdot4\cdot3\cdot2\cdot1}{\underset{7}{14}\cdot13} = \dfrac{6\cdot5\cdot4\cdot3}{7\cdot13} = \dfrac{\mathbf{360}}{\mathbf{91}}$

h. $\dfrac{(7-3)!\,9!}{12!\,(7-2)!} = \dfrac{4!\,9!}{12!\,5!} = \dfrac{4!\,9!}{12\cdot11\cdot10\cdot9!\cdot5\cdot4!} = \dfrac{\cancel{4}!\,\cancel{9}!}{12\cdot11\cdot10\cdot\cancel{9}!\cdot5\cdot\cancel{4}!} = \dfrac{1}{12\cdot11\cdot10\cdot5} = \dfrac{\mathbf{1}}{\mathbf{6600}}$

2. Write the following products in factorial form.

a. $7\cdot6\cdot5\cdot4\cdot3\cdot2\cdot1 = \mathbf{7!}$ b. $10\cdot11\cdot12\cdot13\cdot14\cdot15 = \dfrac{\mathbf{15!}}{\mathbf{9!}}$ c. $22\cdot23\cdot24\cdot25 = \dfrac{\mathbf{25!}}{\mathbf{21!}}$

d. $8\cdot7\cdot6\cdot5\cdot4 = \dfrac{\mathbf{8!}}{\mathbf{3!}}$ e. $4\cdot5\cdot6\cdot7\cdot8\cdot9 = \dfrac{\mathbf{9!}}{\mathbf{3!}}$ f. $35 = \dfrac{\mathbf{35!}}{\mathbf{34!}}$

3. Expand the following factorial expressions.

a. $5(n!) = \mathbf{5[n\,(n-1)(n-2)(n-3)(n-4)(n-5)(n-6)\cdots4\cdot3\cdot2\cdot1]}$

b. $(n-7)! = (n-7)(n-8)(n-9)(n-10)(n-11)(n-12)\cdots 4\cdot 3\cdot 2\cdot 1$

c. $(n+10)! = (n+10)(n+9)(n+8)(n+7)(n+6)(n+5)(n+4)(n+3)(n+2)(n+1)\,n\,(n-1)(n-2)\cdots 4\cdot 3\cdot 2\cdot 1$

d. $(5n-5)! = (5n-5)(5n-6)(5n-7)(5n-8)(5n-9)(5n-10)\cdots 4\cdot 3\cdot 2\cdot 1$

e. $(2n-8)! = (2n-8)(2n-9)(2n-10)(2n-11)(2n-12)(2n-13)\cdots 4\cdot 3\cdot 2\cdot 1$

f. $(2n+6)! = (2n+6)(2n+5)(2n+4)(2n+3)(2n+2)(2n+1)\,2n\,(2n-1)(2n-2)(2n-3)\cdots 4\cdot 3\cdot 2\cdot 1$

g. $(2n-5)! = (2n-5)(2n-6)(2n-7)(2n-8)(2n-9)(2n-10)\cdots 4\cdot 3\cdot 2\cdot 1$

h. $(3n+3)! = (3n+3)(3n+2)(3n+1)\,3n\,(3n-1)(3n-2)(3n-3)\cdots 4\cdot 3\cdot 2\cdot 1$

4. Expand and simplify the following factorial expressions.

a. $\dfrac{(n-2)!}{(n-4)!} = \dfrac{(n-2)!}{(n-4)(n-3)(n-2)!} = \dfrac{(\not n-2)!}{(n-4)(n-3)(\not n-2)!} = \dfrac{1}{(n-4)(n-3)}$

b. $\dfrac{(n+4)!}{n!} = \dfrac{(n+4)(n+3)(n+2)(n+1)\,n!}{n!} = \dfrac{(n+4)(n+3)(n+2)(n+1)\,\not n!}{\not n!} = (n+4)(n+3)(n+2)(n+1)$

c. $\dfrac{(n+5)!}{(n-2)!} = \dfrac{(n+5)(n+4)(n+3)(n+2)(n+1)(n)(n-1)(n-2)!}{(n-2)!} = \dfrac{(n+5)(n+4)(n+3)(n+2)(n+1)(n)(n-1)(\not n-2)!}{(\not n-2)!}$

$= (n+5)(n+4)(n+3)(n+2)(n+1)(n)(n-1)$

d. $\dfrac{(n-1)(n+1)!}{(n+2)!} = \dfrac{(n-1)(n+1)!}{(n+2)(n+1)!} = \dfrac{(n-1)(\not n+1)!}{(n+2)(\not n+1)!} = \dfrac{n-1}{n+2}$

e. $\dfrac{(3n)!(3n-2)!}{(3n+1)!(3n-4)!} = \dfrac{(3n)!(3n-2)!}{(3n+1)(3n)!(3n-4)(3n-3)(3n-2)!} = \dfrac{(3n)!(3n-2)!}{(3n+1)(3\not n)!(3n-4)(3n-3)(3\not n-2)!} = \dfrac{1}{(3n+1)(3n-4)(3n-3)}$

f. $\dfrac{(n-1)!}{(n+2)!\,(n!)^2} = \dfrac{(n-1)!}{(n+2)!\,(n!)(n!)} = \dfrac{(n-1)!}{(n+2)!\,(n!)(n)(n-1)!} = \dfrac{(\not n-1)!}{(n+2)!\,(n!)(n)(\not n-1)!} = \dfrac{1}{(n+2)!\,(n!)(n)}$

g. $\dfrac{(2n-3)!\,2(n!)}{(2n)!\,(n-2)!} = \dfrac{(2n-3)!\,2[(n)(n-1)(n-2)!]}{(2n)(2n-1)(2n-2)(2n-3)!\,(n-2)!} = \dfrac{(2\not n-3)!\,2[(n)(n-1)(\not n-2)!]}{[(2n)(2n-1)(2n-2)(2\not n-3)!]\,(\not n-2)!} = \dfrac{2(n)(n-1)}{(2n)(2n-1)(2n-2)}$

$= \dfrac{(2\not n)(n-1)}{(2\not n)(2n-1)(2n-2)} = \dfrac{n-1}{(2n-1)(2n-2)}$

5. Write the following expressions in factorial notation form. Simplify the answer.

a. $\dbinom{5}{3} = \dfrac{5!}{3!(5-3)!} = \dfrac{5!}{3!2!} = \dfrac{5\cdot 4\cdot 3!}{3!2!} = \dfrac{5\cdot \overset{2}{\not 4}}{1\cdot \not 2} = \dfrac{5\cdot 2}{1} = \dfrac{10}{1} = \mathbf{10}$

b. $\dbinom{10}{6} = \dfrac{10!}{6!(10-6)!} = \dfrac{10!}{6!4!} = \dfrac{10\cdot 9\cdot 8\cdot 7\cdot 6!}{6!4!} = \dfrac{\overset{5}{\not{10}}\cdot \overset{3}{\not 9}\cdot \overset{2}{\not 8}\cdot 7}{4\cdot 3\cdot 2\cdot 1} = \dfrac{5\cdot 3\cdot 2\cdot 7}{1} = \dfrac{210}{1} = \mathbf{210}$

c. $\dbinom{8}{0} = \dfrac{8!}{0!(8-0)!} = \dfrac{8!}{0!8!} = \dfrac{1}{1\cdot 1} = \mathbf{1}$

d. $\dbinom{8}{8} = \dfrac{8!}{8!(8-8)!} = \dfrac{8!}{8!0!} = \dfrac{1}{1\cdot 1} = \mathbf{1}$

e. $\dbinom{6}{3} = \dfrac{6!}{3!(6-3)!} = \dfrac{6!}{3!3!} = \dfrac{6\cdot 5\cdot 4\cdot 3!}{3!3!} = \dfrac{\overset{2}{\not 6}\cdot 5\cdot \overset{2}{\not 4}}{1\cdot 2\cdot \not 3} = \dfrac{2\cdot 5\cdot 2}{1} = \dfrac{20}{1} = \mathbf{20}$

f. $\begin{pmatrix} 5 \\ 1 \end{pmatrix} = \dfrac{5!}{1!\,(5-1)!} = \dfrac{5!}{1!\,4!} = \dfrac{5\cdot 4!}{1!\,4!} = \dfrac{5}{1} = 5$

g. $\begin{pmatrix} n \\ n-5 \end{pmatrix} = \dfrac{n!}{(n-5)!\,[n-(n-5)]!} = \dfrac{n!}{(n-5)!\,(\not{n}-\not{n}+5)!} = \dfrac{n!}{(n-5)!\,5!} = \dfrac{n\cdot(n-1)\cdot(n-2)\cdot(n-3)\cdot(n-4)\cdot(\not{n}-\not{5})!}{(\not{n}-\not{5})!\,5!}$

$= \dfrac{n\cdot(n-1)\cdot(n-2)\cdot(n-3)\cdot(n-4)}{5!} = \dfrac{n\cdot(n-1)\cdot(n-2)\cdot(n-3)\cdot(n-4)}{5\cdot4\cdot3\cdot2\cdot1} = \dfrac{\boldsymbol{n\cdot(n-1)\cdot(n-2)\cdot(n-3)\cdot(n-4)}}{\boldsymbol{120}}$

h. $\begin{pmatrix} 2n \\ 2n-1 \end{pmatrix} = \dfrac{2n!}{(2n-1)!\,[2n-(2n-1)]!} = \dfrac{2n!}{(2n-1)!\,(2\not{n}-2\not{n}+1)!} = \dfrac{2n!}{(2n-1)!\,1!} = \dfrac{2n\cdot(2\not{n}-1)!}{(2\not{n}-1)!} = \boldsymbol{2n}$

i. $\begin{pmatrix} 3n \\ 3n-3 \end{pmatrix} = \dfrac{3n!}{(3n-3)!\,[3n-(3n-3)]!} = \dfrac{3n!}{(3n-3)!\,(3\not{n}-3\not{n}+3)!} = \dfrac{3n!}{(3n-3)!\,3!} = \dfrac{3n\cdot(3n-1)\cdot(3n-2)\cdot(3\not{n}-3)!}{(3\not{n}-3)!\cdot1\cdot2\cdot3}$

$= \dfrac{3n\cdot(3n-1)\cdot(3n-2)}{1\cdot2\cdot3} = \dfrac{\boldsymbol{n\cdot(3n-1)\cdot(3n-2)}}{\boldsymbol{2}}$

j. $\begin{pmatrix} n \\ n-6 \end{pmatrix} = \dfrac{n!}{(n-6)!\,[n-(n-6)]!} = \dfrac{n!}{(n-6)!\,(\not{n}-\not{n}+6)!} = \dfrac{n!}{(n-6)!\,6!} = \dfrac{n\cdot(n-1)\cdot(n-2)\cdot(n-3)\cdot(n-4)\cdot(n-5)\cdot(\not{n}-\not{6})!}{(\not{n}-\not{6})!\,6!}$

$= \dfrac{n\cdot(n-1)\cdot(n-2)\cdot(n-3)\cdot(n-4)\cdot(n-5)}{6\cdot5\cdot4\cdot3\cdot2\cdot1} = \dfrac{\boldsymbol{n\cdot(n-1)\cdot(n-2)\cdot(n-3)\cdot(n-4)\cdot(n-5)}}{\boldsymbol{720}}$

6. Expand the following binomial expressions.

a. $(x-2)^4 = \begin{pmatrix} 4 \\ 0 \end{pmatrix}x^4 + \begin{pmatrix} 4 \\ 1 \end{pmatrix}x^{4-1}\cdot(-2) + \begin{pmatrix} 4 \\ 2 \end{pmatrix}x^{4-2}(-2)^2 + \begin{pmatrix} 4 \\ 3 \end{pmatrix}x^{4-3}(-2)^3 + \begin{pmatrix} 4 \\ 4 \end{pmatrix}x^{4-4}(-2)^4 = \begin{pmatrix} 4 \\ 0 \end{pmatrix}x^4 - 2\begin{pmatrix} 4 \\ 1 \end{pmatrix}x^3 + 4\begin{pmatrix} 4 \\ 2 \end{pmatrix}x^2 - 8\begin{pmatrix} 4 \\ 3 \end{pmatrix}x^1$

$+ 16\begin{pmatrix} 4 \\ 4 \end{pmatrix}x^0 = \dfrac{4!}{0!\,(4-0)!}x^4 - \dfrac{2\cdot4!}{1!\,(4-1)!}x^3 + \dfrac{4\cdot4!}{2!\,(4-2)!}x^2 - \dfrac{8\cdot4!}{3!\,(4-3)!}x + \dfrac{16\cdot4!}{4!\,(4-4)!} = \dfrac{4!}{0!\,4!}x^4 - \dfrac{2\cdot4!}{3!}x^3 + \dfrac{4\cdot4!}{2!\,2!}x^2$

$- \dfrac{8\cdot4!}{3!\,1!}x + \dfrac{16\cdot4!}{4!\,0!} = \dfrac{4!}{0!\,4!}x^4 - \dfrac{2\cdot4\cdot3!}{3!}x^3 + \dfrac{4\cdot4\cdot3\cdot2!}{1\cdot2\cdot2!}x^2 - \dfrac{8\cdot4\cdot3!}{3!\,1!}x + \dfrac{16\cdot4!}{4!\,0!} = x^4 - 8x^3 + 24x^2 - 32x + 16$

b. $(u+2)^7 = \begin{pmatrix} 7 \\ 0 \end{pmatrix}u^7 + \begin{pmatrix} 7 \\ 1 \end{pmatrix}u^{7-1}\cdot2 + \begin{pmatrix} 7 \\ 2 \end{pmatrix}u^{7-2}\cdot2^2 + \begin{pmatrix} 7 \\ 3 \end{pmatrix}u^{7-3}\cdot2^3 + \begin{pmatrix} 7 \\ 4 \end{pmatrix}u^{7-4}\cdot2^4 + \begin{pmatrix} 7 \\ 5 \end{pmatrix}u^{7-5}\cdot2^5 + \begin{pmatrix} 7 \\ 6 \end{pmatrix}u^{7-6}\cdot2^6 + \begin{pmatrix} 7 \\ 7 \end{pmatrix}u^{7-7}\cdot2^7$

$= \begin{pmatrix} 7 \\ 0 \end{pmatrix}u^7 + 2\begin{pmatrix} 7 \\ 1 \end{pmatrix}u^6 + 4\begin{pmatrix} 7 \\ 2 \end{pmatrix}u^5 + 8\begin{pmatrix} 7 \\ 3 \end{pmatrix}u^4 + 16\begin{pmatrix} 7 \\ 4 \end{pmatrix}u^3 + 32\begin{pmatrix} 7 \\ 5 \end{pmatrix}u^2 + 64\begin{pmatrix} 7 \\ 6 \end{pmatrix}u + 128\begin{pmatrix} 7 \\ 7 \end{pmatrix}u^0 = \dfrac{7!}{0!\,(7-0)!}u^7 + \dfrac{2\cdot7!}{1!\,(7-1)!}u^6 + \dfrac{4\cdot7!}{2!\,(7-2)!}u^5$

$+ \dfrac{8\cdot7!}{3!\,(7-3)!}u^4 + \dfrac{16\cdot7!}{4!\,(7-4)!}u^3 + \dfrac{32\cdot7!}{5!\,(7-5)!}u^2 + \dfrac{64\cdot7!}{6!\,(7-6)!}u + \dfrac{128\cdot7!}{7!\,(7-7)!} = \dfrac{7!}{0!\,7!}u^7 + \dfrac{2\cdot7!}{1!\,6!}u^6 + \dfrac{4\cdot7!}{2!\,5!}u^5 + \dfrac{8\cdot7!}{3!\,4!}u^4$

$+ \dfrac{16\cdot7!}{4!\,3!}u^3 + \dfrac{32\cdot7!}{5!\,2!}u^2 + \dfrac{64\cdot7!}{6!\,1!}u + \dfrac{128\cdot7!}{7!\,0!} = u^7 + (2\cdot7)u^6 + (7\cdot12)u^5 + (8\cdot35)u^4 + (16\cdot35)u^3 + (32\cdot21)u^2 + (64\cdot7)u + 128$

$= u^7 + 14u^6 + 84u^5 + 280u^4 + 560u^3 + 672u^2 + 448u + 128$

c. $(y-3)^5 = \begin{pmatrix} 5 \\ 0 \end{pmatrix}y^5 + \begin{pmatrix} 5 \\ 1 \end{pmatrix}y^{5-1}\cdot(-3) + \begin{pmatrix} 5 \\ 2 \end{pmatrix}y^{5-2}(-3)^2 + \begin{pmatrix} 5 \\ 3 \end{pmatrix}y^{5-3}(-3)^3 + \begin{pmatrix} 5 \\ 4 \end{pmatrix}y^{5-4}(-3)^4 + \begin{pmatrix} 5 \\ 5 \end{pmatrix}y^{5-5}(-3)^5 = \begin{pmatrix} 5 \\ 0 \end{pmatrix}y^5 - 3\begin{pmatrix} 5 \\ 1 \end{pmatrix}y^4$

$+ 9\begin{pmatrix} 5 \\ 2 \end{pmatrix}y^3 - 27\begin{pmatrix} 5 \\ 3 \end{pmatrix}y^2 + 81\begin{pmatrix} 5 \\ 4 \end{pmatrix}y^1 - 243\begin{pmatrix} 5 \\ 5 \end{pmatrix}y^0 = \dfrac{5!}{0!\,(5-0)!}y^5 - \dfrac{3\cdot5!}{1!\,(5-1)!}y^4 + \dfrac{9\cdot5!}{2!\,(5-2)!}y^3 - \dfrac{27\cdot5!}{3!\,(5-3)!}y^2 + \dfrac{81\cdot5!}{4!\,(5-4)!}y$

$$-\frac{243\cdot 5!}{5!(5-5)!} = \frac{5!}{0!\,5!}y^5 - \frac{3\cdot 5!}{1!\,4!}y^4 + \frac{9\cdot 5!}{2!\,3!}y^3 - \frac{27\cdot 5!}{3!\,2!}y^2 + \frac{81\cdot 5!}{4!\,1!}y - \frac{243\cdot 5!}{5!\,0!} = y^5 - (3\cdot 5)y^4 + (9\cdot 10)y^3 - [27\cdot 10]y^2$$

$$+(81\cdot 5)y - 243 = y^5 - 15y^4 + 90y^3 - 270y^2 + 405y - 243$$

7. Use the general equation for binomial expansion to solve the following exponential numbers to the nearest hundredth.

a. $(0.95)^5 = (1-0.05)^5$ therefore, $a=1$, $b=-0.05$, $n=5$. Using the general binomial expansion formula

$$(a+b)^n = \binom{n}{0}a^n + \binom{n}{1}a^{n-1}b + \binom{n}{2}a^{n-2}b^2 + \cdots + \binom{n}{r-1}a^{n-r+1}b^{r-1} + \cdots + \binom{n}{n}b^n$$ we obtain the following:

$$(1-0.05)^5 = \binom{5}{0}\cdot 1^5 + \binom{5}{1}\cdot 1^4\cdot(-0.05) + \binom{5}{2}\cdot 1^3\cdot(-0.05)^2 + \binom{5}{3}\cdot 1^2\cdot(-0.05)^3 + \binom{5}{4}\cdot 1^2\cdot(-0.05)^4 + \binom{5}{5}\cdot(-0.05)^5$$

$$= \binom{5}{0} - 0.05\binom{5}{1} + 0.0025\binom{5}{2} - \cdots = \frac{5!}{0!(5-0)!} - 0.05\frac{5!}{1!(5-1)!} + 0.0025\frac{5!}{2!(5-2)!} - \cdots = \frac{5!}{5!} - 0.05\frac{5!}{4!} + 0.0025\frac{5!}{2!3!} - \cdots$$

$$= \frac{5!}{5!} - 0.05\frac{5\cdot 4!}{4!} + 0.0025\frac{5\cdot 4\cdot 3!}{2\cdot 1\cdot 3!}^{2} - \cdots = 1 - 0.25 + 0.025 - \cdots \approx 0.775$$

Therefore, $(0.95)^5$ to the nearest hundredth is equal to **0.78**. (Note that this is an estimate.)

b. $(2.25)^7 = (2+0.25)^7$ therefore, $a=2$, $b=0.25$, $n=7$. Using the general binomial expansion formula

$$(a+b)^n = \binom{n}{0}a^n + \binom{n}{1}a^{n-1}b + \binom{n}{2}a^{n-2}b^2 + \cdots + \binom{n}{r-1}a^{n-r+1}b^{r-1} + \cdots + \binom{n}{n}b^n$$ we obtain the following:

$$(2+0.25)^7 = \binom{7}{0}2^7 + \binom{7}{1}2^6\cdot 0.25 + \binom{7}{2}2^5\cdot 0.25^2 + \binom{7}{3}2^4\cdot 0.25^3 + \binom{7}{4}2^3\cdot 0.25^4 + \binom{7}{5}2^2\cdot 0.25^5 + \binom{7}{6}2\cdot 0.25^6 + \binom{7}{7}0.25^7$$

$$= 128\binom{7}{0} + 16\binom{7}{1} + 2\binom{7}{2} + 0.25\binom{7}{3} + 0.0313\binom{7}{4} + \cdots = \frac{128\cdot 7!}{0!(7-0)!} + \frac{16\cdot 7!}{1!(7-1)!} + \frac{2\cdot 7!}{2!(7-2)!} + \frac{0.25\cdot 7!}{3!(7-3)!} + \frac{0.0313\cdot 7!}{4!(7-4)!} + \cdots$$

$$= \frac{128\cdot 7!}{7!} + \frac{16\cdot 7!}{6!} + \frac{2\cdot 7!}{2!\,5!} + \frac{0.25\cdot 7!}{3!\,4!} + \frac{0.0313\cdot 7!}{4!\,3!} + \cdots = \frac{128\cdot 7!}{7!} + \frac{16\cdot 7\cdot 6!}{6!} + \frac{2\cdot 7\cdot 6\cdot 5!}{2\cdot 1\cdot 5!} + \frac{0.25\cdot 7\cdot 6\cdot 5\cdot 4!}{3\cdot 2\cdot 1\cdot 4!} + \frac{0.0313\cdot 7\cdot 6\cdot 5\cdot 4!}{4!\cdot 3\cdot 2\cdot 1} + \cdots$$

$$= 128 + 112 + 42 + 8.75 + 1.095 + \cdots \approx 291.845$$

Therefore, $(2.25)^7$ to the nearest hundredth is equal to **291.85**. (Note that this is an estimate.)

c. $(1.05)^4 = (1+0.05)^4$ therefore, $a=1$, $b=0.05$, $n=4$. Using the general binomial expansion formula

$$(a+b)^n = \binom{n}{0}a^n + \binom{n}{1}a^{n-1}b + \binom{n}{2}a^{n-2}b^2 + \cdots + \binom{n}{r-1}a^{n-r+1}b^{r-1} + \cdots + \binom{n}{n}b^n$$ we obtain the following:

$$(1+0.05)^4 = \binom{4}{0}\cdot 1^4 + \binom{4}{1}\cdot 1^3\cdot 0.05 + \binom{4}{2}\cdot 1^2\cdot 0.05^2 + \binom{4}{3}\cdot 1^1\cdot 0.05^3 + \binom{4}{4}\cdot 0.05^4 = \binom{4}{0} + 0.05\binom{4}{1} + 0.0025\binom{4}{2} + 0.000125\binom{4}{3} + \cdots$$

$$= \frac{4!}{0!(4-0)!} + \frac{0.05\cdot 4!}{1!(4-1)!} + \frac{0.0025\cdot 4!}{2!(4-2)!} + \frac{0.000125\cdot 4!}{3!(4-3)!} + \cdots = \frac{4!}{4!} + \frac{0.05\cdot 4!}{3!} + \frac{0.0025\cdot 4!}{2!\,2!} + \frac{0.000125\cdot 4!}{3!\,1!} + \cdots$$

$$= \frac{4!}{4!} + \frac{0.05\cdot 4\cdot 3!}{3!} + \frac{0.0025\cdot 4\cdot 3\cdot 2!}{2\cdot 1\cdot 2!}^{2} + \frac{0.000125\cdot 4\cdot 3!}{3!} + \cdots = 1 + 0.2 + 0.015 + 0.0005 + \cdots = 1.2155$$

Therefore, $(1.05)^4$ to the nearest hundredth is equal to **1.22**.

8. Find the stated term of the following binomial expressions.

a. To find the eighth term of $(x+3)^{12}$ first identify the a, b, r, and n terms, i.e., $a = x$, $b = 3$, $r = 8$, and $n = 12$.

Then, use the equation $\binom{n}{r-1} a^{n-r+1} b^{r-1} = \dfrac{n!}{(r-1)!\,(n-r+1)!} a^{n-r+1} b^{r-1}$

$\binom{12}{7} x^5 \cdot 3^7 = \dfrac{12!}{7!(12-7)!} x^5 \cdot 3^7 = \dfrac{12!}{7!5!} x^5 \cdot 2187 = 2187 x^5 \cdot \dfrac{12 \cdot 11 \cdot 10 \cdot 9 \cdot 8 \cdot 7!}{7! \cdot 5 \cdot 4 \cdot 3 \cdot 2 \cdot 1} = (2187 \cdot 792) x^5 = \mathbf{1{,}732{,}104 x^5}$

b. To find the ninth term of $(x-y)^{10}$ first identify the a, b, r, and n terms, i.e., $a = x$, $b = -y$, $r = 9$, and $n = 10$.

Then, use the equation $\binom{n}{r-1} a^{n-r+1} b^{r-1} = \dfrac{n!}{(r-1)!\,(n-r+1)!} a^{n-r+1} b^{r-1}$

$\binom{10}{8} x^2 \cdot (-y)^8 = \dfrac{10!}{8!(10-8)!} x^2 \cdot y^8 = \dfrac{10!}{8!2!} x^2 y^8 = \dfrac{10 \cdot 9 \cdot 8!}{8! \cdot 2 \cdot 1} x^2 y^8 = \dfrac{90}{2} x^2 y^8 = \mathbf{45 x^2 y^8}$

c. To find the seventh term of $(u-2a)^{11}$ first identify the a, b, r, and n terms, i.e., $a = u$, $b = -2a$, $r = 7$, and $n = 11$.

Then, use the equation $\binom{n}{r-1} a^{n-r+1} b^{r-1} = \dfrac{n!}{(r-1)!\,(n-r+1)!} a^{n-r+1} b^{r-1}$

$\binom{11}{6} u^5 \cdot (-2a)^6 = \dfrac{11!}{6!(11-6)!} 64 a^6 u^5 = \dfrac{11!}{6!5!} 64 a^6 u^5 = 64 a^6 u^5 \dfrac{11 \cdot \overset{2}{\cancel{10}} \cdot \overset{3}{\cancel{9}} \cdot \overset{2}{\cancel{8}} \cdot 7 \cdot 6!}{6! \cdot 5 \cdot 4 \cdot 3 \cdot 2 \cdot 1} = (64 \cdot 462) a^6 u^5 = \mathbf{29{,}568 a^6 u^5}$

d. To find the twelfth term of $(x-1)^{18}$ first identify the a, b, r, and n terms, i.e., $a = x$, $b = -1$, $r = 12$, and $n = 18$.

Then, use the equation $\binom{n}{r-1} a^{n-r+1} b^{r-1} = \dfrac{n!}{(r-1)!\,(n-r+1)!} a^{n-r+1} b^{r-1}$

$\binom{18}{11} x^7 \cdot (-1)^{11} = -x^7 \dfrac{18!}{11!(18-11)!} = -x^7 \dfrac{18!}{11!9!} = -x^7 \dfrac{18 \cdot 17 \cdot \overset{2}{\cancel{16}} \cdot \overset{3}{\cancel{15}} \cdot \overset{2}{\cancel{14}} \cdot 13 \cdot 12 \cdot 11!}{11! \cdot 9 \cdot 8 \cdot 7 \cdot 6 \cdot 5 \cdot 4 \cdot 3 \cdot 2 \cdot 1} = -x^7 \dfrac{17 \cdot 2 \cdot 3 \cdot 2 \cdot 13}{6} = -(17 \cdot 13 \cdot 2) x^7$

$= \mathbf{-442 x^7}$

Chapter 2 Solutions:

1. Find the derivative of the following functions by using the difference quotient method.

a. Given $f(x) = x^2 - 1$, then $\dfrac{f(x+h) - f(x)}{h} = \dfrac{\left[(x+h)^2 - 1\right] - \left(x^2 - 1\right)}{h} = \dfrac{x^2 + h^2 + 2hx - 1 - x^2 + 1}{h} = \dfrac{h^2 + 2hx}{h} = \dfrac{h(h + 2x)}{h}$

$= h + 2x$. Therefore, $f'(x) = \lim_{h \to 0} \dfrac{f(x+h) - f(x)}{h} = \lim_{h \to 0}(h + 2x) = 0 + 2x = \mathbf{2x}$

b. Given $f(x) = x^3 + 2x - 1$, then $\dfrac{f(x+h) - f(x)}{h} = \dfrac{\left[(x+h)^3 + 2(x+h) - 1\right] - \left[x^3 + 2x - 1\right]}{h}$

$= \dfrac{x^3 + h^3 + 3x^2h + 3xh^2 + 2x + 2h - 1 - x^3 - 2x + 1}{h} = \dfrac{h^3 + 3x^2h + 3xh^2 + 2h}{h} = \dfrac{h\left(h^2 + 3x^2 + 3xh + 2\right)}{h} = h^2 + 3x^2 + 3xh + 2$.

Therefore, $f'(x) = \lim_{h \to 0} \dfrac{f(x+h) - f(x)}{h} = \lim_{h \to 0}\left(h^2 + 3x^2 + 3xh + 2\right) = 0^2 + 3 \cdot x^2 + 3 \cdot 0 \cdot h + 2 = \mathbf{3x^2 + 2}$

c. Given $f(x) = \dfrac{x}{x-1}$, then $\dfrac{f(x+h) - f(x)}{h} = \dfrac{\dfrac{x+h}{x+h-1} - \dfrac{x}{x-1}}{h} = \dfrac{\dfrac{(x+h)(x-1) - x(x+h-1)}{(x+h-1)(x-1)}}{h} = \dfrac{(x+h)(x-1) - x(x+h-1)}{h\left[(x+h-1)(x-1)\right]}$

$= \dfrac{x^2 - x + hx - h - x^2 - hx + x}{h\left(x^2 - x + hx - h - x + 1\right)} = -\dfrac{h}{h\left(x^2 - 2x + hx - h + 1\right)} = -\dfrac{1}{x^2 - 2x + hx - h + 1}$. Therefore,

$f'(x) = \lim_{h \to 0} \dfrac{f(x+h) - f(x)}{h} = \lim_{h \to 0}\left(\dfrac{-1}{x^2 - 2x + hx - h + 1}\right) = \dfrac{-1}{x^2 - 2x + 0 \cdot x - 0 + 1} = \dfrac{-1}{x^2 - 2x + 1} = -\dfrac{1}{(x-1)^2}$

d. Given $f(x) = -\dfrac{1}{x^2}$, then $\dfrac{f(x+h) - f(x)}{h} = \dfrac{-\dfrac{1}{(x+h)^2} + \dfrac{1}{x^2}}{h} = \dfrac{\dfrac{-x^2 + (x+h)^2}{x^2(x+h)^2}}{h} = \dfrac{-x^2 + (x+h)^2}{hx^2(x+h)^2} = \dfrac{-x^2 + \left(x^2 + h^2 + 2hx\right)}{hx^2\left(x^2 + h^2 + 2hx\right)}$

$= \dfrac{-x^2 + x^2 + h^2 + 2hx}{hx^2\left(x^2 + h^2 + 2hx\right)} = \dfrac{h(h + 2x)}{hx^2\left(x^2 + h^2 + 2hx\right)} = \dfrac{h + 2x}{x^4 + h^2x^2 + 2hx^3}$. Therefore,

$f'(x) = \lim_{h \to 0} \dfrac{f(x+h) - f(x)}{h} = \lim_{h \to 0}\left(\dfrac{h + 2x}{x^4 + h^2x^2 + 2hx^3}\right) = \dfrac{0 + 2x}{x^4 + 0^2 \cdot x^2 + 2 \cdot 0 \cdot x^3} = \dfrac{2x}{x^{4=3}} = \dfrac{2}{x^3}$

e. Given $f(x) = 20x^2 - 3$, then $\dfrac{f(x+h) - f(x)}{h} = \dfrac{\left[20(x+h)^2 - 3\right] - \left[20x^2 - 3\right]}{h} = \dfrac{\left[20\left(x^2 + 2hx + h^2\right) - 3\right] - \left[20x^2 - 3\right]}{h}$

$= \dfrac{20x^2 + 40hx + 20h^2 - 3 - 20x^2 + 3}{h} = \dfrac{40hx + 20h^2}{h} = \dfrac{h(40x + 20h)}{h} = 40x + 20h$

Therefore, $f'(x) = \lim_{h \to 0} \dfrac{f(x+h) - f(x)}{h} = \lim_{h \to 0}(40x + 20h) = 40x + (20 \cdot 0) = 40x + 0 = \mathbf{40x}$

f. Given $f(x) = \sqrt{x^3}$, then $\dfrac{f(x+h) - f(x)}{h} = \dfrac{\sqrt{(x+h)^3} - \sqrt{x^3}}{h} = \dfrac{\sqrt{(x+h)^3} - \sqrt{x^3}}{h} \cdot \dfrac{\sqrt{(x+h)^3} + \sqrt{x^3}}{\sqrt{(x+h)^3} + \sqrt{x^3}} = \dfrac{(x+h)^3 - x^3}{h\sqrt{(x+h)^3} + \sqrt{x^3}}$

$= \dfrac{x^3 + h^3 + 3x^2h + 3xh^2 - x^3}{h\sqrt{x^3 + h^3 + 3x^2h + 3xh^2} + \sqrt{x^3}} = \dfrac{h\left(h^2 + 3x^2 + 3xh\right)}{h\sqrt{x^3 + h^3 + 3x^2h + 3xh^2} + \sqrt{x^3}} = \dfrac{h^2 + 3x^2 + 3xh}{\sqrt{x^3 + h^3 + 3x^2h + 3xh^2} + \sqrt{x^3}}$. Therefore,

$$f'(x) = \lim_{h \to 0} \frac{f(x+h)-f(x)}{h} = \lim_{h \to 0}\left(\frac{h^2+3x^2+3xh}{\sqrt{x^3+h^3+3x^2h+3xh^2}+\sqrt{x^3}}\right) = \frac{0^2+3x^2+3x\cdot 0}{\sqrt{x^3+0^3+3x^2\cdot 0+3x\cdot 0^2}+\sqrt{x^3}}$$

$$= \frac{3x^2}{\sqrt{x^3}+\sqrt{x^3}} = \frac{3x^2}{2\sqrt{x^3}} = \frac{3x^2}{2x^{\frac{3}{2}}} = \frac{3}{2}x^2\cdot x^{-\frac{3}{2}} = \frac{3}{2}x^{2-\frac{3}{2}} = \frac{3}{2}x^{\frac{4-3}{2}} = \frac{3}{2}x^{\frac{1}{2}} = \frac{3}{2}\sqrt{x}$$

g. Given $f(x)=\dfrac{10}{\sqrt{x-5}}$, then $\dfrac{f(x+h)-f(x)}{h} = \dfrac{\dfrac{10}{\sqrt{x+h-5}}-\dfrac{10}{\sqrt{x-5}}}{h} = \dfrac{\dfrac{10\sqrt{x-5}-10\sqrt{x+h-5}}{\sqrt{x+h-5}\cdot\sqrt{x-5}}}{h} = \dfrac{10\sqrt{x-5}-10\sqrt{x+h-5}}{h\cdot\sqrt{x+h-5}\cdot\sqrt{x-5}}$

$$= 10\cdot\frac{\sqrt{x-5}-\sqrt{x+h-5}}{h\cdot\sqrt{x+h-5}\cdot\sqrt{x-5}}\cdot\frac{\sqrt{x-5}+\sqrt{x+h-5}}{\sqrt{x-5}+\sqrt{x+h-5}} = 10\cdot\frac{x-5-x-h+5}{\left(h\cdot\sqrt{x+h-5}\cdot\sqrt{x-5}\right)\cdot\left(\sqrt{x-5}+\sqrt{x+h-5}\right)}$$

$$= 10\cdot\frac{-h}{\left(h\cdot\sqrt{x+h-5}\cdot\sqrt{x-5}\right)\cdot\left(\sqrt{x-5}+\sqrt{x+h-5}\right)} = \frac{-10}{\left(\sqrt{x+h-5}\cdot\sqrt{x-5}\right)\cdot\left(\sqrt{x-5}+\sqrt{x+h-5}\right)} \text{. Therefore,}$$

$$f'(x) = \lim_{h \to 0}\frac{f(x+h)-f(x)}{h} = \lim_{h \to 0}\left(\frac{-10}{\left(\sqrt{x+h-5}\cdot\sqrt{x-5}\right)\cdot\left(\sqrt{x-5}+\sqrt{x+h-5}\right)}\right) = \frac{-10}{\left(\sqrt{x+0-5}\cdot\sqrt{x-5}\right)\cdot\left(\sqrt{x-5}+\sqrt{x+0-5}\right)}$$

$$= \frac{-10}{\left(\sqrt{x-5}\cdot\sqrt{x-5}\right)\cdot\left(\sqrt{x-5}+\sqrt{x-5}\right)} = \frac{-10}{(x-5)\cdot\left(2\sqrt{x-5}\right)} = \frac{-10}{2(x-5)\left(\sqrt{x-5}\right)} = \frac{-5}{(x-5)^{1+\frac{1}{2}}} = \frac{-5}{(x-5)^{\frac{3}{2}}} = -\frac{5}{\sqrt{(x-5)^3}}$$

h. Given $f(x)=\dfrac{ax+b}{cx}$, then $\dfrac{f(x+h)-f(x)}{h} = \dfrac{\left[\dfrac{a(x+h)+b}{c(x+h)}\right]-\left[\dfrac{ax+b}{cx}\right]}{h} = \dfrac{\dfrac{cx(ax+ah+b)-(cx+ch)(ax+b)}{cx(cx+ch)}}{h}$

$$= \frac{acx^2+achx+bcx-acx^2-bcx-achx-bch}{chx(cx+ch)} = \frac{acx^2+achx+bcx-acx^2-bcx-achx-bch}{chx(cx+ch)} = \frac{-bch}{chx(cx+ch)} = \frac{-b}{cx^2+chx}\text{.}$$

Therefore, $f'(x) = \lim_{h \to 0}\dfrac{f(x+h)-f(x)}{h} = \lim_{h \to 0}\left(\dfrac{-b}{cx^2+chx}\right) = -\dfrac{b}{cx^2+c\cdot 0\cdot x} = -\dfrac{b}{cx^2}$

2. Compute $f'(x)$ for the specified values by using the difference quotient equation as the $\lim_{h \to 0}$.

a. Given $f(x)=x^3$, then using $f'(x) = \lim_{h \to 0}\dfrac{f(x+h)-f(x)}{h}$ we obtain $f'(x) = \lim_{h \to 0}\dfrac{(x+h)^3-x^3}{h}$ at $x=1$

$$f'(1) = \lim_{h \to 0}\frac{(1+h)^3-1^3}{h} = \lim_{h \to 0}\frac{1+h^3+3h+3h^2-1}{h} = \lim_{h \to 0}\frac{h\left(h^2+3+3h\right)}{h} = \lim_{h \to 0}h^2+3+3h = 0^2+3+3\cdot 0 = 3$$

b. Given $f(x)=1+2x$, then using $f'(x) = \lim_{h \to 0}\dfrac{f(x+h)-f(x)}{h}$ we obtain $f'(x) = \lim_{h \to 0}\dfrac{1+2(x+h)-(1+2x)}{h}$ at $x=0$

$$f'(0) = \lim_{h \to 0}\frac{1+2(0+h)-(1+2\cdot 0)}{h} = \lim_{h \to 0}\frac{1+2h-1}{h} = \lim_{h \to 0}\frac{2h}{h} = \lim_{h \to 0}2 = 2$$

c. Given $f(x)=x^3+1$, then using $f'(x) = \lim_{h \to 0}\dfrac{f(x+h)-f(x)}{h}$ we obtain $f'(x) = \lim_{h \to 0}\dfrac{\left[(x+h)^3+1\right]-\left(x^3+1\right)}{h}$ at $x=-1$

$$f'(-1) = \lim_{h \to 0}\frac{\left[(-1+h)^3+1\right]-\left[(-1)^3+1\right]}{h} = \lim_{h \to 0}\frac{-1+h^3+3h+3h^2+1-(-1+1)}{h} = \lim_{h \to 0}\frac{\left(h^3+3h+3h^2\right)-0}{h}$$

$$= \lim_{h \to 0}\frac{h\left(h^2+3+3h\right)}{h} = \lim_{h \to 0}\left(h^2+3+3h\right) = 0^2+3+3\cdot 0 = 3$$

d. Given $f(x)=x^2(x+2)=x^3+2x^2$, then using $f'(x)=\lim_{h\to0}\dfrac{f(x+h)-f(x)}{h}$ we obtain

$$f'(x)=\lim_{h\to0}\frac{\left[(x+h)^3+2(x+h)^2\right]-\left(x^3+2x^2\right)}{h} \quad \text{at } x=2 \quad f'(2)=\lim_{h\to0}\frac{\left[(2+h)^3+2(2+h)^2\right]-\left(2^3+2\cdot2^2\right)}{h}$$

$$=\lim_{h\to0}\frac{\left[8+h^3+12h+6h^2+8+2h^2+8h\right]-16}{h}=\lim_{h\to0}\frac{16+h^3+20h+8h^2-16}{h}=\lim_{h\to0}\frac{h\left(h^2+8h+20\right)}{h}$$

$$=\lim_{h\to0}\left(h^2+8h+20\right)=0^2+8\cdot0+20=\mathbf{20}$$

e. Given $f(x)=x^{-2}+x^{-1}+1=\dfrac{1}{x^2}+\dfrac{1}{x}+1=\dfrac{x+x^2}{x^3}+1=\dfrac{x+x^2+x^3}{x^3}=\dfrac{\cancel{x}\left(1+x+x^2\right)}{x^{3=2}}=\dfrac{1+x+x^2}{x^2}$, then using

$$f'(x)=\lim_{h\to0}\frac{f(x+h)-f(x)}{h} \text{ we obtain } f'(x)=\lim_{h\to0}\frac{\dfrac{1+x+h+(x+h)^2}{(x+h)^2}-\dfrac{1+x+x^2}{x^2}}{h} \text{ at } x=1$$

$$f'(1)=\lim_{h\to0}\frac{\dfrac{1+1+h+(1+h)^2}{(1+h)^2}-\dfrac{1+1+1^2}{1^2}}{h}=\lim_{h\to0}\frac{\dfrac{2+h+1+h^2+2h}{1+h^2+2h}-3}{h}=\lim_{h\to0}\frac{3+h^2+3h-3\left(1+h^2+2h\right)}{h\left(1+h^2+2h\right)}$$

$$=\lim_{h\to0}\frac{3+h^2+3h-3-3h^2-6h}{h\left(1+h^2+2h\right)}=\lim_{h\to0}\frac{-2h^2-3h}{h\left(1+h^2+2h\right)}=\lim_{h\to0}\frac{\cancel{h}\left(-2h-3\right)}{\cancel{h}\left(1+h^2+2h\right)}=\lim_{h\to0}\frac{-2h-3}{1+h^2+2h}$$

$$=\frac{(-2\cdot0)-3}{1+0^2+2\cdot0}=\frac{-3}{1}=\mathbf{-3}$$

f. Given $f(x)=\sqrt{x}+2$, then using $f'(x)=\lim_{h\to0}\dfrac{f(x+h)-f(x)}{h}$ we obtain $f'(x)=\lim_{h\to0}\dfrac{\left(\sqrt{x+h}+2\right)-\left(\sqrt{x}+2\right)}{h}$

at $x=10$ $f'(10)=\lim_{h\to0}\dfrac{\left(\sqrt{10+h}+2\right)-\left(\sqrt{10}+2\right)}{h}=\lim_{h\to0}\dfrac{\sqrt{10+h}+\cancel{2}-\sqrt{10}+\cancel{2}}{h}=\lim_{h\to0}\dfrac{\sqrt{10+h}-\sqrt{10}}{h}$

$$=\lim_{h\to0}\frac{\left(\sqrt{10+h}\right)-\sqrt{10}}{h}\cdot\frac{\left(\sqrt{10+h}\right)+\sqrt{10}}{\left(\sqrt{10+h}\right)+\sqrt{10}}=\lim_{h\to0}\frac{\cancel{10}+h-\cancel{10}}{h\cdot\left(\sqrt{10+h}+\sqrt{10}\right)}=\lim_{h\to0}\frac{\cancel{h}}{\cancel{h}\cdot\left(\sqrt{10+h}+\sqrt{10}\right)}=\frac{1}{\sqrt{10+0}+\sqrt{10}}$$

$$=\frac{1}{\sqrt{10}+\sqrt{10}}=\frac{1}{2\sqrt{10}}=\frac{1}{2\cdot3.16}=\frac{1}{6.32}=\mathbf{0.158}$$

Section 2.2 Solutions - Differentiation Rules Using the Prime Notation

1. Find the derivative of the following functions. Compare your answers with the practice problem number one in Section 2.1.

a. Given $f(x)=x^2-1$, then $f'(x)=2x^{2-1}-0=\mathbf{2x}$

b. Given $f(x)=x^3+2x-1$, then $f'(x)=3x^{3-1}+2\cdot1x^{1-1}-0=3x^2+2x^0=\mathbf{3x^2+2}$

c. Given $f(x)=\dfrac{x}{x-1}$, then $f'(x)=\dfrac{\left[1\cdot(x-1)\right]-\left[1\cdot x\right]}{(x-1)^2}=\dfrac{\cancel{x}-1-\cancel{x}}{(x-1)^2}=-\dfrac{1}{(x-1)^2}$

d. Given $f(x)=-\dfrac{1}{x^2}$, then $f'(x)=-\dfrac{\left(0\cdot x^2\right)-(2x\cdot1)}{x^4}=-\dfrac{0-2x}{x^4}=\dfrac{2\cancel{x}}{x^{4=3}}=\dfrac{2}{x^3}$

e. Given $f(x)=20x^2-3$, then $f'(x)=(20\cdot2)x^{2-1}-0=\mathbf{40x}$

f. Given $f(x)=\sqrt{x^3}=x^{\frac{3}{2}}$, then $f'(x)=\dfrac{3}{2}x^{\frac{3}{2}-1}=\dfrac{3}{2}x^{\frac{3-2}{2}}=\dfrac{3}{2}x^{\frac{1}{2}}=\dfrac{3}{2}\sqrt{x}$

g. Given $f(x)=\dfrac{10}{\sqrt{x-5}}=\dfrac{10}{(x-5)^{\frac{1}{2}}}$, then $f'(x) = \dfrac{\left[0\cdot(x-5)^{\frac{1}{2}}\right]-\left[10\cdot\frac{1}{2}(x-5)^{-\frac{1}{2}}\right]}{x-5} = \dfrac{0-5(x-5)^{-\frac{1}{2}}}{x-5} = \dfrac{-5}{(x-5)^{\frac{1}{2}}\cdot(x-5)}$

$= \dfrac{-5}{(x-5)^{\frac{1}{2}+1}} = \dfrac{-5}{(x-5)^{\frac{3}{2}}} = \dfrac{-5}{\sqrt{(x-5)^3}} = -\dfrac{5}{(x-5)\sqrt{x-5}}$

h. Given $f(x)=\dfrac{ax+b}{cx}$, then $f'(x) = \dfrac{a\cdot cx-c\cdot(ax+b)}{(cx)^2} = \dfrac{acx-acx-bc}{(cx)^2} = -\dfrac{b\cancel{c}}{c^{2=1}x^2} = -\dfrac{b}{cx^2}$

2. Differentiate the following functions:

a. Given $f(x)=x^2+10x+1$, then $f'(x) = 2x^{2-1}+10x^{1-1}+0 = 2x+10x^0 = \mathbf{2x+10}$

b. Given $f(x)=x^8+3x^2-1$, then $f'(x) = 8x^{8-1}+(3\cdot2)x^{2-1}-0 = \mathbf{8x^7+6x}$

c. Given $f(x)=3x^4-2x^2+5$, then $f'(x) = (3\cdot4)x^{4-1}-(2\cdot2)x^{2-1}+0 = \mathbf{12x^3-4x}$

d. Given $f(x)=2(x^5+10x^4+5x)=2x^5+20x^4+10x$, then $f'(x) = (2\cdot5)x^{5-1}+(20\cdot4)x^{4-1}+(10\cdot1)x^{1-1} = 10x^4+80x^3+10x^0$

$= \mathbf{10x^4+80x^3+10}$

e. Given $f(x)=a^2x^3+b^2x+c^2$, then $f'(x) = (3\cdot a^2)x^{3-1}+(1\cdot b^2)x^{1-1}+0 = 3a^2x^2+b^2x^0 = \mathbf{3a^2x^2+b^2}$

f. Given $f(x)=x^2(x-1)+3x=x^3-x^2+3x$, then $f'(x) = 3x^{3-1}-2x^{2-1}+3x^{1-1} = 3x^2-2x+3x^0 = \mathbf{3x^2-2x+3}$

g. Given $f(x)=(x^3+1)(x^2-5)$, then $f'(x) = \left[3x^2(x^2-5)\right]+\left[2x(x^3+1)\right] = 3x^4-15x^2+2x^4+2x = \mathbf{5x^4-15x^2+2x}$

h. Given $f(x)=(3x^2+x-1)(x-1)$, then $f'(x) = \left[(6x+1)\cdot(x-1)\right]+\left[1\cdot(3x^2+x-1)\right] = 6x^2-6x+x-1+3x^2+x-1$

$= (6+3)x^2+(-6+1+1)x+(-1-1) = \mathbf{9x^2-4x-2}$

i. Given $f(x)=x(x^3+5x^2)-4x=x^4+5x^3-4x$, then $f'(x) = 4x^{4-1}+(5\cdot3)x^{3-1}-(4\cdot1)x^{1-1} = 4x^3+15x^2-4x^0$

$= \mathbf{4x^3+15x^2-4}$

j. Given $f(x)=\dfrac{x^3+1}{x}$, then $f'(x) = \dfrac{\left[(3x^2+0)\cdot x\right]-\left[1\cdot(x^3+1)\right]}{x^2} = \dfrac{3x^3-x^3-1}{x^2} = \dfrac{2x^3-1}{x^2}$

k. Given $f(x)=\dfrac{x^5+2x^2-1}{3x^2}$, then $f'(x) = \dfrac{\left[(5x^4+4x)\cdot3x^2\right]-\left[6x\cdot(x^5+2x^2-1)\right]}{(3x^2)^2} = \dfrac{15x^6+12x^3-6x^6-12x^3+6x}{9x^4}$

$= \dfrac{9x^6+6x}{9x^4} = \dfrac{3x(3x^5+2)}{3=9x^{4=3}} = \dfrac{3x^5+2}{3x^3}$

l. Given $f(x)=\dfrac{x^2}{(x-1)+3x}=\dfrac{x^2}{4x-1}$, then $f'(x) = \dfrac{\left[2x\cdot(4x-1)\right]-\left[4\cdot x^2\right]}{(4x-1)^2} = \dfrac{8x^2-2x-4x^2}{(4x-1)^2} = \dfrac{4x^2-2x}{(4x-1)^2} = \dfrac{2x(2x-1)}{(4x-1)^2}$

m. Given $f(x)=x^2\left(2+\dfrac{1}{x}\right)=2x^2+\dfrac{x^2}{x}=2x^2+x$, then $f'(x) = (2\cdot2)x^{2-1}+x^{1-1} = 4x+x^0 = \mathbf{4x+1}$

n. Given $f(x)=(x+1)\cdot\dfrac{2x}{x-1}=\dfrac{2x^2+2x}{x-1}$, then $f'(x) = \dfrac{\left[(4x+2)\cdot(x-1)\right]-\left[1\cdot(2x^2+2x)\right]}{(x-1)^2} = \dfrac{4x^2-4x+2x-2-2x^2-2x}{(x-1)^2}$

$= \dfrac{(4-2)x^2-4x-2}{(x-1)^2} = \dfrac{2x^2-4x-2}{(x-1)^2}$

o. Given $f(x)=\dfrac{x^3+3x-1}{x^4}$, then $f'(x) = \dfrac{\left[(3x^2+3)\cdot x^4\right]-\left[4x^3\cdot(x^3+3x-1)\right]}{x^8} = \dfrac{3x^6+3x^4-4x^6-12x^4+4x^3}{x^8}$

$$= \frac{-x^6 - 9x^4 + 4x^3}{x^8} = \frac{x^3\left(-x^3 - 9x + 4\right)}{x^{8-5}} = -\frac{x^3 + 9x - 4}{x^5}$$

p. Given $f(x) = \left(x^2 - 1\right)\left(\frac{2x^3 + 5}{x}\right)$, then $f'(x) = 2x \cdot \left(\frac{2x^3 + 5}{x}\right) + \left(\frac{\left(6x^2 \cdot x\right) - \left(2x^3 + 5\right)}{x^2}\right) \cdot \left(x^2 - 1\right) = 4x^3 + 10$

$+ \left(\frac{6x^3 - 2x^3 - 5}{x^2}\right) \cdot \left(x^2 - 1\right) = 4x^3 + 10 + \left(\frac{4x^3 - 5}{x^2}\right)\left(x^2 - 1\right) = 4x^3 + 10 + \frac{4x^5 - 4x^3 - 5x^2 + 5}{x^2} = \frac{8x^5 - 4x^3 + 5x^2 + 5}{x^2}$

q. Given $f(x) = \frac{3x^4 + x^2 + 2}{x - 1}$, then $f'(x) = \frac{\left[\left(12x^3 + 2x\right) \cdot (x - 1)\right] - \left[1 \cdot \left(3x^4 + x^2 + 2\right)\right]}{(x-1)^2} = \frac{12x^4 - 12x^3 + 2x^2 - 2x - 3x^4 - x^2 - 2}{(x-1)^2}$

$= \frac{9x^4 - 12x^3 + x^2 - 2x - 2}{(x-1)^2}$

r. Given $f(x) = x^{-1} + \frac{1}{x^{-2}} = x^{-1} + x^2$, then $f'(x) = -x^{-1-1} + 2x^{2-1} = -x^{-2} + 2x = -\frac{1}{x^2} + 2x$

3. Compute $f'(x)$ at the specified value of x. Compare your answers with the practice problem number two in Section 2.1.

a. Given $f(x) = x^3$, then $f'(x) = 3x^{3-1} = 3x^2$ at $x = 1$ $f'(x) = 3 \cdot 1^2 = \mathbf{3}$

b. Given $f(x) = 1 + 2x$, then $f'(x) = 0 + (2 \cdot 1)x^{1-1} = 2x^0 = 2$ at $x = 0$ $f'(x) = \mathbf{2}$

c. Given $f(x) = x^3 + 1$, then $f'(x) = 3x^{3-1} + 0 = 3x^2$ at $x = -1$ $f'(x) = 3 \cdot (-1)^2 = \mathbf{3}$

d. Given $f(x) = x^2(x + 2) = x^3 + 2x^2$, then $f'(x) = 3x^{3-1} + (2 \cdot 2)x^{2-1} = 3x^2 + 4x$

    at $x = 2$ $f'(x) = 3 \cdot 2^2 + 4 \cdot 2 = 3 \cdot 4 + 8 = 12 + 8 = \mathbf{20}$

e. Given $f(x) = x^{-2} + x^{-1} + 1$, then $f'(x) = -2x^{-2-1} - x^{-1-1} + 0 = -2x^{-3} - x^{-2} = -\frac{2}{x^3} - \frac{1}{x^2}$

    at $x = 1$ $f'(x) = -\frac{2}{1^3} - \frac{1}{1^2} = -2 - 1 = \mathbf{-3}$

f. Given $f(x) = \sqrt{x} + 2 = x^{\frac{1}{2}} + 2$, then $f'(x) = \frac{1}{2}x^{\frac{1}{2}-1} + 0 = \frac{1}{2}x^{\frac{1-2}{2}} = \frac{1}{2}x^{-\frac{1}{2}} = \frac{1}{2x^{\frac{1}{2}}} = \frac{1}{2\sqrt{x}}$

    at $x = 10$ $f'(x) = \frac{1}{2\sqrt{10}} = \frac{1}{2 \cdot 3.16} = \frac{1}{6.32} = \mathbf{0.158}$

4. Find $f'(0)$ and $f'(2)$ for the following functions:

a. Given $f(x) = x^3 - 3x^2 + 5$, then $f'(x) = 3x^{3-1} - (3 \cdot 2)x^{2-1} + 0 = 3x^2 - 6x$

    Therefore, $f'(0) = \left(3 \cdot 0^2\right) - (6 \cdot 0) = 0 - 0 = \mathbf{0}$ and $f'(2) = \left(3 \cdot 2^2\right) - (6 \cdot 2) = 12 - 12 = \mathbf{0}$

b. Given $f(x) = \left(x^3 + 1\right)(x - 1)$, then $f'(x) = \left[3x^2 \cdot (x - 1)\right] + \left[1 \cdot \left(x^3 + 1\right)\right] = 3x^3 - 3x^2 + x^3 + 1 = 4x^3 - 3x^2 + 1$

    Therefore, $f'(0) = \left(4 \cdot 0^3\right) - \left(3 \cdot 0^2\right) + 1 = 0 - 0 + 1 = \mathbf{1}$ and $f'(2) = \left(4 \cdot 2^3\right) - \left(3 \cdot 2^2\right) + 1 = 32 - 12 + 1 = \mathbf{21}$

c. Given $f(x) = x\left(x^2 + 1\right) = x^3 + x$, then $f'(x) = 3x^{2-1} + x^{1-1} = 3x + x^0 = 3x^2 + 1$

    Therefore, $f'(0) = \left(3 \cdot 0^2\right) + 1 = 0 + 1 = \mathbf{1}$ and $f'(2) = \left(3 \cdot 2^2\right) + 1 = 12 + 1 = \mathbf{13}$

d. Given $f(x) = 2x^5 + 10x^4 - 4x$, then $f'(x) = (2 \cdot 5)x^{5-1} + (10 \cdot 4)x^{4-1} - 4x^{1-1} = 10x^4 + 40x^3 - 4x^0 = 10x^4 + 40x^3 - 4$

    Therefore, $f'(0) = \left(10 \cdot 0^4\right) + \left(40 \cdot 0^3\right) - 4 = \mathbf{-4}$ and $f'(2) = \left(10 \cdot 2^4\right) + \left(40 \cdot 2^3\right) - 4 = 160 + 320 - 4 = \mathbf{476}$

e. Given $f(x)=2x^{-2}-3x^{-1}+5x$, then $f'(x) = (2\cdot-2)x^{-2-1}+(-3\cdot-1)x^{-1-1}+5x^{1-1} = -4x^{-3}+3x^{-2}+5x^0$

$= -4x^{-3}+3x^{-2}+5 = -\dfrac{4}{x^3}+\dfrac{3}{x^2}+5$. Therefore, $f'(0) = -\dfrac{4}{0^3}+\dfrac{3}{0^2}+5$ **which is undefined due to division by zero** and

$f'(2) = -\dfrac{4}{2^3}+\dfrac{3}{2^2}+5 = -\dfrac{4}{8}+\dfrac{3}{4}+5 = -0.5+0.75+5 = \mathbf{5.25}$

f. Given $f(x)=x^{-2}(x^5-x^3)+x=x^3-x+x=x^3$, then $f'(x) = 3x^{3-1} = 3x^2$

Therefore, $f'(0) = 3\cdot0^2 = \mathbf{0}$ and $f'(2) = 3\cdot2^2 = \mathbf{12}$

g. Given $f(x)=\dfrac{x}{1+x^2}$, then $f'(x) = \dfrac{[1\cdot(1+x^2)]-[2x\cdot x]}{(1+x^2)^2} = \dfrac{1+x^2-2x^2}{(1+x^2)^2} = \dfrac{1-x^2}{(1+x^2)^2}$

Therefore, $f'(0) = \dfrac{1-0^2}{(1+0^2)^2} = \mathbf{1}$ and $f'(2) = \dfrac{1-2^2}{(1+2^2)^2} = \dfrac{1-4}{(1+4)^2} = \dfrac{-3}{5^2} = -\dfrac{3}{25}$

h. Given $f(x)=\dfrac{1}{x}+x^3$, then $f'(x) = \dfrac{(0\cdot x)-(1\cdot1)}{x^2}+3x^{3-1} = -\dfrac{1}{x^2}+3x^2$

Therefore, $f'(0) = -\dfrac{1}{0^2}+(3\cdot0^2)$ **which is undefined due to division by zero** and $f'(2) = -\dfrac{1}{2^2}+(3\cdot2^2) = -\dfrac{1}{4}+12 = \mathbf{11.75}$

i. Given $f(x)=\dfrac{ax^2+bx}{cx-d}$, then $f'(x) = \dfrac{[(2ax+b)\cdot(cx-d)]-[c\cdot(ax^2+bx)]}{(cx-d)^2} = \dfrac{2acx^2-2adx+bcx-bd-acx^2-bcx}{(cx-d)^2}$

$= \dfrac{acx^2-2adx-bd}{(cx-d)^2}$. Therefore, $f'(0) = \dfrac{(ac\cdot0^2)-(2ad\cdot0)-bd}{(c\cdot0-d)^2} = \dfrac{0-0-bd}{(0-d)^2} = \dfrac{-bd}{d^{2=1}} = -\dfrac{b}{d}$ and

$f'(2) = \dfrac{(ac\cdot2^2)-(2ad\cdot2)-bd}{(c\cdot2-d)^2} = \dfrac{\mathbf{4ac-4ad-bd}}{\mathbf{(2c-d)^2}}$

5. Given $f(x)=x^2+1$ and $g(x)=2x-5$ find $h(x)$ and $h'(x)$.

a. Given $h(x)=x^3f(x)$ where $f(x)=x^2+1$, then $h(x) = x^3(x^2+1) = \mathbf{x^5+x^3}$ and $h'(x) = \mathbf{5x^4+3x^2}$

b. Given $f(x)=3+h(x)$ where $f(x)=x^2+1$, then $h(x) = f(x)-3 = (x^2+1)-3 = \mathbf{x^2-2}$ and $h'(x) = \mathbf{2x}$

c. Given $2g(x)=h(x)-1$ where $g(x)=2x-5$, then $h(x) = 2g(x)+1 = 2(2x-5)+1 = 4x-10+1 = \mathbf{4x-9}$ and $h'(x) = \mathbf{4}$

d. Given $3h(x)=2xg(x)-1$ where $g(x)=2x-5$, then $h(x) = \dfrac{2xg(x)-1}{3} = \dfrac{2x(2x-5)-1}{3} = \dfrac{4x^2-10x-1}{3}$ and

$h'(x) = \dfrac{[(8x-10)\cdot3]-[0\cdot(4x^2-10x-1)]}{3^2} = \dfrac{3(8x-10)}{9} = \dfrac{\mathbf{8x-10}}{\mathbf{3}}$

e. Given $3[f(x)]^2-2h(x)=1$ where $f(x)=x^2+1$, then $h(x) = \dfrac{1}{2}(-1+3[f(x)]^2) = -\dfrac{1}{2}+\dfrac{3}{2}[f(x)]^2 = -\dfrac{1}{2}+\dfrac{3}{2}(x^2+1)^2$ and

$h'(x) = 3(x^2+1)\cdot2x = \mathbf{6x^3+6x}$

f. Given $h(x)=g(x)\cdot3f(x)$ where $f(x)=x^2+1$ and $g(x)=2x-5$, then $h(x) = (2x-5)\cdot3(x^2+1) = (2x-5)(3x^2+3)$

$= \mathbf{6x^3-15x^2+6x-15}$ and $h'(x) = \mathbf{18x^2-30x+6}$

g. Given $3h(x)-f(x)=0$ where $f(x)=x^2+1$, then $h(x) = \dfrac{f(x)}{3} = \dfrac{x^2+1}{3}$ and $h'(x) = \dfrac{2}{3}x$

h. Given $2g(x)+h(x)=f(x)$ where $f(x)=x^2+1$ and $g(x)=2x-5$, then $h(x)=f(x)-2g(x)=\left(x^2+1\right)-2(2x-5)$

$=x^2+1-4x+10=\textbf{\textit{x}}^2-\textbf{4\textit{x}}+\textbf{11}$ and $h'(x)=\textbf{2\textit{x}}-\textbf{4}$

i. Given $f(x)=x^3+5x^2+h(x)$ where $f(x)=x^2+1$, then $h(x)=f(x)-x^3-5x^2=\left(x^2+1\right)-x^3-5x^2=-x^3+(-5+1)x^2+1$

$=-\textbf{\textit{x}}^3-\textbf{4\textit{x}}^2+\textbf{1}$ and $h'(x)=-\textbf{3\textit{x}}^2-\textbf{8\textit{x}}$

j. Given $h(x)=\dfrac{x^3+1}{x}-f(x)$ where $f(x)=x^2+1$, then $h(x)=\dfrac{x^3+1}{x}-\left(x^2+1\right)$ and $h'(x)=\dfrac{\left[3x^2\cdot x\right]-\left[1\cdot\left(x^3+1\right)\right]}{x^2}-2x$

$=\dfrac{3x^3-x^3-1}{x^2}-2x=\dfrac{\textbf{2\textit{x}}^3-\textbf{1}}{\textbf{\textit{x}}^2}-\textbf{2\textit{x}}$

k. Given $h(x)=2f(x)+g(x)$ where $f(x)=x^2+1$ and $g(x)=2x-5$, then $h(x)=2\left(x^2+1\right)+(2x-5)=2x^2+2+2x-5$

$=\textbf{2\textit{x}}^2+\textbf{2\textit{x}}-\textbf{3}$ and $h'(x)=\textbf{4\textit{x}}+\textbf{2}$

l. Given $[h(x)]^2-f(x)=10$ where $f(x)=x^2+1$, then $[h(x)]^2=10+f(x)$; $h(x)=\sqrt{10+f(x)}=\left[10+\left(x^2+1\right)\right]^{\frac{1}{2}}$

$=\left(x^2+11\right)^{\frac{1}{2}}$ and $h'(x)=\dfrac{1}{2}\left(x^2+11\right)^{\frac{1}{2}-1}\cdot2x=\textbf{\textit{x}}\left(\textbf{\textit{x}}^2+\textbf{11}\right)^{-\frac{1}{2}}$

m. Given $f(x)=\dfrac{2g(x)}{h(x)}$ where $f(x)=x^2+1$ and $g(x)=2x-5$, then $h(x)=\dfrac{2g(x)}{f(x)}=\dfrac{2(2x-5)}{x^2+1}=\dfrac{4x-10}{x^2+1}$ and $h'(x)$

$=\dfrac{\left[4\cdot\left(x^2+1\right)\right]-\left[2x\cdot(4x-10)\right]}{\left(x^2+1\right)^2}=\dfrac{4x^2+4-8x^2+20x}{\left(x^2+1\right)^2}=\dfrac{-\textbf{4\textit{x}}^2+\textbf{20\textit{x}}+\textbf{4}}{\left(\textbf{\textit{x}}^2+\textbf{1}\right)^2}$

n. Given $\dfrac{3f(x)}{h(x)}=\dfrac{1}{x}$ where $f(x)=x^2+1$, then $h(x)=3x\,f(x)=3x\left(x^2+1\right)=\textbf{3\textit{x}}^3+\textbf{3\textit{x}}$ and $h'(x)=\textbf{9\textit{x}}^2+\textbf{3}$

o. Given $f(x)=\dfrac{1}{h(x)+4}$, which is equivalent to $\dfrac{f(x)}{1}=\dfrac{1}{h(x)+4}$; $f(x)\cdot[h(x)+4]=1\cdot1$; $f(x)h(x)+4f(x)=1$

; $f(x)h(x)=1-4f(x)$, and $f(x)=x^2+1$, then $h(x)=\dfrac{1-4f(x)}{f(x)}=\dfrac{1}{f(x)}-\dfrac{4f(x)}{f(x)}=\dfrac{1}{f(x)}-4=\dfrac{1}{x^2+1}-\textbf{4}$ and $h'(x)$

$=\dfrac{\left[0\cdot\left(x^2+1\right)\right]-[2x\cdot1]}{\left(x^2+1\right)^2}-0=\dfrac{0-2x}{\left(x^2+1\right)^2}=-\dfrac{\textbf{2\textit{x}}}{\left(\textbf{\textit{x}}^2+\textbf{1}\right)^2}$

Section 2.3 Solutions - Differentiation Rules Using the $\frac{d}{dx}$ Notation

1. Find $\dfrac{dy}{dx}$ for the following functions:

a. Given $y=x^5+3x^2+1$, then $\dfrac{dy}{dx}=\dfrac{d}{dx}\left(x^5+3x^2+1\right)=\dfrac{d}{dx}x^5+\dfrac{d}{dx}3x^2+\dfrac{d}{dx}1=5x^4+(3\cdot2)x+0=\textbf{5\textit{x}}^4+\textbf{6\textit{x}}$

b. Given $y=3x^2+5$, then $\dfrac{dy}{dx}=\dfrac{d}{dx}\left(3x^2+5\right)=\dfrac{d}{dx}3x^2+\dfrac{d}{dx}5=(3\cdot2)x+0=\textbf{6\textit{x}}$

c. Given $y=x^3-\dfrac{1}{x}$, then $\dfrac{dy}{dx}=\dfrac{d}{dx}\left(x^3-\dfrac{1}{x}\right)=\dfrac{d}{dx}\left(x^3-x^{-1}\right)=\dfrac{d}{dx}x^3-\dfrac{d}{dx}x^{-1}=3x^2+x^{-1-1}=3x^2+x^{-2}=\textbf{3\textit{x}}^2+\dfrac{\textbf{1}}{\textbf{\textit{x}}^2}$

d. Given $y=\dfrac{x^2}{1-x^3}$, then $\dfrac{dy}{dx}=\dfrac{d}{dx}\left(\dfrac{x^2}{1-x^3}\right)=\dfrac{\left[\left(1-x^3\right)\dfrac{d}{dx}x^2\right]-\left[x^2\dfrac{d}{dx}\left(1-x^3\right)\right]}{\left(1-x^3\right)^2}=\dfrac{\left[\left(1-x^3\right)\cdot2x\right]-\left[x^2\cdot\left(-3x^2\right)\right]}{\left(1-x^3\right)^2}$

$=\dfrac{2x-2x^4+3x^4}{\left(1-x^3\right)^2}=\dfrac{x^4-2x}{\left(1-x^3\right)^2}=\dfrac{\textbf{\textit{x}}\left(\textbf{\textit{x}}^3-\textbf{2}\right)}{\left(\textbf{1}-\textbf{\textit{x}}^3\right)^2}$

e. Given $y = 4x^2 + \dfrac{1}{x-1}$, then $\dfrac{dy}{dx} = \dfrac{d}{dx}\left(4x^2 + \dfrac{1}{x-1}\right) = \dfrac{d}{dx}\left(4x^2\right) + \dfrac{d}{dx}\left(\dfrac{1}{x-1}\right) = 8x + \dfrac{\left[(x-1)\cdot\dfrac{d}{dx}1\right] - \left[1\cdot\dfrac{d}{dx}(x-1)\right]}{(x-1)^2}$

$= 8x + \dfrac{0-1}{(x-1)^2} = \mathbf{8x - \dfrac{1}{(x-1)^2}}$

f. Given $y = \dfrac{x^2+2x}{x^3+1}$, then $\dfrac{dy}{dx} = \dfrac{d}{dx}\left(\dfrac{x^2+2x}{x^3+1}\right) = \dfrac{\left[(x^3+1)\dfrac{d}{dx}(x^2+2x)\right] - \left[(x^2+2x)\dfrac{d}{dx}(x^3+1)\right]}{(x^3+1)^2} = \dfrac{\left[(x^3+1)\cdot(2x+2)\right]}{(x^3+1)^2}$

$\dfrac{-\left[(x^2+2x)\cdot 3x^2\right]}{(x^3+1)^2} = \dfrac{2x^4+2x^3+2x+2-3x^4-6x^3}{(x^3+1)^2} = \dfrac{-x^4-4x^3+2x+2}{(x^3+1)^2}$

g. Given $y = x^3(x^2+5x-2) = x^5+5x^4-2x^3$, then $\dfrac{dy}{dx} = \dfrac{d}{dx}(x^5+5x^4-2x^3) = \dfrac{d}{dx}x^5 + \dfrac{d}{dx}5x^4 - \dfrac{d}{dx}2x^3 = \mathbf{5x^4 + 20x^3 - 6x^2}$

h. Given $y = x^2(x+3)(x-1) = (x^3+3x^2)(x-1)$, then $\dfrac{dy}{dx} = \dfrac{d}{dx}\left[(x^3+3x^2)(x-1)\right] = \left[(x-1)\dfrac{d}{dx}(x^3+3x^2)\right] + \left[(x^3+3x^2)\dfrac{d}{dx}(x-1)\right]$

$= \left[(x-1)\cdot(3x^2+6x)\right] + \left[(x^3+3x^2)\cdot 1\right] = 3x^3+6x^2-3x^2-6x+x^3+3x^2 = 4x^3+6x^2-6x = \mathbf{2x(2x^2+3x-3)}$

i. Given $y = 5x - \dfrac{1}{x^3}$, then $\dfrac{dy}{dx} = \dfrac{d}{dx}\left(5x - \dfrac{1}{x^3}\right) = \dfrac{d}{dx}(5x - x^{-3}) = \dfrac{d}{dx}5x - \dfrac{d}{dx}x^{-3} = 5 + 3x^{-3-1} = 5 + 3x^{-4} = \mathbf{5 + \dfrac{3}{x^4}}$

j. Given $y = \dfrac{(x-1)(x+3)}{x^2} = \dfrac{x^2+3x-x-3}{x^2} = \dfrac{x^2+(3-1)x-3}{x^2} = \dfrac{x^2+2x-3}{x^2}$, then $\dfrac{dy}{dx} = \dfrac{d}{dx}\left[\dfrac{x^2+2x-3}{x^2}\right]$

$= \dfrac{\left[x^2\dfrac{d}{dx}(x^2+2x-3)\right] - \left[(x^2+2x-3)\dfrac{d}{dx}x^2\right]}{x^4} = \dfrac{\left[x^2\cdot(2x+2)\right] - \left[(x^2+2x-3)\cdot 2x\right]}{x^4} = \dfrac{2x^3+2x^2-2x^3-4x^2+6x}{x^4}$

$= \dfrac{-2x^2+6x}{x^4} = \dfrac{-2x(x-3)}{x^{4=3}} = -\dfrac{\mathbf{2(x-3)}}{\mathbf{x^3}}$

k. Given $y = x\left(\dfrac{x-1}{3}\right) = \dfrac{x^2-x}{3}$, then $\dfrac{dy}{dx} = \dfrac{d}{dx}\left(\dfrac{x^2-x}{3}\right) = \dfrac{\left[3\cdot\dfrac{d}{dx}(x^2-x)\right] - \left[(x^2-x)\cdot\dfrac{d}{dx}3\right]}{3^2} = \dfrac{[3\cdot(2x-1)] - [(x^2-x)\cdot 0]}{9}$

$= \dfrac{3(2x-1)-0}{9} = \dfrac{3(2x-1)}{9=3} = \dfrac{\mathbf{2x-1}}{\mathbf{3}}$

l. Given $y = x^2(x+3)^{-1}$, then $\dfrac{dy}{dx} = \dfrac{d}{dx}\left[x^2(x+3)^{-1}\right] = \left[(x+3)^{-1}\dfrac{d}{dx}x^2\right] + \left[x^2\dfrac{d}{dx}(x+3)^{-1}\right] = \left[(x+3)^{-1}\cdot 2x\right] + \left[x^2\cdot\left(-(x+3)^{-2}\right)\right]$

$= 2x(x+3)^{-1} - x^2(x+3)^{-2} = \dfrac{\mathbf{2x}}{\mathbf{x+3}} - \dfrac{\mathbf{x^2}}{\mathbf{(x+3)^2}}$

m. Given $y = \left(\dfrac{x}{1+x}\right)\left(\dfrac{x-3}{5}\right)$, then $\dfrac{dy}{dx} = \left(\dfrac{x-3}{5}\right)\dfrac{d}{dx}\left(\dfrac{x}{1+x}\right) + \left(\dfrac{x}{1+x}\right)\dfrac{d}{dy}\left(\dfrac{x-3}{5}\right) = \dfrac{x-3}{5}\cdot\dfrac{(1+x)\dfrac{d}{dx}x - x\dfrac{d}{dx}(1+x)}{(1+x)^2}$

$+ \dfrac{x}{1+x}\cdot\dfrac{5\dfrac{d}{dx}(x-3) - (x-3)\dfrac{d}{dx}5}{5^2} = \dfrac{x-3}{5}\cdot\dfrac{[(1+x)\cdot 1]-[x\cdot 1]}{(1+x)^2} + \dfrac{x}{1+x}\cdot\dfrac{[5\cdot 1]-[(x-3)\cdot 0]}{25} = \dfrac{x-3}{5}\cdot\dfrac{\cancel{x}+1-\cancel{x}}{(1+x)^2} + \dfrac{x}{1+x}\cdot\dfrac{\cancel{5}}{\cancel{25}_5}$

$= \dfrac{x-3}{5}\cdot\dfrac{1}{(1+x)^2} + \dfrac{x}{1+x}\cdot\dfrac{1}{5} = \dfrac{\mathbf{x-3}}{\mathbf{5(1+x)^2}} + \dfrac{\mathbf{x}}{\mathbf{5(1+x)}}$

n. Given $y = x^3\left(1 + \dfrac{1}{x-1}\right) = x^3 + \dfrac{x^3}{x-1}$, then $\dfrac{dy}{dx} = \dfrac{d}{dx}x^3 + \dfrac{d}{dx}\left(\dfrac{x^3}{x-1}\right) = 3x^2 + \dfrac{\left[(x-1)\dfrac{d}{dx}x^3\right] - \left[x^3\dfrac{d}{dx}(x-1)\right]}{(x-1)^2}$

$= 3x^2 + \dfrac{\left[(x-1)\cdot 3x^2\right] - \left[x^3 \cdot 1\right]}{(x-1)^2} = 3x^2 + \dfrac{3x^3 - 3x^2 - x^3}{(x-1)^2} = 3x^2 + \dfrac{2x^3 - 3x^2}{(x-1)^2} = \mathbf{3x^2 + \dfrac{x^2(2x-3)}{(x-1)^2}}$

o. Given $y = \dfrac{1}{x}\left(\dfrac{2x-1}{3x+1}\right) = \dfrac{2x-1}{3x^2 + x}$, then $\dfrac{dy}{dx} = \dfrac{d}{dx}\left(\dfrac{2x-1}{3x^2 + x}\right) = \dfrac{\left[(3x^2 + x)\dfrac{d}{dx}(2x-1)\right] - \left[(2x-1)\dfrac{d}{dx}(3x^2 + x)\right]}{(3x^2 + x)^2}$

$= \dfrac{\left[(3x^2 + x)\cdot 2\right] - \left[(2x-1)\cdot(6x+1)\right]}{(3x^2 + x)^2} = \dfrac{6x^2 + 2x - (12x^2 + 2x - 6x - 1)}{(3x^2 + x)^2} = \dfrac{6x^2 + 2x - 12x^2 - 2x + 6x + 1}{(3x^2 + x)^2} = \dfrac{\mathbf{-6x^2 + 6x + 1}}{\mathbf{(3x^2 + x)^2}}$

p. Given $y = \dfrac{ax^2 + bx + c}{bx}$, then $\dfrac{dy}{dx} = \dfrac{d}{dx}\left(\dfrac{ax^2 + bx + c}{bx}\right) = \dfrac{\left[bx\dfrac{d}{dx}(ax^2 + bx + c)\right] - \left[(ax^2 + bx + c)\dfrac{d}{dx}bx\right]}{(bx)^2} = \dfrac{bx\cdot(2ax+b)}{(bx)^2}$

$\dfrac{-(ax^2 + bx + c)\cdot b}{(bx)^2} = \dfrac{2abx^2 + b^2x - abx^2 - b^2x - bc}{b^2 x^2} = \dfrac{abx^2 - bc}{b^2 x^2} = \dfrac{b(ax^2 - c)}{b^{2=1} x^2} = \dfrac{\mathbf{ax^2 - c}}{\mathbf{bx^2}}$

q. Given $y = \dfrac{x^3 - 2}{x^4 - 3}$, then $\dfrac{dy}{dx} = \dfrac{d}{dx}\left(\dfrac{x^3 - 2}{x^4 - 3}\right) = \dfrac{\left[(x^4 - 3)\dfrac{d}{dx}(x^3 - 2)\right] - \left[(x^3 - 2)\dfrac{d}{dx}(x^4 - 3)\right]}{(x^4 - 3)^2} = \dfrac{\left[(x^4 - 3)\cdot 3x^2\right] - \left[(x^3 - 2)\cdot 4x^3\right]}{(x^4 - 3)^2}$

$= \dfrac{3x^6 - 9x^2 - 4x^6 + 8x^3}{(x^4 - 3)^2} = \dfrac{\mathbf{-x^6 + 8x^3 - 9x^2}}{\mathbf{(x^4 - 3)^2}}$

r. Given $y = \dfrac{5x}{(1+x)^2}$, then $\dfrac{dy}{dx} = \dfrac{d}{dx}\left(\dfrac{5x}{(1+x)^2}\right) = \dfrac{\left[(1+x)^2 \dfrac{d}{dx}5x\right] - \left[5x\dfrac{d}{dx}(1+x)^2\right]}{(1+x)^4} = \dfrac{\left[(1+x)^2 \cdot 5\right] - \left[5x\cdot 2(1+x)\right]}{(1+x)^4}$

$= \dfrac{5(x^2 + 2x + 1) - 10x(1+x)}{(1+x)^4} = \dfrac{5x^2 + 10x + 5 - 10x - 10x^2}{(1+x)^4} = \dfrac{\mathbf{-5x^2 + 5}}{\mathbf{(1+x)^4}}$

2. Find the derivative of the following functions:

a. $\dfrac{d}{dt}(3t^2 + 5t) = \dfrac{d}{dt}(3t^2) + \dfrac{d}{dt}(5t) = (3\cdot 2)t^{2-1} + (5\cdot 1)t^{1-1} = 6t + 5t^0 = \mathbf{6t + 5}$

b. $\dfrac{d}{dx}(6x^3 + 5x - 2) = \dfrac{d}{dx}(6x^3) + \dfrac{d}{dx}(5x) + \dfrac{d}{dx}(-2) = (6\cdot 3)x^{3-1} + (5\cdot 1)x^{1-1} + 0 = 18x^2 + 5x^0 = \mathbf{18x^2 + 5}$

c. $\dfrac{d}{du}(u^3 + 2u^2 + 5) = \dfrac{d}{du}(u^3) + \dfrac{d}{du}(2u^2) + \dfrac{d}{du}5 = 3u^{3-1} + (2\cdot 2)u^{2-1} + 0 = \mathbf{3u^2 + 4u}$

d. $\dfrac{d}{dt}\left(\dfrac{t^2 + 2t}{5}\right) = \dfrac{\left[5\cdot\dfrac{d}{dt}(t^2 + 2t)\right] - \left[(t^2 + 2t)\cdot\dfrac{d}{dt}5\right]}{5^2} = \dfrac{[5\cdot(2t+2)] - [(t^2 + 2t)\cdot 0]}{25} = \dfrac{5(2t+2) - 0}{25} = \dfrac{\cancel{5}(2t+2)}{\underset{5}{\cancel{25}}} = \dfrac{\mathbf{2t+2}}{\mathbf{5}}$ or,

$\dfrac{d}{dt}\left(\dfrac{t^2 + 2t}{5}\right) = \dfrac{1}{5}\dfrac{d}{dt}(t^2 + 2t) = \dfrac{1}{5}\left(\dfrac{d}{dt}t^2 + \dfrac{d}{dt}2t\right) = \dfrac{1}{5}(2t+2) = \dfrac{\mathbf{2t+2}}{\mathbf{5}}$

e. $\dfrac{d}{ds}\left(\dfrac{s^3 + 3s - 1}{s^2}\right) = \dfrac{\left[s^2 \cdot \dfrac{d}{dt}(s^3 + 3s - 1)\right] - \left[(s^3 + 3s - 1)\cdot\dfrac{d}{dt}s^2\right]}{s^4} = \dfrac{\left[s^2 \cdot(3s^2 + 3)\right] - \left[(s^3 + 3s - 1)\cdot 2s\right]}{s^4} = \dfrac{\mathbf{3s^4 + 3s^2 - 2s^4 - 6s^2 + 2s}}{\mathbf{s^4}}$

$$\frac{s^4 - 3s^2 + 2s}{s^4} = \frac{s(s^3 - 3s + 2)}{s^{4=3}} = \frac{s^3 - 3s + 2}{s^3}$$

f. $\dfrac{d}{dw}\left(w^3 + \dfrac{w^2}{1+w}\right) = \dfrac{d}{dw}w^3 + \dfrac{d}{dw}\left(\dfrac{w^2}{1+w}\right) = 3w^2 + \dfrac{\left[(1+w)\cdot\frac{d}{dw}w^2\right] - \left[w^2\cdot\frac{d}{dt}(1+w)\right]}{(1+w)^2} = 3w^2 + \dfrac{[(1+w)\cdot 2w] - [w^2\cdot 1]}{(1+w)^2}$

$= 3w^2 + \dfrac{2w + 2w^2 - w^2}{(1+w)^2} = \mathbf{3w^2 + \dfrac{w^2 + 2w}{(1+w)^2}}$

g. $\dfrac{d}{dt}\left[t^2(t+1)(t^2-3)\right] = \dfrac{d}{dt}\left[(t^3+t^2)(t^2-3)\right] = (t^2-3)\dfrac{d}{dt}(t^3+t^2) + (t^3+t^2)\dfrac{d}{dt}(t^2-3) = (t^2-3)\cdot(3t^2+2t) + (t^3+t^2)\cdot 2t$

$= 3t^4 + 2t^3 - 9t^2 - 6t + 2t^4 + 2t^3 = \mathbf{5t^4 + 4t^3 - 9t^2 - 6t}$, or

$\dfrac{d}{dt}\left[t^2(t+1)(t^2-3)\right] = \dfrac{d}{dt}\left[(t^3+t^2)(t^2-3)\right] = \dfrac{d}{dt}\left(t^5 + t^4 - 3t^3 - 3t^2\right) = \dfrac{d}{dt}t^5 + \dfrac{d}{dt}t^4 - 3\dfrac{d}{dt}t^3 - 3\dfrac{d}{dt}t^2 = \mathbf{5t^4 + 4t^3 - 9t^2 - 6t}$

h. $\dfrac{d}{dx}\left[(x+1)(x^2+5)\right] = \left[(x^2+5)\dfrac{d}{dx}(x+1)\right] + \left[(x+1)\dfrac{d}{dx}(x^2+5)\right] = \left[(x^2+5)\cdot 1\right] + [(x+1)\cdot 2x] = x^2 + 5 + 2x^2 + 2x = \mathbf{3x^2 + 2x + 5}$

or, $\dfrac{d}{dx}\left[(x+1)(x^2+5)\right] = \dfrac{d}{dx}(x^3 + x^2 + 5x + 5) = \dfrac{d}{dx}x^3 + \dfrac{d}{dx}x^2 + \dfrac{d}{dx}5x + \dfrac{d}{dx}5 = \mathbf{3x^2 + 2x + 5}$

i. $\dfrac{d}{du}\left[\dfrac{u^2}{1-u} - \dfrac{u}{1+u}\right] = \dfrac{d}{du}\left(\dfrac{u^2}{1-u}\right) - \dfrac{d}{du}\left(\dfrac{u}{1+u}\right) = \dfrac{\left[(1-u)\cdot\frac{d}{du}u^2\right] - \left[u^2\cdot\frac{d}{du}(1-u)\right]}{(1-u)^2} - \dfrac{\left[(1+u)\cdot\frac{d}{du}u\right] - \left[u\cdot\frac{d}{du}(1+u)\right]}{(1+u)^2}$

$= \dfrac{[(1-u)\cdot 2u] - [u^2\cdot -1]}{(1-u)^2} - \dfrac{[(1+u)\cdot 1] - [u\cdot 1]}{(1+u)^2} = \dfrac{2u - 2u^2 + u^2}{(1-u)^2} - \dfrac{1 + u - u}{(1+u)^2} = \dfrac{2u - u^2}{(1-u)^2} - \dfrac{1}{(1+u)^2} = \mathbf{\dfrac{u(2-u)}{(1-u)^2} - \dfrac{1}{(1+u)^2}}$

j. $\dfrac{d}{dr}\left(\dfrac{3r^3 - 2r^2 + 1}{r}\right) = \dfrac{\left[r\cdot\frac{d}{dr}(3r^3 - 2r^2 + 1)\right] - \left[(3r^3 - 2r^2 + 1)\cdot\frac{d}{dr}r\right]}{r^2} = \dfrac{[r\cdot(9r^2 - 4r)] - [(3r^3 - 2r^2 + 1)\cdot 1]}{r^2}$

$= \dfrac{9r^3 - 4r^2 - 3r^3 + 2r^2 - 1}{r^2} = \mathbf{\dfrac{6r^3 - 2r^2 - 1}{r^2}}$

k. $\dfrac{d}{ds}\left[\dfrac{3s^2}{s^3+1} - \dfrac{1}{s^2}\right] = \dfrac{d}{ds}\left(\dfrac{3s^2}{s^3+1}\right) - \dfrac{d}{ds}\left(\dfrac{1}{s^2}\right) = \dfrac{\left[(s^3+1)\cdot\frac{d}{ds}3s^2\right] - \left[3s^2\cdot\frac{d}{ds}(s^3+1)\right]}{(s^3+1)^2} - \dfrac{\left[s^2\cdot\frac{d}{ds}1\right] - \left[1\cdot\frac{d}{ds}s^2\right]}{s^4}$

$= \dfrac{[(s^3+1)\cdot 6s] - [3s^2\cdot 3s^2]}{(s^3+1)^2} - \dfrac{(s^2\cdot 0) - (1\cdot 2s)}{s^4} = \dfrac{6s^4 + 6s - 9s^4}{(s^3+1)^2} - \dfrac{0 - 2s}{s^4} = \dfrac{-3s^4 + 6s}{(s^3+1)^2} + \dfrac{2s}{s^{4=3}} = \mathbf{-\dfrac{3s(s^3-2)}{(s^3+1)^2} + \dfrac{2}{s^3}}$

l. $\dfrac{d}{du}\left[\dfrac{u^3}{1-u} - \dfrac{u+1}{u^2}\right] = \dfrac{d}{du}\left(\dfrac{u^3}{1-u}\right) - \dfrac{d}{du}\left(\dfrac{u+1}{u^2}\right) = \dfrac{\left[(1-u)\cdot\frac{d}{du}u^3\right] - \left[u^3\cdot\frac{d}{du}(1-u)\right]}{(1-u)^2} - \dfrac{\left[u^2\cdot\frac{d}{du}(u+1)\right] - \left[(u+1)\cdot\frac{d}{du}u^2\right]}{u^4}$

$= \dfrac{[(1-u)\cdot 3u^2] - [u^3\cdot -1]}{(1-u)^2} - \dfrac{[u^2\cdot 1] - [(u+1)\cdot 2u]}{u^4} = \dfrac{3u^2 - 3u^3 + u^3}{(1-u)^2} - \dfrac{u^2 - 2u^2 - 2u}{u^4} = \dfrac{-2u^3 + 3u^2}{(1-u)^2} - \dfrac{-u^2 - 2u}{u^4}$

$= \dfrac{u^2(-2u+3)}{(1-u)^2} + \dfrac{u(u+2)}{u^{4=3}} = \mathbf{-\dfrac{u^2(2u-3)}{(1-u)^2} + \dfrac{u+2}{u^3}}$

3. Find the derivative of the following functions at the specified value.

a. $\dfrac{d}{dx}(x^3 + 3x^2 + 1) = \dfrac{d}{dx}(x^3) + \dfrac{d}{dx}(3x^2) + \dfrac{d}{dx}(1) = 3x^{3-1} + (3\cdot 2)x^{2-1} + 0 = \mathbf{3x^2 + 6x}$

at $x = 2$ $\dfrac{d}{dx}\left(x^3 + 3x^2 + 1\right) = 3 \cdot 2^2 + 6 \cdot 2 = 12 + 12 = \mathbf{24}$

b. $\dfrac{d}{dx}\left[(x+1)\left(x^2 - 1\right)\right] = \left(x^2 - 1\right)\dfrac{d}{dx}(x+1) + (x+1)\dfrac{d}{dx}\left(x^2 - 1\right) = \left(x^2 - 1\right) \cdot 1 + (x+1) \cdot 2x = x^2 - 1 + 2x^2 + 2x = 3x^2 + 2x - 1$

at $x = 1$ $\dfrac{d}{dx}\left[(x+1)\left(x^2 - 1\right)\right] = \left(3 \cdot 1^2\right) + (2 \cdot 1) - 1 = 3 + 2 - 1 = \mathbf{4}$

c. $\dfrac{d}{ds}\left[3s^2(s-1)\right] = \dfrac{d}{ds}\left(3s^3 - 3s^2\right) = \dfrac{d}{ds}\left(3s^3\right) + \dfrac{d}{ds}\left(-3s^2\right) = 9s^2 - 6s$

at $s = 0$ $\dfrac{d}{ds}\left[3s^2(s-1)\right] = \left(9 \cdot 0^2\right) - (6 \cdot 0) = \mathbf{0}$

d. $\dfrac{d}{dt}\left[\dfrac{t^2 + 1}{t - 1}\right] = \dfrac{\left[(t-1)\dfrac{d}{dt}\left(t^2 + 1\right)\right] - \left[\left(t^2 + 1\right)\dfrac{d}{dt}(t-1)\right]}{(t-1)^2} = \dfrac{\left[(t-1) \cdot 2t\right] - \left[\left(t^2 + 1\right) \cdot 1\right]}{(t-1)^2} = \dfrac{2t^2 - 2t - t^2 - 1}{(t-1)^2} = \dfrac{t^2 - 2t - 1}{(t-1)^2}$

at $t = -1$ $\dfrac{d}{dt}\left[\dfrac{t^2 + 1}{t - 1}\right] = \dfrac{(-1)^2 + (-2 \cdot -1) - 1}{(-1-1)^2} = \dfrac{1 + 2 - 1}{4} = \dfrac{2}{4} = \dfrac{\mathbf{1}}{\mathbf{2}}$

e. $\dfrac{d}{du}\left[\dfrac{u^3}{(u+1)^2}\right] = \dfrac{\left[(u+1)^2 \dfrac{d}{du}u^3\right] - \left[u^3 \dfrac{d}{du}(u+1)^2\right]}{(u+1)^4} = \dfrac{\left[(u+1)^2 \cdot 3u^2\right] - \left[u^3 \cdot 2(u+1)\right]}{(u+1)^4} = \dfrac{\left[\left(u^2 + 2u + 1\right) \cdot 3u^2\right] - \left[2u^4 + 2u^3\right]}{(u+1)^4}$

$= \dfrac{3u^4 + 6u^3 + 3u^2 - 2u^4 - 2u^3}{(u+1)^4} = \dfrac{u^4 + 4u^3 + 3u^2}{(u+1)^4}$ at $u = 1$ $\dfrac{d}{du}\left[\dfrac{u^3}{(u+1)^2}\right] = \dfrac{1^4 + \left(4 \cdot 1^3\right) + \left(3 \cdot 1^2\right)}{(1+1)^4} = \dfrac{1 + 4 + 3}{2^4} = \dfrac{8}{16} = \dfrac{\mathbf{1}}{\mathbf{2}}$

f. $\dfrac{d}{dw}\left[\dfrac{w\left(w^2 + 1\right)}{3w^2}\right] = \dfrac{d}{dw}\left[\dfrac{w\left(w^2 + 1\right)}{3w^{2=1}}\right] = \dfrac{d}{dw}\left[\dfrac{w^2 + 1}{3w}\right] = \dfrac{\left[3w \dfrac{d}{du}\left(w^2 + 1\right)\right] - \left[\left(w^2 + 1\right)\dfrac{d}{du}3w\right]}{(3w)^2} = \dfrac{[3w \cdot 2w] - \left[\left(w^2 + 1\right) \cdot 3\right]}{9w^2}$

$= \dfrac{6w^2 - 3w^2 - 3}{9w^2} = \dfrac{3\left(w^2 - 1\right)}{9w^2} = \dfrac{w^2 - 1}{3w^2}$ at $w = 2$ $\dfrac{d}{dw}\left[\dfrac{w\left(w^2 + 1\right)}{3w^2}\right] = \dfrac{2^2 - 1}{\left(3 \cdot 2^2\right)} = \dfrac{3}{12} = \dfrac{\mathbf{1}}{\mathbf{4}}$

g. $\dfrac{d}{dv}\left[\left(v^2 + 1\right)v^3\right] = \dfrac{d}{dv}\left(v^5 + v^3\right) = \dfrac{d}{dv}v^5 + \dfrac{d}{dv}v^3 = 5v^4 + 3v^2$

at $v = -2$ $\dfrac{d}{dv}\left[\left(v^2 + 1\right)v^3\right] = 5 \cdot (-2)^4 + 3 \cdot (-2)^2 = (5 \cdot 16) + (3 \cdot 4) = 80 + 12 = \mathbf{92}$

h. $\dfrac{d}{dx}\left(\dfrac{x^3}{x^2 + 1}\right) = \dfrac{\left[\left(x^2 + 1\right)\dfrac{d}{dx}x^3\right] - \left[x^3 \dfrac{d}{dx}\left(x^2 + 1\right)\right]}{\left(x^2 + 1\right)^2} = \dfrac{\left[\left(x^2 + 1\right) \cdot 3x^2\right] - \left[x^3 \cdot 2x\right]}{\left(x^2 + 1\right)^2} = \dfrac{3x^4 + 3x^2 - 2x^4}{\left(x^2 + 1\right)^2} = \dfrac{x^4 + 3x^2}{\left(x^2 + 1\right)^2}$

at $x = 0$ $\dfrac{d}{dx}\left(\dfrac{x^3}{x^2 + 1}\right) = \dfrac{0^4 + \left(3 \cdot 0^2\right)}{\left(0^2 + 1\right)^2} = \dfrac{0 + 0}{1^2} = \dfrac{0}{1} = \mathbf{0}$

i. $\dfrac{d}{du}\left[u^3\left(\dfrac{u^2}{1-u}\right)\right] = \dfrac{d}{du}\left(\dfrac{u^5}{1-u}\right) = \dfrac{\left[(1-u)\dfrac{d}{du}u^5\right] - \left[u^5 \dfrac{d}{du}(1-u)\right]}{(1-u)^2} = \dfrac{\left[(1-u) \cdot 5u^4\right] - \left[u^5 \cdot -1\right]}{(1-u)^2} = \dfrac{5u^4 - 5u^5 + u^5}{(1-u)^2} = \dfrac{-4u^5 + 5u^4}{(1-u)^2}$

at $u = 0$ $\dfrac{d}{du}\left[u^3\left(\dfrac{u^2}{1-u}\right)\right] = \dfrac{\left(-4 \cdot 0^5\right) + \left(5 \cdot 0^4\right)}{(1-0)^2} = \dfrac{0 + 0}{1} = \dfrac{0}{1} = \mathbf{0}$

4. Given the functions below find their derivatives at the specified value.

a. $\dfrac{ds}{dt}$ given $s = \left(t^2 - 1\right) + (3t + 2)^2$, then $\dfrac{ds}{dt} = \dfrac{ds}{dt}\left(t^2 - 1\right) + \dfrac{ds}{dt}(3t + 2)^2 = 2t + 2(3t + 2)^{2-1} \cdot \dfrac{ds}{dt}(3t + 2) = 2t + 2(3t + 2) \cdot 3$

$$= 2t + 18t + 12 = 20t + 12 \qquad\qquad \text{at } t = 2 \quad \frac{ds}{dt} = (20 \cdot 2) + 12 = 40 + 12 = \mathbf{52}$$

b. $\frac{dy}{dt}$ given $y = \frac{t^3 + 3t^2 + 1}{2t}$, then $\frac{dy}{dt} = \frac{2t \frac{d}{dt}\left(t^3 + 3t^2 + 1\right) - \left(t^3 + 3t^2 + 1\right)\frac{d}{dt} 2t}{(2t)^2} = \frac{2t \cdot \left(3t^2 + 6t\right) - \left(t^3 + 3t^2 + 1\right) \cdot 2}{4t^2}$

$$= \frac{6t^3 + 12t^2 - 2t^3 - 6t^2 - 2}{4t^2} = \frac{4t^3 + 6t^2 - 2}{4t^2} = \frac{2\left(2t^3 + 3t^2 - 1\right)}{\underset{2}{4t^2}} = \frac{2t^3 + 3t^2 - 1}{2t^2}$$

at $t = 1$ $\quad \frac{dy}{dt} = \frac{2 \cdot 1^3 + 3 \cdot 1^2 - 1}{2 \cdot 1^2} = \frac{2 + 3 - 1}{2} = \frac{4}{2} = \mathbf{2}$

c. $\frac{dw}{dx}$ given $w = \left(x^2 + 1\right)^2 + 3x$, then $\frac{dw}{dx} = \frac{d}{dx}\left(x^2 + 1\right)^2 + 3x = 2\left(x^2 + 1\right)^{2-1} \frac{d}{dx}\left(x^2 + 1\right) + \frac{d}{dx} 3x = 2\left(x^2 + 1\right) \cdot 2x + 3$

$$= 4x^3 + 4x + 3 \qquad\qquad \text{at } x = -1 \quad \frac{dw}{dx} = 4 \cdot (-1)^3 + (4 \cdot -1) + 3 = -4 - 4 + 3 = \mathbf{-5}$$

d. $\frac{dy}{dx}$ given $y = x^2\left(x^3 + 2x + 1\right)^2 + 3x$, then $\frac{dy}{dx} = \frac{d}{dx}\left[x^2\left(x^3 + 2x + 1\right)^2 + 3x\right] = \frac{d}{dx}\left[x^2\left(x^3 + 2x + 1\right)^2\right] + \frac{d}{dx} 3x$

$$= \left(x^3 + 2x + 1\right)^2 \frac{d}{dx}x^2 + x^2 \frac{d}{dx}\left(x^3 + 2x + 1\right)^2 + 3 = \left[\left(x^3 + 2x + 1\right)^2 \cdot 2x\right] + \left[x^2 \cdot 2\left(x^3 + 2x + 1\right)^{2-1} \cdot \frac{d}{dx}\left(x^3 + 2x + 1\right)\right] + 3$$

$$= 2x\left(x^3 + 2x + 1\right)^2 + \left[2x^2\left(x^3 + 2x + 1\right)\left(3x^2 + 2\right)\right] + 3 \qquad \text{at } x = 0 \quad \frac{dy}{dx} = 0 + 0 + 3 = \mathbf{3}$$

Section 2.4 Solutions - The Chain Rule

1. Find the derivative of the following functions. Do not simplify the answer to its lowest term.

a. Given $y = \left(x^2 + 2\right)^3$, then $y' = 3\left(x^2 + 2\right)^{3-1} \cdot 2x = \mathbf{6x\left(x^2 + 2\right)^2}$

b. Given $y = \left(x^2 + 1\right)^{-2}$, then $y' = -2\left(x^2 + 1\right)^{-2-1} \cdot 2x = \mathbf{-4x\left(x^2 + 1\right)^{-3}}$

c. Given $y = \left(x^3 - 1\right)^5$, then $y' = 5\left(x^3 - 1\right)^{5-1} \cdot 3x^2 = \mathbf{15x^2\left(x^3 - 1\right)^4}$

d. Given $y = \left(1 - \frac{1}{x^2}\right)^2$, then $y' = 2\left(1 - \frac{1}{x^2}\right)^{2-1} \cdot \frac{2x}{x^4} = 4\left(1 - \frac{1}{x^2}\right) \cdot \frac{\cancel{x}}{x^{4=3}} = \mathbf{\frac{4}{x^3}\left(1 - \frac{1}{x^2}\right)}$

e. Given $y = 2x^3 + \frac{1}{3x^2}$, then $y' = (2 \cdot 3)x^{3-1} + \frac{\left(0 \cdot 3x^2\right) - (6x \cdot 1)}{\left(3x^2\right)^2} = 6x^2 + \frac{0 - 6x}{9x^4} = 6x^2 - \frac{6\cancel{x}}{9x^{4=3}} = \mathbf{6x^2 - \frac{2}{3x^3}}$

f. Given $y = \left(\frac{1+x^2}{x^3}\right)^4$, then $y' = 4\left(\frac{1+x^2}{x^3}\right)^{4-1} \cdot \frac{\left[2x \cdot x^3\right] - \left[3x^2\left(1 + x^2\right)\right]}{x^6} = 4\left(\frac{1+x^2}{x^3}\right)^3 \cdot \frac{2x^4 - 3x^2 - 3x^4}{x^6} = 4\left(\frac{1+x^2}{x^3}\right)^3$

$$\cdot \frac{-x^4 - 3x^2}{x^6} = 4\left(\frac{1+x^2}{x^3}\right)^3 \cdot \frac{-x^2\left(x^2 + 3\right)}{x^{6=4}} = \mathbf{-4\left(\frac{1+x^2}{x^3}\right)^3\left(\frac{x^2 + 3}{x^4}\right)}$$

g. Given $y = x^2\left(\frac{x+1}{3}\right)^3$, then $y' = \left[2x^{2-1} \cdot \left(\frac{x+1}{3}\right)^3\right] + \left[3\left(\frac{x+1}{3}\right)^{3-1} \cdot \frac{1}{3} \cdot x^2\right] = \mathbf{2x\left(\frac{x+1}{3}\right)^3 + x^2\left(\frac{x+1}{3}\right)^2}$

h. Given $y = \left[x(x+1)^2 + 2x\right]^3 = \left[x\left(x^2 + 2x + 1\right) + 2x\right]^3 = \left(x^3 + 2x^2 + x + 2x\right)^3 = \left(x^3 + 2x^2 + 3x\right)^3$, then $y' = 3\left(x^3 + 2x^2 + 3x\right)^{3-1}$

$$\cdot \left(3x^2 + 4x + 3\right) = \mathbf{3\left(x^3 + 2x^2 + 3x\right)^2\left(3x^2 + 4x + 3\right)}$$

i. Given $y = \left(\frac{x}{3} - 2x^3\right)^{-1}$, then $y' = -\left(\frac{x}{3} - 2x^3\right)^{-1-1} \cdot \left(\frac{1}{3} - (2 \cdot 3)x^{3-1}\right) = \mathbf{-\left(\frac{x}{3} - 2x^3\right)^{-2}\left(\frac{1}{3} - 6x^2\right)}$

j. Given $y = \left(x^3 + 3x^2 + 1\right)^4$, then $y' = 4\left(x^3 + 3x^2 + 1\right)^{4-1} \cdot \left(3x^{3-1} + (3 \cdot 2)x^{2-1} + 0\right) = 4\left(x^3 + 3x^2 + 1\right)^3\left(3x^2 + 6x\right)$

k. Given $y = \left(\dfrac{t^2}{1+t^2}\right)^3$, then $y' = 3\left(\dfrac{t^2}{1+t^2}\right)^{3-1} \cdot \dfrac{\left[2t \cdot \left(1+t^2\right)\right] - \left[2t \cdot t^2\right]}{\left(1+t^2\right)^2} = 3\left(\dfrac{t^2}{1+t^2}\right)^2 \cdot \dfrac{2t + 2t^3 - 2t^3}{\left(1+t^2\right)^2} = \dfrac{3t^4}{\left(1+t^2\right)^2} \cdot \dfrac{2t}{\left(1+t^2\right)^2}$

$= \dfrac{3t^4 \cdot 2t}{\left(1+t^2\right)^2 \cdot \left(1+t^2\right)^2} = \dfrac{6t^{4+1}}{\left(1+t^2\right)^{2+2}} = \dfrac{6t^5}{\left(1+t^2\right)^4}$

l. Given $y = \left(1+x^{-2}\right)^{-1}$, then $y' = -\left(1+x^{-2}\right)^{-1-1} \cdot -2x^{-2-1} = 2x^{-3}\left(1+x^{-2}\right)^{-2} = \dfrac{2}{x^3\left(1+x^{-2}\right)^2}$

m. Given $y = \dfrac{(x+1)^{-2}}{x^3}$, then $y' = \dfrac{\left[-2(x+1)^{-2-1} \cdot 1 \cdot x^3\right] - \left[3x^{3-1} \cdot (x+1)^{-2}\right]}{x^6} = \dfrac{\left[-2x^3(x+1)^{-3}\right] - \left[3x^2(x+1)^{-2}\right]}{x^6}$

n. Given $y = \left(\dfrac{1}{1-x^3}\right)^2 + \dfrac{1}{x}$, then $y' = \left[2\left(\dfrac{1}{1-x^3}\right)^{2-1} \cdot \dfrac{3x^2}{\left(1-x^3\right)^2}\right] - \dfrac{1}{x^2} = \left[\left(\dfrac{1}{1-x^3}\right) \cdot \dfrac{6x^2}{\left(1-x^3\right)^2}\right] - \dfrac{1}{x^2} = \dfrac{6x^2}{\left(1-x^3\right)^3} - \dfrac{1}{x^2}$

o. Given $y = \dfrac{x^3}{x^3+2} - x^2$, then $y' = \dfrac{\left[3x^{3-1} \cdot \left(x^3+2\right)\right] - \left[3x^2 \cdot x^3\right]}{\left(x^3+2\right)^2} - 2x^{2-1} = \dfrac{3x^2\left(x^3+2\right) - 3x^5}{\left(x^3+2\right)^2} - 2x = \dfrac{3x^5 + 6x^2 - 3x^5}{\left(x^3+2\right)^2} - 2x$

$= \dfrac{6x^2}{\left(x^3+2\right)^2} - \dfrac{2x}{1} = \dfrac{6x^2 - 2x\left(x^3+2\right)^2}{\left(x^3+2\right)^2}$

2. Find the derivative of the following functions at $x=0$, $x=1$, and $x=-1$.

a. Given $y = \left(x^3+1\right)^5$, then $y' = 5\left(x^3+1\right)^{5-1} \cdot 3x^2 = 15x^2\left(x^3+1\right)^4$. Therefore,

$y'(0) = \left(15 \cdot 0^2\right)\left(0^3+1\right)^4 = 0 \cdot 1^4 = 0 \cdot 1 = \mathbf{0}$

$y'(1) = \left(15 \cdot 1^2\right)\left(1^3+1\right)^4 = 15 \cdot 2^4 = 15 \cdot 2^4 = 15 \cdot 16 = \mathbf{240}$ and

$y'(-1) = \left[15 \cdot (-1)^2\right]\left[(-1)^3+1\right]^4 = 15 \cdot (-1+1)^4 = 15 \cdot 0^4 = 15 \cdot 0 = \mathbf{0}$

b. Given $y = \left(x^3+3x^2-1\right)^4$, then $y' = 4\left(x^3+3x^2-1\right)^{4-1}\left(3x^2+6x\right) = 12x\left(x^3+3x^2-1\right)^3(x+2)$. Therefore,

$y'(0) = (12 \cdot 0)\left(0^3+3 \cdot 0^2-1\right)^3(0+2) = 0.(-1)^3 \cdot 2 = \mathbf{0}$

$y'(1) = (12 \cdot 1)\left(1^3+3 \cdot 1^2-1\right)^3(1+2) = 12 \cdot (1+3-1)^3 \cdot 3 = 12 \cdot 27 \cdot 3 = \mathbf{972}$ and

$y'(-1) = (12 \cdot -1)\left[(-1)^3+3 \cdot (-1)^2-1\right]^3(-1+2) = -12(-1+3-1)^3 = -12(-2+3)^3 = -12 \cdot 1^3 = -12 \cdot 1 = \mathbf{-12}$

c. Given $y = \left(\dfrac{x}{x+1}\right)^2$, then $y' = 2\left(\dfrac{x}{x+1}\right)^{2-1} \cdot \dfrac{1 \cdot (x+1) - 1 \cdot x}{(x+1)^2} = 2\left(\dfrac{x}{x+1}\right) \cdot \dfrac{x+1-x}{(x+1)^2} = 2\left(\dfrac{x}{x+1}\right) \cdot \dfrac{1}{(x+1)^2} = \dfrac{2x}{(x+1)^3}$. Thus,

$y'(0) = \dfrac{2 \cdot 0}{(0+1)^3} = \dfrac{0}{1^3} = \dfrac{0}{1} = \mathbf{0}$

$y'(1) = \dfrac{2 \cdot 1}{(1+1)^3} = \dfrac{2}{2^3} = \dfrac{2}{8} = \dfrac{1}{4} = \mathbf{0.25}$ and

$y'(-1) = \dfrac{2 \cdot -1}{(-1+1)^3} = \dfrac{-2}{0^3} = -\dfrac{2}{0}$ **which is undefined due to division by zero**

d. Given $y = x\left(x^2+1\right)^2$, then $y' = \left[1 \cdot \left(x^2+1\right)^2\right] + \left[2\left(x^2+1\right)^{2-1} \cdot 2x\right] = \left(x^2+1\right)^2 + 4x\left(x^2+1\right)$. Therefore,

$y'(0) = \left(0^2+1\right)^2 + (4 \cdot 0)\left(0^2+1\right) = 1^2 + 0 = \mathbf{1}$

$y'(1) = \left(1^2+1\right)^2 + (4\cdot 1)\left(1^2+1\right) = 2^2 + 4\cdot 2 = 4+8 = \mathbf{12}$ and

$y'(-1) = \left[(-1)^2+1\right]^2 + (4\cdot -1)\left[(-1)^2+1\right] = (1+1)^2 - 4\cdot(1+1) = 2^2 - 4\cdot 2 = 4-8 = \mathbf{-4}$

e. Given $y = x^3 + 2\left(x^2+1\right)^3$, then $y' = 3x^{3-1} + 2\cdot 3\left(x^2+1\right)^{3-1}\cdot 2x = 3x^2 + 12x\left(x^2+1\right)^2$. Therefore,

$y'(0) = 3\cdot 0^2 + (12\cdot 0)\left(0^2+1\right)^2 = 3\cdot 0 + 0\cdot 1^2 = 0+0 = \mathbf{0}$

$y'(1) = 3\cdot 1^2 + (12\cdot 1)\left(1^2+1\right)^2 = 3+12\cdot 2^2 = 3+12\cdot 4 = 3+48 = \mathbf{51}$ and

$y'(-1) = 3\cdot(-1)^2 + (12\cdot -1)\left[(-1)^2+1\right]^2 = 3\cdot 1 - 12\cdot(1+1)^2 = 3-12\cdot 2^2 = 3-12\cdot 4 = 3-48 = \mathbf{-45}$

f. Given $y = \left(\dfrac{x^2}{1+x^2}\right)^3$, then $y' = 3\left(\dfrac{x^2}{1+x^2}\right)^{3-1}\cdot \dfrac{\left[2x\cdot\left(1+x^2\right)\right]-\left[2x\cdot x^2\right]}{\left(1+x^2\right)^2} = 3\left(\dfrac{x^2}{1+x^2}\right)^2\cdot \dfrac{2x+2x^3-2x^3}{\left(1+x^2\right)^2}$

$= \dfrac{3x^4}{\left(1+x^2\right)^2}\cdot \dfrac{2x}{\left(1+x^2\right)^2} = \dfrac{(3\cdot 2)x^{4+1}}{\left(1+x^2\right)^{2+2}} = \dfrac{6x^5}{\left(1+x^2\right)^4}$. Therefore,

$y'(0) = \dfrac{6\cdot 0^5}{\left(1+0^2\right)^4} = \dfrac{0}{1^4} = \dfrac{0}{1} = \mathbf{0}$

$y'(1) = \dfrac{6\cdot 1^5}{\left(1+1^2\right)^4} = \dfrac{6}{2^4} = \dfrac{6}{16} = \dfrac{3}{8} = \mathbf{0.375}$ and

$y'(-1) = \dfrac{6\cdot(-1)^5}{\left[1+(-1)^2\right]^4} = \dfrac{6\cdot -1}{(1+1)^4} = -\dfrac{6}{2^4} = -\dfrac{6}{16} = -\dfrac{3}{8} = \mathbf{-0.375}$

g. Given $y = \left(\dfrac{x}{x^2+1}\right)^5$, then $y' = 5\left(\dfrac{x}{x^2+1}\right)^{5-1}\cdot \dfrac{\left[1\cdot\left(x^2+1\right)\right]-[2x\cdot x]}{\left(x^2+1\right)^2} = 5\left(\dfrac{x}{x^2+1}\right)^4\cdot \dfrac{x^2+1-2x^2}{\left(x^2+1\right)^2} = \dfrac{5x^4}{\left(x^2+1\right)^4}\cdot \dfrac{1-x^2}{\left(x^2+1\right)^2}$

$= \dfrac{5x^4\left(1-x^2\right)}{\left(x^2+1\right)^{4+2}} = \dfrac{5x^4\left(1-x^2\right)}{\left(x^2+1\right)^6}$. Therefore,

$y'(0) = \dfrac{5\cdot 0^4\cdot\left(1-0^2\right)}{\left(0^2+1\right)^6} = \dfrac{0\cdot 1}{1^6} = \dfrac{0}{1} = \mathbf{0}$

$y'(1) = \dfrac{5\cdot 1^4\cdot\left(1-1^2\right)}{\left(1^2+1\right)^6} = \dfrac{5\cdot(1-1)}{2^6} = \dfrac{5\cdot 0}{64} = \dfrac{0}{64} = \mathbf{0}$ and

$y'(-1) = \dfrac{5\cdot(-1)^4\cdot\left(1-(-1)^2\right)}{\left((-1)^2+1\right)^6} = \dfrac{5\cdot 1\cdot(1-1)}{(1+1)^6} = \dfrac{5\cdot 0}{2^6} = \dfrac{0}{64} = \mathbf{0}$

h. Given $y = \left(x^2+1\right)^3\cdot \dfrac{1}{x^2} = y = \dfrac{\left(x^2+1\right)^3}{x^2}$, then $y' = \dfrac{\left[3\left(x^2+1\right)^{3-1}\cdot 2x\cdot x^2\right]-\left[2x\cdot\left(x^2+1\right)^3\right]}{x^4} = \dfrac{6x^3\left(x^2+1\right)^2 - 2x\left(x^2+1\right)^3}{x^4}$

$= \dfrac{2x\left(x^2+1\right)^2\left[3x^2-\left(x^2+1\right)\right]}{x^{4=3}} = \dfrac{2\left(x^2+1\right)^2\left(3x^2-x^2-1\right)}{x^3} = \dfrac{2\left(x^2+1\right)^2\left(2x^2-1\right)}{x^3}$. Therefore,

$y'(0) = \dfrac{2\left(0^2+1\right)^2\cdot\left(2\cdot 0^2-1\right)}{0^3} = \dfrac{2\cdot 1^2\cdot(0-1)}{0} = \dfrac{2\cdot -1}{0} = -\dfrac{2}{0}$ **which is undefined due to division by zero**

$y'(1) = \dfrac{2\left(1^2+1\right)^2\cdot\left(2\cdot 1^2-1\right)}{1^3} = \dfrac{2\cdot 2^2\cdot(2-1)}{1} = \dfrac{2\cdot 4\cdot 1}{1} = \dfrac{8}{1} = \mathbf{8}$ and

$y'(-1) = \dfrac{2\left[(-1)^2+1\right]^2\cdot\left[2\cdot(-1)^2-1\right]}{(-1)^3} = \dfrac{2(1+1)^2\cdot(2\cdot 1-1)}{-1} = \dfrac{2\cdot 2^2\cdot(2-1)}{-1} = \dfrac{2\cdot 4\cdot 1}{-1} = -\dfrac{8}{1} = \mathbf{-8}$

i. Given $y = \left(\dfrac{x^3}{x-1}\right)^2 + 5x$, then $y' = 2\left(\dfrac{x^3}{x-1}\right)^{2-1} \cdot \dfrac{\left[3x^2 \cdot (x-1)\right] - \left[1 \cdot x^3\right]}{(x-1)^2} + 5 = 2\left(\dfrac{x^3}{x-1}\right)\cdot\left(\dfrac{3x^3 - 3x^2 - x^3}{(x-1)^2}\right) + 5$

$= \dfrac{2x^3}{x-1}\cdot\dfrac{2x^3 - 3x^2}{(x-1)^2} + 5 = \dfrac{2x^3}{x-1}\cdot\dfrac{x^2(2x-3)}{(x-1)^2} + 5 = \dfrac{2x^3 \cdot x^2\left(2x^3 - 3x^2\right)}{(x-1)(x-1)^2} + 5 = \dfrac{2x^{3+2}(2x-3)}{(x-1)^{1+2}} + 5 = \dfrac{2x^5(2x-3)}{(x-1)^3} + 5$. Therefore,

$y'(0) = \dfrac{2\cdot 0^5 \cdot(2\cdot 0 - 3)}{(0-1)^3} + 5 = \dfrac{0\cdot(0-3)}{1^3} + 5 = \dfrac{0}{1} + 5 = 0 + 5 = \mathbf{5}$

$y'(1) = \dfrac{2\cdot 1^5 \cdot(2\cdot 1 - 3)}{(1-1)^3} + 5 = \dfrac{2\cdot(2-3)}{0^3} + 5 = \dfrac{2\cdot -1}{0} + 5 = -\dfrac{\mathbf{2}}{\mathbf{0}} + \mathbf{5}$ **which is undefined due to division by zero** and

$y'(-1) = \dfrac{2\cdot(-1)^5 \cdot\left[(2\cdot -1) - 3\right]}{(-1-1)^3} + 5 = \dfrac{2\cdot -1\cdot[-2-3]}{(-2)^3} + 5 = \dfrac{-2\cdot -5}{-8} + 5 = \dfrac{10}{8} + 5 = \dfrac{5}{4} + 5 = 1.25 + 5 = \mathbf{6.25}$

3. Use the chain rule to differentiate the following functions.

a. $\dfrac{d}{dt}\left[\dfrac{(t+1)^3}{t^2}\right] = \dfrac{\left[t^2 \cdot \frac{d}{dt}(t+1)^3\right] - \left[(t+1)^3 \cdot \frac{d}{dt}t^2\right]}{t^4} = \dfrac{\left[t^2\cdot 3(t+1)^2\right] - \left[(t+1)^3\cdot 2t\right]}{t^4} = \dfrac{3t^2(t+1)^2 - 2t(t+1)^3}{t^4} = \dfrac{t(t+1)^2\left[3t - 2(t+1)\right]}{t^{4=3}}$

$= \dfrac{(t+1)^2(3t - 2t - 2)}{t^3} = \dfrac{(t+1)^2(t-2)}{t^3}$

b. $\dfrac{d}{du}\left[\dfrac{(u^2+1)^3}{3u^4}\right] = \dfrac{\left[3u^4\frac{d}{du}(u^2+1)^3\right] - \left[(u^2+1)^3\frac{d}{du}3u^4\right]}{(3u^4)^2} = \dfrac{\left[3u^4\cdot 3(u^2+1)^{3-1}\cdot 2u\right] - \left[(u^2+1)^3\cdot(3\cdot 4)u^{4-1}\cdot 1\right]}{9u^8}$

$= \dfrac{18u^5(u^2+1)^2 - 12u^3(u^2+1)^3}{9u^8} = \dfrac{6u^3(u^2+1)^2\left[3u^2 - 2(u^2+1)\right]}{9u^{8=5}} = \dfrac{2(u^2+1)^2(3u^2 - 2u^2 - 2)}{3u^5} = \dfrac{2(u^2+1)^2(u^2-2)}{3u^5}$

c. $\dfrac{d}{dx}\left[\dfrac{(2x+1)^3}{(1-x)^2}\right] = \dfrac{\left[(1-x)^2\frac{d}{dx}(2x+1)^3\right] - \left[(2x+1)^3\frac{d}{dx}(1-x)^2\right]}{(1-x)^4} = \dfrac{\left[(1-x)^2\cdot 3(2x+1)^{3-1}\cdot 2\right] - \left[(2x+1)^3\cdot 2(1-x)^{2-1}\cdot -1\right]}{(1-x)^4}$

$= \dfrac{6(1-x)^2(2x+1)^2 + 2(2x+1)^3(1-x)}{(1-x)^4} = \dfrac{2(1-x)(2x+1)^2\left[3(1-x) + (2x+1)\right]}{(1-x)^{4=3}} = \dfrac{2(2x+1)^2(4-x)}{(1-x)^3}$

d. $\dfrac{d}{dx}\left[(x^3-1)^2(2x+1)^3\right] = \left[(2x+1)^3\frac{d}{dx}(x^3-1)^2\right] + \left[(x^3-1)^2\frac{d}{dx}(2x+1)^3\right] = \left[(2x+1)^3\cdot 2(x^3-1)^{2-1}\cdot 3x^2\right]$

$+ \left[(x^3-1)^2\cdot 3(2x+1)^{3-1}\cdot 2\right] = \left[6x^2(2x+1)^3(x^3-1)\right] + \left[6(x^3-1)^2(2x+1)^2\right] = 6(2x+1)^2(x^3-1)\left[x^2(2x+1) + (x^3-1)\right]$

$= 6(2x+1)^2(x^3-1)(2x^3 + x^2 + x^3 - 1) = \mathbf{6(2x+1)^2(x^3-1)(3x^3 + x^2 - 1)}$

e. $\dfrac{d}{ds}\left[s^3 - \dfrac{1}{s^2+6}\right]^2 = 2\left[s^3 - \dfrac{1}{s^2+6}\right]^{2-1}\cdot\dfrac{d}{ds}\left[s^3 - \dfrac{1}{s^2+6}\right] = 2\left[s^3 - \dfrac{1}{s^2+6}\right]\cdot\left[\dfrac{d}{ds}s^3 - \dfrac{d}{ds}\dfrac{1}{s^2+6}\right] = 2\left[s^3 - \dfrac{1}{s^2+6}\right]$

$\cdot\left[3s^{3-1} - \dfrac{(s^2+6)\cdot\frac{d}{ds}(1) - 1\cdot\frac{d}{ds}(s^2+6)}{(s^2+6)^2}\right] = 2\left[s^3 - \dfrac{1}{s^2+6}\right]\cdot\left[3s^2 - \dfrac{0-2s}{(s^2+6)^2}\right] = \mathbf{2}\left[s^3 - \dfrac{1}{s^2+6}\right]\cdot\left[3s^2 + \dfrac{2s}{(s^2+6)^2}\right]$

f. $\dfrac{d}{dt}\left[\dfrac{(t^2-1)^3}{t^2+1}\right] = \dfrac{\left[(t^2+1)\frac{d}{dt}(t^2-1)^3\right] - \left[(t^2-1)^3\frac{d}{dt}(t^2+1)\right]}{(t^2+1)^2} = \dfrac{\left[(t^2+1)\cdot 3(t^2-1)^{3-1}\cdot 2t\right] - \left[(t^2-1)^3\cdot 2t\right]}{(t^2+1)^2}$

$$= \frac{\left[6t\left(t^2+1\right)\left(t^2-1\right)^2\right]-\left[2t\left(t^2-1\right)^3\right]}{\left(t^2+1\right)^2} = \frac{2t\left(t^2-1\right)^2\left[3\left(t^2+1\right)-\left(t^2-1\right)\right]}{\left(t^2+1\right)^2} = \frac{2t\left(t^2-1\right)^2\left[3t^2+3-t^2+1\right]}{\left(t^2+1\right)^2} = \frac{4t\left(t^2-1\right)^2\left(t^2+2\right)}{\left(t^2+1\right)^2}$$

g. $\dfrac{d}{du}\left[\left(u^2+1\right)^3\left(\dfrac{1}{u+1}\right)^2\right] = \dfrac{d}{du}\left[\left(u^2+1\right)^3\dfrac{1}{(u+1)^2}\right] = \dfrac{d}{du}\left[\dfrac{\left(u^2+1\right)^3}{(u+1)^2}\right] = \dfrac{\left[(u+1)^2\frac{d}{du}\left(u^2+1\right)^3\right]-\left[\left(u^2+1\right)^3\frac{d}{du}(u+1)^2\right]}{(u+1)^4}$

$$= \frac{\left[(u+1)^2\cdot 3\left(u^2+1\right)^{3-1}\cdot 2u\right]-\left[\left(u^2+1\right)^3\cdot 2(u+1)^{2-1}\right]}{(u+1)^4} = \frac{\left[6u(u+1)^2\left(u^2+1\right)^2\right]-\left[2\left(u^2+1\right)^3(u+1)\right]}{(u+1)^4}$$

$$= \frac{2(u+1)\left(u^2+1\right)^2\left[3u(u+1)-\left(u^2+1\right)\right]}{(u+1)^{4=3}} = \frac{2\left(u^2+1\right)^2\left(3u^2+3u-u^2-1\right)}{(u+1)^3} = \frac{2\left(u^2+1\right)^2\left(2u^2+3u-1\right)}{(u+1)^3}$$

h. $\dfrac{d}{d\theta}\left[\dfrac{\theta^2+3}{(\theta-1)^3}\right]^2 = 2\left[\dfrac{\theta^2+3}{(\theta-1)^3}\right]^{2-1}\cdot\dfrac{d}{d\theta}\left[\dfrac{\theta^2+3}{(\theta-1)^3}\right] = 2\left[\dfrac{\theta^2+3}{(\theta-1)^3}\right]\cdot\dfrac{\left[(\theta-1)^3\frac{d}{d\theta}\left(\theta^2+3\right)\right]-\left[\left(\theta^2+3\right)\frac{d}{d\theta}(\theta-1)^3\right]}{(\theta-1)^6}$

$$= \frac{2\left(\theta^2+3\right)}{(\theta-1)^3}\cdot\frac{\left[(\theta-1)^3\cdot 2\theta\right]-\left[\left(\theta^2+3\right)\cdot 3(\theta-1)^2\right]}{(\theta-1)^6} = \frac{2\left(\theta^2+3\right)}{(\theta-1)^3}\cdot\frac{2\theta(\theta-1)^3-3\left(\theta^2+3\right)(\theta-1)^2}{(\theta-1)^6} = \frac{2\left(\theta^2+3\right)}{(\theta-1)^3}\cdot\frac{(\theta-1)^2\left[2\theta(\theta-1)-3\left(\theta^2+3\right)\right]}{(\theta-1)^{6=4}}$$

$$= \frac{2\left(\theta^2+3\right)}{(\theta-1)^3}\cdot\frac{\left[2\theta(\theta-1)-3\left(\theta^2+3\right)\right]}{(\theta-1)^4} = \frac{2\left(\theta^2+3\right)\cdot\left(2\theta^2-2\theta-3\theta^2-9\right)}{(\theta-1)^{3+4}} = \frac{-2\left(\theta^2+3\right)\cdot\left(\theta^2+2\theta+9\right)}{(\theta-1)^7}$$

i. $\dfrac{d}{dr}\left[\dfrac{r^7}{\left(r^2+2r\right)^3}\right] = \dfrac{\left(r^2+2r\right)^3\frac{d}{dr}r^7-r^7\frac{d}{dr}\left(r^2+2r\right)^3}{\left(r^2+2r\right)^6} = \dfrac{\left[\left(r^2+2r\right)^3\cdot 7r^{7-1}\right]-\left[r^7\cdot 3\left(r^2+2r\right)^{2-1}\cdot\frac{d}{dr}\left(r^2+2r\right)\right]}{\left(r^2+2r\right)^6}$

$$= \frac{7r^6\left(r^2+2r\right)^3-\left[r^7\cdot 3\left(r^2+2r\right)\cdot(2r+2)\right]}{\left(r^2+2r\right)^6} = \frac{7r^6\left(r^2+2r\right)^3-\left[r^7\cdot 3r(r+2)\cdot 2(r+1)\right]}{\left(r^2+2r\right)^6} = \frac{7r^6\left(r^2+2r\right)^3-6r^8(r+2)(r+1)}{\left(r^2+2r\right)^6}$$

4. Given the following y functions in terms of u, find y'.

a. Given $y=2u^2-1$ and $u=x-1$, then $y=2(x-1)^2-1$ and $y' = 2\cdot 2(x-1)^{2-1}-0 = 4(x-1)$

b. Given $y=\dfrac{u}{u-1}$ and $u=x^3$, then $y=\dfrac{x^3}{x^3-1}$ and $y' = \dfrac{\left[3x^{3-1}\cdot\left(x^3-1\right)\right]-\left[3x^{3-1}\cdot x^3\right]}{\left(x^3-1\right)^2} = \dfrac{\left[3x^2\left(x^3-1\right)\right]-\left[3x^2\cdot x^3\right]}{\left(x^3-1\right)^2}$

$$= \frac{3x^5-3x^2-3x^5}{\left(x^3-1\right)^2} = -\frac{3x^2}{\left(x^3-1\right)^2}$$

c. Given $y=\dfrac{u}{1+u^2}$ and $u=x^2+1$, then $y=\dfrac{x^2+1}{1+\left(x^2+1\right)^2}$ and $y' = \dfrac{2x\cdot\left[1+\left(x^2+1\right)^2\right]-\left[2\left(x^2+1\right)^{2-1}\cdot 2x\cdot\left(x^2+1\right)\right]}{\left[1+\left(x^2+1\right)^2\right]^2}$

$$= \frac{\left[2x+2x\left(x^2+1\right)^2\right]-\left[4x\left(x^2+1\right)\cdot\left(x^2+1\right)\right]}{\left[1+\left(x^2+1\right)^2\right]^2} = \frac{2x+2x\left(x^2+1\right)^2-4x\left(x^2+1\right)^2}{\left[1+\left(x^2+1\right)^2\right]^2} = \frac{2x-2x\left(x^2+1\right)^2}{\left[1+\left(x^2+1\right)^2\right]^2} = \frac{2x\left[1-\left(x^2+1\right)^2\right]}{\left[1+\left(x^2+1\right)^2\right]^2}$$

d. Given $y=u^2-\dfrac{1}{2}$ and $u=x^4$, then $y=x^8-\dfrac{1}{2}$ and $y' = 8x^{8-1}-0 = 8x^7$

e. Given $y = u^4$ and $u = \dfrac{1}{1-x^2}$, then $y = \dfrac{1}{\left(1-x^2\right)^4}$ and $y' = \dfrac{\left[0 \cdot \left(1-x^2\right)^4\right] - \left[4\left(1-x^2\right)^{4-1} \cdot -2x \cdot 1\right]}{\left(1-x^2\right)^8} = \dfrac{0 + 8x\left(1-x^2\right)^3}{\left(1-x^2\right)^8}$

$= \dfrac{8x\left(1-x^2\right)^3}{\left(1-x^2\right)^{8=5}} = \dfrac{\boldsymbol{8x}}{\left(\boldsymbol{1-x^2}\right)^{\boldsymbol{5}}}$

f. Given $y = \dfrac{u^2}{(u+1)^3}$ and $u = x-1$, then $y = \dfrac{(x-1)^2}{(x-1+1)^3} = \dfrac{(x-1)^2}{x^3}$ and $y' = \dfrac{\left[2(x-1)^{2-1} \cdot x^3\right] - \left[3x^2 \cdot (x-1)^2\right]}{x^6}$

$= \dfrac{2x^3(x-1) - 3x^2(x-1)^2}{x^6} = \dfrac{x^2(x-1)\left[2x - 3(x-1)\right]}{x^{6=4}} = \dfrac{(x-1)(2x-3x+3)}{x^4} = \dfrac{(\boldsymbol{x-1})(\boldsymbol{-x+3})}{\boldsymbol{x^4}}$

Section 2.5 Solutions - Implicit Differentiation

Use implicit differentiation method to solve the following functions.

a. Given $x^2 y + x = y$, then $\dfrac{d}{dx}\left(x^2 y + x\right) = \dfrac{d}{dx}(y)$; $\left(2x \cdot y + x^2 \cdot y'\right) + 1 = y'$; $2xy + 1 = y' - x^2 y'$; $2xy + 1 = y'\left(1 - x^2\right)$

; $\boldsymbol{y' = \dfrac{2xy+1}{1-x^2}}$

b. Given $xy - 3x^2 + y = 0$, then $\dfrac{d}{dx}\left(xy - 3x^2 + y\right) = \dfrac{d}{dx}(0)$; $(1 \cdot y + x \cdot y') - 6x + y' = 0$; $y - 6x = -xy' - y'$; $y - 6x = -y'(x+1)$

; $\dfrac{y-6x}{x+1} = -y'$; $y' = -\dfrac{y-6x}{x+1}$; $\boldsymbol{y' = \dfrac{6x-y}{x+1}}$

c. Given $x^2 y^2 + y = 3y^3$, then $\dfrac{d}{dx}\left(x^2 y^2 + y\right) = \dfrac{d}{dx}\left(3y^3\right)$; $\left(2x \cdot y^2 + 2y y' \cdot x^2\right) + y' = 9y^2 y'$; $2xy^2 + 2x^2 y y' = 9y^2 y' - y'$

; $2xy^2 = 9y^2 y' - y' - 2x^2 y y'$; $2xy^2 = y'\left(9y^2 - 1 - 2x^2 y\right)$; $\boldsymbol{y' = \dfrac{2xy^2}{9y^2 - 1 - 2x^2 y}}$

d. Given $xy + y^3 = 5x$, then $\dfrac{d}{dx}\left(xy + y^3\right) = \dfrac{d}{dx}(5x)$; $(1 \cdot y + y' \cdot x) + 3y^2 \cdot y' = 5$; $y + y'x + 3y^2 y' = 5$; $xy' + 3y^2 y' = 5 - y$

; $y'\left(x + 3y^2\right) = 5 - y$; $\boldsymbol{y' = \dfrac{5-y}{x+3y^2}}$

e. Given $4x^4 y^4 + 2y^2 = y - 1$, then $\dfrac{d}{dx}\left(4x^4 y^4 + 2y^2\right) = \dfrac{d}{dx}(y-1)$; $4\left(4x^3 \cdot y^4 + 4y^3 y' \cdot x^4\right) + 4y y' = y'$; $16x^3 y^4 + 16x^4 y^3 y'$

$+ 4y y' = y'$; $16x^3 y^4 = y' - 16x^4 y^3 y' - 4y y'$; $16x^3 y^4 = y'\left(1 - 16x^4 y^3 - 4y\right)$; $\boldsymbol{y' = \dfrac{16x^3 y^4}{1 - 16x^4 y^3 - 4y}}$

f. Given $xy + x^2 y^2 - 10 = 0$, then $\dfrac{d}{dx}\left(xy + x^2 y^2 - 10\right) = \dfrac{d}{dx}(0)$; $(1 \cdot y + y' \cdot x) + \left(2x \cdot y^2 + 2y y' \cdot x^2\right) = 0$; $y + y'x + 2xy^2$

$+ 2x^2 y y' = 0$; $y + 2xy^2 = -y'x - 2x^2 y y'$; $y + 2xy^2 = -y'\left(x + 2x^2 y\right)$; $\boldsymbol{y' = -\dfrac{y + 2xy^2}{x + 2x^2 y}}$

g. Given $xy^2 + y = x^2$, then $\dfrac{d}{dx}\left(xy^2 + y\right) = \dfrac{d}{dx}\left(x^2\right)$; $\left(1 \cdot y^2 + 2y y' \cdot x\right) + y' = 2x$; $2y y'x + y' = 2x - y^2$; $\boldsymbol{y' = \dfrac{2x - y^2}{2xy + 1}}$

h. Given $xy^3 + x^3 y = x$, then $\dfrac{d}{dx}\left(xy^3 + x^3 y\right) = \dfrac{d}{dx}(x)$; $\left(1 \cdot y^3 + 3y^2 y' \cdot x\right) + \left(3x^2 \cdot y + y' \cdot x^3\right) = 1$; $y^3 + 3xy^2 y' + 3x^2 y + x^3 y' = 1$

; $3xy^2 y' + x^3 y' = 1 - y^3 - 3x^2 y$; $y'\left(3xy^2 + x^3\right) = 1 - y^3 - 3x^2 y$; $\boldsymbol{y' = \dfrac{1 - y^3 - 3x^2 y}{3xy^2 + x^3}}$

i. Given $y^{\frac{1}{2}} + x^2 y = x$, then $\dfrac{d}{dx}\left(y^{\frac{1}{2}} + x^2 y\right) = \dfrac{d}{dx}(x)$; $\dfrac{1}{2} y^{\frac{1}{2}-1} \cdot y' + \left(2x \cdot y + y' \cdot x^2\right) = 1$; $\dfrac{1}{2} y^{-\frac{1}{2}} y' + 2xy + x^2 y' = 1$

$; \ y'\left(\dfrac{1}{2}y^{-\frac{1}{2}}+x^2\right)=1-2xy \ ; \ y'\left(\dfrac{1}{2y^{\frac{1}{2}}}+x^2\right)=1-2xy \ ; \ y'=\dfrac{1-2xy}{\dfrac{1}{2\sqrt{y}}+x^2}$

j. Given $x^2y+y^2=y^{\frac{1}{4}}$, then $\dfrac{d}{dx}\left(x^2y+y^2\right)=\dfrac{d}{dx}\left(y^{\frac{1}{4}}\right)$; $\left(2x\cdot y+y'\cdot x^2\right)+2yy'=\dfrac{1}{4}y^{\frac{1}{4}-1}y'$; $2xy+x^2y'+2yy'=\dfrac{1}{4}y^{-\frac{3}{4}}y'$

$; \ x^2y'+2yy'-\dfrac{1}{4}y^{-\frac{3}{4}}y'=-2xy \ ; \ y'\left(x^2+2y-\dfrac{1}{4}y^{-\frac{3}{4}}\right)=-2xy \ ; \ y'=\dfrac{-2xy}{x^2+2y-\dfrac{1}{4}y^{-\frac{3}{4}}} \ ; \ y'=\dfrac{-2xy}{x^2+2y-\dfrac{1}{4\sqrt[4]{y^3}}}$

k. Given $x+y^2=x^2-3$, then $\dfrac{d}{dx}\left(x+y^2\right)=\dfrac{d}{dx}\left(x^2-3\right)$; $1+2yy'=2x$; $2yy'=2x-1$; $y'=\dfrac{2x-1}{2y}$

l. Given $x^4y^2+y=-3$, then $\dfrac{d}{dx}\left(x^4y^2+y\right)=\dfrac{d}{dx}(-3)$; $\left(4x^3\cdot y^2+2yy'\cdot x^4\right)+y'=0$; $4x^3y^2+2x^4yy'+y'=0$

$; \ 2x^4yy'+y'=-4x^3y^2 \ ; \ y'\left(2x^4y+1\right)=-4x^3y^2 \ ; \ y'=\dfrac{-4x^3y^2}{2x^4y+1}$

m. Given $y^7-x^2y^4-x=8$, then $\dfrac{d}{dx}\left(y^7-x^2y^4-x\right)=\dfrac{d}{dx}(8)$; $7y^6y'-\left(2x\cdot y^4+4y^3y'\cdot x^2\right)-1=0$; $7y^6y'-2xy^4-4x^2y^3y'=1$

$; \ 7y^6y'-4x^2y^3y'=1+2xy^4 \ ; \ y'\left(7y^6-4x^2y^3\right)=1+2xy^4 \ ; \ y'=\dfrac{1+2xy^4}{7y^6-4x^2y^3}$

n. Given $(x+3)^2=y^2-x$, then $\dfrac{d}{dx}\left[(x+3)^2\right]=\dfrac{d}{dx}\left(y^2-x\right)$; $2(x+3)=2yy'-1$; $2x+6+1=2yy'$; $y'=\dfrac{2x+7}{2y}$

o. Given $3x^2y^5+y^2=-x$, then $\dfrac{d}{dx}\left(3x^2y^5+y^2\right)=\dfrac{d}{dx}(-x)$; $3\left(2x\cdot y^5+5y^4y'\cdot x^2\right)+2yy'=-1$; $6xy^5+15x^2y^4y'+2yy'=-1$

$; \ 15x^2y^4y'+2yy'=-1-6xy^5 \ ; \ y'\left(15x^2y^4+2y\right)=-1-6xy^5 \ ; \ y'=-\dfrac{1+6xy^5}{15x^2y^4+2y}$

Section 2.6 Solutions - The Derivative of Functions with Fractional Exponent

1. Find the derivative of the following exponential expressions.

a. Given $y=x^{\frac{1}{5}}$, then $y'=\dfrac{1}{5}x^{\frac{1}{5}-1}=\dfrac{1}{5}x^{\frac{1-5}{5}}=\dfrac{1}{5}x^{-\frac{4}{5}}$

b. Given $y=\left(4x^3\right)^{\frac{1}{2}}=4x^{\frac{3}{2}}$, then $y'=\dfrac{3}{2}\cdot 4x^{\frac{3}{2}-1}=\dfrac{3}{2}\cdot 4x^{\frac{3-2}{2}}=6x^{\frac{1}{2}}$

c. Given $y=(2x+1)^{\frac{1}{3}}$, then $y'=\dfrac{1}{3}(2x+1)^{\frac{1}{3}-1}\cdot 2=\dfrac{2}{3}(2x+1)^{\frac{1-3}{3}}=\dfrac{2}{3}(2x+1)^{-\frac{2}{3}}$

d. Given $y=\left(2x^2+1\right)^{\frac{1}{8}}$, then $y'=\dfrac{1}{8}\left(2x^2+1\right)^{\frac{1}{8}-1}\cdot 4x=\dfrac{4x}{8}\left(2x^2+1\right)^{\frac{1-8}{8}}=\dfrac{x}{2}\left(2x^2+1\right)^{-\frac{7}{8}}$

e. Given $y=\left(2x^3+3x\right)^{\frac{3}{5}}$, then $y'=\dfrac{3}{5}\left(2x^3+3x\right)^{\frac{3}{5}-1}\cdot\left(6x^2+3\right)=\dfrac{3}{5}\left(2x^3+3x\right)^{-\frac{2}{5}}\cdot 3\left(2x^2+1\right)=\dfrac{9}{5}\left(2x^2+1\right)\left(2x^3+3x\right)^{-\frac{2}{5}}$

f. Given $y=\left(x^3+8\right)^{\frac{2}{3}}$, then $y'=\dfrac{2}{3}\left(x^3+8\right)^{\frac{2}{3}-1}\cdot 3x^2=\dfrac{2}{3}\left(x^3+8\right)^{\frac{2-3}{3}}\cdot 3x^2=2x^2\left(x^3+8\right)^{-\frac{1}{3}}$

g. Given $y=\left(x^3\right)^{\frac{1}{2}}-(3x-1)^{\frac{1}{3}}=x^{\frac{3}{2}}-(3x-1)^{\frac{1}{3}}$, then $y'=\dfrac{3}{2}x^{\frac{3}{2}-1}-\dfrac{1}{3}(3x-1)^{\frac{1}{3}-1}\cdot 3=\dfrac{3}{2}x^{\frac{3-2}{2}}-\dfrac{1}{3}(3x-1)^{\frac{1-3}{3}}\cdot 3$

$=\dfrac{3}{2}x^{\frac{1}{2}}-(3x-1)^{-\frac{2}{3}}$

h. Given $y=x^2(x+1)^{\frac{1}{8}}$, then $y'=2x\cdot(x+1)^{\frac{1}{8}}+\dfrac{1}{8}(x+1)^{\frac{1}{8}-1}\cdot x^2=2x(x+1)^{\frac{1}{8}}+\dfrac{x^2}{8}(x+1)^{\frac{1-8}{8}}=2x(x+1)^{\frac{1}{8}}+\dfrac{x^2}{8}(x+1)^{-\frac{7}{8}}$

i. Given $y = (x^3+1)^{\frac{2}{5}} + x^{\frac{1}{2}}$, then $y' = \frac{2}{5}(x^3+1)^{\frac{2}{5}-1} \cdot 3x^2 + \frac{1}{2}x^{\frac{1}{2}-1} = \frac{6}{5}x^2(x^3+1)^{\frac{2-5}{5}} + \frac{1}{2}x^{\frac{1-2}{2}} = \frac{6}{5}x^2(x^3+1)^{-\frac{3}{5}} + \frac{1}{2}x^{-\frac{1}{2}}$

j. Given $y = \dfrac{x+1}{x^{\frac{2}{3}}}$, then $y' = \dfrac{\left[1 \cdot x^{\frac{2}{3}}\right] - \left[\frac{2}{3}x^{\frac{2}{3}-1} \cdot (x+1)\right]}{x^{\frac{4}{3}}} = \dfrac{x^{\frac{2}{3}} - \frac{2}{3}x^{\frac{2-3}{3}}(x+1)}{x^{\frac{4}{3}}} = \dfrac{x^{\frac{2}{3}} - \frac{2}{3}x^{-\frac{1}{3}}(x+1)}{x^{\frac{4}{3}}}$

k. Given $y = \dfrac{(x^2+1)^{\frac{1}{2}}}{x^2}$, then $y' = \dfrac{\left[\frac{1}{2}(x^2+1)^{\frac{1}{2}-1} \cdot 2x \cdot x^2\right] - \left[2x \cdot (x^2+1)^{\frac{1}{2}}\right]}{x^4} = \dfrac{\left[\frac{2x^3}{2}(x^2+1)^{\frac{1-2}{2}}\right] - \left[2x \cdot (x^2+1)^{\frac{1}{2}}\right]}{x^4}$

$= \dfrac{x^3(x^2+1)^{-\frac{1}{2}} - 2x(x^2+1)^{\frac{1}{2}}}{x^4}$

l. Given $y = \dfrac{(x+1)^2}{x^{\frac{1}{3}}}$, then $y' = \dfrac{\left[2(x+1)^{2-1} \cdot x^{\frac{1}{3}}\right] - \left[\frac{1}{3}x^{\frac{1}{3}-1} \cdot (x+1)^2\right]}{x^{\frac{2}{3}}} = \dfrac{2x^{\frac{1}{3}}(x+1) - \frac{1}{3}x^{-\frac{2}{3}}(x+1)^2}{x^{\frac{2}{3}}}$

2. Use the $\dfrac{d}{dx}$ notation to find the derivative of the following exponential expressions.

a. $\dfrac{d}{dx}\left(x^{\frac{1}{5}}\right)^2 = 2\left(x^{\frac{1}{5}}\right)^{2-1} \cdot \dfrac{d}{dx}x^{\frac{1}{5}} = 2x^{\frac{1}{5}} \cdot \frac{1}{5}x^{\frac{1}{5}-1} = \frac{2}{5}x^{\frac{1}{5}} \cdot x^{-\frac{4}{5}} = \frac{2}{5}x^{\frac{1}{5}-\frac{4}{5}} = \frac{2}{5}x^{-\frac{3}{5}}$, or

$\dfrac{d}{dx}\left(x^{\frac{1}{5}}\right)^2 = \dfrac{d}{dx}x^{\frac{2}{5}} = \frac{2}{5}x^{\frac{2}{5}-1} = \frac{2}{5}x^{\frac{2-5}{5}} = \frac{2}{5}x^{-\frac{3}{5}}$

b. $\dfrac{d}{dx}(x-1)^{\frac{1}{2}} = \frac{1}{2}(x-1)^{\frac{1}{2}-1} \cdot \dfrac{d}{dx}(x-1) = \frac{1}{2}(x-1)^{-\frac{1}{2}} \cdot 1 = \frac{1}{2}(x-1)^{-\frac{1}{2}}$

c. $\dfrac{d}{dx}(x^2+1)^{\frac{1}{3}} = \frac{1}{3}(x^2+1)^{\frac{1}{3}-1} \cdot \dfrac{d}{dx}(x^2+1) = \frac{1}{3}(x^2+1)^{\frac{1-3}{3}} \cdot 2x = \frac{2x}{3}(x^2+1)^{-\frac{2}{3}}$

d. $\dfrac{d}{dx}(x^3+1)^{-\frac{1}{4}} = -\frac{1}{4}(x^3+1)^{-\frac{1}{4}-1} \cdot \dfrac{d}{dx}(x^3+1) = -\frac{1}{4}(x^3+1)^{\frac{-1-4}{4}} \cdot 3x^2 = -\frac{3x^2}{4}(x^3+1)^{-\frac{5}{4}}$

e. $\dfrac{d}{dx}\left[\dfrac{(x-1)^{\frac{1}{2}}}{x^2}\right] = \dfrac{\left[x^2\frac{d}{dx}(x-1)^{\frac{1}{2}}\right] - \left[(x-1)^{\frac{1}{2}}\frac{d}{dx}x^2\right]}{x^4} = \dfrac{\left[x^2 \cdot \frac{1}{2}(x-1)^{\frac{1}{2}-1}\right] - \left[(x-1)^{\frac{1}{2}} \cdot 2x\right]}{x^4} = \dfrac{\frac{x^2}{2}(x-1)^{-\frac{1}{2}} - 2x(x-1)^{\frac{1}{2}}}{x^4}$

f. $\dfrac{d}{dx}(x^3+2x)^{\frac{1}{8}} = \frac{1}{8}(x^3+2x)^{\frac{1}{8}-1} \cdot \dfrac{d}{dx}(x^3+2x) = \frac{1}{8}(x^3+2x)^{-\frac{7}{8}} \cdot (3x^2+2) = \left(\dfrac{3x^2+2}{8}\right)(x^3+2x)^{-\frac{7}{8}}$

g. $\dfrac{d}{dx}\left[(x^3+1)(x^2)^{\frac{1}{3}}\right] = \dfrac{d}{dx}\left[(x^3+1)x^{\frac{2}{3}}\right] = \left[x^{\frac{2}{3}} \cdot \dfrac{d}{dx}(x^3+1)\right] + \left[(x^3+1) \cdot \dfrac{d}{dx}x^{\frac{2}{3}}\right] = \left[x^{\frac{2}{3}} \cdot 3x^2\right] + \left[(x^3+1) \cdot \frac{2}{3}x^{\frac{2}{3}-1}\right]$

$= 3x^{2+\frac{2}{3}} + \frac{2}{3}(x^3+1) \cdot x^{-\frac{1}{3}} = 3x^{\frac{8}{3}} + \dfrac{2x^{-\frac{1}{3}}}{3} \cdot (x^3+1)$

h. $\dfrac{d}{dx}\left[x^3 \cdot \dfrac{1}{(x^2+1)^{\frac{1}{2}}}\right] = \dfrac{d}{dx}\left[\dfrac{x^3}{(x^2+1)^{\frac{1}{2}}}\right] = \dfrac{\left[(x^2+1)^{\frac{1}{2}} \cdot \frac{d}{dx}x^3\right] - \left[x^3 \cdot \frac{d}{dx}(x^2+1)^{\frac{1}{2}}\right]}{x^2+1} = \dfrac{\left[(x^2+1)^{\frac{1}{2}} \cdot 3x^2\right] - \left[x^3 \cdot \frac{1}{2}(x^2+1)^{\frac{1}{2}-1} \cdot 2x\right]}{x^2+1}$

$= \dfrac{\left[3x^2(x^2+1)^{\frac{1}{2}}\right] - \left[x^4(x^2+1)^{\frac{1-2}{2}}\right]}{x^2+1} = \dfrac{\left[3x^2(x^2+1)^{\frac{1}{2}}\right] - \left[x^4(x^2+1)^{-\frac{1}{2}}\right]}{x^2+1}$

i. $\dfrac{d}{dx}\left[\dfrac{x^5}{\left(x^3+1\right)^{\frac{2}{3}}}\right] = \dfrac{\left[\left(x^3+1\right)^{\frac{2}{3}}\cdot\frac{d}{dx}x^5\right]-\left[x^5\cdot\frac{d}{dx}\left(x^3+1\right)^{\frac{2}{3}}\right]}{\left(x^3+1\right)^{\frac{4}{3}}} = \dfrac{\left[\left(x^3+1\right)^{\frac{2}{3}}\cdot 5x^4\right]-\left[x^5\cdot\frac{2}{3}\left(x^3+1\right)^{\frac{2}{3}-1}\cdot 3x^2\right]}{\left(x^3+1\right)^{\frac{4}{3}}}$

$= \dfrac{\left[5x^4\left(x^3+1\right)^{\frac{2}{3}}\right]-\left[\frac{2x^7}{3}\left(x^3+1\right)^{\frac{2-3}{3}}\right]}{\left(x^3+1\right)^{\frac{4}{3}}} = \dfrac{\left[5x^4\left(x^3+1\right)^{\frac{2}{3}}\right]-\left[2x^7\left(x^3+1\right)^{-\frac{1}{3}}\right]}{\left(x^3+1\right)^{\frac{4}{3}}}$

j. $\dfrac{d}{dx}\left[(x-1)^{\frac{1}{2}}(x+1)^{\frac{1}{3}}\right] = \left[(x+1)^{\frac{1}{3}}\frac{d}{dx}(x-1)^{\frac{1}{2}}\right]+\left[(x-1)^{\frac{1}{2}}\frac{d}{dx}(x+1)^{\frac{1}{3}}\right] = \left[(x+1)^{\frac{1}{3}}\cdot\frac{1}{2}(x-1)^{\frac{1}{2}-1}\right]+\left[(x-1)^{\frac{1}{2}}\cdot\frac{1}{3}(x+1)^{\frac{1}{3}-1}\right]$

$= \left[\frac{1}{2}(x+1)^{\frac{1}{3}}\cdot(x-1)^{\frac{1-2}{2}}\right]+\left[\frac{1}{3}(x-1)^{\frac{1}{2}}\cdot(x+1)^{\frac{1-3}{3}}\right] = \left[\frac{1}{2}(x+1)^{\frac{1}{3}}(x-1)^{-\frac{1}{2}}\right]+\left[\frac{1}{3}(x-1)^{\frac{1}{2}}(x+1)^{-\frac{2}{3}}\right]$

k. $\dfrac{d}{dx}\left[x^3\left(x^2+1\right)^{\frac{1}{2}}\right] = \left[\left(x^2+1\right)^{\frac{1}{2}}\frac{d}{dx}x^3\right]+\left[x^3\frac{d}{dx}\left(x^2+1\right)^{\frac{1}{2}}\right] = \left[\left(x^2+1\right)^{\frac{1}{2}}\cdot 3x^2\right]+\left[x^3\cdot\frac{1}{2}\left(x^2+1\right)^{\frac{1}{2}-1}\cdot 2x\right]$

$= \left[3x^2\left(x^2+1\right)^{\frac{1}{2}}\right]+\left[x^4\left(x^2+1\right)^{\frac{1-2}{2}}\right] = 3x^2\left(x^2+1\right)^{\frac{1}{2}}+x^4\left(x^2+1\right)^{-\frac{1}{2}}$

l. $\dfrac{d}{dx}\left[x^3\left(x^2+1\right)^{-\frac{1}{3}}\right] = \left[\left(x^2+1\right)^{-\frac{1}{3}}\frac{d}{dx}x^3\right]+\left[x^3\frac{d}{dx}\left(x^2+1\right)^{-\frac{1}{3}}\right] = \left[\left(x^2+1\right)^{-\frac{1}{3}}\cdot 3x^2\right]+\left[x^3\cdot-\frac{1}{3}\left(x^2+1\right)^{-\frac{1}{3}-1}\cdot 2x\right]$

$= \left[3x^2\left(x^2+1\right)^{-\frac{1}{3}}\right]+\left[-\frac{2x^4}{3}\left(x^2+1\right)^{\frac{-1-3}{3}}\right] = 3x^2\left(x^2+1\right)^{-\frac{1}{3}}-\frac{2x^4}{3}\left(x^2+1\right)^{-\frac{4}{3}}$

Section 2.7 Solutions - The Derivative of Radical Functions

1. Find the derivative of the following radical expressions. Do not simplify the answer to its lowest term.

a. Given $y=\sqrt{x^2+1}=\left(x^2+1\right)^{\frac{1}{2}}$, then $y'=\frac{1}{2}\left(x^2+1\right)^{\frac{1}{2}-1}\cdot 2x=\frac{2x}{2}\left(x^2+1\right)^{-\frac{1}{2}}=\dfrac{x}{\left(x^2+1\right)^{\frac{1}{2}}}$

b. Given $y=\sqrt{x^3+3x-5}=\left(x^3+3x-5\right)^{\frac{1}{2}}$, then $y'=\frac{1}{2}\left(x^3+3x-5\right)^{\frac{1}{2}-1}\cdot\left(3x^2+3\right)=\frac{1}{2}\left(x^3+3x-5\right)^{-\frac{1}{2}}\left(3x^2+3\right)=\dfrac{3\left(x^2+1\right)}{2\left(x^3+3x-5\right)^{\frac{1}{2}}}$

c. Given $y=x^2+\sqrt{x-1}=x^2+(x-1)^{\frac{1}{2}}$, then $y'=2x^{2-1}+(x-1)^{\frac{1}{2}-1}=2x+(x-1)^{-\frac{1}{2}}=2x+\dfrac{1}{(x-1)^{\frac{1}{2}}}$

d. Given $y=\dfrac{\sqrt{x+1}}{x}=\dfrac{(x+1)^{\frac{1}{2}}}{x}$, then $y'=\dfrac{\left[\frac{1}{2}(x+1)^{\frac{1}{2}-1}\cdot x\right]-\left[1\cdot(x+1)^{\frac{1}{2}}\right]}{x^2}=\dfrac{\frac{x}{2}(x+1)^{-\frac{1}{2}}-(x+1)^{\frac{1}{2}}}{x^2}$

e. Given $y=\dfrac{x^2}{\sqrt{x^2-1}}=\dfrac{x^2}{\left(x^2-1\right)^{\frac{1}{2}}}$, then $y'=\dfrac{\left[2x\cdot\left(x^2-1\right)^{\frac{1}{2}}\right]-\left[\frac{1}{2}\left(x^2-1\right)^{\frac{1}{2}-1}\cdot 2x\cdot x^2\right]}{x^2-1}=\dfrac{2x\left(x^2-1\right)^{\frac{1}{2}}-x^3\left(x^2-1\right)^{-\frac{1}{2}}}{x^2-1}$

f. Given $y=\sqrt{x^3}+3x^2=\left(x^3\right)^{\frac{1}{2}}+3x^2=x^{\frac{3}{2}}+3x^2$, then $y'=\frac{3}{2}x^{\frac{3}{2}-1}+(3\cdot 2)x^{2-1}=\frac{3}{2}x^{\frac{3-2}{2}}+6x=\frac{3}{2}x^{\frac{1}{2}}+6x$

g. Given $y=\dfrac{\sqrt{x^2+3}}{\sqrt{x+1}}=\sqrt{\dfrac{x^2+3}{x+1}}=\left(\dfrac{x^2+3}{x+1}\right)^{\frac{1}{2}}$, then $y'=\frac{1}{2}\left(\dfrac{x^2+3}{x+1}\right)^{\frac{1}{2}-1}\cdot\dfrac{2x\cdot(x+1)-1\cdot\left(x^2+3\right)}{(x+1)^2}=\frac{1}{2}\left(\dfrac{x^2+3}{x+1}\right)^{-\frac{1}{2}}\cdot\dfrac{x^2+2x-3}{(x+1)^2}$

h. Given $y = \dfrac{\sqrt[4]{x^3-1}}{\sqrt{x}} = \dfrac{\left(x^3-1\right)^{\frac{1}{4}}}{x^{\frac{1}{2}}}$, then $y' = \dfrac{\left[\frac{1}{4}\left(x^3-1\right)^{\frac{1}{4}-1}\cdot 3x^2\cdot x^{\frac{1}{2}}\right]-\left[\frac{1}{2}x^{\frac{1}{2}-1}\cdot\left(x^3-1\right)^{\frac{1}{4}}\right]}{x} = \dfrac{\frac{3x^{\frac{5}{2}}}{4}\left(x^3-1\right)^{-\frac{3}{4}}-\frac{1}{2}x^{-\frac{1}{2}}\left(x^3-1\right)^{\frac{1}{4}}}{x}$

i. Given $y = \dfrac{x^3}{x^2\sqrt{x}} = \dfrac{x^{3=1}}{x^2\cdot x^{\frac{1}{2}}} = \dfrac{x}{x^{\frac{1}{2}}} = x\cdot x^{-\frac{1}{2}} = x^{1-\frac{1}{2}} = x^{\frac{1}{2}}$, then $y' = \dfrac{1}{2}x^{\frac{1}{2}-1} = \dfrac{1}{2}x^{\frac{1}{2}-\frac{1}{1}} = \dfrac{1}{2}x^{\frac{1-2}{2}} = \mathbf{\dfrac{1}{2}x^{-\frac{1}{2}}}$

2. Use the $\dfrac{d}{dx}$ notation to find the derivative of the following radical expressions.

a. $\dfrac{d}{dx}\left(\sqrt{x^2}+\dfrac{1}{x}\right) = \dfrac{d}{dx}\left(x+\dfrac{1}{x}\right) = \dfrac{d}{dx}x+\dfrac{d}{dx}\left(\dfrac{1}{x}\right) = 1+\dfrac{(0\cdot x)-(1\cdot 1)}{x^2} = \mathbf{1-\dfrac{1}{x^2}}$

b. $\dfrac{d}{dx}\left(\sqrt{\dfrac{x}{x-1}}\right) = \dfrac{d}{dx}\left(\dfrac{x}{x-1}\right)^{\frac{1}{2}} = \dfrac{1}{2}\left(\dfrac{x}{x-1}\right)^{\frac{1}{2}-1}\cdot\dfrac{\left[(x-1)\frac{d}{dx}x\right]-\left[x\frac{d}{dx}(x-1)\right]}{(x-1)^2} = \dfrac{1}{2}\left(\dfrac{x}{x-1}\right)^{\frac{1}{2}-1}\cdot\dfrac{[1\cdot(x-1)]-[1\cdot x]}{(x-1)^2}$

$= \dfrac{1}{2}\left(\dfrac{x}{x-1}\right)^{\frac{1-2}{2}}\cdot\dfrac{\cancel{x}-1-\cancel{x}}{(x-1)^2} = \mathbf{-\dfrac{1}{2}\left(\dfrac{x}{x-1}\right)^{-\frac{1}{2}}\cdot\dfrac{1}{(x-1)^2}}$

c. $\dfrac{d}{dx}\left(\dfrac{x^3}{\sqrt{x+1}}\right) = \dfrac{d}{dx}\left(\dfrac{x^3}{(x+1)^{\frac{1}{2}}}\right) = \dfrac{\left[(x+1)^{\frac{1}{2}}\frac{d}{dx}x^3\right]-\left[x^3\frac{d}{dx}(x+1)^{\frac{1}{2}}\right]}{x+1} = \dfrac{\left[(x+1)^{\frac{1}{2}}\cdot 3x^{3-1}\right]-\left[x^3\cdot\frac{1}{2}(x+1)^{\frac{1}{2}-1}\right]}{x+1}$

$= \dfrac{\left[3(x+1)^{\frac{1}{2}}x^2\right]-\left[\frac{x^3}{2}(x+1)^{\frac{1-2}{2}}\right]}{x+1} = \dfrac{\left[3x^2(x+1)^{\frac{1}{2}}\right]-\left[\frac{x^3}{2}(x+1)^{-\frac{1}{2}}\right]}{x+1}$

d. $\dfrac{d}{dx}\left(\dfrac{\sqrt{x+5}}{x}\right) = \dfrac{d}{dx}\left(\dfrac{(x+5)^{\frac{1}{2}}}{x}\right) = \dfrac{\left[x\frac{d}{dx}(x+5)^{\frac{1}{2}}\right]-\left[(x+5)^{\frac{1}{2}}\frac{d}{dx}x\right]}{x^2} = \dfrac{\left[x\cdot\frac{1}{2}(x+5)^{\frac{1}{2}-1}\right]-\left[(x+5)^{\frac{1}{2}}\cdot 1\right]}{x^2} = \dfrac{\frac{x}{2}(x+5)^{-\frac{1}{2}}-(x+5)^{\frac{1}{2}}}{x^2}$

e. $\dfrac{d}{dx}\left(x^3+\dfrac{\sqrt{x}}{x}\right) = \dfrac{d}{dx}\left(x^3+\dfrac{x^{\frac{1}{2}}}{x}\right) = \dfrac{d}{dx}\left(x^3+x^{\frac{1}{2}}\cdot x^{-1}\right) = \dfrac{d}{dx}\left(x^3+x^{\frac{1}{2}-1}\right) = \dfrac{d}{dx}\left(x^3+x^{-\frac{1}{2}}\right) = \dfrac{d}{dx}x^3+\dfrac{d}{dx}x^{-\frac{1}{2}}$

$= 3x^{3-1}-\dfrac{1}{2}x^{-\frac{1}{2}-1} = 3x^2-\dfrac{1}{2}x^{\frac{-1-2}{2}} = \mathbf{3x^2-\dfrac{1}{2}x^{-\frac{3}{2}}}$

f. $\dfrac{d}{dx}\left(1+\dfrac{2\sqrt{x}}{x^3}\right) = \dfrac{d}{dx}\left(1+\dfrac{2x^{\frac{1}{2}}}{x^3}\right) = \dfrac{d}{dx}\left(1+2x^{\frac{1}{2}}x^{-3}\right) = \dfrac{d}{dx}\left(1+2x^{\frac{1}{2}-3}\right) = \dfrac{d}{dx}\left(1+2x^{\frac{1-6}{2}}\right) = \dfrac{d}{dx}\left(1+2x^{-\frac{5}{2}}\right)$

$= \dfrac{d}{dx}(1)+\dfrac{d}{dx}\left(2x^{-\frac{5}{2}}\right) = 0-\dfrac{5}{2}x^{-\frac{5}{2}-1}\cdot 2 = -\dfrac{5}{2}x^{\frac{-5-2}{2}}\cdot 2 = \mathbf{-5x^{-\frac{7}{2}}}$

3. Find the derivative of the following radical expressions.

a. $\dfrac{d}{dx}\left(\sqrt{x^3}+\sqrt{y}\right) = \dfrac{d}{dx}(x)$; $\dfrac{d}{dx}\left(x^{\frac{3}{2}}+y^{\frac{1}{2}}\right)=1$; $\dfrac{d}{dx}x^{\frac{3}{2}}+\dfrac{d}{dx}y^{\frac{1}{2}}=1$; $\dfrac{3}{2}x^{\frac{3}{2}-1}+\dfrac{1}{2}y^{\frac{1}{2}-1}y'=1$; $\dfrac{3}{2}x^{\frac{3-2}{2}}+\dfrac{1}{2}y^{-\frac{1}{2}}y'=1$

; $\dfrac{3}{2}x^{\frac{1}{2}}+\dfrac{1}{2}y^{-\frac{1}{2}}y'=1$; $\dfrac{1}{2}y^{-\frac{1}{2}}y'=1-\dfrac{3}{2}x^{\frac{1}{2}}$; $\dfrac{y'}{2y^{\frac{1}{2}}}=1-\dfrac{3}{2}x^{\frac{1}{2}}$; $y'=2y^{\frac{1}{2}}\left(1-\dfrac{3}{2}x^{\frac{1}{2}}\right)$; $\mathbf{y'=2y^{\frac{1}{2}}-3(xy)^{\frac{1}{2}}}$

b. $\dfrac{d}{dx}\left(\sqrt{x}+y^3\right) = \dfrac{d}{dx}(2)$; $\dfrac{d}{dx}\left(x^{\frac{1}{2}}+y^3\right)=0$; $\dfrac{d}{dx}x^{\frac{1}{2}}+\dfrac{d}{dx}y^3=0$; $\dfrac{1}{2}x^{\frac{1}{2}-1}+3y^2y'=0$; $\dfrac{1}{2}x^{-\frac{1}{2}}+3y^2y'=0$; $3y^2y'=-\dfrac{1}{2}x^{-\frac{1}{2}}$

$; 3y^2 y' = -\dfrac{1}{2x^{\frac{1}{2}}} \;; y' = -\dfrac{1}{6x^{\frac{1}{2}}y^2}$

c. $\dfrac{d}{dx}(x\,y) = \dfrac{d}{dx}(\sqrt{x}) \;; \dfrac{d}{dx}(x\,y) = \dfrac{d}{dx}x^{\frac{1}{2}} \;; y\dfrac{d}{dx}x + x\dfrac{d}{dx}y = \dfrac{d}{dx}x^{\frac{1}{2}} \;; y + xy' = \dfrac{1}{2}x^{\frac{1}{2}-1} \;; xy' = \dfrac{1}{2}x^{-\frac{1}{2}} - y \;; y' = \dfrac{1}{x}\left(\dfrac{1}{2x^{\frac{1}{2}}} - y\right)$

d. $\dfrac{d}{dx}\left(\sqrt{y}+x^3\right)=0 \;; \dfrac{d}{dx}\left(y^{\frac{1}{2}}+x^3\right)=0 \;; \dfrac{d}{dx}y^{\frac{1}{2}}+\dfrac{d}{dx}x^3 = 0 \;; \dfrac{1}{2}y^{\frac{1}{2}-1}y'+3x^2 = 0 \;; \dfrac{1}{2}y^{-\frac{1}{2}}y' = -3x^2 \;; \dfrac{y'}{2y^{\frac{1}{2}}} = -3x^2$

$; \dfrac{y'}{2y^{\frac{1}{2}}} = \dfrac{-3x^2}{1} \;; y' = -6x^2 y^{\frac{1}{2}}$

e. $\dfrac{d}{dx}\left(\sqrt{x^4}+y^2\right)=\dfrac{d}{dx}(x) \;; \dfrac{d}{dx}\left(x^2+y^2\right)=1 \;; \dfrac{d}{dx}x^2+\dfrac{d}{dx}y^2 = 1 \;; 2x+2y\,y' = 1 \;; 2y\,y' = 1-2x \;; y' = \dfrac{1-2x}{2y}$

f. $\dfrac{d}{dx}\left(\sqrt{x+1}\right)=\dfrac{d}{dx}\left(y^3\right) \;; \dfrac{d}{dx}(x+1)^{\frac{1}{2}} = 3y^2\dfrac{d}{dx}y \;; \dfrac{1}{2}(x+1)^{\frac{1}{2}-1} = 3y^2 y' \;; \dfrac{1}{2}(x+1)^{-\frac{1}{2}} = 3y^2 y' \;; \dfrac{1}{2(x+1)^{\frac{1}{2}}} = \dfrac{3y^2 y'}{1}$

$; 6y^2 y'(x+1)^{\frac{1}{2}} = 1 \;; y' = \dfrac{1}{6y^2(x+1)^{\frac{1}{2}}}$

g. $\dfrac{d}{dx}\left(x\,y^2+\sqrt{x}\right)=\dfrac{d}{dx}(2) \;; \dfrac{d}{dx}\left(x\,y^2+x^{\frac{1}{2}}\right)=0 \;; y^2\dfrac{d}{dx}x+x\dfrac{d}{dx}y^2+\dfrac{1}{2}x^{\frac{1}{2}-1} = 0 \;; y^2 \cdot 1 + x \cdot 2y\,y' + \dfrac{1}{2}x^{-\frac{1}{2}} = 0$

$; 2xy\,y' = -\dfrac{1}{2}x^{-\frac{1}{2}} - y^2 \;; y' = \dfrac{1}{2xy}\left(-\dfrac{1}{2}x^{-\frac{1}{2}} - y^2\right) \;; y' = -\dfrac{x^{-\frac{1}{2}}}{4xy} - \dfrac{y^2}{2xy} \;; y' = -\dfrac{1}{4x^{1+\frac{1}{2}}y} - \dfrac{y^2}{2xy} \;; y' = -\dfrac{1}{4x^{\frac{3}{2}}y} - \dfrac{y}{2x}$

h. $\dfrac{d}{dx}\left(\sqrt{x^3}\right)+\dfrac{d}{dx}(x\,y)=0 \;; \dfrac{d}{dx}x^{\frac{3}{2}}+\dfrac{d}{dx}(x\,y)=0 \;; \dfrac{3}{2}x^{\frac{3}{2}-1}+\left(y\dfrac{d}{dx}x+x\dfrac{d}{dx}y\right)=0 \;; \dfrac{3}{2}x^{\frac{1}{2}}+(y+xy')=0 \;; xy' = -\dfrac{3}{2}x^{\frac{1}{2}} - y$

$; y' = \dfrac{1}{x}\left(-\dfrac{3}{2}x^{\frac{1}{2}} - y\right) \;; y' = -\dfrac{3x^{\frac{1}{2}}}{2x} - \dfrac{y}{x} \;; y' = -\dfrac{3}{2x\cdot x^{-\frac{1}{2}}} - \dfrac{y}{x} \;; y' = -\dfrac{3}{2x^{1-\frac{1}{2}}} - \dfrac{y}{x} \;; y' = -\dfrac{3}{2x^{\frac{1}{2}}} - \dfrac{y}{x}$

i. $\dfrac{d}{dx}\left(\sqrt{x}+3y\right)=\dfrac{d}{dx}(y) \;; \dfrac{d}{dx}\left(x^{\frac{1}{2}}+3y\right)=y' \;; \dfrac{d}{dx}x^{\frac{1}{2}}+\dfrac{d}{dx}3y = y' \;; \dfrac{1}{2}x^{\frac{1}{2}-1}+3y' = y' \;; \dfrac{1}{2}x^{-\frac{1}{2}}+3y'-y' = 0 \;; \dfrac{1}{2}x^{-\frac{1}{2}}+2y' = 0$

$; 2y' = -\dfrac{1}{2}x^{-\frac{1}{2}} \;; 2y' = -\dfrac{1}{2x^{\frac{1}{2}}} \;; y' = -\dfrac{1}{4x^{\frac{1}{2}}}$

4. Evaluate the derivative of the following radical expressions for the specified value of x.

a. Given $y = \sqrt{3x^3}+x^2 = \left(3x^3\right)^{\frac{1}{2}}+x^2 = 3^{\frac{1}{2}}x^{\frac{3}{2}}+x^2$, then $y' = 3^{\frac{1}{2}}\cdot\dfrac{3}{2}x^{\frac{3}{2}-1}+2x^{2-1} = \dfrac{3^{\frac{3}{2}}}{2}x^{\frac{3-2}{2}}+2x = \dfrac{3^{\frac{3}{2}}}{2}x^{\frac{1}{2}}+2x$

at $x=1$ $y' = \dfrac{3^{\frac{3}{2}}}{2}\cdot 1^{\frac{1}{2}}+(2\cdot 1) = \dfrac{3^{\frac{3}{2}}}{2}+2 = \dfrac{5.196}{2}+2 = \textbf{4.598}$

b. Given $y = \left(x^2+1\right)\sqrt{x} = \left(x^2+1\right)x^{\frac{1}{2}} = x^{2+\frac{1}{2}}+x^{\frac{1}{2}} = x^{\frac{4+1}{2}}+x^{\frac{1}{2}} = x^{\frac{5}{2}}+x^{\frac{1}{2}}$, then $y' = \dfrac{5}{2}x^{\frac{5}{2}-1}+\dfrac{1}{2}x^{\frac{1}{2}-1} = \dfrac{5}{2}x^{\frac{5-2}{2}}+\dfrac{1}{2}x^{\frac{1-2}{2}}$

$= \dfrac{5}{2}x^{\frac{3}{2}}+\dfrac{1}{2}x^{-\frac{1}{2}} = \dfrac{5}{2}x^{\frac{3}{2}}+\dfrac{1}{2x^{\frac{1}{2}}}$ at $x=0$ $y' = \dfrac{5}{2}\cdot 0^{\frac{3}{2}}+\dfrac{1}{2\cdot 0^{\frac{1}{2}}} = 0+\dfrac{1}{0}$ **is undefined due to division by zero**

c. Given $y = \dfrac{x^2-1}{\sqrt{4x^2}} = \dfrac{x^2-1}{2x}$, then $y' = \dfrac{[2x\cdot 2x]-\left[2\cdot\left(x^2-1\right)\right]}{(2x)^2} = \dfrac{4x^2-2x^2+2}{4x^2} = \dfrac{2x^2+2}{4x^2} = \dfrac{2\left(x^2+1\right)}{4x^2} = \dfrac{x^2+1}{2x^2}$

at $x=2$ $y' = \dfrac{2^2+1}{2\cdot 2^2} = \dfrac{4+1}{2\cdot 4} = \dfrac{5}{8} = \mathbf{0.625}$

d. Given $y = \sqrt{\dfrac{x}{x^2+1}} = \left(\dfrac{x}{x^2+1}\right)^{\frac{1}{2}}$, then $y' = \dfrac{1}{2}\left(\dfrac{x}{x^2+1}\right)^{\frac{1}{2}-1}\cdot\dfrac{\left[1\cdot\left(x^2+1\right)\right]-\left[2x\cdot x\right]}{\left(x^2+1\right)^2} = \dfrac{1}{2}\left(\dfrac{x}{x^2+1}\right)^{-\frac{1}{2}}\cdot\dfrac{-x^2+1}{\left(x^2+1\right)^2}$

at $x=1$ $y' = \dfrac{1}{2}\left(\dfrac{1}{1^2+1}\right)^{-\frac{1}{2}}\cdot\dfrac{-1^2+1}{\left(1^2+1\right)^2} = \dfrac{1}{2}\left(\dfrac{1}{2}\right)^{-\frac{1}{2}}\cdot\dfrac{-1+1}{2^2} = \dfrac{1}{2}\left(\dfrac{1}{2}\right)^{-\frac{1}{2}}\cdot 0 = \mathbf{0}$

e. Given $y = \sqrt{x^3+1}+4x^3 = \left(x^3+1\right)^{\frac{1}{2}}+4x^3$, then $y' = \dfrac{1}{2}\left(x^3+1\right)^{\frac{1}{2}-1}\cdot 3x^2+(4\cdot 3)x^{3-1} = \dfrac{3x^2}{2}\left(x^3+1\right)^{-\frac{1}{2}}+12x^2$

at $x=0$ $y' = \dfrac{3\cdot 0^2}{2}\cdot\left(0^3+1\right)^{-\frac{1}{2}}+12\cdot 0^2 = \dfrac{3\cdot 0^2}{2}\cdot\dfrac{1}{\left(0^3+1\right)^{\frac{1}{2}}}+12\cdot 0^2 = 0\cdot\dfrac{1}{1^{\frac{1}{2}}}+0 = 0+0 = \mathbf{0}$

f. Given $y = \dfrac{x^2+1}{\sqrt{x^3}} = \dfrac{x^2+1}{\left(x^3\right)^{\frac{1}{2}}} = \dfrac{x^2+1}{x^{\frac{3}{2}}}$, then $y' = \dfrac{\left[2x\cdot x^{\frac{3}{2}}\right]-\left[\dfrac{3}{2}x^{\frac{3}{2}-1}\cdot\left(x^2+1\right)\right]}{x^3} = \dfrac{2x^{\frac{5}{2}}-\dfrac{3}{2}x^{\frac{1}{2}}\left(x^2+1\right)}{x^3}$

at $x=3$ $y' = \dfrac{2\cdot 3^{\frac{5}{2}}-\dfrac{3}{2}\cdot 3^{\frac{1}{2}}\left(3^2+1\right)}{3^3} = \dfrac{(2\cdot 15.58)-(1.5\cdot 1.732\cdot 10)}{27} = \dfrac{31.18-25.98}{27} = \mathbf{0.193}$

Section 2.8 Solutions - Higher Order Derivatives

1. Find the second derivative of the following functions.

a. Given $y = x^3+3x^2+5x-1$, then $y' = 3x^{3-1}+(3\cdot 2)x^{2-1}+5x^{1-1}-0 = 3x^2+6x+5x^0 = 3x^2+6x+5$ and

$y'' = (3\cdot 2)x^{2-1}+6x^{1-1}+0 = 6x+6x^0 = \mathbf{6x+6}$

b. Given $y = x^2(x+1)^2$, then $y' = 2x\cdot(x+1)^2+2(x+1)^{2-1}\cdot x^2 = 2x(x+1)^2+2x^2(x+1) = 2x(x+1)^2+2x^3+2x^2$ and

$y'' = \left[2\cdot(x+1)^2+2(x+1)^{2-1}\cdot 2x\right]+(2\cdot 3)x^{3-1}+(2\cdot 2)x^{2-1} = 2(x+1)^2+4x(x+1)+6x^2+4x = 2\left(x^2+2x+1\right)+4x^2+4xx$

$+6x^2+4x = 2x^2+4x+2+10x^2+8x = 12x^2+12x+2 = \mathbf{2\left(6x^2+6x+1\right)}$

c. Given $y = 3x^3+50x$, then $y' = (3\cdot 3)x^{3-1}+50x^{1-1} = 9x^2+50$ and $y'' = (9\cdot 2)x^{2-1}+0 = \mathbf{18x}$

d. Given $y = x^5+\dfrac{1}{x^2} = x^5+x^{-2}$, then $y' = 5x^{5-1}-2x^{-2-1} = 5x^4-2x^{-3}$ and $y'' = (5\cdot 4)x^{4-1}+(-2\cdot -3)x^{-3-1} = \mathbf{20x^3+6x^{-4}}$

e. Given $y = \dfrac{x^3}{x+1}-5x^2$, then $y' = \dfrac{\left[3x^2\cdot(x+1)\right]-\left[1\cdot x^3\right]}{(x+1)^2}-(5\cdot 2)x^{2-1} = \dfrac{3x^3+3x^2-x^3}{(x+1)^2}-10x = \dfrac{2x^3+3x^2}{(x+1)^2}-10x$ and

$y'' = \dfrac{\left[(6x^2+6x)\cdot(x+1)^2\right]-\left[2(x+1)\cdot\left(2x^3+3x^2\right)\right]}{(x+1)^4}-10x^{1-1} = \dfrac{\left[6x(x+1)(x+1)^2\right]-\left[2(x+1)\left(2x^3+3x^2\right)\right]}{(x+1)^4}-10$

$= \dfrac{(x+1)\left[6x(x+1)^2-2\left(2x^3+3x^2\right)\right]}{(x+1)^{4=3}}-10 = \dfrac{6x(x+1)^2-2x^2(2x+3)}{(x+1)^3}-10$

f. Given $y = x^3\left(x^2-1\right) = x^5-x^3$, then $y' = 5x^{5-1}-3x^{3-1} = 5x^4-3x^2$ and $y'' = (5\cdot 4)x^{4-1}-(3\cdot 2)x^{2-1} = \mathbf{20x^3-6x}$

g. Given $y = x^4+\dfrac{x^8-7x^5+5x}{10}$, then $y' = 4x^{4-1}+\dfrac{1}{10}\left(8x^{8-1}+(-7\cdot 5)x^{5-1}+5x^{1-1}\right) = 4x^3+\dfrac{1}{10}\left(8x^7-35x^4+5\right)$ and

$y'' = (4\cdot 3)x^{3-1}+\dfrac{1}{10}\left[(8\cdot 7)x^{7-1}+(-35\cdot 4)x^{4-1}+0\right] = 12x^2+\dfrac{1}{10}\left(56x^6-140x^3\right) = \mathbf{5.6x^6-14x^3+12x^2}$

h. Given $y = x^2 - \dfrac{1}{x+1} = x^2 - (x+1)^{-1}$, then $y' = 2x^{2-1} + (x+1)^{-1-1} = 2x + (x+1)^{-2}$ and $y'' = 2x^{1-1} - 2(x+1)^{-2-1}$

$= 2x^0 - 2(x+1)^{-3} = 2 - 2(x+1)^{-3} = 2 - \dfrac{2}{(x+1)^3}$

i. Given $y = \dfrac{1}{x^2} - 3x = x^{-2} - 3x$, then $y' = -2x^{-2-1} - 3x^{1-1} = -2x^{-3} - 3$ and $y'' = (-2 \cdot -3)x^{-3-1} - 0 = 6x^{-4}$

2. Find y''' for the following functions.

   a. Given $y = x^5 + 6x^3 + 10$, then $y' = 5x^{5-1} + (6 \cdot 3)x^{3-1} + 0 = 5x^4 + 18x^2$, $y'' = (5 \cdot 4)x^{4-1} + (18 \cdot 2)x^{2-1} = 20x^3 + 36x$,

   and $y''' = (20 \cdot 3)x^{3-1} + 36x^{1-1} = 60x^2 + 36$

   b. Given $y = x^2 + \dfrac{1}{x} = x^2 + x^{-1}$, then $y' = 2x^{2-1} - x^{-1-1} = 2x - x^{-2}$, $y'' = 2x^{1-1} + 2x^{-2-1} = 2 + 2x^{-3}$, and $y''' = -6x^{-4}$

   c. Given $y = 4x^3(x-1)^2$, then $y' = \left[(4 \cdot 3)x^{3-1} \cdot (x-1)^2\right] + \left[2(x-1)^{2-1} \cdot 4x^3\right] = 12x^2(x-1)^2 + 8x^3(x-1) = 12x^2(x^2 - 2x + 1)$

   $+ 8x^4 - 8x^3 = 12x^4 - 24x^3 + 12x^2 + 8x^4 - 8x^3 = 20x^4 - 32x^3 + 12x^2$ $y'' = (20 \cdot 4)x^{4-1} - (32 \cdot 3)x^{3-1} + (12 \cdot 2)x^{2-1}$

   $= 80x^3 - 96x^2 + 24x$ and $y''' = (80 \cdot 3)x^{3-1} - (96 \cdot 2)x^{2-1} + 24x^{1-1} = 240x^2 - 192x + 24x^0 = 240x^2 - 192x + 24$

   d. Given $y = \dfrac{x}{x+1}$, then $y' = \dfrac{[1 \cdot (x+1)] - [1 \cdot x]}{(x+1)^2} = \dfrac{x+1-x}{(x+1)^2} = \dfrac{1}{(x+1)^2} = (x+1)^{-2}$, $y'' = -2(x+1)^{-2-1} = -2(x+1)^{-3}$ and

   $y''' = (-2 \cdot -3)(x+1)^{-3-1} = 6(x+1)^{-4} = \dfrac{6}{(x+1)^4}$

   e. Given $y = x^8 - 10x^5 + 5x - 10$, then $y' = 8x^{8-1} + (-10 \cdot 5)x^{5-1} + 5x^{1-1} - 0 = 8x^7 - 50x^4 + 5x^0 = 8x^7 - 50x^4 + 5$,

   $y'' = (8 \cdot 7)x^{7-1} - (50 \cdot 4)x^{4-1} + 0 = 56x^6 - 200x^3$ and $y''' = (56 \cdot 6)x^{6-1} - (200 \cdot 3)x^{3-1} = 336x^5 - 600x^2$

   f. Given $y = \dfrac{x-1}{x^2} + 5x^3$, then $y' = \dfrac{[1 \cdot x^2] - [2x \cdot (x-1)]}{x^4} + (5 \cdot 3)x^{3-1} = \dfrac{x^2 - 2x^2 + 2x}{x^4} + 15x^2 = \dfrac{-x^2 + 2x}{x^4} + 15x^2$

   $= \dfrac{x(-x+2)}{x^{4=3}} + 15x^2 = \dfrac{-x+2}{x^3} + 15x^2$, $y'' = \dfrac{[-1 \cdot x^3] - [3x^2 \cdot (-x+2)]}{x^6} + (15 \cdot 2)x^{2-1} = \dfrac{-x^3 + 3x^3 - 6x^2}{x^6} + 30x$

   $= \dfrac{2x^3 - 6x^2}{x^6} + 30x = \dfrac{x^2(2x-6)}{x^{6=4}} + 30x = \dfrac{2x-6}{x^4} + 30x$ and $y''' = \dfrac{[2 \cdot x^4] - [4x^3 \cdot (2x-6)]}{x^8} + 30x^{1-1} = \dfrac{2x^4 - 8x^4 + 24x^3}{x^8} + 30$

   $= \dfrac{-6x^4 + 24x^3}{x^8} + 30 = \dfrac{-6x^3(x-4)}{x^{8=5}} + 30 = -\dfrac{6(x-4)}{x^5} + 30$

3. Find $f''(0)$ and $f''(1)$ for the following functions.

   a. Given $f(x) = 6x^5 + 3x^3 + 5$, then $f'(x) = (6 \cdot 5)x^{5-1} + (3 \cdot 3)x^{3-1} + 0 = 30x^4 + 9x^2$ and $f''(x) = (30 \cdot 4)x^{4-1} + (9 \cdot 2)x^{2-1}$

   $= 120x^3 + 18x$. Therefore, $f''(0) = 120 \cdot 0^3 + 18 \cdot 0 = 0$ and $f''(1) = 120 \cdot 1^3 + 18 \cdot 1 = 120 + 18 = 138$

   b. Given $f(x) = x^3(x+1)^2$, then $f'(x) = \left[3x^{3-1} \cdot (x+1)^2\right] + \left[2(x+1)^{2-1} \cdot x^3\right] = 3x^2(x+1)^2 + 2x^3(x+1) = 3x^2(x^2 + 2x + 1)$

   $+ 2x^4 + 2x^3 = 3x^4 + 6x^3 + 3x^2 + 2x^4 + 2x^3 = 5x^4 + 8x^3 + 3x^2$ and $f''(x) = (5 \cdot 4)x^{4-1} + (8 \cdot 3)x^{3-1} + (3 \cdot 2)x^{2-1}$

   $= 20x^3 + 24x^2 + 6x$. Therefore, $f''(0) = (20 \cdot 0^3) + (24 \cdot 0^2) + (6 \cdot 0) = 0$ and $f''(1) = (20 \cdot 1^3) + (24 \cdot 1^2) + (6 \cdot 1) = 50$

c. Given $f(x)=x+(x-1)^2$, then $f'(x) = 1+2(x-1)^{2-1} = 1+2(x-1) = 1+2x-2 = 2x-1$ and $f''(x) = 2x^{1-1}-0 = 2$.

   Therefore, $f''(0) = 2$ and $f''(1) = 2$

d. Given $f(x)=(x-1)^{-3}$, then $f'(x) = -3(x-1)^{-3-1} = -3(x-1)^{-4}$ and $f''(x) = (-3\cdot-4)(x-1)^{-4-1} = 12(x-1)^{-5} = \dfrac{12}{(x-1)^5}$

   Therefore, $f''(0) = \dfrac{12}{(0-1)^5} = \dfrac{12}{-1} = -12$ and $f''(1) = \dfrac{12}{(1-1)^5} = \dfrac{12}{0}$ **which is undefined due to division by zero.**

e. Given $f(x)=(x-1)(x^2+1)$, then $f'(x) = \left[1\cdot(x^2+1)\right]+\left[2x\cdot(x-1)\right] = x^2+1+2x^2-2x = 3x^2-2x+1$ and

   $f''(x) = (3\cdot2)x^{2-1}-2x^{1-1}+0 = 6x-2$. Therefore, $f''(0) = (6\cdot0)-2 = 0-2 = -2$ and $f''(1) = (6\cdot1)-2 = 6-2 = 4$

f. Given $f(x)=(x^3-1)^2+\sqrt{2x} = (x^3-1)^2+(2x)^{\frac{1}{2}}$, then $f'(x) = 2(x^3-1)^{2-1}\cdot3x^2+\dfrac{1}{2}(2x)^{\frac{1}{2}-1}\cdot2 = 6x^2(x^3-1)+(2x)^{-\frac{1}{2}}$

   and $f''(x) = \left[12x\cdot(x^3-1)+3x^2\cdot6x^2\right]-\dfrac{1}{2}(2x)^{-\frac{1}{2}-1}\cdot2 = 12x^4-12x+18x^4-(2x)^{-\frac{3}{2}} = 30x^4-12x-(2x)^{-\frac{3}{2}}$. Thus, $f''(0)$

   $= 30\cdot0^4-12\cdot0-\dfrac{1}{(2\cdot0)^{\frac{3}{2}}} = -\dfrac{1}{0}$ **which is undefined**, and $f''(1) = 30\cdot1^4-12\cdot1-\dfrac{1}{(2\cdot1)^{\frac{3}{2}}} = 30-12-0.35 = \mathbf{17.65}$

Chapter 3 Solutions:

1. Find the derivative of the following trigonometric functions:

a. Given $y = \sin(3x+1)$, then $\dfrac{dy}{dx} = \dfrac{d}{dx}[\sin(3x+1)] = \cos(3x+1)\cdot\dfrac{d}{dx}(3x+1) = \cos(3x+1)\cdot 3 = \mathbf{3\cos(3x+1)}$

b. Given $y = 5\cos x^3$, then $\dfrac{dy}{dx} = \dfrac{d}{dx}5\cos x^3 = -5\sin x^3\cdot\dfrac{d}{dx}x^3 = -5\sin x^3\cdot 3x^2 = \mathbf{-15x^2\sin x^3}$

c. Given $y = x^3\cos x^2$, then $\dfrac{dy}{dx} = \dfrac{d}{dx}\left(x^3\cos x^2\right) = \cos x^2\cdot\dfrac{d}{dx}x^3 + x^3\cdot\dfrac{d}{dx}\cos x^2 = \cos x^2\cdot 3x^2 + x^3\cdot -\sin x^2\cdot\dfrac{d}{dx}x^2$

$= \cos x^2\cdot 3x^2 - x^3\cdot\sin x^2\cdot 2x = \mathbf{3x^2\cos x^2 - 2x^4\sin x^2}$

d. Given $y = \sin 5x\cdot\tan 3x$, then $\dfrac{dy}{dx} = \dfrac{d}{dx}\left(\sin 5x\cdot\tan 3x\right) = \tan 3x\cdot\dfrac{d}{dx}\sin 5x + \sin 5x\cdot\dfrac{d}{dx}\tan 3x = \tan 3x\cdot\cos 5x\cdot\dfrac{d}{dx}5x$

$+ \sin 5x\cdot\sec^2 3x\cdot\dfrac{d}{dx}3x = \tan 3x\cdot\cos 5x\cdot 5 + \sin 5x\cdot\sec^2 3x\cdot 3 = \mathbf{5\tan 3x\cos 5x + 3\sin 5x\sec^2 3x}$

e. Given $y = \tan^2 x^3 = \left(\tan x^3\right)^2$, then $\dfrac{dy}{dx} = \dfrac{d}{dx}\left(\tan x^3\right)^2 = 2\tan x^3\cdot\dfrac{d}{dx}\tan x^3 = 2\tan x^3\cdot\sec^2 x^3\cdot\dfrac{d}{dx}x^3 = 2\tan x^3\cdot\sec^2 x^3\cdot 3x^2$

$= \mathbf{6x^2\tan x^3\sec^2 x^3}$

f. Given $y = \cot(x+3)^3$, then $\dfrac{dy}{dx} = \dfrac{d}{dx}\left[\cot(x+3)^3\right] = -\csc^2(x+3)^3\cdot\dfrac{d}{dx}(x+3)^3 = -\csc^2(x+3)^3\cdot 3(x+3)^2\cdot\dfrac{d}{dx}(x+3)$

$= -\csc^2(x+3)^3\cdot 3(x+3)^2\cdot 1 = \mathbf{-3\csc^2(x+3)^3(x+3)^2}$

g. Given $y = x^5\cos x^7$, then $\dfrac{dy}{dx} = \dfrac{d}{dx}\left(x^5\cos x^7\right) = \cos x^7\cdot\dfrac{d}{dx}x^5 + x^5\cdot\dfrac{d}{dx}\cos x^7 = \cos x^7\cdot 5x^4 + x^5\cdot -\sin x^7\cdot\dfrac{d}{dx}x^7$

$= \cos x^7\cdot 5x^4 - x^5\cdot\sin x^7\cdot 7x^6 = \mathbf{5x^4\cos x^7 - 7x^{11}\sin x^7}$

h. Given $y = \sec^4\sqrt{x^3} = \sec^4 x^{\frac{3}{2}} = \left(\sec x^{\frac{3}{2}}\right)^4$, then $\dfrac{dy}{dx} = \dfrac{d}{dx}\left(\sec x^{\frac{3}{2}}\right)^4 = 4\left(\sec x^{\frac{3}{2}}\right)^3\cdot\dfrac{d}{dx}\left(\sec x^{\frac{3}{2}}\right) = 4\sec^3 x^{\frac{3}{2}}\cdot\sec x^{\frac{3}{2}}$

$\times\tan x^{\frac{3}{2}}\cdot\dfrac{d}{dx}x^{\frac{3}{2}} = 4\sec^3 x^{\frac{3}{2}}\cdot\sec x^{\frac{3}{2}}\tan x^{\frac{3}{2}}\cdot\dfrac{3}{2}x^{\frac{3}{2}-1} = \mathbf{6x^{\frac{1}{2}}\sec^3 x^{\frac{3}{2}}\sec x^{\frac{3}{2}}\tan x^{\frac{3}{2}}}$

i. Given $y = \sec 3x^2 + x^3$, then $\dfrac{dy}{dx} = \dfrac{d}{dx}\left(\sec 3x^2 + x^3\right) = \dfrac{d}{dx}\sec 3x^2 + \dfrac{d}{dx}x^3 = \sec 3x^2\cdot\tan 3x^2\cdot\dfrac{d}{dx}\left(3x^2\right) + 3x^2$

$= \sec 3x^2\cdot\tan 3x^2\cdot 6x + 3x^2 = \mathbf{6x\sec 3x^2\tan 3x^2 + 3x^2}$

j. Given $y = \tan x^5$, then $\dfrac{dy}{dx} = \dfrac{d}{dx}\left(\tan x^5\right) = \sec^2 x^5\cdot\dfrac{d}{dx}\left(x^5\right) = \sec^2 x^5\cdot 5x^4 = \mathbf{5x^4\sec^2 x^5}$

k. Given $y = \tan^5 x$, then $\dfrac{dy}{dx} = \dfrac{d}{dx}\left(\tan^5 x\right) = \dfrac{d}{dx}(\tan x)^5 = 5\tan^4 x\cdot\dfrac{d}{dx}\tan x = 5\tan^4 x\cdot\sec^2 x\cdot\dfrac{d}{dx}x = \mathbf{5\tan^4 x\sec^2 x}$

l. Given $y = \csc\left(x^3+1\right)$, then $\dfrac{dy}{dx} = \dfrac{d}{dx}\csc\left(x^3+1\right) = -\csc\left(x^3+1\right)\cot\left(x^3+1\right)\cdot\dfrac{d}{dx}\left(x^3+1\right) = \mathbf{-3x^2\csc\left(x^3+1\right)\cot\left(x^3+1\right)}$

2. Find the derivative of the following trigonometric functions:

a. Given $y = \dfrac{\tan x}{\csc x}$, then $\dfrac{dy}{dx} = \dfrac{d}{dx}\left(\dfrac{\tan x}{\csc x}\right) = \dfrac{\left(\csc x\cdot\dfrac{d}{dx}\tan x\right) - \left(\tan x\cdot\dfrac{d}{dx}\csc x\right)}{\csc^2 x} = \dfrac{\left(\csc x\cdot\sec^2 x\right) - \left(\tan x\cdot -\csc x\cot x\right)}{\csc^2 x}$

$$= \frac{\csc x \left(\sec^2 x + \tan x \cot x\right)}{\csc^2 x} = \frac{\sec^2 x + \tan x \cot x}{\csc x}$$

b. Given $y = \dfrac{\sin\left(x^3+1\right)}{x^2}$, then $\dfrac{dy}{dx} = \dfrac{d}{dx}\dfrac{\sin\left(x^3+1\right)}{x^2} = \dfrac{\left[x^2 \cdot \frac{d}{dx}\sin\left(x^3+1\right)\right] - \left[\sin\left(x^3+1\right)\cdot\frac{d}{dx}x^2\right]}{x^4} = \dfrac{\left[x^2 \cdot \cos\left(x^3+1\right)\cdot\frac{d}{dx}\left(x^3+1\right)\right]}{x^4}$

$$-\frac{\left[\sin\left(x^3+1\right)\cdot 2x\right]}{x^4} = \frac{\left[x^2 \cdot \cos\left(x^3+1\right)\cdot 3x^2\right]-\left[\sin\left(x^3+1\right)\cdot 2x\right]}{x^4} = \frac{3x^4\cos\left(x^3+1\right)-2x\sin\left(x^3+1\right)}{x^4} = \frac{3x^3\cos\left(x^3+1\right)-2\sin\left(x^3+1\right)}{x^3}$$

c. Given $y = \dfrac{\sec x}{\csc x^3}$, then $\dfrac{dy}{dx} = \dfrac{d}{dx}\left(\dfrac{\sec x}{\csc x^3}\right) = \dfrac{\left(\csc x^3 \cdot \frac{d}{dx}\sec x\right) - \left(\sec x \cdot \frac{d}{dx}\csc x^3\right)}{\csc^2 x^3} = \dfrac{\left(\csc x^3 \cdot \sec x \tan x\right)}{\csc^2 x^3}$

$$-\frac{\left(\sec x \cdot -\csc x^3 \cot x^3 \cdot 3x^2\right)}{\csc^2 x^3} = \frac{\left(\csc x^3 \sec x \tan x\right) + \left(3x^2 \sec x \csc x^3 \cot x^3\right)}{\csc^2 x^3}$$

d. Given $y = x^5 \tan x^3$, then $\dfrac{dy}{dx} = \dfrac{d}{dx}\left(x^5 \tan x^3\right) = \tan x^3 \cdot \dfrac{d}{dx}x^5 + x^5 \cdot \dfrac{d}{dx}\tan x^3 = \tan x^3 \cdot 5x^4 + x^5 \cdot \sec^2 x^3 \cdot \dfrac{d}{dx}x^3$

$$= \tan x^3 \cdot 5x^4 + x^5 \cdot \sec^2 x^3 \cdot 3x^2 = 5x^4 \tan x^3 + 3x^7 \sec^2 x^3 = x^4\left(5\tan x^3 + 3x^3 \sec^2 x^3\right)$$

e. Given $y = x^5 \sin x^2$, then $\dfrac{dy}{dx} = \dfrac{d}{dx}\left(x^5 \sin x^2\right) = \sin x^2 \cdot \dfrac{d}{dx}x^5 + x^5 \cdot \dfrac{d}{dx}\sin x^2 = \sin x^2 \cdot 5x^4 + x^5 \cdot \cos x^2 \cdot \dfrac{d}{dx}x^2$

$$= \sin x^2 \cdot 5x^4 + x^5 \cdot \cos x^2 \cdot 2x = 5x^4 \sin x^2 + 2x^6 \cos x^2 = x^4\left(5\sin x^2 + 2x^2 \cos x^2\right)$$

f. Given $y = (x+5)^2\cos x$, then $\dfrac{dy}{dx} = \dfrac{d}{dx}\left[(x+5)^2\cos x\right] = \cos x \cdot \dfrac{d}{dx}(x+5)^2 + (x+5)^2 \cdot \dfrac{d}{dx}\cos x$

$$= \cos x \cdot 2(x+5) \cdot \dfrac{d}{dx}(x+5) + (x+5)^2 \cdot -\sin x = \cos x \cdot 2(x+5)\cdot 1 - (x+5)^2 \cdot \sin x = 2(x+5)\cos x - (x+5)^2\sin x$$

g. Given $y = x^2 \tan^3 x^5$, then $\dfrac{dy}{dx} = \dfrac{d}{dx}\left(x^2 \tan^3 x^5\right) = \tan^3 x^5 \cdot \dfrac{d}{dx}x^2 + x^2 \cdot \dfrac{d}{dx}\tan^3 x^5 = \tan^3 x^5 \cdot 2x + x^2 \cdot 3\tan^2 x^5 \cdot \dfrac{d}{dx}\tan x^5$

$$= \tan^3 x^5 \cdot 2x + x^2 \cdot 3\tan^2 x^5 \cdot \sec^2 x^5 \cdot \dfrac{d}{dx}x^5 = 2x \tan^3 x^5 + x^2 \cdot 3\tan^2 x^5 \cdot \sec^2 x^5 \cdot 5x^4 = 2x \tan^3 x^5 + 15x^6 \tan^2 x^5 \sec^2 x^5$$

h. Given $y = \sqrt{x + \sin x^3}$, then $\dfrac{dy}{dx} = \dfrac{d}{dx}\left(\sqrt{x+\sin x^3}\right) = \dfrac{d}{dx}\left(x + \sin x^3\right)^{\frac{1}{2}} = \dfrac{1}{2}\left(x + \sin x^3\right)^{\frac{1}{2}-1} \cdot \dfrac{d}{dx}\left(x + \sin x^3\right)$

$$= \dfrac{1}{2}\left(x + \sin x^3\right)^{-\frac{1}{2}} \cdot \left(\dfrac{d}{dx}x + \dfrac{d}{dx}\sin x^3\right) = \dfrac{1}{2}\left(x + \sin x^3\right)^{-\frac{1}{2}} \cdot \left(1 + 3x^2 \cos x^3\right) = \dfrac{1 + 3x^2 \cos x^3}{2\left(x + \sin x^3\right)^{\frac{1}{2}}} = \dfrac{1 + 3x^2 \cos x^3}{2\sqrt{x + \sin x^3}}$$

i. Given $y = \sin\left(1 + x^5\right)$, then $\dfrac{dy}{dx} = \dfrac{d}{dx}\sin\left(1 + x^5\right) = \cos\left(1 + x^5\right)\cdot\dfrac{d}{dx}\left(1 + x^5\right) = \cos\left(1 + x^5\right)\cdot 5x^4 = 5x^4 \cos\left(1 + x^5\right)$

j. Given $y = \cot^2 x^3$, then $\dfrac{dy}{dx} = \dfrac{d}{dx}\left(\cot^2 x^3\right) = \dfrac{d}{dx}\left(\cot x^3\right)^2 = 2\cot x^3 \cdot \dfrac{d}{dx}\cot x^3 = 2\cot x^3 \cdot -\csc^2 x^3 \cdot \dfrac{d}{dx}x^3$

$$= 2\cot x^3 \cdot -\csc^2 x^3 \cdot 3x^2 = -6x^2 \cot x^3 \csc^2 x^3$$

k. Given $y = \sin^3\left(1 + 5x\right)$, then $\dfrac{dy}{dx} = \dfrac{d}{dx}\left[\sin\left(1 + 5x\right)\right]^3 = 3\left[\sin\left(1 + 5x\right)\right]^2 \cdot \dfrac{d}{dx}\sin\left(1 + 5x\right) = 3\left[\sin\left(1 + 5x\right)\right]^2 \cdot \cos\left(1 + 5x\right)$

$$\times \dfrac{d}{dx}\left(1 + 5x\right) = 3\left[\sin\left(1 + 5x\right)\right]^2 \cdot \cos\left(1 + 5x\right)\cdot 5 = 15\sin^2\left(1 + 5x\right)\cos\left(1 + 5x\right)$$

l. Given $y = x^5 \csc x^3$, then $\dfrac{dy}{dx} = \dfrac{d}{dx}\left(x^5 \csc x^3\right) = \csc x^3 \cdot \dfrac{d}{dx}x^5 + x^5 \cdot \dfrac{d}{dx}\csc x^3 = \csc x^3 \cdot 5x^4 + x^5 \cdot -\csc x^3 \cot x^3 \cdot \dfrac{d}{dx}x^3$

$$= \csc x^3 \cdot 5x^4 - x^5 \cdot \csc x^3 \cot x^3 \cdot 3x^2 = 5x^4 \csc x^3 - 3x^7 \csc x^3 \cot x^3$$

3. Find the derivative of the following trigonometric functions:

a. Given $y = \sin 7x$, then $\dfrac{dy}{dx} = \dfrac{d}{dx}(\sin 7x) = \cos 7x \cdot \dfrac{d}{dx}(7x) = \cos 7x \cdot 7 = \mathbf{7\cos 7x}$

b. Given $y = \cos x^3$, then $\dfrac{dy}{dx} = \dfrac{d}{dx}\cos x^3 = -\sin x^3 \cdot \dfrac{d}{dx}x^3 = -\sin x^3 \cdot 3x^2 = \mathbf{-3x^2 \sin x^3}$

c. Given $y = \cot x^3$, then $\dfrac{dy}{dx} = \dfrac{d}{dx}\left(\cot x^3\right) = -\csc^2 x^3 \cdot \dfrac{d}{dx}\left(x^3\right) = -\csc^2 x^3 \cdot 3x^2 = \mathbf{-3x^2 \csc^2 x^3}$

d. Given $y = x^3 \tan x^2$, then $\dfrac{dy}{dx} = \dfrac{d}{dx}\left(x^3 \tan x^2\right) = \tan x^2 \cdot \dfrac{d}{dx}x^3 + x^3 \cdot \dfrac{d}{dx}\tan x^2 = \tan x^2 \cdot 3x^2 + x^3 \cdot \sec^2 x^2 \cdot \dfrac{d}{dx}x^2$

$= \tan x^2 \cdot 3x^2 + x^3 \cdot \sec^2 x^2 \cdot 2x = 3x^2 \tan x^2 + 2x^4 \sec^2 x^2 = \mathbf{x^2\left(3\tan x^2 + 2x^2 \sec^2 x^2\right)}$

e. Given $y = \cot (x+9)$, then $\dfrac{dy}{dx} = \dfrac{d}{dx}\left[\cot (x+9)\right] = -\csc^2(x+9) \cdot \dfrac{d}{dx}(x+9) = -\csc^2(x+9) \cdot 1 = \mathbf{-\csc^2(x+9)}$

f. Given $y = \sin^2\left(x^3 + 5x + 2\right)$, then $\dfrac{dy}{dx} = \dfrac{d}{dx}\sin^2\left(x^3 + 5x + 2\right) = \dfrac{d}{dx}\left(\sin x^3 + 5x + 2\right)^2 = 2\sin\left(x^3 + 5x + 2\right) \cdot \dfrac{d}{dx}\sin\left(x^3 + 5x + 2\right)$

$= 2\sin\left(x^3 + 5x + 2\right) \cdot \cos\left(x^3 + 5x + 2\right) \cdot \dfrac{d}{dx}\left(x^3 + 5x + 2\right) = \mathbf{2\left(3x^2 + 5\right)\sin\left(x^3 + 5x + 2\right)\cos\left(x^3 + 5x + 2\right)}$

g. Given $y = \sin \sqrt{x+3} = \sin (x+3)^{\frac{1}{2}}$, then $\dfrac{dy}{dx} = \dfrac{d}{dx}\sin (x+3)^{\frac{1}{2}} = \dfrac{1}{2}\sin (x+3)^{-\frac{1}{2}} \cdot \dfrac{d}{dx}\sin (x+3) = \dfrac{1}{2}\sin (x+3)^{-\frac{1}{2}}$

$\times \cos (x+3) \cdot \dfrac{d}{dx}(x+3) = \dfrac{1}{2}\sin (x+3)^{-\frac{1}{2}} \cdot \cos (x+3) \cdot 1 = \dfrac{1}{2}\dfrac{\cos (x+3)}{\sin (x+3)^{\frac{1}{2}}} = \dfrac{\cos (x+3)}{2\sqrt{\sin (x+3)}}$

h. Given $y = \sin x^2 + \cos x^3$, then $\dfrac{dy}{dx} = \dfrac{d}{dx}\left(\sin x^2 + \cos x^3\right) = \dfrac{d}{dx}\sin x^2 + \dfrac{d}{dx}\cos x^3 = \cos x^2 \cdot \dfrac{d}{dx}x^2 - \sin x^3 \cdot \dfrac{d}{dx}x^3$

$= \cos x^2 \cdot 2x - \sin x^3 \cdot 3x^2 = \mathbf{2x\cos x^2 - 3x^2 \sin x^3}$

i. Given $y = x^2 \sin x^3$, then $\dfrac{dy}{dx} = \dfrac{d}{dx}\left(x^2 \sin x^3\right) = \sin x^3 \cdot \dfrac{d}{dx}x^2 + x^2 \cdot \dfrac{d}{dx}\sin x^3 = \sin x^3 \cdot 2x + x^2 \cdot \cos x^3 \cdot \dfrac{d}{dx}x^3$

$= 2x\sin x^3 + x^2 \cdot \cos x^3 \cdot 3x^2 = \mathbf{2x\sin x^3 + 3x^4 \cos x^3}$

Section 3.2 Practice Problems – Differentiation of Inverse Trigonometric Functions

1. Find the derivative of the following inverse trigonometric functions:

a. Given $y = \sin^{-1} 3x$, then $\dfrac{dy}{dx} = \dfrac{d}{dx}\text{arc} \sin 3x = \dfrac{1}{\sqrt{1-(3x)^2}} \cdot \dfrac{d}{dx}3x = \dfrac{3}{\sqrt{1-9x^2}}$

b. Given $y = x^3 + \text{arc} \cos 5x$, then $\dfrac{dy}{dx} = \dfrac{d}{dx}\left(x^3 + \text{arc} \cos 5x\right) = 3x^2 - \dfrac{1}{\sqrt{1-(5x)^2}} \cdot \dfrac{d}{dx}5x = 3x^2 - \dfrac{5}{\sqrt{1-25x^2}}$

c. Given $y = x^5 \text{arc} \tan x^4$, then $\dfrac{dy}{dx} = \dfrac{d}{dx}\left(x^5 \text{arc} \tan x^4\right) = \text{arc} \tan x^4 \cdot \dfrac{d}{dx}x^5 + x^5 \cdot \dfrac{d}{dx}\text{arc} \tan x^4 = \text{arc} \tan x^4 \cdot 5x^4$

$+ x^5 \cdot \dfrac{1}{1+x^8} \cdot \dfrac{d}{dx}x^4 = 5x^4 \text{arc} \tan x^4 + x^5 \cdot \dfrac{1}{1+x^8} \cdot 4x^3 = \mathbf{5x^4 \text{arc} \tan x^4 + \dfrac{4x^8}{1+x^8}}$

d. Given $y = \text{arc} \sin \left(x^3 + 2\right)$, then $\dfrac{dy}{dx} = \dfrac{d}{dx}\left[\text{arc} \sin \left(x^3 + 2\right)\right] = \dfrac{1}{\sqrt{1-(x^3+2)^2}} \cdot \dfrac{d}{dx}\left(x^3 + 2\right) = \dfrac{3x^2}{\sqrt{1-(x^3+2)^2}}$

e. Given $y = \text{arc} \cot \dfrac{1}{x^3}$, then $\dfrac{dy}{dx} = \dfrac{d}{dx}\text{arc} \cot \dfrac{1}{x^3} = -\dfrac{1}{1+\left(\dfrac{1}{x^3}\right)^2} \cdot \dfrac{d}{dx}\dfrac{1}{x^3} = -\dfrac{1}{1+\dfrac{1}{x^6}} \cdot \dfrac{-3}{x^4} = \dfrac{3}{x^4\left(1+\dfrac{1}{x^6}\right)}$

f. Given $y = \dfrac{arc \sin 3x}{x^2}$, then $\dfrac{dy}{dx} = \dfrac{d}{dx}\left(\dfrac{arc \sin 3x}{x^2}\right) = \dfrac{\left(x^2 \cdot \frac{d}{dx} arc \sin 3x\right) - \left(arc \sin 3x \cdot \frac{d}{dx} x^2\right)}{x^4}$

$= \dfrac{x^2 \cdot \frac{1}{\sqrt{1-9x^2}} - (arc \sin 3x \cdot 2x)}{x^4} = \dfrac{\frac{x^2}{\sqrt{1-9x^2}} - 2x\, arc \sin 3x}{x^4} = \dfrac{x^2 - 2x\, arc \sin 3x \sqrt{1-9x^2}}{x^4\sqrt{1-9x^2}} = \dfrac{x - 2arc\, sin\, 3x\,\sqrt{1-9x^2}}{x^3\sqrt{1-9x^2}}$

g. Given $y = \dfrac{\tan^{-1} x}{x}$, then $\dfrac{dy}{dx} = \dfrac{d}{dx}\left(\dfrac{\tan^{-1} x}{x}\right) = \dfrac{\left(x \cdot \frac{d}{dx}\tan^{-1} x\right) - \left(\tan^{-1} x \cdot \frac{d}{dx} x\right)}{x^2} = \dfrac{\left(x \cdot \frac{1}{1+x^2}\right) - \left(\tan^{-1} x \cdot 1\right)}{x^2} = \dfrac{\frac{x}{1+x^2} - \tan^{-1} x}{x^2}$

$= \dfrac{\frac{x - \left(1+x^2\right)\tan^{-1} x}{1+x^2}}{x^2} = \dfrac{x - \left(1+x^2\right)\tan^{-1} x}{x^2\left(1+x^2\right)}$

h. Given $y = \dfrac{x^3}{x+5} - arc \sin x$, then $\dfrac{dy}{dx} = \dfrac{d}{dx}\left(\dfrac{x^3}{x+5} - arc \sin x\right) = \dfrac{\left[\frac{d}{dx} x^3 \cdot (x+5)\right] - \left[x^3 \cdot \frac{d}{dx}(x+5)\right]}{(x+5)^2} - \dfrac{1}{\sqrt{1-x^2}}$

$= \dfrac{\left[3x^2 \cdot (x+5)\right] - \left[x^3 \cdot 1\right]}{(x+5)^2} - \dfrac{1}{\sqrt{1-x^2}} = \dfrac{3x^3 + 15x^2 - x^3}{(x+5)^2} - \dfrac{1}{\sqrt{1-x^2}} = \dfrac{2x^3 + 15x^2}{(x+5)^2} - \dfrac{1}{\sqrt{1-x^2}}$

i. Given $y = x + \cos^{-1} x$, then $\dfrac{dy}{dx} = \dfrac{d}{dx}\left(x + \cos^{-1} x\right) = \dfrac{d}{dx} x + \dfrac{d}{dx}\cos^{-1} x = 1 - \dfrac{1}{\sqrt{1-x^2}} \cdot \dfrac{d}{dx} x = 1 - \dfrac{1}{\sqrt{1-x^2}}$

2. Find the first and second derivative of the following inverse trigonometric functions:

a. Given $y = x^2 + arc \sin ax$, then $\dfrac{dy}{dx} = \dfrac{d}{dx}\left(x^2 + arc \sin ax\right) = 2x + \dfrac{1}{\sqrt{1-(ax)^2}} \cdot \dfrac{d}{dx} ax = 2x + \dfrac{a}{\sqrt{1-a^2 x^2}}$

b. Given $y = \cos^{-1} 6x$, then $\dfrac{dy}{dx} = \dfrac{d}{dx}\cos^{-1} 6x = -\dfrac{1}{\sqrt{1-(6x)^2}} \cdot \dfrac{d}{dx} 6x = -\dfrac{1}{\sqrt{1-36x^2}} \cdot 6 = -\dfrac{6}{\sqrt{1-36x^2}}$

c. Given $y = x + arc \sin x^3$, then $\dfrac{dy}{dx} = \dfrac{d}{dx}\left(x + arc \sin x^3\right) = 1 + \dfrac{1}{\sqrt{1-x^6}} \cdot \dfrac{d}{dx} x^3 = 1 + \dfrac{1}{\sqrt{1-x^6}} \cdot 3x^2 = 1 + \dfrac{3x^2}{\sqrt{1-x^6}}$

d. Given $y = arc \tan^2 x$, then $\dfrac{dy}{dx} = \dfrac{d}{dx}\left(arc \tan^2 x\right) = \dfrac{d}{dx}\left(\tan^{-1} x\right)^2 = 2\tan^{-1} x \cdot \dfrac{d}{dx}\tan^{-1} x = 2\tan^{-1} x \cdot \dfrac{1}{1+x^2} = \dfrac{2\tan^{-1} x}{1+x^2}$

e. Given $y = x - arc \tan x^2$, then $\dfrac{dy}{dx} = \dfrac{d}{dx}\left(x - \tan^{-1} x^2\right) = \dfrac{d}{dx} x + \dfrac{d}{dx}\tan^{-1} x^2 = 1 + \dfrac{1}{1+x^4} \cdot \dfrac{d}{dx} x^2 = 1 + \dfrac{2x}{1+x^4}$

f. Given $y = x^2 - arc \sin x$, then $\dfrac{dy}{dx} = \dfrac{d}{dx}\left(x^2 - arc \sin x\right) = \dfrac{d}{dx} x^2 - \dfrac{d}{dx} arc \sin x = 2x - \dfrac{1}{\sqrt{1-x^2}} \cdot \dfrac{d}{dx} x = 2x - \dfrac{1}{\sqrt{1-x^2}}$

g. Given $y = \tan^{-1} x^3$, then $\dfrac{dy}{dx} = \dfrac{d}{dx}\tan^{-1} x^3 = \dfrac{1}{1+\left(x^3\right)^2} \cdot \dfrac{d}{dx} x^3 = \dfrac{1}{1+x^6} \cdot 3x^2 = \dfrac{3x^2}{1+x^6}$

h. Given $y = \tan^{-1} 5x$, then $\dfrac{dy}{dx} = \dfrac{d}{dx}\tan^{-1} 5x = \dfrac{1}{1+(5x)^2} \cdot \dfrac{d}{dx} 5x = \dfrac{1}{1+25x^2} \cdot 5 = \dfrac{5}{1+25x^2}$

i. Given $y = arc \sin 5x^2$, then $\dfrac{dy}{dx} = \dfrac{d}{dx}\left(arc \sin 5x^2\right) = \dfrac{1}{\sqrt{1-25x^4}} \cdot \dfrac{d}{dx} 5x^2 = \dfrac{1}{\sqrt{1-25x^4}} \cdot 10x = \dfrac{10x}{\sqrt{1-25x^4}}$

1. Find the derivative of the following logarithmic functions:

a. Given $y = \ln 10x$, then $\dfrac{dy}{dx} = \dfrac{d}{dx}(\ln 10x) = \dfrac{1}{10x} \cdot \dfrac{d}{dx}(10x) = \dfrac{1}{10x} \cdot 10 = \dfrac{10}{10x} = \dfrac{1}{x}$

b. Given $y = 10\ln 5x^3$, then $\dfrac{dy}{dx} = \dfrac{d}{dx}\left(10\ln 5x^3\right) = 10 \cdot \dfrac{1}{5x^3} \cdot \dfrac{d}{dx}\left(5x^3\right) = \dfrac{10}{5x^3} \cdot 15x^2 = \dfrac{150x^2}{5x^3} = \dfrac{30}{x}$

c. Given $y = \ln\left(x^2 + 3\right)$, then $\dfrac{dy}{dx} = \dfrac{d}{dx}\left[\ln\left(x^2 + 3\right)\right] = \dfrac{1}{x^2+3} \cdot \dfrac{d}{dx}\left(x^2 + 3\right) = \dfrac{2x}{x^2 + 3}$

d. Given $y = x^3 \ln x$, then $\dfrac{dy}{dx} = \dfrac{d}{dx}\left(x^3 \ln x\right) = \left(\ln x \cdot \dfrac{d}{dx}x^3\right) + \left(x^3 \cdot \dfrac{d}{dx}\ln x\right) = \ln x \cdot 3x^2 + x^3 \cdot \dfrac{1}{x} = 3x^2 \ln x + x^2 = x^2(3\ln x + 1)$

e. Given $y = x \ln x - 5x^2$, then $\dfrac{dy}{dx} = \dfrac{d}{dx}\left(x \ln x - 5x^2\right) = \dfrac{d}{dx}(x \ln x) - \dfrac{d}{dx}5x^2 = \left[x \cdot \dfrac{d}{dx}(\ln x) + \ln x \dfrac{d}{dx}(x)\right] - 10x$

$= \left[x \cdot \dfrac{1}{x} \cdot \dfrac{d}{dx}(x) + \ln x \cdot 1\right] - 10x = \left[\dfrac{\cancel{x}}{\cancel{x}} + \ln x\right] - 10x = 1 + \ln x - 10x = \mathbf{\ln x - 10x + 1}$

f. Given $y = \left(x^3 + 2\right)\ln x^2$, then $\dfrac{dy}{dx} = \dfrac{d}{dx}\left[\left(x^3 + 2\right)\ln x^2\right] = \dfrac{d}{dx}\left(x^3 \ln x^2\right) + 2\dfrac{d}{dx}\left(\ln x^2\right) = \left[\ln x^2 \cdot \dfrac{d}{dx}\left(x^3\right) + x^3 \cdot \dfrac{d}{dx}\left(\ln x^2\right)\right]$

$+ \dfrac{2}{x^2} \cdot \dfrac{d}{dx}x^2 = \left[\ln x^2 \cdot 3x^2 + x^3 \cdot \dfrac{1}{x^2} \cdot 2x\right] + \dfrac{2 \cdot 2x}{x^2} = \mathbf{3x^2 \ln x^2 + 2x^2 + \dfrac{4}{x}}$

g. Given $y = \sin\left(\ln x^2\right)$, then $\dfrac{dy}{dx} = \dfrac{d}{dx}\left[\sin\left(\ln x^2\right)\right] = \cos\left(\ln x^2\right) \cdot \dfrac{1}{x^2} \cdot \dfrac{d}{dx}x^2 = \cos\left(\ln x^2\right) \cdot \dfrac{1}{x^2} \cdot 2x = \mathbf{\dfrac{2}{x}\cos\left(\ln x^2\right)}$

h. Given $y = \ln \csc x$, then $\dfrac{d}{dx}(\ln \csc x) = \dfrac{1}{\csc x} \cdot \dfrac{d}{dx}\csc x = \dfrac{1}{\csc x} \cdot -\csc x \cdot \cot x = -\dfrac{\csc x \cdot \cot x}{\csc x} = \mathbf{-\cot x}$

i. Given $y = \ln \sqrt{x^3} = \ln x^{\frac{3}{2}}$, then $\dfrac{dy}{dx} = \dfrac{d}{dx}\ln x^{\frac{3}{2}} = \dfrac{1}{x^{\frac{3}{2}}} \cdot \dfrac{d}{dx}x^{\frac{3}{2}} = \dfrac{1}{x^{\frac{3}{2}}} \cdot \dfrac{3}{2}x^{\frac{1}{2}} = \dfrac{3}{2}x^{\frac{1}{2}} \cdot x^{-\frac{3}{2}} = \dfrac{3}{2}x^{\frac{1}{2}-\frac{3}{2}} = \dfrac{3}{2}x^{-1} = \mathbf{\dfrac{3}{2} \cdot \dfrac{1}{x}}$

j. Given $y = \cos\left(\ln x^2\right)$, then $\dfrac{dy}{dx} = \dfrac{d}{dx}\left[\cos\left(\ln x^2\right)\right] = -\sin\left(\ln x^2\right) \cdot \dfrac{1}{x^2} \cdot \dfrac{d}{dx}x^2 = -\sin\left(\ln x^2\right) \cdot \dfrac{1}{x^2} \cdot 2x = \mathbf{-\dfrac{2}{x}\sin\left(\ln x^2\right)}$

k. Given $y = \ln\dfrac{x+1}{x-1}$, then $\dfrac{dy}{dx} = \dfrac{d}{dx}\left(\ln\dfrac{x+1}{x-1}\right) = \dfrac{x-1}{x+1} \cdot \dfrac{d}{dx}\dfrac{x+1}{x-1} = \dfrac{x-1}{x+1} \cdot \dfrac{(x-1)-(x+1)}{(x-1)^2} = \dfrac{x-1}{x+1} \cdot \dfrac{-2}{(x-1)^2} = \mathbf{-\dfrac{2}{x^2-1}}$

l. Given $y = x^3 \ln x + 5x$, then $\dfrac{dy}{dx} = \dfrac{d}{dx}\left[x^3 \ln x + 5x\right] = \dfrac{d}{dx}x^3 \ln x + \dfrac{d}{dx}5x = \left(\ln x \cdot \dfrac{d}{dx}x^3\right) + \left(x^3 \cdot \dfrac{d}{dx}\ln x\right) + 5$

$= \left(\ln x \cdot 3x^2\right) + \left(x^3 \cdot \dfrac{1}{x}\right) + 5 = \mathbf{3x^2 \ln x + x^2 + 5}$

2. Find the derivative of the following exponential functions:

a. Given $y = x^3 e^{2x}$, then $\dfrac{dy}{dx} = \dfrac{d}{dx}\left(x^3 e^{2x}\right) = e^{2x} \cdot \dfrac{d}{dx}x^3 + x^3 \cdot \dfrac{d}{dx}e^{2x} = e^{2x} \cdot 3x^2 + x^3 \cdot e^{2x} \cdot \dfrac{d}{dx}2x = 3x^2 e^{2x} + x^3 \cdot e^{2x} \cdot 2$

$= 3x^2 e^{2x} + 2x^3 e^{2x} = \mathbf{x^2 e^{2x}(3 + 2x)}$

b. Given $y = \left(x^2 + 3\right)e^{3x}$, then $\dfrac{dy}{dx} = \dfrac{d}{dx}\left[\left(x^2 + 3\right)e^{3x}\right] = \left[e^{3x} \cdot \dfrac{d}{dx}\left(x^2 + 3\right)\right] + \left[\left(x^2 + 3\right) \cdot \dfrac{d}{dx}e^{3x}\right] = e^{3x} \cdot 2x + \left(x^2 + 3\right) \cdot e^{3x} \cdot \dfrac{d}{dx}3x$

$= e^{3x} \cdot 2x + \left(x^2 + 3\right) \cdot e^{3x} \cdot 3 = 2x e^{3x} + 3e^{3x}\left(x^2 + 3\right) = e^{3x}\left[2x + 3\left(x^2 + 3\right)\right] = \mathbf{e^{3x}\left(3x^2 + 2x + 9\right)}$

c. Given $y = e^{\sin 3x}$, then $\dfrac{dy}{dx} = \dfrac{d}{dx}\left(e^{\sin 3x}\right) = e^{\sin 3x} \cdot \dfrac{d}{dx}(\sin 3x) = e^{\sin 3x} \cdot 3\cos 3x = \mathbf{3\cos 3x\, e^{\sin 3x}}$

d. Given $y = e^{\ln x^2}$, then $\dfrac{dy}{dx} = \dfrac{d}{dx}e^{\ln x^2} = e^{\ln x^2} \cdot \dfrac{1}{x^2} \cdot \dfrac{d}{dx}x^2 = e^{\ln x^2} \cdot \dfrac{1}{x^2} \cdot 2x = \dfrac{2}{x}e^{\ln x^2}$

e. Given $y = e^{9x}\sin 5x$, then $\dfrac{dy}{dx} = \dfrac{d}{dx}\left(e^{9x}\sin 5x\right) = \left[\sin 5x \cdot \dfrac{d}{dx}\left(e^{9x}\right)\right] + \left[e^{9x} \cdot \dfrac{d}{dx}(\sin 5x)\right] = \left[\sin 5x \cdot e^{9x} \cdot \dfrac{d}{dx}9x\right]$

$+ \left[e^{9x}\cdot\cos 5x \cdot \dfrac{d}{dx}5x\right] = \left[\sin 5x \cdot e^{9x}\cdot 9\right] + \left[e^{9x}\cdot\cos 5x\cdot 5\right] = e^{9x}\left(9\sin 5x + 5\cos 5x\right)$

f. Given $y = e^{2x}arc\sin x$, then $\dfrac{dy}{dx} = \dfrac{d}{dx}\left(e^{2x}arc\sin x\right) = \left[arc\sin x \cdot \dfrac{d}{dx}\left(e^{2x}\right)\right] + \left[e^{2x}\cdot\dfrac{d}{dx}\left(arc\sin x\right)\right]$

$= arc\sin x \cdot e^{2x}\cdot\dfrac{d}{dx}2x + e^{2x}\cdot\dfrac{1}{\sqrt{1-x^2}} = arc\sin x \cdot e^{2x}\cdot 2 + \dfrac{e^{2x}}{\sqrt{1-x^2}} = e^{2x}\left(2arc\sin x + \dfrac{1}{\sqrt{1-x^2}}\right)$

g. Given $y = (x-5)e^{-x}$, then $\dfrac{dy}{dx} = \dfrac{d}{dx}(x-5)e^{-x} = \left[(x-5)\cdot\dfrac{d}{dx}e^{-x}\right] + \left[e^{-x}\cdot\dfrac{d}{dx}(x-5)\right] = \left[(x-5)\cdot -e^{-x}\right] + \left[e^{-x}\cdot 1\right]$

$= -e^{-x}(x-5) + e^{-x} = -xe^{-x} + 5e^{-x} + e^{-x} = -xe^{-x} + 6e^{-x} = e^{-x}(-x+6)$

h. Given $y = e^{\ln(x+1)}$, then $\dfrac{dy}{dx} = \dfrac{d}{dx}e^{\ln(x+1)} = e^{\ln(x+1)}\cdot\dfrac{d}{dx}\ln(x+1) = e^{\ln(x+1)}\cdot\dfrac{1}{x+1}\cdot\dfrac{d}{dx}(x+1) = \dfrac{e^{\ln(x+1)}}{x+1}$

i. Given $y = \dfrac{e^x}{\tan x}$, then $\dfrac{dy}{dx} = \dfrac{d}{dx}\left(\dfrac{e^x}{\tan x}\right) = \dfrac{\left[\tan x \cdot \dfrac{d}{dx}\left(e^x\right)\right] - \left[e^x\cdot\dfrac{d}{dx}(\tan x)\right]}{\tan^2 x} = \dfrac{\tan x \cdot e^x - e^x\cdot\sec^2 x}{\tan^2 x} = \dfrac{e^x\left(\tan x - \sec^2 x\right)}{\tan^2 x}$

j. Given $y = 3^x\cdot e^x$, then $\dfrac{dy}{dx} = \dfrac{d}{dx}3^x\cdot e^x = e^x\cdot\dfrac{d}{dx}3^x + 3^x\cdot\dfrac{d}{dx}e^x = e^x\cdot 3^x\cdot\ln 3\cdot\dfrac{d}{dx}x + 3^x\cdot e^x\cdot\dfrac{d}{dx}x = e^x\cdot 3^x\cdot\ln 3\cdot 1 + 3^x\cdot e^x\cdot 1$

$= e^x\cdot 3^x\cdot\ln 3 + 3^x\cdot e^x = \left(\ln 3 + 1\right)3^x e^x$

k. Given $y = x^3 arc\sin x$, then $\dfrac{dy}{dx} = \dfrac{d}{dx}\left(x^3 arc\sin x\right) = arc\sin x\cdot\dfrac{d}{dx}x^3 + x^3\cdot\dfrac{d}{dx}arc\sin x = 3x^2 arc\sin x + \dfrac{x^3}{\sqrt{1-x^2}}$

l. Given $y = e^{5x}\cos(5x+1)$, then $\dfrac{dy}{dx} = \dfrac{d}{dx}\left[e^{5x}\cos(5x+1)\right] = \left[\cos(5x+1)\cdot\dfrac{d}{dx}\left(e^{5x}\right)\right] + \left[e^{5x}\cdot\dfrac{d}{dx}\cos(5x+1)\right]$

$= \left(\cos(5x+1)\cdot e^{5x}\cdot\dfrac{d}{dx}5x\right) + \left(e^{5x}\cdot -\sin(5x+1)\cdot\dfrac{d}{dx}(5x+1)\right) = \left[\cos(5x+1)\cdot e^{5x}\cdot 5\right] + \left[e^{5x}\cdot -\sin(5x+1)\cdot 5\right]$

$= 5e^{5x}\left[\cos(5x+1) - \sin(5x+1)\right]$

Section 3.4 Practice Problems – Differentiation of Hyperbolic Functions

Find the derivative of the following hyperbolic functions:

a. Given $y = \cosh\left(x^3+5\right)$, then $\dfrac{dy}{dx} = \dfrac{d}{dx}\cosh\left(x^3+5\right) = \sinh\left(x^3+5\right)\cdot\dfrac{d}{dx}\left(x^3+5\right) = \sinh\left(x^3+5\right)\cdot 3x^2 = 3x^2\sinh\left(x^3+5\right)$

b. Given $y = x^3\sinh x$, then $\dfrac{dy}{dx} = \dfrac{d}{dx}\left(x^3\sinh x\right) = \sinh x\cdot\dfrac{d}{dx}x^3 + x^3\cdot\dfrac{d}{dx}\sinh x = \sinh x\cdot 3x^2 + x^3\cdot\cosh x = 3x^2\sinh x + x^3\cosh x$

c. Given $y = (x+6)\sinh x^3$, then $\dfrac{dy}{dx} = \dfrac{d}{dx}\left[(x+6)\sinh x^2\right] = \sinh x^3\cdot\dfrac{d}{dx}(x+6) + (x+6)\cdot\dfrac{d}{dx}\sinh x^3 = \sinh x^3\cdot 1$

$+ (x+6)\cdot\cosh x^3\cdot\dfrac{d}{dx}x^3 = \sinh x^3 + (x+6)\cdot\cosh x^3\cdot 3x^2 = \sinh x^3 + 3x^2(x+6)\cosh x^3$

d. Given $y = \ln(\sinh x)$, then $\dfrac{dy}{dx} = \dfrac{d}{dx}\left(\ln\sinh x\right) = \dfrac{1}{\sinh x}\cdot\dfrac{d}{dx}\left(\sinh x\right) = \dfrac{1}{\sinh x}\cdot\cosh x\cdot\dfrac{d}{dx}x = \dfrac{\cosh x}{\sinh x} = \coth x$

e. Given $y = e^{\sinh x}$, then $\dfrac{dy}{dx} = \dfrac{d}{dx}\left(e^{\sinh x}\right) = e^{\sinh x} \cdot \dfrac{d}{dx}(\sinh x) = e^{\sinh x} \cdot \cosh x \cdot \dfrac{d}{dx} x = e^{\sinh x} \cdot \cosh x \cdot 1 = e^{\sinh x} \cosh x$

f. Given $y = \cosh 3x^5$, then $\dfrac{dy}{dx} = \dfrac{d}{dx}\left(\cosh 3x^5\right) = \sinh 3x^5 \cdot \dfrac{d}{dx} 3x^5 = \sinh 3x^5 \cdot 15x^4 = 15x^4 \sinh 3x^5$

g. Given $y = \dfrac{\tanh^2 x}{x}$, then $\dfrac{dy}{dx} = \dfrac{d}{dx}\left(\dfrac{\tanh^2 x}{x}\right) = \dfrac{\left[x \cdot \frac{d}{dx}\tanh^2 x\right] - \left[\tanh^2 x \cdot \frac{d}{dx} x\right]}{x^2} = \dfrac{\left[x \cdot 2\tanh x \cdot \sec h^2 x\right] - \left[\tanh^2 x \cdot 1\right]}{x^2}$

$= \dfrac{2x\tanh x \sec h^2 x - \tanh^2 x}{x^2} = \dfrac{\tanh x\left(2x \sec h^2 x - \tanh x\right)}{x^2}$

h. Given $y = \coth \dfrac{1}{x^3}$, then $\dfrac{dy}{dx} = \dfrac{d}{dx}\left(\coth \dfrac{1}{x^3}\right) = -\csc h^2 \dfrac{1}{x^3} \cdot \dfrac{d}{dx}\dfrac{1}{x^3} = -\csc h^2 \dfrac{1}{x^3} \cdot \dfrac{-3}{x^4} = \dfrac{3}{x^4}\csc h^2 \dfrac{1}{x^3}$

i. Given $y = \left(x^2 + 9\right)\tanh x$, then $\dfrac{dy}{dx} = \dfrac{d}{dx}\left[\left(x^2 + 9\right)\tanh x\right] = \tanh x \cdot \dfrac{d}{dx}\left(x^2 + 9\right) + \left(x^2 + 9\right) \cdot \dfrac{d}{dx}\tanh x = \tanh x \cdot 2x$

$+\left(x^2 + 9\right) \cdot \sec h^2 x = 2x \tanh x + \left(x^2 + 9\right)\sec h^2 x$

j. Given $y = \sinh^3 x^2$, then $\dfrac{dy}{dx} = \dfrac{d}{dx}\left(\sinh^3 x^2\right) = \dfrac{d}{dx}\left(\sinh x^2\right)^3 = 3\sinh^2 x^2 \cdot \dfrac{d}{dx}\sinh x^2 = 3\sinh^2 x^2 \cdot \cosh x^2 \cdot \dfrac{d}{dx} x^2$

$= 3\sinh^2 x^2 \cdot \cosh x^2 \cdot 2x = 6x \sinh^2 x^2 \cosh x^2$

k. Given $y = \tanh^5 x$, then $\dfrac{dy}{dx} = \dfrac{d}{dx}\left(\tanh^5 x\right) = 5\tanh^4 x \cdot \dfrac{d}{dx}\tanh x = 5\tanh^4 x \sec h^2 x$

l. Given $y = x^5 \coth\left(x^3 + 1\right)$, then $\dfrac{dy}{dx} = \dfrac{d}{dx}\left[x^5 \coth\left(x^3 + 1\right)\right] = \coth\left(x^3 + 1\right) \cdot \dfrac{d}{dx} x^5 + x^5 \cdot \dfrac{d}{dx}\coth\left(x^3 + 1\right) = \coth\left(x^3 + 1\right) \cdot 5x^4$

$+ x^5 \cdot -\csc h^2\left(x^3 + 1\right) \cdot \dfrac{d}{dx}\left(x^3 + 1\right) = \coth\left(x^3 + 1\right) \cdot 5x^4 - x^5 \cdot \csc h^2\left(x^3 + 1\right) \cdot 3x^2 = 5x^4 \coth\left(x^3 + 1\right) - 3x^7 \csc h^2\left(x^3 + 1\right)$

Section 3.5 Practice Problems – Differentiation of Inverse Hyperbolic Functions

Find the derivative of the following inverse hyperbolic functions:

a. Given $y = x \sinh^{-1} 3x$, then $\dfrac{dy}{dx} = \dfrac{d}{dx}\left(x \sinh^{-1} 3x\right) = \sinh^{-1} 3x \cdot \dfrac{d}{dx} x + x \cdot \dfrac{d}{dx}\sinh^{-1} 3x = \sinh^{-1} 3x \cdot 1 + x \cdot \dfrac{1}{\sqrt{1 + 9x^2}} \cdot \dfrac{d}{dx} 3x$

$= \sinh^{-1} 3x + x \cdot \dfrac{1}{\sqrt{1 + 9x^2}} \cdot 3 = \sinh^{-1} 3x + \dfrac{3x}{\sqrt{1 + 9x^2}}$

b. Given $y = \cosh^{-1}(x + 5)$, then $\dfrac{dy}{dx} = \dfrac{d}{dx}\cosh^{-1}(x + 5) = \dfrac{1}{\sqrt{(x + 5)^2 - 1}} \cdot \dfrac{d}{dx}(x + 5) = \dfrac{1}{\sqrt{(x + 5)^2 - 1}} \cdot 1 = \dfrac{1}{\sqrt{(x + 5)^2 - 1}}$

c. Given $y = \tanh^{-1}\sqrt{x}$, then $\dfrac{dy}{dx} = \dfrac{d}{dx}\left(\tanh^{-1}\sqrt{x}\right) = \dfrac{1}{1 - \left(\sqrt{x}\right)^2} \cdot \dfrac{d}{dx}\sqrt{x} = \dfrac{1}{1 - x} \cdot \dfrac{1}{\sqrt{x}} = \dfrac{1}{\sqrt{x}\,(1 - x)}$

d. Given $y = \dfrac{\sinh^{-1} x}{x^3}$, then $\dfrac{dy}{dx} = \dfrac{d}{dx}\left(\dfrac{\sinh^{-1} x}{x^3}\right) = \dfrac{x^3 \cdot \frac{d}{dx}\sinh^{-1} x - \sinh^{-1} x \cdot \frac{d}{dx} x^3}{x^6} = \dfrac{x^3 \cdot \frac{1}{\sqrt{x^2 + 1}} - \sinh^{-1} x \cdot 3x^2}{x^6}$

$= \dfrac{\frac{x^3}{\sqrt{x^2 + 1}} - 3x^2 \sinh^{-1} x}{x^6} = \dfrac{x^3 - 3x^2\sqrt{x^2 + 1}\,\sinh^{-1} x}{x^6\sqrt{x^2 + 1}} = \dfrac{x - 3\sqrt{x^2 + 1}\,\sinh^{-1} x}{x^4\sqrt{x^2 + 1}}$

e. Given $y = \dfrac{\sinh^{-1} x}{\cosh 3x}$, then $\dfrac{dy}{dx} = \dfrac{d}{dx}\left(\dfrac{\sinh^{-1} x}{\cosh 3x}\right) = \dfrac{\cosh 3x \cdot \frac{d}{dx}\sinh^{-1} x - \sinh^{-1} x \cdot \frac{d}{dx}\cosh 3x}{\cosh^2 3x} = \dfrac{\frac{\cosh 3x}{\sqrt{x^2 + 1}} - 3\sinh^{-1} x \sinh 3x}{\cosh^2 3x}$

$$= \frac{\cosh 3x - 3\sqrt{x^2+1}\,\sinh^{-1}x\,\sinh 3x}{\cosh^2 3x\,\sqrt{x^2+1}}$$

f. Given $y = \dfrac{\cosh^{-1}x}{x^5}$, then $\dfrac{dy}{dx} = \dfrac{d}{dx}\left(\dfrac{\cosh^{-1}x}{x^5}\right) = \dfrac{x^5\cdot\frac{d}{dx}\cosh^{-1}x - \cosh^{-1}x\cdot\frac{d}{dx}x^5}{x^{10}} = \dfrac{x^5\cdot\frac{1}{\sqrt{x^2-1}} - \cosh^{-1}x\cdot 5x^4}{x^{10}}$

$$= \frac{\frac{x^5}{\sqrt{x^2-1}} - 5x^4\cosh^{-1}x}{x^{10}} = \frac{x^5 - 5x^4\sqrt{x^2-1}\,\cosh^{-1}x}{x^{10}\sqrt{x^2-1}} = \frac{x - 5\sqrt{x^2-1}\,\cosh^{-1}x}{x^6\sqrt{x^2-1}}$$

g. Given $y = \left(x^2+3\right)\coth^{-1}x$, then $\dfrac{dy}{dx} = \dfrac{d}{dx}\left(x^2+3\right)\coth^{-1}x = \coth^{-1}x\cdot\dfrac{d}{dx}\left(x^2+3\right) + \left(x^2+3\right)\cdot\dfrac{d}{dx}\coth^{-1}x = \coth^{-1}x\cdot 2x$

$$+\left(x^2+3\right)\cdot\frac{1}{1-x^2} = 2x\coth^{-1}x + \frac{x^2+3}{1-x^2}$$

h. Given $y = e^{3x}\cosh^{-1}x$, then $\dfrac{dy}{dx} = \dfrac{d}{dx}\left(e^{3x}\cosh^{-1}x\right) = \cosh^{-1}x\cdot\dfrac{d}{dx}e^{3x} + e^{3x}\cdot\dfrac{d}{dx}\cosh^{-1}x = \cosh^{-1}x\cdot e^{3x}\cdot\dfrac{d}{dx}3x$

$$+e^{3x}\cdot\frac{1}{\sqrt{x^2-1}} = \cosh^{-1}x\cdot e^{3x}\cdot 3 + \frac{e^{3x}}{\sqrt{x^2-1}} = 3e^{3x}\cosh^{-1}x + \frac{e^{3x}}{\sqrt{x^2-1}} = \frac{e^{3x}\left(3\sqrt{x^2-1}\,\cosh^{-1}x + 1\right)}{\sqrt{x^2-1}}$$

i. Given $y = x^3 + \tanh^{-1}x^5$, then $\dfrac{dy}{dx} = \dfrac{d}{dx}\left(x^3 + \tanh^{-1}x^5\right) = \dfrac{d}{dx}x^3 + \dfrac{d}{dx}\tanh^{-1}x^5 = 3x^2 + \dfrac{1}{1-x^{10}}\cdot\dfrac{d}{dx}x^5 = 3x^2 + \dfrac{5x^4}{1-x^{10}}$

j. Given $y = \sinh^{-1}7x + \ln e^{x^2}$, then $\dfrac{dy}{dx} = \dfrac{d}{dx}\left(\sinh^{-1}7x + \ln e^{x^2}\right) = \dfrac{d}{dx}\sinh^{-1}7x + \dfrac{d}{dx}\ln e^{x^2} = \dfrac{1}{\sqrt{1+(7x)^2}}\cdot\dfrac{d}{dx}7x$

$$+\frac{1}{e^{x^2}}\cdot\frac{d}{dx}e^{x^2} = \frac{1}{\sqrt{1+49x^2}}\cdot 7 + \frac{1}{e^{x^2}}\cdot e^{x^2}\cdot\frac{d}{dx}x^2 = \frac{7}{\sqrt{1+49x^2}} + \frac{1}{e^{x^2}}\cdot e^{x^2}\cdot 2x = \frac{7}{\sqrt{1+49x^2}} + 2x$$

k. Given $y = \tanh^{-1}\left(e^{2x}\right)$, then $\dfrac{dy}{dx} = \dfrac{d}{dx}\tanh^{-1}\left(e^{2x}\right) = \dfrac{1}{1-e^{4x}}\cdot\dfrac{d}{dx}e^{2x} = \dfrac{1}{1-e^{4x}}\cdot e^{2x}\dfrac{d}{dx}2x = \dfrac{1}{1-e^{4x}}\cdot e^{2x}\cdot 2 = \dfrac{2e^{2x}}{1-e^{4x}}$

l. Given $y = \coth^{-1}\left(3x+5\right)$, then $\dfrac{dy}{dx} = \dfrac{d}{dx}\left[\coth^{-1}\left(3x+5\right)\right] = \dfrac{1}{1-(3x+5)^2}\cdot\dfrac{d}{dx}\left(3x+5\right) = \dfrac{3}{1-\left(3x+5\right)^2}$

Section 3.6 Practice Problems - Evaluation of Indeterminate Forms Using L'Hopital's Rule

Evaluate the limit for the following functions by applying the L'Hopital's rule, if needed:

a. $\lim_{x\to\infty}\dfrac{\ln x}{x^2} = \dfrac{\ln\infty}{\infty^2} = \dfrac{\infty}{\infty}$ Apply the L'Hopital's rule $\lim_{x\to\infty}\dfrac{\ln x}{x^2} = \lim_{x\to\infty}\dfrac{\frac{d}{dx}\ln x}{\frac{d}{dx}x^2} = \lim_{x\to\infty}\dfrac{\frac{1}{x}\cdot\frac{d}{dx}x}{2x}$

$$= \lim_{x\to\infty}\frac{\frac{1}{x}\cdot 1}{2x} = \lim_{x\to\infty}\frac{\frac{1}{x}}{2x} = \lim_{x\to\infty}\frac{1}{2x^2} = \frac{1}{2\cdot\infty^2} = \frac{1}{\infty} = 0$$

b. $\lim_{x\to 0}\dfrac{\sin x}{e^x-1} = \dfrac{\sin 0}{e^0-1} = \dfrac{0}{1-1} = \dfrac{0}{0}$ Apply the L'Hopital's rule $\lim_{x\to 0}\dfrac{\sin x}{e^x-1} = \lim_{x\to 0}\dfrac{\frac{d}{dx}\sin x}{\frac{d}{dx}\left(e^x-1\right)}$

$$= \lim_{x\to 0}\frac{\frac{d}{dx}\sin x}{\frac{d}{dx}e^x - \frac{d}{dx}1} = \lim_{x\to 0}\frac{\cos x}{e^x-0} = \lim_{x\to 0}\frac{\cos x}{e^x} = \frac{\cos 0}{e^0} = \frac{1}{1} = 1$$

c. $\lim_{x\to 0}\dfrac{1-\cos x}{x^3} = \lim_{x\to 0}\dfrac{1-\cos x}{x^3} = \dfrac{1-\cos 0}{0^3} = \dfrac{1-1}{0} = \dfrac{0}{0}$ Apply the L'Hopital's rule $\lim_{x\to 0}\dfrac{1-\cos x}{x^3}$

$$= \lim_{x \to 0} \frac{\frac{d}{dx}(1-\cos x)}{\frac{d}{dx}x^3} = \lim_{x \to 0} \frac{\frac{d}{dx}1 - \frac{d}{dx}\cos x}{\frac{d}{dx}x^3} = \lim_{x \to 0} \frac{0+\sin x}{3x^2} = \lim_{x \to 0} \frac{\sin x}{3x^2} = \frac{\sin 0}{3 \cdot 0^2} = \frac{0}{0} \quad \text{Apply the L'Hopital's}$$

$$\text{rule again } \lim_{x \to 0} \frac{\sin x}{3x^2} = \lim_{x \to 0} \frac{\frac{d}{dx}\sin x}{\frac{d}{dx}3x^2} = \lim_{x \to 0} \frac{\cos x}{6x} = \frac{\cos 0}{6 \cdot 0} = \frac{1}{0} = \infty$$

d. $\lim_{t \to 2} \dfrac{t-2}{t^2+2t-1} = \dfrac{2-2}{2^2+2 \cdot 2 -1} = \dfrac{2-2}{4+4-1} = \dfrac{0}{7} = \mathbf{0}$

e. $\lim_{x \to \frac{\pi}{2}} \dfrac{\cos x}{x-\frac{\pi}{2}} = \lim_{x \to \frac{\pi}{2}} \dfrac{\cos \frac{\pi}{2}}{\frac{\pi}{2}-\frac{\pi}{2}} = \dfrac{0}{0} \quad \text{Apply the L'Hopital's rule } \lim_{x \to \frac{\pi}{2}} \dfrac{\cos x}{x-\frac{\pi}{2}} = \lim_{x \to \frac{\pi}{2}} \dfrac{\frac{d}{dx}\cos x}{\frac{d}{dx}\left(x-\frac{\pi}{2}\right)} = \lim_{x \to \frac{\pi}{2}} \dfrac{-\sin x}{1}$

$$= -\lim_{x \to \frac{\pi}{2}} \sin x = -\sin \frac{\pi}{2} = \mathbf{-1}$$

f. $\lim_{t \to 0} \dfrac{t \cos t}{2\sin t} = \dfrac{0 \cdot \cos 0}{2\sin 0} = \dfrac{0 \cdot 1}{2 \cdot 0} = \dfrac{0}{0} \quad \text{Apply the L'Hopital's rule } \lim_{t \to 0} \dfrac{t \cos t}{2\sin t} = \lim_{t \to 0} \dfrac{\frac{d}{dt}t \cos t}{2 \cdot \frac{d}{dx}\sin t}$

$$= \lim_{t \to 0} \dfrac{\cos t \cdot \frac{d}{dt}t + t \cdot \frac{d}{dt}\cos t}{2 \cdot \frac{d}{dt}\sin t} = \lim_{t \to 0} \dfrac{\cos t \cdot 1 + t \cdot -\sin t}{2 \cdot \cos t} = \lim_{t \to 0} \dfrac{\cos t - t \sin t}{2\cos t} = \dfrac{\cos 0 - 0 \cdot \sin 0}{2\cos 0} = \dfrac{1-0 \cdot 0}{2 \cdot 1} = \dfrac{1}{2}$$

g. $\lim_{x \to 8} \dfrac{x-8}{x^2-64} = \dfrac{8-8}{8^2-64} = \dfrac{8-8}{64-64} = \dfrac{0}{0} \quad \text{Apply the L'Hopital's rule } \lim_{x \to 8} \dfrac{x-8}{x^2-64} = \lim_{x \to 8} \dfrac{\frac{d}{dx}(x-8)}{\frac{d}{dx}(x^2-64)}$

$$= \lim_{x \to 8} \dfrac{\frac{d}{dx}x - \frac{d}{dx}8}{\frac{d}{dx}x^2 - \frac{d}{dx}64} = \lim_{x \to 8} \dfrac{1-0}{2x-0} = \lim_{x \to 8} \dfrac{1}{2x} = \dfrac{1}{2 \cdot 8} = \dfrac{1}{16}$$

h. $\lim_{x \to \pi} \dfrac{\sin x}{\pi - x} = \dfrac{\sin 0}{\pi - \pi} = \dfrac{0}{0-0} = \dfrac{0}{0} \quad \text{Apply the L'Hopital's rule } \lim_{x \to \pi} \dfrac{\sin x}{\pi - x} = \lim_{x \to \pi} \dfrac{\frac{d}{dx}\sin x}{\frac{d}{dx}(\pi - x)}$

$$= \lim_{x \to \pi} \dfrac{\frac{d}{dx}\sin x}{\frac{d}{dx}\pi - \frac{d}{dx}x} = \lim_{x \to \pi} \dfrac{\cos x}{0-1} = \lim_{x \to \pi} \dfrac{\cos x}{-1} = -\lim_{x \to \pi} \cos x = -\cos \pi = \mathbf{1}$$

i. $\lim_{x \to 0} \dfrac{\sin 8x}{3x} = \dfrac{\sin(8 \cdot 0)}{3 \cdot 0} = \dfrac{\sin 0}{0} = \dfrac{0}{0} \quad \text{Apply the L'Hopital's rule } \lim_{x \to 0} \dfrac{\sin 8x}{3x} = \lim_{x \to 0} \dfrac{\frac{d}{dx}\sin 8x}{\frac{d}{dx}3x}$

$$= \lim_{x \to 0} \dfrac{\cos 8x \cdot \frac{d}{dx}8x}{\frac{d}{dx}3x} = \lim_{x \to 0} \dfrac{\cos 8x \cdot 8}{3} = \lim_{x \to 0} \dfrac{8\cos 8x}{3} = \lim_{x \to 0} \dfrac{8 \cdot \cos(8 \cdot 0)}{3} = \lim_{x \to 0} \dfrac{8 \cdot \cos 0}{3} = \dfrac{8 \cdot 1}{3} = \dfrac{8}{3}$$

j. $\lim_{t \to \infty} \dfrac{8t+3}{4t-2} = \dfrac{\infty+3}{\infty-2} = \dfrac{\infty}{\infty} \quad \text{Apply the L'Hopital's rule } \lim_{t \to \infty} \dfrac{8t+3}{4t-2} = \lim_{t \to \infty} \dfrac{\frac{d}{dt}(8t+3)}{\frac{d}{dt}(4t-2)} = \lim_{t \to \infty} \dfrac{\frac{d}{dt}8t + \frac{d}{dt}3}{\frac{d}{dt}4t - \frac{d}{dt}2}$

$$= \dfrac{8+0}{4-0} = \dfrac{8}{4} = \mathbf{2}$$

k. $\lim_{x \to 0} \dfrac{\cos x - 1}{x^2} = \dfrac{\cos 0 - 1}{0^2} = \dfrac{1-1}{0} = \dfrac{0}{0} \quad \text{Apply the L'Hopital's rule } \lim_{x \to 0} \dfrac{\cos x - 1}{x^2} = \lim_{x \to 0} \dfrac{\frac{d}{dx}(\cos x - 1)}{\frac{d}{dx}x^2}$

$$= \lim_{x \to 0} \dfrac{\frac{d}{dx}\cos x - \frac{d}{dx}1}{\frac{d}{dx}x^2} = \lim_{x \to 0} \dfrac{-\sin x - 0}{2x} = -\lim_{x \to 0} \dfrac{\sin x}{2x} = -\dfrac{\sin 0}{2 \cdot 0} = -\dfrac{0}{0} \quad \text{Apply the L'Hopital's rule again}$$

$$-\lim_{x \to 0} \dfrac{\sin x}{2x} = -\lim_{x \to 0} \dfrac{\frac{d}{dx}\sin x}{\frac{d}{dx}2x} = -\lim_{x \to 0} \dfrac{\cos x}{2} = -\dfrac{\cos 0}{2} = -\dfrac{1}{2}$$

1. $\lim_{x \to \frac{\pi}{2}} \frac{1 - \sin x}{1 + \cos 2x} = \frac{1 - \sin \frac{\pi}{2}}{1 + \cos 2 \cdot \frac{\pi}{2}} = \frac{1 - \sin \frac{\pi}{2}}{1 + \cos \pi} = \frac{1 - 1}{1 - 1} = \frac{0}{0}$ Apply the L'Hopital's rule $\lim_{x \to \frac{\pi}{2}} \frac{1 - \sin x}{1 + \cos 2x}$

$= \lim_{x \to \frac{\pi}{2}} \frac{\frac{d}{dx}(1 - \sin x)}{\frac{d}{dx}(1 + \cos 2 x)} = \lim_{x \to \frac{\pi}{2}} \frac{\frac{d}{dx} 1 - \frac{d}{dx} \sin x}{\frac{d}{dx} 1 + \frac{d}{dx} \cos 2x} = \lim_{x \to \frac{\pi}{2}} \frac{0 - \cos x}{0 - 2\sin 2x} = \frac{1}{2} \lim_{x \to \frac{\pi}{2}} \frac{\cos x}{\sin 2x} = \frac{1}{2} \cdot \frac{\cos \frac{\pi}{2}}{\sin 2 \cdot \frac{\pi}{2}}$

$= \frac{1}{2} \cdot \frac{\cos \frac{\pi}{2}}{\sin \pi} = \frac{1}{2} \cdot \frac{0}{0} = \frac{0}{0}$ Apply the L'Hopital's rule again $\frac{1}{2} \lim_{x \to \frac{\pi}{2}} \frac{\cos x}{\sin 2x} = \frac{1}{2} \lim_{x \to \frac{\pi}{2}} \frac{\frac{d}{dx} \cos x}{\frac{d}{dx} \sin 2x}$

$= \frac{1}{2} \lim_{x \to \frac{\pi}{2}} \frac{-\sin x}{2\cos 2x} = -\frac{1}{4} \lim_{x \to \frac{\pi}{2}} \frac{\sin x}{\cos 2x} = -\frac{1}{4} \cdot \frac{\sin \frac{\pi}{2}}{\cos 2 \cdot \frac{\pi}{2}} = -\frac{1}{4} \cdot \frac{\sin \frac{\pi}{2}}{\cos \pi} = -\frac{1}{4} \cdot \frac{1}{-1} = \frac{1}{4}$

Chapter 4 Solutions:

1. Evaluate the following indefinite integrals:

a. $\int -3\,dx = -3\int dx = -3\int x^0 dx = -\dfrac{3}{0+1}x^{0+1}+c = \boldsymbol{-3x+c}$

Check: Let $y = -3x+c$, then $y' = -3x^{1-1}+0 = -3x^0 = -3\cdot 1 = -3$

b. $\int 5k\,dx = 5k\int dx = 5k\int x^0 dx = \dfrac{5k}{0+1}x^{0+1}+c = \boldsymbol{5k\,x+c}$

Check: Let $y = 5k\,x+c$, then $y' = 5k\,x^{1-1}+0 = 5k\,x^0 = 5k\cdot 1 = 5k$

c. $\int \dfrac{2}{3}x\,dx = \dfrac{2}{3}\int x\,dx = \dfrac{2}{3}\int x^1 dx = \dfrac{2}{3}\cdot\dfrac{1}{1+1}x^{1+1}+c = \dfrac{2}{3}\cdot\dfrac{1}{2}x^2+c = \boldsymbol{\dfrac{1}{3}x^2+c}$

Check: Let $y = \dfrac{1}{3}x^2+c$, then $y' = \dfrac{1}{3}\cdot 2x^{2-1}+0 = \dfrac{2}{3}x$

d. $\int x^5 dx = \dfrac{1}{5+1}x^{5+1}+c = \boldsymbol{\dfrac{1}{6}x^6+c}$

Check: Let $y = \dfrac{1}{6}x^6+c$, then $y' = \dfrac{1}{6}\cdot 6x^{6-1}+0 = x^5$

e. $\int ax^7 dx = a\int x^7 dx = a\cdot\dfrac{1}{7+1}x^{7+1}+c = \boldsymbol{\dfrac{a}{8}x^8+c}$

Check: Let $y = \dfrac{a}{8}x^8+c$, then $y' = a\cdot\dfrac{1}{8}\cdot 8x^{8-1}+0 = ax^7$

f. $\int\left(x^4+x^3\right)dx = \int x^4 dx+\int x^3 dx = \dfrac{1}{1+4}x^{1+4}+\dfrac{1}{3+1}x^{3+1}+c = \boldsymbol{\dfrac{1}{5}x^5+\dfrac{1}{4}x^4+c}$

Check: Let $y = \dfrac{1}{5}x^5+\dfrac{1}{4}x^4+c$, then $y' = \dfrac{1}{5}\cdot 5x^{5-1}+\dfrac{1}{4}\cdot 4x^{4-1}+0 = x^4+x^3$

g. $\int\dfrac{dx}{x^4} = \int x^{-4}dx = \dfrac{1}{-4+1}x^{-4+1}+c = -\dfrac{1}{3}x^{-3}+c = \boldsymbol{-\dfrac{1}{3x^3}+c}$

Check: Let $y = -\dfrac{1}{3}x^{-3}+c$, then $y' = -\dfrac{1}{3}\cdot -3x^{-3-1}+0 = x^{-4} = \dfrac{1}{x^4}$

h. $\int\left(\dfrac{1}{x^4}-\dfrac{1}{x^2}\right)dx = \int\dfrac{1}{x^4}dx-\int\dfrac{1}{x^2}dx = \int x^{-4}dx-\int x^{-2}dx = \dfrac{1}{-4+1}x^{-4+1}-\dfrac{1}{-2+1}x^{-2+1} = -\dfrac{1}{3}x^{-3}+x^{-1}+c = \boldsymbol{-\dfrac{1}{3x^3}+\dfrac{1}{x}+c}$

Check: Let $y = -\dfrac{1}{3}x^{-3}+x^{-1}+c$, then $y' = -\dfrac{1}{3}\cdot -3x^{-3-1}-x^{-1-1}+0 = x^{-4}-x^{-2} = \dfrac{1}{x^4}-\dfrac{1}{x^2}$

i. $\int\sqrt[3]{x^6}\,dx = \int x^{\frac{6}{3}}dx = \int x^2 dx = \dfrac{1}{1+2}x^{2+1}+c = \boldsymbol{\dfrac{1}{3}x^3+c}$

Check: Let $y = \dfrac{1}{3}x^3+c$, then $y' = \dfrac{1}{3}\cdot 3x^{3-1}+0 = x^2$

2. Evaluate the following indefinite integrals:

a. $\int\left(\sqrt[3]{x^2}+x\right)dx = \int x^{\frac{2}{3}}dx+\int x\,dx = \dfrac{1}{1+\frac{2}{3}}x^{\frac{2}{3}+1}+\dfrac{1}{1+1}x^{1+1}+c = \dfrac{1}{\frac{3+2}{3}}x^{\frac{2+3}{3}}+\dfrac{1}{2}x^2+c = \boldsymbol{\dfrac{3}{5}x^{\frac{5}{3}}+\dfrac{1}{2}x^2+c}$

Check: Let $y = \dfrac{3}{5}x^{\frac{5}{3}} + \dfrac{1}{2}x^2 + c$, then $y' = \dfrac{3}{5}\cdot\dfrac{5}{3}x^{\frac{5}{3}-1} + \dfrac{1}{2}\cdot 2x^{2-1} + 0 = x^{\frac{5-3}{3}} + x = x^{\frac{2}{3}} + x = \sqrt[3]{x^2} + x$

b. $\displaystyle\int -\dfrac{1}{2\sqrt{t}}\,dt = -\dfrac{1}{2}\int \dfrac{1}{t^{\frac{1}{2}}}\,dt = -\dfrac{1}{2}\int t^{-\frac{1}{2}}dt = -\dfrac{1}{2}\cdot\dfrac{1}{1-\frac{1}{2}}t^{1-\frac{1}{2}} + c = -\dfrac{1}{2}\cdot\dfrac{1}{\frac{2-1}{2}}t^{\frac{2-1}{2}} + c = -\dfrac{1}{2}\cdot 2t^{\frac{1}{2}} + c = -\sqrt{t} + c$

Check: Let $y = -t^{\frac{1}{2}} + c$, then $y' = -\dfrac{1}{2}t^{\frac{1}{2}-1} + 0 = -\dfrac{1}{2}t^{\frac{1-2}{2}} = -\dfrac{1}{2}t^{-\frac{1}{2}} = -\dfrac{1}{2t^{\frac{1}{2}}} = -\dfrac{1}{2\sqrt{t}}$

c. $\displaystyle\int \dfrac{1}{\sqrt[7]{x^2}}\,dx = \int \dfrac{1}{x^{\frac{2}{7}}}\,dx = \int x^{-\frac{2}{7}}dx = \dfrac{1}{1-\frac{2}{7}}x^{1-\frac{2}{7}} + c = \dfrac{1}{\frac{7-2}{7}}x^{\frac{7-2}{7}} + c = \dfrac{7}{5}x^{\frac{5}{7}} + c = \dfrac{7}{5}\sqrt[7]{x^5} + c$

Check: Let $y = \dfrac{7}{5}x^{\frac{5}{7}} + c$, then $y' = \dfrac{7}{5}\cdot\dfrac{5}{7}x^{\frac{5}{7}-1} + 0 = x^{\frac{5-7}{7}} = x^{-\frac{2}{7}} = \dfrac{1}{x^{\frac{2}{7}}} = \dfrac{1}{\sqrt[7]{x^2}}$

d. $\displaystyle\int (1+x)\sqrt{x}\,dx = \int\left(\sqrt{x} + x\sqrt{x}\right)dx = \int\sqrt{x}\,dx + \int x\sqrt{x}\,dx = \int x^{\frac{1}{2}}\,dx + \int x\, x^{\frac{1}{2}}\,dx = \int x^{\frac{1}{2}}\,dx + \int x^{1+\frac{1}{2}}dx$

$= \displaystyle\int x^{\frac{1}{2}}\,dx + \int x^{\frac{3}{2}}dx = \dfrac{1}{1+\frac{1}{2}}x^{\frac{1}{2}+1} + \dfrac{1}{1+\frac{3}{2}}x^{\frac{3}{2}+1} + c = \dfrac{1}{\frac{2+1}{2}}x^{\frac{1+2}{2}} + \dfrac{1}{\frac{2+3}{2}}x^{\frac{3+2}{2}} + c = \dfrac{2}{3}x^{\frac{3}{2}} + \dfrac{2}{5}x^{\frac{5}{2}} + c = \dfrac{2}{3}\sqrt{x^3} + \dfrac{2}{5}\sqrt{x^5} + c$

$= \dfrac{2}{3}\sqrt{x^2\cdot x} + \dfrac{2}{5}\sqrt{x^2\cdot x^2\cdot x} + c = \dfrac{2}{3}x\sqrt{x} + \dfrac{2}{5}x^2\sqrt{x} + c$

Check: Let $y = \dfrac{2}{3}x^{\frac{3}{2}} + \dfrac{2}{5}x^{\frac{5}{2}} + c$, then $y' = \dfrac{2}{3}\cdot\dfrac{3}{2}x^{\frac{3}{2}-1} + \dfrac{2}{5}\cdot\dfrac{5}{2}x^{\frac{5}{2}-1} + 0 = x^{\frac{3-2}{2}} + x^{\frac{5-2}{2}} = x^{\frac{1}{2}} + x^{\frac{3}{2}}$

e. $\displaystyle\int (2x^2+1)(x-1)\,dx = \int\left(2x^3 - 2x^2 + x - 1\right)dx = \int 2x^3 dx - \int 2x^2 dx + \int x\,dx - \int dx = 2\int x^3 dx - 2\int x^2 dx + \int x\,dx - \int x^0 dx$

$= 2\cdot\dfrac{1}{1+3}x^{3+1} - 2\cdot\dfrac{1}{1+2}x^{2+1} + \dfrac{1}{1+1}x^{1+1} - \dfrac{1}{1+0}x^{0+1} + c = \dfrac{1}{2}x^4 - \dfrac{2}{3}x^3 + \dfrac{1}{2}x^2 - x + c$

Check: Let $y = \dfrac{1}{2}x^4 - \dfrac{2}{3}x^3 + \dfrac{1}{2}x^2 - x + c$, then $y' = \dfrac{1}{2}\cdot 4x^{4-1} - \dfrac{2}{3}\cdot 3x^{3-1} + \dfrac{1}{2}\cdot 2x^{2-1} - 1 + 0 = 2x^3 - 2x^2 + x - 1$

f. $\displaystyle\int x(3x-1)^2 dx = \int x\left(9x^2 - 6x + 1\right)dx = \int\left(9x^3 - 6x^2 + x\right)dx = \int 9x^3 dx - \int 6x^2 dx + \int x\,dx = \dfrac{9}{3+1}x^{3+1} - \dfrac{6}{2+1}x^{2+1}$

$+ \dfrac{1}{1+1}x^{1+1} + c = \dfrac{9}{4}x^4 - \dfrac{6}{3}x^3 + \dfrac{1}{2}x^2 + c = \dfrac{9}{4}x^4 - 2x^3 + \dfrac{1}{2}x^2 + c$

Check: Let $y = \dfrac{9}{4}x^4 - 2x^3 + \dfrac{1}{2}x^2 + c$, then $y' = \dfrac{9}{4}\cdot 4x^{4-1} - 2\cdot 3x^{3-1} + \dfrac{1}{2}\cdot 2x^{2-1} + 0 = 9x^3 - 6x^2 + x$

g. $\displaystyle\int (2+x)^3 dx = \int\left(2^3 + x^3 + 3\cdot 2^2 x + 3\cdot 2x^2\right)dx = \int\left(x^3 + 6x^2 + 12x + 8\right)dx = \int x^3 dx + \int 6x^2 dx + \int 12x\,dx + \int 8\,dx$

$= \dfrac{1}{3+1}x^{3+1} + \dfrac{6}{2+1}x^{2+1} + \dfrac{12}{1+1}x^{1+1} + \dfrac{8}{0+1}x^{0+1} + c = \dfrac{1}{4}x^4 + 2x^3 + 6x^2 + 8x + c$

Check: Let $y = \dfrac{1}{4}x^4 + 2x^3 + 6x^2 + 8x + c$, then $y' = \dfrac{1}{4}\cdot 4x^{4-1} + 2\cdot 3x^{3-1} + 6\cdot 2x^{2-1} + 8x^{1-1} + 0 = x^3 + 6x^2 + 12x + 8x^0$

$= x^3 + 6x^2 + 12x + 8 = (2+x)^3$

h. $\displaystyle\int (2-x)^3 dx = \int\left(2^3 - x^3 + 3\cdot 2^2 x - 3\cdot 2x^2\right)dx = \int\left(-x^3 - 6x^2 + 12x + 8\right)dx = -\int x^3 dx - \int 6x^2 dx + \int 12x\,dx + \int 8\,dx$

$= -\dfrac{1}{3+1}x^{3+1} - \dfrac{6}{2+1}x^{2+1} + \dfrac{12}{1+1}x^{1+1} + \dfrac{8}{0+1}x^{0+1} + c = -\dfrac{1}{4}x^4 - 2x^3 + 6x^2 + 8x + c$

Check: Let $y = -\dfrac{1}{4}x^4 - 2x^3 + 6x^2 + 8x + c$, then $y' = -\dfrac{1}{4}\cdot 4x^{4-1} - 2\cdot 3x^{3-1} + 6\cdot 2x^{2-1} + 8x^{1-1} + 0 = -x^3 - 6x^2 + 12x$

$+ 8x^0 = -x^3 - 6x^2 + 12x + 8 = (2-x)^3$

i. $\int \frac{y^5+4y^2}{y^2}\,dy = \int \left(\frac{y^5}{y^2}+\frac{4y^2}{y^2}\right)dy = \int \left(y^3+4\right)dy = \int y^3 dy + \int 4dy = \frac{1}{3+1}y^{3+1}+\frac{4}{0+1}y^{0+1}+c = \frac{1}{4}y^4+4y+c$

Check: Let $w=\frac{1}{4}y^4+4y+c$, then $w' = \frac{1}{4}\cdot 4y^{4-1}+4\cdot 1y^{1-1}+0 = y^3+4y^0 = y^3+4$

---

1. Evaluate the following indefinite integrals:

a. Given $\int t^2\left(1+t^3\right)dt$ let $u=1+t^3$, then $\frac{du}{dt} = \frac{d}{dt}\left(1+t^3\right) = 3t^2$ which implies that $du=3t^2dt$ and $dt = \frac{du}{3t^2}$.

Substituting the u and dx values back into the original integral we obtain

$\int t^2\left(1+t^3\right)dt = \int t^2\cdot u\cdot\frac{du}{3t^2} = \frac{1}{3}\int u\,du = \frac{1}{3}\cdot\frac{1}{1+1}u^{1+1}+c = \frac{1}{3}\cdot\frac{1}{2}u^2+c = \frac{1}{6}\left(1+t^3\right)^2+c$

Check: Let $y=\frac{1}{6}\left(1+t^3\right)^2+c$, then $y' = \frac{1}{6}\cdot 2\left(1+t^3\right)^{2-1}\cdot 3t^2+0 = \frac{1}{3}\left(1+t^3\right)\cdot 3t^2 = t^2\left(1+t^3\right)$

b. Given $\int 18x^2\sqrt{6x^3-5}\,dx$ let $u=6x^3-5$, then $\frac{du}{dx} = \frac{d}{dx}\left(6x^3-5\right) = 18x^2$ which implies that $du=18x^2dx$ and

$dx = \frac{du}{18x^2}$. Substituting the u and dx values back into the original integral we obtain

$\int 18x^2\sqrt{6x^3-5}\,dx = \int 18x^2\cdot\sqrt{u}\cdot\frac{du}{18x^2} = \int u^{\frac{1}{2}}du = \frac{1}{\frac{1}{2}+1}u^{\frac{1}{2}+1}+c = \frac{2}{3}u^{\frac{3}{2}}+c = \frac{2}{3}u^{\frac{3}{2}}+c = \frac{2}{3}\left(6x^3-5\right)^{\frac{3}{2}}+c$

Check: Let $y=\frac{2}{3}\left(6x^3-5\right)^{\frac{3}{2}}+c$, then $y' = \frac{2}{3}\cdot\frac{3}{2}\left(6x^3-5\right)^{\frac{3}{2}-1}\cdot 18x^2+0 = \left(6x^3-5\right)^{\frac{1}{2}}\cdot 18x^2 = 18x^2\sqrt{6x^3-5}$

c. Given $\int\frac{5}{\sqrt{x+5}}\,dx$ let $u=x+5$, then $\frac{du}{dx} = \frac{d}{dx}(x+5) = 1$ which implies that $du=dx$. Substituting the u and dx

values back into the original integral we obtain

$\int\frac{5}{\sqrt{x+5}}\,dx = \int\frac{5}{\sqrt{u}}\cdot du = 5\int u^{-\frac{1}{2}}du = 5\cdot\frac{1}{1-\frac{1}{2}}u^{1-\frac{1}{2}}+c = 5\cdot\frac{1}{\frac{2-1}{2}}u^{\frac{2-1}{2}}+c = 10\,u^{\frac{1}{2}}+c = 10(x+5)^{\frac{1}{2}}+c$

Check: Let $y=10(x+5)^{\frac{1}{2}}+c$, then $y' = 10\cdot\frac{1}{2}(x+5)^{\frac{1}{2}-1}+0 = 5(x+5)^{-\frac{1}{2}} = \frac{5}{(x+5)^{\frac{1}{2}}} = \frac{5}{\sqrt{x+5}}$

d. Given $\int x\left(x^2-2\right)^{\frac{1}{5}}dx$ let $u=x^2-2$, then $\frac{du}{dx} = \frac{d}{dx}\left(x^2-2\right) = 2x$ which implies that $du=2x\,dx$ and $dx = \frac{du}{2x}$.

Substituting the u and dx values back into the original integral we obtain

$\int x\left(x^2-2\right)^{\frac{1}{5}}dx = \int x\cdot u^{\frac{1}{5}}\cdot\frac{du}{2x} = \frac{1}{2}\int u^{\frac{1}{5}}du = \frac{1}{2}\cdot\frac{1}{\frac{1}{5}+1}u^{\frac{1}{5}+1}+c = \frac{1}{2}\cdot\frac{5}{6}u^{\frac{6}{5}}+c = \frac{5}{12}\left(x^2-2\right)^{\frac{6}{5}}+c$

Check: Let $\frac{5}{12}\left(x^2-2\right)^{\frac{6}{5}}+c$, then $y' = \frac{5}{12}\cdot\frac{6}{5}\left(x^2-2\right)^{\frac{6}{5}-1}\cdot 2x+0 = x\left(x^2-2\right)^{\frac{1}{5}}$

e. Given $\int\frac{3x}{\sqrt{x^2+3}}\,dx$ let $u=x^2+3$, then $\frac{du}{dx} = \frac{d}{dx}\left(x^2+3\right) = 2x$ which implies that $du=2x\,dx$ and $dx = \frac{du}{2x}$.

Substituting the u and dx values back into the original integral we obtain

$\int\frac{3x}{\sqrt{x^2+3}}\,dx = \int\frac{3x}{\sqrt{u}}\cdot\frac{du}{2x} = \frac{3}{2}\int u^{-\frac{1}{2}}du = \frac{3}{2}\cdot\frac{1}{1-\frac{1}{2}}u^{1-\frac{1}{2}}+c = \frac{3}{2}\cdot\frac{1}{\frac{2-1}{2}}u^{\frac{2-1}{2}}+c = \frac{3}{2}\cdot 2\,u^{\frac{1}{2}}+c = 3u^{\frac{1}{2}}+c = 3\left(x^2+3\right)^{\frac{1}{2}}+c$

Check: Let $y = 3\left(x^2 + 3\right)^{\frac{1}{2}} + c$, then $y' = \frac{3}{2}\left(x^2 + 3\right)^{\frac{1}{2} - 1} \cdot 2x + 0 = \frac{3}{2}\left(x^2 + 3\right)^{-\frac{1}{2}} \cdot 2x = \frac{3x}{\left(x^2 + 3\right)^{\frac{1}{2}}} = \frac{3x}{\sqrt{x^2 + 3}}$

f. Given $\int \frac{t}{2}\sqrt{1 - t^2}\,dt$ let $u = 1 - t^2$, then $\frac{du}{dt} = \frac{d}{dt}\left(1 - t^2\right) = -2t$ which implies that $du = -2t\,dt$ and $dt = \frac{du}{-2t}$.

Substituting the u and dt values back into the original integral we obtain

$$\int \frac{t}{2}\sqrt{1 - t^2}\,dt = \int \frac{t}{2}\cdot\sqrt{u}\cdot\frac{du}{-2t} = -\frac{1}{4}\int u^{\frac{1}{2}}du = -\frac{1}{4}\cdot\frac{1}{\frac{1}{2}+1}u^{\frac{1}{2}+1} + c = -\frac{1}{4}\cdot\frac{2}{3}u^{\frac{3}{2}} + c = -\frac{1}{6}u^{\frac{3}{2}} + c = -\frac{1}{6}\left(1 - t^2\right)^{\frac{3}{2}} + c$$

Check: Let $y = -\frac{1}{6}\left(1 - t^2\right)^{\frac{3}{2}} + c$, then $y' = -\frac{1}{6}\cdot\frac{3}{2}\left(1 - t^2\right)^{\frac{3}{2} - 1}\cdot -2t + 0 = \frac{1}{2}\cdot t\left(1 - t^2\right)^{\frac{1}{2}} = \frac{t}{2}\sqrt{1 - t^2}$

g. Given $\int x^2\left(1 - x^3\right)^2 dx$ let $u = 1 - x^3$, then $\frac{du}{dx} = \frac{d}{dx}\left(1 - x^3\right) = -3x^2$ which implies that $du = -3x^2 dx$ and $dx = \frac{du}{-3x^2}$.

Substituting the u and dx values back into the original integral we obtain

$$\int x^2\left(1 - x^3\right)^2 dx = \int x^2\cdot u^2\cdot\frac{du}{-3x^2} = -\frac{1}{3}\int u^2 du = -\frac{1}{3}\cdot\frac{1}{2+1}u^{2+1} + c = -\frac{1}{3}\cdot\frac{1}{3}u^3 + c = -\frac{1}{9}\left(1 - x^3\right)^3 + c$$

Check: Let $y = -\frac{1}{9}\left(1 - x^3\right)^3 + c$, then $y' = -\frac{1}{9}\cdot 3\left(1 - x^3\right)^{3-1}\cdot -3x^2 + 0 = \frac{3}{3}\left(1 - x^3\right)^{3-1}\cdot x^2 = x^2\left(1 - x^3\right)^2$

h. Given $\int \frac{x^3}{\sqrt{x^4 + 3}}\,dx$ let $u = x^4 + 3$, then $\frac{du}{dx} = \frac{d}{dx}\left(x^4 + 3\right) = 4x^3$ which implies that $du = 4x^3 dx$ and $dx = \frac{du}{4x^3}$.

Substituting the u and dx values back into the original integral we obtain

$$\int \frac{x^3}{\sqrt{x^4 + 3}}\,dx = \int \frac{x^3}{\sqrt{u}}\cdot\frac{du}{4x^3} = \frac{1}{4}\int u^{-\frac{1}{2}}du = \frac{1}{4}\cdot\frac{1}{1-\frac{1}{2}}u^{1-\frac{1}{2}} + c = \frac{1}{4}\cdot 2\,u^{\frac{1}{2}} + c = \frac{1}{2}u^{\frac{1}{2}} + c = \frac{1}{2}\left(x^4 + 3\right)^{\frac{1}{2}} + c$$

Check: Let $y = \frac{1}{2}\left(x^4 + 3\right)^{\frac{1}{2}} + c$, then $y' = \frac{1}{4}\left(x^4 + 3\right)^{\frac{1}{2} - 1}\cdot 4x^3 + 0 = \frac{1}{4}\left(x^4 + 3\right)^{-\frac{1}{2}}\cdot 4x^3 = \frac{x^3}{\left(x^4 + 3\right)^{\frac{1}{2}}} = \frac{x^3}{\sqrt{x^4 + 3}}$

i. Given $\int x^8\left(2x^9 + 1\right)^2 dx$ let $u = 2x^9 + 1$, then $\frac{du}{dx} = \frac{d}{dx}\left(2x^9 + 1\right) = 18x^8$ which implies that $du = 18x^8 dx$ and $dx = \frac{du}{18x^8}$.

Substituting the u and dx values back into the original integral we obtain

$$\int x^8\left(2x^9 + 1\right)^2 dx = \int x^8\cdot u^2\cdot\frac{du}{18x^8} = \frac{1}{18}\int u^2 du = \frac{1}{18}\cdot\frac{1}{2+1}u^{2+1} + c = \frac{1}{54}u^3 + c = \frac{1}{54}\left(2x^9 + 1\right)^3 + c$$

Check: Let $y = \frac{1}{54}\left(2x^9 + 1\right)^3 + c$, then $y' = \frac{1}{54}\cdot 3\left(2x^9 + 1\right)^{3-1}\cdot 18x^8 + 0 = \left(2x^9 + 1\right)^2\cdot x^8 = x^8\left(2x^9 + 1\right)^2$

2. Evaluate the following indefinite integrals:

a. Given $\int 6x^2\left(2x^3 - 1\right)dx$ let $u = 2x^3 - 1$, then $\frac{du}{dx} = \frac{d}{dx}\left(2x^3 - 1\right) = 6x^2$ which implies that $du = 6x^2 dx$ and $dx = \frac{du}{6x^2}$.

Substituting the u and dx values back into the original integral we obtain

$$\int 6x^2\left(2x^3 - 1\right)dx = \int 6x^2\cdot u\cdot\frac{du}{6x^2} = \int u\,du = \frac{1}{1+1}u^{1+1} + c = \frac{1}{2}u^2 + c = \frac{1}{2}\left(2x^3 - 1\right)^2 + c$$

Check: Let $y = \frac{1}{2}\left(2x^3 - 1\right)^2 + c$, then $y' = \frac{1}{2}\cdot 2\left(2x^3 - 1\right)^{2-1}\cdot 6x^2 + 0 = 6x^2\left(2x^3 - 1\right)$

b. Given $\int x\sqrt{1 + x^2}\,dx$ let $u = 1 + x^2$, then $\frac{du}{dx} = \frac{d}{dx}\left(1 + x^2\right) = 2x$ which implies that $du = 2x\,dx$ and $dx = \frac{du}{2x}$.

Substituting the u and dx values back into the original integral we obtain

$$\int x\sqrt{1+x^2}\,dx = \int x\cdot\sqrt{u}\cdot\frac{du}{2x} = \frac{1}{2}\int u^{\frac{1}{2}}du = \frac{1}{2}\cdot\frac{1}{\frac{1}{2}+1}u^{\frac{1}{2}+1}+c = \frac{1}{2}\cdot\frac{2}{3}u^{\frac{3}{2}}+c = \frac{1}{3}u^{\frac{3}{2}}+c = \frac{1}{3}\left(1+x^2\right)^{\frac{3}{2}}+c$$

Check: Let $y = \frac{1}{3}\left(1+x^2\right)^{\frac{3}{2}}+c$, then $y' = \frac{1}{3}\cdot\frac{3}{2}\left(1+x^2\right)^{\frac{3}{2}-1}\cdot 2x+0 = \frac{1}{2}\cdot 2x\left(1+x^2\right)^{\frac{1}{2}} = x\sqrt{1+x^2}$

c. Given $\int\sqrt[5]{7x+1}\,dx$ let $u = 7x+1$, then $\frac{du}{dx} = \frac{d}{d}(7x+1) = 7$ which implies that $du = 7\,dx$ and $dx = \frac{du}{7}$. Substituting

the u and dx values back into the original integral we obtain

$$\int\sqrt[5]{7x+1}\,dx = \int\sqrt[5]{u}\cdot\frac{du}{7} = \frac{1}{7}\int u^{\frac{1}{5}}du = \frac{1}{7}\cdot\frac{1}{\frac{1}{5}+1}u^{\frac{1}{5}+1}+c = \frac{1}{7}\cdot\frac{5}{6}u^{\frac{6}{5}}+c = \frac{5}{42}u^{\frac{6}{5}}+c = \frac{5}{42}\left(7x+1\right)^{\frac{6}{5}}+c$$

Check: Let $y = \frac{5}{42}(7x+1)^{\frac{6}{5}}+c$, then $y' = \frac{5}{42}\cdot\frac{6}{5}(7x+1)^{\frac{6}{5}-1}\cdot 7+0 = \frac{1}{7}(7x+1)^{\frac{1}{5}}\cdot 7 = (7x+1)^{\frac{1}{5}} = \sqrt[5]{7x+1}$

d. Given $\int x\sqrt[3]{3x^2-1}\,dx$ let $u = 3x^2-1$ then $\frac{du}{dx} = \frac{d}{dx}\left(3x^2-1\right) = 6x$ which implies that $du = 6x\,dx$ and $dx = \frac{du}{6x}$.

Substituting the u and dx values back into the original integral we obtain

$$\int x\sqrt[3]{3x^2-1}\,dx = \int x\cdot\sqrt[3]{u}\cdot\frac{du}{6x} = \frac{1}{6}\int u^{\frac{1}{3}}du = \frac{1}{6}\cdot\frac{1}{\frac{1}{3}+1}u^{\frac{1}{3}+1}+c = \frac{1}{6}\cdot\frac{3}{4}u^{\frac{4}{3}}+c = \frac{1}{8}u^{\frac{4}{3}}+c = \frac{1}{8}\left(3x^2-1\right)^{\frac{4}{3}}+c$$

Check: Let $y = \frac{1}{8}\left(3x^2-1\right)^{\frac{4}{3}}+c$, then $y' = \frac{1}{8}\cdot\frac{4}{3}\left(3x^2-1\right)^{\frac{4}{3}-1}\cdot 6x+0 = \frac{1}{6}\left(3x^2-1\right)^{\frac{1}{3}}\cdot 6x = x\left(3x^2-1\right)^{\frac{1}{3}} = x\sqrt[3]{3x^2-1}$

e. Given $\int\frac{x}{\sqrt{x^2+1}}\,dx$ let $u = x^2+1$, then $\frac{du}{dx} = \frac{d}{dx}\left(x^2+1\right) = 2x$ which implies that $du = 2x\,dx$ and $dx = \frac{du}{2x}$.

Substituting the u and dx values back into the original integral we obtain

$$\int\frac{x}{\sqrt{x^2+1}}\,dx = \int\frac{x}{\sqrt{u}}\cdot\frac{du}{2x} = \frac{1}{2}\int u^{-\frac{1}{2}}du = \frac{1}{2}\cdot\frac{1}{1-\frac{1}{2}}u^{1-\frac{1}{2}}+c = \frac{1}{2}\cdot\frac{1}{\frac{2-1}{2}}u^{\frac{2-1}{2}}+c = \frac{1}{2}\cdot 2u^{\frac{1}{2}}+c = u^{\frac{1}{2}}+c = \left(x^2+1\right)^{\frac{1}{2}}+c$$

Check: Let $y = \left(x^2+1\right)^{\frac{1}{2}}+c$, then $y' = \frac{1}{2}\left(x^2+1\right)^{\frac{1}{2}-1}\cdot 2x+0 = \left(x^2+1\right)^{-\frac{1}{2}}\cdot 2x = \frac{x}{\left(x^2+1\right)^{\frac{1}{2}}} = \frac{x}{\sqrt{x^2+1}}$

f. Given $\int x\left(1-x^2\right)^2\,dx$ let $u = 1-x^2$, then $\frac{du}{dx} = \frac{d}{dx}\left(1-x^2\right) = -2x$ which implies that $du = -2x\,dx$ and $dx = \frac{du}{-2x}$.

Substituting the u and dx values back into the original integral we obtain

$$\int x\left(1-x^2\right)^2\,dx = \int x\cdot u^2\cdot\frac{du}{-2x} = -\frac{1}{2}\int u^2 du = -\frac{1}{2}\cdot\frac{1}{2+1}u^{2+1}+c = -\frac{1}{6}u^3+c = -\frac{1}{6}\left(1-x^2\right)^3+c$$

Check: Let $y = -\frac{1}{6}\left(1-x^2\right)^3+c$, then $y' = -\frac{1}{6}\cdot 3\left(1-x^2\right)^{3-1}\cdot-2x+0 = \frac{6}{6}\left(1-x^2\right)^2\cdot x = x\left(1-x^2\right)^2$

g. Given $\int\frac{x^5}{\sqrt{x^6+3}}\,dx$ let $u = x^6+3$, then $\frac{du}{dx} = \frac{d}{dx}\left(x^6+3\right) = 6x^5$ which implies that $du = 6x^5 dx$ and $dx = \frac{du}{6x^5}$.

Substituting the u and dx values back into the original integral we obtain

$$\int\frac{x^5}{\sqrt{x^6+3}}\,dx = \int\frac{x^5}{\sqrt{u}}\cdot\frac{du}{6x^5} = \frac{1}{6}\int\frac{1}{\sqrt{u}}\cdot du = \frac{1}{6}\int u^{-\frac{1}{2}}du = \frac{1}{6}\cdot\frac{1}{1-\frac{1}{2}}u^{1-\frac{1}{2}}+c = \frac{1}{3}u^{\frac{1}{2}}+c = \frac{1}{3}\left(x^6+3\right)^{\frac{1}{2}}+c$$

Check: Let $y = \frac{1}{3}(x^6+3)^{\frac{1}{2}} + c$, then $y' = \frac{1}{3}\cdot\frac{1}{2}(x^6+3)^{\frac{1}{2}-1}\cdot 6x^5 + 0 = x^5(x^6+3)^{-\frac{1}{2}} = \dfrac{x^5}{\sqrt{x^6+3}}$

h. Given $\displaystyle\int \frac{3x}{\sqrt{x^2-1}}\,dx$ let $u = x^2-1$, then $\frac{du}{dx} = \frac{d}{dx}(x^2-1) = 2x$ which implies that $du = 2x\,dx$ and $dx = \frac{du}{2x}$.

Substituting the u and dx values back into the original integral we obtain

$\displaystyle\int \frac{3x}{\sqrt{x^2-1}}\,dx = \int \frac{3x}{\sqrt{u}}\cdot\frac{du}{2x} = \frac{3}{2}\int u^{-\frac{1}{2}}\,du = \frac{3}{2}\cdot\frac{1}{1-\frac{1}{2}}u^{1-\frac{1}{2}} + c = \frac{3}{2}\cdot\frac{1}{\frac{2-1}{2}}u^{\frac{2-1}{2}} + c = 3u^{\frac{1}{2}} + c = 3(x^2-1)^{\frac{1}{2}} + c$

Check: Let $y = 3(x^2-1)^{\frac{1}{2}} + c$, then $y' = \frac{3}{2}(x^2-1)^{\frac{1}{2}-1}\cdot 2x + 0 = 3x(x^2-1)^{-\frac{1}{2}} = \dfrac{3x}{(x^2-1)^{\frac{1}{2}}} = \dfrac{3x}{\sqrt{x^2-1}}$

i. Given $\displaystyle\int \frac{5x^4+6x}{\sqrt{x^5+3x^2+1}}\,dx$ let $u = x^5+3x^2+1$, then $\frac{du}{dx} = \frac{d}{dx}(x^5+3x^2+1) = 5x^4+6x$ which implies that

$du = (5x^4+6x)dx$ and $dx = \dfrac{du}{5x^4+6x}$. Substituting the u and dx values back into the original integral we obtain

$\displaystyle\int \frac{5x^4+6x}{\sqrt{x^5+3x^2+1}}\,dx = \int \frac{5x^4+6x}{\sqrt{u}}\cdot\frac{du}{5x^4+6x} = \int \frac{du}{\sqrt{u}} = \int u^{-\frac{1}{2}}\,du = \frac{1}{1-\frac{1}{2}}u^{1-\frac{1}{2}} + c = \frac{1}{\frac{2-1}{2}}u^{\frac{2-1}{2}} + c = 2u^{\frac{1}{2}} + c$

$= 2(x^5+3x^2+1)^{\frac{1}{2}} + c = \mathbf{2\sqrt{x^5+3x^2+1} + c}$

Check: Let $y = 2(x^5+3x^2+1)^{\frac{1}{2}} + c$, then $y' = 2\cdot\frac{1}{2}(x^5+3x^2+1)^{\frac{1}{2}-1}(5x^4+6x) + 0 = (x^5+3x^2+1)^{-\frac{1}{2}}(5x^4+6x)$

$= \dfrac{5x^4+6x}{(x^5+3x^2+1)^{\frac{1}{2}}} = \dfrac{5x^4+6x}{\sqrt{x^5+3x^2+1}}$

Section 4.3 Solutions – Integration of Trigonometric Functions

1. Evaluate the following integrals.

a. Given $\displaystyle\int \sin\frac{3x}{5}\,dx$ let $u = \frac{3x}{5}$, then $\frac{du}{dx} = \frac{d}{dx}\frac{3x}{5} = \frac{3}{5}$ which implies $dx = \frac{5}{3}\,du$. Therefore,

$\displaystyle\int \sin\frac{3x}{5}\,dx = \int \sin u\cdot\frac{5}{3}\,du = \frac{5}{3}\int \sin u\,du = -\frac{5}{3}\cos u + c = -\frac{5}{3}\cos\frac{3x}{5} + c$

Check: Let $y = -\frac{5}{3}\cos\frac{3x}{5} + c$, then $y' = -\frac{5}{3}\cdot\frac{d}{dx}\cos\frac{3x}{5} + \frac{d}{dx}c = -\frac{5}{3}\cdot -\sin\frac{3x}{5}\cdot\frac{d}{dx}\frac{3x}{5} + 0 = -\frac{5}{3}\cdot -\sin\frac{3x}{5}\cdot\frac{3}{5} = \sin\frac{3x}{5}$

b. Given $\displaystyle\int 3\cos 2x\,dx$ let $u = 2x$, then $\frac{du}{dx} = \frac{d}{dx}2x = 2$ which implies $dx = \frac{du}{2}$. Therefore,

$\displaystyle\int 3\cos 2x\,dx = 3\int \cos u\cdot\frac{du}{2} = \frac{3}{2}\int \cos u\,du = \frac{3}{2}\sin u + c = \mathbf{\frac{3}{2}\sin 2x + c}$

Check: Let $y = \frac{3}{2}\sin 2x + c$, then $y' = \frac{3}{2}\cdot\frac{d}{dx}\sin 2x + \frac{d}{dx}c = \frac{3}{2}\cdot\cos 2x\cdot\frac{d}{dx}2x + 0 = \frac{3}{2}\cdot\cos 2x\cdot 2 = \frac{6}{2}\cdot\cos 2x = 3\cos 2x$

c. Given $\displaystyle\int (\sin 5x - \cos 7x)\,dx = \int \sin 5x\,dx - \int \cos 7x\,dx$ let:

a. $u = 5x$, then $\frac{du}{dx} = \frac{d}{dx}5x$; $\frac{du}{dx} = 5$; $du = 5dx$; $dx = \frac{du}{5}$ and

b. $v = 7x$, then $\dfrac{dv}{dx} = \dfrac{d}{dx} 7x$; $\dfrac{dv}{dx} = 7$; $dv = 7dx$; $dx = \dfrac{dv}{7}$.

Therefore, $\displaystyle\int \sin 5x \, dx - \int \cos 7x \, dx \ = \ \int \sin u \cdot \dfrac{du}{5} - \int \cos v \cdot \dfrac{dv}{7} \ = \ \dfrac{1}{5}\int \sin u \, du - \dfrac{1}{7}\int \cos v \, dv \ = \ -\dfrac{1}{5}\cos u + c_1 - \dfrac{1}{7}\sin v + c_2$

$= \ -\dfrac{1}{5}\cos 5x - \dfrac{1}{7}\sin 7x + c_1 + c_2 \ = \ -\dfrac{1}{5}\mathbf{\cos 5x} - \dfrac{1}{7}\mathbf{\sin 7x} + \mathbf{c}$

Check: Let $y = -\dfrac{1}{5}\cos 5x - \dfrac{1}{7}\sin 7x + c$ then $y' \ = \ -\dfrac{1}{5}\cdot\dfrac{d}{dx}\cos 5x - \dfrac{1}{7}\cdot\dfrac{d}{dx}\sin 7x + \dfrac{d}{dx}c \ = \ \dfrac{1}{5}\cdot\sin 5x \cdot \dfrac{d}{dx}5x$

$-\dfrac{1}{7}\cdot\cos 7x \cdot \dfrac{d}{dx}7x + 0 = \dfrac{5}{5}\cdot\sin 5x - \dfrac{7}{7}\cdot\cos 7x \ = \ \sin 5x - \cos 7x$

d. Given $\displaystyle\int 2\csc^2 3x \, dx$ let $u = 3x$, then $\dfrac{du}{dx} = \dfrac{d}{dx}3x$; $\dfrac{du}{dx} = 3$; $du = 3dx$; $dx = \dfrac{du}{3}$. Therefore,

$\displaystyle\int 2\csc^2 3x \, dx \ = \ 2\int \csc^2 u \cdot \dfrac{du}{3} \ = \ \dfrac{2}{3}\int \csc^2 u \, du \ = \ -\dfrac{2}{3}\cot u + c \ = \ -\dfrac{2}{3}\mathbf{\cot 3x} + \mathbf{c}$

Check: Let $y = -\dfrac{2}{3}\cot 3x + c$, then $y' \ = \ -\dfrac{2}{3}\cdot\dfrac{d}{dx}\cot 3x + \dfrac{d}{dx}c \ = \ -\dfrac{2}{3}\cdot-\csc^2 3x \cdot \dfrac{d}{dx}3x + 0 \ = \ \dfrac{2}{3}\cdot\csc^2 3x \cdot 3 \ = \ 2\csc^2 3x$

e. Given $\displaystyle\int x \sec^2 x^2 \, dx$ let $u = x^2$, then $\dfrac{du}{dx} = \dfrac{d}{dx}x^2$; $\dfrac{du}{dx} = 2x$; $du = 2x \, dx$; $dx = \dfrac{du}{2x}$. Therefore,

$\displaystyle\int x \sec^2 x^2 \, dx \ = \ \int x \sec^2 u \cdot \dfrac{du}{2x} \ = \ \dfrac{1}{2}\int \sec^2 u \, du \ = \ \dfrac{1}{2}\tan u + c \ = \ \dfrac{1}{2}\mathbf{\tan} \, \mathbf{x^2} + \mathbf{c}$

Check: Let $y = \dfrac{1}{2}\tan x^2 + c$, then $y' \ = \ \dfrac{1}{2}\cdot\dfrac{d}{dx}\tan x^2 + \dfrac{d}{dx}c \ = \ \dfrac{1}{2}\cdot\sec^2 x^2 \cdot \dfrac{d}{dx}x^2 + 0 \ = \ \dfrac{1}{2}\cdot\sec^2 x^2 \cdot 2x \ = \ x \sec^2 x^2$

f. Given $\displaystyle\int 8\sec 5x \, dx$ let $u = 5x$, then $\dfrac{du}{dx} = \dfrac{d}{dx}5x$; $\dfrac{du}{dx} = 5$; $du = 5 \, dx$; $dx = \dfrac{du}{5}$. Therefore,

$\displaystyle\int 8\sec 5x \, dx \ = \ 8\int \sec u \cdot \dfrac{du}{5} \ = \ \dfrac{8}{5}\int \sec u \, du \ = \ \dfrac{8}{5}\ln\left| \sec u + \tan u \right| + c \ = \ \dfrac{8}{5}\mathbf{\ln}\left| \mathbf{\sec 5x + \tan 5x} \right| + \mathbf{c}$

Check: Let $y = \dfrac{8}{5}\ln\left| \sec 5x + \tan 5x \right| + c$, then $y' \ = \ \dfrac{8}{5}\cdot\dfrac{1}{\sec 5x + \tan 5x}\cdot 5\sec 5x \tan 5x + 5\sec^2 5x + 0$

$= \ \dfrac{8}{5}\cdot\dfrac{5\sec 5x \left(\sec 5x + \tan 5x \right)}{\sec 5x + \tan 5x} \ = \ 8\sec 5x$

g. Given $\displaystyle\int \sin^3 x \cos x \, dx$ let $u = \sin x$, then $\dfrac{du}{dx} = \dfrac{d}{dx}\sin x$; $\dfrac{du}{dx} = \cos x$; $du = \cos x \cdot dx$; $dx = \dfrac{du}{\cos x}$. Therefore,

$\displaystyle\int \sin^3 x \cos x \, dx \ = \ \int u^3 \cos x \cdot \dfrac{du}{\cos x} \ = \ \int u^3 du \ = \ \dfrac{1}{4}u^4 + c \ = \ \dfrac{1}{4}\mathbf{\sin^4 x} + \mathbf{c}$

Check: Let $y = \dfrac{1}{4}\sin^4 x + c$, then $y' \ = \ \dfrac{1}{4}\cdot 4\sin^3 x \cdot \cos x + 0 \ = \ \sin^3 x \cos x$

h. Given $\displaystyle\int \cos^3 x \sin x \, dx$ let $u = \cos x$, then $\dfrac{du}{dx} = \dfrac{d}{dx}\cos x$; $\dfrac{du}{dx} = -\sin x$; $dx = \dfrac{du}{-\sin x}$. Therefore,

$\displaystyle\int \cos^3 x \sin x \, dx \ = \ \int u^3 \sin x \cdot \dfrac{du}{-\sin x} \ = \ -\int u^3 du \ = \ -\dfrac{1}{4}u^4 + c \ = \ -\dfrac{1}{4}\mathbf{\cos^4 x} + \mathbf{c}$

Check: Let $y = -\dfrac{1}{4}\cos^4 x + c$, then $y' \ = \ -\dfrac{1}{4}\cdot 4\cos^3 x \cdot -\sin x + 0 \ = \ \dfrac{4}{4}\cos^3 x \sin x \ = \ \cos^3 x \sin x$

i. Given $\displaystyle\int \dfrac{x}{2}\cos x^2 \, dx$ let $u = x^2$, then $\dfrac{du}{dx} = \dfrac{d}{dx}x^2$; $\dfrac{du}{dx} = 2x$; $du = 2x \cdot dx$; $dx = \dfrac{du}{2x}$. Therefore,

$\displaystyle\int \dfrac{x}{2}\cos x^2 \, dx \ = \ \int \dfrac{x}{2}\cos u \cdot \dfrac{du}{2x} \ = \ \dfrac{1}{4}\int \cos u \, du \ = \ \dfrac{1}{4}\cdot\sin u + c \ = \ \dfrac{1}{4}\mathbf{\sin} \, \mathbf{x^2} + \mathbf{c}$

Check: Let $y = \dfrac{1}{4}\sin x^2 + c$, then $y' = \dfrac{1}{4}\cdot\cos x^2 \cdot 2x + c = \dfrac{2}{4}x\cos x^2 + 0 = \dfrac{x}{2}\cos x^2$

2. Evaluate the following integrals.

a. Given $\displaystyle\int e^{3x}\sec e^{3x}\,dx$ let $u = e^{3x}$, then $\dfrac{du}{dx} = \dfrac{d}{dx}e^{3x}$; $\dfrac{du}{dx} = 3e^{3x}$; $du = 3e^{3x}\cdot dx$; $dx = \dfrac{du}{3e^{3x}}$. Therefore,

$$\int e^{3x}\sec e^{3x}\,dx = \int e^{3x}\sec u\cdot\frac{du}{3e^{3x}} = \frac{1}{3}\int\sec u\,du = \frac{1}{3}\ln\big|\sec u + \tan u\big| + c = \mathbf{\frac{1}{3}\ln\big|\sec e^{3x} + \tan e^{3x}\big| + c}$$

Check: Let $y = \dfrac{1}{3}\ln\big|\sec e^{3x} + \tan e^{3x}\big| + c$, then $y' = \dfrac{1}{3}\cdot\dfrac{1}{\sec e^{3x} + \tan e^{3x}}\cdot 3e^{3x}\sec e^{3x}\tan e^{3x} + 3e^{3x}\sec^2 e^{3x} + 0$

$$= \frac{1}{3}\cdot\frac{3e^{3x}\sec e^{3x}\left(\sec e^{3x} + \tan e^{3x}\right)}{\sec e^{3x} + \tan e^{3x}} = e^{3x}\sec e^{3x}$$

b. $\displaystyle\int\tan^9 x\sec^2 x\,dx$ let $u = \tan x$, then $\dfrac{du}{dx} = \dfrac{d}{dx}\tan x$; $\dfrac{du}{dx} = \sec^2 x$; $du = \sec^2 x\,dx$; $dx = \dfrac{du}{\sec^2 x}$. Therefore,

$$\int\tan^9 x\sec^2 x\,dx = \int u^9\cdot\sec^2 x\cdot\frac{du}{\sec^2 x} = \int u^9\,du = \frac{1}{9+1}u^{9+1} + c = \frac{1}{10}u^{10} + c = \mathbf{\frac{1}{10}\tan^{10} x + c}$$

Check: Let $y = \dfrac{1}{10}\tan^{10} x + c$, then $y' = \dfrac{1}{10}\cdot 10(\tan x)^{10-1}\cdot\sec^2 x + 0 = (\tan x)^9\sec^2 x = \tan^9 x\sec^2 x$

c. Given $\displaystyle\int\cot^5 x\csc^2 x\,dx$ let $u = \cot x$, then $\dfrac{du}{dx} = \dfrac{d}{dx}\cot x$; $\dfrac{du}{dx} = -\csc^2 x\,c$; $du = -\csc^2 x\,dx$; $dx = -\dfrac{du}{\csc^2 x}$. Thus,

$$\int\cot^5 x\csc^2 x\,dx = \int u^5\cdot\csc^2 x\cdot\frac{-du}{\csc^2 x} = -\int u^5\,du = \frac{-1}{5+1}u^{5+1} + c = -\frac{1}{6}u^6 + c = \mathbf{-\frac{1}{6}\cot^6 x + c}$$

Check: Let $y = -\dfrac{1}{6}\cot^6 x + c$, then $y' = -\dfrac{1}{6}\cdot 6(\cot x)^{6-1}\cdot -\csc^2 x + 0 = \cot^5 x\csc^2 x$

d. Given $\displaystyle\int\sec 2x\tan 2x\,dx$ let $u = 2x$, then $\dfrac{du}{dx} = \dfrac{d}{dx}2x$; $\dfrac{du}{dx} = 2$; $du = 2dx$; $dx = \dfrac{du}{2}$. Therefore,

$$\int\sec 2x\tan 2x\,dx = \int\sec u\cdot\tan u\cdot\frac{du}{2} = \frac{1}{2}\int\sec u\tan u\,du = \frac{1}{2}\sec u + c = \mathbf{\frac{1}{2}\sec 2x + c}$$

Check: Let $y = \dfrac{1}{2}\sec 2x + c$ then $y' = \dfrac{1}{2}\cdot\sec 2x\tan 2x\cdot 2 + 0 = \dfrac{2}{2}\cdot\sec 2x\tan 2x = \sec 2x\tan 2x$

e. Given $\displaystyle\int x\cot x^2\,dx$ let $u = x^2$, then $\dfrac{du}{dx} = \dfrac{d}{dx}x^2$; $\dfrac{du}{dx} = 2x$; $du = 2x\,dx$; $dx = \dfrac{du}{2x}$. Therefore,

$$\int x\cot x^2\,dx = \int x\cdot\cot u\cdot\frac{du}{2x} = \frac{1}{2}\int\cot u\cdot du = \frac{1}{2}\ln\big|\sin u\big| + c = \mathbf{\frac{1}{2}\ln\big|\sin x^2\big| + c}$$

Check: Let $y = \dfrac{1}{2}\ln\big|\sin x^2\big| + c$, then $y' = \dfrac{1}{2}\cdot\dfrac{1}{\sin x^2}\cdot\cos x^2\cdot 2x + 0 = \dfrac{2x}{2}\cdot\dfrac{\cos x^2}{\sin x^2} = x\cot x^2$

f. Given $\displaystyle\int\frac{1}{3}x^2\csc x^3\,dx$ let $u = x^3$, then $\dfrac{du}{dx} = \dfrac{d}{dx}x^3$; $\dfrac{du}{dx} = 3x^2$; $du = 3x^2 dx$; $dx = \dfrac{du}{3x^2}$. Therefore,

$$\int\frac{1}{3}x^2\csc x^3\,dx = \frac{1}{3}\int x^2\cdot\csc u\cdot\frac{du}{3x^2} = \frac{1}{9}\int\csc u\cdot du = \frac{1}{9}\ln\big|\csc u - \cot u\big| + c = \mathbf{\frac{1}{9}\ln\big|\csc x^3 - \cot x^3\big| + c}$$

Check: Let $y = \dfrac{1}{9}\ln\big|\csc x^3 - \cot x^3\big| + c$, then $y' = \dfrac{-\csc x^3\cot x^3\cdot 3x^2 + \csc^2 x^3\cdot 3x^2}{9\left(\csc x^3 - \cot x^3\right)} + 0 = \dfrac{3x^2\csc x^3\left(\csc x^3 - \cot x^3\right)}{9\left(\csc x^3 - \cot x^3\right)}$

$$= \frac{3x^2\csc x^3}{9} = \frac{x^2}{3}\cdot\csc x^3 = \frac{1}{3}x^2\csc x^3$$

g. $\int (\sin 5x \csc 5x)\,dx = \int \sin 5x \cdot \dfrac{1}{\sin 5x}\,dx = \int \dfrac{\sin 5x}{\sin 5x}\,dx = \int dx = x + c$

Check: Let $y = x + c$, then $y' = x^{1-1} + 0 = x^0 = 1$

h. $\int \left(\cos 5t \sec 5t + 3t^2 + t \right)dt = \int \cos 5t \cdot \dfrac{1}{\cos 5t}\,dt + 3\int t^2\,dt + \int t\,dt = \int dt + 3\int t^2\,dt + \int t\,dt = t^3 + \dfrac{1}{2}t^2 + t + c$

Check: Let $y = t^3 + \dfrac{1}{2}t^2 + t + c$, then $y' = 3t^{3-1} + \dfrac{1}{2}\cdot 2t^{2-1} + 1 + 0 = 3t^2 + t + 1$

i. Given $\int e^{\cot x} \csc^2 x\,dx$ let $u = \cot x$, then $\dfrac{du}{dx} = \dfrac{d}{dx}\cot x$; $\dfrac{du}{dx} = -\csc^2 x$; $du = -\csc^2 x\,dx$; $dx = -\dfrac{du}{\csc^2 x}$. Therefore,

$\int e^{\cot x} \csc^2 x\,dx = \int e^u \csc^2 x \cdot \dfrac{-du}{\csc^2 x} = -\int e^u\,du = -e^u + c = -e^{\cot x} + c$

Check: Let $y = -e^{\cot x} + c$, then $y' = -e^{\cot x}\cdot -\csc^2 x \cdot 1 + 0 = e^{\cot x}\csc^2 x$

3. Evaluate the following integrals.

a. $\int \sin 3x \cos 5x\,dx = \int \dfrac{1}{2}\left[\sin(3-5)x + \sin(3+5)x\right]dx = \int \dfrac{1}{2}\left[\sin(-2x) + \sin(8x)\right]dx = \dfrac{1}{2}\int (\sin 8x - \sin 2x)\,dx$

$= \dfrac{1}{2}\int \sin 8x\,dx - \dfrac{1}{2}\int \sin 2x\,dx = \dfrac{1}{2}\cdot -\dfrac{1}{8}\cos 8x + \dfrac{1}{2}\cdot\dfrac{1}{2}\cos 2x + c = -\dfrac{1}{16}\cos 8x + \dfrac{1}{4}\cos 2x + c$

Check: Let $y = -\dfrac{1}{16}\cos 8x + \dfrac{1}{4}\cos 2x + c$, then $y' = -\dfrac{1}{16}\cdot -8\sin 8x + \dfrac{1}{4}\cdot -2\sin 2x + 0 = \dfrac{8}{16}\sin 8x - \dfrac{2}{4}\sin 2x$

$= \dfrac{1}{2}\sin 8x - \dfrac{1}{2}\sin 2x = \dfrac{1}{2}\sin 8x + \dfrac{1}{2}\sin(-2x) = \dfrac{1}{2}\sin(3+5)x + \dfrac{1}{2}\sin(3-5)x = \sin 3x \cos 5x$

b. $\int \cos 6x \cos 4x\,dx = \int \dfrac{1}{2}\left[\cos(6-4)x + \cos(6+4)x\right]dx = \int \dfrac{1}{2}(\cos 2x + \cos 10x)\,dx = \dfrac{1}{2}\int \cos 2x\,dx + \dfrac{1}{2}\int \cos 10x\,dx$

$= \dfrac{1}{2}\cdot\dfrac{1}{2}\cdot\sin 2x + \dfrac{1}{2}\cdot\dfrac{1}{10}\sin 10x + c = \dfrac{1}{4}\sin 2x + \dfrac{1}{20}\sin 10x + c$

Check: Let $y = \dfrac{1}{4}\sin 2x + \dfrac{1}{20}\sin 10x + c$, then $y' = \dfrac{1}{4}\cdot 2\cos 2x + \dfrac{1}{20}\cdot 10\cos 10x + 0 = \dfrac{1}{2}\cos 2x + \dfrac{1}{2}\cos 10x$

$= \dfrac{1}{2}\cos 2x + \dfrac{1}{2}\cos 10x = \dfrac{1}{2}\cos(6-4)x + \dfrac{1}{2}\cos(6+4)x = \cos 6x \cos 4x$

c. $\int \sin^5 3x\,dx = \int \sin^4 3x \sin 3x\,dx = \int \left(\sin^2 3x\right)^2 \sin 3x\,dx = \int \left(1 - \cos^2 3x\right)^2 \sin 3x\,dx$. Let $u = \cos 3x$, then

$\dfrac{du}{dx} = \dfrac{d}{dx}\cos 3x$; $\dfrac{du}{dx} = -3\sin 3x$; $du = -3\sin 3x\,dx$; $dx = -\dfrac{du}{3\sin 3x}$. Therefore,

$\int \sin^5 3x\,dx = \int \left(1 - \cos^2 3x\right)^2 \sin 3x\,dx = \int \left(1 - u^2\right)^2 \sin 3x\,dx = \int \left(u^4 - 2u^2 + 1\right)\sin 3x\cdot\dfrac{du}{-3\sin 3x} = -\dfrac{1}{3}\int\left(u^4 - 2u^2 + 1\right)du$

$= -\dfrac{1}{3}\int u^4\,du + \dfrac{2}{3}\int u^2\,du - \dfrac{1}{3}\int du = -\dfrac{1}{3}\cdot\dfrac{1}{5}u^5 + \dfrac{2}{3}\cdot\dfrac{1}{3}u^3 - \dfrac{1}{3}u + c = -\dfrac{1}{15}\cos^5 3x + \dfrac{2}{9}\cos^3 3x - \dfrac{1}{3}\cos 3x + c$

Check: Let $y = -\dfrac{1}{15}\cos^5 3x + \dfrac{2}{9}\cos^3 3x - \dfrac{1}{3}\cos 3x + c$, then $y' = -\dfrac{1}{15}\cdot 15\cos^4 3x\cdot -\sin 3x + \dfrac{2}{9}\cdot 9\cos^2 3x\cdot -\sin 3x + \dfrac{1}{3}\cdot 3\sin 3x$

$= \sin 3x \cos^4 3x - 2\sin 3x \cos^2 3x + \sin 3x = \sin 3x\left(\cos^4 3x - 2\cos^2 3x + 1\right) = \sin 3x\left(1 - \cos^2 3x\right)^2$

$= \sin 3x\left(\sin^2 3x\right)^2 = \sin 3x \sin^4 3x = \sin^5 3x$

d. $\int \tan^4 2x\,dx = \int \tan^2 2x \tan^2 2x\,dx = \int \tan^2 2x\left(\sec^2 2x - 1\right)dx = \int \tan^2 2x \sec^2 2x\,dx - \int \tan^2 2x\,dx$

$= \int \tan^2 2x \sec^2 2x \, dx - \int \left(\sec^2 2x - 1 \right) dx \; = \; \int \tan^2 2x \sec^2 2x \, dx - \int \sec^2 2x \, dx + \int dx$. To solve the first integral let

$u = \tan 2x$, then $\dfrac{du}{dx} = \dfrac{d}{dx} \tan 2x$; $\dfrac{du}{dx} = 2 \sec^2 2x$; $du = 2 \sec^2 2x \, dx$; $dx = \dfrac{du}{2 \sec^2 2x}$. Therefore,

$\int \tan^2 2x \sec^2 2x \, dx \; = \; \int u^2 \sec^2 2x \cdot \dfrac{du}{2 \sec^2 2x} \; = \; \dfrac{1}{2} \int u^2 du \; = \; \dfrac{1}{2} \cdot \dfrac{1}{3} u^3 \; = \; \dfrac{1}{6} u^3 \; = \; \dfrac{1}{6} \tan^3 2x$. Grouping the terms we obtain

$\int \tan^4 2x \, dx \; = \; \int \tan^2 2x \sec^2 2x \, dx - \int \sec^2 2x \, dx + \int dx \; = \; \dfrac{1}{6} \tan^3 2x - \int \sec^2 2x \, dx + \int dx \; = \; \boldsymbol{\dfrac{1}{6} \tan^3 2x - \dfrac{1}{2} \tan 2x + x + c}$

Check: Let $y = \dfrac{1}{6} \tan^3 2x - \dfrac{1}{2} \tan 2x + x + c$, then $y' = \dfrac{6}{6} \tan^2 2x \cdot \sec^2 2x - \dfrac{2}{2} \sec^2 2x + 1 + 0 \; = \; \tan^2 2x \cdot \sec^2 2x - \sec^2 2x + 1$

$= \; \sec^2 2x \left(\tan^2 2x - 1 \right) + 1 \; = \; \left(\tan^2 2x + 1 \right) \left(\tan^2 2x - 1 \right) + 1 \; = \; \tan^4 2x - \tan^2 2x + \tan^2 2x - 1 + 1 \; = \; \tan^4 2x$

e. $\int \cos^2 ax \, dx \; = \; \int \left(1 - \sin^2 ax \right) dx \; = \; \int dx - \int \sin^2 ax \, dx \; = \; \int dx - \int \dfrac{1}{2} \left(1 - \cos 2ax \right) dx \; = \; x - \dfrac{1}{2} \int dx + \dfrac{1}{2} \int \cos 2ax \, dx$

$= \; x - \dfrac{x}{2} + \dfrac{1}{2} \cdot \dfrac{1}{2a} \sin 2ax + c \; = \; x \left(1 - \dfrac{1}{2} \right) + \dfrac{1}{4a} \sin 2ax + c \; = \; \boldsymbol{\dfrac{x}{2} + \dfrac{\sin 2ax}{4a} + c}$

Check: Let $y = \dfrac{x}{2} + \dfrac{\sin 2ax}{4a} + c$, then $y' = \dfrac{1}{2} + \dfrac{1}{4a} \cdot \cos 2ax \cdot 2a + 0 \; = \; \dfrac{1}{2} + \dfrac{2a}{4a} \cdot \cos 2ax \; = \; \dfrac{1}{2} + \dfrac{1}{2} \cos 2ax$

$= \; \left(1 - \dfrac{1}{2} \right) + \dfrac{1}{2} \cos 2ax \; = \; 1 - \left(\dfrac{1}{2} - \dfrac{1}{2} \cos 2ax \right) \; = \; 1 - \dfrac{1}{2} \left(1 - \cos 2ax \right) \; = \; 1 - \sin^2 ax \; = \; \cos^2 ax$

f. $\int \tan^3 5x \, dx \; = \; \int \tan^2 5x \tan 5x \, dx \; = \; \int \left(\sec^2 5x - 1 \right) \tan 5x \, dx \; = \; \int \sec^2 5x \tan 5x \, dx - \int \tan 5x \, dx$. To solve the first

integral let $u = \tan 5x$, then $\dfrac{du}{dx} = \dfrac{d}{dx} \tan 5x$; $\dfrac{du}{dx} = 5 \sec^2 5x$; $du = 5 \sec^2 5x \, dx$; $dx = \dfrac{du}{5 \sec^2 5x}$. Therefore,

$\int \sec^2 5x \tan 5x \, dx \; = \; \int \sec^2 5x \cdot u \cdot \dfrac{du}{5 \sec^2 5x} \; = \; \dfrac{1}{5} \int u \, du \; = \; \dfrac{1}{5} \cdot \dfrac{1}{2} u^2 \; = \; \dfrac{1}{10} u^2 \; = \; \dfrac{1}{10} \tan^2 5x$. Combining the term we obtain

$\int \tan^3 5x \, dx \; = \; \int \tan^2 5x \tan 5x \, dx \; = \; \int \left(\sec^2 5x - 1 \right) \tan 5x \, dx \; = \; \int \left(\sec^2 5x \tan 5x - \tan 5x \right) dx \; = \; \int \sec^2 5x \tan 5x \, dx$

$- \int \tan 5x \, dx \; = \; \dfrac{1}{10} \tan^2 5x - \int \tan 5x \, dx \; = \; \boldsymbol{\dfrac{1}{10} \tan^2 5x - \dfrac{1}{5} \ln \left| \sec 5x \right| + c}$

Check: Let $y = \dfrac{1}{10} \tan^2 5x - \dfrac{1}{5} \ln \left| \sec 5x \right| + c$, then $y' = \dfrac{1}{10} \cdot 10 \tan 5x \cdot \sec^2 5x - \dfrac{1}{5} \cdot \dfrac{1}{\sec 5x} \cdot 5 \sec 5x \tan 5x + 0$

$= \; \tan 5x \sec^2 5x - \dfrac{\sec 5x \tan 5x}{\sec 5x} \; = \; \tan 5x \sec^2 5x - \tan 5x \; = \; \tan 5x \left(\sec^2 5x - 1 \right) \; = \; \tan 5x \tan^2 5x \; = \; \tan^3 5x$

g. $\int \cot^4 3x \, dx \; = \; \int \cot^2 3x \cot^2 3x \, dx \; = \; \int \cot^2 3x \left(\csc^2 3x - 1 \right) dx \; = \; \int \cot^2 3x \csc^2 3x \, dx - \int \cot^2 3x \, dx$

$= \; \int \cot^2 3x \csc^2 3x \, dx - \int \left(\csc^2 3x - 1 \right) dx \; = \; \int \cot^2 3x \csc^2 3x \, dx - \int \csc^2 3x \, dx + \int dx$. To solve the first integral let

$u = \cot 3x$, then $\dfrac{du}{dx} = \dfrac{d}{dx} \cot 3x$; $\dfrac{du}{dx} = -3 \csc^2 3x$; $du = -3 \csc^2 3x \, dx$; $dx = - \dfrac{du}{3 \csc^2 3x}$. Therefore,

$\int \cot^2 3x \csc^2 3x \, dx \; = \; \int u^2 \csc^2 3x \cdot - \dfrac{du}{3 \csc^2 3x} \; = \; -\dfrac{1}{3} \int u^2 du \; = \; -\dfrac{1}{9} u^3 \; = \; -\dfrac{1}{9} \cot^3 3x$. Grouping the terms we obtain

$\int \cot^4 3x \, dx \; = \; \int \cot^2 3x \csc^2 3x \, dx - \int \csc^2 3x \, dx + \int dx \; = \; -\dfrac{1}{9} \cot^3 3x - \int \csc^2 3x \, dx + \int dx \; = \; \boldsymbol{-\dfrac{1}{9} \cot^3 3x + \dfrac{1}{3} \cot 3x + x + c}$

Check: Let $y = -\dfrac{1}{9} \cot^3 3x + \dfrac{1}{3} \cot 3x + x + c$, then $y' = -\dfrac{9}{9} \cot^2 3x \cdot -\csc^2 3x - \dfrac{3}{3} \csc^2 3x + 1 \; = \; \cot^2 3x \csc^2 3x - \csc^2 3x + 1$

$= \; \csc^2 3x \left(\cot^2 3x - 1 \right) + 1 \; = \; \left(\cot^2 3x + 1 \right) \left(\cot^2 3x - 1 \right) + 1 \; = \; \cot^4 3x - \cot^2 3x + \cot^2 3x - 1 + 1 \; = \; \cot^4 3x$

h. $\int \cot^3 2x\, dx = \int \cot^2 2x \cot 2x\, dx = \int \left(\csc^2 2x - 1 \right) \cot 2x\, dx = \int \csc^2 2x \cot 2x\, dx - \int \cot 2x\, dx$. To solve the first

integral let $u = \cot 2x$, then $\dfrac{du}{dx} = \dfrac{d}{dx} \cot 2x$; $\dfrac{du}{dx} = -2\csc^2 2x$; $du = -2\csc^2 2x\, dx$; $dx = -\dfrac{du}{2\csc^2 2x}$. Therefore,

$\int \csc^2 2x \cot 2x\, dx = \int \csc^2 2x \cdot u \cdot - \dfrac{du}{2\csc^2 2x} = -\dfrac{1}{2} \int u\, du = -\dfrac{1}{4} u^2 = -\dfrac{1}{4} \cot^2 2x$. Combining the term we obtain

$\int \cot^3 2x\, dx = \int \cot^2 2x \cot 2x\, dx = \int \left(\csc^2 2x - 1 \right) \cot 2x\, dx = \int \left(\csc^2 2x \cdot \cot 2x - \cot 2x \right) dx = \int \csc^2 2x \cot 2x\, dx$

$-\int \cot 2x\, dx = -\dfrac{1}{4} \cot^2 2x - \int \cot 2x\, dx = -\dfrac{1}{4} \cot^2 2x - \dfrac{1}{2} \ln \left| \sin 2x \right| + c$

Check: Let $y = -\dfrac{1}{4} \cot^2 2x - \dfrac{1}{2} \ln \left| \sin 2x \right| + c$, then $y' = -\dfrac{1}{4} \cdot 4\cot 2x \cdot -\csc^2 2x - \dfrac{1}{2} \cdot \dfrac{1}{\sin 2x} \cdot 2\cos 2x + 0$

$= \cot 2x \csc^2 2x - \dfrac{\cos 2x}{\sin 2x} = \cot 2x \csc^2 2x - \cot 2x = \cot 2x \left(\csc^2 2x - 1 \right) = \cot 2x \cot^2 2x = \cot^3 2x$

i. $\int \cot^6 3x\, dx = \int \cot^4 3x \cot^2 3x\, dx = \int \cot^4 3x \left(\csc^2 3x - 1 \right) dx = \int \left(\cot^4 3x \csc^2 3x - \cot^4 3x \right) dx = \int \cot^4 3x \csc^2 3x\, dx$

$-\int \cot^4 3x\, dx$. From the problem g above we know that $\int \cot^4 3x\, dx = -\dfrac{1}{9} \cot^3 3x + \dfrac{1}{3} \cot 3x + x + c$. Therefore,

$\int \cot^6 3x\, dx = \int \cot^4 3x \cot^2 3x\, dx = \int \cot^4 3x \left(\csc^2 3x - 1 \right) dx = \int \cot^4 3x \csc^2 3x\, dx - \int \cot^4 3x\, dx$

$= \int \cot^4 3x \csc^2 3x\, dx - \left(-\dfrac{1}{9} \cot^3 3x + \dfrac{1}{3} \cot 3x + x + c \right) = -\dfrac{1}{15} \cot^5 3x + \dfrac{1}{9} \cot^3 3x - \dfrac{1}{3} \cot 3x - x + c$

Check: Let $y = -\dfrac{1}{15} \cot^5 3x + \dfrac{1}{9} \cot^3 3x - \dfrac{1}{3} \cot 3x - x + c$, then $y' = -\dfrac{15 \cdot \cot^4 3x \cdot -\csc^2 3x}{15} + \dfrac{9 \cdot \cot^2 3x \cdot -\csc^2 3x}{9}$

$+ \dfrac{3\csc^2 3x}{3} - 1 = \cot^4 3x \cdot \csc^2 3x - \cot^2 3x \cdot \csc^2 3x + \csc^2 3x - 1 = \csc^2 3x \left(\cot^4 3x - \cot^2 3x + 1 \right) - 1$

$= \left(1 + \cot^2 3x \right) \left(\cot^4 3x - \cot^2 3x + 1 \right) - 1 = \cot^4 3x - \cot^2 3x + 1 + \cot^6 3x - \cot^4 3x + \cot^2 3x - 1 = \cot^6 3x$

Section 4.4 Solutions – Integration of Expressions Resulting in Inverse Trigonometric Functions

1. Evaluate the following indefinite integrals.

a. **First** - Write the given integral in its standard form $\int \dfrac{dx}{\sqrt{a^2 - x^2}} = arc\, \sin \dfrac{x}{a} + c$, i.e., $\int \dfrac{dx}{\sqrt{25 - 9x^2}} = \int \dfrac{dx}{\sqrt{5^2 - (3x)^2}}$

Second - Use substitution method by letting $u = 3x$, then $\dfrac{du}{dx} = \dfrac{d}{dx} 3x = 3x$ which implies $du = 3x\, dx$; $dx = \dfrac{du}{3}$.

Therefore, $\int \dfrac{dx}{\sqrt{25 - 9x^2}} = \int \dfrac{1}{\sqrt{5^2 - u^2}} \cdot \dfrac{du}{3} = \dfrac{1}{3} \int \dfrac{du}{\sqrt{5^2 - u^2}} = \dfrac{1}{3} arc\, \sin \dfrac{u}{5}$

Third - Write the answer in terms of the original variable, i.e., x. $\dfrac{1}{3} arc\, \sin \dfrac{u}{5} + c = \dfrac{1}{3} arc\, \sin \dfrac{3x}{5} + c$

Fourth - Check the answer by differentiating the solution, i.e., let $y = \dfrac{1}{3} arc\, \sin \dfrac{3x}{5} + c$ then

$y' = \dfrac{1}{3} \dfrac{1}{\sqrt{1 - \left(\frac{3x}{5} \right)^2}} \cdot \dfrac{d}{dx} \dfrac{3x}{5} + 0 = \dfrac{1}{3} \dfrac{1}{\sqrt{1 - \frac{9x^2}{25}}} \cdot \dfrac{3}{5} = \dfrac{3}{15} \dfrac{1}{\sqrt{\frac{25 - 9x^2}{25}}} = \dfrac{1}{5} \dfrac{5}{\sqrt{25 - 9x^2}} = \dfrac{1}{\sqrt{25 - 9x^2}}$

b. **First** - Write the given integral in its standard form $\int \dfrac{dx}{\sqrt{a^2 - x^2}} = arc\, \sin \dfrac{x}{a} + c$, i.e., $\int \dfrac{dx}{\sqrt{4 - x^2}} = \int \dfrac{dx}{\sqrt{2^2 - x^2}}$

Second - Use substitution method by letting $u = x$, then $\dfrac{du}{dx} = \dfrac{d}{dx} x = 1$ which implies $du = dx$. Therefore,

$$\int \frac{dx}{\sqrt{4-x^2}} = \int \frac{du}{\sqrt{2^2 - u^2}} = arc \sin \frac{u}{2} + c$$

Third - Write the answer in terms of the original variable, i.e., x. $arc \sin \dfrac{u}{2} + c = \boldsymbol{arc \sin \dfrac{x}{2} + c}$

Fourth - Check the answer by differentiating the solution, i.e., let $y = arc \sin \dfrac{x}{2} + c$ then

$$y' = \frac{1}{\sqrt{1-\left(\frac{x}{2}\right)^2}} \cdot \frac{d}{dx} \frac{x}{2} + 0 = \frac{1}{\sqrt{1-\frac{x^2}{4}}} \cdot \frac{1}{2} = \frac{1}{2} \frac{1}{\sqrt{\frac{4-x^2}{4}}} = \frac{1}{2} \frac{2}{\sqrt{4-x^2}} = \frac{1}{\sqrt{4-x^2}}$$

c. **First** - Write the given integral in its standard form $\int \dfrac{dx}{\sqrt{a^2 - x^2}} = arc \sin \dfrac{x}{a} + c$, i.e., $\int \dfrac{x^2 dx}{\sqrt{25-x^6}} = \int \dfrac{x^2 dx}{\sqrt{5^2 - \left(x^3\right)^2}}$

Second - Use substitution method by letting $u = x^3$, then $\dfrac{du}{dx} = \dfrac{d}{dx} x^3 = 3x^2$ which implies $du = 3x^2 dx$; $dx = \dfrac{du}{3x^2}$. Thus,

$$\int \frac{x^2 dx}{\sqrt{25-x^6}} = \int \frac{x^2 dx}{\sqrt{5^2 - \left(x^3\right)^2}} = \int \frac{x^2}{\sqrt{5^2 - u^2}} \cdot \frac{du}{3x^2} = \frac{1}{3} \int \frac{du}{\sqrt{5^2 - u^2}} = \frac{1}{3} arc \sin \frac{u}{5} + c$$

Third - Write the answer in terms of the original variable, i.e., x. $\dfrac{1}{3} arc \sin \dfrac{u}{5} + c = \boldsymbol{\dfrac{1}{3} arc \sin \dfrac{x^3}{5} + c}$

Fourth - Check the answer by differentiating the solution, i.e., let $y = \dfrac{1}{3} arc \sin \dfrac{x^3}{5} + c$ then

$$y' = \frac{1}{3} \frac{1}{\sqrt{1-\left(\frac{x^3}{5}\right)^2}} \cdot \frac{d}{dx} \frac{x^3}{5} + 0 = \frac{1}{3} \frac{1}{\sqrt{1-\frac{x^6}{25}}} \cdot \frac{3x^2}{5} = \frac{1}{5} \frac{x^2}{\sqrt{\frac{25-x^6}{25}}} = \frac{1}{5} \frac{5x^2}{\sqrt{25-x^6}} = \frac{x^2}{\sqrt{25-x^6}}$$

d. **First** - Write the given integral in its standard form $\int \dfrac{dx}{a^2 + x^2} = \dfrac{1}{a} arc \tan \dfrac{x}{a} + c$, i.e., $\int \dfrac{dx}{9x^2 + 16} = \int \dfrac{dx}{16 + 9x^2}$

$$= \int \frac{dx}{9\left(\frac{16}{9} + x^2\right)} = \frac{1}{9} \int \frac{dx}{\frac{16}{9} + x^2} = \frac{1}{9} \int \frac{dx}{\left(\frac{4}{3}\right)^2 + x^2}$$

Second - Use substitution method by letting $u = x$, then $\dfrac{du}{dx} = \dfrac{d}{dx} x = 1$ which implies $du = dx$. Therefore,

$$\frac{1}{9} \int \frac{dx}{\left(\frac{4}{3}\right)^2 + x^2} = \frac{1}{9} \int \frac{dx}{\left(\frac{4}{3}\right)^2 + u^2} = \frac{1}{9} \cdot \frac{1}{\frac{4}{3}} arc \tan \frac{u}{\frac{4}{3}} + c = \frac{1}{9} \cdot \frac{3}{4} arc \tan \frac{3u}{4} + c = \frac{1}{12} arc \tan \frac{3u}{4} + c$$

Third - Write the answer in terms of the original variable, i.e., x. $\dfrac{1}{12} arc \tan \dfrac{3u}{4} + c = \boldsymbol{\dfrac{1}{12} arc \tan \dfrac{3x}{4} + c}$

Fourth - Check the answer by differentiating the solution, i.e., let $y = \dfrac{1}{12} arc \tan \dfrac{3x}{4} + c$ then

$$y' = \frac{1}{12} \cdot \frac{1}{1+\left(\frac{3x}{4}\right)^2} \cdot \frac{d}{dx} \frac{3x}{4} + 0 = \frac{1}{12} \cdot \frac{1}{1+\frac{9x^2}{16}} \cdot \frac{3}{4} = \frac{3}{48} \frac{1}{\frac{16+9x^2}{16}} = \frac{1}{16} \cdot \frac{16}{16+9x^2} = \frac{1}{16+9x^2}$$

Note that another way of solving this class of problems is by rewriting the integral in the following way:

$$\int \frac{dx}{9x^2 + 16} = \int \frac{dx}{16 + 9x^2} = \int \frac{dx}{4^2 + \left(3x\right)^2}$$. Now, let $u = 3x$, then $\dfrac{du}{dx} = \dfrac{d}{dx} 3x = 3$ which implies $du = 3dx$; $dx = \dfrac{du}{3}$.

Thus, $\displaystyle \int \frac{dx}{4^2 + \left(3x\right)^2} = \int \frac{1}{4^2 + u^2} \cdot \frac{du}{3} = \frac{1}{3} \int \frac{du}{4^2 + u^2} = \frac{1}{3} \cdot \frac{1}{4} arc \tan \frac{u}{4} + c = \frac{1}{12} arc \tan \frac{u}{4} + c = \boldsymbol{\frac{1}{12} arc \tan \frac{3x}{4} + c}$

e. **First** - Write the given integral in its standard form $\int \dfrac{dx}{a^2+x^2} = \dfrac{1}{a} \, arc \, \tan \dfrac{x}{a} + c$, i.e., $\int \dfrac{x^2 dx}{7+9x^6} = \int \dfrac{x^2 dx}{9\left(\frac{7}{9}+x^6\right)}$

$= \dfrac{1}{9}\int \dfrac{x^2 dx}{\frac{7}{9}+x^6} = \dfrac{1}{9}\int \dfrac{x^2 dx}{\left(\frac{\sqrt{7}}{3}\right)^2+\left(x^3\right)^2}$

Second - Use substitution method by letting $u = x^3$, then $\dfrac{du}{dx} = \dfrac{d}{dx}x^3 = 3x^2$ which implies $du = 3x^2 dx$; $dx = \dfrac{du}{3x^2}$.

Therefore, $\dfrac{1}{9}\int \dfrac{x^2 dx}{\left(\frac{\sqrt{7}}{3}\right)^2+\left(x^3\right)^2} = \dfrac{1}{9}\int \dfrac{x^2}{\left(\frac{\sqrt{7}}{3}\right)^2+u^2} \cdot \dfrac{du}{3x^2} = \dfrac{1}{27}\int \dfrac{du}{\left(\frac{\sqrt{7}}{3}\right)^2+u^2} = \dfrac{1}{27}\cdot\dfrac{1}{\frac{\sqrt{7}}{3}} arc \, \tan \dfrac{u}{\frac{\sqrt{7}}{3}} + c$

$= \dfrac{1}{27}\cdot\dfrac{3}{\sqrt{7}} arc \, \tan \dfrac{3u}{\sqrt{7}} + c = \dfrac{1}{9\sqrt{7}} arc \, \tan \dfrac{3u}{\sqrt{7}} + c$

Third - Write the answer in terms of the x variable, i.e., $\dfrac{1}{9\sqrt{7}} arc \, \tan \dfrac{3u}{\sqrt{7}} + c = \dfrac{1}{9\sqrt{7}} arc \, \tan \dfrac{3x^3}{\sqrt{7}} + c$

Fourth - Check the answer by differentiating the solution, i.e., let $y = \dfrac{1}{9\sqrt{7}} arc \, \tan \dfrac{3x^3}{\sqrt{7}} + c$ then

$y' = \dfrac{1}{9\sqrt{7}}\cdot\dfrac{1}{1+\left(\frac{3x^3}{\sqrt{7}}\right)^2}\cdot\dfrac{d}{dx}\dfrac{3x^3}{\sqrt{7}}+0 = \dfrac{1}{9\sqrt{7}}\cdot\dfrac{1}{1+\frac{9x^6}{7}}\cdot\dfrac{9x^2}{\sqrt{7}} = \dfrac{1}{7}\cdot\dfrac{x^2}{\frac{7+9x^6}{7}} = \dfrac{1}{7}\cdot\dfrac{7x^2}{7+9x^6} = \dfrac{x^2}{7+9x^6}$

or, the alternative approach would be to rearrange the integral in the following way:

$\int \dfrac{x^2 dx}{7+9x^6} = \int \dfrac{x^2 dx}{\left(\sqrt{7}\right)^2+\left(3x^3\right)^2}$. Now, let $u = 3x^3$, then $\dfrac{du}{dx} = \dfrac{d}{dx}3x^3 = 9x^2$ which implies $du = 9x^2 dx$; $dx = \dfrac{du}{9x^2}$.

Therefore, $\int \dfrac{x^2 dx}{\left(\sqrt{7}\right)^2+\left(3x^3\right)^2} = \int \dfrac{x^2}{\left(\sqrt{7}\right)^2+u^2} \cdot \dfrac{du}{9x^2} = \dfrac{1}{9}\int \dfrac{du}{\left(\sqrt{7}\right)^2+u^2} = \dfrac{1}{9}\cdot\dfrac{1}{\sqrt{7}} arc \, \tan \dfrac{u}{\sqrt{7}} + c = \dfrac{1}{9\sqrt{7}} arc \, \tan \dfrac{3x^3}{\sqrt{7}} + c$

f. **First** - Write the given integral in its standard form $\int \dfrac{dx}{x\sqrt{x^2-a^2}} = \dfrac{1}{a} \, arc \, \sec \dfrac{x}{a} + c$, i.e., $\int \dfrac{dx}{x\sqrt{x^4-25}} = \int \dfrac{dx}{x\sqrt{\left(x^2\right)^2-5^2}}$

Second - Use substitution method by letting $u = x^2$, then $\dfrac{du}{dx} = \dfrac{d}{dx}x^2 = 2x$ which implies $du = 2x \, dx$; $dx = \dfrac{du}{2x}$.

Therefore, $\int \dfrac{dx}{x\sqrt{\left(x^2\right)^2-5^2}} = \int \dfrac{1}{x\sqrt{u^2-5^2}} \cdot \dfrac{du}{2x} = \dfrac{1}{2}\int \dfrac{du}{x^2\sqrt{u^2-5^2}} = \dfrac{1}{2}\int \dfrac{du}{u\sqrt{u^2-5^2}} = \dfrac{1}{10} arc \, \sec \dfrac{u}{5} + c$

Third - Write the answer in terms of the original variable, i.e., x . $\dfrac{1}{10} arc \, \sec \dfrac{u}{5} + c = \dfrac{1}{10} arc \, \sec \dfrac{x^2}{5} + c$

Fourth - Check the answer by differentiating the solution, i.e., let $y = \dfrac{1}{10} arc \, \sec \dfrac{x^2}{5} + c$ then

$y' = \dfrac{1}{10}\cdot\dfrac{1}{\frac{x^2}{5}\sqrt{\left(\frac{x^2}{5}\right)^2-1}}\cdot\dfrac{d}{dx}\dfrac{x^2}{5}+0 = \dfrac{1}{10}\cdot\dfrac{1}{\frac{x^2}{5}\sqrt{\frac{x^4}{25}-1}}\cdot\dfrac{2x}{5} = \dfrac{2}{50}\cdot\dfrac{x}{\frac{x^2}{5}\sqrt{\frac{x^4-25}{25}}} = \dfrac{1}{25}\cdot\dfrac{25x}{x^2\sqrt{x^4-25}} = \dfrac{1}{x\sqrt{x^4-25}}$

2. Evaluate the following indefinite integrals.

a. **First** - Write the given integral in its standard form $\int \dfrac{dx}{x\sqrt{x^2-a^2}} = \dfrac{1}{a} \, arc \, \sec \dfrac{x}{a} + c$, i.e., $\int \dfrac{dx}{x\sqrt{x^2-16}} = \int \dfrac{dx}{x\sqrt{x^2-4^2}}$

Second - Use substitution method by letting $u = x$, then $\dfrac{du}{dx} = \dfrac{d}{dx}x = 1$ which implies $du = dx$. Therefore,

$$\int \frac{dx}{x\sqrt{x^2-4^2}} = \int \frac{du}{u\sqrt{u^2-4^2}} = \frac{1}{4} arc\sec\frac{u}{4} + c$$

Third - Write the answer in terms of the original variable, i.e., x . $\dfrac{1}{4} arc\sec\dfrac{u}{4} + c = \dfrac{1}{4} \boldsymbol{arc\sec\dfrac{x}{4}} + \boldsymbol{c}$

Fourth - Check the answer by differentiating the solution, i.e., let $y = \dfrac{1}{4} arc\sec\dfrac{x}{4} + c$ then

$$y' = \frac{1}{4}\cdot\frac{1}{\frac{x}{4}\sqrt{\left(\frac{x}{4}\right)^2-1}}\cdot\frac{d}{dx}\frac{x}{4}+0 = \frac{1}{4}\cdot\frac{1}{\frac{x}{4}\sqrt{\frac{x^2}{16}-1}}\cdot\frac{1}{4} = \frac{1}{16}\frac{4}{x\sqrt{\frac{x^2-16}{16}}} = \frac{1}{16}\frac{16}{x\sqrt{x^2-16}} = \frac{1}{x\sqrt{x^2-16}}$$

b. **First** - Write the given integral in its standard form $\int \dfrac{dx}{x\sqrt{x^2-a^2}} = \dfrac{1}{a} arc\sec\dfrac{x}{a}+c$, i.e., $\int \dfrac{dx}{x\sqrt{7x^2-4}} = \int \dfrac{dx}{\sqrt{7}x\sqrt{x^2-\frac{4}{7}}}$

$$= \int \frac{dx}{\sqrt{7}x\sqrt{x^2-\left(\frac{2}{\sqrt{7}}\right)^2}} = \frac{1}{\sqrt{7}}\int \frac{dx}{x\sqrt{x^2-\left(\frac{2}{\sqrt{7}}\right)^2}}$$

Second - Use substitution method by letting $u = x$, then $\dfrac{du}{dx} = \dfrac{d}{dx}x = 1$ which implies $du = dx$. Therefore,

$$\frac{1}{\sqrt{7}}\int \frac{dx}{x\sqrt{x^2-\left(\frac{2}{\sqrt{7}}\right)^2}} = \frac{1}{\sqrt{7}}\int \frac{du}{u\sqrt{u^2-\left(\frac{2}{\sqrt{7}}\right)^2}} = \frac{1}{\sqrt{7}}\cdot\frac{1}{\frac{2}{\sqrt{7}}} arc\sec\frac{u}{\frac{2}{\sqrt{7}}}+c = \frac{1}{2} arc\sec\frac{\sqrt{7}u}{2}+c$$

Third - Write the answer in terms of the original variable, i.e., x . $\dfrac{1}{2} arc\sec\dfrac{\sqrt{7}u}{2}+c = \dfrac{1}{2}\boldsymbol{arc\sec\dfrac{\sqrt{7}x}{2}}+\boldsymbol{c}$

Fourth - Check the answer by differentiating the solution, i.e., let $y = \dfrac{1}{2} arc\sec\dfrac{\sqrt{7}x}{2}+c$ then y'

$$= \frac{1}{2}\frac{1}{\frac{\sqrt{7}x}{2}\sqrt{\left(\frac{\sqrt{7}x}{2}\right)^2-1}}\cdot\frac{d}{dx}\frac{\sqrt{7}x}{2}+0 = \frac{1}{2}\cdot\frac{1}{\frac{\sqrt{7}x}{2}\sqrt{\frac{7x^2}{4}-1}}\cdot\frac{\sqrt{7}}{2} = \frac{\sqrt{7}}{4}\cdot\frac{1}{\frac{\sqrt{7}x}{2}\sqrt{\frac{7x^2-4}{4}}} = \frac{\sqrt{7}}{4}\cdot\frac{4}{\sqrt{7}x\sqrt{7x^2-4}} = \frac{1}{x\sqrt{7x^2-4}}$$

or, the alternative approach would be to rearrange the integral in the following way:

$$\int \frac{dx}{x\sqrt{7x^2-4}} = \int \frac{dx}{x\sqrt{\left(\sqrt{7}x\right)^2-2^2}}$$. Now, let $u = \sqrt{7}x$, then $\dfrac{du}{dx} = \dfrac{d}{dx}\sqrt{7}x = \sqrt{7}$ which implies $du = \sqrt{7}\,dx$; $dx = \dfrac{du}{\sqrt{7}}$.

Therefore, $\displaystyle\int \frac{dx}{x\sqrt{\left(\sqrt{7}x\right)^2-2^2}} = \int \frac{1}{\frac{u}{\sqrt{7}}\sqrt{u^2-2^2}}\cdot\frac{du}{\sqrt{7}} = \frac{1}{\sqrt{7}}\int \frac{\sqrt{7}\,du}{u\sqrt{u^2-2^2}} = \int \frac{du}{u\sqrt{u^2-2^2}}$

$$= \frac{1}{2} arc\sec\frac{u}{2}+c = \frac{1}{2}\boldsymbol{arc\sec\frac{\sqrt{7}x}{2}}+\boldsymbol{c}$$

c. **First** - Write the given integral in its standard form $\int \dfrac{dx}{a^2+x^2} = \dfrac{1}{a} arc\tan\dfrac{x}{a}+c$, i.e., $\int \dfrac{e^x}{4\,e^{2x}+9}\,dx = \int \dfrac{e^x}{9+4\,e^{2x}}\,dx$

$$= \int \frac{e^x dx}{4\left(\frac{9}{4}+e^{2x}\right)} = \frac{1}{4}\int \frac{e^x dx}{\frac{9}{4}+e^{2x}} = \frac{1}{4}\int \frac{e^x dx}{\left(\frac{3}{2}\right)^2+\left(e^x\right)^2}$$

Second - Use substitution method by letting $u = e^x$, then $\dfrac{du}{dx} = \dfrac{d}{dx}e^x = e^x$ which implies $du = e^x dx$; $dx = \dfrac{du}{e^x}$.

Therefore, $\dfrac{1}{4}\displaystyle\int \dfrac{e^x dx}{\left(\frac{3}{2}\right)^2+\left(e^x\right)^2} = \dfrac{1}{4}\int \dfrac{e^x}{\left(\frac{3}{2}\right)^2+u^2}\cdot\dfrac{du}{e^x} = \dfrac{1}{4}\int \dfrac{du}{\left(\frac{3}{2}\right)^2+u^2} = \dfrac{1}{4}\cdot\dfrac{1}{\frac{3}{2}} arc\tan\dfrac{u}{\frac{3}{2}}+c = \dfrac{1}{6} arc\tan\dfrac{2u}{3}+c$

Third - Write the answer in terms of the original variable, i.e., x . $\dfrac{1}{6}arc\,\tan\dfrac{2u}{3}+c \;=\; \dfrac{1}{6}\boldsymbol{arc\,\tan}\dfrac{2e^x}{3}+c$

Fourth - Check the answer by differentiating the solution, i.e., let $y=\dfrac{1}{6}arc\,\tan\dfrac{2e^x}{3}+c$ then

$$y' = \dfrac{1}{6}\cdot\dfrac{1}{1+\left(\frac{2e^x}{3}\right)^2}\cdot\dfrac{d}{dx}\dfrac{2e^x}{3}+0 = \dfrac{1}{6}\cdot\dfrac{1}{1+\frac{4e^{2x}}{9}}\cdot\dfrac{2e^x}{3} = \dfrac{2}{18}\cdot\dfrac{e^x}{\frac{9+4e^{2x}}{9}} = \dfrac{1}{9}\cdot\dfrac{9e^x}{9+4e^{2x}} = \dfrac{e^x}{9+4e^{2x}} = \dfrac{e^x}{4e^{2x}+9}$$

or, the alternative approach would be to rearrange the integral in the following way:

$$\int\dfrac{e^x}{9+4\,e^{2x}}\,dx = \int\dfrac{e^x dx}{3^2+\left(2e^x\right)^2}. \text{ Now, let } u=2e^x, \text{ then } \dfrac{du}{dx}=\dfrac{d}{dx}2e^x=2e^x \text{ which implies } du=2e^x dx\;;\;dx=\dfrac{du}{2e^x}.$$

Thus, $\int\dfrac{e^x dx}{3^2+\left(2e^x\right)^2} = \int\dfrac{e^x}{3^2+u^2}\cdot\dfrac{du}{2e^x} = \dfrac{1}{2}\int\dfrac{du}{3^2+u^2} = \dfrac{1}{2}\cdot\dfrac{1}{3}arc\,\tan\dfrac{u}{3}+c = \dfrac{1}{6}arc\,\tan\dfrac{u}{3}+c = \dfrac{1}{6}\boldsymbol{arc\,\tan}\dfrac{2e^x}{3}+c$

d. Write the given integral $\int\dfrac{dx}{\sqrt{25-(x-3)^2}}$ in its standard form $\int\dfrac{dx}{\sqrt{a^2-x^2}}=arc\,\sin\dfrac{x}{a}+c$ by letting $u=x-3$. Thus,

$$\int\dfrac{dx}{\sqrt{25-(x-3)^2}} = \int\dfrac{dx}{\sqrt{5^2-u^2}} = arc\,\sin\dfrac{u}{5}+c = \boldsymbol{arc\,\sin}\dfrac{x-3}{5}+c$$

Check: Let $y=arc\,\sin\dfrac{x-3}{5}+c$ then $y' = \dfrac{1}{\sqrt{1-\left(\frac{x-3}{5}\right)^2}}\cdot\dfrac{d}{dx}\dfrac{x-3}{5}+0 = \dfrac{1}{\sqrt{1-\frac{(x-3)^2}{25}}}\cdot\dfrac{1}{5} = \dfrac{1}{5}\dfrac{1}{\sqrt{\frac{25-(x-3)^2}{25}}}$

$$= \dfrac{1}{5}\dfrac{5}{\sqrt{25-(x-3)^2}} = \dfrac{1}{\sqrt{25-(x-3)^2}}$$

e. **First** - Write the given integral in its standard form $\int\dfrac{dx}{a^2+x^2}=\dfrac{1}{a}arc\,\tan\dfrac{x}{a}+c$, i.e., $\int\dfrac{dx}{25+(x+4)^2}$

Second - Use substitution method by letting $u=x+4$, then $\dfrac{du}{dx}=\dfrac{d}{dx}(x+4)=1$ which implies $du=dx$. Therefore,

$$\int\dfrac{dx}{25+(x+4)^2} = \int\dfrac{du}{5^2+u^2} = \dfrac{1}{5}arc\,\tan\dfrac{u}{5}+c$$

Third - Write the answer in terms of the x variable, i.e., $\dfrac{1}{5}arc\,\tan\dfrac{u}{5}+c = \dfrac{1}{5}\boldsymbol{arc\,\tan}\dfrac{x+4}{5}+c$

Fourth - Check the answer by differentiating the solution, i.e., let $y=\dfrac{1}{5}arc\,\tan\dfrac{x+4}{5}+c$ then

$$y' = \dfrac{1}{5}\cdot\dfrac{1}{1+\left(\frac{x+4}{5}\right)^2}\cdot\dfrac{d}{dx}\left(\dfrac{x+4}{5}\right)+0 = \dfrac{1}{5}\cdot\dfrac{1}{1+\frac{(x+4)^2}{25}}\cdot\dfrac{1}{5} = \dfrac{1}{25}\cdot\dfrac{25}{25+(x+4)^2} = \dfrac{1}{25+(x+4)^2}$$

f. **First** - Write the integral in its standard form $\int\dfrac{dx}{a^2+x^2}=\dfrac{1}{a}arc\,\tan\dfrac{x}{a}+c$, i.e., $\int\dfrac{dx}{x^2-10x+26} = \int\dfrac{dx}{\left(x^2-10x+25\right)+1}$

$$= \int\dfrac{dx}{(x-5)^2+1} = \int\dfrac{dx}{1+(x-5)^2}$$

Second - Use substitution method by letting $u=x-5$, then $\dfrac{du}{dx}=\dfrac{d}{dx}(x-5)=1$ which implies $du=dx$. Therefore,

$$\int\dfrac{dx}{1+(x-5)^2} = \int\dfrac{du}{1+u^2} = arc\,\tan u+c$$

Third - Write the answer in terms of the x variable, i.e., $arc\,\tan u+c = \boldsymbol{arc\,\tan}\left(x-5\right)+c$

Fourth - Check the answer by differentiating the solution, i.e., let $y=arc\,\tan\left(x-5\right)+c$ then

$$y' = \frac{1}{1+(x-5)^2} \cdot \frac{d}{dx}(x-5) + 0 = \frac{1}{1+(x-5)^2} \cdot 1 = \frac{1}{1+(x-5)^2}$$

Section 4.5 Solutions – Integration of Expressions Resulting in Exponential or Logarithmic Functions

1. Evaluate the following indefinite integrals.

a. Given $\int \frac{dx}{2x+1}$ let $u = 2x+1$, then $\frac{du}{dx} = \frac{d}{dx}(2x+1)$; $\frac{du}{dx} = 2$; $du = 2\,dx$; $dx = \frac{du}{2}$. Therefore,

$$\int \frac{dx}{2x+1} = \int \frac{1}{u} \cdot \frac{du}{2} = \frac{1}{2}\int \frac{1}{u}du = \frac{1}{2}\ln|u| + c = \frac{1}{2}\ln|2x+1| + c$$

Check: Let $y = \frac{1}{2}\ln|2x+1| + c$, then $y' = \frac{1}{2} \cdot \frac{1}{2x+1} \cdot 2 + 0 = \frac{2}{2} \cdot \frac{1}{2x+1} = \frac{1}{2x+1}$

b. Given $\int \frac{x}{x^2-a}dx$ let $u = x^2 - a$, then $\frac{du}{dx} = \frac{d}{dx}(x^2-a)$; $\frac{du}{dx} = 2x$; $du = 2x\,dx$; $dx = \frac{du}{2x}$. Therefore,

$$\int \frac{x}{x^2-a}dx = \int \frac{x}{u} \cdot \frac{du}{2x} = \frac{1}{2}\int \frac{1}{u}du = \frac{1}{2}\ln|u| + c = \frac{1}{2}\ln|x^2-a| + c$$

Check: Let $y = \frac{1}{2}\ln|x^2-a| + c$, then $y' = \frac{1}{2} \cdot \frac{1}{x^2-a} \cdot 2x + 0 = \frac{2}{2} \cdot \frac{x}{x^2-a} = \frac{x}{x^2-a}$

c. Given $\int \frac{x^3}{x^4-1}dx$ let $u = x^4 - 1$, then $\frac{du}{dx} = \frac{d}{dx}(x^4-1)$; $\frac{du}{dx} = x^4 - 1$; $du = 4x^3\,dx$; $dx = \frac{du}{4x^3}$. Therefore,

$$\int \frac{x^3}{x^4-1}dx = \int \frac{x^3}{u} \cdot \frac{du}{4x^3} = \frac{1}{4}\int \frac{1}{u}du = \frac{1}{4}\ln|u| + c = \frac{1}{4}\ln|x^4-1| + c$$

Check: Let $y = \frac{1}{4}\ln|x^4-1| + c$, then $y' = \frac{1}{4} \cdot \frac{1}{x^4-1} \cdot 4x^3 + 0 = \frac{x^3}{x^4-1}$

d. Given $\int \left(\frac{1}{x+3} + \frac{3}{x-5}\right)dx = \int \frac{1}{x+3}dx + \int \frac{3}{x-5}dx$ let $u_1 = x+3$ and $u_2 = x-5$ respectively. Therefore,

$$\int \frac{1}{x+3}dx + \int \frac{3}{x-5}dx = \int \frac{1}{u_1}du_1 + 3\int \frac{1}{u_2}du_2 = \ln|u_1| + 3\ln|u_2| + c = \ln|x+3| + 3\ln|x-5| + c$$

Check: Let $y = \ln|x+3| + 3\ln|x-5| + c$, then $y' = \frac{1}{x+3} + \frac{3}{x-5} + 0 = \frac{1}{x+3} + \frac{3}{x-5}$

e. Given $\int xe^{3x^2}dx$ let $u = 3x^2$, then $\frac{du}{dx} = \frac{d}{dx}3x^2$; $\frac{du}{dx} = 6x$; $du = 6x\,dx$; $dx = \frac{du}{6x}$. Therefore,

$$\int xe^{3x^2}dx = \int xe^u \cdot \frac{du}{6x} = \frac{1}{6}\int e^u du = \frac{1}{6}e^u + c = \frac{1}{6}e^{3x^2} + c$$

Check: Let $y = \frac{1}{6}e^{3x^2} + c$, then $y' = \frac{1}{6} \cdot e^{3x^2} \cdot 6x + 0 = \frac{6}{6} \cdot xe^{3x^2} = xe^{3x^2}$

f. Given $\int \frac{e^{5x}}{1-e^{5x}}dx$ let $u = 1 - e^{5x}$, then $\frac{du}{dx} = \frac{d}{dx}(1-e^{5x})$; $\frac{du}{dx} = -5e^{5x}$; $du = -5e^{5x}dx$; $dx = -\frac{du}{5e^{5x}}$. Therefore,

$$\int \frac{e^{5x}}{1-e^{5x}}dx = \int \frac{e^{5x}}{1-e^{5x}} \cdot -\frac{du}{5e^{5x}} = -\frac{1}{5}\int \frac{1}{u}du = -\frac{1}{5}\ln|u| + c = -\frac{1}{5}\ln|1-e^{5x}| + c$$

Check: Let $y = -\frac{1}{5}\ln|1-e^{5x}| + c$, then $y' = -\frac{1}{5} \cdot \frac{1}{1-e^{5x}} \cdot -5e^{5x} + 0 = \frac{e^{5x}}{1-e^{5x}}$

g. Given $\int 3e^{-ax}dx$ let $u = -ax$, then $\frac{du}{dx} = \frac{d}{dx}(-ax)$; $\frac{du}{dx} = -a$; $du = -a\,dx$; $dx = -\frac{du}{a}$. Therefore,

$$\int 3e^{-ax}\,dx = 3\int e^u \cdot -\frac{du}{a} = -\frac{3}{a}\int e^u\,du = -\frac{3}{a}e^u + c = -\frac{3}{a}e^{-ax} + c$$

Check: Let $y = -\frac{3}{a}e^{-ax} + c$, then $y' = -\frac{3}{a}e^{-ax}\cdot -a + 0 = \frac{3a}{a}e^{-ax} = 3e^{-ax}$

h. $\int\left(x^3 + e^{2x} - e^{-5x}\right)dx = \int x^3\,dx + \int e^{2x}\,dx - \int e^{-5x}\,dx = \frac{1}{4}x^4 + \frac{1}{2}e^{2x} - \frac{1}{-5}e^{-5x} + c = \frac{1}{4}x^4 + \frac{1}{2}e^{2x} + \frac{1}{5}e^{-5x} + c$

Check: Let $y = \frac{1}{4}x^4 + \frac{1}{2}e^{2x} + \frac{1}{5}e^{-5x} + c$ then $y' = \frac{1}{4}\cdot 4x^3 + \frac{1}{2}\cdot e^{2x}\cdot 2 + \frac{1}{5}\cdot e^{-5x}\cdot -5 + 0 = x^3 + e^{2x} - e^{-5x}$

i. $\int \dfrac{e^{\frac{1}{x^3}}}{5x^4}\,dx$ let $u = \dfrac{1}{x^3}$, then $\dfrac{du}{dx} = \dfrac{d}{dx}\dfrac{1}{x^3}$; $\dfrac{du}{dx} = -\dfrac{3}{x^4}$; $x^4\,du = -3\,dx$; $dx = -\dfrac{x^4}{3}\,du$. Therefore,

$$\int \dfrac{e^{\frac{1}{x^3}}}{5x^4}\,dx = \int \dfrac{e^u}{5x^4}\cdot -\dfrac{x^4}{3}\,du = -\dfrac{1}{15}\int e^u\,du = -\dfrac{1}{15}e^u + c = -\dfrac{1}{15}e^{\frac{1}{x^3}} + c$$

Check: Let $y = -\dfrac{1}{15}e^{\frac{1}{x^3}} + c$, then $y' = -\dfrac{1}{15}\cdot e^{\frac{1}{x^3}}\cdot -\dfrac{3}{x^4} + 0 = \dfrac{3}{15}\cdot e^{\frac{1}{x^3}}\cdot\dfrac{1}{x^4} = \dfrac{e^{\frac{1}{x^3}}}{5x^4}$

2. Evaluate the following indefinite integrals.

a. Given $\int\left(3e^{5x} + 5\right)e^{5x}\,dx$ let $u = 3e^{5x} + 5$, then $\dfrac{du}{dx} = \dfrac{d}{dx}\left(3e^{5x} + 5\right)$; $\dfrac{du}{dx} = 15e^{5x}$; $du = 15e^{5x}\,dx$; $dx = \dfrac{du}{15e^{5x}}$. Thus,

$$\int\left(3e^{5x} + 5\right)e^{5x}\,dx = \int u\cdot e^{5x}\cdot\dfrac{du}{15e^{5x}} = \dfrac{1}{15}\int u\,du = \dfrac{1}{15}\cdot\dfrac{1}{2}u^2 + c = \dfrac{1}{30}\left(3e^{5x} + 5\right)^2 + c$$

Check: Let $y = \dfrac{1}{30}\left(3e^{5x} + 5\right)^2 + c$, then $y' = \dfrac{1}{30}\cdot 2\left(3e^{5x} + 5\right)\cdot 15e^{5x} + 0 = \dfrac{30}{30}\left(3e^{5x} + 5\right)e^{5x} = \left(3e^{5x} + 5\right)e^{5x}$

b. Given $\int\left(e^x - 1\right)^5 e^x\,dx$ let $u = e^x - 1$, then $\dfrac{du}{dx} = \dfrac{d}{dx}\left(e^x - 1\right)$; $\dfrac{du}{dx} = e^x$; $du = e^x\,dx$; $dx = \dfrac{du}{e^x}$. Thus,

$$\int\left(e^x - 1\right)^5 e^x\,dx = \int u^5\cdot e^x\cdot\dfrac{du}{e^x} = \int u^5\,du = \dfrac{1}{6}u^6 + c = \dfrac{1}{6}\left(e^x - 1\right)^6 + c$$

Check: Let $y = \dfrac{1}{6}\left(e^x - 1\right)^6 + c$, then $y' = \dfrac{1}{6}\cdot 6\left(e^x - 1\right)^5\cdot e^x + 0 = \dfrac{6}{6}\left(e^x - 1\right)^5 e^x = \left(e^x - 1\right)^5 e^x$

c. Given $\int\dfrac{x^2}{1+x^3}\,dx$ let $u = 1 + x^3$, then $\dfrac{du}{dx} = \dfrac{d}{dx}\left(1 + x^3\right)$; $\dfrac{du}{dx} = 3x^2$; $dx = \dfrac{du}{3x^2}$. Therefore,

$$\int\dfrac{x^2}{1+x^3}\,dx = \int\dfrac{x^2}{u}\cdot\dfrac{du}{3x^2} = \dfrac{1}{3}\int\dfrac{1}{u}\,du = \dfrac{1}{3}\ln|u| + c = \dfrac{1}{3}\ln\left|1 + x^3\right| + c$$

Check: Let $y = \dfrac{1}{3}\ln\left|1 + x^3\right| + c$, then $y' = \dfrac{1}{3}\cdot\dfrac{1}{1+x^3}\cdot 3x^2 + 0 = \dfrac{3}{3}\cdot\dfrac{x^2}{1+x^3} = \dfrac{x^2}{1+x^3}$

d. Given $\int\dfrac{x^4}{1+x^5}\,dx$ let $u = 1 + x^5$, then $\dfrac{du}{dx} = \dfrac{d}{dx}\left(1 + x^5\right)$; $\dfrac{du}{dx} = 5x^4$; $dx = \dfrac{du}{5x^4}$. Therefore,

$$\int\dfrac{x^4}{1+x^5}\,dx = \int\dfrac{x^4}{u}\cdot\dfrac{du}{5x^4} = \dfrac{1}{5}\int\dfrac{1}{u}\,du = \dfrac{1}{5}\ln|u| + c = \dfrac{1}{5}\ln\left|1 + x^5\right| + c$$

Check: Let $y = \dfrac{1}{5}\ln\left|1 + x^5\right| + c$, then $y' = \dfrac{1}{5}\cdot\dfrac{1}{1+x^5}\cdot 5x^4 + 0 = \dfrac{5}{5}\cdot\dfrac{x^4}{1+x^5} = \dfrac{x^4}{1+x^5}$

e. $\int\dfrac{x+7}{x+6}\,dx = \int\dfrac{(x+6)+1}{x+6}\,dx = \int\dfrac{x+6}{x+6}\,dx + \int\dfrac{1}{x+6}\,dx = \int dx + \int\dfrac{1}{x+6}\,dx = x + \ln|x+6| + c$

Check: Let $y = x + \ln|x+6| + c$, then $y' = 1 + \dfrac{1}{x+6} \cdot 1 + 0 = 1 + \dfrac{1}{x+6} = \dfrac{x+6+1}{x+6} = \dfrac{x+7}{x+6}$

f. $\displaystyle\int \dfrac{x+9}{x+5}\,dx = \int \dfrac{(x+5)+4}{x+5}\,dx = \int \dfrac{x+5}{x+5}\,dx + \int \dfrac{4}{x+5}\,dx = \int dx + \int \dfrac{4}{x+5}\,dx = \boldsymbol{x + 4\ln|x+5| + c}$

Check: Let $y = x + 4\ln|x+5| + c$, then $y' = 1 + \dfrac{4}{x+5} \cdot 1 + 0 = 1 + \dfrac{4}{x+5} = \dfrac{x+5+4}{x+5} = \dfrac{x+9}{x+5}$

g. Given $\displaystyle\int a^{x^2+k} x\,dx$ let $u = x^2 + k$, then $\dfrac{du}{dx} = \dfrac{d}{dx}\left(x^2 + k\right)$; $\dfrac{du}{dx} = 2x$; $du = 2x\,dx$; $dx = \dfrac{du}{2x}$. Therefore,

$\displaystyle\int a^{x^2+k} x\,dx = \int a^u \cdot x \cdot \dfrac{du}{2x} = \dfrac{1}{2}\int a^u\,du = \dfrac{1}{2}\dfrac{a^u}{\ln a} + c = \boldsymbol{\dfrac{1}{2}\dfrac{a^{x^2+k}}{\ln a} + c}$

Check: Let $y = \dfrac{1}{2}\dfrac{a^{x^2+k}}{\ln a} + c$, then $y' = \dfrac{1}{2\ln a} \cdot a^{x^2+k} \ln a \cdot 2x + 0 = \dfrac{2\ln a}{2\ln a} \cdot a^{x^2+k} \cdot x = a^{x^2+k} x$

h. Given $\displaystyle\int \dfrac{2}{3} a^{x^3+5} x^2\,dx$ let $u = x^3 + 5$, then $\dfrac{du}{dx} = \dfrac{d}{dx}\left(x^3 + 5\right)$; $\dfrac{du}{dx} = 3x^2$; $du = 3x^2\,dx$; $dx = \dfrac{du}{3x^2}$. Therefore,

$\displaystyle\int \dfrac{2}{3} a^{x^3+5} x^2\,dx = \dfrac{2}{3}\int a^u \cdot x^2 \cdot \dfrac{du}{3x^2} = \dfrac{2}{9}\int a^u\,du = \dfrac{2}{9}\cdot\dfrac{a^u}{\ln a} + c = \boldsymbol{\dfrac{2}{9}\cdot\dfrac{a^{x^3+5}}{\ln a} + c}$

Check: Let $y = \dfrac{2}{9}\cdot\dfrac{a^{x^3+5}}{\ln a} + c$, then $y' = \dfrac{2}{9}\cdot\dfrac{1}{\ln a} \cdot a^{x^3+5} \ln a \cdot 3x^2 + 0 = \dfrac{6}{9}\dfrac{\ln a}{\ln a} \cdot a^{x^3+5} \cdot x^2 = \dfrac{2}{3} a^{x^3+5} x^2$

i. Given $\displaystyle\int \left(e^{2x} + 3\right)^3 e^{2x}\,dx$ let $u = e^{2x} + 3$, then $\dfrac{du}{dx} = \dfrac{d}{dx}\left(e^{2x} + 3\right)$; $\dfrac{du}{dx} = 2e^{2x}$; $du = 2e^{2x}\,dx$; $dx = \dfrac{du}{2e^{2x}}$. Therefore,

$\displaystyle\int \left(e^{2x} + 3\right)^3 e^{2x}\,dx = \int u^3 \cdot e^{2x} \cdot \dfrac{du}{2e^{2x}} = \dfrac{1}{2}\int u^3\,du = \dfrac{1}{2}\cdot\dfrac{1}{4}u^4 + c = \dfrac{1}{8}u^4 + c = \boldsymbol{\dfrac{1}{8}\left(e^{2x} + 3\right)^4 + c}$

Check: Let $y = \dfrac{1}{8}\left(e^{2x} + 3\right)^4 + c$, then $y' = \dfrac{1}{8} \cdot 4\left(e^{2x} + 3\right)^3 \cdot 2e^{2x} + 0 = \dfrac{8}{8}\left(e^{2x} + 3\right)^3 e^{2x} = \left(e^{2x} + 3\right)^3 e^{2x}$

j. $\displaystyle\int \dfrac{1}{(x+1)^2 - 25}\,dx = \int \dfrac{1}{(x+1)^2 - 5^2}\,dx = \dfrac{1}{2\cdot5}\ln\left|\dfrac{(x+1)-5}{(x+1)+5}\right| + c = \boldsymbol{\dfrac{1}{10}\ln\left|\dfrac{x-4}{x+6}\right| + c}$

Check: Let $y = \dfrac{1}{10}\ln\left|\dfrac{x-4}{x+6}\right| + c$ then $y' = \dfrac{1}{10}\cdot\dfrac{1}{\frac{x-4}{x+6}}\cdot\dfrac{1\cdot(x+6) - 1\cdot(x-4)}{(x+6)^2} + 0 = \dfrac{1}{10}\cdot\dfrac{x+6}{x-4}\cdot\dfrac{x+6-x+4}{(x+6)^2}$

$= \dfrac{1}{10}\cdot\dfrac{1}{x-4}\cdot\dfrac{10}{x+6} = \dfrac{1}{(x-4)(x+6)} = \dfrac{1}{x^2 + 6x - 4x - 24} = \dfrac{1}{x^2 + 2x - 24} = \dfrac{1}{(x+1)^2 - 25}$

k. $\displaystyle\int \dfrac{1}{x^2 + 6x + 8}\,dx = \int \dfrac{1}{\left(x^2 + 6x + 9\right) - 1}\,dx = \int \dfrac{1}{(x+3)^2 - 1}\,dx = \dfrac{1}{2\cdot1}\ln\left|\dfrac{(x+3)-1}{(x+3)+1}\right| + c = \boldsymbol{\dfrac{1}{2}\ln\left|\dfrac{x+2}{x+4}\right| + c}$

Check: Let $y = \dfrac{1}{2}\ln\left|\dfrac{x+2}{x+4}\right| + c$ then $y' = \dfrac{1}{2}\cdot\dfrac{1}{\frac{x+2}{x+4}}\cdot\dfrac{1\cdot(x+4) - 1\cdot(x+2)}{(x+4)^2} + 0 = \dfrac{1}{2}\cdot\dfrac{x+4}{x+2}\cdot\dfrac{x+4-x-2}{(x+4)^2} = \dfrac{1}{2}\cdot\dfrac{1}{x+2}\cdot\dfrac{2}{x+4}$

$= \dfrac{1}{(x+2)(x+4)} = \dfrac{1}{x^2 + 4x + 2x + 8} = \dfrac{1}{x^2 + 6x + 8}$

l. $\displaystyle\int \dfrac{1}{9 - (x-1)^2}\,dx = \int \dfrac{1}{3^2 - (x-1)^2}\,dx = \dfrac{1}{2\cdot3}\ln\left|\dfrac{3+(x-1)}{3-(x-1)}\right| + c = \dfrac{1}{6}\ln\left|\dfrac{3+x-1}{3-x+1}\right| + c = \boldsymbol{\dfrac{1}{6}\ln\left|\dfrac{2+x}{4-x}\right| + c}$

Check: Let $y = \dfrac{1}{6}\ln\left|\dfrac{2+x}{4-x}\right| + c$ then $y' = \dfrac{1}{6}\cdot\dfrac{1}{\frac{2+x}{4-x}}\cdot\dfrac{1\cdot(4-x) + 1\cdot(2+x)}{(4-x)^2} + 0 = \dfrac{1}{6}\cdot\dfrac{4-x}{2+x}\cdot\dfrac{4-x+2+x}{(4-x)^2} = \dfrac{1}{6}\cdot\dfrac{1}{2+x}\cdot\dfrac{6}{4-x}$

$= \dfrac{1}{(2+x)(4-x)} = \dfrac{1}{8 - 2x + 4x - x^2} = \dfrac{1}{8 + 2x - x^2} = \dfrac{1}{9 - (x-1)^2}$

Chapter 5 Solutions:

1. Evaluate the following integrals using the integration by parts method.

a. Given $\int xe^{4x}dx$ let $u=x$ and $dv=e^{4x}dx$ then $du=dx$ and $\int dv=\int e^{4x}dx$ which implies $v=\frac{1}{4}e^{4x}$. Using the integration

by parts formula $\int u\,dv=u\,v-\int v\,du$ we obtain

$$\int xe^{4x}dx \;=\; \frac{1}{4}xe^{4x}-\frac{1}{4}\int e^{4x}dx \;=\; \frac{1}{4}xe^{4x}-\frac{1}{16}e^{4x}+c \;=\; \frac{1}{4}e^{4x}\left(x-\frac{1}{4}\right)+c$$

Check: Let $y=\frac{1}{4}e^{4x}\left(x-\frac{1}{4}\right)+c$, then $y'=\frac{4e^{4x}}{4}\cdot\left(x-\frac{1}{4}\right)+1\cdot\frac{1}{4}e^{4x}+0 = e^{4x}\cdot\left(x-\frac{1}{4}\right)+\frac{1}{4}e^{4x} = xe^{4x}-\frac{e^{4x}}{4}+\frac{e^{4x}}{4} = xe^{4x}$

b. Given $\int \frac{x}{2}\cos x\,dx$ let $u=x$ and $dv=\cos x\,dx$ then $du=dx$ and $\int dv=\int\cos x\,dx$ which implies $v=\sin x$. Using the

integration by parts formula $\int u\,dv=u\,v-\int v\,du$ we obtain

$$\int \frac{x}{2}\cos x\,dx \;=\; \frac{1}{2}x\cdot\sin x-\frac{1}{2}\int\sin x\,dx \;=\; \frac{1}{2}x\sin x+\frac{1}{2}\cos x+c$$

Check: Let $y=\frac{1}{2}x\sin x+\frac{1}{2}\cos x+c$, then $y'=\frac{1}{2}(1\cdot\sin x+\cos x\cdot x)-\frac{1}{2}\sin x+0 = \frac{1}{2}\sin x+\frac{1}{2}x\cos x-\frac{1}{2}\sin x = \frac{1}{2}x\cos x$

c. Given $\int (5-x)e^{5x}dx$ let $u=5-x$ and $dv=e^{5x}dx$ then $du=-dx$ and $\int dv=\int e^{5x}dx$ which implies $v=\frac{e^{5x}}{5}$. Using

the integration by parts formula $\int u\,dv=u\,v-\int v\,du$ we obtain

$$\int (5-x)e^{5x}dx \;=\; (5-x)\frac{e^{5x}}{5}+\int\frac{e^{5x}}{5}dx \;=\; e^{5x}-\frac{1}{5}xe^{5x}+\frac{1}{25}e^{5x}+c$$

Check: Let $y=e^{5x}-\frac{1}{5}xe^{5x}+\frac{1}{25}e^{5x}+c$, then $y'=5e^{5x}-\frac{1}{5}\left(e^{5x}+5xe^{5x}\right)+\frac{5}{25}\cdot e^{5x}+0 = 5e^{5x}-\frac{1}{5}e^{5x}-xe^{5x}+\frac{1}{5}e^{5x}$

$$= 5e^{5x}-xe^{5x} = (5-x)e^{5x}$$

d. Given $\int x\sin 5x\,dx$ let $u=x$ and $dv=\sin 5x\,dx$ then $du=dx$ and $\int dv=\int\sin 5x\,dx$ which implies $v=-\frac{1}{5}\cos 5x$.

Using the integration by parts formula $\int u\,dv=u\,v-\int v\,du$ we obtain

$$\int x\sin 5x\,dx \;=\; x\cdot-\frac{1}{5}\cos 5x+\frac{1}{5}\int\cos 5x\,dx \;=\; -\frac{1}{5}x\cos 5x+\frac{1}{5}\sin 5x+c$$

Check: Let $y=-\frac{1}{5}x\cos 5x+\frac{1}{5}\sin 5x+c$, then $y'=-\frac{1}{5}(1\cdot\cos 5x-\sin 5x\cdot 5\cdot x)+\frac{1}{5}\cos 5x+0 = -\frac{1}{5}\cos 5x+\frac{5}{5}x\sin 5x$

$$+\frac{1}{5}\cos 5x = \frac{5}{5}x\sin 5x = x\sin 5x$$

e. Given $\int x\sqrt{3-x}\,dx$ let $u=x$ and $dv=\sqrt{3-x}\,dx$ then $du=dx$ and $\int dv=\int\sqrt{3-x}\,dx$ which implies $v=-\frac{2}{3}(3-x)^{\frac{3}{2}}$.

Using the integration by parts formula $\int u\,dv=u\,v-\int v\,du$ we obtain

$$\int x\sqrt{3-x}\,dx \;=\; x\cdot-\frac{2}{3}(3-x)^{\frac{3}{2}}+\int \frac{2}{3}(3-x)^{\frac{3}{2}}dx \;=\; -\frac{2}{3}x(3-x)^{\frac{3}{2}}-\frac{2}{3}\cdot\frac{1}{1+\frac{3}{2}}(3-x)^{\frac{3}{2}+1}+c \;=\; -\frac{2}{3}x(3-x)^{\frac{3}{2}}-\frac{2}{3}\cdot\frac{2}{5}(3-x)^{\frac{5}{2}}+c$$

$$=\; -\frac{2}{3}x(3-x)^{\frac{3}{2}}-\frac{4}{15}(3-x)^{\frac{5}{2}}+c$$

Check: Let $y=-\frac{2}{3}x(3-x)^{\frac{3}{2}}-\frac{4}{15}(3-x)^{\frac{5}{2}}+c$, then $y'=-\frac{2}{3}(3-x)^{\frac{3}{2}}+\frac{2}{3}\cdot\frac{3}{2}x(3-x)^{\frac{1}{2}}+\frac{4}{15}\cdot\frac{5}{2}(3-x)^{\frac{3}{2}}+0$

$$=\; -\frac{2}{3}(3-x)^{\frac{3}{2}}+x(3-x)^{\frac{1}{2}}+\frac{2}{3}(3-x)^{\frac{3}{2}} \;=\; x(3-x)^{\frac{1}{2}} \;=\; x\sqrt{3-x}$$

f. Given $\int x^3 e^{3x}dx$ let $u=x^3$ and $dv=e^{3x}dx$ then $du=3x^2dx$ and $\int dv=\int e^{3x}dx$ which implies $v=\frac{1}{3}e^{3x}$. Using the

integration by parts formula $\int u\,dv=u\,v-\int v\,du$ we obtain

$$\int x^3 e^{3x}dx \;=\; x^3\cdot\frac{1}{3}e^{3x}-\frac{1}{3}\int e^{3x}\cdot 3x^2 dx \;=\; \frac{x^3 e^{3x}}{3}-\int x^2 e^{3x}dx$$

In example 5.1-1, problem letter b, we showed that $\int x^2 e^{3x}dx=\frac{1}{3}x^2 e^{3x}-\frac{2}{9}xe^{3x}+\frac{2}{27}e^{3x}+c$. Therefore,

$$\int x^3 e^{3x}dx \;=\; \frac{x^3 e^{3x}}{3}-\left(\frac{1}{3}x^2 e^{3x}-\frac{2}{9}xe^{3x}+\frac{2}{27}e^{3x}\right)+c \;=\; \frac{x^3 e^{3x}}{3}-\frac{1}{3}x^2 e^{3x}+\frac{2}{9}xe^{3x}-\frac{2}{27}e^{3x}+c$$

Check: Let $y=\frac{x^3 e^{3x}}{3}-\frac{1}{3}x^2 e^{3x}+\frac{2}{9}xe^{3x}-\frac{2}{27}e^{3x}+c$, then $y'=\frac{1}{3}\left(3x^2\cdot e^{3x}+3e^{3x}\cdot x^3\right)-\frac{1}{3}\left(2x\cdot e^{3x}+3e^{3x}\cdot x^2\right)$

$$+\frac{2}{9}\left(1\cdot e^{3x}+3e^{3x}\cdot x\right)-\frac{2}{27}\cdot 3e^{3x}+0 \;=\; x^2 e^{3x}+x^3 e^{3x}-\frac{2}{3}xe^{3x}-x^2 e^{3x}+\frac{2}{9}e^{3x}+\frac{2}{3}xe^{3x}-\frac{2}{9}e^{3x} \;=\; x^3 e^{3x}$$

g. Given $\int \cos(\ln x)\,dx$ let $u=\cos(\ln x)$ and $dv=dx$ then $du=\dfrac{-\sin(\ln x)}{x}dx$ and $\int dv=\int dx$ which implies $v=x$.

Using the integration by parts formula $\int u\,dv=u\,v-\int v\,du$ we obtain

$$\int \cos(\ln x)\,dx \;=\; \cos(\ln x)\cdot x+\int x\cdot\frac{\sin(\ln x)}{x}dx \;=\; x\cos(\ln x)+\int \sin(\ln x)\,dx \qquad (1)$$

To integrate $\int \sin(\ln x)\,dx$ use the integration by parts formula again, i.e., let $u=\sin(\ln x)$ and $dv=dx$ then

$du=\dfrac{\cos(\ln x)}{x}dx$ and $\int dv=\int dx$ which implies $v=x$. Therefore,

$$\int \sin(\ln x)\,dx \;=\; \sin(\ln x)\cdot x-\int x\cdot\frac{\cos(\ln x)}{x}dx \;=\; x\sin(\ln x)-\int \cos(\ln x)\,dx \qquad (2)$$

Combining equations (1) and (2) together we have

$$\int \cos(\ln x)\,dx \;=\; x\cos(\ln x)-\int \sin(\ln x)\,dx \;=\; x\cos(\ln x)+x\sin(\ln x)-\int \cos(\ln x)\,dx$$

Taking the integral $-\int \cos(\ln x)\,dx$ from the right hand side of the equation to the left hand side

we obtain $\int \cos(\ln x)\,dx+\int \cos(\ln x)\,dx=x\cos(\ln x)+x\sin(\ln x)$ Therefore,

$$2\int \cos(\ln x)\,dx=x\cos(\ln x)+x\sin(\ln x)+c \text{ and thus } \int \cos(\ln x)\,dx=\frac{x}{2}\cos(\ln x)+\frac{x}{2}\sin(\ln x)+c$$

Check: Let $y=\frac{x}{2}\cos(\ln x)+\frac{x}{2}\sin(\ln x)+c$, then $y'=\dfrac{\cos(\ln x)}{2}-\dfrac{x\sin(\ln x)}{2x}+\dfrac{\sin(\ln x)}{2}+\dfrac{x\cos(\ln x)}{2x}+0$

$$=\; \frac{\cos(\ln x)}{2}-\frac{\sin(\ln x)}{2}+\frac{\sin(\ln x)}{2}+\frac{\cos(\ln x)}{2} \;=\; \frac{\cos(\ln x)}{2}+\frac{\cos(\ln x)}{2} \;=\; \cos(\ln x)$$

h. Given $\int \frac{x}{3}\tan^{-1}x\, dx$ let $u = \tan^{-1}x$ and $dv = \frac{x}{3}dx$ then $du = \frac{1}{1+x^2}dx$ and $\int dv = \int \frac{x}{3}dx$ which implies $v = \frac{1}{6}x^2$.

Using the integration by parts formula $\int u\, dv = u\,v - \int v\, du$ we obtain

$$\int \frac{x}{3}\tan^{-1}2x\, dx = \tan^{-1}x \cdot \frac{x^2}{6} - \frac{1}{6}\int x^2 \cdot \frac{dx}{1+x^2} = \frac{1}{6}x^2\tan^{-1}x - \frac{1}{6}\int \frac{x^2}{1+x^2}dx = \frac{1}{6}x^2\tan^{-1}x - \frac{1}{6}\int \left(1 - \frac{1}{1+x^2}\right)dx$$

$$= \frac{1}{6}x^2\tan^{-1}x - \frac{1}{6}\int dx + \frac{1}{6}\int \frac{1}{1+x^2}dx = \frac{1}{6}x^2\tan^{-1}x - \frac{1}{6}x + \frac{1}{6}\tan^{-1}x + c$$

Check: Let $y = \frac{1}{6}x^2\tan^{-1}x - \frac{1}{6}x + \frac{1}{6}\tan^{-1}x + c$, then $y' = \frac{1}{6} \cdot 2x \cdot \tan^{-1}x + \frac{1}{6}\frac{x^2}{1+x^2} - \frac{1}{6} + \frac{1}{6} \cdot \frac{1}{1+x^2} + 0$

$$\frac{1}{3} \cdot x\tan^{-1}x + \frac{1}{6} \cdot \frac{x^2}{1+x^2} + \frac{1}{6} \cdot \frac{1}{1+x^2} - \frac{1}{6} = \frac{1}{3}x\tan^{-1}x + \frac{1}{6} \cdot \frac{x^2+1}{1+x^2} - \frac{1}{6} = \frac{1}{3}x\tan^{-1}x + \frac{1}{6} - \frac{1}{6} = \frac{1}{3}x\tan^{-1}x$$

i. Given $\int \ln x^5 dx$ let $u = \ln x^5$ and $dv = dx$ then $du = \frac{1}{x^5} \cdot 5x^4 dx = \frac{5}{x}dx$ and $\int dv = \int dx$ which implies $v = x$. Using the

integration by parts formula $\int u\, dv = u\,v - \int v\, du$ we obtain

$$\int \ln x^5 dx = \ln x^5 \cdot x - \int x \cdot \frac{5}{x}dx = x\ln x^5 - 5\int dx = x\ln x^5 - 5x + c$$

Check: Let $y = x\ln x^5 - 5x + c$, then $y' = \left(1 \cdot \ln x^5 + \frac{1}{x^5} \cdot 5x^4 \cdot x\right) - 5 + 0 = \ln x^5 + \frac{5x^5}{x^5} - 5 = \ln x^5 + 5 - 5 = \ln x^5$

j. Given $\int x\,e^{-ax}dx$ let $u = x$ and $dv = e^{-ax}dx$ then $du = dx$ and $\int dv = \int e^{-ax}dx$ which implies $v = -\frac{1}{a}e^{-ax}$. Using the

integration by parts formula $\int u\, dv = u\,v - \int v\, du$ we obtain

$$\int x\,e^{-ax}dx = x \cdot -\frac{1}{a}e^{-ax} + \frac{1}{a}\int e^{-ax}dx = -\frac{1}{a}x\,e^{-ax} + \frac{1}{a}\int e^{-ax}dx = -\frac{1}{a}x\,e^{-ax} - \frac{1}{a^2}e^{-ax} + c$$

Check: Let $y = -\frac{1}{a}x\,e^{-ax} - \frac{1}{a^2}e^{-ax} + c$, then $y' = -\frac{1}{a}\left(1 \cdot e^{-ax} - ae^{-ax} \cdot x\right) - \frac{1}{a^2} \cdot -2e^{-ax} + 0 = -\frac{1}{a}e^{-ax} + \frac{axe^{-ax}}{a}$

$$+\frac{a}{a^2} \cdot e^{-ax} = -\frac{1}{a}e^{-ax} + xe^{-ax} + \frac{1}{a}e^{-ax} = xe^{-ax}$$

k. Given $\int e^x \sin 3x\, dx$ let $u = e^x$ and $dv = \sin 3x\, dx$ then $du = e^x dx$ and $\int dv = \int \sin 3x\, dx$ which implies $v = -\frac{\cos 3x}{3}$.

Using the integration by parts formula $\int u\, dv = u\,v - \int v\, du$ we obtain

$$\int e^x \sin 3x\, dx = e^x \cdot -\frac{\cos 3x}{3} - \int -\frac{\cos 3x}{3} \cdot e^x dx = -\frac{1}{3}e^x \cos 3x + \frac{1}{3}\int e^x \cos 3x\, dx \qquad (1)$$

To integrate $\int e^x \cos 3x\, dx$ let $u = e^x$ and $dv = \cos 3x\, dx$ then $du = e^x dx$ and $\int dv = \int \cos 3x\, dx$ which implies

$v = \frac{\sin 3x}{3}$. Thus, $\int e^x \cos 3x\, dx = e^x \cdot \frac{\sin 3x}{3} - \int \frac{\sin 3x}{3} \cdot e^x dx = \frac{1}{3}e^x \sin 3x - \frac{1}{3}\int e^x \sin 3x\, dx \qquad (2)$

Combining equations (1) and (2) together we obtain:

$$\int e^x \sin 3x\, dx = -\frac{1}{3}e^x \cos 3x + \frac{1}{3}\int e^x \cos 3x\, dx = -\frac{1}{3}e^x \cos 3x + \frac{1}{9}e^x \sin 3x - \frac{1}{9}\int e^x \sin 3x\, dx$$

Taking the $-\frac{1}{9}\int e^x \sin 3x\, dx$ from the right hand side of the equation to the left hand side we obtain

$\int e^x \sin 3x \, dx + \dfrac{1}{9}\int e^x \sin 3x \, dx = -\dfrac{1}{3}e^x \cos 3x + \dfrac{1}{9}e^x \sin 3x$ which implies $\dfrac{10}{9}\int e^x \sin 3x \, dx = -\dfrac{1}{3}e^x \cos 3x + \dfrac{1}{9}e^x \sin 3x$

and $\int e^x \sin 3x \, dx = \dfrac{9}{10}\left(-\dfrac{1}{3}e^x \cos 3x + \dfrac{1}{9}e^x \sin 3x \right) = -\dfrac{\mathbf{3}}{\mathbf{10}}\boldsymbol{e^x \cos 3x} + \dfrac{\mathbf{1}}{\mathbf{10}}\boldsymbol{e^x \sin 3x}$

Check: Let $y = -\dfrac{3}{10}e^x \cos 3x + \dfrac{1}{10}e^x \sin 3x$, then $y' = -\dfrac{3}{10}e^x \cdot \cos 3x + \dfrac{3}{10}\sin 3x \cdot 3 \cdot e^x + \dfrac{1}{10}e^x \cdot \sin 3x + \dfrac{1}{10}\cos 3x \cdot 3 \cdot e^x$

$= -\dfrac{3}{10}e^x \cos 3x + \dfrac{9}{10}e^x \sin 3x + \dfrac{1}{10}e^x \sin 3x + \dfrac{3}{10}e^x \cos 3x = \dfrac{9}{10}e^x \sin 3x + \dfrac{1}{10}e^x \sin 3x = \dfrac{10}{10}e^x \sin 3x = e^x \sin 3x$

1. Given $\int e^x \cos 5x \, dx$ let $u = e^x$ and $dv = \cos 5x \, dx$ then $du = e^x dx$ and $\int dv = \int \cos 5x \, dx$ which implies $v = \dfrac{1}{5}\sin 5x$.

Using the integration by parts formula $\int u \, dv = u \, v - \int v \, du$ we obtain

$$\int e^x \cos 5x \, dx = e^x \cdot \dfrac{1}{5}\sin 5x - \dfrac{1}{5}\int \sin 5x \cdot e^x dx = \dfrac{e^x \sin 5x}{5} - \dfrac{1}{5}\int e^x \sin 5x \, dx \qquad (1)$$

To integrate $\int e^x \sin 5x \, dx$ let $u = e^x$ and $dv = \sin 5x \, dx$ then $du = e^x dx$ and $\int dv = \int \sin 5x \, dx$ which implies $v = -\dfrac{1}{5}\cos 5x$.

Thus, $\int e^x \sin 5x \, dx = e^x \cdot -\dfrac{1}{5}\cos 5x + \dfrac{1}{5}\int \cos 5x \cdot e^x dx = -\dfrac{e^x \cos 5x}{5} + \dfrac{1}{5}\int \cos 5x \cdot e^x dx \qquad (2)$

Combining equations (1) and (2) together we obtain:

$$\int e^x \cos 5x \, dx = \dfrac{e^x \sin 5x}{5} - \dfrac{1}{5}\int e^x \sin 5x \, dx = \dfrac{e^x \sin 5x}{5} - \dfrac{1}{5}\left(-\dfrac{e^x \cos 5x}{5} + \dfrac{1}{5}\int e^x \cos 5x \, dx \right) = \dfrac{e^x \sin 5x}{5} + \dfrac{e^x \cos 5x}{25}$$

$-\dfrac{1}{25}\int e^x \cos 5x \, dx$. Taking the $-\dfrac{1}{25}\int e^x \cos 5x \, dx$ from the right hand side of the equation to the left hand side we obtain

$\int e^x \cos 5x \, dx + \dfrac{1}{25}\int e^x \cos 5x \, dx = \dfrac{e^x \sin 5x}{5} + \dfrac{e^x \cos 5x}{25}$ which implies $\dfrac{26}{25}\int e^x \cos 5x \, dx = \dfrac{e^x \sin 5x}{5} + \dfrac{e^x \cos 5x}{25}$ and

$\int e^x \cos 5x \, dx = \dfrac{25}{26}\left(\dfrac{e^x \sin 5x}{5} + \dfrac{e^x \cos 5x}{25} \right) = \dfrac{\mathbf{5}}{\mathbf{26}}\boldsymbol{e^x \sin 5x} + \dfrac{\mathbf{1}}{\mathbf{26}}\boldsymbol{e^x \cos 5x}$

Check: Let $y = \dfrac{5}{26}e^x \sin 5x + \dfrac{1}{26}e^x \cos 5x$, then $y' = \dfrac{5}{26}e^x \cdot \sin 5x + \dfrac{5}{26}\cos 5x \cdot 5 \cdot e^x + \dfrac{1}{26}e^x \cdot \cos 5x - \dfrac{1}{26}\sin 5x \cdot 5 \cdot e^x + 0$

$= \dfrac{5}{26}e^x \sin 5x + \dfrac{25}{26}e^x \cos 5x + \dfrac{1}{26}e^x \cos 5x - \dfrac{5}{26}e^x \sin 5x = \dfrac{25}{26}e^x \cos 5x + \dfrac{1}{26}e^x \cos 5x = \dfrac{26}{26}e^x \cos 5x = e^x \cos 5x$

2. Evaluate the following integrals using the integration by parts method.

a. Given $\int x \sec^2 x \, dx$ let $u = x$ and $dv = \sec^2 x \, dx$ then $du = dx$ and $\int dv = \int \sec^2 x \, dx$ which implies $v = \tan x$. Using

the integration by parts formula $\int u \, dv = u \, v - \int v \, du$ we obtain

$\int x \sec^2 x \, dx = x \cdot \tan x - \int \tan x \, dx = \boldsymbol{x\tan x - \ln|\sec x| + c}$

Check: Let $y = x\tan x - \ln|\sec x| + c$, then $y' = \left(1 \cdot \tan x + \sec^2 x \cdot x\right) - \dfrac{\sec x \tan x}{\sec x} + 0 = \tan x + x\sec^2 x - \tan x = x\sec^2 x$

b. Given $\int arc\sin 3y \, dy$ let $u = arc\sin 3y$ and $dv = dy$ then $du = \dfrac{3dy}{\sqrt{1-9y^2}}$ and $\int dv = \int dy$ which implies $v = y$. Using

the integration by parts formula $\int u \, dv = u \, v - \int v \, du$ we obtain

$$\int arc\sin 3y \, dy \;=\; arc\sin 3y \cdot y - \int y \cdot \frac{3dy}{\sqrt{1-9y^2}} \;=\; y \, arc\sin 3y - 3\int \frac{y\,dy}{\sqrt{1-9y^2}} \qquad (1)$$

To integrate $\int \dfrac{3y\,dy}{\sqrt{1-9y^2}}$ use the substitution method by letting $w = 1-9y^2$ then $\dfrac{dw}{dy} = -18y$ and $dy = -\dfrac{dw}{18y}$. Therefore,

$$\int \frac{3y\,dy}{\sqrt{1-9y^2}} \;=\; \int \frac{3y}{\sqrt{w}} \cdot \frac{dw}{-18y} \;=\; -\frac{1}{6}\int \frac{dw}{\sqrt{w}} \;=\; -\frac{1}{6}\int w^{-\frac{1}{2}}dw \;=\; -\frac{1}{6}\cdot\frac{1}{1-\frac{1}{2}}w^{1-\frac{1}{2}} \;=\; -\frac{1}{6}\cdot\frac{1}{\frac{2-1}{2}}w^{\frac{2-1}{2}} \;=\; -\frac{1}{6}\cdot\frac{2}{1}w^{\frac{1}{2}}$$

$$=\; -\frac{1}{3}w^{\frac{1}{2}} \;=\; -\frac{1}{3}\left(1-9y^2\right)^{\frac{1}{2}} \qquad (2)$$

Combining equations (1) and (2) together we obtain:

$$\int arc\sin 3y \, dy \;=\; y\, arc\sin 3y - \int \frac{3y\,dy}{\sqrt{1-9y^2}} \;=\; \boldsymbol{y \, arc\sin 3y + \frac{1}{3}\left(1-9y^2\right)^{\frac{1}{2}} + c}$$

Check: Let $w = y\, arc\sin 3y + \frac{1}{3}\left(1-9y^2\right)^{\frac{1}{2}} + c$, then $w' = arc\sin 3y + \dfrac{3y}{\sqrt{1-9y^2}} - \dfrac{1}{3}\cdot\dfrac{1}{2}\cdot\dfrac{18y}{\sqrt{1-9y^2}} + 0 = arc\sin 3y$

$$+\frac{3y}{\sqrt{1-9y^2}} - \frac{1}{6}\cdot\frac{18y}{\sqrt{1-9y^2}} \;=\; arc\sin 3y + \frac{3y}{\sqrt{1-9y^2}} - \frac{3y}{\sqrt{1-9y^2}} \;=\; arc\sin 3y$$

c. Given $\int arc\tan x \, dx$ let $u = arc\tan x$ and $dv = dx$ then $du = \dfrac{dx}{1+x^2}$ and $\int dv = \int dx$ which implies $v = x$. Using the

integration by parts formula $\int u \, dv = u\,v - \int v \, du$ we obtain

$$\int arc\tan x \, dx \;=\; arc\tan x \cdot x - \int x \cdot \frac{dx}{1+x^2} \;=\; x\, arc\tan x - \int \frac{x\,dx}{1+x^2} \qquad (1)$$

To integrate $\int \dfrac{x\,dx}{1+x^2}$ use the substitution method by letting $w = 1+x^2$ then $\dfrac{dw}{dx} = 2x$ And $dx = \dfrac{dw}{2x}$. Therefore,

$$\int \frac{x\,dx}{1+x^2} \;=\; \int \frac{x}{w}\cdot\frac{dw}{2x} \;=\; \frac{1}{2}\int \frac{dw}{w} \;=\; \frac{1}{2}\ln|w| \;=\; \frac{1}{2}\ln\left|1+x^2\right| \qquad (2)$$

Combining equations (1) and (2) together we obtain:

$$\int arc\tan x \, dx \;=\; x\, arc\tan x - \int \frac{x\,dx}{1+x^2} \;=\; \boldsymbol{x\, arc\tan x - \frac{1}{2}\ln\left|1+x^2\right| + c}$$

Check: Let $y = x\,arc\tan x - \frac{1}{2}\ln\left|1+x^2\right| + c$, then $y' = arc\tan x + \dfrac{x}{1+x^2} - \dfrac{1}{2}\dfrac{2x}{1+x^2} + 0 \;\; arc\tan x + \dfrac{x}{1+x^2} - \dfrac{x}{1+x^2} = arc\tan x$

d. Given $\int \sin^3 5x \, dx = \int \sin^2 5x \cdot \sin 5x \, dx$ let $u = \sin^2 5x$ and $dv = \sin 5x \, dx$ then $du = 10\sin 5x \cos 5x \, dx$ and

$\int dv = \int \sin 5x \, dx \;\; \int dv = \int \sin x \, dx$ which implies $v = -\dfrac{1}{5}\cos 5x$. Using the integration by parts formula $\int u \, dv = u\,v - \int v \, du$

we obtain $\int \sin^3 5x\,dx \;=\; \sin^2 5x \cdot -\dfrac{\cos 5x}{5} + \dfrac{1}{5}\int \cos 5x \cdot 10\sin 5x \cos 5x \, dx \;=\; -\dfrac{1}{5}\sin^2 5x \cos 5x + 2\int \cos^2 5x \sin 5x\,dx \quad (1)$

To integrate $\int \cos^2 5x \sin 5x \, dx$ use the integration by parts method again, i.e., let $u = \cos^2 5x$ and $dv = \sin 5x$ then

$du = -10\sin 5x \cos 5x \, dx$ and $\int dv = \int \sin 5x \, dx$ which implies $v = -\dfrac{1}{5}\cos 5x$. Therefore,

$\int \cos^2 5x \sin 5x \, dx \;=\; \cos^2 5x \cdot -\dfrac{1}{5}\cos 5x - \dfrac{1}{5}\int \cos 5x \cdot 10\sin 5x \cos 5x \, dx \;=\; -\dfrac{1}{5}\cos^3 5x - 2\int \cos^2 5x \sin 5x \, dx$. Taking the integral $-2\int \cos^2 5x \sin 5x\, dx$ from the right hand side of the equation to the left side we obtain

$$\int \cos^2 5x \sin 5x\, dx + 2\int \cos^2 5x \sin 5x\, dx \;=\; -\frac{1}{5}\cos^3 5x \;.\; \text{Therefore,} \int \cos^2 5x \sin 5x\, dx \;=\; -\frac{1}{15}\cos^3 5x \qquad (2)$$

Combining equations (1) and (2) together we have

$$\int \sin^3 5x\, dx \;=\; -\frac{1}{5}\sin^2 5x \cos 5x + 2\int \cos^2 5x \sin 5x\, dx \;=\; -\frac{1}{5}\sin^2 5x \cos 5x + 2\cdot -\frac{1}{15}\cos^3 5x \;=\; -\frac{\sin^2 5x \cos 5x}{5} - \frac{2}{15}\cos^3 5x$$

$$-\frac{1}{5}\left(1 - \cos^2 5x\right)\cos 5x - \frac{2}{15}\cos^3 5x + c \;=\; -\frac{1}{5}\cos 5x + \left(\frac{1}{5}\cos^3 5x - \frac{2}{15}\cos^3 5x\right) + c \;=\; \frac{1}{15}\cos^3 5x - \frac{1}{5}\cos 5x + c$$

Note that another method of solving the above problem is in the following way:

$$\int \sin^3 5x\, dx \;=\; \int \sin^2 5x \cdot \sin 5x\, dx \;=\; \int \left(1 - \cos^2 5x\right)\cdot \sin 5x\, dx \ \text{ let } u = \cos 5x \,, \text{ then } \frac{du}{dx} = -5\sin 5x \ \text{ and } \ dx = -\frac{du}{5\sin 5x}\;.$$

Therefore, $\int \sin^3 5x\, dx \;=\; \int \sin^2 5x \cdot \sin 5x\, dx \;=\; \int \left(1 - \cos^2 5x\right)\cdot \sin 5x\, dx \;=\; \int \left(1 - u^2\right)\cdot \sin 5x \cdot -\frac{du}{5\sin 5x}$

$$=\; -\frac{1}{5}\int \left(1 - u^2\right) du \;=\; \int \left(\frac{1}{5}u^2 - \frac{1}{5}\right) du \;=\; \frac{1}{15}u^3 - \frac{1}{5}u + c \;=\; \frac{1}{15}\cos^3 5x - \frac{1}{5}\cos 5x + c$$

Check: Let $y = \frac{1}{15}\cos^3 5x - \frac{1}{5}\cos 5x + c$, then $y' \;=\; \frac{1}{15}\cdot 3\cos^2 5x \cdot -\sin 5x \cdot 5 + \frac{1}{5}\cdot 5\sin 5x + 0 \;=\; -\cos^2 5x \cdot \sin 5x + \sin 5x$

$$=\; \sin 5x\left(1 - \cos^2 5x\right) \;=\; \sin 5x \sin^2 5x \;=\; \sin^3 5x$$

e. Given $\int x^2 \cos x\, dx$ let $u = x^2$ and $dv = \cos x\, dx$ then $du = 2x\, dx$ and $\int dv = \int \cos x\, dx$ which implies $v = \sin x$. Using

the integration by parts formula $\int u\, dv = u\,v - \int v\, du$ we obtain

$$\int x^2 \cos x\, dx \;=\; x^2 \cdot \sin x - \int \sin x \cdot 2x\, dx \;=\; x^2 \sin x - 2\int x \sin x\, dx \qquad (1)$$

To integrate $\int x \sin x\, dx$ use the integration by parts formula again, i.e., let $u = x$ and $dv = \sin x\, dx$ then $du = dx$ and

$\int dv = \int \sin x\, dx$ which implies $v = -\cos x$. Using the integration by parts formula $\int u\, dv = u\,v - \int v\, du$ we obtain

$$\int x \sin x\, dx \;=\; x\cdot -\cos x + \int \cos x \cdot dx \;=\; -x \cos x + \sin x \qquad (2)$$

Combining equations (1) and (2) together we have

$$\int x^2 \cos x\, dx \;=\; x^2 \sin x - 2\int x \sin x\, dx \;=\; x^2 \sin x - 2(-x \cos x + \sin x) \;=\; x^2 \sin x + 2x \cos x - 2\sin x$$

Check: Let $y = x^2 \sin x + 2x \cos x - 2\sin x$, then $y' \;=\; \left(2x \sin x + x^2 \cos x\right) + 2(\cos x - x\sin x) - 2\cos x$

$$2x \sin x + x^2 \cos x + 2\cos x - 2x\sin x - 2\cos x \;=\; x^2 \cos x$$

f. Given $\int e^{-2x} \cos 3x\, dx$ let $u = \cos 3x$ and $dv = e^{-2x}dx$ then $du = -3\sin 3x\, dx$ and $\int dv = \int e^{-2x}dx$ which implies

$v = -\frac{1}{2}e^{-2x}$. Using the integration by parts formula $\int u\, dv = u\,v - \int v\, du$ we obtain

$$\int e^{-2x} \cos 3x\, dx \;=\; \cos 3x \cdot -\frac{1}{2}e^{-2x} - \frac{3}{2}\int e^{-2x}\cdot \sin 3x\, dx \;=\; -\frac{1}{2}e^{-2x}\cos 3x - \frac{3}{2}\int e^{-2x} \sin 3x\, dx \qquad (1)$$

To integrate $\int e^{-2x} \sin 3x\, dx$ use the integration by parts formula again, i.e., let $u = \sin 3x$ and $dv = e^{-2x}dx$ then

$du = 3\cos 3x\, dx$ and $\int dv = \int e^{-2x}dx$ which implies $v = -\frac{1}{2}e^{-2x}$. Therefore,

$$\int e^{-2x}\sin 3x\,dx \;=\; \sin 3x \cdot -\frac{1}{2}e^{-2x}+\frac{1}{2}\int e^{-2x}\cdot 3\cos 3x\,dx \;=\; -\frac{1}{2}e^{-2x}\sin 3x+\frac{3}{2}\int e^{-2x}\cos 3x\,dx \qquad (2)$$

Combining equations (1) and (2) together we have

$$\int e^{-2x}\cos 3x\,dx \;=\; -\frac{1}{2}e^{-2x}\cos 3x-\frac{3}{2}\int e^{-2x}\sin 3x\,dx \;=\; -\frac{1}{2}e^{-2x}\cos 3x+\frac{3}{4}e^{-2x}\sin 3x-\frac{9}{4}\int e^{-2x}\cos 3x\,dx$$

Taking the integral $\int e^{-2x}\cos 3x\,dx$ from the right hand side of the equation to the left hand side

we obtain $\int e^{-2x}\cos 3x\,dx+\dfrac{9}{4}\int e^{-2x}\cos 3x\,dx \;=\; -\dfrac{1}{2}e^{-2x}\cos 3x+\dfrac{3}{4}e^{-2x}\sin 3x$. Therefore,

$$\frac{13}{4}\int e^{-2x}\cos 3x\,dx \;=\; -\frac{1}{2}e^{-2x}\cos 3x+\frac{3}{4}e^{-2x}\sin 3x \text{ and thus } \int e^{-2x}\cos 3x\,dx \;=\; -\frac{2}{13}e^{-2x}\cos 3x+\frac{3}{13}e^{-2x}\sin 3x+c$$

Check: Let $y=-\dfrac{2e^{-2x}\cos 3x}{13}+\dfrac{3e^{-2x}\sin 3x}{13}+c$, then $y'\;=\;-\dfrac{2}{13}\Big(-2e^{-2x}\cos 3x-3e^{-2x}\sin 3x\Big)+\dfrac{3}{13}\Big(-2e^{-2x}\sin 3x+3e^{-2x}\cos 3x\Big)$

$$=\;\frac{4}{13}e^{-2x}\cos 3x+\frac{6}{13}e^{-2x}\sin 3x-\frac{6}{13}e^{-2x}\sin 3x+\frac{9}{13}e^{-2x}\cos 3x \;=\; \frac{4}{13}e^{-2x}\cos 3x+\frac{9}{13}e^{-2x}\cos 3x \;=\; e^{-2x}\cos 3x$$

g. Given $\int x(5x-1)^3\,dx$ let $u=x$ and $dv=(5x-1)^3\,dx$ then $du=dx$ and $\int dv=\int(5x-1)^3\,dx$ which implies $v=\dfrac{1}{20}(5x-1)^4$.

Using the integration by parts formula $\int u\,dv=u\,v-\int v\,du$ we obtain

$$\int x(5x-1)^3\,dx \;=\; x\cdot\frac{(5x-1)^4}{20}-\frac{1}{20}\int(5x-1)^4\,dx \;=\; \frac{x(5x-1)^4}{20}-\frac{1}{20}\cdot\frac{1}{25}(5x-1)^{4+1}+c \;=\; \frac{x(5x-1)^4}{20}-\frac{1}{500}(5x-1)^5+c$$

Check: Let $y=\dfrac{x(5x-1)^4}{20}-\dfrac{1}{500}(5x-1)^5+c$, then $y'\;=\;\dfrac{1}{20}\Big[(5x-1)^4+20x(5x-1)^3\Big]-\dfrac{25}{500}(5x-1)^4+0$

$$=\;\frac{1}{20}(5x-1)^4+x(5x-1)^3-\frac{1}{20}(5x-1)^4 \;=\; x(5x-1)^3$$

h. Given $\int x\csc^2 x\,dx$ let $u=x$ and $dv=\csc^2 x\,dx$ then $du=dx$ and $\int dv=\int\csc^2 x\,dx$ which implies $v=-\cot x$. Using

the integration by parts formula $\int u\,dv=u\,v-\int v\,du$ we obtain

$$\int x\csc^2 x\,dx \;=\; x\cdot-\cot x+\int\cot x\,dx \;=\; -x\cot x+\ln\big|\sin x\big|+c$$

Check: Let $y=-x\cot x+\ln\big|\sin x\big|+c$, then $y'\;=\;-\Big(\cot x-x\csc^2 x\Big)+\dfrac{\cos x}{\sin x}+0\;=\;-\cot x+x\csc^2 x+\cot x\;=\;x\csc^2 x$

i. Given $\int\dfrac{2}{3}\cos^{-1}5x\,dx$ let $u=\cos^{-1}5x$ and $dv=dx$ then $du=-\dfrac{5}{\sqrt{1-25x^2}}\,dx$ and $\int dv=\int dx$ which implies $v=x$.

Using the integration by parts formula $\int u\,dv=u\,v-\int v\,du$ we obtain

$$\int\frac{2}{3}\cos^{-1}5x\,dx \;=\; \frac{2}{3}\cos^{-1}5x\cdot x+\frac{2}{3}\int x\cdot\frac{5\,dx}{\sqrt{1-25x^2}} \;=\; \frac{2}{3}x\cos^{-1}5x+\frac{2}{3}\int\frac{5x}{\sqrt{1-25x^2}}\,dx$$

To integrate $\int\dfrac{5x}{\sqrt{1-25x^2}}\,dx$ use the substitution method by letting $w=1-25x^2$ then $\dfrac{dw}{dx}=-50x$ which implies $dx=-\dfrac{dw}{50x}$.

Thus, $\displaystyle\int\frac{5x}{\sqrt{1-25x^2}}\,dx \;=\; \int\frac{5x}{\sqrt{w}}\cdot-\frac{dw}{50x} \;=\; -\frac{1}{10}\int\frac{dw}{\sqrt{w}} \;=\; -\frac{1}{10}\int\frac{1}{w^{\frac{1}{2}}}\,dw \;=\; -\frac{1}{10}\int w^{-\frac{1}{2}}\,dw \;=\; -\frac{1}{10}\cdot 2w^{\frac{1}{2}} \;=\; -\frac{\sqrt{1-25x^2}}{5}$

and $\int \frac{2}{3} \cos^{-1} 5x \, dx = \frac{2}{3} x \cos^{-1} 5x + \frac{2}{3} \int \frac{5x}{\sqrt{1 - 25x^2}} \, dx = \frac{2}{3} x \cos^{-1} 5x - \frac{2\sqrt{1 - 25x^2}}{15} + c$

Check: Let $y = \frac{2}{3} x \cos^{-1} 5x - \frac{2\sqrt{1 - 25x^2}}{15} + c$, then $y' = \frac{2}{3} \cos^{-1} 5x - \frac{2}{3} \cdot \frac{5x}{\sqrt{1 - 25x^2}} - \frac{2}{15} \cdot \frac{-50x}{2\sqrt{1 - 25x^2}} + 0$

$$= \frac{2}{3} \cos^{-1} 5x - \frac{10x}{3\sqrt{1 - 25x^2}} + \frac{10x}{3\sqrt{1 - 25x^2}} = \frac{2}{3} \cos^{-1} 5x$$

j. Given $\int \sinh^{-1} x \, dx$ let $u = \sinh^{-1} x$ and $dv = dx$ then $du = \frac{1}{\sqrt{1 + x^2}} \, dx$ and $\int dv = \int x \, dx$ which implies $v = x$. Using

the integration by parts formula $\int u \, dv = u \, v - \int v \, du$ we obtain

$$\int \sinh^{-1} x \, dx = \sinh^{-1} x \cdot x - \int x \cdot \frac{dx}{\sqrt{1 + x^2}} = x \sinh^{-1} x - \int \frac{x \, dx}{\sqrt{1 + x^2}} \qquad (1)$$

To get the integral of $\int \frac{x \, dx}{\sqrt{1 + x^2}}$ use the substitution method by letting $w = 1 + x^2$ then $dw = 2x \, dx$ which implies

$dx = \frac{dw}{2x}$. Therefore, $\int \frac{x \, dx}{\sqrt{1 + x^2}} = \int \frac{x}{\sqrt{w}} \cdot \frac{dw}{2x} = \frac{1}{2} \int \frac{1}{\sqrt{w}} \, dw = \frac{1}{2} \int \frac{1}{w^{\frac{1}{2}}} \, dw = \frac{1}{2} \int w^{-\frac{1}{2}} \, dw = \frac{1}{2} \cdot \frac{1}{1 - \frac{1}{2}} w^{1 - \frac{1}{2}} = \frac{2}{2} w^{\frac{1}{2}}$

$$= w^{\frac{1}{2}} = \left(1 + x^2\right)^{\frac{1}{2}} = \sqrt{1 + x^2} \qquad (2)$$

Combining equations (1) and (2) together we have

$$\int \sinh^{-1} x \, dx = x \sinh^{-1} x - \int \frac{x \, dx}{\sqrt{1 + x^2}} = x \sinh^{-1} x - \left(1 + x^2\right)^{\frac{1}{2}} + c$$

Check: Let $y = x \sinh^{-1} x - \left(1 + x^2\right)^{\frac{1}{2}} + c$, then $y' = \sinh^{-1} x + \frac{x}{\sqrt{1 + x^2}} - \frac{2x}{2\sqrt{1 + x^2}} + 0 = \sinh^{-1} x$

k. Given $\int x \sec^2 10x \, dx$ let $u = x$ and $dv = \sec^2 10x \, dx$ then $du = dx$ and $\int dv = \int \sec^2 10x \, dx$ which implies $v = \frac{\tan 10x}{10}$.

Using the integration by parts formula $\int u \, dv = u \, v - \int v \, du$ we obtain

$$\int x \sec^2 10x \, dx = x \cdot \frac{\tan 10x}{10} - \frac{1}{10} \int \tan 10x \, dx = \frac{1}{10} x \tan 10x - \frac{1}{100} \ln \left| \sec 10x \right| + c$$

Check: Let $y = \frac{1}{10} x \tan 10x - \frac{1}{100} \ln \left| \sec 10x \right| + c$, then $y' = \frac{1}{10} \tan 10x + x \sec^2 10x - \frac{1}{100} \frac{\sec 10x \tan 10x \cdot 10}{\sec 10x} + 0$

$$= \frac{1}{10} \tan 10x + x \sec^2 10x - \frac{1}{10} \tan 10x = x \sec^2 10x$$

l. Given $\int \frac{x}{5} \sinh 7x \, dx$ let $u = \frac{x}{5}$ and $dv = \sinh 7x \, dx$ then $du = \frac{dx}{5}$ and $\int dv = \int \sinh 7x \, dx$ which implies $v = \frac{1}{7} \cosh 7x \, dx$.

Using the integration by parts formula $\int u \, dv = u \, v - \int v \, du$ we obtain

$$\int \frac{x}{5} \sinh 7x\, dx = \frac{x}{5} \cdot \frac{1}{7} \cosh 7x - \int \frac{1}{7} \cosh 7x \cdot \frac{dx}{5} = \frac{1}{35} x \cosh 7x - \frac{1}{35} \int \cosh 7x \cdot dx = \frac{1}{35} x \cosh 7x - \frac{1}{245} \sinh 7x + c$$

Check: Let $y = \frac{1}{35} x \cosh 7x - \frac{1}{245} \sinh 7x + c$, then $y' = \frac{1}{35} \cosh 7x + \frac{1}{35} \cdot 7x \sinh 7x - \frac{1}{245} \cdot 7 \cosh 7x + 0 = \frac{1}{35} \cosh 7x$

$$+ \frac{1}{5} x \sinh 7x - \frac{1}{35} \cosh 7x = \frac{1}{5} x \sinh 7x$$

Section 5.2 Solutions – Integration Using Trigonometric Substitution

Evaluate the following indefinite integrals.

a. Given $\int \dfrac{dx}{x^2 \sqrt{16 - x^2}}$ let $x = 4 \sin t$, then $dx = 4 \cos t\, dt$ and $\sqrt{16 - x^2} = \sqrt{16 - 16 \sin^2 t} = \sqrt{16\left(1 - \sin^2 t\right)}$

$= \sqrt{16 \cos^2 t} = 4 \cos t$. Substituting these values back into the original integral we obtain:

$$\int \frac{dx}{x^2 \sqrt{16 - x^2}} = \int \frac{4 \cos t\, dt}{(4 \sin t)^2 \cdot 4 \cos t} = \int \frac{4 \cos t}{16 \sin^2 t \cdot 4 \cos t}\, dt = \int \frac{1}{16 \sin^2 t}\, dt = \frac{1}{16} \int \csc^2 t\, dt = -\frac{1}{16} \cot t + c$$

$$= -\frac{1}{16} \frac{\cos t}{\sin t} + c = -\frac{1}{16} \frac{\frac{\sqrt{16-x^2}}{4}}{\frac{x}{4}} + c = -\frac{1}{16} \cdot \frac{4 \cdot \sqrt{16 - x^2}}{4 \cdot x} + c = -\frac{\sqrt{16 - x^2}}{16x} + c$$

Check: Let $y = -\dfrac{\sqrt{16 - x^2}}{16x} + c$, then $y' = -\dfrac{\frac{-2x}{2\sqrt{16-x^2}} \cdot x - 1 \cdot \sqrt{16 - x^2}}{16x^2} + 0 = \dfrac{\frac{-2x^2}{2\sqrt{16-x^2}} - \frac{\sqrt{16-x^2}}{1}}{16x^2} = -\dfrac{\frac{-x^2 - \sqrt{16-x^2} \cdot \sqrt{16-x^2}}{\sqrt{16-x^2}}}{16x^2}$

$$= -\frac{\frac{-x^2 - \left(16 - x^2\right)}{\sqrt{16 - x^2}}}{16x^2} = -\frac{-x^2 - 16 + x^2}{16x^2 \sqrt{16 - x^2}} = \frac{16}{16x^2 \sqrt{16 - x^2}} = \frac{1}{x^2 \sqrt{16 - x^2}}$$

b. Given $\int \dfrac{x^2}{\sqrt{9 - x^2}}\, dx$ let $x = 3 \sin t$, then $dx = 3 \cos t\, dt$ and $\sqrt{9 - x^2} = \sqrt{9 - 9 \sin^2 t} = \sqrt{9\left(1 - \sin^2 t\right)} = \sqrt{9 \cos^2 t}$

$= 3 \cos t$. Substituting these values back into the original integral we obtain:

$$\int \frac{x^2}{\sqrt{9 - x^2}}\, dx = \int \frac{9 \sin^2 t \cdot 3 \cos t\, dt}{3 \cos t} = \int 9 \sin^2 t\, dt = 9 \int \frac{1 - \cos 2t}{2}\, dt = \frac{9}{2} \int \left(1 - \cos 2t\right) dt = \frac{9}{2}\left(t - \frac{1}{2} \sin 2t\right) + c$$

$$= \frac{9}{2} t - \frac{9}{2} \sin t \cos t + c = \frac{9}{2} \cdot \sin^{-1} \frac{x}{3} - \frac{9}{2} \cdot \frac{x}{3} \cdot \frac{\sqrt{9 - x^2}}{3} + c = \frac{9}{2} \sin^{-1} \frac{x}{3} - \frac{x}{2} \sqrt{9 - x^2} + c$$

Check: Let $y = \dfrac{9}{2} \sin^{-1} \dfrac{x}{3} - \dfrac{x}{2} \sqrt{9 - x^2} + c$, then $y' = \dfrac{9}{2} \dfrac{1}{\sqrt{1 - \frac{x^2}{9}}} \cdot \dfrac{1}{3} - \left[\dfrac{1}{2} \sqrt{9 - x^2} + \dfrac{-2x}{2\sqrt{9 - x^2}} \cdot \dfrac{x}{2}\right]$

$$= \frac{9}{2} \frac{3}{\sqrt{9 - x^2}} \cdot \frac{1}{3} - \left(\frac{\sqrt{9 - x^2}}{2} - \frac{x^2}{2\sqrt{9 - x^2}}\right) = \frac{9}{2\sqrt{9 - x^2}} - \left(\frac{2\left(9 - x^2\right) - 2x^2}{4\sqrt{9 - x^2}}\right) = \frac{9}{2\sqrt{9 - x^2}} - \left(\frac{18 - 4x^2}{4\sqrt{9 - x^2}}\right)$$

$$= \frac{9}{2\sqrt{9 - x^2}} - \frac{2\left(9 - 2x^2\right)}{4\sqrt{9 - x^2}} = \frac{9}{2\sqrt{9 - x^2}} - \frac{9 - 2x^2}{2\sqrt{9 - x^2}} = \frac{9 - 9 + 2x^2}{2\sqrt{9 - x^2}} = \frac{2x^2}{2\sqrt{9 - x^2}} = \frac{x^2}{\sqrt{9 - x^2}}$$

c. Given $\int \dfrac{dx}{x\sqrt{9 + 4x^2}}$ let $x = \dfrac{3}{2} \tan t$, then $dx = \dfrac{3}{2} \sec^2 t\, dt$ and $\sqrt{9 + 4x^2} = \sqrt{9 + 4\left(\frac{3}{2} \tan t\right)^2} = \sqrt{9 + 4 \cdot \frac{9}{4} \tan^2 t}$

$$= \sqrt{9+9\tan^2 t} = \sqrt{9\left(1+\tan^2 t\right)} = \sqrt{9\sec^2 t} = 3\sec t . \text{ Therefore,}$$

$$\int \frac{dx}{x\sqrt{9+4x^2}} = \int \frac{\frac{3}{2}\sec^2 t}{\frac{3}{2}\tan t \cdot 3\sec t} dt = \frac{1}{3}\int \frac{\sec^2 t}{\tan t \cdot 3\sec t} dt = \frac{1}{3}\int \frac{\sec t}{\tan t} dt = \frac{1}{3}\int \frac{1}{\tan t}\cdot \sec t \; dt = \frac{1}{3}\int \frac{\cos t}{\sin t}\cdot\frac{1}{\cos t} dt$$

$$= \frac{1}{3}\int \frac{1}{\sin t} dt = \frac{1}{3}\int \csc t \; dt = \frac{1}{3}\ln\left|\csc t - \cot t\right|+c = \frac{1}{3}\ln\left|\frac{\sqrt{9+4x^2}}{2x} - \frac{3}{2x}\right|+c = \frac{1}{3}\ln\left|\frac{\sqrt{9+4x^2}-3}{2x}\right|+c$$

Check: Let $y = \frac{1}{3}\ln\left|\frac{\sqrt{9+4x^2}-3}{2x}\right|+c$, then $y' = \frac{1}{3}\cdot \frac{2x}{\sqrt{9+4x^2}-3}\cdot \frac{2x\cdot \frac{1}{2\sqrt{9+4x^2}}\cdot 8x - 2\cdot\left(\sqrt{9+4x^2}-3\right)}{4x^2}$

$$= \frac{1}{3}\cdot\frac{2x}{\sqrt{9+4x^2}-3}\cdot\frac{\frac{8x^2}{\sqrt{9+4x^2}}-2\cdot\left(\sqrt{9+4x^2}-3\right)}{4x^2} = \frac{1}{3}\cdot\frac{2x}{\sqrt{9+4x^2}-3}\cdot\frac{8x^2-2\cdot\left(\sqrt{9+4x^2}-3\right)\sqrt{9+4x^2}}{4x^2\sqrt{9+4x^2}}$$

$$= \frac{1}{3}\cdot\frac{2x}{\sqrt{9+4x^2}-3}\cdot\frac{8x^2-2\cdot\left(9+4x^2\right)+6\sqrt{9+4x^2}}{4x^2\sqrt{9+4x^2}} = \frac{1}{3}\cdot\frac{2x}{\sqrt{9+4x^2}-3}\cdot\frac{8x^2-18+8x^2+6\sqrt{9+4x^2}}{4x^2\sqrt{9+4x^2}}$$

$$= \frac{1}{3}\cdot\frac{2x}{\sqrt{9+4x^2}-3}\cdot\frac{6\left(\sqrt{9+4x^2}-3\right)}{4x^2\sqrt{9+4x^2}} = \frac{1}{3}\cdot\frac{2x}{1}\cdot\frac{6}{4x^2\sqrt{9+4x^2}} = \frac{12x}{12x^2\sqrt{9+4x^2}} = \frac{1}{x\sqrt{9+4x^2}}$$

d. Given $\int \frac{1}{\left(49+x^2\right)^2} dx$ let $x = 7\tan t$, then $dx = 7\sec^2 t \; dt$ and $49+x^2 = 49+\left(7\tan t\right)^2 = 49+49\tan^2 t = 49\left(1+\tan^2 t\right)$

$= 49\sec^2 t$. Substituting these values back into the original integral we obtain

$$\int \frac{1}{\left(49+x^2\right)^2} dx = \int \frac{7\sec^2 t \; dt}{\left(49\sec^2 t\right)^2} = \int \frac{7\sec^2 t \; dt}{2401\sec^2 t \sec^2 t} = \frac{7}{2401}\int \frac{dt}{\sec^2 t} = \frac{1}{343}\int \cos^2 t \; dt = \frac{1}{343}\cdot\frac{1}{2}\int \left(1+\cos 2t\right) dt$$

$$= \frac{1}{686}\left(t+\frac{1}{2}\sin 2t\right)+c = \frac{1}{686}\left(t+\sin t \cos t\right)+c = \frac{1}{686}\left(\tan^{-1}\frac{x}{7}+\frac{x}{\sqrt{49+x^2}}\cdot\frac{7}{\sqrt{49+x^2}}\right)+c = \frac{1}{686}\left(\tan^{-1}\frac{x}{7}+\frac{7x}{49+x^2}\right)+c$$

Check: Let $y = \frac{1}{686}\left(\tan^{-1}\frac{x}{7}+\frac{7x}{49+x^2}\right)+c$ then $y' = \frac{1}{686}\cdot\frac{1}{7\left(1+\frac{x^2}{49}\right)}+\frac{1}{686}\cdot\frac{7\left(49+x^2\right)-2x\cdot 7x}{\left(49+x^2\right)^2} = \frac{1}{686}\cdot\frac{49}{7\left(49+x^2\right)}$

$$+\frac{1}{686}\cdot\frac{343+7x^2-14x^2}{\left(49+x^2\right)^2} = \frac{1}{686}\cdot\frac{7}{\left(49+x^2\right)}+\frac{1}{686}\cdot\frac{343-7x^2}{\left(49+x^2\right)^2} = \frac{1}{98\left(49+x^2\right)}+\frac{7\left(49-x^2\right)}{686\left(49+x^2\right)^2} = \frac{1}{98\left(49+x^2\right)}$$

$$+\frac{\left(49-x^2\right)}{98\left(49+x^2\right)^2} = \frac{49+x^2+49-x^2}{98\left(49+x^2\right)^2} = \frac{98}{98\left(49+x^2\right)^2} = \frac{1}{\left(49+x^2\right)^2}$$

e. $\int \left(\frac{x^2}{\sqrt{x^2-1}}+5x\right) dx = \int \frac{x^2}{\sqrt{x^2-1}} dx + \int 5x \; dx$. In Example 5.2-1, problem letter e, the solution to the first integral was:

$\int \frac{x^2}{\sqrt{x^2-1}} dx = \frac{1}{2}x\sqrt{x^2-1}+\frac{1}{2}\ln\left|x+\sqrt{x^2-1}\right|+c$. Therefore, combining the two integrals we have

$$\int \frac{x^2}{\sqrt{x^2-1}}\,dx + \int 5x\,dx \;=\; \frac{1}{2}x\sqrt{x^2-1}+\frac{1}{2}\ln\left|x+\sqrt{x^2-1}\right|+\frac{5}{2}x^2+c$$

Check: Let $y=\frac{1}{2}x\sqrt{x^2-1}+\frac{1}{2}\ln\left|x+\sqrt{x^2-1}\right|+\frac{5}{2}x^2+c$, then $y'=\frac{1}{2}\left(\sqrt{x^2-1}+\frac{2x^2}{2\sqrt{x^2-1}}\right)+\frac{1}{2}\frac{1}{x+\sqrt{x^2-1}}$

$$\times\left(1+\frac{2x}{2\sqrt{x^2-1}}\right)+\frac{5}{2}\cdot 2x = \frac{1}{2}\frac{2\left(x^2-1\right)+2x^2}{2\sqrt{x^2-1}}+\frac{1}{2}\frac{1}{x+\sqrt{x^2-1}}\frac{2\left(x+\sqrt{x^2-1}\right)}{2\sqrt{x^2-1}}+5x = \frac{1}{2}\left(\frac{4x^2-2}{2\sqrt{x^2-1}}+\frac{1}{\sqrt{x^2-1}}\right)+5x$$

$$=\frac{1}{2}\frac{4x^2-2+2}{2\sqrt{x^2-1}}+5x = \frac{4x^2}{4\sqrt{x^2-1}}+5x = \frac{x^2}{\sqrt{x^2-1}}+5x$$

f. Given $\int\sqrt{x^2-25}\,dx$ let $x=5\sec t$, then $dx=5\sec t\tan t\,dt$ and $\sqrt{x^2-25}=\sqrt{25\sec^2 t-25}=\sqrt{25\tan^2 t}=5\tan t$. Thus,

$$\int\sqrt{x^2-25}\,dx = \int 5\tan t\cdot 5\sec t\tan t\,dt = \int 25\sec t\tan^2 t\,dt = 25\int\sec t\left(\sec^2 t-1\right)dt = 25\int\sec^3 t\,dt - 25\int\sec t\,dt$$

$$=\frac{25}{2}\left(\tan t\sec t+\ln\left|\sec t+\tan t\right|\right)-25\ln\left|\sec t+\tan t\right|+c = \frac{25}{2}\left(\tan t\sec t-\ln\left|\sec t+\tan t\right|\right)+c$$

$$=\frac{25}{2}\left(\frac{x}{5}\cdot\frac{\sqrt{x^2-25}}{5}-\ln\left|\frac{x}{5}+\frac{\sqrt{x^2-25}}{5}\right|\right)+c = \frac{x}{2}\sqrt{x^2-25}-\frac{25}{2}\ln\left|\frac{x+\sqrt{x^2-25}}{5}\right|+c = \frac{x}{2}\sqrt{x^2-25}-\frac{25}{2}\ln\left|x+\sqrt{x^2-25}\right|$$

$$+\frac{25}{2}\ln 5+c = \frac{x}{2}\sqrt{x^2-25}-\frac{25}{2}\ln\left|x+\sqrt{x^2-25}\right|+c \quad\text{Note: }\frac{25}{2}\ln 5 \text{ is a constant which can be included in the constant } c.$$

Check: Let $y=\frac{x}{2}\sqrt{x^2-25}-\frac{25}{2}\ln\left|x+\sqrt{x^2-25}\right|+c$, then $y'=\frac{1}{2}\left(\sqrt{x^2-25}+\frac{2x^2}{2\sqrt{x^2-25}}\right)-\frac{25}{2}\cdot\frac{1}{x+\sqrt{x^2-25}}$

$$\times\left(1+\frac{2x}{2\sqrt{x^2-25}}\right)+0 = \frac{x^2-25+x^2}{2\sqrt{x^2-25}}-\frac{25}{2}\cdot\frac{1}{x+\sqrt{x^2-25}}\cdot\frac{x+\sqrt{x^2-25}}{\sqrt{x^2-25}} = \frac{x^2-25+x^2}{2\sqrt{x^2-25}}-\frac{25}{2\sqrt{x^2-25}} = \frac{2x^2-50}{2\sqrt{x^2-25}}$$

$$=\frac{2\left(x^2-25\right)}{2\sqrt{x^2-25}} = \frac{x^2-25}{\sqrt{x^2-25}} = \frac{x^2-25}{\sqrt{x^2-25}}\times\frac{\sqrt{x^2-25}}{\sqrt{x^2-25}} = \frac{\left(x^2-25\right)\sqrt{x^2-25}}{\left(x^2-25\right)} = \sqrt{x^2-25}$$

g. Given $\int\sqrt{36-x^2}\,dx$ let $x=6\sin t$, then $dx=6\cos t\,dt$ and $\sqrt{36-x^2}=\sqrt{36-36\sin^2 t}=\sqrt{36\cos^2 t}=6\cos t$. Thus,

$$\int\sqrt{36-x^2}\,dx = \int 6\cos t\cdot 6\cos t\,dt = \int 36\cos^2 t\,dt = \frac{36}{2}\int(1+\cos 2t)\,dt = \frac{36}{2}\left(t+\frac{1}{2}\sin 2t\right)+c = \frac{36}{2}(t+\sin t\cos t)+c$$

$$=\frac{36}{2}\left(\sin^{-1}\frac{x}{6}+\frac{x}{6}\cdot\frac{\sqrt{36-x^2}}{6}\right)+c = \frac{36}{2}\sin^{-1}\frac{x}{6}+\frac{36x\sqrt{36-x^2}}{2\cdot 36}+c = \frac{36}{2}\sin^{-1}\frac{x}{6}+\frac{x\sqrt{36-x^2}}{2}+c$$

Check: Let $y=\frac{36}{2}\sin^{-1}\frac{x}{6}+\frac{x\sqrt{36-x^2}}{2}+c$, then $y'=\frac{36}{2}\cdot\frac{1}{\sqrt{1-\left(\frac{x}{6}\right)^2}}\cdot\frac{1}{6}+\frac{1}{2}\sqrt{36-x^2}+\frac{-2x^2}{2\sqrt{36-x^2}} = \frac{36}{2}\cdot\frac{6}{6\sqrt{36-x^2}}$

$$+\frac{36-x^2-x^2}{2\sqrt{36-x^2}} = \frac{36}{2\sqrt{36-x^2}}+\frac{36-2x^2}{2\sqrt{36-x^2}} = \frac{72-2x^2}{2\sqrt{36-x^2}} = \frac{2\left(36-x^2\right)}{2\sqrt{36-x^2}} = \frac{36-x^2}{\sqrt{36-x^2}} = \frac{36-x^2}{\sqrt{36-x^2}}\cdot\frac{\sqrt{36-x^2}}{\sqrt{36-x^2}}$$

$$=\frac{\left(36-x^2\right)\sqrt{36-x^2}}{36-x^2} = \sqrt{36-x^2}$$

h. Given $\int \dfrac{dx}{\left(9+36x^2\right)^{\frac{3}{2}}} = \int \dfrac{dx}{\left(\frac{9}{36}+x^2\right)^{\frac{3}{2}}}$ let $x = \dfrac{3}{6}\tan t = \dfrac{1}{2}\tan t$, then $dx = \dfrac{1}{2}\sec^2 t\, dt$ and $9+36x^2 = 9+36\cdot\left(\dfrac{1}{2}\tan t\right)^2$

$= 9+36\cdot\dfrac{1}{4}\tan^2 t = 9+9\tan^2 t = 9\left(1+\tan^2 t\right) = 9\sec^2 t$. Therefore,

$$\int \dfrac{dx}{\left(9+36x^2\right)^{\frac{3}{2}}} = \int \dfrac{\frac{1}{2}\sec^2 dt}{\left(9\sec^2\right)^{\frac{3}{2}}} = \int \dfrac{\frac{1}{2}\sec^2 dt}{9^{\frac{3}{2}}\sec^{2\times\frac{3}{2}}t} = \int \dfrac{\frac{1}{2}\sec^2 dt}{\sqrt[2]{9^3}\ \sec^3 t} = \dfrac{1}{2}\int \dfrac{dt}{\sqrt[2]{729}\sec t} = \dfrac{1}{2}\int \dfrac{dt}{27\sec t} = \dfrac{1}{54}\int \dfrac{dt}{\sec t} = \dfrac{1}{54}\int \cos t\, dt$$

$$= \dfrac{1}{54}\sin t + c = \dfrac{1}{54}\dfrac{6x}{\sqrt{9+36x^2}} + c = \dfrac{x}{9\sqrt{9+36x^2}} + c$$

Check: Let $y = \dfrac{x}{9\sqrt{9+36x^2}} + c$, then $y' = \dfrac{1\cdot 9\sqrt{9+36x^2} - 9\cdot\dfrac{72x}{2\sqrt{9+36x^2}}\cdot x}{81\left(9+36x^2\right)} = \dfrac{\dfrac{9\left(9+36x^2\right)-324x^2}{\sqrt{9+36x^2}}}{81\left(9+36x^2\right)} = \dfrac{\dfrac{81+324x^2-324x^2}{\sqrt{9+36x^2}}}{81\left(9+36x^2\right)}$

$= \dfrac{81}{81\left(9+36x^2\right)\cdot\sqrt{9+36x^2}} = \dfrac{1}{\left(9+36x^2\right)\cdot\left(9+36x^2\right)^{\frac{1}{2}}} = \dfrac{1}{\left(9+36x^2\right)^{1+\frac{1}{2}}} = \dfrac{1}{\left(9+36x^2\right)^{\frac{3}{2}}}$

i. Given $\int \dfrac{\sqrt{9-4x^2}}{x}\, dx$ let $x = \dfrac{3}{2}\sin t$, then $dx = \dfrac{3}{2}\cos t\, dt$ and $\sqrt{9-4x^2} = \sqrt{9-4\left(\frac{3}{2}\sin t\right)^2} = \sqrt{9-4\cdot\frac{9}{4}\sin^2 t}$

$= \sqrt{9-9\sin^2 t} = \sqrt{9\left(1-\sin^2 t\right)} = \sqrt{9\cos^2 t} = 3\cos t$. Therefore,

$$\int \dfrac{\sqrt{9-4x^2}}{x}\, dx = \int \dfrac{3\cos t}{\frac{3}{2}\sin t}\cdot\dfrac{3}{2}\cos t\, dt = \dfrac{6}{3}\cdot\dfrac{3}{2}\int \dfrac{\cos^2 t}{\sin t}\, dt = 3\int \dfrac{1-\sin^2 t}{\sin t}\, dt = 3\int \dfrac{1}{\sin t}\, dt - 3\int \dfrac{\sin^2 t}{\sin t}\, dt = 3\int \dfrac{1}{\sin t}\, dt$$

$-3\int \sin t\, dt = 3\int \csc t\, dt - 3\int \sin t\, dt = 3\ln\left|\csc t - \cot t\right| + 3\cos t + c = 3\ln\left|\dfrac{3}{2x} - \dfrac{\sqrt{9-4x^2}}{2x}\right| + 3\cdot\dfrac{\sqrt{9-4x^2}}{3} + c$

$= 3\ln\left|\dfrac{3-\sqrt{9-4x^2}}{2x}\right| + \sqrt{9-4x^2} + c$

Check: Let $y = 3\ln\left|\dfrac{3-\sqrt{9-4x^2}}{2x}\right| + \sqrt{9-4x^2} + c$, then $y' = 3\cdot\dfrac{2x}{3-\sqrt{9-4x^2}}\cdot\dfrac{8x\cdot\dfrac{1}{2\sqrt{9-4x^2}}\cdot 2x - 2\cdot\left(3-\sqrt{9-4x^2}\right)}{4x^2}$

$-\dfrac{8x}{2\sqrt{9-4x^2}} = \dfrac{6x}{3-\sqrt{9-4x^2}}\cdot\dfrac{8x^2 - 2\cdot\left(3-\sqrt{9-4x^2}\right)\cdot\sqrt{9-4x^2}}{4x^2\sqrt{9-4x^2}} - \dfrac{4x}{\sqrt{9-4x^2}} = \dfrac{6x}{3-\sqrt{9-4x^2}}$

$\times\dfrac{8x^2-6\sqrt{9-4x^2}+2\left(9-4x^2\right)}{4x^2\sqrt{9-4x^2}} - \dfrac{4x}{\sqrt{9-4x^2}} = \dfrac{6x\left(8x^2-6\sqrt{9-4x^2}+18-8x^2\right)}{3-\sqrt{9-4x^2}\ \cdot 4x^2\sqrt{9-4x^2}} - \dfrac{4x}{\sqrt{9-4x^2}}$

$= \dfrac{3\left(-6\sqrt{9-4x^2}+18\right)}{3-\sqrt{9-4x^2}\ \cdot 2x\sqrt{9-4x^2}} - \dfrac{4x}{\sqrt{9-4x^2}} = \dfrac{18\left(3-\sqrt{9-4x^2}\right)}{\left(3-\sqrt{9-4x^2}\right)\cdot 2x\sqrt{9-4x^2}} - \dfrac{4x}{\sqrt{9-4x^2}} = \dfrac{9}{x\sqrt{9-4x^2}}$

$-\dfrac{4x}{\sqrt{9-4x^2}} = \dfrac{9-4x^2}{x\sqrt{9-4x^2}} = -\dfrac{9-4x^2}{x\sqrt{9-4x^2}}\times\dfrac{\sqrt{9-4x^2}}{\sqrt{9-4x^2}} = \dfrac{\left(9-4x^2\right)\sqrt{9-4x^2}}{x\left(9-4x^2\right)} = \dfrac{\sqrt{9-4x^2}}{x}$

a. Evaluate the integral $\int \dfrac{dx}{x^2+5x+6}$.

First - Check to see if the integrand is a proper or an improper rational fraction. If the integrand is an improper rational fraction use synthetic division (long division) to reduce the rational fraction to the sum of a polynomial and a proper rational fraction.

Second - Factor the denominator x^2+5x+6 into $(x+2)(x+3)$.

Third - Write the linear factors in partial fraction form. Since each linear factor in the denominator is occurring only once, the integrand can be represented in the following way:

$$\frac{1}{x^2+5x+6} = \frac{1}{(x+2)(x+3)} = \frac{A}{x+2} + \frac{B}{x+3}$$

Fourth - Solve for the constants A and B by equating coefficients of the like powers.

$$\frac{1}{x^2+5x+6} = \frac{A(x+3)+B(x+2)}{(x+2)(x+3)}$$

$$1 = A(x+3)+B(x+2) = Ax+3A+Bx+2B$$

$$1 = (A+B)x+(3A+2B) \text{ therefore,}$$

$$A+B=0 \qquad\qquad 3A+2B=1$$

which result in having $A=1$ and $B=-1$

Fifth - Rewrite the integral in its equivalent partial fraction form by substituting the constants with their specific values.

$$\int \frac{dx}{x^2+5x+6} = \int \frac{A}{x+2}dx + \int \frac{B}{x+3}dx = \int \frac{1}{x+2}dx - \int \frac{1}{x+3}dx$$

Sixth - Integrate each integral individually using integration methods learned in previous sections.

$$\int \frac{1}{x+2}dx - \int \frac{1}{x+3}dx = \ln|x+2| - \ln|x+3| + c$$

Seventh - Check the answer by differentiating the solution. The result should match the integrand.

Let $y = \ln|x+2| - \ln|x+3| + c$, then $y' = \dfrac{1}{x+2}\cdot 1 - \dfrac{1}{x+3}\cdot 1 + 0 = \dfrac{(x+3)-(x+2)}{(x+2)(x+3)} = \dfrac{x+3-x-2}{x^2+3x+2x+6} = \dfrac{1}{x^2+5x+6}$

b. Evaluate the integral $\int \dfrac{x^2+1}{x^3-4x} dx$.

First - Check to see if the integrand is a proper or an improper rational fraction. If the integrand is an improper rational fraction use synthetic division (long division) to reduce the rational fraction to the sum of a polynomial and a proper rational fraction.

Second - Factor the denominator x^3-4x into $x(x^2-4) = x(x-2)(x+2)$.

Third - Write the linear factors in partial fraction form. Since each linear factor in the denominator is occurring only once, the integrand can be represented in the following way:

$$\frac{x^2+1}{x^3-4x} = \frac{x^2+1}{x(x-2)(x+2)} = \frac{A}{x} + \frac{B}{x-2} + \frac{C}{x+2}$$

Fourth - Solve for the constants A, B, and C by equating coefficients of the like powers.

$$\frac{x^2+1}{x^3-4x} = \frac{A(x-2)(x+2)+Bx(x+2)+Cx(x-2)}{x(x-2)(x+2)}$$

$$x^2+1 = A(x^2+2x-2x-4)+B(x^2+2x)+C(x^2-2x) = Ax^2-4A+Bx^2+2Bx+Cx^2-2Cx$$

$$x^2+1 = (A+B+C)x^2+(2B-2C)x-4A \text{ therefore,}$$

$$A+B+C=1 \qquad\qquad 2B-2C=0 \qquad\qquad -4A=1$$

which result in having $A = -\dfrac{1}{4}$, $B = \dfrac{5}{8}$, and $C = \dfrac{5}{8}$.

Fifth - Rewrite the integral in its equivalent partial fraction form by substituting the constants with their specific values.

$$\int \frac{x^2+1}{x^3-4x}\,dx = \int \frac{A}{x}\,dx + \int \frac{B}{x-2}\,dx + \int \frac{C}{x+2}\,dx = -\frac{1}{4}\int \frac{1}{x}\,dx + \frac{5}{8}\int \frac{1}{x-2}\,dx + \frac{5}{8}\int \frac{1}{x+2}\,dx$$

Sixth - Integrate each integral individually using integration methods learned in previous sections.

$$-\frac{1}{4}\int \frac{1}{x}\,dx + \frac{5}{8}\int \frac{1}{x-2}\,dx + \frac{5}{8}\int \frac{1}{x+2}\,dx = -\frac{1}{4}\ln|x| + \frac{5}{8}\ln|x-2| + \frac{5}{8}\ln|x+2| + c$$

Seventh - Check the answer by differentiating the solution. The result should match the integrand.

Let $y = -\dfrac{1}{4}\ln|x| + \dfrac{5}{8}\ln|x-2| + \dfrac{5}{8}\ln|x+2| + c$, then $y' = -\dfrac{1}{4x} + \dfrac{5}{8(x-2)} + \dfrac{5}{8(x+2)} = \dfrac{-2(x-2)(x+2) + 5x(x+2) + 5x(x-2)}{8x(x-2)(x+2)}$

$$= \frac{-2x^2+8+5x^2+10x+5x^2-10x}{8x(x^2-4)} = \frac{8x^2+8}{8x(x^2-4)} = \frac{8(x^2+1)}{8x(x^2-4)} = \frac{x^2+1}{x^3-4x}$$

c. Evaluate the integral $\displaystyle\int \frac{1}{36-x^2}\,dx$.

First - Check to see if the integrand is a proper or an improper rational fraction. If the integrand is an improper rational fraction use synthetic division (long division) to reduce the rational fraction to the sum of a polynomial and a proper rational fraction.

Second - Factor the denominator $36-x^2$ into $(6-x)(6+x)$.

Third - Write the linear factors in partial fraction form. Since each linear factor in the denominator is occurring only once, the integrand can be represented in the following way:

$$\frac{1}{36-x^2} = \frac{1}{(6-x)(6+x)} = \frac{A}{6-x} + \frac{B}{6+x}$$

Fourth - Solve for the constants A and B by equating coefficients of the like powers.

$$\frac{1}{36-x^2} = \frac{A(6+x)+B(6-x)}{(6-x)(6+x)}$$

$$1 = A(6+x)+B(6-x) = 6A + Ax + 6B - Bx$$

$$1 = (A-B)x + (6A+6B) \text{ therefore,}$$

$$6A+6B = 1 \qquad\qquad A-B = 0$$

which result in having $A = \dfrac{1}{12}$, and $B = \dfrac{1}{12}$

Fifth - Rewrite the integral in its equivalent partial fraction form by substituting the constants with their specific values.

$$\int \frac{1}{36-x^2}\,dx = \int \frac{A}{6-x}\,dx + \int \frac{B}{6+x}\,dx = \frac{1}{12}\int \frac{1}{6-x}\,dx + \frac{1}{12}\int \frac{1}{6+x}\,dx$$

Sixth - Integrate each integral individually using integration methods learned in previous sections.

$$\frac{1}{12}\int \frac{1}{6-x}\,dx + \frac{1}{12}\int \frac{1}{6+x}\,dx = \frac{1}{12}\ln|6-x| + \frac{1}{12}\ln|6+x| + c$$

Seventh - Check the answer by differentiating the solution. The result should match the integrand.

Let $y = \dfrac{1}{12}\ln|6-x| + \dfrac{1}{12}\ln|6+x| + c$, then $y' = \dfrac{1}{12}\cdot\dfrac{1}{6-x} + \dfrac{1}{12}\cdot\dfrac{1}{6+x} + 0 = \dfrac{6+x+6-x}{12(6-x)(6+x)} = \dfrac{1}{36-x^2}$

d. Evaluate the integral $\displaystyle\int \frac{x+5}{x^3+2x^2+x}\,dx$.

First - Check to see if the integrand is a proper or an improper rational fraction. If the integrand is an improper rational fraction use synthetic division (long division) to reduce the rational fraction to the sum of a polynomial and a proper rational fraction.

Second - Factor the denominator $x^3 + 2x^2 + x$ into $x(x^2 + 2x + 1) = x(x+1)^2$.

Third - Write the linear factors in partial fraction form. Since one of the factors in the denominator is repeated, the integrand can be represented in the following way:

$$\frac{x+5}{x^3 + 2x^2 + x} = \frac{x+5}{x(x^2 + 2x + 1)} = \frac{x+5}{x(x+1)^2} = \frac{A}{x} + \frac{B}{x+1} + \frac{C}{(x+1)^2}$$

Fourth - Solve for the constants A, B, and C by equating coefficients of the like powers.

$$\frac{x+5}{x^3 + 2x^2 + x} = \frac{A(x+1)^2 + Bx(x+1) + Cx}{x(x+1)(x+1)^2}$$

$$x+5 = A(x^2 + 2x + 1) + B(x^2 + x) + Cx = Ax^2 + 2Ax + A + Bx^2 + Bx + Cx$$

$$x+5 = (A+B)x^2 + (2A+B+C)x + A \quad \text{therefore,}$$

$$A+B = 0 \qquad\qquad 2A+B+C = 1 \qquad\qquad A = 5$$

which result in having $A = 5$, $B = -5$, and $C = -4$

Fifth - Rewrite the integral in its equivalent partial fraction form by substituting the constants with their specific values.

$$\int \frac{x+5}{x^3 + 2x^2 + x} dx = \int \frac{A}{x} dx + \int \frac{B}{x+1} dx + \int \frac{C}{(x+1)^2} dx = 5\int \frac{1}{x} dx - 5\int \frac{1}{x+1} dx - 4\int \frac{1}{(x+1)^2} dx$$

Sixth - Integrate each integral individually using integration methods learned in previous sections.

$$5\int \frac{1}{x} dx - 5\int \frac{1}{x+1} dx - 4\int \frac{1}{(x+1)^2} dx = 5\ln|x| - 5\ln|x+1| + \frac{4}{x+1} + c$$

Seventh - Check the answer by differentiating the solution. The result should match the integrand.

Let $y = 5\ln|x| - 5\ln|x+1| + \frac{4}{x+1} + c$, then $y' = 5\cdot\frac{1}{x} - 5\cdot\frac{1}{x+1} - \frac{4}{(x+1)^2} + 0 = \frac{5(x+1)^2 - 5x(x+1) - 4x}{x(x+1)^2}$

$$= \frac{5x^2 + 10x + 5 - 5x^2 - 5x - 4x}{x^3 + 2x^2 + x} = \frac{x+5}{x^3 + 2x^2 + x}$$

e. Evaluate the integral $\int \frac{1}{x^3 - 2x^2 + x} dx$.

First - Check to see if the integrand is a proper or an improper rational fraction. If the integrand is an improper rational fraction use synthetic division (long division) to reduce the rational fraction to the sum of a polynomial and a proper rational fraction.

Second – Factor the denominator $x^3 - 2x^2 + x$ into $x(x-1)^2$.

Third - Write the linear factors in partial fraction form. Since one of the factors in the denominator is repeated, the integrand can be represented in the following way:

$$\frac{1}{x(x-1)^2} = \frac{A}{x} + \frac{B}{x-1} + \frac{C}{(x-1)^2}$$

Fourth - Solve for the constants A, B, and C by equating coefficients of the like powers.

$$\frac{1}{x^3 - 2x^2 + x} = \frac{A(x-1)^2 + Bx(x-1) + Cx}{x(x-1)^2}$$

$$1 = A(x^2 - 2x + 1) + B(x^2 - x) + Cx = Ax^2 - 2Ax + A + Bx^2 - Bx + Cx$$

$$1 = (A+B)x^2 + (-2A-B+C)x + A \quad \text{therefore,}$$

$$A+B = 0 \qquad\qquad -2A-B+C = 0 \qquad\qquad A = 1$$

which result in having $A = 1$, $B = -1$, and $C = 1$

Fifth - Rewrite the integral in its equivalent partial fraction form by substituting the constants with their specific values.

$$\int \frac{dx}{x^3 - 2x^2 + x} = \int \frac{A}{x}dx + \int \frac{B}{x-1}dx + \int \frac{C}{(x-1)^2}dx = \int \frac{1}{x}dx - \int \frac{1}{x-1}dx + \int \frac{1}{(x-1)^2}dx$$

Sixth - Integrate each integral individually using integration methods learned in previous sections.

$$\int \frac{1}{x}dx - \int \frac{1}{x-1}dx + \int \frac{1}{(x-1)^2}dx = \ln|x| - \ln|x-1| - \frac{1}{x-1} + c$$

Seventh - Check the answer by differentiating the solution. The result should match the integrand.

Let $y = \ln|x| - \ln|x-1| - \frac{1}{x-1} + c$, then $y' = \frac{1}{x} - \frac{1}{x-1} + \frac{1}{(x-1)^2} + 0 = \frac{(x-1)^2 - x(x-1) + x}{x(x-1)^2} = \frac{x^2 - 2x + 1 - x^2 + x + x}{x^3 - 2x^2 + x}$

$$= \frac{x^2 - x^2 - 2x + 2x + 1}{x^3 - 2x^2 + x} = \frac{1}{x^3 - 2x^2 + x}$$

f. Evaluate the integral $\int \frac{x^2 + 3}{x^2 - 1}dx$.

First – Rewrite the integral in the following form: $\int \frac{x^2+3}{x^2-1}dx = \int \frac{(x^2-1)+4}{x^2-1}dx = \int \left(\frac{x^2-1}{x^2-1} + \frac{4}{x^2-1} \right)dx = \int \left(1 + \frac{4}{x^2-1} \right)dx$.

Then, check to see if the integrand of the second integral is a proper or an improper rational fraction. If the integrand is an improper rational fraction use synthetic division (long division) to reduce the rational fraction to the sum of a polynomial and a proper rational fraction.

Second - Factor the denominator $x^2 - 1$ into $(x-1)(x+1)$.

Third - Write the linear factors in partial fraction form. Since one of the factors in the denominator is repeated, the integrand can be represented in the following way:

$$\frac{4}{x^2-1} = \frac{4}{(x-1)(x+1)} = \frac{A}{x-1} + \frac{B}{x+1}$$

Fourth - Solve for the constants A and B by equating coefficients of the like powers.

$$\frac{4}{x^2-1} = \frac{A(x+1) + B(x-1)}{(x-1)(x+1)}$$

$$4 = A(x+1) + B(x-1) = Ax + A + Bx - B$$

$$4 = (A+B)x + (A-B) \quad \text{therefore,}$$

$$A+B = 0 \qquad A-B = 4$$

which result in having $A = 2$ and $B = -2$

Fifth - Rewrite the integral in its equivalent partial fraction form by substituting the constants with their specific values.

$$\int \frac{x^2+3}{x^2-1}dx = \int \left(1 + \frac{4}{x^2-1} \right)dx = \int dx + \int \frac{4}{x^2-1}dx = \int dx + \int \frac{A}{x-1}dx + \int \frac{B}{x+1}dx = \int dx + 2\int \frac{1}{x-1}dx - 2\int \frac{1}{x+1}dx$$

Sixth - Integrate each integral individually using integration methods learned in previous sections.

$$\int dx + 2\int \frac{1}{x-1}dx - 2\int \frac{1}{x+1}dx = x + 2\ln|x-1| - 2\ln|x+1| + c$$

Seventh - Check the answer by differentiating the solution. The result should match the integrand.

Let $y = x + 2\ln|x-1| - 2\ln|x+1| + c$, then $y' = 1 + 2 \cdot \frac{1}{x-1} - 2 \cdot \frac{1}{x+1} + 0 = \frac{(x-1)(x+1) + 2(x+1) - 2(x-1)}{(x-1)(x+1)}$

$$= \frac{(x^2 + x - x - 1) + 2x + 2 - 2x + 2}{(x^2 + x - x - 1)} = \frac{x^2 - 1 + 4}{x^2 - 1} = \frac{x^2 + 3}{x^2 - 1}$$

g. Evaluate the integral $\int \frac{1}{x^3 - 1}dx$.

First - Check to see if the integrand is a proper or an improper rational fraction. If the integrand is an improper rational fraction use synthetic division (long division) to reduce the rational fraction to the sum of a polynomial and a proper rational fraction.

Second - Factor the denominator $x^3 - 8$ into $(x-1)(x^2 + x + 1)$.

Third - Write the factors in partial fraction form. Since one of the factors in the denominator is in quadratic form, the integrand can be represented in the following way:

$$\frac{1}{x^3 - 1} = \frac{1}{(x-1)(x^2 + x + 1)} = \frac{A}{x-1} + \frac{Bx+C}{x^2+x+1}$$

Fourth - Solve for the constants A, B and C by equating coefficients of the like powers.

$$\frac{1}{x^3 - 1} = \frac{A(x^2 + x + 1) + (Bx + C)(x-1)}{(x-1)(x^2 + x + 1)}$$

$$1 = A(x^2 + x + 1) + (Bx + C)(x - 1) = Ax^2 + Ax + A + Bx^2 - Bx + Cx - C$$

$$1 = (A+B)x^2 + (A - B + C)x + (A - C) \text{ therefore,}$$

$$A + B = 0 \qquad\qquad A - B + C = 0 \qquad\qquad A - C = 1$$

which result in having $A = \dfrac{1}{3}$, $B = -\dfrac{1}{3}$, and $C = -\dfrac{2}{3}$

Fifth - Rewrite the integral in its equivalent partial fraction form by substituting the constants with their specific values.

$$\int \frac{1}{x^3 - 1} dx = \int \frac{A}{x-1} dx + \int \frac{Bx+C}{x^2+x+1} dx = \frac{1}{3}\int \frac{1}{x-1} dx + \int \frac{-\frac{1}{3}x - \frac{2}{3}}{x^2 + x + 1} dx = \frac{1}{3}\int \frac{1}{x-1} dx - \frac{1}{3}\int \frac{x+2}{x^2+x+1} dx$$

Sixth - Integrate each integral individually using integration methods learned in previous

sections. To solve the second integral let $u = x^2 + x + 1$, then $\dfrac{du}{dx} = 2x + 1$ and $dx = \dfrac{du}{2x+1}$. Also, $x + 2$ can be rewritten

as $x + 2 = \dfrac{1}{2}(2x + 1) + \dfrac{3}{2}$. Therefore,

$$\frac{1}{3}\int \frac{1}{x-1} dx - \frac{1}{3}\int \frac{x+2}{x^2+x+1} dx = \frac{1}{3}\int \frac{1}{x-1} dx - \frac{1}{3}\int \frac{\frac{1}{2}(2x+1) + \frac{3}{2}}{x^2+x+1} dx = \frac{1}{3}\int \frac{1}{x-1} dx - \frac{1}{6}\int \frac{2x+1}{x^2+x+1} dx - \frac{1}{3}\int \frac{\frac{3}{2}}{x^2+x+1} dx$$

$$= \frac{1}{3}\int \frac{1}{x-1} dx - \frac{1}{6}\int \frac{2x+1}{x^2+x+1} dx - \frac{1}{2}\int \frac{1}{x^2+x+1} dx = \frac{1}{3}\int \frac{1}{x-1} dx - \frac{1}{6}\int \frac{2x+1}{u} \cdot \frac{du}{2x+1} - \frac{1}{2}\int \frac{1}{\left(x+\frac{1}{2}\right)^2 + \frac{3}{4}} dx$$

$$= \frac{1}{3}\ln|x-1| - \frac{1}{6}\ln|u| - \frac{1}{2} \cdot \frac{1}{\frac{\sqrt{3}}{2}} \tan^{-1}\frac{\frac{2x+1}{2}}{\frac{\sqrt{3}}{2}} + c = \frac{1}{3}\ln|x-1| - \frac{1}{6}\ln|x^2 + x + 1| - \frac{1}{2} \cdot \frac{2}{\sqrt{3}} \tan^{-1}\frac{2(2x+1)}{2\sqrt{3}} + c$$

$$= \frac{1}{3}\ln|x-1| - \frac{1}{6}\ln|x^2 + x + 1| - \frac{1}{\sqrt{3}} \tan^{-1}\frac{2x+1}{\sqrt{3}} + c = \frac{1}{3}\ln|x-1| - \frac{1}{6}\ln|x^2 + x + 1| - \frac{\sqrt{3}}{3}\tan^{-1}\frac{2x+1}{\sqrt{3}} + c$$

Seventh - Check the answer by differentiating the solution. The result should match the integrand.

Let $y = \dfrac{1}{3}\ln|x-1| - \dfrac{1}{6}\ln|x^2 + x + 1| - \dfrac{\sqrt{3}}{3}\tan^{-1}\dfrac{2x+1}{\sqrt{3}} + c$, then

$$y' = \frac{1}{3} \cdot \frac{1}{x-1} - \frac{1}{6} \cdot \frac{1}{x^2+x+1} \cdot (2x+1) - \frac{\sqrt{3}}{3} \cdot \frac{1}{1 + \left(\frac{2x+1}{\sqrt{3}}\right)^2} \cdot \frac{\left(2 \cdot \sqrt{3}\right) - 0 \cdot (2x+1)}{\left(\sqrt{3}\right)^2} + 0 = \frac{1}{3(x-1)} - \frac{2x+1}{6\left(x^2 + x + 1\right)}$$

$$- \frac{\sqrt{3}}{3} \cdot \frac{3}{3 + (2x+1)^2} \cdot \frac{2\sqrt{3}}{3} = \frac{1}{3(x-1)} - \frac{2x+1}{6\left(x^2+x+1\right)} - \frac{2}{3 + (2x+1)^2} = \frac{6x^2 + 6x + 6 - (2x+1)\cdot(3x-3)}{18(x-1)\left(x^2+x+1\right)} - \frac{2}{3 + 4x^2 + 4x + 1}$$

$$= \frac{6x^2+6x+6-6x^2+6x-3x+3}{18\left(x^3-1\right)} - \frac{2}{4x^2+4x+4} = \frac{9x+9}{18\left(x^3-1\right)} - \frac{1}{4\left(x^2+x+1\right)} = \frac{x+1}{2\left(x^3-1\right)} - \frac{1}{2\left(x^2+x+1\right)}$$

$$= \frac{(x+1)\left(x^2+x+1\right)-\left(x^3-1\right)}{2\left(x^3-1\right)\left(x^2+x+1\right)} = \frac{x^3+x^2+x+x^2+x+1-x^3+1}{\left(x^3-1\right)\left(2x^2+2x+2\right)} = \frac{2x^2+2x+2}{\left(x^3-1\right)\left(2x^2+2x+2\right)} = \frac{1}{x^3-1}$$

h. Evaluate the integral $\int \frac{1}{x^4-1}dx$.

First - Check to see if the integrand is a proper or an improper rational fraction. If the integrand is an improper rational fraction use synthetic division (long division) to reduce the rational fraction to the sum of a polynomial and a proper rational fraction.

Second - Factor the denominator x^4-1 into $\left(x^2-1\right)\left(x^2+1\right) = (x-1)(x+1)\left(x^2+1\right)$.

Third - Write the factors in partial fraction form. Since one of the factors in the denominator is in quadratic form, the integrand can be represented in the following way:

$$\frac{1}{x^4-1} = \frac{1}{(x-1)(x+1)\left(x^2+1\right)} = \frac{A}{x-1} + \frac{B}{x+1} + \frac{Cx+D}{x^2+1}$$

Fourth - Solve for the constants A, B, C and D by equating coefficients of the like powers.

$$\frac{1}{x^4-1} = \frac{A(x+1)\left(x^2+1\right)+B(x-1)\left(x^2+1\right)+(x-1)(x+1)(Cx+D)}{(x-1)(x+1)\left(x^2+1\right)}$$

$$1 = A(x+1)\left(x^2+1\right)+B(x-1)\left(x^2+1\right)+(x-1)(x+1)(Cx+D)$$

$$1 = Ax^3+Ax^2+Ax+A+Bx^3-Bx^2+Bx-B+Cx^3+Dx^2-Cx-D$$

$$1 = (A+B+C)x^3+(A-B+D)x^2+(A+B-C)x+(A-B-D) \text{ therefore,}$$

$$A+B+C=0 \qquad A-B+D=0 \qquad A+B-C=0 \qquad A-B-D=1$$

which result in having $A=\frac{1}{4}$, $B=-\frac{1}{4}$, $C=0$, and $D=-\frac{1}{2}$

Fifth - Rewrite the integral in its equivalent partial fraction form by substituting the constants with their specific values.

$$\int \frac{1}{x^4-1}dx = \int \frac{A}{x-1}dx + \int \frac{B}{x+1}dx + \int \frac{Cx+D}{x^2+1}dx = \frac{1}{4}\int \frac{1}{x-1}dx - \frac{1}{4}\int \frac{1}{x+1}dx - \frac{1}{2}\int \frac{1}{x^2+1}dx$$

Sixth - Integrate each integral individually using integration methods learned in previous sections.

$$\frac{1}{4}\int \frac{1}{x-1}dx - \frac{1}{4}\int \frac{1}{x+1}dx - \frac{1}{2}\int \frac{1}{x^2+1}dx = \frac{1}{4}\ln|x-1| - \frac{1}{4}\ln|x+1| - \frac{1}{2}\tan^{-1}x + c$$

Seventh - Check the answer by differentiating the solution. The result should match the integrand.

Let $y = \frac{1}{4}\ln|x-1| - \frac{1}{4}\ln|x+1| - \frac{1}{2}\tan^{-1}x + c$, then $y' = \frac{1}{4}\cdot\frac{1}{x-1}\cdot 1 - \frac{1}{4}\cdot\frac{1}{x+1}\cdot 1 - \frac{1}{2}\cdot\frac{1}{x^2+1}\cdot 1 + 0$

$$= \frac{1}{4(x-1)} - \frac{1}{4(x+1)} - \frac{1}{2\left(x^2+1\right)} = \frac{(x+1)\left(x^2+1\right)-(x-1)\left(x^2+1\right)-2(x-1)(x+1)}{4(x-1)(x+1)\left(x^2+1\right)} = \frac{x^3+x+x^2+1-x^3-x+x^2+1}{4\left(x^2-1\right)\left(4+x^2\right)}$$

$$\frac{-2x^2-2x+2x+2}{4\left(x^2-1\right)\left(4+x^2\right)} = \frac{4}{4\left(x^4-1\right)} = \frac{1}{x^4-1}$$

i. Evaluate the integral $\int \frac{1}{x^3+64}dx$.

First - Check to see if the integrand is a proper or an improper rational fraction. If the integrand is a rational fraction use synthetic division (long division) to reduce the rational fraction to the sum of a polynomial and a proper rational fraction.

Second - Factor the denominator x^3+64 into $(x+4)\left(x^2-4x+16\right)$.

Third - Write the factors in partial fraction form. Since one of the factors in the denominator is in quadratic form, the integrand can be represented in the following way:

$$\frac{1}{x^3+64} = \frac{1}{(x+4)(x^2-4x+16)} = \frac{A}{x+4} + \frac{Bx+C}{x^2-4x+16}$$

Fourth - Solve for the constants A, B, and C by equating coefficients of the like powers.

$$\frac{1}{x^3+64} = \frac{A(x^2-4x+16)+(Bx+C)(x+4)}{(x+4)(x^2-4x+16)}$$

$$1 = A(x^2-4x+16)+(Bx+C)(x+4) = Ax^2-4Ax+16A+Bx^2+4Bx+Cx+4C$$

$$1 = (A+B)x^2+(-4A+4B+C)x+(16A+4C) \quad \text{therefore,}$$

$$A+B=0 \qquad\qquad -4A+4B+C=0 \qquad\qquad 16A+4C=1$$

which result in having $A=\dfrac{1}{48}$, $B=-\dfrac{1}{48}$, and $C=\dfrac{1}{6}$

Fifth - Rewrite the integral in its equivalent partial fraction form by substituting the constants with their specific values.

$$\int\frac{1}{x^3+64}dx = \int\frac{A}{x+4}dx+\int\frac{Bx+C}{x^2-4x+16}dx = \frac{1}{48}\int\frac{1}{x+4}dx+\int\frac{-\frac{1}{48}x+\frac{1}{6}}{x^2-4x+16}dx = \frac{1}{48}\int\frac{1}{x+4}dx-\frac{1}{48}\int\frac{x-8}{x^2-4x+16}dx$$

Sixth - Integrate each integral individually using integration methods learned in previous

sections. To solve the second integral let $u=x^2-4x+16$, then $\dfrac{du}{dx}=2x-4$ and $dx=\dfrac{du}{2x-4}$. Also, $x-8$ can be

rewritten as $x-8 = (x-2)-6 = \dfrac{1}{2}(2x-4)-6$. Therefore,

$$\frac{1}{48}\int\frac{1}{x+4}dx-\frac{1}{48}\int\frac{x-8}{x^2-4x+16}dx = \frac{1}{48}\int\frac{1}{x+4}dx-\frac{1}{48}\int\frac{(x-2)-6}{x^2-4x+16}dx = \frac{1}{48}\int\frac{1}{x+4}dx-\frac{1}{48}\int\frac{\frac{1}{2}(2x-4)-6}{x^2-4x+16}dx$$

$$= \frac{1}{48}\int\frac{1}{x+4}dx-\frac{1}{96}\int\frac{2x-4}{x^2-4x+16}dx+\frac{1}{48}\int\frac{6}{x^2-4x+16}dx = \frac{1}{48}\int\frac{1}{x+4}dx-\frac{1}{96}\int\frac{2x-4}{u}\cdot\frac{du}{2x-4}+\frac{6}{48}\int\frac{1}{(x-2)^2+12}dx$$

$$= \frac{1}{48}\int\frac{1}{x+4}dx-\frac{1}{96}\int\frac{1}{u}du+\frac{1}{8}\int\frac{1}{(x-2)^2+12}dx = \frac{1}{48}\ln|x+4|-\frac{1}{96}\ln|u|+\frac{1}{8\sqrt{12}}\tan^{-1}\frac{x-2}{\sqrt{12}}+c$$

$$= \frac{1}{48}\ln|x+4|-\frac{1}{96}\ln\left|x^2-4x+16\right|+\frac{\sqrt{12}}{8\cdot12}\tan^{-1}\frac{x-2}{\sqrt{12}}+c = \frac{1}{48}\ln|x+4|-\frac{1}{96}\ln\left|x^2-4x+16\right|-\frac{\sqrt{12}}{96}\tan^{-1}\frac{x-2}{\sqrt{12}}+c$$

Seventh - Check the answer by differentiating the solution. The result should match the integrand.

Let $y = \dfrac{1}{48}\ln|x+4|-\dfrac{1}{96}\ln\left|x^2-4x+16\right|+\dfrac{\sqrt{12}}{96}\tan^{-1}\dfrac{x-2}{\sqrt{12}}+c$, then

$$y' = \frac{1}{48}\cdot\frac{1}{x+4}-\frac{1}{96}\cdot\frac{1}{x^2-4x+16}\cdot(2x-4)+\frac{\sqrt{12}}{96}\cdot\frac{1}{1+\left(\frac{x-2}{\sqrt{12}}\right)^2}\cdot\frac{\left(1\cdot\sqrt{12}\right)-0\cdot(x-2)}{\left(\sqrt{12}\right)^2}+0 = \frac{1}{48(x+4)}-\frac{x-2}{48\left(x^2-4x+16\right)}$$

$$+\frac{\sqrt{12}}{96}\cdot\frac{12}{12+(x-2)^2}\cdot\frac{\sqrt{12}}{12} = \frac{1}{48(x+4)}-\frac{x-2}{48\left(x^2-4x+16\right)}+\frac{6}{48\left(x^2-4x+16\right)} = \frac{1}{48(x+4)}+\frac{-x+2+6}{48\left(x^2+4x+16\right)}$$

$$= \frac{1}{48(x+4)}+\frac{-x+8}{48\left(x^2-4x+16\right)} = \frac{\left(x^2-4x+16\right)+(-x+8)(x+4)}{48(x+4)\left(x^2-4x+16\right)} = \frac{x^2-4x+16-x^2-4x+8x+32}{48(x+4)\left(x^2-4x+16\right)}$$

$$= \frac{16+32}{48(x+4)\left(x^2-4x+16\right)} = \frac{48}{48(x+4)\left(x^2-4x+16\right)} = \frac{1}{x^3-4x^2+16x+4x^2-16x+64} = \frac{1}{x^3+64}$$

Section 5.4 Practice Problems – Integration of Hyperbolic Functions

---

1. Evaluate the following integrals:

a. Given $\int \cosh 3x \, dx$ let $u = 3x$, then $\dfrac{du}{dx} = \dfrac{d}{dx} 3x = 3$ which implies $dx = \dfrac{du}{3}$. Therefore,

$$\int \cosh 3x \, dx \;=\; \int \cosh u \cdot \frac{du}{3} \;=\; \frac{1}{3}\int \cosh u \, du \;=\; \frac{1}{3}\sinh u + c \;=\; \mathbf{\frac{1}{3}\sinh 3x + c}$$

Check: Let $y = \dfrac{1}{3}\sinh 3x + c$, then $y' = \dfrac{1}{3}\cdot\dfrac{d}{dx}\sinh 3x + \dfrac{d}{dx}c = \dfrac{1}{3}\cdot\cosh 3x\cdot\dfrac{d}{dx}3x + 0 = \dfrac{1}{3}\cdot\cosh 3x\cdot 3 = \cosh 3x$

b. Given $\int\left(\sinh 2x - e^{3x}\right)dx = \int \sinh 2x\, dx + \int e^{3x}dx$ let:

a. $u = 2x$, then $\dfrac{du}{dx} = \dfrac{d}{dx}2x$; $\dfrac{du}{dx} = 2$; $du = 2dx$; $dx = \dfrac{du}{2}$ and

b. $v = 3x$, then $\dfrac{dv}{dx} = \dfrac{d}{dx}3x$; $\dfrac{dv}{dx} = 3$; $dv = 3dx$; $dx = \dfrac{dv}{3}$.

Therefore, $\int \sinh 2x\,dx + \int e^{3x}dx = \int \sinh u \cdot \dfrac{du}{2} + \int e^{3x}\cdot\dfrac{dv}{3} = \dfrac{1}{2}\int \sinh u\,du + \dfrac{1}{3}\int e^{v}dv = \dfrac{1}{2}\cosh u + c_1 + \dfrac{1}{3}e^{v} + c_2$

$= \dfrac{1}{2}\cosh 2x + \dfrac{1}{3}e^{3x} + c_1 + c_2 = \mathbf{\dfrac{1}{2}\cosh 2x + \dfrac{1}{3}e^{3x} + c}$

Check: Let $y = \dfrac{1}{2}\cosh 2x + \dfrac{1}{3}e^{3x} + c$ then $y' = \dfrac{1}{2}\cdot\dfrac{d}{dx}\cosh 2x + \dfrac{1}{3}\cdot\dfrac{d}{dx}e^{3x} + \dfrac{d}{dx}c = \dfrac{1}{2}\cdot\sinh 2x\cdot\dfrac{d}{dx}2x + \dfrac{1}{3}\cdot e^{3x}\cdot\dfrac{d}{dx}3x + 0$

$\qquad = \dfrac{1}{2}\cdot\sinh 2x\cdot 2 + \dfrac{1}{3}\cdot e^{3x}\cdot 3 = \dfrac{2}{2}\cdot\sinh 2x + \dfrac{3}{3}\cdot e^{3x} = \sinh 2x + e^{3x}$

c. Given $\int \operatorname{csch} 5x \, dx$ let $u = 5x$, then $\dfrac{du}{dx} = \dfrac{d}{dx}5x = 5$ which implies $du = 5dx$; $dx = \dfrac{du}{5}$. Therefore,

$$\int \operatorname{csch} 5x \, dx \;=\; \int \operatorname{csch} u \cdot \frac{du}{5} \;=\; \frac{1}{5}\int \operatorname{csch} u \, du \;=\; \frac{1}{5}\ln\left|\tanh\frac{u}{2}\right| + c \;=\; \mathbf{\frac{1}{5}\ln\left|\tanh\frac{5x}{2}\right| + c}$$

Check: Let $y = \dfrac{1}{5}\ln\left|\tanh\dfrac{5x}{2}\right| + c$, then $y' = \dfrac{1}{5}\cdot\dfrac{d}{dx}\ln\left|\tanh\dfrac{5x}{2}\right| + \dfrac{d}{dx}c = \dfrac{1}{5}\cdot\dfrac{1}{\tanh\frac{5x}{2}}\cdot\dfrac{d}{dx}\left(\tanh\dfrac{5x}{2}\right) + 0$

$= \dfrac{1}{5}\cdot\dfrac{1}{\tanh\frac{5x}{2}}\cdot\left(\operatorname{sech}^2\dfrac{5x}{2}\cdot\dfrac{5}{2}\right) + 0 = \dfrac{1}{10}\cdot\dfrac{5\operatorname{sech}^2\frac{5x}{2}}{\tanh\frac{5x}{2}} = \dfrac{1}{2}\cdot\dfrac{\operatorname{sech}^2\frac{5x}{2}}{\tanh\frac{5x}{2}} = \dfrac{1}{2}\cdot\dfrac{1-\tanh^2\frac{5x}{2}}{\tanh\frac{5x}{2}} = \dfrac{1}{2}\cdot\dfrac{1-\frac{\sinh^2\frac{5x}{2}}{\cosh^2\frac{5x}{2}}}{\frac{\sinh\frac{5x}{2}}{\cosh\frac{5x}{2}}}$

$= \dfrac{\frac{\cosh^2\frac{5x}{2}-\sinh^2\frac{5x}{2}}{\cosh^2\frac{5x}{2}}}{2\cdot\frac{\sinh\frac{5x}{2}}{\cosh\frac{5x}{2}}} = \dfrac{\frac{1}{\cosh^2\frac{5x}{2}}}{2\cdot\frac{\sinh\frac{5x}{2}}{\cosh\frac{5x}{2}}} = \dfrac{1}{2}\cdot\dfrac{\cosh\frac{5x}{2}}{\cosh^2\frac{5x}{2}\cdot\sinh\frac{5x}{2}} = \dfrac{1}{2\cosh\frac{5x}{2}\cdot\sinh\frac{5x}{2}} = \dfrac{1}{\sinh 2\cdot\frac{5x}{2}} = \dfrac{1}{\sinh 5x} = \operatorname{csch} 5x$

d. Given $\int x^2 \operatorname{sech}^2 x^3 dx$ let $u = x^3$, then $\dfrac{du}{dx} = \dfrac{d}{dx}x^3$; $\dfrac{du}{dx} = 3x^2$; $du = 3x^2 dx$; $dx = \dfrac{du}{3x^2}$. Therefore,

$$\int x^2 \operatorname{sech}^2 x^3 dx \;=\; \int x^2 \operatorname{sech}^2 u \cdot \frac{du}{3x^2} \;=\; \frac{1}{3}\int \operatorname{sech}^2 u \, du \;=\; \frac{1}{3}\tanh u + c \;=\; \mathbf{\frac{1}{3}\tanh x^3 + c}$$

Check: Let $y = \dfrac{1}{3}\tanh x^3 + c$, then $y' = \dfrac{1}{3}\cdot\dfrac{d}{dx}\tanh x^3 + \dfrac{d}{dx}c = \dfrac{1}{3}\cdot\operatorname{sech}^2 x^3\cdot\dfrac{d}{dx}x^3 + 0 = \dfrac{1}{3}\cdot\operatorname{sech}^2 x^3\cdot 3x^2 = x^2 \operatorname{sech}^2 x^3$

e. Given $\int \dfrac{2}{3}x^3 \operatorname{csch}^2\left(x^4+1\right)dx$ let $u = x^4 + 1$, then $\dfrac{du}{dx} = \dfrac{d}{dx}\left(x^4+1\right)$; $\dfrac{du}{dx} = 4x^3$; $du = 4x^3 dx$; $dx = \dfrac{du}{4x^3}$. Therefore,

$$\int \frac{2}{3} x^3 \csc h^2 \left(x^4 +1\right) dx \;=\; \frac{2}{3} \int x^3 \csc h^2 u \cdot \frac{du}{4x^3} \;=\; \frac{1}{6} \int \csc h^2 u \; du \;=\; -\frac{1}{6} \coth u + c \;=\; -\frac{1}{6} \coth \left(x^4 +1\right) + c$$

Check: Let $y = -\frac{1}{6} \coth \left(x^4 +1\right) + c$, then $y' = -\frac{1}{6} \cdot \frac{d}{dx} \coth \left(x^4 +1\right) + \frac{d}{dx} c = -\frac{1}{6} \cdot -\csc h^2 \left(x^4 +1\right) \cdot \frac{d}{dx} \left(x^4 +1\right) + 0$

$$= \frac{1}{6} \cdot \csc h^2 \left(x^4 +1\right) \cdot 4x^3 = \frac{4x^3}{6} \cdot \csc h^2 \left(x^4 +1\right) = \frac{2}{3} x^3 \csc h^2 \left(x^4 +1\right)$$

f. Given $\int x^3 \csc h \left(2x^4 +5\right) dx$ let $u = 2x^4 +5$, then $\frac{du}{dx} = \frac{d}{dx} \left(2x^4 +5\right);\; \frac{du}{dx} = \left(2x^4 +5\right);\; du = 8x^3 dx \,;\, dx = \frac{du}{8x^3}$. Therefore,

$$\int x^3 \csc h \left(2x^4 +5\right) dx \;=\; \int x^3 \csc h\, u \cdot \frac{du}{8x^3} \;=\; \frac{1}{8} \int \csc h\, u \; du \;=\; \frac{1}{8} \ln \left| \tanh \frac{u}{2} \right| + c \;=\; \frac{1}{8} \ln \left| \tanh \frac{2x^4 +5}{2} \right| + c$$

Check: Let $y = \frac{1}{8} \ln \left| \tanh \frac{2x^4 +5}{2} \right| + c$, then $y' = \frac{1}{8} \cdot \frac{d}{dx} \ln \left| \tanh \frac{2x^4 +5}{2} \right| + \frac{d}{dx} c = \frac{1}{8} \cdot \frac{1}{\tanh \frac{2x^4+5}{2}} \cdot \frac{d}{dx} \tanh \frac{2x^4 +5}{2} + 0$

$$= \frac{1}{8 \tanh \frac{2x^4+5}{2}} \cdot \sec h^2 \frac{2x^4 +5}{2} \cdot \frac{d}{dx} \frac{2x^4 +5}{2} + 0 = \frac{\sec h^2 \frac{2x^4+5}{2}}{8 \tanh \frac{2x^4+5}{2}} \cdot \frac{8x^3}{2} = \frac{x^3}{2} \cdot \frac{1 - \tanh^2 \frac{2x^4+5}{2}}{\tanh \frac{2x^4+5}{2}} = \frac{x^3}{2} \cdot \frac{1 - \frac{\sinh^2 \frac{2x^4+5}{2}}{\cosh^2 \frac{2x^4+5}{2}}}{\frac{\sinh \frac{2x^4+5}{2}}{\cosh \frac{2x^4+5}{2}}}$$

$$= \frac{x^3}{2} \cdot \frac{\frac{\cosh^2 \frac{2x^4+5}{2} - \sinh^2 \frac{2x^4+5}{2}}{\cosh^2 \frac{2x^4+5}{2}}}{\frac{\sinh \frac{2x^4+5}{2}}{\cosh \frac{2x^4+5}{2}}} = \frac{x^3}{2} \cdot \frac{\frac{1}{\cosh^2 \frac{2x^4+5}{2}}}{\frac{\sinh \frac{2x^4+5}{2}}{\cosh \frac{2x^4+5}{2}}} = \frac{x^3}{2} \cdot \frac{\cosh \frac{2x^4+5}{2}}{\cosh^2 \frac{2x^4+5}{2} \cdot \sinh \frac{2x^4+5}{2}} = \frac{x^3}{2 \cosh \frac{2x^4+5}{2} \cdot \sinh \frac{2x^4+5}{2}}$$

$$= \frac{x^3}{\sinh 2 \cdot \frac{2x^4+5}{2}} = \frac{x^3}{\sinh \left(2x^4 +5\right)} = x^3 \csc h \left(2x^4 +5\right)$$

g. Given $\int \cosh^7 (x+1) \sinh (x+1) dx$ let $u = \cosh (x+1)$, then $\frac{du}{dx} = \frac{d}{dx} \cosh (x+1);\; \frac{du}{dx} = \sinh (x+1) \,;\; dx = \frac{du}{\sinh (x+1)}$. Thus,

$$\int \cosh^7 (x+1) \sinh (x+1) dx \;=\; \int u^7 \sinh (x+1) \cdot \frac{du}{\sinh (x+1)} \;=\; \int u^7 du \;=\; \frac{1}{8} u^8 + c \;=\; \frac{1}{8} \cosh^8 (x+1) + c$$

Check: Let $y = \frac{1}{8} \cosh^8 (x+1) + c$, then $y' = \frac{1}{8} \cdot 8 \cosh^7 (x+1) \cdot \sinh (x+1) + 0 = \cosh^7 (x+1) \sinh (x+1)$

h. Given $\int \csc h (5x+3) \coth (5x+3) dx$ let $u = 5x+3$, then $\frac{du}{dx} = \frac{d}{dx} (5x+3);\; \frac{du}{dx} = 5 \,;\; du = 5dx \,;\, dx = \frac{du}{5}$. Therefore,

$$\int \csc h (5x+3) \coth (5x+3) dx \;=\; \int \csc h\, u \coth u \, \frac{du}{5} \;=\; \frac{1}{5} \int \csc h\, u \coth u \; du \;=\; -\frac{1}{5} \csc h\, u + c \;=\; -\frac{1}{5} \csc h (5x+3) + c$$

Check: Let $y = -\frac{1}{5} \csc h (5x+3) + c$, then $y' = -\frac{1}{5} \cdot -\csc h (5x+3) \coth (5x+3) \cdot \frac{d}{dx} (5x+3) + 0 = \frac{\csc h (5x+3) \coth (5x+3)}{5} \cdot 5$

$$= \frac{5}{5} \csc h (5x+3) \coth (5x+3) = \csc h (5x+3) \coth (5x+3)$$

i. Given $\int e^{x+1} \sec h\, e^{x+1} \, dx$ let $u = e^{x+1}$, then $\frac{du}{dx} = \frac{d}{dx} e^{x+1} \,;\; \frac{du}{dx} = e^{x+1} \,;\, du = e^{x+1} \cdot dx \,;\, dx = \frac{du}{e^{x+1}}$. Therefore,

$$\int e^{x+1} \sec h\, e^{x+1} \, dx \;=\; \int e^{x+1} \sec h\, u \cdot \frac{du}{e^{x+1}} \;=\; \int \sec h\, u \; du \;=\; \sin^{-1} (\tanh u) + c \;=\; \sin^{-1} \left(\tanh e^{x+1} \right) + c$$

Check: Let $y = \sin^{-1} \left(\tanh e^{x+1} \right) + c$, then $y' = \frac{1}{\sqrt{1 - \tanh^2 e^{x+1}}} \cdot \frac{d}{dx} \tanh e^{x+1} + 0 = \frac{\sec h^2 e^{x+1}}{\sqrt{\sec h^2 e^{x+1}}} \cdot \frac{d}{dx} e^{x+1}$

$$= \frac{\sec h^2 e^{x+1}}{\sec h \; e^{x+1}} \cdot e^{x+1} \; = \; e^{x+1} \sec h \; e^{x+1}$$

2. Evaluate the following integrals:

a. $\int \tanh^5 x \sec h^2 x \, dx$ let $u = \tanh x$, then $\dfrac{du}{dx} = \dfrac{d}{dx}\tanh x$; $\dfrac{du}{dx} = \sec h^2 x$; $du = \sec h^2 x \, dx$; $dx = \dfrac{du}{\sec h^2 x}$. Thus,

$$\int \tanh^5 x \sec h^2 x \, dx \; = \; \int u^5 \cdot \sec h^2 x \cdot \frac{du}{\sec h^2 x} \; = \; \int u^5 du \; = \; \frac{1}{5+1} u^{5+1} + c \; = \; \frac{1}{6} u^6 + c \; = \; \frac{1}{6}\mathbf{tanh^6}\, x + c$$

Check: Let $y = \dfrac{1}{6}\tanh^6 x + c$ then $y' = \dfrac{1}{6} \cdot 6(\tanh x)^{6-1} \cdot \sec h^2 x + 0 \; = \; (\tanh x)^5 \sec h^2 x \; = \; \tanh^5 x \sec h^2 x$

b. Given $\int \coth^6(x+1) \csc h^2(x+1)\, dx$ let $u = \coth(x+1)$, then $\dfrac{du}{dx} = \dfrac{d}{dx}\coth(x+1)$; $\dfrac{du}{dx} = -\csc h^2(x+1)\mathrm{c}$

; $du = -\csc h^2(x+1)\, dx$; $dx = -\dfrac{du}{\csc h^2(x+1)}$. Thus, $\int \coth^6(x+1)\csc h^2(x+1)\, dx \; = \; \int u^6 \cdot \csc h^2(x+1) \cdot \dfrac{-du}{\csc h^2(x+1)}$

$$= \; -\int u^6 du \; = \; -\frac{1}{7} u^7 + c \; = \; -\frac{1}{7}\mathbf{coth^7}(x+1) + c$$

Check: Let $y = -\dfrac{1}{7}\coth^7(x+1) + c$ then $y' = -\dfrac{1}{7} \cdot 7\big[\coth(x+1) \big]^{7-1} \cdot -\csc h^2(x+1) + 0 \; = \; \coth^6(x+1)\csc h^2(x+1)$

c. Given $\int e^{3x} \tanh e^{3x} dx$ let $u = e^{3x}$, then $\dfrac{du}{dx} = \dfrac{d}{dx} e^{3x}$; $\dfrac{du}{dx} = 3e^{3x}$; $du = 3e^{3x} dx$; $dx = \dfrac{du}{3e^{3x}}$. Therefore,

$$\int e^{3x} \tanh e^{3x} dx \; = \; \int e^{3x} \tanh u \cdot \frac{du}{3e^{3x}} \; = \; \frac{1}{3}\int \tanh u \, du \; = \; \frac{1}{3}\ln\cosh u + c \; = \; \frac{1}{3}\mathbf{ln\cosh}\, e^{3x} + c$$

Check: Let $y = \dfrac{1}{3}\ln\cosh e^{3x} + c$, then $y' = \dfrac{1}{3} \cdot \dfrac{1}{\cosh e^{3x}} \cdot \sinh e^{3x} \cdot 3e^{3x} + 0 \; = \; \dfrac{3}{3} e^{3x} \tanh e^{3x} \; = \; e^{3x}\tanh e^{3x}$

d. Given $\int x^3 \sec h\left(x^4 + 1\right) dx$ let $u = \left(x^4 + 1\right)$, then $\dfrac{du}{dx} = \dfrac{d}{dx}\left(x^4 + 1\right)$; $\dfrac{du}{dx} = 4x^3$; $du = 4x^3 dx$; $dx = \dfrac{du}{4x^3}$. Thus,

$$\int x^3 \sec h\left(x^4 + 1\right) dx \; = \; \int x^3 \cdot \sec h\, u \cdot \frac{du}{4x^3} \; = \; \frac{1}{4}\int \sec h\, u \cdot du \; = \; \frac{1}{4}\sin^{-1}(\tanh u) + c \; = \; \frac{1}{4}\mathbf{sin^{-1}}\left[\mathbf{tanh}\left(x^4 + 1\right)\right] + c$$

Check: Let $y = \dfrac{1}{4}\sin^{-1}\left[\tanh\left(x^4 + 1\right)\right] + c$ then $y' = \dfrac{1}{4\sqrt{1 - \tanh^2\left(x^4 + 1\right)}} \cdot \dfrac{d}{dx}\tanh\left(x^4 + 1\right) + 0 \; = \; \dfrac{\sec h^2\left(x^4 + 1\right)}{4\sqrt{\sec h^2\left(x^4 + 1\right)}}$

$$\times \frac{d}{dx}\left(x^4 + 1\right) = \frac{\sec h^2\left(x^4 + 1\right)}{4\sec h\left(x^4 + 1\right)} \cdot 4x^3 \; = \; \frac{4x^3}{4} \cdot \sec h\left(x^4 + 1\right) \; = \; x^3 \sec h\left(x^4 + 1\right)$$

e. Given $\int \sec h(3x + 2)\, dx$ let $u = 3x + 2$, then $\dfrac{du}{dx} = \dfrac{d}{dx}(3x + 2)$; $\dfrac{du}{dx} = 3$; $du = 3dx$; $dx = \dfrac{du}{3}$. Thus,

$$\int \sec h(3x + 2)\, dx \; = \; \int \sec h\, u \cdot \frac{du}{3} \; = \; \frac{1}{3}\int \sec h\, u \, du \; = \; \frac{1}{3}\sin^{-1}(\tanh u) + c \; = \; \frac{1}{3}\mathbf{sin^{-1}}\left[\mathbf{tanh}\,(3x + 2)\right] + c$$

Check: Let $y = \dfrac{1}{3}\sin^{-1}\left[\tanh(3x + 2)\right] + c$, then $y' = \dfrac{1}{3\sqrt{1 - \tanh^2(3x + 2)}} \cdot \dfrac{d}{dx}\tanh(3x + 2) + 0 \; = \; \dfrac{\sec h^2(3x + 2)}{3\sqrt{\sec h^2(3x + 2)}}$

$$\times \frac{d}{dx}(3x + 2) = \frac{\sec h^2(3x + 2)}{3\sec h\,(3x + 2)} \cdot 3 \; = \; \frac{3}{3} \cdot \frac{\sec h^2(3x + 2)}{\sec h\,(3x + 2)} \; = \; \sec h\,(3x + 2)$$

f. Given $\int e^{\cosh(3x+5)}\sinh(3x + 5)\, dx$ let $u = \cosh(3x + 5)$, then $\dfrac{du}{dx} = \dfrac{d}{dx}\cosh(3x + 5)$; $\dfrac{du}{dx} = \sinh(3x + 5) \cdot \dfrac{d}{dx}(3x + 5)$

$\dfrac{du}{dx} = \sinh(3x+5)\cdot 3$; $dx = \dfrac{du}{3\sinh(3x+5)}$. Therefore,

$$\int e^{\cosh(3x+5)}\sinh(3x+5)\,dx = \int e^u \sinh(3x+5)\cdot \dfrac{du}{3\sinh(3x+5)} = \dfrac{1}{3}\int e^u\,du = \dfrac{1}{3}e^u + c = \dfrac{1}{3}e^{\cosh(3x+5)} + c$$

Check: Let $y = \dfrac{1}{3}e^{\cosh(3x+5)} + c$, then $y' = \dfrac{1}{3}\cdot e^{\cosh(3x+5)}\cdot \sinh(3x+5)\cdot \dfrac{d}{dx}(3x+5) + 0 = \dfrac{3}{3}\cdot e^{\cosh(3x+5)}\cdot \sinh(3x+5)$

$\qquad = e^{\cosh(3x+5)}\sinh(3x+5)$

g. $\displaystyle\int \tanh^5 x\,dx = \int \tanh^3 x\tanh^2 x\,dx = \int \tanh^3 x\left(1 - \sec h^2 x\right)dx = -\int \tanh^3 x\sec h^2 x\,dx + \int \tanh^3 x\,dx$. To solve the first

integral let $u = \tanh x$, then $\dfrac{du}{dx} = \dfrac{d}{dx}\tanh x$; $\dfrac{du}{dx} = \sec h^2 x$; $du = \sec h^2 x\,dx$; $dx = \dfrac{du}{\sec h^2 x}$. Therefore,

$-\displaystyle\int \tanh^3 x\sec h^2 x\,dx = -\int u^3 \sec h^2 x\cdot \dfrac{du}{\sec h^2 x} = -\int u^3\,du = -\dfrac{1}{4}u^4 + c = -\dfrac{1}{4}\tanh^4 x + c$. In Example 5.4-6, problem

letter d, we found that $\displaystyle\int \tanh^3 x\,dx = -\dfrac{1}{2}\tanh^2 x + \ln\cosh x + c$. Therefore,

$$\int \tanh^5 x\,dx = \int \tanh^3 x\tanh^2 x\,dx = \int \tanh^3 x\left(1 - \sec h^2 x\right)dx = -\int \tanh^3 x\sec h^2 x\,dx + \int \tanh^3 x\,dx$$

$= -\dfrac{1}{4}\tanh^4 x + \displaystyle\int \tanh^3 x\,dx = -\dfrac{1}{4}\tanh^4 x + \left(-\dfrac{1}{2}\tanh^2 x + \ln\cosh x + c\right) = -\dfrac{1}{4}\tanh^4 x - \dfrac{1}{2}\tanh^2 x + \ln\cosh x + c$

Check: Let $y = -\dfrac{1}{4}\tanh^4 x - \dfrac{1}{2}\tanh^2 x + \ln\cosh x + c$, then $y' = -\dfrac{4}{4}\tanh^3 x\cdot\sec h^2 x - \dfrac{2}{2}\tanh x\cdot\sec h^2 x + \dfrac{\sinh x}{\cosh x} + 0$

$\qquad = -\tanh^3 x\sec h^2 x - \tanh x\sec h^2 x + \tanh x = -\sec h^2 x\left(\tanh^3 x + \tanh x\right) + \tanh x = -\left(1 - \tanh^2 x\right)\left(\tanh^3 x + \tanh x\right)$

$\qquad + \tanh x = -\left(\tanh^3 x + \tanh x - \tanh^5 x - \tanh^3\right) + \tanh x = -\tanh^3 x - \tanh x + \tanh^5 x + \tanh^3 + \tanh x = \tanh^5 x$

h. $\displaystyle\int \coth^5 x\,dx = \int \coth^3 x\coth^2 x\,dx = \int \coth^3 x\left(1 + \csc h^2 x\right)dx = \int \coth^3 x\csc h^2 x\,dx + \int \coth^3 x\,dx$. To solve the first

integral let $u = \coth x$, then $\dfrac{du}{dx} = \dfrac{d}{dx}\coth x$; $\dfrac{du}{dx} = -\csc h^2 x$; $du = -\csc h^2 x\,dx$; $dx = -\dfrac{du}{\csc h^2 x}$. Thus,

$\displaystyle\int \coth^3 x\csc h^2 x\,dx = \int u^3 \csc h^2 x\cdot -\dfrac{du}{\csc h^2 x} = -\int u^3\,du = -\dfrac{1}{4}u^4 + c = -\dfrac{1}{4}\coth^4 x + c$. In example 5.4-6, problem

letter g, we found that $\displaystyle\int \coth^3 x\,dx = -\dfrac{1}{2}\coth^2 x + \ln\left|\sinh x\right| + c$. Grouping the terms together we have

$$\int \coth^5 x\,dx = \int \coth^3 x\coth^2 x\,dx = \int \coth^3 x\left(1 + \csc h^2 x\right)dx = \int \coth^3 x\csc h^2 x\,dx + \int \coth^3 x\,dx = -\dfrac{1}{4}\coth^4 x + \int \coth^3 x\,dx$$

$= -\dfrac{1}{4}\coth^4 x + \left(-\dfrac{1}{2}\cot^2 x + \ln\left|\sinh x\right| + c\right) = -\dfrac{1}{4}\coth^4 x - \dfrac{1}{2}\coth^2 x + \ln\left|\sinh x\right| + c$

Check: Let $y = -\dfrac{1}{4}\coth^4 x - \dfrac{1}{2}\coth^2 x + \ln\left|\sinh x\right| + c$, then $y' = \dfrac{-4\cdot\coth^3 x\cdot -\csc h^2 x}{4} - \dfrac{2\cdot\coth x\cdot -\csc h^2 x}{2} + \dfrac{\cosh x}{\sinh x} + 0$

$\qquad = \coth^3 x\csc h^2 x + \coth x\csc h^2 x + \coth x = \csc h^2 x\left(\coth^3 x + \coth x\right) + \coth x = \left(\coth^2 x - 1\right)\left(\coth^3 x + \coth x\right) + \coth x$

$\qquad = \coth^5 x + \coth^3 x - \coth^3 x - \coth x + \coth x = \coth^5 x$

i. $\displaystyle\int \coth^6 x\,dx = \int \coth^4 x\coth^2 x\,dx = \int \coth^4 x\left(1 + \csc h^2 x\right)dx = \int \left(\coth^4 x\csc h^2 x + \coth^4 x\right)dx = \int \coth^4 x\csc h^2 x\,dx$.

$+\displaystyle\int \coth x\,dx$. In example 5.4-6, problem letter e, we found that $\int \coth^4 x\,dx = -\dfrac{1}{3}\coth^3 x - \coth x + x + c$. Therefore,

$$\int \coth^6 x \, dx = \int \coth^4 x \coth^2 x \, dx = \int \coth^4 x \left(1 + \csc h^2 x\right) dx = \int \coth^4 x \, \csc h^2 x \, dx + \int \coth^4 x \, dx = \int \coth^4 x \, \csc h^2 x \, dx$$

$$+ \left(-\frac{1}{3}\coth^3 x - \coth x + x + c\right) = -\frac{1}{5}\coth^5 x - \frac{1}{3}\coth^3 x - \coth x + x + c$$

Check: Let $y = -\frac{1}{5}\coth^5 x - \frac{1}{3}\coth^3 x - \coth x + x + c$, then $y' = -\frac{5 \cdot \coth^4 x \cdot -\csc h^2 x}{5} - \frac{3 \cdot \coth^2 x \cdot -\csc h^2 x}{3}$

$$+ \csc h^2 x + 1 + 0 = \coth^4 x \cdot \csc h^2 x + \coth^2 x \cdot \csc h^2 x + \csc h^2 x + 1 = \csc h^2 x \left(\coth^4 x + \coth^2 x + 1\right) + 1$$

$$= \left(\coth^2 x - 1\right)\left(\coth^4 x + \coth^2 x + 1\right) + 1 = \coth^6 x + \coth^4 x + \coth^2 x - \coth^4 x - \coth^2 x - 1 + 1 = \coth^6 x$$

Index

A

Adjacent, *308*
Antiderivative, *213*
Arithmetic sequence, *13-16*
Arithmetic series, *16-19*

B

Binomial coefficient, 45
Binomial expansion, *47-52*

C

Chain rule, *82-96*
Common difference, *13*
Common multiplier, *20*
Common ratio, *20*
Constant function, *59*
Constant of integration, *213*
Converge, *31*
Cosecant, *233*
Cosecant hyperbolic, *351*
Cosine, *233*
Cosine hyperbolic, *351*
Cotangent, *233*
Cotangent hyperbolic, *351*

D

Derivative
    higher order, *124-138*
    of fractions with fractional exponents, *102-108*
    of radical functions, *109-123*
Difference quotient, *54-58*
Differentiation
    of hyperbolic functions, *181-186*
    of indeterminate forms using L'Hopital's rule, *193-211*
    of inverse hyperbolic functions, *187-192*
    of inverse trigonometric functions, *158-165*
    of logarithmic and exponential functions, *166-180*
    of trigonometric functions, *140-157*
Differentiation rules
    using $\frac{d}{dx}$ notation, *71-81*
    using prime notation, *59-70*

D (continued)

Diverge, *31*

E

Even functions, *233*
Even power, *244, 365*

F

Factorial notation, *42-52*
Finite sequence function, *2*
Formulas,
    addition, *232,350*
    double angle, *232,350*
    half angle, *232,350*
    unit, *232,350*

G

General term of a sequence, *2*
Geometric sequence, *20-25*
Geometric series, *25-29*

H

Hyperbolic functions
    addition formulas for, *350*
    double angle formulas for, *350*
    half angle formulas for, *350*
    unit formulas for, *350*
Hypotenuse, *308*

I

Identity function, *59*
Implicit differentiation, *97-101*
Indefinite integral, *213*
Indeterminate forms, *140, 193, 207, 209*
Index of summation, *7*
Infinite sequence function, *2*
Integral sign, *213*
Integrand, *213*
Integration
    by partial fractions, *320-349*
    by parts, *287-307*
    of hyperbolic functions, *350-378*
    of trigonometric functions, *232-258*
    resulting in exponential or logarithmic functions, *273-285*

About the Author

Said Hamilton received his B.S. degree in Electrical Engineering from Oklahoma State University and Master's degree, also in Electrical Engineering from the University of Texas at Austin. He has taught a number of math and engineering courses as a visiting lecturer at the University of Oklahoma, Department of Mathematics, and as a faculty member at Rose State College, Department of Engineering Technology, at Midwest City, Oklahoma. He is currently working in the field of aerospace technology and has published several books and numerous technical papers.

Other Hamilton Education Guides Publications:

- Mastering Fractions
- Mastering Algebra: An Introduction
- Mastering Algebra: Intermediate Level
- Mastering Algebra: Advanced Level

To order please call: 1-800-209-8186
or write to: Hamilton Education Guides
                           P.O. Box 681
                           Vienna, VA 22183

~ Notes ~